40th World Congress of Vine and Wine 2017

BIO Web of Conferences Volume 9 (2017)

Sofia, Bulgaria
29 May - 2 June 2017

Editor:

Jean-Marie Aurand

ISBN: 978-1-5108-4435-3

Printed from e-media with permission by:

Curran Associates, Inc.
57 Morehouse Lane
Red Hook, NY 12571

Some format issues inherent in the e-media version may also appear in this print version.

Printed by Curran Associates, Inc. (2017)

For additional information, please contact EDP Sciences – Web of Conferences
at the address below.

EDP Sciences – Web of Conferences
17, Avenue du Hoggar
Parc d'Activité de Courtabœuf
BP 112
F-91944 Les Ulis Cedex A
France

Phone: +33 (0) 1 69 18 75 75
Fax: +33 (0) 1 69 28 84 91

contact-edps@webofconferences.org

Additional copies of this publication are available from:

Curran Associates, Inc.
57 Morehouse Lane
Red Hook, NY 12571 USA
Phone: 845-758-0400
Fax: 845-758-2633
Email: curran@proceedings.com
Web: www.proceedings.com

TABLE OF CONTENTS

Statement of Peer review

In submitting conference proceedings to BIO Web of Conferences, I certify to the Publisher that I adhere to the **Policy on Publishing Integrity** of the journal in order to safeguard good scientific practice in publishing.

1. All articles have been subjected to peer review administered by the proceedings editors.
2. Reviews have been conducted by expert referees, who have been requested to provide unbiased and constructive comments aimed, whenever possible, at improving the work.
3. Proceedings editors have taken all reasonable steps to ensure the quality of the materials they publish and their decision to accept or reject a paper for publication has been based only on the merits of the work and the relevance to the journal.

Title, date and place of the conference

OIV 40th WORLD CONGRESS OF VINE AND WINE
May 29th – June 2nd 2017, Sofia, Bulgaria

Proceedings editor(s):

MR JEAN-MARIE AURAND
OIV Director General

Date and editor's signature

Paris 16 May 2017

Viticulture in Bulgaria – condition and prospects

Kiril Popov[1], Slavcho Pandeliev[2], Ludmil Angelov[2], and Marko Mihaylov[1]

[1] University of Forestry, 10 Kliment Ohridski Blvd, 1797 Sofia, Bulgaria
[2] Agrarian University, 12 Mendeleev Blvd, 4000 Plovdiv, Bulgaria

Abstract. As industrial culture the vine is grown successfully between 25–50 ° N and 30–50 ° S. with certain deviations depending on the orographic features of the terrain. Most suitable conditions for the development of the vine there are in moderately warm regions, located between 34–45 ° north latitude. Precisely in this most favorable for development of vine culture zone is located the territory of the Republic of Bulgaria – between 41° 14′ and 44° 12′ N / Fig. 1 /.

The wonderful soils and climatic conditions for growth and development of the vine and accumulated in Antiquity and the Middle Ages production experience, handed down from generation to generation, were being important prerequisite for the development of viticulture in our lands.

Depending on the climatic diversity in different parts of the country in 1960 are divided and legalized 4 viticultural areas: north, south, east and southwest [5] / Fig. 2 /.

North vine region. The climate in this region is typical continental – with cold winter and hot summer. The amount of active temperature for the area bordering the Danube part are from 3800 to 3980 °C, and for foothill – from 3600 to 3760 °C. The average monthly temperature in July is between 22.4 and 24.2 °C, and the average of the absolute minimum temperatures is from −18 to −20 °C. The length of growing season ranges from 180 to 210 days. The annual rainfall is between 532 and 600 mm for the area bordering the Danubian part and between 680 and 850 mm in the foothills.

The soils in the northern region are deep and with good nutrient and mode air. In the streak along Danube they are represented by carbonate and typical, often times eroded chernozems, in the middle – mostly leached and degraded chernozems, and in the foothills meets mostly gray and dark gray forest soils.

Natural conditions in the Northern region are favorable for the production of table and wine grapes for table and quality red wines. Area-specific varieties are Gamza and Vratsa Muscat.

East vine region. The amount of active temperatures fluctuated from 3500 to 3700 °C for the northern part and from 3700 to 4000 °C for the southern part. The length of growing season varies between 180 and 240 days. For the region are not typical very low temperatures – the average of the absolute annual minimum varies from −12 °C for the seaside streak to −16 °C inside. The average monthly temperature in July is between 21.8 and 23.5 °C. The annual rainfall is between 470 (Pomorie) and 630 (Shumen) mm. High atmospheric humidity in the coastal area largely offset the deficit of soil moisture.

Soils are deep, well humid and with suitable food regime. They are presented all subtypes of chernozemic and eroded cinnamon forest soils. In the regions of Burgas and Pomorie meet vertisols and in region Varna - gray forest soils overlapped with sea sands.

The area is suitable for cultivation of dessert and wine for the table and quality red and white wines and wine materials for naturally sparkling wines, spirits and aperitifs. Area-specific varieties are Dimiat, Ugni Blanc and Muscat red.

South vine region. The amount of active temperatures from 3500 to 3700 °C for the first subregion and from 3700 to 4000 °C for the second. The average of the absolute annual minimum temperatures are of −14 to −16 °C, but in lowland fluctuated around −18 °C and here there is a danger of frost on the vines. The average monthly temperature in July is between 22.6 and 24.5 °C. The length of the vegetation period ranges from 180 to 200 days. The annual rainfalls fluctuates from 650 to 560 mm. The soils are represented by the diversity of cinnamon forest soils near Chirpan meet humus-carbonate, and in the Rhodope collar - and talus soils. In most places they are enough deep with good water, air and diet rejime. In the higher parts of Srednogorie and Rhodope depth of root dwell soil layer is insufficient.

In the area are cultivated vinous varieties for red and white table quality wines. Area-specific varieties are Mavrud, Muscat and red Pamid.

Southwest vine region. By territory is the smallest – It covers the hills along the Struma and Mesta. In the climatic attitude differs sharply from other parts of the country. It belongs in the zone of transient continental Mediterranean climate. Low temperatures below −16 °C happens rarely and for a short time. The amount of active temperatures is above 4000 °C. The average monthly temperature for July is 23.5 to 25.1 °C. The length of the vegetation period ranges from 180 to 220 days or more. The annual amount of rainfalls is small (550 mm) with a pronounced maximum in autumn-winter period.

The terrain of the area is hilly and the vineyards are planted exclusively only on slopes. The soils are mainly

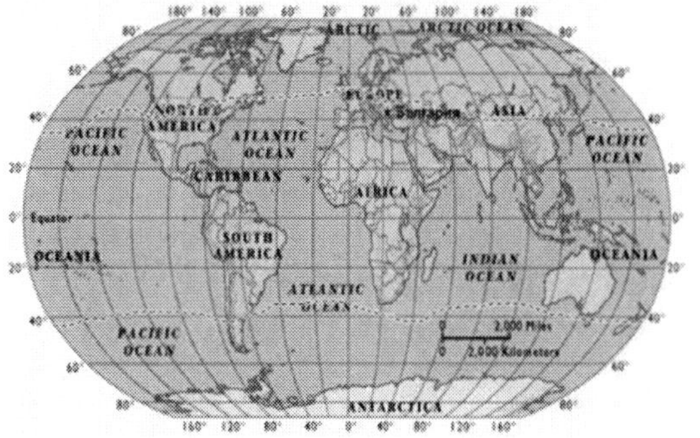

Figure 1. Location of Bulgaria in the area of distribution of vine culture in the world [1].

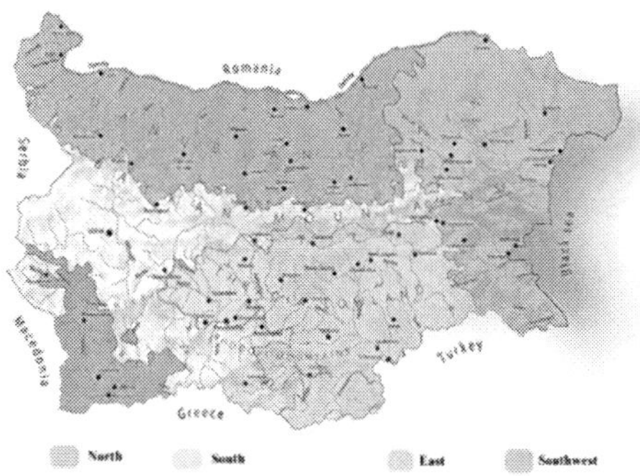

Figure 2. Wine-growing regions in Bulgaria [3].

cinnamon forest, middle to strongly leached In many places, shallow.

The area is suitable for growing a table and vinous varieties for red wines for producing table and quality red dry and table wines. Specific varieties for area are Shiroka melnishka loza and elniê 55.

Depending mainly of the total temperature each of the four areas is divided into three sub-regions with relevant soil and climatic characteristics and varietal composition.

The first subregion covers the hottest places in the country with total temperature sum during the vegetation period of 3700 to 4000 °C. The second subregion with a total temperature sum from 3500 to 3700 °C covers the hilly and partly semi hilly places of the four described areas.

The third sub-region with a total temperature sum from 3100 to 3500 °C covers semi mountains places in the vine-growing areas.

Natural conditions in Bulgaria are suitable as for growing wine varieties / with predominance of red /, and for and the varieties for fresh consumption with different time of ripening grapes. the varietal composition is diverse and includes both local, and introduced varieties vines. Prevailing by acreage from local wine varieties are: Pamid/ 10% / Muscat red / 7% / Dimiat / 5% / Mavrud / 2% / Melnik / 2% / Gamza / 1% /, of the introduced – Merlot / 11% / Cabernet Sauvignon / 10% / Rkatziteli / 9% / Muscat Ottonel / 7% / Sauvignon Blanc / 2% /, Syrah / 2% / Ugni Blanc / 1% /.

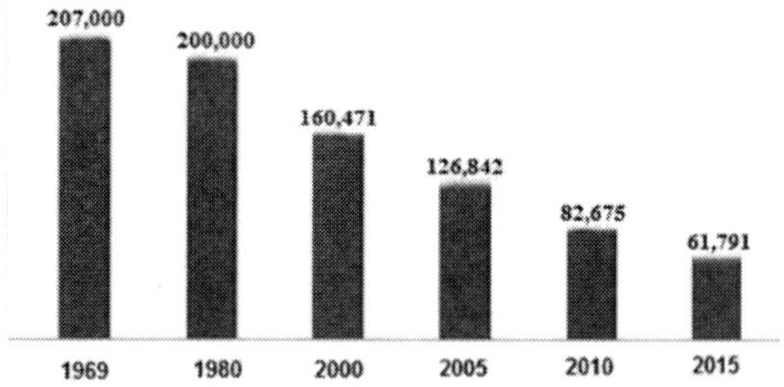

Figure 3. Amendment of areas under vines (ha).

Table 1. Area on harvested vines bystatistical areas and species in 2015.

Statistical zones	Wine varieties			Table variety	Total
	Red	White	Total		
North and Southeast Bulgaria	11319	10058	21377	1004	22381
Southwest and South Central Bulgaria	12718	2363	15081	1250	16331
Total area	24037	12421	36458	2254	38712

In our country have been proven two technologies of growing vines – low ground and stem / medium and high stem /. low ground fledge is imposed only in areas where absolute minimum temperatures fall at least once every 10 years under −15, −18 °C. These technologies are applied to small areas in the personal farm and collections. Stem formations are with height of the stems of the vines up to 100 and over 100 cm with unilateral or bilateral branches and with formed on their short or mixed fruit units. Planting distances are depending on the biological characteristics of the variety, shape and the soil. Prevailing are the distances from 2 to 2.50 m between the lines and from 1–1.20 m inside the lines vinous and from 1.20 to 1.50 m for table wines. The retaining structures are built from wooden, metal or ferroconcrete pillar.

By area of vineyards Bulgaria has been always between the top 10 countries in the world. In 1969 the area of vineyards in the country amounted to 207 thousand ha, by which 149 thousand ha are occupied by wine varieties and 51,7 thousand ha with table wines /Fig. 3/.

In subsequent years, the area of vineyards gradually decreases, as in last 15 years these changes are essential. As a result, and occurred socio-economic changes in the country, the area of vineyards decreased with 62%.

One of the reason for keeping a relatively high production of grapes till 2000 is good preserved juridical status of holdings. In the four areas, physical persons grown about 76% of the vineyards, and the agricultural cooperatives around 20%. Other manufacturers grows 4% of vineyards.

The harvested area of vineyards by species to 2015./ Table 1/ indicates that the structure is good. The greatest area is with red, followed by white wine varieties and at least table varieties.

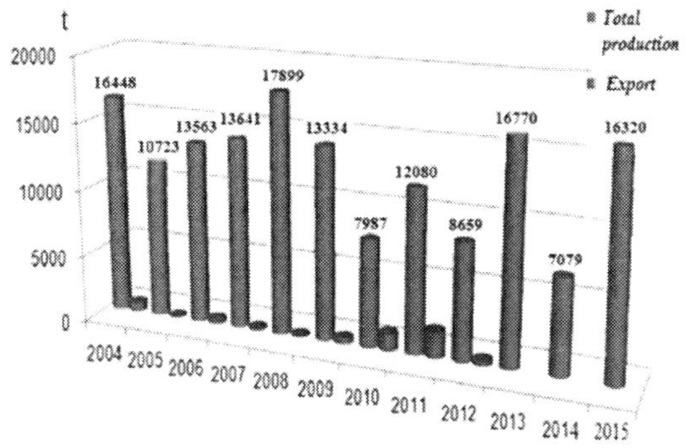

Figure 4. Dynamics of production and export of table grapes for the period 2004–2015.

Table 2. Production of grapes in Bulgaria by statistical areas and species in 2015, t.

Statisticall zones	Grapes from vineyards, t			Grapes from liana vines t	Total produced grapes t
	Vinous	Table vines	Total		
North and Southeast Bulgaria	156327	8163	164490	-	-
Southwest and South Central Bulgaria	88030	8157	96187	-	-
Total for Bulgaria	244357	16320	260677	1143	261820

The average yield of grapes for the country of wine varieties is 6702 Kg/ha, and for table variety – 7240 sc kg/ha.

The total production of grapes in 2015 amounted to 261 820 tons /Table 2/. [4]

From 261,820 tons produced grape, 242,936 tons is processed, as 195,860 tons is vinified in wineries, and 42 176 tons outside them and 4900 tons is processed in other products.

Processing of grapes is concentrated in more than 250 wineries in Bulgaria [4].

The table grape production in the country till 1980 is relatively well developed. The first export of table grapes was made in 1926 from 155 tons. Twelve years later /1938./, Bulgaria exports 12 000 tons mainly variety Bolgar and ranks first in the world in the export of table grapes. In 1965 Bulgaria is again first by exporting grapes in the world. Today, Bulgaria almost not export table grapes /Fig. 4/ [4].

Sharp decline in the consumption of table grapes in the country from 25–30 kg in 1960. on 2–3 kg per capita in 2016. Average yields from table grapes vary from 2640 to 7305 kg/ha for the period 2000 – 2015 [6].

The reasons for reduced production of table grapes in the country are a few:

1.1 Accelerated abandon and poor maintenance of the table vineyards.
1.2 Giving the vineyards for temporary use by different farmers.
1.3 Breaking the retaining construction of a large part of the plantations.
1.4 Losses of markets led to a deterioration of economic efficiency on vineyard and accelerating the eradication of most of them.
1.5 Improper and ineffective land reform.
1.6 Insufficient manpower and low qualification.
1.7 Impaired link between science and production.
1.8 Insufficient using of available resources for various programs.

The controls in the sector is regulated by law wine and spirit drinks(decree N0283/16.09.1999.), with which are created two structures:

1. Executive Agency on Vine and Wine (EAVW). It is a state body supervising the observance of the law on wine and spirits onto vineyard plantations, grapes intended for wine making, grape musts, products of grapes and wines. EAVW is a legal entity with headquarters city Sofia and 9 territorial links in Montana, Pleven, Ruse, Varna, Burgas, Sliven, Haskovo, Plovdiv and Blagoevgrad;
2. National Vine and Wine Chamber (NVWC). It is a professional branch organization, which supports,controls and protects the interests of its members. NVWC is legal entity with headquarters city Sofia and carries out its activities through 5 regional vine-wines chambers with headquarters in Pleven, Varna, Haskovo (since 2001 Moved in Sliven), Plovdiv and Blagoevgrad. In July 2006 launched a reform of the vine-wine sector in the EU to prioritize the following prospects:

- Expanding markets;
- Increasing competition in the European winemaking;
- Saving the best traditions of Europe in the area of winemaking.

Are recommended the following measures in national envelopes:

- Popularization of wine in third countries;
- Restructuring and conversion of vine plantations;
- Financial support for green harvest;
- New measures to manage crises - crop insurance and setting up mutual funds.

The state of viticulture in Bulgaria gives rise to outline the following prospects:

- Adjustment of production structure to the requirements of the international market;
- Introduction of varieties with different ripening period to overcome the seasonality and expanding production on large fruit seedless and seed varieties for consumption in fresh condition and juices from table varieties.
- Increasing acreage, mostly with quality red and white grape varieties and production of quality

wines with pronounced environmental with assertive individuality.

- Using of certified branch planting material;
- Lighter and cheaper supporting structures and replacing them with simpler forms of cultivation;
- Improving production technologies, through maximum mechanization on technological processes;
- Creation of vine arrays with a greater number of vines per unit area (4000–5000 vines/ ha);
- Correctly territorial location, specialization and concentration on vine arrays depending on the biological requirements of varieties to soil and climatic conditions of the region;
- Investments for increasing the quality of human potential through applying the systems of education and qualification.

References

[1] `http://www.thecityedition.com/2012/GPS.html`
[2] National Strategy for Development of viticulture and winemaking in the Republic of Bulgaria 2005– 2025
[3] National Vine and Wine Chamber
[4] MAF, department "Agrostatistics", "Production of grapes and wine" – newsletters from (2000) to (2016)
[5] D. Pantaleeva. "Condition, opportunities and prospects for development of viticulture in Bulgaria", Econ. Mgmt of Agriculture, J. E. **5**, 14–22 (2010)
[6] D. Toteva., P. Kirovski. Impact of the Common Agricultural Policy on the competitiveness, opportunities and prospects for development on the vinous viticulture in Bulgaria, Econ. Mgmt of Agriculture, J. E. **3–4**, 33–40 (2011)

DOI: 10.1051/bioconf/20170901002

ECA&D and E-OBS: High-resolution datasets for monitoring climate change and effects on viticulture in Europe

Christiana Photiadou[1], *Natacha* Fontes[2], *Antonio* Rocha Graça[2], and *Gerard* van der Schrier[1,a]

[1] Royal Netherlands Meteorological Institute (KNMI), De Bilt, The Netherlands
[2] Investigação e Desenvolvimento Sogrape Vinhos, SA, Avintes, Portugal

Abstract. Climate change and climate variability profoundly affect the production of wine. When facing a changing climate, the characteristics of wine produced in each region will change while the natural year-to-year variations in climate will increase variability of income for wine businesses and therefore affect profitability and economic resilience. The challenge posed to the viticulture community is thus to closely monitor these changes to be able to adapt business practices. The European Climate Assessment and Dataset (ECA&D) and its gridded version E-OBS are tools to monitor the changing climatic conditions over Europe, with an emphasis on changes in extreme climatic conditions. In this paper, the potential of ECA&D and E-OBS for viticulture is demonstrated by a few examples. The examples include the changing areal suitable for Chardonnay cultivation, showing an expansion to areas which have been too cold only a few decades ago and retraction of optimal conditions from areas which have been suitable in the recent past. Other examples show the change in the diurnal temperature range in the latter stages of the ripening process of grapes and the variability in heavy precipitation events. Finally, first results of a new dataset for South America are presented.

1. Introduction

Climate is of paramount importance to the production of wine with critical moments during the growing season that the weather needs to be just right to attain a level of excellent quality. Climate is thus a driver of value for producers of both grapes and wines. High-quality grapes may be produced at a fairly narrow range of average air temperatures during the growing season, while regional differences in the diurnal temperature range, frost frequency in Spring, the duration of heat waves or cold snaps affect the production of quality compounds in grapes (sugars, acids, polyphenols, aroma precursors, etc.). Changes in the moisture regime and specifically on the timing of rains, the amount of moisture available from rains, and the atmospheric evaporative demand (atmospheric 'thirst'!), all influence grape composition in such a way that wine taste, colour and aroma is and will most likely will be impacted in the near future.

When facing a changing climate, with its persistent changes in temperature and precipitation patterns, the characteristics of wine produced in each region will change. These changes impact the quality of grapes and wines and may generate higher year-to-year variability in the production, which in return hamper the sustainability and economic resilience of wine businesses. Winegrowers face a big challenge in managing the increased variability between heavy rainfall on the one hand, and relatively high temperatures and drought on the other; a scenery that central Europe is becoming familiar with during summer seasons. Future business practise will require adaptation to overcome these changing and challenging conditions; a monitoring tool is thus essential to understand how these changing develop through time and how they are affecting viticulture and winemaking. The aim of this paper is to introduce an observation-based climate monitoring tool which specifically addresses variability and trends in wind and vine relevant indices.

The *European Climate Assessment and Dataset (ECA&D)* and its gridded version *E-OBS* are tools to monitor changing climatic conditions over Europe - the geographical area responsible for 45% of all wine-growing areas, 65% of global production and 70% of global exports of wine. ECA&D is a dataset with daily station data for over 10.000 stations in Europe, the Middle East and Northern Africa, which includes 12 elements – among them temperature, precipitation, and sunshine duration. The station data are provided by National Meteorological Services and other sources such as the climate-aware private company *Sogrape Vinhos, S.A.* Based on these station data, the high-resolution E-OBS maps of daily gridded data are made available for scientific research. *Climate Information Bulletins* are released once every year and provide a retrospective outlook of the past year with extreme events. In this study, we will discuss how E-OBS can be used as a monitoring tool in the grape and wine industry. Some limitations of E-OBS, which the user needs to be aware of are presented here as well.

Both ECA&D and E-OBS enjoy strong support from the wider scientific community, with over 1250 citations to the E-OBS introducing study (status January 2017), and within the community focusing on viticulture to quantify climatic trends and climatic variability across Europe and in Europe's subregions (e.g., [1–4]). Following results of

[a] e-mail: `schrier@knmi.nl`

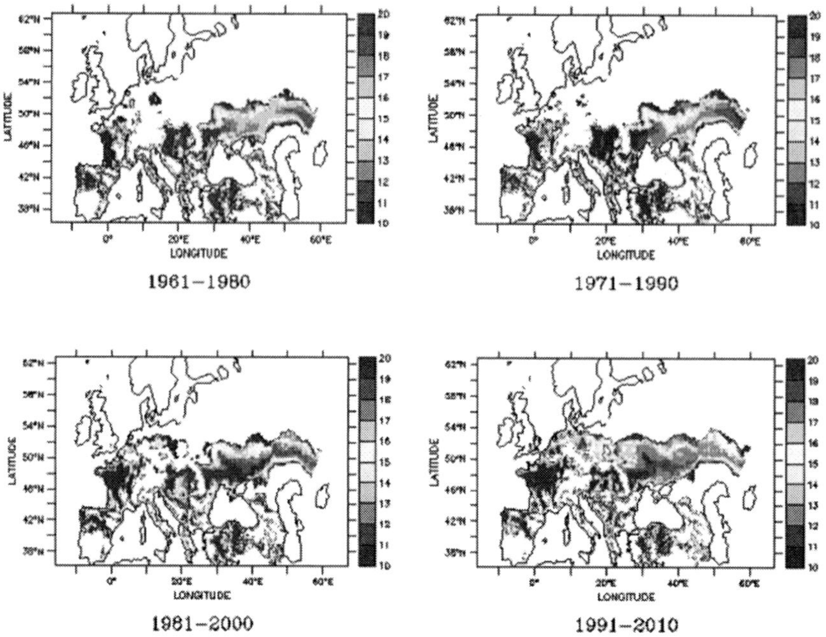

Figure 1. Number of seasons in 20-year overlapping periods which have optimal average temperatures for Chardonnay following the climate-maturity groupings [8].

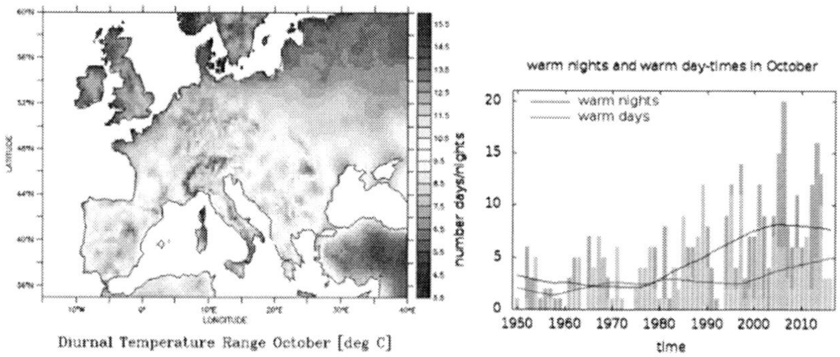

Figure 2. The 1981–2010 average Diurnal Temperature Range for October (left) and the evolution of the number of warm nights (blue) and day-times (red) in October for Bordeaux (France). The red and blue lines denote smoothed values; actual October values are given in the light and dark grey bars for warm day-times and warm nights respectively.

these earlier studies with some new analysis, the benefits of using E-OBS in understanding the nexus between climate, farming vineyards, wine quality, yields, and economics will be quantified. This will be done by discussing trends and variability in *Climate Impact Indicators* (CII, both on station and gridded data) which are widely used in the viticulture scientific community such as the Huglin Index and of more common descent, like the number of frost days in Spring, the number of rainy days and the number of warm days for specific seasons. Finally, an outlook will be provided to the LA-OBS dataset with daily temperature and precipitation focused on South America, using similar geostatistical techniques as its European counterpart. A similar tool is set to be made available for the benefit of South American wine regions, their grape-growers, winemakers, and local businesses, as well as national and regional regulatory bodies.

2. Materials and methods

2.1. Description of ECA&D dataset

The ECA&D data used in this study (`http://www.ecad.eu`) receives data from 68 participants from 64 countries, mainly National Meteorological and Hydrological Services (NMHSs). ECA&D is a collection of daily station observations of currently 12 elements, the Essential Climate Variables, from around 10500 stations throughout Europe and the Mediterranean. The daily station data are subject to data policy limitations of the NMHSs, some of which allow distribution of derived data (like averages or gridded data products) only.

Although most data provided to ECA&D is validated by the responsible NMHS, they first go through a series of basic quality checks to remove obvious problems and

 DOI: 10.1051/bioconf/20170901002

outliers. These tests are described elsewhere [5]. No attempt is made to correct data flagged as suspect and this data is not used in further analysis.

2.2. E-OBS dataset

Hofstra et al. [6] compared several interpolation methods for daily temperature and precipitation sums and concluded that the method which involves kriging using a geographically independent variogram is the most optimal interpolation method for the daily dataset. This method was described in detail by Haylock et al. [7], a brief summary is given here.

The gridding procedure adopts a three-step methodology of interpolating the daily data: interpolating the monthly mean using thin-plate splines to define the underlying spatial structure of the data; kriging the daily anomalies with regard to the monthly mean; and applying the interpolated daily anomaly to the interpolated monthly mean to create the final result. In such a large and complex region as Europe, thin-plate splines are a more appropriate method for trend estimation than other alternatives. For the gridding of the monthly mean temperature and monthly sums of precipitation, a three-dimensional spline is used, taking into account the station elevation.

The ECA&D station data are first gridded onto a high-resolution 0.1° × 0.1° grid, followed by a regridding to a slightly coarser 0.25° × 0.25° regular longitude-latitude grid. Other grids are produces as well for e.g., regional climate modelling applications. The motivation for the regridding is to provide area-averaged values of over the grid squares, which is of importance for precipitation given the low spatial coherency in precipitation.

2.3. Climate Impact Indicators

Climate Impact Indicators (CII) are currently widely used as tools to capture the extremes and the impact of climate change. The ECA&D, through a continuous collaboration with the World Meteorological Organisation (WMO) Expert Teams on Climate Change Detection and Indices (ETCCDI) and on Sector-specific Climate Indices (ET-SCI), provides CII based on station data and on E-OBS gridded data. The later provide an immediate visualization of the spatial variability throughout Europe and, for this occasion, in winemaking regions. ECA&D provides online around 70 CII for describing different regimes – from cold, drought, to extreme rain etc. Here we present a small sample of what is available and how the CII can be used in the viticulture management.

3. Results

3.1. Change in climate-maturity grouping for Chardonnay

The effect of the rise of temperatures on the viability of vineyards and the area where vine cultivation is possible, is illustrated by analysing the climate-maturity ripening potential for quality wine varieties. Figure 1 shows such an analysis for Chardonnay, using the climate maturity grouping based on [8] and for overlapping 20-year periods. In this figure, we show the number of growing seasons (defined as the April-October period) in this 20-year period for which the climate is suited to grow Chardonnay. The most optimal conditions are found in the areas where all growing seasons have temperatures in the optimal temperature range, whereas areas with only about half of the years with averaged temperatures in the optimal temperature range, are marginally suited. Figure 1 shows the gradual northward expansion of the area with suitable conditions to cultivate Chardonnay as time progresses, with south eastern England and Holland now marginally suited for this variety. The expansion of the area in France where climatic conditions are suitable for Chardonnay cultivation is interesting. The area centred around 20°E, 46°N, covering Hungary, northern Serbia and Croatia and Austria's Burgenland and Niederösterreich regions saw an initial improvement of conditions into the 1971–1990 period, followed by a deterioration in the following decades. The conditions to produce high-quality wines based on the Chardonnay grape in this area are deteriorating in the most recent period because of too high temperatures.

On the eastern side of the Carpathian mountain range, the total area suitable for Chardonnay cultivation has generally increased over time, but areas can be distinguished where the climate has become too hot in the most recent period for optimal ripening conditions.

3.2. Warm nights

The climatic influences on wine production are complex, but in an ideal situation, the grower would like to have in the period from *veraison* to harvest little to no rainfall with the vines ripening into the early fall and a large Diurnal Temperature Range (DTR) [8]. The climate change signal can be detected in the DTR as well, especially in the balance between the number of warm day-times and the number of warm nights. The former is based on the daily maximum temperature where a warm day-time is defined as a day with the temperature above the 90th percentile (based in its turn on the 1961–1990 period). The latter is based on the daily minimum temperature using a threshold based on the 90th percentile of daily minimum temperatures. Figure 2 shows a map with the 1981–2010 mean DTR for October and a graph showing the evolution of the number of warm day-times and warm nights in that month for Bordeaux (France). October is chosen to focus on the period around harvest. The graph shows a steep increase in the number of warm nights from the late 1970s onwards which is not present in the number of warm day-times. The latter increases since the late 1990s. Averaged over October, the trend in DTR is slightly negative but Fig. 2 shows that changes in the extremes in day-time and night temperatures can be more pronounced.

3.3. Extreme precipitation indicators

Apart from warmer conditions over the growing seasons, climate change also brings changes to the hydrologic cycle. While the Mediterranean region sees generally drier conditions in the summer half year, areas in Europe further north are seeing generally wetter conditions. Nevertheless, an increase in more extreme precipitation is generally observed over all of Europe. The implications of more frequent and severe downpours are that organic vineyards run a larger risk of crop damage or crop loss. Organic farmers are limited to use non-synthetic chemicals to fight

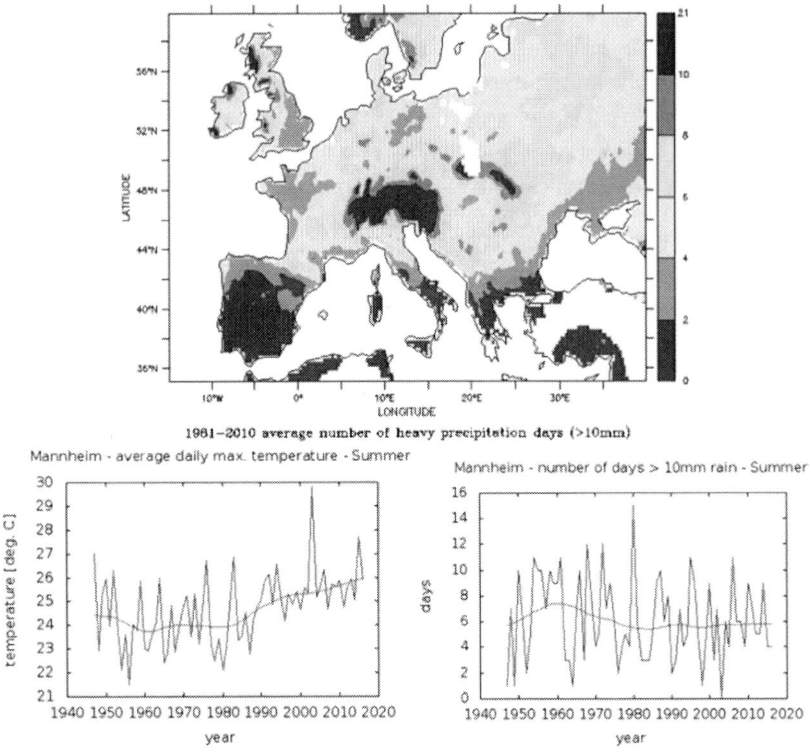

Figure 3. The average number of days with heavy precipitation (> 10 mm/day) in summer for the 1981–2010 period in Europe (top panel). The bottom panels show the change in time of the average daily maximum temperature in summer for Mannheim, in the German Rhine valley (left). The bottom right panel shows the number of days in summer with heavy precipitation (>10 mm/day) for Mannheim.

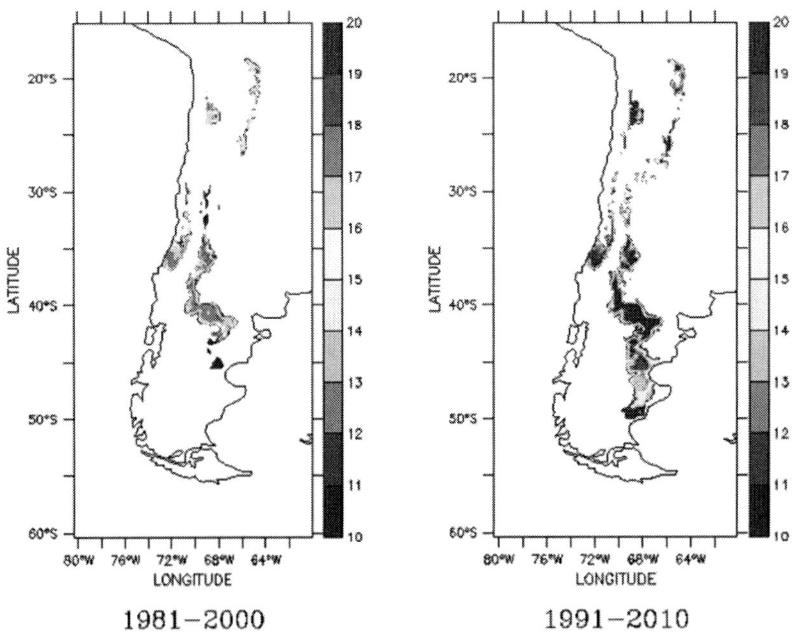

Figure 4. Number of seasons in 20-year overlapping periods which have optimal average temperatures for Chardonnay following the climate-maturity groupings (Jones, 2007).

fungal infections like downy mildew. Heavy rains wash the anti-fungal solution off the vines' leaves while warm conditions fuel the growth of the mildew.

Figure 3 shows a map with the climatological mean value of the number of days in summer with heavy precipitation (> 10 mm/day) and two graphs for Mannheim with the change over time for the summer daily maximum temperature and the number of days with heavy precipitation. Mannheim is in the German Rhine valley, north in the Baden wine district. The graphs show a distinct trend towards higher day-time temperatures (with a peak for the 2003 heat wave). For the days with heavy precipitation, a trend cannot be clearly distinguished for this station. Nevertheless, the large year-to-year variability of the number of days with downpours, which can be observed from the graph, makes it clear that one harvest may be more affected than others.

4. A new dataset with daily temperature and precipitation data for South America

Using a similar set-up as in the ECA&D, a station-based dataset for daily temperature and precipitation has been developed in collaboration with National Meteorological and Hydrological Services of South America. This activity is hosted by the CIIFEN institute in Ecuador and is coined the *Latin American Climate Assessment and Dataset (LACA&D)*†.

Based on the station dataset, a preliminary gridded dataset with daily temporal resolution and a spatial resolution of 0.25° has been developed [9] and will be made available for scientific research under the name LA-OBS. Currently, this dataset spans the period 1981/01/01 to 2014/12/31. Based on the LA-OBS dataset, a similar exercise as for Fig. 1 is done for the wine producing regions in Chili and Argentina. Figure 4 shows the number of growing seasons (defined as the October – April period) in overlapping 20-year periods where climatic conditions are optimal to grow Chardonnay. Figure 4 shows that the area where climatic conditions are optimal for this variety has increased, and it shows that the areas which were already well suited to grow Chardonnay have become excellent areas in the more recent period.

5. Conclusion

The quality and quantity of wine produced in each region is affected by climate change, with its gradual and persistent change in climatic zones and precipitation patterns. The more random natural year-to-year variability profoundly affects viticulture management and general profitability of vineyard and winery operations as well. In this paper, a tool to monitor climate change and climate variability is presented by showing a few basic examples. It is shown, using simple indices relevant for the climate maturity grouping of Chardonnay, that the area where cultivation is adequate for quality wine spreads northward and eastward. It is also shown that some areas already became too hot to produce high-quality wines on every growing season. This observation is under the reservation that the use of hard-limit thresholds will insufficiently reflect the ability of the grapes to grow under stressed conditions. Ancient grapevine varieties such as Chardonnay have a high-level of intra-varietal genetic diversity allowing for high adaptive potential, if vineyards are not planted eugenic, i.e., using a single clone (plants multiplied from a single original mother-plant).

In conclusion, while the station-based dataset ECA&D and its gridded counterpart E-OBS can be used to monitor changes in climate, some caution is required in interpreting trend analysis. Although data have been quality checked, there are issues with the homogeneity of some of the series. When a meteorological station is moved from one place to the other (because of e.g., an expanding city), a discontinuity in the temperature and/or precipitation will be introduced in the series. A simple trend analysis will then confuse this discontinuity with climate change. It is ongoing work to adjust the series for these discontinuities which are still present in the current version of E-OBS.

References

[1] Martínez-Lüscher, J., Sánchez-Díaz, M., Delrot, S., Aguirreolea, J., Pascual, I., & Gomès, E. Australian journal of grape and wine research **22**, 87–95 (2016)
[2] Fraga, H., Malheiro, A. C., Moutinho-Pereira, J., Jones, G. V., Alves, F., Pinto, J. G., & Santos, J. A. Regional Environmental Change **14**, 295–306 (2014)
[3] Fraga, H., Malheiro, A. C., Moutinho-Pereira, J., & Santos, J. A. International Journal of Biometeorology **57**, 909–925 (2013)
[4] Malheiro, A. C., Santos, J. A., Pinto, J. G., & Jones, G. V. Bulletin de l'OIV **85**, 973 (2012)
[5] ECA&D Team. European Climate Assessment & Dataset Algorithm Theoretical Basis Document (ATBD). Royal Netherlands Meteorological Institute KNMI, De Bilt, NL, version 10.5 (2012)
[6] Hofstra, N., Haylock, M. R., New, M., Jones, P. D. & Frei, C. Journal of Geophysical Research: Atmospheres, 113: D22220, doi:10.1029/2008JD010100 (2008)
[7] Haylock, M. R., Hofstra, N., Klein Tank, A. M. G., Klok, E. J., Jones, P. D., & New, M. (2008). Journal of Geophysical Research: Atmospheres **113**(D20) (2008)
[8] Jones, G. V. Proceedings of the Thirteenth Australian Wine Industry Technical Conference, The Australian Wine Research Institute (2007)
[9] Cornes, R. van den Besselaar, E. J. M., van der Schrier, G. Klein Tank, A. M. G. (in preparation) (2017)

† `http://lacad.ciifen.org/`

DOI: 10.1051/bioconf/20170901003

Effects of hot water treatments on dormant grapevine propagation materials used for grafted vine production

Oguzhan Soltekin[1,a] and Ahmet Altindisli[2]

[1] Manisa Viticultural Research Institute, Ataturk St., Yunusemre, PO Box 12, Manisa, Turkey
[2] Ege University, Faculty of Agriculture, Department of Horticulture, Izmir, Turkey

Abstract. *Agrobacterium vitis* is responsible for the crown gall disease of grapevine which breaks the grapevine trunk vascular system. Nutrient flow is prevented by crown gall and it leads to weak growth and death of the plants. It can be destructive disease often encountered in vineyards and it can be spread in cuttings for propagation. Thermotherapy treatment is an alternative method for eradicating *A. vitis* from grapevine cuttings but effects of thermotherapy treatments on dormant vine tissue, bud vitality, rooting and shooting of the propagation materials are not yet fully understood. In this research, it is aimed to determine the effects of thermotherapy treatment (Hot water treatment) on callus formation (at the basal part and grafting point), grafted vine quality (shoot length, shoot width, root number, shooting and rooting development, fresh and dry weight of shoots and roots) and final take in the grafted vine production. Experiment was conducted in the nursery of Manisa Viticultural Research Institute. Rootstocks (Kober 5BB, Couderc 1613 and 41B) and scions (Sultan 7 and Manisa sultanı) were hot-water treated at 50°C for 30 minutes which is the most common technique against *Agrobacterium vitis*. After thermotherapy treatment, all rootstocks were grafted with Sultan 7 and Manisa sultanıvarieties. They were kept for 22 days in callusing room for callus development and then they were planted in polyethlyene bags for rooting. At the end of the study, significant treatment x rootstock interaction were observed for the final take of Sultan 7 variety. Thermotherapy treated of 1613C/Sultan 7 combinations had more final take than the control (untreated) group. For instance, hot water treated cuttings of 1613C/Sultan 7 combinations had 75% final take while the control group had the 70%. Also there were not observed any adverse effects of HWT on bud and tissue vitality.

1. Introduction

Viticulture constitutes an important part of agricultural production in the world and it is common between 20–52° latitudes in the Northern Hemisphere and 20–40° latitudes for Southern Hemisphere. Turkey, which has a strong viticultural potential, is one of the important gene centers of the vine. Today, 74 million tons of world grape yield (production) that is 4.1 million tons yielded in Turkey. In addition, Turkey ranks 5th in the world with 467.093 ha grape harvested area [1] and that the grape yield per hectare is about 8.9 tons. It can be observed that this value is considerably less than the advanced countries of viticulture. This situation is due to various negativities encountered during the period planting to the marketing. Especially, some of the pests and diseases can affect adversely the grape yield per unit area in our country.

Crown gall, caused by the soil-borne bacterium (*Agrobacterium vitis*), is also the most important disease in our country's vineyards. The most remarkable signs are galls and cracks that appear on grapevine trunks [2]. Galls are more common on the lower trunk but they can develop in canes and at graft unions. Crown gall can kill young vines, reduce growing potential and require establishment of new trunks [3].

There are not any effective chemical treatments for grape crown gall control yet [4]. Therefore its control focuses on prevention of injuries and using of pathogen free propagation materials [5].

In our country, grafted vine producers may select the infected cuttings in the vineyard for grafting process without being aware of it. Therefore the possibility of using infected propagation materials with *Agrobacterium vitis* while establishment of the vineyards, is one of the enhancing dissemination and transmission of disease. Furthermore if *Agrobacterium vitis* identify in herbal products, it is among the organisms subject to quarantine which is prohibited to circulate in our country. Therefore it is determined that *A. vitis* has an extremely important precaution for our country, according to the Regulation on "Plant passport system and registration of operators" published in the official gazette dated 2nd January 2011, numbered 27813 [6].

Hot water treatment (Thermotherapy) is an efficient, environmentaly safe and commercially viable method for sterilization of *Agrobacterium vitis* [7]. The most common hot water treatment for dormant propagation materials (rootstocks and vine cuttings) is submerging them in a water bath at 50° C for 30 min [3,7,8]. However the most common problems in HWT studies are generally related its affects on bud and tissue vitality, growth in the nursery and the effects on grafting of vines for different varieties [3,9–13].

In this study, it was aimed to determine the effects of hot-water treatment (thermotherapy) on propagation

[a] e-mail: oguz.soltekin@tarim.gov.tr

 DOI: 10.1051/bioconf/20170901003

materials used for grafted vine production. In this way, new registered grape varieties (Sultan 7 and Manisa sultanı) were used as plant materials and it was observed responses of these varieties to the thermotherapy treatments.

2. Materials and methods

2.1. Plant material

This research was carried out in the nursery of Viticultural Research Institute in Manisa province. Five rootstock cultivars Kober 5BB, Couderc 1613, 41B and two scion cultivars Sultan 7, Manisa Sultanıwere used as plant material. Sultan 7 is a seedless variety and it is the most common raisin variety in Turkey. It matures in second half of the August. It has high drying efficiency and yield potential. Manisa Sultanıis a seedless variety growing for primarily table grape. Berries are elliptic, green-yellow color. It matures in second half of the August. All these propagation materials were collected in the dormant season (November and February) from the vineyards of the Institute. Then they were placed in cold storage at 2–4° C until grafting. Sultan 7 and Manisa Sultanıgrape varieties are newly registered varieties. Therefore, in this research, it was observed responses of these varieties to the hot water treatment (HWT) first time.

2.2. Methods

In the research, all propagation materials (rootstocks and scions) were kept in the room temperature overnight before the HWT. Then they were immersed in hot-water at 50° C for 30 minutes which is the most common technique against *Agrobacterium vitis*. Hot water tank had 1000 L water capacity with a recirculation pump and temperature was monitored via probes during the treatment process. After HWT, all propagation materials were left to drying at room temperature. Also there were not any hot water treatments for control group.

One bud scions of the Sultan 7 and Manisa sultanıgrape varieties were omega shaped bench-grafted onto three rootstocks [14,15]. Grafted cuttings were dipped into paraffin wax (55–60° C) to cover the grafting point and placed in plastic boxes filled up with sawdust and coarse chip (3:1). All boxes were kept in callusing room (25–27° C, 90–95% humidity) for three weeks and then transferred to the outdoor conditions for adaptation in two days. When the callus formation was completed in the callusing room, rooting rates (%) and sprouting rates (%) were determined. Then, callusing level at the grafting point and at the basal part were measured according to the 0–4 scale (0-no callus development, 1- weak callus development, 2- crescent shaped callus development, 3- partially interrupted callus development, 4- complete callus development).

After these measurements, grafted cuttings were planted in polyethlyene bags for rooting over 6–7 weeks at 25–27 °C, 70–75% humidity in greenhouse. At the end of the this growing period in greenhouse, following parameters were measured; Shooting and rooting development level (0-4 scale), shoot length (cm), shoot width (mm), root number, fresh and dry weight of shoots and roots (g) and final take (%) were measured. All parameters were examined to determine the effects of thermotherapy treatments on propagation materials (rootstocks and scions) for grafted vine production.

The experimental design was randomized plots with three replications and there were 30 cuttings in each replication. Analysis of variance is implemented to research data by using SPSS statistical analysis software package on computer, and in order to determine the differences among averages, LSD test is implemented.

3. Results and discussion

According to the measurements of after callus formation in the callusing room, effect of hot water treatment on rooting and sprouting rates was found non-significant. However hot water treated cuttings of all rootstocks x Manisa Sultanıcombinations had higher values than control group in terms of rooting rates. Also the highest values were obtained from the hot water treated (64.4%) and untreated (56.7%) 1613C x Manisa Sultanıcombinations respectively (Table 1 and Table 2). Besides that [16] reported similar results in their study. For instance, they observed different sprouting ratios during the callusing development for the all rootstock x variety combinations. On the other hand the highest rooting ratio was obtained from the 1613C x variety combinations (54.6%) in their study.

The Table 3 and Table 4 show that effects of HWT on callusing level at the basal part and grafting point, respectively. Treatment x rootstocks interactions for the Sultan 7 variety were found statistically significant in terms of callusing level at the basal part. The highest value (2.65) was observed in hot water treated vines of 41B/Sultan 7 combination. Also [17] reported that thermotherapy treatments stimulated callus formation at the basal part in most cases.

When measuring the callus formation at the grafting point, the lowest level were observed in hot water treated vines of 5BB/Sultan 7 combination (2.30). On the other hand the highest values were observed in 1613C/Sultan7 combinations. As can be seen from the results, callus formation was affected adversely by hot water treatment on 5BB/Sultan 7 and 5BB/ Manisa Sultanıcombinations. However other combinations were placed in the highest level group statistically (Table 4). When the callus formation was considered as one of the important criteria determining the success in the production of grafted vines [18], 41B and 1613C rootstocks come into prominence for the both varieties. Hot water treated cuttings of 41B/Manisa Sultanıcombinations was found 6.5% more callus formation than untreated group. In a study hot water treated and grafted cuttings were found 5–10% better callus formation than the control cuttings [3].

Significant treatment x rootstock interactions were observed for the Sultan 7 variety in terms of shooting and rooting development levels (Table 5 and Table 6). Hot water treated vines of 1613C/Sultan7 and 41B/Sultan 7 combinations had higher values than each control groups for the both shooting and rooting development levels. As can be seen from the results of all rootstocks x Sultan 7 combinations, hot water treatments induced more shooting and rooting development except 5BB/Sultan 7 combination. These findings was found parallel with the previous studies [11,17,19].

Table 1. Effects of HWT on rooting rates (%).

Manisa Sultanı				*Sultan 7*			
Rootstocks	Control	HWT	*Rootstock Mean*	Rootstocks	Control	HWT	*Rootstock Mean*
5BB	32.400	33.333	32,867b	5BB	51.297	34.167	42.732
41B	43.380	49.417	46,398ab	41B	41.733	54.373	48.053
1613C	56.783	64.430	60,607a	1613C	61.03	52.943	56.987
Treatment Mean	44.188	49.060		*Treatment Mean*	51.353	47.161	
LSD Treatment: ns				LSD Treatment: ns			
LSD Rootstock: 15,219**				LSD Rootstock: ns			
LSD Treatment x Rootstock: ns				LSD Treatment x Rootstock: ns			
*p < 0.05, **p < 0.01, ns: non-significant							

Table 2. Effects of HWT on sprouting rates (%).

Manisa Sultanı				*Sultan 7*			
Rootstocks	Control	HWT	*Rootstock Mean*	Rootstocks	Control	HWT	*Rootstock Mean*
5BB	30.070	30.640	30.355	5BB	40.977	36.113	38.545
41B	35.417	35.417	35.417	41B	31.25	41.667	36.458
1613C	37.693	37.190	37.442	1613C	37.447	28.677	33.062
Treatment Mean	34.393	34.416		*Treatment Mean*	36.558	35.486	
LSD Treatment: ns				LSD Treatment: ns			
LSD Rootstock: ns				LSD Rootstock: ns			
LSD Treatment x Rootstock: ns				LSD Treatment x Rootstock: ns			
*p < 0.05, **p < 0.01, ns: non-significant							

Table 3. Callusing level at the basal part (0–4).

Manisa Sultanı				*Sultan 7*			
Rootstocks	Control	HWT	*Rootstock Mean*	Rootstocks	Control	HWT	*Rootstock Mean*
5BB	2.70	1.10	1,900a	5BB	2,393a	0,973b	1,683b
41B	2.97	1.84	2,407a	41B	2,580a	2,650a	2,615a
1613C	0.59	0.59	0,593b	1613C	0,647b	0,753b	0,700c
Treatment Mean	2,088a	1,179b		*Treatment Mean*	1,873a	1,459b	
LSD Treatment: 0,835**				LSD Treatment: 0,399*			
LSD Rootstock: 1,022**				LSD Rootstock: 0,685**			
LSD Treatment x Rootstock: ns				LSD Treatment x Rootstock: 0,969**			
*p < 0.05, **p < 0.01, ns: non-significant							

Table 4. Callusing level at the grafting point (0–4).

Manisa Sultanı				*Sultan 7*			
Rootstocks	Control	HWT	*Rootstock Mean*	Rootstocks	Control	HWT	*Rootstock Mean*
5BB	3,777a	3,333b	3,555b	5BB	3,807a	2,303b	3,055c
41B	3,697a	3,940a	3,818a	41B	3,593a	3,623a	3,608b
1613C	3,867a	3,867a	3,867a	1613C	4,000a	3,960a	3,980a
Treatment Mean	3.780	3.713		*Treatment Mean*	3,800a	3,296b	
LSD Treatment: ns				LSD Treatment: 0,242**			
LSD Rootstock: 0,206*				LSD Rootstock: 0,296**			
LSD Treatment x Rootstock: 0,291*				LSD Treatment x Rootstock: 0,419**			
*p < 0.05, **p < 0.01, ns: non-significant							

Table 5. Effects of HWT on shooting development level (0–4).

Manisa Sultanı				*Sultan 7*			
Rootstocks	Control	HWT	*Rootstock Mean*	Rootstocks	Control	HWT	*Rootstock Mean*
5BB	1.900	1.633	1.767b	5BB	2.657 ab	1.733 d	2.195b
41B	2.410	2.247	2.328ab	41B	1.913 cd	2.243 bcd	2.078b
1613C	2.330	3.020	2.677a	1613C	2.553 abc	3.007 a	2.780a
Treatment Mean	2.214	2.300		*Treatment Mean*	2.374	2.328	
LSD Treatment: ns				LSD Treatment: ns			
LSD Rootstock: 0.575*				LSD Rootstock: 0.496**			
LSD Treatment x Rootstock: ns				LSD Treatment x Rootstock: 0.702**			
*p < 0.05, **p < 0.01, ns: non-significant							

Table 6. Effects of HWT on rooting development level (0–4).

Manisa Sultanı				*Sultan 7*			
Rootstocks	Control	HWT	*Rootstock Mean*	Rootstocks	Control	HWT	*Rootstock Mean*
5BB	1.833	1.750	1,792b	5BB	2,500b	1,500c	2,000b
41B	2.833	2.333	2,583a	41B	1,500c	3,000ab	2,250b
1613C	2.500	2.500	2,500a	1613C	2,830b	3,830a	3,333a
Treatment Mean	2.389	2.194		*Treatment Mean*	2,278b	2,778a	
LSD Treatment: ns				LSD Treatment: 0,401*			
LSD Rootstock: 0.529*				LSD Rootstock: 0,689**			
LSD Treatment x Rootstock: ns				LSD Treatment x Rootstock: 0,975**			
*p < 0.05, **p < 0.01, ns: non-significant							

Table 7. Effects of HWT on shoot length (cm).

Manisa Sultanı				*Sultan 7*			
Rootstocks	Control	HWT	*Rootstock Mean*	Rootstocks	Control	HWT	*Rootstock Mean*
5BB	9.897	7.667	8,782b	5BB	14,783a	9,933b	12,358ab
41B	13.210	12.333	12,772ab	41B	9,440b	10,540b	9,990b
1613C	12.610	17.903	15,257a	1613C	13.200ab	15,887a	14,543a
Treatment Mean	11.906	12.634		*Treatment Mean*	12.474	12.120	
LSD Treatment: ns				LSD Treatment: ns			
LSD Rootstock: 4,550**				LSD Rootstock: 2,914**			
LSD Treatment x Rootstock: ns				LSD Treatment x Rootstock: 4,121**			
*p < 0.05, **p < 0.01, ns: non-significant							

Table 8. Effects of HWT on shoot width (mm).

Manisa Sultanı				*Sultan 7*			
Rootstocks	Control	HWT	*Rootstock Mean*	Rootstocks	Control	HWT	*Rootstock Mean*
5BB	1.830	1.507	1.668	5BB	1.973	1.620	1.797
41B	2.010	1.643	1.827	41B	2.143	5.330	3.737
1613C	1.950	2.187	2.068	1613C	1.847	1.920	1.883
Treatment Mean	1.930	1.779		*Treatment Mean*	1.988	2.957	
LSD Treatment: ns				LSD Treatment: ns			
LSD Rootstock: ns				LSD Rootstock: ns			
LSD Treatment x Rootstock: ns				LSD Treatment x Rootstock: ns			
*p < 0.05, **p < 0.01, ns: non-significant							

Table 9. Effects of HWT on root number.

Manisa Sultanı				*Sultan 7*			
Rootstocks	Control	HWT	*Rootstock Mean*	Rootstocks	Control	HWT	*Rootstock Mean*
5BB	12,083c	8,500c	10,292b	5BB	14,000cd	11,183cd	12,592b
41B	27,333ab	13,000bc	20,167b	41B	8,000d	18,667c	13,333b
1613C	27,333ab	41,667a	34,500a	1613C	36,000b	49,667a	42,833a
Treatment Mean	22.250	21.056		*Treatment Mean*	19,333b	26,506a	
LSD Treatment: ns				LSD Treatment: 5,099**			
LSD Rootstock: 10,209**				LSD Rootstock: 6,245**			
LSD Treatment x Rootstock: 14,438**				LSD Treatment x Rootstock: 8,832**			
*p < 0.05, **p < 0.01, ns: non-significant							

The Table 7 shows that effects of hot water treatments on shoot length while the Table 8 shows the effects of hot water treatments on shoot width. Significant treatment x rootstock interactions were observed for the shoot length of Sultan 7 variety. The highest values were observed in hot water treated vines of 1613C/Sultan 7 (15.89 cm) combinations. Also hot water treated cuttings was found 20.3% and 11.7% more shoot length than the untreated ones in 1613C/Sultan 7 and 41B/Sultan 7 combinations respectively. The effect of hot water treatment on shoot width was statistically non-significant for both varieties. However the highest value was observed in hot water treated vines of 41B/Sultan 7 (5.33 mm) and 1613C/Manisa Sultanı(2.19 mm) combinations. Also in previous studies, it was observed that the effects of thermotherapy treatments were varied in terms of shoot length and shoot width on different rootstock x variety combinations [13,20]. In a study carried out by [3], hot water treated and grafted vines of K51-40/Zante Currant and Ramsey/Zante Currant combinations had longer shoots than the control group.

For both varieties, significant treatment x rootstock interaction were observed in terms of root number. Hot water treated vines of 1613C/Manisa sultanıand

DOI: 10.1051/bioconf/20170901003

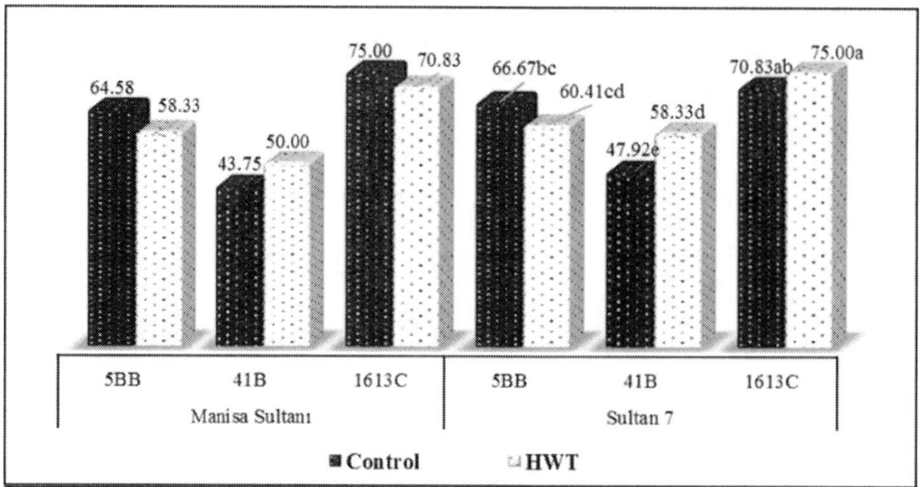

Manisa Sultanı	Sultan 7
LSD Treatment: ns	LSD Treatment: ns
LSD Rootstock: 9,001**	LSD Rootstock: 5,810**
LSD Treatment x Rootstock: ns	LSD Treatment x Rootstock: 8,217**
*p < 0.05, **p < 0.01, ns: non-significant	

Figure 1. Effects of HWT on Final Take value (%).

Table 10. Effects of HWT on fresh weight of shoots (g).

Manisa Sultanı				*Sultan 7*			
Rootstocks	Control	HWT	*Rootstock Mean*	Rootstocks	Control	HWT	*Rootstock Mean*
5BB	45,710c	44,253cd	44,982b	5BB	51,143b	37,330c	44,237b
41B	42,533d	46,363c	44,448b	41B	39,050c	45,143bc	42,097b
1613C	54,640b	57,590a	56,115a	1613C	51,877b	64,267a	58,072a
Treatment Mean	47,628b	49,402a		*Treatment Mean*	47.357	48.913	
LSD Treatment: 1,599 *				LSD Treatment: ns			
LSD Rootstock: 2,747**				LSD Rootstock: 6,132**			
LSD Treatment x Rootstock: 2,770*				LSD Treatment x Rootstock: 8,672**			
*p < 0.05, **p < 0.01, ns: non-significant							

Table 11. Effects of HWT on dry weight of shoots (g).

Manisa Sultanı				*Sultan 7*			
Rootstocks	Control	HWT	*Rootstock Mean*	Rootstocks	Control	HWT	*Rootstock Mean*
5BB	10.050	10.620	10,335b	5BB	11,077b	11,190b	11.133
41B	11.400	11.013	11,207b	41B	10,273b	12,870a	11.572
1613C	14.220	13.890	14,055a	1613C	11,630ab	11,797ab	11.713
Treatment Mean	11.890	11.841		*Treatment Mean*	10,993b	11,952a	
LSD Treatment: ns				LSD Treatment: 0.943**			
LSD Rootstock: 0,972**				LSD Rootstock: ns			
LSD Treatment x Rootstock: ns				LSD Treatment x Rootstock: 1.633**			
*p < 0.05, **p < 0.01, ns: non-significant							

1613C/Sultan7 combinations had higher values compare to the control vines. Root number of grafted vines for the hot water treated 1613C/Manisa sultanıcombination was 41.67 while 1613C/Sultan 7 combination was 49.67 (Table 9). Also it was observed more root numbers with the hot water treatments in 41B/Sulltan7, 1613C/Sultan7 and 1613C/Manisa sultanıcombinations. In a study conducted by [13], there was an increment in the root numbers with the hot water treatments.

According to the measurements of final take, it was found that significant treatment x rootstock interaction for the Sultan 7 variety (Fig. 1). Hot water treated group of 41B/Sultan 7 and 1613C/Sultan 7 combinations had 21.7% and 5.9% more final take than the control group. On the other hand hot water treated vines of 5BB/Sultan 7 combinations had 9.4% lower final take compare to the control group. Generally hot water treatment affected the final take positively in the rootstocks x Sultan 7

DOI: 10.1051/bioconf/20170901003

Table 12. Effects of HWT on fresh weight of roots (g).

Manisa Sultanı				*Sultan 7*			
Rootstocks	Control	HWT	*Rootstock Mean*	Rootstocks	Control	HWT	*Rootstock Mean*
5BB	57.823	55.587	56,705b	5BB	70,187a	63,340ab	66,763a
41B	64.520	64.663	64,592a	41B	42,430e	50,710d	46,570c
1613C	67.047	62.157	64,602a	1613C	52,033cd	59,733bc	55,883b
Treatment Mean	63.130	60.802		*Treatment Mean*	54.883	57.928	
LSD Treatment: ns				LSD Treatment: ns			
LSD Rootstock: 4,172**				LSD Rootstock: 5,607**			
LSD Treatment x Rootstock: ns				LSD Treatment x Rootstock: 7,930**			
*p < 0.05, **p < 0.01, ns: non-significant							

Table 13. Effects of HWT on dry weight of roots (g).

Manisa Sultanı				*Sultan 7*			
Rootstocks	Control	HWT	*Rootstock Mean*	Rootstocks	Control	HWT	*Rootstock Mean*
5BB	6,587c	8,720bc	7,653c	5BB	8,237b	9,300ab	8,237a
41B	9,813b	10,247b	10,030b	41B	4,063d	10,927a	4,063c
1613C	13,140a	10,147b	11,643a	1613C	6,347c	9,697ab	6,347b
Treatment Mean	9.847	9.704		*Treatment Mean*	6,216b	9,974a	
LSD Treatment: ns				LSD Treatment: 1,089**			
LSD Rootstock: 1,568**				LSD Rootstock: 0,951*			
LSD Treatment x Rootstock: 2,218**				LSD Treatment x Rootstock: 1,885**			
*p < 0.05, **p < 0.01, ns: non-significant							

combinations. However it was observed that 5BB/variety combination was more sensitive than the others and this combination was affected adversely by thermotherapy treatment. Additionally it was observed that thermotherapy treatments effects can be changed by the varieties. These findings were similar to the previous studies [8,11,13,19,21].

According to the fresh shoot weight values, significant treatment x rootstock interaction were observed for the Manisa Sultanıand Sultan 7 varieties (Table 10). Hot water treated vines of all combinations, except 5BB/Manisa sultanıand 5BB/Sultan 7 combinations had higher fresh shoot weight values compare to the control group. Additionally hot water treated vines of 1613C/Manisa Sultanıand 1613C/Sultan 7 combinations had the highest fresh shoot weight values for each variety, 57.59 g and 64.27 g respectively. Also significant treatment x rootstock interaction was found for the Sultan 7 variety in terms of dry shoot weight (Table 11). The highest and lowest value was obtained from hot water treated (12.87 g) and untreated (10.27 g) vines of 41B/Sultan 7 combination respectively.

The Table 12 shows that effect of HWT on fresh weight of roots. In Sultan 7 variety, it was determined that significant treatment x rootstock interaction. The highest value was observed in hot water treated vine of 1613C/Sultan 7 combination while the lowest value was obtained from untreated vine of the 41B/Sultan 7 combination. Moreover there were significant treatment x rootstock interactions for the both Manisa Sultanıand Sultan 7 varieties in terms of dry root weight. The highest (10.93g) and lowest value (4.06g) was observed in hot water treated vines and untreated vines of 41B/Sultan 7 combination.

According to the fresh and dry weights of the shoots and roots, generally hot water treatments had positive effects except 5BB x variety combinations. It was observed that 5BB rootstock x variety combination was the most sensitive combination in this experiment. On the other hand there were not determined any adverse effects of thermotherapy treatments on bud and tissue vitality. Also similar results were found in previous studies [3,7,17,22,23].

4. Conclusion

Thermotherapy treatment at 50°C for 30 minutes was found an effective method against *Agrobacterium vitis* in previous studies. Therefore this technique was applied to the dormant grapevine propagation materials before the grafting process in an attempt to examine the growing period of different rootstocks x variety combinations.

It is known that the tolerance of plant material to the hot water treatment is an important matter. Overall the findings reveal that in this study there were no any adverse effects on bud and tissue vitality in Sultan 7 and Manisa Sultanıgrafted onto 5BB, 41B and 1613C rootstocks. On the other hand it was observed that combination of 5BB rootstock with the varieties was more sensitive than the others against to the thermotherapy treatment. Therefore growers should pay attention to this relation between Kober 5BB rootstock and thermotherapy treatment.

Moreover hot water treated vines of 1613C/Sultan 7 and 41B/Sultan 7 combinations became prominent in terms of final take. As can be seen from the results hot water treated cuttings of Sultan 7, which is a raisin variety, has stood out with its high final take values for all rootstocks.

In addition, considering that 41B is a hard-rooted rootstock and final take values are generally low, it is so important that the thermotherapy has a positive effect on final take of 41B rootstock and variety combinations.

As a result, in order to grow healthy grafted vines and provide economic technique against *Agrobacterium vitis* in Sultan 7 and Manisa sultanıvarieties - newly registered varieties which were obtained from the Sultani Çekirdeksiz - thermotherapy treatment needs to be widespread.

References

[1] FAO, 2014. Accessed date: 20.04.2017
[2] C. Bauer, T.F. Schulz, D. Lorenz, K.W. Eichhorn, R. Plapp, Vitis **33**, 25–29 (1994)
[3] K. Ophel and A. Kerr, Internat. Journ of Systematic and Evolutionary Microbiology **40**, 236–241 (1990)
[4] T.J. Burr, and B.H. Katz, *Phytopathology* **73**, 163–165 (1983)
[5] T. Martinson and T. Burr, Research news from Cornell's Viticulture and Enology Program, Research Focus **1** (2012)
[6] Anonymous, Official gazette, (Accessed date: 17 Ocak 2016) (2011)
[7] T.J. Burr, K. Ophel, B.H. Katz, A. Kerr, Plant Dis. **73**, 242–245 (1989)
[8] C. Bazzi, E. Stefani, R. Gozzi, T.J. Burr, C.L. Moore, A. Anaclerio, Vitis **30**, 177–187 (1991)
[9] T.J. Burr, C.L. Reid, D.F. Splittstoesser, M. Yoshimura, Am. J. Enol. Vitic. **47**, 119–123 (1996)
[10] Caudwell, J. Larrue, E. Boundon-Padieu, G.D. Mclean. Aust. J. Grape Wine Res. **3**, 21–25 (1997)
[11] C. Ilgın, and Y.Z. Gürsoy, TAGEM, 6.Türkiye Bağcılık Semp. **I**, 114–120 (2005)
[12] F. Halleen, P.H. Fourie, P.W. Crous, Plant Pathol. **56**, 637–645 (2007)
[13] E. Kacar, B. Işçi, A. Altindişli, Bulletin de l'OIV **85**, 974 (2012)
[14] D. Le Roux, Farming in South Africa, pamphlet VORI, 212 (1988)
[15] İ. Korkutal, G. Kaygusuz, S. Bayram, Afr. J. Biotech. **10**, 15123–15129 (2011)
[16] N. Sivritepe and C. Türkben, Uludağ Üniv. Ziraat Fak. Derg. **15**, 47–58 (2001)
[17] H. Mahmoodzadeh, A. Nazemieh, I. Majidi, I. Paygami and A. Khalighi, J. Phytopathology. **151**, 481–484 (2003)
[18] İ. Yavaş and Y. Fidan, Türkiye 1. Fidancılık Sempozyumu. Ankara. 79–84 (1991)
[19] O. Soltekin, Y. Savaş, E.T. Özcan, E. Kacar, Turkish Journal of Agricultural and Natural Science **4**(1), 30–39 (2017)
[20] K. Ophel, T.J. Burr, P.A. Magarey, A. Kerr, Australasian Plant Pathol. **17,** 61–6 (1988)
[21] R.L. Wample, A. Bary, T.J. Burr, Am. J. Enol. Viticult. **42**, 67–72 (1991)
[22] P.G. Goussard, Vitis **16**, 272–278 (1977)
[23] C.J. Orffer and P.G. Goussard, Vitis. **19**, 1–3 (1980)

Chitosan and grape secondary metabolites: A proteomics and metabolomics approach

Luigi Bavaresco[1,a], Maurizio Zamboni[1], Cecilia Squeri[1], Shuying Xu[1], Anna Abramowicz[1], and Luigi Lucini[2]

[1] Dept. of Sustainable Crop Production, Pomology and Viticulture section, Università Cattolica del Sacro Cuore, Via Emilia Parmense, 84, 29122 Piacenza, Italy
[2] Institute of Agricultural and Environmental Chemistry, Università Cattolica del Sacro Cuore, Via Emilia Parmense, 84, 29122 Piacenza, Italy

Abstract. Chitosan is a polysaccharide obtained by deacetylation of chitin, and it is involved in defence mechanisms of plants toward diseases. In the present work, *V. vinifera* L. cv. Ortrugo, grafted on 420A rootstock was grown in pot and treated, at veraison, by 0.03% chitosan solution at cluster level. Just before the treatment (T0) and 24 hours (T1), 48 hours (T2), 72 hours (T3) and 10 days (T4) later, the concentration of stilbenic compounds was detected, and at T1 proteomics and metabolomics analyses were done. Proteomics relies on the analysis of the complete set of proteins existing in a given substrate, while metabolomics relies on the analyses of the complete set of metabolites in a given substrate. The treatment improved the stilbene concentration over the control at T1. Proteomic analysis showed that superoxide dismutase (SOD) and phenylalanine ammonia-lyase (PAL) were overexpressed in the treated grapes. SOD is known to be an enzyme active against reactive oxygen species (ROS) while PAL is a key enzyme in the phenylpropanoids pathway. Metabolomics analysis highlighted the positive role of the treatment in improving the triperpenoid concentration (betulin, erythrodiol, uvaol, oleanolate); these compounds are known to be effective against microbes, insects and fungi.

1. Introduction

The importance of chitosaccharides, as plant growth promoters and disease control agents has been recently emphasized [1,2]. Chitosan, a linear D-(1,4)-glucosamine polymer produced by deacetylation of chitin and an important structural component of several plant fungi cell walls, can be applied to plants as antimicrobial agent (against fungi, bacteria and viruses) and as elicitor of plant defence mechanisms [3,4]. As concerning the latter, evidences have been obtained on the stimulation of phenylalanine ammonia lyase (PAL), peroxidase and lopoxygenase activities as well as the accumulation of phytoalexins and PR proteins [5–7]. Their biological activity is most likely resulting from the binding to membrane receptors and it depends on the molecular weight and degree of N-acelylation of the molecule [8,9]; chitosan oligomers of low molecular weight were shown to be more effective in inducing defence responses than those of higher molecular weight. Moreover chitosan has been reported to enhance the systemic resistance [1].

V. vinifera plantlets and leaves treated *in vitro* by chitosan and infected by *Botrytis cinerea* and *Plasmopara viticola*, did not show any disease symptom [10,11]. Chitosan showed to be effective during table grape post harvest storage, by replacing the SO_2 treatments, resulting in the control of fungal diseases [12,13].

The aim of the research was to investigate the role of cluster treatments with chitosan on the plant (grapes) response in terms of enzymes regulation and secondary metabolites production, by applying new high performing technologies such as proteomics and metabolomics.

2. Materials and methods

Plant material. *V. vinifera* L. cv. Ortrugo clone PC ORT 81 (white berry color) grafted on *V. berlandieri* x *V. riparia* 420A rootstock was grown on pot (70 L volume) containing sandy soil. The vines were cane pruned and trained vertical shoot positioning (VSP) with 10 buds/vine. The pots were placed outdoor on a platform covered by a hail protection net, basic fertilizer supply was provided at bud burst and the soil was kept near field capacity by drip irrigation.

Treatments. Chitosan from shrimp shell, low viscosity, purchased from Sigma-Aldrich (50494-100G-F), was utilized for the spray treatments. After the preparation of the mother solution with acetic acid, which was kept overnight at −20° C, a 0.03% chitosan solution was sprayed at cluster level at veraison, by treating all the clusters of 8 vines; untreated vines (8) were kept as control, by spraying the clusters only with 0.1% acetic acid.

Tests. The following tests were performed, on both control and treated clusters, at time 0 (corresponding to veraison, just before the treatment), time 1 (24 h after the treatment), time 2 (48 h after the treatment), time 3 (72 h after the treatment), time 4 (10 days after the treatment): total soluble solids (° Brix), pH, stilbenic compounds; only at T1 proteomics and metabolomics analyses were done.

[a] e-mail: luigi.bavaresco@unicatt.it

Stilbenes were extracted from frozen berries by the method of Bavaresco et al. [14] by using methanol 95% and phase partitioning; a 1200 series liquid chromatograph system, equipped with a quaternary pump and an electrospray ionization system, and coupled to a G6410A triple quadrupole mass spectrometer detector (Agilent, Santa Clara, CA, USA) was used. Chromatographic separation was performed by a Synergy Fusion C 18 column from Phenomenex. The LC mobile phase A consisted of 0.05% ammonium formate in water, while mobile phase B was 0.05% ammonium formate in methanol. The gradient started with 50% B, to be increased till 75% at 5 minutes, 80% B at 7.5 minutes, 85% B till 20 minutes; the flow was 0.2 mL/min, at a temperature of 45° C. The electrospray source was set using N at 340° C, flowing at 10 L/min, and operating in the negative ionization mode (Vcap = 4000 V).

Proteomics. After extraction of the proteins from the berry skins using TCA/acetone, followed by reduction and alkylation, the analysis of tryptic peptides was performed by using a shotgun MS/MS approach, by a hybrid quadrupole-time-of-flight (Q-TOF) mass spectrometer coupled to a nano LC Chip Cube source. MS/MS spectra analysis of peptide identification allowed protein inference, against the proteome of *V. vinifera*, using Spectrum Mill MS Proteomics Workbench.

Metabolomics. An untargeted screening was performed via high-resolution mass spectrometry by using a hybrid Q-TOF spectrometer coupled to an UHPLC chromatographic system. MS acquisition was performed in positive mode, in the range 100–1200 m/z and compounds identification carried out according to isotopic patterns (accurate mass, isotopic spacing, isotopic ratio).

Partial Least Square-discriminant analysis (PLS-DA) was the statistical approach used for both proteomics and metabolomics, and differential metabolites/proteins were then exported from covariance structures.

Table 1a. Resveratrol and resveratroloside in grapes, depending on the sampling time and the treatment.

Sampling Time	Tr.	Resveratrol (μg/ kg f.w.)	Resveratroloside (μg/ kg f.w.)
T0	C.	25 a	344 a
	T.	12 a	224 a
T1	C.	16 a	223 b
	T.	18 a	452 a
T2	C.	17 a	360 a
	T.	15 a	370 a
T3	C.	24 a	397 a
	T.	13 a	426 a
T4	C.	16 a	352 a
	T.	13 a	371 a

Tr.: Treatment; C.: Control; T.: Treated. Letters represent comparison between control and treated grapes per each sampling time. Values followed by a different letter are significantly different at $p < 0.05$.

Table 1b. Piceid and Pterostilbene in grapes, depending on the sampling time and the treatment.

Sampling Time	Tr.	Piceid (μg/ kg f.w.)	Pterostilbene (μg/ kg f.w.)
T0	C.	1417 a	83 a
	T.	1405 a	90 a
T1	C.	1430 a	92 a
	T.	1480 a	74 a
T2	C.	1332 a	68 a
	C.	1238 a	102 a
T3	T.	1482 a	73 a
	C.	1259 a	67 a
T4	T.	1298 a	65 a
	C.	738 a	64 a

Tr.: Treatment; C.: Control; T.: Treated. Letters represent comparison between control and treated grapes per each sampling time. Values followed by a different letter are significantly different at $p < 0.05$.

3. Results and discussion

TSS increased from 11.4° Brix at T0 till 16.2° Brix at T4, while pH changed from 2.78 (T0) to 3.08 (T4), as expected (data not shown).

As concerning the stilbenes, resveratrol, resveratroloside, piceid and pterostilbene were detected, and only at T1 the treatment induced a significantly higher concentration of resveratroloside, while no statistical differences were calculated at the other sampling times and for the other compounds (Table 1a, Table 1b).

Proteomics results are reported in Table 2, where proteins expressed at a score (WC0) above 0.2 and below −0.2 are listed since they are considered the most prominent for class prediction. The proteins with WC0 above 0.2 were more expressed in the control (untreated) samples, whilst the proteins with WC0 below −0.2 were more expressed in the treated samples. According to Table 2 the following proteins were more expressed under chitosan spray treatment: superoxide dismutase, proteasome subunit beta type, nascent polypeptide-associated complex subunit beta, ATP-dependent Clp protease proteolytic subunit, phenylalanine ammonia-lyase (PAL). Each of those is related to biochemical function (pathways) which can be identified by screening the Uniprot database. For instance the treatment was involved in oxidative stresses (superoxide dismutase increasing, providing an action against ROS) and in the phenylpropanoid pathway through PAL enzyme. Phenylpropanoid pathway is involved in the biosynthesis of flavonoids and stilbenes.

The main metabolomics results are reported in Table 3, where it is shown that some triterpenoids (botulin, erythrodiol, uvaol, oleanolate) increased very much under the chitosan treatment (up to 65 fold change). Triterpenoids including steroids are a highly diverse group of natural products widely distributed in plants, including grapevine [15]. Plants often accumulate these compounds in their glycosylated form – saponin, which are considered defensive compounds against pathogenic microbes, insects or fungi [16]. Triterpenes are synthesized via mevalonate pathway [17] and their accumulation is known to be triggered by biotic and abiotic stresses [18].

4. Conclusions

Chitosan spray treatment at cluster level was able to enhance phenylpropanoid and mevalonate pathways, while further investigations are needed to check the role of the metabolites produced from those pathways on the defence reactions of the vine toward diseases.

Table 2. List of differential proteins.

PLS-DA important proteins		
Compound	**WC0**	**Mass**
Peroxidase 4	0.217184	34516.2
Superoxide dismutase	−0.22816	31569.7
Pathogenesis-related protein 10	0.266685	17356.5
Cytochrome b-cl 7 complex subunit 7	0.211336	14706.9
Chalcone-flavonone isomerase 1	0.214391	25310.5
Aspartate aminotransferase	0.271999	45286.1
Proteasome subunit beta type	−0.23107	23205.3
Nascent polypeptide-associated complex subunit beta	−0.2325	17173.2
ATP-dependent Clp protease proteolytic subunit	−0.24473	31202.3
Sucrose synthase	0.269306	92937.3
Phenylalanine ammonia-lyase	−0.22497	78591.9
Carotenoid cleavage dioxygenase 1	0.262366	61965.1
Chalcone-flavonone isomerase family protein	0.244305	27747.5
Ribulose bisphosphate carboxylase small chain	0.244345	20713.5
Serine/threonine-protein kinase	0.24413	98336.8

Table 3. Enhanced accumulation of triterpenoids shows fold change in treatment samples as compared to control.

Fold-change	
Compound	**FC (abs) ([treated] vs [control]): Raw**
Betulin	64.67152
Erythrodiol	64.67152
Uvaol	64.67152
Oleanolate	11.47518

The work is a part of Master theses (International Master of Science Erasmus Mundus VINTAGE) developed by Shuying Xu and Anna Abramowicz.

References

[1] G. Agrawal, R. Rakwal, S. Tamogami, M. Yonekura, A. Kubo, H. Saji, Stilbene content of mature *Vitis vinifera* berries in response to UV-C elicitation. J. Agric. Food Chem. **48** 6103–6105 (2000)

[2] E. At Barka, P. Eullaffroy, C. Clément, G. Vernet, Chitosan improves development and protects *Vitis vinifera* L. against Botrytis cinerea. Plant cell rep. **22** 608–614 (2004)

[3] E. I. Rabaea, M. E. T. Badway, C. Stevens, G. Smagghe, W. Steurbaut, Chitosan as antimicrobial agent: applications and mode of action. Biomacromolecules **4**, 1457–1465 (2003)

[4] M. Iriti, M. Sironi, S. Gomarasca, A. P. Casazza, C. Soave, F. Faoro, Cell death-mediated antiviral effect of chitosan in tobacco. Plant Physiol. Biochem. **44**, 893–900 (2006)

[5] M. M. Chang, L.A., D. Horowitz, Molecular characterization of a pea beta-1,3-glucanase induced by *Fusarium solon* and chitosan challenge. Plant Pathol. **44**, 289–299 (1992)

[6] L.A. Hadwiger, P. Jolles, R. A. A. Muzzare, Chitin and Chitosanes. Birkhauser Verlag, Basel, Switzerland, pp. 185–200 (1999)

[7] S. H. Doares, T. Syrovets, E. Weiler, C. A. Ryan, Oligalacturonides and chitosan activate plant defensive genes through the octadecanoid pathway. Proc. Natl. Acad. Sci. USA **92**, 4095–4098 (1995)

[8] H. Kauss, W. Jeblick, A. Domard, The degrees of polymerization and N-acetylation of chitosan determine its ability to elicit callose formation in suspension cells and protoplasts of Catharanthus roseus. Planta **178**, 385–392 (1989)

[9] P. Vander, K. M. Varum, Comparison of the ability of partially N-acetylated chitosans and chitooligosaccharides to elicit resistance reactions in wheat leaves. Plant Physiol. **118**, 1353–1359 (1998).

[10] A. Bézier, B. Lambert, F. Baillieul, Study of defense-related gene expression in grapevine leaves and berries infected with *Botrytis* cinerea. Eur. J. Plant Pathol. **108**, 111–120 (2002)

[11] A. Aziz, P. Trotel-Aziz, P. Dhuicq, Chitosan oligomers and coper sulfate induce grapevine defense reactions and resistance to gray mold and downy mildew. Phytopathology **96**, 1188–1194 (2006)

[12] G. Romanassi, F. Nigro, A. Ippolito, D. Di Venere, M. Salerno, Effects of pre and post-harvest chitosan tratments to control storage grey mould of table grapes. Journal of Food Science **67**, 1862–1867 (2002)

[13] G. Romanazzi, E. Feliziani, S. Bautista Baños, D. Sivakumar, Shelf life ectension of fresh fruit and vegetables by chitosan treatment. Critical Reviews in Food Science and Nutrition **57**, 579–601 (2017)

[14] L. Bavaresco, D. Pettegolli, E. Cantù, M. Fregoni, G. Chiusa, M. Trevisan, Elicitation and accumulation of stilbene phytoalexins in grapevine berries infected by Botrytis cinerea. Vitis **36**, 77–83 (1997)

[15] F. Pensec, A. Szakiel, C. Paczkowski, A. Woźniak, M. Grabarczyk, Ch. Bertsch, J. Marc, C. Fisher, J. Chong, Characterization of triterpenoid profiles and triterpene synthase expression in the leaves of eight *Vitis* vinifera cultivars grown in the Upper Rhine Valley. J Plant Res **129**, 499–512 (2016)

[16] J. M. Augustin, V. Kunzina, S. B. Andersen, B. Bak, Molecular activities, biosynthesis and evolution of triterpenoid saponins. Phytochemistry **72**, 435–457 (2011)

[17] J. Chappell, The genetics and molecular genetics of terpene and sterol origami. Curr. Opin. Plant Biol. **5**, 151–157 (2002)

[18] R. Thimmappa, K. Geisler, T. Louveau, P. O'Maille, A. Osbourn, Triterpene biosynthesis in plants. Annu. Rev. Plant Biol. **65**, 225–257 (2014)

Renewable energy and hydrogen on-site generation for irrigation and mobility in vineyards / energía renovable e hidrógeno generados in situ para riego y movilidad en viñedos

Javier Carroquino[1], Nieves García-Casarejos[2], Pilar gargallo[2], Francisco-Javier García-Ramos[3], y Jesús Yago[4]

[1] Universidad de Zaragoza, Departamento de Ingeniería Eléctrica, María de Luna 3, 50018 Zaragoza, España
[2] Universidad de Zaragoza, Facultad de Economía y Empresa, Doctor Cerrada 1-3, 50005 Zaragoza, España
[3] Universidad de Zaragoza, Departamento de Ciencias Agrarias y del Medio Natural, Escuela politécnica Superior, Carretera de Cuarte s/n, 22071 Huesca, España
[4] Intergia energía sostenible S.L., Tomás Higuera 19, 50013 Zaragoza, España

Abstract. The energy used in agriculture is a source of greenhouse gases. In order to reduce the emissions produced by the use of diesel and electricity, the European project LIFE REWIND addresses the supply from renewable resources. Firstly, the energy demand of the vineyard and winery have been characterized. Secondly, the optimal design of generation has been sought, both from an economic and emission reduction point of view. A prototype with photovoltaic generation has been installed in a vineyard, including a set of solar panels floating on an irrigation pond. This avoids the use of terrain, reduces the loss of water from the pond and improves the performance of photovoltaic panels. The electricity produced feeds the wastewater treatment plant of the cellar and the pumps for drip irrigation of the vineyard. With the surplus energy, an electrolyser produces hydrogen by electrolysis of the water, which is used by an off-road vehicle with a fuel cell. The results show the feasibility of renewable energy systems in the vineyards. This avoids both the indiscriminate extensions of the electric grid and the use of diesel generator sets. Furthermore, the ability of replacing diesel for electricity or hydrogen produced on the farm itself is shown.

Resumen. La energía utilizada en agricultura es fuente de emisiones de gases de efecto invernadero. Para reducir las producidas por el uso de gasóleo y electricidad, el proyecto europeo LIFE REWIND aborda el suministro desde fuentes renovables. Para ello, se ha caracterizado la demanda energética y recursos renovables en viñedo y bodega. Se ha buscado el diseño óptimo económico y medioambiental de la generación. Se ha instalado un prototipo en una explotación vitivinícola, con generación fotovoltaica, incluyendo un conjunto flotante sobre una balsa. Esto evita el uso de terreno, reduce la pérdida de agua de la balsa y mejora el rendimiento de los paneles fotovoltaicos. La electricidad producida alimenta la depuración de agua y el bombeo de riego por goteo del viñedo. Con la energía sobrante, un electrolizador produce hidrógeno mediante electrólisis del agua, que es utilizado por un vehículo todoterreno con pila de combustible. Los resultados muestran la viabilidad técnica y económica de los sistemas autónomos de energía renovable para alimentar el riego estacional en los viñedos. Así se evitan las extensiones indiscriminadas de la red y los grupos electrógenos diésel. Adicionalmente se muestra la posibilidad de reemplazar el gasóleo por electricidad o hidrógeno producido en la propia explotación.

1. Introducción

1.1. El cambio climático y el sector vitivinícola

La búsqueda de sostenibilidad es una tendencia necesaria para consolidar el futuro de cualquier actividad económica, lo que requiere un enfoque transversal y multidisciplinar [1]. La viticultura y en general la industria del vino están empezando a verse afectadas por el cambio climático, cuyos impactos pueden requerir en un futuro próximo importantes medidas de adaptación [2]. En línea con la diversidad de regiones vitivinícolas, con sus particulares condiciones y diferenciados vinos producidos, los impactos pueden ser diferentes y merecen ser estudiados [3–6]. Sin embargo, probablemente pocos productos sean tan sensibles como el vino a las variaciones del clima y por ello es conveniente que las empresas del sector no sólo estén preparadas para su adaptación, sino que también contribuyan en lo posible en las acciones de mitigación, tales como la reducción de emisiones de gases de efecto invernadero. Estas acciones pueden abordarse de diversas formas, requiriéndose un conocimiento que debería ser aportado y compartido dentro del sector [7], ya que su aplicación internacional beneficiaría al conjunto del mismo [8].

Otra consideración a tener en cuenta es el paulatino aumento de la importancia de la sostenibilidad del producto en la decisión de compra de los consumidores [9], máxime en un producto como el vino.

Por otra parte, además de los muchos trabajos acerca de suministro de energía renovable para bombeos de riego, se ha avanzado en su estudio concreto para los cultivos mediterráneos, entre los cuales se encuentra la vid [10].

1.2. El proyecto LIFE REWIND

La energía que se utiliza en agricultura es una fuente de emisiones de gases de efecto invernadero. Para reducir esas emisiones, el proyecto europeo LIFE REWIND aborda el suministro de esta demanda energética desde fuentes renovables. Su nombre completo es "Sistemas rentables de energía renovable de pequeña escala en la industria agroalimentaria y las áreas rurales: una demostración en el sector vitivinícola". Su acrónimo procede del nombre corto en inglés, *"Renewable Energy in the Wine INDustry"*, dando como resultado "REWIND". El programa LIFE es el instrumento financiero de la Unión Europea de apoyo a los proyectos medioambientales, de conservación de la naturaleza y de acción por el clima. Concretamente, LIFE REWIND tiene una duración de 37 meses y un presupuesto de 1.562.994 €, cofinanciado por la Comisión Europea.

Aunque el proyecto es de aplicación más amplia en el sector agrario, se ha tomado como caso de estudio y demostrador el sector vitivinícola. Su gran extensión e importancia en la Europa meridional, así como el ser la vid uno de los más importantes cultivos mediterráneos, hace de la viticultura un excelente modelo. Más aún dada la sensibilidad de la vid al cambio climático [11]. Así, el objetivo del proyecto LIFE REWIND es demostrar que el uso de energías renovables es viable técnica, medioambiental y económicamente, así como promover su uso. De esta forma, contribuye al cumplimiento de los objetivos medioambientales de la Unión Europea.

2. Metodología

2.1. Recogida de datos del sector

La obtención de datos agregados del sector se ha hecho mediante consultas a fuentes estadísticas, europeas y españolas. Para la obtención de datos específicos para su tratamiento en el ámbito del proyecto, se ha elaborado un cuestionario destinado a propietarios y operadores de instalaciones vitivinícolas, que se ha cumplimentado por una muestra representativa de empresas del sector vitivinícola, pertenecientes a todas las regiones españolas. Igualmente, aunque en menor número, se han mantenido entrevistas personales y se han obtenido datos históricos detallados de costes energéticos, así como datos cualitativos de interés. También se han efectuado mediciones in situ en diversas instalaciones y ubicaciones.

En particular, las mediciones y tomas de datos más detalladas se efectuaron en las instalaciones donde se desarrollaría la acción demostradora y se ubicarían los prototipos.

2.1.1. Demanda energética

Para afrontar el estudio del suministro de energía renovable al sector, es necesario caracterizar su demanda energética. Con ese objetivo se han recabado datos de diferentes orígenes, indicados en la Tabla 1.

Desde un primer momento se identificaron los dos ámbitos, viñedo y bodega, donde se produce no sólo la actividad, sino también una demanda de energía bien diferenciada. En consecuencia de ello, todos los análisis y estudios posteriores se hicieron de forma diferenciada para ambos entornos.

Tabla 1. Datos de demanda.

Origen	Tipo de dato
Facturas del suministro eléctrico	Históricos de consumo eléctrico
	Coste de la electricidad
	Estacionalidad de la demanda
Albaranes y anotaciones del usuario	Históricos de consumo de gasóleo
	Costes de mantenimiento
Precios de mercado	Coste de la instalación
Entrevistas y encuestas	Gestionabilidad de la demanda
	Criterios de gestión
Mediciones in situ	Potencias y consumos

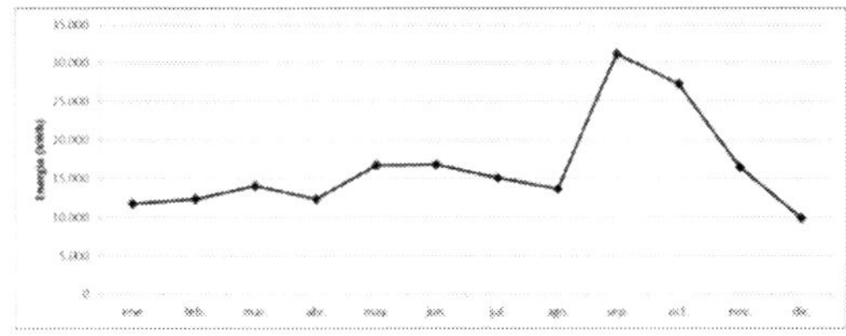

Figura 1. Consumo eléctrico anual de una de las bodegas estudiadas.

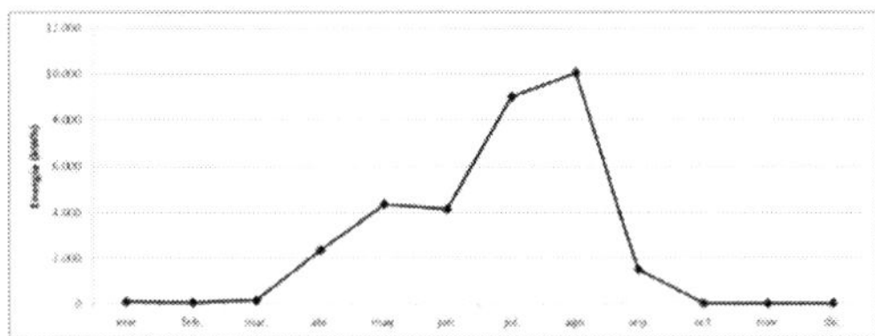

Figura 2. Consumo eléctrico anual de uno de los bombeos de riego estudiados.

La demanda de electricidad en las bodegas ha resultado ser relativamente estable a lo largo del año, con la excepción de un período próximo y posterior a la vendimia y procesos de refrigeración. Adicionalmente se han encontrado variaciones derivadas de consumos de calefacción o climatización de zonas de trabajo y oficinas, en aquellos casos que dichos consumos se alimentaban de electricidad. La Fig. 1 muestra el consumo eléctrico anual, mes a mes, de una de las bodegas estudiadas.

La demanda de electricidad en los viñedos sólo se ha considerado en los casos en que había bombeos de agua para riego, que en todos los casos era por goteo. Estos bombeos podían ser de extracción/elevación, de de presurización o de ambos tipos. Su demanda ha resultado ser muy estacional, nula durante una parte del año y concentrada obviamente en la temporada de riego. La Fig. 2 muestra el consumo eléctrico anual, mes a mes, de uno de los sistemas de riego de viñedo estudiados.

Un dato relevante a tener en cuenta en los sistemas de riego es la existencia o no de una balsa de acumulación de agua. En caso de existir es frecuente la existencia de dos bombeos, uno para alimentar agua a la balsa y otro para impulsar el agua desde la balsa al sistema de riego presurizado. En este caso, la existencia de dos bombeos y una balsa de acumulación permite un alto grado de gestionabilidad, accionando una bomba, la otra o ambas de forma relativamente independiente.

DOI: 10.1051/bioconf/20170901005

Figura 3. Estación de medida autnoma ubicada en viñedo.

2.1.2. Recursos renovables

Una vez identificadas las áreas geográficas con presencia de explotaciones vitícolas, se obtuvieron los datos de recurso eólico y solar disponibles, desde las fuentes indicadas en la Tabla 2.

Tabla 2. Datos de recursos renovables.

Origen	Tipo de dato
Sistemas de información geográfica fotovoltaica	Recurso solar
Bases de datos bibliográficas	Recurso eólico
Campañas de mediciones con sensores in situ	Recurso solar
	Recurso eólico
	Temperatura

En varios de los casos de estudio se llevó a cabo una campaña de medición in situ de los recursos eólico y solar, mediante la instalación de estaciones de medida, dotadas de dos anemómetros, veleta y piranómetro. Tres de las estaciones (Fig. 3) eran completamente autónomas, ya que se alimentaban de energía solar y enviaban los datos a un servidor central mediante comunicación GSM. El resto de las estaciones requerían visitas periódicas, ya que estaban alimentadas a pilas y grababan la información en tarjetas de memoria.

Los anemómetros fueron colocados para la medición del viento a baja altura, ya que por el orden de magnitud de la energía necesaria, la posible generación eólica a instalar sería del tipo minieólico, muy diferente de los grandes aerogeneradores para inyección a red. La campaña de medición se prolongó durante más de tres años.

Los datos de recurso eólico resultaron ser muy diferentes (Fig. 4) de una ubicación a otra, aun a distancias menores de 150 km, lo que es coherente con la gran influencia que ejercen el relieve, la rugosidad del suelo y los obstáculos del terreno. También se observó una gran variación de un año a otro, así como la existencia de períodos de calma, con ausencia de viento a veces durante más de una semana.

En cuanto al recurso solar, resultó ser muy similar (Fig. 5) entre las ubicaciones de las diversas estaciones de medida, aún a distancias de hasta 150 km. Esto es coherente con que el recurso solar depende principalmente de la latitud geográfica y en menor medida de otros factores, como la altitud, la claridad atmosférica y el albedo del terreno.

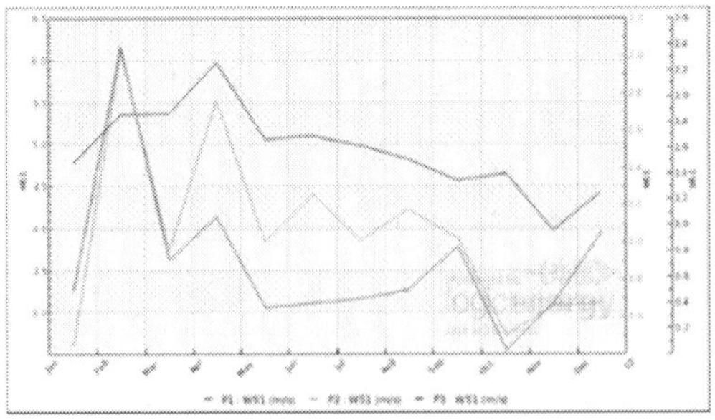

Figura 4. Recurso eólico en gráfica anual, medido en tres estaciones distantes entre sí menos de 150 km.

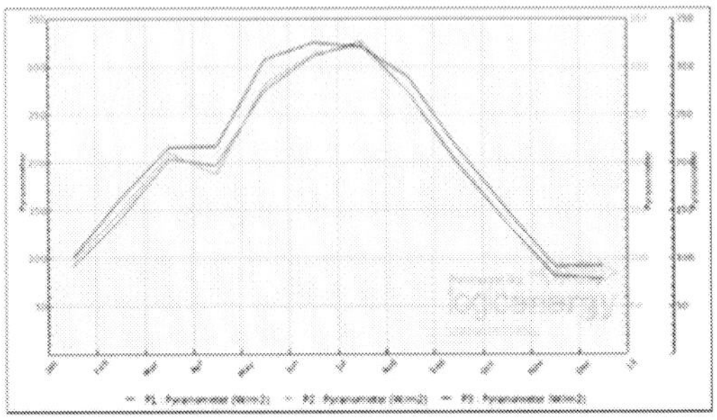

Figura 5. Recurso solar en gráfica anual, medido en tres estaciones distantes entre sí menos de 150 km.

El ámbito de estudio de los recursos renovables se limitó al de las zonas europeas de mayor presencia del sector vitivinícola, donde consecuentemente con el clima y la latitud geográfica se encontró una gran variedad de recurso eólico y una relativa uniformidad del recurso solar.

2.2. Búsqueda de soluciones óptimas

Con los datos obtenidos de demanda energética de bodegas y viñedos, se realizaron mediante ordenador numerosas simulaciones de sistemas de generación renovable. La búsqueda de soluciones óptimas se hizo por métodos heurísticos, empleando el software de simulación y optimización iHOGA [12]. Este programa utiliza algoritmos genéticos para determinar el dimensionado de sistemas de generación renovable que minimicen el coste de la energía, las emisiones de CO_2 o ambos criterios.

En el proceso de optimización mono objetivo, se buscan los mínimos de una única función, como puede ser el coste de la energía. En la Fig. 6 se muestra una gráfica de iHOGA al término de una optimización mono objetivo, con las soluciones ordenadas por su coste de energía (línea roja). El coste de la energía se ha comparado según su LCE (*Levelized Cost of Energy*) [13].

También pueden buscarse las configuraciones de sistemas de energía con mínimas de emisiones de CO_2. Sin embargo, si no se tiene en cuenta su coste pueden resultar soluciones no competitivas económicamente y por lo tanto de difícil aceptación en la práctica. Para evitar esto, se recurre a la optimización multiobjetivo, teniendo en cuenta ambos criterios: coste de la energía y emisiones de CO_2 [14]. Este proceso da lugar a un número limitado de soluciones que pueden ser mostradas en un diagrama de Pareto (Fig. 7), eligiendo luego de entre ellas aquella que se considere mejor compromiso entre costes y emisiones.

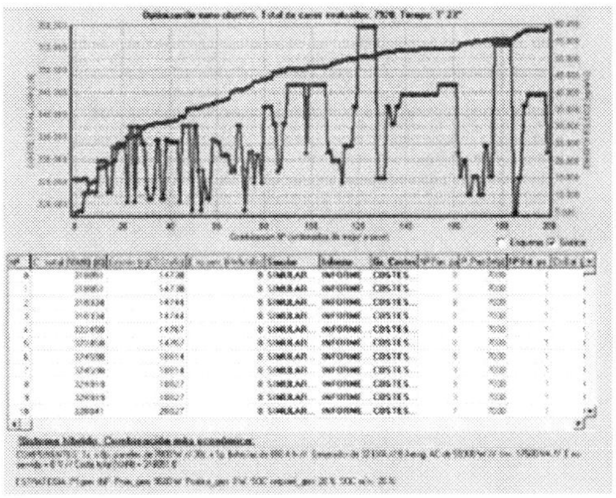

Figura 6. Gráfica de soluciones de iHOGA al término de una optimización mono objetivo (coste de energía).

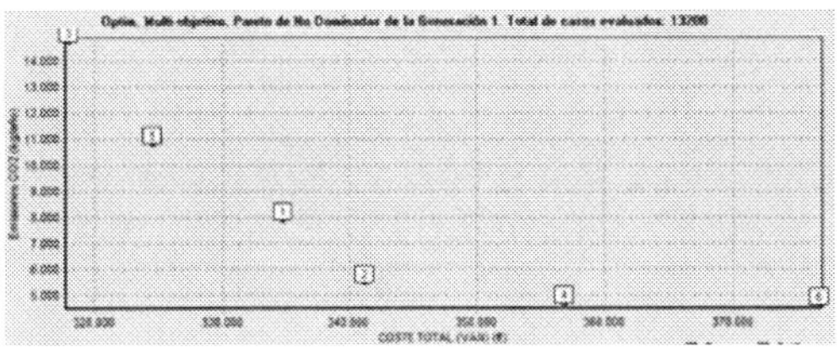

Figura 7. Diagrama de Pareto obtenido en optimización multiobjetivo con iHOGA.

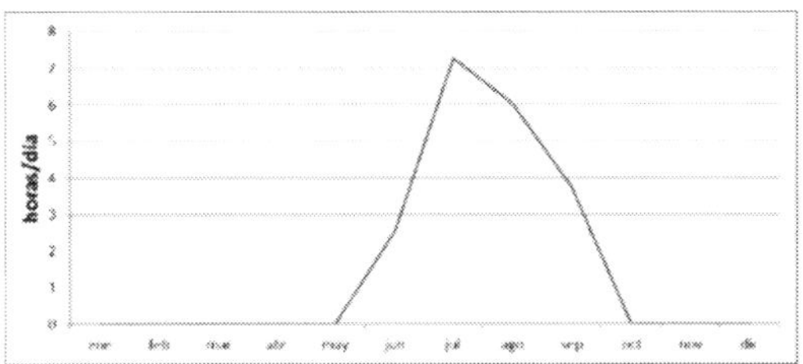

Figura 8. Perfil anual de la demanda hídrica del riego del viñedo.

Con los procedimientos descritos se han estudiado los dimensionados y configuraciones de diferentes casos de estudio y en particular el caso donde se han instalado los prototipos.

2.3. Implementación de los prototipos

Los prototipos se han definido con una función demostrativa. Conforme a lo anteriormente expresado, se ha estudiado la demanda de energía eléctrica de la bodega y viñedos de Viñas del Vero, situados en la región vitivinícola del Somontano en Barbastro (España).

La bodega, como sucede en la mayor parte de los casos, obtiene su electricidad de la red de distribución eléctrica. En un caso general, la configuración adoptada en bodega hubiera consistido en un conjunto de generación fotovoltaica en autoconsumo conectado a la red, de tal forma que aportase una fracción del total de la electricidad consumida por la bodega. Sin embargo, para un mayor efecto demostrador y también a causa de la inapropiada normativa española sobre autoconsumo, se optó por identificar un consumo concreto y alimentarlo al 100% con energía renovable, sin conexión a la red. La instalación

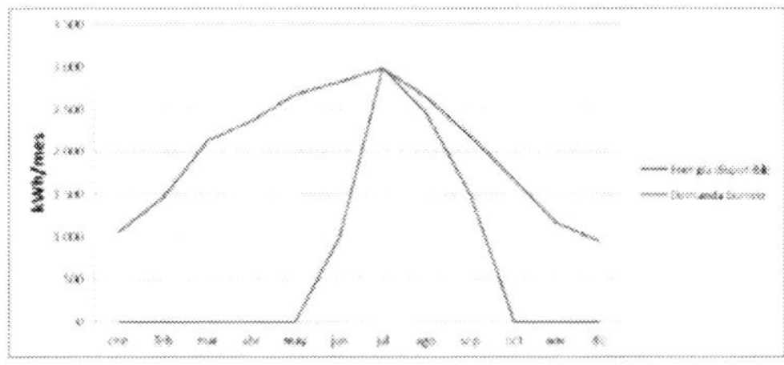

Figura 9. Gráficas anuales de energía disponible y energía demandada.

escogida fue la estación que depura el agua utilizada en la propia bodega, para su posterior uso en riego, cerrando así el ciclo de sostenibilidad del agua.

De entre los diversos sistemas de riego por goteo existentes en los viñedos se escogió el más próximo a la bodega, que toma el agua de la ya mencionada estación depuradora e incluye un bombeo de elevación hasta la balsa de riego y otro de presurización al sistema de goteo.

La demanda en horas de funcionamiento de este tipo de riegos (Fig. 8) es muy estacional, máxima en verano y nula en invierno. La Fig. 9 compara mes a mes a lo largo del año la energía que puede producir un generador fotovoltaico con la demanda de energía del sistema de riego. Así, existe una energía excedentaria, producida en períodos en que no existe demanda.

En una aplicación comercial se evitaría esto mediante la utilización de un generador híbrido fotovoltaico-diésel, que corresponde a la solución económicamente óptima. Sin embargo, para el prototipo se optó por mostrar una novedosa posibilidad de aprovechamiento de la energía excedentaria. Como no resulta viable el almacenamiento a largo plazo de electricidad en baterías, se optó por almacenar esa energía excedentaria en forma de hidrógeno. En el prototipo, un hidrolizador utiliza la electricidad excedentaria para producir hidrógeno por medio de electrolisis del agua, que posteriormente es comprimido y almacenado a presión. El hidrógeno puede utilizarse para satisfacer la otra demanda de energía en el viñedo: la maquinaria agrícola y la movilidad del personal.

Aunque el hidrógeno podría sustituir al gasóleo utilizándose de forma similar en motores de combustión interna, es mucho más eficiente y limpio si en el propio vehículo vuelve a transformarse en electricidad, mediante una pila de combustible, para alimentar motores de tracción eléctricos. Para mostrar esta posibilidad, el prototipo incluye un vehículo todo terreno eléctrico (Fig. 10) que se modificó incorporándole una pila de combustible, los depósitos de hidrógeno y los sistemas de control necesarios.

En su conjunto, los prototipos consisten en un sistema de generación de origen fotovoltaico, que produce electricidad de iguales características que la proporcionada por la red eléctrica trifásica. Además de los paneles fotovoltaicos, incorpora inversores, baterías, la mencionada producción de hidrógeno y diversos elementos auxiliares, de medida y de control.

Para la ubicación de los campos fotovoltaicos se ha estudiado el horizonte de las posibles ubicaciones, mediante fotografías a 360°, la trayectoria anual del sol y los obstáculos del entorno. En cuanto a la colocación de los paneles fotovoltaicos en los prototipos, se han

 DOI: 10.1051/bioconf/20170901005

Figura 10. Vehículo todo terreno con pila de combustible y alimentado por hidrógeno.

Figura 11. Conjunto de paneles en estructura metálica sobre suelo.

Figura 12. Conjunto fotovoltaico flotante, durante su montaje.

utilizado tres tipos de soporte, a efectos comparativos y demostrativos. Uno de ellos es un seguidor solar a dos ejes, semejante a los utilizados en los parques solares de generación para venta a red.

El segundo conjunto de paneles (Fig. 11) está montado en un soporte de estructura metálica fija sobre el terreno. En este caso, la fijación al suelo se ha hecho mediante contrapesos prefabricados, lo que permite un montaje sin obra civil y un desmontaje sin residuos al final de su ciclo de vida. La posición de los paneles es ajustable manualmente, bien sea para maximizar la producción en la temporada de riego o para la demanda de bodega.

El tercer conjunto de paneles (Fig. 12) constituye un prototipo en sí mismo y es un sistema fotovoltaico flotante especialmente diseñado para balsas de riego. Respecto a otros soportes flotantes, se diferencia en que está pensado para adaptarse a las fluctuaciones de nivel producidas por el llenado y vaciado completos de la balsa y en que la posición de los paneles fotovoltaicos se ha definido para la temporada de riego.

La colocación de los paneles sobre la balsa evita la utilización, preparación y vallado de terreno, esperándose también obtener efectos positivos como la reducción de pérdida de agua de la balsa por evaporación y la mejora del comportamiento térmico de los paneles fotovoltaicos.

En los sistemas de generación aislados, no conectados a la red, es clave que la energía se gestione de tal forma que se eviten tanto períodos donde falte energía como otros en los que se desaproveche. Por ello, otra de las innovaciones incorporadas a los prototipos es un avanzado

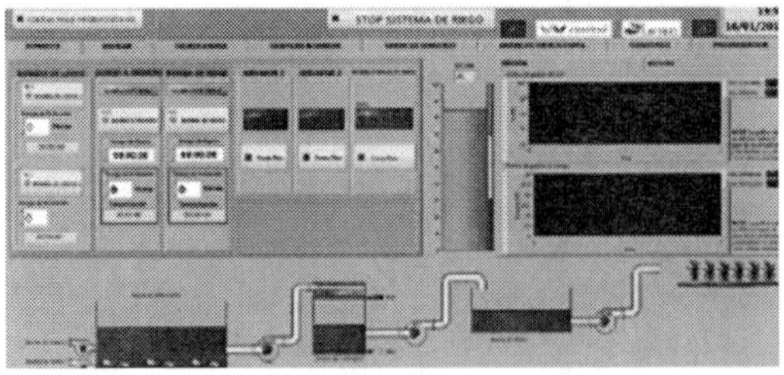

Figura 13. Una de las pantallas del sistema de control.

software de control y gestión de la energía (Fig. 13). Algunos de los consumos relacionados con la depuración se gestionan de forma automática, mientras que otros, como el riego, se han dejado a criterio del usuario, de forma manual o programable. El sistema es capaz de prever el estado de energía para las próximas horas para permitir gestionarlo en consecuencia. Su manejo y supervisión puede efectuarse de forma remota con un ordenador, una tableta o un teléfono móvil conectado a internet. También a través de internet puede vigilarse o mostrarse la instalación, gracias a la existencia de unas cámaras web motorizadas.

2.4. Análisis de impactos

2.4.1. Impacto medioambiental

La medición del impacto medioambiental de los prototipos se efectúa comparando la situación inicial, es decir, el uso de la red eléctrica, un grupo electrógeno y un vehículo diésel, con la nueva situación que los ha sustituido por un sistema de generación fotovoltaico y un vehículo con pila de combustible de hidrógeno. Para ello, se identifica y caracteriza la situación ex ante y ex post a la instalación de los prototipos, en lo relativo a sus impactos ambientales, tanto en bodega (red eléctrica) como en viñedo (generación diésel) y movilidad (vehículo todo terreno). Se establecen tres grupos de parámetros de control medioambiental: el primero incluye las emisiones de CO2, SO2 y NO? por kWh; el segundo, la reducción anual de consumo de lubricantes y filtros; el tercero, los niveles de ruido en el entorno durante la operación del sistema.

Se identifican los coeficientes de paso de los vectores energéticos: electricidad de la red y gasóleo. Se calculan los valores iniciales de los parámetros de control y las emisiones en función de los coeficientes de paso anteriores. Finalmente se calculan las emisiones evitadas y la reducción de ruido y se calculan las emisiones producidas durante la fabricación de los módulos FV.

2.4.2. Impacto socioeconómico

Se desea evaluar el impacto socioeconómico sobre la población y sociedad local en los territorios implicados en el proyecto. Se ha explorado la metodología existente, revisando diversas opciones de entre las utilizadas para la evaluación de impactos socioeconómicos [15], no existiendo una alternativa única consolidada y resultando algunas no idóneas.

Para evaluar el impacto en los grupos de interés se ha determinado que la metodología apropiada es el método de valoración contingente. Este método se incluye entre

Tabla 3. Grupos de interés para cuantificación del impacto socioeconómico.

	Composición	
Personas	Habitantes de la comarca	
	Visitantes	
	Trabajadores de la comarca	
	Consumidores de productos de la empresa	
Empresas	Clientes	Bodegas vitícolas en la zona
		Otras empresas susceptibles de instalar energías renovables
	Proveedores	Servicios: alojamientos, restauración, comercio...
		Materiales para sistemas energéticos renovables
Instituciones	Públicas	Diputación, comarca, ayuntamientos
	Privadas	Asociaciones: empresarios, consumidores
	Comunidad Científica	Centros de formación
		Centros de investigación

Figura 14. Una de las pantallas de la aplicación de diagnóstico de viabilidad.

las formas de valoración directa, en concreto por encuesta, e intenta medir en unidades monetarias los cambios en el nivel de bienestar de las personas y la satisfacción de las empresas e instituciones implicadas, debidos a la puesta en marcha de sistemas más respetuosos con el medio ambiente, en nuestro caso, el proyecto LIFE REWIND. La Tabla 1 muestra los grupos de interés identificados, a los que se va a consultar.

2.5. Métodos de ayuda a la replicación

Uno de los objetivos que se desea conseguir es la replicación de las soluciones propuestas. Para ello se han programado numerosas acciones de difusión del proyecto y de sus resultados.

Los prototipos se conciben para una utilización demostrativa: son visitables, dotados de paneles informativos y con entrega de folletos explicativos a los visitantes. Además, desde internet es posible la visita remota accediendo a las cámaras web motorizadas de alta definición, al ordenador de control y al servidor que alberga datos de su funcionamiento y de sus sensores, incluidos gráficos y estadísticas.

Para facilitar la replicación, se crean dos herramientas informáticas. Una primera que permite realizar un primer diagnóstico de viabilidad de incorporación de energía renovable a cada instalación concreta, sean bodegas, riegos, granjas, etc. La Fig. 14 muestra la pantalla en

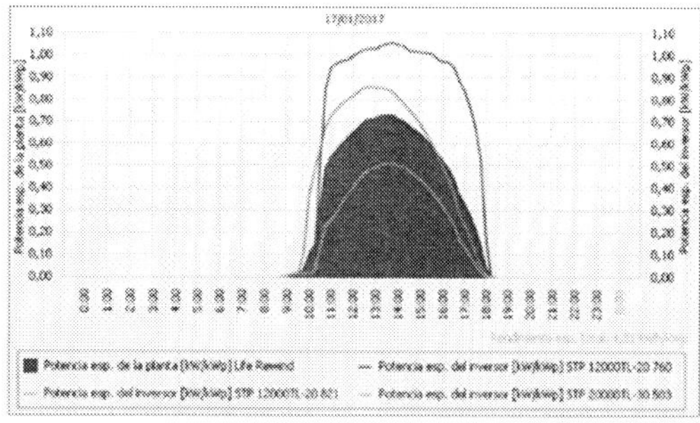

Figura 15. Rendimientos de los tres diferentes campos fotovoltaicos.

que se introduce la ubicación del caso a considerar. Una segunda herramienta informática se ha desarrollado para dimensionado técnico, que facilita la replicación del diseño propuesto, adaptándolo a cada caso concreto.

3. Resultados y discusión

Por no estar aún disponibles los resultados definitivos, así como por su previsible extensión, su publicación completa será objeto de varias publicaciones futuras de los ámbitos energético, medioambiental y socioeconómico. Sin embargo, se pueden mencionar algunos resultados cualitativos que aparecen claramente en los datos ya disponibles.

De los datos obtenidos se observa que las demandas energéticas de diferentes bodegas comparten ciertas características comunes, aún en las diferentes regiones y climas estudiados. La gran mayoría obtienen la electricidad de la red eléctrica, bien sea por estar ubicadas o muy próximas a núcleos urbanos o bien porque se llevó hasta allí la red en el momento de su construcción. Los costes que pagan las bodegas por la electricidad dependen del contrato de suministro y de las tarifas eléctricas vigentes, por lo que pueden variar considerablemente incluso de un país a otro.

En cuanto a los viñedos, también presentan características compartidas, especialmente las derivadas de la estacionalidad del riego (si lo hay) y de los trabajos con maquinaria agrícola. Existen meses de demanda prácticamente nula, sin embargo el almacenamiento en baterías a largo plazo no es viable. El perfil estacional del recurso solar guarda considerable semejanza con el de la demanda de energía de las instalaciones de riego estudiadas, lo que favorece el uso de generación fotovoltaica. Por el contrario, el recurso eólico presenta un perfil estacional muy diferente del de la demanda. Debido a la moderada cantidad de energía demandada para los usos estudiados, los aerogeneradores que serían apropiados corresponden al segmento de minieólica, con bujes a poca altura sobre el suelo, donde la velocidad del viento es muy escasa. A ello se añade que, en las zonas estudiadas de viñedo, el recurso eólico es escaso o moderado. Todo ello dificulta el uso de generación eólica.

La existencia en el prototipo de tres campos fotovoltaicos ubicados muy próximos pero montados sobre diferentes clases de soporte permite la toma de datos y su estudio comparativo. El mayor rendimiento y la curva de producción diaria más aplanada se obtiene con el seguidor

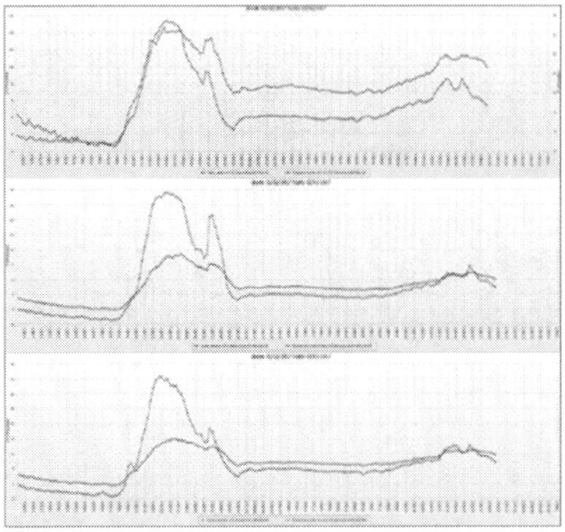

Figura 16. Gráficas de temperatura ambiente y de mdulo de los tres campos fotovoltaicos (balsa, seguidor y suelo).

solar, como era de esperar. La gráfica de rendimientos específicos mostrada en la Fig. 15 y correspondiente a un día de invierno, muestra que el segundo mejor corresponde al soporte en suelo, cuya posición es ajustable y está adaptada al período posterior a la vendimia, mientras que el peor rendimiento se obtiene en el campo de la balsa, diseñado para la época veraniega. La misma gráfica en un día de la temporada de riego mostraría un rendimiento mejor en el campo de la balsa y peor en el campo de suelo, a no ser que en éste se reajustase la posición de los paneles para dicha temporada.

Continuando con la comparativa de los tres tipos de soporte para los paneles fotovoltaicos, la Fig. 16 muestra para cada conjunto de paneles la gráfica de temperatura ambiente (en color azul) y temperatura interna del panel (en color rojo). Esta gráfica abarca dos días consecutivos, el primero soleado y el segundo nublado. En el primero se observa el aumento de la temperatura interna debido a la producción de electricidad, mientras que el día nublado no existe ese aumento y ambas temperaturas (líneas roja y azul) se mantienen próximas. En la gráfica superior, correspondiente a los paneles sobre la balsa, se observa un comportamiento térmico muy diferente a las otras dos, manteniéndose más fríos los paneles, incluso por debajo de la temperatura ambiente. Esto requiere un estudio exhaustivo de la colección de datos ya disponible, aunque apunta a un comportamiento térmico claramente diferenciado.

4. Conclusiones

Igualmente a lo indicado en el apartado de resultados, al no estar aún éstos analizados completamente, aquí sólo pueden avanzarse algunas conclusiones provisionales. Su descripción completa se hará en publicaciones futuras de ámbito energético, medioambiental y socioeconómico.

En el sector vitivinícola, las soluciones propuestas permitirán reducir las emisiones de CO_2 asociadas al proceso y al producto. Al intervenir únicamente en el origen de la energía utilizada, no resulta necesaria ninguna adaptación o modificación del proceso de trabajo.

La principal conclusión del estudio es que la incorporación de energía renovable en el sector vitivinícola es posible de forma rentable, reduciendo las emisiones asociadas al proceso y por lo tanto también las imputables a las empresas y al producto. Cada instalación requiere un estudio específico. Los mejores candidatos son los bombeos para riego por goteo actualmente alimentados por generadores diésel. En segundo lugar están las bodegas.

4.1. Sistemas aislados: Viñedo

Para alimentar instalaciones no conectadas a la red eléctrica, como los sistemas de riego, en muchos casos los sistemas de generación híbridos fotovoltaico-diésel pueden ser la solución óptima, tanto por el coste de la energía como por la reducción de emisiones de CO_2. Esta opción de suministro debería ser tenida en cuenta tanto para nuevas instalaciones como para la renovación de las ya existentes, efectuando una comparativa económica y medioambiental en competencia con las soluciones convencionales.

Los sistemas aislados, no conectados a la red, 100% fotovoltaicos, presentan una considerable fracción de energía excedentaria que puede ser utilizada para otros fines. Mientras que los sistemas fotovoltaicos o híbridos pueden ser rentables según los casos, la producción y uso de hidrógeno en maquinaria agrícola y movilidad es técnicamente viable pero todavía económicamente inadecuada para aplicación comercial.

En la revisión de instalaciones que se ha realizado, se ha detectado que en muchos casos existe un potencial de reducción de la demanda y del gasto energético mediante la incorporación de medidas de eficiencia como la compensación de reactiva, la incorporación de variadores de frecuencia, etc. En algunas instalaciones de riego, el dimensionado hidráulico no es adecuado, lo que habría que cuidar especialmente en el diseño de nuevas instalaciones, incluso teniendo en cuenta el coste y origen de la energía que van a utilizar.

En el caso de bodegas situadas en zonas alejadas del suministro, es de aplicación parte de lo dicho en este apartado sobre sistemas aislados. La opción aislada híbrida debería ser tenida en cuenta, efectuando una comparativa económica y medioambiental en competencia con la construcción de una extensión de la red eléctrica.

4.2. Sistemas conectados a la red: Bodegas

En general, la instalación de una fracción de energía renovable en bodegas puede proporcionar importantes ahorros en el suministro eléctrico, especialmente si se dimensiona de forma adecuada.

Al igual que se puede construir una bodega utilizando técnicas de arquitectura bioclimática se reduce la demanda térmica de la instalación, la integración arquitectónica de generación fotovoltaica puede reducir drásticamente la demanda total de energía eléctrica.

En forma más sencilla, los paneles fotovoltaicos pueden instalarse en las cubiertas, aparcamientos o en otras superficies disponibles con orientación adecuada. Esto permite integrar la generación en las instalaciones existentes, sin causar ninguna afección a sus actividades.

DOI: 10.1051/bioconf/20170901005

4.3. Futuros trabajos

Varios estudios se están preparando como consecuencia de los datos obtenidos a lo largo del desarrollo del proyecto LIFE REWIND, relativos a los resultados cuantitativos económicos, medioambientales y socioeconómicos. Igualmente respecto a los efectos de la colocación de los paneles sobre la balsa, al posicionamiento de los paneles para adecuarse a la demanda estacional y a la utilización del hidrógeno en maquinaria y movilidad agrícola.

Los autores agradecen al programa LIFE de la Comisión Europea su soporte financiero al proyecto LIFE REWIND (LIFE13/ENV/ES/000280) y también a Viñas del Vero S.A por su colaboración.

Referencias

[1] P. Aigrain, F. Brugière, E. Duchêne, I. Garcia de Cortazar, J. Gautier, E. Giraud Héraud, H. Hannin, N. Ollat, J. M. Touzard, BIO Web Conf. **7**, 3015 (2016)
[2] M. R. Mozell, L. Thach, Wine Econ. Policy **3**, 81–89 (2014)
[3] I. Bernetti, S. Menghini, N. Marinelli, S. Sacchelli, and V. A. Sottini, Wine Econ. Policy **1**, 73–86 (2012)
[4] G. Benedetto, Wine Econ. Policy **2**, 33–41 (2013)
[5] S. Penavayre,V. Lempereur, A. P. Huet, S. Gillet, C. Pernet, A. Besnier, L. Farrant, F. Jolibert, BIO Web Conf. **7**, 01009 (2016)
[6] F. G. Rosner, W. Dobritzhofer, A. Geyrhofer, J. Glatt, S. Großauer, S. Pchtrager, F. Regner, N. Sauberer, H. Scheiblhofer, G. Sellner, G. Soja, K. Vogl, F. Zehetner, BIO Web Conf. **5**, 1022 (2015)
[7] J. Galbreath, Int. J. Wine Bus. Res. **27**, 220–230 (2015)
[8] S. Sabbado Flores, R. M. Vieira Medeiros, BIO Web Conf. **7**, 3018 (2016)
[9] D. Borra, A. Viberti, S. Massaglia, A. Del Vecchio, BIO Web Conf. **3**, 03003 (2014)
[10] J. Carroquino, R. Dufo-Lpez, J. L. Bernal-Agustn, Renew. Energy **76**, 566–574 (2015)
[11] J. Rochard, BIO Web Conf. **7**, 1010 (2016)
[12] R. Dufo-Lpez, J. L. Bernal-Agustn, Sol. Energy **79**, 33–46 (2005)
[13] W. Short, D. Packey, T. Holt, *A manual for the economic evaluation of energy efficiency and renewable energy technologies, Univ. Press Pacific*, **2**, 120 (2005)
[14] R. Dufo-Lpez, J. L. Bernal-Agustn, J. M. Yusta-Loyo, J. A. Domnguez-Navarro, I. J. Ramrez-Rosado, J. Lujano, I. Aso, Appl. Energy **88**, 4033–4041 (2011)
[15] C. Beber, M. Padilla, M. Razès, F. Fort, J. L. Rastoin, BIO Web Conf. **7**, 3009 (2016)

DOI: 10.1051/bioconf/20170901006

La richesse de la flore comme levier pour maintenir la biodiversité dans le vignoble ? / Flora richness as a trigger to maintain biodiversity in viticulture?

Chantal Rabolin[1,a], Christophe Schneider[1], Marie Thiollet-Scholtus[2], et Christian Bockstaller[1]

[1] LAE, INRA, Université de Lorraine, 28, rue de Herrlisheim, 68000 Colmar, France
[2] ATSER INRA, SAD - UR-0055, 28, rue de Herrlisheim, 68000 Colmar, France

Abstract. The intensification of crop management have led to an impoverishment of the biological diversity, resulting in the loss of ecosystem services. In viticulture, flora could play an interesting role because it can compensate the negative impacts of vine monoculture. The flora biodiversity favours fauna biodiversity, sustaining especially the presence of pollinating insects in the grapevine inter row sown with grass or legumes and area around grapevine plots. These aspects have been poorly studied until now. A grapevine experiment set in Alsace at INRA Colmar in the frame of the DEPHY-EXPE PEPSVI project was used to address these questions.

Between 2014 and 2016, floristic survey was carried out on 4 sites, within an area of 500 m^2 within each plot. On those sites, different grapevine systems based on integrated, biological, or biodynamic production were tested. The "presence/absence" method implemented to characterize flora composition and richness on the grapevine row, the inter-row sown with grass, and the tilled inter-row.

Results showed that Asteraceae, Fabaceae and Poaceae were the most represented flora families. Asteraceae and Fabaceae may provide pollination ecosystem services if they are well managed and allowed to flower. Moreover, Fabaceae family increases soil fertility by symbiotic fixation.

The role of environmental factors, i.e., semi-natural area like hedge or forest edge close to vineyard, and anthropic factors, i.e., soil cover management, were also characterized and shown as impacted on flora composition and richness under grapevine rows and between inter- rows.

Flora characterization will continue in 2017 and 2018 to validate the present results and to assess the evolution of flora composition and richness in different climatic conditions.

1. Introduction

La viticulture est une culture phare en France. Elle est considérée comme étant «intensive, grande consommatrice de produits phytosanitaires et peu intéressante pour la biodiversité» (Pain, G., 2010). En effet, elle est une des cultures nécessitant le plus de produits phytosanitaires avec un indice de fréquence de traitement (IFT) de 12,5, tandis que l'IFT d'un blé est de 4,1. De plus, la viticulture consomme 20 % de la quantité totale des produits phytosanitaires utilisés en France, alors qu'elle n'occupe que 4 % de la surface agricole utile (Butault, J.P., 2010). Ces produits phytosanitaires entrainent des conséquences importantes sur l'environnement: (i) la pollution des sols et des eaux (Mailly, F., 2015) ; (ii) le déclin de la biodiversité et en particulier l'érosion de la diversité floristique (Clavien, Y., Delabays, N., 2006) ; (iii) des conséquences sur la santé humaine. Une des clés permettant de maintenir la réussite de la viticulture française est d'assurer la durabilité des agrosystèmes viticoles. (IFV, 2016). Pour ce faire, l'Union Européenne a lancé des mesures d'augmentation et de conservation de la biodiversité dans les paysages agricoles et viticoles, afin de contrer sa diminution probable. (Bruggisser, O.T., Schmidt-Entling, M.H., Bacher, S., 2010.)

2. Les systèmes de production viticoles

Différents systèmes de production viticoles peuvent contribuer au maintien de la durabilité de la viticulture: la viticulture biologique, dont la biodynamie et la viticulture en production intégrée. Celle-ci cherche à remplacer certains intrants issus de l'industrie agrochimique par des produits de régulation naturelle (Chambre d'Agriculture des Ardennes, 2016). La viticulture biologique se rapproche des critères d'une viticulture durable, car elle n'a pas recourt à des produits de synthèse. La viticulture biodynamique est une déclinaison de viticulture biologique qui vise entre autre à améliorer la diversité végétale et animale dans le vignoble et repose sur une approche holistique (Koepf, H.H., 1990 in Reeve, J., 2005).

Dans un objectif de durabilité de la viticulture, des pratiques viticoles comme l'enherbement des inter-rangs sont mises en place, en raison de leurs nombreux avantages environnementaux et agronomiques. En effet, dans le vignoble, maintenir ou implanter des couverts dans l'inter-rang et voire dans le rang (cavaillon) permettant de préserver cette biodiversité floristique, représente un des leviers environnemental fondamental.

[a] e-mail: `chantal.rabolin-meinrad@inra.fr`

Tableau 1. *Caractérisation des sites PESVI : cépages système, surfaces et type de sol* (Abréviations: *P.I. : Production Intégrée - AB : viticulture biologique - Biody : Viticulture biodynamique*).

Commune	Nomenclature	Cépage	Syst.	Surface	Type de sol
Rouffach (EPLEFPA Rouffach-Wintzenheim)	Rouff_PI	Riesling	P.I	42 ares	Limon-argileux
Rouffach (EPLEFPA Rouffach-Wintzenheim)	Rouff_PIopti	Pinot gris			Limon-argileux
Rouffach (EPLEFPA Rouffach-Wintzenheim)	Rouff_PIPODmild				Limon-argileux
Ribeauvillé (INRA)	Ribeau_PI	Riesling	P.I	36,9 ares	Limono-argileux caillouteux sur marnes calcaires
Ribeauvillé (INRA)	Ribeau_AB		AB		Limono-argileux caillouteux sur marnes calcaires
Châtenois (OPABA)	Châten_AB	Riesling	AB - *Biody*	25 ares	Sol brun sur arène argilo-granitique
Ingersheim (OPABA)	Inger_AB	Riesling	AB - *Biody*	1,6 ha	Alluvions granitiques

Les objectifs de cette étude, sont de suivre l'évolution de la flore dans les différents systèmes de production viticole évoquée, puis de mettre en évidence les facteurs favorisants ou non la richesse spécifique floristique. La connaissance de cette richesse spécifique nous permet de calculer différents indicateurs. Nous allons ici nous focaliser sur **la valeur pollinisatrice** (Ricou, C., 2014) des différentes espèces trouvées.

3. Matériel et méthode

Pour évaluer la biodiversité floristique en milieu agricole, des observations de terrains par relevés floristiques et des calculs d'indicateurs sont nécessaires.

Les relevés floristiques sont effectués sur 7 parcelles du projet PEPSVI (Réf.) réparties sur 4 communes viticoles, parcelles conduites en viticulture biologique, biodynamique ou intégrée (Tableau 1). Un relevé floristique représente l'inventaire des espèces végétales présentes sur la parcelle viticole.

Le relevé floristique doit satisfaire à quatre critères qualitatifs : i) la représentativité, il doit donner une image de la ou des communauté(s) présente(nt) sur le terrain la plus fidèle possible (Delpech, 2006). Pour cela, nous avons effectué nos relevés sur des surfaces de 500 m^2, surface qui permet d'avoir une représentation fidèle de la richesse présente. En effet, la courbe aire-espèce ne présente plus qu'une augmentation limitée du nombre d'espèce observée au-delàs de 500 m^2 (Clavien, Y., 2005). Enfin, un relevé doit être ii) exhaustif, iii) reproductible (qualité essentielle pour un suivi) et iv) faisable (Chevalier, R., Gautier, G., Archaux, F., 2010).

Des observations de «présence-absence»et le recouvrement sont réalisées pour caractériser respectivement la richesse et la composition floristique, son abondance, ce qui permet d'évaluer la diversité flore présente.

La méthode «présence-absence» consiste à identifier si une espèce est présente ou non. Le codage 0 ou 1, qu'on appelle binaire ou booléen (Vanpeene-Bruhier S., 2010) a été utilisé : le chiffre «1» marque la présence d'une espèce et «0» son absence.

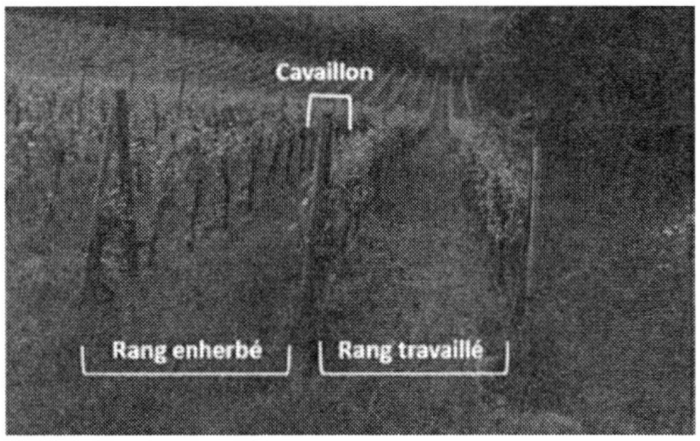

Figure 1. Localisation des différentes zones de relevé dans une parcelle viticole (Copyright © : C. Rabolin, INRA- Colmar).

La méthode «présence-absence»permet d'obtenir la richesse spécifique de chaque parcelle par le simple calcul du nombre total ou moyen d'espèces présentes par unité de surface (Clavien, Y., 2005 ; Grall, J., Coïc, N., 2005). La structure taxonomique du peuplement par les espèces peut être analysée grâce à cet indice (Grall, J. et Coïc, N., 2005). La richesse spécifique représente, au niveau de la gestion de l'espace, un outil intéressant pour la prise en compte de la biodiversité (Vanpeene-Bruhier S., 2010). Cet indicateur de biodiversité est couramment utilisé en raison de sa simplicité mais présente des limites: le fonctionnement de l'écosystème ne peut pas être appréhendé par celle-ci. (Chevalier, R., Gautier, G., Archaux, F., 2010). L'estimation du recouvrement de chaque espèce en permet d'avoir des informations sur la dominance des espèces.

Pour les relevés floristiques, des fiches distinctes sont utilisées afin évaluer la présence-absence et les recouvrements.

La flore est relevée sur différentes positions dans les parcelles: rang enherbé, rang travaillé et cavaillon (Fig. 1).

Trois campagnes de relevés espacées de six semaines les unes des autres ont été programmées aux dates suivantes: mars, mai, fin juin. Durant cette large période de relevés, on considère que la majorité des espèces

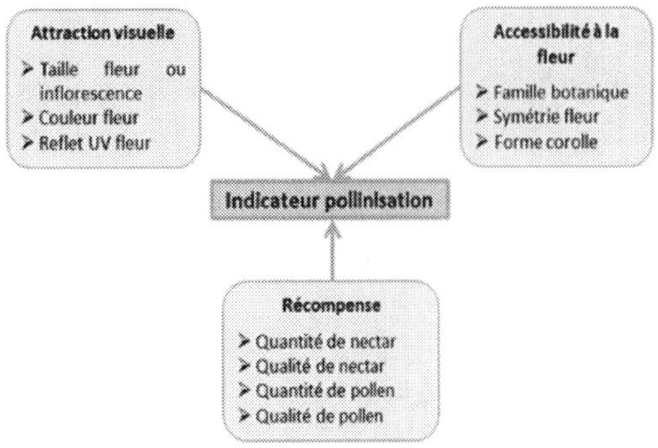

Figure 2. Schéma des 3 compartiments pris en compte dans le calcul de l'indicateur.

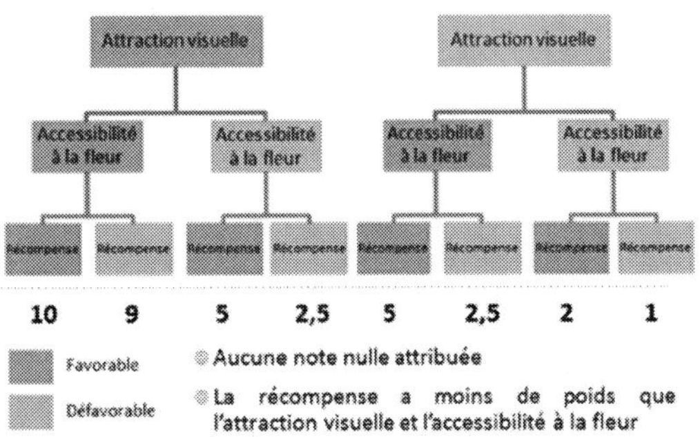

Figure 3. Arbre de décision global permettant le calcul de l'indicateur.

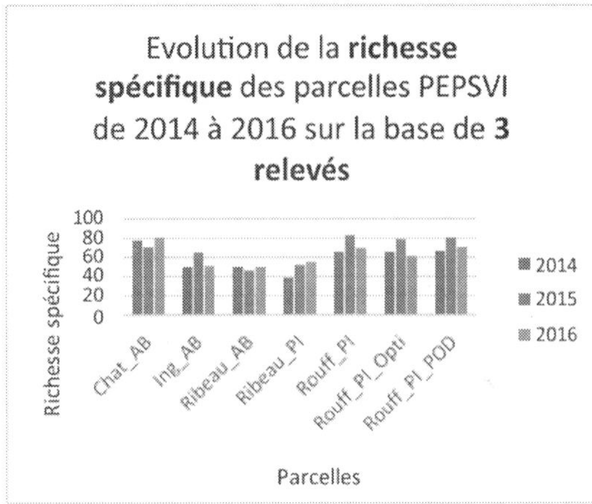

Figure 4. Evolution de la richesse spécifique des parcelles de 2014 à 2016 sur la base de 3 relevés.

Lotier corniculé
(*Lotus corniculatus* L.)

Luzerne cultivée
(*Medicago sativa*)

Figure 5. Répartition des familles botaniques en 2016 sur la base de 3 relevés.

printanières pré-estivales, estivales sont représentées (Boudjedjou, L., Fenni, M., 2011). Cette méthode permet un relevé exhaustif de la biodiversité présente. Les relevés sont réalisés aux mêmes périodes chaque année.

L'identification des espèces végétales est réalisée dans le meilleur des cas sur les parcelles, par des personnes formées en botanique. Lorsque l'identification est incomplète, des photos sont prises et des échantillons prélevés afin de les identifier en laboratoire. Quand l'espèce végétale est au stade plantule, l'identification reste peu fiable, l'annotation par famille est alors utilisée. (Espèce .sp. // Famille)

Après avoir intégrées les données obtenues, des traitements statistiques sont réalisés sur les résultats. Pour cela, nous avons eu recours au logiciel «R» (version 3.1.3 ; R Development Core Team, 2008 ; packages «agricolae», «RVAideMemoire», «pgirmess»et «ez»).

Différents indicateurs peuvent être calculés, I-Phy (indicateur phytosanitaire) I-N (indicateur Azote). Les indicateurs sont de plusieurs ordres.

3.1. Les indicateurs

Plusieurs types d'indicateurs (Bockstaller et al, [2008]) peuvent être calculés à partir des données prélevées et des itinéraires techniques:

- Indicateurs simples : une variable ou combinaison de variables obtenues par sondages, bases de données : pas directement mesurées. Exemple : pourcentage de couverture du sol nu en hiver. Evaluation indirecte des impacts environnementaux, faible qualité de prédiction.
- Indicateurs basés sur le calcul : utilisation de plus d'un type de facteur. Grande diversité d'indicateurs. Simulation de modèles ou facteurs d'émission. Exemple : évaluation des émissions de polluants.
- Indicateurs basés sur une ou plus mesures : exemples : indicateurs de biodiversité. Utilisés quand il n'y a pas de modèle précis disponible. Coûts parfois élevés, relations cause/effet non précises.

Ici, nous calculerons un indicateur de biodiversité, la valeur pollinisatrice : Ipol.

3.2. L'indicateur de pollinisation

Valeur pollinisatrice = capacité de la fleur à attirer des insectes pollinisateurs et leur fournir du pollen et du nectar pendant une période de l'année. Le calcul de la valeur pollinisatrice dépend de différents critères:

4. Valeur pollinisatrice

Elle est exprimée sur une échelle entre 0 à 10 (10 : meilleur valeur pollinisatrice), pour chaque plante. Elle est calculée en fonction du type de pollinisateur (abeille,

DOI: 10.1051/bioconf/20170901006

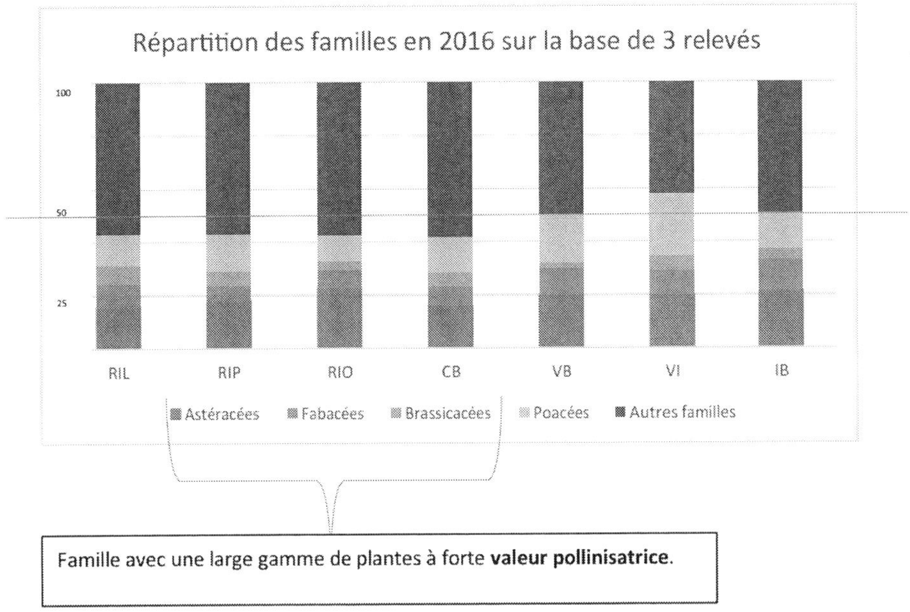

A partir de la composition floristique observée sur les différentes parcelles, nous avons calculé la valeur pollinisatrice :

	Attraction visuelle	Accessibilité à la fleur	Récompense	**Indicateur**	2014	2015	2016
Vicia sativa L.	7,33	8,35	4,58	**7,13**	Ribeau - Ing	Rouff - Chât - Ribeau - Ing	Ribeau
Picris hieracioides L.	5,70	10,00	5,50	**7,02**		Rouff - Ribeau - Ing	Rouff - Chât - Ribeau - Ing
Sonchus oleraceus L.	5,70	10,00	5,31	**7,00**		Rouff - Ribeau	Rouff - Ribeau - Ing
Tragopogon pratensis L.	5,67	10,00	5,50	**7,00**		Rouff - Ribeau	Ribeau
Sonchus asper (L.) Hill	5,66	10,00	5,50	**7,00**		Rouff	Rouff - Chât - Ribeau
Taraxacum officinale Weber	6,21	8,35	4,77	**6,61**	Rouff - Ribeau - Ing	Rouff - Chât - Ribeau	Rouff - Chât - Ribeau - Ing
Carduus nutans L.	7,94	6,68	4,58	**6,50**	Rouff - Chât - Ribeau - Ing	Chât - Ribeau - Ing	Rouff - Chât - Ribeau - Ing
Lamium purpureum L.	5,80	8,35	4,58	**6,43**	Rouff - Chât - Ribeau - Ing	Rouff - Chât - Ribeau - Ing	Rouff - Chât - Ribeau - Ing
Bellis perennis L.	5,29	8,35	5,50	**6,30**		Rouff - Chât - Ribeau - Ing	Rouff
Crepis capillaris (L.) Wallr.	5,40	8,35	4,58	**6,27**		Rouff	Rouff

Figure 6. Exemple de valeur pollinisatrice pour 10 espèces observées.

bourdon, syrphe…), en fonction des caractéristiques de chaque plante (taille de la fleur, couleur, reflet UV, famille botanique, symétrie, forme de la corolle, quantité et qualité de nectar, quantité et qualité de pollen).

Pour exemple, la phacélie (Phacelia tanacetifolia) et le mélilot blanc (Melilotus albus) comptent parmi les plantes préférées des abeilles et «les abeilles visitent plus souvent le mélilot que la phacélie» (Teittinen, 1980 ; Balzekas, 1978)

5. Résultats de l'étude

En 2014, 124 espèces ont été recensées, en 2015, 178 espèces, et 212 en 2016. Ce nombre important d'espèces permet de mettre en évidence la capacité du vignoble à maintenir une diversité floristique, malgré les interventions effectuées sur le vignoble.

Les valeurs pollinisatrices issues de l'indicateur : Phacélie : 6,5 et Mélilot : 7.

Les bourdons visitent préférentiellement les espèces des familles de Fabaceae, Lamiaceae et Asteraceae (Fussell et Corbet, 1992 ; Goulson et Darvill, 2004).

Les fleurs les plus visitées par les bourdons: bleuet (Centaurea cyanus), centaurée des prés (Centaurea thuillieri), mélilot jaune (Mélilotus officinalis) et sainfoin (Onobrychis viciifolia).

A partir de la composition floristique observée sur les différentes parcelles, nous avons calculé la valeur pollinisatrice.

6. Discussion

Cette étude a permis de mettre en avant la capacité du vignoble à maintenir une certaine diversité floristique. Ceci est essentiel dans le cadre de l'utilisation de celui-ci en tant que corridor écologique entre des milieux d'intérêts différents. Une meilleure identification des plantes auxiliaires mais aussi des espèces hôtes de vecteurs indésirables en proximité de parcelle permettra une meilleure gestion de celle-ci. Cette connaissance des plantes, liée à une gestion adaptée maintiendrait une réelle diversité et n'affecterait pas le potentiel du vignoble pour sa production. Dans le contexte alsacien, et sa localisation sur le piemont des Vosges, (lien entre la forêt et plaine d'alsace), il permettra d'augmenter la biodiversité de cette étendue et d'enrichir le patrimoine floristique à grande échelle.

DOI: 10.1051/bioconf/20170901006

Concernant les facteurs environnementaux, de nouvelles espèces sont apparues en 2015. Leur apparition est souvent liée à la proximité de milieux boisés. L'environnement a donc un impact important sur la flore présente dans le vignoble (publication en cours).

L'étude de la répartition des familles sur 2014, 2015 et 2016 a permis aussi de mettre en évidence le développement de certaines familles indésirables (ex : poaceae), le brome stérile (*Bromus sterilis*) a une valeur pollinisatrice de 1, induisant une diminution de la richesse spécifique des autres familles à forte valeur pollinisatrice. Il est donc nécessaire de suivre l'évolution de la flore présente sur les sites afin de valoriser au mieux les avantages apportés par une grande richesse spécifique, et de limiter le développement des graminées. Les espèces invasives comme le brome stérile (*Bromus sterilis*) récemment apparues réduisent également la diversité floristique, créent une concurrence avec la vigne et de fait nécessite une surveillance et un contrôle. L'ensemble des sites présentent des inter-rangs enherbés, il est donc indispensable de parfaitement maîtriser l'enherbement, soit par un ITK adapté, soit par un choix réfléchis des espèces semées. Les dates de fauche seront essentielles pour suivre les espèces jusqu'à la floraison.

Les viticulteurs pourront favoriser différentes familles comme les fabacées et les astéracées pour maintenir une diversité au sein de leur vignoble. Certains mélanges locaux peuvent être implantés, avec une composition faible en poacées. Des études sont en cours.

L'étude de la valeur pollinisatrice à travers la flore présente au sein des sites PEPSVI n'a débutée qu'en 2014. Il serait donc intéressant de poursuivre ce travail afin de confirmer les tendances observées. Il faudrait néanmoins homogénéiser les interventions d'une année à l'autre afin d'analyser réellement l'influence des différentes pratiques culturales de chaque système sur la flore présente. Une meilleure connaissance de la flore viticole permettrait également aux viticulteurs d'orienter leurs pratiques vers la promotion de la biodiversité, tout en tenant compte des intérêts agronomiques.

Nous remercions les personnes ayant participées au projet :

Julie Buchmann : Caractérisation de la flore, Influence des différents facteurs anthropiques et environnementaux de 2014 à 2016 sur un réseau de parcelles expérimentales – Stage master 2 – (INRA Colmar 2016).

Chloé Schneller : Développement et validation d'un indicateur de la valeur pollinisatrice des bordures de champs en grandes cultures (2011).

Mots clés : Biodiversité, viticulture, flore, pollinisation, durabilité, fabacées, zones non productives, évaluation environnementale.

Références

Bockstaller, C., L. Guichard, D. Makowski, A. Aveline, P. Girardin, and S. Plantureux. 2008. Agri-environmental indicators to assess cropping and farming systems. A review. **28**, 139–149

Clavien, Y. and Delabays, N. (2006). "Inventaire floristique des vignes de Suisse romande: connaître la flore pour mieux la gérer." Revue suisse Vitic. Arboric. Hortic. **38**(6), 335–341

Delabays, N., Spring, J-L., Mermillod, G., (2006). "Essai d'enherbement de la vigne avec des espèces peu concurrentielles: aspects botaniques et malherbologiques." Revue suisse Vitic. Arboric. Hortic. **38**(6), 343–352

Delabays, N., Wirth, J., et Vaz, C., (2009). "Nouveaux enjeux dans la gestion de la flore des vignobles". Revue suisse Vitic. Arboric. Hortic. **41**(4), 207–211

Mailly, F., et al. (2015). "Impact of soil and canopy management practices on pesticide use in viticulture in french regions" In: *IPM Innovation in Europe*: *book of abstracts* (p. 149)

Pain, G., et al. (2010). «La préservation des espaces viticoles en matière d'agriculture durable, de gestion des espaces naturels et du maintien de la biodiversité-Biodiversité en territoire viticole: exemple.» (Disponible en ligne: `http://www.biodivine.eu`) 26p.

Ricou, C., C. Schneller, B. Amiaud, S. Plantureux, and C. Bockstaller. 2014. A vegetation-based indicator to assess the pollination value of field margin flora. Ecol. indic. **45**, 320–331

Thiollet-Scholtus, M., Ley,L., (2016) Good quality of grapes and small yield: Tension between no use of pesticides and use of organic fertilizers?, [ESA (2016)]

 DOI: 10.1051/bioconf/20170901007

Le guide de l'agroécologie en viticulture, un outil pour la filière / Guide for agroecology in viticulture, a tool for the sector

Valérie Lempereur[1], Jacques Gautier[2], et Carine Herbin[1]

[1] IFV - Institut Français de la Vigne et du Vin, France
[2] INAO - Institut national de l'Origine et de la Qualité, France

Abstract. In 2014, the French public authorities launched the agroecological project for France. This project aimed at starting the transition of French agriculture towards new production systems that are more efficient in the economic, social and environmental dimensions. In 2016, the National Institute of Origin and Quality (INAO), in charge of French geographical indications (GI), validated the integration of agrienvironmental measures into specifications of French GIs. A work involving the French Institute of Vine and Wine (IFV) and INAO resulted in the publication of a *"Guide for agroecology in viticulture"* in early 2017. This guide describes agroecological measures that can be integrated into the production rules of French GIs. Those measures are classified according to the five themes initially identified by INAO. An agrienvironmental measure is the implementation of a strategic orientation articulated on the five themes of agroecology. Each measure is constructed on the basis of the knowledge of sustainable viticulture, the evolution of practices, examples already existing in the regions, including the specifications of the GIs. Accessible to all, the ambition of this guide is to provide technical benchmarks for winegrowers who wish to develop their practices individually or collectively in order to initiate or deepen an agrienvironmental approach.

1. Une politique publique française tournée vers l'agroécologie et la préservation de l'environnement

En 2014, les Pouvoirs Publics français ont lancé le projet agroécologique pour la France [1], projet ayant l'objectif d'engager la transition de l'agriculture française vers de nouveaux systèmes de production plus performants dans les dimensions économique, sociale et environnementale.

Les dispositions de la Loi d'avenir Agricole promulguée le 14 octobre 2014 [2] confirment cette orientation, notamment en retenant la définition de l'agroécologie :

«Les systèmes de production agroécologiques privilégient l'autonomie des exploitations agricoles et l'amélioration de leur compétitivité, tout en maintenant ou en augmentant leur rentabilité économique, en améliorant la valeur ajoutée des productions et en réduisant la consommation d'énergie, d'eau, d'engrais, de produits phytopharmaceutiques et de médicaments vétérinaires, en particulier les antibiotiques. Ces systèmes de production agroécologiques sont fondés sur les interactions biologiques et l'utilisation des services écosystémiques et des potentiels offerts par les ressources naturelles, en particulier les ressources en eau, la biodiversité, la photosynthèse, les sols et l'air, en maintenant leur capacité de renouvellement du point de vue qualitatif et quantitatif. Ils contribuent à l'atténuation et à l'adaptation aux effets du changement climatique.»

Les Pouvoirs Publics français ont également mis en place le plan Ecophyto, débuté en 2008 et remplacé par la suite par le plan Ecophyto II [3], qui vise à instaurer un cadre d'action pour parvenir à une réduction de la consommation des pesticides compatible avec le développement durable.

Par ailleurs l'Etat a développé depuis 2012 un système de certification environnementale des exploitations agricoles [4], basé sur les thématiques Biodiversité, Stratégie phytosanitaire, Gestion de la fertilisation et Gestion de la ressource en eau. Ce système, conçu sur une logique de certification progressive de l'ensemble de l'exploitation, présente trois niveaux : le premier correspond à la maitrise de la réglementation environnementale de la conditionnalité des aides de la politique agricole commune, le second correspond à des obligations de moyens et permet de mettre en œuvre des axes de progrès environnementaux, le troisième (Haute Valeur Environnementale) s'appuyant sur des niveaux d'indicateurs à atteindre dans les exploitations et traduit donc une obligation de résultats.

Les différentes orientations prises nationalement visent donc à favoriser l'implantation effective de projets environnementaux dans les différents systèmes agricoles français, tout en s'assurant qu'ils bénéficient également d'une réelle rentabilité économique.

2. Les orientations de l'INAO dans les modes de production des indications géographiques viticoles

2.1. Depuis 1935, le système des indications géographiques françaises, AOC et IGP, repose sur un système original de cogestion

Depuis la création du système des AOC en 1935, la gestion des indications géographiques a été confiée à l'Institut national de l'origine et de la qualité (INAO), organisme public dont une des missions est de valider les propositions de condition de production dont le respect permet au produit d'acquérir les qualités substantielles ayant permis la reconnaissance de l'AOC.

La revendication d'une indication géographique par une exploitation entraine de facto l'adhésion de celle-ci à une structure collective, longtemps dénommée syndicat de défense de l'appellation. Cette structure est reconnue depuis 2007 par l'INAO comme étant «l'organisme de défense et de gestion» (ODG) de l'indication géographique concernée [5].

L'ensemble des propositions d'évolution des conditions de production inscrites dans les cahiers des charges des indications géographiques est de la seule compétence de l'ODG et de ses adhérents, et donc des producteurs eux-mêmes.

Une fois le dossier transmis à l'INAO, l'examen et la validation des propositions des ODG est assurée par le comité national, structure majoritairement composée d'acteurs professionnels (notamment producteurs et négociants) de la filière viticole française concernée, AOC ou IGP.

Les décisions prises par le comité national de l'INAO sont ensuite transmises au Ministère français de l'Agriculture, pour publication du cahier des charges de l'AOC au Journal Officiel. Par la suite interviendra la procédure d'enregistrement auprès des autorités communautaires.

Le système de cogestion mis en place depuis 80 ans pour les AOC a donc associé les producteurs (via leur ODG), les Pouvoirs Publics et l'INAO, le pouvoir de proposition dans l'établissement des conditions de production de leur indication géographique étant de la seule responsabilité des ODG : qui connait le mieux les règles de production vitivinicoles permettant d'aboutir à un produit d'exception, si ce n'est les producteurs eux-mêmes ?

Ce même système a été étendu aux IGP viticoles depuis leur mise en place en 2010.

2.2. Une évolution de la réglementation des indications géographiques françaises

2.2.1. Les dispositions législatives

Outre les orientations relatives à l'agroécologie décrites ci-dessus, la Loi d'Avenir Agricole d'octobre 2014 a également acté la possibilité pour les ODG des indications géographiques d'intégrer dans les cahiers des charges des mesures visant à la préservation de leur terroir.

Cette disposition législative concrétise une demande exprimée par la filière viticole depuis plus de dix ans, et permet aux ODG des indications géographiques de pouvoir prendre des mesures de préservation de leur principal outil de travail, le terroir.

Pour mémoire l'ensemble des pays membres de l'OIV votait en 2010 la définition suivante du terroir : « Le *terroir* vitivinicole est un concept qui se réfère à un espace sur lequel se développe un savoir collectif des interactions entre un milieu physique et biologique identifiable et les pratiques vitivinicoles appliquées, qui confèrent des caractéristiques distinctives aux produits originaires de cet espace. Le *terroir* inclut des caractéristiques spécifiques du sol, de la topographie, du climat, du paysage et de la biodiversité » [6].

2.2.2. Des orientations fortes de l'INAO vers l'intégration de l'agroécologie

Début 2016, les différentes structures décisionnelles de l'INAO ont souhaité intégrer les principes de l'agroécologie dans les modes de production des Signes Officiels de la Qualité et de l'Origine (SIQO). Cette décision, prise en accord avec le Ministre de l'Agriculture, confortait ainsi les initiatives précédentes des professionnels, et été inscrite dans le contrat d'objectif de l'INAO [7].

La méthode retenue est basée sur le volontariat de chaque ODG et sur une mise en place progressive des dispositions agroécologiques dans les règles de production des indications géographiques.

Il a également été retenu d'établir une liste de *mesures-type agroenvironnementales*, mesures qui seraient intégrées de manière rapide et simplifiée dans les règles des IG, après demande de l'ODG.

Pour la viticulture, cinq thématiques agroécologiques ont été retenues par l'INAO :

1. Préservation et développement de la biodiversité : par exemple l'implantation de haies et de bandes enherbées, la réduction voire interdiction de l'utilisation d'herbicides, le développement de l'agroforesterie, ...
2. Maîtrise et réduction de la fertilisation, et notamment la fertilisation azotée : limiter le recours aux engrais azotés de synthèse, favoriser l'introduction de légumineuses, favoriser la culture d'engrais verts, ...
3. Diminution de l'usage des produits phytopharmaceutiques et développement du biocontrôle : améliorer l'efficience des quantités de produits phytopharmaceutiques utilisées en utilisant un matériel de pulvérisation efficient limitant notamment la dérive aérienne, favoriser l'utilisation de produits de biocontrôle comme la confusion sexuelle, ...
4. Recherche d'une meilleure gestion de l'eau par les exploitations : respect de bonnes pratiques d'irrigation avec notamment un matériel et des pratiques efficientes, réflexion sur l'enherbement des parcelles, sur la présence de haies et de talus limitant la vitesse des eaux de ruissellement, ...
5. Recours à un matériel végétal plus adapté aux enjeux de l'agroécologie ; réfléchir à la réintroduction de certaines variétés locales pouvant présenter une meilleure adaptation aux caractéristiques du milieu, réflexion sur la possible introduction de variétés plus résistantes à la sécheresse ou à des conditions climatiques extrêmes ou à des certaines maladies comme le mildiou ou l'oïdium, ...

La Commission nationale environnement de l'INAO a travaillé avec l'Institut Français de la Vigne et du Vin (IFV) – à titre d'expertise pour la filière - à l'élaboration des mesures agroenvironnementales-types illustrant les thématiques agroécologiques principales. Ces mesures-types ont été adoptées fin 2016 au sein des comités nationaux de l'INAO.

3. Le guide de l'agroécologie en viticulture de l'IFV et de l'INAO

Il a été décidé d'aller plus loin et d'accompagner la filière viticole et tout particulièrement les ODG, vers

Figure 1. Couverture du guide de l'agroécologie en viticulture.

l'agroécologie au moyen – dans un premier temps – d'un guide de l'agroécologie en viticulture permettant le développement des mesures agroenvironnementales (Fig. 1).

Les objectifs sont d'accompagner la profession dans ses réflexions, de valoriser les démarches existantes en région, d'inventorier les documents et outils de références, et de centraliser les informations sur l'agroécologie dans un seul et unique support.

Le guide a pour objet de fournir des points de repère techniques pour tous les vignerons qui souhaitent, sur le plan individuel ou dans un cadre collectif, faire évoluer leurs pratiques, afin d'initier ou d'approfondir une démarche agroenvironnementale.

Le projet a reçu le soutien financier de l'ONEMA-Agence Française pour la Biodiversité dans le cadre de l'appel à projet national Ecophyto-2 de 2016.

Un travail associant les équipes techniques de l'Institut Français de la Vigne et du Vin (IFV), les services de l'INAO, a abouti à la publication de ce guide début 2017. Un comité de pilotage, constitué de l'IFV, l'INAO, la Direction Générale de la Performance Economique et Environnementales des Entreprises (DGPE) du Ministère de l'Agriculture, la Confédération des vins IGP, la Confédération nationale des producteurs de vins et eaux-de-vie de vin à appellations d'origine contrôlées (CNAOC), l'ONEMA-Agence Française de la Biodiversité, a validé les orientations, les étapes et les résultats du projet.

Le guide de l'agroécologie en viticulture a été lancé officiellement au salon international de l'agriculture en présence du ministre de l'agriculture début mars 2017.

3.1. Principes de l'approche agroécologique

3.1.1. L'agroécologie, une approche systémique ancrée dans une dynamique de territoire

L'agroécologie implique de repenser son système d'exploitation en utilisant au maximum les fonctionnalités offertes par la nature [8].

L'objectif est double : amplifier les écosystèmes tout en visant à diminuer les pressions sur l'environnement en réduisant, par exemple, les émissions de gaz à effet de serre ou en limitant au maximum le recours aux produits phytosanitaires.

Il s'agit aussi de préserver les ressources naturelles telles que l'eau, l'énergie ou les éléments minéraux. La nature est utilisée comme facteur de production en réintroduisant de la biodiversité dans une dynamique de paysage.

3.1.2. L'agroécologie, une approche systémique de l'exploitation agricole

L'agroécologie implique le recours à un ensemble de techniques en synergie et ne peut pas être assimilée à une seule technique particulière. Elle considère l'exploitation dans son ensemble. C'est grâce à cette approche systémique que les résultats techniques et économiques peuvent être maintenus ou améliorés tout en améliorant les performances environnementales...

L'agronomie est au centre des systèmes de production agroécologiques.

De solides connaissances dans ce domaine sont indispensables, tant pour les agriculteurs que pour leurs conseillers. Par conséquent, les références technicoéconomiques pour la conduite des cultures doivent être adaptées aux parcelles, en particulier à travers une série d'expérimentations dans ses propres champs.

Ces démarches d'expérimentations peuvent être conduites collectivement, avec d'autres agriculteurs-viticulteurs et/ou conseillers, en s'appuyant notamment sur l'expertise des acteurs en fonction de leur milieu.

3.1.3. L'agroécologie, une évolution raisonnée en fonction du territoire

Chaque évolution vers un système de production agroécologique doit être raisonnée au cas par cas, en fonction du territoire : conditions pédoclimatiques, tissu socio-économique et possibilités de débouchés...mais aussi en fonction des objectifs de l'exploitant quant à la qualité de vie.

Le passage à l'agroécologie doit aussi être pensé à l'échelle des territoires. En effet, les dynamiques des eaux, les cycles bio-géochimiques, les épidémies ou les attaques de ravageurs sont liés à des échelles plus vastes que celles des parcelles cultivées.

La bonne gestion des fonctionnalités écologiques nécessite l'existence d'infrastructures agroécologiques. Si ces infrastructures n'existent plus, suite par exemple au remembrement, il s'agira de réfléchir à leur réaménagement.

Ces deux niveaux d'organisation, la parcelle et celui des territoires, sont à intégrer de façon cohérente.

 DOI: 10.1051/bioconf/20170901007

Figure 2. Paysage viticole-type.

3.2. Organisation des leviers d'action agroécologiques sur le système viticole

Sur un paysage viticole-type (Fig. 2), le «système» viticole est un ensemble de motifs (éléments), de structures, d'actions, de productions et d'effets, interagissant plus ou moins et de façon plus ou moins contrôlée, dans le temps, aux échelles imbriquées de l'exploitation, du territoire et du paysage.

Les leviers d'action agroécologiques s'organisent selon les 5 thématiques arrêtées pour la viticulture :

- Préservation et développement de la biodiversité,
- Maîtrise/réduction de la fertilisation,
- Diminution de l'usage de produits phytosanitaires & développement du biocontrôle,
- Recherche d'une meilleure gestion de l'eau,
- Recours à un matériel végétal plus adapté.

Les 5 thématiques sont déclinées en orientations principales, lesquelles sont traduites en mesures agroenvironnementales règlementaires, envisageables - dont les mesures-types INAO - ; les mesures agissant entre elles et sur une combinaison d'enjeux.

3.3. Construction des mesures agroenvironnementales

Ce guide décline des mesures agroenvironnementales pouvant être mises en œuvre dans la production viticole, classées selon les cinq thématiques retenues initialement par l'INAO citées ci-dessus, et qui sont la traduction d'orientations stratégiques, elles-mêmes décrites dans le sommaire du guide.

Chaque mesure est construite sur la base des acquis de la viticulture durable, de l'évolution des pratiques, des exemples déjà existants en région – notamment les cahiers des charges des indications géographiques et les Groupements d'intérêts économiques et environnementaux –, des connaissances techniques et des règlementations, actualisées par les groupes d'experts de l'IFV.

Ce guide comprend également de nombreux exemples d'applications simples relevés dans les différentes régions viticoles françaises par le réseau des ingénieurs de l'IFV implantés dans les différentes zones de production. Même si la liste des exemples n'est bien évidemment pas exhaustive, ces exemples soulignent la volonté de pouvoir indiquer au lecteur que les différentes propositions ont déjà été mises en œuvre avec des exemples concrets et accessibles.

Les mesures proposées dans le guide comprennent en particulier une première série de huit *mesures-type agroenvironnementales* jugées prioritaires par les instances professionnelles viticoles de l'INAO, et qui pourront faire l'objet d'une intégration rapide et simplifiée dans les règles de production des indications géographiques françaises :

1. Mesures concernant l'enherbement des contours des parcelles
2. Mesures interdisant le désherbage chimique en plein des parcelles de vigne
3. Mesures relatives à l'enherbement des vignes
4. Mesures visant à l'amélioration de l'efficience du matériel de pulvérisation : différentes mesures progressives sont décrites, allant de l'interdiction de l'utilisation de certains matériels ne maitrisant pas la dérive aérienne, au recours obligatoire à certains types de matériels, (jet porté, buses à jet projeté, panneaux récupérateurs)
5. Mesures de réduction des quantités de produits phytosanitaires
6. Mesures de limitation des apports d'azote minéral de synthèse
7. Mesures de préservation des murets, bosquets, terrasses, ...
8. Mesures visant au respect de la séquence morphologique originelle des sols.

Le guide de 52 pages, téléchargeable sur les sites de l'IFV (`www.vignevin.com`), de l'INAO (`www.inao.gouv.fr`) et du Ministère de l'Agriculture français (`www.agriculture.gouv.fr`), constitue un document de référence accessible à tous et qui sera mis à jour régulièrement en fonction des évolutions techniques et réglementaires.

En accompagnement du guide, il est également prévu dans un terme proche un outil en ligne comprenant un volet pédagogique et un volet d'orientation agroécologique. Celui-ci permettra d'affiner sa stratégie environnementale sur l'exploitation dans un contexte local et une dynamique de territoire, et de choisir son orientation agroécologique et les mesures agroenvironnementales à développer et/ou à introduire dans les cahiers des charges de production.

4. Conclusion

Afin d'accompagner la filière viticole française dans sa volonté d'évolution avec l'intégration de mesures environnementales dans ses règles de production, le guide de l'agroécologie en viticulture de l'IFV et de l'INAO apporte un certain nombre d'informations, portant notamment sur la réglementation existante, des propositions de mesures à mettre en œuvre avec le descriptif de leur intérêt et les modalités de leur réalisation, des exemples déjà appliqués dans différentes régions, ...

Ce guide doit aider la filière à concrétiser ses ambitions en matière de stratégie environnementale, tout en apportant une réponse aux attentes sociétales en matière

DOI: 10.1051/bioconf/20170901007

de protection de l'environnement, notamment en décrivant des mesures de limitation des consommations d'intrants (fertilisants, produits phytosanitaires, eau notamment).

Références

[1] Ministère de l'agriculture, *Projet agro-écologique pour la France*, `http://agriculture.gouv.fr/sites/minagri/files/documents/projet-agroecologique-2013_cle43b56c-1.pdf` (2014)
[2] Assemblée nationale et Sénat, *Loi 2014-1170 du 13 avril 2014 d'Avenir pour l'agriculture, l'alimentation et la forêt*, Journal officiel, n° 238 (2014)
[3] Ministère de l'agriculture, ministère de l'écologie, *plan Ecophyto II*, `http://agriculture.gouv.fr/sites/minagri/files/151022_ecophyto.pdf` (2015)
[4] Ministère de l'agriculture, *une certification environnementale pour les exploitations agricoles* `http://agriculture.gouv.fr/sites/minagri/files/1506-ae-dep-certificationenvironnement-bd-11.pdf` (2015)
[5] Assemblée nationale, *LOI n° 2007-1821 du 24 décembre 2007 ratifiant l'ordonnance n° 2006-1547 du 7 décembre 2006 relative à la valorisation des produits agricoles, forestiers ou alimentaires et des produits de la mer*, JO du 27/12/2007 texte : 0300;1 page **21210** (2007)
[6] OIV, *Définition du terroir vitivinicole*, Résolution OIV /VITI 333/2010 (2010)
[7] INAO, Intégrer les principes de l'agro-écologie dans les signes de l'origine et de la qualité (2016)
[8] Ministère de l'agriculture, l'agroécologie, une approche systémique ancrée dans une dynamique de territoire, `http://agriculture.gouv.fr/lagro-ecologie-une-approche-systemique- ancree-dans-une-dynamique-de-territoire` (2013)

DOI: 10.1051/bioconf/20170901008

Cycle, productivity and must chemical characteristics of varieties of white wine grape

Mara Fernandes Moura[1,a], Juliana Rocha de Souza[2,b], Lenon Romano Modesto[2,b], José Luiz Hernandes[1,a], and Taiane Silva Santos[2,b]

[1] Scientific reseacher, Centro APTA de Frutas, Instituto Agronômico, Av. Luiz Pereira dos Santos, 1500, Corrupira, Jundiaí, SP, Brasil, CEP: 13214-820
[2] Postgraduate student inTropical and Subtropical Agriculture of Instituto Agronômico. Av. Barão de Itapura, 1481 Botafogo, Campinas, SP, Brasil, CEP 13020-902

Abstract. The objective of this work was to evaluate five varieties of white grapes destined to winemaking, grafted on the two main rootstocks used in viticulture in São Paulo, Brazil, 'IAC 766 Campinas' and 'IAC 572 Jales'. The experiment was performed on the experimental area of Centro APTA de Frutas in the Jundiaí city, São Paulo state, Brazil. The experimental design was a randomized block with split plot with five replications, the plots represented by varieties and subplots by rootstocks. We evaluated the number of days to budding (NDB), number of days to flowering (NDF) and number of days to maturation (NDM), productivity (Prod) (ton.ha^{-1}), content of soluble solids (SS), titratable acidity (TA) and maturation index (MI). There was no difference between rootstocks, and there was interaction among rootstocks and varieties for SS and TA. The varieties did not show significant differences for yield, but there was a significant difference for the other variables. We can concluded that among the assessed varieties on both rootstocks, in Jundiaí / SP, the variety SR 501-17 distinguished itself due to its precocity, productivity and chemical quality of the must.

1. Introduction

Brazilian viticulture is in increasingly expanding, always willing to change in order to improve the vineyards, and consequently, obtaining grapes and wines of a higher quality [1–3].

Brazil presents countless difficulties in producing quality wines, especially those related to environmental factors, such as soil and climate [4]. In addition to it, the production of quality grapes proper for wine production is closely related to the choice of canopies and rootstocks. The interaction of all of these factors, besides directly affecting productivity and some chemical characteristics of the berry, such as pH, soluble solids and titratable acidity, can also affect other important quality parameters such as nutrient absorption, accumulation of phenolic compounds and anthocyanin content [5].

Cultivation of grapes destined to the production of wines and its derivatives comprises around 1,536,100 grapevines and is mainly concentrated in the regions surrounding Sorocaba, which includes the towns of São Roque, Capão Bonito and São Miguel Arcanjo; Campinas, comprising the towns of Serra Negra, Vinhedo, Jundiaí, Itupeva, Itatiba, Jarinu, Jaguariúna and Santo Antônio do Jardim; and Araçatuba, and the towns of Florida Paulista and Lucélia [6].

[a] e-mail: mouram@iac.sp.gov.br;jlhernandes@iac.sp.gov.br
[b] e-mail: julianarocha_agrofer@hotmail.com;lenon_modesto@hotmail.com; ss.taiane@gmail.com

In the state of São Paulo, the main grapes cultivars proper for wine production are: Seibel-2, Isabel, Bordô, Niagara Branca, Niagara Rosada, IAC 138-22 'Máximo' and Moscatel. We can notice that there has been a slight increase in the farming of grapes, such as Cabernet Sauvignon, Merlot and Syrah, and the highest growth of hybrid cultivars Máximo and Moscatel. However, a good alternative to wine producers is the white grapes cultivars, because the quality of their wines, in addition to not depending on phenolic maturation, accepts higher levels of total acidity. Little to no information is available in literature regarding the behavior of new white grape varieties in different Brazilian regions [7].

During the last years, we verified through grape producers along the region of Jundiaí-SP, that there is a demand for studies regarding cultivars of grapes proper for wine production, since this is an alternative for the verticalization of grape production, allowing the their processing and the elaboration of beverages, thus increasing the production value [8].

With the phenological study of grapevines we aim to describe and correlate the course of events and their variability with climate and other phenotype events for each and every kind, in order to determine the acclimation of the grapevines to the farming region.

The main phenological stages of a grapevine's vegetative and reproductive cycle are budding, flowering, early maturation and complete maturation (harvest), visually identified when 50% of the buds, flowers and berries reach each event [10].

Determining the physico-chemical properties of wines makes it possible to rate their quality [11]. The berries' composition at the time of harvest, such as the sugars, acids, tannins, non-oxidizable polyphenols content, their antioxidant capacity, scent, oxidoreductive enzymes and microelements are essential for wine quality. All of these composition factors guarantee the distinctive character, quality and is directly related to the cultivar's genetic material, cultural techniques adopted in the vineyard and with the viticulture ecosystem itself. Important elements that compose the grape are absorbed by the root system, for that reason, besides cultivar, climate and soil; the rootstock can play an important role in the grape's composition, determining its quality, as well as the wine's final quality [5, 12].

Based on what we presented, the objective of this study was to evaluate the phenological behavior, productivity and quality of must of five grape varieties for white wine in the eastern region of the São Paulo state.

2. Material e methods

We carried out the experiment in the experimental area of Centro APTA de Frutas located in the city of Jundiaí-SP. Which is found at 23°17″ S and 46°9″ O, with an altitude of 700 to 900 m, with annual rainfall averages of 1,400 mm, an average temperature of 19.5°C and a relative humidity of 70.6%. The experimental area's soil type is the Red Cambrian Distrophic [13]. Regarding the number of cold hours, the altitude compensates for the latitude along these regions, thus allowing the farming of temperate climate viticulture. Winter is mild, but subject to rimes, and with low rainfall rates; summer is hot and dry, making fungal diseases, such as anthracnose, mildew, leaf spot, rust, powdery mildew and clump rot, possible threats.

The support system we used was the espalier one, in spaces of 2.0 × 1.1 m. These two rootstock varieties seeding was carried out on September 2013 and the canopy varieties grafting was done on July 2014. The experimental design was a randomized block design with five replications.

The treatments consisted of combinations from hybrid varieties such as SR 0.501-17, Moscato Embrapa, Moscatel de Jundiaí, BRS Lorena and IAC Madalena, grafted onto the rootstocks IAC 572 'Jales' e IAC 766' Campinas'.

The SR 0.501-17 variety, is a hybrid cultivar, that presents a medium maturity cycle, scent and flavor noticeably Moscatel, a good productivity and tolerance to major fungal diseases [14].

The Moscato Embrapa variety originates wines with a muscatel flavor, typically aromatic. It is resistant to bunch rot, presents a moderate acidity, and it's very well accepted among Brazilian consumers [15].

The Moscatel of Jundiaí variety has vigorous, productive plants, but it's vulnerable to mildew, and presents large, long, winged clusters [14].

The BRS Lorena variety presents a must rich in sugar (soluble solids content around 20 to 22° Brix) and its acidity is relatively high [16].

The IAC Madalena variety presents medium / late cycles, from its grapes we obtain white, aromatic muscat wines with good acidity, and also, being fit for the production of sparkling wines, both alone and in parts with neutral white ones [14].

The rootstock IAC 572 'Jales' is vigorous, with leaves resistant to major diseases, branches that lignify later and hardly ever has its leaves falling, presenting great rooting and attachment properties [17].

The rootstock IAC 766 'Campinas' is vigorous, perfectly adaptable to the São Paulo's environmental conditions, its leaves are very resistant to diseases [17].

We used an experimental plot composed by five plants. Pruning was performed on August 1st, 2016, when we left two gems per branch.

During the experiment's maintenance, we adopted every cultivation technique practiced by wine producers of the region. With the start of budding, we carried out shoots topping, we then tied them to wires, and lastly, the innotation was performed.

We determined the plant's phenology through weekly visual observations performed after pruning using Eichor Lorentz's methodology [18]. Afterwards we determined the number of days for budding (NDB), the number of days for flowering (NDF), and the number of days for maturity (NDM) in according to the classification proposed by Baillod & Baggiolini [19].

Harvests were then carried out from December 28th, 2016 to January 23rd, 2017. We identified the grapes harvesting point from monitoring the maturation curve, taking into account a maximum pH of 3.5.

We carried out weekly analyzes at the Centro APTA de Frutas Laboratory, using a sample of 30 random berries per treatment.

At the time of harvesting, all of the clusters from each and every plot were weighed and the value obtained, we divided by the number of plants, thus determining the average production rate per plant, expressed in kg.planta^{-1}.

In order to estimate the average productivity per hectare, we multiplied the average production per plant from each plot by the number of plants per hectare; the result has been expressed in t ha^{-1}.

We determined the must's chemical composition within a sample composed of 100 berries per experimental plot. The must was obtained through the berries manual maceration and then it was homogenized. We used an Atago® digital refractometer to perform a direct refractometry in order to determine the soluble solids (SS) contents, which, in turn, was expressed in ° Brix; the titratable acidity (AT), expressed as g of tartaric acid 100 g-1 of pulp; And pH, was found using Digimed®DM-22 pH measurer. We obtained the maturation index from the relationship between the soluble solids content and the titratable acidity, according to the method described by the Institute of Food Technology [20].

At last, we submitted the obtained results to a variance analysis (test F) whilst the average values, we compared them through the Tukey test, at a 5% probability.

3. Results and discussion

There was no significant difference among the rootstocks, and they did not present any interaction with the different varieties, except regarding the soluble solids content and index maturation variables. In addition to the rootstock influence, along the same region, other variables may

Table 1. Means of number of days for budding (NDB), number of days for flowering (NDF), number of days for ripening (NDM), yield and titratable acidity (AT) in cultivars of white wine grapes on two rootstocks, Jundiaí-SP, 2016.

Varieties	NDB (days)	NDF (days)	NDM (days)	Prod. (ton.$^{-1}$)	AT (% tartaric acid)
IAC Madalena	18,50a	54,0b	157,0d	14,1	0,70b
Moscatel de Jundiaí	17,0bc	51,2c	171,0b	17,1	0,80a
Moscato Embrapa	18,3ab	60,0a	164,0c	15,4	0,58c
SR –501	16,8c	51,2c	150,0e	16,1	0,67 b
BRS Lorena	17,8abc	54,80b	176,0a	14,3	0,70 b
Porta-enxerto					
766	17,91	54,64	163,60	15,39	0,67b
572–1	17,65	54,56	163,60	13,49	0,71a
			P Value		
Varieties	0,003	<0.001	—	0,712	<0.001
Rootstocks	0,308	0,271	—	0,049	0,044
Varieties x Rootstocks	0,941	0,239	—	0,742	0,482
CV (%)	6,19	3,73	—	34,62	8,35

Means followed by the same lowercase letter in the column, do not present significant difference between them (Tukey ≤ 0,05).

Table 2. Mean values of the soluble solids (SS) and maturation index (MI) variable in cultivars of white wine grapes on two rootstocks, Jundiaí-SP, Brazil, 2016.

	SS (° Brix)		MI	
Cultivares	IAC 766	IAC 572	IAC 766	IAC 572
SR 501-17	16.80abB	18.32aA	26.05bA	26.70aA
Madalena	15.10cA	14.26cA	22.48bcA	20.24bA
Lorena	17.98cA	18.90aA	26.20bA	16.85aA
Moscatel de JD	15.5bcA	15.88bbA	19.78cA	19.50bA
Moscato Embrapa	16.78abA	16.36bA	31.72aA	26.18aB
		P Value		
Varietiies	<0.001		<0.001	
Rootstocks	0.29		0.06	
Varieties x Rootstocks	0.03*		0.04	
CV %	5.21		9.99	

Means followed by the same letters in the column and in the line, do not present significant difference between them (Tukey ≤ 0,05).

interfere in the productive behavior of the cultivars, such as the climate conditions in the farming year, pests and/or diseases incidence on the clusters, directly affecting the production, or on the leaves, interfering with the photosynthetic rate.

Grape varieties are characterized by their high sensitivity to biological adversities [21] and the great influence from environmental conditions on physiology and productive behavior [22]. Several factors that affect its phenology ought to be identified in each and every producing region [23].

Although there was no significant difference among them for productivity, the Moscatel de Jundiaí cultivars presented the highest productivity average (17.1 ton.ha^{-1}), directly followed by the SR 501-17 (16.1 ton.ha^{-1}), Moscato Embrapa (15.4 ton.ha^{-1}) and BRS Lorena cultivars (14.3 ton.ha^{-1}). The IAC Madalena variety presented the lowest one (14.1 ton.ha^{-1}).

The length of the main grapevines phenological stages during the 2016 cycle are presented in the Table 1. Buds occurred in between 16 and 19 days after pruning. In which, the SR 501-17 variety presented a lower budding and flowering number of days, not that different from the Moscatel de Jundiaí variety.

The classification based on the budding season is of extreme value to the grape producers, once it allows the use of early budding varieties in regions with low risks of late rimes and late budding varieties in regions prone to this phenomenon [24].

Another important factor is to know the grapevine maturation potential in a certain region, its accumulation of sugars and acidity. Colder regions can prolong this period, also adding more polyphenols, which, in turn, allows the elaboration of high quality fine wines [25–27].

Grape maturation started occurring in between 150 and 176 days, with an earlier ripening observed on the SR-501 variety (150 days) and a later one on the BRS Lorena variety (176 days). In São Paulo, earlier ripening varieties are required, once there is a possibility of two cycles a year when using early maturation varieties.

For the quality of the must, there were interactions between the varieties and the rootstock regarding the soluble solids contents variable (SS) and maturation index (MI) (Table 2). For both IAC 766 and IAC 572 rootstocks, the BRS Lorena variety presented higher averages (17.98° Brix and 18.9° Brix, respectively) not so distant from the SR 501-17 variety (16.8° Brix and 18.32° Brix, Respectively). All in all, the soluble solids values did not differ among rootstocks, except for the SR 501-17 variety that presented a higher average (18.32° Brix) when grafted on IAC 572 rootstocks.

The appropriate level of SST for quality winemaking should be between 18 and 22° Brix for white grapes and above 21° Brix for red ones [28]. In general, the SS levels we obtained from wine intended grape cultivars in the region of Jundiaí are low. Studies carried out in Jundiaí showed that the 'Moscatel de Jundiaí', 'IAC 138-22 Máximo', 'IAC 21-14 Madalena', 'Seibel 10096' and 'Isabel' grape varieties presented content levels of 17.2, 17.0, 16 , 2, 17.2 and 15.9° Brix, respectively [29]. According to the authors, these low SS levels were found due to the fact that the maturation period of the grapes occurred under conditions of too much water and high day and night temperatures.

Regarding the titratable acidity, the lowest average value we obtained was from the Moscato Embrapa variety (0.58% tartaric acid), directly influencing its maturation index (28.95), but this variety did not present differences from the maturation index of SR 501 −17 (26.38) and BRS Lorena varieties (26.53). These values were found due to their own high soluble solids content, whilst a high maturation index depending on soluble solids content is expected. The recommended must's titratable acidity contents are somewhere in between 0.4 and 0.6% for the production of a quality wine [5]. However, these values should not be considered high because during the winemaking process there is a considerable reduction on the titratable acidity, mainly due to salitification, of tartaric acid precipitation and malolactic fermentation [30]. In addition to it, there are several potential wine styles, for which there will be different levels for optimum acidity, keeping in mind the technical limits already established by oenological experience.

In relation to MI, higher mean values were found in the Moscato Embrapa variety (Table 2). However, this variety when grafted on the rootstock IAC 572, did not present significant difference from the varied SR 501-17 and BRS Lorena, being due the high soluble solids content of the varieties SR 501-17 and BRS Lorena, being more desirable while low acidity is not recommended to elaboration of a quality white wine [5].

4. Conclusion

We can concluded that among the assessed varieties on both rootstocks, in Jundiaí / SP, the variety SR 501-17 distinguished itself due to its precocity, productivity and chemical quality of the must.

Fundação de Amparo à Pesquisa do Estado de São Paulo – FAPESP.

References

[1] F.A. Brito Panorama e perspectives da vitivinicultura. In: 7° Seminário Nacional sobre Fruticultura de Clima Temperado. Resumos. São Joaquim, 7–11 (2006)

[2] L. M. R. Mello Vitivinicultura brasileira: Panorama 2009. Bento Gonçalves: 2010, 4p. (Artigo Técnico)

[3] L. I. Malinovski, L. J. Welter, A. F. Brighenti, H. J. Vieira, M. P. Guerra, A. L. Silva Highlands of Santa Catarina/Brazil: A region with high potential for wine production. Acta Horticulturae **931**, 433–440 (2012) `http://dx.doi.org/10.17660/ActaHortic.2012.931.51`

[4] J. P. P. Gardin et al. Ácido abscísico e etefom: influência sobre a maturação e qualidade das uvas Cabernet Sauvingnon. Revista Brasileira de Fruticultura, Jaboticabal **34**(2), p. 321–327 (2012)

[5] R.V. da Mota, C. R. de Souza, A. C. Favero, C. P. C. Silva, E. L. do Carmo, A. R. Fonseca, M. de A. Regina Produtividade e composição físicoquímica de bagas de cultivares de uva em distintos portaenxertos. Pesquisa Agropecuária Brasileira **44**, 576–582 (2009). doi: 10.1590/S0100204X2009000600005

[6] S. C. S. Bueno, et al. Vinhedo paulista. Campinas: Coordenação de Assistência Técnica Integral, 2010, 256 p.

[7] M. A. Regina, et al. Influência da altitude na qualidade das uvas 'Chardonnay' e 'Pinot Noir' em Minas Gerais. Revista Brasileira de Fruticultura, Jaboticabal **32**(1), 143–150 (2010)

[8] L. C. Silva, et al. Níveis de produção em vinhedos de altitude da cv. Malbec e seus efeitos sobre os compostos fenólicos. Revista Brasileira de Fruticultura, Jaboticabal **30**(3), 675–680 (2008)

[9] G. Jones, R. Davis Climate influences on grapevine phenology, grape composition, and wine production and quality for Bordeaux, France. American Journal of Enology and Viticulture, Davis **51**, 249–261 (2000)

[10] D. H. Lorenz, K.W. Eichhorn, H. Bleiholder, R. Klose, U. Meier, E. Weber Phenological growth stages of the grapevine (Vitis vinifera L. ssp. vinifera) – codes and descriptions according to the extended BBHC scale. Australian Journal of Grape and Wine Research **1**, 100–103 (1995)

[11] M. B. M Castilhos, V. L. Del Bianchi Caracterização físico-química e sensorial de vinhos brancos na Região Noroeste de São Paulo. Holos, Natal **4**, 148–158 (2011)

[12] L. A. Rizzon, M. B. G. Salvador, A. Miele Teores de cátions dos vinhos da Serra Gaúcha. Ciência e Tecnologia de Alimentos, **28**, 635–641 (2008). doi: 10.1590/S010120612008000300020

[13] EMBRAPA. Sistema brasileiro de classificação do solo. Brasília: Embrapa Produção de Informação; Rio de Janeiro: Embrapa Solo **42** (1999)

[14] J. S. I. de Souza, MARTINS, F. Picarelli, **Viticultura Brasileira:** Principais Variedades e suas Características. Piracicaba: FEALQ **368** (2002)

[15] U. A. Camargo, M. C. Zanuz Embrapa 131-Moscato Embrapa Ð Nova cultivar para a elaboração de vinho branco. Bento Gonçalves: Embrapa Uva e Vinho, 1997. 4p. (Embrapa Uva e Vinho. Comunicado Técnico, 24)

[16] U. A. Camargo, C. C. Guerra BRS Lorena: cultivar para a elaboração de vinhos aromáticos. Bento Gonçalves: Embrapa Uva e Vinho, ago. 2001. 4p. (Embrapa Uva e Vinho. Comunicado Técnico, 39)

[17] F. M. Pereira, H. F. Leitão Filho Caracterização botânica de porta-enxertos de videira. Campinas: Instituto Agronômico **19** (1973) (Boletim técnico, 7)

[18] K.W. Eichorn, D. H. Lorenz **Phaenologische Entwicklungstadien der Rede**. European and Mediterranean Plant Protection Organization, Paris **14**(2), 295–298 (1984)

[19] M. Baillod, M. Baggiolini Les stades répères de la vigne. Revue Suisse de Viticulture Arboriculture Horticulture, Lausanne **28**, 7–9 (1993)

[20] Instituto de Tecnologia de Alimentos (1990) Análises químicas dos alimentos. Manual técnico. Campinas, Ital, 120p.

[21] P. R. Merz, T. Moser, J. Hool, A. Kortekamp, G. Buchholz, E. Zyprian, J. Bogs The transcription factor VvWRKY33 is involved in the regulation of grapevine (Vitis vinifera) defense against the oomycete pathogen Plasmopara viticola. Physiologia Plantarum **153**, p. 365–380, doi: 10.1111/ppl.12251 (2014)

[22] A. M. Rosa, et al. Fertilidade e reservas de carbono e nitrogênio em gemas de ramos das viníferas 'Cabernet Sauvignon' e 'Nebiolo'. Revista Brasileira de Fruticultura **36**, 576–585 (2014). doi: 10.1590/01002945198/13

[23] T. C. Tomazetti, et al. Fenologia e acúmulo térmico em videiras viníferas na região da Fronteira Oeste do Rio Grande do Sul. Pesq. Agropec. Bras., [s.l.], **50**(11), p.1033-1041 (2015) WINKLER, A.J. Viticultura. 6.ed. México: Editorial Continental,1 980, 791p.

[24] F. Mandelli, M. A. Berlato, J. Tonietto, H. Bergamaschi Fenologia da videira na Serra Gaúcha Pesquisa Agropecuária Gaúcha **9**, 129–144 (2003)

[25] E. Brighenti, J. Tonietto, O clima de São Joaquim para a viticultura de vinhos finos: classificação pelo sistema CCM geovitícola. In: CONGRESSO BRASILEIRO DE FRUTICULTURA, 8., 2004, Florianópolis, SC. Anais eletrônicos... Florianópolis: SBF (2004) 4p. (CD-ROM)

[26] E. F. Gris, V. M. Burin, E. Brighenti, H. Vieira, M. T. Bordignon-Luiz Phenology and ripening of *Vitis vinifera L.* grape varieties in São Joaquim, southern Brazil: a new South American wine growing region. Ciencia e Investigación Agraria **37**, 61–75 (2010)

[27] Borghezan, Marcelo et al. Comportamento vegetativo e produtivo da videira e composição da uva em São Joaquim, Santa Catarina. Pesquisa Agropecuária Brasileira, Brasília **46**(4), 398–405 (2011)

[28] Brighenti, Alberto Fontanella et al. Desempenho vitícola de variedades autóctones italianas em condição de elevada altitude no Sul do Brasil. Pesq. Agropec. Bras. **49**(6), 465–474 (2014) J. L. Hernandes, P. Júnior, M. J, A. O. Santos, M. A. Tecchio Fenologia e produção de cultivares americanas e híbridas de uvas para vinho, em Jundiaí-SP. Revista Brasileira Fruticultura, **32**, 135–142 (2010). doi: 10.1590/ S0100-29452010005000040

[29] L. A. Rizzon, J. Meneguzzo, L. Manfroi, Processamento de uva, vinho tinto, graspa e vinagre (2004)

[30] T. das G. S. Orlando, M. J. Pedro Júnior, A. O. Santos, J. L. Hernades, Ciênc. Agrotec. **32**(3) (2008)

Diez portainjertos de vid: Efectos sobre el crecimiento, la producción y la composición de la uva del cv. Sauvignon blanc en la denominación de origen rueda (España) / Ten grapevine rootstocks: Effects on vegetative growth, production and grape composition of cv. Sauvignon Blanc in the Appellation of Origin Rueda (Spain)

Jesus Yuste[1], Alejandro Vicente[2], Enrique Barajas[1], y Maria Alburquerque[2]

[1] Instituto Tecnologico Agrario de Castilla y Leon, 47071 Valladolid, Spain
[2] At present: external viticulture activity, Spain

Abstract. The knowledge of the behaviour of different grapevine rootstocks is basic to achieve a good adaptation of vine to its growing area. With the objective of knowing the agronomic and qualitative response of cv. Sauvignon Blanc to the use of several rootstocks, a trial was established in 2006 in the D.O. Rueda. The vines are vertical trellised, with bilateral Royat cordon pruning, and the vine spacing is 2.60 m × 1.25 m. The row orientation is NNW (N-25°). The experimental design consisted of randomized blocks with 4 replications and elemental plot of 60 vines. Over the period 2010–2014 it has been developed the study of the following 10 rootstocks (treatments): 110R, 101-14M, 420A, 3309C, 41B, 161-49C, 196-17C, Fercal, Gravesac and RGM. The rootstocks 420A and 41B showed a production higher than 50% with respect to 196-17C and 161-49C, and higher to 100% than RGM, due to the number of clusters per vine and, in greater amount, to the cluster weight. Fercal and Gravesac showed an increase of pruning wood weight of 24% with respect to 196-17C and 161-49C, and 90% with respect to RGM, mainly due to the shoot weight. The sugar concentration increased with 101-14M, 196-17C and Fercal, and became reduced with 161-49C, 41B and RGM. The pH of must was reduced with Fercal whereas the titratable acidity increased, which also showed increase with Gravesac and 161-49C. The tartaric acid hardly increased slightly with Fercal and 161-49C, whereas the malic acid increased with Gravesac and Fercal, and was reduced with 41B, 3309C, RGM and 101-14M. The potassium concentration increased with 196-17C, Gravesac and Fercal, and was reduced with 41B, 161-49C, 420A and 3309C. The effects observed show alternatives for rootstock election according to the growing conditions and objectives of the vineyard.

1. Introducción

La utilización del portainjerto en el cultivo de la vid tuvo su origen en la lucha contra la filoxera. Sin embargo, actualmente los portainjertos se emplean también para mejorar la adaptación de la vid al terreno, como por ejemplo en los suelos calizos, y para controlar ciertas plagas del suelo, además de la filoxera, como los nemátodos [1].

Los primeros trabajos de mejora de portainjertos en el mundo estuvieron orientados a conseguir resistencia a la caliza y a la filoxera, pero progresivamente se han ido obteniendo portainjertos que presentan además otras cualidades. Dichas cualidades suponen criterios añadidos a considerar en la elección de portainjerto, como son el vigor que éste confiere a la variedad, la influencia sobre el ciclo vegetativo del injerto, la facilidad de estaquillado y de injerto, la adaptación a condiciones específicas del lugar (sequía, humedad, salinidad, etc.) y la calidad de la uva [2].

El uso de distintos portainjertos conlleva respuestas variables, de las diferentes variedades de vid, en las diversas regiones y zonas productoras de vino del mundo, las cuales dependen de las condiciones edafoclimáticas del lugar de cultivo [3]. Así, la elección adecuada del portainjerto, en función de las características del lugar de cultivo y los objetivos productivos, puede permitir el incremento de la eficiencia productiva del viñedo, y por consiguiente el aumento de la rentabilidad del viticultor [4]. El análisis de los factores de producción y los criterios para la elección de portainjerto deben ser considerados como aspectos preferentes en términos tanto de experimentación como de aplicación práctica en la plantación de viñedo. En definitiva, se trata de un factor de producción que lleva implícita una enorme repercusión en la rentabilidad del viñedo, en el mismo sentido que ocurre, por ejemplo, con la mejora de la eficiencia hídrica y fisiológica del viñedo [5], las técnicas de riego subterráneo, el sistema RDI (*regulated deficit irrigation*) o el PRD (*partial root-zone drying*), la mecanización integral u otras operaciones vitivinícolas alternativas. Muchas de las citadas técnicas suelen estar encaminadas a paliar el déficit hídrico severo o el exceso de vigor y diversos inconvenientes de conducción del viñedo. En este ámbito, el portainjerto puede proporcionar respuestas permanentes que ayuden a resolver este tipo de problemas, sobre todo los relacionados con la escasez de agua o con el control del desarrollo vegetativo del viñedo [6].

En España, a pesar de ser el país con mayor superficie de cultivo de viñedo para vinificación del mundo, cercana a un millón de hectáreas, se utiliza mayoritariamente un número reducido de portainjertos, la mayoría de los cuales pertenece a especies de origen americano del género *Vitis*, concretamente a cruzamientos entre dichas especies, *V. riparia*, *V. rupestris* y *V. berlandieri*, así como a su cruzamiento con *V. vinifera*. De entre ellos destacan los portainjertos 110 Richter, sobre todo, y 41B, empleado específicamente en zonas con suelos calizos [7]. Sin embargo, la existencia de zonas productoras con diferentes características edafoclimáticas sugiere la conveniencia de conocer el funcionamiento de portainjertos alternativos que puedan mejorar la adaptación del viñedo dependiendo de las condiciones de la zona de cultivo [8]. Esta situación se contempla también en la región de Castilla y León, donde ya se han llevado a cabo algunos trabajos con portainjertos, entre los que cabe mencionar los realizados con Tempranillo en la D.O. Rueda [7], en la D.O. Toro [6,8] y en la D.O. Cigales [9], con Sauvignon Blanc en la D.O. Rueda [10], con Verdejo en la D.O. Rueda [11] y con Mencía en la D.O. Bierzo [12].

Por tanto, se considera conveniente profundizar en el estudio de portainjertos, entre los cuales habría que incluir, además de los citados 110R y 41B, otros de diferentes características, como 140Ru, 1103P, 101-14M, 420A, 5BB, 161-49C, 333EM y SO4, para conocer su influencia en las diversas zonas de Castilla y León.

Con el ánimo de conocer los efectos del portainjerto en el comportamiento agronómico del viñedo, se ha planteado un ensayo experimental con un grupo de 10 portainjertos de vid, encaminado a evaluar la influencia de los mismos en el cv. Sauvignon Blanc, a través de la cuantificación de parámetros vegetativos, productivos y de calidad de la uva, en la Denominación de Origen Rueda, ubicada en el centro de Castilla y León.

2. Material y métodos

El estudio del ensayo se ha realizado, a lo largo del periodo 2010–2014, en la localidad de Medina del Campo (Valladolid, España), en un viñedo perteneciente a las bodegas Grupo Yllera S.L., ubicado dentro de la D.O. Rueda, en el centro de Castilla y León. Las coordenadas del ensayo son 41° 21′02″ N y 4° 56′15″ O.

Las cepas del viñedo experimental, plantadas en 2006, son del cv. Sauvignon Blanc y están injertadas sobre un grupo de distintos portainjertos, que constituyen los tratamientos experimentales objeto de estudio, que son: 110R, 101-14M, 420A, 3309C, 41B, 161-49C, 196-17C, Fercal, Gravesac y Riparia Gloire de Montpellier (RGM).

Las cepas han sido conducidas en espaldera, con formación en cordón Royat bilateral y posicionamiento vertical de la vegetación (*VSP*). La orientación de filas es NNO (N-25°). El marco de plantación es de 2,60 m × 1,25 m, que corresponde a un área de suelo de 3,25 m^2/cepa y a una densidad de 3.077 cepas/ha. La carga de poda ha sido de 16 yemas por cepa, en 8 pulgares de 2 yemas. Se ha aplicado anualmente una operación de poda en verde, tras el período de riesgo de helada primaveral, para el ajuste de la carga a 16 pámpanos por cepa.

El diseño experimental es en bloques al azar, con 4 repeticiones de 60 cepas y parcela elemental de 20 cepas de control.

El suelo del ensayo es suficientemente profundo para el desarrollo del viñedo, con un horizonte superior de textura arcillosa, y presenta una suave pendiente homogénea y decreciente en el sentido este-oeste. El régimen hídrico de cultivo durante el estudio se ha mantenido a través de riego deficitario, con una aplicación de agua del 25% de ETo, desde parada de crecimiento vegetativo de pámpanos principales hasta vendimia.

El análisis estadístico de los resultados se ha realizado mediante análisis de varianza (ANOVA), con el programa STATISTICA 7.0.

Los datos medios mensuales de pluviometría y temperatura, correspondientes al periodo 2010-2014, se detallan en la tabla siguiente.

Tabla 1. Datos termo-pluviométricos medios del periodo 2010–2014 (octubre-2009 / septiembre-2014), registrados en Rueda (Valladolid). Tm: temperatura media (°C), Tmax: temperatura media de máximas (°C), Tmin: temperatura media de mínimas (°C), P: precipitación (mm).

	Tm	Tmax	Tmin	P
Oct	13,0	20,3	6,5	37,1
Nov	7,6	12,4	3,3	19,3
Dic	3,9	8,4	–0,2	41,3
Ene	4,2	8,4	0,2	32,4
Feb	4,2	9,8	–1,1	25,7
Mar	8,1	14,5	2,0	44,6
Abr	11,1	17,1	4,9	35,6
May	14,8	21,7	7,6	17,4
Jun	19,1	26,7	10,9	14,8
Jul	22,1	30,3	13,3	6,4
Ago	21,9	30,2	13,4	2,5
Sep	18,2	26,0	10,9	28,0
Año	12,3	18,8	6,0	305

3. Resultados y discusión

3.1. Producción de uva

El rendimiento en uva ha mostrado la influencia del empleo de los diferentes portainjertos, con diferencias estadísticamente significativas observadas a favor fundamentalmente de 420A y 41B, que han propiciado una mayor producción media de uva, en torno a 2,2 kg/cepa, con proximidad de los portainjertos 110R y 3309C, en torno a 1,9 kg/cepa (Tabla 2). Los tratamientos RGM, con menos de 1 kg/cepa, y 196-17C y 161-49C, con 1,4 kg/cepa, han mostrado un rendimiento claramente inferior al resto. Los portainjertos 420A y 41B han mostrado un incremento medio de rendimiento superior al 50% respecto a los tratamientos 196-17C y 161-49C.

El número de racimos por cepa ha estado estrechamente relacionado con la producción de uva, destacando significativamente 420A y 41B, con promedio de 21,6 racimos, frente a 196-17C y 161-49C, con 17,9 racimos, y, sobre todo, RGM, con 15,0 racimos. El incremento de 17,9 a 21,6 racimos supone un 20%. Las diferencias de rendimiento han sido debidas en mayor

Tabla 2. Número de racimos, N.rac.; Producción de uva, Prod. (kg/cepa); Peso de racimo, P.rac. (g); Peso de baya, P.bay. (g); Número de bayas del racimo, N.bay.; de los 10 portainjertos (PI). Promedio del periodo 2010-2014. Significación estadística * = $p < 5\%$; letras distintas indican valores estadísticos significativamente diferentes (para todas las tablas).

PI	N.rac.	Prod.	P.rac.	P.bay.	N.bay.
110R	21,2^{ab}	1,95^{ab}	91,7^{ab}	1,22^{ab}	76,7^{ab}
101-14M	19,2^{bc}	1,63^{bc}	84,7^{bc}	1,11^{b}	78,3^{ab}
420A	21,7^{a}	2,29^{a}	104,9^{a}	1,27^{a}	83,9^{a}
3309C	20,9^{ab}	1,91^{ab}	91,3^{ab}	1,21^{ab}	78,4^{ab}
41B	21,5^{a}	2,11^{a}	97,5^{ab}	1,20^{ab}	82,4^{a}
161-49C	18,1^{c}	1,42^{c}	78,6^{c}	1,21^{ab}	66,9^{bc}
196-17C	17,7^{c}	1,41^{c}	78,8^{c}	1,16^{ab}	68,9^{bc}
Fercal	20,3^{ab}	1,81^{b}	89,0^{bc}	1,18^{ab}	76,7^{ab}
Gravesac	18,8^{bc}	1,59^{bc}	83,7^{bc}	1,19^{ab}	71,2^{b}
RGM	15,0^{d}	0,93^{d}	62,8^{d}	1,07^{b}	59,7^{c}
Sig.	*	*	*	*	*

Tabla 3. Número de sarmientos francos, Franc.; Número de chupones, Chup.; y Número total de sarmientos, S.tot., por cepa; Peso de madera de poda, PMP (g/cepa); Peso de sarmiento, P.sar. (g); Indice de Ravaz, I.Rz.; de los 10 portainjertos (PI). Promedio del periodo 2010–2014.

PI	Franc.	Chup.	S.tot.	PMP	P.sar.	I.Rz.
110R	14,69	1,67^{b}	16,0	970^{ab}	63,0^{ab}	2,01^{ab}
101-14M	14,35	1,93^{ab}	15,9	932^{b}	58,9^{b}	1,76^{b}
420A	14,69	1,92^{ab}	16,3	993^{ab}	61,5^{ab}	2,31^{a}
3309C	14,72	1,68^{b}	16,1	885^{bc}	55,5^{c}	2,17^{a}
41B	14,75	2,13^{a}	16,5	924^{b}	56,6^{bc}	2,30^{a}
161-49C	14,81	1,13^{c}	15,6	846^{c}	54,1^{c}	1,68^{b}
196-17C	14,63	1,90^{ab}	16,2	820^{c}	51,4^{c}	1,72^{b}
Fercal	14,94	1,62^{b}	16,1	1049^{a}	65,3^{a}	1,72^{b}
Gravesac	14,17	1,59^{b}	15,4	1017^{a}	66,8^{a}	1,57^{b}
RGM	14,77	1,16^{c}	15,5	541^{d}	34,9^{d}	1,75^{b}
Sig.	–	*	–	*	*	*

medida a la variación del peso del racimo, que en los dos portainjertos más productivos ha mostrado un promedio de 101 g y en los menos productivos citados ha sido de 79 g; o sea, en torno a un 30% mayor en los primeros, lo que ha supuesto que las diferencias entre ambos grupos fuesen estadísticamente significativas. Evidentemente, el portainjerto menos productivo, RGM, mostró asimismo un peso de racimo significativamente muy inferior, de solo 63 g.

El peso de baya también ha variado entre los distintos tratamientos, aunque las diferencias estadísticamente significativas han sido menores. El portainjerto 420A ha mostrado una baya de tamaño significativamente mayor, con 1,27 g, que RGM y 101-14, con 1,07 y 1,11 g respectivamente. Las diferencias en peso de baya entre los tratamientos restantes no resultaron estadísticamente significativas. El número de bayas del racimo fue significativamente superior en los dos portainjertos más productivos, 420A y 41B, con 83 bayas, que en 161-49C y 196-17C, con 68 bayas, y sobre todo que en RGM, con apenas 60 bayas por racimo. Por tanto, el número de bayas del racimo ha contribuido generalmente en mayor medida que el peso de baya a las diferencias en el peso del racimo entre tratamientos.

En resumen, los portainjertos utilizados han provocado claras diferencias en el rendimiento en uva, destacando 420A y 41B como más productivos, y RGM, 196-17C y 161-49C como menos productivos, en relación más estrecha con el tamaño del racimo que con el número de racimos.

3.2. Desarrollo vegetativo

El número total de sarmientos por cepa ha sido muy similar en todos los tratamientos, oscilando entre 15,4 y 16,5, gracias al ajuste de carga de brotes que se aplicó en las semanas posteriores a la fecha de brotación (Tabla 3). Las escasas diferencias observadas entre algunos portainjertos, que no han sido estadísticamente significativas, se debieron a las diferencias en el desarrollo de chupones, que resultaron significativamente favorables a 41B y desfavorables a 161-49C y RGM, puesto que el número de sarmientos francos resultó muy similar en todos los tratamientos.

El peso de madera de poda fue significativamente superior en los tratamientos Fercal y Gravesac, con más de 1 kg/cepa, que en 196-17C y 161-49C, con menos de 850 g/cepa, y sobre todo que en RGM, con 540 g/cepa. Los portainjertos Fercal y Gravesac han mostrado un incremento medio del peso de madera de poda del 24% con respecto a los tratamientos 196-17C y 161-49C, y del 90% con respecto a RGM.

La tendencia del vigor del sarmiento ha sido muy similar a la del peso de madera de poda, destacando nuevamente Gravesac y Fercal, con 66 g/sarmiento, frente a 196-17C, 161-49C y 3309C, con valores en torno a 53 g/sarmiento, y sobre todo frente a RGM, con un sarmiento de 35 g, muy significativamente inferior.

El índice de Ravaz también ha mostrado diferencias significativas entre tratamientos, favorables a los portainjertos más productivos, 420A y 41B, y a 3309C, con respecto al resto de tratamientos, entre los cuales no se han observado diferencias estadísticamente significativas. Los bajos valores de este índice observados indican que se trata de unas condiciones de cultivo del viñedo con un nivel productivo bastante moderado.

Atendiendo a los resultados obtenidos, los portainjertos Fercal y Gravesac han mostrado un mayor desarrollo vegetativo, mientras 196-17C, 161-49C y, sobre todo, RGM han mostrado un desarrollo claramente inferior al resto de portainjertos.

3.3. Composición de la uva

La concentración de azúcares se ha visto ligeramente afectada por los portainjertos, con diferencias de escasa magnitud (Tabla 4). Los tratamientos 101-14M, 196-17C y Fercal han mostrado un valor significativamente superior, en torno a 23,8 ° brix, que 161-49C, 41B y RGM, en torno a 23,2 ° brix.

El pH del mosto ha mostrado algunas diferencias estadísticamente significativas entre los portainjertos 196-17C, 41B y RGM, con valores medios cercanos a 3,60, y Fercal, con un valor medio de 3,44. La acidez titulable ha mostrado diferencias estadísticamente significativas entre los portainjertos Fercal, Gravesac y 161-49C, con valores superiores a 6,0 g/L, y los portainjertos RGM, 3309C, 101-14M, 41B y 196-17C, con valores por debajo de 5,6 g/L.

El ácido tartárico en mosto apenas mostró diferencias entre tratamientos, aunque se observó una tendencia

DOI: 10.1051/bioconf/20170901009

Tabla 4. Sólidos solubles totales, SST (° brix); pH; Acidez titulable, A.T. (g TH_2/L); Acido tartárico, TH_2 (g/L); Acido málico, MH_2 (g/L); Potasio, K (mg/L); de los 10 portainjertos (PI). Promedio del periodo 2010–2014.

PI	SST	pH	A.T.	TH_2	MH_2	K
110R	23,6^{ab}	3,50^{ab}	5,90^{ab}	8,39	1,73^{ab}	1778^{ab}
101-14M	23,9^{a}	3,54^{ab}	5,56^{b}	8,27	1,69^{b}	1803^{ab}
420A	23,4^{ab}	3,49^{ab}	5,80^{ab}	8,35	1,78^{ab}	1759^{b}
3309C	23,5^{ab}	3,56^{a}	5,56^{b}	8,23	1,64^{b}	1760^{b}
41B	23,3^{b}	3,58^{a}	5,60^{b}	8,30	1,61^{b}	1713^{b}
161-49C	23,2^{b}	3,52^{ab}	6,05^{a}	8,69	1,84^{ab}	1757^{b}
196-17C	23,7^{a}	3,60^{a}	5,62^{b}	8,35	1,81^{ab}	1838^{a}
Fercal	23,7^{a}	3,44^{b}	6,19^{a}	8,70	1,99^{a}	1827^{a}
Gravesac	23,5^{ab}	3,49^{ab}	6,07^{a}	8,17	2,08^{a}	1829^{a}
RGM	23,3^{b}	3,57^{a}	5,52^{b}	8,22	1,67^{b}	1783^{ab}
Sig	*	*	*	–	*	*

a ser más elevado en Fercal y 161-49C, con valores medios en torno a 8,7 g/L, mientras que en el resto de portainjertos estuvo entre 8,2 y 8,4 g/L. El ácido málico fue responsable en gran medida de la tendencia observada en la acidez total, ya que mostró diferencias estadísticamente significativas entre algunos tratamientos. Así, los portainjertos Gravesac y Fercal presentaron valores medios claramente más elevados, en torno a 2,04 g/L, que 41B, 3309C, RGM y 101-14M, con valores en torno a 1,65 g/L. El contenido en potasio del mosto mostró diferencias estadísticamente significativas favorables a los portainjertos 196-17C, Gravesac y Fercal, con valores en torno a 1830 mg/L, con respecto a 41B, 161-49C, 420A y 3309C, con valores en torno a 1750 mg/L.

4. Conclusiones

El empleo de distintos portainjertos de vid afectó al desarrollo vegetativo-productivo y a la calidad de la uva del cv. Sauvignon Blanc de forma variable. Los portainjertos 420A y 41B mostraron un potencial productivo medio más del 50% superior que 196-17C y 161-49C, y más del 100% superior que RGM. Las diferencias en el rendimiento fueron debidas al número de racimos por cepa y, en mayor cuantía, al peso del racimo. Los portainjertos Fercal y Gravesac presentaron incremento medio del peso de madera de poda del 24% con respecto a 196-17C y 161-49C, y del 90% con respecto a RGM. Las diferencias en la madera de poda fueron debidas mayoritariamente al vigor del sarmiento.

En cuanto a calidad de la uva, la concentración de azúcares se vio favorecida por 101-14M, 196-17C y Fercal y perjudicada por 161-49C, 41B y RGM. El pH del mosto fue favorecido por 196-17C, 41B y RGM y desfavorecida por Fercal. La acidez titulable fue incrementada por Fercal, Gravesac y 161-49C y disminuida por RGM, 3309C, 101-14M, 41B y 196-17C. El ácido tartárico apenas se vio afectado, aunque Fercal y 161-49C mostraron un valor medio superior. El ácido málico aumentó con Gravesac y Fercal y se redujo con 41B, 3309C, RGM y 101-14M. El contenido en Potasio se incrementó con 196-17C, Gravesac y Fercal y se vio reducido con 41B, 161-49C, 420A y 3309C.

En resumen, los efectos observados en los distintos parámetros analizados muestran la posibilidad de distintas alternativas de elección de portainjerto para el cultivo de Sauvignon Blanc en función de las condiciones medioambientales y los objetivos productivos del viñedo.

El desarrollo de este trabajo ha sido posible gracias a la colaboración del Grupo Yllera, propietario del viñedo, así como a la aportación analítica de la Estación Enológica de Castilla y León (Rueda, Valladolid), la contribución financiera de la Junta de Castilla y León y del FEDER, y por supuesto, la ayuda del personal del Grupo de Viticultura del ITACyL.

Referencias

[1] P. May. "Using grapevine rootstocks. The Australian perspective". Winetitles, Cowandilla SA (Australia), 62 p. (1994)

[2] P. Galet. "Grape varieties and rootstocks varieties". Oenoplurimédia, Chaintré (France), 315 p. (1998)

[3] A. Reynier. "Manual de viticultura". Ed. Mundi-Prensa, Madrid (España), 497 p. (2002)

[4] N. Dry. "Grapevine rootstocks. Selection and management for South Australian vineyards". Hyde Park Press, Adelaide (Australia), 85 p. (2007)

[5] T. Sampaio, C. Vasconcelos. "Optimizing water status, gas-exchange, fuit yield and composition using rootstocks". 14th GiESCO Symposium, Geisenheim (Germany), August 23–27 (2005)

[6] M.V. Alburquerque, F.J. Castaño, J. Yuste. "Influencia de diez portainjertos sobre el comportamiento de la variedad Tempranillo. Desarrollo productivo y vegetativo y calidad de la uva". Vida Rural **305**, 52–56 (2010)

[7] J. Yuste, M.V. Alburquerque. "Influence of 9 grapevine rootstocks on vegetative-productive development and grape quality of Tempranillo variety in the A.O. Rueda (Spain)". Le progrès agricole et viticole: Special 17th GiESCO Symposium, Asti-Alba (Italy), August 29 - September 2, pp: 455–458 (2011)

[8] J. Yuste, M.V. Alburquerque. "Influencia de 10 portainjertos de vid sobre el desarrollo productivo y vegetativo y la calidad de la uva de cv. Tempranillo en la denominación de origen Toro (España)". Enoviticultura **22**, 6–15 (2013)

[9] J. Yuste, M.V. Alburquerque. "Effects of 9 grapevine rootstocks on vegetative development, production and grape quality of cv. Tempranillo in the A.O. Cigales (Spain)". X Int. Terroir Congress, Budapest (Hungary), July 7–10 (2014)

[10] J. Yuste, E. Barajas, M.V. Alburquerque. "Influencia de 10 portainjertos de vid sobre el desarrollo productivo, la producción y la calidad de la uva del cv. Sauvignon blanc en la DO Rueda". Enoviticultura **25**, 7–13 (2013)

[11] J. Yuste, M.V. Alburquerque. "Rootstocks of grapevines: effects on vegetative development, production and grape quality of cv. Verdejo in the D.O. Rueda (Spain)". 19th GiESCO Symposium, Gruissan (France), June 1–5 (2015)

[12] J. Yuste, R. Yuste, M.V. Alburquerque. "Ten grapevine rootstocks: effects on vegetative development, production and grape quality of cv. Mencia in the D.O. Bierzo (Spain)". XI Int. Terroir Congress, McMinnville, OR (USA), July 10–14 (2016)

 DOI: 10.1051/bioconf/20170901010

Evolution des pratiques viticoles françaises : de la viticulture raisonnée à l'agroécologie / Evolution of the French vineyard practices: From "controlled vine growing" to agro-ecology

Joël Rochard

IFV (Institut Français de la Vigne et du Vin), 17 rue Jean Chandon Moët, 51200 Epernay, France

Résumé. Après la deuxième guerre mondiale, l'agriculture et la viticulture s'inscrivaient dans un contexte de défi alimentaire. La priorité de l'époque était avant tout d'intégrer les progrès de la science dans les itinéraires de production pour augmenter la productivité et les rendements. En complément des effets sur le milieu naturel, l'utilisation massive des produits phytosanitaires a pu conduire à la présence de résidus dans le raisin. Progressivement, la lutte puis la production intégrée ont émergé en France avec un souci de limitation des impacts sur les auxiliaires de la vigne. Parallèlement à ces enjeux territoriaux, la notion de développement durable a mis l'accent sur l'impact planétaire de l'homme, et vis-à-vis des générations futures. Ainsi, la biodiversité, l'utilisation des énergies fossiles, la production de déchet et de gaz à effet de serre, s'immiscent progressivement dans les contraintes viticoles. Parallèlement aux aspects techniques, la certification des démarches s'est appuyée sur un processus d'évaluation et de contrôle. L'IFV a accompagné cette évolution, notamment par de nombreuses recherches (modélisation des maladies, pulvérisation, mécanismes de résistance des cépages, dépérissements, etc.) et par l'élaboration de supports de vulgarisation (publications, colloques, plaquette, site Internet) ainsi que par la création d'outils d'aide à la décision.

Abstract. After the Second World War, agriculture and the viticulture fell under a food context of challenge. The priority of the time was above all to integrate the advances in knowledge in the routes of production to increase the productivity and the outputs. In complement of the effects on the natural environment, the massive use of the plant health products could lead to the presence of residues in the grape. Gradually, the fight then the integrated viticulture emerged in France with a preoccupation with a limitation of the impacts on the auxiliaries of the vine. Parallel to these territorial challenges, the concept of sustainable development focused on the worldwide impact of the man, and with respect to the future generations. Thus, the biodiversity, the use of fossil energies, the production of waste and greenhouse gas, are involved gradually in the wine constraints. Parallel to the technical sides, the certification of the approachs was based on a process of rating and check out. The IFV accompanied this evolution, in particular by many search (modeling of the diseases, pulverization, mechanisms of resistance of type of vines, deteriorations, etc.) and by the development of supports of popularization (publications, conferences, plate, and website) like by the creation of tools of decision-making aid.

1. Introduction

La viticulture a longtemps été fondée sur un système de culture ancestral pour lequel l'énergie humaine et animale conservait un rôle prépondérant. Mais progressivement le développement des moyens de transport a privé le vignoble d'une partie de sa main-d'œuvre, en quête d'une vie plus «facile» en ville. Après la deuxième guerre mondiale, l'agriculture et la viticulture s'inscrivaient dans un contexte de défi alimentaire. La priorité de l'époque était avant tout d'intégrer les progrès de la science dans les itinéraires de production pour augmenter la productivité et les rendements. Au cours de cette période, les vignerons cultivaient l'espoir d'un progrès définitif notamment vis-à-vis de la protection de la vigne contre les parasites. Mais ils durent déchanter devant les pullulations d'acariens et l'apparition de plus en plus fréquente de résistance des bio-agresseurs et de mauvaises herbes [1].

Parallèlement aux phénomènes de résistance, l'utilisation des produits agro-pharmaceutiques a pu conduire, par la modification de l'équilibre biologique et la destruction des auxiliaires, à l'émergence de nouveaux parasites. Tel fut le cas pour le désherbage avec la pullulation de mauvaises herbes insensibles aux nouveaux herbicides. De même, l'utilisation des insecticides, conduisit à une diminution de la population des auxiliaires et notamment les typhlodromes, aboutissant au développement des acariens.

Parallèlement aux effets sur le milieu naturel, l'utilisation massive des produits phytosanitaires a pu conduire à la présence de résidus dans le raisin, parfois à l'origine d'une perturbation de la fermentation alcoolique et, dans certains cas, d'un dépassement des seuils de résidus fixés pour les vins dans le cadre notamment des échanges internationaux.

DOI: 10.1051/bioconf/20170901010

Écophyto est un plan national qui vise à réduire progressivement l'utilisation des produits phytosanitaires en France tout en maintenant une agriculture économiquement performante (Fig. 1).

Pour y parvenir, toute une batterie d'outils a été mise en place comme par exemple :

- la formation des agriculteurs à une utilisation responsable des pesticides : le certiphyto (certificat individuel produits phytopharmaceutiques),
- la création d'un vaste réseau de fermes pilotes pour mutualiser les bonnes pratiques,
- la mise en ligne dans chaque région, de bulletins de santé du végétal qui alertent les producteurs sur l'arrivée des parasites,
- un programme de contrôle de tous les pulvérisateurs qui sont utilisés pour l'application des produits phytosanitaires.

écophyto2018

Réduire et améliorer l'utilisation des phytos :
moins, c'est mieux

Figure 1. Logo de la démarche Ecophyto 2018.

Si les préoccupations environnementales du grand public ont concerné dans un premier temps majoritairement les industries chimiques et métallurgiques, l'agriculture et la viticulture furent progressivement associées aux attentes sociétales relayées par les médias. De nombreuses émissions firent état des risques environnementaux liés à la santé des opérateurs et consommateurs. La dimension culturelle de la vigne et du vin, la forte consommation de produits de protection des plantes pour la protection contre les maladies cryptographiques ainsi que la proximité de cette culture vis-à-vis des zones d'habitation, ont contribué à privilégier le secteur viticole dans les supports médiatiques dénonçant les méfaits des pesticides (Fig. 2).

Au-delà des préoccupations légitimes du consommateur par rapport à l'environnement et à la sécurité alimentaire, les vignerons ont progressivement pris conscience que l'image de leur vin est liée, bien sûr, à sa qualité intrinsèque, mais aussi aux conditions dans lesquelles il est produit, un environnement naturel étend considéré comme le bien commun de tous les citoyens.

Sur le plan législatif, de nouvelles exigences environnementales se sont traduites par une réglementation de plus en plus stricte vis-à-vis de l'utilisation des produits de protection des plantes :

- Restriction, voire interdiction, d'utilisation de certains produits détectés dans les eaux.
- Renforcement de l'impact environnemental dans la procédure d'homologation des produits.
- Procédure d'agrément des distributeurs et des applicateurs des produits agro-pharmaceutiques.

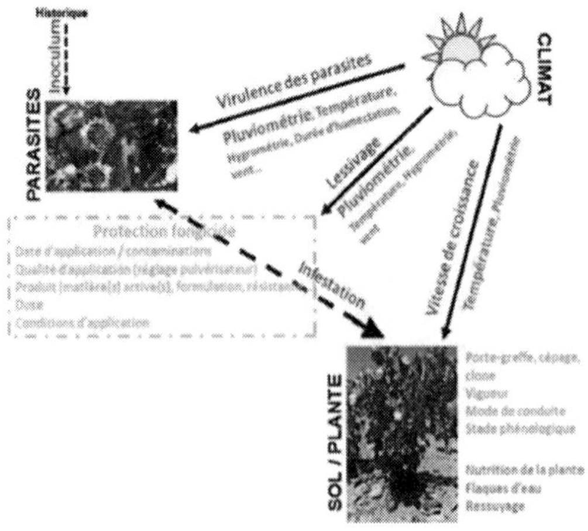

Figure 2. Schéma simplifié de l'efficacité d'une protection phytosanitaire (M. Raynal, A. Davy et coll., IFV).

- Création des zones tampons (Zones Non Traitées) à proximité des points d'eau.
- Renforcement de la protection des manipulateurs de produits phytosanitaires.
- Optimisation de la pulvérisation (contrôle des pulvérisateurs, conception du local de stockage, gestion du rinçage).

2. De la production intégrée au développement durable

Dès les années 1970, des scientifiques de plusieurs pays se regroupèrent au sein de l'OILB (Organisation Internationale de Lutte Biologique et Intégrée). Les agrobiologistes de l'OILB avaient pour objectif la Lutte Intégrée qui repose sur l'association de toutes les solutions s agronomiques permettant de préserver l'équilibre de l'agroécosystème et la santé globale du végétal cultivé (Fig. 3).

Cette approche, limitée au départ à la protection des vergers, puis des vignobles, a été élargie à une notion de Production Intégrée, lorsque les connaissances sur les interactions entre la conduite du végétal, sa nutrition et son état sanitaire ont pu être intégrées dans un raisonnement global, incluant des objectifs qualitatifs pour la production.

L'OILB publia en 1996 un guide qui inspira l'élaboration de référentiels des cahiers des charges institutionnelles ou privés. Parallèlement, l'agriculture biologique et bio dynamique, fondée notamment sur un rejet de l'utilisation des « intrants chimiques », se développa progressivement dans le secteur viticole (Fig. 4).

La fin des années 1970 et début des années 1980 marquaient le terme des «30 glorieuses», caractérisées par l'intensification des méthodes de production qui touche également le monde agricole. Elle permit de réduire la pénibilité des travaux, de régulariser les rendements et de proposer aux consommateurs des denrées abondantes, variées et bon marché. Mais si elle a apporté une réponse à la fois sociale et économique au problème posé par la population au sortir de la seconde guerre mondiale, cette mutation s'est accompagné dans le même temps de multiples effets collatéraux sur l'environnement.

 DOI: 10.1051/bioconf/20170901010

IOBC-WPRS

Figure 3. Logo de l'OILB/IOBC www.iobc-wprs.org.

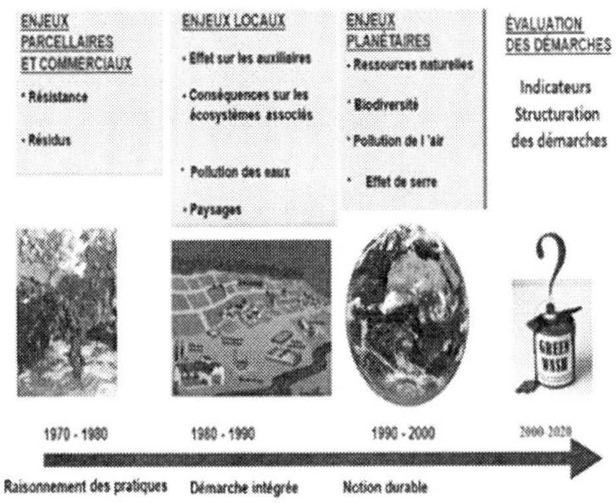

Figure 4. Evolution de la viticulture : du concept raisonné à la viticulture durable (Source J. Rochard) [2].

Dès les années 1970, cette absence de prise en compte des équilibres naturels couplée à la première crise pétrolière déclencha l'émergence de nouveaux courants de pensée (résistance à la société de consommation, prise de conscience des limites de la planète) (Schéma numéro 4). En 1972, le rapport Meadows commandé par le Club de Rome, «Les limites de la croissance», envisageait un effondrement brutal des économies au cours du 21e siècle suite aux effets conjugués de l'évolution démographique, des pressions environnementales et de l'épuisement des ressources.

Depuis la Conférence des Nations Unies de Stockholm en 1972, le problème du rapport entre le développement économique et la détérioration de l'environnement s'est inscrit au centre des préoccupations internationales.

En 1983, L'ONU créa la Commission Mondiale pour l'Environnement et le Développement qui fit apparaître clairement que les effets secondaires du développement industriel posaient de graves problèmes de survie, notamment aux pays en développement (Fig. 5). La Commission élabora alors le concept de développement durable, en opposition avec celui d'une croissance économique sans limites : *«Répondre aux besoins du présent sans compromettre la capacité des générations futures de satisfaire leurs propres besoins »*. En 1992, 172 états étaient réunis à la conférence des Nation Unies à Rio, au cours de laquelle, cinq accords importants furent adoptés, visant à infléchir la démarche traditionnelle du développement.

PNUE

Figure 5. Logo du Programme des Nations Unies pour l'Environnement.

Le programme aborda des problèmes urgents et présente des propositions d'actions dans les domaines sociaux et économiques (lutte contre la pauvreté, modification des modes de production et des schémas de consommation, études des tendances démographiques) et aborde des stratégies de conservation et de gestion des ressources naturelles indispensables à la vie (protection de l'atmosphère, des océans et de la biodiversité, lutte contre le déboisement, promotion d'une agriculture durable, etc.).

La notion de développement durable a mis l'accent sur l'impact de l'homme, par son activité domestique, agricole ou industrielle, sur le devenir des générations futures. Ainsi, la biodiversité, l'utilisation des énergies fossiles, la production de déchet et de gaz à effet de serre, se sont progressivement immiscés dans les itinéraires viticoles. Cette orientation a initié de nombreuses démarches dont la crédibilité s'appuie de plus en plus sur des notions d'indicateurs en lien notamment avec des méthodologies de type Analyse de Cycle de Vie.

En 1996 apparu la norme ISO 14001 sur le management environnemental. Elle repose sur le principe d'amélioration continue de la performance environnementale par la maîtrise des impacts liés à l'activité de l'entreprise (Fig. 8).

Le respect de ses exigences nécessite de mettre en place un système de management environnemental qui inclut :

- Une analyse environnementale permettant de dresser un état des lieux des activités, de la réglementation applicable à ces dernières et des impacts environnementaux qu'elles induisent.
- Une politique environnementale comportant un engagement d'amélioration continue et de prévention de la pollution, de conformité à la législation et à la réglementation environnementale applicable et aux autres exigences auxquelles l'organisme a souscrit.
- La structure organisationnelle, les activités de planification, les responsabilités, les pratiques, les procédures, les procédés et les ressources pour élaborer, mettre

Stratégie de réduction de réduction de l'impact «effet de serre» de la filière viticole française.

Le Bilan Carbone ®développé par l'ADEME a permis de formaliser les démarches de réduction de l'impact sur l'effet de serre de nombreuses structures viticoles françaises avec des plans climats établis notamment dans le bordelais, la Bourgogne, la Champagne (Fig. 7). En complément les projets AGRIBALYSE pour la production de raisins et ACYVIA pour l'élaboration des vins ont pour objectif d'établir une base de données d'Inventaires du Cycle de Vie qui requiert un certain nombre de règles méthodologiques, ainsi que la cohérence entre ces jeux de données (Figs. 6–7).

Figure 6. Logo de la méthode de calcul de l'impact effet de serre Bilan Carbone®.

Figure 7. Logo des méthodes de calcul de l'impact environnemental bilan Carbone, Agribalyse et Acyvia.

Figure 8. Logo de la démarche ISO 14001.

en œuvre, réaliser, passer en revue et maintenir la politique environnementale de la structure.

Si les démarches de développement durable ont surtout concerné dans un premier temps le volet environnemental, l'aspect social et plus globalement sociétal s'intègre de plus en plus dans les réflexions avec l'émergence de la norme ISO 26 000.

La Responsabilité Sociétale des Entreprises (RSE) est un «concept dans lequel les entreprises intègrent les préoccupations sociales, environnementales, et économiques dans leurs activités et dans leurs interactions avec leurs parties prenantes sur une base volontaire ». La norme ISO 26000 contient des lignes directrices et non des exigences. Elle ne se prête donc pas à la certification, contrairement à d'autres normes très connues de l'ISO (Fig. 10). Elle permet en revanche de clarifier la notion de responsabilité sociétale, d'aider les entreprises et les organisations à traduire les principes en actes concrets, et de faire connaître les meilleures pratiques en matière de responsabilité sociétale, dans le monde entier. Elle vise les organisations de tous types, quelle que soit leur activité, leur taille ou leur localisation.

Management environnemental collectif : l'exemple du bordelais

Un Systeme de Management Environnemental collectif au service des performances individuelles a été mis en place par l'interprofession bordelaise (CIVB) (Fig. 9). Il est basé sur des groupes d'entreprises qui avancent ensemble mais chacune a leur rythme. Les fondements de cette demarche reposent sur le partage d'experience et la mutualisation de moyens. En 2011 la première association comptait 28 membres, aujourd'hui, plus de 250 entreprises réparties sur l'ensemble du vignoble bordelais sont engagees dans la demarche du SME soient environ 10 % de la surface du vignoble.

Figure 9. Logo du système de management environnemental du vin de Bordeaux.

Figure 10. Logo de la norme ISO 26 000.

Plusieurs régions françaises bénéficie de ce label (Fig. 12):

- Juridiction de Saint-Émilion
- Val de Loire entre Sully-sur-Loire et Chalonnes
- Coteaux, maisons et caves de Champagne
- Climats du vignoble de Bourgogne.

Dans le prolongement du développement durable, une approche patrimoniale des terroirs en lien notamment avec les paysages à progressivement émergé. C'est ainsi que le classement patrimoine mondial Unesco a concerné une quinzaine de régions au niveau mondial et de nombreuses régions viticoles ont établi des chartes paysagères en

La filière vin française dispose désormais d'un guide de la norme internationale ISO 26000 (responsabilité sociétale des entreprises – RSE) adapté à ses spécificités.

Impulsé par Y. Chabin au sein d'Inter Oc, Interprofession des Vins Pays d'Oc IGP, il a été élaboré collectivement, sous l'égide d'AFNOR, par une quarantaine d'organismes professionnels très engagés pour que chaque intervenant, du raisin à la bouteille, contribue à la préservation des territoires. Les démarches de développement durable se multiplient dans la filière française du vin, en réponse aux attentes croissantes exprimées par les acheteurs français et étrangers. Elles traduisent les actions des entreprises pour préserver l'environnement (comme la préservation des sols, la gestion de l'eau, le traitement des déchets) mais aussi leur contribution au développement local et en particulier leur implantation dans les territoires (Fig. 11).

Figure 11. Accompagnement de l'AFNOR dans les démarches de responsabilité sociétale et environnementale ISO 26 000, `www.afnor.org`.

Figure 12. Label patrimoine mondial Unesco.

lien pour certaines avec le Réseau International Paysages Viticole (Figs. 13–14). La gouvernance territoriale imposée par cette approche, au-delà de l'oeno-tourisme, a souvent également conduit à une réflexion sur les enjeux environnementaux écologiques avec la mise en place de stratégies à l'échelle des communes vis-à-vis des problèmes de ruissellement et d'érosion, la mise en place d'une lutte par confusion sexuelle et parfois des

Figure 13. Logo du réseau international paysages viticole.

Figure 14. Régions françaises qui bénéficient du label Réseau International Paysages Viticoles.

approches collectives de gestion des effluents de cave et de pulvérisation.

Le label Réseau International Paysages Viticoles, établit à partir de la Charte de Fontevraud, témoigne de la qualité des démarches volontaires et concertées sur des terroirs viticoles, où tous les acteurs du territoire (collectivités locales, syndicats de producteurs, opérateurs économiques, culturels et touristiques, universités et laboratoires) sont engagés, à partir d'un diagnostic paysager et un plan d'action.

En lien avec la vision paysagère des terroirs, l'agro-écologie intègre les services écosystémiques fournis par l'environnement autour et dans la parcelle agricole et les utilise notamment pour privilégier les auxiliaires et ainsi limiter l'utilisation des pesticides. Les espaces viticoles étant souvent composés de parcelles de dimensions réduites, les nombreux espaces interstitiels sont propices à la restauration de milieux favorables à la biodiversité, notamment des haies, des murets ou encore des bandes

 DOI: 10.1051/bioconf/20170901010

herbacées et fleuries. La FAO a organisé en 2014 à Rome un symposium international sur l'agroécologie pour la sécurité alimentaire et la nutrition. Elle considère que l'agro-écologie est un moyen viable d'assurer la sécurité alimentaire mondiale, en particulier dans un contexte de changements climatiques.

«Guide de l'agro-écologie en viticulture »,
Elaboré par l'IFV et l'INAO, il a pour objectif de fournir des points de repère techniques pour tous les vignerons qui souhaitent, sur le plan individuel ou dans un cadre collectif, faire évoluer leurs pratiques, afin d'initier ou d'approfondir une démarche agroenvironnementale. Conçu de façon très pratique, ce guide regroupe des mesures d'application simples pour accompagner la filière vers des systèmes de production durables et performants (Fig. 15) [3,4].

Figure 15. Guide de l'agroécologie – `www.vignevin.com`.

3. Principales démarches viticoles initiées en France

A l'initiative de groupes de viticulteurs, de relais en amont ou en aval de la filière de structures professionnelles ou des autorités publiques, différentes démarches ont été engagées dans les régions viticoles françaises.

3.1. Viticulture biologique

En France, la loi d'orientation agricole de juillet 1980, a reconnu complétée par le décret de mars 1981, officiellement l'agriculture biologique en tant que mode de production.

Un logo national (la marque AB) fut proposé en 1985 (Fig. 16). Les cahiers des charges, jusqu'alors privés, ont été homologués à l'échelon national. La même année, le livre vert, document de réflexion élaboré par la Commission Européenne, indiquait de façon claire que la fonction de l'agriculture ne peut se limiter à la fourniture alimentaire mais qu'elle doit intégrer la fonction de protection de l'environnement (Fig. 15).

Figure 16. Logo national agriculture biologique.

Figure 17. Le logo européen «agriculture biologique», encore appelé «Eurofeuille», a pour objet d'aider les consommateurs à repérer les produits biologiques. Sa présence sur l'étiquetage assure le respect du règlement sur l'agriculture biologique de l'Union européenne.

En 2009, les réglementations européennes et nationales relatives à l'agriculture biologique furent abrogées et remplacées par un nouveau règlement, qui permit d'apposer le logo vin biologiques «Eurofeuille» en intégrant un cahier des charges sur le volet élaboration (Fig. 17).

3.2. Agriculture raisonnée

L'agriculture raisonnée est une démarche, en France, qui s'applique aux productions agricoles prenant en compte la protection de l'environnement, la santé et le bien-être animal. Le principe central de l'agriculture raisonnée est d'optimiser le résultat économique en maitrisant les quantités d'intrants utilisées. Le concept a été notamment promu par le Forum des agriculteurs responsables respectueux de l'environnement (FARRE). En France, de 2002 à 2013, le concept était réglementé par les pouvoirs publics (ministères de l'Agriculture et de l'Écologie) par une certification.

Le référentiel porte notamment sur le respect de l'environnement, la maîtrise des risques sanitaires, la santé et la sécurité au travail et le bien-être des animaux.

- Agri Confiance® : Cette démarche est née en 1992 à l'initiative de Coop de France pour l'ensemble du secteur agricole afin initialement d'organiser la mesure de la qualité et structurer la traçabilité dans les structures agricoles. Progressivement elle a intégré le volet environnement en s'appuyant notamment sur la norme Afnor NF V01-007 : «système de management de la qualité et de l'environnement de la

Figure 18. Logo de la démarche Agri confiance.

Figure 19. Logo Terra Vitis®.

production agricole ») Cette certification se conjugue avec l'engagement de chaque agriculteur adhérent à sa coopérative, d'aménager et d'entretenir son territoire, son terroirs, de les préserver, les protéger, par amour de la terre et de la nature, pour léguer aux générations futures des sols fertiles et des environnements sains et authentiques (Fig. 18) [5].

- Terra Vitis® : Née dans la région du Beaujolais en 1998, cette association s'est progressivement développée dans d'autres régions viticoles françaises (Val de Loire, Rhône-Méditerranée, Bordeaux, etc.). L'originalité de la démarche de viticulture intégrée repose sur un double contrôle) qui permet de vérifier le respect du cahier des charges par l'exploitant :
  - Contrôle interne par un technicien du groupement ou un conseiller viticole.
  - Contrôle externe par un organisme indépendant des exploitations et des structures de coordination.

Par cette organisation, l'association gère une marque privée TERRA VITIS valorisée par le biais de macarons, étiquettes, vignettes (Fig. 19).

- Vignerons en Développement Durable® : Concernées par l'avenir de la Terre, la préservation des terroirs et la durabilité du métier de viticulteur, 17 caves coopératives de plusieurs régions (Bourgogne, Rhône Alpes, Provence Alpes Côte d'Azur, Languedoc Roussillon, Pays de la Loire) se sont réunies autour d'une charte collective spécifique au secteur viti-vinicole reposant sur 37 engagements comprenant quatre axes :
  - Gouvernance et Pratiques Managériales
  - Performance économique

Figure 20. Logo Vignerons en Développement Durable®.

  - Responsabilité sociale et territoriale
  - Responsabilité environnementale.

Face aux défis du monde agricole, chacune a su mettre en place des actions garantissant la qualité des produits dans le respect des personnes et de l'environnement (Fig. 20).

- Certification environnementale : La certification environnementale des exploitations agricoles mises en place en 2011, par le Ministère de l'Agriculture, est une démarche volontaire qui vise à identifier et valoriser les pratiques plus particulièrement respectueuses de l'environnement mises en œuvre par les agriculteurs. Elle porte sur quatre thématiques clés : la préservation de la biodiversité, la stratégie phytosanitaire, la gestion de la fertilisation et la gestion de la ressource en eau.

  Afin d'engager l'ensemble de la filière agricole dans une démarche de progrès environnemental, elle est conçue selon une logique de certification de l'ensemble de l'exploitation en trois niveaux.

- Niveau 1 : respect des exigences environnementales de la conditionnalité des aides Pac et réalisation d'un autodiagnostic, vérifié par un organisme habilité, montrant que l'exploitation s'inscrit dans une démarche de progrès pour atteindre les niveaux 2 et/ou 3 ;
- Niveau 2 : respect de 16 exigences rassemblées dans un référentiel. Ce niveau correspond à des démarches existantes (par exemple l'agriculture raisonnée). Le référentiel comporte des obligations de moyens autour de quatre thématiques : biodiversité, utilisation des produits phytosanitaires, fertilisation, gestion de l'eau.
- Niveau 3 : fondé sur des indicateur de résultats relatifs à la biodiversité, la stratégie phytosanitaire, la gestion de la fertilisation et de l'irrigation. C'est le niveau qui permet d'obtenir le label de «haute valeur environnementale ». qui garantit un haut niveau de performance environnementale de l'exploitation agricole. Le retour d'expérience montre que l'engagement des agriculteurs dans le dispositif est beaucoup plus efficace lorsque les agriculteurs sont accompagnés par une structure (coopérative, syndicat, collectivité territoriale…)» ce qui a motivé la mise en place de certification collective par le Ministère de l'Agriculture

 DOI: 10.1051/bioconf/20170901010

Figure 21. Logo de la certification « Haute Valeur Environnementale ».

Figure 22. Logo de la démarche «Viticulture durable en champagne». L'interprofession champenoise a développé depuis plusieurs années une dynamique de viticulture durable, dont le référentiel, dans le cadre d'une gestion collective, permet de bénéficier de la certification «Haute Valeur Environnementale» [6].

avec néanmoins des indicateurs de performance environnementale calculés au niveau de l'exploitation individuelle (Fig. 21).

- Initiatives professionnelles régionales : Conscientes des enjeux environnementaux, certaines structures professionnelles régionales (syndicats, interprofessions, coopérative, etc.) ont souhaité établir un cadre relatif à la production raisonnée. Ces travaux sont bien souvent l'aboutissement de suivis d'exploitations pilotes qui ont permis de cerner les points critiques et d'apprécier les contraintes de mise en œuvre. Certaines ont pu bénéficier du label «Haute Valeur Environnementale» dans le cadre de démarches collectives (Fig. 22).

4. Conclusion

La France, comme beaucoup de pays viticoles du monde a connu au cours des dernières décennies une évolution significative de ses pratiques. Depuis l'après-guerre et les «30 glorieuses» marqués par l'introduction de la mécanisation et l'utilisation intensive des produits phytosanitaires, la viticulture a progressivement évolué vers des pratiques environnementales plus durables. Si la protection du vignoble a été en premier concernée par cette évolution, l'ensemble des itinéraires techniques (gestion des sols, amendement, système de conduite etc.) s'est progressivement intégré dans les cahiers des charges. Ainsi les concepts de viticulture raisonnée, intégrée, durable et pour certain biologique, se sont progressivement imposés dans les exploitations soutenues par les démarches régionales ou nationales et encouragée par les pouvoirs publics. Parallèlement les paysages, la biodiversité et dans le prolongement l'agro-écologie, s'intègrent également dans une vision durable à l'échelle des terroirs et des territoires viticoles. L'enseignement, le développement, la recherche accompagnent cette évolution qui devrait se poursuivre dans un contexte évolutif, notamment en lien avec les changements climatiques.

«L'évolution de l'agriculture vers l'agroécologie doit désormais être un élément structurant des réflexions de tous ceux qui participent à construire l'avenir de l'agriculture. Une réalité s'impose : il est possible de faire coexister les enjeux environnementaux et économiques. Il nous appartient collectivement de dépasser ce qui pouvait ressembler hier encore à des contradictions, des oppositions et surtout d'en tirer profit».

Stéphane LE FOLL

Ministre de l'Agriculture, de l'Agroalimentaire et de la Forêt, Source : Guide de l'agro écologie, IFV/INAO, 2017.

Références

[1] I. Pierot, J. Rochard, *Développement durable vitivinicole : état des lieux international stratégique et règlementaire*, congrès OIV, Bucarest, Roumanie, (2013)

[2] J. Rochard, *Traité de viticulture et d'oenologie durables*, Editions Avenir Oenologie (2005)

[3] IFV/INAO, *Guide de l'agro écologie*, IFV INAO, (2017), www.vignevin.com

[4] J. Rochard, C. Herbin, V. Lempereur, *L'agro-écologie des terroirs: concept, application viticole, exemple de la France,* Congres Brésil, Bento Gonçalves OIV (2016)

[5] Y. Chabin, J. Rochard, *Les coopératives vitivinicoles : quelles voies de durabilité ?* Colloque diversité et durabilité des modèles agricoles coopératifs, Paris, 6–7 novembre (2012)

[6] A. Descôtes, D. Moncomble, *Champagne, environnement et développement durable : 30 ans d'histoire*, Le vigneron champenois, novembre (2013)

 DOI: 10.1051/bioconf/20170901011

Analysis of climate change indices in relation to wine production: A case study in the Douro region (Portugal)

Daniel Blanco-Ward[1,a], Alexandra Monteiro[1], Myriam Lopes[1], Carlos Borrego[1], Carlos Silveira[1], Carolina Viceto[2], Alfredo Rocha[2], António Ribeiro[3], João Andrade[3], Manuel Feliciano[3], João Castro[3], David Barreales[3], Cristina Carlos[4], Carlos Peixoto[5], and Ana Miranda[1]

[1] Department of Environment and Planning (DAO) & CESAM, Aveiro University, 3810-193 Aveiro, Portugal
[2] Physics Department & CESAM, Aveiro University, 3810-193 Aveiro, Portugal
[3] Mountain Research Centre (CIMO), School of Agriculture, Polytechnic Institute of Bragança, Campus de Santa Apolónia, 5300-253 Bragança, Portugal
[4] Association for the Development of Viticulture in the Douro Region (ADVID), 5050 106 Peso da Régua, Portugal
[5] Casa Ramos Pinto, 5150 338 Vila Nova de Foz Côa, Portugal

Abstract. Climate change is of major relevance to wine production as most of the wine-growing regions of the world, in particular the Douro region, are located within relatively narrow latitudinal bands with average growing season temperatures limited to 13–21°C. This study focuses on the incidence of climate variables and indices that are relevant both for climate change detection and for grape production with particular emphasis on extreme events (e.g. cold waves, storms, heat waves). Dynamical downscaling of MPI-ESM-LR global data forced with RCP8.5 climatic scenario is performed with the Weather Research and Forecast (WRF) model to a regional scale including the Douro valley of Portugal for recent-past (1986–2005) and future periods (2046–2065; 2081–2100). The number, duration and intensity of events are superimposed over critical phenological phases of the vine (dormancy, bud burst, flowering, *véraison*, and maturity) in order to assess their positive or negative implications on wine production in the region. An assessment on the statistical significance of climatic indices, their differences between the recent-past and the future scenarios and the potential impact on wine production is performed. Preliminary results indicate increased climatic stress on the Douro region wine production and increased vulnerability of its vine varieties. These results will provide evidence for future strategies aimed to preserve the high-quality wines in the region and their typicality in a sustainable way.

1. Introduction

There is a general acceptance by the scientific community of the reality of climate change in relation to human activities, especially concerning greenhouse gases (GHGs) emissions. Depending on the GHGs emissions scenario, it is expected an increase in global mean surface temperature from 1°C to 3.7 °C by the end of the century when compared to the reference period 1986–2005 [1]. The influence of climate is critical in viticulture and wine production. Year-to-year meteorological variations affect the yield and the optimal environmental conditions for the grape to ripe and, therefore, whether wine typicity for a given 'terroir' or grapevine growing region will be correctly expressed to achieve its full potential. This is known as the "vintage effect" with climate having a greater impact on yield and quality than other environmental factors such as soil type or grapevine variety. In this sense, it is already known that an increase of temperature produces an advance in phenology, which in turn, could have advantageous effects in northern or Atlantic conditions but be detrimental under Mediterranean conditions. As the grapes are exposed to higher temperatures due both to climate change and advanced phenology, the supply of metabolites to the grape is altered, generally causing greater sugar accumulation and higher alcohol levels, lower acidity and variable effects on different aromas and secondary metabolites [2].

Most of the wine-growing regions of the world, in particular the Douro region, are located within relatively narrow latitudinal bands with average growing season temperatures limited to 13–21°C. Therefore, small changes in temperature could end affecting the typicity and style of the wine produced in them, and even produce shifts in their potential for viticulture, making it too warm or better suited to produce quality wines than before [3].

The Multicriteria Climatic Classification System (MCC System) for grape growing regions worlwide [4] along with other bioclimatic indices (e.g. Winkler index, Hydrothermic index of Branas) have been already used to assess the impact of climate change on the suitability for wine production across Europe. The indices were calculated from climate variables obtained from simulations with a grid size of about 25 km performed with the regional climate model COSMO-CLM for the recent-past climate (1960–2000) and for the 21st century (2011–2040, 2041–2070 and 2071–2100) under the Intergovernmental Panel on Climate Change Special Report on Emission Scenarios (IPCC-SRES) B1 and A1B. The A1B scenario corresponds to a balance across all

[a] e-mail: dblancoward@ua.pt

energy sources, whereas in the B1 scenario the emphasis is on environmental sustainability. All simulations were forced by ECHAM5/MPI-OM1 boundary conditions. The results of this study indicated an increased soil water deficit and cumulative thermal effects during the growing season in southern Europe, which could imply negative effects for wine production for these areas unless suitable adaptation measurements (e.g. rootstock and variety selection, training system, irrigation) are taken. In contrast, western and central Europe could benefit with higher quality potential for the grape and even new potential areas for wine production [5].

In another study, the atmospheric variables taken from coupled global and regional climate models (GCM-RCMs) simulations in combination with the MCC system were also used to assess present and future scenarios for the Portuguese grapevine growing regions. For present scenarios (1950–2000), the WorldClim project 1 km high resolution dataset [6], validated with E-OBS observational data [7] was used. To assess the impacts of climate change on future viticulture suitability in Portugal, the period 2014–2070 under the IPCC-SRES A1B was chosen. An ensemble of 13 RCM simulations driven by 3 different GCMs produced by the ENSEMBLES project [8] was selected. To pass from the coarse grid RCM resolution $0.22° \times 0.22°$ (~ 25 km) to the 1 km grid size of the WorldClim baseline dataset, a bi-cubical interpolation (pattern downscaling or 'delta' method) was used. The final results illustrated significant changes in the current bioclimatic viticultural Portuguese zones as they depict a lower bioclimatic diversity and a more homogenous warm and dry climate for most Portuguese wine regions [9].

With regards to specific studies for the Portuguese Douro valley region, a multivariate linear regression analysis related a long wine production series (1932–2010) collected by Porto and Douro Wines Institute (IVDP), with atmospheric variables (namely, temperature and precipitation) selected from the Vila Real station available for the period 1941–2010. Results indicated that high rainfall and cool temperatures during bud burst, shoot and inflorescence development (February-March) and warm temperatures during flowering and berry development (May) are generally favourable to high production. A logistic regression was also developed to predict three vintage yield categories (low, normal and high). Both regression approaches were then applied to climate parameters computed from the atmospheric variables provided by 16 regional climate model experiments following the SRES A1B scenario to estimate recent-past and future changes for the period 2001–2099. The datasets ranged from 25 to 18 km resolution and originated from the ENSEMBLES project [8]. Results indicated an overall positive impact under that SRES scenario with an increase of 10% in production by the end of the 21st century and an increase of high production years from 25% to over 60%. However, this study also cautions about the fact that the modelling set-up used does not take yet into account possible changes during the ripening period (e.g. rising heat stress).

There is also another pertinent statistical work relating wine production to climate in the Portuguese Douro Demarcated Region (DDR). A series of climate parameters and bioclimatic indices derived from weather stations from the Portuguese National Meteorological Service present in the Douro valley area were related with yield and quality vintage records. Temperature was introduced in the model as the successive time length periods required to reach a given threshold of heat accumulation or thermal time to pass from a given grapevine phenological phase (e.g. bud burst) to the next one (e.g. flowering). The model takes into account four key grapevine phenological phases (bud burst, flowering, '*véraison*' or berry onset, and maturity). Precipitation entered the model as accumulated totals from January to March, April to June and July to September. Finally, the growing season mean temperature from March to September was taken into account as well. The results of this study illustrated that weather characteristics (e.g. growing season temperatures above the mean, warm winters, cool temperatures during ripening) are strongly associated with better quality vintages [10, 11].

Most recently, a grape production model (GPM) has been developed that relates daily historic temperature and precipitation from the E-OBS observational dataset [7] with the grape production of three wineries within the Douro valley region. The model compares thermal/hydric conditions in a given year against the average conditions for high and low yield vintages. Results indicate that relatively cool pre-flowering temperatures and relatively warm conditions during berry development favour higher yields. Higher production is also associated with years with precipitation above the mean conditions before the flowering stage [12].

The present research makes use of an ensemble of Weather and Research Forecasting (WRF) model simulations driven by two forcings, namely the European Reanalysis (ERA) Interim and the Max Planck Institute Earth System - low resolution model (MPI-ESM-LR) GCM. The simulations were performed for 20-year periods as adopted by the IPCC 5th Assessment Report [13], namely 1986–2005 for the recent-past, 2046–2065 for the mid-future and 2081–2100 for the long-future climate using the Representative Concentration Pathway (RCP) 8.5 new emission scenario [14]. More details on the climatological modelling can be found in the Data and methods section.

The purpose of this study is to evaluate whether the ERA-interim driven WRF simulations for the recent-past time period (1986–2005) can be used successfully to assess the influence of climate parameters, and bioclimatic and climate change indices on vintage yield and quality records concerning the Portuguese Douro Demarcated Region (DDR). The GCM MPI-ESM-LR driven simulations are used to assess the possible impacts of the estimated climate changes under RCP8.5 for the study area in the light of the relationships found with the ERA-interim driven WRF dataset. Compared with other work previously done, this study relates regional high-resolution 9×9 km WRF simulations directly to DRR vintage yield and quality records evaluating the usefulness of computing climate parameters, and bioclimatic and climate change indices, taking into account key grapevine phenological stages (bud burst, flowering, *véraison*, maturity) based on specific heat requirements of local varieties in the region.

2. The study area

The Portuguese Douro Demarcated Region (DDR) runs along both margins of the Douro River from its midcourse

DOI: 10.1051/bioconf/20170901011

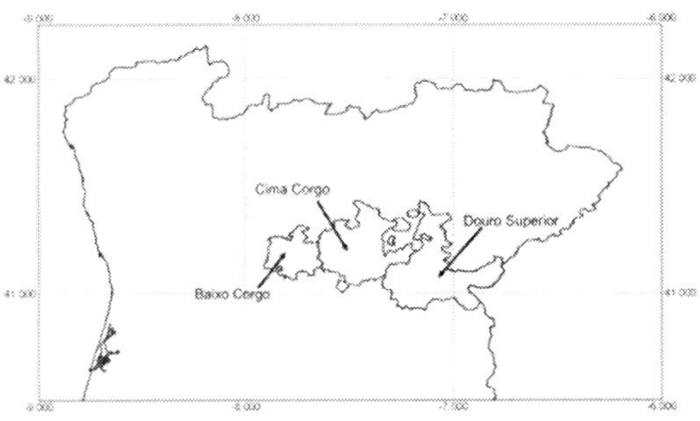

Figure 1. Demarcated Region of Douro (Portugal).

in the East up to the border with Spain in the West. The western most area of the region is located 70 km from the Atlantic Ocean. The Douro Valley extends along 90 km in the West-East direction and along 50 km in the North-South direction (Fig. 1). The landscape is characterized by mountainous terrain, rising above the Douro River and its tributaries, with moderate to steep slopes and varying exposures. The geology of the Douro Valley is dominated by schistose-layered rock, oriented nearly vertical, with some outcrops of granite. The average elevation over the entire region is 443 m, but ranges from a low near 40 m to a high of just over 1,400 m [15]. The Region covers approximately 250,000 hectares with vineyard area representing roughly 43,480 hectares, 17.4% of the total land area. The DDR is divided into three sub-regions: Baixo Corgo, Cima Corgo, and Douro Superior. The Baixo Corgo covers the smallest area (45,000 ha) with the Cima Corgo the next largest (95,000 ha) and the Douro Superior the largest sub-region (110,000 ha). The vineyard figures for these sub-regions are 13,368 ha, 20,270 ha and 9,842 ha respectively [16].

The region is characterized topographically by sloping vineyards arranged in various terraced configurations. These terraces were created and perfected throughout the centuries enabling man to cultivate vines on the steepest slopes. After the phylloxera epidemic, the devastated small terraces were abandoned and new, wider and steeper terraces were built, with or without supporting walls and allowing for a greater planting density (approximately 6,000 vines per hectare). It was then, too, that vineyards were planted according to the natural slope of the land. The introduction of mechanization to the region at the end of the 60's and early 70's, led to the appearance of a new cultivation system composed by horizontal terraces with earth supporting walls, each bearing 1–2 rows of vines and with a low planting density of some 3,000 to 3,500 vines/ha. More recently, as an alternative to the wide terraces, vines are being planted in vertical rows rising up the steeper hillsides. With a planting density similar to that of traditional vineyards (about 4,500 to 5,000 plants per/ha), this system is better adapted to small plots of land with up to 40% slopes and can be worked by mechanical means [17].

The DRR has a Mediterranean climate, with highly variable rainfall events, concentrated in winter months, and hot summers. It is sheltered from Atlantic wet and cold winds by two mountain ranges, Marão and Montemuro, located at its western border. Temperature increases and precipitation decreases from West to East. The westernmost sub-region inside the Douro Valley (Baixo Corgo) is nearer to the Atlantic Ocean and therefore more affected by the moist maritime winds. The eastern most regions within the Douro Superior sub-area are more distant from the Atlantic Ocean therefore having a more continental climate influence. The region is classified as a warm temperate climate (Köppen Csb), with average annual temperatures during 1980–2009 of 15.4°C, average daily minima temperatures (T min) in the coldest month dropping to 2.7°C, and average daily maxima temperatures (T max) in the warmest month reaching 32.1°C [10]. Mean growing season temperature (GST) from April to September for the same climatological period is 20.6 °C. Growing season precipitation (GSP) has a mean value of 193 mm, representing 30% of the annual total (624 mm). The average precipitation of the driest month (July) is just 11.2 mm. Low precipitation values along with high temperatures and high radiation exposure give rise to situations of intense summer plant-soil-water stress, particularly in the Cima Corgo and Douro Superior sub-regions [3]. The Huglin Index (HI) for the 1980–2009 period averaged 2,740 °C d^{-1} whereas the cool night index (CI) 13.6 °C and the dryness index (DI) −126 mm. In the Geoviticulture MCC System [4] the DDR climate is currently classified as HI+2/DI+2/CI+1 (Warm/Very dry/Cool nights) [18].

3. Data and methods

3.1. High resolution 9 x 9 km WRF climate simulations

Marta-Almeida et al. [19] performed an ensemble of WRF high-resolution climate simulations driven by two forcings, namely ERA-Interim reanalysis and the Max Planck Institute Earth System - low resolution model (MPI-ESM-LR). The reanalysis data were obtained from the European Centre for Medium-Range Weather Forecasts (ECMWF) through the ERA-Interim project, with a horizontal resolution of approximately 79 km [20]. The MPI-ESM-LR is a global Earth System Model developed by the Max-Planck Institute, with a 1.9°horizontal resolution [21] which corresponds to about 160 km horizontal resolution. This model participated in the Coupled Model Intercomparison Project Phase 5 (CMIP5), which uses new emission scenarios, namely the RCPs.

Currently there is a new set of RCPs associated to future concentrations of GHGs in the atmosphere. In this study we used the RCP8.5, defined by a radiative forcing of 8.5 W m^{-2} by 2100 and a continuous increase after this year [22]. The RCP 8.5 provides an updated and revised quantification of the original IPCC A2 SRES scenario [23].

The WRF high-resolution climate simulations were performed for 20-year periods as adopted by the IPCC 5th Assessment Report [24], namely 1986–2005 for the recent-past, 2046–2065 for the mid-future and 2081-2100 for the long-future climate. These simulations were implemented for three nested domains with increasing horizontal resolutions, namely 81, 27 and 9 km (Fig. 2). Validation of the WRF recent-past simulations driven by the two forcings with observational temperature and precipitation datasets for the Iberian Peninsula provided

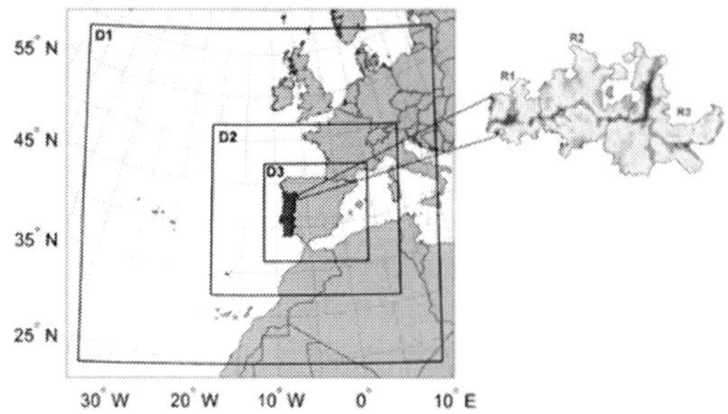

Figure 2. Nested model domains used in the regional WRF implementation with resolutions of 81 (D1), 27 (D2) and 9 (D3) km^2.

acceptable comparisons of the probability distributions of temperature and precipitation for both models, with WRF-ERA providing better results most of the times although there are occasions where WRF-MPI performed better [14].

3.2. Phenological modelling

The concept of 'thermal time' as the daily temperature summation above a temperature base required for a given crop to complete a specific phenological stage was established by Monteith [25]:

$$\theta = \int (T - Tb)dt \tag{1}$$

Where θ(°C day) represents the thermic duration or thermic time for a specific phenological stage, T is the daily average temperature and Tb is a threshold above which there is plant development within that phenological stage. Models based on growing degree days (GDD, °C day) have been extensively used to model the phenology of the grapevine [26,27] or as grapevine zoning bioclimatic indices [28–30].

A specific phenological squeme partitioning the annual development cycle of the grapevine into three major phases (b – bud burst to flowering, f – flowering to *véraison*, and v – *véraison* or berry onset to maturity) has been already proposed for the study area. A.C.Real et al. [10,11] used a mixed grapevine variety arrangement as the 'Tinta Roriz' variety was used to define the required GDDs to reach bud burst, flowering and *véraison* and the 'Touriga Franca' variety was used as the set-up for the required GGDs to reach maturity in the DDR. This was done in order to simulate an intermediate varietal behaviour (neither an early nor a late grapevine variety class) for all of those phenological stages using two varieties present extensively in the Douro valley region.

This work makes use of the mixed 'Tinta Roriz-Touriga Franca' local grapevine varietal phenological model to compute the timings of the key grapevine phenological timings (b-bud burst, f-flowering and v-*véraison*) for the four climate modelling set-ups presented in Sect. 3.1 (WRF-ERA recent-past, WRF-MPI recent-past, WRF-MPI mid-future and WRF-MPI long-future) based on the specific varietal GDD requirements as observed at the Portuguese National Ampelographic Collection [31]. Results are contrasted with recently-made available specific phenological observations for the Douro valley region [32] in the Results section.

3.3. Climate parameters and indices

Using the daily maximum and minimum temperatures and the daily total precipitation from the four different regional WRF modelling set-ups (WRF-ERA recent-past, WRF-MPI recent-past, WRF-MPI mid-future and WRF-MPI long future), a total of 26 climatic parameters, 4 grapevine bioclimatic indices, and 28 climate change indices were calculated.

3.3.1. Climate parameters

The climatic parameters consisted of computing the average of daily maximum temperatures, daily minimum temperatures and daily mean temperatures (as T max + T min / 2) and the total precipitation (P) for each one of the yearly phenological stages based on the conventional monthly scheme for the northern hemisphere (bud burst usually occurs between April and May, flowering relates more often with June and July and *véraison* usually happens between August and September). Therefore, a set of twelve climatic parameters was calculated following the conventional scheme (e.g. T max Apr-May, T min Apr-May, T avg Apr-May, P Apr-May etc.). Another set of twelve climatic parameters was computed taking into account the yearly dates according to the phenological modelling arrangement (T max-b, T min-b, T avg-b, P-b etc.). Finally, two other climatic parameters were computed following the conventional scheme (months from April to September) to summarize information for intercomparison purposes: growing season average temperature (GST) and growing season total precipitation (GSP).

3.3.2. Bioclimatic indices

Four specific grapevine bioclimatic indices were selected: Winkler index (WI), the Heliothermal index of Huglin (HI), the cool night index (CI) and the dryness index (DI). These indices are commonly used to characterize grapevine growing areas worldwide in an standardized way and as an agronomic zoning tool [4], [33].

The Winkler index is the sum of the daily average temperatures above a threshold temperature of 10 °C considered the active grapevine temperature (the temperature above which it activates its vegetative cycle) along the growing season. The index is usually calculated from monthly data:

$$WI = \sum_{April1st}^{October31st} (Tavg - 10°C), Tavg \geq 10°C$$
$$Tavg = \frac{T\max + T\min}{2} \tag{2}$$

The HI provides information regarding heliothermal and sugar potential. It is very much correlated with the Thermal Index of Winkler ($r^2 = 0.98$ over 97 grape-growing regions worldwide) but, according to Tonietto et al. [4], is more pertinent to the qualitative factors (e.g. berry sugar potential):

$$HI = \sum_{April1st}^{September30th} \frac{(T\max - 10°C) + (Tavg - 10°C)}{2} d$$
$$Tavg = \frac{T\max + T\min}{2} \tag{3}$$

Where again, Tavg is the mean air temperature (°C), Tmax is the maximum air temperature (°C), d is length of day coefficient ranging from 1.02 to 1.06 between 40° and 50° of latitude. A value of 1.02 was assumed for a latitude between 40°01′ and 42°00′. This index is usually also calculated from monthly climatic means.

The purpose of CI is to improve the assessment on the grape qualitative potentials, notably in relation to secondary metabolites (polyphenols, aromas) in grapes [4]:

$$CI = \sum_{September\,1st}^{September\,30st} \frac{T\min}{30} \tag{4}$$

Finally, DI indicates the potential water availability in the soil, related to the level of dryness in a region. It is also related with the level of grape ripening and wine quality [4].

$$W = W_0 - P - T_v - E_s \tag{5}$$

For intercomparison reasons, it is also calculated on a monthly basis during the same period used for HI (April 1st to September 30th), which is acceptable for most of grape growing regions in the northern hemisphere. W is the estimate of soil water reserve at the end of the April 1st–September 30th modelled growing season period, W_0 is the initial soil water reserve, which can be accessed by the vine roots, P total monthly precipitation, T_v the potential transpiration of the vineyard and E_s the direct evaporation from the soil. To compute T_v and E_s is also necessary to compute the monthly total potential evapotranspiration. This is usually done by the Penman-Monteith method but as we only worked with temperature and precipitation records, it was approximated by the Hargreaves method which produces comparable results in arid and semiarid environments and requires temperature data only [34]. The result is mm of water in the soil. The initial W_0 is usually taken as 200 mm [4].

3.3.3. Climate change indices

A group of six climate change indices from the group of indices developed by the Expert Team (ET) on Climate Change Detection and Indices (CCDI) [35] related with the incidence of cold waves, storms and heat waves, which could proof to be relevant for the grapevine, were calculated on a yearly basis. Namely, these indices were:

a. SU25, the number of summer days with T max >25 °C.
b. SU35, the number of very hot or stressful days with T max >35 °C.
c. CSDI, cool spell duration index or total number of days being part of cool spells longer than 6 consecutive days in duration.
d. WSDI, warm spell duration index or heatwave index, or the total number of days being part of warm spells longer than 6 consecutive days in duration.
e. R10, the number of days with heavy precipitation (daily precipitation >10 mm).
f. CWD, maximum number of consecutive wet days where daily precipitation >1 mm.

Table 1. Sources of yearly quality ratings for Port vintages.

Source	Acronym	Rating
(1) Berry Bros & Rudd	BBR	1–10
(2) Decanter	DC	1–5
(3) Instituto dos Vinhos do Douro e do Porto	IVDP	0–1
(4) Michael Broadbent	MB	0–5
(5) Sotheby's Wine Encyclopaedia Wine Encyclopaedia	SWE	0–100
(6) Vintages	VT	0–10
(7) Wine Advocate	WA	50–100
(8) Wine Enthusiast	WE	50–100
(9) Wine Spectator	WS	50–100

Table 2. Original vintage scores for 1986–2005 according to the vintage chart rating systems presented in Table 1 and a consensus ranking (CR).

Vintage	CR	1	2	3	4	5	6	7	8	9
86	18			0					84	
87	11		3	1	3		8		85	88
88	19			0					83	
89	10			1	3				86	
90	17			0	3				85	
91	5	7	4	1	4	95	9		92	93
92	6	8	4	1	4	85	9		93	94
93	20			0						
94	1	9	5	1	4	95	10		96	99
95	9		1	0	3	88	9		91	92
96	16		2	0	3				85	
97	4	8	4	1	4	90	10		93	96
98	13	6	3	0	3	80			87	
99	15			0	3	75			86	
00	2	9	5	1	5	95	10		90	97
01	12		4	0	3	86	8		84	
02	14		2	0	3	70			84	
03	3	8	4	1	5	94	10		96	98
04	8		4	0	4	88	9		90	
05	7		5	0	5	80	8		91	

A specific index for the grapevine, SU33, the number of very hot or stressful days for the grapevine (T max >33 °C) was added to the previous set. The final seven climate change indices were computed for each phenological stage as well (bud burst, flowering and *véraison*) yielding another 21 climate change indices computed on a yearly basis.

3.4. Porto vintage yield and quality data

The average yield (hl/ha) of all types of wines produced in the Douro valley region for the recent-past 1986-2005 period was considered. Data were available from [18] consisting on a compilation of data from the Portuguese Office for National Statistics (INE) taking into account the increase of planted areas since 1982.

Data on Port vintage quality were available from [18] and [10] consisting of ratings from six different sources, as illustrated in Table 1 where the highest values correspond to the best Port vintages.

The original vintage scores for the recent-past period (1986–2005) along with a derived consensus ranking (CR) can be found in Table 2 as compiled in [10].

DOI: 10.1051/bioconf/20170901011

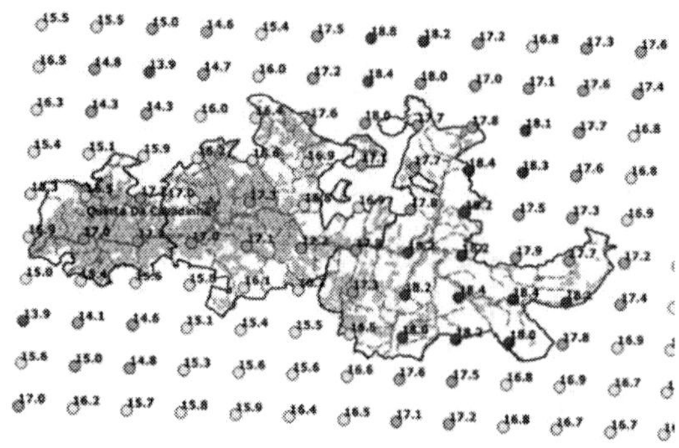

Figure 3. Growing Season Average Temperature (GST) as calculated by the 1986–2005 WRF-ERA simulation. The circles represent 9 × 9 km cell centres. Each WRF cell is associated with its GST value (°C). WRF cells with highest GST values are coloured with red tones whereas cooler areas are depicted with bluish tones. The DRR limits are represented with its main subdivisions (Baixo Corgo, Cima Corgo, and Douro Superior), main river courses and grapevine cultivated areas.

3.5. WRF cell selection and statistics

One 9 × 9 km cell representative for each of the three subdivisions (Baixo Corgo, Cima Corgo and Douro Superior) present in the DDR region was selected to compute the timing of different key phenological stages (bud burst, blooming, *véraison* and maturity) for each year belonging to the simulated 1986–2005 recent-past period. This was done by taking into account the GST values for each WRF-ERA cell within the DRR limits at the same time they were overlain on mapped vineyard areas as provided by the CORINE 2009 dataset [36] in a Geographic Information System (GIS) set-up as illustrated in Fig. 3 of the Results section.

Data on normalized vintage yield and quality scores, climate parameters, bioclimatic and climate change indices were submitted to a Pearson correlation test using the data visualization and statistical programming R software. Tests for the significance of differences in means and variances [37] of the statistical distributions of climate parameters and indices between the different 20-year periods (recent-past, mid-future, long-future) as modelled by WRF-MPI were also implemented in R.

4. Results

4.1. WRF-ERA data validation

The high-resolution WRF simulations forced by ERA-Interim reanalysis resulted in realistic patterns of surface atmospheric variables. Figure 3 shows the spatial pattern of the grapevine growing season average temperature (GST) over the Douro valley as modelled for the final nested highest resolution (9 × 9 km) domain. Within the DRR, the GST values portray an East-West trend resulting from the orientation of the main valley itself and the increasing distance from the sea.

Table 3 portrays the resulting mean *véraison* dates along with the WRF-ERA GST median statistics obtained for each one of the DRR sub-regions. The GST median values are close to those reported for the area by [3] based on the WorldClim 1 km × 1 km 1950–2000 database [6]. Only the cell selected for Douro Superior sub-region presents an average *véraison* date within the 95% confidence limits of the timings reported for the area [32]. These rather delayed phenological figures could be attributed to an important difference between the average height of the selected WRF-ERA cells (434 m for the Baixo Corgo cell, 481 m for the Cima Corgo cell and 417 m for the Douro Superior cell) and that of the field plot where the phenological dates were observed (Quinta da Cavadinha, 205 m). In fact, the average height of the selected WRF-ERA cells corresponds more to transitional sub-plateau locations than to sheltered valley areas where the Mediterranean climate characteristics are stronger.

A closer inspection of the ombrothermic chart (Fig. 4) for the Douro superior WRF-ERA cell against climatological observations reported by an equivalent sub-plateau location in that sub-region, reveals a quite close match for precipitation (except for February and March). However, temperatures are significantly lower for all months (−1.8 °C on average), which could be in part explained by the difficulty to reproduce the specific mesoclimates associated to the varied topography and shelter effects present in the area. WorldClim figures for annual precipitation in the Douro Superior sub-region are significantly higher (median 832 mm) than those reported by the literature for a sub-plateau location (in the order of 500 mm) [3,38].

Table 3. GST and *Véraison* statistics for the three DDR sub-regions.

1986–2005	GST medians		*Véraison*date	
Region	Model	WorldClim	Model	Field
Baixo Corgo	17.1	17.5	234	175–229
Cima Corgo	17.0	17.5	234	
D. Superior	18.0	18.0	223	

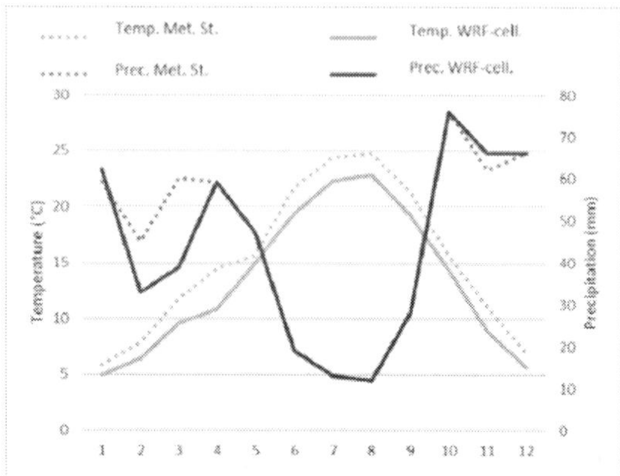

Figure 4. Ombrothermic chart comparing monthly average temperatures and total precipitations from the WRF-ERA cell selected within the Douro Superior region compared with literature sources [38]. Solid lines are for WRF-ERA cell data and dashed lines are for meteorological observations from the literature.

Table 4. Correlation coefficients for the different vintage rating systems (* 95% significance level, ** 99% significance level).

	WE	WS	SWE	MB	BBR
WS	0.85**				
SWE	0.73**				
VT	0.76**	0.96**	0.79**		
MB	0.74**	0.78*			
DC				0.76**	0.94**

Table 5. Correlations among climate parameters and indices, grapevine yield and vintage quality (* 95% significance level, ** 99% significance level).

	Yield	WE	MBR	DC
GDDgs	0.56**			
HI	0.52*			
T max-b			0.51*	
T max-v	0.52*			
SU33-f		0.49*		
SU35-f		0.64**		
WSDI				–0.62*

4.2. Vintage yield and quality in relation to climate

Table 4 illustrates the significant correlations found among the normalized scores of the different vintage rating systems applied to vintage Port used in this study. All of these rating systems have high and significant Pearson's correlation coefficients with at least other two within the 1986–2005 period with the exception of BBR which presents only one significant correlation and WA with none (thus, not shown in the table). As the number of Port vintages specifically evaluated was higher in WE (19), MB (17) and DC (14) compared to the other rating systems, these were the final vintage chart systems selected to analyse the relationship between Port wine vintage quality and climate. It is worthwhile noticing that yield presented significant negative correlations with two of the rating systems used in this study (–0.74* WS, –0.72* VT). This suggests an inverse relationship between yield and quality.

The significant correlations found among the climate parameters and indices, as computed from the daily temperature and precipitation records from the Douro Superior sub-region WRF-ERA cell, and the corresponding vintage yield and quality records for the recent-past 1986–2005 time period were all of moderate type from a statistical point of view, as it can be observed in Table 5. However, it is worth noticing that, within a viticulture context, correlation values as low as 0.54 have been considered highly significant in an study relating water deficit with vintage quality [2]. These correlations are higher than those found in another study at a close latitude (Galicia, Northwest Spain) for the case of T max-b and T max-v [39] showing the benefit of using phenological timings based on accumulated heat for some of the climate indices and parameters (e.g. T max-b, T max-v, SU33-f, SU35-f) rather than using the conventional monthly schemes.

pcElevated heat sums (GDDgs, HI) along with elevated mean maximum temperatures from*véraison* to maturity time (T max-v) are related with high grapevine yields in the study area within the 1986–2005 time period, which is consistent with other studies performed in the area [10,12,40]. Port vintage quality ratings are positively related to the total incidence of days with T max >35°C from the flowering to the *véraison* stage (SU35-f), and to the mean maximum temperatures achieved from the bud burst to the flowering stage (T max-b), but they are inversely related to the occurrence of heat waves (WSDI). In one hand, these results could relate with the fact that the temperate range that favours vegetative development –and therefore yield- is slightly higher (23-25 °C) than the one which is considered best for fructification (22–22 °C) [41]. On the other hand, a moderate incidence of non-contiguous, stressful, scalding days between flowering and *véraison* (SU35-f) could result in a smaller number of grapes developed, which would result in a greater chance of a sufficient supply of photoassimilates to them [42]. At this point, it can be noted that another positive relation exists between vintage quality and the total incidence of days with T max > 33 °C from flowering to *véraison*, SU33-f, but the correlation coefficient was lower and less significant than that of SU35-f. Finally, prolonged periods of scalding days or heatwaves (WSDI) appear negatively related with vintage quality which might have to do with excessive plant damage or too unfavourable conditions for a balanced supply of metabolites (polyphenols, aromas) to the grape [4].

Table 6. Top quality and regular vintages from the climate perspective: positive factors are displayed in green cells whereas negative ones in orange.

Year	CI	V	T max-v	SU35-f	WSDI	CWD-v
Outstanding vintages						
95	9.9	229	23.3	4	0	3
00	12.7	228	23.3	3	0	3
03	12.8	217	29.3	7	14	3
97	15.2	217	28.0	5	6	2
91	14.4	225	29.0	6	0	3
Not outstanding vintages						
96	11.1	225	23.3	0	0	3
90	14.8	216	30.6	6	0	1
86	15.2	235	24.4	0	0	3
88	11.7	236	24.0	0	8	3
93	11.9	225	20.4	8	6	12
Avrg	12.7	223	26.4	4	7	

Table 6 portrays the five highest and the five lowest quality Port vintages for the 1986–2005 recent-past time period as scored according to a consensus ranking (CR in Table 2) taking into account all vintage rating systems.

When contrasted with the values of climate parameters and indices as computed from the Douro Superior sub-region WRF-ERA cell, it can be observed that, in one hand, all of the outstanding vintages had a moderate incidence of non-contiguous, isolated, 'extremely hot' or 'scalding' days from flowering to *véraison* (SU35-f). Despite this fact, the two top quality vintages present below-average *véraison* to maturity mean maximum temperatures (T max-v) and cool to very cool September mean minimum temperatures (CI), which are conditions of slow maturation and high qualitative potential for the grape. On the other hand, three of the worse classified vintages (96, 86 and 88) totally lack any incidence of 'scalding days' from flowering to *véraison* time (SU35-f) with two of them portraying a considerable delay on the onset of *véraison*. Vintage 90 displays incidence of a suitable number of SU35-f days but the average maximum temperature from *véraison* to maturity (T max-v) was considerable higher

DOI: 10.1051/bioconf/20170901011

Table 7. Future scenarios for some key grape growing parameters (* 95% significance level, ** 99% significance level).

		MPIr	MPIm	MPIl
GST (°C)	μ	15.6**	18.0**	20.1**
	σ^2	0.8	0.9	0.8
Véraison **onset**	μ	236**	213**	198**
	σ^2	159	84	57
SU35 (°C)	μ	2**	15**	42**
	σ^2	12**	71*	176*

(30.6 °C). GGDs for that vintage were also quite elevated (1664 GGDs against an average of 1525 GDDs) being the vintage of highest yield of the 1986–2005 recent-past period. Finally, vintage 93 clearly shows too low mean maximum temperatures from *véraison* to maturity (T max-v=20.4 °C) accompanied with a too prolonged contiguous rain period from *véraison* to maturity time (CWD-v = 12) which caused quite unfavourable conditions both for vintage yield (19th ranking position regarding yield) and quality (last position not being scored for any of the vintage charting systems).

4.3. Future scenarios

The GST, the *véraison* timing and the SUF35 ETCCDI index were calculated for the previously selected Douro superior 9 × 9 km cell according to the daily temperature records provided by the WRF-MPI 1986–2005 recent-past (MPIr), 2046–2065 mid-future (MPIm) and 2081–2100 long-term (MPIl) RCP8.5 modelling scenarios. Tests for stability of means (t-test) and variances (F-test) (Table 7) illustrate a significant increase of GST means (2.4 °C for the mid-future, and 2.1 °C more for the long-term future) not accompanied by significant differences in GST variance. Regarding timing of *véraison*, there is a significant advancement in the means (22 days for the mid-future and other 15 more days in advance for the long-future compared with the recent-past period) without significant differences in the variances. Finally, the mean number of 'scalding days' or too hot days, SU35, increases very significantly (from 2 days to 15 days in the mid-future and 27 days more in the long-term future) with a significant increase in their variances too.

Although the change of GST provided by these modelling scenarios remains within the general 13–21 °C interval considered as suitable to grape growing, it can be clearly observed that an ETCCDI index such as SU35 will pass from tolerable levels (a mean value of 2 incidences per year according to WRF-MPIr and of 7 according to WRF-ERA) to much higher frequencies covering, on average, half a month for the mid-tem future scenario, and almost a month and a half for the long-term future scenario. These results suggest a strong likelihood of present-baseline heatwaves and water deficits.

5. Conclusions

Several conclusions can be drawn from this study at the research level:

a. Climate parameters and indices calculated from daily climate variables from the WRF-ERA climate simulation for the recent-past time period (1985–2005) can be successfully related with vintage yield and quality scores records.
b. Results confirm previous studies on the positive current relation between higher growing season heat accumulations and greater vintage yields.
c. A moderate incidence of 'scalding' days from the flowering to the *véraison* stage has been found positively related with vintage quality.
d. The negative relation between vintage quality and the incidence of heat waves also confirms previous studies and much background work regarding berry ripening.
1. Correlations found by using a phenological set-up are usually higher than those found using a regular monthly set-up.
e. The changes in climate estimated under the RCP8.5 scenario are triggering advancement on phenology in a very significant way with the consequent risk of overexposing the grapes to too warm conditions for balanced ripening.
f. Although GST remains within range for quality wine production, a very significant increase of 'scalding' days is forecasted for both mid and long-term future scenarios, which also increases the likelihood of incidence of present-baseline heatwaves. These conditions can impair the ripening of the berries with regards the high quality standards and the wine typicity of the region.

As the future modelled WRF-MPI climate scenarios illustrate and advancement on vine phenology and changes in temperature conditions in the DRR, where grape ripening can be significantly affected with regards to the balance of sugars, acidity, secondary metabolites and aromas, several actions can be proposed to preserve as much as possible DRR wine typicity taking current state-of-the-art knowledge [2] and previous work done in the study area [3]:

a. Too early ripeness can be delayed in the order of seven to ten days by means of clonal selection and use of suitable rootstocks. Longer delays can be achieved by using late-ripening varieties or carefully selected non-local varieties.
b. Training systems and late pruning can also be useful to delay phenology. Training systems can also have an impact on water deficit. Care should be taken to maintain an appropriate leaf area/fruit weight ratio.
c. Increase the water-holding capacity (SWHC) of soils or look for soils with greater SWHC and select rootstocks resistant to water deficits.
d. Irrigation of a drought-resistant plant such as the vine should be considered a last option as it has an economic, environmental and social cost.
e. Move the vineyards to higher altitudes.

The authors wish to thank the financial support of the DOUROZONE project (PTDC/AAG-MAA/3335/2014; POCI-01-0145-FEDER-016778) through the Project 3599 – Promoting the scientific production and the technological development, and thematic networks (3599-PPCDT) and through FEDER.

DOI: 10.1051/bioconf/20170901011

List of abbreviations

1. **DDR:** Douro Demarcated Region.
2. **MCC:** Multicriteria Climatic Classification.
3. **ERA-Interim:** European Reanalysis-Interim forcing.
4. **MPI-EMS-LR:** Max Plank Institute Earth System low resolution model forcing.
5. **WRF:** Weather and Research Forecasting model.
6. **T max:** daily maximum temperature.
7. **T min:** daily minimum temperature.
8. **T avg:** daily mean temperature as (T max + Tmin)/2
9. **P:** daily total precipitation.
10. **b:** bud burst to flowering phenological stage.
11. **f:** flowering to *véraison* phenological stage
12. **v:** *véraison* to maturity phenological stage.
13. **T max Apr-May:** mean of daily maximum temperatures from April to May.
14. **T max-f:** mean of daily maximum temperatures from the flowering to the *véraison* phenological stage.
15. **GDD:** growing degree days.
16. **gs:** growing season.
17. **GST:** growing season mean temperature.
18. **GSP:** growing season total precipitation.
19. **WI:** Winkler index.
20. **HI:** Huglin index.
21. **CI:** cool night index.
22. **DI:** dryness index.
23. **SU25:** number of summer days per year.
24. **SU33:** number of 'scalding' days for the vine per year.
25. **SU35:** number of very hot days per year.
26. **CSDI:** cool spell duration index.
27. **WSDI:** warm or heatwave duration index.
28. **R10:** number of days with heavy precipitation per year.
29. **CWD:** maximum number of wet days per year.
30. **SU33-f:** yearly SU33 from flowering to *véraison*.
31. **SU35-f:** yearly SU35 from flowering to *véraison*.

References

[1] IPPC (International Panel on Climate Change), *Climate Change 2014: Impacts, Adaptation and Vulnerability. Synthesis report* (2014)

[2] C. Van Leeuwen and P. Darriet, J. Wine Econ. **11**, 1, 150–167 (2016)

[3] G. V. Jones and F. Alves, Int. J. Glob. Warm. **4**, 3/4, 383–406 (2012)

[4] J. Tonietto and A. Carbonneau, Agric. For. Meteorol. **124**, 81–97 (2004)

[5] A. Malheiro, J. Santos, H. Fraga, and J. Pinto, Clim. Res. **43**, 163–177 (2010)

[6] R. J. Hijmans, S. E. Cameron, J. L. Parra, P. G. Jones, and A. Jarvis, Int. J. Climatol. **25**, 15, 1965–1978 (2005)

[7] E. J. M. Van Den Besselaar, M. R. Haylock, G. Van Der Schrier, A. M. G. Klein Tank, and A. M. G. Klein J. Geophys. Res **116** (2011)

[8] P. Van Der Linden, J. F. B. Mitchell, P. Gilbert, and P. Tulkens, *Summary of research and results from the ENSEMBLES project ENSEMBLES Climate change and its impacts at seasonal, decadal and centennial timescales*, Exeter, UK (2009)

[9] H. Fraga et al., Reg. Environ. Chang. **14**, 1, 295–306 (2014)

[10] A. C. Real, J. Borges, J. S. Cabral, and G. V. Jones, Int. J. Climatol. (2016)

[11] A. C. Real, J. Borges, J. S. Cabral, and G. V. Jones, Int. J. Biometeorol. **59**, 8, 1045–1059 (2015)

[12] H. Fraga and J. A. Santos, Aust. J. Grape Wine Res. **Apr**, 1–9 (2017)

[13] IPCC (International Panel on Climate Change), "Climate Change 2013: the Physical Science Basis. Summary for Polycymakers (2013)

[14] J. Sillmann, V. V. Kharin, F. W. Zwiers, X. Zhang, and D. Bronaugh, J. Geophys. Res. Atmos. **118**, 6, 2473–2493 (2013)

[15] G. Jones, *Uma Avaliação do Clima para a Região Demarcada do Douro?: Uma análise das condições climáticas do passado, presente e futuro para a produção de vinho*, ADVID (2013)

[16] Porto and Douro Wines Institute (IVDP), *Statistics. Área de vinha e sua composição.* Ano 2016. [Online]. Available: `https://www.ivdp.pt/estatisticas_novo2.php?codIdioma=0&codEstatistica=2&codIdioma=0&periodos=461.` [Accessed: 27-Apr-2017]

[17] Porto and Douro Wines Institute (IVDP), *Statistics*, 2017. [Online]. Available: `https://www.ivdp.pt/` (accessed 18/04/2017)

[18] A. C. de Sousa, *Analyzing the influence of the Douro valley weather on the quality and yield of vintage Port.*, PhD Dissertation, University of Porto (2014)

[19] M. Marta-Almeida, J. C. Teixeira, M. J. Carvalho, P. Melo-Gonçalves, and A. M. Rocha, Phys. Chem. Earth **94**, 94–105 (2016)

[20] D. P. Dee et al., Q. J. R. Meteorol. Soc. Q. J. R. Meteorol. Soc **137**, 553–597 (2011)

[21] M. A. Giorgetta et al., J. Adv. Model. Earth Syst. **5**(3), 572–597 (2013)

[22] Z. M. Moss et al., *Towards new scenarios for analysis of emissions, climate change, impacts, and response strategies*. IPCC Expert meeting report (2008)

[23] K. Riahi et al., Clim. Change **109**(1–2), 33–57 (2011)

[24] IPPC (International Panel on climate Change), *Climate Change 2013. The Physical Science Basis.*, (2013) [Online]. Available: `http://www.ipcc.ch/pdf/assessment-report/ar5/wg1/WG1AR5_TS_FINAL.pdf.` [Accessed: 27-Apr-2017]

[25] J.L. Monteith, Climate in P.T.Alvim, T.T. Kozlowski (eds.). Ecophysiology of Tropical Crops. New York: Academic Press (1977)

[26] L. Hidalgo, Comun. INIA. Ser. Prod. Veg. **29**, 5–255 (1980)

[27] A. Carbonneau, C. Riou, and D. Guyon, *Agrometeorologie de la vigne en France*, Luxembourg (1992)

[28] A.J. Winkler, J.A.Cook, W.M. Kliiewer, and L.A. Lider, *General viticulture*, 4th ed. Berkeley: University of California Press (1974)

[29] P. Huglin, *Biologie et Ecologie de la Vigne*. Lausanne (1986)

[30] C. Riou, *Le determinisme climatique de la maturation du raisin?: application au zonage de la teneur en sucre dans la Communaute europeenne*, Bordeaux, France (1994)

[31] J. Lopes et al., Ciência Tèc. Vitiv. **23**(1), 61–71 (2008)
[32] F. Alves et al., Int. Symp. GIESCO, Porto, Port. 7-11 (2013)
[33] D. Blanco-Ward, J. M. G. Queijeiro, and G. V Jones, Vitis **46**, 2, 63–70 (2007)
[34] G. H. Hargreaves, F. Asce, and R. G. Allen, J. Irrig. Drain Eng. **129**(1), 53–63 (2003)
[35] ETCCDI/CRD Climate change indices [Online]. Available: http://etccdi.pacificclimate.org [Accessed: 24-Apr-2017]
[36] A. N. M- Caetano, V. Nunes, *Corine Land Cover 2006 for Continental Portugal, Relatorio técnico*, Instituto Geográfico Portugués (2009)
[37] E. R. Dahmen and M. J. Hall, *Screening of hydrological data: Tests for stationarity and relative consistency* (1990)
[38] J. A. Ribeiro, *Caracterização genérica da região vinhateira do Alto Douro*, Douro - Estudos & Documentos **5**, 11–29 (2000)
[39] M. N. Lorenzo, J. J. Taboada, J. F. Lorenzo, and A M Ramos, Reg. Environ. Chang. (2012)
[40] J. A. Santos, A. C. Malheiro, M. K. Karremann, and J. G. Pinto, Int J Biometeorol. **55**, 119–131 (2011)
[41] J. Gladstones, *Wine, Terroir and Climate Change*. Kent Town, South Australia: Wakefield Press (2011)
[42] G. Soja, *et al.*, Atmos. Environ. **38**, 2313–2321 (2004)

DOI: 10.1051/bioconf/20170901012

WineSeq®: A new tool for the study of the functional biodiversity of soils, and its use as a biomarker and guide for vitiviniculture practices

Ignacio Belda[1,2], Antonio Palacios[1,3], Javier Fresno[1], Hector Ortega[1], and Alberto Acedo[1]

[1] WineSeq, Biome Makers, 48011 Valladolid, Spain / 94107 San Francisco, USA
[2] Department of Microbiology, Complutense University of Madrid, 28040 Madrid, Spain
[3] Excell Ibérica Laboratories, 26006 Logroño, Spain

Abstract. Apart from their explicit viticulture value, vineyards are natural reservoirs of biological diversity, constituting a complex, but interesting, anthropogenic ecosystem. The functional biodiversity found in vineyard soils is determinant not only for the physical-chemical and nutritional properties of these soils, but also for vine health and grape yield and quality. Diseases affecting vine health or grape quality cause significant economic losses for wineries, specially in the case of old and unique vineyards. In a precision viticulture context, the use of rational phytosanitary treatments and the application of adequate agronomical practices is the only way to maintain the biodiversity required by the fields to keep their stability and resilience. In this context, we have developed WineSeq® (`https://wineseq.com/`), as the first online portal created to support the management of vine health and yield and to prevent or to diagnose vine diseases. The portal (`https://portal.wineseq.com/`) shows filtered and interpreted results of the global microbial population analyzed by metagenomics (DNA Next Generation Sequencing) from soil, wood, grapes or fermentation samples. WineSeq® technology is based on machine learning and cloud computing to integrate microbiome information with climatic, edaphic and agronomical information of interest. Thus, we make easier and understandable the analysis of the microbial variable for vinegrowers and winemakers.

1. Introduction

In 2015, the total world area under vines reached 7.534.000 ha. More than a half are located in the European continent, with Spain topping the list (1.021.000 ha), followed by France (786.000 ha) and Italy (682.000 ha). Outside Europe, China (830.000 ha), Turkey (497.000 ha) and USA (419.000 ha) are the biggest vineyards; the planted surface in Argentina (225.000 ha), Chile (211.000 ha) and Australia (149.000 ha) should also be highlighted. Regarding wine production, it reached 274.400 hl worldwide in 2015, standing out Italy (49.500 hl), France (47.500 hl), Spain (37.200 hl) and US (22.100 hl) [1].

Different countries establish national legislations to guarantee the sustainability of vineyard plantations and replacements, and also, together with regulatory councils, they establish quality requirements in both viticulture and enology practices with the aim of defining quality standards. In this regard, wines with distinctive autochthonous peculiarities have a great acceptance among consumers, causing important economic consequences, and Apellations of Origin are looking for distinctive markers of their terroir. However, the main heritage of wine regions are their old and genuine vineyards, in which a lot of time and money have been spent by vine growers to develop high quality grapes that produce recognizable and high-quality wines [2, 3].

Thus, preserving the health of vineyards is the only way to guarantee their future life, and for that purpose it is necessary to understand the basis of vine diseases and to control and prevent their development in the field. Two groups of diseases are in the spotlight due to their economic consequences: vine trunk diseases and grape rots, being the former those that threaten the vineyard heritage and the later the ones that cause short-term losses in wine quality and production.

We can define 33 key vine diseases with microbial origin [4], where those affecting long-time organs (trunk, roots, etc.), like Esca, eutyposis or *Botryosphaeria* infection are the main ones related with trunk and could affect both old and new plantations. Other diseases such as Black foot or Petri diseases are more specific for young vineyards, threatening their longevity and productivity. Altogether, trunk diseases cause economic losses of about 1.500 million dollars per year [5, 6].

The substitution of the affected vines with new plantations, together with an adequate disinfection treatment of soils, used to be the most definitive way to stop the spread of an infection; however, it implies substantial investments of money [7]. Table 1 shows that, compared with the cost of keeping rigorous diagnostics and preventive and palliative actions in vineyard management (estimated in 600–1.000€ per Ha and year), the economic consequences of a diseased vineyard could overcome 300.000€ considering the cost

 DOI: 10.1051/bioconf/20170901012

Table 1. Mean calculation of the economic losses associated with a vineyard death and replantation.

Cost of replacement of diseased vines		
Vine cost (original and replacement)	1,20€/vine	2,40 €
Planting and maintenance costs (first 5 years)	4€/year	20 €
Total cost per vine		**22,40 €**
Vines/Ha	1.500	
Total cost per Ha		**33.600,00 €**
Economic losses in wine production		
Kg of grapes per vine	4	
Kg of grapes per Ha	6.000	
Bottles per Ha	6.000	
Cost per bottle	9,00 €	
Indirect losses per year		**54.000,00 €**
Years until ideal productivity	5	
Total losses		**270.000,00 €**

of vines replacement and the losses from wine miss-production.

Due to the evident economic consequences of vine diseases, different solutions have been proposed to monitor and diagnose the phytosanitary conditions of vineyards. As an example, the use of multispectral aerial images, taken by satellite or by drones, are in the spotlight of these approaches. However, the difficulty of the treatment of vine trunk diseases when external symptoms appear, do not allow the use of these image-based technologies with early diagnosis objectives.

WineSeq® is the first online platform designed to prevent and diagnose vine diseases, based on the study of the complete vineyard microbiology. It reports the phytosanitary state of the sampled vineyard, with the precise detection of any microbial pathogen of vine or grape, estimating its severity or risk level. WineSeq® technology uses Next Generation DNA Sequencing to identify the entire microbial population (bacteria, filamentous fungi and yeasts) of different samples (soils, wood, grapes, musts, wine, etc.). The vast amount of microbial information is processed by an informatics algorithm based on machine learning to integrate it with climatic and Geografical Information to create a robust model that makes it possible to define the epidemiology of diseases, the effectivity of agronomical practices, the existence of cross-contaminations (fecal, landfill, etc.), and to assist in zonification works.

Based on this technology, we developed WineSeq® Portal (`https://portal.wineseq.com`) as the online platform where winegrowers and winemakers can explore the microbiome of their vineyards. This Portal has been created to easily understand the origin of vine diseases or potential wine inocula.

2. Material and methods

2.1. Sample collection

2.500 soils and 635 grapes were sampled from wineries of 18 countries during 2015 and 2016 (Fig. 1).

All the samples were collected according to a methodology standardized with bibliography and our own

Figure 1. Worldwide distribution of soil samples contained in WineSeq database.

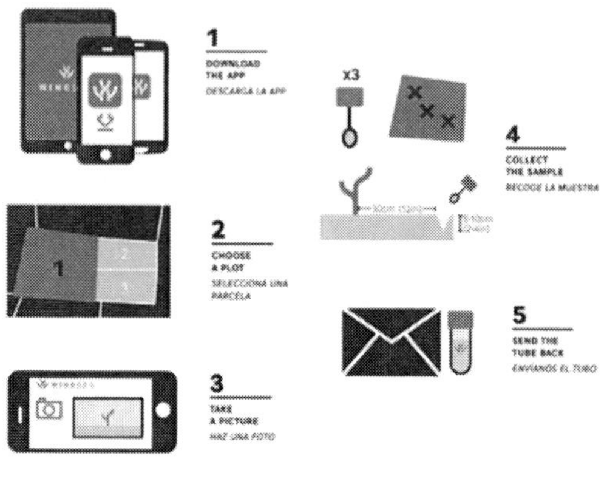

Figure 2. Sample collection instructions for WineSeq analysis.

experience [8], and with the aim of making the work for the winegrower simple (Fig. 2):

- Plot selection: a plot is defined as a vineyard surface with homogeneous soil properties, the same grape variety and managed with the same agronomical practices. If a plot is located in a slope higher than 7%, it should be divided in 3 zones: upper, middle and lower.
- Sample collection: winegrowers received a labeled sterile tube that is scanned with WineSeq App® available for Android and iOS mobile devices. Three spoons of soil should be collected from 3 points randomly selected in the plot. The exact point for soil collection should be at 30 cm from the vine (onto the lane) and at 5–10 cm deep from the first untilled fraction of soil.
- Metadata: information of interest, such as the location of the vineyard, is collected by taking a picture of the sampling point, using WineSeq App®. Additional data (grape variety, agronomical practices, etc) are collected by a digital form completed by WineSeq App® or online Portal.
- Samples are sent by regular courier service and received at WineSeq laboratories within 24 h.

2.2. Microbial population analysis

Nucleic acids are extracted directly from samples, using our own extraction methodology that avoid biases,

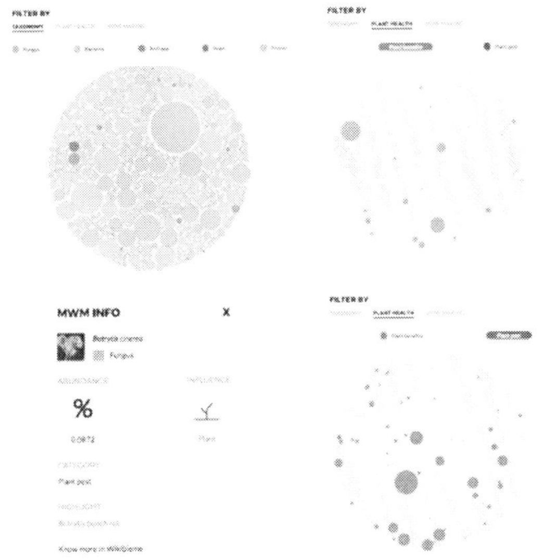

Figure 3. Microbiome exploration tool available at WineSeq portal. Different images are included: complete taxonomical composition of microbiome (upper left); information of beneficial microorganisms for vine health (upper right); summary information of a specific pathogen detected (*Botrytis cinerea*) (bottom left); information about detrimental microorganisms for vine health (plant pests) (bottom right).

allowing us to detect the maximum possible diversity values. Next, taxonomic-marker genes for both prokaryote and eukaryote microorganisms are amplified using oligonucleotides designed and patented by Biome Makers Inc. Each amplicon sequence obtained from the PCR is sequenced using Next Generation Sequencing (Illumina technology).

As a mean, 300.000 reads of DNA are obtained per sample that are then analyzed using a meticulous bioinformatics pipeline, also protected under patent. In brief, DNA reads are clustered by their sequence similitude in Operational Taxonomic Units (OTUs). Each OTU is compared with a taxonomic database, developed and curated from international databases (such as SILVA, GreenGenes, etc.), making possible the later taxonomic assignment to each sequence cluster. The abundance of each taxonomic group is calculated as a percentage of the total amount of OTUs detected, and the distribution and functional diversity of the population is represented in an intuitive tool, available at Wineseq® Portal (Fig. 3).

2.3. VERITAS

Genomic data and the metadata associated to each sample are collected in an intelligent database composed by 119 different information boxes. For example, the 'climatic box' contains several parameters collected by connecting to the climatic station closest to the sample point. Every data, both genomic and metadata, are integrated and interconnected with machine learning algorithms, generating significant correlations and risk estimations within the disease prevention model. As an example of the connection of algorithms, the presence of powdery mildew, downy mildew and botrytis pathogens are crossed with the previously known algorithms of climatic risk for these diseases.

2.4. WineSeq® portal

WineSeq® portal is the online platform that translates the information generated by VERITAS (coming from the previous microbiome analysis) in simple, understandable and intuitive results and conclusions of vitiviniculture relevance. This portal is divided into a private zone, composed by the reports of the samples analyzed (regarding vine health, fermentative microbiota, and microbiome exploration tools) and a public community zone, with open access to the registered users of the 'WineSeq® community' (`https://community.wineseq.com`). This section contains a WikiBiome, as a repository of biological and viticulture/enological information of relevant species for vine health and nutrition and for the winemaking process. WineSeq® community, is supported by a group of international experts (from industry and academia) and it is conceived as a dynamic source of information coming from the questions and answers shared in the open forum by the users of the portal (always moderated by experts).

3. Results

The study of the microbiome of more than 2.500 vineyards, allowed us to extract some relevant conclusions about the frequency and virulence (that is, about the epidemiology) of diseases affecting vine health and grape quality. Soils are the main reservoir of vineyards' microbiome [8] and their biodiversity interacts positively and negatively with the plant and the later fermentative processes [9]. Thus, analyzing the entire microbiome of the soils makes it possible to detect the presence of pathogenic bacteria and fungi (as the basis of the precise application of phytosanitaries), but also to unveil their nutritional potential by identifying microbes related with nitrogen or phosphorous bioavailability and their future stability when detecting biocontrol agents or elicitor-producing organisms.

Figure 4 shows a comparison of the relative abundance of some vines and grapes diseases detected in soil and grape samples. In the case of trunk diseases, all of them are detectable in soil samples, but in certain cases (such as crown galls) they are not found in any grape samples. This fact, together with a precise detection of grape diseases/rots, males soil samples the most adequate matrix to develop general studies of the vineyard microbiome. In addition, and being aware that soil is a dynamic ecosystem, it could be considered as the most stable part of the vineyard along the year, since the development of periodical analyses confirms the consistence of the study of microbiome in soils.

Thus, the analysis of soil microbiomes, has emerged as a useful tool for the monitoring of the health status of vineyards, making possible the implementation of precision viticultural practices focused on the functional biodiversity of soils.

An early detection of microbial pathogens allows for an efficient and targeted intervention to detain the development of the specific infectious agents risking the health of the vineyard. Figure 5 shows a comparison of

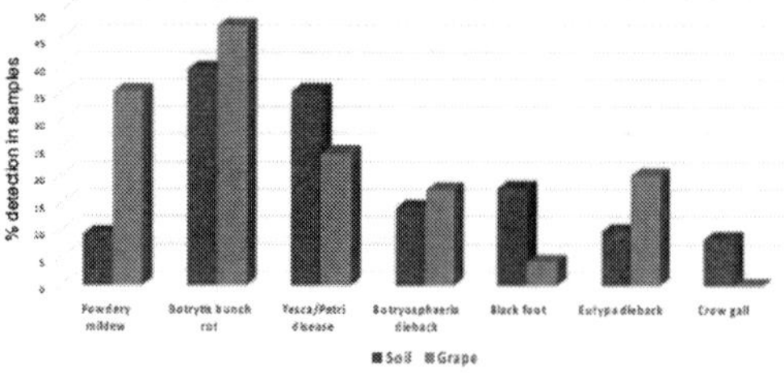

Figure 4. Frequency of detection of vine and grape diseases in soil vs. grape samples.

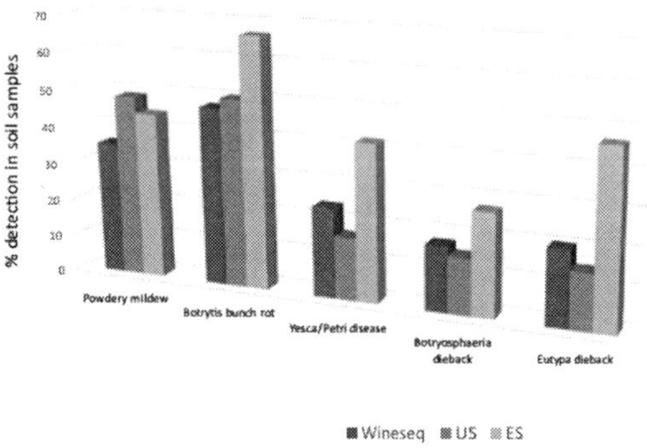

Figure 5. Frequency of detection of vine and grape diseases in soil samples from US, Spain (ES) and from the worldwide WineSeq database.

the frequency of diseases in vineyard soils from Spain and USA (Napa Valley). The main conclusion of this analysis is that, regarding trunk diseases, Spanish vineyards appear to be more susceptible. This fact could be determined by the different viticulture strategies that are commonly applied in new-world *vs* old-world vineyards, and also by the differential susceptibility of the grape varieties and clones used in both continents. These conclusions are enough evidence to start the establishment of a global strategy against trunk disease, based in the knowledge of the epidemiology trends, the edapho-climatic determinants and the biodiversity context of these diseases, with special focus on saving the future of genuine *terroirs* and old vineyards.

Figure 6 shows a partial visualization of the summary report from a soil microbiome analysis in WineSeq® Portal. Within the vine health report, there is simple information about the detection or not of the main 33 vine/grape diseases (Table 2) and, in the case of a positive detection, three additional pieces of information are reported: severity (estimated by pondering the abundance of the pathogenic microorganism with the mean abundance level of this pathogen in healthy samples and with the virulence of the specific microorganism detected causing this disease), frequency (as the global frequency of detection of this disease in our database) and the responsible microorganism (as specific information on the pathogenic agent detected in a sample among the diversity of microorganisms involved in a single disease).

The severity level of a disease detected ranges from Mild to Aggressive, being the complete absence of the pathogens causing a disease informed as "not detected". Furthermore, for every reported disease, there are available

Figure 6. Disease risk test report (WineSeq Portal) from a model sample containing three diseases: Eutypa dieback (with aggressive presence); Black foot disease (with Medium risk); and *Botrytis* brunch rot (with a high risk). It is also shown the specific brief report for the Eutypa dieback detected, including the severity grade, the worldwide frequency of this disease and the number of microorganisms causing this disease in the sample. In addition, a brief description of the disease and its symptoms is included in the bottom part of the image.

Figure 7. Information available at WineSeq Portal for a disease detected in a sample analyzed by WineSeq.

images of the symptoms, making easier their identification in plants or fruits. For each disease present in a sample, WineSeq® Portal contains information and recommendations about treatments organized in sections depending on their origin: 'inorganic compounds', 'organic chemical inducers/natural extracts' and 'biocontrol agents' (Fig. 7).

As a mean, each vine disease can be caused by 5 different microbial species. Some important diseases with a single origin are downy mildew *(Plasmopara viticola)*, botritis infection *(Botrytis cinerea)* or Pierce disease *(Xylella fastidiosa)*. On the other hand, there are diseases with a complex microbial origin, such as esca (up to 40 related species) or the black foot disease (up to 13 related species). In these cases, the specific species involved in an infective process can cause different symptomatology, infection virulence and spreading, and thus, the treatment should be selected and applied with precision.

Table 2. List of important vine/grape diseases with microbial origin, and summary of the microbial groups with enological relevance.

Main vine/grape diseases	Microbial groups with enological relevance
- Alternaria rot	- Lactic acid bacteria
- Brenner Rot	- Acetic acid bacteria
- Angular leaf spot	- Conventional starters (*Saccharomyces* and *Oenococcus*)
- Anthracnose	- non-*Saccharomyces* yeasts
- Armillaria root rot	- *Brettanomyces*
- Grape rots (*Aspergillus, Botrytis*)	- Fermentation problems
- Bacterial blight	•Refermentations (*Zygosaccharomyces*)
- Bitter rot	•Fermentation stucks (*Lactobacillus kunkeei*)
- Black foot disease	
- Black rot	
- *Botryosphaeria* dieback	
- Crown gall tumors	
- *Dematophora* root rot)	
- Downy/powdery mildew	
- *Cladosporium* leaf spot	
- Esca / Petri disease	

Apart from the vine health report, Wineseq® provides an additional report with information about 'potential fermentative microorganisms' found in the vineyard. This report contains information about relevant bacteria and yeast species for the winemaking process (Table 2). Since the correlation between soil, grape and must microbiomes have been demonstrated by robust scientific works [8], with the aim to foresee the wine microbiome from vineyard samples we have developed an algorithm based in the survival rate of microbial species detected along entire the production line (soil-grape-must-wine) in both organic and conventional vineyards. The information contained in this report could be used to guide the individual management of grapes from each vineyard in cellar, developing high-risk (but appreciated by consumers) enological practices such as spontaneous fermentations or low sulphite wines with a previous knowledge about fermentative risks and enhancers found in different vineyards.

4. Conclusions

The health status of vineyards should be controlled to guarantee the quality of the later winemaking process. Precision viticulture pursues the control of all the variables determining the quality of grapes, and thus the microbial determinants of vine health and wine quality should be understood. This knowledge will make it possible to develop precise and eco-logical viticulture and agronomical practices (nutrition, tillage, phytosanitary addition…) depending on the microbial potential and risks of vineyard soils. In addition, developing routine microbiome analyses such as WineSeq® that allows for the detection of vine diseases (before symptoms appearance and pathogen spreading) can save high amounts of money derived from the uncontrolled use of phytosanitary products, and even from huge losses caused by vines death and vineyard replacement (Table 1). In this sense, WineSeq® can also be useful to evaluate the quality of vineyards in purchasing processes, accrediting their sanitary conditions or their functional biodiversity.

Furthermore, the vineyard microbiome, as a reflect and biomarker of terroir, can be used as a tool for plots zonification and for the definition of differential management zones or vineyards of special consideration (such as the Spanish "Vinos de Pago").

Massive DNA sequencing technology for soil samples, such as WineSeq®, is the only way to explore the entire microbial community of vineyards. The complexity of the raw results obtained with this avant-garde technology requires a screening and interpretation to extract relevant and understandable information for winegrowers and winemakers. In the case of WineSeq® Portal, it shows a complete but intuitive report with information about diseases affecting vines and grapes and with an estimation of the fermentative potential of the grapes analyzed from different vineyards. Finally, WineSeq® Community has been created as the first worldwide online community to share information, questions and advice about vines health, vineyard management and winemaking trends.

References

[1] OIV, *State of the Vitiviniculture World Market* (OIV, Paris, 2016)
[2] J. Moulard, B.J. Babin, M. Griffin, IJWBR **27** (1), 61–78 (2015)
[3] S. Caple, M. Thyne, (2016) `http://academyofwinebusiness.com/wp-content/uploads/2014/07/Co04_Caple_Sue.pdf`
[4] WineSeq, (2017) `https://portal.wineseq.com/wikibiome`
[5] V. Hofstetter, B. Buyck, D. Croll, O. Viret, A. Couloux, K. Gindro, Fungal Divers. **4**, 51-67 (2012)
[6] D. Gramaje, J. Armengol, Plant Dis. **95** (9), 1040-1055 (2011)
[7] G. Bruno, L. Sparapano, Physiol. Mol. Plant P. **71**, 210–229 (2007)
[8] I. Zarraonaindia, S.M. Owens, P. Weisenhorn, K. West, J. Hampton-Marcell, S. Lax, et al. MBio **6**(2), e02527-02514 (2015)
[9] I. Belda, I. Zarraonaindia, M. Persisin, A. Palacios, A. Acedo, Front. Microbiol. **8**, 821 (2017)

DOI: 10.1051/bioconf/20170901013

Amélioration des performances environnementales des pratiques viticoles et analyse des risques pour la qualité du raisin : étape vers l'écoqualiconception©

Improvement of the environmental performance of winegrowing practices and risk analysis for grape quality: Step towards ecoqualiconception©

Sandra Beauchet[1], Marie Thiollet-Scholtus[2], Christel Renaud-Gentié[1], et Frédérique Jourjon[1]

[1] Unité de Recherche GRAPPE, Univ Bretagne Loire, Ecole Supérieure d'Agricultures (ESA)-INRA, 55 rue Rabelais, BP. 30748, 49007 Angers Cedex, France
[2] INRA - SAD - UR-0055-ASTER, 28, rue de Herrlisheim 68000 Colmar France. DP Sciences, Service Editorial, 91944 Les Ulis Cedex A, France

Résumé. L'évolution des itinéraires techniques viticoles (ITKv) vers un plus grand respect de l'environnement passe par une évaluation de leurs performances environnementales. Dans un contexte de gestion durable, plusieurs autres facteurs sont à prendre en compte, notamment le risque de modification de la qualité des raisins issus de ces ITKv. Les résultats d'évaluation environnementale par la méthode d'Analyse du Cycle de Vie (ACV) ainsi que l'évaluation de la qualité du raisin s'appuient sur beaucoup données. Ce travail montre l'intérêt d'une méthode d'évaluation multicritère conjointe «environnement et qualité», dans le processus de changement de pratiques pour une amélioration des performances. Les résultats obtenus sur 5 ITKv montrent des différences de notes entre les 2 millésimes étudiés et des pratiques ont été identifiées pour leurs forts impacts sur les résultats ACV. Des propositions théoriques de changement de pratiques sont testées sur l'un des ITKv. Ces propositions améliorent les notes environnementales finales de +0,15 points à +0,77 points (/10) quand elles sont mises en œuvre séparément et +2,29 points (/10) en étant combinées. Un modèle explicatif de la qualité du raisin permet de vérifier que les nouvelles pratiques ne peuvent potentiellement pas altérer la qualité du raisin. La méthode d'évaluation conjointe CONTRA-QUALENVIC adaptée à la viticulture montre ici qu'elle peut être une aide à l'amélioration des ITKv.

Abstract. The evolution of technical management routes (TMRs) towards greater respect for the environment requires an assessment of environmental performances. In the context of sustainable management other factors must be taken into account, as the risk of grape quality changes. Environmental assessment results using the Life Cycle Assessment (LCA) method and the assessment of grape quality are based on numerous data and needs multicriteria decision-making analyzes. This work aims to show the interest of a joint evaluation method "environment and quality", in the process of change of practices to improve performances. The results obtained on 5 TMRs show a great difference of score between the 2 vintages studied and some practices were identified as having the biggest impact difference. Proposals for modifying practices were tested on a TMR. These theoretical proposals for changing practices improved the final environmental scores from +0.15 points to +0.77 points (/ 10) while studied independently and +2.29 points (/ 10) while combined. A model explaining the quality of grapes using viticultural practices verifies that the proposed new practices don't present a potential risk for grape quality. The method of joint evaluation CONTRA-QUALENVIC adapted to viticulture shows here that it can be a tool to help improving TMRs.

1. Introduction

Les filières agricoles sont confrontées aujourd'hui à des exigences de pluri-performance de leur système de production : économique, environnementale, sociale en cohérence avec les attentes des consommateurs et des acheteurs.

La prise en compte de l'environnement dans le secteur viticole mondial est de plus en plus prégnante, particulièrement en France où des programmes de réduction de l'utilisation des pesticides de 50 % entre 2008 et 2018 ont été mis en œuvre. Cette volonté de réduction des impacts rejoint les enjeux économiques de la filière que ce soit vis-à-vis de la réduction du coût des intrants ou pour s'adapter à la demande croissante de vins produits de manière respectueuse de l'environnement de la part des metteurs en marché, et de certains consommateurs [1].

L'évolution des itinéraires techniques viticoles vers un plus grand respect de l'environnement passe par une évaluation des performances environnementales. Dans un contexte de gestion durable des itinéraires techniques viticoles, plusieurs autres facteurs sont à prendre en compte, notamment le risque de modification de la qualité des raisins issus de ces itinéraires techniques. En effet, une adaptation des pratiques viticoles pour une amélioration des performances environnementales ne peut se faire qu'à condition de maintenir le niveau de rendement et de qualité du raisin visé.

Les résultats d'évaluation environnementale par la méthode d'Analyse du Cycle de Vie (ACV) (ISO 14040-14044) [2,3] ainsi que l'évaluation de la qualité du raisin s'appuient sur de nombreuses données et sur des analyses multicritères rendant complexe la prise de décision de changement de pratiques quant à ses conséquences sur la performance globale du système viticole. Les méthodes d'évaluations multicritère représentent une aide dans ce type de situation afin d'évaluer conjointement plusieurs évaluations [4]. Ce travail vise à montrer l'intérêt d'une méthode d'évaluation multicritère conjointe «environnement et qualité», agrégeant ces résultats complexes, dans le processus de changement de pratiques à des fins d'amélioration de performances.

L'objectif de ce travail est de tester l'utilisation de la démarche d'évaluation multicritère (environnement/ qualité) sur un itinéraire technique viticole afin d'en améliorer ses performances environnementales sans risquer de nuire à la qualité du raisin. En testant cette démarche cela permettra de construire des scénarios d'amélioration de performances environnementales à partir de changement de pratiques et de simuler les conséquences de modifications de pratiques sur la qualité potentielle du raisin via l'utilisation d'un modèle d'explication de la qualité du raisin [5].

2. Matériels et méthodes

La démarche mise en place s'articule en 5 étapes. i) Une évaluation environnementale par ACV et une évaluation multicritère de la qualité du raisin à la vendange réalisées sur l'itinéraire technique viticole réel ii) L'agrégation des résultats des deux évaluations selon la méthode multicritère CONTRA-QUALENVIC [4,6] aboutissant à une note de performance environnementale et une note de qualité du raisin. iii) L'identification des pratiques viticoles les plus impactantes sur un plan environnemental. iv) La simulation de changement de pratiques et ses conséquences sur un plan performance environnementale v) L'estimation des risques potentiels du changement pratiques sur la qualité du raisin en s'appuyant sur un modèle explicatif de la qualité du raisin construit à partir des pratiques viticoles et permettant de vérifier que les nouvelles pratiques proposées n'engendrent pas d'importantes modifications «théoriques» de la qualité du raisin.

Les évaluations environnementales par ACV ainsi que les évaluations multicritères de la qualité du raisin à la vendange ont été réalisées sur 5 itinéraires techniques viticoles (appelés dans le texte Technical Management Routes (TMRs)) au cours de 2 millésimes. Ces 5 TMRs sont localisés en zone d'appellation d'Origine Contrôlée Anjou-Saumur (Moyenne Vallée de la Loire, France) et produisent des vins blancs secs) à partir du cépage Chenin Blanc. Ils ont été sélectionnés car ils représentent la diversité des itinéraires techniques que l'on peut rencontrer dans la région pour le cépage concerné [7]. Les 2 années étudiées (2011 et 2013) ont été choisies pour leurs climats contrastés : 2011 se caractérise par une année chaude et sèche avec peu de pression parasitaire (année dite favorable) et 2013 par une année froide et pluvieuse avec une forte pression parasitaire (année dite défavorable) [8].

L'évaluation environnementale par la méthode ACV s'appuie sur 14 catégories d'impacts environnementaux issues de la méthode ReCiPe [9] regroupées autour de 4 compartiments environnementaux : l'air avec les catégories d'impacts «Changement climatique (kg CO_2 eq.)», «Amincissement de la couche d'ozone (kg CFC-11 eq.)», «Formation d'oxydants photochimiques (kg NMVOC eq.)», «Formation de particules (kg PM_{10} eq.)» ; l'eau avec les catégories d'impacts «Eutrophisation des eaux douces (kg P eq.)», «Eutrophisation des eaux de mers (kg N eq.)», «Écotoxicité des eaux douces (kg 1,4-DB eq.)», «Écotoxicité des eaux de mer (kg 1,4-DB eq.)» ; le sol avec les catégories d'impacts «Acidification des sols (kg SO_2 eq.)», «Écotoxicité des sols (kg 1,4-DB eq.)» ; et enfin les ressources avec les catégories d'impacts «Occupation des terres agricoles (m2a)», «Diminution des ressources en eau (m^3)», «Appauvrissement des ressources en métal (kg Fe eq.)» et «Appauvrissement en ressources fossiles (kg oil eq.)». L'évaluation de la qualité du raisin est réalisée à partir de 7 critères également regroupés en plusieurs catégories ; elle a été définie en fonction des types de production visées : vin blanc sec de printemps ou vin blanc sec de garde [6].

Les résultats des évaluations environnementales et de qualité du raisin des 5 TMRs sur 2011 et 2013 ont été agrégés par la méthode CONTRA-QUALENVIC [6]. Cette méthode d'évaluation multicritère est adaptée à l'évaluation environnementale et l'évaluation de la qualité du raisin. Elle permet à partir de fonctions d'appartenances associées à la logique floue de fournir des notes finales d'évaluation environnementale d'une part et de qualité du raisin d'autre part pour chacun des TMRs et chacune des années étudiées. Ainsi le TMR numéro 4 étudié pour l'année 2013 sera identifié : 13TMR4.

A partir de l'ensemble des résultats d'évaluation environnementale et qualité réalisés sur les 5 TMR étudiés, nous avons choisi de travailler sur le TMR qui présentait le plus d'écarts de performance entre les deux années étudiées. Il s'agit du TMR 4 (ITK de type agriculture biologique modérée) caractérisé par un mode de production en culture biologique selon le cahier des charges de l'AOC Savennières. En effet, pour les 5 TMR, les écarts entre les notes finales d'évaluation environnementales vont de 1,5 à 6,7/10 points et les écarts entre les notes finales d'évaluation de la qualité du raisin vont de 1,8 à 6,75/10 points.

En 2011, cet itinéraire technique permet d'obtenir un bon compromis entre environnement et qualité avec une note de 6,85/10 en environnement et de 7,37/10 en qualité du raisin pour la production d'un vin de garde. En 2013, les notes de qualité et d'environnement diminuent fortement avec, des notes de 0,18/10 en environnement et de 0,62/10 en qualité du raisin pour la production de vin de garde.

Le TMR4 semble donc un bon exemple pour illustrer les possibilités d'adaptation des pratiques viticoles pour une amélioration de ses performances environnementales en veillant à assurer une bonne qualité du raisin produit.

DOI: 10.1051/bioconf/20170901013

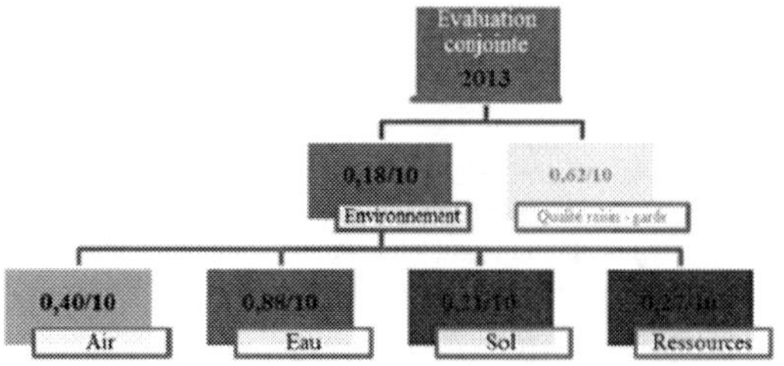

Figure 1. Représentation des notes obtenues par le TMR4 en 2013 dans CONTRA pour son évaluation environnementale.

Il s'agira dans un premier temps de revenir à l'origine de la mauvaise note obtenue en environnement en 2013. Ainsi les résultats agrégés et intermédiaires obtenus par la méthode CONTRA-QUALENVIC pour l'évaluation environnementale puis les résultats bruts obtenus par la méthode ReCiPe [9] pour l'Analyse du Cycle de Vie (ACV) seront analysés afin de remonter aux pratiques viticoles responsables des impacts environnementaux. Par la suite, des propositions théoriques de changements de pratiques seront évaluées d'un point de vue environnemental pour être ensuite évaluées par la méthode CONTRA. Si ces pratiques améliorent la performance environnementale du TMR4 en 2013, elles seront ensuite analysées quant à leur effet potentiel sur la qualité du raisin en s'appuyant sur le modèle explicatif de la qualité du raisin à partir des pratiques viticoles [5].

3. Résultats

3.1. Diagnostic de l'origine des impacts environnementaux

La Fig. 1 montre les notes obtenues par le TMR4 en 2013 via la méthode CONTRA-QUALENVIC.

On peut observer que les mauvaises notes de performance environnementale sont généralisées à tous les compartiments de l'environnement (air, eau, sol et ressources). Les impacts sont particulièrement importants pour les compartiments sol (0,21/10) et ressources (0,27/10). A partir des notes finales obtenues, il faut rechercher la source de l'impact, en remontant dans l'évaluation jusqu'aux résultats ACV, pour cibler la ou les pratiques à améliorer en priorité.

Pour le compartiment sol, l'impact provient principalement de la catégorie d'impact «écotoxicité terrestre» où l'impact observé est proche de la limite défavorable utilisée dans le modèle CONTRA avec un impact de 37,23 kg 1,4-DB eq. en 2013 contre 5,25 kg 1,4-DB eq. en 2011. Cela explique la mauvaise note du TMR4 pour ce compartiment.

Pour le compartiment ressources, c'est la consommation de métal (313 kg Fe eq.) et de ressources fossiles (652 kg de pétrole eq.) qui engendre un impact important sur l'environnement.

Pour le compartiment eau, ce sont les catégories «écotoxicité aquatique» (7,50 kg 1,4-DB eq. pour l'écotoxicité aquatique en eau douce) et «eutrophisation» (0,61 kg P eq. pour l'eutrophisation en eau douce) qui expliquent les mauvais résultats. Il est aussi à noter que pour le compartiment air, l'impact des pratiques viticoles mises en œuvre sur le TMR4 sur le changement climatique est aussi important avec une valeur de 1 992 kg CO2 eq. en 2013 contre 1289 kg CO2 eq. en 2011.

Le nombre important de traitements phytosanitaires effectués sur la parcelle est à l'origine d'une partie des impacts. Ces traitements nécessitent une consommation importante de produits phytosanitaires et d'eau ainsi qu'une consommation importante de carburant et une usure du matériel agricole. Pour le compartiment sol, cela se traduit principalement sur les catégories d'impact écotoxicité terrestre et acidification. Pour le compartiment air, l'impact se localise sur la catégorie changement climatique. Pour le compartiment ressources, c'est la consommation de ressources fossiles qui est la plus touchée par ce nombre important de traitements. Et enfin pour le compartiment eau, c'est la catégorie écotoxicité terrestre qui est la plus touchée.

Aux traitements phytosanitaires s'ajoutent le nombre important de désherbages mécaniques et de travaux mécaniques réalisés, généralement, dans la parcelle (passage de lames, buttage, débuttage). Ces pratiques engendrent une consommation de carburant et une usure du matériel agricole. Les impacts de ces pratiques mécanisées sont ciblés sur le compartiment air pour leur impact sur le changement climatique et sur le compartiment ressources pour leur impact sur la consommation de ressources fossiles.

3.2. Propositions théoriques de changements de pratiques

L'objectif visé est d'améliorer la note d'évaluation environnementale globale. Pour cela, trois propositions théoriques d'amélioration des performances environnementales sont faites ; elles sont basées sur le la recherche dans les autres itinéraires techniques étudiés pour la même année, la «meilleure pratique» (celle présentant la meilleure note environnementale) et de l'appliquer sur l'itinéraire étudié que l'on cherche à améliorer (TMR4). Trois pratiques sont ainsi identifiées et testées dans notre démarche. Elles seront distinguées au travers de leur nom.

3.2.1. 13TMR4EcoPhyt (économie de produits phytosanitaires pour le TMR4 en 2013)

Nous remplaçons le programme de traitements du TMR4 par celui du TMR5 qui respecte le même cahier des charges (agriculture biologique). Le programme de traitements du TMR5 permet de diminuer presque de moitié le nombre de traitements. Cela aura un impact sur plusieurs catégories d'impact environnemental (écotoxicité terrestre, acidification des sols, changement climatique, consommation des ressources fossiles et écotoxicité aquatique).

3.2.2. 13TMR4EcoTract (Economie d'émission du tracteur en prenant un quad pour le TMR4 en 2013)

A partir de l'exemple du TMR3, l'utilisation d'un quad à la place du tracteur pour réaliser les traitements sera intégrée. L'utilisation d'un quad permettra de diminuer les impacts pour les mêmes catégories d'impact environnementaux que celles ciblées en changeant le programme de traitements.

*Variable Importance in the Projection.

DOI: 10.1051/bioconf/20170901013

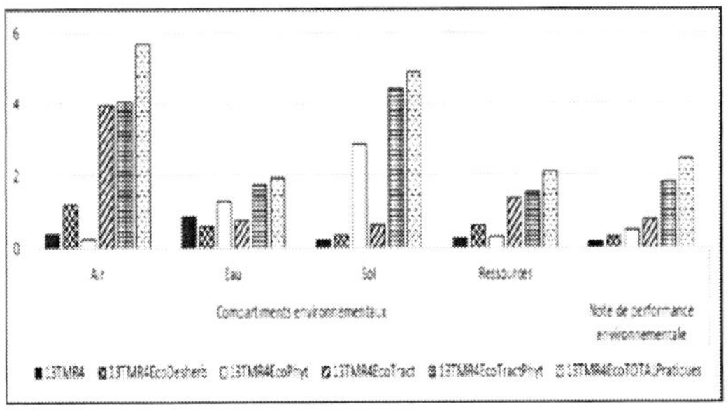

Figure 2. Représentation des notes obtenues par les propositions théoriques de changement de pratiques sur le TMR4 en 2013 pour chaque compartiment de l'environnement ainsi que pour l'étude de la note environnementale finale par rapport aux notes originelles du TMR4 en 2013 (les notes obtenues sont /10).

3.2.3. 13TMR4EcoDesherb (économie de passage de lames pour le désherbage du TMR4 en 2013)

Enfin des modifications sur le travail mécanique effectué sur la parcelle se feront par une diminution du temps de passage des lames passant de 6h/ha à 3h/ha comme c'est le cas du TMR2.

3.3. Résultats CONTRA-environnement avec les nouvelles pratiques mises en œuvre sur TMR4

Les résultats sont représentés pour chaque compartiment de l'environnement (Fig. 2). Les notes obtenues sont représentées Tableau 1 ainsi que la note finale environnementale. Il est à noter que les propositions théoriques d'amélioration de pratiques sont testées dans un premier temps indépendamment les unes des autres.

On observe que globalement chacune des propositions théoriques de changement de pratiques proposées améliore chacune des notes de compartiment. La note de compartiment air est principalement améliorée par l'utilisation d'un quad à la place d'un tracteur pour effectuer les traitements phytosanitaires. On passe d'une note de 0,40/10 pour le TMR4 original à 3,95/10 pour le TMR4 qui utilise un quad. La note de compartiment eau est globalement la note la plus difficilement améliorée avec ces changements de pratiques.

Néanmoins on observe que la diminution de la quantité de produits phytosanitaires utilisée permet d'améliorer la performance environnementale de la parcelle pour le compartiment eau en passant de 0,18/10 pour le TMR4 original à 0,48 pour le TMR4 modifié. La note du compartiment sol est nettement améliorée par la diminution de la quantité de produits phytosanitaires permettant de passer de 0,21/10 à 2,86/10. La note du compartiment ressource est principalement améliorée par l'utilisation du quad à la place du tracteur : on passe d'une note de 0,27/10 pour le TMR4 originel contre 1,37/10 pour l'épandage des produits phytosanitaires avec un quad. La note environnementale globale est ainsi améliorée quelle que soit la proposition théorique de changement de pratiques. Néanmoins cette amélioration n'est que partielle. Il faudrait combiner l'amélioration de plusieurs pratiques pour obtenir une importante amélioration de la note globale environnementale. C'est pour cela que l'amélioration environnementale doit être réfléchie à l'échelle de l'itinéraire technique. Il est possible d'améliorer de façon individuelle les pratiques les plus impactantes. Mais étant donné qu'elles sont en relations avec d'autres pratiques, il s'agit de penser l'amélioration de plusieurs pratiques en même temps et de repenser l'itinéraire technique viticole dans sa globalité.

Ainsi associer la modification du programme de traitement phytosanitaire tout en utilisant un quad à la place d'un tracteur permet d'améliorer la note finale environnementale de 1,63 points (/10) en passant de 0,18/10 à 1,81/10. Enfin, en associant les trois propositions théoriques de changement de pratiques étudiées, on peut augmenter la note finale de 2,29 points (/10) pour arriver à une note finale de performance environnementale de 2,47/10. Cette amélioration est surtout visible au niveau des notes obtenues dans le compartiment air avec une augmentation de 5,30 points pour ce compartiment.

3.4. Quelles conséquences ces changements peuvent-ils provoquer sur la qualité du raisin ?

Plusieurs pratiques parmi celles proposées ci-dessus n'ont a priori aucune incidence directe sur la qualité du raisin, que l'on considère la bibliographie ou le modèle explicatif développé dans notre travail

Ainsi concernant la réduction du temps de passage des lames, notre modèle explicatif de la qualité du raisin n'identifie aucune relation significative entre la pratique et les critères de qualité étudiés (sucre, acidité totale, acide malique et tartrique, azote assimilable, pH et taux de pourriture).

D'après la bibliographie et notre modèle explicatif, un lien positif existe entre l'utilisation de fongicides qui représente une importante partie des produits phytosanitaires utilisés dans ce TMR et les concentrations des baies en acidité totale (VIP*=1,040) et en acide malique (VIP=0,835). Une diminution du nombre de traitements fongicides, si la parcelle est sensible aux attaques fongiques, risque en effet d'affecter l'état sanitaire du feuillage de la vigne et des baies de raisin et la maturité du raisin. De plus, la diminution d'utilisation d'insecticides pourrait diminuer la concentration des baies en sucre (VIP=1,256). Ce changement risque néanmoins de faire augmenter le pH du raisin (VIP=1,774) car l'utilisation d'insecticides apparait liée négativement au pH.

Enfin, le modèle explicatif de la qualité ne montre aucune relation significative entre le type de matériel de traction (tracteur ou quad) et les critères de qualité étudiés. Il est donc probable que l'utilisation d'un quad à la place d'un tracteur pour l'épandage des produits phytosanitaires sur la parcelle n'aura aucune conséquence sur la qualité du raisin.

Aux vues de l'utilisation du modèle explicatif de la qualité du raisin à partir des pratiques viticoles et des facteurs du milieu et des travaux bibliographiques nous pouvons considérer que le changement de pratiques tel que proposé ici pour TMR4 en 2013 pourrait permettre d'augmenter la note globale de qualité du raisin bien qu'elle engendrera probablement des modifications de notes intermédiaires de plusieurs critères de qualité (acidité totale, acide malique, sucre et pH).

*Variable Importance in the Projection.

 DOI: 10.1051/bioconf/20170901013

Tableau 1. Notes obtenues suite aux changements théoriques de pratiques sur le TMR4 en 2013 pour chaque compartiment de l'environnement ainsi que pour l'étude de la note environnementale finale par rapport aux notes originelles du TMR4 en 2013 (les notes obtenues sont /10).

	Compartiments environnementaux				**Note de performance environnementale**
	Air	Eau	Sol	Ressources	
13TMR4	0.40	0.88	0.21	0.27	0.18
13TMR4EcoDesherb	1.19	0.60	0.35	0.62	0.30
13TMR4EcoPhyt	0.24	1.28	2.86	0.31	0.48
13TMR4EcoTract	3.95	0.78	0.65	1.37	0.79
13TMR4EcoTractPhyt	4.06	1.76	4.42	1.52	1.81
13TMR4EcoTOTALPratiques	5.70	1.93	4.88	2.11	2.47

4. Discussion

Les simulations effectuées montrent dans un premier temps que la méthode CONTRA-QUALENVIC permet de simuler et évaluer les effets d'un changement de pratiques. Il est ainsi possible d'envisager d'utiliser cette méthode pour travailler de manière pédagogique avec le viticulteur et son conseiller afin de trouver les pratiques viticoles qui apportent le meilleur compromis entre performance environnementale et performance de la qualité du raisin.

La démarche mise en œuvre montre tout l'intérêt du cumul des pratiques pour améliorer la performance environnementale globale de l'itinéraire technique et pour encourager à raisonner de manière globale et systémique le raisonnement de son itinéraire technique. La simulation réalisée sur le TMR4 montre clairement que la révision de l'itinéraire technique dans son ensemble est plus prometteuse pour une amélioration des performances environnementales qu'une amélioration des pratiques prises de manière isolée. Changer le quad ou modifier les traitements phytosanitaires ou faire un passage de lames plus léger permet d'améliorer les performances environnementales du TMR4 d'au minimum 0,15 points. Et si l'on prend la solution de remplacer le tracteur par un quad pour les traitements phytosanitaires, le TMR4 obtient de meilleures performances environnementales que le TMR5 pour la même année. Il faut combiner plusieurs améliorations pour que le changement soit important. La méthode CONTRA-QUALENVIC permet de calculer la note globale environnementale de ce nouvel itinéraire technique théorique «amélioré» par le cumul de toutes les pratiques modifiées et permet ainsi de revoir l'itinéraire technique viticole dans son ensemble.

Cependant, les propositions théoriques de changements de pratiques ne sont pas toutes envisageables dans la réalité. En effet, le quad qui est l'amélioration présentant les meilleurs résultats ne peut être associé à un pulvérisateur classique pour l'épandage des produits phytosanitaires, il demande d'y associer un pulvérisateur spécifique motorisé qui consomme lui aussi du carburant, ce qui n'a pu être pris en compte ici faute de données. Cette proposition de changement de pratiques n'est donc peut-être pas réalisable en réalité.

L'amélioration des résultats via CONTRA-QUALENVIC ne peut se faire pour le moment qu'en souhaitant améliorer principalement la performance environnementale des TMRs pour ensuite analyser les éventuelles conséquences que cette amélioration pourrait causer sur la qualité du raisin.

La démarche mise en œuvre ne peut se faire actuellement que dans ce sens «amélioration environnementale vs. qualité du raisin» la réciproque n'étant pas possible faute de modèle prédictif de la qualité du raisin à partir des pratiques viticoles. Pour le moment, le modèle développé dans notre travail peut uniquement déterminer s'il y a un lien potentiel entre la pratique identifiée et l'un des critères de qualité du raisin étudié ; si cette influence est forte et si elle pourrait risquer de faire augmenter ou diminuer la valeur du critère de qualité. De plus, ce modèle n'est pour le moment utilisable que pour la production de vin blanc sec à partir de Chenin en moyenne vallée de la Loire.

5. Conclusion

Nos travaux ont permis de démontrer que concilier performance environnementale et qualité des raisins est possible ; le TMR4 en est la meilleure illustration.

Ensuite nous avons montré qu'il est possible d'améliorer les pratiques viticoles d'un point de vue environnemental tout en veillant au maintien de la bonne qualité du raisin produit. A travers l'analyse de nos cinq cas d'étude et des simulations de changements de pratiques réalisées sur un cas «théorique» nous avons pu proposer des améliorations des performances environnementales tout en validant que les nouvelles pratiques proposées n'entrainaient pas de conséquences négatives sur la qualité du raisin souhaité.

Par cette étude, la méthode d'évaluation conjointe environnement/qualité CONTRA-QUALENVIC a pu être éprouvée. Le paramétrage de la méthode adaptée à la viticulture est sensible à une modification de pratique viticole et aux nouveaux impacts qu'elle engendre. L'application de la méthode CONTRA-QUALENVIC a permis de faire ressortir les pratiques qui présentaient un impact important sur les performances environnementales de l'itinéraire technique viticole étudié. Il est ainsi possible de remonter dans les notes d'évaluations obtenues jusqu'à identifier les substances responsables des impacts principaux causés, puis de proposer de nouvelles pratiques amélioratrices Le modèle explicatif des liens entre pratiques, facteurs du milieu et critères de qualité du raisin a permis d'étudier l'influence potentielle des nouvelles pratiques proposées à des fins d'amélioration des performances environnementales, sur les différents critères de qualité du raisin.

L'enjeu est désormais de s'appuyer sur cette méthode d'évaluation conjointe pour engager des démarches d'amélioration continue d'itinéraires techniques viticoles à des fins de conception d'itinéraires respectueux de l'environnement et intégrant les exigences attendues de qualité du raisin. On parle alors d'Ecoconception et d'Ecoqualiconception©). Pour être complet, cette

DOI: 10.1051/bioconf/20170901013

méthode d'évaluation devra aussi être enrichie d'un pilier économique et social ; ceci fait l'objet de travaux en cours au sein de l'équipe de recherche.

Références

[1] Agence-Bio, Agence française pour le développement et la promotion de l'agriculture biologique (2013)
[2] ISO (International Standards Organization) (2006a)
[3] ISO (International Standards Organization) (2006b)
[4] C. Bockstaller, L. Guichard, O. Keichinger, P. Girardin, M. B. Galan, and G. Gaillard. Sustainable Agriculture, pp. 769-784 (2009)
[5] S. Beauchet, V. Cariou, C. Renaud-Gentié, M. Meunier, R. Siret, M. Thiollet-Scholtus, and F. Jourjon, Oeno One
[6] S. Beauchet, *thèse de doctorat*, L'Université Bretagne Loire (2016)
[7] C. Renaud-Gentié, S. Burgos, M. Benoît, Europ. Jour. of Agron. **56**, 19–36 (2014a)
[8] C. Renaud, C. Renaud-Gentié, S. Beauchet, and F. Jourjon, *37th OIV Congress, 9–14 November 2014, Mendoza, Argentina* (2014b)
[9] M. Goedkoop, R. Heijungs, M. Huijbregts, A. De Schryver, J. Struijs, R. van Zelm, ReCiPe 1 (2009)

Ricerche fondamentali per scrivere della viticoltura la "Carta della Sostenibilità Universale 4.1C o Sostenibilità MetaEtica 4.1C", anche, secondo la "Rivoluzione Produttiva 4.1C" della "MetaEtica 2.1C" e della "Grande Filiera MetaEtica 4.1C" del Conegliano Campus 5.1C

Fundamental researches to write of the Viticulture a "Chard of the Universal or MetaEthical Sustainability 4.1C", also, according to the "Productive Revolution 4.1C" the "Metaethics 2.1C" and the "Great Chain MetaEthics 4.1C" and the "Great Chain MetaEthics 4.1C" of Conegliano Campus 5.1C

Giovanni Cargnello[a]

Conegliano Campus 5.1C (Treviso-I)-Port, Italy

Riassunto. Facendo seguito a quanto, anche da noi, pubblicato sulla sostenibilità, in questo lavoro verranno esposte le ricerche di base condotte per poter scrivere, come è stata scritta, la "Carta della Sostenibilità Universale 4.1C della Viticoltura" o "Carta della Sostenibilità MetaEtica 4.1C della Viticoltura", anche secondo la "Rivoluzione Produttiva della Viticoltura 4.1C e non 4.0" del Conegliano Campus 5.1C. Cioè di una sostenibilità che considera "tutti" i fattori coefficentati materiali ed immateriali non solo economici, ambientali, sociali, ma anche quelli tecnici, esistenziali, (esistenziali per l'uomo e per tutti gli altri esseri viventi e non viventi, per il territorio, per la madre terra, per la litosfera, per gli universi, ...), etici settoriali ed "Etici Universali 2.1C e 4.1C o MetaEtici 2.1C e 4.1C" in armonica filiera, (così come nella "Rivoluzione Produttiva della Viticoltura 4.1C"), secondo la nota filosofia e metodologia, il noto algoritmo della "Grande Filiera MetaEtica 4.1C" del Conegliano Campus 5.1C. Queste attività che sono state esposte in questo lavoro hanno compreso, tra l'altro, ricerche su varie sostenibilità: ONU, UNESCO, CE, OIV, Conegliano Campus, GiESCO, Equalitas, ecc. e sulla rivoluzione della produzione così detta 4.0 o "Rivoluzione 4.0". E tutto ciò è stato fatto, anche, per poter praticamente certificare in viticoltura, non una insoddisfacente sostenibilità settoriale, limitata, ma una "Sostenibilità Universale 4.1C" o "Sostenibilità MetaEtica 4.1C" o "Sostenibilità 4.1C", cioè una sostenibilità che va oltre a quelle settoriali, le quali come tali, anche, secondo queste ricerche sono risultate limitate, insoddisfacenti secondo una visione universale come quella, nota, della "Grande Filiera MetaEtica 4.1C" del Conegliano Campus 5.1C, fondamenta, anche, per scrivere le carte e relative certificazioni e garanzie sulle "Sostenibilità 2.1C e 4.1C". **Parole chiave**-viticoltura, viticultura, rivoluzione 4.0, rivoluzione 2.1C, 4.1C e 5.1C, carta della sostenibilità universale 2.1C, 4.1C, carta della sostenibilità metaetica 2.1C, 4.1C , conegliano campus 5.1C, vite, vino

Abstract. Following previously published research, also from us, on sustainability, in this paper presents the basic research conducted in order to write the "Charter of the Universal Sustainability 4.1C Viticulture" or "Charter of Sustainability MetaEthics 4.1C Viticulture", according to the "Productive Revolution Viticulture 4.1C" (not 4.0), of the Conegliano Campus 5.1C. Sustainability in this paper considers "all" descriptors with coefficient, not only economic, environmental, social, but also technical, existential (existential for humans and all other living and non-living beings, for the territory, for mother earth, to the lithosphere, to the universes), ethical, sectorial and "Universal Ethics 2.1C and 4.1C or MetaEthics 2.1C and 4.1C" in the harmonic chain, as well as in the "Productive Revolution 4.1C, and of Viticulture", according to the methods and the well-known algorithm of the "Great Chain MetaEthics 4.1C" of Conegliano Campus 5.1C. These activities w presented in this work include, among other things, research on various types of sustainability: UN, UNESCO, EC, OIV, Conegliano Campus, GiESCO, Equalitas, etc. and the production of the so-called "Revolution 4.0". The aim of all of this is to certify in viticulture, not an unsatisfactory sectorial limited sustainability, but a "Universal Sustainability 4.1C" or "Meta-Ethical Sustainability 4.1C" or "Sustainability 4.1C,"; in other words, a sustainability that goes beyond those that are sectorial, which as such, also, according to this research are limited and unsatisfactory according to a universal view, such as noted in the "Great Chain MetaEthics 4.1C" of Conegliano Campus 5.1C foundation.

[a] e-mail: giovannicargnello@gmail.com

 DOI: 10.1051/bioconf/20170901014

1. Premessa

Premesso che dopo una specifica ricerca è emerso che "durabilità" sta per duraturo indefinitamente, mentre la "sostenibilità" implica, anche, sostenere ciò che intendiamo duri indefinitamente. (Devoto, Oli 2016; Zingarelli, 2016; Pittano 2016; Wikipedia, 2017). Pertanto appare che "Duraturo" e "Sostenibile" non sono sinonimi e neanche equivalenti e analoghi e quindi noi così li useremo.

Della sostenibilità, cioè della caratteristica di un processo o di uno stato di essere mantenuto tale indefinitamente, soprattutto negli ultimi cinquanta anni si è parlato e si sta parlando moltissimo relativamente all'ambito ambientale, economico e sociale, mentre secondo la nota filosofia, la metodologia della "MetaEtica 2.1C" e della "Grande Filiera MetaEtica 4.1C" algoritmi del Conegliano Campus 5.1C, e quindi per noi come oltre vedremo, la "sostenibilità", alla fine la "Sostenibilità Universale 4.1C e 5.1C[1]" o "Sostenibilità MetaEtica 4.1C e 5.1C" o "Sostenibilità 4.1C e 5.1C" considera "tutti" i fattori materiali ed immateriali coefficentati tecnici, economici, ambientali, sociali.occupazionali, esistenziali, etici settoriali, "Etici Universali 4.1C o MetaEtici 4.1C".

Ciò premesso è fondamentale aggiungere che a differenza della "Rivoluzione Produttiva 4.0" o "Rivoluzione dell'Industria 4.0" o "Industria 4.0" la "Rivoluzione Produttiva Globale, 4.1C e 5.1C, Viticola e VitiVinicola 4.1C e 5.1C" comprese, così come le "Sostenibilità e le relative carte e certificazioni 2.1C, 4.1C e 5.1C", considerano oltre all'aspetto "produttivo-economico" anche "tutti" gli altri fattori coefficentati materiali ed immateriali tecnici, ambientali, sociali.occupazionali, esistenziali, (esistenziali per l'uomo e per tutti gli altri esseri viventi e non viventi, per il territorio, per la madre terra, per la litosfera, per gli universi, ...), etici settoriali, "Etici Universali 4.1C o MetaEtici 4.1C" in armonica filiera, secondo la nota filosofia, metodologia della "Grande Filiera MetaEtica 4.1C", nostro algoritmo del Conegliano Campus 5.1C esposto più volte sin dalla fine del secolo scorso, e pure, all'OIV dall' 2000. [1–34]**. Merita, inoltre ricordare che la "Rivoluzione 4.0" o "Rivoluzione Industriale 4.0" o "Industria 4.0" è nata in Germania nel ottobre del 2012 e soprattutto recentemente, da noi, della "Rivoluzione 4.0" ci si riempie la bocca senza gli indispensabili fondamentali approfondimenti, così come un tempo non vennero fatti, ad esempio, per: terroir, territoir, zonazioni, qualità, mode, sostenibilità, valori, mezzi, obiettivi, aspetti: tecnici, economici, ambientali, sociali.occupazionali, esistenziali, etici, "MetaEtici", nonché sulla "MetaEtica", sulla "MetaEtica CC, 2C, 2.1C, 5.1C", sulla: "Grande Filiera", "Grande Filiera MetaEtica", "Grande Filiera MetaEtica: CC, 2C, 2.0, 2.1C, 4C, 4.1C, 5.1C"**, sulle scelte di "Politica" produttiva, sulla "Rivoluzione Produttiva in generale, dell'Industria , della Viticoltura, della Vitivinicoltura comprese", sugli acronimi: CC, 2C, 4C, 2.0, 4.0, 2.1C, 4.1C, 5.1C, ecc. Ed allora considerando che l'applicazione della "Rivoluzione Produttiva 4.0" sta già creando, anche da noi, problemi, alcuni molto gravi, una visione non a compartimenti stagni e non settoriale insoddisfacente, ma una visione armonicamente globale relativamente a "tutti" i fattori indicizzati, (indicizzati in funzione della realtà in cui si opera), della "Grande Filiera 4C", del Conegliano Campus 5.1C, come quella degli algoritmi della "Rivoluzione Produttiva 4.1C", della "MetaEtica 2.1C" e della più articolata, interdisciplinarmente globale, olistica 4.1C della "Grande Filiera MetaEtica 4.1C" rappresentano un'opportunità, un mezzo fondamentale per realizzare una rivoluzione, anche produttiva, non settoriale insoddisfacente, o addirittura catastrofica, ma una visione, una rivoluzione produttiva, così detta, "Universale 4.1C" o "MetaEtica 4.1C", cioè una rivoluzione, un'attività, un agricultura.agricoltura, una viticultura.viticoltura, una vitivinicoltura. vitivinicoltura "Universalmente Sostenibile 4.1C" o "MetaEticamente Sostenibile 4.1C base fondamentale per scrivere, come sono state scritte, anche, della viticoltura la "Carta della Sostenibilità Universale 4.1C" o "Carta della Sostenibilità MetaEtica 4.1C" e le relative certificazioni e "Garanzie MetaEtiche" delle certificazini. (lc).

Pertanto per noi, come vedremo, la "MetaEtica 2.1C" del Conegliano Campus 5.1C, non va confusa con la "Grande Filiera MetaEtica 4.1C", sempre del Conegliano Campus 5.1C. Infatti, sempre per noi, la "MetaEtica 2.1C", come noto, per pensare, progettare, realizzare, giudicare tutto e tutti e quindi anche relativamente alle "rivoluzioni", alle "viticolture", alle "sostenibilità", alle "carte", alle "certificazioni", ecc., considera in modo "universale" ("Universale 4.1C o MetaEtico 4.1C) singolarmente o per gruppi i fattori relativi gli aspetti tecnici, economici, ambientali, sociali.occupazionali, esistenziali, etici secondo la "Grande Filiera MetaEtica 4.1C" del Conegliano Campus 5.1C mentre la "Grande Filiera MetaEtica 4.1C" sempre del Conegliano Campus 5.1C per pensare, progettare, realizzare, giudicare tutto e tutti, sempre, si rifà all'algoritmo che considera in armonica filiera "tutti" i fattori coefficentati materiali ed immateriali tecnici, economici, ambientali, sociali.occupazionali, esistenziali, etici settoriali, "Etici Universali 4.1C o MetaEtici 4.1C" della "Grande Filiera MetaEtica 4.1C", oltre ad altri fondamentali aspetti che successivamente verranno esposti. (lc). Ed allora se l'universo "uomo" viene considerato, a livello operativo, "MetaEtico 4.1C" tutto quello che da lui discende, ad esempio relativamente ai fattori indicizzati materiali ed immateriali tecnici, economici, ambientali, sociali occupazionali, esistenziali, etici è "MetaEtico 2.1C".

Si evidenzia che allo stato attuale delle cose questa visione, l'applicazione della "MetaEtica 2.1C" del Conegliano Campus 5.1C per tutti i singoli fattori o per gruppi di essi, ha rappresentato e rappresenta una base importante ad esempio, per scrivere, come sono state scritte, le relative carte e certificazione su una loro "sostenibilità universale 2.1C" e la metodologia della "Grande Filiera MetaEtica 4.1C" del Conegliano Campus 5.1C ha rappresentato e rappresenta, pure, le fondamenta per scrivere, come sono state scritte, le relative carte e certificazioni su una loro "sostenibilità universale 4.1C": "Carta dell'Agricoltura, della Viticoltura e della Vitivinicoltura Universalmente o MetaEticamente Sostenibile 4.1C" o "Carta Universale della Sostenibilità 4.1C" o "Carta della Sostenibilità MetaEtica 4.1C" secondo la "Grande Filiera MetaEtica 4.1C" del Conegliano Campus 5.1C. Pertanto come a suo tempo da noi previsto e allo stato attuale delle cose, quanto prima , bisogna cercare di acquisire una cultura, una forma mentis interdisciplinarmente globale, fondamentale olistica 4.1C come è quella della metodologia dall'algoritmo della

 DOI: 10.1051/bioconf/20170901014

"Grande Filiera MetaEtica 4.1C" per poi, in pratica agire semplificando al massimo, in questo caso, "Metaeticamente 2.1C" ad imbuto relativamente ai descrittori o gruppi di essi da prendere in considerazione, così come abbiamo predisposto per la relazione per l'OIV del 4 aprile scorso a Parigi, relazione esposta dal Prof. Carbonneau.

La filosofia applicata, la metodologia, l'algoritmo della "Grande Filiera MetaEtica 4.1C" del Conegliano Campus 5.1C per pensare, progettare, realizzare, condurre, produrre, comunicare, valorizzare, vendere e incassare correttamente, recuperare "MetaEticamente 4.1C" i crediti, valutare, giudicare ogni produzione, ogni attività e ogni cosa, e quindi anche la "Sostenibilità", la "Qualità", la "Viticultura-Viticoltura" comprese e per scrivere le relative carte e certificazioni delle "Sostenibilità Universali 4.1C o MetaEtiche 4.1C" essenzialmente prevede: 1-di considerare, dal basso verso l'alto, in armonica filiera "tutti" i fattori materiali ed immateriali coefficentati tecnici, economici, ambientali, sociali.occupazionali, esistenziali, (esistenziali per uomo, gli altri esseri viventi e non viventi, per il territorio, per la terra madre, per la litosfera, per il globo e per gli universi, ...), etici settoriali e "Etici Universali 4.1C o MetaEtici 4.1C" secondo la "Grande Filiera MetaEtica 4.1C" del Conegliano Campus 5.1C, 2-di definire in modo univoco il significato, la semantica storica ed attuale, "un passo indietro verso il futuro 4.1C®", dei termini che si usano e di usarli in modo appropriato sempre e tassativamente nel mondo della ricerca e dell'Università (ad esempio quando si parla di: "Grande" e "MataEtica" (tra virgolette e con la "G", la "M" e la "E" maiuscole), CdiC; CC; 2C, 3C; 4C; 2.1C, 4.1C, 5.1C, nonché di: cultura, antropizzazione, paesaggio, qualità, maturità, "Smar.t.C-Smar'C-SmarC", sostenibilità, mode, valori, terroir, territoir, mezzi, obiettivi, etica, "MetaEtica", "MetaEtica 2.1C", filiera: "piccola filiera", "Grande Filiera", "Grande Filiera MetaEtica", "Grande Filiera MetaEtica 2C, 3C, 4C", "Grande Filiera MetaEtica 4.1C", nonché di "Rivoluzione Produttiva Generale o dell'Industria o della Viticoltura o della Vitivinicoltura, ecc.: 4.0, 2.1C, 4.1C, 5.1C, ecc.), 3-di chiarire inequivocabilmente quali sono gli obiettivi e quali sono i mezzi (qualità sensoriale "classica" compresa, la quale rappresenta un mezzo per poter raggiungere al "meglio", cioè "MetaEticamente 4.1C SmarC" secondo la "Grande Filiera MetaEtica 4.1C", i vari obiettivi sopra esposti), 4-di collegare armonicamente le tecniche, i mezzi agli obiettivi e quindi i mezzi, le tecniche impiegate devono discendere dagli obiettivi secondo la "Grande Filiera MetaEtica 4.1C"; pertanto la vitivinicoltura deve discendere dell'enologia e l'enologia dal consumatore e questo deve discendere "MetaEticamente 4.1C", cioè secondo la"Grande Filiera MetaEtica 4.1C", 5-di considerare armonicamente tra loro "tutti" i fattori coefficentati materiali e immateriali relativi agli aspetti sopra esposti, 6-di operare sempre metodologicamente partendo non dal basso e fermarsi lì ai mezzi, agli aspetti tecnici, o ai singoli obiettivi sopra esposti o a gruppi di essi, ma, quanto meno a livello culturale, a livello di forma mentis, per operare partire dall'alto di questa "Grande" filiera, ("Grande 4.1C filiera), cioè partire dai "Descrittori Universali 4.1C o MetaEtici 4.1C" per poi scendere a considerare singoli fattori o gruppi di essi o armonicamente tra loro tutti i fattori coefficentati materiali e immateriali tecnici, economici, ambientali, sociali.occupazionali, esistenziali, etici settoriali e "Etici Universali 4.1C o MetaEtici 4.1C" per poi, "fare quello che si può", ma in questa "MetaEtica 4.1C-SmarC ottica", cioè con l'obiettivo di operare, quanto prima, secondo la "Grande Filiera MetaEtica 4.1C" del Conegliano Campus 5.1C per pensare, progettare, realizzare e giudicare ogni cosa, ogni attività compresa quella viticola e quella vitivinicola.

Quindi la "MetaEtica 2.1C" non è la "Grande Filiera MetaEtica 4.1C" e, come noto, praticamente rappresenta un algoritmo importante relativamente alla "Grande Filiera MetaEtica 4.1C" del Conegliano Campus 5.1C, la quale "Grande Filiera MetaEtica 4.1C" ha rappresentato, rappresenta e rappresenterà, tra l'altro, la base fondamentale per una più articolata, interdisciplinarmente globale, olistica 4.1C, finale "Universale Sostenibilità 4.1C e 5.1C SmarC" o "MetaEtica Sostenibilità 4.1C e 5.1C SmarC", nonché: della "Rivoluzione Produttiva 4.1C, anche, della Viticoltura 4.1C" o "Rivoluzione Viticola 4.1C", del "BIOBIOSOS4.1C", "BIOINtegrato 4.1C", del' "l'AltraViticultura-l'Altraviticoltura 4.1C", delle "Viticulture-viticolture 4.1C smarc" di "un passo indietro verso il futuro 4.1C®", delle viticolture, cosi dette, non "industriali", ma anche di quelle "industriali" e di "rapina", nonché quelle scelte come "stili di vita-autosufficenti", storiche, culturali, sociali.occupazionali, esistenziali, integrate e valorizzanti le risorse del territorio, concorsuali: "MetaEticamente 2C" e "MetaEticamente 4.1C", nonché per scrivere, come è stata scritta, una "Carta della Viticoltura e della Vitivinicoltura Sostenibile MetaEticamente 4.1C o Carta Universale della Sostenibilità 4.1C o Carta della Sostenibilità MetaEtica 4.1C" secondo la "Grande Filiera MetaEtica 4.1C" del Conegliano Campus 5.1C. (lc). Un esempio di "MetaEtica 2.1C": la meccanizzazione della vendemmia, (così come per le altre tecniche colturali, e non solo), la si può analizzare come risolutrice, solo e solamente, di un problema di tecnica colturale oppure la si può considerare, come dovrebbe essere considerata sempre, in un contesto "universale" secondo l'algoritmo della "MetaEtica 2.1C" che la analizza in fase progettuale, operativa e di giudizio finale considerando dal basso verso l'alto, in armonica filiera "MetaEtica 4.1C" "tutti" gli altri fattori materiali ed immateriali coefficentati tecnici, economici, ambientali, sociali.occupazionali, esistenziali, etici relativi alla "vendemmia meccanica", così come può e deve essere fatto singolarmente o per gruppi di essi per tutti gli altri fattori tecnici, economici, ambientali, sociali.occupazionali, esistenziali, etici della "Grande Filiera 4C" e come ad esempio per la "quantità", la "qualità", per l'uomo "solitamente considerato" e per gli altri mezzi della "Charter of sustainable viticulture the G*i*ESCO" (46) mentre la "Grande Filiera MetaEtica 4.1C" del Conegliano Campus 5.1C, oltre ha rappresentare il palinsesto base, anche per la "MetaEtica 2C" prevede in fase progettuale, operativa e di giudizio finale di considerare i fattori indicizzati non settorialmente o per gruppi di essi, ma di considerare "tutti" complessivamente nella globalità interdisciplinare olistica 4.1C i fattori materiali ed immateriali coefficentati tecnici ad esempio operativi, nonché "tutti" gli altri fattori economici, ambientali, sociali.occupazionali, esistenziali, (esistenziali per uomo, gli altri esseri viventi e non viventi, per il territorio, per la terra madre, per la litosfera, per il globo e per gli universi), etici settoriali ed "Etici Universali 4.1C o MetaEtici 4.1C" per pensare, progettare, realizzare, condurre, produrre, comunicare, valorizzare, vendere e incassare correttamente, recuperare "MetaEticamente 4.1C" i crediti, valutare, giudicare ogni produzione, ogni attività e ogni cosa. (lc).

Pertanto si risottolinea che la "MetaEtica 2.1C" e la "Grande Filiera MetaEtica 4.1C", (come altrove viene fatto), dovrebbe rappresentare, sempre e sempre più, la premessa e la conclusione di tutto e per tutto: convegni, corsi, incontri vari, ecc, nonché l'unità di misura di tutto, in quanto "Universale 4.1C o MetaEtica 4.1C".

Ed allora se questo modo di far fare e di fare, non solo viticoltura, sortito da attività e ricerche condotte e pubblicate in epoche non sospette fosse allora stato applicato si sarebbero potute evitare le note realtà negative o addirittura catastrofiche, anche da noi, agli occhi di tutti a livello tecnico, economico, ambientale, paesaggistico, dell'antropizzazione, sociale, occupazionale, esistenziale, etico, "MetaEtico 4.1C".

Quindi queste attività e ricerche sono risultate fondamentali per poter scrivere, come sono stati scritti gli algoritmi delle sostenibilità universale 4.1C o sostenibilità MetaEtica 4.1C: globale ideale, quella da realizzare quanto prima e quella realizzabile attualmente, anche, per la viticoltura e per la vitivinicoltura. Algoritmi fondamentali per poter scrivere, come sono state scritte le relative "Carte e certificazioni della Sostenibilità Universale 4.1C o Sostenibilità MetaEtica 4.1C", anche, secondo la "Grande Filiera MetaEtica 4.1C" e la "MetaEtica 2.1C" generali e quelle che l'azienda realmente, ("MetaEticamente 4.1C), desidera perseguire e quindi "MetaEticamente 4.1C auto-certificare e se lo desidera, "MetaEticamente 4.1C, far certificare, anche, da enti terzi pubblici, privati o misti, nonché le carte e le certificazioni relative alla "Rivoluzione Produttiva 4.1C", la quale, come noto è "Universalmente 4.1C o MetaEticamente 4.1C" fondamentale per prevenire i problemi, alcuni nolto gravi determinati dalla "Rivoluzione Produttiva 4.0".

2. Materiali, metodi

Dopo un breve periodo, (fine anni sessanta, inizio anni settanta) nel quale, per decisione di altri, l'attività e le ricerche riguardavano solo aspetti tecnici viticoli settoriali e a se stanti, sono passato a collegare tra loro gli aspetti tecnici viticoli a quelli economici, secondo la nostra "prima piccola filiera C" di Conegliano, per poi collegare questi aspetti a quelli ambientali e sociali secondo la nostra "seconda piccola filiera 2C" del Campus di Conegliano per poi collegare questi aspetti a quelli occupazionali ed esistenziali, etici settoriali secondo la nostra "terza piccola filiera 3C" del Campus di Conegliano 3C pensando così di essere arrivato alla fine della filiera e del relativo algoritmo detto della "piccola filiera 3C". Ma approfondimenti, soprattutto, sull'"etica" e sui relativi descrittori ci hanno portato alla "MetaEtica" e quindi alla filosofia applicata, alla metodologia, all'algoritmo della così detta "Grande Filiera MetaEtica 4C" del Conegliano Campus 4C e in seguito alla inaccettabile "Quarta Rivoluzione industriale" o "Rivoluzione Industriale 4.0" o "Industria 4.0" o "Rivoluzione 4.0" del 2012 siamo passati alla "Grande Filiera MetaEtica 4.1C" del Conegliano Campus 5.1C. Qui si aggiunge che le nostre "Sostenibilità 4.1C", le relative carte e certificazioni sono state ispirate, anche, da quelle dell'ONU, dell'UNESCO, della CE, dell'OIV, del GiESCO, dell' Equalitas, ecc, ecc., ma scritte secondo la "Grande Filiera MetaEtica 4.1C" del Conegliano Campus 5.1C. In particolare alla luce delle definizioni delle sostenibilità, si è proceduto ad imbuto:1-al decimo sotto-raggruppamento e coefficentazione dei descrittori o fattori materiali ed immateriali (oltre 1000 dagli oltre 20000 iniziali): tecnici, economici, ambientali, sociali.occupazionali, esistenziali, etici settoriali e "Etici Universali 4.1C o MetaEtici 4.1C" secondo la "Grande Filiera MetaEtica 4.1C" del Conegliano Campus 5.1C relativi alle sostenibilità, andando così oltre i descrittori delle altre sostenibilità, 2-alla definizione delle soglie minime per essere "Sostenibili 4.1C" relativamente ai vari descrittori, le quali sono risultate 5: 100-30-10-5-1 su una scala da 1 a 100 e all'attribuzione degli aggettivi di "sostenibilità di merito" se la valutazione del descrittore supera quella minima di sostenibilità e cioè: se supera dall' "1" al "4" %: "sostenibilità di merito": "Più che sufficiente", dal "5" al "25" %: "Buona", dall' "26" al "50" %: "Molto Buona", dall' "51" al "75" %: "Ottima", dall' "76" al "100" %: "Eccellente", superiore al "101%": "Esemplare" e le approssimazioni sono per difetto, 3-e come primo approccio all'individuazione e alla relativa coefficentazione dei "sotto gruppi omogenei" di appartenenza dei vari fattori che sempre ad imbuto sono risultati 6 "sotto gruppi di descrittori tecnici, economici, ambientali, sociali, esistenziali, etici" secondo la "Grande Filiera MetaEtica 4.1C" del Conegliano Campus 5.1C; tale approccio "MetaEtico 2.1C" può essere esteso ai singoli descrittori,

Sempre dalla "MetaEtica 2.1C" e dalla "Grande Filiera MetaEtica 4.1C" del Conegliano Campus 5.1C, sono derivate, tra l'altro, le carte e le certificazioni: del "BIO-BIOSOS4.1" (acronimo di "Biologico" e "Biodinamico" "Universalmente o MetaEticamente Sostenibili 4.1C"), della "Rivoluzione Produttiva della Viticoltura 4.1C" o "Rivoluzione Viticola 4.1C", della "Rivoluzione Produttiva della Vitivinicoltura 4.1C" o "Rivoluzione Vitivinicola 4.1C", della "Qualità Universalmente Sostenibile 4.1C" o "Qualità Sostenibile MetaEticamente 4.1C" o "Qualità MetaEtica 4.1C", ecc. L'originale "BIOBIOSOS4.1C", prevede che in realtà particolari e/o straordinarie, "MetaEticamente 4.1C", si possono applicate interventi non previsti dal "Biologico" e dal "Biodinamico" "classici", ma interventi di assoluto rispetto, tutela, meglio se anche migliorativi dell'ambiente inteso, questo, secondo la "Grande Filiera MetaEtica 4.1C". Interventi ufficialmente autorizzati dal mondo della scienza "pubblica", mentre la garanzia della relativa certificazione viene "Garantita MetaEticamente 4.1C e 5.1C" dal produttore e in caso di necessità di costui da enti terzi pubblici o privati o misti. Merita evidenziare che dopo vari incontri e discussioni con illustri, stimati e titolati personaggi, attività datate, anche, fine dell'altro secolo, costoro pur condividendo la filosofia, le metodologie della "MetaEtica" e della "Grande Filiera MetaEtica" e le relative carte e certificazioni delle "Sostenibilità 4.1C" del Conegliano Campus 5.1C, ritenevano, ma ora le cose stanno completamente cambiando, che l'adozione della "Sostenibilità 4.1C" avrebbe enormemente complicato le cose in sede di ridefinizione internazionale, ad esempio all'OIV, della "Sostenibilità". (51).

Pertanto, per non perdere tempo prezioso, a ragione si proponeva di urgentemente procedere per tappe: I^a tappa: di considerare i fattori esistenziali, etici, "MetaEtici" della "Grande Filiera MetaEtica 4.1C" del Conegliano Campus 5.1C come prerequisiti della "Sostenibilità" ad esempio dell'OIV, II^a tappa: di aggiungere agli aspetti ambientali, economici, sociali, anche quelli esistenziali ed etici settoriali, nonché quelli tecnici, III^a tappa: di considerare tutti i fattori, etici settoriali compresi, in modo "Universale 2.1C o MetaEtico 2.1C", IV^a tappa: di applicare la filosofia, le metodologie, l'algoritmo della "Grande Filiera MetaEtica", V^a tappa: di dare corso, dopo "adeguato confronto con chi lo desiderasse", subito ad una "Sostenibilità Universale 2.1C" o "Sostenibilità MetaEtica 2.1C" e ad una "Sostenibilità Universale 4.1C" o "Sostenibilità MetaEtica 4.1C", VI^a tappa: alta formazione e alta informazione. (lc)

E tutto ciò relativamente ad una "Carta della Sostenibilità Universale 2.1C" o "Carta della

Sostenibilità MetaEtica 2.1C" e ad una "Carta della Sostenibilità Universale 4.1C" o "Carta della Sostenibilità MetaEtica 4.1C", anche, secondo la "Rivoluzione Produttiva 4.1C" del Conegliano Campus 5.1C.

3. Risultati e discussioni

Per ovvi motivi per tutti i risultati si rimanda alle relative specifiche numerose pubblicazioni nazionali ed internazionali datate, anche, l'altro secolo alcune delle quali sono riportate in bibliografia. (lc).

Qui ci limitiamo a riportare che queste attività e queste ricerche ci hanno permesso di evidenziare che: 1-la "MetaEtica 2C" per singoli fattori o per gruppi di essi non va confusa con la "Grande Filiera MetaEtica 4.1C" e praticamente può rappresentare una base per una più articolata, interdisciplinarmente globale, olistica 4.1C, finale "Universale Sostenibilità 4.1C e 5.1C" o "MetaEtica Sostenibilità 4.1C e 5.1C " secondo la "Grande Filiera MetaEtica 4.1C" del Conegliano Campus 5.1C, 2-la "MetaEtica 2.1C" del "Conegliano Campus 5.1C" (47.48.50) era "MetaEtica 2C" del Conegliano Campus 5.1C durante la preparazione all' incontro con l'OIV del 4 aprile 2017 (51) e prima ancora era "MetaEtica" non ancora codificata, 3-la "Grande Filiera MetaEtica 4.1C" (abbreviando "MetaEtica 4.1C") del Conegliano Campus 5.1C all'origine era "Grande Filiera" di Conegliano" divenuta successivamente "Grande Filiera Etica 2C" del "Campus di Conegliano 2C", divenuta poi "Grande Filiera MetaEtica 4C" del "Conegliano Campus 4C" divenuta infine, dopo la rivoluzione dell' "Industria 4.0", "Grande Filiera MetaEtica 4.1C" del "Conegliano Campus 5.1C" (acronimo "MetaEtica 4.1C") del Conegliano Campus 5.1C. Perché "1" e non "0": per andare oltre l'innovazione, la rivoluzione settoriale produttiva 4.0 che ha creato e sta creando, da noi e in tutto il mondo gravi problemi; problemi che si sarebbero potuti evitare e razionalmente porre rimedio se si fosse operato secondo l'allora "MetaEtica 2C" e "Grande Filiera MetaEtica 4C". Quindi considerando che lo "0" è "0", lo "0" del 4.0 grazie al precedente algoritmo della "Grande Filiera MetaEtica 4C" lo "0" è stato sostituito con un "1" che significa oltre lo "0" per cui "Grande Filiera MetaEtica 4.1C"; pertanto questi nostri algoritmi universali, grazie alla "Grande Filiera MetaEtica 4C" sono tutti con un "1": "MetaEtica 2.1C", "MetaEtica 4.1C" e "Grande Filiera MetaEtica 4.1C" e 5.1C quando la "MetaEtica 4.1C", la "Grande Filiera MetaEtica 4.1C", la "Sostenibilità 4.1C", ecc. sono applicati.

4. Conclusioni

Queste attività, queste ricerche iniziate l'altro secolo sulla "MetaEtica", sulla "Grande Filiera", sulla "MetaEtica 2.1C" e sulla "Grande Filiera MetaEtica 4.1C" o "MetaEtica 4.1C" sono risultate le fondamenta della filosofie applicate, delle metodologie, degli algoritmi della "Rivoluzione Produttiva 4.1C" generale, della "Rivoluzione Viticola 4.1C", della "Rivoluzione VitiVinicola 4.1C", delle "Viticoltura MetaEtica 2.1C", della "Viticoltura MetaEtica 4.1C" o "Viticoltura della Grande Filiera MetaEtica 4.1C" del "Conegliano Campus 5.1C Smar'C"; filosofie applicate, metodologie, algoritmi che ci hanno permesso di scrivere, come abbiamo scritto, le "Carte della Sostenibilità Universale 2.1C o Carte della Sostenibilità MetaEtica 2.1C" e le "Carte della Sostenibilità Universale 4.1C o Carte della Sostenibilità MetaEtica 4.1C" generale, viticola e vitivinicola e delle relative "MetaEtiche 4.1C" certificazioni e garanzie delle certificazioni del "Conegliano Campus 5.1C".

Inoltre dall'attività e dalle ricerche svolte è emerso che la "MetaEtica 2.1C" e la "Grande Filiera MetaEtica 4.1C" del Conegliano Campus 5.1C sortite in epoche non sospette, anche, in questo particolare momento storico della realtà locale, nazionale, europea e soprattutto internazionale sono risultate di fondamentale attualità e interesse non solo per porre rimedio e per evitare il ripetersi delle note conseguenze negative o addirittura catastrofiche, anche da noi, agli occhi di tutti a livello tecnico, economico, ambientale, paesaggistico, sociale, occupazionale, esistenziale, etico secondo la "MetaEtica 2.1C" e secondo la "Grande Filiera MetaEtica 4.1C" del Conegliano Campus 5.1C, ma anche, tra l'altro, queste attività e ricerche sono risultate fondamentali per scrivere, come sono state scritte, le carte e le relative certificazioni e garanzie "MetaEtiche 2.1C" e "MetaEtiche 4.1C", ad esempio, delle "Sostenibilità": generali, del vigneto, della cantina, della vitivinicoltura, del "Biologico 4.1C", del "Biodinamico 4.1C", del "BIOBIOSOS4.1C e 5.1C", de "l'AltraViticultura-l'Altraviticoltura 5.1C", delle "Qualità 4.1C e 5.1C", ecc. e quindi, anche per avere sempre più e sempre meglio una "Viticultura-viticoltura MeteEtica Sostenibile 4.1C" secondo la "Grande Filiera MetaEtica 4.1C" del Conegliano Campus 5.1C.

[1]L'acronimo "5.1C" lo possiamo già usare in quanto siamo già nella fase applicativa.

Per il determante contributo dato da d. Zuliani A., Prestamburgo M., Fregoni M., Eynard I., Intrieri C., Amati A., Favaretti G., Cescon P., Milotc A., Persuric G., Forno S. e dato da d. Frilli F., Girolami V., Boatto V., Galletto L., Giacomini C., Scalabrelli GC., Ubigli M., Duso C., Gaeta D., Bavaresco L., De Marco D., Spellini, Dalla Torre D., d. Dassie M., Floriani L., Milanese R., Tonon C., Doimo G., Camatta T., Di Gaetano R., Zironi R., Giulivo C., Silvestroni O., Di Lorenzo R., Novello V., Tonutti P., Bonghi C., Paliotti A., Bravi M., Scaggiante S., D'Onofrio C., Piazza G., Scalacci R., Bisoffi S., Zinzani G., Gerbi V., Celotti E., Stefanini M., Raifer B., Morandel A., Streifeneder T., Boni G., Castaldi R., Belvini P., Tarzariol F., Cecchetto G., Lunardelli R..

 DOI: 10.1051/bioconf/20170901014

e da semplici, anche "analfabeti", ma illuminati, saggi cittadini, montanari e contadini ai quali tutti devo, dobbiamo molto, cosi come al GiESCO e IAVW, all'OIV, al CONAVI, al "Conegliano Campus", all' Istituto Sperimentale per la Viticoltura-CREA, all' UNIPD e altre Università e Istituti di Ricerca, all'Accademia Italiana della Vite e del Vino, al "Centro Culturale Humanitas", CIA, Coldiretti, Confagricoltura, UIV, Federvini e dato da altri numerosi signore e signori e soprattutto dato da Gianni Teo e da Alain Carbonneau.

Bibliografia

[1] Carbonneau A., Cargnello G., 2003. Architectures de la Vigne et Systèmes de Conduite. Dunod Ed., pp188
[2] Cargnello G. et Carbonneau A. (2007a): Méthode de la "Grande Filiére" appliquée au management d'un modéle productif de vignoble. Proceedings XVth International Symposium GESCO Porec-Croatia 20-23 june, pp. 16-33
[3] Carbonneau A. 2012. Métaéthique: ésotérisme inutile ou véritable enjeu ? Véritable enjeu pour la société et la durabilità! Progr. Agric. Vitic., **129**(13-14), 319-323. A.A. 2012. La "MetaEtica" non è un esoterismo inutile, ma una vera importante e fondamentale posta in gioco per la società per cui urgentemente va considerata, dibattuta, comunicata ed applicata ovunque. Traduzione dell'articolo del prof. Alain Carbonneau su PAV. **129**(13-14), 319-323
[4] Carbonneau A. and CARGNELLO G. (2013m): "MetaEthics": a social challenge and a process of sustainable viticulture. Acts 18 th International Symposium GiESCO from the 7-11 July 2013. Porto (Portugal). Pp. 1123-1128
[5] Cargnello G. (1986): Recherches de nouveaux systèmes de conduite pour une viticulture économique et de qualité dans le Nord de l'Italie. Thèse. Université de Montpellier (France), pp. 1-332
[6] Cargnello G. (1994b): social, existential, ethical and economic aspects concerning new models of an increasingly profitable viticulture in researches on "global reasoned viticulture". XXIV International horticultural congress, Kyoto (Japan)
[7] Cargnello G. (1996a): la qualité économique, l'économie de la qualité et la qualité économique des préférences: différentes considérations. Compte-rendu n° 9 GESCO, Budapest (Hongrie), 21 -23 août, pp. 379-384
[8] Cargnello G. (1999b): Evaluation "globale" de processus productifs, d'aspects techniques et scientifiques divers, etc ... : considérations générales relatives à la "Grande Filière" viti-vinicole. GESCO Sicilie, Italie, 6-12 June, pp. 362-371
[9] Cargnello G. (2000): Evaluation "globale" de les "qualité" du produit par le biais de l'idée d'une "nouvelle" méthode dénommée C.IM.C., aussi pour contribuer à rendre les systèmes productifs vitivinicoles et leurs itinéraires techniques durables. Actes O.I.V. Paris 19-23 Juin, Section I pp. 207-213
[10] Cargnello G. avec Cielo M., Michieli M., Dalla Torre D., Pezza L., Spera G., Milotic A., Persuric G., Sladonja Barbara, Staver M. (2005b):"Grande Filière": universalité de la philosophie méthodologique base de la "Grande Filière" dans un contexte d'equitable soutenabilité économique, de temps, humain, technique et scientifique pour la lecture, l'évaluation, la valorisation, la communication et le developpement d'activités variées viticoles aussi, et application pratique de la "grande filière".Proceedings Vol. 2 XIVthInternational Symposium GESCO, Geisenheim, 23-27 August, pp. 779-786
[11] Cargnello G. (2007c): "Innovations" to make "universal" "sustainable" the development of viticolture. Various researchs and considerations. International symposium on Viticolture and Enology. Yangling, Shaanxi (CN). 20-22 april, pp. 1-34
[12] Cargnello G. (2008a): "Grande Filiera" - "Viticoltura Etica": urgente necessità di andare oltre la "qualità.profitto" tecnico ed economico puntando sulla viticoltura "etica". Considerazioni varie. Atti 2°Convegno Nazionale di Viticoltura (CONAVI), Marsala 14-19 luglio pp. 1-12
[13] Cargnello G. (2008c): Gli universi delle "qualità" e delle "preferenze" (oltre novanta) secondo la "Grande Filiera Base" (GFB) e la "Grande Filiera Super Corta" (GFSC)quali fondamenta per l'"innovativa" e"sostenibile" realizzazione, preservazione,miglioramento,comunicazione, marketing e valorizzazione dell' originalità e della tipicità delle nostre produzioni locali dei "terroir", dei "territoir" e del nostro "mosaico paesistico".XIIIConvegno internazionale interdisciplinare. Aquileia(UD) 17-18/9
[14] Boatto V., Galletto L., Scaggiante S., Teo G., Barisan L., Cargnello G., Pezza L., Gallo G., Di Gaetano R., Camatta Teresa, Bonghi C., Bianchini Federica, Veilleux Lissa, Negro V, †Curtolo C., Pancot M., Cecchetto G. (2009c): Research in the technical, socio-economical, and ethical aspects for Veneto vitivinicolture (north est Italy). 16 th International Symposium GIESCO 2009 Uc-Davis (USA - CA). Pp. 283-287
[15] Cargnello G. (2009d): "Great chain: urgent necessity the focusing on the "metaethical" viticulture or "great" viticulture: research and various considerations. 16 th International Symposium GIESCO 2009 Uc-Davis (USA - CA). Pp. 31-35
[16] Cargnello G. (2011b): Ancora sull'"ethical e sul "MetaEtical" wine & food & farming & living, ... Ricerche pluriannuali e considerazioni varie. Taste-Vin n. 4. pp. 25-29
[17] Cargnello G. (2011d): Conclusioni sull'attività e sull'incontro multidisciplinare del "Conegliano Campus" sulla "Sostenibilità", sui "Modelli Viticoli Made in Italy", sui "Modelli Viticoli Made in Italy del Conegliano Campus" e sui "Modelli Viticoli Non Made in Italy" relativi, in questo periodo, ad applicativi fondamentali più significativi, per l'Italia, modelli viticoli e sulla loro valutazione tecnica-economica-sociale-esistenziale-etica armonicamente complessiva e "MetaEtica" e sul "MEVIEXPO". "Conegliano Campus" e MEVIEXPO
[18] Cargnello G. with 645 coll. (2012a): Research on new sustainable viticultural models which allow to safeguard an equi profit for viticultural and winemaking businesses increasing occupation levels and the necessary resources and costs. 35th World

Congress of Vine and Wine 9th General Assembly of the O.I.V. Izmir, Turkey, 18-22 June

[19] Cargnello G., Carbonneau A., Schultz H., Murisier F., Queiroz J. (2012f): Researches on traditional and new models of cultivation systems for the development of mountain or steep slope viticulture. IV International Congress on Viticulture in Mountain. Lyon (France) 7-9 novembre 2012

[20] Cargnello G. (2012p): È la "Grande Filiera" e non la "MetaEtica" la vera posta in gioco per il futuro degli universi, della terra, dell'umanità, delle varie identità socio-culturali-paesagistiche-produttive, nonché di tutte le produzioni, prodotti ed attività , per la terra madre e per l'universo. "Conegliano Campus" e MEVIEXPO

[21] Cargnello G., Carbonneau A. (2013b): Researches and considerations about mountain sustainable viticulture, with the dimension of 'MetaEthics'. Acts 18 th International Symposium GiESCO from the 7-11 July 2013. Porto (Portugal). Pp. 247-250

[22] Cargnello G. avec 53 Col. (2013c): Recherches conclues conduites pour definir les mode de conduite "Made in Italy", "Made in Italy du Conegliano Campus", "Made non in Italy" applicatifs actuellement les plus significatifs et soutenables pour l'Italie et pour definir l'argorithme pour leur evaluation technique, economique, sociale, existentielle, ethique harmonieusement d'ensemble et "metaethique". Acts 18 th International Symposium GiESCO from the 7-11 July 2013. Porto (Portugal). Pp. 598-604

[23] Cargnello G. (2013h): Nouvelles perspectives relatives à la soutenabilité pour une viticulture durable en vue d'une activité "MétaEthique" selon la "Grande Filière MétaEthique" du "Conegliano Campus". Acts 18 th International Symposium GiESCO from the 7-11 July 2013. Porto (Portugal). Pp. 1072-1077

[24] Cargnello G., avec Queiroz J., Carbonneau A., De Castro R., Schulz H., Murisier F. (2013i): Viticultures difficiles et heroïques: recherches sur des modèles de viticolture " vieux-nouveaux" et convenables au niveau economiques, socio environmental, esistentiel, ethique de façon " MetaEthique ". Acts 18 th International Symposium GiESCO from the 7-11 July 2013. Porto (Portugal). Pp. 1078-1086

[25] Cargnello G. with 39 Col. (2013n): Research aimed to reduce production costs. Results obtained at "Conegliano Campus" in 42 years of innovative research on vineyard full mecchanization and its effects on MetaEthical" "quality-profit". Poceedings 36th World OIV Congress of Vine and Wine. Bucharest – Romania 2-7 June 2013

[26] Cargnello G. with 47 Col. (2013o): Research aimed at conserving or increasing occupation levels and the necessary resources to safeguard an equiprofit for viticultural enterprise. Research on technical, socio-economic, existential and "MetaEthical" aspects according to the "MetaEthical Great Chain" production model. Third contribute. Poceedings 36th World OIV Congress of vine and wine.Bucharest.Romania 2-7 June

[27] Cargnello G. (2014c): In "Sustainability" ("Great or Universal Sustainability") is essential and urgent to consider in a "MetaEthical" even existential, ethical and other social descriptors as well as economic, environmental and social ones. OIV: 37th World Congress of Vine and Wine. November 9-14.2014 Mendoza (ARG)

[28] Cargnello G., Teo G., Lunardelli R., Bertuzzo P., Franzin G., Lunardelli S., Bonghi C., Tonielli Marialuisa and Veilleux Lissa (2014d): Innovative and natural techniques applied pre and postharvest for the production of wines of high sensorial, competitive, and socio.economic.MetaEthic quality that are produced in a reality dedicated to the production of "core of selection" wines. OIV: 37th World Congress of Vine and Wine. November 9th - 14th 2014 Mendoza (Argentina)

[29] Cargnello G. et Carbonneau A., avec : Teo G., Forno S., Spera G. et plus de 2500 personnalités (2015a): Les éléments essentiels issus de la philosophie et de la méthodologie d' évaluation de la «Grande Filière», à la base de l'algorithme de la viticulture durable «Universelle-MétaEthique». Essential elements provided by philosophy and the evaluation methodology of the «Great Chain», used in the algorithm of the sustainable «Universal – MetaEthic» viticulture. 19 th International Symposium GiESCO from the 31 May - 5 Juine 2015. Pech Rouge-Montpellier (France). **2**, pp. 770-776

[30] Carbonneau A., Cargnello G. et col. (2015b): Proposition d'une charte de viticulture durable suite a une enquête au sein du GiESCO. Proposition of a charter of sustainable vitticulture following a survey inside the GiESCO. 19 th International Symposium GiESCO from the 31 May - 5 Juine 2015. Pech Rouge-Montpellier (France). **3**, pp. 60-67

[31] Cargnello G., 2016a. La "Grande Filiera MetaEtica 4.1C", la "Rivoluzione 4.1C dell 'Agricoltura" viticoltura compresa, la "Sostenibilità Universale 4.1C e 5.1C" e la "BIOBIOSOS4.1C e 5.1C". ARGAV, 25-11. Pp 1-38.

[32] Cargnello G., 2016b. la "Grande Filiera MetaEtica 4.1C" algoritmo fondamentale per "un passo indietro verso il futuro metaetico 4.1C" viticoltura compresa, per il "Progresso Universalmente Sostenibile 4.1C", della "Rivoluzione dell'Agricoltura 4.1c" (Viticoltura e Vitivinicoltura comprese), della "Carta della Sostenibilità Universale o MetaEtica 4.1C ", della "BIOBIOSOS 4.1C" o "Biologico e Biodinamico Universalmente Sostenibili 4.1C", BIOBIOSOS DISTRETTO 4.1C, de "l' AltraViticoltura-Viticultura 4.1C E 5.1C", della "Internazionale Scuola alle Scuole sull'innovazione 4.1C, su: modelli viticoli, potature, gestione del vigneto e post vigneto, tecniche varie ..." del Conegliano Campus 4,1C. Seminario, A.L.S.- Bolzano, 29-11. Pp1-79

[33] Cargnello G. and Col. 2017. The "Great Chain MetaEthics 4.1C" fundamental basis for a "Universal Sustainability 4.1C and 5.1C" or "MetaEthics Sustainability 4.1C and 5.1C Smarc", as well as the "Revolution 4.1C", the"Organic 4.1C", the "Biodynamic 4.1C" and also of viticulture and of the viticulture, extreme heroic, for concours, innovative, integrated of mountain, land in strong and very strong slope. Acts 5th International Congress of the

DOI: 10.1051/bioconf/20170901014

Mountain and Steep Slope Viticulture. 29th march to 1st april. Conegliano-Italy
[34] Carbonneau A., Cargnello G. and Col. 2017. Travaux du G*i*ESCO sur la viticulture durable: Charte générale (2015) Priorités; Proposition of a charter of sustainable viticulture from a survey inside the G*i*ESCO. OIV, Parigi 4 aprile 2017

** Purtroppo per i lavori cancellati e per gli acronimi non potuti illustrare: Cell. +39 3477191342

Plataforma comparativa para consumos de energía, agua y huella de carbono en bodegas de vinos / Forecasting water and carbon footprint and utilities consumption in Chilean vineyards

Patricio Núñez[1], Paulina Flores[2], y Lucía Preller[1]

[1] Departamento de Ingeniería Química, Universidad Técnica Federico Santa María, Chile
[2] Consorcio I+D de Vinos de Chile, Chile

Abstract. This work, relates the wine production with efficient use of water, energy and fuel, and the greenhouse gases generation. A web platform was developed, based on the results of a set of six test runs in vineyards, which production are in the range of 3 to 25 million of liters of wine per year. The following stages was considered: 1. Winemaking Stage. 2. Reception of grape in hoppers or selection tables. 3. Alcohol Fermentation, Maceration and Reassembly. 4. Take-off and pressing. 5. Malolactic fermentation. 6. Stabilization and Filtered. 7. Bottling. The process of white wine is quite similar, except for the order of the stages and the absence of maceration. The platform results are based on the experimental data, which was used to formulate a model among the wine production geographic location and utility consumption: water, fuel and electricity. At the time, the platform is based on excel and visual basic, so it can be run on any standard computer. Main results of the web platform are expressed in kWh/liters of wine, or liters of water/liters of wine, for example, parameters that are useful for the generation of opportunities to improve best practices in the industry of wine.

1. Introducción

La plataforma web es un instrumento de apoyo elaborado para que las viñas pertenecientes al Consorcio I+D de Vinos de Chile puedan gestionar de mejor manera la información disponible sobre consumo eléctrico, de combustible e hídrico, permitiendo detectar a futuro oportunidades de mejora dentro de cada una de ellas. En la actualidad muchas de las viñas no gestionan este tipo de información, y algunas de ellas lo hacen pero no en sistemas en línea, como la herramienta propuesta.

Esta plataforma es uno de los resultados del proyecto "Herramientas expertas de apoyo para la toma de decisiones en tecnologías de energía y GEI en empresas vitivinícolas", del Consorcio I+D de Vinos de Chile (Consorcio), ejecutado por la Universidad Técnica Federico Santa María (UTFSM), y en donde participaron 6 viñas piloto con diversos tamaños de producción, como se muestra en la Tabla 1.

El proceso de levantamiento de información, se inició con la presentación del proyecto a las viñas piloto seleccionadas, con el fin de generar el contacto entre el equipo ejecutor (UTFSM) y éstas.

En una siguiente etapa se realizaron diversas actividades para recopilar información a través de distintas fuentes, como por ejemplo seminarios, talleres, capacitaciones y visitas.

1.1. Seminario de inicio de proyecto

El lanzamiento oficial del proyecto, se llevó a cabo en un seminario en donde fueron invitadas todas las viñas pertenecientes al Consorcio. En esta actividad se realizó la presentación del proyecto, destacando la importancia de disminuir los consumos energéticos, huella de carbono e hídrica para mejorar las ventajas competitivas del vino chileno en el mercado internacional.

1.2. Capacitaciones

Se llevó a cabo una capacitación en donde se presentó la Plataforma Web de Gestión desarrollada para el proyecto, en donde asistieron las viñas piloto. En esa oportunidad se presentaron avances de esta plataforma, y se entregaron los usuarios y claves para que cada viña tuviese acceso a ella. Además, se realizó una capacitación sobre cómo utilizar esta herramienta.

1.3. Visitas técnicas

Se llevaron a cabo visitas técnicas a las viñas pilotos, con el objetivo de conocer sus procesos productivos y particularidades, identificando de forma preliminar y general eventuales alternativas de ahorro energético, además de consolidar la relación con las viñas involucradas. La primera tanda de visitas técnicas se realizó en noviembre del 2015, en donde se logró una participación activa por parte de las viñas pilotos. Una nueva ronda de visitas se llevó a cabo en agosto de 2016, para finiquitar los temas pendientes que existieran en el caso de información no entregada.

1.4. Eficiencia energética

La eficiencia energética es un concepto que nace bajo la necesidad de reducir el consumo energético en la industria,

Table 1. Tamaño (según capacidad de producción) viñas piloto.

Viña Piloto	Capacidad de producción [kg uva]
Viña 1	2.613.413
Viña 2	4.249.760
Viña 3	6.599.085
Viña 4	15.432.576
Viña 5	22.198.558
Viña 6	27.766.330

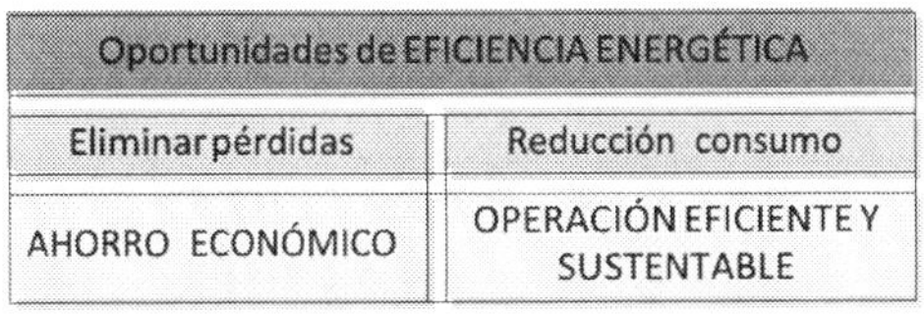

Figura 1. Oportunidades de eficiencia energética.

optimizando el uso de ésta. Para lo anterior, se debe realizar una optimización de los procesos productivos, a modo de obtener los mismos o mejores resultados, consumiendo lo mismo o menos energía, ya que la eficiencia energética apunta al ahorro energético sin que este represente un impacto negativo en el desarrollo de las actividades.

El uso correcto y optimizado de la energía, implica realizar el mínimo consumo posible de ésta, no emplearla en actividades innecesarias, mejorar la eficiencia de los equipos utilizados, entre otras actividades que se determinan a partir de la exploración de nuevas tecnologías y el análisis de la línea base bajo la cual opera una empresa.

Las organizaciones que consumen energía pueden disminuir el consumo de ésta con el fin de reducir costos y paralelamente mejorar su desempeño sostenible, es decir, perfilarse como una empresa que desarrolla mejores prácticas en los ejes social y ambiental, además del económico. Esto último además permite que el producto ofrecido tenga un sello verde, lo cual es tomado en cuenta en la actualidad por los consumidores. Hay estudios que indican que el mercado de productos ecológicos, incluyendo los vinos, en un año ha crecido 8,8% [1].

A nivel mundial y también nacional, se han llevado a cabo estudios que han permitido encontrar oportunidades de mejora en eficiencia energética, por ejemplo en el caso del consumo eléctrico se pueden implementar mejoras en el bombeo, en el sistema de climatización, en el sistema de iluminación, en el sistema eléctrico, mejorar el factor de potencia de diversos equipos, entre otros.

La implementación de herramientas que ayuden a tomar decisiones de eficiencia energética y que permitan reducir el consumo, otorgaran además de un beneficio económico a la empresa, una imagen sustentable que traerá consigo otro tipo de beneficios, como cercanía con los clientes y a la larga un aporte al medio ambiente, como se observa en el diagrama de la Fig. 1 [2].

La sustentabilidad ambiental en el sector del vino es actualmente una prioridad, razón por la cual la eco-innovación, la huella de carbono y agua, además de la eficiencia energética, son fundamentales para alcanzar altos estándares de sustentabilidad en el proceso productivo [3].

Tanto a nivel nacional como internacional, la preocupación por desarrollar prácticas sostenibles desde el viñedo hasta la distribución del vino, han generado la necesidad de investigar y estudiar distintas variables que permitan mejorar el proceso de producción del vino.

Diversas organizaciones e instituciones se han sumado a esta labor, patrocinando diversos estudios en distintas áreas de interés para el desarrollo sostenible de la industria.

En nuestro país, el Consorcio I+D de Vinos de Chile tiene como objetivo desarrollar Investigación y Desarrollo a largo plazo, centrado en las necesidades de la industria y con el fin de mejorar la competitividad del producto chileno en el mercado internacional [4].

Para este estudio, se revisaron investigaciones y herramientas, como por ejemplo la herramienta Benchmarking and Energy and water Savings Tool Winery (BEST Winery), elaborada en California, que nace debido a que no todas las viñas tenían la oportunidad de realizar una auditoría de sus operaciones, por lo tanto mediante este instrumento se puede llevar a cabo una autoevaluación en la industria del vino [5]. El principal objetivo de esta herramienta es comparar la eficiencia energética e hídrica de una bodega, respecto de otra que posee mejores prácticas en esas áreas.

Otras organizaciones como The Wine Council Ontario, también son patrocinadores de numerosos estudios al respecto. Uno de los documentos más reconocidos es el Sustainable Winemaking Ontario, que entre los tópicos desarrollados toma en cuenta el crecimiento ecológico de la vid, mejores prácticas para viñedos, mejores prácticas en energía, mejores prácticas de recurso hídrico, entre otros [6].

Sumado a lo anterior, hay una serie de normas ISO (International Organization for Standardization), las cuales son reconocidas y validadas tanto nacional como internacionalmente, y que son útiles para guiar la operación de viñas (e industrias en general) a una más eficiente y sustentable. Entre las más reconocidas en este ámbito se encuentran:

- ISO 50.001: Facilita una herramienta para mejorar el desempeño energético y los costos financieros asociados, además de las emisiones GEI [7].
- ISO 14.044: Se utiliza para calcular el ciclo de vida (LCA por sus siglas en ingles) de los productos, lo que permite a una industria conocer mejor su proceso e impactos generados, anticipándose a los riesgos y necesidades [8].
- ISO 14.067: Norma para medir la huella de carbono de productos, como una forma de cuantificar el impacto medioambiental de actividades, productos y servicios consumidos [9].
- ISO 14.064: Norma para la verificación y contabilización de GEI, proporciona un conjunto de herramientas para elaborar programas destinados a disminuir las emisiones de gases de efecto invernadero [10].
- ISO 14.046: Evalúa el impacto de productos, servicios u organizaciones en el agua, desde una perspectiva integral [11].

2. Metodología

La plataforma web es una herramienta elaborada exclusivamente para este proyecto, como un apoyo para la gestión de datos de las viñas. Mediante el estudio

Table 2. Rango de tamaño en bodegas.

Bodega grande: producción sobre 5.000.000 litros
Bodega mediana: producción entre 2.500.000 y 5.000.000 litros
Bodega pequeña: producción bajo 2.500.000 litros

de otras herramientas desarrolladas, y de acuerdo a la información disponible con la que se contaba, se propuso crear un instrumento en donde las viñas puedan ingresar información relevante a su proceso productivo (relacionado al consumo energético, hídrico y emisiones GEI) y éste pueda entregarles de vuelta el análisis de sus datos, comparándolos a valores referenciales estándar para determinar si la viña tiene un buen rendimiento en los diversos ámbitos.

2.1. Ingreso

El ingreso a la plataforma web es unificado, lo que significa que primero se ingresa a la Base de Datos para cargar información y posteriormente se pasa directamente a la Plataforma de Gestión de Datos, para evaluar resultados (o bien se ingresa directamente a la segunda, si ya están cargados los datos). Para lo anterior, cada viña cuenta con un usuario y clave asignados.

Además, se implementará una limitación de acceso, además de la seguridad usuario/contraseña, con pendrives de seguridad que permitirá acceder a la plataforma sólo si el dispositivo se encuentra conectado al computador.

2.2. Estructura

Para la elaboración de la plataforma de gestión de datos, se realizó una división de todos los aspectos a evaluar, por lo tanto, se llevó a cabo una división entre Vino Tinto y Vino Blanco, y en cada uno se consideraron los siguientes puntos: Consumo eléctrico, Consumo combustible, Consumo energético global, Consumo hídrico y Huella de Carbono.

2.3. Escenarios de evaluación

Se consideran inicialmente entre los escenarios de evaluación la zona de vinificación, ya que con esto varía la temperatura ambiental y por ende hay una incidencia en el balance térmico del proceso de vinificación. Se evaluó el impacto que tiene considerar los cambios térmicos según el lugar de emplazamiento de una viña, y se determinó que no es relevante, por lo tanto, se dejó de lado.

Posteriormente se consideró el tamaño de producción de la bodega, ya que esto tiene una relación directa con los datos de entrada. Se evalúa la sensibilidad del tamaño de la bodega, considerando los rangos indicados en la Tabla 2.

Para cada uno de estos tamaños, se contrastan los resultados generados por la herramienta para cada caso, con datos reales.

2.4. Base de datos

La base de datos consiste en la primera sección de la plataforma, en donde las viñas cargan información relevante a la producción y consumos dentro de ella. Esta data será utilizada en la segunda sección, de análisis y reporte de datos.

La base de datos se encuentra normalizada para realizar una recopilación de la información más eficiente y permite realizar un análisis rápido al optimizar la velocidad de las consultas a ésta. Se basa en el motor de base de datos de código libre PostgreSQL, el cual se encuentra exento del pago de licencias, posee un desarrollo activo y un muy buen rendimiento.

2.5. Plataforma de gestión de datos

Posterior al ingreso de información en la plataforma web, esta es utilizada para realizar una serie de cálculos que se exponen mediante tablas y gráficas. Se exponen aquí los resultados de consumo energético, consumo hídrico y huella de carbono, además de comparar esto con valores referenciales promedio, a fin de comprender si la viña se encuentra dentro de rangos esperados de estos parámetros. A partir de estos resultados, se espera que a futuro las viñas del Consorcio puedan utilizar la herramienta como puntapié en la detección de ineficiencias energéticas, de consumo de agua o emisiones GEI, con el fin de mitigarlas y generar mejoras operacionales que permitan aumentar su rendimiento y competitividad.

2.6. Desarrollo web

Para la creación de la aplicación que permite subir los datos de forma accesible y estructurada, se utilizó Django como *backend*, permitiendo el uso de un *frontend* editable "*on page*" basado en programación javascript.

Utilizando un sistema basado en *plugins*, ya no es necesaria la modificación del código para realizar cambios. Los gráficos se adaptan al API implementado para la realización de consultas, además de poder añadir diferentes tipos en implementaciones futuras sin afectar el resto de la plataforma (de forma modular). Lo anterior se traduce a que, cualquier usuario con permisos de administración puede modificar la Plataforma de Gestión de Datos, sin necesidad de modificar el código. Esto implica agregar/sacar gráficos, cálculo de indicadores, información relevante, fotografías, etc.

Se implementó el uso de un API con permisos y restricciones configurables para cada entidad, con lo que se abre la posibilidad de una comunicación de forma segura con aplicaciones Android/IOS para el despliegue de la información en estos dispositivos, así como la utilización de sensores y procesamiento digital de imágenes usando la cámara de estos y enviando el resultado a la base de datos.

2.7. Consumo energético

La información solicitada para llevar a cabo el cálculo de consumo energético de la viña, corresponde a los datos de producción de vino anual, listado de equipos eléctricos y térmicos y sus consumos anuales, con el fin de estimar los kWh utilizados en ese período. Además se solicitó indicar la etapa a la cual pertenece cada equipo para determinar cuál de ellas es la que posee un mayor consumo energético dentro del proceso. El consumo se estima como indican las ecuaciones (1) y (2):

$$Consumo\ Eléctrico\ [kWh/año] = Potencia\ Instalada\ [kW] * Uso\ [hr/año] \quad (1)$$

$$Consumo\ Térmico\ [kWh/año] = Cantidad\ combustible\ [L] * Capacidad\ Calorífica\ [kWh/L]^a \quad (2)$$

2.8. Consumo hídrico

El reporte de información de consumo hídrico consiste en datos de producción de vino anual y los consumos hídricos en cada una de las etapas de producción. De este modo se estima el consumo de acuerdo a la ecuación (3):

$$Consumo\ Hídrico\ [L\ agua/L\ vino] = Litros\ de\ agua\ total\ N/Litros\ de\ vino\ N$$

En donde N corresponde a la etapa en cuestión, o bien al total del proceso.

2.9. Huella de carbono

El primer paso para el cálculo de la huella de carbono es seleccionar el año base de cálculo, identificar las fuentes emisoras y determinar con cuales se trabajará para recopilar la información necesaria para su cálculo. Ésta considera tres alcances:

- Alcance 1: Considera fuentes que son propiedad de la empresa o son directamente controladas por ella. Para fines de este proyecto se tomó en cuenta el uso de combustibles tanto para el proceso productivo (calderas por ejemplo), como para transporte de materiales o productos, o uso de vehículos como tractores.
- Alcance 2: Este alcance considera fuentes indirectas de GEI, asociadas al consumo eléctrico. Debido a lo anterior, se cuantifica el uso de energía eléctrica consumida en el proceso productivo.
- Alcance 3: Corresponde a otras emisiones indirectas, lo cual se refiere a actividades de la empresa pero que ocurren en fuentes que no son controladas por esta, por lo tanto es una categoría opcional de medir. Para este proyecto se consideraron envases y embalajes, viajes nacionales de empleados y transporte de insumos y productos.

La metodología consiste en calcular las emisiones directas e indirectas de GEI producidas como resultado de las actividades de la compañía durante un año. Las emisiones directas se refieren a las emisiones que se producen en la propia empresa, mientras que las emisiones indirectas tienen lugar en otra parte y la empresa no puede ejercer cambios sobre ellas. Para obtener el total de emisiones de CO_2 equivalentes producidos por una empresa, se lleva a cabo el cálculo de la fórmula (3), en donde FE es el factor de emisión [12].

$$Cantidad\ de\ combustible * FE = toneladas\ CO_2 \quad (3)$$

Finalmente con los datos productivos reportados, se pueden informar *Key Performance Indicators (KPI)* como por ejemplo consumo eléctrico por litro de vino, consumo hídrico por litro de vino, etc.

[a] La unidad dependerá del tipo de combustible. En este caso para ejemplificar se utilizó litros.

Table 3. Resultados de consumo energético (globales).

Viña piloto	Potencia [kW]	Consumo energético [kWh/año]
Viña 1	729	2.719.078
Viña 2	581	955.793
Viña 3	1.128	4.477.516
Viña 4	962	1.424.510
Viña 5	4.823	12.937.703
Viña 6	3.612	7.865.759

Table 4. Valores referenciales de indicadores de consumo energético [13, 14].

Razón energía eléctrica kWh/HL	11
Razón energía térmica kWh/HL	1

3. Resultados y análisis

3.1. Línea base

Una vez levantada la información, de cada viña se obtiene el diagrama de flujo, datos de producción y consumos, principalmente. Lo anterior permite analizar como llevan a cabo la producción del vino y detectar posibles zonas o equipos en donde implementar mejoras en el proceso mismo, en el equipamiento, en las prácticas operacionales, entre otras.

3.2. Consumo energético

Se realizó un análisis segregado del consumo eléctrico y consumo térmico dentro de las viñas piloto, identificando cada uno de los equipos que las componen.

En base a lo anterior se conoció la potencia instalada total, y el consumo energético total y por etapa, en kWh, con el fin de comparar estos valores a valores referenciales y detectar si los consumos son excesivos o se encuentran dentro de un rango aceptable. Algunos de estos resultados se exponen en la Tabla 3.

A partir de este valor, se pueden calcular indicadores con el fin de compararlos con valores referenciales para identificar oportunidades de mejora. Por ejemplo, de acuerdo a la literatura, valores referenciales de diversos indicadores se muestran en la Tabla 4.

Estos parámetros se calculan conociendo el consumo total versus la producción de vino de la empresa. Se pueden calcular diversos tipos de indicadores.

3.3. Consumo hídrico

Se repite la metodología anterior, expuestA en el punto 2.8 en relación al consumo hídrico, en donde el valor referencial corresponde a 700 litros de agua por litros de vino producido [15, 16]. En esta oportunidad los valores son notoriamente menores a lo esperado, lo cual se debe a que la información reportada respecto del consumo hídrico en la mayoría de los casos fue incompleta.

3.4. Huella de carbono

Respecto al cálculo de la huella de carbono, para fines de este trabajo se consideró el análisis del Alcance 1 y Alcance 2, que corresponden aproximadamente al 10% del total de la huella, alcanzando un valor de 0,22 kg CO2 equivalente/litro de vino [17, 18]. En relación a los valores

DOI: 10.1051/bioconf/20170901015

obtenidos, se debe verificar la información reportada por parte de las viñas pilotos ya que son muy bajos en comparación a los referencias lo que hace inferir que la información se encuentra incompleta.

3.5. Plataforma web

La implementación de la plataforma web se encuentra en vías de estar operativa y cada viña piloto cuenta con acceso usuario/clave, además del pendrive de seguridad, que permite ingresar seguramente y resguardar la data.

El ingreso de información se realiza en la primera etapa, base de datos, en donde posteriormente en la plataforma de gestión de datos, estos son divididos en vino tinto y vino blanco, y respectivamente en sus etapas productivas, a fin de detectar en cada una de ellas alguna oportunidad de mejora en los distintos ámbitos analizados.

La ventaja de la plataforma web, además de reportar resultados relevantes, es que lo hace de manera dinámica y automatizada, es amigable con el usuario, y permite que el administrador pueda realizar las modificaciones necesarias para la presentación de resultados. Por otro lado, permite comparar visualmente los resultados obtenidos con parámetros estándar tanto nacionales como internacionales, con lo cual se detectan oportunidades de mejorar los indicadores obtenidos.

4. Conclusiones

A través del estudio llevado a cabo, se logra diseñar y establecer una metodología para el análisis de la información presentadas por las viñas pilotos.

Por otro lado, la elaboración de la Plataforma Web constituye un importante paso en el apoyo a las viñas en evaluar su operación, con el fin de identificar puntos en donde se pueden implementar mejoras, y convertir su proceso en uno sustentable y de mayor eficiencia. Con esta herramienta es posible comparar el desempeño del viñedo con valores referenciales estándar, y detectar oportunidades de implementar mejoras en el proceso productivo, desde la cosecha de la uva, hasta el empaquetado del producto.

Referencias

[1] Laure Cuilhé y Carmen Valor Martínez. *Vino Ecológico: comercialización en España y Francia*, **Pág45** (2013)

[2] Consorcio I+D de Vinos de Chile. Código Nacional de Sustentabilidad, Objetivos, `http://www.consorciovinosdechile.cl/htm/codigo-nacional-de-sustentabilidad.php`

[3] Consorcio I+D de Vinos de Chile. *Sustentavid: Certified Sustainable Wine of Chile*, `http://www.sustentavid.org/codigo/1/`

[4] Consorcio I+D de Vinos de Chile, `http://www.winesofchile.org/`

[5] *BEST Winery Guidebook*: Benchmarking and Energy and water Savings Tool for the wine industry. Lawrence Berkeley National Laboratory (October 2005)

[6] Narelle Martin and Two Hemispheres Environmental Consulting Inc. Wine Council of Ontario. *Sustainable Winemaking Ontario: A Newcomer's Primer-The Environmental and the wine industry in Ontario* (2007)

[7] International Organization for Standardization (ISO). *ISO50001/2011: Energy management systems – Requirements with guidance for use*

[8] International Organization for Standardization (ISO). *ISO14044/2006: Environmental management – Life cycle assessment*

[9] International Organization for Standardization (ISO). *ISO14067/2013: Greenhouse gases – Carbon footprint of products – Requirements and guidelines for quantification and communication*

[10] International Organization for Standardization (ISO). *ISO14064/2006: Greenhouse gases – Part 1: Specification with guidance at the organization level for quantification and reporting of greenhouse gas emissions and removals*

[11] International Organization for Standardization (ISO). *ISO14046/2014: Environmental management – Water footprint – Principles, requirements and guidelines*

[12] World Resources Institute (WRI) and the World Business Council for Sustainable Development (WBCSD) *GHG Protocol*

[13] Cooperativa Agro-alimentaria, Aragón, España. *Manual de ahorro y eficiencia energética del sector*

[14] Universidad Politécnica de Madrid. *Manual de Eficiencia Energética en Bodegas, Transfering Energy Save Laid on Agroindustry*

[15] Anu Kumar, Penny Frost, Ray Correll and Darren Oemcke. CSIRO Land and Water Report. *Winery wastewater generation, treatment and disposal: A survey of Australian practice* (March 2009)

[16] Grape and Wine research and development corporation, Adelaide. *Winery wastewater management and recycling: Operational Guidelines* (2011)

[17] Navarro, A., Puig, R., Kilic, E., Penavayre, S.Y., Fullana-i-Pañmer, P., Journal of Cleaner Production. Eco-innovation and benchmarking of carbon footprint data for vineyards and wineries in Spain and France **142**, 1661–1671 (2017)

[18] Wine Institute and the California Association of Winegrape Growers. *California Wine's Carbon Footprint* (2002)

Genetic analysis of loose cluster architecture in grapevine

Robert Richter[1], Susanne Rossmann[2], Reinhard Töpfer[1], Klaus Theres[2], and Eva Zyprian[1]

[1] Julius Kühn-Institute, Institute for Grapevine Breeding Geilweilerhof, Siebeldingen
[2] Max Planck Institute for Plant Breeding, Research Dept. of Plant Breeding and Genetics, Köln

Abstract. Loose cluster architecture is a well known trait supporting *Botrytis* resilience by permitting a faster drying of bunches. Furthermore, a loose bunch enables a better application of fungicides into the cluster. The analysis of 150 F1 plants of the superior breeding line GF.GA-47-42 ('Bacchus' x 'Seyval blanc') crossed with ?Villard blanc' segregating for compactness of the cluster was used for QTL analysis. Plenty of QTL were identified reproducibly for two years, QTLs stable over three growing seasons were identified for rachis length, peduncle length, and pedicel length. In a second approach 'Pinot noir' clones showing variation for cluster architecture were analyzed for differential gene expression. Grown in three different German viticultural areas, loose versus compact clustered 'Pinot noir' clones showed in gene expression experiments a candidate gene expressed fivefold higher in loosely clustered clones between stages BBCH57 and BBCH71.

1. Introduction

The phyto-sanitary status of grapevine fruits is positively influenced by a loose architecture of bunches (Thoma 1995). The extended distance between the berries leads to reduced physical stress on the berry cuticula. The effect of berries squeezing each other is avoided. In the epidermis of berries the formation of micro cracks induced by droppable water is reduced. Loose cluster architecture allows an accelerated air exchange and supports efficient and fast drying-off in the succession of high humidity or rain fall. In this way a physical barrier is established against humidity-affected diseases like *Botryotinia fuckeliana*, teleomorph of *Botrytis cinerea*. In addition, loose cluster architecture allows a better application of fungicides into the cluster.

Several options exist to achieve a loose cluster architecture. Shoot trimming (e.g., Bondada et al. 2016) is one option to be applied in existing vineyards. For new plantations, however, loosely clustered clones of a particular cultivar are a less labour consuming option. Several 'Pinot noir' clones are available and represent an excellent example for variation within clonal material (Thoma 1995). Consequently, during the selection of new grapevine varieties breeders evaluate cluster architecture and select individuals showing loose clusters. A very new example is the breeding strain Gf.1993-22-6 (denominated as 'Calardis blanc', `https://www.julius-kuehn.de/media/ Veroeffentlichungen/Flyer/ Rebsorte-Calardisblanc_Gf.1993-22-6_pdf`) which shows very good *Botrytis* resilience due to loose clusters and tight berry skin. Calardis blanc is a descendent of the selection GF.GA-47-42 bred by Gerhardt Alleweldt in 1964 which is used in this study (see below). To systematically improve the selection efficiency in breeding programmes we set to investigate the genetic determinants of loosely clustered grapevine phenotypes. For this purpose two different approaches are followed: (1) The cross of breeding line GF.GA-47-42 x 'Villard blanc' resulted in 150 F1 individuals which segregate for the trait of cluster architecture. (2) Differential gene expression experiments with loosely and compactly clustered 'Pinot' clones grown in three different German viticultural areas.

2. Material and methods

150 F1-plants of GF.GA-47-42 crossed with? Villard blanc' was used for phenotyping. Gf.GA-47-42 is an elite breeding line originating from 'Bacchus' x 'Seyval blanc' (Zyprian et al. 2016) showing loose clusters and excellent wine quality. F1-plants and the parents grown at Geilweilerhof were evaluated for cluster traits determining various length, size and volume parameter (Richter et al. in preparation). Evaluations were done for at least two years.

Clonal material as indicated in figure 1 of 'Pinot noir' was used. In all cases 125 AA was used as the rootstock and plants were at least 10 years old. In order to identify effects independent of the location and the season 'Pinot noir' clones grown at three places and for the seasons, 2015, 2016 were investigated. First location was at the vineyards of the JKI Institute for Grapevine Breeding Geilweilerhof (Siebeldingen, Pfalz), second location was at the vineyards of Antes Rebenverdelung (Heppenheim, Hessische Bergstraße), and third location was at Reben Sibbus GmbH (Sasbach-Jechtingen, Baden).

3. Results and discussion

'Pinot noir' clones showing differences in their cluster architecture grown at three different places in Germany as indicated in Fig. 2 were studied. The phenology was carefully monitored during the years 2015 and 2016. In parallel measurements were done to determine the length of rachis, peduncle pedicel, and to determine further subtraits of cluster architecture. It became evident that the growth gain measured in weekly intervals was higher in clones that show a loose cluster architecture (Fig. 3). In

DOI: 10.1051/bioconf/20170901016

Figure 1. 'Pinot noir' clones used in this study and examples for their cluster architecture. The variation range is from compact clusters (left), to mixed sized berries (middle), and loose cluster architecture (right).

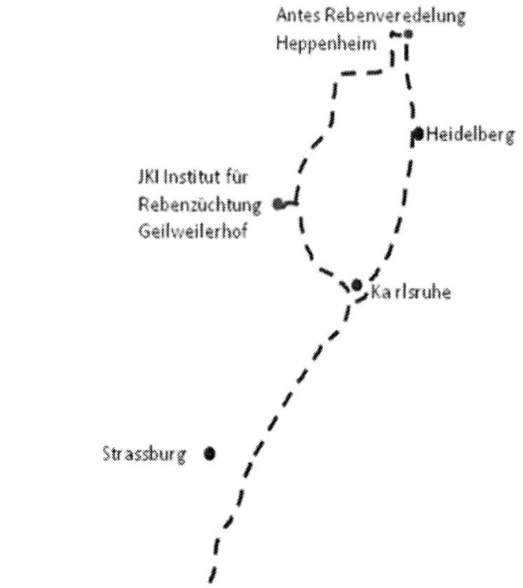

Figure 2. Locations (Antes, JKI, Sibbus) for sampling 'Pinot noir' clonal plant material. The distance between Heppenheim and Sasbach-Jechningen is around 200 km.

2015 the rachis during June grew 40 to 45 mm within three weeks while in compact variants the growth gain was 20 to 25 mm. In the following year the rachis grew about 35 mm within three weeks in loose cluster clones while the compact variants again showed a growth rate of about 25 mm in three weeks. Thus, in both seasons weekly measurements of the increase in length of the rachis revealed a reduced growth in compact and mixed berry size clones versus Pinot clones showing loose clusters (Fig. 3).

In a concomitant RNAseq approach (Rossmann et al. in preparation) experiments were performed on the 'Pinot noir' clones differing in their cluster architecture. Several differentially expressed genes were identified. One of the genes was investigated in more detail during the seasons 2015 and 2016. Elongation of the rachis was recorded as a function of time. The comparison of gene expression of the candidate gene identified in RNAseq studies at BBCH71 in reference to glycerolaldehydephosphate dehydrogenase (GAPDH) reproducibly revealed an altered expression pattern. The candidate gene was expressed higher in loose clusters compared to either compact ones or those that showed the mixed berry sized phenotype. Despite of the correlation further evidence needs to be generated to prove the causal link between candidate gene and phenotype. In a complementary approach a genetic analysis was initiated. For that purpose a F1 population GF.GA-47-42 x 'Villard blanc' was used (Zyprian et al. 2016). Several sub-traits of cluster architecture were phenotypically characterized each year in this population.

The phenotypic data were used for the computation of QTLs in the genetic map for GF.GA-47-42 x 'Villard blanc' which consists of 543 loci. So far more than 20 QTLs related to sub-traits of cluster architecture were reproducibly found. These QTLs were stable over two years and they are dispersed throughout the genome. QTLs

DOI: 10.1051/bioconf/20170901016

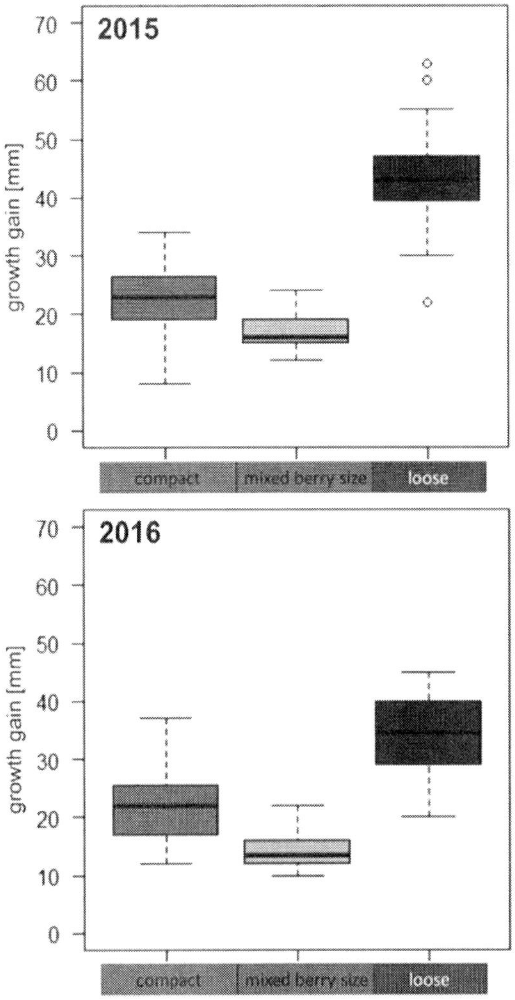

Figure 3. Growth gain of the rachis during June 2015 and June 2016, respectively. The rachis in loose clusters grows faster per week than that of compact clusters. As a consequence their rachis becomes longer than that of compact clustered clones.

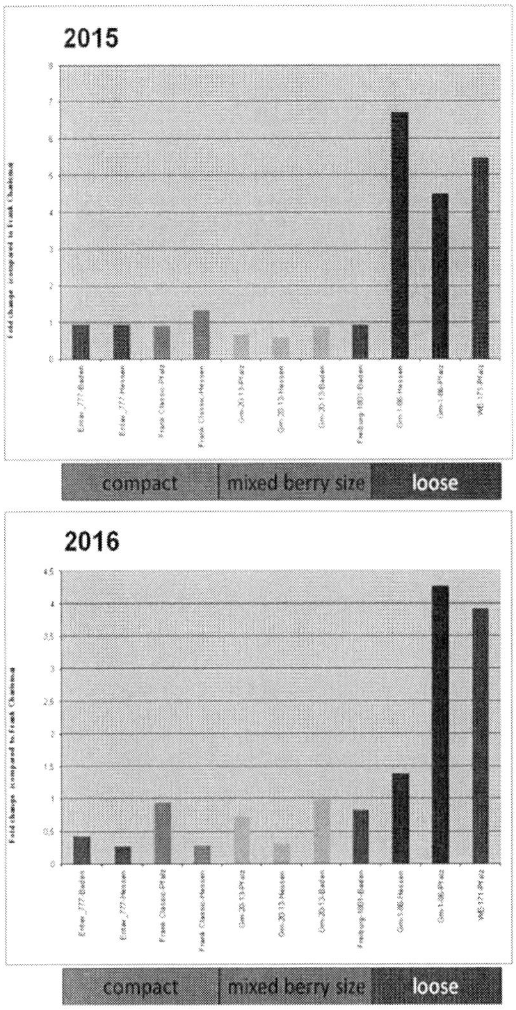

Figure 4. Expression pattern of a candidate gene at BBCH 71 in compact, mixed berry size and loosely clustered 'Pinot noir' clones. The candidate gene follows the gouping and is stronger expressed in loose clustered clones. It needs to be figured out whether or not there is a causal link between candidate gene and phenotype.

stable over three growing seasons could be identified for the sub-traits rachis length, peduncle length, and pedicel length indicating the importance of these subtraits for cluster architecture in 'Pinot noir' clonal material.

The molecular markers flanking QTL will be used to project the corresponding genetic regions onto the physical map of the grapevine reference genome. These regions of the reference genome will further be screened for candidate genes.

References

[1] Bondada, B., Covarrubias, J. I., Tessarin, P., Boliani, A. C., Marodin, G., Rombola, A. D. (2016) Postveraison shoot trimming reduces cluster compactness without compromising fruit quality attributes in organically grown Sangiovese grapevines. American Journal of Enology and Viticulture **67**, 206–211

[2] bib2 Thoma K (1995) 'Pinot noir': Less *Botrytis* attack using 'L-clones'. Der Badische Winzer, Freiburg **20**, 31–34

[3] Zyprian, E., Ochßner, I., Schwander, F., Simon, S., Hausmann, L., Bonow-Rex, M., Moreno-Sanz, P., Grando, M. S., Wiedemann-Merdinoglu, S., Merdinoglu, D., Eibach, R., Töpfer, R. (2016) Quantitative trait loci affecting pathogen resistance and ripening of grapevines. Molecular Genetics and Genomics **291**, 1573–1594

DOI: 10.1051/bioconf/20170901017

Intensity of Anthracnose in resistant varieties (PIWI) in the high altitude regions of southern Brazil

Bruno Bonin[1], Betina de Bem[1], Alberto Brighenti[2], Douglas Würz[1], Ricardo Allebrandt[1], Emilio Brighenti[2], Leonardo Araujo[2], Felipe Augusto Moretti Ferreira Pinto[2], and Amauri Bogo[1]

[1] Santa Catarina State University, College of Agriculture and Life Science, Av. Luiz de Camões, 2090, Lages, SC, Brazil
[2] Santa Catarina State Agricultural Research and Rural Extension Agency (EPAGRI), Rua João Araújo Lima, 102, São Joaquim, SC, Brazil

Abstract. The objective of this study was to evaluate the different degrees of tolerance to Anthracnose infection in three PIWI varieties (Cabernet Cortis, Regent and Bronner) compared to the Cabenert Sauvignon variety. The experiment was conducted at EPAGRI Experimental Station, located in the city of São Joaquim, Santa Catarina State (28°17′39″ S; 49°55′56″ W, altitude 1,415 m a.s.l.), during 2016/2017 growing season. The evaluation of the disease was performed every 15 days, applying a scale of notes, from the onset of symptoms, and lasted for a period of two months, under the condition of infection Natural. The intensity of the disease was compared by epidemiological measures: time to reach maximum disease severity (TRMDS), maximum disease severity (Smax), and area under the severity disease progress curve (AUDSPC). Maturity indices of the varieties were obtained from juice physical-chemical analysis at the harvest. Although the PIWI varieties were resistant to mildew, they were susceptible to anthracnose. Bronner and Cabernet Cortis showed on average the same intensity of disease in the leaves and branches as the Cabernet Sauvignon variety, the Regent variety showed greater susceptibility to the pathogen, differing statistically from the other varieties evaluated.

1. Introduction

Anthracnose (*Elsinoe ampelina*) is one of the major fungal diseases in southern Brazil. Rainy springs, fog, relative humidity above 90% and cold winds are ideal conditions for the development of the disease, its control is usually carried out with systematic applications of fungicides.

Anthracnose, caused by the fungus Ascomycetes *Elsinoe ampelina*, sexed form of *Sphaceloma ampelinum*, is one of the main fungal diseases of the grapevine in humid regions. It originates from the European continent and is reported in all the grape producing areas of the world, however, it causes damages in regions of high humidity and high temperature. Due to the characteristic symptom in the berries, this disease is also known as "bird's eye" [1]. Recently, some studies in India and China have associated species of the genus *Colletotrichum* as causal agent of the anthracnose of the vine [2].

The fungus can infect all the aerial organs of the plant, however, the young tissues are more susceptible. In the leaves, the initial symptoms are small, dark brown, slightly depressed, circular patches. Usually, the lesions are very numerous and can coalesce into a small hole. In the petiole and in the veins the lesions are elongated and cause the unequal development of the foliar tissues, causing the folding and shading of the leaves. In the branches, the disease causes the appearance of cancers with irregular shapes of gray coloration in the center and black borders. In the berries there are circular patches of gray color in the center and black on the edges [3].

Infection can occur at temperatures ranging from 2 °C to 32 °C, but temperatures between 24 °C and 26 °C, associated with rainy springs, fogs or locks, relative humidity of over 90% and cold winds are conditions Ideal for the development of the pathogen and disease. In addition, there is need for at least 12 hours of liquid water on the vegetable tissue. Under favorable conditions, the incubation of the pathogen (period between infection and onset of symptoms) occurs around seven days [1].

The sources of resistance to anthracnose of Asian origin are more promising for genetic improvement purposes, because there is an indication that the inheritance for this character is dominant. Among the Asian species, V. amurensis is most used in the genetic improvement of the vine. Genetic resistance to anthracnose is found in both American and Asian species.

The varieties with resistance genes to mildew and powdery mildew, such as PIWI (Pilzwiderstandsfähige) varieties, are an alternative to reduce the number of conventional fungicide applications and reduce production costs for grapevine planting in southern Brazil. However, few studies have been conducted with anthracnose for PIWI varieties. The objective of this study was to evaluate the different degrees of tolerance to Anthracnose infection in three PIWI varieties (Cabernet Cortis, Regent and Bronner) compared to the Cabenert Sauvignon variety.

DOI: 10.1051/bioconf/20170901017

Table 1. Scale of severity notes in the leaves and branches, used to evaluate the anthracnose in vine (PEDRO JUNIOR et al., 1998).

Nota	Anthracnose intensity
0	0
0,1	One leaf or branch with injury
0,2	Injury in up to 5 leaves or branches
0,5	Injury in 6 to 10 leaves or branch
0,7	Injury in 11 to 15 leaves or branches
1	25% of leaves or branches of the plant with lesions
2	50% of leaves or branches of the plant with lesions
3	75% of leaves or branches of the plant with lesions
4	100% of leaves or branches of the plant with lesions

2. Materials and methods

The experiments were conducted in São Joaquim Experimental Station (EPAGRI) Vineyards (28°17′39″ S; 49°55′56″ W, altitude 1,415 m a.s.l), in 2016/2017 cycle. The evaluated varieties were the PIWI varieties (Cabernet Cortis, Regent and Bronner) compared to the Cabenert Sauvignon variety. The vineyard was planted in 2013, the plants were grafted on Paulsen 1103, vines were trained in vertical shooting positioning trellis and spacing 3.0 × 1.5 m. The meteorological data were obtained from a meteorological station located at vineyard. The daily data of precipitation (mm), air relative humidity (%) and mean temperature (°C) were obtained from CIRAM (Center of Environmental Resources Information and Hydrometeorology of Santa Catarina).

The evaluation of the disease was performed every 15 days from the onset of symptoms, which occurred on September 25, approximately 15 days after sprouting, and lasted for a period of two and a half months, under the condition of infection Natural. The experiments followed a randomized completed-block design with three replications per variety and all branches and leaves of each plant were evaluated. The severity was evaluated according to the methodology proposed by Pedro Júnior et al. [5] applying the scale of notes (Table 1). The severity data were transformed into a percentage to perform the statistical analysis by the Tukey test ($P < 0.05$) ; where 0, 0.1, 0.2, 0.5 and 0.7 were equivalent to the values 0%, 2.5%, 5% , 12.5% and 17.5%, respectively, and the values of 1, 2, 3 and 4 are pre-set in the Table as 25%, 50%, 75% and 100%, respectively.

Anthracnose disease progress curves were constructed and the epidemic were compared using four epidemiological measures: time to reach maximum disease severity (TRMDS), maximum disease severity (Smax), and area under the severity disease progress curve (AUDSPC). AUDPC was calculated according the equation 1.

$$AUDPC = \sum((Yi + Yi + 1)/2)(ti + 1 - ti). \quad (1)$$

Where: Y =disease intensity (incidence or severity), t= unit of time, and i = cumulative number of evaluations. This area represented thetrapezoidal integration value of severity [6]. Tests for significance were performed using an analysis of variance and post-hoc comparisons performed using a Tukey test ($\alpha = 0.05$). Sis Var software, was used for the data analysis.

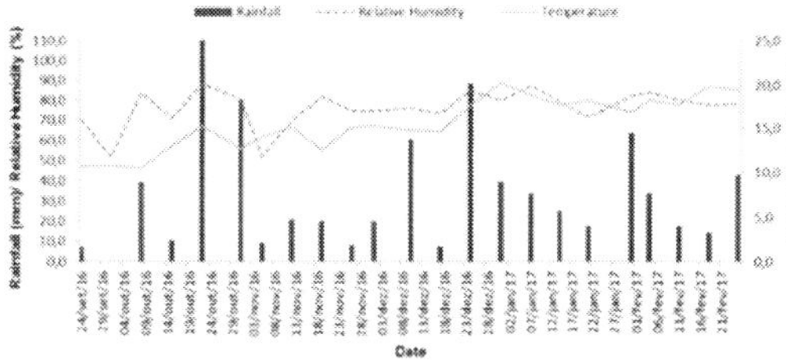

Figure 1. Average monthly accumulated precipitation (mm), air relative humidity (%) and mean temperature (°C) at Sao Joaquim, Santa Catarina State, Brazil, across 2016/2017 growing season.

At harvest time the productivity were evaluated by clusters weight (kg. plant^{-1}) of pre-selected plants. The estimated productivity (ton.ha^{-1}) were obtained by the plant density per hectare and by plants production. At that time, 300 berries of each cultivar were randomly collected and sent to perform physical-chemical and phenolic analysis. From grape must were performed analyses of total soluble solids (°Brix), total acidity (meq L^{-1}) and pH, according to the methodology proposed by OIV [7].

The soluble solids (°Brix) were measured using an optical refractometer (model Instrutherm – RTD-45) with temperature correction. The pH was measured with a pH meter (model MP 220 Metler-Toledo). Total acidity was measured by titration method with a 10 mL aliquot of juice with standardized 0.1 N NaOH. Total polyphenols content was quantified by the Folin-Ciocalteu method, with absorbance readings at 760 nm [8].

3. Results and discussion

Mean temperature, relative humidity and total precipitation between the first and the last evaluation were, respectively, 13.4 °C, 73% and 429,6 mm (Fig. 1). The beginning of symptoms appearance occurred at the end of September, in this period October (277,9 mm) and December (87,4 mm) were the months with higher rainfall. A combination of frequent rain and sustained humidity, particularly in spring and summer, provided conditions for leaves and branches remain wet, enabling infections by *Elsinoe ampelina*. Previous studies have shown that for *Elsinoe ampelina*, temperatures above 15 °C C are favorable to the growth of the fungus [1,3].

The spring temperature influences the onset of the disease [8], an occurrence of higher temperatures and higher precipitation in the spring of 2016/2017 cycle provided that as different variants initiated the symptoms at similar times and also resulted in a more homogeneous disease occurrence.

The PIWI varieties took the same time to reach maximum severity of disease (TRMSD) (Table 2), but the variety Cabernet Sauvignon reached the maximum severity of anthracnose (TRMSD) on average 30 days earlier than the resistant varieties, differing statistically from them.

The resistance in plants can be classified in horizontal or vertical by delaying the onset of the epidemic by reducing the initial infections, or making it slower after it is start, by reducing the rate of infection or progress (r) [9]. The Cabernet Cortiz and Regent varieties started first the beginning of the epidemic at September 29.

Table 2. Characteristics of anthracnose disease, time to reach the maximum disease severity (TRMDS), maximum value of disease severity (Smax) and area under the disease severity (AUDSPC). Disease was evaluated in three tolerant varieties and one susceptible during 2016–2017 growing seasons at Sao Joaquim, Santa Catarina State, Brazil.

Variety / Parameter	TRMDS (days)	Smax(%)	AUDSPS
C.Cortiz	66 a	17,5 a	1041,43 ab
Regent	71 a	41,66 b	1347,09 b
Bronner	71 a	15,83 a	899,28 a
C. Sauvignon	36 b	17,5 a	1073,28 ab
Average	61	23,12	1090,27
C.V(%)	16,39	32,83	24,40

C.V. – Coefficient of variance. *Means followed by the same letter in the same row are not significant different (Tukey test, $P < 0.05$). a Percentage of disease by a diagrammatic scale of Pedro Junior et al. [5]. b Area calculated by trapezoidal integration value according to Campbell and Madden [6].

The Regent variety presented the maximum severity (Smax) (Table 2), 41,66% on average, superior from other varieties, differing statistically from them. Cabernet Cortiz, Bronner and Cabernet Sauvignon, presented statistically, the same maximum value of disease severity.

In the comparative epidemiology, the parameter used to compare different degrees of susceptibility of plants to pathogens is the disease progress rate. Quantification of a variable that expresses the severity (intensity) of the disease is important to describe the progress of epidemics over time and its relationship with the weather or with different forms of management [10].

In comparative epidemiology, the parameter used to differentiate the susceptibility of plants under different conditions is the rate of disease progression. The quantification of a variable that expresses the incidence and severity (intensity) of the disease is important to describe the progress of epidemics over time and their relation to the climate or with different forms of management [11].

There were significant differences between the varieties by area under the curve progress severity of disease (AUDSPC) (Table 2). There was a significant increase in AUDSPC of Regent variety (Fig. 2). The AUDSPS of Regent variety with a total 1347,09, was higher than the average of the curves of the disease of Cabernet Cortiz (1041,43) Bronner (899,28) and Cabernet Sauvignon (1073,27).

The primary infection of anthracnose occurred from late September to early October, approximately four weeks after sprouting. Leaves that were not infected in the first four weeks remained anthracnose free and the most severe epidemics occurred at times of short rainfall intervals [12]. In the same way as reported above, in the present study infections started soon after sprouting, but the disease also occurred in the older leaves, but not as in the young leaves.

A significant increase in disease of intensity occurred in Regent variety, on the 70nd day after the first evaluation (Fig. 2). Possibly, the increasing intensity of disease is related to the increase in temperatures and rainfall in the region, and the more susceptibility from this variety to anthracnose.

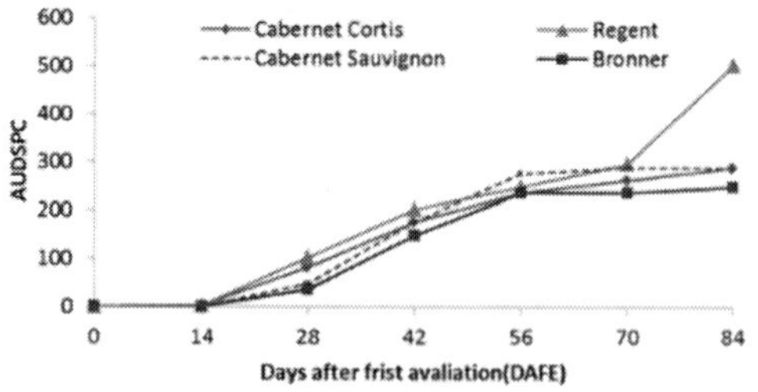

Figure 2. Antrachnose disease progress curves for area under disease severity (AUDSPC) for PIWI varieties Cabernet Cortiz, Regent and Bronner and for Cabernet Sauvignon variety during 2016–2017 growing season in Sao Joaquim, Santa Catarina State, Brazil.

Table 3. Maturity indices for Cabernet Coriz, Regent, Bronner and Cabernet Sauvignon obtained from juice physical-chemical analysis at the harvest on 2016–2017 growing season, in Sao Joaquim, Santa Catarina State, Brazil.

Variety	TSS (° Brix)	TA (meq L^{-1})	pH	Total Polyphenols (mg L^{-1})
C. Cortiz	23,9	104,16	2,98	2045,8
Regent	19,6	69,4	3,45	4048,4
Bronner	20,5	100,3	2,87	256,9
C. Sauvignom	20,6	160,3	2,9	1764,3

Maturity indices obtained by physical-chemical analysis of grape juice from different varieties at harvest, was appropriate for the elaboration of high quality wines in the season 2016/2017 (Table 3). Appropriate parameters for winemaking are pH between 2.9 and 3.6; titratable acidity above 100 meq L^{-1}, total soluble solids above 20° Brix.

The white variety Bronner presented a quantity of polyphenols suitable for this type of grape. Polyphenols have the function of protecting plants from physical (such as the sun UV radiation) and biological (fungi, viruses, bacteria) attacks. Polyphenols from the group of non-flavonoids, as the stilbenes are phytoalexins, i.e., compounds synthesized by vine in response to a stressful situation, such as pathogen attack [12]. The high contents of polyphenols in red varieties, especially on Regent berries are a good alternative for the elaboration of red wines with complexity and quality.

4. Conclusion

Although the PIWI varieties were resistant to mildew, they were susceptible to anthracnose. Bronner and Cabernet Cortis showed on average the same intensity of disease in the leaves and branches as the Cabernet Sauvignon variety, which presented 17.5% of average severity.

The Cabernet Cortis variety had an average severity of 17.5%, and Bronner presented 15.6%. The Regent variety showed greater susceptibility to the pathogen, differing statistically from the other varieties evaluated, with an average of 41.6% of leaf and branch severity.

Therefore, the mildew resistant varieties evaluated had the same degree or greater susceptibility to Anthracnose compared to the Cabernet Sauvignon variety under the climatic conditions of southern Brazil.

 DOI: 10.1051/bioconf/20170901017

References

[1] L. Amorim.; H. Kuniyuki, Diseases of the vine. (Ed.) *Manual of phytopathology: diseases of cultivated plants*. 3.ed. São Paulo: Agronômica Ceres, 1997 **2**, p. 736–757

[2] S. Sawant, Indu et al. Emergence of Colletotrichum gloeosporioides sensu lato as the dominant pathogen of anthracnose disease of grapes in India as evidenced by cultural, morphological and molecular data. Australasian Plant Pathology **41**(5), p. 493–504 (2012)

[3] P.J. Brook, Epidemiology of grapevine anthracnose, caused by Elsinoe ampelina. New Zealand Journal of Agricultural Research (1973)

[4] M. A. Ellis, & O. Erincik, Anthracnose of Grape. The Ohio State University Extension (2008)

[5] M. J, Pedro Júnior; I.J. Ribeiro; A. Martins, F.P. Microclimate conditioning by the removal of leaves and occurrence of anthracnose, downy mildew and 49 leaf spot on the 'Niagara Rosada' vine. Summa Phytopathologica, Jaguariúna **24**(2), p. 151–156 (1998)

[6] C.L. Campbell, L.V. Madden. Introduction to Plant Disease Epidemiology (Wiley, New York, 532, 1990)

[7] OIV – Organization Internationale de la Vigne et du Vin. Compendium of International Methods of Wine and Must Analysis (OIV, Paris, 504, 2009)

[8] C.R.C. Buffara, F. Angelotti, F.A. Vieira, A. Bogo, D.J. Tessmann, B.P. de Bem. Elaboration and validation of a diagrammatic scale to assess downy mildew severity in grapevine. Cienc. Rural **44**, 1384–1391 (2014)

[9] J.E. Van der Plank. Plant Diseases: epidemics and control (Academic Press, New York, 349, 1963)

[10] M.B. Spósito, Temporal and spatial dynamics of the black spot (Guignardia citricarpa) and quantification of the damage caused to the citrus crop. Thesis (Doctorate). School of Agriculture Luiz de Queiroz, Piracicaba (2003)

[11] P.J. Brook Epidemiology of grapevine anthracnose and downy mildew in Auckland. New Zealand vineyards, New Zealand Journal of Horticulture Science **20**, 37–49 (1992)

[12] N.F.S. Vaccari, M.C.H. Soccol, G.M. Ide. Phenolic compounds in wines and their antioxidant effects in disease prevention. Revista de Ciências Agroveterinárias **8**, 710–83 (2009)

DOI: 10.1051/bioconf/20170901018

Effect of genotype x environment interactions of grapevine hybrids characteristics

Dragan Nikolić[1], Zorica Ranković-Vasić[1], Aleksandar Petrović[1], Ivana Radojević[2], Tatjana Jovanović-Cvetković[3], and Branislava Sivčev[1]

[1] University of Belgrade, Faculty of Agriculture, Belgrade, Serbia
[2] Centre of Viticulture and Eenology, Niš, Serbia
[3] University of Banja Luka, Faculty of Agriculture, Banja Luka, Republika Srpska - BiH

Abstract. Research in this paper was performed at two different locations: Radmilovac and Vršac in Serbia. Four new interspecific hybrids (9846, 9896, 19574 and 20506) which are intended for table consumption were used as a material. Grape yield per unit area, the properties of the bunch (bunch weight, bunch length, bunch width and number of berries in bunch), the properties of berry (berry weight, berry length and berry width), as well as the characteristic of grape quality (sugar content and total acids in the must) were studied in selected hybrids.

The highest yield per unit area in the localities Radmilovac and Vršac had a hybrid 9896 (14 998 kg/ha; 11 365 kg/ha). Analysis of variance results showed for the bunch weight, bunch width and number of berries in bunch, berry weight and berry length significant differences among the genotypes. Significant differences between investigated localities were determined for the bunch length and all the berry characters. The interaction between genotype and localities showed significant differences for bunch length, berry length and berry width.

Since the genotypes in the initial yielding (third year after planting), they are showed satisfactory results in relation to the objectives of selection.

1. Introduction

Grapevine (*Vitis vinifera* L.) is of great economic importance in worldwide [15]. It is a very divergent species from the genetic standpoint. Since its distribution range is very wide, and its cultivation history very long, over time, a rich genetic diversity has arisen within the species. The differences arising from genetic modifications due to adaptations to various environmental conditions or crossings (spontaneous or anthropogenic) have resulted in the emergence of numerous cultivars [4]. Cultivars are differ from each other according to the morphological and physiological characteristics as well as economic and technological properties. The approximate number of different cultivars and germplasm collections held is 10,000 [14].

Serbia has very favorable agro-ecological conditions for growing a large number of grapevine cultivars, from early to very late table cultivars and wine cultivars. Grapevine has been grow in Serbia at about 25,000 ha. Wine cultivars are represented at about 75.7%, and cultivars which grape is intended for consumption in the fresh state are grown at about 24.3% of the total area [2].

It is well known that grape production depends on cultivar selection. Thus, breeding could improve the cultivar genetic potentials [5]. Today, interspecies hybridization is the most important and most widespread method for creating new grapevine cultivars in many breeding programs in the world [1,7].

The Faculty of Agriculture in Belgrade for many years working on the creation of new grapevine cultivars. Until now, it has created 23 cultivars (15 for fresh consumption and 8 for wine production) [8]. Also, from the different crossing combinations has been obtained a large number of the hybrids which are investigating in relation to the purpose of selection. For each of them is study the most important agrobiological and economic-technological characteristics. After the first stage of the selection single out best genotypes that are vegetatively multiplied. They can be further examined at different locations before or during the procedure for recognition of new cultivars.

Most commercially important traits of grapevine, such as berry ripening time, berry weight, soluble solids concentration, titratable acidity, and berry flesh texture are quantitative, and fluctuate depending on environmental factors. Therefore, it is important for grape breeders to obtain information on the contribution of genetic and environmental factors to phenotypic expression [11]. Methodologies for determining genotypic stability range from conventional analysis of variance to regression analysis, as well as genetic correlations [9].

The importance of grapevine grown lies in the fact that it is one of the most humble plant species because thrives in very different ecological conditions. On the overall grape quality of some cultivar is greatly influenced by external factors such as climate, soil composition, weather conditions during the growing season, agrotechnical practices, choice of rootstock and tradition of cultivation

DOI: 10.1051/bioconf/20170901018

Figure 1. Hybrid 9846.

in some region. In so heterogeneous localities choice of appropriate cultivars or clones has very great importance for obtaining optimal grape quality [13].

The aim of this study was to investigate the economic and technological characteristics of four different interspecific hybrids of grapevine at two localities in Serbia.

2. Material and methods

2.1. Materials

2.1.1. Experimental site

The examinations were conducted in two grape growing regions: Belgrade region – Grocka vineyards and South Banat region – Vršac vineyards. Grocka vineyards located in the hilly area on the right side of the Danube, downstream from Belgrade. The vineyards are located at 496.12 ha. There dominate table cultivars, which are grown in 83.7% of the area of all the vineyards [2]. Area of Radmilovac belonging to this vineyards. Vršac vineyards has about 2,000 ha. The vineyards stretch along the southern and northern sides of the Vršac Mountains at an altitude of 50–200 m.

2.1.2. Examined grapevine hybrids

Four new interspecific hybrids, which are intended for table consumption: 9846, 9896, 19574 and 20506 were used as a material in this paper. All hybrids were grafted on *V. berlandieri* x *V. riparia* Kober 5BB rootstock and planted at a distance of 3.0 x 1.0 m. The training system was double Guyot. Table 1 shows the hybrids and their parental partners.

Table 1. Labels and cross combinations of investigated hybrids.

Hybrid	Crossing combination
9846	Muscat Hamburg × SV 12375
9896	Muscat Hamburg × SV 12375
19574	SV 18315 × Muscat Hamburg
20506	Seedling 113 × Muscat Hamburg

Figures 1, 2, 3 and 4 shows the bunche appearance of investigated hybrids.

Figure 2. Hybrid 9896.

Figure 3. Hybrid 19574.

Figure 4. Hybrid 20506.

2.2. Methods

Grape yield per unit area, the properties of the bunch (bunch weight, bunch length, bunch width and number of berries in bunch), the properties of berry (berry weight, berry length and berry width), as well as the characteristic of grape quality (sugar content and total acids in the must) were studied in selected hybrids. Grape yield per unit area was determined by calculation, based on the total number of vines per hectare and the average grape yield per vine. The bunch weight was established by the measurement of the bunch on "CAS-Shollex type SHRE-122" scale. Length and width of bunch were determined by measuring with a ruler. Number of berries in bunch was established by counting the total number of berries in bunch. The berry weight is determined by measuring on analytical ballance "Tecator-6110 Balance". Length and width of berry were determined by caliper measurements. Sugar content in

 DOI: 10.1051/bioconf/20170901018

Table 2. Properties of the bunch of investigated hybrids in the localities Radmilovac and Vršac.

Hybrid	Locality		Bunch weight (g)	Bunch length (cm)	Bunch width (cm)	Number of berries in bunch
9846	Radmilovac		239.0	17.8	11.7	56.6
	Vršac		222.6	14.2	10.9	62.4
9896	Radmilovac		262.4	16.6	11.9	71.2
	Vršac		270.6	14.6	11.4	69.8
19574	Radmilovac		185.8	15.9	9.7	50.0
	Vršac		169.0	14.8	9.3	49.4
20506	Radmilovac		217.2	16.5	11.8	53.0
	Vršac		192.0	14.7	10.3	52.6
LSD Hybrid (H)		0.005	25.22	–	1.12	7.0
		0.001	33.96	–	1.51	9.43
LSD Locality (L)		0.005	–	0.61	–	–
		0.001	–	0.82	–	–
LSD H x L		0.005	–	1.22	–	–
		0.001	–	1.65	–	–

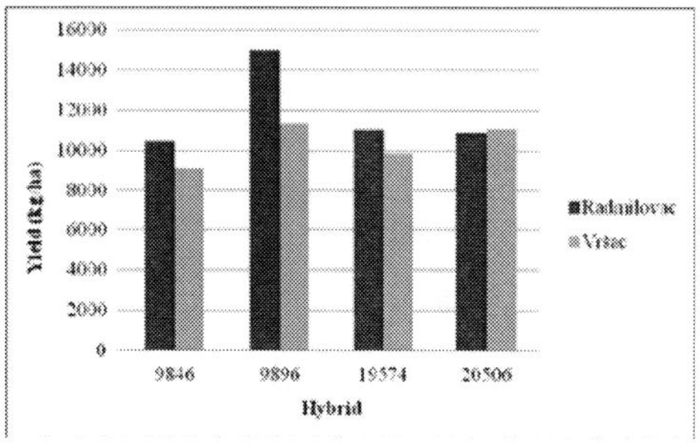

Figure 5. Grape yield per unit area of investigated hybrids in the localities Radmilovac and Vršac.

the must was established using a digital refractometer (PocketPal - 1, Atago, Japan). The total acid content in the must was determined by the method of titration with n/4 NaOH.

2.3. Statistical analysis

Statistical analysis was done by the method of analysis of variance (ANOVA) two-factorial design, using the F test at $P < 0.05$ and $P < 0.01$. In cases where the F test was significant, the significance of differences between mean values was determined using a LSD test for the significance level $P < 0.05$ and $P < 0.01$. Data analysis was performed using the statistical software package Statistica, Version 8 (StatSoft, Inc., Tulsa, Oklahoma, USA).

3. Results and discussion

On the basis of the total number of vines per unit area and the average yield of grape per vine, it was found that the highest yield per hectare had the hybrid 9896 (14998 kg) at Radmilovac, and the smallest hybrid 9846 (9099 kg) at Vršac (Fig. 5). In Serbia, [12] examining the production and technological properties of some promising grapevine seedlings at Radmilovac found that the greatest differences between genotypes were determined in the yield.

Table 2 shows the results of the mechanical analysis of the bunch at studied hybrids grown on Radmilovac and Vršac, as well as the results of statistical analysis of the parameters analyzed. The average largest bunch weight on both localities was achieved by hybrid 9896 (262.4 g; 270.6 g), and the smallest hybrid 19574 (185.8 g; 169.0 g). The smallest bunch length had the hybrid 19574 (15.9 cm), and the highest hybrid 9846 (17.8 cm) at the Radmilovac. At the locality of Vršac the smallest bunch length had the hybrid 9846 (14.2 cm), and the highest hybrid 19574 (14.8 cm). The highest bunch width and the number of berries in bunch at both localities was achieved by hybrid 9896 (11.9 cm; 11.4 cm; 71.2; 69.8), while the smallest bunch width and the number of berries in bunch had the hybrid 19574 (9.7 cm; 9.3 cm; 50.0; 49.4). Analysis of variance showed significant differences between investigated hybrids for the bunch weight, bunch width and number of berries in bunch. Significant differences between studied localities and the interactions between hybrid and localities were determined only for the bunch length.

The average largest berry weight on both localities was achieved by hybrid 20506 (3.6 g, 3.1 g), while the lowest berry weight at the locality Radmilovac had the hybrid 9896 (2.9 g), and at the locality Vršac the hybrid 19574 (2.6 g). At both localities, the average largest berry length was achieved by hybrid 9846 (19.7 mm; 17.1 mm), while the smallest berry length had the hybrid 9896 (16.1 mm) at the locality Radmilovac, and at locality Vršac the hybrid 19574 (16.0 mm). The average largest berry width had the hybrid 9846 (16.4 mm), and the lowest hybrid 9896 (15.6 mm) at the locality Radmilovac. At the locality Vršac, average largest berry width had the hybrid 9896 (15.4 mm), and the lowest hybrid 9846 (13.0 mm). Analysis of variance showed significant differences among the hybrids for the berry weight and berry length. Significant differences between localities were detected for all three characteristics. The interaction between hybrids and localities showed significant differences in the length and width of berry.

Significant differences between genotypes, localities and their interaction for some of bunch and berry properties are shown by [10], examining the characterization and adaptation of some Pinot Noir clones of Radmilovac and Vršac (Table 3).

The content of sugar and total acids in the must is an important factor in the quality of the grape. At the degree

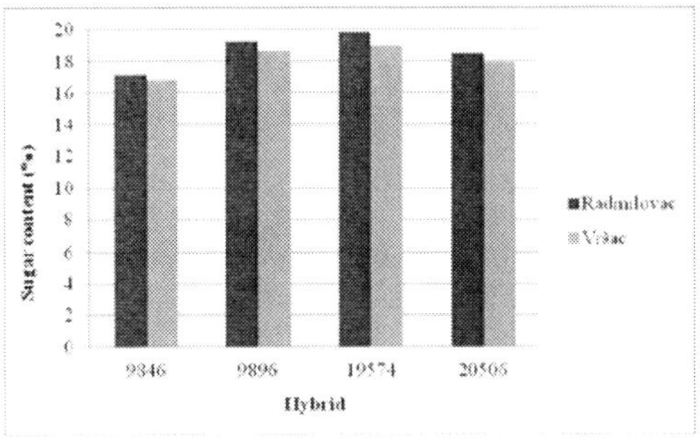

Figure 6. Sugar content in the must of investigated hybrids in the localities Radmilovac and Vršac.

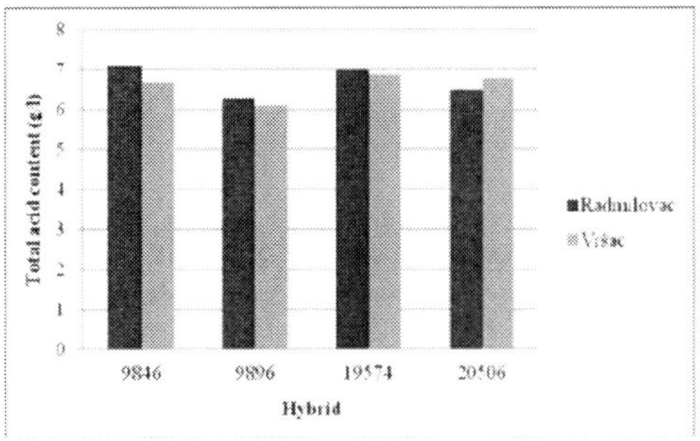

Figure 7. Total acids in the must of investigated hybrids in the localities Radmilovac and Vršac.

Table 3. Properties of berry of investigated hybrids in the localities Radmilovac and Vršac.

Hybrid	Locality		Berry weight (g)	Berry length (mm)	Berry width (mm)
9846	Radmilovac		3.3	19.7	16.4
	Vršac		2.7	17.1	13.0
9896	Radmilovac		2.9	16.1	15.6
	Vršac		2.8	16.4	15.4
19574	Radmilovac		3.1	18.3	16.3
	Vršac		2.6	16.0	14.0
20506	Radmilovac		3.6	18.0	16.3
	Vršac		3.1	16.6	14.7
LSD Hybrid (H)		0.005	0.39	0.75	–
		0.001	0.52	1.02	–
LSD Locality (L)		0.005	0.28	0.53	0.77
		0.001	0.38	0.71	1.04
LSD H x L		0.005	–	1.08	1.51
		0.001	–	1.46	2.03

of sweetness of grape juice influences the content and ratio of the amount of acids and sugars. The high sugar content and a high acids content is advantageous of ratio. Too low acidity adversely affects the taste, and therefore the quality of the grape [3].

At both localities the largest sugar content in must had the hybrid 19574 (19.8%; 19.0%), and the smallest hybrid 9846 (17.1%; 16.8%) (Figure 6). Nikolić [6], examining of interspecific hybrids at the locality Radmilovac also was determined the largest sugar content in hybrid 19574 (21.2%). The highest content of the total acids in the must at the locality Radmilovac had the hybrid 9846 (7.1 g/l), and at the locality Vršac hybrid 19574 (6.9 g/l). Based on Figure 7 it can be seen that at both localities, the smallest total acids content in the must had the hybrid 9896 (6.3 g/l; 6.1 g/l).

4. Conclusions

Based on examination of four different hybrids (9846, 9896, 19574 and 20506) which have been grown on two different localities (Radmilovac and Vršac) the following can be concluded:

- For most of the studied traits were determined the differences between investigated hybrids and between investigated localities.
- The hybrid 9896 had the highest yield per hectare and the largest bunch weight at both localities.
- The average largest berry weight had the hybrid 20506 at both localities, and the smallest hybrid 19574 from locality of Vršac.
- The highest sugar content had the hybrid 19574, and the lowest hybrid 9846 at both localities.
- The highest content of total acids in the must had the hybrid 9846 from the locality Radmilovac, and the smallest hybrid 9896 from both localities.
- Hybrid 9896 showed the best characteristics at both localities and is considered to be the most promising hybrid for the introduction of further production.

Ths study was supported by and Ministry of Education, Science and Technological Development of the Republic of Serbia, Project TR 31063.

References

[1] Gray, D.J., Li, Z.T., Dhekney, S.A. Precision breeding of grapevine (*Vitis vinifera* L.) for improved traits. Plant Sci. **228**, 3–10 (2014)
[2] Ivanišević, D., Jakšić, D., Korać, N. Popis poljoprivrede 2012. Poljoprivreda u Republici Srbiji. Vinogradarski atlas. Republički zavod za statistiku, Beograd (2015)
[3] Liu, H.F., Wu, B.H., Fan, P.G., Xu, H.Y., Li, S.H. Inheritance of sugars and acids in berries of grape (*Vitis vinifera* L.). Euphytica **153**(1), 99–107 (2007)
[4] Maletić, E., Karoglan-Kontić, J., Pejić, I., Preiner, D., Zdunić, G. Bubola, M., Stupić, D. Andabaka, Ž., Marković, Z., Šimon, S., Žulj-Mihaljević, M., Ilijaš, I., Marković, D. Zelena knjiga: Hrvatske izvorne sorte vinove loze. Državni zavod za zaštitu prirode, Zagreb (2015)
[5] Nikolić, D. Oplemenjivanje vinove loze. Monografija, Fleš, Zemun (2012)
[6] Nikolić, D. Properties of two interspecies grapevine hybrids from Serbia. Acta Hortic. **1082**, 141–147 (2015)
[7] Nikolić, D., Milutinović, M., Rakonjac, V., Fotirić, M. Evaluation of resistance to low temperatures in promising interspecies grapevine hybrids. Acta Hortic. **827**, 461–464 (2009)
[8] Nikolić, D., Ranković-Vasić, Z., Atanacković, Z. New Serbian grapevine genotypes for red wine production. Vitis **54**, 165–168 (2015)
[9] Onokpise, O.U., Mortensen, A. Genotype x environment interaction in *Vitis rotundifolia* Michaux. Am. J. Enol. Vitic. **39**(2), 121–124 (1988)

[10] Ranković-Vasić, Z, Nikolić, D., Atanacković, Z., Sivčev, B., Ruml, M. Characterization and adaptation of some 'Pinot Noir' clones to the environmental conditions of Serbian grape growing regions. Vitis **54**, 147–149 (2015)
[11] Sato, A., Yamada, M., Iwanami, H., Hirakawa, N. Optimal spatial and temporal measurement repetition for reducing environmental variation of berry traits in grape breeding. Sci. Hort. **85**(1-2), 75–83 (2000)
[12] Sivčev, B., Žunić, D., Jović, S. Production-technological characteristics of some promising grapevine seedlings in the condition of experimental station "Radmilovac". J. Agric. Sci. **45**(2), 93–99 (2000)
[13] Sivčev, B., Ruml, M., Sivčev, I., Ranković-Vasić, Z. Organska proizvodnja grožđ a. Monografija, Poljoprivredni fakultet, Beograd (2015)
[14] This, P., Lacombe, T., Thomas, M.R. Historical origins and genetic diversity of wine grapes. Trends Genet. **22**(9), 511–519 (2006)
[15] Viana, A.P., Riaz, S., Walker, M.A. Evaluating genetic diversity and optimizing parental selections in a segregating table-grape population. Am. J. Enol. Vitic. **62**(3), 285–290 (2011)

Evaluación de la diversidad genética de la Monastrell, una variedad antigua en la provincia de Alicante (España) mediante Genotipado por Secuenciación (GBS) / Genotyping by Sequencing (GBS) to evaluate the genetic diversity of Monastrell, an old grapevine variety in the province of Alicante (Spain)

C. Esteras[1], D. López-Lluch[2], S. Derdak[3,4], B. Picó[1], y J.J. Ruiz[5]

[1] Universitat Politécnica de Valencia, UPV-COMAV, Spain
[2] Dpto. Economía Agroambiental, Ing. Cartográfica y Expresión Gráfica en la Ingeniería, Escuela Politécnica Superior de Orihuela, UMH, Spain
[3] CNAG-CRG, Centre for Genomic Regulation (CRG), Barcelona Institute of Science and Technology (BIST), Baldiri i Reixac 4, 08028 Barcelona, Spain
[4] Universitat Pompeu Fabra (UPF), Barcelona, Spain
[5] Dpto. Biología Aplicada, Escuela Politécnica Superior de Orihuela, UMH, Spain

Abstract. Monastrell variety, also known as Mourvèdre, is cultivated mainly in the southeast of Spain in more than 45,000 ha. It is an ancient variety, whose origins are not known with certainty, highly adapted to the dry and warm climate of the southeast Spanish. The existence of a great clonal diversity is well known but, although some clonal selection programs have been carried out (Murcia, Bandol, etc.), the genetic diversity has not been evaluated to date. The measure of this diversity is urgent, since very old vines have been plucked in recent years. We have initiated a project to estimate the genetic diversity of the conserved clones of this variety through GBS (Genotyping by Sequencing). To carry out this analysis, a survey was carried out covering the whole area of Monastrell cultivation in the Alicante Denomination of Origin (D.O.). We selected 46 vines from 46 different vineyards, always prioritizing the oldest vines at each sampling site. The GBS has allowed us to obtain between 0.54 and 1.17 million of reads per sample, of which between 73 and 78% have been mapped to the reference genome developed by the French-Italian consortium (vvinifera.IGGP 12x) with a mapping quality above threshold. More than 100,000 SNPs have been identified, ranging from 14,000 to 43,000 per sample, with a coverage greater than 10X. A high percentage of these variants are polymorphic among Monastrell clones, confirming the existence of significant levels of genetic diversity. In addition, this analysis provides a high number of high quality SNPs well distributed across the genome, suitable for clone genotyping, which will allow the design of strategies to optimize its conservation and use.

1. Introducción

La variedad Monastrell se cultiva mayoritariamente en el sureste de España, en más de 45.000 ha (Fig. 1). Se trata de una variedad antigua, cuyos orígenes no se conocen con certeza, muy adaptada al clima seco y cálido del sureste español.

Se tiene conocimiento de la existencia de una gran diversidad clonal, pero, aunque se han llevado a cabo algunos programas de selección clonal (Murcia, Bandol, etc.), no se ha evaluado de forma adecuada la diversidad genética existente. La medida de esta diversidad es urgente, por cuanto se han venido arrancando en los últimos años cepas muy antiguas. En la Denominación de Origen Protegida Vinos de Alicante (DOP Alicante) se cultivan 6.284 ha de Monastrell, que suponen un 66,1% de la superficie cultivada. Hemos iniciado recientemente un proyecto para estimar la diversidad genética existente en los clones conservados de esta variedad mediante Genotipado por secuenciación (GBS, Genotyping by Sequencing). En este trabajo se presentan los resultados preliminares obtenidos de un muestreo llevado a cabo en las comarcas del Alto y Medio Vinalopó, en Alicante.

2. Material y métodos

2.1. Muestras evaluadas

Para llevar a cabo este análisis, durante el mes de junio de 2016 se realizó una toma de muestras, recorriendo las principales zonas de cultivo de la variedad Monastrell en la DOP Alicante, fundamentalmente campos de cultivo de las comarcas del Alto y Medio Vinalopó (Fig. 2).

Procedentes del muestreo anteriormente citado, se han seleccionado 46 cepas individuales, procedentes de 46 viñedos diferentes, priorizando siempre las cepas más antiguas en cada uno de los sitios de muestreo. No se ha tratado de obtener muestras representativas de todo el

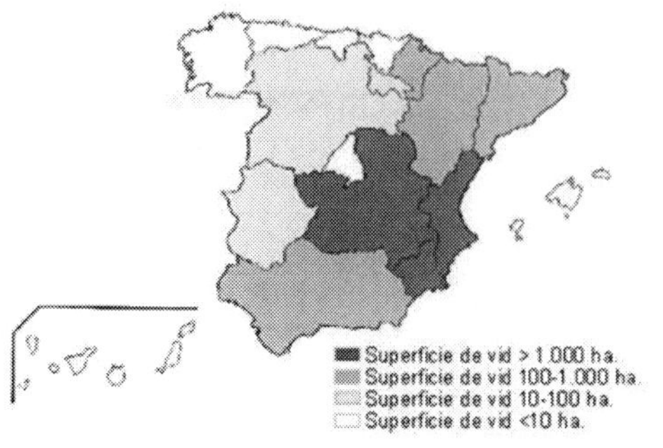

Figura 1. Distribución geográfica de la variedad Monastrell en España. (Ministerio de Agricultura y Pesca, Alimentación y Medio Ambiente, Gobierno de España.)

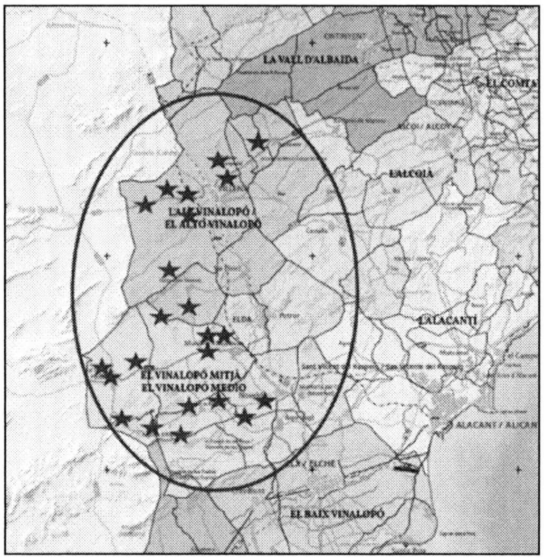

Figura 2. Distribución aproximada de las muestras de Monastrell analizadas. En cada localidad representada se ha muestreado uno o más viñedos, obteniéndose hasta un total de 46 muestras.

territorio cultivado de la variedad Monastrell, sino que, basándose en el conocimiento de los técnicos de la D.O. y de bodegueros y viticultores locales, se ha buscado siempre obtener muestras de las cepas más antiguas y que pudieran tener posibles características diferenciales.

Cada una de las 46 cepas seleccionadas fue posicionada en un mapa físico además de ser georeferenciada. Se obtuvieron de dos a tres muestras de brotes jóvenes para el posterior análisis de ADN.

2.2. Genotipado

2.3. Extracción de ADN

El ADN se extrajo de hojas jóvenes de cada una de los genotipos evaluados. La extracción se llevó a cabo empleando el kit DNeasy Plant Mini Kit (Qiagen, Hilden, Germany). Las concentraciones de ADN se midieron mediante un espectofotómetro Nanodrop ND-1000 v.3.5, seleccionando las muestras con al menos una concentración de 50 ng/μl. Finalmente, 46 muestras con ADN de alta calidad fueron elegidas para la secuenciación, incluyendo además 6 genotipos duplicados como control. Las muestras se enviaron para su genotipado al CNAG (Centro Nacional de Análisis Genómico, Centre de Regulació Genómica, Barcelona, Spain). Se empleó un mínimo de 500 ng de ADN para el genotipado de SNPs (Polimorfismos de nucleótido simple).

2.4. Genotipado por secuenciación (GBS)

Para la identificación de SNPs entre los genotipos evaluados se empleó la estrategia de Genotipado por secuenciación (Genotyping-by-sequencing, GBS) [1]). Se ensayaron en una prueba piloto con 6 muestras dos endonucleasas (ApeKI y PstI). La enzima ApeKI fue finalmente seleccionada, ya que produjo más fragmentos genómicos secuenciables, que fueron más reproducibles en las muestras evaluadas. Además, los fragmentos obtenidos con ApeKI fueron más diversos y representaron una mayor parte del genoma, dando por tanto lugar a un proceso de identificación de variantes más robusto que el obtenido con la enzima PstI.

Las librerías GBS de las 46 muestras se prepararon empleando la endonucleasa ApeKI (sitio de reconocimiento: G/CWGC) y se secuenciaron mediante la plataforma Illumina HiSeq 2000 (Illumina Inc, San Diego, CA, USA). Las lecturas obtenidas se procesaron, eliminando las secuencias de los adaptadores y se analizaron por su calidad. Las lecturas con menos de 30 pares de bases y que no superaron los umbrales de calidad fueron eliminadas antes de proceder al mapado e identificación de SNPs. El mapado se realizó empleando como genoma de referencia la secuencia de *Vitis vinifera* producida por el consorcio público franco-italiano (secuencia de Pinot Noir PN40024, recubrimiento 12X) [2] empleando el programa Bowtie 2 [3]). La identificación de SNPs se realizó empleando el programa GATK (`https://software.broadinstitute.org/gatk/`). Se empleó un umbral de calidad de mapado de 30 y una cobertura mínima de 5. Los SNPs resultantes se seleccionarán, además, descartando aquellos con más de un 70% de datos faltantes, los no bialélicos, y exigiendo una frecuencia alélica mínima superior al 10%.

3. Resultados y discusión

3.1. Pérdida de diversidad en la variedad Monastrell

Durante la toma de muestras en campo, recorriendo las distintas zonas de cultivo del Alto y Medio Vinalopó, se constató que en los últimos años está teniendo lugar un acelerado proceso de pérdida de cepas antiguas de la variedad Monastrell. Muchos de los viñedos más antiguos habían sido arrancados muy recientemente, siendo ya relativamente infrecuente encontrar cepas como las de la Fig. 3, con una edad superior a 60–70 años. Algunos de los viticultores consultados refieren la existencia de viñedos con cepas muy antiguas que no han podido ser mantenidas en producción, pero que hubieran sido de interés para su caracterización genética y para su posible inclusión en programas de selección clonal de la variedad o en bancos de germoplasma.

Figura 3. Cepa de Monastrell, con una edad estimada de aproximadamente 80 años (cerca de Pinoso, Alicante).

3.2. Genotipado por secuenciación

En el presente ensayo, la técnica de genotipado por secuenciación (GBS) ha permitido obtener entre 0.54 y 1.17 millones de lecturas por muestra, de las cuales entre un 73 y un 78% han podido maparse al genoma de referencia desarrollado por el consorcio franco-italiano (vvinifera.IGGP_12x) [2] con una calidad de mapado superior al umbral.

Entre los resultados preliminares del ensayo destaca el número total de SNPs identificados, que ha sido superior a 100.000. Se obtuvieron entre 14.000 y 43.000 SNPs por muestra, con una cobertura mayor a 10X. Un elevado porcentaje de las variantes detectadas han sido polimórficas entre las cepas de Monastrell evaluadas, lo que confirma la existencia de importantes niveles de diversidad genética. El estudio detallado de la diversidad y estructura genética de todas las muestras analizadas nos proporcionará en breve plazo información de interés para determinar la importancia de estos niveles de diversidad, la necesidad de ampliar las muestras y zonas de cultivo a analizar, así como posibilitará la comparación con la diversidad genética existente en otras áreas de cultivo de la variedad Monastrell.

Además, este análisis preliminar nos ha proporcionado un elevado número de SNPs de alta calidad y distribuidos por el genoma. Una vez depurados y seleccionados teniendo en cuenta las dificultades asociadas a las especies altamente heterocigotas como la vid [4,5], serán adecuados para el genotipado de clones de esta variedad, lo que permitirá en el futuro diseñar estrategias para optimizar su conservación y uso, por ejemplo, en programas de selección clonal.

La presente investigación está siendo financiada por la Cátedra Cajamar de Economía y Desarrollo Agroambiental de la Universidad Miguel Hernández de Elche. Queremos agradecer la colaboración prestada por la Denominación de Origen Protegida Vinos de Alicante, y en especial la colaboración de Omar Amorós, técnico de la D.O.

References

[1] R.J. Elshire, J.C. Glaubitz, Q. Sun, J.A. Poland, K. Kawamoto, E.S. Buckler et al. PLoS ONE **6**, 5 (2011)

[2] URGI. 12X.2 version of the grapevine reference genome sequence from The French-Italian Public Consortium (PN40024) 2014 [cited 2016]. Available from: `https://urgi.versailles.inra.fr/Species/Vitis/DataSequences/Genome-sequences`

[3] B. Langmead, S. Salzberg, Nat. Methods **9**, 357 (2012)

[4] K.E. Hyma, P. Barba, M. Wang, J.P. Londo, C.B. Acharya, S.E. Mitchell, L. Cadle-Davidson, PLoS ONE **10**, 8 (2015)

[5] C. Kim, H. Guo, W. Kong, R. Chandnani, L.S. Shuang, A.H. Paterson, Plant Sci. **242**, 14 (2016)

BIO Web of Conferences **9**, 01020 (2017) DOI: 10.1051/bioconf/20170901020
40th World Congress of Vine and Wine

Genetic identification and characterization of Armenian grapevine cultivars

Anna Nebish[1,2,3,a], Iris Ochssner[4], Erika Maul[4], Reinhard Töpfer[4], Ludger Hausmann[4], Anahit Hovhannisyan[5], Hakob Devejyan[2], Gagik Melyan[3], and Rouben Aroutiounian[1,2]

[1] Department of Genetics and Cytology, Yerevan State University, 1 Alex Manoogian, 0025 Yerevan, Armenia
[2] Research group of Plant genetics and Immunology, Institute of Molecular Biology of National Academy of Sciences RA, 7 Hasratyan, 0014 Yerevan, Armenia
[3] Scientific Center of Viticulture, Fruit-Growing and Wine-Making of the Armenian National Agrarian University, 1139 Merdzavan, Armenia
[4] Institute for Grapevine Breeding, Julius Kühn-Institut (JKI), Geilweilerhof, 76833 Siebeldingen, Germany
[5] Laboratory of Ethnogenomics, Institute of Molecular Biology of National Academy of Sciences RA, 7 Hasratyan, 0014 Yerevan, Armenia

Abstract. Armenia is an important area of crop origins including grapes. The objectives of this study were the molecular characterization of 38 Armenian grape genotypes and the detection of the genetic relationships between the accessions. In total, 164 alleles were obtained at 18 SSRs loci. The number of alleles per locus ranged from 5 to 15 with a mean number of 9.17 alleles per locus. The expected heterozygosity ranged from 0.514 (VVIN73) to 0.90 (VVS2) and the observed heterozygosity from 0.417 (VVIV37) to 0.947 (VVMD7 and VMC1B11). In our research the highest values of polymorphic information content (PIC) were obtained using markers VVS2, VRZAG62 and VRZAG79, while the least informative ones were VVIN73 and VVIB01. A high level of gene diversity was observed between ancient and new bred cultivars. Higher values were recorded for the new grape varieties, (He = 0.792) and lower values for ancient cultivars (He = 0.739). The mean number of alleles (MNA) for all loci per population ranged from 7.39 in ancient to 8,00 in new crosses. A clear separation was observed for the groups of ancient aboriginal cultivars, *Vitis vinifera* L. x*Vitis vinifera* L. and *Vitis vinifera* L. x *Vitis amurensis* Rupr .crosses, which were bred in Armenia and seedless cultivars. Very close relationships with high similarity were determined for 5 pairs of cultivars. Two cases of possible homonymy were also detected. Our results are the basis for future MAS selection and target breeding.

1. Introduction

Armenia has a long history of viticulture and winemaking. Numerous wild grapevine (*Vitis vinifera* L. subsp. *sylvestris)* populations are thriving in the country and a rich biodiversity of local grapevine genetic resources including more than 450 traditional autochthonous cultivars exist [1]. The earliest traces of grape domestication and cultivation were found in the Armenian highlands dating to the ancient Neolithic settlements (6,000-5,000 BC) [2–4]. The oldest evidence of wine-production (Areni-1 cave) was found in southeastern Armenia dating to 4,000 BC [5]. Armenia is a mountainous country with a dry and continental climate with hot summers and cold winters. Climatic conditions are favorable for the successful cultivation of grapes for different usage, e.g., differing in ripening time and yield. However, very low temperatures in winter as low as −35 °C can sometimes cause great damage to the grape plants. Thus the winter frost tolerant species *Vitis amurensis* Rupr. was widely used in breeding programs at the Armenian Scientific Research Institute of Viticulture, Fruit-Growing and Winemaking.

Nowadays, two ampelographic field collections are present in Armenia:

1. The collection of the Scientific Center for Viticulture, Fruit Growing and Wine-Making (Nalbandyan, Armavir province);
2. The new ampelographic collection at the Scientific Center of Agriculture (Echmiadzin, Armavir province) established in 2016.

The classification of Armenian grapevine cultivars is traditionally based on classical ampelographic characteristics according to OIV descriptors. However, the common morphological identification has different limitations and does sometimes not provide enough evidence for the correct identification of the accessions [6].

Having already been studied by the methods of ampelography, Armenian grapevine biodiversity still needs to be investigated on a molecular level in accordance with modern requirements. During the last few years, eighty Armenian cultivars have already been identified and characterized according to OIV phenotypic descriptors [7,8], but their genotypic characteristics are yet to be identified. The genetic analysis of Armenian grapevine cultivars is essential for the correct description of

[a] e-mail: anita.nebish@gmail.com

genetic resources and the ascertainment of genetic relationships between grapevine genotypes and their pedigrees.

DNA microsatellite markers, or simple sequence repeats (SSRs), are widely used in the molecular characterization of grapevine germplasm and cultivar identification. Due to their high degree of polymorphism, co-dominance and stability they are used as a complementary tool to traditional ampelographic methods, as well as for the evaluation of genetic diversity and clarification of nomenclature [9–12]. SSR markers are thus in use for DNA fingerprinting and also for molecular marker-assisted selection (MAS) in breeding programs [13,14] as well as for studying the historical origin of grape [15].

The objectives of this study were the molecular identification of Armenian grape genotypes including ancient as well as new cultivars using SSR analysis and the detection of the genetic relationships between the investigated accessions.

2. Material and methods

Plant material. Thirty-eight accessions were chosen from (1) the Armenian National Grape collection (WIEWS-Institute code: ARM011, Armenian Academy of Viticulture and Winemaking, Nalbandyan, Armenia; 31 genotypes) and (2) the German national grape collection (WIEWS-Institute code DEU098, Julius Kühn-Institut, Institute for Grapevine Breeding Geilweilerhof, JKI-IRZ; 7 genotypes). The material included 19 ancient Armenian and 10 newly bred *V. vinifera L.* cultivars as well as 9 interspecific crosses (*V.vinifera.* x *V.amurensis*). Table 1 lists the analyzed cultivars, the institute code of the grapevine collection, the species and the pedigree given by the breeder. In our study 'Karmrahyut' (*V.vinifera* x *V.amurensis* interspecific cross) is the parental genotype of the cultivars 'Charentsi', 'Nalbandyani', 'Agaraki', 'Anushahyut', 'Nerkarat' and possibly 'Csarensi'. The two *V. vinifera* cultivars 'Cabernet franc' and 'Muscat à petits grains blancs' were used as references for the standardization of allele sizes and for the comparison of genetic profiles with microsatellite databases.

DNA isolation and microsatellite genotyping. Samples of 8–10 young leaves per genotype were collected and dried in silica gel from the ARM011 collection. Samples from the DEU098 collection were collected and stored at −20 °C until extraction of genomic DNA. DNA was extracted from all leaves using the DNeasy Plant Maxi Kit (Qiagen) following the manufacturer's instructions. DNA concentration and purity was estimated using the Thermo Scientific NanoDrop 2000 spectrophotometer. For the genetic characterization of the Armenian cultivars the collected material was genotyped in multiplexes at the following 18 microsatellite loci: VVS2 [16], VVMD5, VVMD7, VVMD24, VVMD25, VVMD32 [17,18], VRZAG62, VRZAG79 [19], VVIB01, VVIH54, VVIN16, VVIN73, VVIP31, VVIP60, VVIV37, VVIV67 [20], VMC1B11 [21] and VMC4f3.1 [22]. The majority of these SSR markers is commonly used for the assessment of trueness to type in grapevine collections and the search of putative relationships. (European project GENRES081, `http://www.genres.de/vitis`) [23–26].

One primer of each marker was fluorescently labeled at the 5′ end with FAM (blue), TAMRA (yellow), HEX (green), ROX or NED (red) (Metabion, Martinsried, Germany). Up to 6 markers with different labels and diverse fragment lengths were combined in multiplexes and polymerase chain reaction (PCR) was performed using 5 μl reaction mixtures containing master mix, 100 pmol/μl of each primer and 1 ng/μl of template DNA using a GeneAmp PCR system 9700 thermal cycler (Applied Biosystems, Darmstadt, Germany). Amplification was started with a 15 min initial denaturation at 95 °C, followed by 30 cycles with denaturation at 95 °C for 30 s, annealing at 60 °C for 90 s and extension at 72 °C for 60 s. A final extension was performed at 72 °C for 7 min. 1 μl of the PCR product was used for fragment length determination using an ABI 3130 $\times$ *l* Genetic Analyzer (Applied Biosystems, Foster City, USA) at the Institute for Grapevine Breeding Geilweilerhof at the Julius Kühn Institute (Siebeldingen, Germany). The results were processed with GeneMapper software version 4.0 (Applied Biosystems, Darmstadt, Germany).

Data analysis. The genotypic data obtained for the 38 Armenian grapevine accessions were processed using the genetic analysis software Cervus© 3.0.7 (Copyright Tristan Marshall 1998–2014) [27]. Homozygous genotypes show a single peak at the investigated locus. Cervus software was used to calculate allele frequency and number of alleles per locus, estimated frequency of null alleles and probability of identity per locus, expected and observed heterozygosity [28], probability of identity per locus and the polymorphic information content for diversity analysis. The Arlequin 3.5 software was used to calculate gene diversity and observed heterozygosity per population for two groups of cultivars: (1) ancient cultivars and (2) newly bred cultivars including those obtained by open pollination [29]. Genotypic results were converted to a similarity matrix and a dendrogram was constructed with the algorithm UPGMA (unweighted pair-group method with arithmetic mean) using the NTSYS-pc software (Numerical Taxonomy and Multivariate Analysis System, ver. 2.1) [30].

Results and discussion. The analysis of the genetic diversity of the Armenian grape accessions of different origin included the average number of alleles per locus (Na), *Ra – range of alleles sizes (bp)*, allele frequency, estimated heterozyosity (He) and observed heterozygosity (Ho), probability of identity (PI), frequency of null alleles (Fo) and polymorphic information content (PIC) (Table 2).

In our research a total of 164 alleles were obtained at the 18 SSRs loci analyzed. The number of alleles per locus ranged from 5 (VVIN16, VVIB01, VVIN73) to 15 (VVS2) with a mean number of 9.17 alleles per locus. According to other authors the average number of alleles per locus for grape accessions of different geographical origins ranged from 8.17 [31] to 24.4 [32].

The expected heterozygosity of the investigated cultivars ranged from 0.514 (VVIN73) to 0.90 (VVS2) with a mean 0.789. This value is less than data obtained by Martinez et al. (2006) (0.810) and Martin et al. (2003) (0.806) [33,34]. The lowest observed heterozygosity (0.417) was detected at the VVIV37 locus and the highest at the VVMD7 and VMC1B11 loci (0.947). The high heterozygosity at some of the loci may be explained by the high number of crosses in the cultivar set. Twenty from the 38 investigated cultivars are new, including 11 *V. vinifera*

Table 1. List of grape accessions evaluated in this study.

ID	Cultivar name	Short name	Institute code	Species	Pedigree
1	Areni	ARE	ARM011	*V. vinifera L.*	Ancient
2	Areni chernyi	ACH	DEU098	*V. vinifera L.*	Ancient
3	Ararati	ARA	ARM011	*V. vinifera L.*	Ancient
4	Arevar	ARV	ARM011	*V. vinifera L.*	Ancient
5	Chilar	CHL	ARM011	*V. vinifera L.*	Ancient
6	Garan dmak	GDM	ARM011	*V. vinifera L.*	Ancient
7	Eraskheni	ERA	ARM011	*V. vinifera L.*	Ancient
8	Itsaptuk	ITS	ARM011	*V. vinifera L.*	Ancient
9	Hastamashk	HMK	ARM011	*V. vinifera L.*	Ancient
10	Kakhet	KKH	ARM011	*V. vinifera L.*	Ancient
11	Marmari	MAR	ARM011	*V. vinifera L.*	Ancient
12	Mskhali	MSK	ARM011	*V. vinifera L.*	Ancient
13	Mskhali clone	MSC	ARM011	*V. vinifera L.*	Ancient
14	Parvana	PAR	ARM011	*V. vinifera L.*	Ancient
15	Spitak Arakseni	SPA	ARM011	*V. vinifera L.*	Ancient
16	Spitak kaqavik	SPK	ARM011	*V. vinifera L.*	Ancient
17	Vanqapatkan	VQP	ARM011	*V. vinifera L.*	Ancient
18	Voskehat	VOS	ARM011	*V. vinifera L.*	Ancient
19	Armenia	ARM	ARM011	*V. vinifera L.*	**Itsaptuk** (synonymKhusaine Belyi)**X Sateni Chernyi**
20	Shahumyani	SHA	ARM011	*V. vinifera L.*	**Itsaptuk** (synonymKhusaine Belyi), Open pollination
21	Anush	ANU	ARM011	*V. vinifera L.*	**Armenia** *(Khusaine Belyi X Sateni Chernyi)* **X KishmishRozovyi**
22	Karmir kakhani	KAH	DEU098	*V. vinifera L.*	**Tana Kouzy XSevan**
23	Muscat Tskha	MTS	ARM011	*V. vinifera L.*	**Madeleine Angevine** X **Muskat Desertnyi**
24	Muscat haykakan	MHA	ARM011	*V. vinifera L.*	**Muskat Chernyi** (free pollination) x **Muscat a Petits Grains Noirs**
25	Muscat Yerevani	MUY	ARM011	*V. vinifera L.*	**C 484 X Rizamat**
26	Merdzavani vaghahas	MEV	ARM011	*V. vinifera L.*	**C 484** (Madeleine Angevine X Chasselas Musque) **X Kishmish Khishrau** (Nimrang X Kishmish Noir)
27	Nazeli	NAZ	ARM011	*V. vinifera L.*	**Sultanine** X**Blanc D'iran**
28	Tokun	TOK	ARM011	*V. vinifera L.*	**Spitak Arakseni,** open pollination
29	Agaraki	AGA	DEU098	*V.* **vinifera L.x** *V. amurensis Rupr.*	*V. vinifera L.* (NO.1563/1+21) **X Karmrahyut**
30	Anushahyut	ANH	DEU098	*V. vinifera L.x V.amurensis Rupr.*	**C 1262** (*V. amurensis Rupr.* **X Perle Von) Csaba) X Karmrahyut**
31	Burmunk	BUR	ARM011	*V. vinifera L.x V.amurensis Rupr.*	*V.amurensis Rupr.* X **Muscat Fleur D'Oranger**
32	Charentsi	CHA	ARM011	*V. vinifera L.x V.amurensis Rupr. V.amurensis Rupr.*	**Seyanets C 1262** (*V.amurensis Rupr.* X Zhemchug Saba) X **Karmrahyut**
33	Csarenci	CSA	DEU098	*V. vinifera L.x V.amurensis Rupr.*	*V. vinifera L.x V.amurensis Rupr.*
34	Karin	KAR	DEU098	*V. vinifera L.x V.amurensis Rupr.*	**Lernatu X Fioletovyi Rannii** *V.amurensis Rupr xV. vinifera L..)*
35	Nalbandyani	NAB	ARM011	*V. vinifera L.x V.amurensis Rupr.*	**Karmrahyut X Seyanets 1563/1+21**
36	Nerkarat	NRA	DEU098	*V. vinifera L.x V.amurensis Rupr.*	**C 1262** (*V.amurensis Rupr.* X Perle Von X Csaba) **X Karmrahyut**
37	Nerkeni	NER	ARM011	*V. vinifera L.x V.amurensis Rupr.*	**Saperavi X**("Garan Dmak X) Kikhet)
38	Karmrahyut	KRH	ARM011	*V. vinifera L.x V. V.amurensis Rupr.*	**Hadisi X No 15-7-1** (*V.amurensis Rupr.* X X Sladkii Chernyi)

and 9 *V. vinifera* x *V. amurensis* crosses. The probability of identity (PI) values for different loci ranged from 0.023 (VVS2) to 0.291 (VVIN73).

Polymorphic information content (PIC) for each locus was calculated for the evaluation of the effectiveness of each marker (Table 2). In our research PIC values were lower than the values of expected heterozygosity and higher than 0.7 for the majority of loci meaning that these loci are highly informative. The most informative loci were VVS2 (0.899), VRZAG62 (0.862) and VRZAG79 (0.857), while the least informative were the VVIN73 and VVIB01 loci with a PIC value of 0.459 and 0.575,

Table 2. Genetic parameters for SSR loci in Armenian accessions.

Locus	Na	Ra	He	Ho	PI	Fo	PIC
VVMD5	8	226–248	0.825	0.730	0.059	0.060	0.789
VVMD7	9	233–255	0.817	0.947	0.064	−0.087	0.780
VVMD24	7	206–219	0.809	0.868	0.069	−0.046	0.770
VVMD25	7	235–255	0.737	0.658	0.120	0.045	0.682
VVMD32	12	240–277	0.773	0.667	0.083	0.047	0.736
VVIV37	9	152–182	0.799	0.417	0.073	0.310	0.760
VVIV67	11	352–383	0.809	0.611	0.061	0.143	0.777
VVIN16	5	147–157	0.671	0.763	0.166	−0.076	0.628
VVIN73	5	260–272	0.514	0.595	0.291	−0.099	0.459
VVIH54	10	139–177	0.856	0.632	0.042	0.142	0.828
VVIB01	5	294–312	0.653	0.763	0.196	−0.100	0.575
VVIP31	11	173–196	0.847	0.895	0.045	−0.034	0.818
VVIP60	9	306–332	0.755	0.789	0.100	−0.041	0.710
VMC1B11	8	168–191	0.809	0.947	0.068	−0.090	0.771
VMC4f3.1	12	165–208	0.874	0.868	0.032	−0.009	0.848
VVS2	15	124–161	0.901	0.871	0.023	0.021	0.876
VRZAG62	10	184–204	0.878	0.758	0.033	0.065	0.876
VRZAG79	12	237–261	0.881	0.676	0.030	0.130	0.856
Mean	9.17	–	0.789	0.748	0.086	0.021	0.752

Table 3. The most frequent alleles per locus and their frequency.

SSR-marker	VVMD5	VVMD7	VVMD24	VVMD25	VVMD32	VVIV37	VVIV67	VVIN16	VVIN73
Most frequent allele	238	239	208	241	272	154	360	151	268
Major allele frequency, %	27.03	28.95	28.95	38.16	40.28	34.72	37.50	48.68	66.22
SSR-marker	**VVIH54**	**VVIB01**	**VVIP31**	**VVIP60**	**VMC1B11**	**VMC4f3.1**	**VVS2**	**VRZAG62**	**VRZAG79**
Most frequent allele	163	294	183	322	168	181	137	186	247
Major allele frequency, %	25.0	43.42	30.26	38.16	31.58	21.05	19.35	21.21	21.62

Table 4. Genetic variability within the studied population: observed and expected heterozygosity (Ho and He), mean number of alleles (MNA), averaged over all loci.

Population	$H_o \pm$ s.d.	$H_e \pm$ s.d.	MNA
Ancient cultivars	0.753 ± 0.161	0.779 ± 0.082	7.39
New crosses	0.739 ± 0.171	0.792 ± 0.118	8.00

S.d. - *standard deviations.*

respectively. According to literature data, VRZAG62 [33] and VRZAG79 [35] were the most informative ones both with PIC values of 0.88 which comes very close to our data.

The most frequent alleles at each locus and their frequencies are presented in Table 3. The highest frequency of alleles was registered for VVIN73 – 268 (66.22%), VVIN16 – 151 (48.68%), VVIB01 – 294 (43.42%) and VVMD32 – 272 (40.28%). In total, 68 rare alleles (41,46%) for all loci were found with a frequency of occurrence of less than 5%.

In Table 4 the genetic variability of ancient cultivars and the new bred *V. vinifera* x *V. vinifera* and *V. vinifera* x *V. amurensis* varieties is presented. Observed and expected heterozygosity (Ho and He) for all loci was high on average in both populations. The higher variation of gene diversities with (He) 0.792 was obtained for interspecific crosses and a lower value of 0.739 was found in the ancient cultivars. The mean number of alleles (MNA) for all loci per population ranged from 7.39 in ancient to 8.00 in new cultivars.

Genetic relatedness. The dendrogram shown in Fig. 1 was constructed using the UPGMA method for the evaluation of genetic diversity and relatedness between the investigated cultivars. In the dendrogram the investigated cultivars were separated into two main clusters. The smaller cluster included only the two cultivars 'Agaraki' and 'Karin' from the German collection. Both genotypes are interspecific crosses (*V. vinifera* x *V. amurensis*). The ancient cultivar Eraskheni separates from the other two. The large cluster exhibited 4 distinct subclusters labelled S1, S2, S3 and S4 and the interspecific cross 'Burmunk' which is characterized by a very high sugar content in its berries (more than 30%) and very high frost resistance.

Subcluster S1 was separated into two smaller subgroups, with the first one including only interspecific crosses (*V. vinifera* x *V. amurensis*) from both Armenian and German collections: 'Anushahyut', 'Csarensi', 'Nerkarat', 'Charentsi' and 'Nerkeni'. The first four of them are offspring of 'Karmrahyut'. The second subgroup included three ancient cultivars – 'Areni', 'Areni chernyi' and 'Chilar'.

The second subcluster S2 separated into two distinct subgroups. The first one included mainly ancient cultivars and also 'Itsaptuk' (synonym 'Khusaine belyi'), its offspring 'Shahumyani' and 'Armenia' together with

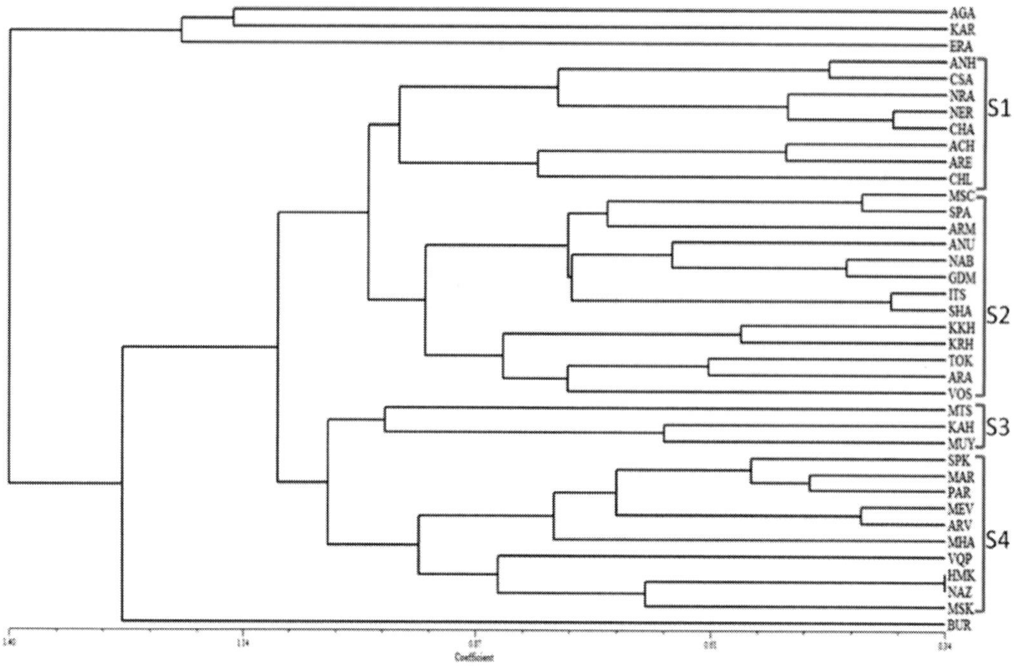

Figure 1. Genetic relationships in Armenian grapevine cultivars based on SSR analysis.

Armenia's offspring 'Anush'. In the second subgroup the ancient cultivar 'Kakhet' was grouped with 'Karmrahyut'. It seems that the name 'Karmrahyut' of this accession is incorrect because it did not group into cluster 1 or subcluster S1 which included its offspring and other *V. vinifera* x *V. amurensis.* crosses, although we did not investigate the second parental genotypes of the crosses.

The second part of the subgroup was formed by the ancient cultivars 'Voskehat' and 'Ararati', and the cultivar 'Tokun' obtained as the result of open pollination of the ancient cultivar 'Spitak Arakseni'.

Subcluster S3 included three *V. vinifera* interspecific cultivars: 'Muscat Tskha', 'Muscat Yerevani' and 'Karmir kakhani'.

The fourth subcluster S4 was made up of ten accessions which were separated into two sub groups and included mainly ancient cultivars. 'Spitak kaqavik', 'Marmari' and 'Parvana' seedless fell into the first subgroup. Unexpectedly, the newly bred cultivar 'Nazeli' fell into the second subgroup together with the ancient cultivars 'Hastamashk', 'Vanqapatkan' and 'Maskhali'.

The closest relationships were detected between the following pairs of genotypes: The ancient cultivar 'Itsaptuk' and its offspring 'Sahumyani' obtained in the result of open pollination; 'Anushahyut' and 'Csarenci', both from the German collection, which are offsprings of the same parent 'Karmrahyut' (*V. vinifera* x *V. amurensis*). The ancient cultivars 'Areni' from Armenia and 'Areni chernyi' from Germany matched except at VVMD32, where heterozygosity versus homozygosity was stated, most likely due to clone variation. 'Charentsi' from Armenia and 'Csarenci' from Germany were not identical but had high similarity and were placed in the same group of *V. vinifera* x *V.amurensis* cultivars). As it was expected, all investigated interspecific crosses were characterized by their high frost resistance, but differed by other phenotypic characteristics. This, however, was not enough for a clear distinction of the investigated genotypes, especially new bred varieties. Thus an application of modern techniques with a genetic approach was necessary for the determination of the genetic relation between the investigated genotypes.

3. Conclusion

Our comprehensive genetic research has shown high polymorphism and heterozygosity in Armenian grapevine cultivars of different origin. The Armenian genetic pool was formed during thousands years of folk selection and later was enriched by hybridizations and mutations.

The molecular characterization and analysis of the genetic structure in Armenian grape germplasm based on allele frequencies and heterozygosity contributes to the knowledge about levels and distribution of genetic diversity of the investigated cultivars. It reveals differences between aboriginal ancient and newly bred cultivars in the existing gene pool of *Vitis*.

By evaluation of the genetic relationship among the investigated cultivars a clear separation between the groups of ancient aboriginal cultivars, the Armenian new *V. vinifera* x*V. vinifera and V. vinifera* x *V. amurensis* crosses and seedless cultivars was detected. Very close relationships with high similarity were determined for five pairs of cultivars with the same pedigree or origin. However, two cases of possible homonymy were also detected. Our results reveal the high effectiveness of using SSR based genetic analysis of the grapevine germplasm and provide correct information about genetic diversity and genetic relations between accessions.

All genetic profiles will be compared with allelic data from the *Vitis* International Variety Catalogue (www.vivc.de) and the European *Vitis Database*, (`www.eu-vitis.de`) [36]. The material of Armenian local cultivars significantly contributes to the understanding of the genetic variation in our grapevine collection and presents an important source for target selection. Special care should be taken to conserve these cultivars in ampelographic collections, because in some areas in Armenia the old vineyards are rapidly disappearing.

This research was partially supported by the DAAD (German Academic Exchange Service) Research visit grant (A/13/03780-326) and the Research project of State committee of science at the Ministry of Education and Science of Armenia (15T-1E232).

References

[1] G. Melyan, S. Gasparyan, *Caucasus and Northern Black Sea Region Ampelography* (Vitis, Germany, 2012)

[2] P.E. McGovern, *Ancient wine: the search of the origin of the viniculture.* (Princeton University Press, New Jersey, 2003)

[3] F. Haroutunian, *Reflections on the history of viticulture and winemaking in ancient and medieval Armenia.* (Printinfo, Yerevan, 2005)

[4] R. Arroyo-García, L. Ruiz-García, L. Bolling, R. Ocete, M.A. López, C. Arnold, A. Ergul, G. Söylemezoğlu, H.I. Uzun, F. Cabello, J. Ibáñez, M.K. Aradhya, A. Atanassov, I. Atanassov, S. Balint, J.L. Cenis, L. Costantini, S. Goris-Lavets, M.S. Grando, B.Y. Klein, P.E. McGovern, D. Merdinoglu, I. Pejic, F. Pelsy, N.Primikirios, V. Risovannaya, K.A. Roubelakis-Angelakis, H. Snoussi, P. Sotiri, S. Tamhankar, P. This, L. Troshin, J.M. Malpica, F. Lefort, J.M. Martinez-Zapater, Mol. Ecol. **15**, 8 (2006)

[5] H. Barnard, A. Dooley, G. Areshian, B. Gasparian, K. Faull, J Archaeol Sci **38**, 8 (2010)

[6] J. Smith, O. Smith. Advances in agronomy **47**, 56 (1992)

[7] R. Aroutiounian, A. Nebish, K. Margaryan, G. Melyan, Vitis **54**, 4 (2015a)

[8] R. Aroutiounian, A. Nebish, G. Melyan, K. Margaryan, BIO Web of Conferences **5**, 5 (2015b)

[9] M.R. Thomas, N.S. Scott, Theor. Appl. Genet. **86**, 985–990 (1993)

[10] H. Lin, M.A. Walker, Am. J. Enol. Vitic **49**, 5 (1998)

[11] J. Ibanez, M.T., De Andres, A. Molino, J. Borrego, Am. J. Enol. Vitic. **54**, 9 (2003)

[12] E. Maletic, I. Pejic, J.K. Kontic, G.S. Dangl, A. Vokurka,T. Lacombe, N. Mirosevic, C.P. Meredith, Am. J. Enol. Vitic. **55**, 7 (2004)

[13] R. Eibach, E. Zyprian, L. Welter, R. Töpfer, Vitis **46**, 5 (2007)

[14] C. Bergamini, C., M.F. Cardone, A. Anaclerio, R. Perniola, A. Pichierri, R. Genghi, V. Alba, L.R. Forleo, A.R. Caputo, C. Montemurro, A. Blanco, D. Antonacci, Mol. Biotechnol. **54**, 10 (2013)

[15] P. This, T. Lacombe, M.R. Thomas. Trends Genet. **22**, 9 (2006)

[16] M.R. Thomas, N.S. Scott, Theor. Appl. Genet. **86**, 6 (1993)

[17] J.E. Bowers, G.S. Dangl, R. Vignani, C.P. Meredith, Genome **39**, 6 (1996)

[18] J.E. Bowers, G.S. Dangl, C.P. Meredith, Am. J. Enol. Viticult. **50**, 8 (1999)

[19] K.M. Sefc, F. Regner, E. Turetschek, J. Glossl, H. Steinkellner, Genome **42**, 7 (1999)

[20] D. Merdinoglu, G. Butterlin, L. Bevilacqua, V. Chiquet, A. Adam-Blondon, S. Decroocq, Mol. Breed. **15**, 18 (2005)

[21] E. Zyprian, R. Topfer, *National Center For Biotechnology Information NCBI, Genbank* (Bethesda, Maryland, USA, 2005)

[22] G. Di Gaspero, E. Peterlunger, R.Testolin, K. Edwards, G. Cipriani, Theor. Appl. Genet. **101**, 8 (2000)

[23] P. This, A. Jung, P. Boccacci, J. Borrego, R. Botta, L. Costantini, M. Crespan, G.S. Dangl, C. Eisenheld, F. Ferreira-Monteiro , S. Grando , J. Ibáñez, T. Lacombe, V. Laucou, R. Magalhães, C.P. Meredith, N. Milani, E. Peterlunger, F. Regner, L. Zulini, E. Maul,Theor. Appl. Genet. **109**, 11 (2004)

[24] E. Maul, K.N. Sudharma, S. Kecke, G. Marx, C. Müller, L. Audeguin et al., Vitis **51**, 7 (2012)

[25] V. Laucou, T. Lacombe, F. Dechesne, R. Siret, J.P. Bruno, M. Dessup, T. Dessup, P. Ortigosa, P. Parra, C. Roux, S. Santoni, D. Vares, J.P. Peros, J.M. Boursiquot, P.This, Theor. Appl. Genet. **122**, 13 (2011)

[26] T. Lacombe, J.M. Boursiquot, V. Laucou, M. Di Vecchi-Staraz, J.P. Péros, P. This, Theor. Appl. Genet. **126**, 14 (2013)

[27] S.T. Kalinowski, M.L. Taper, T.C. Marshall, Mol. Ecol. **16**, 8 (2007)

[28] M. Nei, *Analysis of gene diversity in subdivided populations.* Proceedings of the National Academy of Sciences USA **70**, 3 (1973)

[29] L. Excoffier, H.E.L. Lischer, Mol. Ecol. Res. **10**, 8 (2010)

[30] F.J. Rohlf, *NTSYS-PC numerical taxonomy and multivariate analysis system* (Exeter Publishing, New York, NY, USA, 1988)

[31] M.C. Almadanim, M.M. Baleiras-Couto, H.S. Pereira, Carneiro L.C., P. Fevereiro, J.E. Eira-Dias, L. Moraiscecilio, W. Viegas, M.M. Veloso, Vitis **46**, 4 (2007)

[32] W.F. Lamboy, C.G. Alpha, Journal of the American Society of Horticultural Science **123**, 7 (1998)

[33] L.E. Martinez, P.F. Cavagnaro, R.W. Masuelli, M. Zuniga, Plant Science **170**, 9 (2006)

[34] J. Martín, J. Borrego, F. Cabello, J. Ortiz, Genome **46**, 9 (2003)

[35] P. C. de S. Leao, C.D. Cruz, S.Y.Motoike, Rev. Bras. Frutic. **35**, 10 (2013)

[36] R. Töpfer, K.N. Sudharma, S. Kecke, G. Marx, R. Eibach, D. Maghradze, E. Maul, Bulletin de l'OIV, **11** (2009)

BIO Web of Conferences **9**, 01021 (2017)
40th World Congress of Vine and Wine
DOI: 10.1051/bioconf/20170901021

Grapevine powdery mildew (Uncinula necator (Schw.) Burr.) – a permanent issue concerning the health status of grapes cenosis in Bulgaria

Mariana B. Nakova[1], Boris K. Nakov[1], and Miroslav Tityanov[2]

[1] Department of phytopathology, Agriculture University Plovdiv, Plovdiv, Bulgaria
[2] SAB Sumi-Agro, Bulgaria

Abstract. Grapes powdery mildew appeared permanently in the grapes growing regions of Bulgaria. The disease was reported in our country by Kostov in 1900. More than a century the powdery mildew disease has been causing less or greater yield losses depending on the climate. In the past when the vineyards were small and scattered, the disease was not an economic problem. Later, during the second half of the 20th century, the grapes growing areas consolidated and enlarged, and the concentration and intensification of production took place. The pathogen spread permanently to epidemic in the vineyards and the disease became destructive and economically important. During that period commercial varieties were grown susceptible to the causal agent of the powdery mildew, and organic fungicides were applied in the disease control system. Mistakes in the grapes growing technology are observed and the disease management strategies applied are not always scientifically proved. The statements that in Bulgaria there exist conditions suitable for the appearance and development of grapes powdery mildew only in the Black sea region are disproved. A new research is necessary to be done to answer the questions about: the sources of primary inoculum, the influence of the ecological conditions on the appearance and spread of the powdery mildew, and the timing for the disease control. In the survey the overwintering of the pathogen, the appearance of the first symptoms and the dynamics of the disease spread have been discussed. As for the ecological conditions in Bulgaria, it is considered that the fungus mainly survives as mycelia in the buds and on the shoots of the vines serving as a source of a permanent infection background. The studies carried through during 1994–2002 proved that the pathogen influenced by the ecological conditions could also form cleistothecia and they could be the source of the initial infection. The effect of the leaves removal around the clusters on the spread of the grapes powdery mildew and the field resistance of some varieties were studied. Critical stages of the disease appearance and spread are defined based on the research data. Management strategies are proposed including alternative approach practices.

The grapes powdery mildew permanently appears in the grapes cenosis in Bulgaria. More than a century it has been causing less or greater yield losses. The disease was reported by Kostov in 1900. In the past, when the vineyards were small and scattered, the disease was not an economic problem. But after the consolidation of the areas, the concentration and intensification of the production during the second half of the 20th century, the development of the disease became permanent and, not rarely, epiphytotic and economically important. The problem is due to the growing of vine varieties sensitive to the pathogen causing the disease, to the introduction of organic fungicides in the production, to the mistakes done in the technology applied and to methods for control not always scientifically based.

In Bulgaria the studies of the biological characteristics, the epidemiology, the prediction and the control are limited and not unidirectional. Debatable are the questions concerning the overwintering and the source of the primary inoculum; the influence of the ecological conditions, as well as the development dynamics and the prediction of the powdery mildew; and questions concerning the moments and methods of control.

The history data show that the powdery mildew is the oldest grapevine disease known to science and practice. It comes from North America where the disease has always existed [1], but the grown grapevine varieties were comparatively resistant, with low yield losses, so the disease did not attract the attention of the specialists.

As far as Europe is concerned the fungus was first brought to England in 1845. In 1847 Tucker and Berkeley report independently a new cryptogamic parasite on grapes. Berkeley identifies it as *Oidium tuckeri*.

Leclerc (1853) [2] reports the oidium to be noticed in France in 1847 and only for two years it had managed to spread throughout the country. In 1850 the vineyards around Paris were destructed, i.e. the fungus caused the first epiphytothic after being transferred to Europe.

In Italy the disease was found in 1849 [3]. During 1851–1852 the powdery mildew spread to devastating proportions in several European and Mediterranean countries. Within 1853–1856 the disease caused serious economic problems: eradicated vineyards; selling-off lands; resettlement of the poorest people; raising of the wine prices; 2 to 3 times crop reduction in comparison with the good years [4].

 DOI: 10.1051/bioconf/20170901021

Thus, by the end of the 19th century the grapevine powdery mildew and its causal agent had become a worldwide issue. The outbreak of the epidemic drew the scientists' attention resulting in an intensive research work on the symptoms, the biological features and the measures for control.

Viala (1853, 1877) [5] is the first to state that the earliest symptoms can be found on the vines during bud break and later the symptoms spread on the whole surface of the new shoots. This he calls "flag shoots". According to him the fungus remains as a micelium in the dormant buds under their scale. The infection happens in the autumn. Appel (1903) [6] confirms Viala's thesis. Amici (1852) [7] and Mohl (1853) [8] report damages caused on the inflorescences. Arnaud (1931) [1], Rendu (1853) [9] and Viala (1887) [10] add some information about infections and serious symptoms on the grape berries up to the berry touch period. Pearson and Gartel (1985) [11] find out that the buds become infected mainly from the 3rd to the 5th internodes.

Jossifovitch (1923) [12], Frezal (1953) [13] and Molnar (1914) [14] also state that in mild winters the micelium stays alive and restores during the following spring. Arnaud (1931) [1] rejects that possibility. Viala (1893) [10] finds the micelium on "the newly opened buds". Wortmann (1900) [15] and Ravaz (1921) [16] confirm the statement of "the presence of the micelium" in the buds and the fact that out of the infected buds come shoots "fully covered with the micelium and the spores of the fungus".

Boublas (1961) [17] finds out that the buds having micelium are located from the 3rd to the 6th node. Lafon (1967) [18], Bulit and Lafon (1978) [19], Sall and Wrysinski (1982) [20] also observe that "the early symptoms of the oidium appear from the infected buds". Pearson and Gartel (1985) [11] report that "the disease happens in large scale from the infected buds during April and May". Rumbolz et al. (2002) [21], Rugner et al. (2002) [22], Ypema and Gubler (2000) [23] also state the same data of the fungus overwintering as a micelium. The tempetatures and the humidity in June favour the mass spread of the oidium.

The slight curl on the periphery of the leaf blades is the first direct symptom of the powdery mildew appearance followed by a rapid invasion of the micelium and the abundant spore formation. The late infections establish conditions for inoculations and the survival of the parasite till the next vegetation season.

There are discussions on the role of the teleomorph in the life cycle of the powdery mildew causal agent. In Europe the teleomorph is reported by Couderc (1893) [24]. Cleistothecia are observed in large numbers in September-October 1893, after a dry and hot summer followed by the drop in temperatures to 4°–8°C [25]. Pacottet (1903) [26] also assumes the preservation of the pathogen via teleomorph. According to Foex (1911) [27] the run out of the nutrient substances is a very important factor for the cleistothecia formation. Laibach (1930) [28] states the role of the "worsened physiological conditions of the host plants, and especially the lack of water to stimulate the teleomorph formation". Jossifovitch (1923) [12] proves that the cleistothecia put on dry leaves or wet sand, during October-February, can survive till March. Nedeltscheff (1924) [29] and Geoffrion (1960) [30] also support the thesis that dry weather favours the formation of the cleistothecia. Schnathorst (1965) [31] reveals the correlation between the environmental conditions and the development of the teleomorph. According to Gadoury and Pearson (1988, 1988) [32,33] the ascospores are discharged during the stages of the bud break and flowering, at 4°–32°C and rain – 2.5 mm. The first spots can be found during the last week of May, when shoots are 7–15 cm high. In certain regions of California [23] the fungus is stated to survive as micelia and cleistothecia. In South Africa the sources of the initial infection are the micelium in the infected buds and the cleistothecia [34]. Gee et al. (2000) [35] conclude that the cleistothecia are the source of the additional inoculum in Australia. Miazzi et al. (2003, 2008) [36,37] announce the existence of A and B genetic groups in South Italy developed in separate agro-cenosis during the vegetation period. They correlate this to the cleistothecia formation. Gadoury et al. (2012) [38] survey the spread of these two groups in the USA and think they appear to be the parental forms of those groups registered in Italy and introduced separately in Europe. A certain additional research should be carried out in order to define the mechanism of the appearance of both groups.

The status of the grapevine powdery mildew issue in Bulgaria

The scientific researches on the biological features of the oidium causal agent in our literature are not enough. The first report is made by Nedeltscheff (1923/24) [39]. The author establishes that cleistothecia are formed comparatively rarely and they do not play an essential role in the life cycle of the the fungus. Atanasov (1934) [40] explains that the question of the overwintering is debatable: cleistothecia or dormant mycelia on the shoots. Hristov (1956) [41] states that the causal agent overwinters with cleistothecia ripening after the first or the second overwintering. According to him the large scale appearance of the disease in Europe can be explained with the formation of the dormant mycelium on the annual shoots in autumn but not with the cleistothecia. That micelium overwinters and causes the disease during the next spring. It is also accepted that there are favourable conditions for the development of the powdery mildew only along the Black sea cost line and in the regions along the Danube River [42]. Perhaps the authors are not aware of the surveys carried out in the end of the 19th and the beginning of the 20th centuries on the overwintering of the fungus in the buds and the early development of the disease [10,11,15,16].

Contradictory are the data in some Bulgarian articles on the proper period of the first spraying. Martinov (1938, 1940) [43,44] suggests three sprayings with sulphur, at 20°C, after the end of the flowering. According to Nedeltscheff and Kondarev (1949) [45] the first treatment with sulphur should be done after the fruit set. The next applications has to be done every other 15 days, if it is necessary. The authors support their thesis explaining the fact that "in Bulgaria the oidium does not appear before the fruit set. Thus the early treatments are useless". Raykov and Nenov (1959) [42] are the first who give in 1953 some data about the damages of the disease in the Black Sea region reaching up to 50% of the yield, or even "the full devastation of the grapes crops". The authors advise how to limit the losses: the first sulphur application should be done in the end of the flowering stage, just at the very beginning of the disease development. "The end of the flowering" means the moment when the last caps fall.

All these instructions how to control the oidium are based on the real observations on the first symptoms of the disease, without considering the studies on the biological features of the pathogen described in the foreign literature, i.e. the early development of the pathogen causing infections with the very beginning of the buds break. The first latent period lasts about 3–4 weeks coinciding with the stages of forming the inflorescence and flowering when the first symptoms are discovered.

After 1960 comes a period of introducing the organic fungicides replacing the traditional copper and sulphur active ingredients. The scientific circles and the practices do not consider the fact that the zineb fungicides show an effect to the downy mildew (*Plasmopara viticola*), and at the same time they strongly stimulate the development of the powdery mildew (*Uncinula necator*). The very first experiments in Bulgaria during 1958–1960, show that with the zineb treated variants, symptoms of powdery mildew are registered on 80% of the leaves and the clusters. While with the copper and sulphur treated variants the registered spots are few [46]. The oidium starts its development in large scales permanently. During 1965–1970 the disease becomes epidemic in some villages in the region of Plovdiv (the villages of Markovo, Belashtitsa, Brestnik), as well in the towns of Nessebar, Pomorie and the entire region of Bourgas (Nakov, not published data). Iliev (1969) [47] writes about great damages of the powdery mildew in the region of Bourgas.

The carried out investigations of the vineyards, greatly affected by the powdery mildew, have established the fact that the first centers of infection appear from the "flag shoots" developed from the infected buds. The microscopic analysis of the infected buds shows that there are micelia and spores right before the buds break, just under the scales. After being cut and coloured with cotton-blue dye the bud tissues show the micelium hyphae and the chains of conidia at 6–8°C. The surveys done in the region of the town of Nessebar show that the infected shoots develop from inoculated buds [48]. The author assumes that spores might be transferred from the southern regions, i.e. from Turkey. According to him the cleistothecia are formed but they can not ripen. The observations and surveys confirm the information in the literature that the infected bud is the source of the initial inoculums. Thus, it is necessary for early treatments to be done, just after the bud break, when the shoots are 0.5 – 2 – 3 cm long.

After 1970–1980 the systemic (therapeutical) fungicides become popular in the plant protection. Their ability to penetrate into the plants' tissues; to go through the organism and kill the pathogen is considered to be a revolutionary period in science [49]. When the treatment is done in the stage of the bud break and the the first leaves unfolded, the systematic fungicides penetrate inside and kill the micelium. The late teatments omit the earliest infections which lead to sporulation and spread of the pathogen.

The authors of this article present the results of their surveys on the influence of certain epidemiological factors on the appearance and development of the teleomorph of the pathogen. The data show the fact that during the last decade of the 20th century there appear some natural phenomena, such as: extremely high temperatures, dry and hot weather during the vegetation season, provoking stress situations and disturbance not only in the host plant physiology but in the fungus life cycle. The damages of the powdery mildew increase. The existent complicated situation requires constant observations on the disease symptoms, the sources of the initial infection and the critical moments furthering its intensive development. The meteorological data are received from the automised system EnviroCaster (Neogen corporation, Michigan, USA), having an installed module forecasting the powdery mildew. The disease development dynamics is registered with the variety Brestovitsa vines, not being treated with fungicides. Every three days they are monitored for the appearance and development of the disease symptoms on the leaves, the shoots and the grapes. The growth of the micelium on the newly formed leaf spots is measured, in mm, every day. Periodically the spots on the shoots, from the 1st to the 8th nodes, are reported in numbers, while the spots on the berries are reported in percentage. Observations of the cleistothecia formation on the infected parts are done regularly. The forming of the ascospores and their ripening is monitored in the experimental field of the Agriculture University Plovdiv and in the laboratory [50].

The development of the powdery mildew is examined with the following varieties too: Bolgar, Super early Bolgar, Rousalka, Hamburgski Misket, Cardinal, Aligote, Dimyat, Cabernet Sauvignon, Mavrud during 1994–1997 on the experimental field of the Agriculture University Plovdiv. These varieties were treated with fungicides and leaf removal around the clusters (under, above, and in contact with cluster), in the stages "end of the flowering to fruit set".

During 1994–1995 vines in the village of Brestnik (the University experimental field) were observed for the cleistothecia formation. During the end of July and the beginning of August the damages of the powdery mildew on the leaves with the Brestovitsa variety reached 63.92%, while the damages on the fruits were 47.32%. The mass formation of the cleistothecia on the leaves, the leaf stalks and the shoots is registered during the second half of September and the beginning of October. This period (1994–1995) had dry and hot summer, with extremely high temperatures reaching 38.6°C, while is certain periods up to 39–41°C. In the afternoons, for 3-4 hours, the relative humidity dropped to 29–33%. There is observed a depression in the plant development; the leaves with the oidium spots grow older fast, and perhaps they cannot serve as an adequate pathogen nurishing source. That makes the pathogen start the developement of the teleomorph. This process is stimulated by the sudden dramatic temperature drop to 8°–12°C. Similar are the characteristics of years 1996, 1997, 1999, 2001, 2002 [50]. During 1994–1995 the development of the cleistothecia is studied in the laboratory – at 4°–8°C, with and without moistering leaves, with cleistothecia, on steriled sand. During 2002 and 2003 the experiments are carried outdoor, on the experimental field of the Agriculture University Plovdiv – the leaves with cleistothecia were placed on the soil surface covered with 1–2 cm sand. With both experiments the development of the teleomorph, the formation of the asci and ascospores, happens during the last decade of October and in November. Ascospores were formed in the end

DOI: 10.1051/bioconf/20170901021

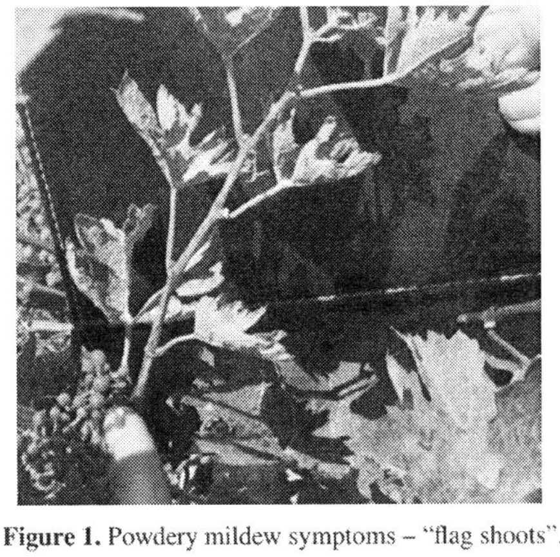

Figure 1. Powdery mildew symptoms – "flag shoots".

of November and the beginning of December. Part of the ascospores put in a wet chamber germinated. The ripened ascospores preserve their fertility, at the variable temperatures (2–8°C) and humidity, till the beginning of the vine vegetation, i.e. till the end of March and during April (1995, 2003). During the inoculation of the young vines with ripened cleistothecia (in pots and on leaves) in a wet chamber, at 26°C and high relative humidity, above 76%, symptoms appear after 6–7 days (May 1995; May 2003). The received data coincide with the data present by Viala (1893) [10], Pacottet (1903) [26], Yossifovitch (1923) [12], Pearson and Gadoury (1987) [51], Geoffrion (1960) [30], Foex (1911) [27], Gee et al., (2000) [35], i.e. cleistothecia can be formed in different regions under unfavourable meteorological conditions and they might be a source of an additional infection on the grapevines cenosis and serve as an initial infection background.

Surveys on the prediction of the powdery mildew development were carried out during 1993–1999 with the help of the EnviroCaster system (Neogen corporation, Michgan, USA). The results show that in the grapevine cenosis there exists a permanent initial infection background. The most strongly infected are the buds from the 3rd to the 6th node, but the infection can also be seen on the 8th node. The flag shoots growing out of the infected buds are covered with mycelia and conidia. At the end of May there appear shoots with smaller young leaves looking distorted and slightly curled up (Fig. 1). On the upper leaf surface light green asymmetrical small spots can be found. These spots resemble "oil spots" and are covered spares mycelia and spores (Fig. 2).

The data coincide with the publications of Gadoury et al. (2012) [38], Halleen and Holz (2000) [34] developing the hypothesis that these are spots caused by the ascospores of the pathogen. The studies on the oidium in Bulgaria show that these symptoms can be seen in the years, when during the last vegetation period, the fungus cleistothecia were massively formed.

During the research work the registered relative humidity is in the following intervals:

April – 45–70–82%;
May – 48–50–85%;
June – 42–50–70–85%;
July – 55–60–70–80%;
August – 40–45–70–85%.

Figure 2. Powdery mildew symptoms – ascospore infections ("oil spots").

Table 1. 'Critical periods' in the powdery mildew development.

Temperatures, °C from ..... to......	Relative humidity, %	Leaf wetness period, in hours
9.2–12.4	62–96	21–27
14.0–15.2	60 –95	11–16
16.0–19.9	61–94	8–10
22.5–26.5	60–86	5–6
26.9–29.1	76–92	9–11

Table 2. Development of the incubation periods, in days.

Temperatures, °C from .... to	Duration of the incubation period, in days
9.8–10.7	24–26
1.2–12.9	18–19
13.5–15.1	14–16
19.8–20.6	8 - 9
26.9–28.1	5–6

During the hot daily hours in the second half of July and in August the humidity drops down to 29%–30%, for one to three hours, at 39°–41°C. The relative humidity data compared to the data published by other authors show that in the grapevine cenosis such humidity favours the powdery mildew development during the vegetation period.

Based on the information about the influence of the temperature, the relative humidity and the leaf wetness period /in hours/ on the pathogen development, several periods of risk (critical periods) are defined. The critical stages could be used to predict the powdery mildew appearance and spread (Tables 1, 2).

During the years 1994–1997 a lot of reseach work was done concerning the effect of the leaf removal around the cluster (under, above and in contact) on the powdery mildew development (Table 3). With the Brestovitsa variety showing the highest sensitivity towards the powdery mildew causal agent, the difference between the treatments with the leaf removal and without the leaf removal is from 10.5 to 20%. With the other varieties, having 'green operations' and sprayings with fungicides applied, the differences in the the powdery mildew development on the clusters are: from 5.5 to 5.67% (variety Bolgar); from 13.4 to 21% (variety Dimyat), growth stage

DOI: 10.1051/bioconf/20170901021

Table 3. The influence of the partial leaf removal around the cluster on the development of *Uncinula necator* with the different varieties in the region of Plovdiv (average data) (1996–1997).

Variety	Disease severity on berries - % of infected berries, growth stages			
	"berries pea-sized", bunches hang		"beginning of berry touch"	
	Without leaf removal	With leaf removal	Without leaf removal	With leaf removal
Brestovitsa	2.32	0.32	77.8–95.3	67.3–75.3
Bolgar	2.36	0.21	17.2–19.1	11.53–13.60
Dimyat	1.80	0.0	22.38–29.99	8.98–8.98
Cabernet Sauvignon	2.43	0.0	1.80–2.29	0.32–1.05
Mavrud	1.10	0.0	2.32-2.32	0.0–1.02
Кардинал	3.42	1.32	2.76–2.76	0.37–0.64
Super early Bolgar	2.78	0.014	4.33–6.49	2.64–3.54
Rousalka	1.40	0.0	1.09–2.07	0.31–0.62
Hamburgski Misket	1.39	0.0	3.26–3.26	0.72–0.72
Aligote	0.5	0.0	0.12–0.55	0.0

beginning of the berry ripening. This tendency is the same with the rest of the varieties.

Conclusion

The conclusions based on the research work on certain biological features of the powdery mildew causal agent (*Uncinula necator*) are as follows:

- The pathogen overwinters mainly as a micelium in the buds, from the 3rd to the 6th node, causing the first early infections on the bud break and the later development of the systemic infection 'flag shoots'.
- Cleistothecia are formed under the influence of certain conditions such as: dry weather and extremely high temperatures in July and August followed by a dramatic drop of temperatures in September–October. Asci and ascospores are formed at the end of November and in December. Part of the ascospores stay alive till the beginning of the next vines vegetation period. The ascospores can cause later infections and serve as an infection background of the agrocenosis.
- Critical periods (periods with high infection risk) during the powdery mildew development are defined, which could be used for powdery mildew prediction and signaling.
- The effect of the leaf removal practices, around the grapes clusters, on the powdery mildew development is studied.

References

[1] G. et M. Arnaud, *Traite de patholgie vegetele*. T 1: 281–318 Lechevalier (1931)
[2] Leclerc L., *Les vignes maladies*. Rapport a M. le Comte de Persigny, Ministre de'Imterieur, sur un voyage d'etude entrepris par ses orders dans les vignobles fraincais pendant l'ete 1852. Paris, Hachette, 81p. (1853)
[3] Targioni-Tozzeti, *Sulla Mallatia delle vine*. Raporto generale della commissione della R. academia dei Georgofili, Firenze, 320 p. (1856)
[4] H. Mares, Note sur le soufrage des vognes. Bull. Soc. Agr. de l'Herault 43me an.: 86–96 (1856)
[5] P. Galet, *Les maladies et les parasites de la vigne*. T.I. Pausan de Midi. Montpelier, France (1977)
[6] O. Appel, Zur Kenntnis der Uberwinterung des *Oidium Tuckeri*. Centralblatt fur Bakt. U parasitenkr. 2 Abt. XI, Kassel **I**: 143–146 (1903)
[7] G.B. Amici, Sulla malattia dell'uva. Att. Acc. Georg. Firence **30**, 454–469 (1852)
[8] H. (Von) Mohl, Ubre die Traubenkrankheit. Bot. Zeitung Berlin **XI**, 585–595 (1853)
[9] V. Rendu, *De la maladie de la vigne dans le midi de la France et le nord de l'Italie*. Paris Impr. Imperilae (1853)
[10] P. Viala, *Le Malades de la vigne*. Coiulet ed., Montpellier, 67-124; 2me ed. 1887: 9-84; 3me ed. 1893: 1–56 (1885)
[11] R.C. Pearson, Gartel W., Occurrence of hyphae of *Uncinula necator* in buds of grapevine. Plant disease **69**, 149–151 (1985)
[12] M. Jossifovitch, *Contriution a l'etude de l' Oidium de la vigne et de son traitement*, These Fac. Sci. Toulouse, Imp. Bonnet, 176 p. et Bull. Soc. Hist. Nat. Toulouse (1923)
[13] P. Frezal, Rapport national (Algerie) sur l'oidium. 7me Cong. Rome O.I.V. 1953 et Bull. O.I.V. 1954 **281**, 51 (1953)
[14] G.Y. Molnar, Die uber Winterung des oidium der Weinrebe. Ampelologiai Intezetblrk. Budapest V an: 100-111 et Bull. Rens. Ag. Riome 1915 **672** (1914)
[15] Wortmann, Des Ascheriges des Weinstoke (*Oidium Tuckeri*). Bericht der Konigl Lehranst. Fur Obst Wein und Gertenbau in Geisenheim fur 1899–1900: 80–82 (1900)
[16] L. Ravaz, Les traitements d'havier de l'oidium. Prog. Agri. Et Vit. **76**, 392–393 (1921)
[17] D. Boublas, Etude des cause de la resistance des Vitacees a l' Oidium de la vigne et de leur mode de transmission hereditaire. Ann. Amelior. Plantes. **II**(4), 401–500 (1961)
[18] J. Lafon, L'oidium. Le Vrai Cognac 214: 2260 et Bull. O.I.V. 1967 **437**, 850 (1967)
[19] J. Bulit, R. Lafon, Powdery mildew of the vine, In: *The powdery mildews* (Spenser D. M. ed.), Academic press, New York, p. 525–548 (1978)
[20] A.M. Sall, J. Wrysinski, Perennation of powdery mildew in buds of grapevines. Plant disease **66**(8), 678–679 (1982)
[21] J. Rumbolz, E.G. LaRedo, W.D. Gubler, Overwintering of *Uncinula necator* in dormant grape buds. Determination of infection window". *Proceedings of the 4th International workshop of powdery and downy mildew in grapevine*, p. 50 (2002)
[22] A. Rügner, J. Rumbolz, B. Huber, G. Bleyer, U. Gisi, H.-H. Kassemeyer, R. Guggenheim, Formation of overwintering structures of *Uncinula necator* and colonization of grapevine under field conditions. Plant pathology **51**, 322–330 (2002)
[23] H.L. Ypema, D. Gubler, The distribution of early season grapevine shoots infected by *Uncinula*

necator from year to year: a case study in two California vineyards. American journal of enology and viticulture **51**(1), 1–6 (2000)

[24] G. Couderc, Sur les perithecés de *l'Uncinula spiralis* en France. C.R. Ac. Sc. Paris **116**, 210-211-Séance du 30 janvier et Bull. Soc. Mycol. De France. IX, 253 (1893)

[25] P. Viala, Sur les peritheces de l'oidium de la vigne. C.R. Ac. Sc. Paris **119**, 411 (1894)

[26] P. Pacottet, Oidium et *Uncinula spiralis*. Rev de Vit. **20**, 685 (1903)

[27] Foex et., Notes sur les modes d'hibernation de l'oidium de la vigne. Congr. Vit. Montpellier, Coulet ed., 8 p., et Prog. Agri et Vit. 1912 **57**, 47 (1911)

[28] F. Laibach, Uber die Bedingungen der Perithezien-bildung bei den Erisipheen. Jahrb. Wissenschaft. Bot. **72**, 106–132 (1930)

[29] N. Nedeltscheff, Perfect stage of oidium tuckeri in Bulgarie. Ann. Univer. Sofia V. Faculte Agron. **II**, 281–291 (1924)

[30] R. Geoffrion, Formation de peritheces de l'oidium de la vigne dans la vallee de la Loire. Phytoma mai **32** (1960)

[31] W.C. Schnathorst, Environmental relationship in the powdery mildew. Ann. Rev. phytopath. **3**, 343–366 (1965)

[32] D.M. Gadoury, R.C. Pearson, Germination of ascospores and infection of Vitis by *Uncinula necator*. Phytopathology **78**, 1538 (1988)

[33] D.M. Gadoury, R.C. Pearson, Initiation, development, dispersal and survival of cleistothecia of *Uncinula necator* in New York vineyards, Phytopathology **78**, 1413–1421 (1988)

[34] P. Halleen, G. Holz, Cleistothecia and flag shoots: sources of primary inoculum for grape powdery mildew in the Western Cape province, South Africa. S. Afr. J. Enol. Vitic. **21**, 66–70 (2000)

[35] L.M. Gee, B.E. Stummer, D.M. Gadoury, L.T. Biggins, E.S. Scott, Maturation of cleistothecia of *Uncinula necator* (powdery mildew) and release of ascospores in Southern Australia. Australian journal of grape and wine research **5**, 13–20 (2000)

[36] M. Miazzi, H. Hajjeh, F. Faretra, Observations on the population biology of the grape powdery mildew fungus *Uncinula necator*. Journal of plant pathology **85**(2), 123–129 (2003)

[37] M. Miazzi, H. Hajjeh, F. Faretra, Occurrence and distribution of two distinct genetic groups in populations of *Erysiphe necator* Schw. in Southern Italy. Journal of plant pathology **90**(3), 563–573 (2008)

[38] D.M. Gadoury, L. Cadle-Dawidson, W.F. Wilcox, I.B. Dry, C.R. Seem, G.M. Milgroom, Grapevine powdery mildew (*Erysiphe necator*): a fascinating system for the study of the biology, ecology and epidemiology of an obligate biotroph. Molecular Plant Pathology **13**(1), 1–16 (2012)

[39] N. Nedeltscheff, The perfect stage of *Oidium tickeri* in Bulgaria, Ann. Reports, Agriculture faculty, Sofia **2** (1923/24)

[40] D. Atanasoff, *Diseases on agriculture plants*, Univ. library, Sofia, 137 (1934)

[41] Al. Hristoff, *Special plant pathology*, Zemizdat, Sofia (1956)

[42] E. Raykov, Sv. Nenov, Scientific works of CNIK on viticulture in Ploven, **T II** (1959)

[43] S.I. Martinov, Comparative experiment with some chemicals for powdrey mildew control, (1938)

[44] S.I. Martinov, *Control of the powdery mildew on grapes*, IPP Sofia (1940)

[45] N. Nedelchev, M. Kondarev, *Pest and diseases and disorders in grapes*. Zemizdat, Sofia (1949)

[46] B. Nakov, R. Raychev, P. Kotezov, Viticulture and enology **2**, 26–29 (1960)

[47] I. Iliev, Plant protection **11**, 24–28 (1969)

[48] G. Valkov, *Thiesis of doctors work*, Sofia (1970)

[49] H.G. Hewitt, *Fungicides in crop protection*. University press Cambridge, UK (1998)

[50] M. Nakova, B. Nakov, Kr. Dimitrov, K. Sakakusheva, Viticulture and enology **2**, 42–45 (2010)

[51] R.C. Pearson, D.M. Gadoury, Cleistothecia, the source of primary inoculum for grape powdery mildew in New York. Phytopathology **77**, 1509–1514 (1987)

 DOI: 10.1051/bioconf/20170901022

Teledetección no es Tele-adivinación: La importancia de las causas en Viticultura de Precisión

Teledetection is not Tele-divination: The importance of the causes in Precision Viticulture

Vicente D. Gómez-Miguel[1,a], Vicente Sotés[1], y María L. González-SanJosé[2]

[1] Departamento de Producción Agraria, Universidad Politécnica de Madrid (UPM), España
[2] Department of Biotechnology and Food, Science, Universidad de Burgos (UBU), España

Resumen. La cartografía realizada a partir de la información procedente de sensores remotos es un producto de actualidad que se ofrece con éxito gracias principalmente a su atractiva presentación y a su bajo coste. La información se obtiene mediante sensores instalados en satélites, aviones y otras aeronaves a alta, media o baja altura, tripuladas o no y en vehículos terrestres autopropulsados o no. Los sensores y elementos asociados son capaces de procesar la radiación de gran parte del espectro electromagnético y convertirla en imágenes que representan la emisión de una determinada zona geográfica. A pesar de que el procedimiento puede llegar a ser muy sofisticado, el resultado que se proporciona al viticultor con mayor o menor detalle (resolución) o más o menos elaborado, es prácticamente siempre el mismo: uno o varios mapas en los que se recoge la respuesta obtenida por el sensor de lo que hay en la parcela en el preciso momento en el que pasa el vehículo en el que se ha instalado. Por lo tanto, este mapa representa la respuesta del estado "electromagnético" de una parcela en un momento determinado, sin embargo, las causas de la variabilidad que se aprecia en el propio mapa se desconocen. Estas causas son muy complejas y se relacionan principalmente con la variabilidad de los elementos del clima, la geología, el relieve, el suelo, la planta y el momento de realización del proceso. Factores todos ellos de gran importancia en la configuración del terroir y que influyen en la calidad del producto. En este trabajo tratamos del alcance real de la teledetección y de la importancia de la cartografía de las causas que inciden en la heterogeneidad de la parcela/explotación para un uso adecuado en Viticultura de Precisión, todo ello dentro de un modelo general de producción vitícola para la obtención de vinos de calidad.

Abstract. The cartography that has been carried out from the information coming from remote sensing is a current issue that is successfully offered mainly due to its attractive presentation and to its low cost. The information is obtained by sensors installed in satellite in/on satellites, planes and other aircraft at high, medium or low altitude – manned or unmanned- and in land vehicles – self-propelled or non-propelled. The sensors and associated elements are able to process the radiation of a great deal of the electromagnetic spectrum and transform it into images representing the emission of a certain geographical area. In spite of the procedure can be really sophisticated, the result given to the winegrower, with more or less detail (resolution) or more or less elaborated, is practically always the same: one or various maps with the answer, obtained by the sensor, of the elements in the field in the precise moment that the vehicle, where it is installed, passes. Therefore, this map represents the response of the "electromagnetic" state of a certain moment however the causes of the variability observed in the same map are unknown. These causes are very complex and they are in relationship mainly with the variability of the elements of climate, geology, relief, soil, plant and the moment of the process execution. All these factors are of a great importance in the configuration of the terroir and they influence the product quality. In this work, we deal with the real range of teledetectión (remote sensing) and the importance of the cartography of the causes that affect the heterogeneity of the field / exploitation in order to get a suitable use for the Precision Viticulture. The whole of it is inside a general model of vine-growing production to obtain quality wines.

1. Introducción

La cartografía realizada a partir de la información procedente de sensores remotos *(remote sensing)* es un producto de actualidad que se ofrece con éxito a los viticultores gracias principalmente a su atractiva presentación y a su bajo coste. La oferta disponible en el mercado promete al usuario un número importante de elementos cartografiados mediante teledetección que son simplemente imposibles de obtener: teledetección no es tele-adivinación.

La información se obtiene mediante sensores instalados en satélites, aviones y otras aeronaves a alta, media o baja altura, tripuladas o no y en vehículos terrestres

[a] e-mail: `vicente.gomez@upm.es`

DOI: 10.1051/bioconf/20170901022

autopropulsados o no. Los sensores y elementos asociados son capaces de procesar la radiación de gran parte del espectro electromagnético y convertirla en imágenes que representan la emisión de una determinada zona geográfica. Las más usuales se corresponden con el espectro visible, el infrarrojo y el radar.

A pesar de que el procedimiento puede llegar a ser muy sofisticado, el resultado que se proporciona al viticultor con mayor o menor detalle (resolución) o más o menos elaborado, es prácticamente siempre el mismo: uno o varios mapas en los que se recoge la respuesta obtenida por el sensor de los elementos que hay en la parcela sólo en el preciso momento en el que pasa el vehículo en el que este se ha instalado. Por lo tanto, este mapa representa la respuesta del estado "electromagnético" de una parcela en un momento determinado, sin embargo, las causas de la variabilidad que se aprecia en el propio mapa se desconocen.

En este trabajo tratamos del alcance real de la teledetección y de la importancia de la cartografía de las causas que inciden en la heterogeneidad de la parcela/explotación para un uso adecuado, tanto en Manejo Convencional o Tradicional (MC) como en Viticultura de Precisión (VP), todo ello dentro de un modelo general de producción vitícola para la obtención de vinos de calidad.

2. Causas de la Variabilidad

Las causas de la variabilidad espacial de los viñedos son muy complejas y se relacionan principalmente con la de los factores naturales como el clima (insolación, temperatura, precipitación,...), la geología (litología, mineralogía,...), el relieve (altitud, pendiente, orientación, exposición...) y el suelo (propiedades físicas, químicas, biológicas, mineralógicas,...); de los factores biológicos (patrón, cultivar, biodiversidad intrínseca, estado fisiológico, plagas, enfermedades,...); y de los factores agronómicos (preparación, protección y laboreo del suelo, fertilización, conducción, sanidad, riego...). Estos factores pueden ser modificados por el hombre mediante lo que se describe frecuentemente con los términos gestión o manejo. Factores todos ellos de gran importancia en la configuración del terroir ambiental y que influyen en la calidad del producto [1].

Estos factores, los procesos enológicos, los elementos socioeconómicos y los relacionados con la presentación y el marketing, e incluso, la tradición son los constituyentes del terroir y la metodología de su estudio constituye la Zonificación de los Terroir (*Terroir Zoning*), o mejor, la Zonificación Integrada del Terroir (ZIT) [2].

A pesar de su importancia, a la hora de establecer el terroir como unidad de análisis (y de manejo) nos encontramos con una barrera que la mayoría de las veces resulta insalvable: la falta de datos enológicos, de mercado y socioeconómicos georreferenciados, es decir, que puedan ser plasmados en un mapa, en definitiva, que sean cartografiables. Este es el motivo principal por el que se utiliza preferentemente el primer grupo de factores para la elaboración del mapa de las causas: el Mapa de Variabilidad del Terroir (MVT) o sus sinonimias: mapa de variabilidad espacial del viñedo, mapa de los terroir o mapa de zonificación del terroir [3].

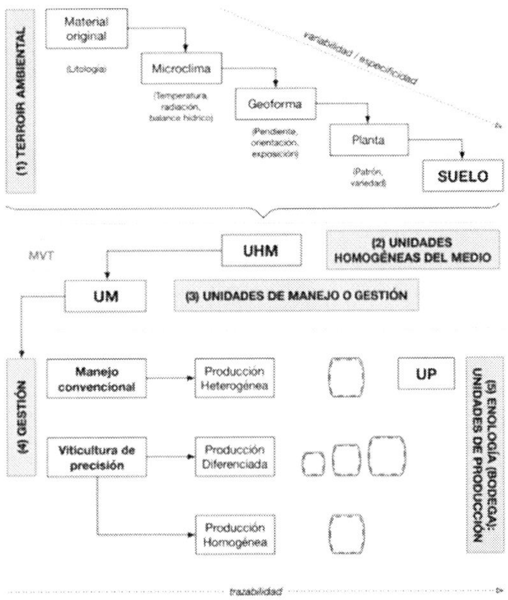

Figura 1. Esquema del modelo de producción vitícola para la obtención de vinos de calidad (para explicación, ver texto).

3. Importancia de la causas en el sistema de gestión del viñedo

En la Fig. 1 se incluye un esquema del modelo de producción vitícola para la obtención de vinos de calidad.

El MVT se sitúa como una necesidad entre el Terroir (ambiental) y la gestión del viñedo y está constituido por el mapa de las Unidades Homogéneas del Medio (en adelante, UHM) y el de las Unidades de Manejo o Gestión (en adelante, UM).

El modelo general de producción vitícola se realiza en cinco etapas que comienzan con la identificación, descripción, caracterización y cartografía de los elementos que constituyen el terroir ambiental (1), todos en una escala de detalle comparable. A continuación se crean las UHM (2) y a partir de ellas, se diseñan las UM (3) que permiten la toma de decisiones sobre el modelo de gestión (4) ya sea como MC o como VP. Finalmente, la relación con las Unidades de Producción (por ejemplo: uno o varios depósitos) localizadas en bodega se realiza a través de estas UM y en función del modelo de gestión previamente definido (5). Las tres fortalezas del sistema son el manejo diferencial (cada terroir se maneja de forma particularizada), el aprovechamiento de la especificidad (cada terroir produce un vino determinado) y la trazabilidad (cada terroir se relaciona con su producto de procedencia).

4. Cartografía de las causas: el mapa de Variabilidad del terroir

La variabilidad del medio que soporta la viña (variabilidad espacial del viñedo o del terroir) y las dificultades que esta conlleva para una gestión adecuada están en el origen de la heterogeneidad del producto (uva) y en definitiva de la calidad del vino.

¿Es posible medir, caracterizar, cuantificar y cartografiar la variabilidad espacial del viñedo? ¿Existen técnicas disponibles para ello? ¿Cuáles son esas técnicas?

Un MVT responde a estas cuestiones y representa la caracterización, cuantificación y cartografía de los sectores de la parcela que son similares y su separación de aquellos que son diferentes, lo que permite un tratamiento deferencial.

¿Por qué es importante el tratamiento diferencial de cada sector de cada parcela/explotación? Lo es, tanto si optamos por un objetivo simple o anual como por un objetivo ambicioso o a medio plazo o permanente.

Nos referimos a un objetivo simple cuando se plantea cada año y está relacionado con una propiedad concreta de un determinado factor o un elemento concreto y previamente detectado y seleccionado y sin interferencias de los demás que puede ser cartografiado utilizando un tipo de recursos adecuado. En Viticultura este tipo de objetivos suele estar relacionado con la mejora anual del producto (en cuanto a calidad y/o producción) y los elementos de que depende y que se pretenden corregir frecuentemente se detectan indirectamente y son, por ejemplo, carencias y toxicidades del cultivo, plagas y enfermedades, vigor, estrés hídrico, exceso o falta de humedad, factores de calidad puntual, maduración y vendimia, clorosis, salinidad, etc. En definitiva, se incide en la detección de la diferenciación del sector (por ejemplo, cambio en el vigor) no de sus causas.

Sin embargo, en la Viticultura moderna es frecuente el planteamiento de un objetivo ambicioso que perdure en el tiempo, a futuro, para lo que es necesario conseguir la optimización y homogeneización estable del producto. Para alcanzar este objetivo es imprescindible el MVT y un sistema de gestión que, como la VP, permita actuar de forma independiente en cada propiedad de cada factor y en cada sector de cada parcela.

Se trata por lo tanto de la identificación de las causas de la variabilidad espacial intraparcela a través de las interrelaciones entre las diferentes propiedades plasmadas en su cartografía para su utilización en VP como sistema de gestión o manejo de la explotación. Dicho de otra manera, se pretende solucionar problemas inmediatos dependientes de propiedades específicas e interrelacionadas y hacer a corto o medio plazo (futuro), mediante la gestión llevada a cabo con VP, que los sectores de la parcela sean lo más homogéneos posible y que de esta manera también el producto lo sea.

El MVT permite y facilita determinadas actuaciones que se llevan a cabo en la gestión del viñedo y que no se podrían realizar si este no existiera (Tabla 1). Estas actuaciones están relacionadas, como no podía ser menos, con la VP, sin embargo no hemos de olvidar que el mapa permite y facilita y la VP ejecuta y gestiona.

No obstante, debe hacerse hincapié en que el mapa de variabilidad espacial del viñedo presenta ciertas limitaciones y para potenciar sus ventajas, es necesario comprender la naturaleza y las interrelaciones entre las propiedades de los diferentes factores que explican la variabilidad de cada sector y es muy conveniente vincular estos factores con las propiedades y la calidad de la uva y el vino. Obviamente las características de cada sector son propias y específicas de él y no deben extrapolarse a otras situaciones por más próximas o semejantes que parezcan.

Tabla 1. Actuaciones (gestión o manejo) que permite y facilita el Mapa de Variabilidad del Terroir (MVT).

Elección del sitio	Microzonificación, MVT, UHM, UM
Elección del Sistema de Gestión	MC; Manejo Mixto; VP
Material vegetal	Variedades, patrones, clones
Conservación de suelos y aguas	Elección de medidas de prevención de la erosión y de conservación de suelos y aguas y detectar el lugar de su aplicación
Preparación del suelo	Elección adecuada de las labores previas a la plantación y detección del lugar de actuación: desfonde, subsolado, labor profunda, topo, volteo, *chisel*, vertedera,...
Diseño de riego	Optimización y localización de la alimentación hídrica y la gestión del agua: tipo/forma; diseño; dosis, caudal, frecuencia; Sectores
Pre-plantación	Enmiendas orgánicas y minerales; Fertilización/abonado de fondo
Material vegetal y diseño de la formación	Selección y localización de portainjertos y variedades; Gestión diferenciada del dosel vegetal y de la poda en los lugares precisos;
Plantación	Optimización del diseño de la plantación: geometría; sistema, tamaño de unidades/módulo; tipo/modo; UM
Mantenimiento del suelo	Optimización y localización del manejo del suelo
Cubiertas	Diseño y manejo de cubiertas específicas para sector: inertes o vegetales; ciclo vegetativo; manejo
Nutrición mineral	Localización de la aplicación de los diferentes insumos en función de las necesidades reales de cada sector: enmiendas; fertilización, fertirrigación; mantenimiento/restitución
Tratamientos	Tratamiento diferencial y localización de plagas y enfermedades
Maduración	Seguimiento del ciclo y de la maduración específico de cada sector
Vendimia	Gestión de la vendimia y elección del momento adecuado en cada sector;
SIG/*GIS*	Seguimiento de la plantación: alimentación hídrica, mineral, vigor, estado sanitario, plagas Investigación (diseños experimentales)
	Control temporal (anual, bianual,...) de los parámetros de variabilidad
Trazabilidad	Seguimiento del producto: UM – UP – Mercado

5. Realización del MVT

El MVT es la respuesta a la pregunta ¿dónde y cómo efectuar las diferentes actuaciones (manejo)? y

 DOI: 10.1051/bioconf/20170901022

las unidades que lo conforman, bien definidas, bien caracterizadas y bien cartografiadas (UHM), son la base de las UM que permiten el diseño de una viticultura de tal detalle que sea consecuente con las modernas ideas de trazabilidad suelo-planta-uva-bodega-mercado.

La calidad del MVT se corresponde con la de las fuentes de información y con las técnicas cartográficas utilizadas para su realización, está en la precisión y exactitud requerida por una escala adecuadamente elegida de acuerdo con los objetivos, en la densidad de observaciones y en las técnicas de muestreo del suelo seleccionados y en la importancia y el número de propiedades, variables o parámetros considerados en las descripciones y los análisis y finalmente en un lenguaje internacionalmente admitido que facilite la transferencia de tecnología.

En los párrafos siguientes se describen brevemente los diferentes procesos manejados en la realización del MVT, en una determinada parcela, imprescindible para el diseño de las UM con la siguiente secuencia: objetivos y escala, metodología y procedimiento, muestreo, información y lenguaje y finalmente, interpretación de resultados.

5.1. Objetivos y escala

La escala del mapa es importante porque está relacionada con la cantidad y la calidad de la información proporcionada al viticultor. En un mundo en el que las herramientas de trabajo que, como los Sistemas de Información Geográfica (SIG), facilitan enormemente el manejo y la presentación de resultados (ampliaciones y reducciones, diseño, colores…) es necesario utilizar criterios de calidad del mapa fijados con propiedades relacionadas con la escala, tales como el objetivo y el muestreo.

En este sentido, si el promotor del trabajo es la Administración o un organismo privado y los objetivos se relacionan con la planificación y la ordenación del cultivo de la viña en una zona geográfica relativamente amplia, la escala adecuada sería 1:25.000, y en casos especiales podría ser inferior o igual a 1:50.000. Sin embargo, para viticultores y empresas vitícolas cuyos objetivos se relacionan con la gestión y el manejo de la parcela que condiciona actuaciones directas sobre ella (Tabla 1), la escala necesaria sería al menos 1:5.000 y mejor aún, para casos específicos, sería más recomendable trabajar con escalas superiores, por ejemplo, 1:2.500 [4].

5.2. Metodología para la identificación y separación de sectores

El problema principal que se plantea es la diferenciación de sectores y su delineación, la obtención de polígonos o recintos, principalmente porque sobre el dibujo resultante se hace el diseño del muestreo (¿dónde y cómo se hacen las observaciones de campo?) y, a partir de él, el MVT. El objetivo principal consiste en que los sectores iguales deben aparecer en el mapa como iguales y los diferentes como diferentes: parece obvio, pero eso es un mapa. Este conjunto de polígonos en el que todavía no hemos identificado o definido, ni suelos, ni terroir, ni UHM, pero si los hemos identificado, lo denominamos "mapa de sectores".

Para conseguir este mapa de sectores se emplean tres tipos de técnicas específicas: la fotointerpretación aérea **(FIA)**, la percepción remota (*remote sensing*) o **teledetección** y el método directo (**MD**).

La FIA consiste en la utilización de la información que procede principalmente de los fotogramas aéreos convencionales para visión estereoscópica. La visibilidad de diferentes elementos en los fotogramas aéreos con visión estereoscópica (básicos como la pendiente..., compuestos como el patrón de drenaje, ... o deducidos como las formas de paisaje, ...) es de gran importancia para la definición de la variabilidad horizontal del suelo, y su tradicional uso en Edafología [5–7] es debido también a que facilita su delineación y por este motivo la FIA es un instrumento metodológico muy efectivo para la realización del mapa de sectores, más aún a escalas medias que a escalas grandes. Sin embargo, fotointerpretación no es foto-adivinación.

La Teledetección es el conjunto de metodologías que permiten el tratamiento de imágenes obtenidas mediante sensores remotos transportados por cualquier medio: satélites, aeronaves o aviones de alta, media o baja altitud, tripulados o no, automóviles o no, … La información de objetos o fenómenos (el medio natural) se trata a partir de imágenes espectrales de alta resolución (IEAR) obtenidas por tales sensores. La resolución es una medida de la habilidad que posee un sistema sensor para distinguir entre respuestas que son semejantes espectralmente o próximas espacialmente. Existen varios tipos de resolución (espacial, espectral, radiométrica, temporal) que permiten la selección de imágenes adecuadas para un objetivo determinado. Hoy día este método de diferenciación de sectores adquiere gran importancia principalmente por su fácil acceso, bajo coste y porque también aumenta la precisión de los límites entre sectores de diferenciación a escalas grandes [8].

La cartografía obtenida por este método recoge la respuesta obtenida por el sensor de los elementos que hay en la parcela sólo en el preciso momento en el que pasa el vehículo en el que este se ha instalado. Por lo tanto, este mapa representa la respuesta del estado "electromagnético" de una parcela en un momento determinado sin entrar en las causas de su variabilidad. Por este motivo los mapas resultantes carecen de estabilidad y son diferentes no sólo en función de las propiedades del terroir, sino también, -y esta es la diferencia esencial- de las propiedades del medio (elementos meteorológicos, estado del suelo, fenología y estado fisiológico de la planta, etc) en el momento de realización del proceso y hasta las diferentes propiedades de la tecnología disponible.

Por lo tanto, también se puede afirmar que teledetección no es tele-adivinación y en el estado actual de nuestros conocimientos no es posible cartografiar el conjunto de las propiedades permanentes del medio y de las causas de su variabilidad únicamente mediante Teledetección. En particular, el suelo no se ve y por ello es imprescindible el uso de otras técnicas, especialmente en los mapas de escalas grandes y muy grandes.

En tales mapas la utilización del MD adquiere mayor relevancia. El MD obtiene la información mediante técnicas cartográficas de muestreo que se aplican frecuentemente en disciplinas relacionadas con el medio natural como la Climatología, la Geología, la Geomorfología, la Edafología [4].

Este método se optimiza con el mejor conocimiento intrínseco de la parcela (diferencias en la pendiente y/o en la exposición; existencia de variaciones estructurales como terrazas o bancales, nivelaciones, desagües o drenajes; variabilidad de la vegetación natural y/o de los cultivos, su crecimiento y su vigor; humedad relativa e hidromorfía; salinidad u otros accidentes negativos; respuesta diferencial al manejo y/o paso de maquinaria; etc), pero con técnicas específicas de muestreo, como la malla fija *(fixed grid)*, no es imprescindible.

En conclusión, la variabilidad sólo puede ser considerada correctamente a partir de una cartografía suficientemente detallada en la que aparezcan delineadas todas y cada una de las unidades, agrupadas las que son iguales y separadas las que son diferentes a causa de alguna propiedad de cualquiera de los factores. Por estos motivos parece lo más sensato, como se ha dicho más arriba, utilizar todo el conjunto de las metodologías disponibles, concretamente el MD, la FIA y la teledetección preferentemente a partir de IEAR, de forma coordinada y complementaria y aprovechando las ventajas de todas estas técnicas para la elaboración adecuada del mapa de sectores, base de los MVT o mapas de UHM con la finalidad última de que a corto o medio plazo (futuro), mediante la gestión llevada a cabo con VP, los sectores de la parcela tengan la máxima homogeneidad posible y así también el producto la tenga.

5.3. Diseño y realización del muestreo

Una vez que se dispone del dibujo en el que aparece cada recinto con su contenido diferenciador (el mapa de sectores) se procede al diseño del muestreo cuyo punto de partida es la determinación del número de observaciones que hemos de realizar. Generalmente el cálculo inicial se efectúa por unidad de superficie: densidad de observaciones.

La densidad de observaciones relacionada con el mapa de suelos está condicionada por la escala elegida y, junto con el número de determinaciones que se han de realizar en el laboratorio de acuerdo con los objetivos previstos, condicionan su calidad y su coste [4].

El objetivo del muestreo es identificar, calificar, caracterizar y determinar la distribución de todos los tipos de terroir. El componente del terroir de onda más corta, de mayor variación, y que por lo tanto influye más en la variabilidad del terroir, es sin duda el suelo. Por este motivo el objetivo preferente es identificar, calificar, caracterizar y determinar la distribución de todos los tipos de suelo y para ello se realizan dos tipos de observaciones: las calicatas y los sondeos.

Las calicatas se utilizan para identificar, calificar y caracterizar los suelos. La calicata se realiza de forma manual o mecánica, mediante una retroexcavadora, y con unas dimensiones aproximadas de 2 m de largo, 0.4 m de ancho y 2 m de profundidad. En la calicata se aprecia el perfil del suelo y constituye la unidad de muestreo.

Los sondeos se emplean para definir correctamente la distribución de los suelos, corrigiendo sus límites y los errores de identificación. El sondeo se realiza con una barrena o sonda, manual o mecánica y en él no se aprecia el perfil del suelo completamente y, debido a su finalidad, la mayoría de las veces es suficiente con describirlo para su correcta identificación con algún tipo de suelo ya caracterizado en una calicata y sólo en ocasiones particulares es necesario el muestreo y el análisis.

La densidad de observaciones es definida por diferentes metodologías [4]. La resolución de la OIV [9] recomienda para los estudios que nos ocupan una densidad igual a una observación por cada cm^2 de mapa lo que significa 4 observaciones/Ha para mapas a escala 1:5.000 y 16 observaciones/Ha para los de escala 1:2.500. En general, el número de calicatas puede suponer un 25/35% del número de observaciones y el de sondeos el resto.

Una vez determinado el número de observaciones se procede a su localización en el mapa de sectores por alguno de los métodos de muestreo disponibles. Estos métodos de muestreo pretenden una distribución de las observaciones previstas de manera que cumplan sus objetivos de la forma menos costosa posible. Hay muestreos en malla fija (*fixed grid*), líneas de muestreo (*simple line*), bandas de muestreo (*simple strip*), sitios o puntos de muestreo (*simple spot*) y áreas de muestreo (*simple área*). Lo más frecuente es que se utilice una combinación de varios de estos métodos en función de las peculiaridades de la zona y el objetivo del estudio.

Para los estudios detallados y muy detallados a los que nos referimos, y que pretenden caracterizar cada sector y detectar la variabilidad dentro de él (intersector), se suele utilizar el muestreo en malla fija (*fixed grid*) o sus variantes dirigidas. En la malla fija las observaciones se sitúan a lo largo de líneas imaginarias, paralelas entre sí, con igual (cuadrado) o diferente (rectángulo) separación que la existente entre las líneas. El error en los límites entre sectores (precisión) queda definido por la mitad de la distancia entre líneas y de la distancia a lo largo de ellas. El muestreo en malla fija es de coste variable (en función de la amplitud de la malla), pero ciertamente costoso, aunque, sin duda, no caro y si el más favorable.

En situaciones, en las que la disponibilidad de recursos es limitada, se puede optar por un muestreo dirigido a sectores definidos por la intersección de los mapas FIA y de teledetección con la malla fija de lado tan amplio como lo permita el número de observaciones calculado. La malla puede ser variable de forma que sea más densa en sectores en los que el conocimiento de la parcela permita inferir que son más heterogéneos y más amplia en aquellos considerados más homogéneos.

Con demasiada frecuencia se encuentran manuales, y tutoriales en la red, que explican dónde y cómo tomar la muestra que incluso califican de "representativa". Las indicaciones sobre la localización se suelen referir al lugar de forma ambigua (bajo o alto, ladera o llanura, encharcable o no, pedregoso o no, etc) y son, como poco, confusas, sólo consideran la ¿variabilidad? de una de las características del sector y, sobre todo, no permiten delimitarlo. De la misma manera, cuando describen la toma de muestras, se refieren frecuentemente a una muestra superficial que además procede de la mezcla de varias muestras obtenidas siguiendo un itinerario más o menos tortuoso. Como parece obvio, después de las consideraciones realizadas en los párrafos anteriores, esto es, cuando menos, incorrecto: las plantas no toman un determinado elemento por su valor medio, si en un sector es deficiente pueden presentar carencias y si es excesivo, toxicidades.

5.4. Descripción de las observaciones

Una vez localizadas las observaciones en el mapa de sectores se procede a su apertura, descripción y muestreo en el campo.

Como ya se ha dicho el suelo es tridimensional y se estudia a partir de su perfil, que está constituido por una o varias capas llamadas horizontes. El conjunto de los horizontes A y B se denomina *solum* y es un buen índice para valorar la profundidad efectiva –en la que se desarrolla la mayor parte del sistema radicular de la planta-, de ahí la importancia en su alimentación hídrica y nutricional.

En el perfil es imprescindible describir todos los horizontes, muestrear cada uno de ellos y analizar cada muestra independientemente (obviamente sin mezclarlas). Esto es importante porque cada horizonte tiene sus propiedades y no es correcto referirse, por ejemplo, a un suelo arcilloso o arenoso, sino a un horizonte arcilloso o arenoso de un suelo determinado; por ejemplo, en los viñedos es posible que el horizonte A sea francoarenoso, el B arcilloso y el C francolimoso ¿cuál es el adjetivo que debemos aplicar al suelo? Esto incrementa sensiblemente el número de muestras que se han de realizar.

Una vez abierta la calicata (o el sondeo) en determinado sector de la parcela, se procede a su estudio. Este estudio comienza con la apreciación y descripción del entorno exterior (Litología, Geomorfología, Vegetación natural y usos) y continúa con la visualización del frente soleado del interior. En esta última fase se comienza con la diferenciación de los horizontes y la descripción de sus propiedades (Tabla 2) y termina con el muestreo de cada uno de ellos: tantas muestras como horizontes (sin mezclarlas) envasadas y convenientemente etiquetadas para su envío al laboratorio donde se realizarán las determinaciones necesarias (& 5.5).

Las propiedades descritas en el campo [12] y las determinaciones realizadas en el laboratorio [13] son las que aportan la información precisa, disponible para el viticultor y, por lo tanto, también establecen la calidad del estudio, por supuesto, su número afecta al presupuesto.

Con demasiada frecuencia determinadas consultoras y laboratorios basan sus servicios de asesoría al viticultor en análisis realizados a una determinada muestra de suelo y en el mejor de los casos añaden otra del "subsuelo", y obvian todas las propiedades específicas del perfil y las morfológicas de sus horizontes; así se pierde la mayor parte de la información. Lo correcto es disponer de una descripción completa realizada a partir de una calicata.

Las propiedades morfológicas y físicas del suelo permiten poner un especial interés en los problemas e impedimentos naturales que se plantean en la elección del sitio y en la planificación de las actuaciones (manejo) tanto previas a la plantación, como durante su desarrollo, lo que permite disminuir, o en su caso eliminar, las dificultades para el desarrollo radicular de la planta y su alimentación hídrica. El ejemplo de la densidad aparente (Da) de cada horizonte es paradigmático. Es imprescindible disponer del valor de la Da en condiciones de campo porque permite ajustar y valorar la importancia de la profundidad efectiva y el peso del suelo natural y ahorrar muchos recursos (frecuentemente más del 60%) relacionados con la fertilización, el agua de riego, etc.

Tabla 2. Propiedades morfológicas a caracterizar para el Mapa de Variabilidad del Terroir (MVT).

PROPIEDADES	FINALIDAD Y UTILIZACIÓN
Clasificación/Serie[1]	Comunicación y transferencia
Clasificación/Horizontes[1]	de tecnología
Colores y moteados[2]	Detección de hidromorfia, encharcamientos y calidad
Concreciones y costras[3]	Limitaciones, manejo
Compacidad/porosidad[3]	Limitaciones, manejo
Consistencia[3]	Laboreo, manejo
Densidad aparente[3,4]	En campo: relaciones volumen-suelo; peso del suelo
Drenaje/desagüe[3]	Limitaciones, manejo y calidad
Elementos gruesos[3]	Laboreo, *mulching* y calidad
Escorrentía[3,4]	Erosión y balance de humedad
Espesor de horizonte [3]	Vigor, fertilidad y calidad
Estructura[3]	Limitaciones y calidad
Limitaciones y panes[3]	Limitaciones y manejo
Límites entre horizontes[3]	Limitaciones, manejo y calidad
Pedregosidad[3]	Limitaciones, manejo
Perfil cultural[5]	Detección de posibles obstáculos para las raíces; estimación del peso-volumen útil
Permeabilidad, CH; infiltración[3]	En campo: relaciones suelo agua, riego
Profundidad (efectiva)[3]	>80/85 cm. En el campo: volumen de suelo útil explorado por las raíces; peso del suelo (Muy importante)
Raíces[3]	Pe, limitaciones, manejo
Rocosidad[3]	Limitaciones, manejo
[1][10]; [2][11]; [3][12]; [4][13]; [5][14, 15]	

5.5. Análisis de las muestras: elección y definición de las determinaciones analíticas

Es frecuente que los responsables de la gestión del viñedo pregunten por la realización al "suelo" de un "análisis completo". Ya se ha señalado que las determinaciones se realizan a cada muestra de cada horizonte y no al "suelo". Cuando esto queda aclarado, la respuesta fácil es dar una referencia en la que figuren la totalidad de las determinaciones que se pueden realizar a una muestra –ver por ejemplo, [13]-, sin embargo esto encarece mucho cualquier estudio y, desde luego, no es lo recomendable.

¿Qué análisis (determinaciones) hay que hacer? Optimizar la respuesta es complicado. Las determinaciones analíticas que se han de realizar a una muestra dependen de muchos factores, unos relacionados con el medio y con la heterogeneidad del sector, otros con la escala, otros con el objetivo y la finalidad del estudio y otros con el tipo de suelo y las características que hemos observado previamente en la descripción morfológica del perfil y en concreto de la propia muestra.

Ni a todas las muestras se le realizan las mismas determinaciones, ni todas se analizan completamente ¿para qué se necesitan, por ejemplo, los datos de materia

Tabla 3. Propiedades físicas, químicas, físico-químicas y mineralógicas a determinar [13] para el Mapa de Variabilidad del Terroir (MVT).

PROPIEDADES		FINALIDAD Y UTILIZACIÓN
Conductividad Eléctrica, CE		Salinidad, valoración inicial del riesgo
pH (1:2,5)		Índice del estado general del suelo
Elementos gruesos		Granulometría de la fracción mineral, propiedades hídricas, minerales…, límite abrupto, contrastes, etc.
Textura USDA		
Materia orgánica, MO		Contenido en MO del suelo
N		Contenido en nitrógeno total del suelo
Relación C/N		Evolución de la materia orgánica
P		Fósforo
N/P		Antagonismo entre el N y el P
CIC		Capacidad de Intercambio Catiónico: Fertilidad potencial
Bases de cambio		Suma de bases, SB; Ca, Mg, Na y K
Acidez de cambio; Al		Suelos ácidos, toxicidad del Al
Relaciones en el Complejo de Cambio	V; PCI,	% de saturación: V=100*SB/CIC Fertilidad actual.
	PMI; PKI;	% de Ca, Mg y K sobre la CIC
	PSI; Ca/Mg;	% Na sobre la CIC; Alcalinidad
	K/Mg	Antagonismos
Caliza total y activa, CA; Fe		Ca, agua, clorosis, patrones; Índice de Poder Clorosante $(IPC= CA/10^4_{*}Fe^2)$
Extracto de Saturación CE>0,2dS/m	Aniones Cationes	Salinidad y tipología de sales. Problemas limitantes y riesgos de salinización del suelo,
	CEe; pHe	de alcalinización (SAR) y de toxicidad de la planta, Cálculo de lavado
	Hs	Humedad de saturación, porosidad
Oligoelementos		Cu, Zn, Mn, B, Fe, Pb,…
Constantes de Humedad		pF a 33 kpa ya 1500 kpa; régimen de humedad y riego
Análisis especiales		Mineralogía de arenas y limos; correlación

orgánica de un horizonte situado a más de 180 cm de profundidad? La casuística es muy grande y es difícil dar recomendaciones generales, pero la mayoría de las veces es suficiente pensar que cada cosa que se haga ha de servir para algo y que es innecesario utilizar recursos inútilmente.

En la Tabla 3 se incluye una relación de las determinaciones analíticas a realizar en el laboratorio con la finalidad de caracterizar correctamente el suelo dentro de la perspectiva del MVT e incluye algunas ideas generales sobre su utilización para la producción de vinos de calidad.

5.6. ¿Qué actuaciones son necesarias? Importancia del lenguaje e Interpretación de resultados

Una vez definido el mapa de sectores, realizado el muestreo y analizadas las muestras, se ha de llevar a cabo una operación importante que se denomina "correlación" y que pretende dar contenido a este mapa en términos de tipo de suelo de forma que podamos crear las UHM.

Para realizar esta correlación es necesario emplear un lenguaje que sea capaz de diferenciar todos los suelos de cada sector en función de las propiedades cuya variabilidad afecte al cultivo de referencia, a la vid. Es posible que esto pueda hacerse de forma personal, creativa y original por parte de cada operador pero hay dos aspectos importantes que hacen que esto no sea ni correcto, ni adecuado, ni útil: el lenguaje debe ser completo y generalizado.

Completo significa que estén previstos todos los suelos, generalizado que lo use la mayor parte de los técnicos (y científicos) para hacer posible la transferencia de tecnología, es decir, el aprovechamiento de la investigación y las experiencias realizadas en suelos de igual nombre, iguales.

Estas dos condiciones sólo las cumple una sistemática, la *Soil Taxonomy* [4,10,16] que en su categoría más baja (la Serie) considera más de 23.000 tipos de suelo. Otras clasificaciones o leyendas, como la de FAO [17] o las categorías altas de la propia *Soil Taxonomy*, están pensadas para escalas pequeñas o muy pequeñas [18] y no sirven en absoluto a nuestro cometido. Se podrá decir que ni el viticultor, ni el gestor de la explotación vitivinícola conocen este lenguaje, y es posible que sea cierto, tampoco el ganadero ha de conocer necesariamente el lenguaje del veterinario, incluso a veces puede ser hasta contraproducente.

Una vez clasificado el suelo al nivel más bajo - de más detalle-, realizada la correlación entre sectores y tipos de suelo y creadas las series se obtienen las UHM, en definitiva, el mapa de UHM o mapa de zonificación o MVT. Con este mapa se puede interpretar el conjunto de elementos y propiedades de las que se dispone de información y asignar los resultados a una determinada UHM dentro de cada UM. Esta operación (la interpretación de resultados) sólo tiene sentido si el resultado se asigna a una determinada zona geográfica de la UM. Como es obvio para hacer esta asignación se han de diseñar las UM. Este diseño depende de múltiples factores, sobre todo de la elección del tipo de gestión (MC o VP), del objetivo y del tamaño de la UP, y no tiene una solución única.

El objetivo para obtener la mejor opción es la optimización de la calidad del producto, lo que implica que algunas UHM, sin posibilidad de ser mejoradas, deben ser rechazadas. En la Tabla 4 se incluyen algunas consideraciones sobre la relación entre las UM (Viticultura) y las UP (Enología) y la calidad del producto considerando criterios de especificidad, trazabilidad y rentabilidad.

Es de gran importancia, hacer una referencia a los métodos analíticos utilizados. La mejor recomendación al respecto es tratar de utilizar métodos que estén convenientemente calibrados con relación al tipo de suelo, al cultivo – portainjerto y variedad- y al objetivo: producción de vinos de calidad y, si existen problemas para

Tabla 4. Relación entre las UM (Viticultura) y las UP (Enología) y la calidad del producto.

UNIDADES DE MANEJO (UM)	UNIDADES DE PRODUCCIÓN (UP)
Manejo Convencional (MC): sin demasiadas posibilidades	
• La superficie de cada UM es variable en la parcela y sólo está condicionada por la mecanización. La geometría del reparto de las UM es de tipo puzle. • Las diferentes UHM pueden estar constituidas por suelos disímiles • La plantación es igual en toda la UM • El manejo en campo de la UM es sencillo • Se realizan las mismas actuaciones en todas la UHM de las UM • Las actuaciones son definidas por el tipo de suelo dominante y/o limitante	• Pueden ser una o varias UP por cada UM • La capacidad de la UP es independiente de la superficie y de la producción de UM • En la vendimia se recoge un producto heterogéneo que se elabora en la UP • Baja calidad del producto • El manejo en bodega es complejo • No permite trazabilidad
Viticultura de Precisión (VP): objetivo simple (cada añada)	
• La superficie de cada UM es variable en la parcela y está determinada por la capacidad del conjunto de las UP y de la superficie de la UHM mayor y de la UP mayor • La plantación es diferencial en la UM según cada UHM • El manejo en campo es complejo, sobre todo de la plantación • Se realizan diferentes actuaciones a cada una de las UHM de la UM • Las actuaciones son definidas por cada tipo de suelo de cada UHM	• Varias UP: una por cada UHM de la UM • Capacidad de cada UP relacionada con la superficie de cada UHM de la UM • En la vendimia se recoge un producto homogéneo • Alta calidad del producto • El manejo en bodega es convencional • Permite trazabilidad
Viticultura de Precisión (VP): objetivo ambicioso (a futuro)	
• Las UHM, al cabo 5-n años, están prácticamente homogeneizadas • La superficie de cada UM es constante en toda la parcela y está determinada por la capacidad de la UP • La plantación es igual en toda la UM • El manejo en campo es sencillo • Todas las actuaciones son diferenciales en toda la UM • Las actuaciones definidas por cada tipo de suelo de cada UHM son cada vez más parecidas	• Hay una UP para cada UM • La capacidad de cada UP está relacionada con la superficie de la UM • En la vendimia se recoge un producto homogéneo • Alta calidad del producto • El manejo en bodega es convencional • Permite trazabilidad

acceder a esta información, lo mejor es dejarse aconsejar por los responsables de los laboratorios con experiencia. El desvío de los resultados obtenidos en el muestreo respecto a los valores de referencia de todas las propiedades debe ser corregido en cada sector y en el caso de que esto no sea posible la UHM debe ser rechazada.

En este apartado, aunque no referido al suelo, pero sí muy relacionado con él y sobre todo con su uso, es necesario dejar una nota suficientemente explícita respecto al agua de riego: es imprescindible tener información detallada sobre ella, tomar correctamente la muestra y analizarla convenientemente. Las relaciones físico-químicas agua-suelo deben preverse y, en su caso prevenir inconvenientes, en el informe de suelos, en caso contrario éste sería incompleto, con graves perjuicios para el futuro de la plantación.

A partir de la correcta interpretación de resultados se diseña el plan de actuaciones para antes de la plantación, para el momento de la plantación y para su vida útil. En este sentido, no sólo se ha de fijar el qué y el dónde, sino también el cómo, el cuánto y el cuándo y definir los controles necesarios.

Por ejemplo, para satisfacer las necesidades nutricionales de la viña es necesario, no sólo que ponga a su disposición los elementos necesarios, sino también que los encuentre en el momento oportuno. Esto ha de conseguirse mediante un plan de abonado adecuado en el que el suelo toma un papel protagonista como almacén o despensa; el correcto desarrollo fisiológico de la planta, en sentido amplio, constituye la garantía de acceso a este almacén a través del sistema radicular de la viña. Este plan de abonado debe tener en cuenta esto y además solucionar los problemas que surjan durante la vida de la plantación, respondiendo a las principales cuestiones que el viticultor se plantea: ¿qué fertilizante? (tipo, composición, formulación), ¿cuánto? (dosis de fertilizante y/o enmienda), ¿cuándo? (momento de la realización del abonado, calendario), ¿cómo? (extensión, localización, mecanización) y ¿dónde? (en cada una de UHM de la UM correspondiente). Las respuestas a las cuestiones no tratadas hasta ahora están fuera del ámbito de este trabajo principalmente por problemas de espacio.

Durante la vida de la plantación, su situación evoluciona de tal forma que se crea una variabilidad importante dentro de cada una de las UM de manera que es necesario corregir los posibles desequilibrios generados. Son por ello necesarios análisis de control al menos cada 4/5 años que nos permitan volver a la situación de equilibrio. Estos controles tienen que ver principalmente con el suelo, la planta y sus relaciones.

En cuanto a la planta son útiles las directrices provisionales sobre el muestreo de la planta [19, 20] y los análisis foliares y/o peciolares en relación con el suelo y los estados fenológicos; y respecto al suelo, son suficientes controles periódicos del estado sanitario y de la nutrición en el horizonte superficial (A), entre 5 y 20 cm (MO, N, CE, pH y P) y en el horizonte subsuperficial (generalmente B) de la fertilidad (Ca, Mg, Na, K y acidez de cambio).

6. Conclusiones

- El Mapa de Variabilidad del Terroir (MVT), Mapa de los Terroir o Mapa de Zonificación del Terroir constituye la cartografía detallada de la parcela y permite conocer la delimitación y la caracterización de los diferentes sectores que constituyen su variabilidad.

- El MVT se sitúa como una necesidad entre el Terroir (ambiental) y la gestión del viñedo y está constituido por el mapa de las UHM y el de las UM. Este mapa facilita y optimiza la gestión independiente de cada uno de los sectores de la parcela de manera localizada en función las características propias de cada sector.
- La información utilizada para obtener el MVT se consigue coordinando los tres tipos de técnicas (**FIA**, **teledetección** y **MD**) y aprovechando todas sus ventajas.
- El MD es la forma usual de conseguir el MVT a escalas grandes. Este método realiza el mapa con eficacia y además aporta información sobre las propiedades de cada terroir, lo que permite diagnosticar las causas de la heterogeneidad, la relación con el producto y la posibilidad de influir sobre ellas de forma adecuada.
- Los mapas obtenidos de IEAR (teledetección) aportan información importante, son muy asequibles, facilitan la precisión de los límites y se obtienen más rápidamente que los realizados a partir del MD. Sin embargo, en el estado actual de nuestros conocimientos no es posible cartografiar el conjunto de las propiedades permanentes del medio y de las causas de su variabilidad únicamente mediante Teledetección.
- En todos los casos la calidad del muestreo es fundamental. La escala, la densidad de observaciones, el número de propiedades implicadas en el estudio y el lenguaje determinan la utilidad del MVT y las directrices, a ser posible siguiendo normas internacionales, deben ser garantía de calidad.
- El MVT así obtenido permite al viticultor abordar el objetivo más ambicioso: la optimización y homogeneización del producto. En concreto se trata de que a corto o medio plazo (futuro) y mediante la gestión llevada a cabo con la Viticultura de Precisión la actuación sobre las propiedades del terroir en los diferentes sectores de la parcela se consiga la mayor homogeneidad posible y así también el producto lo sea.

Referencias

[1] OIV *Definición de "terroir" vitivinícola*. Res. OIV VITI **333/2010** (2010)
[2] V. Gómez-Miguel *Atlas das castas da Península Ibérica: história, terroir, ampelografia*. Dinalivro **3**, 104–153 (2011)
[3] V. Gómez-Miguel y V. Sotés *Delimitación de zonas vitícolas en la DO Ribera del Duero, en la DO Ca Rioja, en la DO Rueda, en la DO Toro, en la DO Bierzo, en la DO Somontano, en la DO Cigales, en la DO Arribes, en la RD Douro y en la DO campo de Borja*. ETSIA-ETSIAAB. Universidad Politécnica de Madrid (1990–2016)
[4] USDA *Soil Survey Manual*. H. No. **18,** 437 (1993)
[5] D. M. Carroll, R. Evans y V. C. Bendelow *Air photo-interpretation for soil mapping* Soil Survey 85 (1977)
[6] FAO *Aerial photo interpretation in Soil Survey*. Soil Bull. **6** (1967)
[7] L. P. White. *Aerial Photography and Remote Sensing for Soil Survey*. Oxford Univ. Press 104 (1977)
[8] V. Gómez-Miguel, Á. Martínez y ML. González-SanJosé *Nuevos Retos y Oportunidades en Viticultura, Enología y Cultura del Vino* 1–19 (2016)
[9] OIV *Líneas directrices sobre metodologías de zonificación vitivinícola a nivel del suelo y del clima*. Res. OIV-VITI **423-2012** (2012)
[10] USDA *Key to Soil Taxonomy* (2014)
[11] Munsell *Munsell soil color charts* (2015)
[12] USDA *Field book for describing and sampling soils* **3.0** (2012)
[13] USDA *Soil Survey Laboratory Methods Manual*. SSIR **42**, v.5.0 (2014)
[14] S. Henin, R. Gras y G. Monnier *Le profil Cultural* Masson (1969)
[15] A. Reynier. *Manuel de viticulture*. Tec &- Doc Lavoisier 587 (2016)
[16] USDA *National Soils Handbook* (2004)
[17] FAO *World reference base for soil resources* No 106 (2014)
[18] V. Gómez-Miguel y D. Badía *The Soils of Spain*. Springer, WSBS 24-98 (2016)
[19] OIV *The OIV guidelines for sampling in vineyards*. (prop. Prel. mod) CI-TECVIT **2014-03 04.WD23** (2015)
[20] R. B. Ferguson y G. H. Hergert *Soil sampling for Precisión Agriculture*. Univ.Nebraska IANR. EC **00–154**, Lincoln (2016)

Interest of seeing Precision Viticulture through two distributed competences: Determination of resources and schemes allowing some practical recommendations

Leslie B. Moreiro

RITM, Université Paris-Sud, Faculté Jean Monnet, 54 Boulevard Desgranges, 92330 Sceaux, France

Abstract. The aim of this article is to better understand resources needed and constraints to avoid in implementation of Precision Viticulture technologies. In this view, we adapt distributed cognition theory to multilevel model of competence in management sciences. We use a qualitative methodology based on semi-structured interviews in 7 cases study. The main results allow us to distinguish the two aspects of Precision Viticulture, artifacts, providers of resources, and their utilization scheme. Furthermore, Precision Viticulture is decomposed here in two tasks. The first task is characterization of heterogeneity. The second is modulation of technical itineraries. These two tasks are complementary. In addition, we present ours results by highlighting resources used by firm using these technologies and constraints that they face. Finally, with these results, we can do some practical recommendations to designers and users of these technologies.

Introduction

Precision Agriculture is the technological revolution in measuring and processing spatial information [1–3] in order to put the right input at the right place and at the right time [4]. Precision Agriculture is based on concept of heterogeneity (of soil or plant), for characterizing it and for modulating technical itineraries regarding it [5]. In this article, we will focus on a part of Precision Agriculture, Precision Viticulture, because of its specificities. By blending several theories in social sciences like distributed cognition theory and multidimensional model of competence, we try to define Precision Viticulture as competences composed by a set of resources coordinated by a scheme. By this theoretical blend we hope to show that precision viticulture isn't only technology and we try to make some practical recommendations as to designers than to users. To this end, we first do a state of the art, before to explain methods and materials used. Finally, we explain the results founded that allow us to do some recommendations.

1. A particular kind of technology: Precision Agriculture

1.1. A brief history of Precision Agriculture

The concept has been more developed since discovery of NDVI (Normalized Difference Vegetation Index) formula in 1973, the most used index for studying vegetation characteristics [6]. Then, a joint venture for developing and commercializing Variable Rate Technology applicator (VRT) was made in United States between CENEX, Farmer Union Central Exchange, and Computer Company Control Data Corporation. The first VRT was used in 1995 [1]. The first yield maps were used in 1993 in United States of America and in 1997 in Europe [2]. But, Precision Agriculture is an evolving concept. It has been called "Specific Farming" or "Spatially Variable Field Operations" [4] or "Farming by Soil Types" [1]. First, from end of 1980's until middle of 1990's, it has been called "Grid Sampling" because of a fixed sampling mapping of plots. From middle to end of 1990's, we had headed toward "Soft Sampling" by the fact that sampling cans evolutes on the time. Since the 2000's, we talk about "Management Zone" because of the interest to isolate several homogeneous zones in order to provide them best input regarding their needs [7]. But we can ask now what the new paradigm is? By the way, Jullien and Huet (2005) [2] differentiate two types of use. There is a proprietary system where farmers are owners of technology and data recorded with it. Instead, there is service delivery where data collection and mapping are made by advisors.

1.2. Precision Agriculture is composed by several steps with different artifacts

Precision Agriculture is breaking down by scientists in several steps. These steps are presented as data collection, data processing with data interpretation and finally application of inputs [5, 8, 9]. Data collection can be made by several kinds of sensor. We don't come back to differences in terms of technological standards. This question has been discussed by some scientists like Dobermann et al. (2004) [10] who explain differences between mechanic sensor, acoustic sensor, pneumatic sensor, electrochemical sensor, optic and radiometric sensor. In other ways, scientists make differences between several kinds of data collection. There is proxy-detection like on-the-go sensors, or remote sensing by several kind of support like drone, aircraft or satellite [2]. There is sensor for measuring soil characteristics or vegetation recovering or meteorological data. Each of these sensors

DOI: 10.1051/bioconf/20170901023

has advantages and inconvenient regarding farmer's interests. The data collected is localized by several kinds of Global Positioning System (GPS). Finally, sensors can give several indicators of soil (pH, composition and acidity of soil...), indicators of plant characteristics (LAI, Leaf Area Index, NDVI...) [6] or others (meteorological...).

The data collected is first cleaned before to be mapped on Geographic Information System (GIS) software. It is interpolated in order to frame management class. Number of class is determinate regarding several criterions like economic feasibility, dimension of agricultural equipment or hardship of work. Classes are eventually validated by testing [11]. The number of management class is linked to spatial resolution of pictures, or more precisely the coverage area of pixel. More high is resolution, more heterogeneity is considered and less coverage area is important [6]. But management class isn't management zone. Class refers to all area with same characteristics and zone is the spatial and continuous area on a plot [11]. This differentiation is interesting for the understanding of spatial repartition of class in geographic zone and its evolutions in the time [12]. This understanding is possible with regularly frequencies of collection data [6]. Data processing is realized with superposition of several information layers on a map [13]. It's possible to do another map with recommendations of technical itineraries regarding heterogeneity [5].

Finally, application of inputs can be made in several kinds of action like seeding, fertilization, irrigation and disease management [5]. Choice of value to measuring depends of the type of action farmer wants to modulate. It's possible to do modulation in real time during data acquisition or from the data of the last year. Thus, it's possible to collect new data during modulation of technical itineraries and reclosing new cycle of Precision Agriculture [12].

But we can wonder if these steps described in an academic environment can be the same that in firms which use Precision Agriculture Technologies. Our principal question is to understand despite of their differences if firms of a same industry can have a same way of use these technologies and can develop the same competences. By the way, we think that it can be useful to understand complementary resources that can be used to facilitate use of these technologies as constraints users can face.

1.3. Limits of Precision Agriculture

Precision Agriculture technologies can have some constraints and limits. They still have a long and uneven development. It's in part due to the fact that they were on the market in a too immature shape [14]. Scientists have listed ten years ago limits in Precision Agriculture. It can be problem of compatibility between several data standards and compatibility of dimensions with other agricultural machines and tractors [15,16]. Furthermore, there is a lack of training and outreach regarding these technologies [15,17]. Analysis requires rare skills and knowledge and need training [18]. For Robert (2000) [15], this have as consequences that famers use service instead of using their own purchased artifacts. That makes them dependents of service providers. By the way, Jullien and Huet (2005) [2] highlight important costs of purchase. And designers have difficulties to provide adapted decision support tools including economics values and factors of temporal heterogeneity [17]. In fact, economic feasibility is important in modulation of technical itineraries and be too precise in modulation can increase cost and error risks [2].

For Arno et al. (2009) [18], lack of adapted decision support tool is particularly relevant in Precision Viticulture. For them, the choice of value to measuring and the way to use artifact is very important for helping farmers. That's why we partly focus on that.

1.4. Particularity of Precision Viticulture

We choose to focus in one specific industry, viticulture, for having a coherent sampling of firms despite of heterogeneity of firm and technologies used. Viticulture has specifics issues. Sensors and decision-support models used in viticulture are not the same that for other crop because of some particularities like specificity of roots system or particular needs of the vine [3]. Contrary to arable crop, vine life cycle is at least 30 years, it's a perennial crop. This perennial aspect of vine enables comparison years after years [3,18]. Furthermore, a particularity of Precision Viticulture is to take into account variability of grape yield and quality [19]. This variability can have many causes like climate and sanitary situation, soil properties, diversity of Woodstock, age and type of vine, topography, etc.

Moreover, viticulture is linked with important site specific knowledge of winemakers. This is due to the fact that viticulture needs more manual activities than other culture. In other part, in some countries like France, knowledge is linked to importance given to the concept of *terroir*. Due to this concept, the diversity of methods used by winemakers has an impact of vine heterogeneity [3]. That's why there are several indicators of climate, vegetation and soil characteristics that can be used in a complementary way. All these things suppose a different way than in other industry for data processing and analyzing. In order to better understand these specific resources needed in the use of these technologies, we analyze Precision Viticulture as technologies AND competences.

2. From competence to technology

2.1. A multidimensional view of competence

Researchers have highlighted 3 competence perspectives regarding countries. First, the job-oriented definition in the UK considers competence as work station attribute. The US worker-oriented definition defines competence as attribute of individual who is the only one responsible of it. In this view competence doesn't include contextual, organizational aspect. Finally, we follow the multidimensional definition used in European countries, like France [20,21]. In this view competence is a combination of several resources, internal and external at the individual, group or firm who develop it. Competence is expressed in action, depending in part of it environment [22]. It as a goal, the "what is to be made", that mean the task described by de Montmollin (1986) [23]. Competence is dynamic and has a life cycle [22,24]. We consider that this approach can be enhanced by distributed cognition theory, in order to better take into account importance of action and external resources presents in the environment.

2.2. Interest of distributed cognition theory

Theory of distributed cognition can be linked to this view and thus adapted to the concept of competence. Distributed cognition is defined as information flux which is distributed in individuals mind and some media in human environment as artifacts and other human minds. This distribution is made in the time and it's coordinated by a structure which defines several steps of action [25, 26]. In order to better understand it, Hutchins (1995) [25] proposes to focus on activity realized in task frame. So we can adapt this theory by saying that competence can be distributed in the time, between people (that's mean socially distributed) and between human and technology, more precisely under artifact form. This view is in harmony with multidimensional model of competence where competence is composed by several resources and not only human resources. We can adapt distributed cognition theory to competence in order to say that competence is distributed. Orlikowski (2002) [27] has already showed that competence is socially distributed. Latour (1992) [28] has supposed that distribution of action between human and artifact can lead to a distributed competence. Then we try to link competence to technology.

2.3. A multidimensional view of competence

Finally, by mixing these two theoretical pans, we can define competence as composed by set of resources and a structure or a scheme. Resources can be internal to humans or included in their environment like artifacts. Structure, although defined as scheme by several scientists, is needed to mobilizing resources in competence [29].

Technology is discussed through several generic terms, as tool, instrument, and machine. We think that the concept of instrument is particularly relevant for our analysis. Rabardel (1995) [30] define instrument as composed by two entities: artifact and scheme. For him, artifact is material or immaterial structure of instrument. This structure may impose restrictions and bring new opportunities by allowing or not some of these actions in the implementation of activities [25,31]. Scheme is an abstract and stable model of action [29], more particularly; utilization scheme is a model of action concerning use of this artifact [30]. The utilization scheme can be common for a group [32]. The utilization scheme can be the way to coordinate the resource distributed in competence, when this competence is linked to the use of instrument. This definition of technology is very interesting because it allow us to overcome typology of technology like Information and Communication Technology, automated process technology, decision technology. By this way, we can define technology in same terms than competence: scheme and resources (provided by artifact).

3. Materials and methods

We based on qualitative methodology with semi-structured interviews. We have anonymized some data about our informants. On one side, we have interviewed 7 experts of Precision Agriculture (designers, researchers and network facilitators) in order to better understand the technologies studied and interactions between users and designers. On the other side; we interviewed 9 people in 7 firms using Precision Viticulture technologies. These 7 firms constitute our 7 case studies situated in several regions of France. The sample is composed by 6 independents wineries and one cooperative.

The first firm, named firm I is a wine-producing holding of 360 ha and employing 123 people. It uses a vigor sensor. The firm J is a wine-producing holding composed by 1150 ha and 1300 employees. It uses the same vigor sensor than firm I. Firm F is a familial vineyard of 41 ha and with 25 employees. It use drone with camera. Firm P is a familial vineyard in a growing phase, with 135 ha and 15 employees. It use meteorological border with 2 hygrometric probes for characterizing water stress. Firm O is a vineyard with 31 ha and 6 employees, using a service providing of vegetation cover mapping for characterizing vegetation heterogeneity.

Firm Y is a cooperative. It include 18 vineyards for 1870 ha and 181 farmers. It has 3 own vineyards in 140 ha and 91 employees. It use service providing of NDVI and vigor mapping for characterizing vegetation heterogeneity of all cooperative area. Maps have been used one year in their 3 own vineyards with an automated spreader with DPAE for modulation of fertilization. Finally, a vintage machine is used in these 3 vineyards in order to do modulated vintage regarding maps of service providing, but also with sensors of the vintage machine. The last, firm A is a familial vineyard composed by 3 partners and one employee on 9 ha. It uses a DPAE (flow rate proportional to the electronic advance) to modulate doses of input.

4. First results

First we briefly describe artifacts used. Even if there are device composed by several artifacts, we describe it as one unified artifact in order to simplify analysis. Then we explain task where they are used and we finish by describe steps of competence scheme.

In firm I sensor is linked with a GPS, an on-board computer and mounted on a tractor equipped with pre-pruner. In addition, a USB stick, simple software programmed under Python and the free software Qgis are used for data processing. This is used in a task of characterization of heterogeneity. This task is clearly oriented by attainment of environmental and quality target. We distinguish several steps of the competence scheme, were use of technologies is made in group. First, sensor is mounted before pre-pruning on a tractor equipped with pre-pruner. In winter during pre-pruning period, each day tractor-drivers, chosen for their experience, use this sensor for data collection. At the end of the day, after data collection, data are stored in USB stick and transferred to members of management team who will transfer data by mail to the individuals who do the processing step. Data processing is made during the same period than data collection. It consists to filter incorrect data, realize repartition graph of the values and mapping the value. The first year of use, it's made by Deputy Management of the wine-growing holding. Years after, it is made by accountant of the firm who is accustomed to use many data management software, even if she hasn't

DOI: 10.1051/bioconf/20170901023

agronomic degree. At the end of the season, sessions of data analysis are made in group between management team, tractor-drivers, experimentation manager and field labors. This analysis is made by collective step back with confrontation of sensor data with knowledge of field labors, experimentation results, soil analysis, grape analysis, visualization by comeback by foot in some plots, etc. It allows doing zoning for creating modulations of technical itineraries. People were trained, sensitized for using these new technologies. This use is strongly linked to global production goals of the firm. By other way, strong links with sensor designer, who is part of a professional organization, allow him training and help them and adapt sensor to their needs. Much technologic watch is made in this firm and people are accustomed to try new artifacts each year.

In firm J, artifacts used are the same than in firm I. The several steps of the scheme are also the same, even if there are some differences in the way to perform it. First, sensor was tested in R&D department of the firm, with few users and just one individual for data collecting. With transfer to operational services, the use is made in group, but not really coordinated. For example, during the first year of transfer, people who do sensor mounting, data collection and transfer aren't identified. Individuals that take part of this process aren't trained and sensitized. People who help them and do the data processing are trainees who aren't the same year after year. Furthermore, the mounting is not the same as in firm I. Transfer of the data is more complicated and implies more displacement for management team. To finish, data analysis isn't also democratic as in firm J and sensor use isn't linked to production goals of the firm. In this firm, sensor use has been unsuccessful. Few reliable data were collected and few maps were realized. But recent and new organizational choices improve results, as choice and training of new users.

Firm O use service providing of mapping by satellite. These maps are used in the task of characterization of heterogeneity. This task is linked to the need of determining date and modulation of vintage. Data collection, data processing and one part of data analysis are made by service providers. But the farmer, who is user, makes himself analysis of the map by matching with other sources of information and checking the data with observations of the plants. Then, he can take decision about zoning plots and vintage date. But, due to his professional experience in multinational sanitary product firm, he has some difficulties to impose his decisions. It's because employees don't consider he has experience in viticulture. Moreover, they are unwilling to practices change.

Firm F uses drone with camera, GPS, SD card and joystick. Camera provides pictures viewable on a computer. Drone is used by managers in the task of characterization of heterogeneity. This characterization is linked with general goals depending of situation. It can be used for characterization of sanitary situation, vintage modulation, etc. For that's concern utilization scheme, data collection can be made by leading drone alone or accompanied by other people. The first visualization of picture is occasion to "processing" data. It's not processing by making a model, but by cleaning and joining pictures. A second visualization, made alone or with colleagues is occasion to better understand picture and compare it with other information. At the end, it's possible to take decision on zoning plots including advice of other people like oenologist. We can highlight that people of this firm make many technological and agronomic watch, they try to regularly test new product and technologies, especially for improve their sustainable approach.

Firm P uses meteorological border with 2 hygrometric probes, composed in part of GPS, sensors SD cards and coupled with use of Internet and computer for having access to data. This device is used by the farmer and is father in characterization of temporal heterogeneity of water stress. Data acquisition is automatically made. Data are automatically processed and are validated by border provider. A first data analysis is made by the farmer or his father. After this analysis, a is taken in order to choose what kind of action is needed to modulate (irrigation, sanitary protection). In this firm, interest of new knowledge by internal research, openness toward innovation, discussion with international farmers and partnership by testing technologies in development are particularly showed as important for firm growing.

Firm A use a DPAE, an on-board computer with GIS mounted on a tractor. It's composed by bus CAN (Controlled Area Network), a sensor on the vessel, speed sensor and linked with electromagnetic valves. This device is used during task of technical itineraries modulation. This task is strongly linked to the willing to avoid exceed doses authorized by biodynamic label. Utilization scheme is decomposed in three main steps. First step is registration of data (setting) in on-board computer. These data, like vine variety or inter-row width, define heterogeneity of plots and choice of technical itineraries. The second step, application of input is in part made automatically by processing of the registered data and in another part in real time with information collected by speed sensor. Application can also be made manually by the drivers of the tractors regarding information visualized on the on-board computer or the tractor environment. The last step is refinement of setting regarding the new collected data. In order to improve his practice, the farmer takes part to collective exchange of practice within group of users constituted by an organizational organization.

Finally, firm Y uses three sets of device. One of this is a service providing of mapping. The first year, data acquisition was made by drone. The second year it was made by aircraft and years after by satellite. Maps and recommendations are transmitted to users by Internet. This service is used in characterization of heterogeneity to understand crops and their needs of input. This understanding is linked to a general goal of the firm, importance for saving input and to have sustainable behavior in input application. Data collection and a part of data processing are made by service providers. A second data processing with Qgis and an analysis of these data are made in the cooperative. Returns of these data are made in recommendations form to each farmer in order to explain them interest of this service. The other device used is a fertilizer spreader, in part composed by DPAE, sensor, SD card, GPS, IDE (Internet Data Exchange) and an on-board computer where maps realized before were stored. This fertilizer spreader is used in modulation of technical itineraries (modulation of fertilizer) for economics and sustainability goals. Utilization scheme is the same that in firm A, with little adaptations, for example spreader user can see the map in on-go computer, that isn't the case in firm A. The other device used by firm Y is a vintage machine. It includes sensor, on-board computer, IDE, SD card and GPS. It's used for modulation of technical itineraries, more precisely vintage modulation, and it's linked to willing to improve quality of wine. Utilization scheme takes the same form that for using fertilizer spreader. For example, the first step is the setting, with inclusion of map realized during characterization of heterogeneity. Grape selection is made regarding this map and with data recorded in real-time by sensor measuring anthocyanin rate. Feedbacks of data recorded in real time and shaped under map allow setting improvement. This can be the opportunity to make another loop of Precision Viticulture by begining a new task of characterization of heterogeneity.

5. Discussion

Regarding these results, we can develop a grid of analyze. Precision Viticulture includes here two main tasks where distributed competences take place. It also includes several kinds of technologies related by utilization scheme and artifacts providers of resources.

Before use or purchase, some reflections are made by users in order to define tasks where technologies are needed. Task is considered here as the goal of action [23] or the aim of use of technologies. This is what de Sanctis and Poole (1994) [33] called spirit of technology. But this spirit has to be interpreted in managerial and production terms, that's Hatchuel and Weil (1992) [34] have called managerial philosophy. Interpretation scheme is a part of the scheme focused on technology aim, interpreting technology spirit [33]. It allow to link spirit to managerial philosophy. It is highlighted when people that we have interviewed explain us that it is important for them to keep in mind their own production objectives during all the steps of the scheme. This is particularly pointed out during data analysis and decision of zoning. These two steps are also called data interpretation and recommendation zoning by several agronomic scientists. It's important to link spirit to managerial philosophy because there are many values that we can measure or evaluate (soil composition and moisture, NDVI, anthocyanin, vigor...). And there are many kinds of technical itineraries that can be modulated (spread of fertilizer, sanitary product, vintage, modulation of grass in inter-row...). The willing to reach production targets impacts the choice of value to measuring or choice of technical itineraries. The only one firm that doesn't have made this link is firm J and it is the one that has many difficulties to record reliable data.

We can say that Precision Viticulture is decomposed in two main tasks: characterization of heterogeneity and modulation of technical itineraries regarding this heterogeneity. These tasks have been intuitively mentioned by de Carvalho Pinto et al. (2007) [5]. We see here that the task made in first is characterization of heterogeneity, were the diagnosis competence take place. The second task, modulation of technical itineraries is place where we encountered scheme and resources of competence of technical itineraries management. Performing diagnosis competence can help to manage technical itineraries. However we have seen the possibility that management of technical itineraries can give opportunity to begin another diagnosis.

But, as we have seen in the first part of this article, in viticulture the concept of *terroir* induces the fact that characterization and modulation are task already made by winemakers with more traditional methods. Characterization is made by visualizing plots during manual activities of winemakers, or by coring the soil or doing grape analysis [2]. Modulations are generally made during manual activities regarding information obtained by characterization. Thus, what is really new in Precision Viticulture concept? As Tisseyre (2012) [3] says, the new technologies used create some revolution of information for more precise information and better automation. Then, we can say that Precision Viticulture is as a way to act than a set of technologies in two kind of competence. There is a traditional Viticulture Precision and a modern one, some "Viticulture Precision +" with added technologies. We have seen that is possible that a firm use Precision Viticulture technologies only for one of the two referred tasks.

We will recap the resources provided by artifacts and the utilization scheme for each competence. A part of competence scheme can be made by a service provider, like data collection and processing in firm O and Y.

Task of characterization of heterogeneity is where is patterned the scheme of the competence we call diagnosis competence. The scheme associated to this competence is variable regarding kind of technology, situation, and firm.

But we can find some common steps between several firms:

step 1: mounting (in firms I and J)
step 2: data collection
step 3: data transfer (important step for data reliability in firms I and J)
step 4: data processing
step 5: data analysis, it can be made separately (firms O, F, P and Y) or associated with step 6 (firms I and J)
step 6: decision of zoning. All studied firm in this step do complementary manual soil and crop sampling or visual observation in order to be sure of Precision Viticulture data. And as Zwaenepoel and le Bars (1993) [4] say, farmers and employees expertise on their plots is essential to take decision that can at first sight to be contradictory to Precision Viticulture data.

In this task, artifacts provide several kinds of resources that we call here "means". We can see in the Table 1 that even if artifacts used are different, they provide the same kind of resources.

The second task, named task of modulation of technical itineraries, is where is patterned the scheme of the second competence. We choose to call it competence of management of technical itineraries. The scheme associated can be broken down as follows:

step 1: setting by data registration
step 2: application of modulation (application of input, vintage) by data processing
step 3: refinement of setting

In this task, artifacts provide similar kind of resources explained in Table 2.

We can see the common resources provided by these technologies, with some particularities for each

Table 1. Resources provided by Precision Viticulture Technologies in task of characterization of heterogeneity.

Means of data process-sing and analyzing	**Means of data transfer**	**Means of data store**	**Means of data visualization**	**Means of locomotion or fixing device**	**Means of data localization**	**Means of data collection**	**Resources**
Desk computer, software	USB stick, IDE, Inter-net	USB stick, computer	On-board soft-ware	Tractor	GPS	Sensor	**Firm I**
Desk computer, software	USB stick, IDE, Inter-net	USB stick, computer	Computer, software	Tractor	GPS	Sensor	**Firm J**
Desk computer, software	Internet	Computer, binder	Computer, software, binder	Satellite	GPS	Sensor	**Firm O**
Desk computer, software	Remo-e control, SD card	SD card, computer	Desk com-puter, soft-ware	Drone and joystick	GPS	Came-a	**Firm F**
Desk computer, software	Internet by WIFI	SD card, computer	Desk com-puter, soft-ware	Housing of border and probes shaft	GPS	Sensor on bor-der and probes	**Firm P**
Desk computer, software	Internet, IDE	SD card, computer	Desk com-puter, soft-ware	Drone, Aircraft, Satellite	GPS	Sensor	**Firm Y**

Table 2. Resources provided by Precision Viticulture Technologies in task of modulation of technical itineraries.

Resources	**Firm A**	**Firm Y for input modulation**	**Firm Y for vintage modulation**
Means of data collection	Speed sensor, storage meter, on-board computing (for pre-registered data)	Speed sensor, storage meter, on-board computing (for pre-registered data)	Sensor, on-board computing (for pre-registered data)
Means of data localization	GPS	GPS	GPS
Means of locomotion or fixing device	Tractor	Tractor	Vintage machine
Means of data visualiza-tion	On-board computer	On-board computer, desk com-puter, software	On-board computer, desk com-puter, software
Means of data store	On-board computer, SD card	On-board computer, SD card	On-board computer, SD card
Means of data transfer	IED, SD card	IED, SD card	IED, SD card
Means of data processing and analyzing	GIS on on-board computer	GIS on on-board computer	GIS on on-board computer
Means of acting on het-erogeneity	Electromagnetic valves	Electromagnetic valves	Harvesting head, sorting table, vessels

task. In task of characterization of heterogeneity, means of data processing and analyzing are more developed than in task of modulation, where they are only used to allow acting in real-time. In modulation task, new means appear: means of acting on heterogeneity. In these two tasks, data analysis or refinement of setting is the place for step back on work. That is a way to improve competence and develop knowledge [29]. Farmers say us that uses of these technologies allow them to go back to fundamental agronomical concepts. Particularly in task of characterization, they are associated with investigating work in order to better understand causes of heterogeneity. By the way, we can see that even users of service do a part of processing and analysis. This work challenge affirmation made by Jullien and Huet (2005) [2] distinguishing proprietary of technologies like actors and pro-actives from "passive" consumer of service.

However, as we consider competence is distributed, there are other resources to take in consideration than only Precision Viticulture resources, like material and financial resources, human resources other organizational resources and extra-organizational resources.

In the interviews we have leaded, several resources appear to facilitate use of technologies and resources they provides. For example, financial resources were mentioned in interview for all cases. They can facilitate adoption of these technologies considered as expensive. Material resources are important because of compatibility needs as for equipment dimension than for data standard in automation. Human resources are particularly relevant. It's a priority for all our informants. They consider that a lack of people who have the good skills and knowledge can dissuade to use these technologies or encourage requesting service provider more than acquire technologies. There is other kind of organizational resources like openness and interest for innovation, which encourage and facilitate to see opportunity bringing by technology. And the learning made by using other innovation and technology can be resources adapted to use of the new one. The management type can allow users to improve their use of technology. Interest for sustainable practices for example, can give clear goals to technology, clarify it spirit. By this way, we can see that the only one firm which has difficulties in using technology is firm J. This is the one which

Table 3. Main resources provided by firm and its environment in task of characterization of heterogeneity.

Human resources	Extra-organizational resources	Other organizational resources	Material and financial resources	Resources or constraints
Deputy Manager, tractor-drivers, field labors, sector leaders, foremen, experimentation manager, accountant	Help and support from designer, exchange with other users	Openness to new technology (by testing and adapting every year new equipment), participative management, resources provided by training and awareness , interest for sustainable practices	Additional material (like pre-pruner), disposition of the equipment, financial resources	**Firm I**
Team managers, tractor-drivers, sector leaders, R&D engineer, engineer trainee (changing every year), Garage manager	Help and support from designer, exchange with other users	Lack of thinking about goals reached by technology use, R&D department, since 2015 resources provided by training and awareness	Additional material, own garage, financial resources	**Firm J**
General Manager	Regular use of advisors, help and support from professional organization, part of the scheme realized by service provider	Lack of resources in terms of openness to innovations, lack of interest of employees and relational difficulties	Additional material, new fermentation tank	**Firm O**
General Manager, Technical manager, head of cultivation, trainee, soil scientist	Training for drone piloting and data processing software	Familial management, interest for sustainable practices, interest for innovations	Additional material, financial resources	**Firm F**
Farmer and his father	Exchange with other users and designers, researchers, maintain of borders	Interest for sustainable practices, interest for innovations, Openness to new technology (watch), experience in R&D project with designers	Other material (like drip irrigation system, simple agricultural equipment), financial resources	**Firm P**
Innovation Manager and his assistant head of cultivation, oenologist	Help, support and part of the scheme realized by service provider, exchange with other users, interest from any famers	Interest for sustainable practices, interest for innovations, experience in R&D project with designers, representation of almost of designation	Additional material, financial resources	**Firm Y**

doesn't have a clear and strong technology spirit. Even if for firm F and P, technology spirit and goal of use are changing depending situation of use (measure sanitary state, anticipate vintage…). A participative management style can help to make collective discussion, even with people who aren't users but who can provide their knowledge in order to better understand data. This is highlighted in firm I by the help of field labor on collective step back in data analysis. Training and awareness made for users and their colleagues are important resources. Training is the opportunity to develop new knowledge and skills and awareness develop motivation of people and eliminate fears linked to technology. It is particularly relevant in firm I where manager too are trained and aware of this technology. Participative management expands awareness to other people like field labor who aren't direct users but participate to data analysis. We can see difference between this firm and firm J. Firm J is the only one firm where training and awareness aren't made. In this firm, lack of skills, knowledge and motivation has consequences on data acquisition. These lacks can concern as users as their managers (sectors leaders and foremen). In general, training can be made by people external to the firm like advisor or designers. In this case, we consider that the resource is extra-organizational. It can be provided with other help and support supplied by designer. Designers can help users when they face problem with technology. And a discussion with users allow designer to improve technology. Service provider realize part of the scheme of action and other users bring some knowledge and exchange "best practices" in some formal or informal users communities.

We choose to name "constraints" the lacks of resources that can be damaging for a good perform of Precision Viticulture competence. These constraints appear in case J, Y and O. We have seen the problem caused by lack of strong technology spirit and the lack of training, awareness, participative management. In firm O, the constraints are lack of openness to innovation of employees and relational difficulties General Manager has with them. That has for consequences a zoning less precise than General Manager wanted. Constraints can be extra-organizational, like regulatory and institutional constraints. Ours informants have highlighted the relative importance of these aspects in the two Precision Viticulture competences. For example, in case Y, difficulties in obtaining European funding delay the development and testing of new spreader more adapted to firm goals. On the contrary, for firm A, regulatory constraints can be understood as resources. In fact, statutory limits for Demeter label (biodynamic agriculture) is an aim of use of DPAE, it constitute a motivation for farmer of firm A.

DOI: 10.1051/bioconf/20170901023

Table 4. Main resources provided by firm and its environment in task of modulation of technical itineraries.

Resources or constraints	Firm A	Firm Y for input and vintage modulation
Material and financial resources	Management and traceability tool, complementary agricultural equipment and repair equipment, financial resources	Complementary agricultural equipment, financial resources
Other organizational resources	Openness to new technology (technology watch), interest for sustainable practices	Interest for sustainable practices, interest for increase wine quality, interest for innovations, experience in R&D project with designers
Extra-organizational resources	Share of users experience, training	Data obtained in task of characterization of modulation, help and support from designers, share of users experience, difficulties to obtain European financial help
Human resources	Farmer	Innovation Manager, tractor-driver

As we have seen, the two studied competences are distributed between several people (social distribution) and entities, as human and technologies. It distribution is made in the time during several step of a scheme coordinating these resources. The analysis of these resources and constraints liked with a model of action, the scheme, allow us to do some practical recommendation to designers and users of Precision Viticulture technologies.

6. Practical recommendations

This work can help users by giving them some practical recommendations in order to improve performance and interest of Precision Viticulture technologies. We begin by recommendation to designers. We showed here interest highlighted by Prost et al. (2012) [35] for designers to take in account users since first step of conception. Spirit of technology has to be clearly identified and fit with production goals, resources and constraints of the farmers. It can be possible for example through participatory research [36], like in case Y. These kinds of design and conception can inform about needs of some particular resources like specific human resources, or data standard compatibility, needs of other equipments and specificities of designations and regions. That helps designers to adapt their technology to several market segments. Many of the technologies studied in our case are designed by several firms, each of them providing a specific expertise. By the way, partnership with other designers can be helpful in order to create peripheral equipments needed that can have been set in relief by partnership with users. These peripherals equipments can be adapted in term of dimension and data standard to Viticulture Precision technologies. In addition, maintain good relationship with users after purchase is helpful for designers. By helping and providing training to users, they can better understand their needs and problems. That can lead to improve technology and develop other complementary tools. Due to the fact that Precision Viticulture is still unknown and there is few training in this domain, this support provided by designer can be a commercial argument to choose this tool instead of another. Furthermore, as one of our informant said, it's important that at least one of firm designer comes from Viticulture industry, in order to better understand its constraints and specificities. Agriculture is a specific sector and more particularly, viticulture hasn't the same characteristics than other kind of crop. If it's not the case, designers had to do a partnership with professional organization of the sector in order to gain credibility. This partnership can include training and support to users made by a professional organization.

We have also some recommendation to users. Our researches show importance for users to define what aims they want to reach with Precision Viticulture technologies and in what kind of task. The use of these technologies can be linked with sustainable practices, interest for increasing quality or developing tractability. This is the occasion to evaluate feasibility of implementation by an audit of resources they need and those they have. Resources can be determinate in term of financial or material resources, human resources, other organizational resources as production routines or openness to innovation, and finally external resources provided by their environment. These external resources can be the link with other users, the designers or the academic community. Human resources (HR) practices and management are other important aspects. We have seen that identifying the good people, training and making awareness to employees is crucial for success of technology implementation and well perform the competence. Awareness allows employees to better identify aim of technology use. With socially distributed aspect of competence, we have seen that these HR practices shouldn't be focused only to direct users, but to others people with whom they interact during all the steps. A participative management helps users to implicate to improvement of the competence. This participative management gives to the floor to participate in collective feedback and to reach a consensus in order to build a common utilization scheme and have the same interpretation of aim of use (technology spirit). That's why others people than direct users are important, like trainer and advisors (who can be internal or external to the user firm), or the gatekeeper. Gatekeeper is individual who is at interface between users and designers. He can be a referee, that's mean the supervisor of technology implementation. The problems in firm J shows that the referee has to be stable in the time in order to create long relationship with users and better understand them. Managers who supervise users' work are other kind of important people. In firm J, unmotivated managers demotivate tractor-driver in data acquisition, that's lead to bad results. Although with motivated managers, good results have been encountered. To have these good examples of success is a way to lean on lead users. These lead users can motivate others users, and transmitting knowledge [37]. They also can be external to the firm, like users community. Eventually, openness to innovation, watching, habits of testing new tools and agronomics experimentation can facilitate new technology uses. It can be interesting for users doing participative research with designers for testing new technology and see if it can be useful for them before a purchase. Once technology purchased, the firms which have the most of people implicated in technology use (firm I, Y and firm J when they try to solve their problem of sensor use) have a particular way to implement technology use. They choose to gradually increase area where used technologies are. We think this way allow to rapidly identify problem of use before too duplicate them and increase bad way of use and demotivation.

7. Conclusion

Our work allows us to highlight that Precision Viticulture can be considered like technologies and through two kind of distributed competences: diagnosis competence and competence of management of technical itineraries. They are composed by resources and scheme. We think that this work can help to better understand what can provide Precision Viticulture technologies and how it can be used. Thus, we hope having produced the illustrate case studies needed for helping adoption of Precision Viticulture [17]. By this way, we have seen that limits emphasized by scientist 10 years ago are in part still relevant. On the other part, we have identified some good practices and some problems that enable us to do some practical recommendations as for designers than for users. By this way, we hope taking the translator role attributed to social scientist [38] in order to facilitate relationship and understanding between specialists and non-specialists, designers and users of Precision Viticulture technologies.

References

[1] P. C. Robert. Precision agriculture: an information revolution in agriculture. Agricultural Outlook Forum (1999)

[2] A. Jullien, P. Huet. Agriculture de précision. *Agricultures et territoires*, Lavoisier, Paris (2005)

[3] B. Tisseyre. Peut-on appliquer le concept d'agriculture de précision à la viticulture. Mémoire d'habilitation à diriger des Recherches (2012)

[4] P. Zwaenepoel, J.-M. Le Bars. L'agriculture de précision. Ingénieries-EAT **12** (1997)

[5] F. D. A. de Carvalho Pinto, D. M. de Queiroz, E. Chartuni, E. Ruz. Agricultura de precisión: nuevas herramientas para mejorar la gestión tecnológica en la empresa agropecuaria. Revista Palmas **28**(4), 29–34 (2007)

[6] D. J. Mulla. Twenty five years of remote sensing in precision agriculture: Key advances and remaining knowledge gaps. Biosystems Engineering **114**(4), 358–371 (2013)

[7] D. Boisgontier. Les nouvelles technologies de l'information aux Etats-Unis Etats des lieux et perspectives vus de la France. Colloque AgriMédia, France (2004)

[8] R. Ortega, L. Flores. Agricultura de Precisión: Introducción al manejo sitio-específico. *Ministerio de Agricultura, Instituto de investigaciones agropecuarias. CRI Quilamapu*. Chile, 13-46 (1999)

[9] T. Catanzarite Torres, A. B. González Fernández, F. J. Benavente Cobo, J. R. Rodríguez Pérez. SIG para el seguimiento vitícola y diferenciación de lotes de vendimia en la DO Bierzo (León-España). (2009)

[10] A. Dobermann, S. Blackmore, S. E. Cook, V. I. Adamchuk. Precision farming: challenges and future directions. 4th International Crop Science Congress. **26**. Brisbane, Australia (2004)

[11] J. A. Taylor, McBratney, A. B., B. M. Whelan. Establishing management classes for broadacre agricultural production. Agronomy Journal **99**(5), 1366–1376 (2007)

[12] E. Jallas, M. Crétenet, M. Aide à la décision en agriculture et gestion des ressources naturelles: Note introductive. Savanes africaines: des espaces en mutation, des acteurs face à de nouveaux défis. *Colloque Cirad-Prasac*, Garoua, Cameroun (2003)

[13] B. Ariaux. Les SIG utilisés en agriculture de précision. *Colloque Agriculture de Précision, Educagri*, Dijon France 55–65 (2000)

[14] J. Lowenberg-DeBoer. Precision farming or convenience agriculture. *11th Australian Agronomy Conference*. February (2003)

[15] P. C. Robert. L'agriculture de précision: les verrous liés à la technologie et à la gestion agronomique. *Colloque Agriculture de Précision, Educagri*, Dijon (2000)

[16] C. Martin, C. Pages. Conditions et moyens d'une gestion informatisée des données pour la mise en oeuvre de l'agriculture raisonnée: exploration stratégique par la méthode d'audit patrimonial. Ingénieries-EAT, 30 (2002)

[17] A. McBratney, B. Whelan, T. Ancev, J. Bouma. Future directions of precision agriculture. Precision agriculture, **6**(1), 7–23 (2005)

[18] J. Arnó Satorra, J. A. Martinez Casasnovas, M. Ribes Dasi, J. R. Rosell Polo. Review. Precision viticulture. Research topics, challenges and opportunities in site-specific vineyard management. Spanish Journal of Agricultural Research **7**(4), 779–790 (2009)

[19] A. Matese, P. Toscano, S. F. di Gennaro, L. Genesio, F. P. Vaccari, J. Primicerio et al. Intercomparison of uav, aircraft and satellite remote sensing platforms for precision viticulture. Remote Sensing **7**(3), 2971–2990 (2015)

[20] T. N. Garavan, D. McGuire. Competencies and workplace learning: some reflections on the rhetoric and the reality. Journal of Workplace learning **13**(4), 144–164 (2001)

[21] F. D. Le Deist, J. Winterton. What is competence?. Human resource development international **8**(1), 27–46 (2005)

[22] E. Rouby, C. Thomas. L'articulation compétences individuelles/compétences stratégiques: vers une solution de gestion intégrée des compétences. Revue Interventions économiques **40** (2009)

[23] M. De Montmollin. L'intelligence de la tâche: *Eléments d'ergonomie cognitive*. P. Lang. (1986)

[24] T. Durand. L'alchimie de la compétence. Revue française de gestion **160**(1), 261–292 (2006)

DOI: 10.1051/bioconf/20170901023

[25] E. Hutchins. *Cognition in the Wild.* MIT press (1995)

[26] E. Hutchin. Distributed cognition, *Internacional Enciclopedia of the Social and Behavioral Sciences* (2000)

[27] W. J. Orlikowski. Knowing in practice: Enacting a collective capability in distributed organizing. Organization science **13**(3), 249–273 (2002)

[28] B. Latour. On interobjectivity. Mind, culture, and activity **3**(4), 228–245 (1996)

[29] P. Pastré, G. Vergnaud. La didactique professionnelle. Revue française de pédagogie **154**(1) (2006)

[30] P. C. Rabardel. Les hommes et les technologies; approche cognitive des instruments contemporains (1995)

[31] J. Hollan, E. Hutchins, D. Kirsh. Distributed cognition: toward a new foundation for human-computer interaction research. ACM Transactions on Computer-Human Interaction (TOCHI) **7**(2), 174–196 (2000)

[32] P. Beguin, P. Rabardel. Concevoir pour les activites instrumentees. Revue d'Intelligence Artificielle **4**(1), 35–54 (2000)

[33] G. de Sanctis, M. S. Poole. Capturing the complexity in advanced technology use: Adaptive structuration theory. Organization science **5**(2), 121–147 (1994)

[34] A. Hatchuel, B. Weil. *L'expert et le système: gestion des savoirs et métamorphose des acteurs dans l'entreprise industrielle; suivi de Quatre histoires de systèmes-experts.* Economica (1992)

[35] L. Prost, M. Cerf, M. H. Jeuffroy. Lack of consideration for end-users during the design of agronomic models. A review. Agronomy for sustainable development **32**(2), 581–594 (2012)

[36] J. N. Pretty. *Regenerating agriculture: policies and practice for sustainability and self-reliance.* Joseph Henry Press (1995)

[37] L. B. Jeppesen, K. Laursen. The role of lead users in knowledge sharing. Research Policy **38**(10), 1582–1589 (2009)

[38] M. Callon. Éléments pour une sociologie de la traduction: la domestication des coquilles Saint-Jacques et des marins-pêcheurs dans la baie de Saint-Brieuc. L'Année sociologique (1940/1948-) **36**, 169–208 (1986)

Total polyphenols contents in different grapevine varieties in highlands of southern brazil

Emilio Brighenti[1], Katia Casagrande[2], Paula Zelindro Cardoso[2], Mateus da Silveira Pasa[1], Marlise Nara Ciotta[1], and Alberto Fontanella Brighenti[1]

[1] Santa Catarina State Agricultural Research and Rural Extension Agency (EPAGRI), João Araújo Lima St., 102, São Joaquim, Santa Catarina State, Brazil
[2] University of South Santa Catarina (UNISUL), José Acácio Moreira Av., 787, Tubarão, Santa Catarina State, Brazil

Abstract. Phenolic compounds are one of the main parameters of wine quality and contribute to the organoleptic characteristics, particularly color, astringency and body. In the highlands of southern Brazil, low temperatures and high accumulation of global solar radiation favor the synthesis of total polyphenols in grapes. The objective of this work was to evaluate the concentration of total polyphenols of 10 white varieties and 13 red varieties produced in high altitude regions of southern Brazil. The vineyard is located in the Experimental Station of Santa Catarina State Agricultural Research and Rural Extension Agency (EPAGRI), in the city of São Joaquim (28° 16′30″S, 49° 56′09″W, Altitude 1,400 m), the evaluations occurred in 2015/2016 growing season. The content of total polyphenols was determined as proposed by Singleton & Rossi (1965), using the Folin-Ciocalteu method, with spectrophotometer readings. Polyphenol content ranged from 283.56 to 1,387.31 mg/L for white varieties, the varieties with the highest concentrations were Greco di Tufo (1,378.31 mg/L), Trebbiano Toscano (995.59 mg/L) and Ribola Gialla (737.48 mg/L). For the red varieties, the total polyphenol content ranged from 523.87 to 4,929.57 mg/L, Ancellotta (4,929.57 mg/L), Uva di Troia (2,722.27 mg/L) and Croatina (2,410 mg/L) stood out for presenting the highest levels.

1. Introduction

The wine production in high altitude regions of Santa Catarina State has less than 15 years of history and is relatively recent, when compared with other producing regions in Brazil. Today in high altitude regions of Santa Catarina State, there are 35 wineries in operation, with a planted area of 600 hectares and over 200 labels of white, red and sparkling wines. In high-altitude vineyards of Santa Catarina, the main varieties planted are Cabernet Sauvignon, Merlot and Sauvignon Blanc [1].

The enlargement of the varietal choice is a powerful tool in the hands of grape growers and winemakers in the highly competitive wine market; however, it should not be repeated the error of relying on few varieties either local or imported. The French varieties, it has to be reminded, are spread all over the viticultural world. In addition, some practices largely adopted in modern winemaking emphasized the globalization of wine taste minimizing on this respect the contribution of the variety and of the environment of production. In such a situation, the most forward thinking wine producers are starting to again look at local varietal heritage, re-examining their enological value and marketing potential [2].

When introducing new grape varieties, the phenology and ripening monitoring are important in determining the ability of a region to produce a crop within the confines of its climate regime [3]. Grape ripening encompasses the period that starts with the color change (véraison) and ends at harvest, and varies according to the variety and the region. Among the modifications that occur in the berries and, therefore, in the composition of the grape during the ripening are pigment accumulation in the skin, synthesis of aromatic substances and flavor modification [4,5].

These modifications are related to the synthesis of phenolic compounds, the grape is one of the major sources of phenolic compounds, which are mainly found in the skin and in the seed, presenting little content in the pulp [6].

Phenolic compounds are important for plant growth and reproduction, also acting as antipathogens (stress conditions such as infections and wounds) and protection against UV radiation [7]. They still contribute to pigmentation, astringency, aromas and oxidative stability. In viticulture these compounds are responsible for the qualitative and organoleptic composition of wines, such as color, body and astringency [8].

The synthesis and accumulation of phenolic compounds are greatly influenced by the environment conditions, including light, air temperature, altitude, soil type, water availability, nutritional status, disease incidence and other developmental processes [9,10].

The objective of this work was to evaluate the concentration of total polyphenols of 10 white varieties and 13 red varieties produced in high altitude regions of southern Brazil.

2. Materials and methods

The experiments was conducted in variety collection of the Experimental Station of Santa Catarina State Agricultural Research and Rural Extension Agency (EPAGRI), located in the city of São Joaquim (28° 16′30″S, 49° 56′09″W,

DOI: 10.1051/bioconf/20170901024

Altitude 1,400 m a.s.l.). The vineyard was planted in 2006, spacing 3.00 m between rows and 1.50 m between plants, the training system adopted was VSP and evaluations occurred in the 2015/2016 growing season.

The climate of the region is characterized as humid subtropical, without dry season, with annual mean temperature and rainfall of 13.5°C and 1,700 mm, respectively. According to the Köppen classification, the climate is Cfb. The soils are classified as Haplumbrept and Udorthent developed from rhyodacite and basalt rock [11].

The evaluated white grape varieties were Riesling Renano, Verdelho, Marsanne, Petit Manseng, Moscato Giallo, Fiano, Chardonnay, Ribola Gialla, Trebbiano Toscano and Greco di Tufo. The evaluated red grape varieties were Negroamaro, Pinot Nero, Canaiolo Nero, Malvasia Nera, Barbera, Cabernet Sauvignon, Marselan, Merlot, Aglianico, Syrah, Croatina, Uva di Troia and Ancellotta.

Climatic data, like precipitation (mm), relative humidity (%), maximum, minimum and mean air temperature (°C) and sunshine (h) were measured on site with an automatic weather station from CIRAM (Center of Environmental Resources Information and Hydrometeorology of Santa Catarina).

For quantification of the grapes phenolic content the extract used was obtained as follows: grape skins were carefully removed from 90 berries and used for the extraction process of the phenolic ripening analysis [12].

To obtain extract solutions, the following skin and extract ratio was used: 50 g of skins were separated manually from berries samples, which were added 20 mL of hydroalcoholic solution of methanol 50% v/v, and maintained At 30 °C (+ 0.5 °C) for 24 hours. After this time, the "hot" extract was separated and the skins were rinsed with 5 mL of the methanol solution. After this, 20 mL of methanol extracting solution were added to the skins, which were then placed at 0 °C (+ 0.5 °C) for further 24 hours. After this extraction, the "cold" extract was homogenized with the "hot" extract, and the skin rinsing was repeated with another 5 mL of methanol solution. The extract solution was filtered at the end of the process [12].

The concentration of skin total polyphenols was quantified in spectrophotometer, using the reagent Folin-Ciocalteu (Vetec) and gallic acid as standard, with absorbance readings at 760 nm [13]. The calibration curve was made using the concentrations of 0, 100, 200, 300, 400, 500, 600 and 1000 mg/L. The results were expressed in mg/L of gallic acid.

Data were submitted to analysis of variance (ANOVA, $p \leq 0.05$), and when variety effects were detected, the Tukey test was performed ($p \leq 0.05$).

3. Results and discussion

The climate is one of the basic aspects in the grape culture and, consequently, for the viticulture. Viticulture for the wine production is developed in many types of climate and soil in the different continents [14]. A survey of world viticultural areas can indicate relationship between climate and quality [4]. Thus, the monitoring of the climatic parameters are important because these influence both grape and wine quality [3].

Grape ripening period occurred between January and April and the climatic parameters for the period are shown in Table 1. Temperatures observed in this phase confirm the potential for wine production in the region, due to temperature reduction and substantial thermal amplitude during fruit ripening. Another authors reported for the region of São Joaquim thermal amplitude around 10 °C, which is favorable for production of quality grapes [3, 15, 16].

Table 1. Climatic parameters obtained with meteorological station, from *véraison* to harvest, during the growing season 2015/2016, in São Joaquim – Santa Catarina State, Brazil.

Climatic Parameters	Jan	Feb	Mar	Apr	Mean
Maximum Temp. (°C)	23.4	23.5	20.7	21.0	22.2
Minimum Temp. (°C)	13.4	14.6	11.2	12.7	13.0
Mean Temp. (°C)	17.8	18.4	15.4	16.1	16.9
Termal Amplitude (°C)	10.0	8.9	9.5	8.3	9.2
Relative Humidity (%)	83.2	85.3	84.2	80.2	83.2
Precipitation (mm)	119.9	177.5	159.4	157.9	614.7
Rainy Days (n)	10	20	12	14	56
Rainy Days (%)	32	69	39	47	47
Sunshine hours (h)	198.3	160.2	180.5	155.4	694.4

The climatic conditions of São Joaquim favored grape phenolic maturation, since the accumulation of anthocyanins during maturation presents negative correlation with high temperatures and positive with low temperatures. The grapes present a good coloration when the thermal amplitude does not exceed 10 °C and during maturation it must be below 15 °C [17].

During grape ripening, the observed amount of precipitation was 614.7 mm. The frequency and distribution of rainfall are climatic elements of great importance, because the free water on the leaves and the fruit is the main factor to trigger the onset of fungal infections on the vine [18]. The high rate and frequency of rainfall is the main limiting climatic factor of vine cultivation in high altitude regions of southern Brazil [19].

The precipitation is an important climatic factor in terms of the formation of phenolic compounds in the skin of the grapes [20]. Polyphenols have the function of protecting plants from physical (such as the sun UV radiation) and biological (fungi, viruses, bacteria) attacks. Typically, in places where there are high rainfall levels, there is stress from the plant to combat, mainly, fungal diseases, inducing the production of phenolic compounds [9, 10, 21].

The accumulation of sunshine hours during grape ripening was higher in São Joaquim (694.4 hours) than in other grapevine producing regions of Santa Catarina State. The regions of Campo Belo do Sul, Bom Retiro and Água Doce showed 546, 571 and 288 sunshine hours, respectively, during ripening of different varieties [16, 22]. In São Joaquim, during the growing seasons of 2005/2006 and 2006/2007 it was reported the average of 424 sunshine hours during the ripening of four *Vitis vinifera* L. varieites [3]. This climatic parameter is important because it is directly related to fruit quality, stimulating mainly the accumulation of phenolic compounds such as anthocyanins [5].

The incidence of solar radiation in grape clusters also contributes to the activation of metabolism and the formation of substances responsible for wine quality, such as phenolic compounds, especially flavonols [9]. This

Table 2. Date of occurrence from véraison to harvest phases for the white varieties evaluated during 2015/2016 growing season, in São Joaquim – Santa Catarina State, Brazil.

Variety	Véraison	Harvest	Ripening Duration	Total cycle (days)
Riesling R.	25-jan	9-mar	44	192
Verdelho	9-feb	28-mar	48	203
Marsanne	14 feb	28-mar	43	196
P. Manseng	2-feb	28-mar	55	194
Moscato G.	10-feb	22-mar	41	189
Fiano	11-feb	28-mar	46	202
Chardonnay	19-jan	16-mar	57	208
Ribola G.	31-jan	28-mar	57	205
Trebbiano T.	15-feb	5-apr	50	196
Greco di Tufo	10-feb	15-apr	55	223

Table 3. Date of occurrence from véraison to harvest phases for the red varieties evaluated during 2015/2016 growing season, in São Joaquim – Santa Catarina State, Brazil.

Variety	Véraison	Harvest	Ripening Duration	Total cycle (days)
Negroamaro	12-feb	5-apr	53	200
Pinot Nero	17-jan	1-mar	44	183
Canaiolo N.	11-feb	28-mar	46	201
Malvasia N.	13-feb	28-mar	44	192
Barbera	6-feb	22-mar	45	190
Cabernet S.	6-feb	22-mar	45	186
Marselan	5-feb	22-mar	46	200
Merlot	4-feb	22-mar	47	186
Aglianico	15-feb	5-apr	50	201
Syrah	7-feb	5-apr	58	211
Croatina	10-feb	28-mar	47	192
Uva di Troia	10-feb	28-mar	47	188
Ancellotta	29-jan	28-mar	59	194

difference can give to the wines high altitude regions of Santa Catarina State more intense coloration, more complex aroma, greater structure and persistence [3,22].

The mean date of occurrence and extent of phenological cycle, between véraison and harvest, are shown in Tables 2 and 3. Greco di Tufo (white) and Syrah (red) were the varieties with the longest cycle with 223 and 211 days respectively. Among white varieties, Chardonnay, Ribola Gialla, Greco di Tufo and Petit Manseng presented the longest duration of grape ripening. While for the red varieties, Ancellotta and Syrah presented the longest ripening duration.

The extension of the phenological cycle is related to climatic conditions of each region. In Brazilian climatic conditions, regions of high altitude (above 900 meters a.s.l.), present longer phenological cycles due to the occurrence of lower temperatures [15,19,23].

Significant differences were observed in polyphenol contents between the different varieties (Figs. 1 and 2). although the environmental conditions under which the grapes were grown have a great influence on the synthesis of polyphenolic compounds, the nature and the different concentration of these compounds follow the determinant genetics of each variety [24].

Many authors have reported that polyphenol composition is due not only to the type of cultivar but also to the location where the grapes are grown and environmental

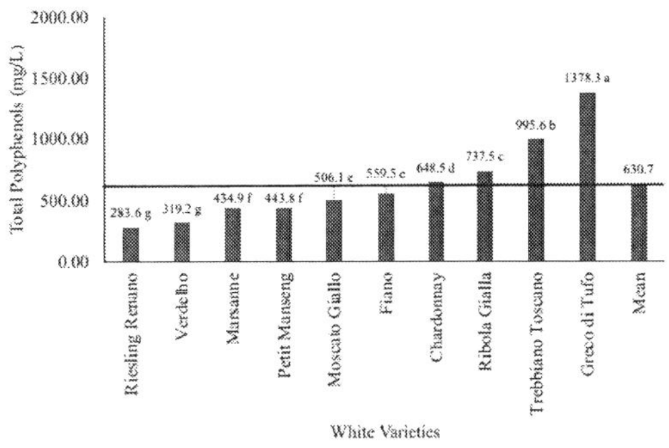

Figure 1. Total polyphenols concenttration (mg/L) in berries of different white varieties grown in São Joaquim – Santa Catarina State, Brazil, during the growing season of 2015/2016.

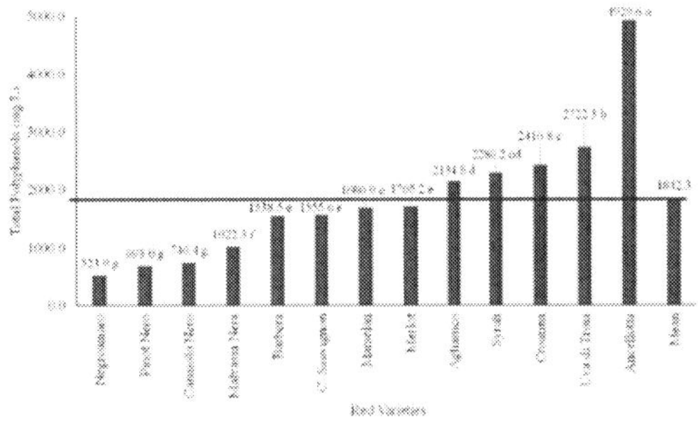

Figure 2. Total polyphenols concenttration (mg/L) in berries of different red varieties grown in São Joaquim – Santa Catarina State, Brazil, during the growing season of 2015/2016.

and management practices, as well as the growing season [4,20,25].

Polyphenol content ranged from 283.56 to 1,387.31 mg/L for white varieties. For the red varieties, the total polyphenol content ranged from 523.87 to 4,929.57 mg/L. The main phenolic compounds found in grapes and wines are classified as: non-flavonoids and flavonoids [26]. Flavonoid compounds represent the largest group of polyphenols found in foods, presenting high antioxidant power [27]. The most common flavonoids present in grapes and red wines include anthocyanins, flavanols and flavonols. These compounds are responsible for color and structure characteristics of wines [9]. Therefore the concentration of total polyphenols is on average higher in the red varieties.

The white varieties with the highest concentrations of total polyphenols were Greco di Tufo (1,378.31 mg/L), Trebbiano Toscano (995.59 mg/L) and Ribola Gialla (737.48 mg/L), while Riesling Renano (283.56 mg/L), Verdelho (319.16 mg/L) and Marsanne (434.87 mg/L) presented the lowest values.

In southern Italy, the average total polyphenols ranged from 491.6 to 536.8 mg/L for different white grape varieties [28]. While in Água Doce – Santa Catarina State, Brazil, the average total polyphenols ranged from 601 to 706 mg/L [16]. In Água Doce, the value obtained for Fiano was 601 mg/L, quite similar to those obtained in this study (559.5 mg/L). While in São Joaquim, it was found values of 474.94 mg/L for Prosecco, 517.32 for Vermentino and 540.04 for Verdicchio [15].

DOI: 10.1051/bioconf/20170901024

Considering red varieties, Ancellotta (4,929.57 mg/L), Uva di Troia (2,722.27 mg/L) and Croatina (2,410 mg/L) stood out for presenting the highest levels of total polyphenols, while Negroamaro (523.87 mg/L), Pinot Nero (692.98 mg/L) and Canaiolo Nero (746.38 mg/L) presented the lowest levels of total polyphenols.

In a study performed in Calabria/Italy, with four different red grape varieties, it was found variations in polyphenol contents from 1,503 to 2,170 mg/L. The diversity of results found, is probably related to several factors, among them: the variety, management practices, methods of extraction and analysis used [29].

While in Água Doce – Santa Catarina State, Brazil, the average total polyphenols ranged from 570 to 2,217 mg/L to four red varieites [16]. In Água Doce, the variety Ancellotta stood out with the highest concentration of total polyphenols (2,217 mg/L), that value was lower the observed in São Joaquim, for the same variety (4,929.57 mg/L). In São Joaquim, during the growing seasons of 2010/2011, 2011/2012 and 2012/2013 it was found values between 734.2 and 2,515.1 mg/L for 9 red Italian varieties, in this work, the varieties Aglianico (1,730.7 mg/L) and Ancellotta (2,515.1 mg/L) stood out with highest total polyphenols concentration [15].

It is believed that high altitude is related to the high concentration of total polyphenols obtained in grapes grown in São Joaquim. Among the environmental factors, climate exerts the greatest influence on the accumulation of polyphenols; the higher the altitude, the lower the temperature and humidity and the higher the incidence of solar radiation, which favors the accumulation of total polyphenols [9,30]. In this case, the causes of the greatest accumulation of total polyphenols may be related to the quantity and quality of solar radiation and thermoperiod, which are effects of altitude [19].

4. Conclusion

Significant differences were observed in polyphenol contents between the different varieties. This fact evidences the interaction between genotype and environment as one of the main factors related to accumulation and concentration of grape polyphenols.

Polyphenol content ranged from 283.56 to 1,387.31 mg/L for white varieties, the varieties with the highest concentrations were Greco di Tufo (1,378.31 mg/L), Trebbiano Toscano (995.59 mg/L) and Ribola Gialla (737.48 mg/L), while Riesling Renano (283.56 mg/L), Verdelho (319.16 mg/L) and Marsanne (434.87 mg/L) presented the lowest values.

For the red varieties, the total polyphenol content ranged from 523.87 to 4,929.57 mg/L, Ancellotta (4,929.57 mg/L), Uva di Troia (2,722.27 mg/L) and Croatina (2,410 mg/L) stood out for presenting the highest levels, while Negroamaro (523.87 mg/L), Pinot Nero (692.98 mg/L) and Canaiolo Nero (746.38 mg/L) presented the lowest levels of total polyphenols.

References

[1] A.F. Brighenti, E. Brighenti and M.S. Pasa. Vitivinicultura de altitude: realidade e perspectivas. Agropecuária Catarinense **29**, 140–146 (2016)

[2] F. Mannini. Italian indigenous grapevine cultivars: guarantee of genetic biodiversity and economic resources. Acta Hort. **652**, 8795 (2004)

[3] E.F. Gris, V.M. Burin, E. Brighenti, H. Vieira and M.T. Bordignon-Luiz. Phenology and ripening of Vitis vinifera L. grape varieties in São Joaquim, southern Brazil: a new South American wine growing region. Cien. Inv. Agr. **37**, 61–75 (2010)

[4] D.I. Jackson and P.B. Lombard. Environmental and management practices affecting grape composition and wine quality. A review. Am. J. Enol. Vitic. **44**, 409–430 (1993)

[5] R.V. Mota, M.A. Regina, D.A. Amorim and A.C. Fávero. Fatores que afetam a maturação e a qualidade da uva para vinificação. Inf. Agrop. **27**, 56–64 (2006)

[6] G. Sanna, S. Ledda, G. Manca and M.A. Franco. Characterization of the content of antioxidant substances in the wines of Sardinia. Journal of Commodity Science, Technology and Quality **47**, 5–25 (2008)

[7] D. Rusjan, R. Veberic and M. Mikulic-Petkovsek. The response of phenolic compounds in grapes of the variety 'Chardonnay' (*Vitis vinifera* L.) to the infection by phytoplasma Bois noir. European journal of plant pathology **133**, 1–10 (2012)

[8] M. Naczk and F. Shahidi. Extraction and analysis of phenolics in food. Journal of Chromatography **1054**, 95–111 (2004)

[9] M.O. Downey, N.K. Dokoozlian and M.P. Krstic. Cultural practice and environmental impacts on the flavonoid composition of grapes and wine: a review of recent research. American Journal of Enology and Viticulture **57**, 257–268 (2006)

[10] Z.W. Dai, N. Ollat, E. Gomès, S. Decroocq, J.P. Tandonnet, L. Bordenave, P. Pieri, G. Hilbert, C. Kappel, C.V. Leeuwen, P. Vivin and S. Delrot. Ecophysiological, Genetic, and Molecular Causes of Variation in Grape Berry Weight and Composition: A Review. American Journal of Enology and Viticulture **62**, 413–425 (2011)

[11] Embrapa. Centro Nacional de Pesquisa de Solos. *Solos do Estado de Santa Catarina* (Rio de Janeiro: Embrapa Solos, Boletim de Pesquisa e Desenvolvimento **46**, 726, 2004)

[12] J.L. Marcon Filho, J.S. Hipólito, T.A. de Macedo, A.A. Kretzschmar and L. Rufato. A survey of clusters thinning in the 'Cabernet Franc' winemaking potential in two seasons. Ciencia Rural **45**, 2150–2156 (2015)

[13] V.L. Singleton and J.A. Rossi. Colorymetry of total phenolics with phosphomolibdic- phosphotungstic acid reagent. Am. J. Enol. Vitic. **16**, 144–158 (1965)

[14] J. Tonietto and A. Carbonneau. A multicriteria climatic classification system for grape growing regions worldwide. Agricultural and Forest Meteorology **124**, 81–97 (2004)

[15] A.F. Brighenti, A.L. Silva, E. Brighenti, D. Porro and M. Stefanini. Viticultural performance of native Italian varieties in high-altitude conditions in Southern Brazil. Pesq. Agropec. Bras. **49**, 465–474 (2014)

[16] L.I. Malinovski, A.F. Brighenti, M. Borghezan, M.P. Guerra, A.L. Silva, D. Porro, M. Stefanini and

 DOI: 10.1051/bioconf/20170901024

H.J. Vieira. Viticultural performance of Italian grapevines in high altitude regions of Santa Catarina State, Brazil. Acta Hort. **1115**, 203210 (2016)
[17] J.M. Ubalde, X. Sort, A. Zayas and R.M. Poch. Effects of Soil and Climatic Conditions on Grape Ripening and Wine Quality of Cabernet Sauvignon. Journal of Wine Research **21**, 1–17 (2010)
[18] G. Chavarria, H.P. Santos, O.R. Sônego, G.A.B. Marodin, H. Bergamaschi and L.S. Cardoso. Incidência de doenças e necessidade de controle em cultivo protegido de videira. Rev. Bras. Frutic. **29**, 477–482 (2007)
[19] A.F. Brighenti, L.I. Malinovski, M. Stefanini, H.J. Vieira and A.L. Silva. Comparison between the wine producing regions of São Joaquim - SC, Brazil and San Michele All'Adige - TN, Italy. Rev. Bras. Frutic. **37**, 281–288 (2015)
[20] G.V. Jones and R.E. Davis. Climate Influences on Grapevine Phenology, Grape Composition, and Wine Production and Quality for Bordeaux, France. American Journal of Enology and Viticulture **51**, 249–261 (2000)
[21] C.R.C. Buffara, F. Angelotti, F.A. Vieira, A. Bogo, D.J. Tessmann and B.P. de Bem. Elaboration and validation of a diagrammatic scale to assess downy mildew severity in grapevine. Cienc. Rural **44**, 1384–1391 (2014)
[22] L.I. Malinovski, L.J. Welter, A.F. Brighenti, H.J. Vieira, M.P. Guerra and A.L. Silva. Highlands of Santa Catarina/Brazil: A region with high potential for wine production. Acta Hort. **931**, 433439 (2012)
[23] A.F. Brighenti, E. Brighenti, V. Bonin and L. Rufato. Phenological characterization and thermic requirement of distinct grapevines varieties in São Joaquim, Santa Catarina – Brazil. Ciência Rural **43**, 11621167 (2013)
[24] A. Calò, D. Tomasi and R. Di Stefano. Relationship between environmental factors ant the dynamics of growth and composition of the grapevine. Acta Hort. **427**, 217–231 (1996)
[25] N. Mateus, J.M. Machado and V. Freitas. Development changes of anthocyanins in *Vitis vinifera* grapes grown in the Douro Valley and concentration in respective wines. J. Scie. Food Agric. **82**, 1689–1695 (2002)
[26] G. González-Neves, G. Gil, F. Guzmán and M. Ferrer. Potencial polifenólico de la uva: índices propuestos y posibles aplicaciones. Comunicata Scientiae **2**, 57–69 (2011)
[27] A. Scalbert and G. Williamson. Dietary intake and bioavailability of polyphenols. Journal of Nutrition **130**, 2073–2085 (2000)
[28] Fanara, C. *Evoluzione dei composti polifenolici in vini bianchi della sardegna durante la conservazione in bottiglia e caratterizzazione aromatica*, Thesis (Sassari, Italy: Dottorato in Biotecnologie Microbiche Agroalimentari, Universitaĺ degli Studi di Sassari, 2011)
[29] A.M. Giuffreì. Polifenoli totali nelle bucce di uve rosse in vitigni calabresi. Industrie delle bevande **227**, 18–21 (2010)
[30] N. Mateus, S. Marques, A.C. Gonçalves, J.M. Machado and V. de Freitas. Proanthocyanidin composition of red *Vitis vinifera* varieties from the Douro valley during ripening: influence of cultivation altitude. American Journal of Enology and Viticulture **52**, 115–121 (2001)

New wine-growing regions of Brazil and their importance in the evolution of Brazilian wine

Douglas André Wurz[a], Betina Pereira de Bem, Ricardo Allebrandt, Bruno Bonin, Luiz Gabriel Dalmolin, Adrielen Tamiris Canossa, Leo Rufato, and Aike Anneliese Kretzschmar

Santa Catarina State University (UDESC), College of Agriculture and Life Science, Av. Luiz de Camões, 2090, CEP 88520-000, Lages, SC, Brazil

Abstract. The objective of this work is to characterize the new Brazilian wine regions, describing their edaphoclimatic and productive characteristics, wine types, and their importance in the evolution of Brazilian wine industry. The Campanha Gaúcha is characterized by flat lands, presents a high number of hours of light, and dry summers, guaranteeing a complete maturation of the grapes. Including the locations The Southeastern Region of Rio Grande do Sul presents pronounced ripples, located in altitudes between 400 and 600 m, presenting dry and sunny summers with cold nights, stony soil; Merlot and Cabernet Franc are the outstanding varieties. In the northern plateau of Rio Grande do Sul, at 1000 m a.s.l., the region of Campos de Cima da Serra has a characteristic high solar incidence, and due to the low nocturnal temperatures. The Altitude Region of Santa Catarina State presents similar characteristics to those found in the Campos de Cima da Serra, with vineyards located between 900 and 1400 m, the slow maturation promotes the preservation of acidity and high levels of aromatic compounds, which confer freshness and typical white wines, respectively, especially those made from the Sauvignon Blanc. Located at medium elevations of 900 to 1100 m, the region of Greater Curitiba with hot days and mild nights, stands out for the production of varieties of short to medium cycle, because it presents humid summers favoring the occurrence of diseases fungal infections. The South of Minas Gerais State presents mean altitudes of 800 and 1000 m, in which the technique of double pruning was adopted, leading the grapes maturation to occur during the winter, when a dry season with mild temperatures is found, making it an ideal place to produce high quality Syrah wines. In São Paulo State, altitudes between 1,000 and 1,300 m a.s.l. are found, where cool nights and excellent sunshine during the day provide thermal amplitude of 10°C at the time of harvest. Add to this a dry, well-drained and granitic soil, especially for growing Syrah and Viogner. The São Francisco River Valley presents flat lands, at an altitude of 400 meters. Located outside the suitable range for growing grapes, this semi-arid region presents low rainfall (less than 500 mm a year) and strong insolation. Its permeable clay soils have proven to be suitable for the acclimatization of grapevines such as Moscatel, Cabernet Sauvignon and Syrah.

1. Introduction

The first vines cultivated in Brazil were of European origin and emerged with the arrival of Portuguese colonizers (1532). In the mid-nineteenth century, Italian immigrants introduced the "Isabel" American grape variety, culminating in the rapid replacement of vine varieties of European varieties, becoming the basis for the development of commercial winemaking in the states of Rio Grande do Sul and São Paulo. From the beginning of the twentieth century, São Paulo viticulture replaced the cultivars of the variety "Isabel" by "Niagara Branca" and "Seibel II". In the same period, with government incentives, Rio Grande do Sul intensified the planting of vine varieties [1].

Since 1875, with the arrival of the Italian immigrants, the winemaking activity is present in the Serra Gaúcha, and over the years, production has been improving and reaching quite satisfactory levels of development, transforming the place into a national wine-growing pole [2].

Until the end of the 1950s, Brazilian commercial viticulture was restricted to the three southern states and the eastern regions of São Paulo and southern Minas Gerais. From there, there was a great enlargement of the wine-growing frontier. In traditional regions, production systems have been modified over the years, depending on market opportunities and requirements [3].

The 1990s presented a scenario of a restructuring and/or reconversion of the profile of the wine industry in the State of Rio Grande do Sul. The need for the wine industry to face the intensification of competition with foreign wines and the progressive drop in wine consumption worldwide have led industry leaders to devise strategies that could redefine the organization of viticulture in South-Rio Grande. The 90's can be considered the first great evolution of Brazilian wine.

However, despite the existence of several improvements that facilitate the development of the wine production process, there is a locational movement in Brazilian winemaking: the growth of wine production outside the traditional region (Serra Gaúcha) [2]. This locational movement of viticulture allowed the discovery

[a] e-mail: douglaswurz@hotmail.com

of new terroirs in Brazil, allowing the elaboration and commercialization of differentiated products. Therefore, it is a determining factor for the evolution of Brazilian winemaking.

In this context, the objective of this work is to characterize the new Brazilian wine-growing regions, describing their edaphoclimatic and productive characteristics, wine types, and their importance in the evolution of Brazilian wine.

2. Materials and methods

The study was developed through a qualitative research, from a descriptive perspective.

For the development of the work scientific works and national publications were compiled referring to the evolution of Brazilian wine, as well as the regions that produce fine wines in Brazil. Consultations were made to the bibliographic collection of the State University of Santa Catarina, as well as consultations in indexed periodicals, with the objective of exploring the following contents: Brazilian winemaking, Brazilian wine evolution, wine producing regions in Brazil, Brazilian winemaking map, campanha gaúcha, Southeastern Region, Campos de Cima da Serra, Altitude Regions in Santa Catarina State, Region of Curitiba in Paraná State, State of São Paulo; South Region of Minas Gerais State and the São Francisco Valley in Pernambuco State.

The data were also collected in institutions responsible for the elaboration and dissemination of data related to winemaking in Brazil, such as: Brazilian Wine Institute, IBRAVIN, Embrapa Uva e Vinho, Ministry of Agriculture and Brazilian Union of Winemaking, UVIBRA.

From the data cataloged, it was possible to elaborate tables and figures that demonstrate the evolution of Brazilian wine, as well as to characterize the new Brazilian wine regions, describing their edaphoclimatic and productive characteristics, wine types, and their importance in the evolution of Brazilian wine.

3. Results and discussion

The Brazilian viticulture presents great diversity. The culture is spread from Rio Grande do Sul, at 31 °S latitude, to Rio Grande do Norte and Ceará, at 05 °S latitude. The variation of altitude is also great, with considerable environmental diversity between production zones, including temperate, subtropical and tropical regions [3].

In the last years the grapevine cultivation in the West Frontier of the State of Rio Grande do Sul has been emphasizing among the producers as an alternative of cultivation, driven by the ease of mechanization, due to its flat relief and mainly by the edaphoclimatic conditions that allow the production of Grapes and the elaboration of fine wines with unique characteristics [4]. The long days, with a great period of luminosity for the plants, and the great temperature variation between the day and the night benefit the cultivation of the vines. Favorable conditions are complemented by soil, rich in granite and limestone.

Including the towns of Pinheiro Machado and Encruzilhada do Sul, in the region of Serra do Sudeste, vitiviniculture gained economic importance, more recently, from investments made by wineries located in the

Table 1. Wine-growing regions and their edaphoclimatic characteristics and altitude above sea level.

Wine-growing	Altitude (m)	Characteristics
Campanha Gaúcha	150	High number of hours of light, and dry summers.
Southeastern Region	400–600	Dry and sunny summers with cold nights, stony ground
Campos de Cima da Serra	900–1300	High solar incidence, longer vegetative cycle, high aromatic intensity and content of phenolic compounds
Altitude Wines	900–1400	Slow ripening, preservation of acidity and high levels of aromatic compounds
Metropolitan Region of Curitiba	900–1100	Hot days and mild nights, stands out for the production of short to medium cycle varieties
South Region of Minas Gerais	800–1000	It was adopted the double pruning, that causes that the cycle of the grapevine changes, with the maturation happening in the winter
São Paulo State	1000–1300	Fresh nights and excellent sunshine that provide a thermal amplitude of 10 °C at harvest time
San Francisco Valley	400	Low rainfall (less than 500 mm) and strong sunshine. Pervious sandy-loam soils

Serra Gaúcha. Vitis vinifera grape varieties are cultivated, predominantly of red grapes. The productivity of the vineyards in the region is between 8 and 12 t/ha, depending on the crop and climatic conditions of the crop [5]. The Serra do Sudeste is a region formed by successive undulations in the terrain, with altitudes varying between 400 and 500 meters. Even though it has no tradition in winemaking activity, it is being reported that this region has been considered as one of the newest terroirs in Rio Grande do Sul state. In the year 2014, the Serra do Sudeste region produced 4.76 million kilos of grapes. The hybrid and American grapes totaled 1.84 million kilos and the grapes 2.92 million kilos, confirming the region's specialization for grapes destined to produce fine wines [6,7]. The mountain range has proven suitable in the cultivation of noble grapes, such as Cabernet Franc and Merlot, white such as Sauvignon Blanc and Malvasia. The current harvest of the region produces enough to produce 500 thousand bottles of red and white, vinified for the time being in facilities of the Serra Gaúcha. It is mentioned as one of the most promising Brazilian producing areas. Curiously, the Serra do Sudeste is home to very few canteens. Most of the grapes are transported, usually at night, to other regions of Rio Grande do Sul, where it is vinified. However, with the growth of its importance in the national oenological scenario and with the emergence

DOI: 10.1051/bioconf/20170901025

of local enterprises focused on grape production, this situation must change in the near future.

The region of Campos de Cima da Serra is located on the border with the state of Santa Catarina, is formed by a set of slopes and huge walls whose altitude varies from 900 to 1,100 meters. This region has a small area that stands out due to the wines and sparkling wines from municipalities such as Bom Jesus, Jaquirana, São Francisco de Paula, Vacaria and Many Capões. In 2014 the region produced 7.19 million kilos of grapes [6,7]. For a long time, the region of Campos de Cima de Serra was in the shadow of the Serra Gaúcha. The predominance of the cultivation of hybrid varieties and the cold and windy climate were seen as obstacles to the development of large vineyards. Today, however, the scenario is the opposite. The low temperature and the constant incidence of the wind have been transformed into differentials, because they allow a longer maturation and conditions for the grapes to present excellent sanity. The initiatives of entrepreneurs who ventured to elaborate wines in the region were rewarded with great labels, now nationally known for their quality. The soil presents characteristics favorable to the cultivation of the vine, such as good structure, effective depth above 1.50 m, very good drainage and low natural fertility.

The vineyard of Santa Catarina has recently stood out in front of the other fine wines producers in Brazil. Grapes produced in regions of altitude above 900 m have their own characteristics and are distinct from those cultivated in other areas of the country, as well as phenolic maturation adequate to the elaboration of wines [8,9]. The climatic potential of the high altitude regions for the production of vinifera grapes (Vitis vinifera L.) in Santa Catarina, has been proven through several surveys. These conditions provide a differentiated climate in relation to the other wine regions of Brazil, interfering mainly in the phenological cycle. Due to the milder air temperatures in these regions, the vegetative and reproductive cycle of the grapevine is more extensive, resulting in slower ripening of the grapes for the production of fine quality wines [10,11]. In addition, it causes maturation in less frequent months (April and May) and amount of rainfall, providing grapes with greater sanity and oenological quality.

The elevation region of the state of São Paulo is characterized by altitudes between 1,000 m and 1,300 m, cool nights and excellent sunshine during the day provide a thermal amplitude between 10 °C and 12 °C at the time of harvest, similar to that of large European regions. Add to this a dry soil – with good drainage – and granitic, which is especially suitable for grapes destined to the production of high quality wines. The result is a great terroir.

In Minas Gerais state, viticulture focused on the elaboration of wines is concentrated in the southern region of the State, mainly in the municipalities of Caldas and Andradas [12]. The cultivation of grapes for the elaboration of fine wines in Minas Gerais Is a recent practice.The Technological Center EPAMIG Uva e Vinho has been active in the search for new farming regions in the State of Minas Gerais guided by the principles that fine wines are produced in regions where the grape harvest is made in dry periods and has (10 °C), above which the vine can grow almost all year round, and thus its cycle is defined as a function of the date of pruning [13]. The first projects to explore these principles were installed in the year 2001 in Três Corações, and the results showed that Syrah grapevines, pruned in January, show a new cycle of normal vegetation and production, with rates of sprouting, fertility Of the buds, maturation and sanity markedly higher than those observed for this same cultivar during the summer [12,14]. The great virtue of inverting the grapevine cycle is that during the winter period, in the South of Minas, the days are dry and hot and the nights cold, with great daily thermal amplitude that helps in the slow process of ripening of the grape, which Leads to the excellent fixation of the aromas and favors the accumulation of polyphenols, which dictate the color intensity of the fruit and the production of tannin, essential substance for a good structure and wine longevity. In addition, avoiding the rainy season increases the possibility of a complete phenolic maturity of the grapes that reaches an optimal sugar concentration.

In the Northeast of Brazil, the São Francisco Valley is located between the 8–9°parallels of the Southern Hemisphere, is a semi-arid tropical climate region, with edaphoclimatic characteristics that make it possible to stagger the production of grapes for wines throughout the year. A vine plant can produce two harvests per year due to high temperatures, high insolation rates and abundant water available for irrigation from the São Francisco River. This fact allows investments in physical structure to be smaller than those carried out in temperate regions, where the harvest of grapes is concentrated in three or four months of the year, depending on the cycle of each variety [15]. Depending on the month of harvest and preparation, the wines present great variations in composition, quality and typicity, mainly due to intra-annual climatic variability [16]. The main varieties used for the production of red wines in the region of the Valley of the Submédio San Francisco are syrah, tempranillo, national touriga, cabernet sauvignon, alicante bouschet, ruby cabernet and petit verdot, and syrah represents about 65% of red wines. In the case of whites, the grapes used are chenin blanc, sauvignon blanc, muscato canelli and viognier, and chenin represents about 60% of the still white wines [17,18].

4. Conclusion

Along the years, the Brazilian wine map has changed, with the beginning of the production of grapes in non-traditional regions, This diversity presents itself as a differential in the quality of the elaborated products, contributing to the evolution of the Brazilian wine.

References

[1] R.V. Botelho, J.P. Pires. Viticultura como opção de desenvolvimento para os Campos gerais. In: Anais Encontro de Fruticultura dos Campos Gerais **1**, 40–54 (2009)

[2] L.S. Kummel, M.A.D. Moraes, P.V. Marques. Dinâmica locacional da vitivinicultura no Rio Grande do Sul e no Vale do São Francisco. Revista de Política Agrícola **18**, 36–49 (2009)

[3] U.A. Camargo, J. Tonietto, A. Hoffmann. Progresso na Viticultura Brasileira. Revista Brasileira de Fruticultura **33**, 144–149 (2011)

[4] G.F. Brixner, C.R. Martins, U. Amaral, L.M. Kopp, D.B. Oliveira. Caracterização fenológica e exigência térmica de videiras Vitis viníferа cultivadas no

município de Uruguaiana, na região da fronteira oeste – RS. Revista da FZVA **17**, 221–233 (2010)
[5] J.F. Protas, U.A. Camargo. Vitivinicultura brasileira: panorama setorial (SEBRAE, Brasília, 2010)
[6] M. Copello. As regiões produtoras do Brasil. Revista Anuário Vinho do Brasil, São Paulo (2015)
[7] IBRAVIN – Instituto Brasileiro do Vinho Dados estatísticos – Produção de uvas no Brasil. (IBRAVIN, Bento Gonçalves, 2015)
[8] L.D. Falcão, E.S. Chaves, V.M. Burin, A.P. Falcão, V. Bonin, M.T. Bordigon-Luiz. Maturity of Cabernet Sauvignon verries from grapevines grown with two differente training systems ina a new grape growning region in Brazil. Ciencia e Investigácion Agraria **35**, 271–282 (2008)
[9] L.C. Silva, L. Rufato, A.A. Kretzschmar, J.L. Marcon Filho. Raleio de cachos em vinhedos de altitude e qualidade do vinho da cultivar Syrag. Pesquisa Agropecuária Brasileira **44**, 148–154 (2009)
[10] E.F. Gris, V.M. Burin, E. Brighenti, H.J. Vieira, M. Bordigon-Luiz. Phenology and ripening of Vitis vinifera grape varieties in São Joaquim, southern Brazil: a new wine growing region. Ciência e Investigácion Agraria **37**, 61–75 (2010)
[11] L.I. Malinovski, L.J. Welter, A.F. Brighenti, H.J. Vieira, M.P. Guerra, A.L. Da Silva. Highlands of Santa Catarina/Brazil: a region with high potencial for wine production. Acta Horticulturae **931**, 433–440 (2012)
[12] A.C. Favero, D.A. Amorim, R.V. Mota, A.M. Soares, M.A. Regina. Viabilidade de produção da videira Surah em ciclo de outono inverno na região Sul de Minas Gerais. Revista Brasileira de Fruticultura **30**, 685–690 (2008)
[13] M.A. Regina, D.A. Amorim, A.C. Favero, R.V. Mota, D.J. Rodrigues. Novos polos vitícola para a produção de vinhos finos em Minas gerais. Informe Agropecuário **27**, 111–118 (2006)
[14] D.A. Amorim, A.C. Favero, M.A. Regina. Produção extemporânea da videira cv. Syrah nas condições do sul de Minas Gerais. Revista Brasileira de Fruticultura **27**, 327–331 (2005)
[15] G.E. Pereira. Os vinhos tropicais em desenvolvimento no Nordeste do Brasil. Com Ciência **149**, 1–3 (2013)
[16] O. Tonietto, A.H. Teixeira. Zonage climatique des périodes viticoles de production l'année em zonage tropicale: application de la methodologie du Systéne CCM Géoviticole. In: Joint International Conference on Viticultural Zoning, Cape Town, South Africa. S.I.: s.n., 193–201 (2004)
[17] G.E. Pereira, A.J.B. Araújo, J. Santos, R. Vanderlinde, L. Lima. Chemical and aromatic characteristics of Brazilian tropical wines. Acta Horticulturae **910**, 135–140 (2011)
[18] U.A. Camargo, G.E. Pereira, C.C. Guerra. Wine grape cultivars adaptation and selection for tropical wines. Acta Horticulturae **191**, 121–129 (2011)

BIO Web of Conferences **9**, 01026 (2017)
40th World Congress of Vine and Wine
DOI: 10.1051/bioconf/20170901026

Modifications in climate suitability for wine production of Romanian wine regions as a result of climate change

Liviu Mihai Irimia[1], Cristian Valeriu Patriche[2], Bogdan Roşca[2], and Valeriu V. Cotea[1]

[1] University of Agriculture Sciences, Faculty of Horticulture, 700505 Iasi, Romania
[2] Romanian Academy, Iaşi Branch, Geography Group, 700505 Iasi, Romania

Abstract. The aim of this study is to reveal the shifts in climate suitability for wine production affecting Romanian wine growing regions. For this, we analyzed the spatial distribution over Romanian territory of the *oenoclimate aptitude index* (IAOe) for the 1961 to 1990 and 1991 to 2013 time periods. The IAOe has been calculated based on gridded data at 10 × 10 km resolution of average daily temperature, precipitation and sunshine duration between 1961–2013, originating from about 150 weather stations across Romanian territory and recorded in the ROCADA database. The study reveals: northward shifts and to higher altitudes of suitability for wine production; the appearance of new areas suitable for wine production; the expansion or the shrinkage of the current areas suitable for wine production; shifts in classes of suitability for wine production in the current wine regions and appearance of premises of replacing their specific varieties and traditional wine type production; the tendency to level climate suitability at regional scale by diminishing suitability for white wines and replacing it with climate suitability for red wines. The study provides a solid support for developing strategies to adapt Romanian viticulture to climate change.

1. Introduction

Climate suitability for wine production determines the possibility of grapevine growing, wine grape variety assortments of wine regions and their types of wine production. Climate suitability is a stable trait of a wine area over time, with minor variations from year to year, and yet unable to modify the specific type of wine production of a wine region. In the climate change context [1] this trait of wine regions changes significantly leading to important consequences: modification of grapevine phenology [2]; earlier maturation of grapes, altering their color and aromatic profile [3]; higher alcoholic potential and lower total acidity of grape juice [4]; changes in the consecrated sensory profile of wines [5].

Changing climate suitability for wine production manifests globally, with detrimental consequences in the warm Mediterranean areas [6] and positive consequences for northern wine regions with cooler climate, which can diversify their types of wine production. Projections based on climate models predict the expansion of suitable area for grapevine growing beyond the northern and southern limits of vine area at global level, as well as shifts to higher altitudes [7–9].

The impact of climate change on viticulture is also of great interest for Romania, a traditional viticultural country with a wine area of about 197,000 ha and an annual wine production of about 6 million Hl [10]. In Romania the wine production has local specificity generated by: autochtonous *vinifera* wine grape varieties; traditional grapevine training systems; specific wine types. These features have strengthened over the centuries and have become authentic elements for Romanian vineyards that are trying to be kept under the climate change. Romania's wine-growing area is structured in eight large wine-growing regions including 141 wine growing centers (vine plantations around a locality). The eight Romanian wine regions have particular environmental characteristics, generated by latitude, topography, proximity to the Carpathian Mountains situated in the center of the country, proximity to the Black Sea bordering Romania to SE, or proximity to the Danube River bordering southern limit of the country.

The manifestation of climate change over Romania territory has been studied by Romanian climatologists in recent years in all its aspects [11–15]. The main developments are: increased summer temperatures; precipitation stability; increasing sunshine duration; increasing air temperatures during January and February; diminishing numbers of days with frosts; increasing frequency and intensity of extreme phenomena.

As a result of these developments climatic parameters specific to Romanian vineyards have changed [16] determining changes in climate suitability for wine production at vineyards level [17]. The impact of climate change on viticulture, still less studied at the national level, indicates: earlier bud-burst; shortening of stages of annual growth cycle; early grapes ripening; diminishing grapes total acidity and increasing sugar content. As consequences of these developments, cool climate vineyards have already introduced wine grape varieties for red wines, while the southern, warmer ones have introduced wine grape varieties that better preserve grapes acidity, such as *Rhine Riesling* or even varieties such as*Syrah*, specific to Mediterranean wine regions.

The purpose of this study is to find out the changes in climate suitability for wine production of Romanian wine

DOI: 10.1051/bioconf/20170901026

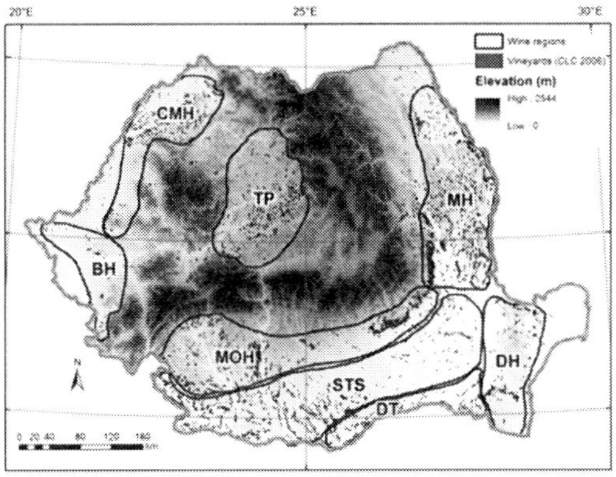

Figure 1. Location of Romanian wine growing regions: TP - Transilvania Plateu; MH - Moldavia Hills; MOH - Muntenia and Oltenia Hills; BH - Banat Hills; CMH - Crişana and Maramureş Hills; DH - Dobrogea Hills; DT - Danube Terraces; STS - Southern Terrains and Sands.

regions as a result of climate change. Climatic suitability is expressed through the IAOe *oenoclimate aptitude index* [18], an index generally used in Romanian viticulture for vineyards climate analysis. It is distinguished from heat summation indices in that it includes, besides the influence of temperature on the grapevine, the influence of sunshine duration and precipitation during the growing season, which provides a much higher accuracy in representation of climate suitability. Its values reveal the conceivable wine type production in an area, especially the possibility of obtaining red wines. This index correlates with the grapes anthocyanins content at the *Cabernet Sauvignon* variety.

2. Material and method

Romania is located in temperate continental climate Dfb and Dfa zone [19], between 43°37′–48°15′ N lat. and 20°15′–29°44′ E long (Fig. 1).

The eight Romanian wine growing areas are: Transylvania Plateu (TP); Moldavia Hills (MH); Muntenia and Oltenia Hills (MOH); Banat Hills (BH); Crişana and Maramureş Hills (CMH); Dobrogea Hills (DH); Danube Terraces (DT); Southern Terrains and Sands (STS). The analysis regards the entire area of each wine growing region, as they were delineated on Romania map in the Fig. 1. The map of wine growing centers and vineyards location within Romanian wine growing regions is extracted from the Corine Land Cover Database 2006 Europe [20].

The climate data base for this study is represented by gridded data at 0.1 degrees' resolution (10 × 10 km) of daily average temperature, precipitation and sunshine duration during the 1961–2013 extracted from the ROCADA database [21]. These parameters have been used to calculate the *sum of active temperatures* (Σ ta, °C), *precipitation* (PP, mm) and *actual sunshine duration* (ASD, hours) during the growing season (April 1st to September 30th), needed to compute the IAOe index for the 1961 to 1990 and for the 1991 to 2013 time periods. The IAO was computed based on [18] formula (1):

$$IAOe = ASD + \Sigma t_a - (PP - 250) \qquad (1)$$

where: *ASD* = *actual sunshine duration* (hours, April 1st to September 30th); Σt_a = the *sum of active temperatures* (sum of daily average temperatures ≥ 10 °C/April 1st to September 30th); *PP* = precipitation (mm, April 1st to September 30th); 250 = (mm) minimum precipitation required for unirrigated grapevine growing. For situations where PP <250 mm, the IAOe is computed as the sum of ASD and Σt_a [18].

Within Romanian wine regions the IAOe values vary between 3793 and >5200 units, with values higher than 5200 characterizing wine areas at the southern limit of the country producing mainly table grapes [22]. In this study, the IAOe is classified according to [23] as follows: IAOe <3793 = unsuitable for grapevine growing (UNS); IAOe = 3793–4300 = suitable for white table wines, sparkling wines, wines for distillates (WTW+SW+WD); IAOe = 4301–4600 = suitable for quality white wines and red table wines (QWW+RTW); IAOe >4600 = suitabile for quality red wines (QRW).

The study involves the analysis at Romanian wine growing regions level of averages and suitability for wine production of the Σt_a, PP, ASD and IAOe for two time periods considered to have different climatic characteristics: 1961–1990 and 1991–2013. In the case of IAOe the changes in suitability and spatial distribution within the eight wine growing regions are analyzed.

3. Results

3.1. Changes in temperature, precipitation and sunshine duration within Romanian wine growing regions between 1961–2013

According to the data in Tables 1 and 2, within the eight Romanian wine growing regions, during the period 1961–2013, there were recorded important increases in temperature and sunshine duration, amid of a relative stability of precipitation during the growing season. According to previous research on grapevine relationships with environmental factors, these are the main determinants of grapes quality [24,25]. Changing their values automatically leads to changes in growing and fruiting conditions for the grapevine with consequences on chemical composition of grapes and implicitly of wines.

The data presented in Table 1 reveals that the Σta (° C) average increased at the Romanian viticulture area level by 198 °C, from 3003 °C between 1961–1990 to 3201 °C between 1991–2013. According to our data the highest increases in Σta occur in the western part of Romania where the CMH and BH wine regions are located and these increases are of +213 °C for the CMH and of +208 °C for the BH. These outcomes are consistent with the data reported by [15] for this part of Romania, but also with the ones revealed by a regional study covering the western part of Romania as well carried out in Hungary by [26]. The lowest temperature increases are recorded in the wine regions from the southern part of Romania (+187 °C for the MOH and 190 °C for the STS), which during the 1961–1990 were the warmest.

Equally important increases are also found in the case of ASD (Table 2), an extremely important factor for anthocyanins accumulation in grapes at the red wine grape varieties [25]. Our data reveal increases of +76 hours to +102 hours for the April 1st to September

Table 1. The sum of active temperatures for Romanian wine regions during the 1961–1990 (61/90) and the 1991–2013 (91/13) time periods. Values computed based on ROCADA climatic data base [21].

Wine region	Σ ta (° C)		Difference (±)	Average for the growing season (° C)	
	61/90	90/13		61/90	90/13
TP	2590	2789	+ 199	14.38	15.49
MH	2961	3154	+ 193	16.45	17.55
MOH	3005	3192	+ 187	16.69	17.73
BH	2999	3207	+ 208	16.66	17.81
CMH	2836	3049	+ 213	15.75	16.93
DH	3109	3311	+ 202	17.27	18.39
DT	3269	3466	+ 197	18.16	19.25
STS	3251	3441	+ 190	18.06	19.11
Averages	**3003**	**3201**	**+198**	**16.67**	**17.43**

Table 2. Precipitation during the growing season (PP) and sunshine duration (ASD) for Romanian wine regions, during the 1961–1990 (61/90) and 1991–2013 (91/13). Values computed based on ROCADA data base [21].

Wine region	PP (mm)		ASD (hours)		Differences	
	61/90	90/13	61/90	90/13	PP	ASD
TP	400	413	1307	1409	+ 13	+ 102
MH	354	354	1418	1515	0	+ 98
MOH	383	392	1433	1509	+ 9	+ 76
BH	382	385	1414	1493	+ 3	+ 79
CMH	377	380	1414	1515	+ 3	+ 101
DH	239	266	1510	1607	+ 27	+ 97
DT	302	320	1518	1609	+ 17	+ 91
STS	313	321	1519	1606	+ 8	+ 87
Averages	**344**	**354**	**1442**	**1533**	**+10**	**+91**

30th in ASD between 1961–2013, with minima in the MOH wine region and maxima in TP wine region, the last one being known as the coldest and poorest in heliothermal resources of all Romanian wine regions [27]. These sunshine increases, with an average of +91 hours at Romanian wine regions level, are suitable for the accumulation of organic compounds in grapes and can be considered to be a factor sustaining grapes quality at red winegrape varieties.

Precipitation dwithin a wide range,uring the growing season (April 1st- September 30th) slightly increased between 1961–2013, with an average of about +10 mm at Romanian wine growing regions level (Table 2). However, the extreme values fall within a wide range, ranging from a minimum of 0 mm for the MH wine region to a maximum of +27 mm for the DH wine region. As compared to grapevine need of precipitation during the growing season [22], one can consider that by the increase of 27 mm, the PP for the DH wine region exceeds the threshold showing the need of grapevine irrigation (≥250 mm). Taking into account the increase of temperatures during the same time period for the DH wine region, and implicitly the intensification of evapotranspiration as a consequence of this phenomenon [28], this increase may not meet the water needs of the grapevine. As for the other wine-growing regions, the increase in the rainfall volume is insignificant. Maintaining the perceived trends is likely to lead to increased in aridity and to the need for grapevine irrigation in the future, especially for the MH, BH, CMH regions characterized by the lowest rainfall

Table 3. Averages of *oenoclimate aptitude index* IAOe for the Romanian wine region, during the 1961–1990 (61/90) and 1991–2013 (91/13). Values computed based on ROCADA climatic data base [21].

Wine region	IAOe					
	Min		Max		Mean	
	61/90	90/13	61/90	90/13	61/90	90/13
TP	2815	3039	4147	4447	3798	4011
MH	2863	3132	4714	5004	4249	4537
MOH	3041	3313	4814	5071	4345	4599
BH	3477	3761	4471	4775	4275	4561
CMH	3089	3371	4383	4702	4134	4446
DH	3870	4125	4829	5106	4625	4896
DT	4480	4761	4884	5144	4736	5005
STS	4469	4722	4937	5203	4708	4978
Averages	**3513**	**3778**	**4647**	**4931**	**4358**	**4629**
Differences	**+265**		**+284**		**+271**	

increases and highest increases in heliothermal resources (Tables 1 and 2).

3.2. Changes in IAOe values within the Romanian wine regions between 1961–2013

Changes in the multi-annual averages of the Σta, PP and ASD between 1961–2013 led to changes in the multiannual averages of the *oenoclimate aptitude index* IAOe, which reveals the suitability of vineyards climate for different types of wine production. During the 1961–2013 at Romanian wine regions level the IAOe values increased by +271 units, with maxima of +312 units for the CMH wine region and minima of +254 units for the MOH wine region. With these increases the average value of the IAOe at Romania wine regions level passes from a value showing the suitability for quality white wines and red table wines to a value (4629 units) revealing the suitability for quality red wines.

Although high, the IAOe Min for the TP, MH, MOH, BH and CMH is still unsuitable for grapevine growing (Table 3), revealing the year-to-year variability and inconsistency of climate suitability for the wine production specific to temperate continental climate which characterizes this part of Europe and which naturally, leads to better or worse vintages. As opposed to the Min for the most wine regions, the IAOe Mean and Max have passed to a higher suitability class for the wine production. In the case of DH, DT, STS, MOH and MH wine regions, where the past IAOe Max was already suitable for red

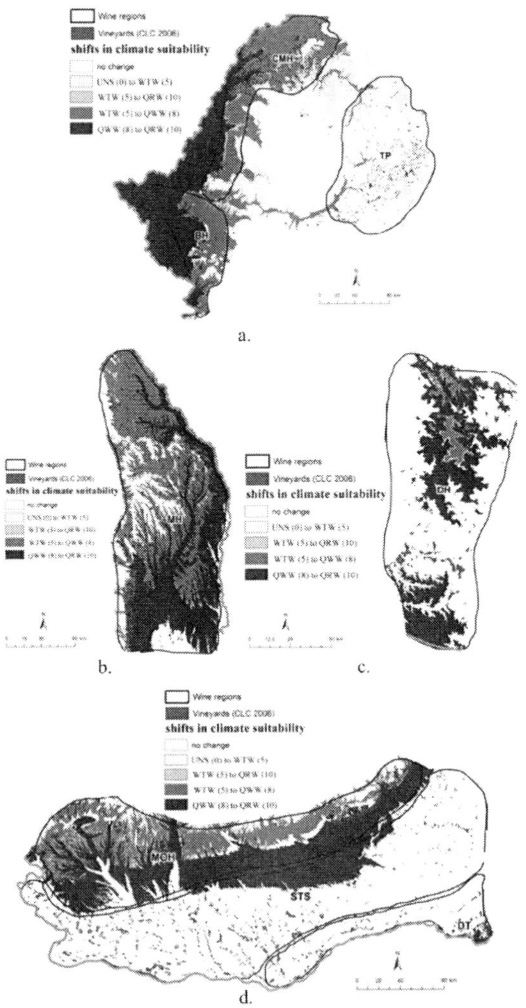

Figure 2. Shifts occurred in the spatial distribution of the IAOe suitability within Romania wine regions between 1961–2013: a. shifts in the BH, CMH and TP wine regions; b. shifts in MH wine region; c. shifts in DH wine region; d. shifts in the MOH, STS and DT wine regions.

wine production, the current Max maintains in the same suitability class but with higher values (Table 4). This indicates a climate evolution towards a predominance of suitability for the red wine production in Romania as well, which corresponds to the predictions of climatologists who studied the perspectives of climate suitability for wine production in the context of climate change for certain regions from the central and eastern part of Europe [26, 29, 30].

3.3. Modifications in climate suitability for wine production within Romanian wine regions between 1961–2013

Analyzing the spatial distribution of changes occurred in climate suitability for wine production of Romanian wine regions, there are major changes in all of them, as well as an expansion of the area suitable for grapevine growing (Fig. 2). Climate suitability massively expands within Transylvania Plateau where the TP wine region is located (Fig. 2a) and to a less extent within Moldavia Plateau, where the MH wine region is located (Fig. 2b). Also in the Transylvania Plateau suitable conditions for the RTW production occur, both in the traditional wine production area and outside it, on the low valleys of the main rivers. In all the other wine regions except the extreme southern ones (DT and STS) there are major shifts in spatial distribution of suitability classes: in the CMH and BH from the western part of Romania, the passage of climate suitability of the traditional areas from QWW to QRW and also the enlargement of this new climate suitability type to the entire Western Plain, far beyond the traditional wine-growing areas can be observed (Fig. 2a). What is more in the CMH wine region on a narrow band on the slopes of the Western Hills, a shift from the climate suitability for WTW (the lowest suitability level) to the suitability for QRW (the highest suitability level) can be observed. This evolution is validated by the initiation of the QRW production in the area during the last decade. In the MH, MOH and DH wine regions, climate suitability for wine production expanded and covered the whole area: the initial climate suitability types passed to an upper suitability type, while the areas unsuitable for the wine production in the past became climatic suitable for WTW production (Figs. 2b, 2c, 2d). These shifts are revealed by the current IAOe averages for the wine growing centers (**WGC**) encompassed by the eight Romanian wine regions (Table 5).

Table 4. Climate suitability types for wine production for Romanian wine regions during the 1961–1990 (61–90) and 1991–2013 (90–13), depending on the IAOe averages for the two time periods*.

Wine region	Climate suitability types according to IAOe values					
	Min		Max		Mean	
	61/90	90/13	61/90	90/13	61/90	90/13
TP	UNS	UNS	WTW	QWW	WTW	4011
MH	UNS	UNS	QRW	QRW	WTW	QWW
MOH	UNS	UNS	QRW	QRW	QWW	QWW
BH	UNS	UNS	QWW	QRW	WTW	QWW
CMH	UNS	UNS	QWW	QRW	WTW	QWW
DH	WTW	WTW	QRW	QRW	QRW	QRW
DT	QWW	QRW	QRW	QRW	QRW	QRW
STS	QWW	QRW	QRW	QRW	QRW	QRW

*UNS = unsuitable for grapevine growing; WTW = suitable for white table wines, sparkling wines, wines for distillates; QWW = suitable for quality white wines and red table wines; QRW = suitable for quality red wines.

Depending on their location on latitude and altitude, the IAOe values for the WGC differ from one another, being in some situations far beyond the region's IAOe average. According to our research (unpublished data) this is due to a specific local climate, characterized by higher or lower precipitation volume than the average one for the region. It is the case of Lechinţa and Bistriţa WGC in the TP where the precipitations diminished as compared to the increases recorded in all other WGC in the same region; or the Murfatlar and Medgidia WGC in the DH, where, on the contrary, rainfall has increased well above the region's average, which is reflected in the particular values of IAOe. This fact is directly related to the spatial variation of climate in the context of climate change. It is

 DOI: 10.1051/bioconf/20170901026

Table 5. Averages and differences for the *oenoclimate aptitude index* (IAOe) for some Romanian wine growing centers (WGC) for the 1961–1990 (61/90) and 1991–2013 (91/13) time periods. Values were computed based on gridded data at 10 × 10 km resolution of temperature, precipitation and sunshine duration extracted from the ROCADA climatic data base [21].

Wine region	WGC	IAOe		Difference
		61/90	91/13	(±)
TH	Lechința	3892	4221	+ 329
	Turda	3894	4172	+ 278
	Jidvei	3983	4269	+ 286
	Blaj	3991	4282	+ 291
MH	Bohotin	4346	4653	+ 307
	Avereşti	4155	4459	+ 304
	Huşi	4436	4741	+ 305
	Panciu	4108	4397	+ 289
	Odobeşti	4333	4614	+ 281
	Coteşti	4244	4519	+ 275
MOH	Pietroasele	4066	4338	+ 272
	Tohani	4156	4401	+ 245
	Urlați	4398	4634	+ 236
	Halânga	4647	4914	+ 267
	Drăgăşani	4560	4819	+ 259
BH	Teremia	4471	4775	+ 304
	Recaş	4339	4626	+ 287
CMH	Sanislău	4250	4548	+ 298
	Valea lui Mihai	4288	4586	+ 298
	Diosig	4365	4671	+ 306
DH	Măcin	4637	4913	+ 276
	Tulcea	4766	5039	+ 273
	Murfatlar	4757	5025	+ 255
TH	Ostrov	4746	5010	+ 264
	Greaca	4668	4933	+ 265
STS	Cetate	4814	5085	+ 271
	Poiana Mare	4896	5166	+ 270

Table 6. Differences between the locations on altitude (m) of climate suitability classes during the 1991–2013 and 1961–1990 time periods at Romania level.

Climate suitability classes	Differences (km)			
	Min	Max	Range	Mean
UNS	+ 96	0	−96	+ 129
WTW	+ 105	+ 182	+ 77	+ 131
QWW	+ 36	+ 181	+ 145	+ 80
QRW	0	+ 130	+ 130	+ 30

also demonstrated by the fact that the climate change does not manifest homogeneously at the regional level, which determines topoclimates with particular suitability for the wine production, higher or lower than the region's average to which they belong. In order to establish the climate suitability type determined by the new climatic context, a fine scale analysis of vineyard climate is required.

Statistical analyse of differences occured in the spatial distribution of climate suitability zones at Romania level during the two time periods reveal their major shifts on altitude and latitude (Tables 6 and 7).

In the case of altitude shifts, the spatial range of the UNS area narrowed by 96 m while its location on altitude increased by 129 m; the spatial range of the WTW, QWW and QRW climate suitability types increased by 77 to 145m, while their location on altitude shifted by 30 to 131 m.

Table 7. Differences between the latitudinal coordinates of the climate suitability classes during the 1991–2013 as compared to 1961–1990.

Climate suitability classes	Differences (km)			
	Min	Max	Range	Mean
UNS	+ 1.8	−3.9	−5.7	−3.92
WTW	+ 1.2	−1.8	−3.0	+ 89.0
QWW	+ 77.0	+ 55.4	−21.6	+ 101.0
QRW	0.0	+ 127.3	+ 127.3	+ 45.0

Table 8. The structure of climate suitability for wine production in Romania during the 1961–1990 and 1991–2013.

Climate suitability class	% in the total area	
	1961–1990	1991–2013
QRW	21.20	36.57
QWW, RTW	16.24	16.13
WTW, SW, WD	25.47	20.28
Unsuitable	37.07	27.01
–	**100**	**100**

Regarding the location on latitude of climate suitability classes, statistical analysis reveals (Table 7): a 5.7 km narrowing of the UNS spatial interval for the recent time period as compared to the 1961–1990 time period and the shift of its minimum altitude limit by 1.8 km to N as compared to 1961–1990; narrowing of the average spatial range of the WTW and QWW climatic zones by 3 to 21.6 km, along with the northward shifts of the minimum latitude by 1.2 to 77 km; widening of the spatial range of the climate-suitability zone for QRW by 127 km and a 45 km northern shift of its average location.

Simultaneously with these major spatial and structural shifts, the structure of area with climate suitable for wine production is changing in Romania in an absolutely advantageous way for the wine production (Table 8). The area with climate unsuitable for wine production decreased by 10%, from 37.07% between 1961–1990 to 27.1% between 1991–2013; the area with climate suitable for WTW, SW and WD diminished from 25.47% to 20.28%; the area with climate suitable for QWW remained quite stable at 16% of the total; and the area with climate suitable for red wine production increased by 15%, from 21.2% between 1961–1990 to 36.57% between 1991–2013.

4. Conclusions

The impact of climate change on Romania's viticulture is a major one. According to the results of our study, the changes in temperature, sunshine duration and precipitation averages in the context of climate change determined the expansion of climatically suitable area for wine grape growing in Romania, the changes in regional climate suitability structure and spatial distribution. New climatically suitable areas for wine grape growing have emerged, most of them in the Transylvanian Plateau, the coldest region of the country, while the climate suitability of traditional wine regions and vineyards has changed in most cases in a positive way. The areas climatically suitable for white table wine production in the past have become climatically suitable for quality white wines production; areas suitable for quality white wines have evolved to ones suitable for red table wine and quality red wines; while areas traditionally suitable for quality red wine production keep their potential, but they are experiencing increases in heliothermal resources that makes them resemble to areas specialized in table grape production. Climatic suitability for wine grape growing shifts on altitude and latitude to cooler areas, respectively to higher elevations and to northern latitudes. The developments revealed by this study are consistent with those revealed or predicted based on climate models up to date. On the other hand, we notice important local variations of suitability for wine grape growing, which manifest in the context of climate change and which are not revealed by regional analysis, requiring a fine scale analysis of suitability of the vineyard climate. The results of this study can be used to develop strategies for Romanian viticulture adaptation to climate change, by adapting vineyard variety assortments and by a judicious placement of the new grapevine plantations.

References

[1] IPCC, *Contribution of working group I to the fifth assessment report of the intergovernmental panel on climate change* (Cambridge University Press, 2013)

[2] L.B. Webb, P.H. Whetton, E.W.R. Barlow, Aust J Grape Wine Res. **13**, 165–175 (2008)

[3] VO Sadras, M.A Moran, M. Bonada, Aust J Grape Wine R **19**, 95–106 (2013)

[4] G.V. Jones, R.E. Davis, Am J Enol Vitic **51**, 249–261 (2000)

[5] R.R. Nemani, M.A. White, D.R. Cayan, G.V. Jones, S.W. Running, J.C. Coughlan, Clim Res, **19**, 25–34, doi: 10.3354/Cr019025 (2001)

[6] G.V. Jones, M.A. White, O.R. Cooper, K. Storchmann, Clim Change **73**, 319–343 (2005)

[7] L. Hannah, P.R. Roehrdanz, M. Ikegami, A.V. Shepard, M.R. Shaw, G. Tabor, L. Zhi, P.A. Marquet, R.J. Hijmans, Proc Natl Acad Sci USA **110**, 6907–6912 (2013)

[8] H. Fraga, A.C. Malheiro, M.J. Mouthino Pereira, J.A. Santos, Int J Biometeorol **57**, 909–925 (2013)

[9] M. Moriondo, G.V. Jones, B. Bois, C. Dibari, R. Ferrise, G. Trombi, M. Bindi, Clim Change **119**, 825–839 (2013)

[10] OIV, *State of the Vitiviniculture World Market* `http://www.oiv.int/en/technical-standards-and-documents/statistical-analysis/state-of-vitiviniculture` (2016)

[11] A.E. Croitoru, A. Piticar, Int J Climatol **33**, 1987–2001 (2013)

[12] M.V. Bîrsan, A. Dumitrescu, DM. Micu et al., Nat Hazards **74**, 1899 (2014)

[13] R. Bojariu, V.M. Bârsan, R. Cică, L. Velea, S. Burcea, Al. Dumitrescu et al., *Schimbările climatice – de la baze fizice la riscuri şi adaptare. Administraţia Naţională de Meteorologie* (Printech Bucureşti, 2015)

[14] Busuioc, A. Dobrinescu, M.V. Bîrsan, A. Dumitrescu, A. Orzan, Int J Climatol **35**, 1278–1300 (2015)

[15] A. Dumitrescu, R. Bojariu, M.V. Bîrsan, L. Marin, A. Manea Theor Appl Climatol **122**, 111–119 (2014)

[16] M. Bucur, L. Dejeu, Bulletin UASMV Horticulture **73**, 126–134 (2016)

[17] L.M. Irimia, C.V. Patriche, H. Quenol, L. Sfâcă, C. Foss, Theor Appl Climatol. doi: 10.1007/s00704-017-2033-9 (2017)

[18] Şt. Teodorescu, A.I. Popa, G. Sandu *Oenoclimatul României* (Scientific and Encyclopedic Press, Bucharest, 1987)

[19] M.C. Peel, B.L. Finlayson, T.A. McMahon, Hydrol Earth Syst Sci **11**, 1633–1644 (2007)

[20] EEA, `http://www.eea.europa.eu/data-and-maps/data/clc-2006-vector-data-version-2` (2006)

[21] A. Dumitrescu, M.V. Bîrsan, Nat Hazards **78**, 1045–1063 (2015)

[22] M. Oşlobeanu, M. Macici, M. Georgescu, V. Stoian *Zonarea soiurilor de viţă de vie în România* (Ceres Press, 1991)

[23] LM Irimia, CV Patriche, H Quenol, J Int Sci Vigne Vin **48**, 145–167 (2014)

[24] W.M. Kliewer, Lider L.A. J Am Soc Hortic Sci **95**, 766–769 (1970)

[25] N.K. Dokoozlian, W.M. Kliewer, J Am Soc Hortic Sci **121**, 869–874 (1996)

[26] Mesterházy, R. Mészáros, Pongrácz R. **118**, 193–206 (2014)

[27] L.M. Irimia, *Biologia, ecologia şi fiziologia viţei de vie* (Ion Ionescu de la Brad Ed, 2012)

[28] E. Duchêne, C. Schneider, Agron Sustain Dev **25**, 93–99 (2005)

[29] M. Stock, F.W. Gerstengarbe, P.C. Werner, Acta Horticulturae **689**, 29-40 (2005)

[30] M. Kryza, M. Szymanovski, M. Blas, Theor Appl Climatol **122**, 207–218 (2015)

Determinations of adaptation level of wine grape varieties in terms of climatic data in wine growing regions of turkey

Fadime Ateş[a] and Hülya Uysal

Manisa Viticulture Research Institute Manisa, Turkey

Abstract. Wine grapes are adapted to a wide range of climate; the best production occurs in regions that meet certain specific climatic conditions. Temperatures during the growing season can affect grape quality and viability. Beneficial climatic conditions will improve the wine's quality. In this study it is aimed that to determine suitable wine grape varieties for the cultivation in some areas of Southeastern Anatolia Region, Eastern Anatolia, Central Anatolia Region, Central Black Sea Region, Aegean Region and Marmara Region in Turkey with related to climate requirements. For this reason, long-term climatic data were collected by meteorological stations including, Diyarbakır (Çermik, Çüngüş), Central-Elazığ, Nevşehir (Central and Ürgüp), Ankara (Kalecik) Tokat (Central, Erbaa, Niksar, İzmir (Seferihisar, Menderes, Urla) ve Denizli (Çal ve Güney), Çanakkale (Bozcaada, Bayramiç), Tekirdağ. In this study heliotermic and hdyrothermic indices were calculated and evaluated for appropriate viticultural practice in this region. It was found that Boğazkere and Öküzgözü in Southeastern Anatolia Region and Eastern Anatolia Region; Kalecik Karası, Dimrit and Narince in Central Anatolia Region and Emir in Central Black Sea Region; Bornova Misketi, Cabernet Sauvignon, Syrah, Alicante Bouschet, Carignane, Kalecik Karası, Merlot, Öküzgözü, Çal Karası, Boğazkere, Sultani Çekirdeksiz in Aegean Region; Karasakız,Karalahana, Vasilaki, Cabernet Sauvignon, Merlot, Syrah, Alicante Bouschet, Semillion, Cinsaut, Yapıncak, Gamay, Merlot, Cabernet Sauvignon can be adapted and grown well in terms of climatic conditions in Marmara Region respectively.

1. Introduction

Turkey has a very rich genetic potential as it is the gene center of grapevine.

Turkey is a major producer of grapes in the world and viticulture is one of the major branches of agriculture with respect to production area and its large share of income in Turkish national economy. Grapevine is grown in almost all parts of Turkey and has been produced commercially in many regions of the country for many years.

Turkey is among the important viticulturist countries with its 478.000 ton of viticulture field and 4, 26 millions of ton of grape production. (5th one for the area and 6th one in the production) 52.9% of table grape, 36.3% to be dried, 10.8% for wort and wine [1].

The climatic conditions have a very important role in the constitution of the maturity, yield and quality values of the wine grape variety. The criterias determining the relations between the Vitis vinifera and the climate and if the substrate is convenient for the grape vine farming cultivation have been presented in the studies [2–7]. Specific temperature data are the basic information for any grape variety. Reaching to the phonologic phases, key for any varieties, is possible when 10 °C heat accumulation is used as base. [8, 8–10]. Each variety of grapes needs a specific heat accumulation starting from the beginning of the vegetation period until the maturity period [9]. Maturity period of grapes is closely connected to the local climate conditions and phenological growth of the variety. Phenological growth is a genetic feature varies from variety to variety [8]. It has been detected that effective heat summation demand is between 1210 °C (Cardinal) and 1500 °C (Müşküle) in Ankara conditions; 1033 °C (Uslu) and 1538 °C (Alphonse L.) in Mediterranean conditions [11]. If a grape variety cannot mature its grape in the demanded level in ecology, it means that it cannot be recommended to be cultivated for hot region.

Oraman [12], has emphasized in the study that he has performed that regular sunshine duration is important as well as the temperature and that the annual sunshine duration of a grape wine should not be less than 1300 hours. According to Çelik et all [13] this value should not be less that 1500–1600 hours in an economical grape wine cultivation and that the vegetation duration must be more that 180 days. Other conditions apart from the vegetation are also important in the grape variety choice of regions (especially in the regions with frost risk). The resistances of the grape varieties to the lower temperatures are very different. Accordingly, it has been detected in the studies held that if the temperature is lower than −20, 5 °C in 3 or less times 10 years, that region is convenient for the grape wine cultivation [13] Not only the heat accumulation but also sunshine and amount of precipitation should also be taken into the consideration for the convenience to the environment during the vegetation period of a grape variety. according to [2–4, 14]. Karantonis [14] the temperature values of the environment is not the unique important factors for the grape cultivation; what is really

[a] e-mail: `fadime.ates@tarim.gov.tr`

important is the balance between the temperature and sunshine. It is stated that the heliothermic proportion (X*12-3/H*10-3) value 1 or higher environments are convenient for the grape cultivation. The method of detecting the varieties that might be convenient in a specific area by studying the relations between Grape varieties and environmental conditions (climate, land) has been applied in many countries of the world.

The grape varieties that might be cultivated according to the climate data of the Aegean Region and Marmara Region have been detected with the studies that have been held in our country. To improve the wine grape cultivation in Aegean Region, the climate factors of the region have seen to have accorded with the region. The studies regarding the detection not only the domestic wine grape but also qualified wine grapes of foreign origins have gained a lot importance in the recent years.

The objective of this study was to determine suitable wine grape varieties for the cultivation in some areas of Southeastern Anatolia Region, Eastern Anatolia, Central Anatolia Region, Central Black Sea Region, Aegean Region and Marmara Region in Turkey with related to climate requirements.

2. Materials and methods

2.1. Materials

In this study was the first group material was provided wine grape growers in Diyarbakır (Çermik, Çüngüş), Central-Elazığ provinces of the Southeastern Region; Nevşehir (Central and Ürgüp), Ankara (Kalecik) provinces of the in Central Anatolia Region; Denizli (Güney and Çal) and İzmir (Menderes, Seferihisar, Urla) provinces of the Aegean Region; Çanakkale (Bozcaada, Bayramiç), Tekirdağ provinces of the Marmara Region. Other materials consist of the climatic data (temperature, rain etc) of long years provided from the meteorological stations of the counties.

2.1.1. Experimental site

Çermik and Çüngüş counties in Diyarbakır provinces of the Southeastern Anatolia region; Central-Elazığ in Elazığ provinces of the Eastern Anatolia region, being the study area has 3681 km^2 of total area. Wine grape area of the examined study area is 53950 decare and total wine grape production amount is 43753 tons (Table 1). Central-Elazığ county has the largest viticulture field (37100 decare) and it is Central-Elazığ county that has the furthest production amount (33245 tons). Average yield of the study area is 749 kg / decare (Table 1).

Central-Nevşehir and Ürgüp counties in Nevşehir provinces; Kalecik counties in Ankara provinces of Central Anatolia Region region; Central-Tokat, Erbaa and Niksar in Tokat provinces of the Eastern Anatolia region, being the study area has 6461 km^2 of total area. Wine grape area of the examined study area is 105803 decare and total wine grape production amount is 72234 tons (Table 1). Central-Nevşehir county has the largest viticulture field (29900 decare) and it is Central-Nevşehir county that has the furthest production amount (23920 tons). Average yield of the study area is 661 kg / decare (Table 1).

Güney and Çal counties in Denizli provinces; Menderes, Seferihisar and Urla counties in İzmir provinces of the Aegean region, being the study area has 329 km^2 of total area. Wine grape area of the examined study area is 87200 decare and total wine grape production amount is 48750 tons. (Table 1) Çal county has the largest viticulture field (400000 decare) and it is Güney county that has the furthest production amount (21250 tons). Average yield of the study area is 562 kg / decare (Table 1).

Bozcaada and Bayramiç counties in Çanakkale provinces; Şarköy counties in Tekirdağ provinces of the Marmara region, being the study area has 1793.6 km^2 of total area. Wine grape area of the examined study area is 42320 decare and total wine grape production amount is 33955 tons (Table 1). Şarköy county has the largest viticulture field (20000 decare) and it is Şarköy county that has the largest production amount (5625 tons). Average yield of the study area is 820 kg / decare (Table 1).

2.2. Methods

Çermik and Çüngüş counties in Diyarbakır provinces of the Southeastern Anatolia region; Central-Elazığ in Elazığ provinces of the Eastern Anatolia region; Central -Nevşehir and Ürgüp counties in Nevşehir provinces; Kalecik counties in Ankara provinces of Central Anatolia Region region; Central-Tokat, Erbaa and Niksar in Tokat provinces of the Eastern Anatolia region; Güney and Çal counties in Denizli provinces; Menderes, Seferihisar and Urla counties in İzmir provinces of the Aegean region; Bozcaada and Bayramiç counties in Çanakkale provinces; Şarköy counties in Tekirdağ provinces of the Marmara region had been the sampling area. The counties have been chosen so as to provide 50% of the wine grape production. 40 Wine grape growers from each county have been interviewed with telic sampling method.

'Effective heat summation (EHS)', being one of the efficient parameters to determine the needs of the wine grape varieties in the specific region, has been calculated. In the calculation of this value expressed as Day-Temperature, 10 °C (threshold temperature), which is accepted as the average temperature when the grape wine growth starts, has been selected as baseline (Çelik and, 1998). The assessments have been formulated as the date and day when the average of many years in several stations for 10 °C, threshold of the grape vine growth, reaches to threshold temperature (end date and day of the vegetation) and the accumulation of the temperatures that the average temperature for each day in this period is higher that the threshold temperature.

$EHS = \sum(T\text{-}Te)$
EHS = accumulation of the effective temperature (°C-day)
T = daily average temperature (°C)
Te: threshold temperature (°C).

Hydrothermic indices was calculated to determine the possibility of supplying the water need of the grape vine from the natural ways. In this calculation it will be determined that if the water consummation related to the temperature change of the varieties in the May-July period and rain, can be provided from the natural ways.

Hydrothermic indices: $(\sum P*10)/\sum T°$
$\sum P$ = Total rain (mm),

Table 1. Surface area of the counties (km^2), viticulture area (decare), production (ton) and yield (kg/ decare) values (Anonym, 2010).

REGIONS	COUNTIES	Surface area of the counties (km^2)	Viticulture area (decare)	Production (ton)	Yield (kg/decare)
SOUTHEASTERN ANATOLİA REGİON AND EASTERN ANATOLİA REGİON	Çermik-DİYARBAKIR	991	6850	6508	950
	Çüngüş-DİYARBAKIR	472	10000	4000	400
	Central-ELAZIG	2218	37100	33245	896
	TOTAL	**3681**	**53950**	**43753**	**749**
CENTRAL ANATOLİA AND CENTRAL BLACK SEA REGİONS	Central – NEVSEHIR	536	29900	23920	800
	Ürgüp-NEVSEHIR	565	31950	19726	**617**
	Kalecik-ANKARA	1341	8500	5525	650
	Central – TOKAT	1924	17910	12537	700
	Erbaa-TOKAT	1177	11493	6896	600
	Niksar-TOKAT	918	6050	3630	600
	TOTAL	**6461**	**105803**	**72234**	**661**
AEGEAN REGION	Menderes-İZMİR	775	96500	7720	800
	Seferihisar-İZMİR	371	32500	2600	800
	Urla-İZMİR	728	8000	2200	150
	Güney-DENIZLİ	534	290000	21250	691
	Çal-DENIZLİ	1521	400000	14800	370
	TOTAL	**3929**	**827000**	**48570**	**562**
MARMARA REGİON	Bozcaada-ÇANAKKALE	37.6	6420	5625	900
	Bayramiç-ÇANAKKALE	1275	15900	11130	700
	Şarköy-TEKİRDAG	481	20000	17200	860
	TOTAL	**1793.6**	**42320**	**33955**	**820**
	TOTAL	**15864,6**	**1029073**	**198512**	**698**

Heliothermic indices have been calculated to determine the balance between the temperature and sunshine for the grape production.

Heliothermic indices: $X*12^{-3}/H*10^{-3}$

X: accumulation of the effective heat temperature during the vegetation period

H: total hours of days (daytime).

3. Results and discussion

Diyarbakır (Çermik and Çüngüş) and Elazığ (Central-Elazığ) two of the cities where important wine grapes are being produced in the Southeastern Anatolia Region and Eastern Anatolıa Region, have been the study area. Terrestial level varies from 29.2 (Central-Elazığ) −30.1 (Çüngüş) according to the geographic and topographic conditions of the study area. The effects of the variety of the terrestial level, annual total rain and altitude difference have been observed. Average temperature value of the study area for many years have been measured between 14.6 °C (Menderes) and 16.5 °C in Çermik (Table 2). Extremely high temperature values are between 42.2 °C (Central-Elazığ) and 46.1 °C (Çermik). Peak value of the low temperatures varies between −23.4 °C (Çermik) and −22.6 °C (Central-Elazığ) (Table 2).

Table 2. Climatic data in the Study Area.

REGİONS	STATIONS	LATITUDE (°N)	LONGITUDE (°E)	ALTITUDE (m)	AVERAGE ANNUA TEMP. (°C)	AVERAGE ANNUAL RAINFALL (kg/m²)	TERRESTIAL LEVEL (°C)	MAX. TEMP. (°C)	MIN. TEMP. (°C)
SOUTHEASTERN ANATOLİA REGİON AND EASTERN ANATOLİA REGİON	Çermik-DİYARBAKIR	38°17'	39°49'	700	16.5	492.7	29.8	46.1	–23.4
	Çüngüş-DİYARBAKIR	38°13'	39°17'	1000	16.2	490.8	30.1	45.8	–22.8
	Central-ELAZIĞ	38°40'	39°13'	1093	14.6	429.6	29.2	42.2	–22.6
CENTRAL ANATOLİA AND CENTRAL BLACK SEA REGİONS	Central-NEVŞEHİR	38°37'	34°42'	1223	13.8	483.4	23.2	39.5	–21.2
	Ürgüp-NEVŞEHİR	38°37'	34°54'	602	14.2	490.0	21.8	38.9	–22.4
	Kalecik-ANKARA	40°05'	33°24'	725	14.8	485.2	22.6	40.8	–21.4
	Central-TOKAT	40°18'	36°33'	649	12.3	446.0	24.0	45.0	–22.1
	Erbaa-TOKAT	40°40'	36°33'	113	12.7	439.2	24.2	45.0	–22.1
	Niksar-TOKAT	40°34'	36°56'	339	12.5	456.4	23.7	45.0	–2.1
AEGEAN REGION	Seferıhısar-İZMİR	38°11'	26°50'	28	16.3	582	18.3	42.4	–6.2
	Menderes-İZMİR	38°16'	27°08'	53	17.8	590	18.6	40	–5.4
	Urla-İZMİR	38°19'	26°45'	75	17.2	535	18.1	41.1	–4.3
	Güney-DENİZLİ	38°09'	29°04'	847	15.3	514	21.7	37.7	–8.7
	Çal-DENİZLİ	38°05'	26°03'	911	14.5	477	22.4	39.4	–10.4
MARMARA REGİON	Bozcaada-ÇANAKKALE	39°42'	26°45'	28	15.4	508	17.3	35.4	–8.2
	Bayramiç-ÇANAKKALE	39°48'	26°37'	72	14.5	655	21.4	39.8	–13.5
	Şarköy-TEKİRDAĞ	40°37'	27°04'	10	14.4	534	20.2	38.4	–11.5

Nevşehir (Central-Nevşehir and Ürgüp), Ankara (Kalecik) and Tokat (Central-Tokat, Erbaa and Niksar) three of the cities where important wine grapes are being produced in Central Anatolia and Central Black Sea Regions, have been the study area. Terrestial level varies from 21.8 (Ürgüp) −24.2 (Erbaa) according to the geographic and topographic conditions of the study area. The effects of the variety of the terrestial level, annual total rain and altitude difference have been observed. Average temperature value of the study area for many years have been measured between 12.3 °C (Central-Tokat) and 14.8 °C in Kalecik (Table 2). Extremely high temperature values are between 38.9 °C (Ürgüp) and 45.0 °C (Central-Tokat, Erbaa and Niksar). Peak value of the low temperatures varies between −22.4 °C (Ürgüp) and −21.2 °C (Central- Nevşehir) (Table 2).

Denizli (Güney and Çal) and İzmir (Menderes, Seferihisar, Urla) two of the cities where important wine grapes are being produced in the Aegean Region, have been the study area. Terrestial level varies from 18.3 (Urla) −22.4 (Çal) according to the geographic and topographic conditions of the study area. The effects of the variety of the terrestial level, annual total rain and altitude difference have been observed. Average temperature value of the study area for many years have been measured between 7.8 °C (Menderes) and 14.5 °C in Çal (Table 2). Extremely high temperature values are between 42.4 °C (Seferihisar) and 39.4 °C (Çal). Peak value of the low temperatures varies between −10.0 °C (Çal) and −4.3 °C (Urla) (Table 2).

Tekirdağ (Şarköy) and Çanakkale (Bozcaada and Bayramiç) two of the cities where important wine grapes are being produced in the Marmara Region, have been the study area. Terrestial level varies from 18.3 (Bozcaada) −21.4 (Bayramiç) according to the geographic and topographic conditions of the study area. The effects of the variety of the terrestial level, annual total rain and altitude difference have been observed. Average temperature value of the study area for many years have been measured between 14.4 °C (Şarköy) and 15.4 °C in Bozcaada (Table 2). Extremely high temperature values are between 39.8 °C (Bayramiç) and 35.4 °C (Bozcaada). Peak value of the low temperatures varies between −13.5 °C (Bayramiç) and −8.2 °C (Bozcaada) (Table 2).

We assess the bio climatic values in the climatic surface in the regard of productions of the wine grape varieties in the Southeastern Anatolia Region and Eastern Anatolia Region (Table 3). It has been detected that the vegetation period starts between 23 March (Çüngüş) and 11 April (Central- ELAZIĞ). It has been understood that the potential vegetation duration was between 220 days (Central- ELAZIĞ) and 245 days (Çermik); and the total temperature was between 4525 °C (Central- ELAZIĞ) and 5605 °C (Çüngüş). Heliothermic indices, presenting the combination of the vegetation days and temperature factor, were high in all counties examined in the scope of the wine grape cultivation and that there were important differences between counties. The heliothermic indices values are between 6.25 (Central- ELAZIĞ) and 8.53 (Çermik) and these values are appropriate for the cultivation of the varieties of the wine grape cultivation in the examined area (according to the vegetation duration and Effective temperature).

The bio climatic values in the climatic surface in the regard of productions of the wine grape varieties in the Central Anatolia and Central Black Sea Regions (Table 3). It has been detected that the vegetation period starts between 09 April (Central- TOKAT and Niksar) and 2 April (Kalecik). It has been understood that the potential vegetation duration was between 199 days (Central-NEVŞEHİR) and 211 (Kalecik); and the total temperature was between 3528 °C (Ürgüp) and 4823 °C (Central-TOKAT). The heliothermic indices values are between 6.15 (Kalecik) and 9.46 (Ürgüp) and these values are appropriate for the cultivation of the varieties of the wine grape cultivation in the examined area.

In the Aegean Region the bio climatic values in the climatic surface in the regard of productions of the wine grape varieties (Table 3) it has been detected that the vegetation period starts between 07 March (Menderes) and 06 April (Çal). It has been understood that the potential vegetation duration was between 205 days (Çal) and 255 days (Seferihisar); and the total temperature was between 4154 °C (Çal) and 6425 °C (Seferihisar). The heliothermic indices values are between 5.79 (Çal) and 9.15 (Seferihisar) and these values are appropriate for the cultivation of the varieties of the wine grape cultivation in the examined area.

The bio climatic values in the climatic surface in the regard of productions of the wine grape varieties in the Marmara Region (Table 3). It has been detected that the vegetation period starts between 25 March (Bozcaada) and 15 April (Şarköy). It has been understood that the potential vegetation duration was between 242 days (Bayramiç) and 280 days (Bozcaada); and the total temperature was between 5087 °C (Bayramiç) and 5414 °C (Bozcaada). The heliothermic indices values are between 6.47 (Bayramiç) and 7.53 (Bozcaaad) and these values are appropriate for the cultivation of the varieties of the wine grape cultivation in the examined area.

A parallelism has been found between the counties of Regions of Turkey in terms of precipitation amount and dispersion in the vegetation period. It is seen that wine grape cultivation is possible in the examined regions without watering in the wine grape growing.

As a result of the assessment made in Southeastern Anatolia Region and Eastern Anatolia Region it was seen that the the effective heat summation (EHS) was between 2315 °C and 2914 °C (Table 3). The temperature need of Boğazkere grape variety was 1252 °C (Table 4) and temperature need of Öküzgözü grape variety was 1542 °C. Total efficient temperature in Çermik county was 2874 °C and EHS of Çüngüş county was 2914 °C in Diyarbakır. The total efficient temperature of Central-ELAZIĞ in Eastern Anatolia Region county was 2315 °C and the total temperature needs of the varieties were as follows: Öküzgözü 1542 °C, Boğazkere 1525 °C (Table 4). The effective heat summation need of these grape varieties demand a lower total temperature than in Southeastern Anatolia Region and Eastern Anatolia Region effective heat summation values.

In the Central Anatolia and Central Black Sea Regions it was seen that the EHS was between 2137 °C and 2961 °C (Table 3). EHS in Central-Nevşehir county was 2874 °C and EHS of Ürgüp county was 2914 °C in Nevşehir. The EHS need of grape varieties were as follows: 1400 °C

DOI: 10.1051/bioconf/20170901027

Table 3. Potential bioclimatic values of the study area in terms of wine grape cultivation (daily average temperature ≥10 °C Period) Hydrothermal.

REGİONS	STATIONS	BUDBURST (DATE)	TIME (DAYS)	∑T (°C)		AVERAGE TEMP. (°C)	RAINFALL (mm)	HELİOTERMIC INDICE	HYDROTERMIC INDICE (May-July)
				TOTAL	EFFICIENT				
SOUTHEASTERN ANATOLİA REGİON AND EASTERN ANATOLIA REGİON	Çermik-DİYARBAKIR	29. 3	245	5245	2874	22.3	203.5	8.53	0.32
	Çüngüş-DİYARBAKIR	23.3	241	5605	2914	23.1	198.9	8.49	0.30
	Central-ELAZIĞ	11.4	220	4525	2315	21.8	168.6	6.25	0.39
	Diferance	**19**	**25**	**2339**	**599**	**1.3**	**34.9**	**2.28**	**0.09**
CENTRAL ANATOLİA AND CENTRAL BLACK SEA REGİONS	Central-NEVŞEHİR	10.4	199	3686	2198	20.1	208.6	9.43	0.27
	Ürgüp-NEVŞEHİR	10.4	204	3528	2137	19.9	215.2	9.46	0.22
	Kalecik-ANKARA	12.4	211	3902	2145	19.8	199.8	6.15	0.35
	Central-TOKAT	09.4	205	4823	2961	21.8	213.9	8.23	0.23
	Erbaa-TOKAT	10.4	202	4798	2872	22.4	224.4	8.27	0.24
	Niksar-TOKAT	09.4	203	4812	2854	20.9	221.6	8.32	0.25
	Diferance	**3**	**12**	**1195**	**824**	**2.6**	**24.6**	**3.31**	**0.13**
AEGEAN REGION	Seferihisar-İZMİR	11.3	255	6425	3875	25.2	207.9	9.15	0.85
	Menderes-İZMİR	07.3	254	6042	3502	23.7	245.2	8.40	0.72
	Urla-İZMİR	10.3	252	6125	3605	24.3	227.3	7.53	0.83
	Güney-DENİZLİ	02.4	207	4460	2390	21.5	210.7	6.49	0.50
	Çal-DENİZLİ	06.4	205	4154	2104	20.2	212.0	5.79	0.45
	Diferance	**22**	**50**	**2271**	**1771**	**50**	**34.5**	**3.36**	**0.40**
MARMARA REGİON	Bozcaada-ÇANAKKALE	20.3	280	5414	3964	24.8	223.9	7.53	0.22
	Bayramiç-ÇANAKKALE	03.4	242	5087	3447	25.4	254.8	6.47	0.35
	Şarköy-TEKİRDAĞ	05.4	250	5151	3533	24.3	236.7	6.66	0.34
	Diferance	**21**	**58**	**327**	**517**	**0.6**	**40.9**	**1.07**	**0.12**

Table 4. Features of some grape wine varieties and their temperature needs (Çelik at all, 1988a).

Wine Grape Varieties	Temperature Needs °C)
Alicante Bouschet	1398
Boğazkere	**1525**
Bornova Misketi	**1250**
Cabernet Sauvignon	1382
Carignan	1547
Cinsaut	1453
Çal Karası	**1395**
Dimrit	**1400**
Emir	**1502**
Gamay	1363
Kalecik Karası	**1421**
Karalahana	**1535**
Karasakız	**1522**
Merlot	1402
Semilion	1382
Narince	**1418**
Sultani Çekirdeksiz	**1380**
Syrah	1399
Vasilaki	**1435**
Yapıncak	**1545**

(Dimrit), 1502 °C (Emir) and 1421 °C (Kalecik Karası) (Table 4). EHS of Kalecik county was 2145 °C. The temperature need of Kalecik Karasıgrape variety was 1421 °C (Table 4). EHS in Central TOKAT county was 2961 °C and EHS of Erbaa county was 2872 °C and EHS of Niksar county in Tokat. The temperature need of Narince grape variety was 1418 °C (Table 4). The effective heat summation need of these grape varieties demand a lower total temperature than effective heat summation values.of in the Central Anatolia and Central Black Sea Regions.

In Aegean region it was seen that the EHS was between 2104 °C and 3875 °C (Table 3) and the temperature need of Alicante Bouschet grape variety was 1398 °C (Table 4) and temperature need of Carignan grape variety was 1547 °C. Total efficient temperature in Urla county was 3605 °C and the needs of grape varieties were as follows: Cabernet Sauvignon 1382 °C, Syrah 1399 °C and Alicante Bouschet 1398 °C (Table 4). EHS of Güney county in Denizli was 2390 °C and the temperature needs of the varieties are as follows: Syrah (1399 °C), Kalecik Karası (1421 °C), Boğazkere (1525 °C), Merlot (1402 °C), Öküzgözü (1542 °C), Cabernet Sauvignon (1382 °C), Çalkarası (1395 °C) and Sultani Çekirdeksiz (1380 °C) and the temperature need of these grape varieties demand a lower total temperature than these values. The total efficient temperature of Çal county was 2104 °C and the total temperature needs of the varieties were as follows: Öküzgözü 1542 °C, Çal Karası (1395 °C), Boğazkere 1525 °C, Merlot 1402 °C and Sultani Çekirdeksiz (1380 °C) (Table 4). The effective heat summation need of these grape varieties demand a lower total temperature than effective heat summation values of Aegean region.

In Marmara region it was seen that the EHS was between 3447 °C and 3964 °C (Table 3). The effective heat summation of Bozcaada county in Çanakkale was 4964 °C and the effective heat summation need of grape varieties were as follows: Alicante Bouschet, 1398 °C, Cabernet Sauvignon 1382 °C, Karalahana 1535 °C, Karasakız 1522 °C, Merlot 1402 °C, Syrah 1399 °C, and Vasikali 1435 °C (Table 4). The effective heat summation of Bayramiç county in Çanakkale was 3447 °C and the effective heat summation need of grape varieties were as follows: Cabernet Sauvignon 1382 °C, Karasakız 1522 °C. The total efficient temperature of Şarköy county in Tekirdağ was 3533 °C and the effective heat summation need of grape varieties were as follows: Cabernet Sauvignon 1382 °C, Cinsaut 1453 °C, Gamay 1363 °C, Merlot 1402 °C Semilion 1382 °C, Yapıncak 1545 °C, (Table 4). The effective heat summation need of these grape varieties demand a lower total temperature than effective heat summation values of Marmara region.

According to the climatic data from the meteorological station and the evaluation carried out with the wine grape growers has been found that Boğazkere, and Öküzgözü in the region of Southeastern Anatolia and Eastern Anatolia Region; Kalecik Karası, Dimrit and Narince in Central Anatolia Region and Emir in the Central Black Sea Region; Bornova Misketi, Cabernet Sauvignon, Syrah, Alicante Bouschet, Carignane, Kalecik Karası, Merlot, Öküzgözü, Çal Karası, Boğazkere, Sultani Çekirdeksiz in Aegean Region; Karasakız, Karalahana, Vasilaki, Cabernet Sauvignon, Merlot, Syrah, Alicante Bouschet, Semillion, Cinsaut, Yapıncak, Gamay, Merlot, Cabernet Sauvignon can be adapt and grow well in terms of climatic conditions in Marmara Region respectively.

References

[1] TUİK, (Turkish Statistical Institute), Agricultural Production Statistics. (www.tuik.gov.tr) (2015)
[2] J., Branas, Viticulture 343–357 (1974)
[3] G.H. Constantinescu, Bulletin de l'OIV **40**, 441 (1967)
[4] P. Huglin, 'Biologie et écologie de la vigne.' (Payot Editions:Lausanne, France) 264–309 (1986)
[5] H. Işık, *Üzüm Çeşidi ve Uyumlu Yer Seçimine ilişkin V. Vinifera L. Üzerine Biyoklimatik Araştırma.* Science Title Thesis. Pleven (1988)
[6] H. Işık, H. Öztürk, Gökçay, E., Kader, S., Ege Bölgesinde Sofralık Üzüm Yetiştiriciliğine İlişkin Biyoklimatik Araştırmalar. Manisa Bağcılık Araştırma Enstitüsü, Final Report Manisa (2001a)
[7] H. Işık, N.Y. Delice, C. Özer., *Sofralık Üzüm Çeşitlerinin Marmara Bölgesi Koşullarına Biyoekolojik uyumu ile Muhafaza ve Pazarlama SorunlarıÜzerinde Araştırmalar.* Final Report Tekirdağ. (2001b)
[8] C. Van Leeuwen, C. Garnier Agut, B. Baculat, G. Barbeau, E. Besnard, B.Bois, J.-M. Boursiquot, I. Chuine, T. Dessup, T. Dufourcq, I. Garcıa-Cortazar, E. Marguerıt, C.Monamy, S. Koundouras, J.-C Payan., A. Parker, V. Renouf, B. Rodriguez-Lovelle, J.-P. Roby, J. Tonıetto, and W. Trambouze,

Proceedings of the 7th International Terroir Congress, Changins, Switzerland (Agroscope Changins-Wädenswil: Switzerland) pp. 222–227 (2008)
[9] A.J. Winkler, J.A. Cooke, M. Kliewer, L.A. Lider, General viticulture, California Uni. Pres. 710 p. (1974)
[10] H., Çelik, Y.S., Y., Ağaoğlu, Fidan, B., Marasalı., G.,Söylemezoğlu, Genel Bağcılık, 253 s. ANKARA (1998)
[11] H. Uzun, Journal of The Faculty of Agriculture, Akdeniz University 9, 40–60 (1996)
[12] M.N. Oraman, 1970. Bağcılık Tekniği I. Ankara university agricultural faculty publications number: 415, Ankara (1970)
[13] H. Çelik, B. Marasalı, İ.Demir, Türkiye III. Bağcılık Sempozyumu, Bursa (1988)
[14] N. Karantonis, 1.er Symposium International su l'Ecologie de la Vigne, Constanta (1978)

Uva da tavola ipocalorica, possibilità di coltivazione e nuove prospettive di mercato / Hypocaloric table grapes: Possibility of cultivation and new market prospects

Antonacci Donato[a], Alba Vittorio[b], Perniola Rocco[c], Roccotelli Sabino, e Genghi Rosalinda[d]

Research Centre for Viticolture and Enology (CREA –VE) seat of Turi-Bari (Italy), Via Casamassima, 148–70010, Italy

Abstract. According to the CODEX STAN 255 (2007), table grapes can be harvested when the refractometric index reaches at least 16° Brix. Grapes with a lower refractometric index are accepted provided the sugar/acid ratio (Total Soluble Solids/Tritatable Acidity – TSS/TA) is at least equal to 20:1 if the Brix level is comprised between 12.5° and 14° Brix, 18:1 if the Brix level is comprised between 14° and 16° Brix. Some table grapes grown in Mediterranean climate can accumulate TSS higher than 16° Brix, with low levels of acidity, thus leading to a TSS/AT ratio even greater than 30: 1. Grapes that accumulate TSS less than 12.5° Brix and low TA levels, hardly reach the recommended ratio TSS/TA of 20:1, rather stopping at lower values.

Based on these considerations in 2012 an experimental trial of four years started on a total of 103 table grape cvs. Phenological phases in relation to the acidic and sugar content were measured in order to assess the TSS/TA ratio and build regression lines with TSS. Almost all cvs. showed a significant linear regression, with R2 values higher than 0.80 and a standard error between 00.18 and 1.74.

TSS values of 103 cvs. at TSS/TA ratio of 20:1, predicted by means of linear regression equations, , were subjected to cluster analysis based on Euclidean Distance. We identified nine clusters, of which the first two with the predicted TSS between 10.04° and 11.77° Brix (7 cvs.) and 12.08° and 12.62° Brix (7 cvs.), respectively.

Some of the cvs. showing TSS values below 12.5° Brix at the TSS/TA of 20: 1, were subjected to a test panel of 24 tasters, who were asked to indicate the acceptability of these grapes on hedonic scale of 9 points.

Preliminary results showed that for some table grapes with TSS below 12.5° Brix, a SST/AT ratio lower than the recommended 20:1 can be proposed, while preserving an acceptability by the consumer. These results suggest a more modern perspective on grapes nutritional intake, which takes into account the real needs of Western countries where diabetes, hyperglycemia, obesity and cardiovascular diseases are the main dismetabolisms arising from a diet too rich in calories. These grapes thus represent a new frontier and an opportunity for producers of table grapes, who want to enterprise a market share on "grape diet", and represent a genetic basis for breeding aimed at obtaining new hypocaloric grape cultivars.

1. Introduzione

L'uva da tavola è un frutto non climaterico, che a maturità ha una bassa attività fisiologica. Le caratteristiche organolettiche ottimali vengono normalmente raggiunte alla maturità commerciale. Il principale indice per definire il momento di raccolta è il contenuto in zuccheri indicato come solidi solubili totali (SST), o° Brix. Come indici di maturità delle uve troviamo anche l'acidità titolabile (TA) ed il rapporto TSS-TA. Varietà diverse da quelle "bianche" hanno anche esigenze minime di maturazione legate al colore tipico della varietà ed identificato come percentuale di bacche del grappolo che presentano una accettabile intensità e copertura del colore. Altri criteri di qualità per le uve da tavola sono una buona apparenza, priva di difetti, acini con buccia sottile, dimensioni grandi, buona consistenza e sapore. Il rachide deve essere turgido e verde (cioè non essiccato e imbrunito). Attualmente i requisiti minimi di maturità variano con la cultivar, le zone di produzione ed i mercati di riferimento; tuttavia, gli standard minimi di maturazione si stanno gradualmente armonizzando per principali mercati di riferimento.

Secondo il CODEX STAN 255 (2007), le uve da tavola possono essere raccolte al raggiungimento di un indice rifrattometrico almeno di 16° Brix. Uve con un indice rifrattometrico più basso sono commercializzabili a condizione che il rapporto zuccheri/acidità totale (Solidi Solubili Totali/ Acidità Titolabile - SST/AT) sia almeno pari a 20:1 se il grado Brix è compreso tra 12,5° e 14° Brix; rapporto che diventa 18:1 se il grado Brix è compreso tra 14° e 16 ° Brix.

Successivamente la Comunità Europea ha recepito e ripreso nel 2011 queste indicazioni, apportando aggiornamenti relativi ad alcune eccezioni nell'apposito paragrafo

[a] L'autore ha ideato ed organizzato la ricerca, coordinando le attività e la redazione dell'articolo.

[b] L'autore ha contribuito con gli approfondimenti di elaborazione statistica dei dati e nella redazione dell'articolo.

[c] L'autore ha contribuito alla effettuazione delle attività di campo, di laboratorio analitico, di valutazione sensoriale e di redazione dell'articolo.

[d] L'autore ha contribuito alla effettuazione delle attività di campo e di laboratorio analitico.

denominato "Requisiti di maturazione" secondo il quale le uve da tavola devono avere un indice rifrattometrico di 12° Brix per le cultivar Alphonse Lavallée, Cardinal e Victoria, 13° Brix per le altre varietà con semi, 14° Brix per tutte le varietà apirene. Mentre, relativamente al rapporto SST/AT il regolamento comunitario non fa rifermento a valori specifici, limitandosi ad autorizzare un rapporto "soddisfacente".

Questa discrasia tra Codex, che definisce anche i rapporti SST/AT, e il regolamento comunitario che tiene conto unicamente dell'indice rifrattometrico, appare naturale conseguenza dell'impossibilità di identificare un parametro univoco per la determinazione del momento di raccolta idoneo alla commercializzazione per tutte le cultivar di uva da tavola nel rispetto della loro accettabilità da parte del consumatore. A seconda della cultivar, del mercato di riferimento, o dell'ambiente di coltivazione, possono variare i criteri di adozione per la definizione dei requisiti minimi di maturazione delle cultivar di uve da tavola, prendendo in considerazione ora solo SST o AT, o piuttosto il loro rapporto [1–4, 6, 7].

Pertanto, sarebbe auspicabile un allineamento tra CODEX e regolamento CE, anche alla luce del fatto che oltre alle tre cultivar considerate commercializzabili al grado rifrattometrico di 12° Brix (Alphonse Lavallée, Cardinal e Victoria) ve ne possono essere altre con le medesime caratteristiche che andrebbero tenute in considerazione.

In tale ambito, l'OIV suggerisce come requisito minimo di commercializzazione per le uve bianche o rosa il raqggiungimento dei 16° Brix e se inferiore a 16° Brix il raggiungimento di un rapporto SST/AT di 20:1. Per le uve rosse, invece, per contenuti in SST da 12,5 a 16° Brix dovranno presentare un rapporto minimo SST/AT di 20:1, mentre uve con un ° Brix inferiore a 12,5 non sono considerate mature.

La ricerca ha inteso verificare, nell'ambito di una buona accettabilità al consumo, sia se esistono e possono essere individuate varietà di uve da tavola con un contenuto di SST inferiore ai 12,5° Brix, sia se esiste la possibilità di proporre un rapporto SST/AT inferiore a quello raccomandato di 20:1 quale standard minimo di commercializzazione delle uve da tavola.

L'importanza di eventuali risultati positivi è del tutto evidente, per una valutazione qualitativa più moderna, che consideri anche l'apporto nutrizionale delle uve consumate nella dieta, che tenga conto delle esigenze reali di paesi occidentali in cui il diabete, l'iperglicemia, l'obesità e le malattie cardiovascolari rappresentano sempre più le principali dismetabolie, derivanti da una dieta eccessivamente ricca di calorie.

Questo potrebbe influenzare significativamente il comportamento del consumatore in relazione all'inserimento nella dieta delle uve da tavola, a seconda dei parametri di qualità e maturità posseduti dall'uva.

Ne deriva che il comportamento del consumatore in relazione alle uve da tavola, logicamente, può variare a seconda dei parametri di qualità e maturità dell'uva.

La possibilità di coltivare e produrre uva da tavola a basso contenuto energetico, da poter mettere a disposizione nella dieta, rappresenterebbe una nuova frontiera e un'opportunità per i consumatori, che disporrebbero di frutta/uva a più basso contenuto calorico. Si otterrebbe quindi il beneficio dell'inserimento nell'alimentazione umana di un frutto importante come l'uva, contraddistinto dai noti effetti benefici, limitando l'apporto glucidico delle varietà tradizionalmente impiegate nel settore.

2. Materiali e metodi

2.1. Valutazione collezione di 103 varietà

La ricerca è stata condotta presso l'Azienda Sperimentale "Lamarossa" del CREA-UTV sita in agro di Rutigliano (BARI) (Lat. 40,956 N, Lon. 17,008 E) nel quadriennio 2012–2015. Le 103 cultivar di uva da tavola oggetto della presente ricerca sono allevate a tendone e potate con 4 capi a frutto. Sono state considerate almeno 3 piante per ogni cultivar e su ciascuna di esse sono state rilevate le fasi fenologiche. Per ogni anno sono stati effettuati diversi campionamenti durante la maturazione dell'uva, valutando il contenuto in Solidi Solubili Totali (SST) e l'Acidità Titolabile (AT) espressa in g/L di acido tartarico, prelevando 20 bacche/grappolo ripetendo su 3 grappoli/pianta, e quindi su 3 piante, per un totale di 180 bacche per campione varietale.

I SST sono stati misurati mediante indice rifrattometrico espresso in ° Brix, mentre l'AT è stata ottenuta mediante titolazione con NaOH (0,1 N) in presenza di blu di bromotimolo come indicatore e i valori espressi in g/L. Tutti i punti di maturazione e il loro relativo giorno giuliano di rilievo derivanti dai vari anni di durata dell'esperimento, sono stati linearizzati in riferimento ad unica scala cronologica annuale.

2.2. Analisi sensoriale

E' stato inoltre condotto un test di assaggio su un totale di 18 cultivar di uva da tavola a diverso grado di maturazione, e quindi di SST, per un totale di 54 campioni. Il test ha coinvolto 24 assaggiatori a cui è stato chiesto di indicare l'accettabilità di queste uve su "scala edonistica" di 9 punti. L'accettabilità gustativa di ogni singolo campione è stata valutata mediante:

- il calcolo della "Mediana" degli scores degli assaggiatori,
- il calcolo della % di accettabilità (%CA) secondo il seguente rapporto: Numero di panelisti che hanno dato un giudizio >5 / n° totale panelisti.

Le analisi statistiche, sia univariate che multivariate, sono state effettuate mediante il software Statgraphics Centurion XV ver. 15.1.02.

3. Risultati e discussione

Per ogni varietà è stato calcolato il rapporto SST/AT al momento del prelievo per la realizzazione di grafici di distribuzione con SST/AT sull'asse delle ascisse e SST su quello delle ordinate.

Sulla base di questi grafici (Figs. 1, 2, 3, 4) è stata calcolata la retta di regressione con la relativa equazione per ognuna delle 103 cultivar. Inoltre, per avvalorare la bontà del "fitting" del modello lineare è stato calcolato il coefficiente di determinazione R2, che è la proporzione di variazione totale della variabile dipendente spiegata da quella indipendente e serve per misurare "quanto" della variabile dipendente Y (SST) sia predetto dalla variabile

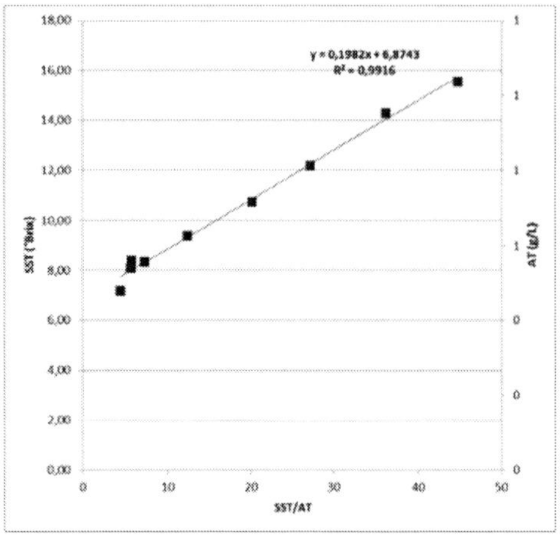

Figura 1. Esempio di distribuzione della cv Victoria.

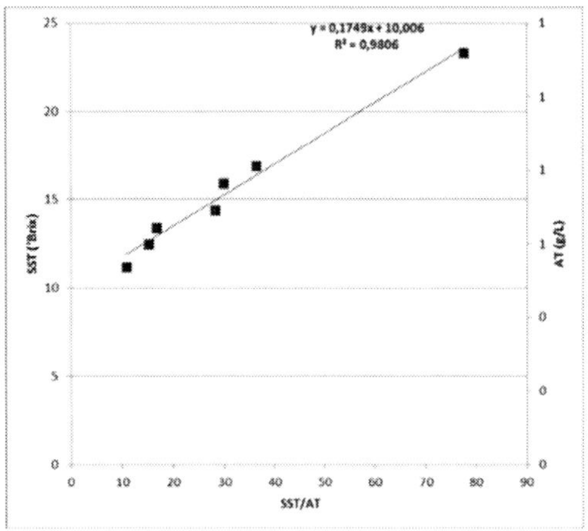

Figura 2. Esempio di distribuzione della cv Sultanina.

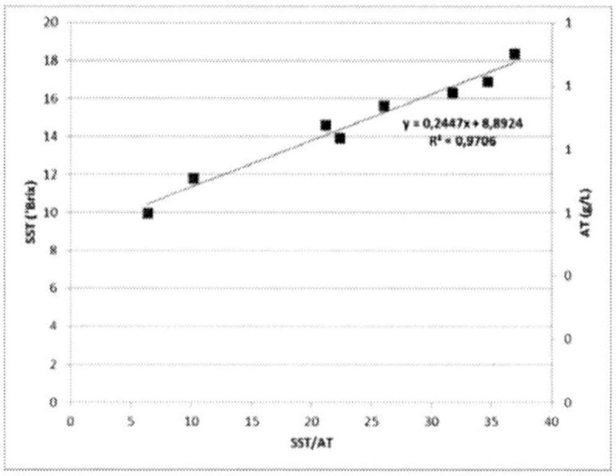

Figura 3. Esempio di distribuzione della cv Italia.

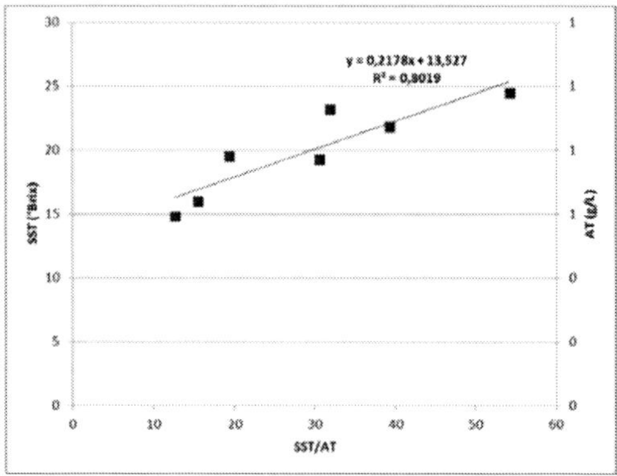

Figura 4. Esempio di distribuzione della cv Red Flame.

indipendente X (SST/AT) e, quindi, per valutare la bontà dell'equazione di regressione ai fini della previsione sui valori della Y.

La significatività della regressione lineare è stata testata mediante il test F di Fisher-Snedecor. I valori plottati nei grafici che sono risultati esterni agli intervalli di confidenza, calcolati con una significatività del 95%, sono stati esclusi dall'analisi di regressione. Le diverse varietà sono state quindi ordniate in funzione del contenuto in solidi solubili totali raggiunti al rapporto SST/AT pari a 20:1 (Tabella 1).

Inoltre, è stato calcolato l'errore standard che indica la deviazione standard dei residui e serve per costruire i limiti di previsione per le nuove osservazioni.

Sulla base delle equazioni di regressione di ogni singola cultivar è stato predetto il valore di SST al raggiungimento di un rapporto SST/AT di 20:1 (da calcolo).

I valori predetti di SST sono stati utilizzati per la suddivisione delle varietà in classi di frequenza (considerando otto classi con intervalli corrispondenti a 10,04–11,50; 11,51–12,50; 12,51–13,50; 13,51–14,50; 14,51–15,50; 15,51–16,50; 16,51–17,50; 17,51–18,50). Tale distribuzione è stata confermata anche da un analisi cluster delle 103 cultivar basata sulla Distanza Euclidea, al fine di evidenziare le differenze fra le cultivar e le possibilità di raggruppamento in funzione delle caratteristiche esaminate (la clusterizzazione è in grado di distinguere 9 nodi - dati non mostrati nel presente lavoro).

Dal grafico della Fig. 5 è possibile rilevare che delle 103 varietà esaminate il 93% è giustificato all'interno dei parametri descritti dalla risoluzione OIV citata in premessa avendo un intervallo compreso fra 12,5 e 18,5 con un rapporto pari a 20:1, evidenziano la fondatezza della stessa risoluzione; mentre per il restante 10,7% delle varietà esaminate la risoluzione pone alcuni limiti in quanto, non potrebbero essere commercializzate, perché aventi un contenuto in SST inferiori a 12,5.

Le analisi sensoriali hanno poi mostrato come per alcune varietà come la Michele Palieri e la Victoria con un contenuto in Solidi solubili totali inferiore a 12,5 e con un rapporto superiore a 20:1, l'accettabilità risultava buona con un %CA superiore alla media.

Per altre varietà come la Maxia, l'Autumn royal e la Crimson seedless, con un contenuto in SST superiore a quello indicato come riferimento per la commercializzazione, e con un rapporto SST/AT superiore a 20:1 è stata evidenziata un'accettabilità inferiore a quella media. In particolare, nel caso della Crimson seedless, il contenuto in SST superiore a 16° Brix non è sufficiente a garantire il raggiungimento di un buon livello di accettabilità.

Tabella 1. Parte a. Delle regressioni lineari per SST/AT pari a 20.

per SST/AT = 20:1	Predicted SST	Standard Error of extimates (±)	Linear equation	R2	Significativity
Poeta Metabon rs.	10,04	0,78	Y = 0,2372 X + 9,3005	0,88	***
Vittoria b.	10,84	0,29	Y = 0,1982 X + 6,8743	0,99	***
M. Palieri n.	10,99	0,45	Y = 0,1484 X + 8,0237	0,97	***
Early victoria b.	11,17	0,86	Y = 0,1512 X + 8,1504	0,87	***
Sublima b.	11,50	1,06	Y = 0,1479 X + 8,5433	0,86	***
Carati b.	11,65	0,88	Y = 0,1442 X + 8,7701	0,93	***
Chasselas rose rs.	11,77	0,85	Y = 0,1381 X + 9,0034	0,91	**
Autumn royal n.	12,08	1,14	Y = 0,119 X + 9,702	0,81	**
Supernova rs.	12,11	0,76	Y = 0,1549 X + 9,0145	0,94	***
Maxia b.	12,21	0,87	Y = 0,1955 X + 8,2953	0,94	***
Zeiruel b.	12,43	0,5	Y = 0,1542 X + 9,3469	0,94	***
Reginone b.	12,56	1,4	Y = 0,1776 X + 9,0034	0,86	***
Inzolia imperiale_Martellata b.	12,58	0,43	Y = 0,2523 X + 7,5326	0,97	***
Perla di Yalova b.	12,62	1,06	Y = 0,0858 X + 10,908	0,83	***
Afrodite b.	12,83	1,36	Y = 0,2568 X + 7,6978	0,82	*
Red globe rs.	12,84	1,14	Y = 0,2002 X + 8,8376	0,86	***
I. Conegliano precoce 218 n.	12,89	1,32	Y = 0,1891 X + 9,1063	0,86	***
Alphonso Lavallée n.	12,90	0,81	Y = 0,2544 X + 7,8539	0,94	***
Nerona n.	12,93	0,7	Y = 0,2605 X + 7,7202	0,95	***
Rubinia n.	12,98	1,24	Y = 0,1588 X + 9,8043	0,83	***
Regina b.	12,99	0,4	Y = 0,204 X + 8,9061	0,97	***
Perlon n.	13,08	1,25	Y = 0,2463 X + 8,1574	0,88	***
Beogradska bessemena b.	13,20	0,8	Y = 0,1642 X + 9,9117	0,88	**
Blush seedless rs	13,20	0,67	Y = 0,2336 X + 8,5293	0,96	***
Imperatrice rs.	13,21	0,93	Y = 0,1913 X + 9,3792	0,86	***
Lord Rothermere b.	13,29	0,23	Y = 0,128 X + 10,73	0,99	***
Doradilla b.	13,34	0,78	Y = 0,061 X + 12,123	0,84	**
Cardinal n.	13,35	1,09	Y = 0,2275 X + 8,8029	0,91	***
Sugraone b.	13,41	1,18	Y = 0,1532 X + 10,346	0,90	***
Pansè precoce b.	13,50	1,03	Y = 0,2742 X + 8,0178	0,95	***
Sultanina n.	13,50	0,61	Y = 0,1749 X + 10,006	0,98	***
Christmas rose rs.	13,51	0,76	Y = 0,2057 X + 9,3985	0,94	***
Pansè rosa di Malaga rs.	13,53	1,3	Y = 0,1013 X + 11,505	0,67	***
Coarna negra n.	13,57	0,74	Y = 0,189 X + 9,7852	0,97	***
Ruby seedless n.	13,60	0,99	Y = 0,2505 X + 8,5941	0,93	***
Rodites rs.	13,62	0,43	Y = 0,2372 X + 8,8742	0,96	***
Alba b.	13,67	1,3	Y = 0,3003 X + 7,6636	0,93	***
Bermestia violacea rs.	13,68	0,82	Y = 0,2669 X + 8,3458	0,88	**
Inzolia b.	13,70	1,65	Y = 0,2998 X + 7,7019	0,87	***
Madeleine royale b.	13,70	0,88	Y = 0,2322 X + 9,0601	0,96	***
Summer royal n.	13,71	1,63	Y = 0,1994 X + 9,7192	0,84	**
Italia b.	13,79	0,51	Y = 0,2449 X + 8,8924	0,97	***
Primus b.	13,82	0,9	Y = 0,1639 X + 10,537	0,96	***
Moscato d'Amburgo n.	13,84	0,93	Y = 0,1559 X + 10,717	0,95	***
Fiorenza b.	13,92	1,32	Y = 0,2123 X + 9,6712	0,86	***
Melissa b.	13,93	1,21	Y = 0,1421 X + 11,089	0,82	***
Black opal n.	13,98	0,68	Y = 0,2516 X + 8,9467	0,95	***
Zagatera n.	14,01	1,12	Y = 0,1519 X + 10,969	0,79	**
Almeria b.	14,04	0,62	Y = 0,1124 X + 11,789	0,86	**
Duraca b.	14,07	0,97	Y = 0,1876 X + 10,316	0,85	***
Baresana b.	14,16	1,08	Y = 0,1356 X + 11,447	0,76	*
Trakya ilkeeren n.	14,19	1,34	Y = 0,2443 X + 9,3029	0,91	***

Tabella 1. Parte b. Delle regressioni lineari per SST/AT pari a 20.

per SST/AT = 20:1	Predicted SST	Standard Error of extimates (±)	Linear equation	R2	Significativity
Huevo de gato rs.	14,23	1,01	Y = 0,2324 X + 9,5	0,87	***
Corinto b.	14,23	1,22	Y = 0,2821 X + 8,5	0,96	***
Prime b.	14,28	0,82	Y = 0,196 X + 10,3	0,92	***
I. Conegliano precoce 199 n.	14,30	1,18	Y = 0,2233 X + 9,8	0,95	***
Olivetta nera n.	14,31	0,61	Y = 0,1025 X + 12,	0,77	***
Sciradzuoli blanc b.	14,32	1,16	Y = 0,0782 X + 12,	0,69	*
I. Alzey real rs.	14,40	0,96	Y = 0,1156 X + 12,	0,83	**
Paula b.	14,43	0,87	Y = 0,2521 X + 9,3	0,96	***
Crimson seedless rs.	14,52	0,18	Y = 0,1684 X + 11,	0,98	***
Red Italia rs.	14,66	0,75	Y = 0,2224 X + 10,	0,89	***
Rossa del Merg rs.	14,67	0,84	Y = 0,2225 X + 10,	0,95	***
Dimiat b.	14,71	0,86	Y = 0,1901 X + 10,	0,89	***
Chasselas lacinie b.	14,74	0,97	Y = 0,1544 X + 11,	0,86	***
I. Bellini n.	14,77	0,97	Y = 0,1971 X + 10,	0,86	**
Autumn seedless b.	14,78	0,86	Y = 0,2297 X + 10,	0,94	***
Pizzutello b.	14,80	1,15	Y = 0,1112 X + 12,	0,90	**
Gros vert b.	14,83	0,44	Y = 0,308 X + 8,66	0,96	***
Olivetta b.	14,85	0,54	Y = 0,3586 X + 7,6	0,96	***
Torbato b.	14,91	1,07	Y = 0,2105 X + 10,	0,87	**
Giada b.	14,92	1,3	Y = 0,2573 X + 9,7	0,91	***
I. Crujdero di Spagna b.	14,93	0,87	Y = 0,2817 X + 9,2	0,94	***
Thompson seedless b.	14,96	0,82	Y = 0,1943 X + 11,	0,96	***
Black pearl n.	14,98	1,15	Y = 0,1524 X + 11,	0,61	n.s.
Centennial b.	15,02	1,72	Y = 0,2618 X + 9,7	0,88	***
Urreti b.	15,03	1,44	Y = 0,2341 X + 10,	0,82	**
Lacrima di Maria b.	15,16	0,55	Y = 0,2778 X + 9,5	0,95	***
Delizia di Vaprio b.	15,20	0,72	Y = 0,2493 X + 10,	0,96	***
Regal seedless b.	15,21	1,33	Y = 0,2151 X + 10,	0,82	***
I. Ceresa n.	15,23	0,83	Y = 0,3295 X + 8,6	0,94	***
Zibibbo b.	15,28	1,29	Y = 0,218 X + 10,9	0,85	**
Perla di Csaba b.	15,31	0,85	Y = 0,2278 X + 10,	0,96	***
Pepin d'Hispanie rs.	15,41	0,73	Y = 0,2046 X + 11,	0,96	***
Argentina rs.	15,41	0,68	Y = 0,2231 X + 10,	0,97	***
Black finger n.	15,43	0,93	Y = 0,2932 X + 9,5	0,87	**
Delight b.	15,46	1,64	Y = 0,2499 X + 10,	0,90	***
Pobjeda b.	15,47	1,14	Y = 0,2415 X + 10,	0,91	***
San Pietro e Paolo n.	15,47	0,97	Y = 0,234 X + 10,7	0,90	***
Perlette b.	15,54	0,23	Y = 0,2906 X + 9,7	1,00	***
Angyal Delzo b.	15,58	1,38	Y = 0,2119 X + 11,	0,84	***
Giovanna n.	15,64	0,65	Y = 0,236 X + 10,9	0,96	***
Aurora b.	15,65	1,39	Y = 0,2582 X + 10,	0,81	*
Patrizia rs.	15,76	1,18	Y = 0,1501 X + 12,	0,75	***
Prunesta n.	15,80	0,65	Y = 0,3215 X + 9,3	0,98	***
Raisin d'Afrique blanc b.	15,82	0,68	Y = 0,3389 X + 9,0	0,97	***
Ciminnita b.	16,03	1,33	Y = 0,132 X + 13,3	0,51	n.s.
Carina b.	16,16	1,71	Y = 0,2153 X + 11,	0,86	***
Chasselas violet b.	16,16	1,47	Y = 0,3938 X + 8,2	0,88	***
Marsigliana nera di Sicilia n.	16,29	1,14	Y = 0,2554 X + 11,	0,87	***
Serna rs.	16,60	0,63	Y = 0,2454 X + 11,	0,98	***
Dawn seedless b.	17,19	1,16	Y = 0,2064 X + 13,	0,91	***
Red flame n.	17,88	1,74	Y = 0,2178 X + 13,	0,80	**

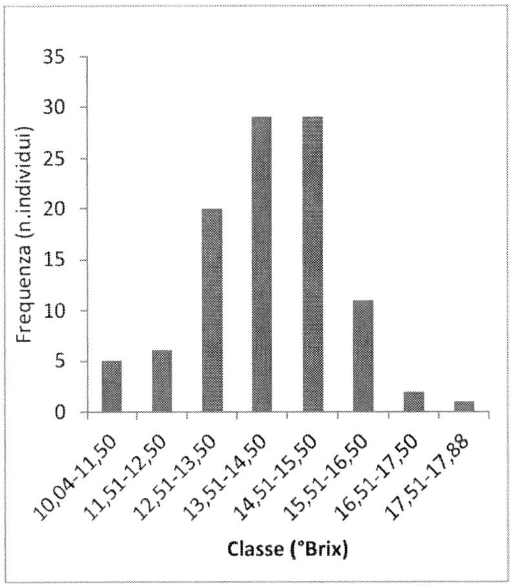

Figura 5. Istogrammi delle frequenze delle varietà.

Tabella 2. Analisi sensoriale.

Varietà	Brix	SST/AT	AT	Mediana	% CA
Palieri 1	11	16	6,9	3,9	31,6%
Palieri 2	11,6	33	3,5	5,1	52,2%
Victoria 1	11,6	39	3	5,1	65,2%
Victoria 2	11,9	22	5,3	3,6	40,0%
Italia 2	12,1	15	8,3	3,9	39,1%
Palieri 3	12,2	27	4,5	5,1	91,3%
I.Gargiulo-1	12,3	13	9,8	4,1	36,8%
Carati 1	12,6	22	5,9	5,4	65,0%
Maxia 2	13	22	5,9	4,5	36,8%
Supernova 1	13,9	26	5,3	7,7	100,0%
Maxia 5	14,1	29	4,9	4,4	36,4%
Redglobe 1	14,1	24	5,9	6,4	94,7%
Afrodite 1	14,3	24	6	6,8	90,0%
Italia 4	14,5	22	6,6	5,4	68,2%
Autumn royal 1	14,7	26	5,6	4,5	47,4%
Sugraone 1	14,9	28	5,4	6,3	90,0%
Carati 3	14,9	52	2,9	6,3	90,5%
Victoria 3	15	57	2,6	5,9	63,6%
Inzolia 1	15,1	22	6,8	6,2	73,7%
Cardinal 1	15,1	31	4,8	5,1	100,0%
Inzolia 5	15,4	27	5,7	6,3	87,0%
Afrodite 2	15,5	49	3,2	5,9	90,9%
Thompson 1	15,6	24	6,5	6,4	86,4%
Redglobe 3	15,7	37	4,2	6,4	89,5%
Autumn royal 2	15,8	36	4,4	6	78,9%
ISV 1	16	30	5,3	5,1	84,2%
Crimson 1	16,2	24	6,8	4,5	42,1%
Regal 1	16,8	25	6,6	5,4	56,5%
Thompson 3	16,9	28	6	6,4	95,5%
Crimson 2	17,1	27	6,5	4,7	47,4%
Thompson 4	17,4	32	5,5	6,4	100,0%
Sugraone 4	17,8	58	3,1	6,3	71,4%
Regal 2	18,7	32	5,9	6,3	82,6%
Supernova 3	19,2	56	3,5	7,7	95,2%
Down seedless 1	22,8	56	4,1	7,7	100,0%

4. Conclusioni

Ci sono diverse varietà che mostrano un rapporto SST/AT di 20:1 con solidi solubili inferiori a 12,5° Brix.

Non tutte le varietà che mostrano rapporti superiori a 20:1 hanno raggiunto un buon livello di accettabilità.

Non tutte le varietà che hanno conseguito 16,0° Brix hanno raggiunto un buon livello di accettabilità.

È possibile individuare varietà e condizioni che possono mostrare una buona apprezzabilità dell'uva con un contenuto in zuccheri inferiore a 12° Brix.

Tali condizioni potrebbero essere spiegate dalla differente composizione degli zuccheri nell'uva. Nell'uva matura, il rapporto glucosio/fruttosio è uguale mediamente a 1,0 ma può variare da 0,7 a 1,3 principalmente in funzione della varietà di uva da tavola.

Pertanto a parità di °Brix, la quantità relativa di glucosio e fruttosio può essere molto diversa. Ad esempio, per uva con 16% di zuccheri:

- Con rapporto pari a 1,0 vuol dire che ci sono 8 g di glucosio e 8 g di fruttosio,
- Con rapporto pari a 0,7 ci sono 6,6 g di glucosio e 9,4 g di fruttosio,
- Con rapporto pari a 1,3 ci sono 9,1 g di glucosio e 6,9 g di fruttosio.

Quindi a parità di zuccheri del 16%, il contenuto di fruttosio può variare in maniera significativa. Ma questo influisce molto sulla sensazione gustativa: infatti, il fruttosio ha un potere dolcificante di 2,5 volte quella del glucosio.

5. Prospettive future

In prospettiva, con la possibilità di commercializzazione di uve a basso contenuto glucidico, i produttori di uve da tavola avrebbero la possibilità di sfruttare una nuova quota di mercato, che potremmo definire delle "diet grape", destinate a consumatori attenti a limitare le "calorie" ingerite con il cibo.

Tali varietà possono anche rappresentare una base importante per approcci di miglioramento genetico mirato all'ottenimento di nuove varietà contraddistinte da minore apporto calorico.

Si ringraziano gli alunni del percorso di alternanza scuola lavoro "Uvamia", responsabile prof.ssa Laura Persico del Liceo scientifico Marconi - Hack di Bari, per la collaborazione alle attività di analisi sensoriale sulle uve da tavola.

References

[1] Combrink, J.C., Ginsburg, L. and Truter, A.B. 1974. Maturity of table grapes. Deciduous Fruit Grower **24**(1), 8–13

[2] Coombe, B.G., Dundon, R.J. and Short, A.W.S. 1980. Indices of sugar acidity as ripeness criteria for wine grapes. J. Sci. Food Agric. **31**(5), 495–502

[3] Crisosto, C.H. and Crisosto, G.M. 2002. Understanding American and Chinese consumer acceptance of

"Redglobe" table grapes. Postharvest Biol. Technol. **24**(2), 155–162

[4] Jayasena, V., Cameron, I. 2008. °Brix/acid ratio as a predictor of consumer acceptability of Crimson Seedless table grapes. Journal of Food Quality **31**, 736–750

[5] M. Maante*, E. Vool, R. Ratsep and K. Karp 2015. The effect of genotype on table grapes soluble solids content. Agronomy Research **13**(1), 141–147

[6] Nelson, K.E., Schutz, H.G., Ahmedullah, M. and Mcpherson, J. 1973. Flavour preferences of supermarket customers for "Thompson Seedless" grapes. Am. J. Enol. Vitic. **24**(1), 31–40

[7] OJ L-157 15/06/2011. Marketing standards for fresh fruits and vegetables. `http://exporthelp.europa.eu/update/requirements/ehir_eu13_02v001/eu/auxi/eu_mktfrveg_annex1b_r543_2011_grapes.pdf`. Accessed 17.02.2017

DOI: 10.1051/bioconf/20170901029

The dynamics of ochratoxigenic fungi contents through different stages of dried grape production

Lusine Hakobyan[1], Karine Grigoryan[2], and Armen Trchounian[1,2]

[1] Department of Biochemistry, Microbiology and Biotechnology, Yerevan State University; 1 Alex Manoogian Str., Yerevan 0025, Republic of Armenia
[2] Research Institute of Biology, Yerevan State University, 1 Alex Manoogian Str., Yerevan 0025, Republic of Armenia

Abstract. Dried vine fruit (raisin, sultana and currant) is the second (after wine) most important product of viticulture. Concerning this, the contamination of dried grape by ochratoxigenic fungi and ochratoxin A (OTA) has attracted much attention. Favorable climatic conditions in countries with well-developed viticulture contribute to the spreading of ochratoxigenic fungi. The aim of this work was to identify the contamination sources of dried vine fruit by ochratoxigenic filamentous fungi and OTA, as well to determine the Critical Control Points (CCP) at different stages of production. Primary contamination of grapes occurred during vegetation, especially maturation period, when the risk of mechanical damages was the highest one. 48 samples of soil and 81 samples of fresh grape berries collected in 4 regions of Armenia were investigated. As a result, 22 micromycetes sp. from 7 genera were isolated. Drying process is one of the main CCP. As the most of dried products is produced by open sun drying method, secondary contamination occurs in plants. In our studies 27 species of filamentous fungi were revealed in 87 samples of dried vine fruit, collected at different stages of production. The samples had quite high contamination level by potential toxigenic *A. niger* and *A. carbonarius* species.

1. Introduction

The quality of dried grapes mainly depends on the parameters of microbiological safety, which include pathogenic and toxigenic microorganisms contaminating dried vine fruit. Dried vine fruits are often contaminated by filamentous fungi, the presence of which in food is a danger to the health of the consumer. Therefore, the content of filamentous fungi and some other microorganisms in food is normalized in SanPiN2.3.2.1078-01 [1].

Dried fruits, including dried vine fruits, are susceptible to contamination by fungi and mycotoxins, since they contain a sufficient amount of moisture and sugars [2]. *Aspergillus, Penicillium, Fusarium* and *Alternaria* are the main genera of filamentous fungi contaminating food including dried vine fruits [3].

The problem of contamination of dried vine fruit by ochratoxin A has been studied in many countries with developed viticulture and winemaking such as Spain [4], France [5], Italy [6], and Australia [7]. In these studies, it is claimed that most of the ochratoxigenic strains belong to the species *A. carbonarius*. A few numbers of strains belonging to *A. niger* species which synthesized OTA. In Hungary among species isolated from raisin samples in addition to *A. niger* and *A. caorbnarius* another ochratoxigenic species of *A. tubingens* dominated [8].

In countries of South America the main contaminants of grapes, raisins and other grape derived products are species of *A. niger, A. ochraceus, A. carbonarius* and *Alternaria alternate* [9]. In Argentina, the most common species was *A. niger*, but only a few number of isolated strains of this species produced OTA. *A. carbonarius* was revealed more rarely but 82.6% of the strains in one case [10] synthesized OTA. Toxigenic species *A. ochraceus* and *A. japonicus*, which synthesized OTA, were also detected.

Studies show that in California the presence of ochratoxin A in raisin samples is also associated mainly with the species *A. carbonarius* [11].

Analyzes conducted in countries of the Middle East region (Saudi Arabia, Yemen and Iraq) have shown the similar results. The species of *A. niger* dominated in all studied samples from three countries. This species was the main ochratoxigenic species in samples from Iraq [12]. *A. carbonarius* also revealed frequently (60.3%) [13] and the frequency of occurrence of *A.niger* was up to 92.3%. Among the species isolated from raisin samples in Saudi Arabia, the fungi from genera *Aspergillus, Alternaria, Cladosporium, Fusarium* and *Rhizopus*, were also extended [14]. In addition to *A.niger* the toxigenic species of *A. flavus, A. fumigatus* and*A. ochraceus* from genus *Aspergillus* [15] were often isolated from raisin samples in Yemen Republic.

In order to reduce the risks of contamination of the final product in dried vine fruit production much attention is paid to the development of technologies and methods of inhibition the growth and development of micromycetes-potential producers of mycotoxins. The practice of drying varies depending on the geographical location and varieties of grapes [16]. There are three main methods that are used in production of dried vine fruits: open sun drying (or open air drying), shadow drying and mechanical drying. Traditional open sun drying method has been used for thousands of years in Asia and other countries around the world. More than 90–95% of dried products produced in all countries of the world are prepared by the open sun drying method. Along with the traditional sun drying method, another drying method on vine (DOV) is used. This method is especially common in Uzbekistan and USA. In order to prevent the development of mold fungi and synthesis of mycotoxins, it is advisable to use solar dryers that accelerate the process and simplify the control of drying conditions [17]. Recently, in America various methods of mechanical drying of grapes have been developed, in which the risk of effluence of unfavorable weather conditions as well contact of workers with raw materials and final products, are reduced [18]. The basis of these methods is the mechanization of all processes: harvesting, drying and collecting of final dried products.

The use of appropriate agrotechnical measures and fungicides in grapes cultivation can significantly prevent the growth of fungi from genus *Aspergillus* and contamination of final product with ochratoxin A [19]. After harvesting, the most effective method for limiting the development of mold fungi is the treatment of raw grape with sulfur dioxide (SO_2) [20]. SO_2 and its derivatives have long been used for treatment of raw grape, as a preservative inhibiting the development of mold-forming fungi and other microorganisms. It also acts as an antioxidant, which prevents oxidation processes and lead to the changes of the color of final product.

2. Materials and methods

2.1. Sampling

87 dried vine fruit samples (white and black varieties of raisin and sultana) were analyzed. 39 samples were collected from different markets and supermarkets of Yerevan, and 48in some manufactures in Armavir region in accordance with GOST 1750–86 [21] and Commission Regulation (EC) No 401 [22]. Samples were collected at different stages of production: raw materials, the pretreated material - after blanching (alkali treatment), after treatment by SO_2, during drying process (2–4 weeks), after drying and during storage.

48 soil samples were taken in vineyards in different depths under vine (5 cm, 15 cm and 25 cm). 81 samples of fresh white and black grape were analyzed. The sampling of soils, vegetative organs of vine and grape fresh grape berries was done during harvesting in four regions of Armenia: Ararat, Armavir, Kotayk and Vayots-Dzor. In total 225 samples were studied.

2.2. Mycological analysis of dried vine fruit, grape berries and soil samples

Mycological analyses of dried vine fruit samples were carried out with direct planting and serial dilution planting methods [23].The dilution includes the following steps: 1) shaking the product suspension for 15 minutes on a shaker; 2) infusion for 10 minutes; 3) preparation of serial dilutions 1:10 and 1: 100; 4) surface plating from dilutions on nutrient mediums. Plates were incubated at 25 °C for 7 days. Following nutrient mediums were used: CYA (Chapek-Yeast Agar medium, HiMedia Ltd.), GYA (Glucose-Yeast Agar medium, HiMedia Ltd.), and MEA (Malt-Extract Agar medium, HiMedia Ltd.).

In order to inhibit the growth of superficial saprotrophic mycobiota and reveal the internal mycoflora of dried vine fruit, fresh grape berries and vegetative organs of vine, the samples were exposed to external sterilization before analysis. For this the samples were treated with 3% solution of sodium hypochlorite for 15 minutes, after which the samples were washed with distilled water.

The aim of this work was to identify the contamination sources of dried vine fruit by ochratoxigenic filamentous fungi and OTA, as well to determine the Critical Control Points (CCP) at different stages of their production.

2.3. Identification of fungi species

Identification of isolated cultures of filamentous fungi was done based on macroscopic and microscopic characteristics by following manuals: [24–27]. For identification of species from section *Aspergillus Nigri* cultures were grown on Czapek Yeast Extract Agar (CYA) at 25 ± 1 °C and 37±1 °C. Fungi species are forming characteristic colonies and retain the ability to generate exudate and pigmentation on this nutrient medium.

2.4. Data analysis

The quantity of microscopic fungi in 1 g food was detected according to NF ISO 7698–91 [28] using the following formula:

$$\omega = \frac{\sum C}{(n_1 + 0.1\,n_2)d} \tag{1}$$

in which
$\sum C$ –sum of colonies of fungi grown in all dishes;
n_1 – number of dishes used for I dilution;
n_2 – number of dishes used for II dilution;
d – dilution coefficient.

The frequency of occurrence of fungi species was determined by the ratio of quantity of samples, in which the species was revealed, to the total quantity of analyzed samples [29]:

$$A = B/C * 100\%, \tag{2}$$

in which
B – number of samples in which species was detected;
C – total number of analyzed samples.

3. Results and discussion

3.1. Filamentous fungi from section *Aspergillus Nigri* contaminating dried vine fruit

Mycological analysis of 87 dried vine fruit samples were carried out. 48 samples were collected immediately in plants and 39 in different markets and supermarkets in Yerevan.

The results showed that the contamination level of dried vine fruits by filamentous fungi (2.2×10^2–4.3×10^4 CFU/g) often exceeds the maximum permissible level of this parameter presented in local and international sanitary rules and regulations.

According to SanPin 2.3.2.1078–01 maximum permissible amount of diaspora of molds in 1 g of dried vine fruit is 5×10^2 CFU/g.

27 species of filamentous fungi belonging to 5 genera: *Mucor, Aspergillus, Penicillium, Trichoderma* and*Alternaria*, were isolated and identified. 13 (48%)

DOI: 10.1051/bioconf/20170901029

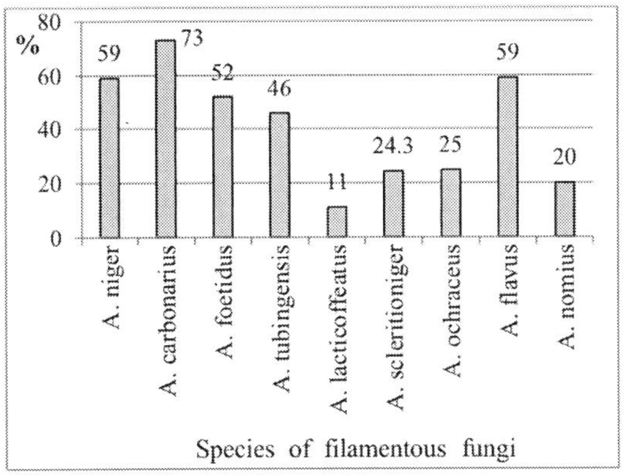

Figure 1. The frequency of occurrences of toxigenic species from genus *Aspergillus* in dried vine fruit.

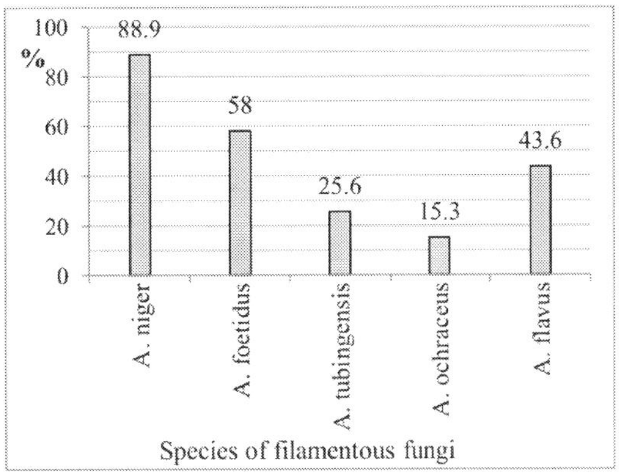

Figure 2. The frequency of occurrences oftoxigenic species isolated from soil samples.

species of the total amount of isolated filamentous fungi belong to the genus *Aspergillus*.

Contamination of dried vine fruits by potentially toxigenic species *A. niger* and *A. carbonarius* is quite high. This species are dominant and hav 59% and 73% of frequency of occurrence respectively. Monopolistic growth of *A. niger* was often observed during mycological analysis of samples. Some strains of this species are characterisied by high antagonistic activity and high energy of growth.

The results of scientific research conducted in a number of countries have shown that the detection of ochratoxin A in vine derived products, including dried vine fruits, is associated first of all with the presence of *A. carbonarius* species [10, 11].

Species *A. sclerotiicarbonarius* (34.7%), *A. tubingensis* (46%), *A. foetidus* (52%) and *A. flavus* (59%) had an average frequency of occurrence (Fig. 1). These species except *A. flavus* belong to section*Aspergillus Nigri*.

A. sclerotioniger and *A. lacticoffeatus* species, known as potential producers of OTA, were revealed on the new substrate - dried vine fruit for the first time in our investigations. According to Samson [27] species *A. sclerotioniger*, *A. lacticoffeatus* and *A. niger* are also potential producers of ochratoxin A. However, only certain strains of these species are capable to produce mycotoxins.

Nine potentially toxigenic fungi species of genus *Aspergillus* were revealed in samples of dried vine fruit. Seven of these species are ochratoxigenic, 6 of which belong to the section *A. Nigri* (*A. niger, A. carbonarius, A. tubingensis, A. foetidus, A. sclerotioniger, A. lacticoffeatus*)and one to the section *A. Circumdati* (*A. ochraceus*). *A. flavus* and *A. nomius* species belong to the section*Aspergillus Flavi*. These fungi are potential producers of aflatoxins.

Investigations of numerous authors [4,6–8] proved that *A. niger, A. carbonarius, A. flavus*and *A. tubingensis* species from genus *Aspergillus* are the specific microflora of dried vine fruit.

3.2. Mycobiota of soil samples from vineyards in different regions in Armenia

The main sources of contamination of dried vine fruit, wine and other grape processing products by species *A. carbonarius* and *A. niger*, is the soil and its residues on grape berries. In order to identify the sources of contamination of Armenian dried vine fruit by toxigenic species, mycological analysis of 48 samples of soil from vineyards in four regions in Armeniawas carried out.

As a result of the study, 22 species of filamentous fungi from 7 genera: *Aspergillus, Penicillium, Alternaria, Fuzarium, Cladosporium, Trichoderma* and*Mucor*, were isolated and identified. 10 species were revealed in analyzed samples of soil belonged to the genus *Aspergillus* and 6 – to the genus *Penicillium*. Species of ochratoxigenic fungi *A. niger* (88.9%), *A. foetidus* (58%) and *A. tubingensis* (25.6%), had high frequency of occurrence (Fig. 2).

A. carbonarius species was not revealed in any of analyzed soil samples. Other species producers of ochratoxin A from section *Aspergillus Nigri* were isolated from the surface layers of soil more often than from deeper layers. High contamination level of surface layers soil by micromycetes - producers of ochratoxin A is revealed during ripening and harvesting. In this period grape berries contain high quantity of glucose and moisture. Implementation of Good Agricultural Practises in vineyards is important to reduce the risks of main contamination of grape berries as the row materials.

3.3. Mycological analysis of fresh grape berries

Contamination of grape by filamentous fungi often occure due to the presence of mechanical damages on the skin of grape berries throuth which micelium of fungi penetrate inside. More favorable conditions for the growth of fungi are created in ripened berries, in which pH value is low and a sugar content is high.

Grape berries in the vegetation season and especially harvesting period are mainly affected by saprotrophic fungi from genera: *Penicillium, Aspergillus, Fuzarium, Cladosporium* and *Alternaria*.

We analyzed 81 samples of white and black varieties of grape cultivated in 4 regions in Republic of Armenia during the harvest season. As a result of the mycological analysis of fresh grape berries 19 species of micromycetes from 5 genera: *Aspergillus, Penicillium, Alternaria, Cladosporium* and *Mucor*, were isolated and identified [30].

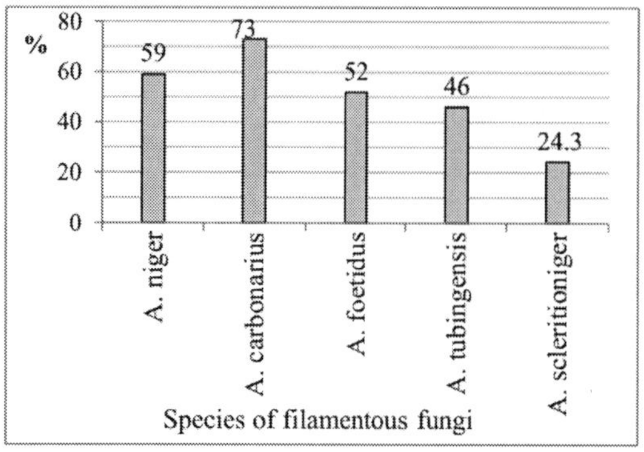

Figure 3. The frequency of occurrence of fungi species from section *Aspergillus Nigri* contaminating Armenian grape.

Table 1. Contamination levels of grape samples collected in four regions in Armenia.

Region	Contamination level in CFU/g	Number of isolated species
Armavir	65–2 × 10^2	13
Ararat	60–95	18
Kotayk	80–2.8 × 10^2	16
Vayotsdzor	40–90	9

Table 2. Fungi species isolated from fresh grape berries.

Species	Frequency of occurrence of fungi (%)			
	Armavir region	Ararat region	Kotayk region	Vayotsdzor region
Aspergillus niger	41	36	49.5	29
A. carbonarius	45.7	24	48.3	8
A. foetidus	36	21	5.4	11
A. sclerotioniger	14	10.4	2	–
A. tubingensis	7.8	6.3	4.3	2.3
A. japonicus	–	4	4	5
A. uvarum	3.9	4.2	2.4	–
A. aculeatus	2	3.4	1.5	–
A. pulverulentum	–	2.6	1.5	3
A. flavus	9	11	17	–

8 species belonged to the section *Aspergillus Nigri*. The frequencies of occurrence of these species are presented in Fig. 3. From potentially ochratoxigenic fungi*A. niger, A. carbonarius, A. foetidus, A. sclerotioniger* and *A. tubingensis* species, were identified.

Species *A. carbonarius, A. niger,* and *A. foetidus* dominated in all analyzed samples. *A. flavus* species had an average frequency of occurrence.

Comparative mycological analysis of grape berries, selected in 4 regions, shown that grape samples from VayotsDzor region had the lowest degree of contamination: 40–90 CFU/g (Table 1).

Only 6 species from genus *Aspergillus* were revealed and identified in these samples (Table 2). And the content of potentially toxigenic fungi was quite low in samples from VayotsDzor region.

The results of mycological analysis grape and soil samples collected in the vineyards allowed us to identify the main sources of contamination of dried vine fruit by *A. carbonarius* and *A. niger* ochratoxigenic species. Thus, grape berries affected by the "black" mold are the main source of contamination of dried grapes by ochratoxigenic species from genus *Aspergillus*.

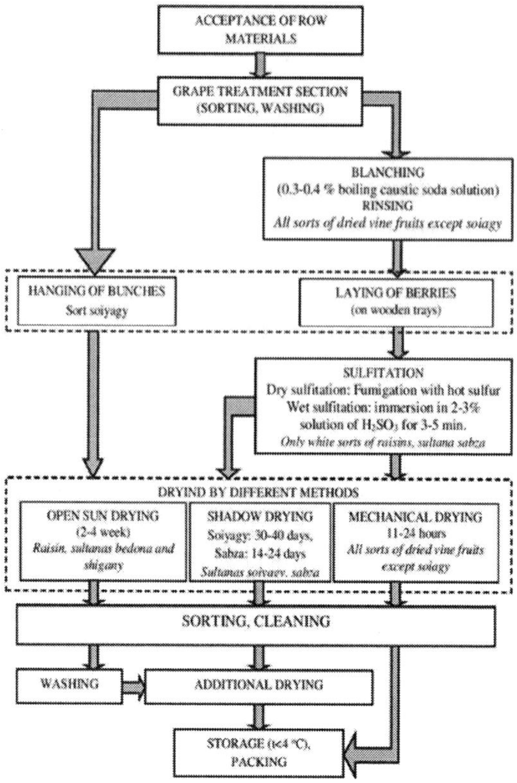

Figure 4. Main technological stages of different dried vine fruit production.

3.4. The influencce of drying technologies on growth and evolution of filamentous fungi in dried vine fruit

Technological regimes of grape drying also significantly affect the contamination level of the final product by fungi and mycotoxins. During open air drying the risk of contamination of dried product by mold fungi and mycotoxins increases. These occurres due to high probabilities of mechanical damage on grape berrie skin. The main schem of dried vine fruit production is given in Fig. 4.

During drying period the contamination degree of dried vine fruits directly depends on the use of sulphitation and the content of SO_2 in the intermediate and final products. The latter is an effective preservative and has a significant inhibitory effect on the development of mycelial fungi, including toxin-producing species, during drying and storage.

Some authors investigated the influence of drying technologies on contamination of the processing and final product by filamentous fungi and mycotoxins [31].

In Armenia, the main part of dried foods produced by traditional open sun drying method. But along with this solar dryers are also used. CCPs are reduced during production of dried vine fruit in this dryers.

 DOI: 10.1051/bioconf/20170901029

Table 3. Contamination level (CFU/G) of dried vine fruit at different stages of production.

Number of analyzed samples	Samples	Raw material	After treatment by sulfur dioxide	During drying	After drying	During storage (at+8°C
12	White raisin	$80–1.4 \times 10^2$	0	$2.4 \times 10^2–3 \times 10^2$	$4.5 \times 10^2–6.8 \times 10^2$	$4 \times 10^2–7 \times 10^2$
12	White sultana	$90–1.3 \times 10^2$	0	$95–2 \times 10^2$	$2.2 \times 10^2–5.2 \times 10^2$	$2 \times 10^2–5.5 \times 10^2$
12	Black raisin	$70–1.8 \times 10^2$	*	$5.2 \times 10^2–2 \times 10^3$	$2.6 \times 10^3–1.8 \times 10^4$	$2.3 \times 10^3–2 \times 10^4$
12	Black sultana	$1.5 \times 10^2–2 \times 10^2$	*	$1.8 \times 10^2–8.5 \times 10^2$	$2.6 \times 10^3–3.4 \times 10^3$	$2.6 \times 10^3–4 \times 10^3$

* Sulfur dioxide treatment is not permissible in manufacture of these dried vine fruit varieties.

We carried out mycological analysis of 48 samples of Armenian dried vine fruit at all stages of manufacturing in order to determine the sources of contamination of the final product by filamentous fungi and identify stages in which the product is most prone to risk of contamination.

Samples were collected in two manufacturs in Armavir region at different stages of production: raw materials, the pretreated material - after blanching (alkali treatment), after treatment by SO_2, during drying process (2–4 weeks), after drying and during storage.

Studies have shown that the contamination level of processing product increased during the entire production process (Table 3). However, no fungi growth was detected in the samples taken immediately after treatment by SO_2, while the high contamination degree of the final product was observed. This proves an inhibitory effect of SO_2 on the growth and evolution of filamentous fungi. Immediately after sulfitation process SO_2 content in the intermediate product is in the maximum level.

As the most quantity of dried vine fruit is produced by the open sun drying method, so a secondary contamination occurs. Analyses of samples taken in this stage of production showed a high contamination level of intermediate product by filamentous fungi. In some cases the value of CFU/g exceeded the maximum permissible (5×10^2). Only contamination levels of the samples dried in closed solar dryers were within the normal range.

As the black varieties of dried vine fruit had not been treated with sulfur dioxide, so their contamination levels were higher than white varieties. Concerning this the blanching stage of these products, which is not followed by sulfitation, becomes a Critical Control Point. The reason of this are microcracks on the skin of grape berries formed due to the blanching.

Contamination of product by toxigenic species increases during grape drying process. The relative proportion of *A. carbonarius* in processing grape increases as the humidity reduces in grape berries (Fig. 5).

In comparison with open air drying method the mechanical drying in closed chambers, with controlled temperature and humidity, provides the mycological safe final product. Grape drying process is often carried out at a temperature of 50°C. Under such a temperature regime, the probability of OTA biosynthesis is minimal, since the growth of *A. carbonarius* is almost completely inhibited [32]. However, a sharp decrease of the content of vitamins, polyphenols, anthocyanins, ensuring the quality and nutritional value of final product, is observed at this temperature.

Vacuum drying method is currently used in dried vine fruits production for obtaining a safe product for the consumer.Independence with the used drying technology,

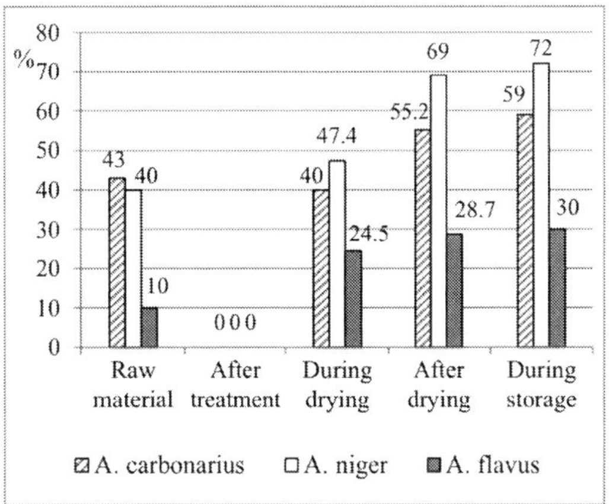

Figure 5. Percentage content of toxigenic species of fungi in different stages of production.

hygienic conditions is essential in all stages of dried vine fruit manufacturing: starting from grape harvest to the drying process, transport, packaging and storage. The absence of these conditions can lead to the secondary contamination and cross contamination of raisins by OTA-producing fungi.

4. Conclusions

Contamination of dried vine fruit by species from *Aspergillus niger* section mainly occurs during the cultivation of grape: in ripening and harvesting stages. Secondary contamination by fungi species can occur in case of bad sorting and drying process, and violation of drying and hygienic conditions in manufactures.

Climatic conditions playe an important role during production by open sun drying method. In rainy weather, the probability of secondary contamination of the intermediate product by toxigenic fungi increases.

The Critical Control Points (CCP) in production of dried vine fruits are raw grape storage, transportation conditions to the manufactur plant and technology of pretreatment of raw materials before drying.

For prediction of contamination degree of dried products by mycotoxins some aspects must be taken into account, such as weather conditions before harvest, types and amounts of fungicides used during vegetation, moisture content of grape berries during harvest and contaminant species of fungi. It is necessary to use unaffected, benign raw materials for reducing the risks of contamination of raw materials by mycotoxigenic species from genus *Aspergillus* and the accumulation of mycotoxins in dried vine fruits. Special attention should be given to the prevention of the contamination of vineyards by ochratoxigenic fungi. This can be done by carrying out hygienic measures in the vineyard areas. It is also necessary to follow the basic principles of Good Agricultural, Industrial and Hygienic Practices (GAP, GMP, GHP).

The drying process is the next important step to prevent the contamination of the final product. Our studies have shown that at this stage of production, the risk of contamination of the product by fungi increases sharply. To ensure the microbiological safety of the dried grape produced, it is important to determine CCP in plants or in storage houses. This will help manufacturers to develop and apply a suitable HACCP system that will eliminate unforeseen sources of contamination of products by mycotoxins. At this stage of production, the Critical Control Points are: pests and insects, drying time, frequency of turning grape berries, hygienic conditions, and weather conditions (temperature, humidity, rain, etc.). The latter is especially important in the production by open air drying method. Control measures should be applied at all critical points. For each identified CCP, the measured parameters must be defined and their critical limits must be established.

References

[1] SanPiN 2.3.2.1078-01, Sanitary-epidemiological rules and regulations. Production row materials and foodstuffs. Hygienic requirements for the safety and nutritional value of food, 180 page (2001)
[2] S.M.S. Alghalibi, A.R.M, Assiut University Bulletin of Environmental Research **7**(2), p. 26–30 (2004)
[3] J.I. Pitt, A.D. Hocking, *Fungi and Food Spoilage*, **3rd** ed., Springer, p. 413–414 (2009)
[4] M. Bau, M.R. Bragulat, M.L. Abarca, S. Minguez, F.J. Cabanes, Int J. Food Microbiol. **98**, p. 125–130 (2005)
[5] L. Sage, S. Krivobok, E. Delbos, F. Seigle-Murandi, E.E. Creppy, J. Agric. Food Chem **50**, p.1306–1311 (2002)
[6] P. Battilani, P. Giorni, T. Bertuzzi, S. Formenti, A. Pietri, Int. J. Food Microbiol. **111**, p. S53–S60 (2006)
[7] S.L. Leong, A.D. Hocking, J.I. Pitt, B.A. Kazi, R.W. Emmett, E.S. Scott, Int. J Food Microbiol. **111**(1), p. S10–S17 (2006)
[8] J. Varga, S. Kocsube, Z. Koncz, J. Teren, Actaalimentaria. **35**(3), p. 289–294 (2006)
[9] S.N. Chilze, C.E. Magnoli, A.M. Dalcero, Int. J. Food Microbiol. **111**(1), S5–9 (2006)
[10] S.M. Romero, R.M. Comerio, G. Larumbe, A. Ritieni, Int. J. Food Microbil. **104**(1) 43–49, (2005)
[11] J.D. Palumbo, T.L. O'Keeffe, S.J. Vasquez, N.E. Mahoney, Let. in Appl. Microbiol. **52**, p. 330–336 (2011)
[12] M.T. Farid, A.L. Neveen Nowzad Rajab, Zan. J. of Pure and Appl. Sci. **26**(4), pp. 49–60 (2014)
[13] A.A.M. Saadullah, S.K. Abdullah, **7**(1), pp. 61–68 (2015)
[14] R.M. Gashgari, Y.M. Shebany, Y.A. Gherbawy, Fborn. Pathog. and Dis. **8**(11), p. 1221–1227 (2011)
[15] S.M.S. Alghalibi, A.R.M. Shater, Assiut Uni. Bulle. Of Envir. Res. **7**(2), p. 26–30 (2004)
[16] C. Domingo, Li. Giralt, Gener. Catalun. Inst. Cat. Vin. **1**, 27 (2006)
[17] D.R. Pangavhane, R.L. Sawhney, P.N. Sarsavadia, Ener. Convers. Manag. **43**(1), p. 45–61 (2002)
[18] R.L., Williams, S.A. Shepard, J.E. Casey, Calif. Agri. Tech. Inst., 74 p (2002)
[19] L. Covarelli, G. Beccari, A. Marini, L. Tosi, Food Cont. **26**, p. 347–356 (2012)
[20] A.D. Hocking, *Xerophilic fungi in intermediate and low moisture foods,* (New York: Marcel Dekker Inc. 3, 1992)
[21] Gosstandart of the USSR, Dried fruits. Acceptance rules, methods of analysis (1986)
[22] Commission Regulation (EC) No 401/2006, Offic. J. EU, L70, p. 12–34 (2006)
[23] JI.Pitt, A.D. Hocking, *Fungi and Food Spoilage*, (2nd Ed. Black. Acad. and Prof., London 1997)
[24] V.I. Bilay, E.Z. Koval, Aspergilles, (Kiev, Naukova Dumka, 1988) (in Russian)
[25] K.B. Raper, D.I. Fennell, *The genus Aspergillus*. (Robert E. Krieger Publ. Co. Inc. Florida, USA, 1977)
[26] R.A. Samson, J.A. Houbraken, F.A. Kuijpers, J.M. Frank, J.C. Frisvad, In Stud. Myc. **50**, 45–61 (2004)
[27] R.A. Samson, P. Noonim, M. Meijer, J. Houbraken, J.C. Frisvad, J. Varga, In Myc. **59**(13), 129–145 (2007)
[28] NF ISO 7698–91.Controle de la qualite des produits alimentaires controle microbiologique. Direktives generales pour le denombrement des levures et moisissures (1993)
[29] I.A. El-Kady, J.N. Abdel-Hafez, S.S. El-Maraghy, In J. Mycopath. **77**, p. 103–109 (1982)
[30] K. Grigoryan, L. Hakobyan, M. Sargsyan, J. Hyg. Engin. Design, p. 97–100 (Ohrid, Republic of Macedonia, 2011)
[31] A.W.S. Kassem, M.A. Al-Sulaiman, A.E.M. Aboukarima, S.S. Kassem, Austr. J. Bas.& Appl. Scien. **5**(6), p. 230–241 (2011)
[32] D. Spadaro, S. Patharajan, A. Lore, M.L. Gullino, Garibaldi A. Phytopathol. Mediterr. **49**, p. 65–73 (2010)

Performance of the new seedless grape 'BRS Isis' grown in subtropical area

Saeed Ahmed[1], Sergio R. Roberto[1,a], Ronan C. Colombo[1], Renata Koyama[1], Muhammad Shahab[1], and Reginaldo T. Souza[2]

[1] Londrina State University, PO Box 6001, Londrina, PR 86051–990, Brazil
[2] Embrapa Grape and Wine, 515 Livramento Rd, Bento Gonçalves, RS 95700–000, Brazil

Abstract. 'BRS Isis' is a new colored seedless table grape tolerant to downy mildew, the main vine disease in subtropical humid areas. This new seedless cultivar is an interspecific hybrid from the crossing of CNPUV 681–29 [Arkansas 1976 X CNPUV 147–3 ('Niagara Branca' x 'Venus')] x 'BRS Linda', and was recently released by Embrapa Grape and Wine, Brazil. The performance of 'BRS Isis' seedless grape was evaluated in an experimental vineyard in 2016 located in a subtropical area at Marialva city, state of Parana, Brazil. The vines were grafted onto 'IAC 766 Campinas' rootstock and trained in an overhead trellising system spaced at 2×5 m. Vines were cane-pruning in late winter of July 2016, and for assessments, 20 representative vines were selected in the area. As 'BRS Isis' is a very fruitful grape, presenting 4 bunches per shoot, a load adjustment was performed after fruitset removing 50% of bunches per shoot, leaving 2 bunches per shoot, equivalent to a density of 10 bunches.m^{-2}. The duration in days of the main phenological stages from pruning to harvest, and the physicochemical and yield characteristics of 'BRS Isis' seedless grape were subjected to evaluation. It was determined that the cycle is 144 days. The means of berry and bunch weight was 6.7 ± 1.0 g and 500.0 ± 0.04 g, respectively, and the color index of berries (CIRG) was 4.3 ± 1.3. The means of total soluble solids (TSS), titratable acidity (TA) and maturity index (TSS/TA) observed were $14.2\pm 0.3°$ Brix, 0.6 ± 0.04% of tartaric acid and 24.1 ± 1.5, respectively, while the yield was 49.0 ± 5.7 tons.ha^{-1}. The yield observed is considered high to keep a sustainable crop over time, thus, in order to obtain grapes of 'BRS Isis' with regular yield, and possibly, with a higher content of soluble solids, it is desirable to keep only 1 bunch per shoot (5 bunches.m^{-2}) after fruitset.

1. Introduction

Viticulture is an important economic activity in Brazil. In the recent years, it has also become important in generating employment in large enterprises for production of table grapes, mainly the seedless types. The worldwide trend for the consumption of seedless grapes has increased the competition between producers, which direct efforts to meet a more demanding consumer market [1,2].

Seedless grapes have certain characteristics that make them a high quality fruit and have better acceptance by consumer. Thus, Embrapa Grape ans Wine has been conducting a breeding program since 1997 aimed at creating cultivars of seedless table grapes well adapted to the Brazilian tropical and subtropical conditions. In this context, 'BRS Isis' seedless grape was released by Embrapa in 2013. It is tolerant to downy mildew and adapts well to the different climates of Brazil. It also stands the high bud fertility and presents firm texture and neutral taste. It is the result of crossing CNPUV 681–29 [Arkansas 1976 X CNPUV 147–3 ('Niagara White' x 'Venus')] x 'BRS Linda', held in 2004 at Embrapa Grape and Wine, Brazil. It presents large and red berries, good adhesion and crunchy texture [3].

The northern region of Parana State is a consolidated area in table grape production, but it is based on seeded grapes such as 'Benitaka', 'Italia', 'Rubi' and 'Brasil'. In this area, due the mild winter and subtropical conditions, two crops of grapes per year are obtained. Thus, seedless grape cultivation, such as 'BRS Isis', could be an alternative to diversify the current production system, opening the possibility of international market. However, there is no information about the performance of 'BRS Isis' grape in the state of Parana, and the results cannot always be extrapolated from one region to another, because these characteristics vary depending on the genotype and climatic conditions of each region, influencing also on the fruit quality [4,5].

2. Materials and methods

This study was conducted in a commercial vineyard of 'BRS Isis' seedless grapes grafted on IAC 766, from 2-year-old vines, located in Marialva, state of Parana (PR), Brazil (23°29′52, 8″S, 51°47′58″0 W, elevation 570 m), in crop season 2016. According to the Köppen classification, the climate is type Cf*a*. The vines were trained on overhead trellis and spaced 2×5 m. Vines were cane-pruning in late winter of July 2016. Pruning was performed to leave two to 3–4 buds per cane. Subsequently, 5% hydrogen cyanamide was applied to the buds to induce and standardize sprouting. During the trials, the standard regional cultivation practices with regard to nutrition, weed control, and pest and disease management were used.

[a] e-mail: sroberto@uel.br

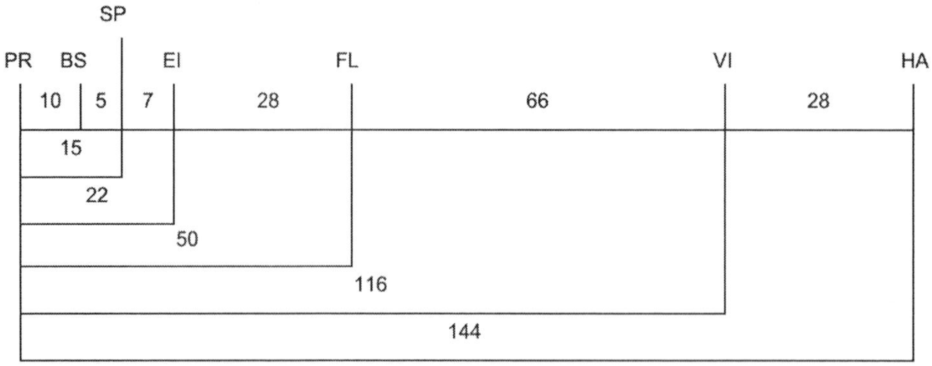

Figure 1. Duration in days of the phenological stages of 'BRS Isis' seedless grape in Marialva, PR. Crop season 2016. Pruning (PR); Bud-swelling (BS); Sprouting (SP); Emergence of inflorescence (EI); Flowering (FL); Veraison (VI); Harvesting (HA).

For assessments, 20 representative vines were selected in the area. As 'BRS Isis' is a very fruitful grape, presenting 4 bunches per shoot, a load adjustment was performed after fruitset removing 50% of bunches per shoot, leaving 2 bunches per shoot, equivalent to a density of 10 bunches.m^{-2}. The duration in days of the following phenological stages were evaluated by means of visual observation, from pruning (PR): a) bud swelling (BS) b) sprouting (SP); c) emergence of inflorescence (EI); d) flowering (FL); e) veraison (VI) and f) harvest (HA) [6].

Then, a diagram was constructed representing the duration in days of each phenological stage of crop season [7].

The grapes were harvested at full maturity and following physical characteristics were analyzed during the study: weight (g) and diameter (mm) of berries; weight (g) and width (cm) of bunches using digital calipers and scales. The number of bunches per plant were also analyzed, and according to the planting density, the production per plant (kg per plant) and per area (tons per ha) were also estimated.

The evolution of the ripening of the grapes was determined by analysis of the chemical characteristics of its berries as total soluble solids (TSS), titratable acidity (TA) and maturity index (TSS/TA).

For anthocyanin analysis 30 berries were collected with two berries taken from the upper, middle, and bottom regions of each marked cluster. The concentrations of total anthocyanins were carried out according to Peppi et al. [8].

3. Results and discussion

The duration in days of the 'BRS Isis' production cycle (PR-HA) was of 144 days, while the duration of the PR-BS, PR-SP, PR-EI, PR-FL and PR-VI was 10; 15; 22; 50 and 116 days, respectively (Fig. 1). Rodrigues [9], working with 'Itália' grapes in Porto Feliz, SP, where the climate is similar to that found in Marialva, PR (Cf*a*), detected a variation of 22 days between the regular and off season cycles, 161 and 139 days, respectively, a value close to the 25 days observed for the 'Black Star' grape in Marialva. The knowledge of the duration of the phenological phases is a requirement of modern viticulture, since it makes possible the rationalization and optimization of the cultural practices, which are indispensable for the cultivation of the vine [10].

Table 1. Physicochemical characteristics of 'BRS Isis' seedless grape. Crop season of 2016.

Physicochemical characteristics	Crop season 2016
Berry weight (g)	6.7 ± 1.0
Berry diameter (mm)	19.0 ± 1.7
Bunch weight (g)	500.0 ± 0.04
Color index of red grapes (CIRG)	4.3 ± 1.3
Soluble solids - SS (o Brix)	14.2 ± 0.3
Titratable acidity – TA (% of tartaric acid)	0.6 ± 0.04
Maturation index (SS/TA)	24.1 ± 1.5
pH	4.3 ± 0.02
Total anthocyanins (mg.100g^{-1})	0.3 ± 0.1
Number of bunches per plant	94.9 ± 11.2
Production per plant (kg/planta)	49.0 ± 5.8
Yield (tons.ha^{-1})	49.0 ± 5.7

The means of production characteristics of the 'BRS Isis' grape such as the diameter and weight of the berry, the number of bunches per plant, the weight of the bunch, and the estimates of production per plant and yield were 19.00 mm, 6.7 g, 94.9, 500.00 g, 49 kg.plant^{-1} and 49 tons.ha^{-1}, respectively, while the color index of berries (CIRG) was 4.3. The means of total soluble solids (TSS), titratable acidity (TA) and maturity index (TSS/TA) observed were 14.2° Brix, 0.6% of tartaric acid and 24.1, respectively (Table 1).

4. Conclusion

The yield of 'BRS Isis' grown in subtropical area is considered high to keep a sustainable crop over time, thus, in order to obtain grapes with regular yield, and possibly, with a higher content of soluble solids, it is desirable to keep only 1 bunch per shoot (5 bunches.m-2) after fruitset.

References

[1] J.C. Nachtigal., U.A. Camargo, J.D.G. Maia, Efeito de reguladores de crescimento em uva apirênica, cv. BRS Clara. Revista Brasileira de Fruticultura, **27**(2), p. 304-307 (2005)

[2] L.M.R. Mello, Vitivinicultura brasileira: Panorama 2009. Bento Gonçalves: Embrapa Uva e

DOI: 10.1051/bioconf/20170901030

Vinho (2010) Disponível em: <http://www.grupocultivar.com.br/artigos/artigo.asp?id=1001>. Acesso em: 20 jul. 2010.

[3] P.S. Ritschel, J.D.G Maia, U.A. Camargo, R.T. De Souza, T.V.M. Fajardo, R. De L. Naves, C.L. Girardi, BRS Isis Nova Cultivar de Uva de Mesa Vermelha, sem Sementes e Tolerante ao Míldio. Bento Gonçalves: Embrapa. p. 6-11, 2013. Comunicado Técnico, 143

[4] A.C. Boliani, F.M. Pereira, Avaliação fenológica de videiras (*Vitis vinifera* L.), cultivares Itália e Rubi, submetidas à poda de renovação na região oeste do estado de São Paulo. Revista Brasileira de Fruticultura **18**(2), p. 193–200 (1996)

[5] P.C.S. Leao, E.E.G. Silva, Caracterização fenológica e requerimento térmico de variedades de uvas sem sementes no vale de São Francisco. Journal of Tropical Fruits **25**(3), p. 379–382 (2003)

[6] M. Baillod, M. Baggiolini, Les stades repères de la vigne. Revue Suisse Viticulture Horticulture **25**(1), p. 7–9 (1993)

[7] A.J. Sato, B.S. Jubileu, C.E. Santos, R. Bertolucci, R.D. Silva, M. Carielo, M.C. Guiraud, I.C.B, Fonseca, S.R. Roberto, Phenology and thermal demand of 'Isabel' and 'Rubea' grapevines on different rootstocks in North of Parana. Semina **29**(2), p. 283–292 (2008)

[8] M.C. Peppi, M.W. Fidelibus, N. Dokoozlian, HortScience **41**, p. 1440–1445 (2006)

[9] Rodrigues A., Desenvolvimento da videira 'Itália' em clima tropical de altitude. (2009) Dissertation, Escola Superior de Agricultura "Luiz de Queiroz"

[10] F. Mandelli, J. Tonietto, U.A. Camargo, A.B.C. Czermainski, phenology and thermal needs of the vine in Serra Gaucha. Annals Brazilian Congress of Fruits, 18, 2004, Florianópolis, SC

Inactivation of microorganisms by UV-treatment of must and wine

Dominik Durner[1,a], Kathrin Diesler[1], Patricia Golombek[1], Lisa Kromm[2], Mario Stahl[2], Karlis Briviba[2], Maren Scharfenberger-Schmeer[1], and Ulrich Fischer[1]

[1] Institute for Viticulture and Enology, Dienstleistungszentrum Ländlicher Raum (DLR) Rheinpfalz, Breitenweg 71, 67435 Neustadt/Weinstrasse, Germany
[2] Department of Physiology and Biochemistry of Nutrition, Max Rubner-Institut, Federal Research Institute of Nutrition and Food, Haid-und-Neu-Straße 9, 76131 Karlsruhe, Germany

Abstract. The objective was to investigate the applicability of UV-C technology to inactivate yeasts and bacteria in must and wine. Experiments were carried out in vintage 2016 with Riesling musts of different quality containing their natural microflora. Yeasts were tested more resistant to UV-C energy than bacteria. *Saccharomyces cerevisiae* showed higher tolerance against UV-C irradiation than *Hanseniaspora uvarum* facilitating new opportunities to control spontaneous fermentations. However, inactivation efficacy was strongly dependent on turbidity of musts and the initial degree of contamination suggesting a shadowing effect of individual germs. Compared with thermal pasteurization, UV-C treatment of must with 1 kJ/L showed similar effects in germ-reduction. While thermal pasteurization significantly decreased aroma precursors in musts, UV-C treatment did not change concentrations of glycosidically-bound C6-alcohols, monoterpenes and C13-norisoprenoids as shown by GC-MS analysis. Applying UV-C technology in wines, it was possible to irreversibly stop ongoing alcoholic fermentation indicating that UV-C treatment is capable to replace SO_2 addition to produce wines with residual sugar. Besides inactivation power, UV-C is known for its ability to form powerful off-flavours such as methional or methanethiol. Sensory analysis revealed that the application of UV-C at doses < 2 kJ/L in must is uncritical. However, applying UV-C after alcoholic fermentation can result in rising concentrations of mercaptans already at doses < 1 kJ/L. In this context, compounds such as caftaric acid, riboflavin and dissolved oxygen are thought to positively contribute to the UV-induced formation of off-flavours in wine.

1. Introduction

Warm temperatures due to climate change decrease acidity in grapes and allow new allochthonous parasites to establish. Higher pH values limit the microbicide effect of SO_2 and enable good living conditions for harmful organisms, which may empower the formation of SO_2 binding metabolites such as pyruvate and acetaldehyde, further reducing the microbicide effect of SO_2. Various physical sterilization processes like UV-C treatment are considered as techniques for the inactivation of microorganisms. While UV-C treatment is a well-established method in other industries such as water purification, only a few publications focused on the effect of UV-C on microorganism in must and wine [1,2]. UV-C induces photochemical reactions such as strand breakage of the DNA. As a result, damaged cells are not able to reproduce. For the inactivation of microorganisms, wavelengths between 250 and 260 nm are particularly suitable, since the DNA absorbs very well in this range. The effect of UV-C in liquids generally depends on the penetration depth of the light waves [3], which are either trapped by the absorption of ingredients or scattered by colloids in the liquid [4]. However, this problem can be circumvented by choosing an appropriate technique such as shown in several UV-C reactors [5]. Application fields for the UV-C technology in wineries are the treatment of musts with microbial contamination, the treatment of young wines to stop fermentation, and treatment of wines for the purpose of stabilization, e.g. prior to bulk wine shipping. The technique might be considered as an alternative for thermal pasteurization. In any case, treated products must not suffer from UV-C irradiation. However, UV-C is known for its ability to form powerful off-favours such as methional or methanethiol [6]. Accordingly, the objective of this study was to strive for the minimal UVC energy to inactivate microorganisms without generating detrimental light-struck flavour.

2. Materials and methods

Vitis vinifera grapes of the variety Riesling were harvested in 2016 from the vineyards of the DLR Rheinpfalz, Neustadt, Germany. Harvesting by hand was conducted in three consecutive weeks on same sites obtaining grapes of different quality and different microflora. Immediately after harvesting, grapes were crushed and pressed obtaining musts which were further processed by means of sedimentation for clarification reasons (< 50 NTU). In order to investigate UV-C efficiency in turbid musts, grape lees were blended into the clarified

[a] e-mail: `dominik.durner@dlr.rlp.de`

juice yielding musts with 100, 500 and 900 NTU. UV-C treatment of musts was achieved by a pilot-scale coiled tube reactor designed and assembled at the Max-Rubner-Institut (Karlsruhe, Germany). The main component is a module which consists of a helically wound quartz glass tube with a diameter of 3.0 mm which is assembled around a 30 W low pressure mercury lamp with maximum peak radiation at 254 nm (UVN 30, UV Technik Speziallampen GmbH, Wümbach, Germany). Between lamp and coiled quartz glass, a water cooling system allows for stable must temperatures during treatment. Musts were pumped through the device at a flow rate of 25 L/h by a peristaltic pump (Pump drive Pd 5206, Heidolph, Schwabach, Germany). Flow was adjusted to 25 L/h after preliminary studies taking into account that turbulent flow is necessary to maximize UV-C effect [7]. In this study, various UV C doses were obtained by recirculating must and wine through the device for multiple times. UV C dose in J/L was determined using a potassium iodide/iodate chemical actinometry method previously described [8]. All experiments were conducted in triplicate.

Microbial analysis was carried out in 10^{-5}-dilutions (in 0.9% NaCl) using YPD medium (with chloramphenicol und kanamycin) for yeasts and LB medium (with cycloheximide) for bacteria. Agar plates were incubated at 30 °C for 24 h, then colonies were counted. GC-MS analysis of aroma precursors in Riesling must was carried out after SPE purification and enzymatic hydrolysis using a 30 m × 0.25 mm i.D. fused silica capillary column ZB-Wax (PEG) as described by Schober [9]. Photometric analysis was conducted according to OIV-MA-AS2-11 procedure using a Varian Cary100 Spectrophotometer. In order to investigate the sensory effect of UV-C treatment, musts were processed by means of a standardized winemaking protocol. Microvinification of UV-C treated and control musts was carried out in 15 L batches at 18 °C using a dry active yeast strain of *Saccharomyces cerevisiae*. After alcoholic fermentation wines were racked off the lees and stored for two months at 15 °C for clarification reasons. Stabilization was achieved by using a commercial solution of ammonium hydrogensulphite yielding 30 mg/L free SO_2. Sensory analysis was performed by 21 judges examining the wines by triangle test according to DIN 10954. Statistical calculations were performed using analysis of variance and Tukey-Kramer post-hoc test, significance level was set at $p \leq 0.05$.

3. Results and discussion

3.1. UV-C inactivation of natural yeast flora in grape must

Harvesting the Riesling on Sep. 24th, Oct. 6th and Oct. 17th yielded musts with total yeast counts of 6.8×10^4, 3.2×10^5 and 2.0×10^7 cfu/ml, respectively. Right after pressing, four yeasts could be clearly identified in the fresh grape musts, in detail *Saccharomyces cerevisiae*, *Hanseniaspora uvarum*, *Schizosaccharomyces pombe* and *Pichia fermentans*. When applying UV-C treatment, cultivable yeast cells decreased immediately in the musts. Regardless of the type of yeast, the inactivation success was strongly depended on the initial cell count. Table 1 shows the inactivation power of UV-C with 2 kJ/l for

Tab. 1. Inactivation power of UV-C treatment with 2 kJ/L for *Schizosaccharomyces pompe* in clarified Riesling grape must. The yeast was inoculated in definite concentrations at after must pasteurization.

Initial cell count (cfu/ml)	Reduction of cells ($\log_{10}$ cfu/ml)
10^7	1
10^6	1.5
10^5	2
10^4	2.5

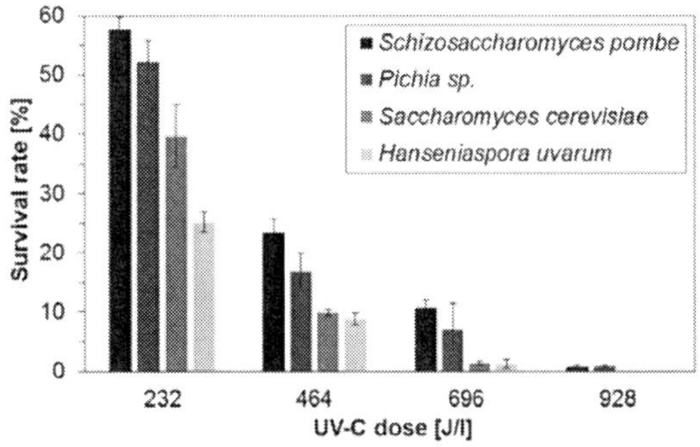

Fig. 1. Survival rates of naturally occurring yeasts in clarified Riesling must after UV-C treatment with different doses. The initial number of total yeast germs counted on YPD medium was 3.2×10^5 cfu/ml.

Schizosaccharomyces pompe in clarified Riesling grape must.

Data in Table 1 clearly reveals that inactivation efficacy of UV-C is dependent on the initial degree of contamination suggesting a shadowing effect of individual germs during treatment. Another factor influencing the inactivation efficacy of UV-C is the type of yeasts present in the grape must. The broken-down view on the four yeast types named earlier allows for detailed assessment of the UV-C sensitivity of different yeasts occurring on grapes. Figure 1 shows the survival rates of naturally occurring yeasts in clarified Riesling must after UV-C treatment with different doses.

Schizosaccharomyces pombe was found to least sensitive to UV-C treatment, followed by *Pichia fermentans*, *Saccharomyces cerevisiae* and *Hanseniaspora uvarum*. While *Saccharomyces* and *Hanseniaspora* were completely killed at doses >1 kJ/l, *Schizosaccharomyces* and *Pichia* still remained in small cell counts. However, at higher doses <1 kJ/l, all occurring yeasts could be inactivated. The technical option to completely or partially inactivate yeasts in grape musts facilitate new opportunities to control spontaneous fermentations.

3.2. UV-C inactivation kinetics of yeast and bacteria in grape must

While facilitating the targeted inactivation of yeasts with UV-C treatment, the question arises as to whether a similar attempt is possible with parallel existence of bacteria and yeast in must and wine. Therefore, UV-C inactivation kinetics have been investigated in clarified Riesling must, which was co-inoculated with *Schizosaccharomyces pompe*, *Brettanomyces bruxellensis* and *Lactobacillus*

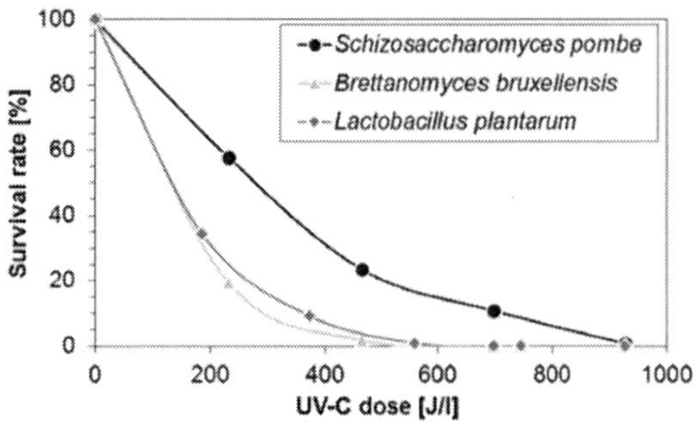

Fig. 2. UV-C inactivation kinetics for *Schizosaccharomyces pompe*, *Brettanomyces bruxellensis* and *Lactobacillus plantarum* in clarified Riesling must. All microorganisms were inoculated at 10^5 cfu/mL after must pasteurization.

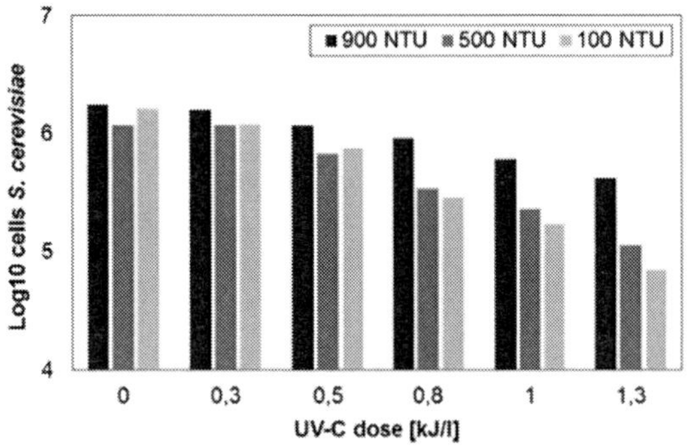

Fig. 3. Influence of turbidity levels on UV-C inactivation of *Saccharomyces cerevisiae* in Riesling must.

plantarum, each at 10^5 cfu/mL, after must pasteurization (Fig. 2).

Brettanomyces bruxellensis and *Lactobacillus plantarum* revealed similar survival rates due to UV-C treatment and were found to be much more sensitive compared to *Schizosaccharomyces pompe*. This finding suggests that targeted inactivation is also possible for mixtures of bacteria and yeasts in must or wine allowing for new possibilities in controlling harmful growth of bacteria, e.g. at the end of alcoholic fermentation. The relatively high sensitivity of *Brettanomyces bruxellensis* towards UV-C might be also considered as an important hint for cellars with *Brettanomyces* problems. However, it was already noted that UV-C efficacy is influenced by anthocyanins and other phenolics, which play a major role in red wines [1].

3.3. Influence of turbidity on UV-C inactivation of yeast in grape must

Turbidity in must and wine is known to reduce UV-C efficacy [1]. Several technical optimization strategies have been proposed to overcome restrictions in UV-C effect in turbid liquids [5]. When employing state-of-the-art UV-C technology, efficacy of the germicidal effect still depended on the turbidity as shown in Fig. 3.

Figure 3 reveals that grape lees attenuate UV-C efficacy with an increasing UV-C dose. At 1 kJ/l, UV-C treatment

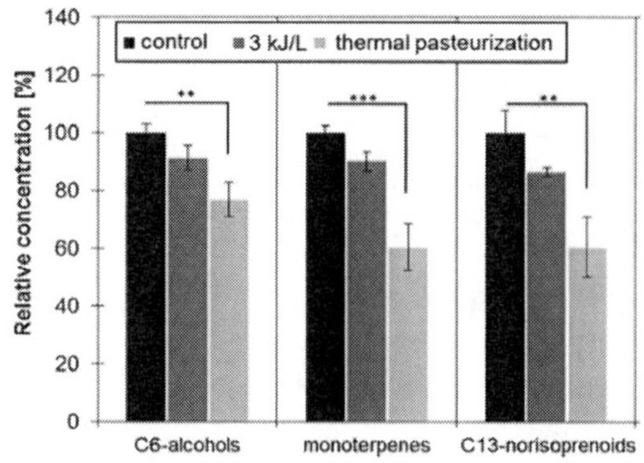

Fig. 4. Influence of UV-C treatment (3 kJ/l) and thermal pasteurization (80 °C, 120 s) on glycosidically bound aroma precursors in Riesling must. The substance classes represent the sum of 2–6 aroma precursor compounds within each class.

resulted in an average $\log_10$ microbial reduction of 0.49, 0.72 and 1.37 for musts with 900, 500 and 100 NTU respectively. While 500 and 900 NTU are unrealistically high turbidity values for fermenting grape musts, these findings indicate that existing UV-C technology can still be improved towards higher efficacy.

3.4. Influence of UV-C treatment on aroma precursors in grape must

The efficacy of UV-C is undisputedly a capable technology to inactivate microorganisms in grape juices. Accordingly, the technology has to be considered as an alternative for thermal pasteurization. Therefore, the effects of UV-C treatment and thermal pasteurization on glycosidically-bound aroma precursors in Riesling must were compared with each other (Fig. 4).

Both, UV-C treatment with 3 kJ/l and thermal pasteurization at 80 °C for 120 s yielded sterile musts. While heat resulted in significantly lower concentrations of glycosidically bound C6-alcohols, monoterpens and C13-norisoprenoids, UV-C did not cause major losses of aroma precursors. It may be concluded, that UV-C with 3 kJ/l is more gently than heating at 80 °C for 120 s.

3.5. Use of UV-C treatment to stop alcoholic fermentation

To obtain natural residual sugar the alcoholic fermentation of wines is typically stopped by the addition of high concentrations of SO_2. In order to investigate the efficacy of UV-C to stop alcoholic fermentation, a fermenting Riesling at 17 g/L residual sugar was treated either with UV-C or 100 mg/L SO_2. The $\log_10$ microbial reduction after these treatments is displayed in Fig. 5.

Although none of the treatments yielded absolute inactivation of *Saccharomyces cerevisiae*, alcoholic fermentation may be considered as interrupted when cells are $<10^2$ cfu/ml. UV-C with 1 kJ/l was not capable to reach that number, however, a treatment with > 2 kJ/l exceeded the effect of 100 mg/L SO_2 by 1.5 $\log_{10}$ counts.

 DOI: 10.1051/bioconf/20170902001

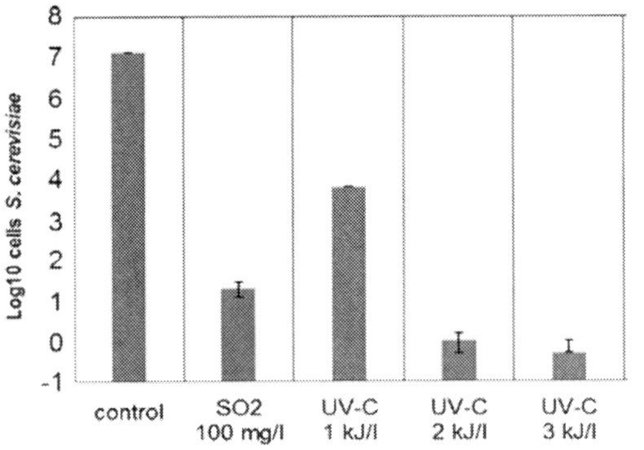

Fig. 5. Efficacy of UV-C treatment to stop alcoholic fermentation of Riesling at 17 g/L residual sugar compared to the use of 100 mg/L SO_2.

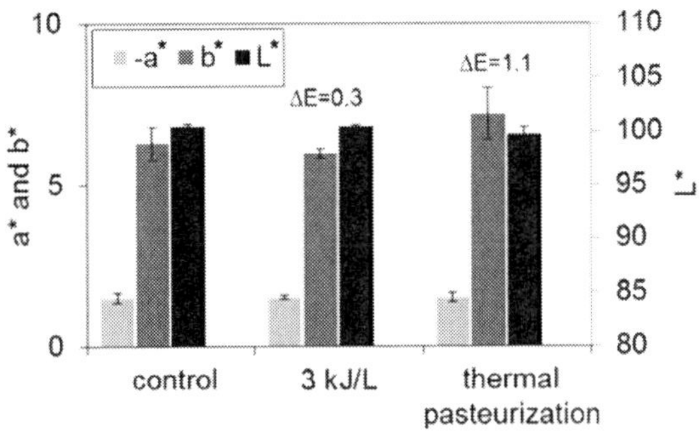

Fig. 6. Chromatic properties of wines made from Riesling musts which were either treated with UV-C (3 kJ/L) or thermal pasteurization (80 °C, 120 s).

3.6. Sensory and colour changes in wine due to UV-C treatment of grape must

UV-C light is known for its ability to form powerful off-favours such as methional or methanethiol [6]. Increasing concentrations of mercaptans due to UV-C treatment of grape must with < 2 kJ/l could not be determined yet. Sensory triangle testing revealed no differences between wines from control an UV-C treated musts. However, higher energies than 2 kJ/l applied in must were tested significantly different in the wine stadium and even low energies <1 kJ/l applied in wines seem to develop off-flavours reminding of cooked cabbage, popcorn and camphor. Other than aroma changes, chromatic properties of wines made from Riesling musts which were either treated with UV-C or thermal pasteurization have already been measured (Fig. 6). A slightly higher yellow colour was observed for wines obtained from heated musts but not due to UV-C treatment.

4. Conclusion and outlook

Outcomes in this ongoing research project support the applicability of UV-C treatment to inactivate microorganisms in must and wine. The efficacy towards different germs is much higher than anticipated and the fact, that UV-C is adjustable in its intensity, opens up new vistas in enology. Different sensitivities of yeasts and bacteria may be utilized to inactivate the unwanted and to let the beneficial ones survive. The current focus lies on the potential off-flavour formation by UV-C. GC-PFPD analysis towards low volatile sulphides and GC-MS analysis of 2-aminoacetophenone are about to be completed shortly. Since high concentrations of caftaric acid, riboflavin and dissolved oxygen seem to positively contribute to UV-C induced sensory defects in wines, this finding will be pursued.

References

[1] I.N. Fredericks, M. du Toit, M. Krügel, Food Microbiol. **28**, 510 (2011)
[2] L. Rizzotti, N. Levav, F. Fracchetti, G.E. Felis, S. Torriani, Food Control **47**, 407 (2015)
[3] R.E. Tucker, *Sterilizer unit for fluid media and process*, U.S. Patent, Patent No.: US4798702 A (1989)
[4] P. Mendes, A. Müller, A. Fernández, M. Stahl, Innov. Food Sci. Emerg. Technol. **21**, 90 (2014)
[5] A. Müller, K. Günther, M. Stahl, R. Greiner, C. Franz, C. Posten, Innov. Food Sci. Emerg. Technol. **29**, 240 (2015)
[6] A. Maujean, N. Sequin, Sci. Aliments **3**, 589 (1983)
[7] M. Huch, A. Müller, J. Strohhacker, S. Vogt, A. Hanak, V. Gräf, M. Stahl, C. Franz, Fruit Processing **20**, 196 (2010)
[8] A. Müller, M. Stahl, R. Greiner, C. Posten, J. Food Eng. **138**, 45 (2014)
[9] D. Schober, *Freisetzung glycosidisch gebundener Aromastoffe während der Wein- und Sektbereitung*, Dissertation, TU Kaiserslautern (2016)

Study of the effectiveness of a strain of *Saccharomyces cerevisiae* selected for the production of wines with higher acidity and lower alcoholic strength

Olga Pascual[1], Pere Pons-Mercadé[1], Jordi Gombau[1], Anne Ortiz-Julien[2], José M. Heras[2], Francesca Fort[1], Joan Miquel Canals[1], and Fernando Zamora[1]

[1] Departamento de Bioquímica y Biotecnología, Facultad de Enología de Tarragona, Universidad Rovira i Virgili, C/ Marcel.lí Domingo, 1, 43007 Tarragona, España
[2] Lallemand Bio S.L. C/ Galileu 303, 1ªplanta, 08028-Barcelona, España

Abstract. Grapes of Tempranillo, Garnacha Tinta and Merlot at very high maturity level were used for red microvinifications using a conventional *Saccharomyces cerevisiae* strain (Lalvin EC 1118®) and new *S. cerevisiae* strain generated using adaptive evolution-based strategies (IONYS™WF). All microvinifications were performed by triplicate and at low (16°C) and at high (27°C) temperatures. The results show that all the wines fermented with IONYS™WF, independently of the fermentation temperature and grape cultivar have significant lower ethanol content (average 0.60 %), higher glycerol content (average of 5.6 g/L), higher titratable acidity (average of 1.3 g of tartaric acid/L) and lower pH (average of 0.1 units) than their corresponding controls. It seems therefore that IONYS™WF strain can be a useful tool to mitigate the excess of ethanol and the lack of acidity that unfortunately many wines present nowadays. Moreover, the high glycerol production can also be an interesting contribution inasmuch as this compound increases mouthfeel and smooth astringency.

1. Introduction

It is a doubtless fact that in recent years the alcoholic strength of wines has increased significantly [1], probably due to climate change and also because winemakers are searching full grape maturity [2,3]. Since high alcohol levels have certain drawbacks, the wine industry is interested in developing techniques to reduce the alcohol content. In this regard, a wide variety of strategies have been proposed with this purpose [3,4], among which we would highlight the selection of yeasts with lower sugar/ethanol transformation ratio [5,6].

According to the stoichiometry of the alcoholic fermentation pathway, 15.45 g/L of sugars are theoretically required to obtain 1% vol. Alcohol [4]. However, besides ethanol yeasts also produce many other compounds such as glycerol, higher alcohols, esters, succinic acid, diacetyl, acetoin, 2,3-butanediol, etc. Moreover, some of the sugars are used by yeasts to increase their biomass, and a percentage of ethanol is evaporated in a greater or lesser extent depending on the CO_2 release rate, the temperature and the dimensions of the tank. Moreover, some winemaking operations such as pumping over can also favor its evaporation [7,8].

It is consequently difficult to know what the real sugar/ethanol transformation ratio (TR) is during alcoholic fermentation. For this reason, the term "Potential Alcohol Content" is used to predict the ethanol content of a wine from the sugar content of the grape juice. OIV considers an average TR of 16.83 g/L. However, winemakers usually use a TR of 16.00 g/L for white wines and 17.00 g/L for red wines. The difference is mainly because white wines are generally fermented at low temperatures and without aeration whereas red wines are fermented at high temperatures and with some operations involving aeration [4,7].

Recently scientists from INRA of Montpellier (France) in collaboration with Lallemand have generated a new S. cerevisiae strain using adaptive evolution-based strategies (100% natural; non-GMO) [9]. Making the yeasts grow continuously and repeatedly under hyperosmotic medium they have generated a S. cerevisiae strain that produces appreciably less alcohol and more glycerol and acidity than usual strains. This yeast is nowadays commercialized with the name IONYS™WF (International Patent N° WO2015/11411).

The aim of this study was to determine if this strain can really reduce the ethanol content improve wine acidity under real winemaking conditions.

2. Materials and methods

Grapes of Tempranillo, Garnacha tinta and Merlot at very high maturity level were destemmed, crushed and sulphited (30 mg/L). Microvinifications were carried out by triplicate at two fermentation temperatures (16 and 27 °C) using submerged cap procedure [10]. Control musts were inoculated with the commercial strain Lalvin EC1118®and the experimental ones with IONYS™WF (both yeasts from Lallemand Inc.). After 15 days of

Table 1. General parameters of Tempranillo wines.

PARAMETER	Temperature: 16 ± 1 °C		Temperature: 27 ± 1 °C	
	EC1118	IONYS™WF	EC1118	IONYS™WF
Ethanol (% v/v)	14.1 ± 0.13 B	13.58 ± 0.08 A	14.22 ± 0.13 B	13.72 ± 0.13 A
Glycerol (g/L)	6.53 ± 0.35 A	10.77 ± 0.42 B	6.73 ± 0.06 A	11.13 ± 0.46 B
Titratable Acidity (g/L)	7.00 ± 0.30 A	8.37 ± 0.12 B	7.20 ± 0.10 A	8.07 ± 0.49 B
Volatile Acidity (g/L)	0.15 ± 0.07 A	0.17 ± 0.03 A	0.17 ± 0.02 A	0.18 ± 0.03 A
pH	3.67 ± 0.01 B	3.56 ± 0.01 A	3.75 ± 0.02 B	3.67 ± 0.05 A
Glucose + Fructose (g/L)	0.17 ± 0.05 A	0.20 ± 0.04 A	0.30 ± 0.23 A	0.37 ± 0.24 A

Results are expressed as mean ± standard deviation of three replicates. Different letters indicate a statistical difference ($p < 0.05$).

Table 2. General parameters of Garnacha Tinta wines.

PARAMETER	Temperature: 16 ± 1 °C		Temperature: 27 ± 1 °C	
	EC1118	IONYS™WF	EC1118	IONYS™WF
Ethanol (% v/v)	16.3 ± 0.05 B	15.40 ± 0.52 A	15.87 ± 0.08 B	15.42 ± 0.15 A
Glycerol (g/L)	6.43 ± 0.42 A	10.93 ± 0.55 B	7.13 ± 0.25 A	13.73 ± 0.49 B
Titratable Acidity (g/L)	5.53 ± 0.12 A	6.87 ± 0.29 B	5.60 ± 0.10 A	7.23 ± 0.23 B
Volatile Acidity (g/L)	0.10 ± 0.01 A	0.15 ± 0.03 B	0.31 ± 0.04 A	0.28 ± 0.03 A
pH	3.58 ± 0.02 B	3.44 ± 0.02 A	3.60 ± 0.01 B	3.47 ± 0.01 A
Glucose + Fructose (g/L)	0.17 ± 0.09 A	0.97 ± 0.67 B	0.23 ± 0.06 A	0.70 ± 0.87 A

Results are expressed as mean ± standard deviation of three replicates. Different letters indicate a statistical difference ($p < 0.05$).

maceration free run wines were separated from the pomace, sulphited to prevent malolactic fermentation (40 mg/L) and stabilized for 30 days at 4 °C. Wines were analyzed one month later.

The analytical methods recommended by the OIV were used to determine the ethanol content, pH and volatile acidity (Organisation Internationale de la Vigne et du Vin, 2014). Glycerol and sugars (D-glucose and D-fructose) were measured using enzymatic kits (R-Biopharm AG., Darmstadt, Germany).

All data are expressed as the arithmetic average ± the standard deviation from three replicates. One-factor analysis of variance (ANOVA) and Tukey's test were carried out with SPSS software (SPSS Inc., Chicago, Illinois, USA).

3. Results and discussion

Tables 1, 2 and 3 show the general parameters of the Tempranillo, Garnacha Tinta and Merlot wines.

Results are expressed as mean ± standard deviation of three replicates. Different letters indicate a statistical difference ($p < 0.05$).

Results are expressed as mean ± standard deviation of three replicates. Different letters indicate a statistical difference ($p < 0.05$).

Results are expressed as mean ± standard deviation of three replicates. Different letters indicate a statistical difference ($p < 0.05$).

All fermentations were correctly performed since the Glucose + Fructose concentration were in all the cases below 1 g/L and the volatile acidity below 0.35 g of acetic acid/L.

In general, no important differences were found between the wines fermented at low (16°C) and high (27°C) temperature for a same yeast in any of the studied parameters. The small differences between the ethanol

Table 3. General parameters of Merlot wines.

PARAMETER	Temperature: 16 ± 1 °C		Temperature: 27 ± 1 °C	
	EC1118	IONYS™WF	EC1118	IONYS™WF
Ethanol (% v/v)	15.5 ± 0.01 B	14.83 ± 0.06 A	15.38 ± 0.08 B	14.73 ± 0.15 A
Glycerol (g/L)	6.07 ± 0.12 A	13.73 ± 0.12 B	6.37 ± 0.21 A	12.80 ± 0.42 B
Titratable Acidity (g/L)	6.10 ± 0.00 A	7.27 ± 0.21 B	6.40 ± 0.10 A	7.73 ± 0.29 B
Volatile Acidity (g/L)	0.27 ± 0.02 A	0.32 ± 0.02 B	0.30 ± 0.01 A	0.32 ± 0.01 A
pH	3.34 ± 0.02 B	3.25 ± 0.02 A	3.39 ± 0.01 B	3.33 ± 0.02 A
Glucose + Fructose (g/L)	0.77 ± 0.50 A	0.83 ± 0.15 A	0.43 ± 0.58 A	0.35 ± 0.23 A

Results are expressed as mean ± standard deviation of three replicates. Different letters indicate a statistical difference ($p < 0.05$).

content at low and at high temperature for a same yeast may be related with the fact that all fermentation were performed with submerged cap which would reduce the temperature effect on ethanol evaporation.

In contrast, the ethanol content was significant lower and glycerol concentration significant higher in all the wines fermented with IONYS™WF strain than in their corresponding controls, independently of the fermentation temperature and grape cultivar. The average difference was of 0.60% for ethanol content and of 5.6 g/L for glycerol concentration. These results clearly indicate that IONYS™WF strain redirect part of the metabolic flow of the sugars towards the production of glycerol.

In addition, all the wines fermented with IONYS™WF strain have significant lower pH and significant higher titratable acidities than their corresponding controls. Specifically, the average difference was of o.1 units for pH and of 1.3 g of tartaric acid/L for titratable acidity.

It can be concluded that this new yeast can be a useful tool to mitigate the excess of ethanol and the lack of acidity that unfortunately many wines present nowadays. Moreover, the high glycerol production can also be an interesting contribution inasmuch as this compound increases mouthfeel and smooth astringency.

This study has been funded by CDTI (CIEN-VINySOST2014 project).

References

[1] P. Godden, R. Muhlack, Aust. NZ. Grapegrow. Winemak. **558**, 47–61 (2010)
[2] H.R. Schultz, G.V. Jones, J. Wine Res. **21**, 137–145 (2010)
[3] N. Kontoudakis, M. Esteruelas, F. Fort, J.M. Canals, F. Zamora, Aust. J. Grape Wine Res. **17**, 230–238 (2011)
[4] F. Zamora, Delacoholised Wines and Low-Alcohol Wines. En "Wine Safety, Consumer Preference, and Human Health". Ed. M.V. Moreno-Arribas & B. Polo. Bartolomé Sualdea, Springer. New York, pp 163–182 (2016)
[5] S. Michnick, J.L. Roustan, F. Remiz, P. Barre, S. Dequin, Yeast **13**, 783–793 (1997)
[6] M. Quirós, V. Rojas, R. Gonzalez, P. Morales, Int. J. Food Microbiol. **181**, 85–91 (2014)
[7] M., Gil, N. Kontoudakis, S. Estévez, E. González-Royo, M. Esteruelas, F. Fort, J.M. Canals, F. Zamora, (2013) Non microbiological strategies to reduce alcohol in wines. En: Alcohol Reduction in Wine; Oneoviti International Network. Vigne et

DOI: 10.1051/bioconf/20170902002

Vin Publications Internationales. Merignac, France, pp. 25–28 (2013)

[8] O. Pascual, E., González-Royo, M. Gil, S. Gómez-Alonso, E. García-Romero, J.M. Canals, I. Hermosín-Gutíerrez, F. Zamora, (2016) J. Agric. Food Chem. **64**, 6555–6566 (2016)

[9] V. Tilloy, A. Ortiz-Julien, S. Dequin, Appl. Environ. Microbiol. **80**, 2623–32 (2014)

[10] T.L. Sampaio, J.A. Kennedy, M.C. Vasconcelos, Am. J. Enol. Vitic. **58**, 534–539 (2007)

[11] Organisation Internationale de la Vigne et du Vin. Methods of analysis of wines and must (2014)

Detection with flash gas chromatography electronic nose of the general influences of glutathione, ascorbic acid, tannin and carbon dioxide treatments on the volatile profiles of white wines of feteasca regala

Arina Oana Antoce[a] and George Adrian Cojocaru[b]

University of Agronomic Sciences and Veterinary Medicine of Bucharest, Faculty of Horticulture, Department of Bioengineering of Horti-Viticultural Systems, 59, Mărăşti Ave., Sector 1, 011464 Bucharest, Romania

Abstract. The present work aimed to determine some of the influences of the treatments with glutathione (GSH), ascorbic acid (AA), catechinic tannin (T) and carbon dioxide (CO_2) during winemaking or bottling on the volatile profiles of white wines of Feteasca Regala. The study is based on the use of a flash gas chromatography electronic nose, which is able to discriminate various clusters of wine samples prepared with various combinations of antioxidants. The treatments that induce enough differences in the volatile profiles of the wine that the electronic nose is able to discriminate are likely to be discriminated also by the consumers. When the electronic nose does not discriminate the clusters of samples with certain treatments it is very likely that those treatments are not sufficient to induce a practically important difference in the aromatic profile of wines. The main detectable influences of the antioxidant treatments are presented and discussed. The treatment of musts with AA (50 mg/l) clearly influences the volatile profile of the wines, reducing their aromatic complexity. The treatment during bottling with catechinic tannin (20 mg/l) does not seem to have an important influence of the profile of the wines produced with the addition of AA and or GSH. The addition of carbon dioxide during bottling does not show much influence one year after bottling, but it shows that in the samples most prone to oxidation, a difference is present. The CO_2 treated wines belonging to the cluster of wines treated with 40 mg/l GSH and 50 mg/l AA have distinct profiles, different from those not protected and closer to the clusters of the less oxidised wines. The treatment of musts with GSH shows clear influences at the dose of 40 mg/l, but for the lower dose of 20 mg/l GSH in Feteasca regala wines it is difficult to identify differences compared to the wines included in the cluster of samples not treated with GSH. This finding is particularly relevant as the dose of 20 mg/l GSH is the maximum allowed at present to be added in must and wines.

1. Introduction

The protection of wine volatiles from oxidation is one of the most desirable effects achieved by the use of the antioxidants or other compounds in winemaking or during bottling. Among the oenological permitted practices are the addition of glutathione (GSH), ascorbic acid (AA) and catechinic tannin (T) in must or wines, as well as the bottling of the wines in the presence of carbon dioxide. For this reason, although not able to replace the sulphur dioxide in spite of many trials [7,10,11,13], these compounds have been widely used and assessed in winemaking in various combinations and doses, both for still [5,7–9,12,16,18–20] and sparkling wines [21, 22]. Taking into account the many possible combinations and reactions in wine, the effect of their application is difficult to assess by sensory analyses or by physico-chemical determinations. One rapid method for evaluating the effects of these compounds on the resulted wines is the comparison of the volatile profiles of wines by using a flash gas chromatography electronic nose. This technology allows for a precise and time-efficient evaluation of clusters of samples grouped based on similar treatments [1], such as combinations of antioxidants, or common product traits, such as origin [14,17] or other [2,4]. The technique does not directly identify the volatile compounds, but rather records non-target signals resulted from the chromatographic separation of the volatile fraction present in the sample headspace and uses them as variables (e-nose sensors) for multivariate statistical analysis.

Although the electronic nose does not tell which volatile profile is better from the viewpoint of the consumer, it can evaluate many samples in a short period of time and show if samples grouped in various clusters in accordance to the undergone treatments can be discriminated or not. If some clusters of samples cannot be discriminated by the electronic nose, most likely, from the viewpoint of the perceived volatile profile, they will not be discriminated by the consumers either. Thus, the presence of clusters which cannot be discriminated would mean that the treatments performed on the wines included in those clusters are not important enough to induce perceivable differences and may not be worth applying in the wine cellar.

[a] e-mail: arina.antoce@horticultura-bucuresti.ro
[b] e-mail: aantoce@yahoo.com

2. Materials and methods

The wines evaluated were obtained in normal wine cellar conditions from Feteasca regala wine grapes cultivated in the experimental field of the University of Agronomic Science and Veterinary Medicine of Bucharest, Department of Viticulture and Enology. The oenological materials used were: L-glutathione reduced (Carl Roth GmbH, purity min. 98% for biochemistry), L(+)-ascorbic acid (Carl Roth GmbH, purity min 99%, for biochemistry), catechinic tannin (Ti Premium, Enologica Vason), carbon dioxide (Buse Gaz, purity min. 99.99%) *Saccharomyces cerevisiae* yeast (Premium blanc 12 V yeast, Enologica Vason), fermentation activator (V Starter TF, Enologica Vason). The free-run must of Feteasca regala had the following parameters: 21.6% Brix; total titratable acidity of 82.7 meq/l; pH 3.33 and 135 mg/l YAN.

The experimental wine was obtained in stainless steel tanks of 50 l volume, with temperature control of fermentation. For the wine variants preparations reduced glutathione (GSH) was added in the musts in doses of 0, 20 and 40 mg/l. The sample variants with 20 and 40 mg/l GSH were prepared either without or with 50 mg/l AA added also before the onset of fermentation. In this way, after fermentation, 5 types of wines were obtained, coded in accordance to the oenological treatments of musts as G0, G20, G20A, G40 and G40A. After racking and cold stabilisation for 3 months in tanks, the variants were bottled in 0.75 l volume glass containers, at this stage dividing each variant in accordance with some supplementary treatments performed during bottling, as follows: addition of 0 or 20 mg/l catechinic tannin (T00 and T20) and addition or not of carbon dioxide in the headspace of the bottle.

Usual dosages of the sulphur dioxide were maintained in the musts and wines, being corrected when necessary, so that the average of the free sulphur dioxide in wines was 26 mg/l and the total sulphur dioxide 132 mg/l. Detailed wine preparation techniques and the complete results for the main wine physico-chemical parameters (measured at the moment of electronic nose assessment) are presented in another paper [3]. The coding for each wine and the treatments made are included in Table 1.

One year from the bottling time the wines were evaluated with an electronic nose working on the principle of gas chromatography (GC E-nose Heracles, Alpha MOS, France). The apparatus used is a flash GC endowed with a Tenax trap and two short chromatographic columns of 2 m length and 0.18 mm diameter – one non-polar (DB5 – 5% diphenyl, 95% dimethylpolysiloxane) and another of low/mid polarity (DB1701 – 14% cyanopropylphenyl, 86% dimethylpolysiloxane) working simultaneously. The separated compounds are detected with flame ionization detectors placed at the end of each chromatographic column. Hydrogen is used both as carrier and combustion gas.

The wines from each variant are introduced in triplicate (repetitions r1–r3) in vials of 10 ml, which are filled with a volume of 4 ml of wine sample and sealed with magnetic caps with silicone septa, specially designed for the autosampler HS 100 (Combi PAL Auto-Sampler System, CTC Analytics AG, Switzerland). The method of sample analysis involves a step of warming up the sample in the autosampler shaking oven for 10 minutes at 60 °C and 500 rpm, extraction with a HS syringe of a 2.5 ml gas sample from the head-space and injection into the GC. The chromatographic program used for the separation on both columns has the following parameters: trap temperature 40 °C, trap desorption temperature 250 °C, trap purge time 50 s, bake-out 50s, initial column temperature 40 °C with an increase rate of 5 °C/s up to 200 °C, FID temperature 220 °C, FID fuel pressure 35 psi. The aquisition time is 46 s, with 5 min between 2 sample analyses. Some other details related to the method and apparatus can be found in other papers [1,2,4]. The data processing was done using the apparatus software, AlphaSoft V12.42.

Table 1. Coding and antioxidant treatments of the wine samples.

Sample code	Glutathione (mg/l)	Ascorbic acid (mg/l)	Catechinic tannin (mg/l)	Carbon diox-ide*
Stage of treatment	Winemaking		Bottling	
G00_A00_T00	0	0	0	no
G00_A00_T00_CO2	0	0	0	yes
G00_A00_T20	0	0	20	no
G00_A00_T20_CO2	0	0	20	yes
G20_A00_T00	20	0	0	no
G20_A00_T00_CO2	20	0	0	yes
G20_A00_T20	20	0	20	no
G20_A00_T20_CO2	20	0	20	yes
G20_A50_T00	20	50	0	no
G20_A50_T00_CO2	20	50	0	yes
G20_A50_T20	20	50	20	no
G20_A50_T20_CO2	20	50	20	yes
G40_A00_T00	40	0	0	no
G40_A00_T00_CO2	40	0	0	yes
G40_A00_T20	40	0	20	no
G40_A00_T20_CO2	40	0	20	yes
G40_A50_T00	40	50	0	no
G40_A50_T00_CO2	40	50	0	yes
G40_A50_T20	40	50	20	no
G40_A50_T20_CO2	40	50	20	yes

*(Bottle headspace filling).

Discriminant factor analysis (DFA) was selected as the most relevant statistical method to apply to separate the groups of samples.

To ensure that no information is lost, unless stated otherwise, for the statistical analysis all sensors (chromatographic peaks) identified by the apparatus were selected and used, irrespective of their discriminant power (Table 2).

As seen in Table 2, the sensors represent the chromatographic peaks determined by the AlphaSoft as being the most discriminant for the analysed samples and they are from both chromatographic columns (suffix 1A for the column DB5, 2A from the column DB1701).

3. Results and discussion

In order to compare wine samples containing various types of antioxidants, the variants were grouped in accordance with the main antioxidants used for their preparation. The variants treated with carbon dioxide were included in the clusters with similar samples not treated with CO_2, as the sensory analyses on the samples showed no perceivable influence [3]. Thus, the following clusters are formed (Table 3).

Table 2. The selection of the sensors (chromatographic peaks) for the DFA analysis.

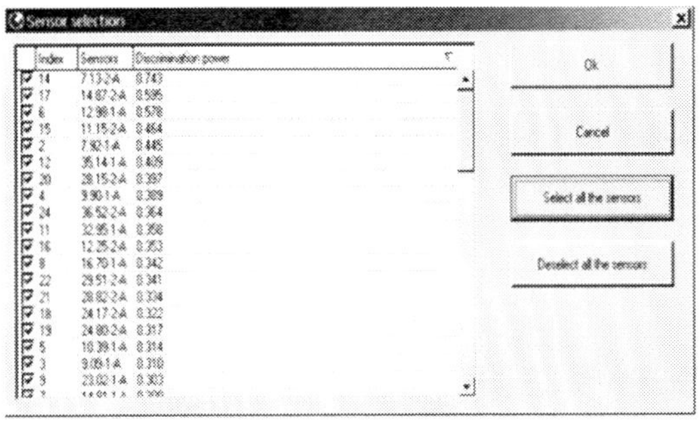

Sensor selection

Index	Sensors	Discrimination power
14	7.13.2A	0.743
17	14.87.2A	0.595
6	12.98.1A	0.578
15	11.15.2A	0.464
2	7.92.1A	0.445
12	35.14.1A	0.409
20	28.15.2A	0.397
4	9.90.1A	0.389
24	36.52.2A	0.364
11	32.95.1A	0.358
16	12.25.2A	0.353
8	16.70.1A	0.342
22	29.51.2A	0.341
21	28.82.2A	0.334
18	24.17.2A	0.322
19	24.80.2A	0.317
5	10.39.1A	0.314
3	9.09.1A	0.310
9	23.02.1A	0.303

Ok
Cancel
Select all the sensors
Deselect all the sensors

Table 3. Clusters of wines grouped in accordance with the main antioxidant treatments.

Cluster name	Sub-cluster name	Sample code	Color in figures
G0	**G0T0**	G00_A00_T00 (r1-r3)	Light
		G00_A00_T00_CO2 (r1-r3)	blue
	G0T20	G00_A00_T20 (r1-r3)	Dark
		G00_A00_T20_CO2 (r1-r3)	blue
G20	**G20**	G20_A00_T00 (r1-r3)	Light
		G20_A00_T00_CO2 (r1-r3)	green
	G20T20	G20_A00_T20 (r1-r3)	Dark
		G20_A00_T20_CO2 (r1-r3)	green
G20A	**G20AT0**	G20_A50_T00 (r1-r3)	Light
		G20_A50_T00_CO2 (r1-r3)	orange
	G20AT20	G20_A50_T20 (r1-r3)	Dark
		G20_A50_T20_CO2 (r1-r3)	orange
G40	**G40T0**	G40_A00_T00 (r1-r3)	Light
		G40_A00_T00_CO2 (r1-r3)	cyan
	G40T20	G40_A00_T20 (r1-r3)	Dark
		G40_A00_T20_CO2 (r1-r3)	cyan
G40A	**G40AT0**	G40_A50_T00 (r1-r3)	Light
		G40_A50_T00_CO2 (r1-r3)	magenta
	G40AT20	G40_A50_T20 (r1-r3)	Dark
		G40_A50_T20_CO2 (r1-r3)	magenta

The clusters take into account only the presence of GSH and AA, irrespective of the tannin treatment, while the sub-clusters take also into account the presence of tannin.

3.1. The influence of glutathione and ascorbic acid treatments

First we tried to discriminate the volatile profiles of the wines treated with various dosages of GSH in the presence or absence of AA. The clusters were formed based on the content of GSH and AA, irrespective of the presence of tannin, as described in Table 3 - first column (clusters G0, G20, G40, G20A, and G40A).

All the chromatographic peaks were selected for the DFA analysis, having discrimination power from 0.060 to 0.709.

As it can be seen in Fig. 1 the electronic nose discriminates the clusters of wines, and most of the variation is included in the discriminant function 1 (87.41%).

The treatment with AA clearly influences the volatile profile of the wines, thus the electronic nose separates in

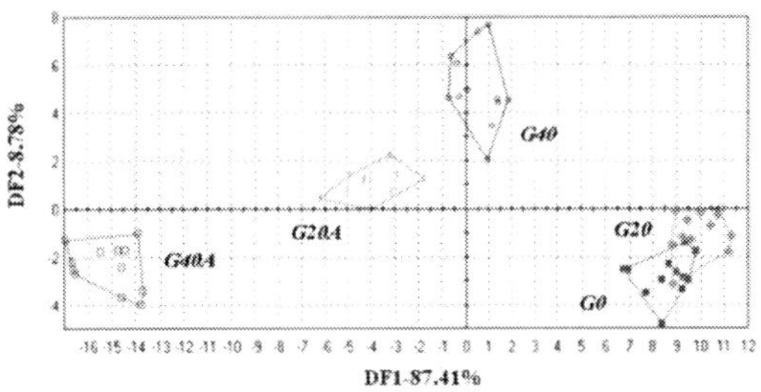

Figure 1. Discrimination of wines grouped in clusters in accordance to the glutathione and ascorbic acid treatments, irrespective of the presence of tannin. (GSH of 0, 20 and 40 mg/l (clusters G0, G20 and G40), GSH of 20 and 40 mg/l plus 50 mg/l AA (clusters G20A and G40A). Cluster colors: G0 - blue, G20 - green, G40 - cyan, G20A - orange, G40A - magenta.

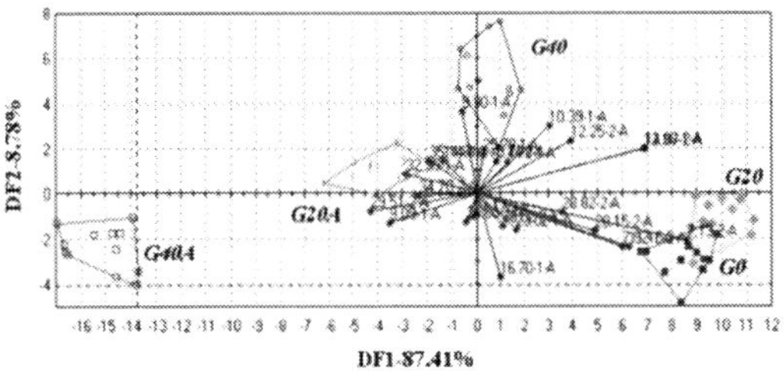

Figure 2. DFA diagram with loadings for the wine clusters discriminated in accordance to the glutathione and ascorbic acid treatments, irrespective of the presence of tannin. (GSH of 0, 20 and 40 mg/l (clusters G0, G20 and G40), GSH of 20 and 40 mg/l plus 50 mg/l AA (clusters G20A and G40A). Cluster colors: G0 - blue, G20 - green, G40 - cyan, G20A - orange, G40A - magenta.

the right part the wines containing only GSH from the wines containing both GSH and AA, which are included in the left part of the diagram (Fig. 1).

The presence of AA reduces the number of discriminative substances (chromatographic peaks) present in the headspace of the wine sample vials, suggesting that the volatile profiles of these wines are less complex. This can be qualitatively demonstrated by the diagram in Fig. 2, where the loadings of the chromatographic peaks are also represented along with the wine clusters. Fewer and shorter loadings determine the differentiation of the clusters G20A and G40A, those that contain samples with AA along with GSH.

Interestingly, the wines treated with only 20 mg/l GSH do not appear to have significantly different volatile profiles compared to the control wines. This can be seen in both Fig. 1 and Fig. 2, where the clusters G0 and G20 overlap. Also, if we take into account the discriminant function 3, same result is obtained (Fig. 3).

This observation is important, as the dose of 20 mg/l GSH is the maximum allowed at present for the treatments of musts and wines in accordance to the International Organisation of Vine and Wine recommendations [15].

The fact that the electronic nose does not discriminate the samples treated with 20 mg/l GSH from those not treated (control samples) means that from the viewpoint of the aromatic profile the consumer will most likely perceive no difference as well.

DOI: 10.1051/bioconf/20170902003

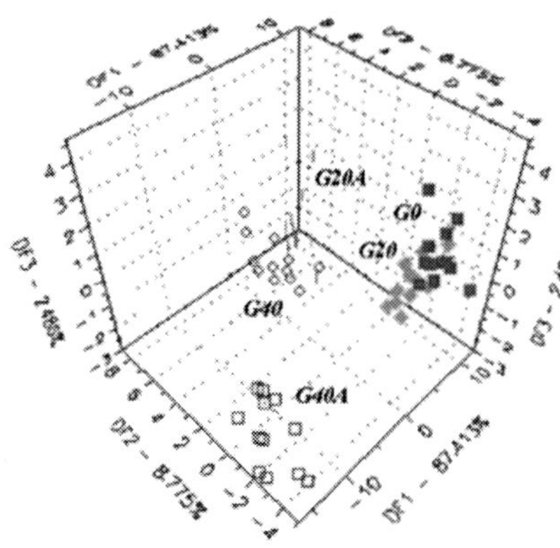

Figure 3. DFA 3D diagram for the wine clusters discriminated in accordance to the glutathione and ascorbic acid treatments, irrespective of the presence of tannin. (GSH of 0, 20 and 40 mg/l (clusters G0, G20 and G40), GSH of 20 and 40 mg/l plus 50 mg/l AA (clusters G20A and G40A.) Cluster colors: G0 - blue, G20 - green, G40 - cyan, G20A - orange, G40A - magenta.

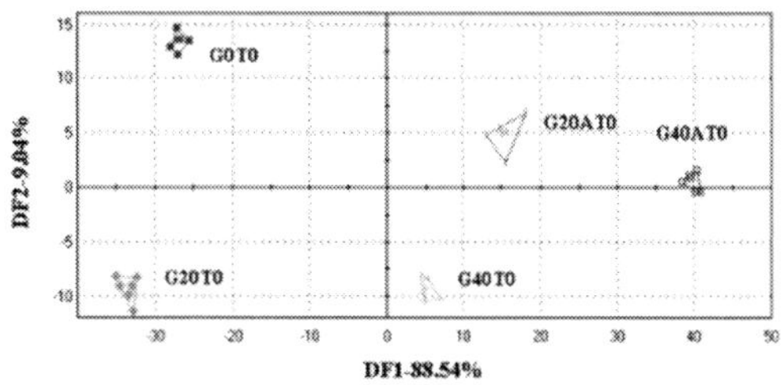

Figure 4. Discrimination of wines grouped in clusters in accordance to the glutathione and ascorbic acid in the absence of tannin. Sub-cluster colors: (G0T0 - light blue, G20T0 and G20T20 - light green, G40T0 - light cyan, G20AT0 - light orange, G40AT0 - light magenta.

This effect may, however, be due to the influence of the tannin present in some of the samples included in the analysed groups.

If we eliminate the effect of tannin, by comparing clusters of wines groups only in accordance with the GSH and AA treatments and with no tannin added, the wines containing 20 mg/l GSH are also separated from the control wines. (Fig. 4), although the discrimination power of the most discriminant sensor was only 0.654 (the selected sensors ranged from 0.654 to 0.218).

Therefore, it is very clear that for the discrimination of the wine samples by the electronic nose and also by the consumers, the complexity of the wine matrix, and the combination of oenological treatments are important.

3.2. The influence of tannin treatment

To determine the influence of the treatment with tannin on the volatile profile of wines already vinified with GSH and AA the DFA analysis was performed by grouping the samples in clusters containing also tannin (0 or 20 mg/l), as described in Table 3 second column (sub-clusters G0T0 and G0T20, G20T0 and G20T20, G40T0 and G40T20, G20AT0 and G20AT20, G40AT0 and G40AT20).

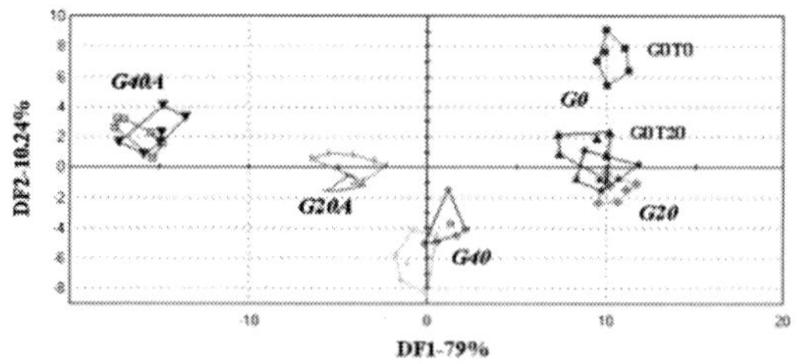

Figure 5. Discrimination of wines grouped in clusters in accordance to the glutathione, ascorbic acid and tannin treatments. GSH of 0, 20 and 40 mg/l (clusters G0, G20 and G40), GSH of 20 and 40 mg/l plus 50 mg/l AA (clusters G20A and G40A). Sub-cluster colors: G0T0 and G0T20 - light and dark blue, G20T0 and G20T20 - light and dark green, G40T0 and G40T20 - light and dark cyan, G20AT0 and G20AT20 - light and dark orange, G40AT0 and G40AT20 - light and dark magenta.

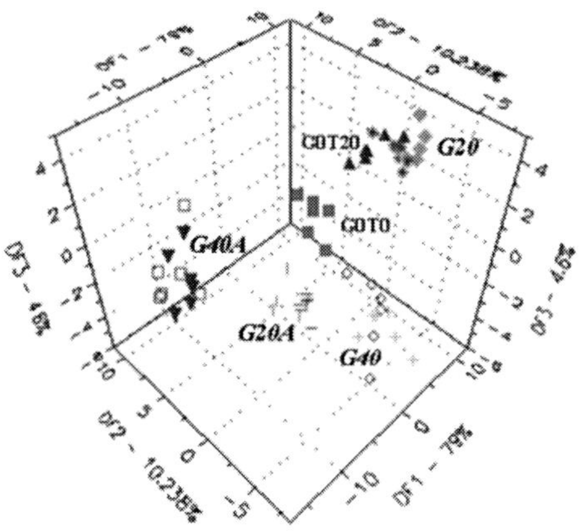

Figure 6. DFA 3D diagram for the wine clusters treated with glutathione, ascorbic acid and tannin treatments.

All the sensors were selected for analysis, with discrimination power ranging from 0.078 to 0.754.

The results showed that the presence of 20 mg/l tannin was not important enough to have a perceivable impact on the on the wine samples, as compared to the GSH or AA treatments.

Figures 5 and 6 show that the clusters with and without tannin overlap in all groups with the exception of the control cluster (blue color) which is split in the G0T0 (light blue) and G0T20 dark blue) sub-clusters.

The fact that in the control cluster we observe a slight discrimination between samples treated with tannin and those not treated suggests that the tannin actually influences the volatile profile of the wines, but when the GSH or AA are also added, its effect is masked by the greater effect induced by these other antioxidants.

The main effect on the volatile profile of the wines is induced by the GSH and the combination of GSH and AA. This is more clearly demonstrated when only the sub-clusters containing tannin are compared (Fig. 7). The sensor selection ranged from 0.817 to 0.222 discrimination power.

In Fig. 7 to the clusters of samples treated with both GSH and AA (right) are separated from the clusters of samples treated only with GSH (left), while the dose

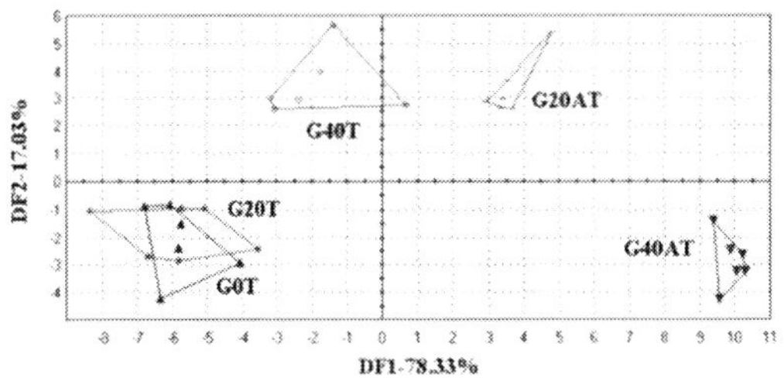

Figure 7. Discrimination of wines treated with tannin grouped in clusters in accordance to the glutathione and treatments. Sub-cluster colors: G0T20 - dark blue, G20T20 - dark green, G40T20 - dark cyan, G20AT20 - dark orange, and G40AT20 - dark magenta.

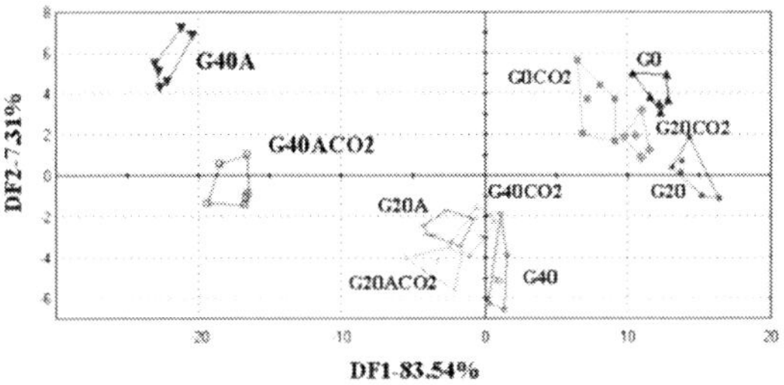

Figure 8. Discrimination of wines grouped in clusters in accordance to the glutathione, ascorbic acid and carbon dioxide treatments. GSH of 0, 20 and 40 mg/l (clusters G0, G20 and G40), GSH of 20 and 40 mg/l plus 50 mg/l AA (clusters G20A and G40A). Sub-cluster colors: G0CO2 and G0 – light and dark blue, G20CO2 and G20 - light and dark green, G40CO2 and G40 – light and dark cyan, G20ACO20 and G20A – light and dark orange, G40ACO2 and G40A– light and dark magenta.

of 20 mg/l GSH is not sufficiently important to ensure discrimination of samples from the control wines, not treated with GSH.

The results obtained under our experimental conditions seem to indicate that at least the treatment with 20 mg/l GSH in Feteasca regala does not lead to any benefit as far as the aromatic profile is concerned. Comparable results were obtained by other authors too [18], the 20 and 40 mg/l GSH treatments of certain white wines (also containing AA from the must treatment) leading to no significant difference in sensory parameters as compared to control wines.

3.3. The influence of carbon dioxide treatment

Although in a previous study the sensory analysis of the samples did not show any perceivable influence of the carbon dioxide on the aroma intensity and quality of these samples [3], the discrimination by electronic nose of the groups of samples with and without carbon dioxide added during bottling was also attempted (Fig. 8). The wines were grouped in the same clusters based on the treatments of GSH and AA, irrespective of the presence of tannin, but then, each cluster was split in sub-clusters with CO_2 (light colored) and without CO_2 (dark colored).

After 1 year from bottling time, the addition of carbon dioxide in the bottles did not influence significantly the aroma profile. All the sub-clusters containing wines with CO_2 overlapped with their corresponding sub-cluster without CO_2, with the exception of the samples with both 40 mg/l GSH and 50 mg/l AA, which, as shown in other studies too [6,23] are the most prone to oxidation. Thus, we can see that after 1 year in bottle, without CO_2 protection, the wines containing a larger dose of GSH and in the presence of AA (group G40A) show different profiles of the volatile substances as compared to all the other wines.

This result suggests that, in time, the effects of preservation with carbon dioxide will be more easily detectable and that the treatment is worth making.

4. Conclusions

The electronic nose can discriminate fast and reliably the samples of wines prepared with various oenological treatments.

Based on the statistical analysis of the discriminant factors (DFA) we were able to evaluate if the antioxidant treatments of white wines of Feteasca regala induced significant differences in the volatile profile 6 month after winemaking and 1 year after bottling.

Thus, in our study, the electronic nose helped to establish the following effects determined by the treatments with the antioxidants on Feteasca regala wines:Generally, the treatment with GSH in the must influences the volatile profile of the final wines only at doses larger than the 20 mg/l allowed at present by the OIV. The cluster of wines resulted from musts treated with 20 mg/l GSH was not discriminated by the electronic nose from the cluster without GSH added. A discrimination was however possible when the effect of tannin added was removed from the analysis, by excluding from the wine sample clusters the samples with added tannin.

The addition before fermentation of 50 mg/l AA in the musts treated with GSH, irrespective of the GSH dosage, reduced the complexity of the volatile profile of wines. The presence of AA in the samples determined the clear discrimination of the clusters containing wines with AA, from those without AA, irrespective of the other treatments.

The treatment with only 20 mg/l catechinic tannin during bottling was generally not important enough to induce perceivable effects on the volatile profile, in the context of addition of AA or/and GSH in the musts, which had a greater influence.

The treatment with carbon dioxide during bottling shows a promising effect in the long run. At 1 year form bottling, the wines mostly prone to oxidation (those included in the cluster of samples treated with 40 mg/l GSH and 50 mg/l AA) are split in sub-clusters of samples with and without CO_2 preservation. The sub-cluster with CO_2 is positioned closer to the samples less oxidised.

Of all the findings, the fact that the treatment of must with 20 mg/l GSH in Feteasca regala wines does not induce major differences in the volatile profiles of the resulted wines from the wines included in the cluster of samples not treated with GSH is particularly relevant as the dose of 20 mg/l GSH is at present the maximum allowed to be added in must and wines.

The authors declare no financial or commercial conflicts of interest.

References

[1] Antoce Arina Oana, Namolosanu C. I., 2011. Rapid and precise discrimination of wines by means of an electronic nose based on gas-chromatography, Revista de Chimie **62**(6), p. 593–595

[2] Antoce A. O., 2012. Influence of Maceration Enzyme Treatment on the Colour and Volatile Profile of Two Red Romanian Wines", Revista de Chimie **63**(9), p. 859–864

[3] Antoce A. O., Cojocaru G. A., 2017. Sensory profile changes induced by the antioxidant treatments of white wines – the case of glutathione, ascorbic acid and tannin treatments on Feteasca regala wines produced in normal cellar conditions, AgroLife Scientific Journal **6**(1) (sent for publication)

[4] Antoce O. A., Stroe M. V., Cojocaru G. A., 2015. Tentative application of an electronic nose to the study of the parentage of Romanian grape varieties Sarba and Alb Aromat, Agriculture and Agricultural Science Procedia, Elsevier **6**, p. 110–117

[5] Bradshaw M. P., Barril C., Clark A. C., Prenzler P. D., and Scollary G. R., 2011. Ascorbic acid: A review of its chemistry and reactivity in relation to a wine environment. Critical Reviews in Food Science and Nutrition **51**(6), 479–498

[6] Cojocaru G. A., Antoce O. A., 2016. Oxygen intake and color evolution in Sauvignon Blanc and Muscat Ottonel wines treated with ascorbic acid and glutathione, AgroLife Scientific Journal **5**(1), pp. 44–50

[7] Comuzzo P., Battistutta F., Vendrame M., Páez M. S., Luisi G. and Zironi R., 2015. Antioxidant properties of different products and additives in white wine. Food Chemistry **168**, 107–114

[8] Danilewicz J. C., Seccombe J. T., Whelan J., 2008. Mechanism of Interaction of Polyphenols, Oxygen and Sulphur Dioxide in Model Wine and Wine. American Journal of Enology and Viticulture, **59**(2), 128–136

[9] El Hosry L., Auezova L., Sakr A., Hajj-Moussa E., 2009. Browning susceptibility of white wine and antioxidant effect of glutathione. Int J Food Sci Technol **44**, p. 2459–2463

[10] Fracassetti D., Gabrielli M., Costa C., Tomas-Barber F. A., Tirelli A., 2016. Characterization and suitability of polyphenols-based formulas to replace sulfur dioxide for storage of sparkling white wine, Food Control **60**, p. 606–614

[11] Guerrero R. F., Cantos-Villar E., 2015. Demonstrating the efficiency of sulphur dioxide replacements in wine: A parameter review, Trends in Food Science & Technology **42**(1), p. 27–43

[12] Kritzinger E. C., Bauer F. F., du Toit W.J., 2013. Role of glutathione in winemaking: a review. J Agric Food Chem **6**, p. 269–277

[13] Li H., Guo A., Wang H., 2008., Mechanisms of oxidative browning of wine, Food Chemistry **108**, p. 1–13

[14] Melucci D., Bendini A., Tesini F., Barbieri S., Zappi A., Vichi S., Conte L., Toschi T.G., 2016. Rapid direct analysis to discriminate geographic origin of extra virgin olive oils by flash gas chromatography electronic nose and chemometrics, Food Chemistry **204**, p. 263–273

[15] OIV, International Organisation of Vine and Wine, 2016. International code of œnological practices. OIV - 18, rue d'Aguesseau - 75008 Paris, `http://www.oiv.int/public/medias/2696/code-2016-en.pdf`, retrieved on January 2017

[16] Panero L., Motta S., Petrozziello M., Guaita M. and Bosso A., 2015. Effect of SO2, reduced glutathione and ellagitannins on the shelf life of bottled white wines. European Food Research Technology **240**, 345–356

[17] Peng Q., Tian R., Chen F., Li B., Gao H., 2015. Discrimination of producing area of Chinese Tongshan kaoliang spirit using electronic nose sensing characteristics combined with the chemometrics methods, Food Chemistry **178**, p. 301–305

[18] Piano F., Petrozziello M., Vaudano E., Bonello F., Ferreira V., Zapata J., HernándezOrte P., 2014. Aroma compounds and sensory characteristics of Arneis Terre Alfieri DOC wines: the concentration of polyfunctional thiols and their evolution in relation to different ageing conditions, Eur Food Res Technol **239**, p. 267–277

[19] Sonni F., Chinnici F., Natali N., Riponi C., 2011a. Pre-fermentative replacement of sulphur dioxide by lysozyme and oenological tannins: Effect on the formation and evolution of volatile compounds during the bottle storage of white wines, Food Chemistry

[20] Sonni F., Moore E. G., Clark A. C., Chinnici F., Riponi C. and Scollary G. R., 2011b. Impact of glutathione on the formation of methylmethine-and carboxymethine-bridged (+)-catechin dimers in a model wine system. Journal of Agricultural and Food Chemistry **59**(13), 7410–7418

[21] Webber V., Dutra S. V., Spinelli F. R., Marcon A. R., Carnieli G. J. and Vanderlinde R., 2014. Effect of glutathione addition in sparkling wine. Food Chemistry **159**, 391–398

[22] Webber V., Dutra S. V., Spinelli F. R., Carnieli G. J., Cardozo A. and Vanderlinde R., 2017. Effect of glutathione during bottle storage of sparkling wine, Food Chemistry **216**, p. 254–259

[23] Wegmann-Herr P., Berner M., Dickescheid C., Wallbrunn C., Nickolaus P., Durner D., 2015. Influence of ascorbic acid, sulphur dioxide and glutathione on oxidation product formation in wine-like system, *BIO Web of Conferences* 5, 02005; `http://dx.doi.org/10.1051/bioconf/20150502005`. **129**, p. 1193–1200

The effect of inactivated yeast-based products on the process of wine aging, phenolic compounds and sensory characteristics of red wine Prokupac

Marko Malićanin[1], Vladislav Rac[2], and Vesna Rakić[2]

[1] Rubin AD, Nade Marković 57, 37000 Kruševac, Serbia
[2] Faculty of Agriculture, University of Belgrade, Nemanjina 6, 11080 Zemun, Serbia

Abstract. Keeping of red wine in bottles is very important for its maturation and quality. However, there are numerous changes that happen during that period, usually caused by oxidative processes and changes in structure and content of polyphenolic compounds. The goal of this study was to determine the effects of Inactivated dry yeast (IDY) products on aging, phenolics content, colour stability and sensory characteristics of red wine Prokupac (Serbian autochthonous variety). The treatment of wine was done by 3 different IDYP (Lallemand, Canada): Optimum White, Opti Less and Noblesse, applied as 0.2 g/L and 0.4 g /L during 15 days. Subsequently, wine clarification was done, followed by filtration and bottling. Untreated wines were used for comparison. Wine was subjected to accelerated aging (10 days at 55°C) and also to normal aging conditions during one year. To determine the influence of IDYP following parameters were monitored: dissolved O_2, free and total SO_2, sugar free extract, content of total phenolics, flavonoids, flavan-3-ols and anthocyanins, percent of polymeric color, color tint, color intensity and sensory characteristics (Panel method). The obtained results clearly show that IDY products are good scavengers of oxygen and have a positive impact on wine quality preservation and its organoleptic characteristics. However, a slight decrease of polyphenols content was detected.

1. Introduction

Inactivate dry yeast (IDY) products are derivatives of yeasts produced from enological variety (Saccharomyces cerevisiae) which have been pre-treated in order to inhibit their fermentative activity. Usually, these products are obtained by growing *Saccharomyces cerevisiae* yeasts in a sugar-rich medium, after which the yeasts are autolysed and dried to obtain final products (in powder form) which, under the generic name of IDY products, may contain inactive yeasts, yeast autolysates, yeast extracts, yeast hulls or walls [1]. In recent years, commercial use of such products in winemaking industry is growing and, according to their composition, IDYP are recommended for various reasons [2,3]. Namely, many bioactive compounds such as peptides, amino acids, gluthation and polysaccharides (mannoproteins) are released during yeast autolysis. Consequently, they can act as: alcoholic and malolactic fermentation enhancers, protective agents to improve active DY rehydration or as organoleptic enhancers. The stabilization of red wine colour is also possible by using mannoprotein-rich IDYPs [3]. Concerning sensory properties and stability, the goal is usually to obtain characteristics similar to those achieved by ageing on lees [2,4].

However, despite the fact that many of these products are currently in the market under different brand names claiming different wine improvements, scientific information about the chemistry behind their application is still insufficient and the mechanisms responsible for the claimed wine's improvements are not completely clear. Moreover, insufficient research on the application of IDYPs in winemaking conditions has been done, since it seems that the application of these products is based mostly on experience acquired during production of wine in small cellar trials. Therefore, well established scientific experiments devoted to the effects that these products induce in wines are needed in order to achieve better understanding of the mechanism of IDYP action.

Having all of previously mentioned in mind, the objective of the present study is to characterise the effect of different products based on yeast autolysates on the wine aging process, phenolic compounds and sensory characteristics of red wine Prokupac during storage in bottles. This work constitutes a primary approach in understanding the action mechanism of IDYP and establishing better criteria for their use in winemaking.

2. Materials and methods

2.1. Materials

Red wine Prokupac (Serbian autochthonous red grape variety), produced from grapes from "Rubin" vineyards, central Serbia wine region and Tri Morave wine sub region, vinified in 2015, was assayed. Catechin, gallic acid and vanillin were from Sigma. Folin–Ciocalteu reagent and sodium bisulfite were from Fluka.

2.2. Methods

Prokupac grapes were harvested at technological maturity, based on indices of sugar content and acidity established

DOI: 10.1051/bioconf/20170902004

Table 1. Wine sample labeling and treatment description.

Sample	IDYP used	Concentration of IDYP
BL	/	/
OW20	Optimum White	0.2 g/L
OW40	Optimum White	0.4 g/L
OL20	Opti Less	0.2 g/L
OL40	Opti Less	0.4 g/L
N20	Noblesse	0.2 g/L
N40	Noblesse	0.4 g/L

by the Vine & Wine Institute (Paris, France), destemmed and crushed. Pomace were inoculated with selected yeast strain (S. cerevisiae var cerevisiae) and fermented under controlled temperature (16–21 °C) to dryness (reducing sugar content below 4 g/L). Maceration lasted nine days. After months of wine care, the treatment of wine was carried out by 3 different products based on inactive yeast (from Lallemand, Canada) which are available on the market under the names Optimum White, Opti Less and Noblesse. These products were applied in two concentrations: 0.2 g/L or 0.4 g /L during a period of 15 days, with stirring every third day. Denotations and treatment descriptions are given in Table 1.

After treatment, wine clarification was done (using bentonite and gelatin), followed by filtration and bottling. Untreated wines were used for comparison ("blank"). After bottling, the wine in some bottles was subjected to the accelerated aging while the wine in other bottles was stored in the stock under normal conditions, in dark place, at 15–20°C during one year.

Titratable acidity, density, alcohol, reducing sugars and extract content, pH and free and total sulphur dioxide were determined using well-established, standard methods [5]. Dissolved oxygen content was measured by polarographic oxygen analyzer (YSI model 51A disolved oxygen meter, Yellow Springs Instruments CO). Total polyphenol concentration was determined with the Folin–Ciocalteu assay [6]. Gallic acid was used as standard and results were expressed as grams per litre of gallic acid equivalents. Flavonoids content was determined using $AlCl_3$ spectrophotometric assay [6]. Catechin was used as standard and results were expressed as milligrams per litre of catechin equivalents. Flavon-3-ols content was measured using vanilin spectrophotometric assay [7]. Catechin was used as standard and results were expressed as milligrams per litre of catechin equivalents. Anthocyanins content was measured according to Ribereau-Gayon and Stonestreet [8]. Percent of polymeric colour was determined using bisulfite bleaching method [9]. Colour intentisty and tint were measured spectrophotometrically [10]. All parameters were monitored after each of the experimental phases (Table 2) and measured in triplicates.

Panel method was used for sensory evaluation. A panel of five judges was recruited comprising "Rubin" staff with extensive experience in wine sensory evaluation, of which four are members of the Commission of Ministry of Agriculture, Republic of Serbia; appointed for the organoleptic assessment of wine and similar products. 19 descriptors were evaluated: smell profile (overall odor, spice, color intensity, vegetable, red fruit berries, black fruit berries, floral, complexity, intensity and typicality), and taste profile (taste, harmony, acidity, astringency, fullness, complexity, duration, structure and typicality).

Table 2. Abbreviations and descriptions of experimental phases.

Phase abbreviation	Phase description
W	Basic wine
TR	After IDYP treatment
CFB	After clarification and filtration - after bottling
ACC	After accelerated aging test
NOR	After one year of normal aging conditions

3. Results and discussion

3.1. Oenological analyses

Basic oenological parameters (acidity, density, alcohol, reducing sugars, extract, pH, free and total sulphur dioxide) of Prokupac wine before and after the IDYP treatment, and after one year of aging are presented in Table 3. The results show that the treatment had not induced significant changes, which was indeed expected, considering the relatively short treatment duration.

3.2. Oxygen content

Oxygen content values measured in all wine samples after each of the experimental phases are presented in Fig. 1. The results indicate that the three implemented commercial IDYPs exhibit good oxygen scavenging properties, since all treated samples contained oxygen quantities lower than those found in untreated wine. This conclusion holds for all phases of the investigation. Especially low oxygen contents were found after the experimental phase which included IDYP treatment, clarification, filtration and bottling (CFB), regardless of IDYP used. After the accelerated aging test, dissolved O_2 concentrations were also found to be lower in all treated wines than in the blank, which was especially pronounced for the sample treated by "Noblesse", using its concentration of 0.4 g/L (0.81 mg/L of oxygen, compared to 1.23 mg/L for blank). After one year in the bottles, there were no significant differences in O_2 content between treated wine samples and blank.

3.3. Total polyphenol content

Figure 2 presents the total polyphenol contents (TPC) of the investigated wine samples. As a general trend, lower polyphenol concentrations were found for all IDYP treated wines. The observed decrease of TPC can be attributed to partial adsorption of phenolic compounds on yeast cells and yeast walls, which has been observed and discussed in literature [1,2]. Also, the treatment mildly influenced the way TPC changed after clarification, filtering and bottling (CFB). For blank wine, only a slight difference in TPC was observed before and after CFB, while for IDYP treated wine samples these differences were more pronounced: lowering of TPC values was observed in the case of treatment by Optimum White and Opti Less, when amounts 0.4 g/L of IDYP were used, while in the case of Noblesse application, this effect was observed for both 0.2 g/L and 0.4 g/L. After one year of wine aging in normal conditions, relative changes of total

Table 3. Oenological parameters of basic wine and wine samples after treatment and normal aging.

Parameter	Basic wine	After one year of normal aging conditions						
		BL	OW 20	OW 40	OL20	OL40	N20	N40
Density D_{20}^{20}	0.99399	0.99353	0.9936	0.99366	0.9936	0.99396	0.99361	0.99364
Alcohol (%vol)	12.2±0.03	12.17±0.02	12.2±0.015	12.07±0.04	12.22±0.035	12.23±0.03	12.09±0.04	12.22±0.02
Total extract (g/L)	26.00±0.24	24.81±0.31	25.09±0.33	24.86±0.26	25.14±0.35	26.09±0.36	24.77±0.28	24.96±0.21
Total acids (as tartaric) (g/L)	5.6±0.13	5.27±0.15	5.43±0.11	5.37±0.21	5.2±0.18	5.2±0.12	5.37±0.21	5.37±0.22
Volatile acids (as acetic) (g/L)	0.63±0.06	0.54±0.04	0.54±0.045	0.51±0.55	0.51±0.062	0.51±0.33	0.54±0.047	0.51±0.055
pH	3.39±0.03	3.42±0.24	3.43±0.022	3.43±0.015	3.43±0.012	3.43±0.03	3.43±0.026	3.42±0.016
Reducing sugars (g/L)	2.20±0.22	2.42±0.15	2.35±0.16	2.31±0.2	2.25±0.14	2.36±0.23	2.3±0.12	2.37±0.17
Sugar free extract (g/L)	24.80±0.68	23.39±0.86	23.74±0.75	23.55±0.7	23.89±0.55	24.73±0.94	23.48±0.85	23.59±0.69
Free SO_2 (mg/L)	26.88±1.28	26.03±1.21	24.75±0.93	24.75±1.08	25.67±1.12	26.03±1.15	23.04±1.24	24.75±1.15
Total SO_2 (mg/L)	56.32±1.66	46.96±1.93	52.05±1.15	50.64±0.99	52.91±1.42	48.6±1.65	50.01±1.33	49.62±1.55

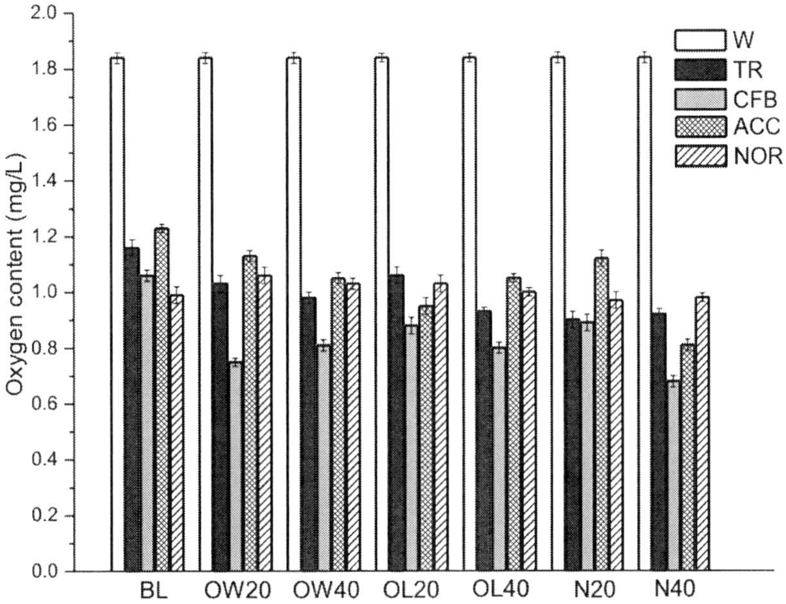

Figure 1. Oxygen content of the investigated wine samples.

polyphenol concentrations (after treatment vs. after aging) were comparable to those observed for blank wine.

3.4. Anthocyanins content

The effects of IDYP treatment on the concentration of anthocyanins in the investigated wines seem to follow a pattern different than the one observed for total polyphenol content. Results presented in Fig. 3 show that similar or only slightly decreased values of anthocyanin contents were detected after the treatment. Again, these changes are probably due to the adsorption of small amounts of anthocyanins on the IDY products used. However, after clarification, filtering and bottling, anthocyanin concentrations decreased to a greater extent in treated wine samples than in blank wine. The largest difference before and after CFB phase was detected when Noblesse IDYP was used. After the accelerated aging phase, values of anthocyanin content in treated wines only slightly differed from the blank wine. However, after one year of aging in normal conditions, higher concentrations were found for the treated samples, compared to the blank, indicating a positive influence of IDYPs on stability of anthocyanins during aging.

3.5. Flavonoids and Flavan-3-ols

The contents of flavonoids and flavan-3-ols in the investigated wines (not presented) showed only slight changes right after treatment (TR) and after clarification, filtering and bottling (CFB). Slightly decreased concentrations were also measured after accelerated aging, for all treated wines. After normal aging, however, noticeably higher values were detected for treated samples, compared to blank wine, which corresponds to smaller relative decreases of flavonoids and flavan-3-ols content during aging, for treated samples.

3.6. Percent of polymeric colour

The percentages of polymeric colour variations in all investigated wine samples, including blank, show a distinct increase after the one year aging period (Fig. 4). This is in accordance with the fact that changes in red wine colour take place as a result of structural changes of the

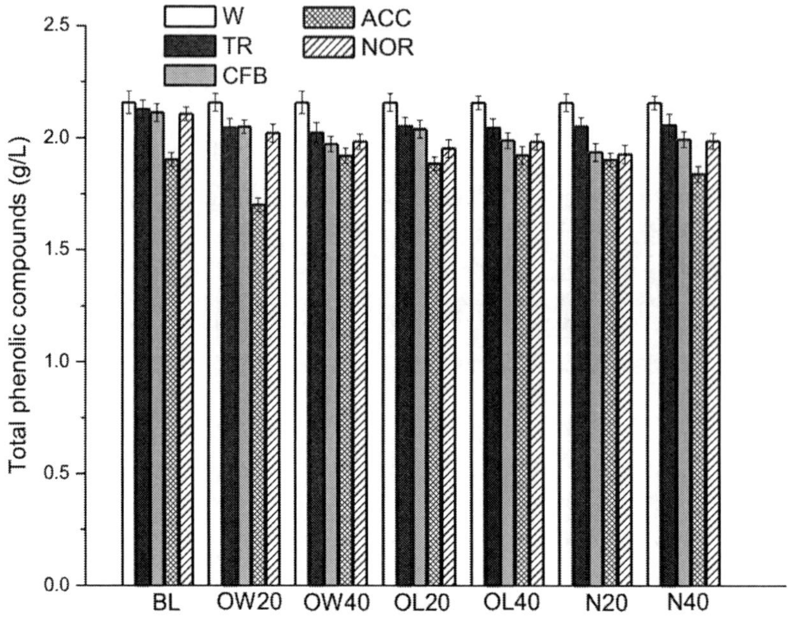

Figure 2. Total polyphenol content of the investigated wine samples.

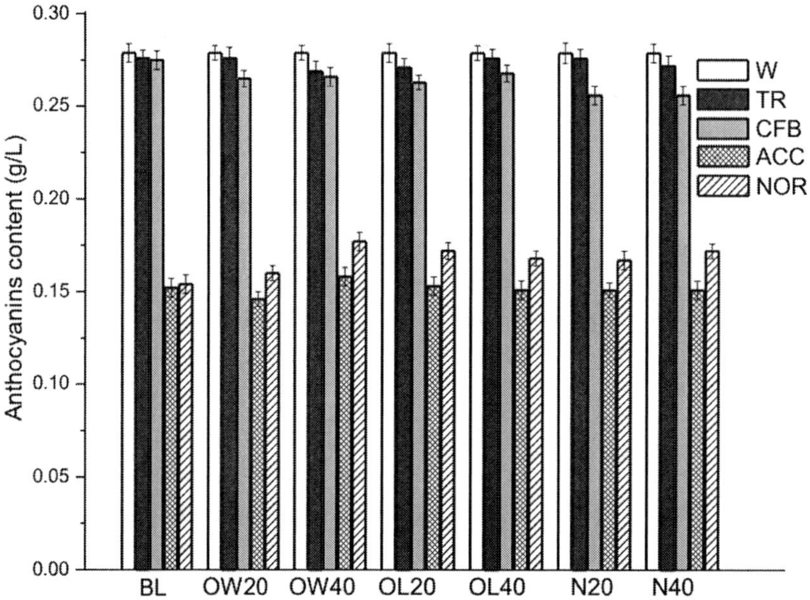

Figure 3. Anthocyanins content of the investigated wine samples.

anthocyanins, from ionized anthocyanins to condensed anthocyanins, which makes the percentage of polymeric colour a parameter of wine ageing. During aging, most of the anthocyanins are subjected to transformations to stable complexes, either with tannins, or between anthocyanins themselves (self-association) or with other colorless cofactors [11] and only a lesser part of them is decomposed under the influence of temperature, light and oxygen or precipitates as colloid fraction. As a result, the absorbance at 520 nm increases in aged wines and, simultaneously, the absorbance at 420 nm decreases, due to the transformation of monomeric anthocyanins to their polymeric forms. The results obtained for IDYP treated samples show that the percentage of polymeric colour slightly increased (compared to blank) after the accelerated aging test and also after normal aging one year period (except for the sample treated with Optimum White at 0.4 g/L). It has been shown [1] that polysaccharides (abundant in IDYPs) can act as protective colloids, preventing precipitation, and therefore lead to an increase in colour stability.

 DOI: 10.1051/bioconf/20170902004

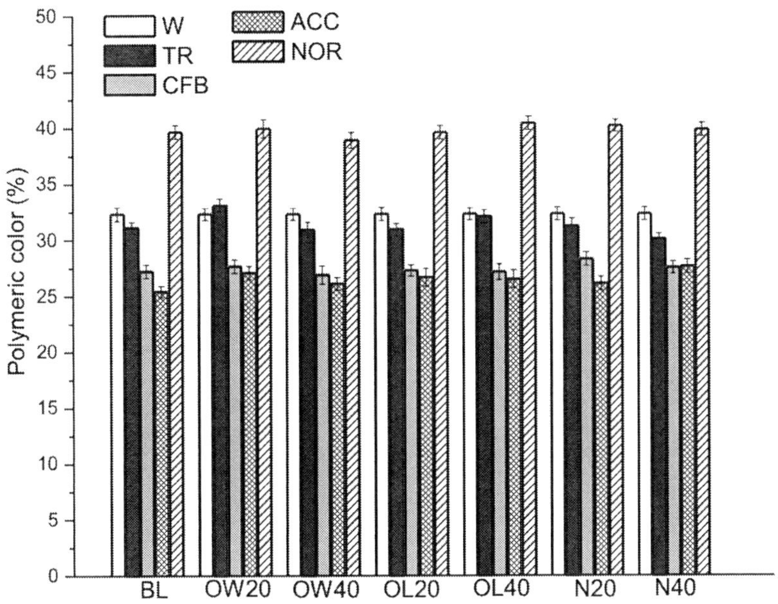

Figure 4. Percent of polymeric colour of the investigated wine samples.

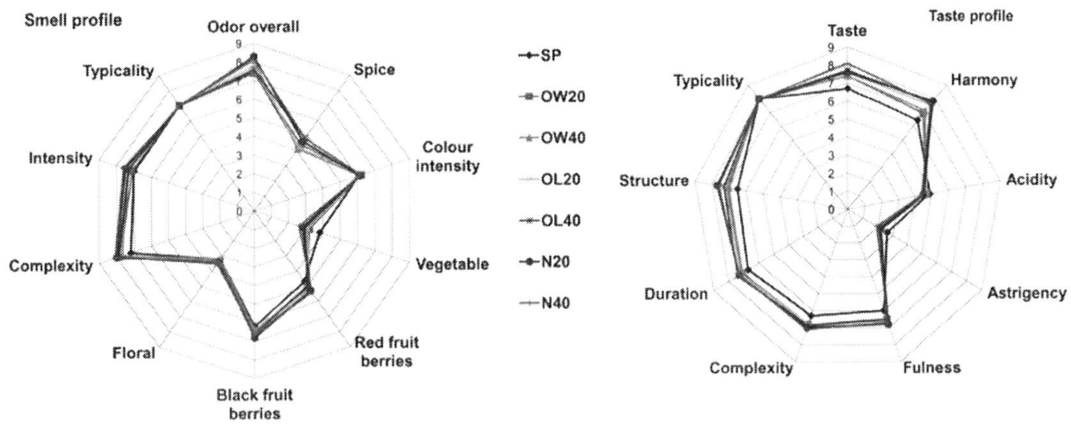

Figure 5. Sensory analysis of the investigated wine samples.

3.7. Colour tint and intensity

The results obtained from colour tint and intensity (not shown) revealed no significant differences between IDYP treated wines and blank

3.8. Sensory analysis

Sensory analysis findings clearly indicate that IDYP treatment had an overall positive influence on the sensory characteristics of the wine. After being bottled and kept in standard storage conditions, the treated wines were found to have improved smell, more complexity, more pronounced black and red fruit berry notes and also less of a vegetable character, which was clearly present in basic wine. Concerning taste, a significant improvement was noticed in terms of fullness, structure, complexity and harmony, accompanied by an astringency decrease.

3.9. Accelerated aging test

Although the evaluation of the accuracy of the accelerated aging test was not amongst primary goals of this investigation, the obtained results should not be left uncommented. Namely, the values of all the monitored parameters presented in Figs. 1–4, measured after the accelerated test and after one year of normal aging, exhibit significant discrepancies, both in terms of absolute and relative values. Hence, our findings indicate that the accelerated aging test results should be used with caution.

4. Conclusion

The results obtained in this investigation show that the use of IDYPs can lead to improved sensory characteristics of red wine after a one year aging period, accompanied by minor changes in wine polyphenol composition and content, decreased oxygen content and slightly

DOI: 10.1051/bioconf/20170902004

enhanced percentage of polymeric colour. However, detailed investigations of treatment parameters, such as quantity of IDYP, temperature and duration of treatment, are yet to be performed in order to reveal the influence of yeast products on wine quality in details.

References

[1] M.Á. Pozo-Bayón, I. Andújar-Ortiz, M.V. Moreno-Arribas, Food Res. Int. **42,** 754 (2009)
[2] J.M. Nguela, N. Sieczkowski, A. Roi, A. Vernhet, J. Agr. Food. Chem. **63**, 660 (2015)
[3] M.A. Pozo-Bayon, I. Andujar-Ortiz, J.M. Alcaide-Hidalgo, P.J. Martin-Alvarez, M.V. Moreno Arribas, J. Agr. Food. Chem. **57**, 10784 (2009)
[4] J.A. Peìrez-Serradilla, M. D. Luque de Castro, Food Chem. **111**, 447 (2008)
[5] International Organization of Vine and Wine, *Compendium of International Methods of Wine and Must Analysis, Vol 1* (OIV, Paris, France, 2017)
[6] V. Ivanova, M. Stefova, F. Chinnici, J. Serb. Chem. Soc. **75**, 45 (2010)
[7] M. Monagas, C. Gomez-Cordoves, B. Bartolome, O. Laureano, J.M.R. Da Silva, J. Agric. Food Chem. **51**, 6475 (2003)
[8] J.C. Rivas-Gonzalo, Y. Gutierrez, E. Hebrero, C. Santos-Buelga, Am. J. Enol. Vitic. **43**, 210 (1992)
[9] T.J. Hager, L.R. Howard, R.L. Prior, J. Agric. Food Chem. **56**, 689 (2008)
[10] M.A. Cliff, M.C. King, J. Schlosser, Food Res. Int. **40**, 92 (2007)
[11] Y. Hayasaka, J.A. Kennedy, Aust. J. Grape Wine Res. **9**, 210 (2003)

Reverse osmosis influence over the content of metals and organic acids in low alcoholic beverages

Mitică Tiberiu Andrieş[1], Cătălin Ioan Zamfir[2], Camelia Elena Luchian[1], Marius Niculaua[2], Cintia Colibaba[1], Gheorghe Odăgeriu[2], Mihai Cristian Focea[1], and Valeriu V. Cotea[1,a]

[1] The University of Agricultural Sciences and Veterinary Medicine, 3 M. Sadoveanu Alley, Iaşi 700490, Romania
[2] Research Centre for Oenology, Romanian Academy - Iaşi Branch, 9 M. Sadoveanu Alley, Iaşi 700490, Romania

Abstract. Wine is defined as an alcoholic beverage resulted from fermentation of grape must, having ethanol content higher than 8.5% (v/v). Wine consumption has health benefits related to the high concentration of polyphenolic compounds with antioxidant activity and cardiovascular protection effects. However, the alcohol content restricts wine consumption, but wines with low-alcohol content can be obtained with the help of the dealcoholisation process, after it was produced through alcoholic fermentation. The purpose of this work is to evaluate the organic acid concentration, metal content and other physical-chemical parameters of low alcoholic beverages obtained from grape must by a process which involves reverse osmosis, mixing in a variable ratio the permeate and concentrate and then fermentation. For the experiments, a Muscat Ottonel grape must from Iaşi vineyard was used. There were ten variants of beverages (wines) with low alcoholic concentration, by mixing known quantities of the two phases resulting from the reverse osmosis process. These beverages (wines) had an alcoholic concentration starting from 2.5% (v/v) in the first variant, up to 7% (v/v) in the tenth variant. Alcoholic concentration varies for each variant by 0.5% (v/v). After fermentation in 50 L stainless steel tanks, the samples were filtered with 0.45 μm sterile membrane and bottled in 0.75 L glass bottles. After 2 months of storage at constant temperature, the beverage samples were analyzed to determine the metal content (AAS method), organic acids concentration (HPLC method), and other physical-chemical characteristics (OIV standard methods). The results obtained indicate that the very complex physical-chemical composition of the low alcoholic beverages analyzed is influenced by the specific chemical composition of a given grape must, as well as by the use of products obtained from reverse osmosis.

1. Introduction

Wine is an ancient beverage spread all over the world and one of the most known alcoholic drinks, which has the benefit to reduce the risk of cardiovascular diseases, thanks to the many compounds playing a role of great importance human health-wise [1].

For the past several years there has been a steady increase of the alcoholic degree of wines because of climate change. Because of global warming, grapes that have a higher sugar concentration are produced. For this reason the cultivation of grape varieties is being moved from Southern to cooler areas. This changes every year, depending on various circumstances like occasionally unfavorable climate conditions [2].

An important aspect of wine production is the process of obtaining a certain lower ethanol concentration, which has gained considerable attention over the last 10 years. Increase consumer demand for wines that are perceived as healthier, changing attitudes of consumers regarding the social consequences of excessive ethanol consumption and more favorable excise rates for lower alcohol beverage are just some of the reasons for lowering ethanol levels in wine. There is a significant consumer demand for the beverages with lower ethanol levels [3].

[a] e-mail: vvcotea@yahoo.com

Metals in wine occur at the mg/L level or less. Though not directly related to the taste of the wine, the metal content should be determined because excess is undesirable, and in some cases prohibited, due to potential toxicity. Lead content in wine, to guarantee consumer health protection is restricted in several states by legislation because toxicity for human health [4].

For reason of direct correlation with soil composition, metals appear to be the best way to identify the geographical origin of the wine [5]. Moreover, knowledge of the metallic ion concentration of wine is of interest because of their influence on the wine-making process where metals such as calcium, potassium, copper and iron can cause cloudiness or precipitates.

The international community set some low acceptance levels for some metals, due to their importance in nutrition and their potentially toxic effects [6].

In grapes, the total acid content normally reaches a maximum content during growth and decreases during ripening. [7] Although having the same genotype, grapes harvested from different climates, have different organic acid contents. [8] During grape ripening, in continuous warm conditions a lower acid content at maturity results mostly due to increasing degradation of malic acid [9]. Malic and tartaric acids are dominant organic acids and account for 90% or more of the total acids in grapes [10].

Table 1. Classical enological parameters of low alcoholic wine obtained by reverse osmosis.

Parameter	VSA 1	VSA 2	VSA 3	VSA 4	VSA 5	VSA 6	VSA 7	VSA 8	VSA 9	VSA 10
Alcoholic strength (% v.a.)	2.5	3.2	3.7	4.1	4.7	5.1	5.7	6.2	6.7	7.2
Sugars (g/L)	0.19	0.58	1.36	0.68	1.32	4.37	3.24	5.21	6.29	6.28
VA (g/L $C_2H_4O_2$)	0.39	0.42	0.42	0.54	0.51	0.45	0.54	0.51	0.51	0.51
TA (g/L $C_4H_6O_6$)	5.15	5.35	5.47	5.69	5.88	6.07	6.19	6.22	6.3	6.45
pH	3.85	3.63	3.52	3.46	3.35	3.24	3.22	3.19	3.10	3.02
Free SO_2(mg/L)	40.96	40.96	38.40	43.52	35.84	38.40	35.84	35.84	35.84	33.28
Total SO_2(mg/L)	53.76	56.32	48.64	51.20	48.64	56.32	51.20	48.64	53.76	48.64
Density	0.9982	0.9982	0.9984	0.9972	0.9973	0.9986	0.9973	0.9978	0.9978	0.9978
NE	4.71	6.82	8.44	7.32	9.28	10.83	10.66	11.49	12.21	13.82

VA - (volatile acidity); TA – (total acidity); NE- (non-reducing extract); VSA-" low alcoholic wine".

Table 2. Instrumental parameters.

Element/ Parameters	Zn	Fe	Cu	Ni	Pb	Mn	Mg	Ca	Na	K
λ(nm)	213.857	371.993	324.754	352.454	405.781	403.076	285.213	396.847	589.592	769.897
Nebulizer Flow (L/min)	0.45	0.65	0.7	0.7	0.75	0.9	0.9	0.6	0.95	0.75
Air Injection Flow Rate	medium	high	high	high	high	high	medium	high	high	high
Read Time (s)	3	3	3	3	3	3	3	3	3	3

All organic acids play an important role in wine, the taste being influenced by the concentration of different organic acids. Wine's organic acids in wine are malic, tartaric, citric, acetic, lactic, succinic and others. Concentration of organic acids varies depending on different factors such as pH, temperature, sulfur dioxide and oxygen concentration. The ratio of tartaric acid to malic acid mainly influences the taste of the wine. When the ratio of these acids is about 2 or less, the wine will not be harmonious and will have a sour aftertaste. The wine with best flavor and bouquet will be obtained at a ratio of tartaric to malic acid equal to 3 and more. For example succinic acid has a salty-bitter taste, citric acid gives freshness to wine and malic acid gives the taste of green apples [11].

A complete analysis of wine composition is a time-consuming process, but it brings necessary information for both obtaining a quality product and its maintenance under proper conditions. Therefore, only a few parameters are checked periodically. Among them, the most common are: alcoholic degree, reducing sugars, pH, sulfur dioxide, total and volatile acidity.

The classical methods used for monitoring the wine composition with high robustness and precision are recognized by the international community as official methods of analysis, even though more recent methods may be based on modern automated instrumental techniques [12].

2. Materials and method

2.1. Grape sample and winemaking

Muscat Ottonel grapes from Iasi vineyard were harvested in 2015 at optimal maturity. The grapes were destemmed and crushed, and must was transferred in stainless steel containers. The grape must was concentrated until 308.3 g/L sugar using a device for reverse osmosis, Bucher Vaslin. The obtained permeate has 8.032 g/L sugar.

There were obtained ten variants of beverages (wines) with low alcoholic concentration, by using known quantities of the two phases resulting from the reverse osmosis process. These beverages (wines) had an alcoholic concentration starting from 2.5% (v/v) in the first variant, up to 7% (v/v) in the tenth variant (VSA1 to VSA10). Alcoholic concentration varies for each variant by 0.5% (v/v). After fermentation in 50 L stainless steel tanks (Zymaflore X16® yeast), the samples were filtered with 0.45 μm sterile membrane, and bottled in 0.75 L glass bottles.

After 2 months of storage at constant temperature, the beverage samples were analyzed to determine the metal content (using an Agilent MP-AES 4200), organic acids concentration (HPLC method), and other physical-chemical characteristics (OIV standard methods).

2.2. Physical-chemical analysis

The main physico-chemical parameters analyzed were alcoholic strength, reducing sugars, total acidity (TA), volatile acidity (VA), real acidity (pH), sulfur dioxide and density (O.I.V., 2016).

The results are presented in Table 1.

2.3. Analysis of the metals content in wines

All analyzed samples were diluted to reduce the salinity of the samples under 1%. If this technique is used, the mineralization and ionization of the sample is not necessary. The measurements were performed using an Agilent MP-AES 4200 instrument equipped with the standard sample introduction system consisting of the OneNeb nebulizer, double pass cyclonic spray chamber and easy fit torch.

The Agilent SPS-3 auto-sampler was used to deliver samples to the instrument allowing automatic operation. Selection of optimal lines depended on wavelengths that were free from spectral interference and matched the appropriate sensitivity. Spectral and background interferences were simultaneously and accurately corrected using the MP software.

Each sample was analyzed for three times and the registration time was 3 seconds. In Table 2 are shown the instrumental parameters and operating conditions of the equipment.

DOI: 10.1051/bioconf/20170902005

Table 3. Zn, Fe, Cu, Ni, Pb, Mn, Mg cations concentration in µg/L and Ca, K, Na cations concentration in mg/L of low alcoholic wine obtained by reverse osmosis.

Parameter	VSA 1	VSA 2	VSA 3	VSA 4	VSA 5	VSA 6	VSA 7	VSA 8	VSA 9	VSA 10
Zn	320.13±0.0001	212.89±0.0001	190.54±0.0001	141.69±0.0002	194.59±0.0003	88.16±0.0003	107.28±0.0003	26.64±0.0011	5.0000±0.0014	2.00±0.0030
Fe	1401.46±0.0017	774.92±0.0014	642.58±0.0016	495.97±0.0005	560.99±0.0003	423.60±0.0012	404.54±0.0001	456.08±0.0011	515.75±0.0007	1028.11±0.0004
Cu	726.66±0.0006	346.60±0.0002	325.89±0.0017	164.28±0.0043	245.75±0.0002	107.11±0.0062	112.81±0.0002	127.49±0.0006	105.94±0.0008	310.81±0.0007
Ni	66.54±0.0051	86.05±0.0052	266.24±0.0048	58.29±0.0067	736.19±0.0036	68.36±0.0043	72.19±0.0014	77.08±0.0004	75.13±0.0090	123.51±0.0079
Pb	66.73±0.0002	79.72±0.0004	99.31±0.0011	101.34±0.0003	138.21±0.0011	94.75±0.0006	109.69±0.0049	117.71±0.0005	133.89±0.0030	547.51±0.0003
Mn	437.78±0.0003	328.38±0.0014	269.97±0.0003	276.27±0.0030	358.06±0.0001	319.31±0.0002	352.68±0.0001	383.39±0.0001	420.06±0.0003	857.29±0.0011
Mg	144.53±0.0002	152.17±0.0008	170.75±0.0062	181.67±0.0007	183.38±0.0006	183.95±0.0043	189.31±0.0002	213.73±0.0017	226.52±0.0002	230.69±0.0006
Ca	37.71±0.0003	12.05±0.0007	18.78±0.0012	20.01±0.0004	18.02±0.0017	25.59±0.0005	14.48±0.0014	24.41±0.0016	18.04±0.0001	18.33±0.0011
K	113.70±0.0020	83.74±0.0018	143.73±0.0007	120.74±0.0009	124.50±0.0006	94.97±0.0002	99.73±0.0012	140.43±0.0002	141.50±0.0009	135.62±0.0006
Na	310.02±0.0036	234.34±0.0090	230.12±0.0043	586.07±0.0079	354.90±0.0051	319.81±0.0067	178.51±0.0052	337.85±0.0048	216.26±0.0014	180.59±0.0004

Table 4. Organic acid content (g/L), (mg/L).

Sample	g/L						mg/L	
	Tartaric acid	Malic acid	Lactic acid	Acetic acid	Citric acid	Succinic acid	Fumaric acid	Shikimic acid
VSA 1	2.39±0.001	1.32±0.001	0.32±0.001	0.36±0.005	0.39±0.002	0.52±0.002	11.3600±0.0008	3.4000±0.0004
VSA 2	2.40±0.002	1.33±0.001	0.32±0.001	0.39±0.004	0.39±0.003	0.50±0.001	9.4800±0.0005	3.4400±0.0006
VSA 3	2.45±0.003	1.28±0.002	0.32±0.001	0.39±0.006	0.40±0.004	0.52±0.004	9.4900±0.0007	3.5100±0.0005
VSA 4	2.45±0.001	1.30±0.001	0.32±0.001	0.42±0.004	0.40±0.006	0.54±0.002	9.5000±0.0005	3.9200±0.0007
VSA 5	2.67±0.002	1.28±0.002	0.33±0.001	0.42±0.005	0.41±0.004	0.54±0.003	9.5100±0.0004	3.8800±0.0006
VSA 6	2.70±0.003	1.30±0.003	0.34±0.002	0.44±0.004	0.41±0.005	0.54±0.001	9.5200±0.0006	4.1500±0.0004
VSA 7	2.72±0.001	1.31±0.001	0.34±0.002	0.46±0.006	0.42±0.004	0.56±0.002	9.5600±0.0007	8.5500±0.0005
VSA 8	2.73±0.003	1.29±0.002	0.34±0.002	0.48±0.004	0.42±0.006	0.57±0.003	9.5700±0.0005	8.4600±0.0007
VSA 9	2.74±0.002	1.31±0.002	0.35±0.002	0.48±0.003	0.43±0.004	0.57±0.003	9.5900±0.0006	8.8200±0.0005
VSA 10	2.84±0.003	1.30±0.001	0.36±0.001	0.48±0.002	0.43±0.005	0.57±0.002	9.6100±0.0004	9.3500±0.0008

2.4. Analysis of the organic acids in wines using HPLC-High-performance liquid chromatography method

For the analysis of the organic acids content a Shimadzu series Proeminence LC20 device was used. The sample was filtered through a 28 mm diameter nylon 0.45 μm

DOI: 10.1051/bioconf/20170902005

syringe cartridge. This method is described in the MA-E-AS313-04-ACIORG by O.I.V. standards.

Using a 5 μL volume of standard, a sample was injected through two analytical columns (YMC-Triart C18 multi-stage hybrid group's 3 μm 150 × 4.6mm 120 Å) at a flow rate of 0.9 mL/min., using a solution of sulphuric acid adjusted to a value of 1.3 pH.

Columns' temperatures were maintained at 45 °C for the entire period of analysis.

3. Results and discussion

Statistical evaluation of the data was performed using Statgraphics Centurion XVI® software, (StatPoint Technologies, Inc, U.S.A.). In this study, a one-way ANOVA procedure was applied.

3.1. Physical-chemical characteristics of wine

For alcoholic strength, there was significant difference between all samples. In the first variant, were obtained wines with 2.5% alcohol degree. Then every sample has an alcoholic degree with 0.5% volume more than the previous one.

Total acidity (TA) also showed similar mean values for all samples, with a minimum in the first sample and with a maximum in the last variant.

All other parameters showed average values with variability to a smaller or larger degree depending on the impact of different quantities of permeate and concentrate used for every wine sample.

All of the wine samples were dry.

3.2. Metals content of wine sample

The values of Zn, Fe, Cu, Ni, Pb, Mn, Mg, Ca, K, Na, cations content identified in samples of low alcoholic wine, as determined by AES, are shown in Table 3.

The measured concentration of zinc, iron, copper, nickel show that these varied with a minimum content in the first sample VSA and then increased gradually in other samples.

Magnesium and calcium concentration in wine have an insignificant change in all analyzed samples.

The lead value progressively increases from the first VSA1 sample to VSA5 with a minimum content of 66.73 μg/L and a maximum content of 138.21 μg/L. This increase is then found again starting with VSA6 with a minimum content of 94.75 μg/L and a maximum content in the 10th VSA10 sample of 547.51 μg/L.

The content of manganese fluctuates in large limits, with a minimum content in VSA3 sample of 269.97 μg/L and a maximum content in VSA10, of 857.29 μg/L.

Potassium cations concentration values present small differences from one sample to another.

Sodium concentrations fluctuates, registering a minimum content in the case of the seventh sample VSA7 178.51 mg/L and a maximum of 586.07 mg/L in the fourth sample VSA4. This increase was not linear since the first sample, but fluctuated.

Table 5. Multiple Range Tests for Malic Acid by Sample.

Contrast	*Sig.*	*Difference*	*+/– Limits*
VSA 1. - VSA 10.		0	0.00295
VSA 1. - VSA 2.	*	–0.03	0.00295
VSA 1. - VSA 3.	*	0.04	0.00295
VSA 1. - VSA 4.	*	0.01	0.00295
VSA 1. - VSA 5.	*	0.02	0.00295
VSA 1. - VSA 6.	*	0.03	0.00295
VSA 1. - VSA 7.	*	–0.01	0.00295
VSA 1. - VSA 8.	*	0.04	0.00295
VSA 1. - VSA 9.	*	–0.01	0.00295
VSA 10. - VSA 2.	*	–0.03	0.00295
VSA 10. - VSA 3.	*	0.04	0.00295
VSA 10. - VSA 4.	*	0.01	0.00295
VSA 10. - VSA 5.	*	0.02	0.00295
VSA 10. - VSA 6.	*	0.03	0.00295
VSA 10. - VSA 7.	*	–0.01	0.00295
VSA 10. - VSA 8.	*	0.04	0.00295
VSA 10. - VSA 9.	*	–0.01	0.00295
VSA 2. - VSA 3.	*	0.07	0.00295
VSA 2. - VSA 4.	*	0.04	0.00295
VSA 2. - VSA 5.	*	0.05	0.00295
VSA 2. - VSA 6.	*	0.06	0.00295
VSA 2. - VSA 7.	*	0.02	0.00295
VSA 2. - VSA 8.	*	0.07	0.00295
VSA 2. - VSA 9.	*	0.02	0.00295
VSA 3. - VSA 4.	*	–0.03	0.00295
VSA 3. - VSA 5.	*	–0.02	0.00295
VSA 3. - VSA 6.	*	–0.01	0.00295
VSA 3. - VSA 7.	*	–0.05	0.00295
VSA 3. - VSA 8.		0	0.00295
VSA 3. - VSA 9.	*	–0.05	0.00295
VSA 4. - VSA 5.	*	0.01	0.00295
VSA 4. - VSA 6.	*	0.02	0.00295
VSA 4. - VSA 7.	*	–0.02	0.00295
VSA 4. - VSA 8.	*	0.03	0.00295
VSA 4. - VSA 9.	*	–0.02	0.00295
VSA 5. - VSA 6.	*	0.01	0.00295
VSA 5. - VSA 7.	*	–0.03	0.00295
VSA 5. - VSA 8.	*	0.02	0.00295
VSA 5. - VSA 9.	*	–0.03	0.00295
VSA 6. - VSA 7.	*	–0.04	0.00295
VSA 6. - VSA 8.	*	0.01	0.00295
VSA 6. - VSA 9.	*	–0.04	0.00295
VSA 7. - VSA 8.	*	0.05	0.00295
VSA 7. - VSA 9.		0	0.00295
VSA 8. - VSA 9.	*	–0.05	0.00295

3.3. Wine organic acids analysis

The descendent pH values of the analyzed samples are concordance whit the weight of each acid, in accordance with the literature [13].

ANOVA table decomposes the variability for all acids identified (Table 4) into contributions due to various factors. Since Type III sums of squares (the default) have been chosen, the contribution of each factor is measured having removed the effects of all other factors.

The P-values test the statistical significance of each of the factors. Since one P-value (0.00) is less than 0.05, this factor has a statistically significant effect on all acids identified at the 95.0% confidence level.

Table 6. The homogenous groups that were identified.

Sample	*Count*	*LS Mean*	*LS Sigma*	*Homogeneous Groups*
VSA 8.	3	1.86	0.001	X
VSA 3.	3	1.86	0.001	X
VSA 6.	3	1.87	0.001	X
VSA 5.	3	1.88	0.001	X
VSA 4.	3	1.89	0.001	X
VSA 10.	3	1.9	0.001	X
VSA 1.	3	1.9	0.001	X
VSA 9.	3	1.91	0.001	X
VSA 7.	3	1.91	0.001	X
VSA 2.	3	1.93	0.001	X

Table 7. Table of Least Squares Means for Malic Acid with 95.0% confidence Intervals.

Level	*Count*	*Mean*	*Stnd. Error*	*Lower Limit*	*Upper Limit*
GRAND MEAN	30	1.891			
Sample					
VSA 1.	3	1.9	0.001	1.89791	1.90209
VSA 10.	3	1.9	0.001	1.89791	1.90209
VSA 2.	3	1.93	0.001	1.92791	1.93209
VSA 3.	3	1.86	0.001	1.85791	1.86209
VSA 4.	3	1.89	0.001	1.88791	1.89209
VSA 5.	3	1.88	0.001	1.87791	1.88209
VSA 6.	3	1.87	0.001	1.86791	1.87209
VSA 7.	3	1.91	0.001	1.90791	1.91209
VSA 8.	3	1.86	0.001	1.85791	1.86209
VSA 9.	3	1.91	0.001	1.90791	1.91209

For example, in Table 5 a multiple comparison procedure was applied to determine which means are significantly different from others in the case of acid malic. The bottom half of the output shows the estimated difference between each pair of means. An asterisk has been placed next to 42 pairs, indicating that these pairs show statistically significant differences at the 95.0% confidence level.

In Table 6, 7 homogenous groups are identified using columns of X's. Within each column, the levels containing X's form a group of means within which there are no statistically significant differences. The method currently being used to discriminate among the means is Fisher's least significant difference (LSD) procedure. With this method, there is a 5.0% risk of calling each pair of means significantly different when the actual difference equals 0.

In Table 7 are shown the mean malic acid for each level of the factors. It also shows the standard error of each mean, which is a measure of its sampling variability. The rightmost two columns show 95.0% confidence intervals for each of the means.

For all other parameters, the main effects are statistically significant at the 95.0 % confidence level. This rule is available for all identified acids in all samples.

4. Conclusion

The results obtained indicate that the very complex physical-chemical composition of the low alcoholic beverages analyzed is influenced by the specific chemical composition of a given grape must, as well as by the use of products obtained from reverse osmosis.

We found moderate correlation between quantities of permeate and concentrate regarding metals content. The content of metals has not increased in large limits.

The study show the influence of used quantities of the two phases resulting from the reverse osmosis process (permeate/concentrate). These had a statistically significant effect on all acids identified at the 95.0% confidence level.

References

[1] L. Liguori, J. Food Chem. **140**, 68–75 (2013)
[2] E. Duchene, C. Schneider, J. Dev. Agron. Sustain. **25**, 93–99 (2005)
[3] B. Saha, J. Oenoviti Int. Net. Fr. **1**, 78–86 (2013)
[4] M. Aceto, O. Abollino, M. C. Bruzzonti, E. Mentasti, C. Sarzanini and M. Malandrino, J. Food Addit. Contam. **19**, 126–133 (2002)
[5] P.R. Ashurst & M. J. Dennis, *Food authentication*, 60–107 (EDP Chapman & Hall, 1996)
[6] H. Eschneauer, Am J Enol Vitic **33**, 226–230 (1982)
[7] N. A. M. Eskin, H. M. Henderson & R. J. Townsend, N. A. M. Eskin *Biochemistry of foods*, 31–63 (EDP Academic Press, 1971)
[8] T. Fuleki, E. Pelayo & R. Palabay, J. AOAC Int. **76**, 591–600 (1993)
[9] Seymour, G.B., Taylor, J.E., Tucker, Gregory A., *Biochemistry of fruit ripening*, 189–220 (EDP Chapman & Hall 1993)
[10] T. Philip & F. E. Nelson. J. Food Sci. **38**, 18–20 (1973)
[11] V. N. Bayraktar, J. Biotech. Acta **6**, 97–106 (2013)
[12] M. D. Luque de Castro, J. González-Rodríguez, P. Pérez-Juan, J.Food Sci. Technol. **21**, 231–265 (2005)
[13] G. Odageriu, *Evaluarea solubilitatii compusilor tartrici din vinuri*, 169–187 (EDP. "Ion Ionescu de la Brad" Iasi, 2006)

The impact of some commercial yeast strains on aroma compounds and sensorial analysis on two white wine varieties made in PGI Dealurile Olteniei, Romania

Liviu-Gabriel Grigorica[1], Marius Niculaua[2], Constantin Bogdan Nechita[2], Alina-Mihaiela Nistor[1], and Valeriu V. Cotea[1]

[1] University of Agricultural Sciences and Veterinary Medicine, 3 Mihail Sadoveanu Alley, Iaşi 700490, Romania
[2] Research Centre for Oenology, Romanian Academy Iasi Branch, 9 Mihail Sadoveanu Alley, Iaşi 700490, Romania

Abstract. Eight wines of Pinot gris and five wines of Sauvignon blanc obtained by fermenting grape juices from PGI Dealurile Olteniei with different commercial selected yeast strains and with indigenous microbiota were compared in terms of sensorial characteristics and their composition in minor volatile aroma compounds. An easy handle methodology with dynamic head-space trap, Gas Chromatography-Mass Spectrometry based, (HS-trap-GC-MS) permits the identification of 49 aroma compounds. The aim of the work was to link these aroma compounds, arranged in different chemical families, with the sensorial analysis and to establish the characteristics of each biotechnology used in the trials so we can find solutions to improve the quality of white wines made in hot climate zone that in time was a traditional area of red wines.

1. Introduction

Ever since oldest times the hill areas of PGI Dealurile Olteniei have proven excellent conditions for winegrowing and especially for the red wine producing grapes varietis as the great savant Ion Ionescu de la Brad mentioned in 1868 in "Agricultura Română din judeţul Mehedinţi": "...the greatest benefits from the vineyards come from Blahniţa and Câmpul areas where the famous vineyards of Oreviţa, Rogova, Drância, Oprişorul are located". In this context it is well known the potential of Golul Drâncei-Mehedinţi for red wine and also for some white wine according to the official task book of the PGI Dealurile Olteniei (2011). The customer preferences for much more fruity-floral style of white wines requires in the last decades improvements in winemaking technology and new approaches in using biotechnologies in winemaking [6, 19].

It is well known that the wine aroma can be classified according it origin into three groups: primary (varietal and pre-fermentative included), secondary (fermentative) and tertiary aroma (wine ageing). As the primary aromas cannot be developed into a very worm climate and the tertiary aroma is not so welcomed for fresh & fruity style white wines the winemakers in the area need to play different strategies in order to achieve as much as possible new and stronger aromas in the wines.

The description of wine aroma is not a simple task for researchers because more than 800 volatile compounds have been identified at a wide range of chemical families and concentration levels [5, 12–14]. The combination of varietal and fermentative aromas are the first quality factor for the differentiation of the white wines obtained from a certain variety and in a certain terroir. In the context of a competitive global market some few international grape varieties in some traditional production areas have been intensely studied [1,2]; nevertheless the characteristics of different indigenous varieties or the potential of international varieties in different East-European areas remain still unknown at the present day.

The differences among wines produced from the same grape variety, same PGI, from the same vintage, at the same ripening period and subjected to the same pre-fermentative treatments, but whose fermentations are carried out with different selected yeasts or with indigenous yeasts should be caused to the effect of inoculated yeast that offers undeniable advantages regard to the control and homogeneity of the fermentations [14, 15]. Despite these assertions, some researchers have not found appreciable differences between fermentations carried out with selected yeasts and those performed with indigenous yeasts. In the opinion of some authors it is advisable to use selected yeasts in new wine-growing areas, from early stages of fermentation, while indigenous yeasts are preferable in well-established areas [4, 8, 11, 16].

In recent years, hyphenated techniques have received ever-increasing attention as the principal means to solve complex analytical problems. This way, the use of Gas Chromatography (GC), coupled to Mass Spectroscopy (MS), in combination with dynamic head-space trap (HS-trap) showed a powerful integrated platform for the analysis of the volatile compounds present in grape, juice or wine samples [2, 3, 18].

This work is focused to the application of modern analytical techniques (HS-GC-MS) and sensorial analysis for the differentiation of wines obtained from the same musts (8 wines of Pinot Gris and 5 wines of Sauvignon blanc) using different starter cultures. The aim is to evaluate the influence of the yeasts on the organoleptic character in order to find the biotechnological solution that can achieve the proper solution for the commercial approach of the wine style.

2. Material and methods

2.1. Experimental wines production

In order to conduct this study, grapes of Pinot gris and Sauvignon blanc variety, healthy, at the technological maturity, manually harvested from Oprişor - Golul Drâncei vineyard - PGI Dealurile Olteniei area, were used. The experiment was conducted three vintages: 2013, 2014 and 2015. The grapes were destemmed and crushed. In the case of Sauvignon Blanc the mash was kept for a 6 hours maceration at 10 °C. The obtained must was homogenized and transferred to 5000 litres stainless steel tanks - 8 tanks (in the case of Pinot Gris) and 5 tanks (in the case of sauvignon Blanc) in equal quantities, for batch fermentation. For each year the same enzymatic products for clarification and selected yeasts for controlled fermentation were used. The difference between the variants consisted in the way the clarification was made (gravitational or enzyme-gravitational) and the use or not of various selected yeast strains for the alcoholic fermentation.

For Pinot gris the experimental variants were the following:

PG0 - control, gravitational clarification of the juice, no enzyme treatment, fermentation 18-20 °C, spontaneous fermentation

PG1 - gravitational clarification of the juice, no enzyme treatment, fermentation 16–18 °C, selected yeast Y1

PG2 - gravitational clarification of the juice, enzyme treatment E1 2 g/hL, fermentation 16–18 °C, selected yeast Y1

PG3 - gravitational clarification of the juice, enzyme treatment E2 1 g/hL, fermentation 16–18 °C, selected yeast Y1

PG4 - gravitational clarification of the juice, enzyme treatment E3 1 g/hL, fermentation 1618 °C, selected yeast Y1

PG5 - gravitational clarification of the juice, enzyme treatment E3 1 g/hL, fermentation 16–18 °C, selected yeast Y2

PG6 - gravitational clarification of the juice, enzyme treatment E3 1 g/hL, fermentation 16–18 °C, selected yeast Y3

PG7 - gravitational clarification of the juice, enzyme treatment E3 1 g/hL, fermentation 16–18 °C, selected yeast Y4

For Sauvignon blanc the experimental variants were the following:

SB0 - control, prefermentative cold maceration at 10 °C, 6 hours, gravitational clarification of the juice, no enzyme treatment, fermentation 18–20 °C, spontaneous fermentation.

SB1 - prefermentative cold maceration at 10 °C, 6 hours, gravitational clarification of the juice, enzyme treatment E1 2 g/hL, fermentation 14–16 °C, selected yeast Y2

SB2 - prefermentative cold maceration at 10 °C, 6 hours, gravitational clarification of the juice, enzyme treatment E1 2 g/hL, fermentation 14–16 °C, selected yeast Y3

SB3 - prefermentative cold maceration at 10 °C, 6 hours, gravitational clarification of the juice, enzyme treatment E1 2 g/hL, fermentation 12–14 °C, selected yeast Y

SB4 - prefermentative cold maceration at 10 °C, 6 hours, gravitational clarification of the juice, enzyme treatment E1 2 g/hL, fermentation 12–14 °C, selected yeast Y6.

All fermentation were made at controlled temperature (12–18 °C) in Oprisor wine cellar of Carl Reh Winery. The commercial yeast strains were encoded from Y1 to Y6. After the end of alcoholic fermentation (when ethanol content remained constant), a sample of 10 L (glass container) was filtered using BVT-20 filter sheets (Omniafiltri, Italy) and stored in cellar conditions (5–10 °C). All wines were analysed two months after filtration.

2.2. Gas-chromatographic analysis of volatile compounds

In order to conduct the determination of the volatiles compounds, all samples were analysed using head space technique with enrichment on a trap material connected to a gas chromatograph with a mass spectrometer detector (Shimadzu HS 20 trap-GC2010plus-MS8040TQ). The system uses helium as carrier gas and the adsorption material in trap as stationary phase is a resin of 2,6-diphenylene oxide (TENAX). At 6 mL of wine were added a mixture of 0.8 g from the following salts NaCl, Na_2SO_4 and KH_2PO_4 in a vial for analysis of vapours (headspace fraction) by the method of salting out. All samples were heated to 70 °C and shacked at 25 rpm for 5 minutes and then the vials were pressurised to 60 KPa with helium to remove condense. The volatile fraction is transferred in to a Tenax trap at −10 °C by a transfer line at 150 °C. Before injection for half a minute the trap is dry purged and then the volatiles are transferred at 280 °C with a split of 1/50 in to the analytic column. The separation column used is a Phenomenex® ZB WAXplus 60 m × 0.25 mm ID ×0.15 μm which is contained in a temperature - controlled oven. Temperature program for separation begins: constant for 3 minute (the total injection time 2 minute) at 82 °C raped with 3 °C/min. to 135 °C, maintained for 1 minute, then 7.5 °C/min. up to 160 °C, maintained for 1 minute and finally raped with 27 °C/min. to 240 °C constant for 8 minute (total temperature program 37 minutes). Separation is carried-out at a constant linear velocity of 35 cm/s (column flow 1.5 mL/min). The 8040 triple quadrupole is utilised in Q3 (third quadrupole) scan mode from 30–400 Da at 0.1 s event time (5000 scan/sec). The ion source interface is maintained at 230 °C and source at 200 °C. The results are processed qualitatively comparing them to different MS databases available: Nist14, Willey 10th, FFNSC and SZTERP. For the quantitative results, all samples are processed by internal standard method relative to 4-methyl-penta-2-ol.

2.3. Sensory evaluation

For each vintage a wine sensory evaluation were performed two months after bottling by an expert panel

Table 1. Analytical parameters of experimental variants, variety Pinot gris, vintages 2013–2015.

Experimental Variant and Vintage	Alcohol %v/v	Total Acidity g/L g/L tartaric ac tartaric ac	Reducing substances g/L	Volatile acidity g/L acetic ac acacid	SO_2 free mg/L	SO_2 total mg/L	pH
PG0–2013	13.3	5.8	7.2	0.48	25	110	3.42
PG0–2014	12.9	6.0	4.1	0.40	26	105	3.29
PG0–2015	13.5	6.2	7.4	0.73	28	118	3.48
PG1–2013	13.7	5.8	2.6	0.32	28	109	3.40
PG1–2014	13.1	6.1	3.7	0.40	26	107	3.33
PG1–2015	13.8	6.0	3.6	0.52	28	120	3.45
PG2–2013	13.6	5.9	2.9	0.28	34	109	3.42
PG2–2014	13.1	6.1	3.5	0.39	24	105	3.27
PG2–2015	13.8	6.2	3.7	0.49	34	120	3.43
PG3–2013	13.7	5.8	2.6	0.29	32	111	3.42
PG3–2014	13.0	6.0	3.4	0.39	27	105	3.31
PG3–2015	13.7	6.1	3.6	0.47	35	117	3.48
PG4–2013	13.7	5.8	2.7	0.26	34	109	3.44
PG4–2014	13.1	6.1	3.4	0.38	28	104	3.30
PG4–2015	13.8	5.9	3.3	0.48	33	122	3.47
PG5–2013	13.7	5.8	2.9	0.28	36	108	3.43
PG5–2014	13.2	6.1	3.1	0.36	32	108	3.32
PG5–2015	13.7	5.9	3.8	0.49	35	120	3.49
PG6–2013	13.6	5.9	3.1	0.26	34	109	3.42
PG6–2014	13.2	6.1	2.8	0.35	30	104	3.30
PG6–2015	13.8	6	3.6	0.50	36	119	3.47
PG7–2013	13.6	5.8	3.0	0.32	34	111	3.44
PG7–2014	13.1	6.1	3.2	0.36	28	106	3.31
PG7–2015	13.7	6.1	3.5	0.54	33	122	3.46

Table 2. Analytical parameters of experimental variants, variety Sauvignon blanc, vintages 2013–2015.

Experimental Variant and Vintage	Alcohol %v/v	Total acidity g/L tartaric ac tartaric ac	Reducing substances g/L	Volatile acidity g/L acetic ac aacid	SO_2 free mg/L	SO_2 total mg/L	pH
SB0 - 2013	12.8	5.9	5.3	0.38	28	111	3.34
SB0 - 2014	11.8	6.1	4.8	0.45	30	115	3.21
SB0 - 2015	13.4	6.6	3.9	0.79	32	125	3.15
SB1 - 2013	13.1	6.0	3.1	0.21	38	109	3.35
SB1 - 2014	12.0	6.1	3.3	0.34	39	113	3.23
SB1 - 2015	13.6	6.5	3.2	0.62	39	122	3.11
SB2 - 2013	13.1	5.9	2.9	0.19	37	108	3.32
SB2 - 2014	12.1	6.2	3.4	0.36	38	112	3.22
SB2 - 2015	13.6	6.7	3.1	0.73	38	124	3.17
SB3 - 2013	13.0	6.0	3.9	0.46	32	110	3.34
SB3 - 2014	11.9	6.0	3.7	0.52	32	115	3.21
SB3 - 2015	13.5	6.5	3.5	0.85	34	125	3.19
SB4 - 2013	13.0	6.0	3.4	0.36	34	109	3.33
SB4 - 2014	12.0	6.1	3.5	0.35	34	114	3.23
SB4 - 2015	13.5	6.5	3.4	0.73	36	123	3.20

of 5 members of Romanian Wine Tasters Association with experience as sensory judges (same members each year). Blind tests were carried out for each grape variety by comparing the control with the variants. Samples were served in ISO glasses and coded with random two digit numbers. Judges were asked to fill two different sheets: one with general attributes of the wines (smell, taste, overall harmony), one with sensory descriptors considered typical for each variety (Pinot Gris: aroma intensity, complexity, fruit aromas, flowers, vegetal, citrus, exotic, mineral; Sauvignon Blanc: aroma intensity and complexity, fruit aromas, flowers, vegetal, citrus, exotic, acacia, elder, green nuts) using marks from 1 to 9 for each sensory descriptor.

3. Results and discussion

3.1. Oenological characteristics

Tables 1 and 2 lists the values for the basic enological chemical analysis for the two varieties technological variants. Both controls (PG0 and SB0) regardless of vintage showed the lowest ethanol content and the highest values

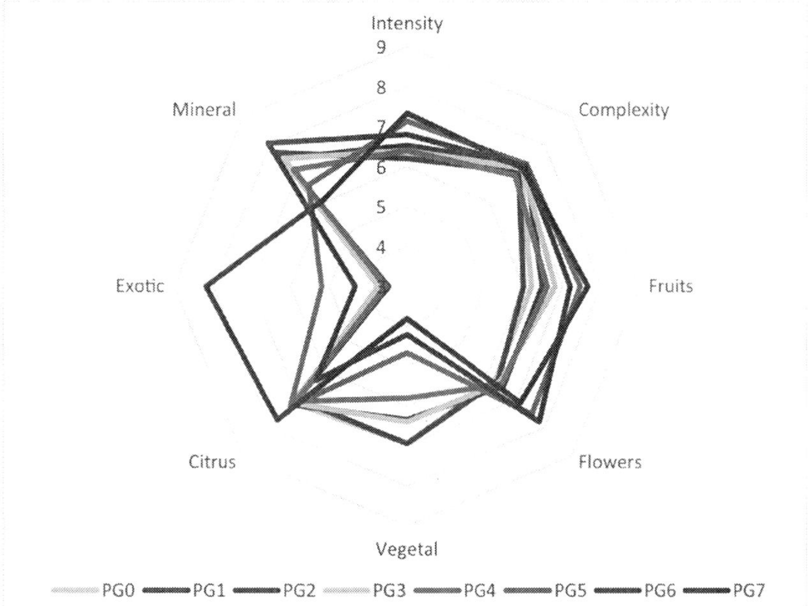

Figure 1. Sensory evaluation of Pinot gris wines fermented with different selected yeasts, PGI Dealurile Olteniei, averages 2013–2015.

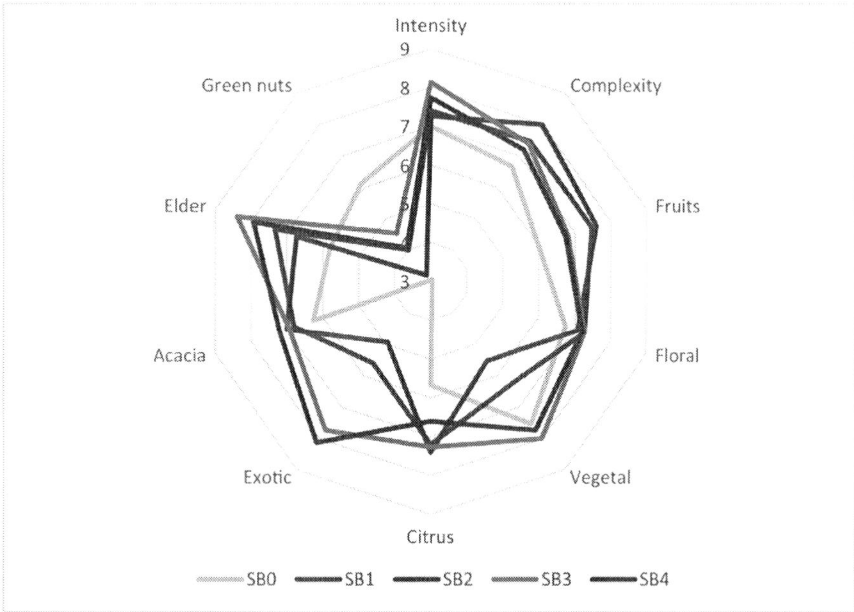

Figure 2. Sensory evaluation of Sauvignon blanc wines fermented with different selected yeasts, PGI Dealurile Olteniei, averages 2013–2015.

of the residual sugars, generally being not considered like dry wines (less than 4 g/L residual sugar). Some selected yeast, Y1 and Y2, were able to ferment every vintage untill less residual sugar. The fermentation carried out by selected yeasts showed slight differences among them with respect to all oenological variables quantified. The volatile acidities were systematically bigger in controls comparing with other variants, unconcerned the variety or vintage.

3.2. Sensory evaluation

The aim of this test was to evaluate if there were significant differences on the organoleptic characteristics among the young wines obtained using indigenous microbiota and different selected yeast with the goal to improve the overall quality of the white wines.

For Pinot gris as it is shown in Fig. 1 the enzyme used in clarification influences some organoleptic

Table 3. Sensory evaluation of Pinot gris wines fermented with different selected yeasts, PGI Dealurile Olteniei, results vintage 2015 compared with average 2013–2015.

Variant	Intensity	Complexity	Fruits	Floral	Vegetal	Citrus	Exotic	Mineral
PGO- 2015	6.20	7.40	6.40	6.60	6.60	7.20	3.80	7.60
AVPGO	6.33	7.13	6.20	6.27	6.93	7.13	3.47	7.67
PG1- 2015	6.20	7.60	6.20	6.80	6.80	6.80	3.20	7.40
AVPG1	6.20	7.07	6.00	6.33	6.93	7.00	3.47	7.73
PG2- 2015	6.40	8.20	6.60	7.00	6.40	7.20	3.60	7.20
AVPG2	6.53	7.33	6.47	6.53	6.33	7.20	3.67	7.53
PG3- 2015	6.20	8.00	6.80	6.80	6.20	7.60	4.20	7.00
AVPG3	6.40	7.20	6.80	6.40	6.40	7.13	3.73	7.53
PG4- 2015	6.20	7.40	6.60	6.20	5.60	7.20	3.60	6.60
AVPG4	6.40	6.93	6.60	6.47	5.80	7.07	3.53	7.13
PG5- 2015	7.20	7.80	7.40	7.60	4.20	7.60	6.20	5.80
AVPG5	7.13	7.33	7.47	7.53	4.67	7.27	5.20	6.60
PG6- 2015	6.60	7.20	7.20	8.00	4.40	6.40	4.80	8.00
AVPG6	6.80	7.27	7.20	7.80	4.20	6.33	4.33	8.07
PG7- 2015	7.20	7.40	8.20	7.20	3.40	7.80	8.60	5.40
AVPG7	7.33	7.20	7.67	7.13	3.80	7.73	8.20	6.07

Table 4. Sensory evaluation of Sauvignon blanc wines fermented with different selected yeasts, PGI Dealurile Olteniei, results vintage 2015 compared with average 2013–2015.

Variant	Intensity	Complexity	Fruits	Floral	Vegetal	Citrus	Exotic	Acacia	Elder	Green nuts
SB0 - 2015	7.00	6.60	6.40	7.00	7.60	5.20	2.60	6.00	5.40	6.20
AV SB0	7.00	6.67	6.00	6.73	7.53	5.67	2.93	6.27	5.80	6.13
SB1 - 2015	7.40	7.80	8.20	7.60	5.40	7.80	6.20	7.00	7.60	4.20
AV SB1	7.40	7.47	7.47	7.27	6.40	7.20	5.60	6.80	7.33	4.00
SB2 - 2015	7.60	8.20	8.60	7.20	4.20	7.00	6.20	6.40	7.20	3.60
AV SB2	7.27	8.00	7.60	7.07	5.53	7.40	4.93	7.00	6.73	3.20
SB3 - 2015	8.40	8.00	7.40	6.80	7.80	7.20	8.40	7.40	8.40	3.80
AV SB3	8.13	7.40	6.80	7.27	8.00	7.27	7.73	6.93	8.40	4.53
SB4 - 2015	7.80	7.40	7.80	7.60	8.20	6.80	8.20	7.80	8.20	4.20
AV SB4	7.73	7.20	6.73	7.20	7.73	6.60	8.13	7.20	7.93	4.07

characteristics (less vegetal, much more fruity/floral aromas) compared with control, but there are not significant differences between different types of enzymes. All controls showed sometimes more aromatic complexity but systematically were less intense compared with the variants fermented with selected yeasts. Among the selected yeasts Y2 and Y4 excels in accentuating some exotic, floral-fruity aromas.

For Sauvignon blanc as it is shown in Fig. 2 the controls are less intense in aromas unconcerned the vintage. Some selected yeasts, Y5 and Y6, showed systematically good marks for exotic aromas, elder and acacia flowers and Y2 were much more consistent in fruit and floral, exotic aromas.

3.3. Identification of free wine volatile compounds and correlation with wine sensory results

Identification of maximum 49 relevant volatile wine compounds was performed by using authentic standards NIST08 and Wiley7 mass spectral databases. We were more interested in esters because they are formed by yeasts during the alcoholic fermentation and they are responsible for the fruity aromas of fresh white wines [4]. The results of GC-MS analysis are in Table 5 for Pinot Gris and in Table 6 for Sauvignon Blanc. Table 5 Volatile aroma compounds identified in Pinot gris wines fermented with different selected yeasts, PGI Dealurile Olteniei, vintage 2015 Table 6 Volatile aroma compounds identified in Sauvignon blanc wines fermented with different selected yeasts, PGI Dealurile Olteniei, vintage 2015.

Regarding the correlations between chemical compounds and the results of sensory analysis we can comment the following: for Pinot gris the high mark of exotic notes for PG7 can be explained by higher concentration of ethyl propionate that is usually responsible for tropical aromas [9]; the higher concentration of methyl octanoate (grass aromas) and isoamyl hexanoate (tomato leaf notes) [17,19] explain the bigger marks of vegetal notes for the control PG0; from the other hand the higher concentration of butyl acetate (sensory attributes – citrus) or isobutyl acetate (passion fruit) [10] for variant PG7 have not any correspondence in the wine tasting marks.

For Sauvignon blanc the higher marks for elder and vegetal notes of SB3 and SB4 can be explained by higher concentration of isobutyl acetate and ethyl nonanoate (sensory attributes - cat urine) [7]; the increasing from SB0 to SB4 of the exotic marks can be explained by higher concentration of ethyl propionate. The vegetal notes in SB0 without exotic and elder high marks can be explained by the higher concentration in isoamyl hexanoate without the support of important concentration of isobutyl acetate or ethyl propionate.

Table 5. Volatile aroma compounds identified in Pinot gris wines fermented with different selected yeasts, PGI Dealurile Olteniei, vintage 2015.

Nr.	Chemical Compound	mvali eq SI							
		Pg0	Pg1	Pg2	Pg3	Pg4	Pg5	Pg6	Pg7
1	1-propanol	6.22	25.46	11.48	18.54	16.43	10.78	40.15	14.69
2	isoamylacetate	110.79	235.11	151.24	143.44	136.31	230.89	112.14	259.66
3	2,6-dimethyl-4-heptanone	19.61	19.16	18.54	22.29	85.77	20.78	32.86	43.89
4	ethyl palmitate	7.46	5.78	6.65	7.88	4.89	5.06	6.07	7.41
5	3-methyl-1-butanol	6043.50	6003.59	5728.61	5942.56	6130.44	5837.36	6121.26	6102.52
6	ethyl nicotinate	3.34	3.82	3.51	4.00	3.12	4.42	5.30	3.23
7	2,3,3 trimethyl 1,7-octadiene	59.83	57.17	64.73	80.94	42.26	45.81	24.51	70.97
8	3-hexen-1-ol, acetate	25.66	4.57	15.54	30.39	582.99	12.06	20.50	19.40
9	3 methyl 1-pentanol	591.68	576.07	564.65	569.52	1286.18	600.23	1028.71	542.58
10	octanoic acid, ethyl ester	0.98	2.62	3.22	2.87	2.98	8.17	2.86	1.42
11	isoamyl hexanoate	42.37	13.31	18.40	22.44	18.94	22.81	23.27	20.93
12	3-hexen-1-ol	2792.63	581.45	4051.92	1451.26	1876.14	2363.82	2536.75	2626.27
13	isobuthyl acetate	0.23	1.98	2.79	2.42	3.48	3.59	1.24	4.98
14	2-propanone, 1-hydroxy-	4.79	12.95	9.65	10.05	8.76	10.24	6.89	2.88
15	3-pentanol	106.51	262.13	133.40	192.55	20.67	81.83	48.91	281.39
16	1-propanol, 3-ethoxy-	2.46	2.45	2.98	1.74	3.25	2.37	1.82	0.98
17	ethyl pelargonate	0.98	3.68	2.76	3.04	2.42	2.98	3.26	3.17
18	1-methoxy-3-2-hydroxyethyl nonane	10.30	7.22	5.37	8.03	6.48	11.10	19.53	5.83
19	1-methoxy-3-hydroxymethylheptane	2.70	1.98	3.29	5.89	7.55	2.57	3.76	2.78
20	2,5-dimethyl-1-hepten-4-ol	7.51	12.43	15.96	14.84	15.47	12.89	13.67	10.45
21	ammonium oxalate	17.99	15.06	18.74	8.05	3.12	11.52	12.54	23.40
22	benzaldehyde	7.45	14.71	17.31	15.70	16.48	12.91	15.72	12.93
23	ammonium acetate	2796.81	2481.46	3679.23	2595.40	810.15	3847.11	3270.45	2463.41
24	ethyl nonanoate	3.47	10.76	12.89	18.21	16.62	12.78	14.90	14.37
25	propanoic acid	45.52	2.97	3.04	27.48	43.02	59.44	50.96	68.28
26	3-methylbutyl ester	1981.13	3176.19	3176.02	2465.39	4348.94	2639.88	2718.06	3175.19
27	n-Butyric acid	23.51	24.80	21.60	22.42	26.68	44.92	55.26	21.89
28	diethyl ester	5.64	9.55	5.78	10.85	26.35	7.90	13.97	28.36
29	isovaleric acid	126.93	78.74	124.85	103.08	66.62	93.97	144.05	112.58
30	hexanoic acid	16.30	7.66	15.32	7.35	2.05	28.24	27.27	8.37
31	hexanal	23.29	20.24	21.16	14.66	18.76	21.42	31.71	21.72
32	2-phenylethyl ester	9.34	3.29	7.55	5.65	2.32	4.56	5.78	5.32
33	ethyl laurate	2.76	5.62	4.60	5.76	4.24	5.20	3.14	2.78
34	pentadecanoic acid 3-methylbutyl ester	90.90	91.29	58.06	94.72	66.43	46.95	74.52	89.45
35	hexanoic acid	269.11	430.30	349.87	316.75	583.85	304.70	312.08	388.74
36	methyl 3-nonenoate	14.79	12.89	16.21	12.65	24.09	16.26	19.04	15.76
37	phenylethyl alcohol	579.40	510.72	520.54	445.13	505.82	659.49	730.59	512.99
38	Ethyl trans-2-hexenoate	3.45	1.45	1.29	2.99	2.64	2.45	1.98	3.25
39	1-dodecanol	369.06	391.57	369.28	312.81	842.49	395.23	407.71	361.96
40	tetradecanoic acid ethyl ester	3.31	4.30	1.98	2.70	3.96	2.76	2.45	2.41
41	ethyl palmitate	1.12	4.49	3.52	2.61	4.12	3.98	4.23	1.34
42	octanoic Acid	4.80	14.92	12.67	13.63	14.88	18.64	12.89	15.98
43	buthyl acetate	5.86	6.90	8.32	9.15	8.17	9.76	6.20	19.92
44	ethyl heptanoate	558.15	250.34	322.62	355.00	495.30	735.49	879.99	441.44
45	ethyl propionate	0.64	4.54	2.84	3.92	2.72	11.58	7.03	14.50
46	hexadecanoic acid, ethyl ester	1.20	0.98	1.22	1.48	3.37	15.18	27.62	13.31
47	n-decanoic acid	3.40	2.25	1.83	2.43	1,12	2.62	35.82	3.56
48	methyl octanoate	33.23	6.35	16.39	21.89	21.43	15.91	16.74	8.77
49	lauric acid	619.54	798.09	489.55	550.84	558.04	579.43	645.93	467.40

4. Conclusions

The HS-GC-MS hyphenated techniques applied to the analysis of wine volatile compounds permits the differentiation of wines obtained from the same must fermented with different selected (commercial) yeasts.

The study showed that it is possible to correlate these results with sensorial analysis marks for both varieties used in this experiment. Not least, during the three vintages study, some selected yeast showed important characteristics from an organoleptic point of view. Y2 and Y4 for Pinot gris and Y2, Y5 or Y6 for Sauvignon blanc can be appropriate solutions for fresh/fruity style wine wines with increased aromas.

The study confirmed that, even in areas with hot climate and not very suitable for good quality white wines,

Table 6. Volatile aroma compounds identified in Sauvignon blanc wines fermented with different selected yeasts, PGI Dealurile Olteniei, vintage 2015.

Nr.	Chemical compound	mvali eq SI				
		SB0	SB1	SB2	SB3	SB4
1	1-propanol	12.34	13.56	12.89	23.56	33.59
2	isoamylacetate	108.86	135.28	144.75	123.50	81.96
3	1-propanol	22.14	14.33	27.18	15.04	31.75
4	2,6-dimethyl-4-heptanone	37.19	16.24	52.54	6.59	44.40
5	1-butanol, 3-methyl-	1422.81	1359.25	1220.57	1167.35	703.37
6	3-hexen-1-ol	11.11	37.65	31.06	4.52	14.49
7	4-methyl-2-hexanol	7.36	7.63	6.20	6.48	7.97
8	1-pentanol, 3-methyl-	6.85	12.09	8.62	6.83	31.63
9	octanoic acid	2372.05	2773.52	5300.48	2493.99	15.74
10	1-hexanol	11.05	25.70	14.66	8.98	40.92
11	isoamyl hexanoate	40.35	17.11	16.38	21.22	15.12
12	propanoic acid, 2-hydroxy-	226.16	515.32	458.92	337.82	496.13
13	isobuthyl acetate	24.33	35.70	28.24	55.66	40.32
14	2-propanone, 1-hydroxy-	10.36	3.37	2.24	2.74	1.42
15	ethyl pelargonate	7.68	5.50	9.01	4.44	2.34
16	2,5-dimethyl-1-hepten-4-ol	11.06	9.49	8.97	8.42	12.28
17	tridecane, 2-methyl-	4.38	8.11	3.78	4.86	6.87
18	4-heptanol, 2-methyl-	3.45	10.29	12.87	12.27	15.59
19	ammonium oxalate	17.35	9.87	4.37	16.25	6.80
20	1-octanol	5.86	7.96	12.93	5.26	8.47
21	ethyl decanoate	2276.79	2591.14	2943.08	1358.20	2399.89
22	propanoic acid, 2-methyl-	21.45	94.98	32.89	35.37	26.57
23	octanoic acid, 3-methylbutyl ester	4.19	5.48	4.95	5.88	6.16
24	ethyl nonanoate	25.88	20.46	24.28	34.88	29.73
25	n-butyric acid	76.54	152.39	112.64	49.02	30.47
26	butanedioic acid, diethyl ester	42.85	80.34	30.95	37.75	18.12
27	isovaleric acid	4.67	47.39	18.56	2.79	12.12
28	acetic acid	89.97	98.62	106.43	47.22	121.02
29	ethyl laurate	93.28	132.86	148.75	68.27	212.93
30	pentadecanoic acid, 3-methylbutyl ester	13.71	11.32	6.04	4.64	3.68
31	hexanoic acid	490.79	849.21	516.30	398.80	396.54
32	benzenemethanol	3.11	10.68	3.71	2.67	4.24
33	phenylethyl alcohol	363.95	608.57	431.66	342.90	476.11
34	methyl octanoate	8.85	6.00	3.93	3.45	3.64
35	ethyl palmitate	6.86	5.99	2.77	1.12	0.96
36	ethyl miristate	2.12	5.04	21.67	31.48	28.41
37	octanoic Acid	1352.91	2435.94	1350.39	1214.42	1033.28
38	hexadecanoic acid, ethyl ester	13.67	41.15	28.05	87.98	75.69
39	ethyl heptanoate	766.85	900.33	357.31	361.38	286.84
40	ethyl 9-hexadecanoate	10.71	16.59	13.88	16.10	62.50
41	1-hexadecanol	5.89	12.77	11.23	14.72	13.42
42	benzoic acid	9.03	0.94	1.91	2.17	0.36
43	buthyl acetate	4.54	9.99	15.57	16.30	14.01
44	lauric acid	24.63	48.14	29.08	28.37	9.82
45	ethyl propionate	8.34	15.41	15.83	18.30	15.98

we can produce good quality fresh/fruity style wines using the correct and adapted biotechnological solution in winemaking.

This work was supported by Doctorate School of USAMV Iasi and the Research Centre for Oenology, Romanian Academy - Iasi Branch. The authors thank Carl Reh Winery for the grapes provided and for the use of the winery for the experiments.

References

[1] G. Antalick, M.-C. Perello & G. de Revel, American Journal of Enology and Viticulture **65**(3), 293–304 (2014)

[2] M. Arbulu, C. Sampedro, A. Sanchez-Ortega, A. Gomez-Caballero, N. Unceta, M.A. Goicole et al. Analytica Chimica Acta **777**, 41–48 (2013)

[3] R. Castro, R. Natera, E. Durán, C. Garcia-Barroso, European Food Research Technology **228**, 1–18 (2008)

[4] D. Dubourdieu, D., Proceedings of the ASEV 50th Anniversary Annual Meeting, 196–203 (2000)

[5] P. X. Etiévant, *Wine* (Ed. H. Maarse, 1991)

[6] L.G. Grigorica, C. Baduca Campeanu, G. Beleniuc, V. Gheorghiu, Lucrari Stiintifice **53**, USAMV Iasi (2010)

DOI: 10.1051/bioconf/20170902006

[7] M.G. Lambrechts, I.S. Pretorius, South African Journal of Enology and Viticulture **21**, 97–129 (2000)
[8] J. Moreno, C. Millan, J.M. Ortega, M. Medina, Journal of Industrial Microbiology **7**, 181–190 (1991)
[9] M.A. Pedroza, A. Zalacain, J.F. Lara, M.R. Salinas, Food Research International **43**, 1003–1008 (2010)
[10] I.S. Pretorius, Yeast **16**, 675–729 (2000)
[11] J.A. Regodon-Mateos, F. Perez-Nevado, M. Ramirez-Fernandez, Enzime and Microbial technology **40**, 151–157 (2006)
[12] P. Reboredo-Rodriguez, C. Gonzales-Barreiro, R. Rial-Otero, B. Cancho-Grande, J. Simal-Gandara, Critical Reviews in Food Science and Nutrition **55**, 202–218 (2014)
[13] P. Reboredo-Rodriguez, C. Gonzales-Barreiro, R. Rial-Otero, B. Cancho-Grande, J. Simal-Gandara, Critical Reviews in Food Science and Nutrition **55**, 1051–1071 (2015)
[14] P. Ribéreau-Gayon, Y. Glories, A. Maujean, D. Dubourdieu, *Handbook of enology* **2**, 51–64 (John Wiley & Sons Ltd., 2006)
[15] P. Romano, C. Fiore, M. Paraggio, M. Caruso, A. Capece, International Journal of Food Microbiology **86**, 169–180 (2003)
[16] B. Suarez, R. Pando, N. Fernández, A. González, R. Rodriguez, LWT **38**, 455–461 (2005)
[17] K. Šuklje, G. Antalick, A. Buica, A.C. Zelmari, J. Brand, L.M. Schmidtke, M. A. Vivier, Food Chemistry **197**, 1073–1084 (2016)
[18] A. Tredoux, A. Villiers, P. Májek, F. Lynen, A. Crouch, P. Sandra, Jour. For Agr. And Food Chem. **56**, 4286–4296
[19] F. Vararu, J. Moreno-Garcia, C.I. Zamfir, V.V. Cotea, J. Moreno, Food Chemistry **197**, 373–381 (2016)

DOI: 10.1051/bioconf/20170902007

Content of metals and organic acids from experimental sparkling white wine

Mihai Cristian Focea[1], Camelia Elena Luchian[1], Ana Maria Moroşanu[1], Marius Niculaua[2], Valeriu V. Cotea[1,a], Gheorghe Odăgeriu[2], and Cătălin Ioan Zamfir[2]

[1] The University of Agricultural Sciences and Veterinary Medicine, Mihail Sadoveanu Alley, 3, Iaşi 700490, Romania
[2] Research Centre for Oenology of Romanian Academy – Iaşi Branch, Mihail Sadoveanu Alley, 9, Iaşi 700490, Romania

Abstract. This work is focused on the study of the influence of different strains of yeasts on the concentration of organic acids, metallic content, and other physical-chemical parameters from experimental sparkling white wines produced by traditional method (bottle fermented). This study was required due to climatic conditions varying from year to year, generating grape harvests with very high alcoholic potential, and very low values for total acidity. In this case, a Muscat Ottonel grape must was used and passed by a reverse osmosis process. The obtained permeate was mixed with a calculated amount of the concentrate to generate a must with a potential of 10.5% (v/v) alcohol, in order o to obtain the base wine for the second fermentation. After fermentation, the wine was treated to get tartaric, protein and microbiological stabilization. For the second fermentation four different strains of yeast species *Saccharomyces cerevisiae* were used. Bottle fermentation and storage was performed at a constant temperature of approximately 12 °C. After six months of storage, sparkling wine samples were analyzed. The metal content was determined using AAS method, and organic acid concentration was determined by a HPLC method. The main physical-chemical characteristics were determined (alcohol concentration, total acidity, volatile acidity, total dry extract, free SO_2, total SO_2, density, pH, conductivity) based on OIV methods. The results obtained indicated significant differences of the analyzed parameters.

1. Introduction

Sparkling wines obtain by the traditional method and here it can be mentioned cava wines (Spain) and champagne wines (France) have been scarcely studied in the literature of the field. The traditional method of obtaining sparkling wines consists of a second fermentation process. The base wine elaborated in a first phase is left to undergo a second alcoholic fermentation, inside a sealed bottle by adding tirage liqueur and yeasts, followed by ageing on yeast lees under anaerobic conditions for several months [1].

The composition of a sparkling wine and in general of a wine is due to many factors related to the specific production area such as: grape variety, soil, climate, yeast, wine making practices and not only. All of this factors have an important influence on the quality of wine and they also have a major role in the characterization and differentiation of wines [2,3].

The mineral content in wines has been extensively studied and reported due to the implications on different levels such as: sensorial, hygienic, dietetic, toxicological [4]. Furthermore this type of studies helps to characterize and to differentiate the wines from the geographical point of view when reliable chemical analyses techniques in combination with modern chemometrical methods are used [5–8].

The levels of organic acids in wines and in general of the volatile component are strongly influenced by the dynamics and the activity of the indigenous or the inoculated yeasts [9]. In wines, organic acids have a major role for the composition, the stability and the organoleptic qualities and their preservative proprieties enhance the microbiological and the physicochemical stability [10].

Reverse osmosis (RO) is a technology that uses a semi permeable membrane to remove ions, molecules and larger particles from different liquids. In reverse osmosis, an applied pressure is used to overcome osmotic pressure a colligative property, which is driven by chemical potential differences of the solvent, a thermodynamic parameter. In the winemaking industry it is desired to obtain a wine from extremely ripe grapes in order to create a wine with a developed sensory palette, but in this case occurs the problem of obtaining wines with a high alcoholic strength [11–14].

Thus, in order to avoid this type of situations reverse osmosis is applied and in this case the level of alcohol is reduced without major changes on the fruit flavors and other elements in the wine.

2. Materials and methods

2.1. Must samples

The present study was carried out on four samples of sparkling wines obtained from Muscat Ottonel juice/must. Muscat Ottonel is a younger and less-distinguished member of the broader Muscat family and it is considered to be a clone obtained from the crossed process of Chasselas and Muscat Ingram [15].

[a] e-mail: vvcotea@yahoo.com

DOI: 10.1051/bioconf/20170902007

The must and implicitly the wines obtained from the Muscat Ottonel must are characterized by a floral aroma with fruity notes such as apricot and quince. Noteworthy is that the juice of Muscat Ottonel are suitable for obtaining both dry and sweet wines. Concerning the present study the raw material - the must was obtained from grapes of Muscat Ottonel that were harvest in 2015 at full maturity from Iaşi vineyard.

2.2. Wine samples

The sparkling wine samples taken in the experiment were produced in the micro-winery belonging to the Oenological department of the University of Agricultural Sciences and Veterinary Medicine of Iasi.

It is a must to emphasize the fact that the samples covered by this scientific study were obtained by applying the classical method. This method consists of a secondary bottle fermentation of the wine raw material, followed by the removal of impurities by riddling and disgorgement operations.

The harvest of grapes along with the primary fermentation proceeds have gone as with any still wine. The still wine produced becomes in this way the base-wine to be used in the next step. It is necessary to mention the fact that a reverse osmosis process was applied to the initial must, then the obtained permeate and concentrate was mixed in a variable ratio to generate a must with a potential of 10.5% (v/v) alcohol, followed by the typical fermentation process.

The second fermentation is induced by adding four different yeasts mixed with a calculated amount of the concentrate must (called "tirage liqueur"). The sugar added with the "tirage liqueur" provides food for the yeast and is entirely consumed during the secondary fermentation and has no effect on the "sweetness" of the final product. At this time the bottle is capped with a crown cap (the same as a beer bottle) and stored in a cellar in the horizontal position to age.

After the completion of the aging process, the bottle is submitted to a removing operation of the dead yeast, this process being known as riddling. At this stage the bottles are placed on special racks at a 45° angle with the cork pointed down and occasionally the bottles are slightly shaken. After, the position of the bottles is changed being pointed straight down with the sediment in the neck of the bottle.

The disgorging operation consists in a frizzing process of the neck bottles that are still pointed down and after that the bottles are turned upright and the cap is removed. In this way the pressure in the bottle will eliminate the lees.

Immediately after disgorging and before final corking, the wine levels from the bottles is completed and the liqueur for dosage is added, in order to obtain certain sensory characteristics and the various types of sparkling wine

2.3. Reagents for the fermentation process

For the fermentation process the most important factor is the type of the yeasts that were used. Thus, the yeasts are becoming the experimental factor helping to differentiate samples regarding the quality of the wine obtained given by: aromatic compounds, metals content, acids content, color and the determination of the amino acids. The samples are: V0- the blank sample, V1- the type 1 of yeast are used, V2- the type 2 of yeast are used, V3- the type 3 of yeast are used, V4- the type 4 of yeast are used. It is also necessarily to underline the fact that the yeasts that were used for this experiment are IOC products.

After the decarbonisation operation, each sample of wine was submitted to the following analyses: sulfur dioxide, volatile acidity, total acidity, alcoholic strength, reducing substances, total dry matter and non reducing substances. The analyses were done according to OIV methods and the specific literature [10].

2.4. The quantification of metals in the wines samples using an Agilent MP-AES 4200 system

The analyzed samples were diluted in order to reduce the salinity of the samples below 1%. The technique employed doesn't need the mineralisation of the samples or the addition of ionisation buffers. The measurements were performed using an Agilent MP-AES 4200 (Microwave Plasma-Atomic Emission Spectrometer) system equipped with the standard sample introduction system consisting of the OneNeb nebulizer, double pass cyclonic spray chamber and easy fit torch.

The Agilent SPS-3 auto sampler was used to deliver samples to the instrument allowing unattended operation. The selection of optimal lines depended on wavelengths that were free from spectral interference and matched the appropriate sensitivity. Concerning the interferences on the background and spectral level can be corrected for using the MP software. The read time for all samples is 3 seconds and all others instrument operating parameters are presented in Table 1.

2.5. The determination of organic acids of the wine samples

The analysis of the organic acids was performed by using a HPLC Shimadzu series 20 system. The samples were filtered through a 28 mm nylon 60.45 μm syringe cartridge according to the modified methodology described in the method MA-E-AS313-04-ACIORG by O.I.V. standards. A 5 μL volume of sample or standard was injected through two analytical columns (YMC-Triart C18 multi-stage hybrid group's 3 μm150 × 4.6 mm 120 Å) at a flow rate of 0.9 mL/min. using a solution of sulphuric acid adjusted to 1.3 pH value. Columns temperatures are maintained to 45 °C for the entire run. The quantification of compounds is done at certain wavelength for each organic acid.

3. Results and discussion

The general physic-chemical parameters of the analyzed sparkling wine samples are presented in Table 2. It is necessarily to mention the fact that in the obtaining process of sparkling wines was used the same raw-material-wine and possible differences could result from using different types of fermentation yeasts.

Comparing the sparkling wine samples, visible differences can be observed between the blank samples and the samples treated with specific yeasts. Thus, the values of the alcoholic strength of the analyzed samples ranged

Table 1. Operating parameters for Agilent MP-AES 4200 system.

Element	Wavelength λ (nm)	Nebulizer Flow (L/min)	Air Injection Flow Rate	Optimal dilution factor of wine	Linear domain for $r^2 \geq 0.995$
Zn	213.857	0.45	medium	1/10	10–200 ppb
Fe	371.993	0.65	high	1/10	3–500 ppb
Cu	324.754	0.7	high	1/10	3–2000 ppb
Ni	352.454	0.7	high	1/10	5–2000 ppb
Pb	405.781	0.75	high	1/10	25–2000 ppb
Mn	403.076	0.9	high	1/10	3–500 ppb
Mg	285.213	0.9	medium	1/100	5–2000 ppb
Ca	396.847	0.6	high	1/100	25–300 ppb
Na	589.592	0.95	high	1/100	0.25–5 ppm
K	769.897	0.75	high	1/100	0.1–5 ppm

Table 2. Quality parameters of the wine raw-material and of the sparkling wines obtained.

Samples	**1**	**2**	**3**	**4**	**5**	**6**	**7**	**8**
Blank sample–V0	0.9921	6.2	0.33	10.3	17.2	70.7	15.2	2.8
Sample V1	0.9905	6.7	0.32	11.6	5.12	56.3	14.5	3.1
Sample V2	0.9908	6.9	0.29	11.3	5.12	48.6	13.3	3.0
Sample V3	0.9906	6.9	0.26	11.6	5.12	51.2	13.5	3.0
Sample V4	0.9907	6.6	0.31	11.3	7.68	64.0	14.3	3.0

1– density; **2**– total acidity (g tartaric acid/L); **3**– volatile acidity (g acetic acid/L);
4– alcohol strength (%); **5**– free SO_2 (mg/L); **6**– total SO_2 (mg/L);
7– Non-reducing extract (g/L); **8**– pH.

Table 3. The alkaline and alkaline-earth metals content in the raw-material wine and in the analyzed sparkling wines.

Samples		**K** (µg/L)	**Ca** (µg/L)	**Mg** (µg/L)	**Na** (µg/L)
Blank sample -**V0**		139,92	28,24	249,02	296,00
6 months	**V1**	553,06	124,77	28,77	37,58
	V2	547,27	126,00	28,82	43,38
	V3	545,52	126,82	28,6	40,23
	V4	556,61	127,27	28,57	38,02
11 months	**V1a**	530,54	126,29	28,87	24,2
	V2a	552,17	124,71	28,64	21,76
	V3a	551,09	124,62	28,36	24,78
	V4a	527,52	124,65	28,17	24,26

between a minimum of 10.34% for the blank sample and a maximum of 11.6% for the samples V1 and V3. The dynamic of the alcoholic strength can result from the fermentation process that was "accelerated" for the samples inoculated in the second fermentation stage with specific yeasts.

The density of the samples is correlated with the sugar content of the wines. Furthermore, observing the sugar levels or the reducing substances, the sparkling wine samples can be included in the category brut.

The total acidity that varied from 6.28 g/L to 6.9 g/L and the volatile acidity that ranged between 0.26 g/L and 0.33 g/L describe the profile of balanced sparkling wine samples. Generally, the real acidity of the wine or the pH has an important impact on almost every aspect: color, stability, flavor and also on the rate of the fermentation process. Referring to the analyzed samples this parameter was situated between 2.8 and 3.1 a range that doesn't allow the growth of bad microorganisms.

Detection of the presence of different metals in the blank sample and also in the samples treated with specific yeasts was carried out using an Agilent MP-AES 4200 system. The values of the metals concentrations that were identified in the sparkling wine samples are listed in Table 2 and Table 3. It is required to mention the fact that the metal analysis of the samples was performed at six months (V1, V2, V3, V4) and at eleven months (V1a, V2a, V3a, V4a).

The presence of metals in wine has a double origin: natural and anthropogenic and their concentrations can play a significant role, affecting the consumption and also the conservation of the wines. Sparkling wines quality and in general wines quality depends also on the metal composition since the metallic ions have important roles in the oxide-reductive reactions resulting in wine browning, turbidity, cloudiness and astringency. Most metals are important for efficient alcoholic fermentation. Ca, Mg, K and Na are implicated in the process of cellular metabolism of yeasts by maintaining the right level of pH and ionic balance.

Observing Table 3 and Fig. 1, it can be noticed a variability concerning the presence of Mg, Ca, K and Na in the blank sample and in the samples analysed. So, the concentrations of K and Ca register a growth after the storage at six and eleven months, but in the case of Mg and Na a considerable decrease is detected.

The composition of metals in wine during fermentation, maturation and storage is not stable. In the fermented

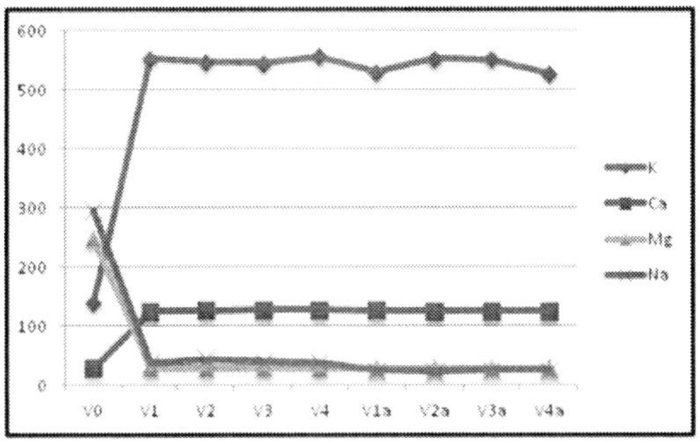

Figure 1. Dynamic of the alkaline and alkaline-earth metals concentrations in the wine raw-material and in the analyzed sparkling wines.

Table 4. Concentrations of Mn, Zn, Fe, Cu and Ni in the wine raw-material and in the analyzed sparkling wines.

Samples		Mn (µg/L)	Zn (µg/L)	Fe (µg/L)	Cu (µg/L)	Ni (µg/L)	Pb (µg/L)
Blank sample – V0		668,84	326,03	1214,65	188,12	128,11	206,23
6 months	V1	708,76	380,09	1619,9	118,02	144,75	449,9
	V2	689,83	377,53	1457,47	124,59	140,74	ND
	V3	696,18	387,35	1476,17	111,3	138,88	ND
	V4	698,66	351,56	1433,72	113,2	145,96	558,74
11 months	V1a	706,65	216,3	1704,21	154,27	143	260,77
	V2a	712,17	361,69	1536,80	152,95	146,86	135,83
	V3a	691,24	315,8	1487,92	118,25	144,69	631,85
	V4a	701,20	212,54	1505,59	118,66	169,14	2,46

musts and wines, yeasts consume Ca, Cu, Fe, K, Mg and Zn, so usually their concentrations decrease [16].

After the fermentation process, racking of yeasts sediments and other precipitates is used to remove these metals. However, referring to the analyzed samples the concentrations of metals such as: Zn, Fe, Cu, Ni, Mn are higher for the samples inoculated with yeasts.

Even more, comparing the concentrations of Fe, Cu, Ni, and Mn in the samples V1, V2, V3, V4 an interesting dynamic is observed (Table 4 and Fig. 2). Thus, the concentrations of these metals have increased with the increasing of the storage period. This phenomenon might suggest the susceptibility of the wine samples to the clouding or sedimentation process as well as flavor deterioration due to the strong catalytic and oxidative proprieties of the cations (especially Fe and Cu). The concentration of Pb in the analyzed samples weren't represented in Fig. 1 due to the fact that in the case of samples V2 and V3 the presence of this metal wasn't detected by the system and in the case of other samples the presence wasn't pure but interfering with the presence of Fe.

The presence of the main organic acids and their concentrations in the analyzed sparkling wines samples are presented in Table 5. As it was mentioned before for the determination of organic acids a HPLC Shimadzu was used. The organic acids that were the subject of this study were: the malic acid, the tartaric acid, the shikimic acid, the lactic acid, the acetic acid, the citric acid, the succinic acid and fumaric acid.

The acids in wine are one of the essential components in both winemaking and the finished product since they

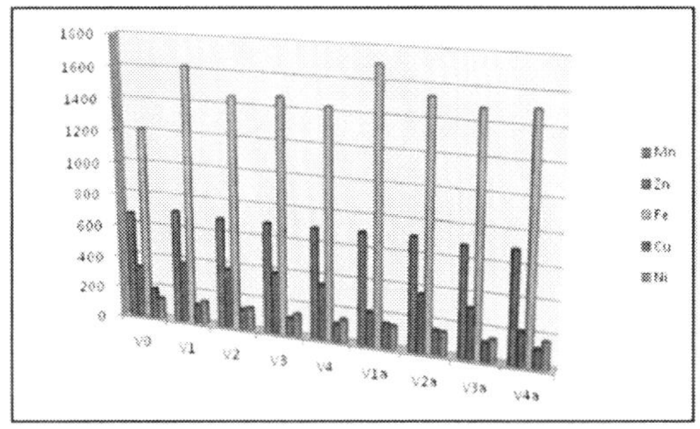

Figure 2. Concentrations of Mn, Zn, Fe, Cu and Ni in the wine raw-material and in the analyzed sparkling wines.

have a direct influence on the color, balance and on the taste. In addition the acids present in wines assure the growth and the vitality of yeast during fermentation and also have a protective role against harmful bacteria. When the presence of the organic acids is at the right level a balanced wine is obtain, a wine characterized by a refreshing organoleptic sensation. On the other hand excessive concentrations give the wine a tart or a sour taste.

Malic acid along with the tartaric acid is one of the principal acids found in grapes and wine. In the analyzed samples the higher value was detected in the blank sample and with the storage of the samples a slight decrease was observed. From winemaking perspective the tartaric acid is one of the most important one due to its prominent role in maintaining the stability of the wine and due to its influence on the color and taste of the finished product. Referring to the results this acid varied from a minimum of 0,99 mg/L for the sample V3a and a maximum of 5,97 mg/L for the blank sample. Actually the same tendency of quantitative decrease was observed for most organic acids determined with some minor exceptions such as: fumaric acid, citric acid and malic acid.

As a minor carboxylic acid, shikimic acid can be identified in a wide range of fruits and it can be a great help in establishing the traceability and the authenticity of a wine. In the analyzed sparkling wines the level of this organic acid has suffer a slight decrease with the time of storage, the higher value being registered for the blank sample V0, respectively 15,97 (μg/L), but a generally decreased was identified at 11 months of storage.

On the other hand observing the dynamic of fumaric acid, an important increase can be noticed with the storage, So, if for the blank sample the level of this acid was somewhere 9,74 mg/L, after six months of storage the quantity of fumaric acid was four times higher and at 11 months a slight decrease was observed.

4. Conclusions

Following the results of the physico-chemical characteristics, the considered samples could be included in the category of "brut" sparkling wines.

The metals analysis revealed important differences between the analyzed samples. So, in the case of K and Ca, the concentrations in the sparkling wines after six months

Table 5. The organic acids identified in the in the wine raw-material and in the analyzed sparkling wines.

Samples		1	2	3	4	5	6	7	8
Blank sample -**V0**		2,84	5,97	15,97	0,36	0,85	0,45	1,07	9,74
6 months	**V1**	6,33	1,39	4,14	0,42	0,38	0,58	0,18	34,54
	V2	6,46	1,42	4,36	0,38	0,39	0,59	0,20	40,43
	V3	6,35	1,40	4,27	0,38	0,38	0,58	0,31	32,31
	V4	6,42	1,41	4,61	0,40	0,39	0,61	0,31	33,34
11 months	**V1a**	6,42	1,39	4,02	0,63	0,39	0,59	0,29	32,05
	V2a	6,63	1,42	4,29	0,60	0,38	0,57	0,30	36,84
	V3a	5,61	0,99	4,68	1,12	0,21	0,51	0,22	15,65
	V4a	6,51	1,45	4,31	0,58	0,37	0,51	0,17	31,77

1– malic acid (g/L); **2**– tartaric acid (g/L); **3**– shikimic acid (mg/L); **4**– lactic acid (g/L); **5**– acetic acid (g/L); **6**– citric acid (g/L); **7**– succinic acid (g/L); **8**– fumaric acid (μg/L).

of storage has increased as compared to the blank sample and after that suffered a slight decrease.

The levels of Mg and Na have registered important decreases at both series of samples, six months and eleven months of storage due to the yeast metabolism. On the other hand, in the case of Mn, Zn, Fe and Ni the analysis showed a marked increase after six months of storage compared with the raw-material wine. At eleven months the levels of Mn and Zn have diminished, but in the case of Fe has augmented probably due to the formation of stable complexes with amino-acids, melanoids.

The level of Cu has decreased in the samples analyzed after the six months of storage, but later an augmentation was observed. However, it should be stated the fact that the variability of the metals can be influenced by the evolution of pH, the activation of prosthetic metallo-enzyme and the ionic balance.

References

[1] F. Torchio, S. Río Segade, V. Gerbi, E. Cagnasso, L. Rolle, Food Research International **44**, 729–738 (2011)
[2] R. Lara, S. Cerutti, J.A. Salonia, R.A. Olsina, L.D. Martínez, Food Chem. Toxicol. **43**, 293–297 (2005)
[3] M. Urbano, M.D. Luque, P.M. de Castro, J.García-Olmo Pérez, M.A. Gómez-Nieto, Food Chem. **97**, 166–175 (2006)
[4] M. López-Artıguez, A.M. Cameán, M. Repetto, J. AOAC Int. **79**, 1191–1197 (1996)
[5] Thiel, K. Danzer, Fresenius J. Anal. Chem. **357**, 553–557 (1997)
[6] L. Sun, K. Danzer, G. Thiel, Fresenius J. Anal. Chem. **359**, 143–149 (1997)
[7] K. Danzer, D. García, G. Thiel, M. Reichenbacher, Am. Lab. 26–34 (1999)
[8] W.U. Kwan, B.R. Kowalski, J. Food Sci. **43**, 1320–1323 (1978)
[9] P. Hernandez-Orte, J. F. Cacho, V. Ferreira, J. Agric. Food Chem. **50**, 2891–2899 (2002)
[10] P. Ribereau-Gayon, Y. Glories, A. Maujeanand, D. Dubourdieu, Handbook of enology. The chemistry of wine: Stabilization and treatments, **2** (EDP Chichester: Wiley, U.K., 2006)
[11] E. Ronald, R. Nina., W. Robert, A. Keith and G Edward, J. Food Sci. **58**(3), 633–637 (1993)
[12] C. Delfini and G. Nicolini, J. Int. Sci. Vigne Vin, **25**(1), 1–35 (1991)
[13] P. Cuenat, D. Kobel and E. Zufferey, Bull. OIV **701–702**, 519–537 (1989)
[14] M. Mietton-Peuchot, V. Milisic and P. Noilet, Desalination **148**, 125–129 (2002)
[15] L. Rotaru, Soiuri de viţă de vie pentru struguri de vin, (EDP "Ion Ionescu de la Brad", Iaşi, 2009
[16] S. Rodriguez Mozaz, A. Garcia Sotro, J. Garrido Segovia, C. Ancin Azpilicueta, Food Research Int. **32**, 683–689 (1999)

 DOI: 10.1051/bioconf/20170902008

Pulsed electric fields (PEF) applications on wine production: A review

Burcu Ozturk[1,a] and Ertan Anli[2,b]

[1] Namık Kemal University Şarköy Vocational School, Wine Production Technology Programme, 59800 Şarköy- Tekirdağ, Turkey
[2] Ankara University Engineering Faculty, Food Engineering Department, 06110 Dışkapı-Ankara, Turkey

Abstract. Novel techniques have been searched in the last decades as a result of increasing demand for high quality food products. Non-thermal processing technologies, such as pulsed electric fields (PEF) have been improved to achieve inhibition of deleterious effects on quality-related compounds. The working principle of PEF is based on the application of pulses of high voltage (typically above 20 kV/cm up to 70 kV/cm) to liquid foods placed between two electrodes. Pulsed electric fields technique has also been studied in winemaking process. Certain positive influences of PEF on vinification have been reported as elimination of pathogenic microorganisms, reduction of maceration time, increase in phenolic compounds extraction , acceleration of wine aging and inactivation of oxidative enzymes. The aim of this review is to summarize the potential applications of PEF in winemaking and to express its effects on quality of wine.

1. Introduction

Non-thermal techniques have been researched widespread aiming at maintaining food qulity. Pulsed electrical fields (PEF) as an alternative to conventional techniques is an innovative process offering improvement of food quality by reducing the cost of energy [1–3]. The PEF technology has various advantages like shorter processing time, lower treatment temperature, and continuous flow when compared with conventional thermal processes [2]. It suggests increase in production efficiency and low processing temperature diminishes the food quality deterioration risk [1].

Winemaking techniques play a crucial role in terms of wine quality and aging stabilitiy of wine. Among them extraction of phenolic compounds has the most powerful effect on both organoleptic quality and stability of wine. On the other hand inactivation of spoilage microorganisms is of crucial significance in order to control the deterioration of wine and reduce the economical losses of wine producers.

Electroporation, which causes increase in the cell membrane permability and conductivity, is a phase that is originated by means of externally applied electric field of adequate strength [3,4]. The electroporation resulted from PEF treatments has indicated effective results for extraction of juice, and other important compounds from grape tissue, nonthermal preservation by microbial inactivation and increasing fermentation activity of yeasts [5,6].

This review is focused on main points of PEF technique, various applications on different wine production stages and the changes in certain wine quality parameters related to these applications.

[a] e-mail: bozturk@nku.edu.tr
[b] e-mail: anli@eng.ankara.edu.tr

2. PEF mechanism

PEF is a novel technology which includes the application of microsecond (μs) pulses of high electric field to a material placed between two electrodes [7]. A classical systems for the treatment of pumpable fluids composed of a PEF generation unit that consists of a high voltage generator and a pulse generator respectively, a treatment chamber, a proper product process system and a set of monitoring and controlling equipment [7,8] (Fig. 1).

With the help of energy transfer to the fluid PEF treatments cause an increase in temperature and electrical conductivity of the treatment medium. The temperature increase changes viscosity and stability of the cell membrane and changed conductivity at constant energy input diminishes the field strength [9].

3. Electroporation strategy

The first step of PEF treatment is electroporation, also named as electropermeabilization, is electrical breakdown of cell membranes via externally applied electrical field of adequate strength yields a raise in cell's plasma membrane conductivity and permeability [4]. This ability of the treatment is utilised for killing microorganisms [8]. Membrane electroporation property has found different field of application to improve other new processes via induction of mechanical, hydrodynamic, osmotic viscoelastic instabilities progressively [10].

There are three possible results of PEF associated with the duration of cell exposure, the local field strength (maximum amount of enery givable to the membrane via external electrical field), and the rate of membrane recovery. When the field strength and exposure time are inadequate, there is no electroporation and no change

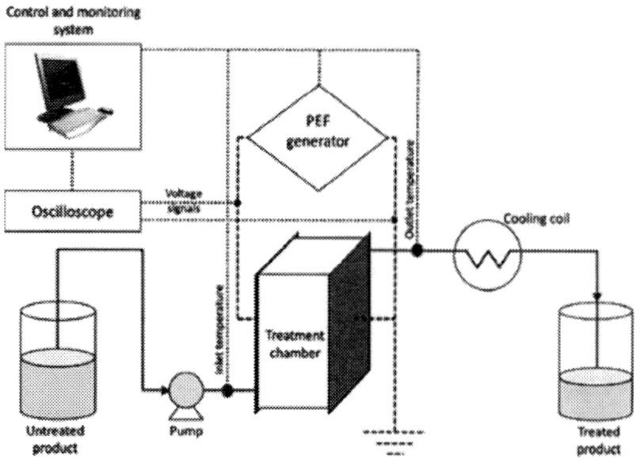

Figure 1. Schematics of a PEF processing system for pumpable products [7].

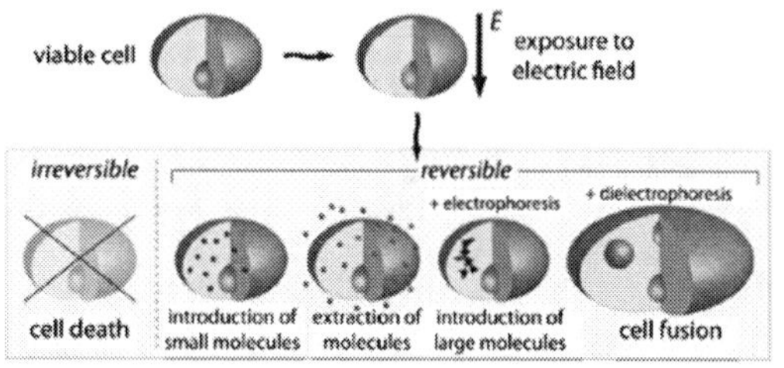

Figure 2. The schematic illustration of possible outcomes of cell electroporation depending on the pulsing dynamic (amplitude, shape, duration of pulses) and extra cell distotion technique (dielectrophoresis) [4, 11].

in cell's viability and permeability. If the field strength goes over the reversable threshold and exposure period is adequate, reversible electroporation takes place (Fig. 2). In other words the membrane permeabilized for a period of time and can turn to its original state thanks to membrane resealing. Irreversible electroporation, which results in loss of cell homoestosis and death of the cell, only occurs providing that the field strength and amount of delivered energy are excessive [4, 11–13].

4. Applications on wine microbiota

The control of spoilage microorganisms like the yeasts of genus *Dekkera/Brettanomyce/Candida* and the lactic acid bacteria of genus *Lactobacillus* in must, wine and wine contact surfaces is of vital importance for winemakers [9, 14]. Thermal treatments, addition of preservatives like sulferdioxide (SO_2), ascorbic acid, sorbic acid and dimetyhl dicarbonate and microfiltration are the most common techniques tried to control or prevent microbial growth [15, 16].

As an alternative to the common method of SO_2 usage PEF is searched to control and inhibit microbial growth. The effect of PEF or high voltage electrical discharge (HVED) was compared for the control of sweet white must fermentation with SO_2 addition by Delsart et al [17]. As a result the inactivation of total yeasts and non-Saccharomyces yeasts was achieved with the influence of PEF (from 4 kV*cm^{-1} to 20 kV*cm^{-1}; from 0,25 ms to 6 ms) and HVED (40 kV/cm; 1 ms or 4 ms). They reported

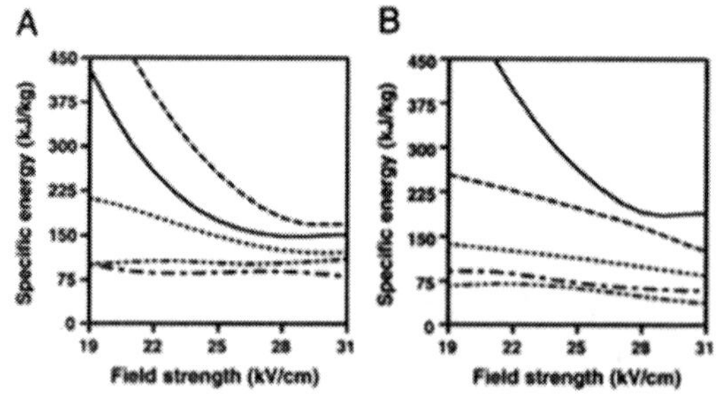

Figure 3. Treatment conditions, field strength and specific enery to achieve 3 log cycles of inactivation in must (A) or wine (B) of red grape, of the different studied microorganisms *D. anomala (. . .), D. bruxellensis (-- --),S. bayanus (-- - -),L. plantarum (—), L. hilgardii (—)* [14].

that the highest inactivation rate with PEF treatment was obtained at E=20 kV/cm, t_{PEF} =4 ms, W=320 kJ*L^{-1}. In addition PEF technology was found more proper for sweet wines than HVED hence it caused no browning problem [17].

Puértolas et al. investigated the lethal effect of PEF on different spoilage microorganisms (*Dekkera/ Brettanomycess and Lactobacillus*) in must and wine. The effect of electrical field strength and the specific energy on inactivation of various microorganisms is illustrated in Fig. 4. The microbial death rate was observed greater during the first period of processing and a progressive decrease was recognised which is referred to as 'tailing behaviour'. Lactic acid bacteria were found more PEF resistant than yeast whichwas explained by their larger size since less intensive field strengths was required to electroporate their cell membranes. Among all microbial species *L. plantarum* was found the most PEF resistant in wine and *L. hilgardii* in must [14]. The higher PEF resistance in wine of *L. plantarum* was related to its high resistance to environmental conditions at raised ethanol concentrations [18].

The efficiency of PEF treatment is reported to be relevant to the temperature since microbial sensitivity rises with temperature [19]. The size, shape, morphological and biochemical features of the cells are incharge of cell resistance to PEF conditions. In addition, other physicochemical factors like pH, alcohol content, water activity, solible solids, electrical conductivity or temperature should also be kept in mind according to sensitivity of the microorganisms [20–23].

PEF technology was also investigated as to the its stimulating effect on microbial flora. It is reported by Mattar et al. (2015) that electroporation was exposed by the advancement of fermentation activity in this way raised yeast metabolism [24]. In accordance with stress response analysis of *S. cerevisiae*, PEF induced expression of the oxidation genes and glutathione played a crucial role in the stres resistance [25]. According to the results of the study performed by Mattar et al. (2015), it is proved that PEF has stimulating influence on synthesis of RNA and ezymes, frequency of cell division events, tolerance to ethanol and fermentation capacity [24].

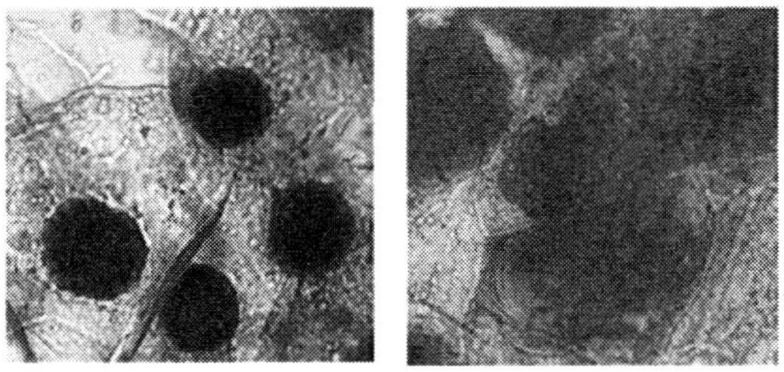

Figure 4. Microscopic photos of peel tissue of Lemberger wine grapes before and after electroporation [29].

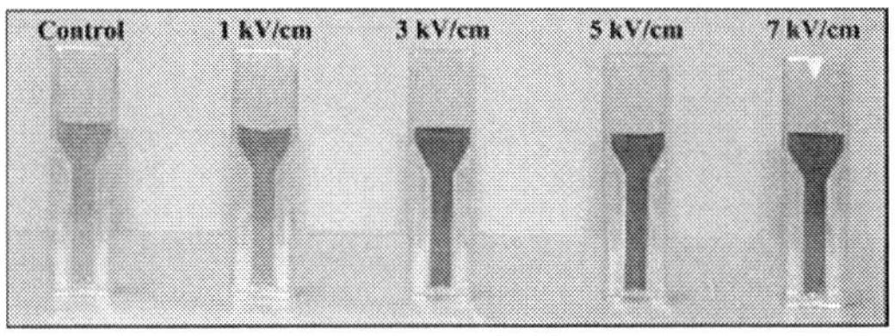

Figure 5. Image of the Garnacha must after 1 h of maceration with grapes untreated and treated by PEF (50 exponential decay pulses; 1–7 kV/cm; 0,4–4,1 kJ/kg) [20].

5. Applications for improvement of oenological characteristics

The extraction of precious substances such as flavouring substances or pigments from wine grapes is of crucial for the wine production industry [26]. The phenolic compounds supply significant organoleptic properties of red wines like colour, bitterness, astringency and mouthfeel. Furthermore via their antimicrobial and antioxidant capacity phenolic substances make essential contribution to aging capacity [27,28]. An adequate extraction is provided by perforating the cells of berry tissue. The novel technique PEF causes formation of pores in the cell membrane via electroporation [29,30]. As illustrated in Fig. 4 after the electric field treatment, cell membrane is charged which causes opening the membranes and vacuoles.

Corrales et al. (2008) reported that the application of PEF treatment (3 kV/cm, 30 pulses, 10 kJ/kg) to the grape skins improved the total phenolic content and anthocyanins in the extraction medium 100% and 17% respectively, compared to the untreated ones [31]. In another study it was proved that phenolic concentration in the extraction medium of Chardonnay grape was doubled after treatment of 1000 μs duration at 1,3 kV/cm for 1 s [32].

The influences of PEF treatment on accelaration and increase of phenolic extraction from grape skins in the course of maceration and fermantation process have been researced extensively. In accordance with the previous studies [3,14,20] performed with a cylindirical batch paralel electrode treatment chamber, with a distance between the electrodes of 1 cm and area of 19,64 cm^2, it has proved that PEF technique enhanced the colour of must in maceration step (Fig. 5).

Effect of PEF treatment during cold maceration and alcoholic fermentation on oenological characteristics of Cabernet Sauvignon red wines have been searched. Moderate and high field strengths (E=0,8 kV/cm and 5 kV/cm) were applied prior and during alcoholic fermentation on red grapes on enhancement of different parameters. The application of PEF utilising moderate strength at different times in the course of cold maceration (0,2, and 4 days) was found more effective during the cold maceration treatment. It was proved that PEF induce membrane electropermeabilization, improving phenolic compounds extraction during cold maceration and raising the resultant colour index [33].

6. Conclusion

In this review, it has been emphasised the innovative technique PEF can be an efficient treatment to improve important onelogical characteristics of wine and to inhibit spoilage microosganisms [20]. In general PEF cause low temperature increase that prevents heat-induced changes in colour, flavour, taste, and nutrient content of food [9,34]. PEF gives rise to local structural changes of cell membrane and degenaration of membrane permeability barrier by means of electroporation. Thus PEF pretreatment can be a promising alternative on microbial inhibition of wine wine without any browning problem. On the other hand the efficiency of the technique is reported to be improved with the utilisation of other electrical pulse shapes, pressurized or degassed wine to avoid electrical discharge problem in gas bubbles and the combination of other technologies with PEF [17].

References

[1] N. Yang, K. Huang, C. Lyu, J. Wang. Pulsed electric field technology in the manufacturing processes of wine, beer, and rice wine: A review. Food Control **61**, 28–38 (2016)

[2] E. Puértolas, N. López, S. Condón, I. Alvarez, and J. Raso J. Potential applications of PEF to improve red wine quality. Trends in Food Science & Technology **21**, 247–255 (2010)

[3] N. López, E. Puértolas, and S. Condon. Application of pulsed electric fields for improving the maceration process during vinification of red wine: influence of grape variety. Eur Food Res Technol. **227**, 1099–1107 (2008)

[4] S. Mahnič-Kalamiza, E. Vorobiev, and D. Miklavčič. Electroporation in food Processing and Biorefinery. J. Membrane Biol. **247**, 1279–1304 (2014)

[5] I. Odriozola-Serrano, I. Aguilo-Aguayo, R. Soliva-Fortuny, and O. Martın-Belloso. Pulsed electric fields processing effects on quality and health-related constituents of plant-based foods. Trends in Food Science&Technology **29**, 98–107 (2013)

[6] S. Y. Leong, D. J. Burritt, and I. Oey. Evaluation of the anthocyanin release and health-promoting properties of Pinot Noir grape juices after pulsed electric fields. Food Chemistry **196**, 833–841 (2016)

[7] R. Soliva-Fortuny, A. Blasa, D. Knorr, and O. Martin-Belloso. Effects of pulsed electric fields on bioactive compounds in foods: a review. Trends in Food Science& Technology **20**, 544–556 (2009)

[8] F. J. Barba, O. Parniakov, S. A. Pereira, A. Wiktor, N. Grimi, N. Boussetta, J. A. Saraiva, J. Raso, O. Martin-Belloso, D. Witrowa-Rajchert. Current applications and new opportunities fort he use of

DOI: 10.1051/bioconf/20170902008

pulsed electric fields in food science and industry. Food Research International **77**, 773–798 (2015)

[9] Q.-X. Ou, M. Nikolic-Jaric, and M. Gänzle. Mechanisms of inactivation of *Candida humilis* and *Saccharomyces cerevisiae* by pulsed electric fields. Bioelectrochemistry **115**, 47–55 (2017)

[10] A. G. Pakhomov, D. Miklavčič, & M. S. Markov, Advanced electroporation techniques in biology and medicine. Boca Raton: CRC Press (2010)

[11] M. Rebersek and D. Miklavčič. Concepts of electroporation pulse generation and overview of electric pulse generators for cell and tissue electroporation. Advanced electroporation techniques in biology and medicine. 323–339 (CRC Press, Boca Raton, 2010)

[12] R. V. Davalos, I. L. M. Mir, B. Rubinsky. Tissue ablation with irreversible electroporation. Ann Biomed Eng. **33**, 223–231 (2005)

[13] P. Y. Phoon, F. G. Galindo, A. Vicente, P. Deimek. Pulsed electric field in combination with vacuum impregnation with trehalose improves the freezing tolerance of spinach leaves. J Food Eng. **88**, 144–148 (2008)

[14] E. Puertolas, N. Lopez, S. Condon, J. Raso, and I. Alvarez. Pulsed electric fields inactivation of wine spoilage yeast and bacteria. International Journal of Food Microbiology **130**, 49–55 (2009)

[15] E. J. Bartowsky. Bacterial spoilage of wine and approaches to minimize it. Lett Appl Microbial **48**(2), 149–156 (2009)

[16] S. Gergely, E. Bekassy-Molnar, and G. Vatai. The use of multiobjective optimization to improve wine filtration. Journal of Food Engineering **58**, 311-316 (2003)

[17] C. Delsart, N. Grimi, N. Boussetta, C. M. Sertier, R. Ghidossi, M. M. Peuchot, and E. Vorobiev. Comparison of the effect of pulsed electric field or high voltage electrical discharge for the control of sweet white must fermentation process with the conventional addition of sulfur dioxide. Food Research International **77**, 718–724 (2015)

[18] E.G. Alegría, I. López, J.I. Ruiz, J. Sáenz, E. Fernández, M. Zarazaga, M. Dizy, C. Torres, F. Ruiz-Larrea High tolerance of wild *Lactobacillus plantarum* and *Oenococcus oeni* strains to lyophilisation and stress environmental conditions of acid pH and ethanol. FEMS Microbiology Letters **230**, 53–61 (2004)

[19] L. Gonzales-Arenzana, J. Portu, R. Lopez, P. Santamaria, T. Garde-Cerdan, and I. Lopez-Alfaro. Inactivation of wine-associated microbiota by continuous pulsed electric field treatments. Innovative Food Science and Emerging Technologies **29**, 187–192 (2015)

[20] E. Puértolas, G. Saldaña, S. Condón, I. Álvarez, and J. Raso. Evolution of polyphenolic compounds in red wine from Cabernet Sauvignon grapes processed by pulsed electric fields during aging in bottle. Food Chemistry **119**, 1063–1070 (2010)

[21] S. Toepfl, V. Heinz, and D. Knorr. High intensity pulsed electric fields applied for food preservation. Chemical Engineering and Processing **46**, 537–546 (2007)

[22] V. Heinz, and D. Knorr. Effect of pH, ethanol addition and high hydrostatic pressure on the inactivation of *Bacillus subtilis* by pulsed electric fields. Innovative Food Science & Emerging Technologies **1**(2), 151–159 (2000)

[23] A. H. El-Hag, S. H. Jayaram, O. R. Gonzales, and M. W. Griffiths. The influence of size and shape of microorganism on pulsed electric field inactivation. IEEE Transactions on Nanobioscience **10**(3), 133–138 (2011)

[24] J. R. Mattar, M. F. Turk, M. Nonus, and N. I. Lebovka. S. cerevisiae fermetation activity after moderate pulsed electric field pre-treatments. Bioelectrochemistry **103**, 92–97 (2015)

[25] T. Tanino, S. Sato, M. Ohshima. Analysis of the stress response of yeast Saccharomyces cerevisae toward pulsed electric field. J. Electrost. **70**, 212–216 (2012)

[26] R. S. Jackson, Chem. Cons. of Gr. and Wine. *Wine Science-Third Edition* 270–331 (San Diego: Academic Press, 2008)

[27] R. B. Boulton, The copigmentation of anthocyanins and its role in the colour of red wine: a critical review. American Journal of Enology and Viticulture **55**, 67–87 (2001)

[28] S. Vidal, L. Francis, A. Noble, M. Kwiatkowski, V. Cheynier & E. Waters. Taste and mouthfeel properties of different types of tannin-like polyphenolic compounds and anthocyanins in wine. Analytica Chimica Acta **513**, 57–65 (2004)

[29] M. Sack, C. Eing, R. Stängle, A. Wolf, G. Müller, J. Sigler, L. Stukenbrock: "Electric Measurement of the Electroporation Efficiency of Mash from Wine Grapes", Trans. DEI **16**(5), 1329–1337 (2009)

[30] M. Sack, J. Sigler, C. Eing, L. Stukenbrock, R. Stängle, A. Wolf, and G. Müller. Operation of an Electroporation Device for Grape Mash IEEE Transactions on Plasma Science **38**(8), 1928–1934 (2010)

[31] M. Corrales, S. Toepfl, P. Butz, D. Knorr, & B. Tauscher. Extraction of anthocyanins from grape by-products assisted by ultrasonics, high hydrostatic pressure or pulsed electric fields: a comparison. Innovative Food Science and Emerging Technologies **9**, 85–91 (2008)

[32] N. Boussetta, N. Lebovka, E. Vorobiev, H. Adenier, C. Bedel- Cloutour, & J. L. Lanoisellé. Electrically assisted extraction of soluble matter from Chardonnay grape skins for polyphenol recovery. Journal of Agricultural and Food Chemistry **57**, 1491–1497 (2009)

[33] N. El Darra, H. N. Rajha, M.-A. Ducasse, M. F. Turk N. Grimi, R. G. Maroun, N. Louka, and E. Vorobiev. Effect of pulsed electric field treatment during cold maceration and alcoholic fermentation on major red wine qualitative and quantitative parameters. Food Chemistry **213**, 352–360 (2016)

[34] A. Angersbach, V. Heinz, and D. Knorr. Effects of pulsed electric fields on cell membranes in real food systems. Innovative Food Science and Emerging Technologies **1**, 135–149 (2000)

 DOI: 10.1051/bioconf/20170902009

Typicité sensorielle de vins de l'AOC Coteaux du Layon et acceptabilité par les acteurs (professionnels et consommateurs) de vins avec des titres d'alcool volumique plus faibles / Sensory typicity of the PDO Coteaux du Layon: Acceptability by the actors (professionals and consumers) of wines with lower alcohol content

Cécile Coulon-Leroy[1], Ronan Symoneaux[1], Philippe Chretien[2], Aurélie Perrin[1], et René Siret[1]

[1] USC 1422 GRAPPE, INRA, Ecole Supérieure d'Agricultures, Univ. Bretagne Loire, SFR 4207 QUASAV, 55 rue Rabelais, Angers, France
[2] Institut Français de la Vigne et du vin, Pôle Val de Loire, Beaucouzé, France

Abstract. In order to maintain an optimum sugar / acid balance in a changing context, winegrowers of the Coteaux du Layon PDO, a sweet wine of the middle Loire valley, wish to know if a final alcohol content (TAV) lower than the minimum imposed by the regulation of the PDO (11% vol. today) would affect the typicality of the wines. To answer this question, a study was carried out in 4 steps. 1) To know the expected typicality of these wines, focus groups were conducted with 26 professionals. 2) In order to assess the acceptability of lower TAV wines, based on a sample of 24 wines (commercial wines and experimental wines with TAV between 8% vol. and 10% vol.), 31 professionals scored levels of exemplarity. 3) A trained panel made a fine sensory description. 4) Finally, a selection of 8 wines from these 24 were evaluated by 163 consumers, they scored levels of preference. The results of steps 1) and 2) show a strong consensus among professionals to assess the exemplarity of wines. Experimental wines obtained exemplary scores above the average of the scale and can therefore be considered as good examples. In step 3), the trained panel generated the sensory attributes to describe the wines: 21 descriptors of odors and aromas and 6 of flavors, which reveal the sensory complexity of the wines. In stage 4), in general, the 8 wines evaluated by the consumers were all appreciated. The results suggest that a sufficiently high amount of residual sugar content would enhance the acceptability of lower TAV wines by both professionals and consumers. The winegrowers of the syndicate of Coteaux du Layon PDO thus have factual elements to update the specification of the PDO (decreasing the minimum TAV threshold), in interaction with INAO (the French national institute of origin and quality).

1. Introduction

Les producteurs de vins de l'AOC Coteaux du Layon sont soumis à un contexte changeant qui les amène à repenser le profil analytique attendu de certains de leurs vins.

1.1. Une concurrence internationale forte

Les vins sous signes de qualité, et notamment les AOC, contribuent fortement à l'économie française en étant le second secteur d'exportation excédentaire après l'aéronautique ou en amenant 10 millions d'oenotouristes en France [1]. Cette production permet le maintien des territoires et paysages et une valorisation satisfaisante pour les vignerons puisqu'ils se basent alors sur une logique de différenciation [2]. Toutefois la concurrence est très présente à l'international avec de plus en plus de zones viticoles revendiquant un lien au terroir.

La revendication d'un lien à l'origine pour les vins, n'est plus une exclusivité des pays producteurs de «l'ancien monde». La concurrence internationale est de plus en plus forte, les vignobles français doivent affirmer leur identité : la typicité, celle-ci est liée à leur origine : le terroir. Les vins de l'AOC Coteaux du Layon n'échappent pas à cette concurrence internationale et à la nécessité de réaffirmer et valoriser leur typicité. Des travaux sur la typicité et le lien avec le terroir ont déjà été entrepris à l'échelle des AOC [3].

1.2. Une production très soumise aux aléas climatiques et à *Botrytis cinerea*

La récolte des raisins destinés à la production de liquoreux est soumise de façon importante aux aléas météorologiques, de plus en plus nombreux. Des pluies trop importantes en automne favorisent le développement de *Botrytis cinerea* sous sa forme pourriture grise qui peut affecter significativement le rendement et la qualité des raisins et ne pas permettre alors d'atteindre, naturellement, le minimum de sucre requis par le cahier des charges de l'AOC. La chaptalisation pour atteindre, si besoin, ce minimum requis, tend, par ailleurs, à être de plus en plus abandonnée.

La forme pourriture noble de *Botrytis cinerea*, quant à elle, produit un antibiotique, la botryticine, qui tend à arrêter naturellement les fermentations. Cela se produit parfois avant le taux minimum d'alcool requis. Il est alors nécessaire de relancer artificiellement la fermentation même si l'équilibre du vin est satisfaisant pour attendre le minimum d'alcool réglementaire.

1.3. Une production actuelle difficilement durable

Selon l'indice de Huglin, la région viticole du Val de Loire a évolué d'un climat frais à un climat tempéré, avec une évolution de +1.6 °C en 50 ans [4]. Ce réchauffement a conduit à une avancée des stades phénologiques et à une évolution importante de la composition des baies. Une diminution de l'acidité des raisins est aussi constatée. Les vins doivent alors tendre à être un peu moins sucrés pour maintenir des équilibres sucres / acides optimaux.

Les tries successives nécessaires pour produire des liquoreux concentrés en sucre, couplés aux faibles rendements, amènent, de plus, à des coûts de production élevés et non rentables pour certaines exploitations.

Par ailleurs, les consommateurs tendent à préférer des produits plus faciles à boire pour une consommation plus fréquente, contrairement aux vins AOC Coteaux du Layon historiquement très concentrés et réservés à quelques occasions particulières.

1.4. Une étude de la typicité des vins de l'AOC Coteaux du Layon pour évoluer vers une segmentation des vins

Une des dimensions clés de la qualité d'un vin AOC est sa «typicité», qui est liée à son origine géographique, son cépage et les techniques de vinification [5]. Elle correspond à la propriété d'appartenance à un type, la typicité liée au terroir est une construction particulière qui concrétise l'effet du terroir pour un produit donné [6]. Elle peut être associée aux valeurs d'origine, d'authenticité et de tradition auxquelles sont de plus en en plus attachés certains consommateurs.

Dans le cadre de ce projet, le focus a été fait sur la typicité sensorielle et non sur la typicité au sens large. La typicité peut être théorique [7] suivant l'appartenance à une AOC ; conceptuelle suivant l'image que se font les connaisseurs du produit ou perceptuelle suite à une dégustation [3]. Cette dernière peut être qualifiée de typicité sensorielle.

Les vignerons de l'AOC des Coteaux du Layon considèrent que l'adaptation au contexte changeant passe par une segmentation des produits :

- Des vins concentrés et riches, avec des arômes propres aux vendanges tardives, avec un travail important du vigneron (plusieurs tris, sélection de cuvée) avec un prix de valorisation plus important que la moyenne de l'AOC. Ces vins sont notamment produits par les vignerons proches de la source du Layon, au sud du département qui ne bénéficie pas de parcelles sur les zones des AOC Coteaux du Layon Villages, Chaume ou Bonnezeaux.

- Des vins plus légers et vifs avec des arômes de fruits frais, aux coûts de production inférieurs, le prix de valorisation étant plus faible que la précédente catégorie. Ces vins permettent une part de risque plus faible (une fois les maturités avancées, il est difficile de «revenir en arrière» en cas d'aléas), facteur essentiel pour une pérennité des exploitations. Ils sont par ailleurs de plus en plus demandés par les clients. Les raisins sont récoltés à un titre d'alcool potientiel (TAP) plus faible.

Pour cette seconde catégorie, les vignerons de l'AOC Coteaux du Layon, souhaitent savoir si, pour ces cuvées à plus faible TAP, un TAV (Titre Alcoométrique Volumique) plus faible que le minimum imposé par la réglementation du cahier des charges de l'AOC (qui est de 11%vol. aujourd'hui) affecterait la typicité des vins.

2. Matériel et méthode

L'objectif global des travaux présentés visait à comprendre comment les professionnels de la filière et les consommateurs acceptent des vins de l'AOC Coteaux du Layon ayant un équilibre 'alcool acquis / sucres résiduels' ne correspondant pas à celui actuellement autorisé par le cahier des charges de l'AOC. Il s'agissait de vérifier si ces nouveaux vins candidats à l'AOC étaient définis tout de même comme de bons exemples de vins AOC Coteaux du Layon par les professionnels de la filière, si leurs caractéristiques sensorielles correspondaient à celles qui était attendues et s'ils n'étaient pas rejetés par les consommateurs. L'étude a été menée entre 2015 et 2017.

2.1. Les vins de l'étude

La sélection des vins de l'étude avait pour objectif de couvrir l'ensemble des profils sensoriels des vins de l'AOC Coteaux du Layon mais se trouvant dans les limites basses du cahier des charges en termes de taux d'alcool. La sélection a été réalisée par le Syndicat de l'AOC Coteaux du Layon. Les critères se sont portés sur des caractéristiques analytiques et techniques (TAV, quantité de sucres résiduels, millésime) et aromatiques. 16 vins commerciaux ont été sélectionnés : 14 de l'AOC Coteaux du Layon, notés «Layon» et 2 hors de l'AOC Coteaux du Layon, mais ayant des caractéristiques proches (un vin blanc de France et un de l'AOC Anjou blanc) notés «NonLayon». L'ensemble des vins sont exclusivement issus du cépage Chenin.

En complément, 8 vins expérimentaux ont été élaborés par l'IFV. Ces vins, notés «Expé», ont un TAV inférieur à ce qui est actuellement autorisé dans le cahier des charges. Ces vins ont été élaborés à partir de raisins de Chenin récoltés dans l'aire de l'AOC Coteaux du Layon avec une concentration en sucre moins importante. Certains vins ont été dilués à l'eau distillée pour atteindre certains TAV fixés.

Dans un second temps, une sous-sélection de 8 vins, 4 vins commerciaux et 4 vins expérimentaux, a été réalisée sur la base des caractérisations sensorielles des 24 vins faites par le panel entrainé et les professionnels. Ces vins ont été évalués par les consommateurs et ont servis de supports de discussion avec les professionnels lors de tables-rondes. Les critères de sélection des 8 vins ont été d'éliminer les vins avec défauts (comme un goût de bouchon), de conserver une large diversité de produits, de focaliser plus strictement sur les limites basses du cahier des charges au niveau du taux d'alcool et de

 DOI: 10.1051/bioconf/20170902009

conserver une répartition équilibrée de vins commerciaux et expérimentaux.

2.2. L'évaluation de la typicité par les connaisseurs et porteurs de mémoire de l'AOC Coteaux du Layon

La **typicité sensorielle conceptuelle** a été évaluée lors de tables-rondes par 26 professionnels de l'AOC Coteaux du Layon (producteurs, œnologues, techniciens et personnes formées aux contrôles organoleptiques).

La **typicité sensorielle perceptuelle** a été étudiée via la notion d'exemplarité lors de dégustations. Cette méthode a été mise au point par Ballester et al. [8] et utilisée par la suite par plusieurs auteurs [7,9,10]. Dans ce cas, l'échelle d'évaluation va de «mauvais exemple» à «bon exemple»

Le test d'exemplarité, sur les 24 vins, a été réalisé avec 31 professionnels de la filière viticole de la région Anjou, connaisseurs des vins de l'AOC Coteaux du Layon dont 18 producteurs, 9 œnologues, 1 ingénieur viticole et 3 personnes formées aux contrôles organoleptiques (25 hommes et 6 femmes, âge moyen de 46 ans). L'objectif était ici de d'évaluer le niveau d'exemplarité des vins de l'étude, d'identifier s'il existait un consensus concernant la typicité des vins AOC Coteaux du Layon et également d'évaluer les caractéristiques qui permettent de les identifier.

Une Analyse en Composantes Principales (ACP) normée aves les juges comme variables et les vins comme individus a été réalisée pour évaluer le consensus entre les juges. La typicité moyenne a également été projetée en variable supplémentaire.

Les professionnels ont ainsi été mobilisés en deux temps : dans un premier temps, lors de dégustation pour réaliser le test d'exemplarité et évaluer les caractéristiques sensorielles des 24 vins ; et dans un second temps, pour participer à des tables-rondes couplant des discussions et des dégustations des 8 vins. Ces deux approches complémentaires ont permis d'obtenir d'une part des informations quantitatives, et d'autre part des informations plus qualitatives.

2.3. La caractérisation sensorielle des vins par un panel entrainé

Le profil conventionnel a été choisi comme étant la méthode la plus appropriée pour l'étude. En effet, une description précise des 24 vins était requise pour pouvoir les différencier. Le panel était composé de 15 personnes âgées de 42 à 65 ans, dont 9 femmes et 6 hommes. Les séances d'évaluation ont eu lieu au sein de la salle d'analyse sensorielle de l'USC Grappe, INRA, ESA. Cette salle est constituée de 20 box standardisés selon la norme AFNOR NF09-105 et équipée du logiciel FIZZ. Le panel a été formé à l'évaluation des vins de l'AOC Coteaux du Layon en s'entrainant sur des références olfactives et gustatives spécifiques aux moelleux et liquoreux.

Dans un premier temps la liste des descripteurs sensoriels a été établie à partir de la littérature existante sur les liquoreux et les vins de l'AOC Coteaux du Layon. Cette liste a ensuite été ajustée par le panel, par consensus lors des entrainements réalisés sur les vins de l'étude. La liste finale contenait 50 termes dont 6 descripteurs de saveurs ou

Tableau 1. Liste des 50 descripteurs sélectionnés par le panel entrainé et P-value associées pour les descripteurs discriminants (>10%).

		Descripteurs	*P-Value* *Profil des 24 vins*		*P-Value* *Profil des 8 vins*	
Saveurs et sensations		Acide	0,000		0,000	
		Sucre	0,000		0,000	
		Amer	0,000		0,000	
		Astringent	0,032			
		Perlant	0,000			
		Brûlant	0,000		0,000	
	Famille		**Odeur**	**Arôme**	**Odeur**	**Arôme**
Odeurs / Arômes (évalués séparemment)	Intensité	Intensité odorante	0,027	0,000	0,086	
	Fruité	Fruits jaunes				
		Agrumes		0,000	0,023	
		Fruits blancs	0,028	0,074	0,013	0,040
		Coing	0,026	0,064	0,013	
		Fruits cuits	0,053	0,012		
		Fruits exotiques		0,066		
		Fruits rouges et noirs				
		Pomme			0,014	
	Animal	Animal	0,000		0,000	
	Végétal	Foin				
		Herbe coupée				
	Fermentaire	Beurre				
		Fermenté	0,002	0,000	0,000	0,000
	Caramélisé	Caramel				
		Miel		0,001		0,001
	Epices	Epices		0,015		
	Empyreumatique	Boisé		0,069		0,012
		Empyreumatique	0,049	0,035		0,051
	Floral	Floral	0,041			0,077
	Alcool	Alcool	0,015	0,000		0,000
	Terreux	Moisi terreux			0,004	0,075

sensations et 22 couples d'arômes et odeurs (Tableau 1). Un entrainement spécifique à la reconnaissance de ces descripteurs a ensuite été réalisé.

Des analyses de la variance (ANOVA) ont été réalisées avec un modèle mixte à trois facteurs. Le facteur «juge» est considéré comme un effet aléatoire, les facteurs «produit» et «répétition» comme des effets fixes. Le modèle prend en compte les interactions entre les facteurs.

Une ACP a été réalisée à partir des descripteurs sensoriels significatifs à 10 % afin d'obtenir une représentation multidimensionnelle des produits en résumant la variabilité entre les vins. Le tableau des moyennes sur le panel a été utilisé et les données ont été normées.

Les 24 vins ont été regroupés par une Classification Ascendante Hiérarchique en prenant en compte l'ensemble des dimensions de l'ACP, six groupes ont été pris en compte pour réaliser la sélection des 8 vins évalués par les consommateurs.

2.4. L'évaluation de l'acceptabilité des consommateurs

L'objectif de cette phase de l'étude était d'obtenir la perception des consommateurs sur les vins. Un test hédonique a été réalisé sur la sous-sélection de 8 vins afin de connaitre leur appréciation. Chaque consommateur devait noter le produit sur une échelle discontinue à 9 points allant de «extrêmement désagréable» à «extrêmement agréable». Il leur était également demandé de s'exprimer sur les qualités et défauts de chaque vin.

Ces tests ont été réalisés auprès de 163 consommateurs, à Angers et Lille, avec un nombre d'individus équilibré selon 3 critères : sexe, âge et ville. Le test s'est déroulé selon le même protocole dans les deux villes, dans des salles d'analyse sensorielle équipées du logiciel FIZZ.

Les vins ont été présentés aux consommateurs de façon monadique séquentielle et selon un plan d'expérience basé sur un carré latin de Williams.

DOI: 10.1051/bioconf/20170902009

Figure 1. Citations des mots utilisés individuellement par les 26 professionnels, lors des tables-rondes, pour décrire un vin d'AOC Coteaux du Layon (la taille est proportionnelle à la fréquence de citation).

3. Résultats et discussion

3.1. La typicité conceptuelle des vins de l'AOC Coteaux du Layon par les professionnels

Le cahier des charges en vigueur (BO 26 juin 2014) indique : «Les vins sont racés et élégants. En bouche, ils allient la suavité à la fraîcheur, la puissance à la finesse et leurs arômes sont complexes. Aptes à un vieillissement prolongé, les vins font le bonheur des dégustateurs avertis». Les vins sont également décrits par l'interprofession comme ayant «une robe claire à profonde, jaune or aux reflets verts, évoluant vers un vieil or aux reflets ambrés ; le nez à des arômes de miel, acacia, fleur, citronnelle, ananas, abricot, coing, fruits frais pour les vins jeunes et notes de fruits confits sur les vins concentrés ou plus évolués ; sa bouche est ronde, onctueuse, équilibrée, franche avec de la fraicheur et vivacité en finale».

Lors des tables-rondes, les Coteaux du Layon ont été décrits individuellement par les 26 professionnels présents au total avec les 3 séances. Les 8 descripteurs qui ont eu les fréquences de citation les plus élevées (>20 %) pour caractériser un vin de l'AOC Coteaux du Layon sont : «fruité» (26 citations), «acidulé» (14 citations), «fraicheur» (12 citations), «moelleux» (9 citations), «gras» (9 citations), «équilibré» (8 citations), «fruits confits» (7 citations) et «jaune or» (7 citations). On retrouve quelques descripteurs du cahier des charges. Ces descriptions sont à relier à la notion de typicité conceptuelle [3].

Les professionnels ajoutent lors des discussions que les vins de l'AOC Coteaux du Layon les plus légers sont reconnaissables à leurs arômes de fruits frais alors que les notes de fruits secs apparaissent, elles, avec des vins plus concentrés. Les professionnels expliquent que les vins AOC Coteaux du Layon sont caractérisés par une certaine vivacité, apportée par le cépage Chenin qui les différencie par rapport aux autres liquoreux. D'ailleurs, l'équilibre est un mot fréquemment cité et notamment l'équilibre entre la quantité de sucre et l'acidité. Les termes «moelleux», «gras» sont également des mots qui reviennent fréquemment. Au niveau aromatique, ce sont des vins pouvant présenter des notes florales, et dont la couleur est jaune/dorée.

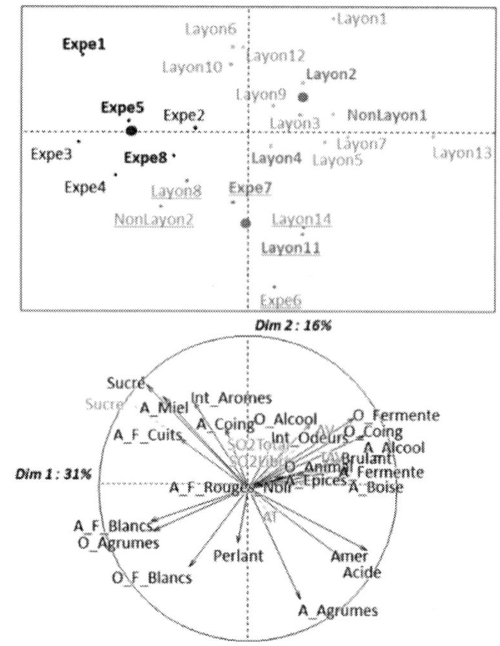

Figure 2. Analyse en Composantes Principales réalisés sur les 24 vins (Expe : vins expérimentaux) et 27 descripteurs discriminants les vins (au risque d'erreur de 10 %, F : fruits, Int : intensité). Les variables en bleu n'ont pas été prises en compte dans l'analyse (AT : acidité totale, AV : acidité volatile). La première composante (abscisse) représente 33 % de l'information et la seconde (ordonné) 15 %. Les groupes en noir, gris et gris souligné correspondent aux groupes optimaux obtenus suite à une Classification Ascendante Hiérarchique (critère de Ward).

3.2. Les caractéristiques sensorielles des vins de l'AOC Coteaux du Layon par le panel entrainé

Sur les 50 descripteurs utilisés, 27 permettent de différencier les 24 vins de l'étude (Tableau 1).

Un groupe est uniquement constitué de vins expérimentaux (à gauche de la Fig. 2). Ces 6 vins se caractérisent par une saveur et une quantité de sucre élevée, ainsi qu'un arôme de «miel» et une odeur de «fruits blancs». Ces vins sont aussi peu alcoolisés. Le groupe en bas du graphique Fig. 2 regroupe 2 vins expérimentaux ainsi que 4 vins commerciaux. Ils se différencient des autres par leurs arômes «agrumes» et une faible saveur «sucré». Enfin, le groupe de droite est composée de 12 vins commerciaux dont un des vins commerciaux qui n'appartient pas à l'AOC. Ces vins se caractérisent par un arôme et une odeur «alcool» qui est associé à un TAV élevé, mais aussi des odeurs «animal» et «coing». De plus, ils présentent une odeur et un arôme «fermenté».

Les 8 vins évalués par les consommateurs ont été à évalués par le panel entrainé au même moment (une année s'étant écoulée, il était nécessaire de revérifier les caractéristiques sensorielles). 6 descripteurs ne sont pas discriminants ni pour le profil des 24 vins, ni pour celui des 8 vins ; avec un taux d'erreur de 10 % (Tableau 1). Globalement les caractérisations sensorielles sont proches.

DOI: 10.1051/bioconf/20170902009

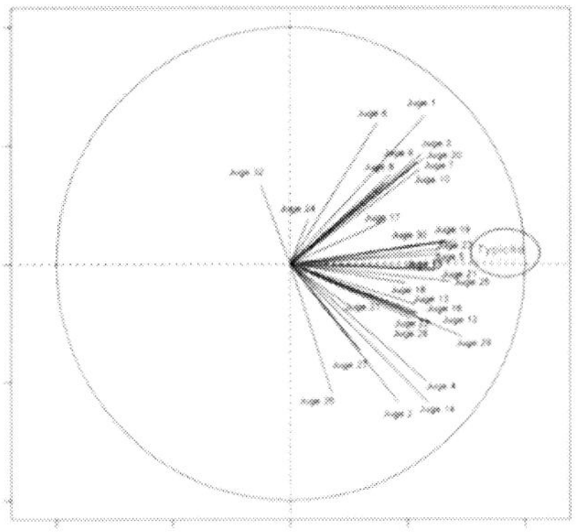

Figure 3. Analyse en Composantes Principales sur les notes d'exemplarité des professionnels. La première composante (abscisse) représente 27 % de l'information et la seconde (ordonné) 12 %.

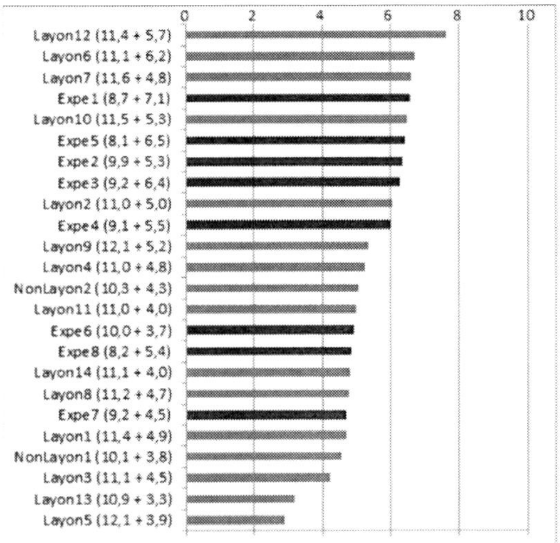

Figure 4. Notes moyennes d'exemplarité et (TAV et taux de sucre en %vol) par produit.

3.3. La typicitéperceptuelle des vins de l'AOC Coteaux du Layon par les professionnels de la filière

L'ACP, avec les juges comme variables, montre qu'il y avait un fort consensus entre les juges (positionnement des juges/variables dans le même demi-cadran) (Fig. 3). Cela nous amène à confirmer qu'il existerait bien une typicité sensorielle des vins de l'AOC Coteaux du Layon, même si cette appellation a une certaine ancienneté et présente une certaine diversité.

La moyenne générale des notes de typicité obtenue est de 5,4 ± 0,70. Deux paliers sont visibles sur la Fig. 5 : un premier après le vin Layon13 et un second après le vin Layon9. Bien que ne respectant pas le cahier des charges de l'AOC, des vins expérimentaux ont des notes d'exemplarité supérieures à la moyenne de l'échelle et peuvent donc être considérés comme de bons exemples. Les vins expérimentaux Expe1 et Expe5 sont proches des vins avec les plus fortes notes moyennes d'exemplarité. Ce résultat montre que des vins à TAV faible peuvent avoir

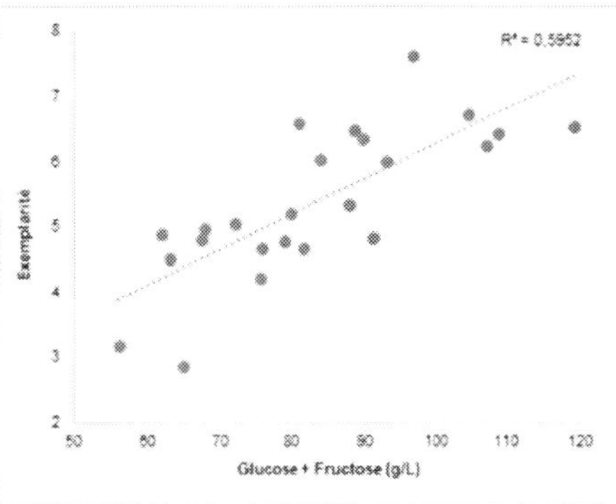

Figure 5. Relation entre les notes d'exemplarité et les quantités de sucres résiduels.

les caractéristiques sensorielles correspondant à la typicité sensorielle attendue des vins de l'AOC Coteaux du Layon.

Parmi les vins les plus typiques, plusieurs vins sont des vins décrits comme des vins aux arômes de «miel», de «fruits exotiques», «sucrés», et aux odeurs de «fruits blancs» selon le panel (Fig. 2). Ces caractéristiques correspondent à celles liées à la typicité conceptuelle.

La note d'exemplarité est fortement liée à la quantité de sucres résiduels (Fig. 4). Alors que le TAV semble, lui, jouer un rôle moindre. Ce résultat montre que les professionnels ont tendance à privilégier le sucre comme facteur de l'exemplarité. Ceci explique également pourquoi les vins Expe6 et Expe7 ont des notes d'exemplarité inférieures aux autres vins expérimentaux, tous deux ont, en comparaison, peu de sucres résiduels par rapport aux autres produits expérimentaux.

Les vins Expe1 et Expe5 sont des vins jugés plus majoritairement que les autres vins comme de «très bons exemples» de l'AOC Coteaux du Layon (Figs. 4 et 7). Les professionnels ont décrit le vin Expe1 comme étant un vin «aromatique» et «concentré», et le vin Expe5 comme un vin «frais» et «équilibré». Le vin Expe5 a cependant était jugé comme étant trop «sucré». Le vin Expe8 est un vin noté comme étant un «bon exemple» par les professionnels et comme étant «frais» et «équilibré»; le vin Expe7 comme un «moins bon exemple» car pas assez «acide». Les vins Layon2 et NonLayon1 ont, eux, été jugés comme de «mauvais exemples». Le vin NonLayon1 est décrit comme étant «plat», «acide» et pas assez «sucré» et le vin Layon2 comme ayant une bonne odeur mais «soufré» et «évolué». Le produit Layon2 crée des désaccords entre professionnels. Il regroupe, en effet, le troisième plus grand nombre de «très bon exemple» après les produits Expe1 et Expe5, mais est aussi cité le plus grand nombre de fois comme «mauvais exemple». Le terme «évolué» a d'ailleurs été utilisé en terme positif ou négatif.

3.4. L'acceptabilitépar les consommateurs des vins avec de nouveaux profils analytiques

Les consommateurs ont d'une manière générale apprécié les vins qui leurs étaient proposés, toutes les moyennes sont supérieures à 5 (Fig. 7). De plus, tous les vins ont été notés comme étant «agréables» par plus de 50 % des consommateurs. Les trois vins les plus appréciés par les consommateurs sont : Expe1, Expe5 et Expe8, avec plus

 DOI: 10.1051/bioconf/20170902009

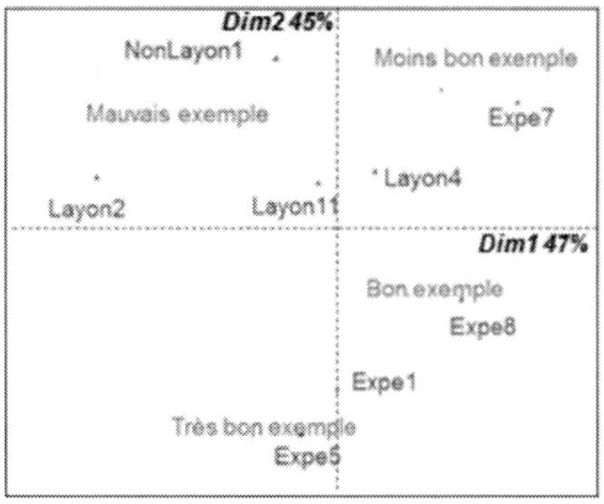

Figure 6. Analyse Factorielle des Correspondances sur la séparation des produits par les professionnels selon leur niveau d'exemplarité. La première dimension (abscisse) représente 47 % de l'information et la seconde (ordonné) 45 %.

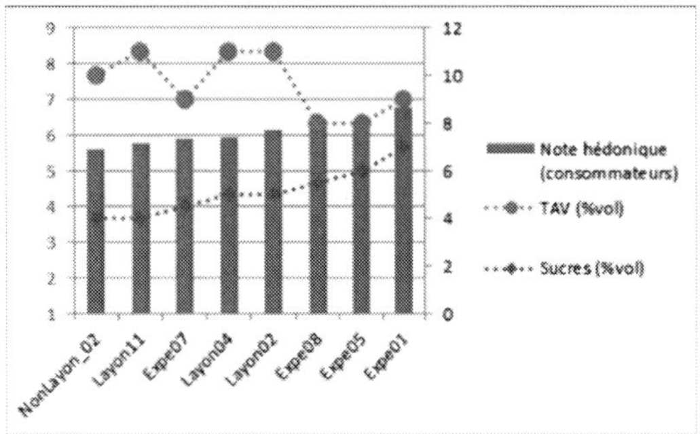

Figure 7. Moyenne de l'appréciation des consommateurs pour chaque vin et niveaux des Titres d'Alcool Volumique (TAV en %vol.) et des quantités de sucres résiduels (exprimées en %vol).

de 85 % des consommateurs qui ont considéré le vin Expe1 comme agréable. Ces trois vins sont ceux ayant le plus de sucres résiduels. Cette corrélation positive entre la moyenne obtenue et le taux de sucre résiduels est d'ailleurs confirmé Fig. 7. Le taux d'alcool semble, lui, ne pas jouer de rôle dans la notation des produits par les consommateurs.

4. Conclusion

Le cahier des charges d'une AOC permet de protéger le savoir-faire des vignerons d'un territoire. Une évolution du cahier des charges ne doit pas amener à modifier le produit. L'aspect innovant de la démarche était d'intégrer l'appréciation du consommateur à l'étude de la typicité mais aussi de coupler des dégustations individuelles, avec les professionnels, pour évaluer et caractériser la typicité, comme proposé par Cadot et al. [3], avec des tables-rondes pour pouvoir échanger sur les résultats et avoir alors des commentaires complémentaires.

Les résultats montrent qu'en considérant un TAV minimal de 10 % vol (soit -1 % vol. par rapport au seuil actuel), des équilibres «alcool/sucre» de 10+4 ou 10+5 sont des équilibres qui semblent conformes à la typicité de l'AOC Coteaux du Layon et à l'appréciation des consommateurs. L'étude montre également que la composante «sucré», et sans doute les équilibres «acide-sucre», ainsi que la composante aromatique sont probablement plus importants pour la construction de l'exemplarité des vins de l'AOC Coteaux du layon que la seule quantité d'alcool présente au final dans le vin. D'ailleurs, il semble également qu'une majorité de consommateurs ne soit pas très sensible à cette question de la quantité d'alcool dans ce type de vins.

La notion d'équilibre «sucre/acide» a été évoquée lors des 3 tables-rondes. Les résultats mettent en avant l'importance de rester avec une quantité de sucre résiduelle assez importante.

Les résultats de l'étude vont permettent aux vignerons du syndicat des Coteaux du Layon en concertation avec la fédération viticole Anjou-Saumur, d'avoir des éléments factuels sur lesquels se reposer pour envisager de faire une demande d'actualisation du cahier des charges en faveur d'un abaissement du TAV minimum, auprès de l'INAO.

Une étude complémentaire sur la définition d'itinéraires techniques permettant d'obtenir et de reproduire des vins avec ces nouveaux équilibres peut aussi être envisagée. En perspectives, il serait aussi pertinent de focaliser une étude sur la notion d'équilibre avec un plan d'expérimentation adapté pour étudier cette question (ce n'était pas l'objectif ici).

Ces travaux ont reçu le soutien financier d'Interloire et du ministère chargé de l'agriculture (CASDAR). La responsabilité du ministère chargé de l'agriculture ne saurait être engagée.

Références

[1] `http://www.vinetsociete.fr/mediatheque#publications`, consulté le 5 avril 2016
[2] P. Colonna, S. Fournier, J.-M Touzard, Chapitre 4. Système alimentaire. Durabilité de l'alimentation face à de nouveaux enjeux. Questions à la recherche (INRA-CIRAD, 2011)
[3] Y. Cadot, S. Caillé, A. Samson, G. Barbeau, V. Cheynier Anal. Chim. Acta **660**, 1–2 (2010)
[4] E. Neethling, G. Barbeau, C. Bonnefoy, H. Quénol, Climate Res **53** (2012)
[5] F. Sauvageot, Sci. aliments **14** (1994)
[6] F. Casabianca, B. Sylvander, Y. Noël, C. Beranger, J.-B. Coulon, G. Giraud, G. Flutet, F. Roncin, E. Vincent In VIth International Terroir Congress (2006)
[7] L. Perrin, J. Pages, J. Sens. Stud. **24**, 5 (2009)
[8] J. Ballester, C. Dacremont, Y.L. Fur, P. Etiévant, Food Qual. Prefer. **16**, 4 (2005)
[9] I. Maitre, R. Symoneaux, F. Jourjon, E. Mehinagic, Food Qual. Prefer. **21**, 7 (2010)
[10] A. Loison-Bou Maroun, R. Symoneaux, P. Deneulin, T. Thomas-Danguin, C. Fant, L. Guérin, Y. Le Fur, Food Qual. Prefer **40**, A (2015)

An assessment of potential applications with pulsed electric field in wines

Foteini Drosou, Enoch Yang, Marina Marinea, Euthalia G. Dourtoglou, Archodoula Chatzilazarou, and Vassilis G. Dourtoglou[a]

Department of Oenology and Beverage Technology, Faculty of Food Technology and Nutrition, Technological Educational Institute of Athens, 12 Ag. Spyridonos St., Egaleo, 122 10 Athens, Greece

Abstract. Pulsed electric fields (PEF) is a non-thermal processing technology that uses instantaneous, pulses of high voltage for a short period in the range of milliseconds to microseconds; the application of high intensity electric field on toasted wood chips leads to a quick diffusion of extractable molecules. Currently most PEF studies, in the field of oenology, have been focusing on the application of PEF as a pretreatment of grape musts by examining the microbial inactivation and the enhancement of polyphenol extraction. In this study a post-treatment of wine is introduced as method to enhance the wood flavor in the wine with a green noninvasive technology. Major phenolic aldehydes that have been identified as the characteristic compounds of oak volatile compounds were selected as markers and were analyzed instrumentally to compare the influence of PEF processing to non-treated samples. PEF treated samples brought about higher concentrations of the examined oak compounds in the samples treated with PEF, which may explain the advantages of its application. The modulation of the intensity of the electric field and the period of pulses influenced the concentrations of the volatile phenols that were leached out. Differences found between the assayed treatments indicate that PEF application could be a potential practice for a rapid extraction of volatile compounds from oak.

1. Introduction

The aging process of wine, in which wine is in direct contact with wood for a specific time, is important for its organoleptic characteristics. During this process, small amounts of air insert into the barrel from its pores. Wine should have good structure and body, together with antioxidant capacity, in order to equilibrate the negative effect of the oxygen (Ortega-Heras et al. 2004).

Although aging conditions, such as the wine cellar temperature, humidity and the dwelling time in barrels affect the characteristic of wine (Arapitsas et al. 2004; Dubois, 1989) the most important factor is the raw material of the oak barrel and its treatment (Arapitsas et al., 2004; Miller & Howell, 1992), as these two factors determine compounds extracted on wine. The traditional way of aging wine in barrels acquires a lot of time, usually a few years, depending on the style of wine.

At present, for cost-effective reasons, alternatives to the oak barrel are being looked at to carry out the wine–ageing process. Therefore, alternative forms have been tested in the past few years to expedite the extraction of aroma compounds from the oak. One of these techniques consists of adding oak chips to wine in order to give a woody aroma (Garde-Cerdan et al., 2006).

Extraction of oak wood compounds into wine is of great importance since they can improve the wine's aroma and mouth feel. Major compounds that are degraded from wood are aromatic aldehydes, also known as volatile oak phenolics. Major degradation product of lignin is vanillin which gives wine a typical vanilla flavor. Other extracted aromatic aldehydes are: syringaldehyde, coniferyladehyde, sinapaldehyde, and furfural (Margalit 2010).

Furfural derives from the degradation of monosaccharides produced by partial hydrolysis of hemicelluloses and celluse, together with Maillard reactions which take place during the interaction with oak (Arapitsas et al. 2004; Ortega-Heras et al., 2006). Oak lactone originates from oak lipids and its odor attributes a woody and coconut character (Arapitsas et al. 2004). Guaiacol is a product of lignin thermal decomposition, and has a smoky flavor.

As the demand of wine market is increasing, new and economically effective methods are conducted to accelerate the extraction of oak aroma compounds (Zhang et al., 2012). Some of the new methods that have been introduced to accelerate wine aging process are ultrasonic waves, gamma irradiation, and electric fields, including pulsed electiric field (PEF) (Zhang et al., 2012).

Pulsed electric field relies on the use of high intensity pulsed electric fields for cell membrane disruption where induced electric fields perforate microbial membranes by electroporation. The high intensity pulsed electric filed processing system is a simple electrical system consisting of a high voltage source, a capacitor bank, a switch, and a treatment chamber. From the electrical point of view, the PEF treatment chamber represents the electrical load consisting of two or more electrodes filled with the liquid substance to be treated. The chamber has to be constructed in such a way that the electrical field acting on the liquid is more or less homogeneous across the entire active region.

[a] e-mail: vdourt@teiath.gr

This goal can be reached in principle with planer, coaxial and axial electrode geometries (Loeffler, 2006).

The application of PEF is primarily used as a non-thermal processing for the inactivation of microorganisms, with the potential of being an alternative for pasteurization of liquid foods (Grahl et al., 1996; Álvarez et al. 2003). Several researchers have introduced the electric field treatment as an effective method to accelerate young wine aging (Zeng et al., 2008). Zhang et al. (2012) examined the influence of PEF on brandy aging in oak barrels and their results showed extraction enhancements of tannins, total phenols, and volatile phenols from the wood.

The study of Zeng et al, (2008) attempted the application of AC high voltage electric field to accelerate wine aging, by examining physiochemical attributes as well as sensory evaluation. Also, Corrales et al. (2008) drew the comparison between ultrasonics and pulsed electric field in the extractability of anthocyannins from grape by-products. Further researches have been conducted in wine processed with pulsed electric fields.

In the traditional way of aging, it has been observed that the bouquet of red and white wines develop after a short period in the barrel. During the aging of red wine various components of color, aroma and flavor develop. The color changes to deeper red and the flavor evolves becoming softer, with less astringency. The rate at which these changes occur is different for each wine, depending on both outside conditions and the wine's specific composition such as oxidative phenomena (O_2 and SO_2), temperature and time and its phenol composition (Ribereau-Gayon et al., 2006).

This study examines the extractability of oak volatiles in wine combined with PEF treatment, in order to expedite the aging process. Oak chips were added to wine samples and were treated with pulsed electric fields in different conditions of time (t_i = 200 ms and 5 sec) and voltage (E=7.7 and 1.2 kV/cm) to examine its effect in the extraction of volatile compounds compared to untreated wine samples. Moreover, organoleptic characteristics are examined of the treated and untreated wine samples in order to conclude whether PEF treatment leads to faster ageing than the traditional barrel ageing.

2. Material and methods

2.1. PEF equipment

The PEF equipment used was a static bench scale system. PEF equipments is consisted of a high voltage power generator (Eisco, India), a digital oscilloscope (model UTD 2062C, ELV Elektronik AG, Germany) and a pulse generator (model UPG 100, ELV Elektronik AG, Germany).

The PEF generator used provides maximum voltage and current 5 kV and 0.8 kV respectively. This generator provides pulses of monopolar rectangular shape. Signals of voltage, current, frequency, and pulse waveform were monitored by the digital oscilloscope.

The treatment chamber (Val-Electronic, Greece) is consisted of three coaxial geometries, as shown in Fig. 1. A bronze cylinder 5 mm thick, 125 mm height and 28 mm diameter. A U shaped cylinder, where the liquid is filled in, 20 mm diameter and 130 mm height, which is inserted in the bronze cylinder. And a stainless steel electrode of

a)
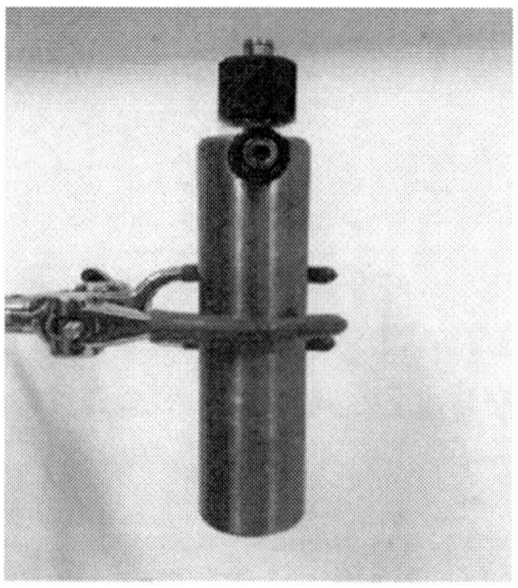

b)
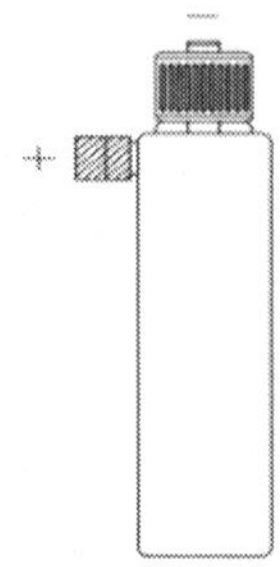

Figure 1. a) PEF treatment chamber, b) schematic aspect of PEF treatment chamber.

5 mm diameter and 120 mm height which screws to the U shaped cylinder. Electric field strength E is evaluated as $E = U/d$, where U is the applied voltage and d is the distance between the electrode and the bronze cylinder, where $d = 0,65$ cm.

For each case the time of treatment is calculated as

$$t = (t_i + t_p) \times P \text{ in (s)}$$

t_i=the pulse duration (s)
t_p=pause time (s)
P=number of pulses.

The treatment conditions used in the experiments are:

- E=7.7 kV/cm, t=1500 s or 25 min (t_i=5 s, t_p=5 s, P=150)
- E=1.2 Kv/cm, t=1800 s or 30 min (t_i=200 ms, t_p=200 ms, infinite pulses).

The detailed treatment procedure is as follows: For each sample 0.2 gr of toasted oak chips were added to 25 ml wine sample in the U shaped cylinder. For the PEF treated samples a voltage of 5 kV and 0.8 kV was applied between the electrodes for 25 minutes.

Under the same experimental procedure and condition, all wine samples treated with 0 kV/cm were used as control for comparison.

2.2. GC/MS analysis

2.2.1. *Ultrasound-assisted extraction*

All samples were immediately Ultrasound-assisted extracted after PEF treatment. 25 mL of wine sample, 10 mL of saturated NaCl solution and 25 mL of Dichloromethane were added in a laboratory flask. The flask was then placed in an ultrasonic bath (Transsonic T570/h, Elma, Germany) for 10 min at room temperature. The samples were centrifuged at 3500 rpm for 5 min to separate the phases. The supernatant was then extracted a second time using the same volume of solvent in the ultrasonic bath for 10 min. The organic layer was then washed with distilled water in a separation funnel. The organic phase was dried on anhydrous sodium sulphate, filtered and then condensed in a vacuum rotary evaporator. Prior to chromatographic analysis, the sample was recovered with 50 μL of chloroform.1 μL was used for GC–MS analysis. To demonstrate replication, all extractions were done in triplicate.

Each sample was subjected to GC coupled with MS analysis, using an Agilent 6890 series GC System, equipped with 5975C VL MSD and a fused silica capillary column, 30 mP × 0, 32 mmi.d. × 0, 25 μm film thickness (HP-%MS 5%, phenylmethylsiloxane, Agilent Technologies). Samples (1.0 μl) injected using split mode with a split ratio of 100:1. The injector temperature hold at 250 °C, the carrier gas was helium at a flow rate 1 ml/min, and the oven temperature programmed as follows: initial 50 °C hold 2.5 °C min to 180 °C, then rate 2 °C/min to 230 °C, then rate 6 °C/min to 250 °C for 5 min and finally rate 5 °C/min to 270°C for 2 min. The above GC conditions translated using the Agilent GC Method Translation to match the mass spectrum retention times. Helium used as carrier gas at 25 cm/s. The temperature of the transfer line fixed at 280°C, the mass spectrometer operated in the electron ionization mode (EI) at an ionization voltage of 70 eV in the mass range 40–550 amu and a manifold temperature of 270°C All data was recorded using the Turbomass 5.0 ChemStation software.

For the quantification of Vanillin, Syringaldehyde, Furfural and Oak Lactone external calibration curves were arranged from standards.

2.2.2. *Wines:*

The wine samples used for this study were from the greek variety Agiorgitiko from the experimental winery of Technological Institute of Athens at the department of Oenology and Beverages Technology (2015).

2.2.3. *Oak chips*

French *Quercus sp.*, untoasted, oak chips were used in this study, purchased by Arabois. These oak chips were extra small in size and heterogeneous in shape.The chips were toasted at 200 °C for 2 hours before being used. In each case 0.2 gr of oak chips were added to 25 ml of sample.

2.2.4. *Chemicals*

Dichloromethane, chloroform, sodium chloride and anhydrous sodium sulphate were purchased from Chem Lab (Zedelgem, Belgium). Vanillin, syringaldehyde, oak lactone and furfural were purchased from. All chemicals are of analytical grade.

2.2.5. *Statistical Analysis*

Every analysis was repeated three times.

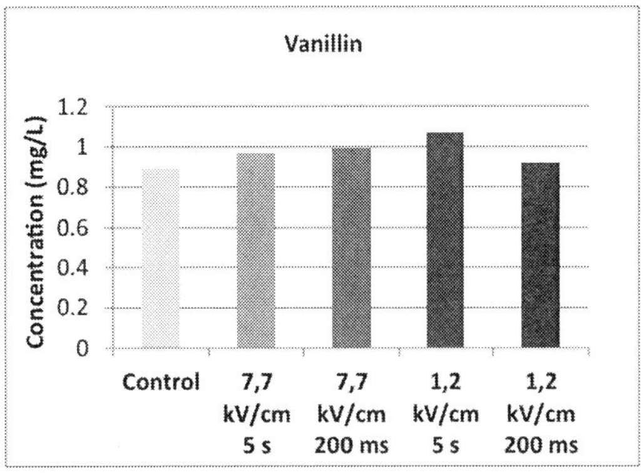

Figure 2a. Concentration of Vanillin for untreated (0 kV/cm) and PEF-treated (7.7 kV/cm for 5 sec and 200 ms and 1.2 kV/cm for 5 sec and 200 ms) red wines.

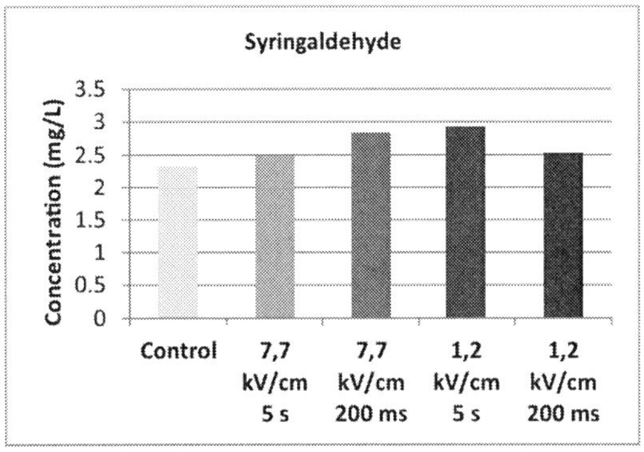

Figure 2b. Concentration of Syringaldehyde for untreated (0 kV/cm) and PEF-treated (7.7 kV/cm for 5 sec and 200 ms and 1.2 kV/cm for 5 sec and 200 ms) red wines.

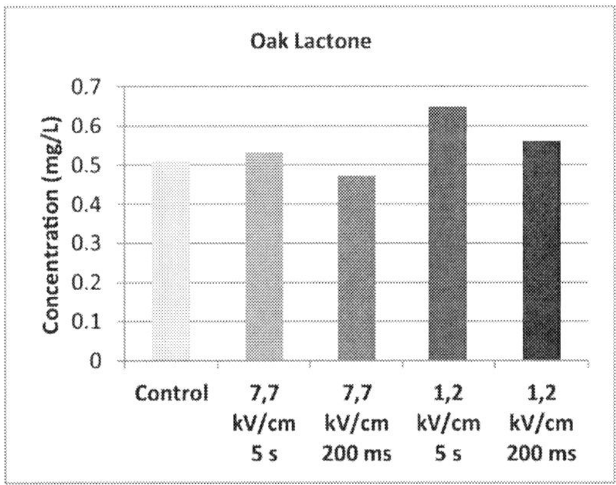

Figure 2c. Concentration of Oak Lactone for untreated (0 kV/cm) and PEF-treated (7.7 kV/cm for 5 sec and 200 ms and 1.2 kV/cm for 5 sec and 200 ms) red wines.

3. Results & discussion

Furanic aldehydes may be reduced during ageing to form their corresponding alcohols (Boidron et al., 1988; Garde-Cerdán & Ancin-Azpilicueta, 2006). Furfural can also be transformed through chemicals mechanisms (Tominaga, Blanchard, Darriet & Dubourdieau, 2000). Other studies (Garde-Cerdán et al., 2002; Gómez-Plaza et al., 2004;

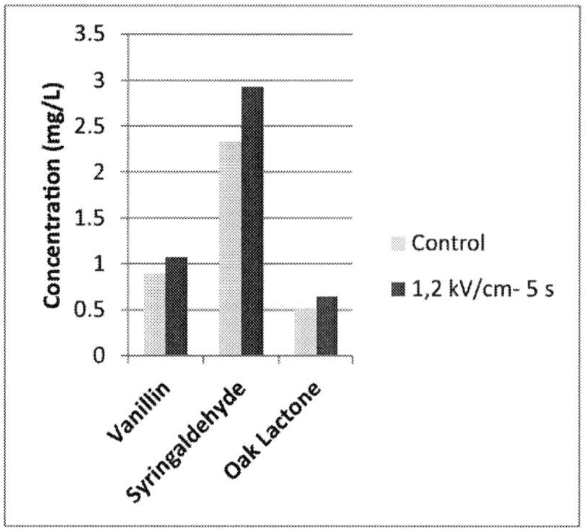

Figure 3. Concentration of Vanillin, Syringaldehyde and Oak Lactone for non-treated and PEF-treated wine samples.

Garde-Cerdán & Ancin-Azpilicueta, 2006) showed that the concentration of phenolic aldehydes is maximal after 10–12 months of ageing. Of this group of compounds most important is vanillin.

The present work studied the extraction rate of the most important volatile compounds of oak wood -vanillin, syringaldehyde, oak lactone and furfural by using oak chips and PEF treatment in order to minimize the time of ageing. In the non- treated wine samples, taking into consideration that the process time was the same for all samples, the compounds being extracted were sufficient enough to be quantified. To be particular, in the non-PEF treated sample, vanillin was extracted in 0.89 mg/L, Syringaldehyde in 2.33 mg/L, oak lactone in 0.51 mg/L and furfural was detected in traces. In Figs. 2a–c, the results of the extraction for the specific 3 volatile compounds in the control and in the samples after treated by PEF are shown. According to Perez-Prieto et al. (2003) after 3, 6 and 9 months of ageing red wine in once used American oak barrels vanillin was detected in the concentration of 140, 165 and 190 μg/L respectively, whereas in our study the extraction rate of the volatile compound of oak wood, vanillin, for PEF-treated sample for 30 min exceeds the 1 mg/lt.

In our study, in general terms, PEF showed the highest efficacy, for the molecular extraction of oak components. For pulse duration 5 sec, 180 pulses and 1,2 kV/cm, the extractability of vanillin and Syringaldehyde reached up to 1.07 mg/L and 2.93 mg/L respectively (Fig. 3). Moreover, oak lactone was increased to 0.65 mg/L, whereas furfural was "found" in traces and had a small extraction after the PEF treatment reaching to 0.03 mg/L.

Taking under advisement these results, different conditions of PEF treatment were examined in order to interpret the best combination of pulse duration and voltage in order to extract high concentration of the volatiles compounds. In order to achieve this, primarily the voltage was increased from 0.8 kV to 5 kV and the time was the same. Likewise, PEF results showed an enhancement in the concentration of the compounds compared with the control samples, but less comparing with the PEF treated samples of the first condition

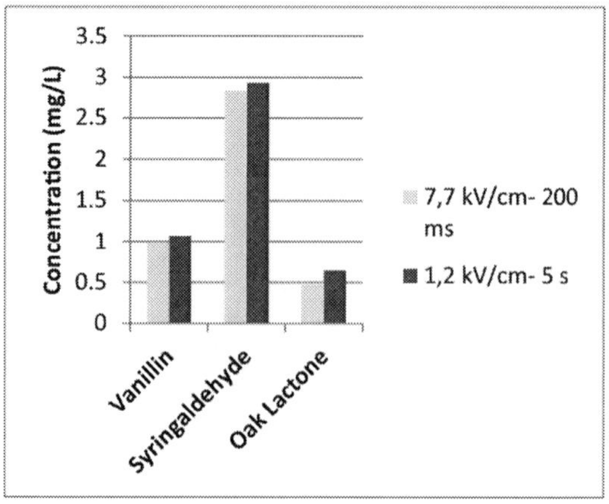

Figure 4. Concentration of Vanillin, Syringaldehyde and Oak Lactone for PEF-treated wine samples in different conditions.

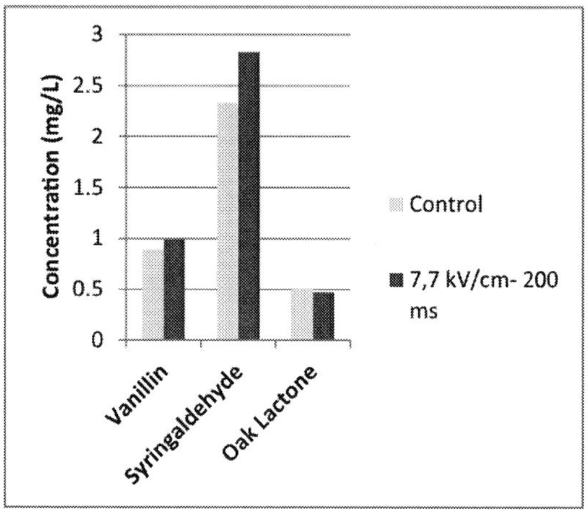

Figure 5. Concentration of Vanillin, Syringaldehyde and Oak Lactone for non-treated and PEF-treated (7.7 kV/cm and pulses 200 ms) red wines.

(1.2 kV/cm), as Vanillin and Syringaldehyde aspired to 0.99 mg/L and 2.83 mg/L respectively, while oak lactone had a smaller increase (Fig. 4). Thus, the best condition, for the fast extraction of volatile compounds is 1.2 kV/cm and pulse duration 5 sec.

Another outcome is that in PEF treatment of 1.2 kV/cm, which considers to be the condition with the greatest impact on the wine samples treated with PEF, as it was mentioned before, the quantity of the extracted compounds differs depending on the pulse duration. As shown in Fig. 5, in the treatment of short pulses (t_i =200 ms) the concentration of all examined compounds were higher than those of long pulses (t_i =5 s). More specific, vanillin's extraction was greater by 0.07% and syringaldehyde's by 1.93%; whereas in oak lactone a reduction was observed in the rate of 0.35%. Remarkably, in the case of long pulses (t_i =5 s) the extractability of furfural augments from traces to 0.03 mg/L.

All the samples, treated and untreated, were evaluated sensorially in order to detect if there was any difference in flavor and mouthfeel. According to Zeng et al. (2008), the results of the wine tasting showed that the

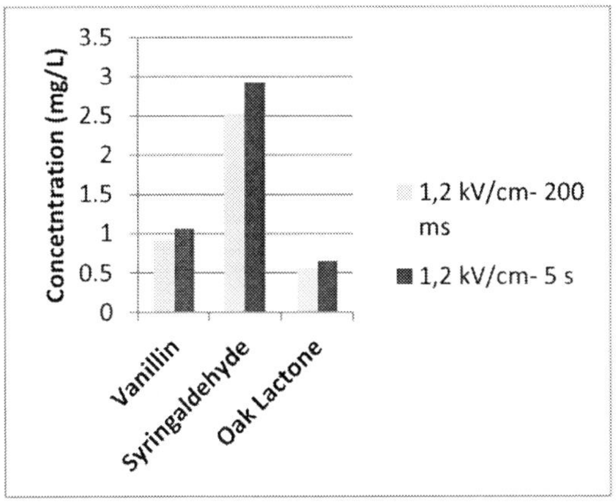

Figure 6. Concentration of the three extracted compounds, PEF-treated with 1.2 KV/cm in different pulse duration.

PEF treated samples had a better sensory quality than the untreated ones, and it was proved that the aging process was accelerated by the application of an AC electric field. In our study, we examined how the parameters (pulse duration, number of pulses and electric field strength) of PEF treatment can evolve the ageing process of wine. PEF treated wines were described by a softer mouthfeel and tannins. The pungent smell of fresh wine was covered by notes of vanilla and wood, as those of an aged wine. Therefore, the combined PEF treatment with oak chips can replace the long-during ageing in oak barrels; since it gives similar outcomes to the traditionally aged wine, but in shorter time.

4. Conclusion

The main aim of the present work was to study the extraction rate of the aromatic compounds –vanillin, Syringaldehyde, oak lactone and furfural from oak wood into wine treated with different conditions of PEF. In this study, it has been demonstrated that the application of PEF treatment to the wine enhanced the extraction of compounds from the wood. The concentration of all the compounds were higher after PEF treated with best condition 1.2 kV/cm and pulse duration 5 sec. Vanillin and Syringaldehyde reached up to 1.07 mg/L and 2.93 mg/L respectively, whereas oak lactone and furfural had a smaller influence reaching to 0.65 mg/L and 0.3 mg/L respectively. Therefore, it can be concluded that the volatile composition of wine was enhanced by PEF application and it can be further investigated in order to minimize the time of aging and the financial costs of this traditional process.

This research has been co-financed by the European Union (European Social Fund – ESF) and Greek national funds through the Operational Program "Education and Lifelong Learning" of the National Strategic Reference Framework (NSRF) - Research Funding Program: ARCHIMEDES III. Investing in knowledge society through the European Social Fund.

References

I. Álvarez, R. Pagán, S. Condón, & J. Raso. The influence of process parameters for the inactivation of Listeria monocytogenes by pulsed electric fields. International Journal of Food Microbiology **87**(1–2), 87–95 (2003)

P. Arapitsas, A. Antonopoulos, E. Stefanou & V.G. Dourtoglou. Artificial aging of wines using oak chips. Food Chemistry **86**, 563–570 (2004)

J.N. Boidron, P. Chatonnet & M. Pons, Effects of wood on aroma compounds of wine. Connaissance de la Vigne et du Vin **22**(4), 275–294 (1988)

M. Corrales, S. Toepfl, P. Butz, D. Knorr, & D. Tauscher. Extraction of anthocyanins from grape by-products assisted by ultrasonics, high hydrostatic pressure or pulsed electric fields: A comparison. Innovative Food Science and Emerging Technologies **9**, 85–91 (2008)

P. Dubois. Apport du fut de chêne neuf a l'arome des vins. Revue Francaise d'œnologie **120,** 19–24 (1989)

T. Garde-Cerdán, & C. Ancín-Azpilicueta. Review in quality factors on wine ageing in oak barrels. Trends in Food Science & Technology **17**, 438–447 (2006)

T. Garde-Cerdán, D. Torrea-Goñi & C. Ancín-Azpilicueta. Changes in the concentration of volatile oak compounds and esters in red wine stored for 18 months in re-used French oak barrels. Australian Journal of Grape and Wine Research **8**(2), 140–145 (2002)

T. Garde-Cerdán, D. Torrea-Goñi, & C. Ancín-Azpilicueta. Accumulation of volatile compounds during ageing of two red wines with different composition. Journal of Food Engineering **65**(3), 349–356 (2004)

E. Gomez-Plaza & P. Rodriguez-Rodriguez. Dependence of Oak-Related Compounds on the Physiochemical Characteristics of Barrel-Aged Wines. Food Technol. Biotechnol. **50**(1), 59–65 (2012)

E. Gomez-Plaza, L.J. Perez-Prieto, J. I. Fernandez-Fernandez & J.M. Lopez-Roca. The effect of successive uses of oak barrels on the extraction of oak-related volatile compounds from wine. International Journal of Food Science and Technology **39**(10), 1069–1078 (2004)

T. Grahl & H. Markl. Killing of microorganisms by pulsed electric fields. Appl. Microbiol. Biotechnol. **45**(1-2), 148–157 (1996)

M. J. Loeffler. Generation and application of high intensity pulsed electric fields. In J. Raso & V. Heinz (Eds.), *Pulsed electric field technology for the food industry: fundamentals and applications* (pp. 27–72), New York: Springer (2006)

D. P. Miller & G. P. Howell. The contents of phenolic acid and aldehyde flavor components of white oak as affected by site and species. American Journal of Enology and Viticulture **43,** 333–338 (1992)

M. Ortega-Heras, C. Gonzalez-Huerta, P. Herrera, & M. L. Gonzalez-Sanjose. Changes in wine volatile compounds of varietal wines during ageing in wood barrels. Analytica Chimica Acta **513**(1), 341–350 (2004)

P. Ribereau-Gayon, Y. Glories, A. Maujean, D. Dubourdieu. *Handbook of Enology Volume* **2** *: The Chemistry of Wine and Stabilization and Treatments*, Chapter 13(411–418), John Wiley & Sons, Ltd (2006)

C. Simer, S. Toepfl & V. Heinz. Inactivation of Bacillus subtilis spores by pulsed electric fields (PEF)

DOI: 10.1051/bioconf/20170902010

in combination with thermal energy – I. Influence of process- and product parameters. Food Control **39**, 163–171 (2014)

S. Spilimbergo, F. Dehghani, A. Bertucco & N.R. Foster. Inactivation of Bacteria and Spores by Pulse Electric Fields and High Pressure CO2 at Low Temperature. Biotechnol. Bioen. **82**, 118–125 (2002)

T. Tominaga, L. Blanchard, P. Darriet & D. Dubourdieu. A powerful aromatic volatile thiol, 2-furanmethanethiol, exhibiting roast coffee aroma in wines made from several Vitis vinifera grape varieties. Journal of Agricultural and Food Chemistry **48**(5), 1799–1802 (2000)

X. A. Zeng, S. J. Yu, L. Zhang & X. D. Chen The effects of AC electric field on wine maturation. Innovative Food Science and Emerging Technologies **9**, 463–468 (2008)

B. Zhang, X. Zeng, W. T. Lin, D. Sun & J. Cai. Effects of electric field treatments on phenol compounds of brandy aging in oak barrels. Innovative Food Science and Emerging Technologies **20**, 106–114 (2013)

B. Zhang, X. A. Zeng, D. Sun, S. J. Yu, M. F. Yang & S. Ma. Effect of Electric Field Treatments on Brandy Aging in Oak Barrels. Food and Bioprocess Technology **6**(7), 1635–1643. doi: 10.1007/s11947-012-0788-7 (2012)

DOI: 10.1051/bioconf/20170902011

Development of a chemical model to predict the doses of calcium sulfate and tartaric acid to acidify musts in Sherry area

J. Gómez[1,a], J.M. Palacios-Santander[2], C. Lasanta[1], L.M. Cubillana-Aguilera[2], R. Arnedo[3], J.A. Casas[3], B. Amilibia[4], and I. LLoret[5]

[1] Department of Chemical Engineering and Food Technology, Faculty of Sciences, University of Cadiz, 11510 Puerto Real, Spain
[2] Department of Analytical Chemistry, Faculty of Sciences, University of Cadiz, 11510 Puerto Real, Spain
[3] Bodegas González-Byass, c/Manuel Maria González, 12, 11403 Jerez, Spain
[4] Grupo Estévez, S.A. Carretera Nacional IV. Km 640, 11408 Jerez, Spain
[5] Certification and Control Authority, Regulation Council of the Denomination of Origin Jerez-Xérèz-Sherry, Avda. Alcalde Álvaro Domecq, 2, 11402 Jerez, Spain

Abstract. Calcium sulfate is normally used as a complementary acidifier combined with tartaric acid. The doses corresponding to each one depend on the desired reduction of pH and on the composition of musts. However, considering that there are several interrelated chemical equilibria implied (tartaric acid dissociation, calcium tartrate and potassium bitartrate precipitation, etc.), it is not easy to predict the effect on pH of a mixed tartaric acid and calcium sulfate addition and to determine the necessary doses to reach the final pH required by the winemaker. In a model previously developed by the authors, the prediction of pH after an acidification was properly achieved. On the contrary, in the same model the prediction of the necessary doses of acidifiers to achieve a desired pH have higher errors due to some parameters, as pH and pK, are found as exponential functions in the equations. This work develops and verify the necessary corrections to the models so that appropriate predictions of the doses are obtained. With these corrections, prediction errors of less than 5% were obtained for all doses of acidifiers, confirming the good comprehension of the chemical equilibria involved in this practice.

1. Introduction

The use of calcium sulfate ($CaSO_4 \cdot 2H_2O$) is authorized in the European Union as a complementary acidifier in liquor wines from Spain provided that the residual sulfate content in the wine does not exceed 2.5 g/L expressed as potassium sulfate [1]. Calcium sulfate is also authorized in the United States for the production of wines aged under yeast veil although residual sulfate cannot exceed 2 g/L [2]. The OIV is also currently considering the approval of calcium sulfate for liquor wines [3]. Gómez et al. [4] recommended a combined acidification with 2 g/L of calcium sulfate and sufficient tartaric acid to achieve a pH of 3.25. In this way, the necessary dose of tartaric acid does not exceed 1.5 g/L, the maximum authorized level in UE, and the final concentration of sulfates is lower than 2.5 g/L. To predict the final pH after an addition of tartaric acid or/and calcium sulfate Moreno and Peinado [5] proposed a simple and easy to apply model. In this model the acidity of wine is considered to be due to a monoprotic acid. This model has been developed and verified in [6,7], obtaining errors in pH predictions lower than 5%. However for winemakers is more interesting to predict the necessary acidifier dose to reach the required pH. In this way the OIV prescribes in [3] that "It is advisable to make previous tests in the laboratory to calculate the doses of calcium sulfate and tartaric acid needed to reach the required pH".

[a] e-mail: juangomez.benitez@uca.es

To predict the dose of acidifiers it is necessary to change the equations proposed in [6] in this way:

For tartaric acid:

$$X = \frac{AAi - TAi \times 10^{pHf - pKi}}{1 + 10^{pHf - pKi}} \tag{1}$$

where X is the dose in meq/L, K_i is the acidity constant, AA is the ash alkalinity and TA is the total acidity. The subscripts "i" and "f" denote initial and final states respectively. In this model it is considered that the medium is saturated in tartrate and, as a consequence, the addition of tartaric acid as an acidifying agent will introduce the common ion HT^- (bitartrate) and this will precipitate naturally or during cold stabilization as potassium bitartrate. In this way, one can consider that all of the HT^- added will precipitate as potassium bitartrate and this does not contribute to the titratable acidity.

For calcium sulfate

$$Y = \frac{AAi - TAi \times 10^{pHf - pKi}}{1 - 10^{pHf - pKi}} \tag{2}$$

and for combined acidification, the final equation would be a combination of both.

Nevertheless, in these equations pH and pK appears as exponential functions what increase the effect of the errors. In consequence, an error of 0.1 units of pH produces an error of 25% in the predicted dose.

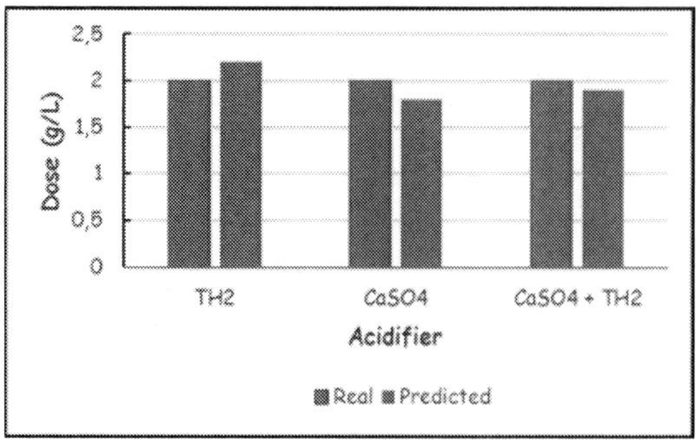

Figure 1. Prediction of the doses of acidifiers.

Table 1. Results of the acidification tests.

Tartaric acid dose (g/L)	Calcium sulfate dose (g(L)	Titratable pH (meq/L)	Buffering acidity (meq/L·u. pH)	power
0	0	3.71	50.38	38.70
1.0	0	3.52	64.78	43.00
0	1.0	3.66	49.14	36.19
1.0	1.0	3.44	65.52	41.03
1.5	0	3.48	69.74	49.02
0	1.5	3.60	50.88	33.68
1.5	1.5	3.36	70.74	46.80
2.0	0	3.41	79.93	55.05
0	2.0	3.63	50.66	31.17
2.0	2.0	3.26	79.68	53.78

Experiences carried out in the harvest of 2016 working with the best laboratory conditions (Fig. 1) and described in [7] showed that the errors in the prediction of the acidification doses cannot be reduced below 10% using the model described in [6].

This fact shows that the simplified model used must be corrected to obtain better prediction results.

The aim of this work consists of designing and verifying some corrections in the model previously used in order to minimize the errors in the prediction of the doses of acidifiers used in the winemaking of Sherry musts.

2. Materials and methods

2.1. Corrections to the model

For Tartaric acid:

Instead of supposing that all added tartaric acid precipates as potassium bitartrate, it is considered that a fraction **f** of the addition does not precipitate because there is not enough time to reach the equilibrium in acidified musts before fermentation. In this case, the dose X of tartaric acid would be obtained by a new equation to be called (3).

For Calcium sulfate

The initial model does not consider the increase of ash alkalinity due to calcium sulfate. In consequence, in this correction it is considered that fraction **g** of added sulfate remains in solution and increases ash alkalinity. The dose Y of calcium sulfate would be obtained by e new equation to be called (4).

For combined acidification

A combination of the equations (3) and (4) is used in combined acidification. The dose X of tartaric acid would be obtained by the Eq. (5) where logically participates the dose Y of calcium sulfate.

2.2. Acidification tests

The Sherry must used in the acidification tests was obtained from *Palomino fino* grapes harvested at optimum maturation state and pressed in a pneumatic press at a pressure lower than 1 Kg/cm^2. Next, 2 g/L of SO_2 was added to the obtained must to allow its preservation until the acidification tests were carried out and to avoid the beginning of alcoholic fermentation. These tests were conducted with 100 mL of must in 250 mL Erlenmeyer flaks, placed on a magnetic stirrer and adding slowly the corresponding dose of acidifier weighed with a precision of 0.1 mg. The flasks were covered with plastic film to avoid SO_2 evaporation. For the first 2 minutes the stirring speed was high to dissolve the acidifiers and later it was maintained until 30 minutes at the lowest possible speed (c.a. 60 r.p.m.) to keep the solids suspended. The doses of tartaric acid and calcium sulfate used were 1.0, 1.5 and 2.0 g/L alone and combined. The samples were centrifuged before analysis and pH, titratable acidity and buffering power were determined in all samples before and after acidification. All tests and analysis were carried out in duplicate indicating the results as the average values.

2.3. Analytical methods

pH, titratable acidity and buffering power were determined as in [6].

2.4. Determination of f and g factors

Iterative calculations using Excel 2016 were carried out to know the values of f and g factors that minimized the errors in the predicted doses of acidifiers obtained by using the Eqs. (3), (4) and (5) in relation to the real doses. In these calculations, f and g factors were gradually changed until obtaining the minimal errors. In the combined acidification, the dose of tartaric acid was fixed and we tried to predict the dose of calcium sulfate. In consequence, couples of f and g factors values minimizing the error in calcium sulfate dose were searched for.

3. Results and discussion

3.1. Acidification tests

The results of the acidification tests carried out in the laboratory are shown in Table 1. In general terms, these results are according with those obtained in previous research works [6].

3.2. Determination of f and g factors

The evolutions of errors with the values of f and g factors for an acidification of 2 g/L are shown in Figs. 1 and 2

 DOI: 10.1051/bioconf/20170902011

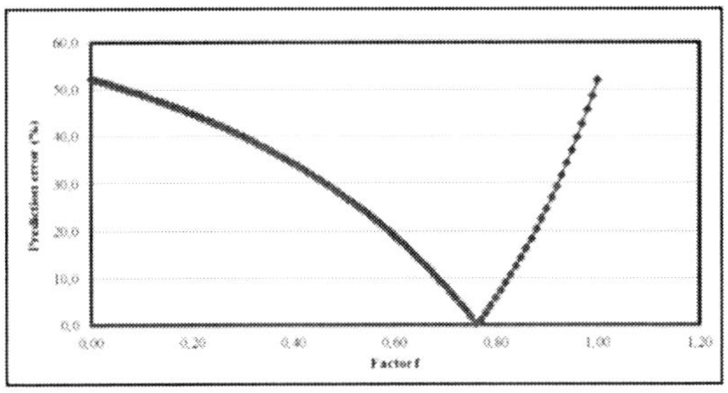

Figure 2. Variation of the prediction error with *f* factor for acidification with 2 g/L of tartaric acid.

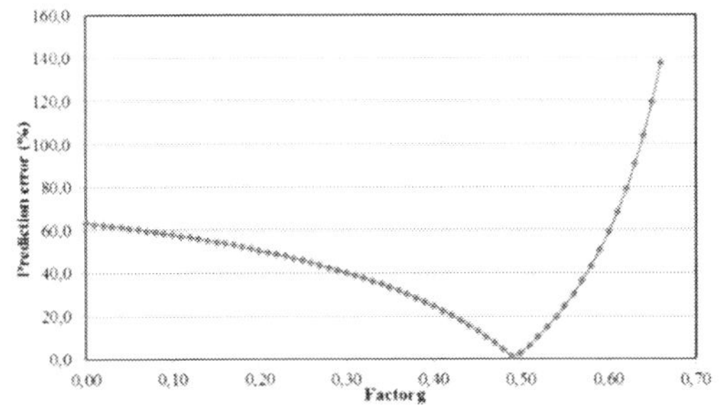

Figure 3. Variation of prediction error with *g* factor for acidification with 2 g/L of calcium sulfate.

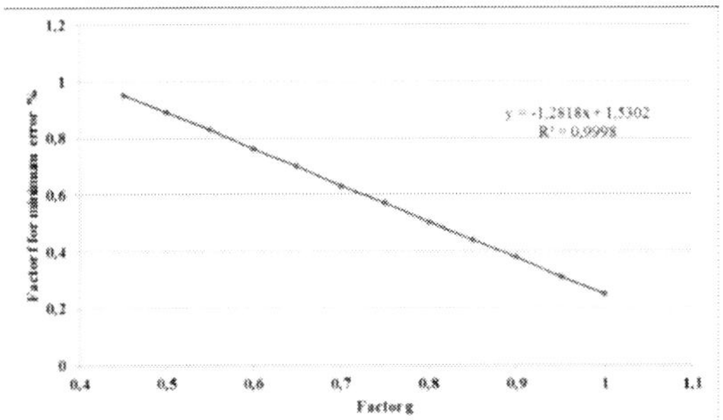

Figure 4. Values of *f* and *g* factors in combined acidification with 2 g/L of tartaric acid and calcium sulfate.

Table 2. *f* and *g* factors for different doses of acidifiers.

Acidifier dose (g/L)	*f* factor (Tartaric acid)	*g* factor (Calcium Sulfate)
2	0.76	0.49
1.5	0.80	0.53
1	0.80	0.55

As it can be seen, both evolution plots show similar shapes, first decreasing the error with the factors until reaching a minimum value, corresponding to certain factor value, and then sharply increasing the prediction errors with the factor values.

Figure 4 shows the values of *f* and *g* factors that minimize the errors in combined acidification. As it can be seen, these factors are very well correlated with a correlation factor of 0.9998.

The procedure for calculating the *f* and *g* values for other acidifier doses was the same and the values obtained are shown in Table 2. As it can be seen, the factors are approximately constants for all considered doses. This fact confirms that the corrections of the model considered are valid regardless of the doses of acidifiers used.

4. Conclusions

The proposed corrections to the chemical model for the acidification equilibria in the Sherry winemaking are able to predict properly the necessary doses to reach the required pH values with very low errors. Hence, the chemical model developed can be a valuable tool for winemakers that can be easily implemented as a spreadsheet in a computer to be used in the laboratory during the harvest.

References

[1] European Union, Commission Regulation (EC) N° 606/2009 laying down the categories of grapevine products, oenological practices and the applicable restrictions, O.J.E.U. **L153**(606/2009), pp. 1–59 (2009)

[2] e-CFR.Electronic Code of Federal Regulations, "Title 27: Alcohol, Tobacco and Firearms PART 24 WINE Subpart L Storage, Treatment and Finishing of Wine §24.246 Materials authorized for the treatment of wine and juice. Available on line at `http://www.ecfr.gov/cgi-bin/text-idx?SID=5a39ba6737562446636`," 2010

[3] OIV, "Resolution OENO-TECHNO 15-583 Step 7. Treatment of musts with calcium sulphate for liqueur wines." 2017

[4] G. J, G. MM, and D. J, "Study of the acidification of Sherry musts with gypsum and tartaric acid," Am. J. Enol. Vitic. **44**(4), pp. 1–5 (1993)

[5] J. Moreno Vigara and R. A. Peinado Amores, *Oenological Chemistry*, 1st ed. San Diego CA. (USA): Academic Press, 2010.

[6] J. Gómez, C. Lasanta, J. M. Palacios-Santander, and L. M. Cubillana-Aguilera, "Chemical modeling for pH prediction of acidified musts with gypsum and tartaric acid in warm regions," Food Chem. **168**, pp. 218–224 (2015)

[7] J. Gomez *et al.*, "Comprehensive chemical study of the acidification of musts in Sherry area with calcium sulphate and tartaric acid.," in *39th World Congress of Vine and Wine* (2016)

New techniques for wine aging

Hatice Kalkan Yıldırım and Ezgi Dündar

Ege University, Department of Food Engineering, 35100 Bornova, Izmir, Turkey

Abstract. The aging of wine requires a long time therefore it can cause loss of time and money. Therefore using of new techniques for wine aging shortens the length of aging time and wines may be placed on the market more quickly. Nowadays, gamma irradiation, ultrasonic waves, AC electric field and micro-oxygenation are the new techniques for wine aging. Gamma irradiation (after fermentation) is accelerated physical maturation method. Gamma irradiation, in a suitable dosage (200 Gy), is a suitable method for improving some wine defects and producing a higher taste quality in wine. The 20 kHz ultrasonic waves aged wine much more quickly than standard aging, with similar quality. The wine treated by 20 kHz ultrasonic waves had a taste equivalent to 1 year aged wine. Wine maturing with AC electric field promises novel process accelerating aging process of fresh wine when suitable conditions are applied. As a result of research, an optimum treatment (electric field 600 V/cm and duration time 3 min) was identified to accelerate wine aging. Harsh and pungent raw wine become harmonious and dainty. This process is equivalent to 6 month aging in oak barrel. Microoxygenation is a very important technique used in aging wines in order to improve their characteristics. The techniques of wine tank aging imply the use of small doses of oxygen (2 ml L^{-1} $month^{-1}$) and the addition of wood pieces of oak to the wine. Studies concerning these new techniques demonstrated that maturation of wines become more quickly than standard maturation procedures with keeping and improving the wine quality.

1. Wine aging

The term "wine aging" included group of reactions occurring after bottling of wines. The "wine maturation" is related to steps of wine storage before bottling [1,2]. The wine maturation can last from 6 to 24 months or many years. At this stage, clarification may be applied taking into account the possibility that the wine will undergo malolactic fermentation. Wine aging process starts with bottling of wine and is also referred to as also reductive aging since aging is carried out without oxygen [3].

During wine aging many reactions occurred that caused significant organoleptical changes in wines. During the process of maturation and aging, the most obvious change occurs in the color of the wine which refers to phenols changes.

The phenolic compounds present in the wines are divided into two groups, flavonoids and phenolic compounds other than flavonoids (Fig. 1). Phenolic compounds containing a hydroxyl (- OH) group attached to a benzene ring in their structure are present in cells forming part of the grape such as shell, nucleus and stem. Flavonoids are in the C6-C3-C6 diphenylpropane structure and the triple carbon bridge between the phenyl groups forms a ring with oxygen. Differences among flavonoids are due to the number of hydroxyl groups attached, the degree of unsaturation and the oxidation level of the triple carbon segment [4].

The changes occurring in the phenols in wines during and after the production are realized by the following mechanisms [5].

1. The destruction of anthocyanins;
2. Reactions of tannins with proteins; Reactions of tannins with polysaccharides;
3. Cation formation of procyanidines;
4. Oxidation reactions of procyanidins;
5. Polymerization reactions of procyanidins;
6. Copigment formation processes of anthocyanins;
7. Reaction of anthocyanins with compounds containing polarized double bonds;
8. Condensation reactions of anthocyanins with tannins;

In the case of young wines "aging" is expected to protect organoleptic properties for long time. In the aging of wines, it is intended to make a specific change in the composition of the wine by changing the organoleptic properties of the wine [2].

Wine aging requires a long time therefore it can cause loss of time and money. Using of new techniques for wine aging by providing high quality wines, could shortens the length of aging time and allow wines be placed on the market more quickly.

2. Wine aging in oak barrels

The oak which used in the time of the ancient Romans, is used for making barrels nowadays, also. Today, oak barrels are used for the wine aging. Oxidation which takes place in a small amount over time in the barrel, allows the wine to produce a variety of unique aromas in the stage of aging [3]. Controversies to this, there are some disadvantages of oak barrels used as traditional

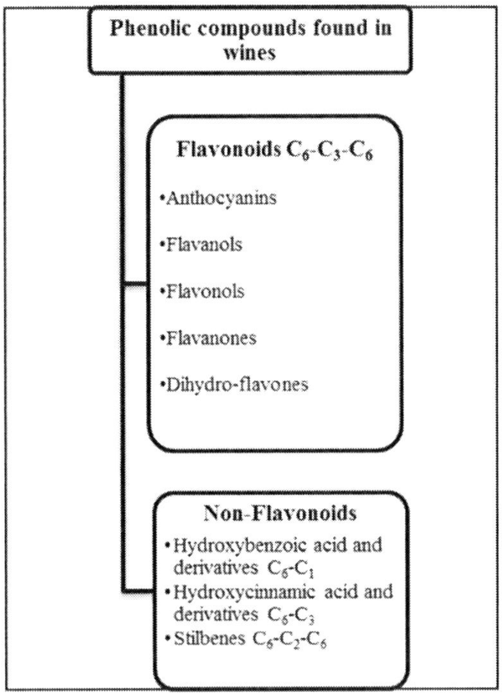

Figure 1. Phenolic compounds found in wines.

aging methods. The aging of wine continues long time so it causes time loss. Additionally, due to the undesired microorganisms such as *Brettanomyces* and *Dekkera* species which may contaminate wine barrel must be renewed over time [6].

Considering all these disadvantages, current studies have focused on new techniques for wine aging which will be an alternative to aging with the barrel that shortens the aging period and the preservation of quality after aging of the wine [6]. These techniques include the use of some physical applications and micro-oxygenation applications of wines. Each technique can contribute to different points for the process of aging the wines [7]. Ultrasonic waves, electric field, gamma irradiation, applications can be used in the aging of wines. In studies conducted using these methods, it is aimed to shorten the period of aging of wine and to offer high quality wines to consumers [7].

3. Physical methods in wine aging

3.1. Ultrasonic waves

The aging process for wines using ultrasound technique is a specific example of ultrasonically developing oxidation [8]. The aging effect of ultrasonic treatment is related to the cavitation created by ultrasonic waves and the growing and collapse of micro sized bubbles. Ultrasonic waves can precipitate these bubbles therefore temperature and pressure may rise. High pressure and temperature can influence the wine aging and the wine could be negatively affected. However, controlled ultrasonic waves can positively affect the chemical reactions in the wine [9].

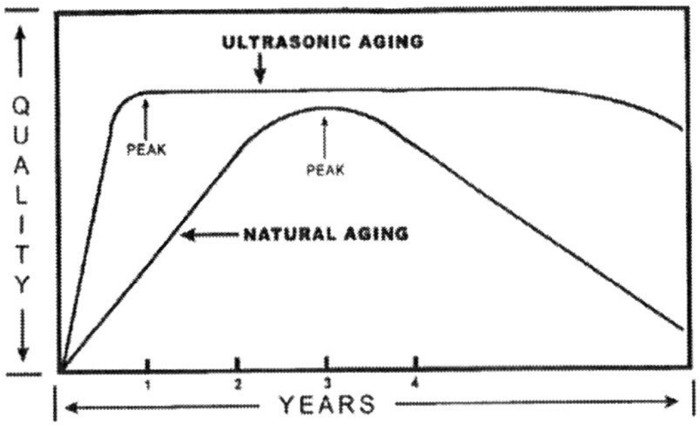

Figure 2. Comparison of qualities in natural aging in wine and ultrasonic aging in wine (adapted from Leonhardt and Morabito, 2007).

The wines reach the 'peak' points after being aged for a certain period of time. The wines are the most fine at the peak points and the changes that occur in the tannins so wine have soft flavor and a complex bouquet [10]. After that point, the quality of the wine starts decreasing, so it is best to consume the wines when they reach their peak. After the application of the ultrasonic waves, the wines can reach their peak quickly, and the wines can retain their peak points and qualities for a longer time than the standard aging process. This relationship has been shown as Fig. 2. Peak points in wines have a very important place in the extension of shelf life [11].

In a study, wines made by using rice and corn were subjected to 20 kHz ultrasonic wave application for one week with purpose of acceleration of wine aging. Concerning rice wines, the 20 kHz ultrasound application allowed the production of wines with the same quality as the standardized aging wines, but in case of corn wine the same quality was not obtained [12] Similar results, related to the differences of used grapes and ultrasound treatment parameters were also determined [13].

3.2. Electric field

Electric field applications cause electroporation of cell membranes. As a result, diffusion of soluble materials and microbial inactivation are increasing [14]. It was determined that the application of pulsed electric field in red wines before oak barrels aging could positively affected the aging process time: the extraction efficiency of polyphenolic compounds could increase and high quality wine could be produced [6].

There is an another study creating an innovative technique to implement on an pilot scale implementing AC electric field to speed up the aging of Cabernet Sauvignon wines. Electricity application of 600 V/cm – 3 min ensures that harsh and pungent wines become harmonious and dainty wines. The results demonstrated that high alcohol and aldehyde content decreased and the number of ester and free amino acids increased slightly in the samples treated with electric field. Thus, the high-voltage electric field could be an alternative for wine aging acceleration for production of wine with high quality [15]. Even this result and similar ones [16,17] the need of more studies are required concerning the clarification of treatment parameters for wines produced from different grape varieties.

3.3. Gamma irradiation

Gamma radiation is one of the three types of natural radioactivity. The other two types of natural radioactivity are alpha and beta radiation, which are in the form of particles. Gamma rays are the most energetic form of electromagnetic radiation, with a very short wavelength of less than one-tenth of a nanometer [18].

The basic technique for its application in foods is to expose to gamma rays at determined conditions [19]. The effects concerning wine is related to accelerating physical maturation.

In a previous study, gamma radiation was applied after fermentation to mature rice wines. Gamma irradiation, in a suitable dosage (200 Gy), is a suitable method for improving some wine defects and producing a higher taste quality in wine, without the presence of irradiation residues [20]. However, more work needs to be done to understand the effects of gamma irradiation on human health [5].

4. Chemical methods in wine aging

4.1. Micro-oxygenation

Microoxygenation works by the addition of oxygen into red wine at a controlled rate and flow to stabilize color and improve astringency and aromatic components of the final wine [21].

It is well known that adding oxygen to wines affects chemical and sensory components of a wine by changing forms of phenolics, sulfur compounds and other reactions related to these compounds. The amounts of used oxygen is of critical importance. In case of too much oxygen, several problems can occur, including the oxygenation of phenolic compounds, increases in astringency, color, mouthfeel, and bacteria populations.

In a study, wines were aged for three months in "new" American oak barrels, "old" American oak barrels and treated with microoxygenation method for three months. Results demonstrated that microoxygenation could improve wine color, phenolic and chromatic characters in similar manner to oak barrels application [22].

Since this technique is used with a lot of purposes as improving the organoleptic properties of the stability of the wine color, reduction of the sulfur-derived smells and to simulating the wine aging [5,23] before application of microoxygenation optimum conditions for wine category, each grape varieties be determined.

5. Concluding remarks

Modern winemaking has discovered many different techniques and technologies to improve upon the quality of wine. Although the traditional method used in years is regarded as creating quality wines, it has been sought for new methods. Using of new techniques for wine aging could shortens the length of aging time and wines may be placed on the market more quickly. More research work is needed to make a definite comment on ultrasonic, electric field and gamma irradiation applications used as an alternative to traditional wine aging. Microoxygenation application is an improved method and it may be recommended to use it in combination with other aging methods.

References

[1] A.L. Waterhouse, G.L. Sacks, D.W. Jeffery *Understanding wine chemistry*. (John Wiley & Sons, 2016)
[2] J. Morena, R. Peinado. *Enological chemistry*. (Academic Press, 2012)
[3] R.S. Jackson, *Wine science: principles and applications*. (Academic press, 2008)
[4] F. Cuyckens, M. Claeys, *Mass spectrometry in the structural analysis of flavonoids*, Journal of Mass Spectrometry **39**, 1–15 (2004)
[5] P. Ribéreau-Gayon, D. Dubourdieu, B. Donèche, A. Lonvaud, (Eds.) *Handbook of enology, the microbiology of wine and vinifications* (Vol. 1). (John Wiley & Sons, 2006)
[6] F.G. Martín, D.W. Sun, *Ultrasound and electric fields as novel techniques for assisting the wine ageing process: The state-of-the-art research*. Trends in Food Science and Technology **33**, 40–53 (2013)
[7] Y. Tao, J.F. Garcia, D.W. Sun, *Advances in wine aging technologies for enhancing wine quality and accelerating wine aging process*, Critical Reviews in Food Science and Nutrition **54**, 817–835 (2014)
[8] T. Mason, *Power ultrasound in food processing—the way*, Ultrasound in Food Processing, 105 (1998)
[9] A.J. Saterlay, R.G. Compton, *Sonoelectroanalysis–an overview*, Fresenius' Journal of Analytical Chemistry **367**, 308–313 (2000)
[10] J. Robinson, J. Harding, *The Oxford Companion to Wine*. (American Chemical Society, 2015)
[11] C.G. Leonhardt, J.A. Morabito, *Wine aging method and system*, US Patent 11/043,121 (2007)
[12] A.C. Chang, F.C. Chen, *The application of 20 kHz ultrasonic waves to accelerate the aging of different wines*. Food Chemistry **79**, 501–506 (2002)
[13] Y. Tao, Z. Zhang, D. Sun, *Experimental and modeling studies of ultrasound-assisted release of phenolics from oak chips into model wine*, Ultrasonics Sonochemistry **21**, 1839–1848 (2014)
[14] L. Picart, J.C. Chetfel, *Pulsed electric fields. Food preservation techniques*. Fla.: CRC Press. 57–68 (2003)
[15] A.X. Zeng, S.J Yu, L. Zhang, X.D Chen, *The effects of AC electric field on wine maturation*, Innovative Food Science and Emerging Technologies **9**, 463–468 (2008)
[16] Z. Lin, X. Zeng, S. Yu, D. Sun, *Enhancement of Ethanol–acetic acid esterification under room temperature and non-catalytic condition via pulsed electric field application*, Food And Bioprocess Technology **5**, 2637–2645 (2011)
[17] B. Zhang, X. Zeng, D. Sun, S. Yu, M. Yang, S. Ma, Effect of Electric Field Treatments on Brandy Aging in Oak Barrels, Food And Bioprocess Technology, **6**, 1635–1643 (2012)
[18] *Gamma Radiation*. [online] Available at: `https://www.ndeed.org/EducationResources/`

 DOI: 10.1051/bioconf/20170902012

`CommunityCollege/Radiography/Physics/gamma.htm` [Accessed 22 Apr. 2017]

[19] J.M. Wetheril, *Gamma Irradiation of Food*, Canadian Journal of Public Health/Revue Canadienne de Sante'e Publique **56**, 521–524 (1965)

[20] A.C. Chang, *The effects of gamma irradiation on rice wine maturation*, Food Chemistry **83**, 323–327 (2003)

[21] *The Effect of Micro-Oxygenation on Wine Quality and Consumer Preferences.* [online] Available at: `http://www.academicwino.com/2013/01/microoxygenation-quality-preferences.html/` [Accessed 22 Apr. 2017]

[22] M. Cano- López, J.M. López, F. Pardo-Mínguez, E. Gómez-Plaza, *Oak barrel maturation vs. micro-oxygenation: Effect on the formation of anthocyanin-derived pigments and wine colour*, Food chemistry **119**, 191–195 (2010)

[23] E. Gómez-Plaza, M. Cano-López,*A review on micro-oxygenation of red wines: Claims, benefits and the underlying chemistry.* Food Chemistry **125**, 1131–1140 (2011)

Impact of different wood chip species (oak, acacia and cherry) on evolution of individual anthocyanins, chromatic characteristics and antioxidant capacity in model wine solutions

António M. Jordão[1,a], Viriginia Lozano[2], Ana C. Correia[1], and María L. González-SanJosé[2]

[1] Instituto Politécnico de Viseu (CI&DETS), Escola Superior Agrária, Estrada de Nelas, Ranhados, 3500-606 Viseu, Portugal
[2] Universidad de Burgos, Facultad de Ciencias, Departamento de Biotecnologia e Ciencia de los Alimentos, Plaza Misael Bañuelos s/n, 09001 Burgos, España

Abstract. The aim of present work was to investigate the impact of different oak (*Q. alba*, *Q. robur* and *Q. pyrenaica*), acacia and cherry wood chips on evolution of individual anthocyanin content, chromatic characteristics and antioxidant capacity by the use of model wine solutions. According to the results obtained, model wine solutions containing acacia and especially cherry wood extracts showed more evident changes of individual anthocyanin composition, especially due to the formation of large number of new direct condensation anthocyanins, than model wines with oak extracts. These results were accompanied also by a higher color degradation and a lower antioxidant potential of solutions containing the different wood chip extract species, compared with solutions containing only anthocyanin alone. This research contributes to a better knowledge about the color evolution of red wines when they are aging in contact with new alternative wood species, especially acacia and cherry.

1. Introduction

During red wine aging in contact with oak wood, phenolic compounds, mainly anthocyanins, undergo several changes that may lead to color loss or stabilization. Previous studied showed that the origin or variety of the oak together with the specific practices of each coopery, in addition to the age of the barrels and the wine aging time in contact with oak wood, play a key role in the formation of "new pigments" which could improve chromatic characteristic of wines [1–3]. Despite the advances made in the evolution of red wine color and composition during wine aging in contact with different oak wood species [4–6], there is a scare knowledge about the potential impact on anthocyanin, chromatic characteristics and antioxidant capacity evolution when red wines are aging in contact with other non-oak wood species. This is an important topic because the high demand of oak wood products has an ecological impact on harvesting of oak trees in forest where the replacement of trees is not guaranteed. Thus, the use of potential alternative woods, such as acacia and cherry, may be an interesting option for potential use in red wine aging process.

The aim of present work was to investigate the impact of oak (French, American and Iberian species), acacia and cherry wood chips on evolution of individual anthocyanin content, chromatic characteristics and antioxidant capacity by the use of model wine solutions.

[a] e-mail: antoniojordao@esav.ipv.pt

2. Material and methods

2.1. Wood chip samples

The wood chip samples used were: acacia (*Acacia pseudorobinia*) chips purchased by SAI company (Paredes, Portugal), cherry (*Prunus avium*) French and American oak (*Quercus petraea* and *Quercus alba*, respectively) chips purchased by AEB Bioquímica company (Viseu, Portugal), and Iberian oak chips from Portugal (*Quercus pyrenaica*) purchased by J.M. Gonçalves company (Palaçoulo, Portugal), with medium toasting (20 min at 160–170 °C), a particle size of 8 mm and submitted to a natural drying.

2.2. Model wine solutions

Aliquots of a red model wine solution (12% ethanol, adjusted to pH 3.5 with tartaric-tartrate buffer and containing grape isolated anthocyanins) was added with different wood chip concentrated extracts (final concentration corresponding to 4 g wood chips/L wine) from oak, acacia and cherry species obtained after 15 and 30 days of extraction. After mixing, the different solutions were stored during 30 days under controlled temperature (around 14 °C) and darkness conditions. All assays were made in duplicate.

2.3. Methods

By the use of HPLC-DAD, the individual anthocyanin content evolution was analyzed followed the methodology described by Pérez and González-Sanjosé [7], while chromatic characteristics evolution (color intensity) was studied by spectrophotometric methodologies [8], and

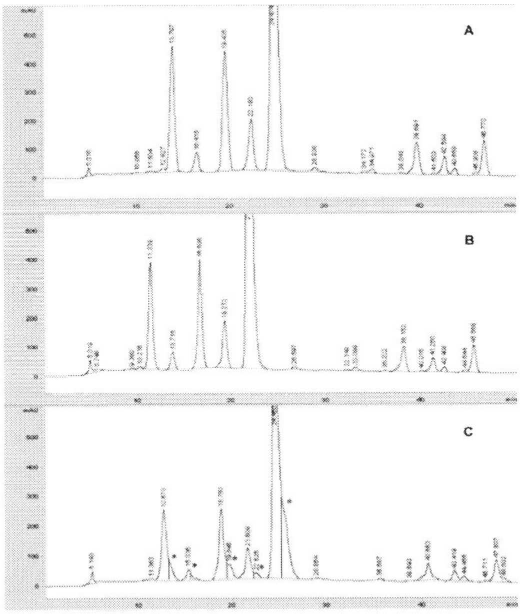

Figure 1. Changes of the individual anthocyanin profile (HPLC) after mixing a model wine solution, containing anthocyanins, with a wood chip extracts. (A) model wine solution containing anthocyanin extract alone; (B) model wine solution containing anthocyanin extract and oak wood chip extract after 15 days of storage; (C) model wine solution containing anthocyanin extract and oak wood chip extract after 30 storage days; (*) new pigments formed during storage.

by the use of CieLAB* coordinates [9]. To distinguish the color more accurately, the color difference was also calculated using the following formula: ($\Delta E = [(\Delta L^*)^2 + (\Delta a^*)^2 + (\Delta b^*)^2]^{1/2}$). Finally, the antioxidant capacity evolution was investigated by the use of ABTS and FRAP assays, applying the conditions optimized by Rivero et al. [10]. All analyses were done in triplicate.

3. Results and discussion

3.1. Individual anthocyanins composition

According to obtained results (Fig. 1), levels of newly formed anthocyanins are notable visible after 30 days of contact of model wine solutions with wood extracts.

During the storage time considered a decrease of free anthocyanins was detected in all model wine solutions (Fig. 1A, 1B and 1C). In addition, new pigments are formed in the solutions containing anthocyanin extract and wood extract (Fig. 1C). These new products formed occurred in all solutions containing anthocyanin extract and different extracts obtained from the different wood chip specie used (oak, cherry and acacia). The number of new products formed (including the anthocyanin pyruvic derivative products) and their concentration increase with the storage time in all model wine solutions containing anthocyanin extract and wood chip extracts. However, the majority of the direct condensation products formed were detected in solutions containing anthocyanin extracts and acacia and cherry wood chip extracts after 15 storage days (data not show). The results obtained in our work are in agreement with the research work previously developed by Sáchez et al. [11] where was analyzed the effect of micro-oxygenation and wood type on the phenolic composition and color parameters of red wines. In addition, Jordão et al. [3] also detected new HPLC peaks corresponding to new degradation products in solutions containing a mixture of malvidin-3-glucoside and ellagic acid extracted from oak wood and also in solutions containing malvidin-3-glucoside and oak wood extract. These new compounds formed increased slightly during 64 storage days.

Thus, our work points out the importance of wood components not only from oak wood species, but also from acacia and cherry wood species in the transformation of anthocyanins, as well as in reactions that occur during the red wine aging in contact with the different wood species. It is also possible that the new products formed could participate in the color changes that usually occurred in red wines aged in wood barrels or in contact with wood chips.

3.2. Chromatic parameters

Color changes are one of the greatest modifications occurring during the wood aging of red wines. Anthocyanins are able to reacts with compounds extracted from wood but also with other compounds formed by redox potential changes, induced by wood components. Under these conditions, anthocyanins transformation may lead to color loss or stabilization, with the corresponding negative or positive effect on wine quality. This fact can be valuated through the measure of color parameters as color intensity, tonality and the CieLAB* parameters.

The color intensity evolution quantified in the model wine solutions containing the different wood chip extract species is illustrated in Fig. 2. After 30 days of storage, a general decrease of color intensity was observed in all the studied mixtures of red wine model solutions and wood extracts. This result was also obtained independently of the extraction time used for obtained the different wood extracts (15 and 30 extraction days). However, quantitative differences were detected among samples by factors, wood extraction time and wood type. Higher decreases of color intensity were always observed in mixtures containing wood extracts obtained after 30 days of chips extractions. So, mixtures with wood extracts obtained after 15 extraction days, showed significant decrease of color intensity only after 30 days of storages. The color reduction detected, may be due to the wood components that were extracted from wood chips that induce the rate of anthocyanin decline and consequently induced a color decrease (especially red color component). Previously Jordão et al. [2] reported a malvidin-3-glucoside declined more rapidly in the presence of oak wood extract than alone and consequently a more evident decrease of red color. In addition, it has been reported that numerous anthocyanin adducts are formed during red wine aging in oak wood barrels [12]. These new compounds presumably have an influence on the chromatic characteristics of red wines.

Another interesting result obtained is related with the impact of the different wood chip extract species on color intensity decrease. In fact, the color decrease was significantly more pronounced for the red model wine solutions mixed with the cherry wood extracts (particularly for the wood extract obtained after 30 extraction days)

DOI: 10.1051/bioconf/20170902013

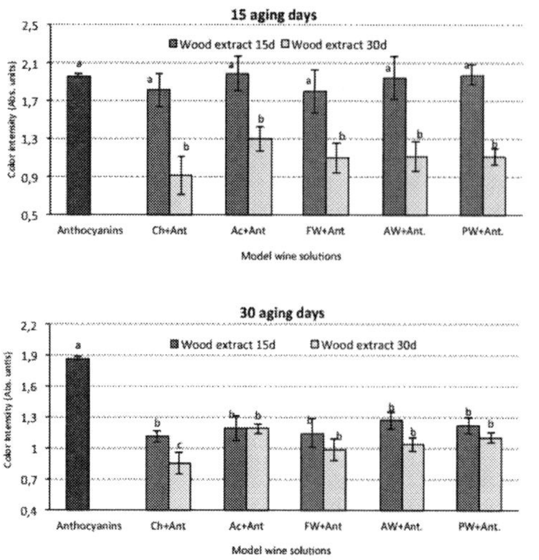

Figure 2. Color intensity quantified in model wine solutions containing different wood chip extracts after 15 and 30 aging days. Ch + Ant = Cherry wood chip extract and anthocyanin extract; Ac + Ant = Acacia wood chip extract and anthocyanin extract; FW + Ant = French wood chip extract and anthocyanin extract; AW + Ant = American wood chip extract and anthocyanin extract; PW + Ant = Portuguese wood chip extract and anthocyanin extract.

and especially after 30 storage days. In wines, several works [13,14] reported that the use of cherry barrels in wine aging induce a faster evolution of wine pigments with a quick augmentation formation of derived and polymeric compounds. In addition the use of this type of wood specie demonstrated to provide an environment favoring oxidative reactions and, thus increases the red color loss, making it less suitable for longer wine aging periods.

Concerning to CieLAB chromatic coordinates evaluation, Fig. 3 shows the L^*, a^* and C^* values obtained in model wine solutions after 30 aging days. For L^* values (lightness), all model wine solutions (independently of the extraction time and wood species) showed higher lightness compared with the initial red model wine. This result is consistent with the decreased of color intensity showed in all red model wine after mixing with wood extracts. In fact, the decrease of color intensity induces a more lightness of the samples.

For a^* values (redness), also a decrease of the red color component values was observed for all the mixtures of wood extracts with the red model wine. This result agrees the results already showed for color intensity. The decrease of red color component was similar for all model wine solutions, except for the model wine solutions containing cherry wood chip and French oak wood extract obtained after 30 extraction days, where the decrease was more evident.

Regarding to the results obtained for c^* values (chroma), similar tendency was also detected. Thus, a tendency for a chroma values decrease was evident for all red model wine solutions mixed with wood chip extracts, corresponding also to a less color intensity.

Finally, the values obtained for color difference (ΔE) among red model wine solution alone and mixed with the different wood chip extracts (Fig. 4) showed that

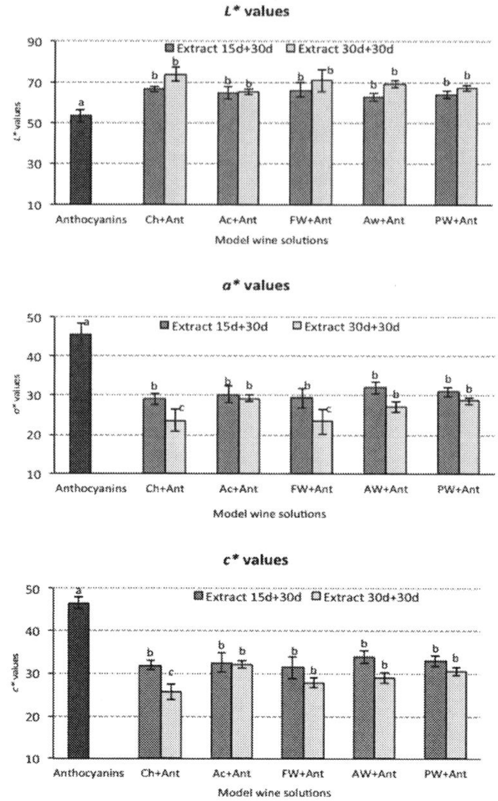

Figure 3. CieLab chromatic coordinates quantified in model wine solutions containing different wood chip extracts after 30 aging days. Ch + Ant = Cherry wood chip extract and anthocyanin extract; Ac + Ant = Acacia wood chip extract and anthocyanin extract; FW + Ant = French wood chip extract and anthocyanin extract; AW + Ant = American wood chip extract and anthocyanin extract; PW + Ant = Portuguese wood chip extract and anthocyanin extract.

all mixture raised differences much higher than three CieLAB units. Thus, taking into account that ΔE values of up to three CieLAB units indicates color differences appreciable to the human eyes [15], it could be affirmed that the contact between anthocyanin extract and wood chip extracts from different species led to color differences visually perceptible between in all model wine solutions. This result confirms the results already show for color intensity and the partial results obtained for CieLAB chromatic coordinates (Figs. 2 and 3). In addition, the high color differences results were independently of the wood chip extracts species. However, the model wine solutions containing wood extracts obtained with longer extraction time (30 days) showed significantly higher color difference values. Recently, Cejudo-Bastante et al. [16], conclude that the addition of oak wood chips (during both fermentation and aging) induced also a visually perceptible color differences from the early months of storage of a Syrah red wine (values among 5 and 12).

3.3. Antioxidant capacity

Figure 5 shows the antioxidant capacity quantified in the different model wine solutions after 30 aging days.

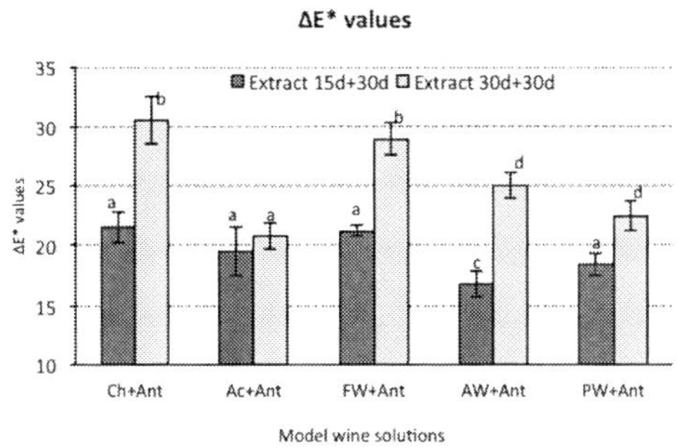

Figure 4. Color difference (ΔE) quantified in model wine solutions containing different wood chip extracts after 30 aging days. Ch + Ant = Cherry wood chip extract and anthocyanin extract; Ac + Ant = Acacia wood chip extract and anthocyanin extract; FW + Ant = French wood chip extract and anthocyanin extract; AW + Ant = American wood chip extract and anthocyanin extract; PW + Ant = Portuguese wood chip extract and anthocyanin extract.

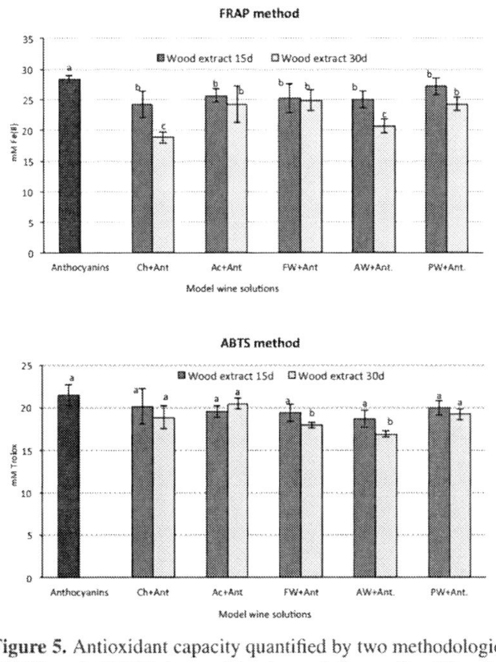

Figure 5. Antioxidant capacity quantified by two methodologies (FRAP and ABTS) in model wine solutions after 30 aging days. Ch + Ant = Cherry wood chip extract and anthocyanin extract; Ac + Ant = Acacia wood chip extract and anthocyanin extract; FW + Ant = French wood chip extract and anthocyanin extract; AW + Ant = American wood chip extract and anthocyanin extract; PW + Ant = Portuguese wood chip extract and anthocyanin extract.

According to the results obtained, red model wine alone showed the highest antioxidant capacity (especially considering the results obtained by the use of FRAP method where the values were significantly higher) in comparison with the results obtained for model wine solutions containing also the different wood chip extracts. According to Rivero et al. [17], glucoside anthocyanins are one of the major anthocyanin group that determine the antioxidant capacity power of red wines, in particular malvidin-3-monoglucoside.

Concerning to the potential impact of the use of different wood chip extract species on antioxidant capacity, the results obtained were not totally distinctive. Thus, a slight tendency for lower antioxidant capacity values were obtained for the red model wine solutions mixed with cherry wood extract and with American oak wood extract (both extracted obtained after 30 extraction days). According to several authors [4,18] the use of oak wood induce an increase of antioxidant capacity of the red wines during the aging process as a result of polyphenol compounds extraction. However, contrary results were obtained by Gortzi et al. [19]. For these authors, after a considerable aging period in contact with oak wood chips, red wines showed a decrease of polyphenolic content and consequently a reduction of antioxidant capacity. In addition, it is important to note that according to Psarra et al. [20] polyphenols extracted from oak wood chips yield almost 6.4 times higher reducing power comparative with acacia chips. Therefore, the antioxidant effects exerted by polyphenols do not depend exclusively on their total amount, but also on the different polyphenolic structures [21]. This fact could help to explain the similar values of antioxidant capacity found in the majority of the model wines solutions studied. Probably the use of a more extended aging period or higher wood chips concentration used to produce the wood chip extracts could be a factor to consider in the future to potentiate eventual differences.

4. Conclusions

In this work it was possible to detected a clear influence of the use of different wood chip species on evolution of the anthocyanin composition and consequently on chromatic parameters. This influence was especially detectable by a decrease of the color intensity and especially on red color component values. In addition, the antioxidant capacity showed also a slightly decrease of the values when anthocyanins were put in contact during 30 days with wood chip extracts. It was also clear that the time used to obtain the wood chip extracts had a particular influence on the results determining an increase of the variation of the parameters analyzed.

This research contributes to a better knowledge about the color evolution of red wines when they are aging in contact with oak wood and in especially when is using new alternative wood species, especially acacia and cherry. This aspect will be important if these wood species may come in the future to be authorized for oenological purposes.

António M. Jordão thanks to FCT (project UID/Multi/04016/2016) and IPV-CI&DETS for their financial support. Authors also appreciated the collaboration of the companies that supplied the wood chip samples used in this study (SAI Enology, AEB Bioquímica and J.M. Gonçalves cooperage).

References

[1] I. Revilla, M.L. González-Sanjosé, M.C. Gómez-Cordovés, Food Sci. Technol. Int. **5**, 177–181 (1999)
[2] A.M. Jordão, J.M. Ricardo-da-Silva, O. Laureano, Am. J. Enol. Vitic. **57**, 377 (2006)

[3] A.M. Jordão, J.M. Ricardo-da-Silva, O. Laureano, W. Mullen, A. Cozier, Aust. J. Grape Wine Res. **14**, 260 (2008)
[4] F.J. Gonçalves, A.M. Jordão, Ital. J. Food Sci. **21**, 51 (2009)
[5] G. Coninck, A.M. Jordão, J.M. Ricardo-da-Silva, O. Laureano, J. Int. Sci. Vigne Vin **40**, 25 (2006)
[6] M. Kyraleou, P.-L. Teissedre, E. Tzanakouli, Y. Kotseridis, N. Proxenia, K. Chira, I. Ligas, S. Kallithraka, Oeno One **50**, 209 (2016)
[7] S. Pérez, M.L. González-Sanjosé, J. Agric. Food Chem. **52**, 1181 (2004)
[8] Y. Glories, Connaiss. Vigne Vin **18**, 253 (1984)
[9] S. Pérez, M.L. González-Sanjosé, Food Chem. **81**, 301 (2003)
[10] M.D. Rivero, P. Muñiz, M.L. González-Sanjosé, J. Agric. Food Chem. **55**, 5476 (2007)
[11] M. Sánchez, M.L. González-Sanjosé, S. Pérez, M. Ortega, C. González, J. Agric. Food Chem. **57**, 11498 (2009)
[12] I. Revilla, M.L. González-Sanjosé, Eur. Food Res. Technol. **213**, 281 (2001)
[13] F. Chinnici, N. Natali, F. Sonni, A. Bellachioma, C. Riponi, J. Agric. Food Chem. **59**, 6575 (2011)
[14] M. De Rosso, A. Panighel, A. Dalla-Vedova, L. Stella, R. Flamini, J. Agric. Food Chem. **57**, 1915 (2009)
[15] J.A. Martínez, M. Melgosa, M.M. Pérez, E. Hita, A.I. Negueruela, Food Sci. Technol. Int. **7**, 439 (2001)
[16] M.J. Cejudo-Bastante, F.J. Rivero-Granados, F.J. Heredia, Eur. Food Res. Technol. **243**, 575 (2017)
[17] M.D. Rivero, P. Muñiz, M.L. González-Sanjosé, Food and Chem. Toxicol. **46**, 2815 (2008)
[18] A.M. Jordão, A.C. Correia, R. Del Campo, M.L. González-Sanjosé, Eur. Food Res. Technol. **235**, 817 (2012)
[19] O. Gortzi, X. Metaxa, G. Mantanis, S. Lalas, Food Chem. **141**, 2887 (2013)
[20] C. Psarra, O. Gortzi, D.P. Makris, J. Inst. Brew. **121**, 207 (2015)
[21] M. Aoun, D.P. Makris, Inter. J. Food Studies **2**, 238 (2013)

Influence of different yeasts on the amino acid pattern of rosé wine

Karin Mandl[a], Karin Silhavy-Richter, Karin Korntheuer, Martin Prinz, Elsa Patzl-Fischerleitner, and Reinhard Eder

Federal College and Research Center of Pomology and Viticulture, Wienerstraße 74, 3400 Klosterneuburg, Austria

Abstract. In an experiment with Rosé wine, 27 different commercial yeasts were tested for their influence on the amino acid pattern of the wine. Amino acids are precursors for aromatic substances; therefore a large variation of the amino acid values in the wine was expected. Blaufränkisch grapes with 20° KMW were matured in the cellar with 27 different commercial yeasts. The fermentation was carried out in 34l vessels. The wines were measured for amino acids using an HP 1200 liquid chromatograph and HP-FLD1100 according to Umagat. The wines showed 13.5% alcohol and little residual sugar. The measurement results of the amino acids of the different wines showed large variations. For example, the amount of the amino acid alanine in wine varied from 17 to 138 mg. In particular, the wines of the yeast Pino Type showed the highest amounts of alanine in comparison to the other fermented wines.

1. Introduction

Amino acids serve as sources of nitrogen for yeasts and as precursors for some aromas [14]. They are building blocks of proteins that are needed for enzymes and the construction of support structures of the cell. For the human body, wine is a good source for undenatured amino acids. Although it is not enough to cover the daily needs, it is a good addition. The various amino acids are incorporated into muscles, tendons, hormones and / or as neurotransmitters [5]. Heated foods [18] make the coverage of daily needs considerably more difficult. Eight amino acids are "essential" for humans; a supply from outside is necessary because they cannot be produced independently. These are: isoleucine, leucine, lysine, methionine, phenylalanine, threonine, tryptophan and valine [5]. Non-essential amino acids, on the other hand, the body is able to produce by itself. But they are not so important. A good source of non-denatured essential amino acids and also good vitamin B suppliers are the yeasts contained in "Sturm" (Austrian partially fermented grape must). Yeasts are generally good nutrient suppliers for humans, since they are rich in vitamins, trace elements and minerals [11]. From the literature it is known that the influence of the yeast results in large differences in the amino acid pattern of the wine [9,10,16]. The amino acid contents of the must are mainly influenced by wine-growing techniques [12], but the wine is influenced by various yeasts. The amino acids are often the basis for the formation of aromas and serve as precursors [1,3,6,7,14]. Since yeasts have mostly been selected for their ability to develop aromas, it is not surprising that there are great differences in the amino acid pattern of the end wine when using different yeasts, despite the same starting must.

In particular, the amino acids leucine, valine, isoleucine and phenylalanine are involved in determining the aroma profile [14]. In principle, amino acids are the building blocks of proteins. They consist chemically of a nitrogen and an acid group. They are enzymes or scaffold proteins that are necessary for different metabolic processes of humans and animals [2].

2. Material and methods

In this experiment, microvinification was carried out with 27 different commercial yeasts.

Yeast list: Uvaferm WAM, Filtraferm C, Lalvin Opal ICV, Zymaflore Alpha X5, Siha Rubin Cru, Zymaflore X5, Prezisio Red and Fruity, Prezisio Arom C, Fermicru AR2, Littolevure Pink, Siha Active Yeast 3, Oenoferm Freddo, Filtraferm Extract, Filtraferm C Fresh, Lalvin EC 1118, Oenoferm Pino Type, Lalvin Rhone 2056, Siha Active Yeast 8, Actiflore Rosé, Preziso White & Fruity, Oenoferm Rosé, Fermicru Rosé, Fermiflor, Lalvin Rhone 4600, Oenoferm Rouge, Oenoferm Tipico, Oenoferm Klosterneuburg.

A must of the Blaufränkisch variety was used for this experiment. The grape material came from the vineyard Götzhof in Langenzersdorf. The enzymatic pretreatment as well as the bentonite treatment were carried out in the HBLA cellar in Klosterneuburg. The must had a gradation of 20 KMW °, a total acidity of 5.1 g / l and a pH of 5.5. The fermentation took place in a double expansion in 34-liter carboys. The fermentation control was carried out using Oenofoss ™(Foss).

The amino acid spectrum was performed from pre-filled wine bottles with a "HP 1200 Liquid" chromatograph and "HP-FLD 1100" according to Umagat [2]. The finished wines showed values of 13.5% vol. alcohol, very little residual sugar (very strongly dependent on the yeast),

[a] e-mail: Karin.Mandl@weinobst.at

 DOI: 10.1051/bioconf/20170902014

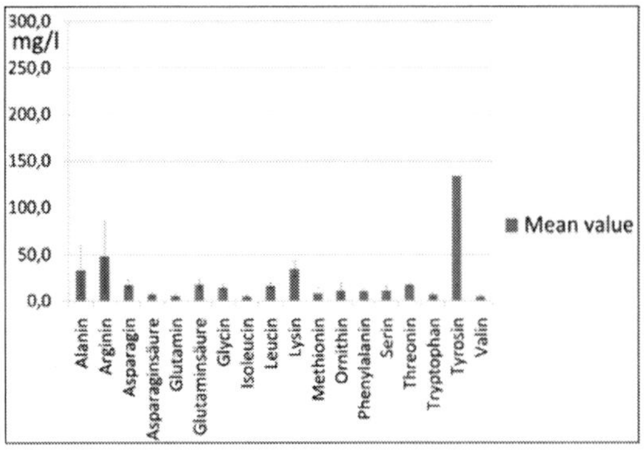

Figure 1. Overview of the investigated amino acids - mean values (n = 54) on all yeasts used (27).

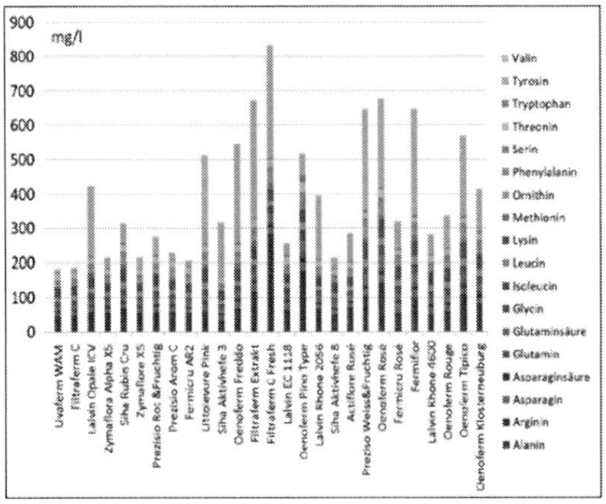

Figure 2. Distribution of the amino acids for the individual commercial yeasts (mean values, n = 2). The highest amounts reached tyrosine, arginine and alanin.

no glucose, 6.3 g/l ± 0.3 g of total acid, a pH value of 3.4 and volatile acid at 0.4 g / l ± 0.1 g.

3. Results and discussion

The amino acids examined in the rosé wines are shown in Fig. 1 (overview with standard deviation from the yeast population) and the more precise distribution of the respective amino acids to the corresponding yeasts is shown in Fig. 2.

Alanine can be produced by the yeast itself [2] when ammonia and pyruvic acid are available. It is not an essential amino acid for humans [5]. In this experiment, the various commercial yeasts showed very wide variations between 17 and 138 mg/l in the wine. The yeasts "Uvaferm WAM", "Filtraferm C", "Zymaflora Alpha", "Prezisio Arom C", "Fermicru AR2", "Littolevure Pink", "Siha Active Yeast 3", "Lalvin EC 1118", "Actiflore Rosé" and "Lalvin Rhone" showed low values. The yeasts "Oenoferm Pino Type" and "Oenoferm Rosé" showed very high contents of the amino acid alanine. In addition, both yeasts had an increased residual sugar value and a detectable lactic acid value.

The measured arginine values varied between 16 and 250 mg/l. The highest levels were found in the yeast "Filtraferm C Fresh". Arginine is the most important source of nitrogen for yeasts in wine. This is a semi-essential amino acid [5], which means that it must be absorbed in some situations with the food and serves as a dietary supplement for certain ailments.

Asparagine is a good source of nitrogen for yeasts [17] and has a negative influence on the formation of succinic acid. It is a non-essential amino acid for humans [5]. The consumption of the yeasts varies. The values ranged from 9 to 27 mg/l. The lowest value was found in the yeast "Uvaferm WAM" and the highest in "Fermicru Rosé".

Aspartic acid is a good source of nitrogen for yeasts [17] and is not an essential amino acid for humans. Values of 4.5 to 14 mg/l were measured. The highest values were found in "Oenoferm Pino Type".

Glutamine is an important amino acid for plants and animals. It is incorporated in high concentrations in the muscle and shows an anabolic effect. Basically, this amino acid can be formed by the human body itself [5]. The highest values were measured with "Oenoferm Pino Type". The values ranged from 2.3 to 18.1 mg/l.

Glutamic acid is the basis for succinic acid formation [2]. In the presence of asparagine, the formation of succinic acid decreases. The glutamate in food is the ester and salt of glutamic acid. The umami taste is a full flavor, which is why glutamate is often used as a food additive. The amino acid can be formed by the body itself. The highest values in the wine were found in "Oenoferm Pino Type". The values ranged from 10.1 to 34.4 mg/l.

Glycine is characterized by its sweet taste. It is contained in all proteins and tissues in the human body and can be produced by itself. The values were 8.4 to 28.3 mg/L and the highest value was found with "Oenoferm Rosé".

Isoleucine is an essential amino acid for the human body. It serves as an endogenous source of energy in hunger phases and regulates some hormones in the body, such as the stimulation of the insulin release and the release of the growth hormone somatotropin [5]. A deficiency of isoleucine leads to muscle weakness and lack of drive.

In wine, isoleucine is a precursor for the fruity fatty acid (2-methylbutyric acid) and the pineapple tone (ethyl 2-methylbutanoate) [14]. The decomposition and conversion is carried out according to the Ehrlich metabolic pathway [4]. The values ranged from 2.8 to 9.8 mg/l. The highest values were produced by "Oenoferm Pino Type".

The amino acid leucine is an essential amino acid and plays a central role in the energy balance of the muscle cell. It is the basis for muscle building and is involved in the healing of wounds [5]. In wine, leucine precursors [14] are fruity, nutty notes (isovaleraldehyde), rancid, rotted fruit and cheese tones (isovaleric acid), apple tones (ethyl isovalerate), fruity banana and pear tones (acetic acid isopentyl esters) and formation of fusel alcohol (isoamyl alcohol). The measured values were 11.6 and 24 mg/ml. The highest value was produced by "Oenoferm Pino Type".

Yeast is a good source for lysine [5]. It is an essential amino acid. It is responsible for the development of collagen. The values in the wine were 22.3 to 54 mg/l, with the highest value for "Preziso White & Fruity".

Methionine is a sulfur-containing essential amino acid. It is involved in the formation of glutathione [5] and provides methyl groups for various chemical processes.

The highest value was found in "Siha Rubin Cru"; the other values were between 2 and 22 mg/l.

Ornithine is not an essential amino acid. The highest value was found in "Preziso White & Fruity", the other values were between 2.5 and 38.8 mg/l. In the human body it shows effects on somatotropin and insulin (endogenous hormones) [5].

Phenylalanine is synthesized in plants via the shikimic acid pathway [2]. The flavonoids from phenylalanine can be biosynthesized via the pathway of cinnamic acid. By 4-hydroxylation from phenylalanine L-tyrosine in the human body is formed. L-tyrosine is involved in the synthesis of hormones such as adrenaline, noradrenaline and melanin. In wine, phenylalanine is a precursor for rose, honey and flower tones (2-phenylacetate) [14]. The values in the wine ranged from 7.1 to 18 mg/l.

Serine is a non-essential amino acid and can be prepared from threonine, glycine or glucose. Serine forms the basic structure of many membranes and is involved in the formation of the amino acids cysteine and tryptophan [5]. In addition, serine is involved in the formation of the neurotransmitter acetylcholine - a substance that plays an important role in the neuronal transmission of nerve impulses [13]. The values in the wine were between 5.9 and 31.5 mg/l.

Threonine is an essential amino acid for the human body. The highest value was produced by "Oenoferm Pino Type". The other wines varied from 12.7 to 29.6 mg/l.

Tryptophan is an essential amino acid and represents the precursor for the happiness hormone serotonin [5]. It is produced by plants and microorganisms and the estimated daily requirement is 3.5 to 6 mg/kg body weight. The values in the wines were from 4 to 23.3 mg/l.

Tyrosine is not an essential amino acid and is produced from phenylalanine. It is the starting substance for the biosynthesis of DOPA (L-3,4-dihydroxyphenylalanine), dopamine, catecholamines, melanin, thyroxine and tyramine (neurotransmitters and thyroid hormones) and a component of all proteins [5]. In the wines, the values were 6.7 to 352 mg/l.

Valine is a precursor in wine for apple flavor (isobutyraldehyde, isobutyric acid), fruity notes (isobutanol) and the banana tone (isobutyl acetate) (Styger et al., 2011). Grapes harvested later have a higher valine content. For the human body, valine is an essential amino acid and an essential building block for proteins and muscle building. The contents in the wines varied between 1.1 and 6.8 mg/l.

Proline is not an essential amino acid and is not metabolized by yeasts [8]. The values were shown for different yeasts from 920 mg to 1,460 mg/l. The amino acid proline can be produced by the human body from glutamate. In the presence of vitamin C, proline is incorporated into collagen as L-hydroxyproline and determines its mechanical properties. Deficiency results in scurvy [5].

It was demonstrated that yeasts have a certain requirement of amino acids. Above all, the essential amino acids are needed as precursors for building aromas. The question is how much can be actively influenced by viticulture treatments and the choice of yeast.

References

[1] T. Bauza, A. Blake, F. Daumas, J.C. Cabanis, Journal of Chromatography A. **707**, 373–379 (1995)
[2] J.B. Berg, J.L. Tymoczko., L Stryer, *Biochemie* 5.Auflage, Spektrum Akademischer Verlag, Heidelberg, Berlin (2003)
[3] R.M Callejón., A.M. Troncoso, M.L. Morales, Talanta **81**, 1143–1152 (2010)
[4] F. Ehrlich, Ber Dtsch Chem Ges **40**, 1027–1047 (1907)
[5] U. Gröber, *Mikronährstoffe*, 3.Auflage, Wissenschaftliche Buchgesellschaft (2010)
[6] P. Hernández-Orte, M.J. Ibarz., J. Cacho., V. Ferreira, Food Chemistry **98**, 300–310 (2006)
[7] M.G. Lambrecht, I.S. Pretorius, S.Afric. J. Enol. Vitic. **21**, 200 (2000)
[8] P. Lehtonen, Am.J.Vitic. **47**, 2 (1996)
[9] A. J. Martínez-Rodríguez, A. V. Carrascosa, P. J. Martín-Álvarez, V. Moreno-Arribas, M. C. Polo., Journal of Industrial Microbiology and Biotechnology **29**, 6, 314–322 (2002)
[10] V. Moreno-Arribas, E M. Pueyo, C. Polo, P. J Martín-Álvarez J., Agric. Food Chem. **46**, 4042–4051 (1998)
[11] J Schormüller. *Alkoholische Genussmittel* (Springer Verlag, Berlin Heidelberg, 1968)
[12] A. Schwab, M. Peternel und E. Grebner, Rebe und Wein, **10**, 25–30 (2003)
[13] S. Silbernagl, *Taschenatlas Physiologie*, (Georg Thieme Verlag, Stuttgart, 2012)
[14] G. Styger., B. Prior, F.F. Bauer, J. Ind. Microbiol. Biotechnol **38**, 1145–1159 (2011)
[15] H. Umagat, P. Kucera, L.F: Wen, J. Chromatogr. **239**, 463–474 (1982)
[16] E. Valero, C. Millán, J.M. Ortega, J. C. Mauricio: Journal of the Science of Food and Agriculture **83**, 830–835 (2003)
[17] G. Würdig, R. Woller, *Chemie des Weines* (Ulmer Verlag, Stuttgart, 1989)
[18] T.Y. Yu, J. D. Morton, S .Clerens, J. M Dyer, Food Science and Food Safety **16**, 141–159 (2017)

DOI: 10.1051/bioconf/20170902015

Linking microbial community on grapes from two Portuguese wine regions to the biogenic amines production in musts

Rita Calisto[1], Ana Patrícia Graça[1,2], Ana Rita Lopes[3], Paula Araújo[4], Eugénia Pinto[2,3], Joana Oliveira[4], José A.M. Catita[5,6], Natacha Fontes[7], António Graça[7], and Olga Maria Lage[1,2]

[1] Department of Biology, Faculty of Sciences, University of Porto, Portugal
[2] Interdisciplinary Centre of Marine and Environmental Research, University of Porto, Portugal
[3] Department of Biological Sciences, Faculty of Pharmacy, University of Porto, Portugal
[4] REQUIMTE- LAQV, Department of Chemistry and Biochemistry, Faculty of Sciences, University of Porto, Portugal
[5] Paralab, SA
[6] CEBIMED - Faculty of Health Sciences; University Fernando Pessoa, Porto, Portugal
[7] Sogrape Vinhos, S.A.

Abstract. Grapevine-associated microbiota influences wine organoleptic properties. Spoilage due to undesired microorganisms and biogenic amines (BAs) presence are two main constrains that must be seriously considered. In wine, BAs can originate from the grape berries or can be produced during fermentation, ageing or storage. This work aimed to understand if the high BAs levels observed in musts can have its origin in the microbial community present on grapes. The following methodologies were done: bacterial and fungal grapes communities' isolation, BAs quantification in grapes and musts and molecular amplification of the genes related to BAs production. For comparative purposes, microbial communities from grapes and musts from Douro (low BAs levels in musts) and Alentejo (high BAs levels in musts) were used. Higher number and diversity of bacteria were observed in Alentejo grapes comparatively to Douro ones. Filamentous fungi were predominant when compared with yeasts and the diversity was higher in Alentejo. BAs levels mainly due to putrescin were about ten times higher in grapes and musts from Alentejo. As bacteria isolated from Alentejo grapes showed a great BAs-production potential, namely putrescin, our results suggest a bacterial grape origin for the high putrescine levels found in fresh musts of this region.

1. Introduction

The production of wine is an important industry with a major economic and social impact in Portugal and in many other countries. Although microorganisms play a fundamental role on the entire process of wine production and strains responsible for wine fermentation are well known and studied, the grape-bunch associated microbial communities are still poorly characterized and their role in wine processing very much unknown [1–3]. Their influence on wine terroir is gaining relevance [3]. However, the microbial community can also induce noxious effects with great impact on this industry. Spoilage microorganisms can grow during fermentations or the storage period [4]. One problem faced during wine making in specific situations is the appearance of high levels of Biogenic Amines (BAs) which may have a microbial origin. BAs are secondary metabolites of organic nitrogenous nature and putrescine, histamine, tyramine and cadaverine are the main BAs associated to wine [4]. High BA levels in wine can be potentially toxic to humans and their effect can even be potentiated by alcohol [5].

The aim of this work was to assess the microbial communities present on grapes from two Portuguese wine producing regions (Douro and Alentejo) with known differences in the levels of must BAs and correlate these levels to the bacterial potential capacity for BAs production.

2. Material and methods

Wine grape sampling in Douro and Alentejo was done one month before (D1 and A1) and at harvesting (D2 and A2). Grapes collected aseptically were transported under cold conditions and processed within 24h. Musts were collected immediately after crushing, under the same conditions in Douro and Alentejo. Isolation was carried out using grape extracts in a non-selective broad medium for heterotrophic bacteria (NA medium pH 5.5 and 7.0: 0.5% peptone, 0.3% yeast extract and 1.5% agar) and in appropriate media for yeasts and filamentous fungi (yeast peptone glucose (YPG) supplemented with biphenyl and chloramphenicol, pH 5.0; Sabouraud dextrose agar (SAB) supplemented with chloramphenicol, pH 5.6; YPG supplemented with biphenyl, chloramphenicol and cycloheximide (YPGciclo), pH 5.0; SAB supplemented with chloramphenicol and cycloheximide (SABciclo), pH 6.9). BAs levels were quantified in the juice extracted from sampled grapes and in musts immediately after harvesting [6]. Bacterial isolates were phylogenetically identified based on the analysis of the 16S rRNA gene [7] and screened for the presence of the genes involved in the production of BAs: cadaverine, putrescine, tyramine and

DOI: 10.1051/bioconf/20170902015

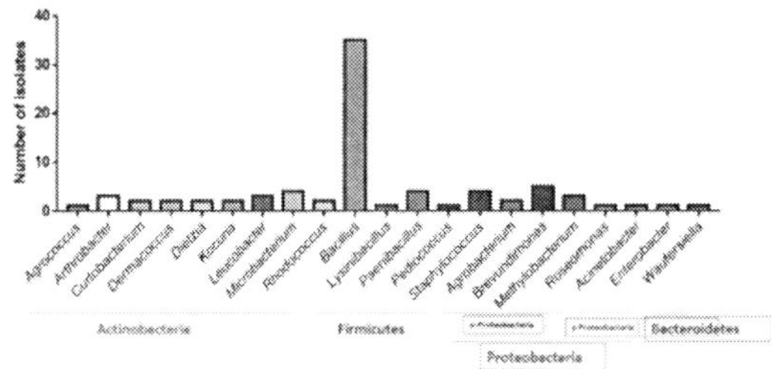

Figure 1. Affiliation of the isolates to corresponding genus and phyla or class. Although the *Firmicutes Bacillus* was the most abundant genus, *Actinobacteria* was de most diversified group (9 different genera).

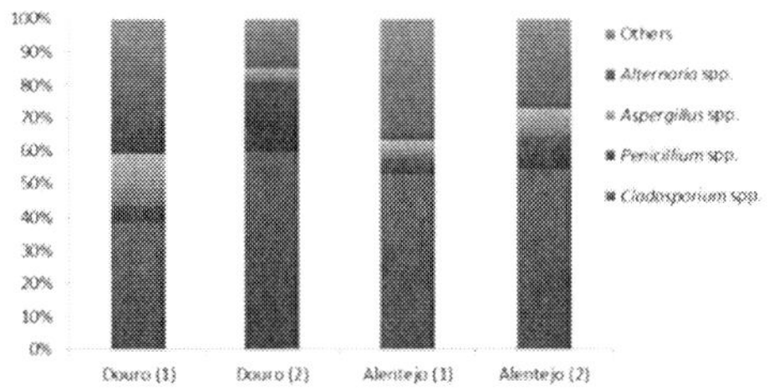

Figure 2. Filamentous fungi isolated from grapes sampled in Douro and Alentejo. Douro (1) – Douro one month before harvesting; Douro (2) – Douro at harvesting; Alentejo (1) – Alentejo one month before harvesting; Alentejo (2) – Alentejo at harvesting.

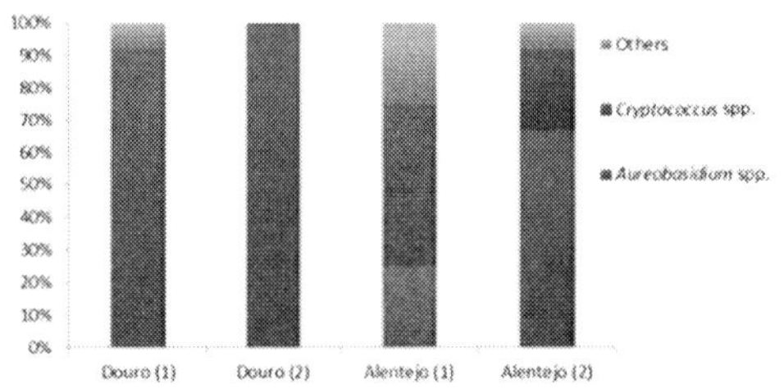

Figure 3. Yeasts isolated from grapes sampled in Douro and Alentejo. Douro (1) – Douro one month before harvesting; Douro (2) – Douro at harvesting; Alentejo (1) – Alentejo one month before harvesting; Alentejo (2) – Alentejo at harvesting.

histamine [8,9]. Yeasts and filamentous fungi isolates were identified by morphological, biochemical and molecular (Internal transcribed spacer gene-ITS) methods [10].

3. Results and discussion

A semi-quantitative analysis allowed concluding that bacterial number was higher in Alentejo isolations comparatively to the Douro ones. However, a total number of fungi was approximately the same in both regions at harvesting but one month before was greater in Douro than in Alentejo. Furthermore, and under the same experimental conditions, a much higher number of bacterial morphotypes that correspond to a higher diversity was retrieved in Alentejo (77%) comparatively to Douro (23%). The most abundant bacterial phylum obtained was *Firmicutes* followed by *Actinobacteria* and *Proteobacteria*

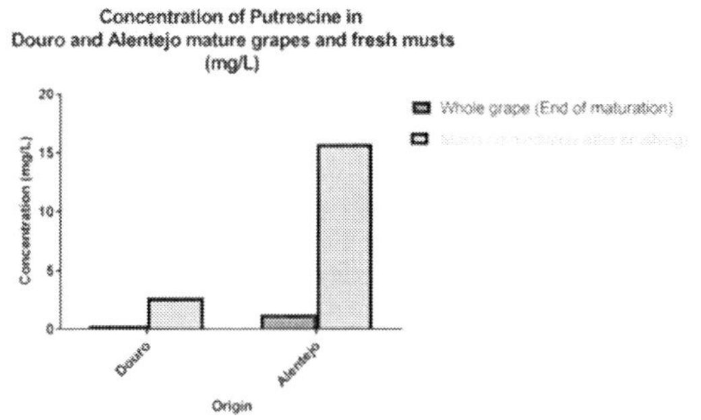

Figure 4. Concentration (mg/L) of putrescine and tyramine in grapes and musts of Alentejo and Douro, quantified through LC-MS technique.

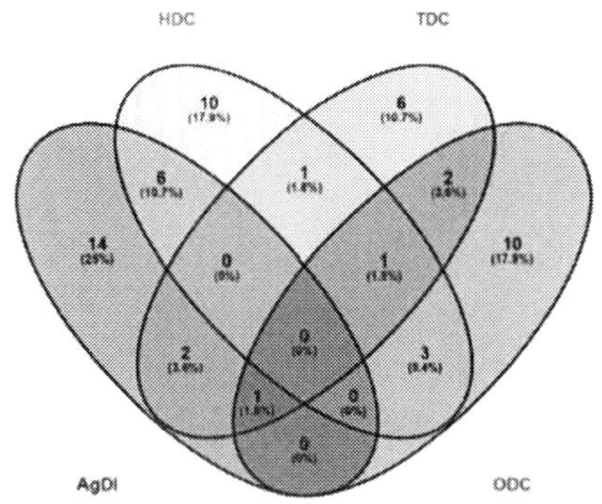

Figure 5. Number of isolates that have the potential to produce one or more BA enzymes (ODC – ornithine decarboxylase, AgDI – agmatine deaminase, HDC – histidine decarboxylase and TDC – tyrosine decarboxylase), by screening of the correspondent gene. 52% of the studied isolates amplified only the *agdi* gene, 17.9% only the *odc* gene and the same percentage the *hdc* gene while only 6 (10.7%) amplified the *tyrd* gene. Regarding isolates that amplified three genes only 1.8% amplified for *odc*, *hdc* and *tyrd* and 1.8% for the *agdi*, *tyrd* and *odc*. Amplification of *agdi* and *odc* was achieved by 3.6% of isolates, 1.8% for the *hdc* and *tyrd*, 10.7% for *agdi* and *hdc*, 5.4% for *odc* and *hdc* and 3.6% for *odc* and *tyrd*. None of the isolates amplified the four genes

(*Alpha* and *Gamma* classes) but the highest diversity (9 genera) was obtained in the *Actinobacteria* (Fig. 1).

Considering grape fungal communities, filamentous fungi were predominant when compared to yeasts.

Fungal species of the genus *Cladosporium* were shown to be prevalent followed by *Aspergillus*, *Penicillium* and *Alternaria* (Fig. 2). *Cryptococcus* spp. were only found in Alentejo while *Aureobasidium* was found both in Douro, where it dominated, and in Alentejo where the number increased with maturation (Fig. 3).

Chemical analysis showed that the levels of putrescine detected in mature grapes were higher in Alentejo compared to Douro. A rapid and relevant increase of the levels of putrescine (more than 10 times) was observed especially in Alentejo from mature grapes to fresh musts.

As relevant differences existed between the bacterial grapes communities of Alentejo and Douro, contrarily to what was observed in the fungal communities, bacteria were screened for potential BAs-production aiming to establish a correlation between bacteria and the high putrescine levels in musts from Alentejo.

The isolated bacteria revealed a great potential for BA-production (57% of the isolates) especially for putrescine (56% of the isolates) and lower levels of amplifications were obtained for histamine, tyramine and cadaverine (Fig. 5). Isolates with gene amplicons for tyramine-production were only detected in Douro region.

Due to the (1) high bacterial number, (2) high bacterial diversity and (3) great potential for putrescine production observed in Alentejo grape bacterial community, our results suggest a bacterial grape origin for the high putrescine levels found in fresh musts of this region.

References

[1] E. Kecskeméti, B. Berkelmann-Löhnertz, A. Reineke, PLoS ONE **11**, e0160852 (2016)

[2] V. Loureiro, M. M. Ferreira, S. Monteiro, R. B. Ferreira, *In* H. Gerós, M. Chaves, S. Delrot (eds) The Biochemistry of the Grape Berry, pp. 241–268 (2012)

[3] N. A. Bokulich, T. S. Collins, C, G. Allen, H. Heymann, S. E. Ebeler, D. A. Mills, mBio **7**(3), e00631–16 (2016)

[4] A. Y. Smit, W. J. du Toit, M. du Toit, South Africa Journal of Enology and Viticulture **29**, 109–126 (2008)

[5] EFSA, European Food Safety Authority (EFSA) The EFSA Journal **9**, 2393. [93 pp.] (2011)

[6] N. Gouveia, Desenvolvimento de uma metodologia analítica para determinação de aminas biogénicas em tunídeos. Master thesis. Funchal: Universidade da Madeira (2009)

[7] A. P. Graça, F. Viana, J. Bondoso, M. I. Correia, L. A. Gomes, M. Humanes, A. Reis, J. Xavier, H. Gaspar, O. M. Lage, Front. Microbiol. **6**, 389 (2015)

[8] M. Coton, A. Romano, G. Spano, K. Ziegler, C. Vetrana, C. Desmarais, A. Lonvaud-Funel, P. Lucas, E. Coton, Food Microbiology **27**, 1078–1085 (2010)

[9] J. M. Landete, S. Ferrer, I. Pardo, Food Control **18**, 1569–1574 (2007)

[10] T. J. White, T. Bruns, S. Lee, J. Taylor. PCR Protocols: a Guide to Methods and Applications **38**, 315–22 (1990)

Influence of different commercial yeasts on the phenol pattern of rosé wine

Karin Mandl[a], Karin Silhavy-Richter, Silvia Wendelin, Martin Prinz, Elsa Patzl-Fischerleitner, and Reinhard Eder

Federal College and Research Center of Pomology and Viticulture

Abstract. In this experiment, 27 different commercial yeasts were tested for their influence on the phenols of a rosé wine. The phenols depend primarily on the vine variety, cultivation and cellar processing. This experiment sought to clarify how great the influence of the yeast on the phenols is. Blaufränkisch grapes with 20° KMW were matured in the cellar with 27 different commercial yeasts. The vinification was carried out in 34 l vessels. An HPLC Agilent Technology 1220 (RRLC) with a column of ZORBAX SB-C18 15 × 2.1 mm (1.8 μm) was used for the measurement. 5 μl of the sample were injected directly. The column oven temperature was 40 °C. The transient was 0.5% formic acid (pH = 2.3) and a gradient of methanol (3%, 4%, 14%, 30% and 70%). The detection took place at 280 and 320 nm. It was shown that the effect of yeast on phenols is low. There was a great influence on the chemical substance tyrosol. This is a phenol that can be formed by the yeast itself. The values varied between 5.4 mg and 13.2 mg/l. The already described influence of the duration of fermentation on tyrosol formation could not be confirmed.

1. Introduction

In this experiment with rosé wine, 27 different commercial yeasts were examined for the influence of phenols in the same starting must. The question arises if differences in the phenol pattern result from the influence of the yeast.

In principle, phenols can be subdivided into flavonoids, phenolic carboxylic acids, volatile phenols, stilbenes, polymeric phenols, and condensed and hydrolyzable tannins. More than 800 different substances have already been described [3] Their effect is mainly in the defense of the plant, in attracting animals by means of colorants and odorants to spread the seeds and as a free radical scavenger [1,4,8]. The phenols of the wine have a great effect on human health. The positive effects of the "French paradox" are well known [6].

2. Material and methods

Must: Grapes from the Blaufränkisch variety from the vineyard Götzhof, Langenzersdorf were used for the production of trial wine. The must had a gradation of 20° KMW, a total acid of 5.1 g/l. The fermentation was carried out in 34 l glass containers in the cellar department.

27 yeast: Uvaferm WAM, Filtraferm C, Lalvin Opal ICV, Zymaflore Alpha X5, Siha Rubin Cru, Zymaflore X5, Prezisio Red & Fruity, Prezisio Arom C, Fermicru AR2, Littolevure Pink, Siha Active Yeast 3, Oenoferm Freddo, Filtraferm Extract, Filtraferm C Fresh, Lalvin EC 1118, Oenoferm Pino Type, Lalvin Rhone 2056, Siha Active Yeast 8, Actiflore Rosé, Preziso White & Fruity, Oenoferm Rosé, Fermicru Rosé, Fermiflor, Lalvin Rhone 4600, Oenoferm Rouge, Oenoferm Tipico, Oenoferm Klosterneuburg Phenol Analysis: An HPLC Agilent Technology 1220 (RRLC) with a column of ZORBAX SB-C18 15 × 2.1 mm (1.8 μm) was used. 5 μl of the sample were injected directly. The column oven temperature was 40 °C. The transient was 0.5% formic acid (pH = 2.3) and a gradient of methanol (3%, 4%, 14%, 30% and 70%). The detection took place at 280 and 320 nm.

3. Results and discussion

Gallic acid, a tannin which naturally occurs in plants and/or can be formed from tannins, was in a range of 0.6 mg/l± 0.2 mg. The measurements also fluctuated within this range during the repetitions. In grapes, gallic acid, a 3,4,5-trihydroxybenzoic acid, is present only to a small extent, and it is very susceptible to oxidation and tastes generally tart and sour. In the industry, this substance is often used as an antioxidant for cosmetics and foodstuffs. "The alkyl esters of gallic acid, besides their anti-oxidative action, have a microbicidal effect against *Saccharomyces cerevisiae*. It increases with increasing chain length of the alkyl radical, that is, with increasing fat solubility of the esters. The effect of gallates on yeast first manifests itself in a clear fermentation inhibition. This inhibition is not specific. The metabolism of the yeast is also impaired in a different way. For example, under the influence of fermentation-inhibiting gallate concentrations, the cells formed were always smaller than without gallate. Under certain conditions, however, increased numbers of cells were found in gallate-containing fermentations. Fermentation inhibition by gallic acid esters ultimately results from the death of the yeast cells. The yeast cell absorbs gallate until a lethal concentration is reached, which leads to cell death. Since this material intake is relatively slow, yeast can even multiply under the influence of fermentation-inhibiting

[a] e-mail: Karin.Mandl@weinobst.at

DOI: 10.1051/bioconf/20170902016

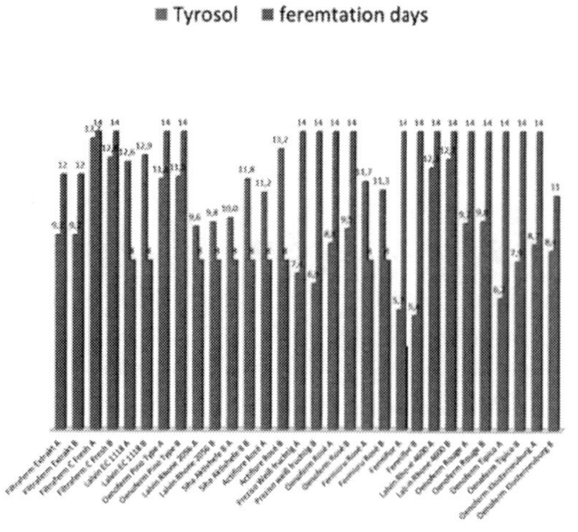

Figure 1. Illustration of the different tyrosol values depending on the fermentation time.

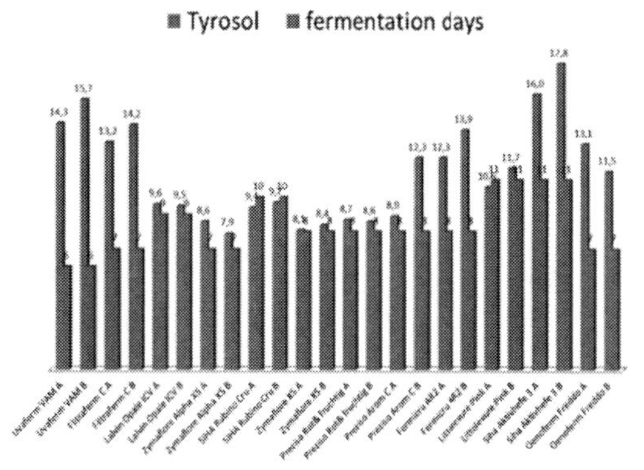

Figure 2. Illustration of the different tyrosol values depending on the fermentation period.

gallate concentrations. The short yeast production time prevents the accumulation of a lethal gallate concentration in the cell. This lethal, intracellular gallate concentration can only arise when the stationary cell number is reached. In optimal substrates, which allow the yeast to grow quickly, the fermentation-inhibiting concentrations of gallate are not sufficient to suppress growth since they are too slow. Only the cell mass already formed is subsequently killed. The killing effect of the gallic acid esters is probably due to the denaturation of cell protein, since the yeast cells become smaller under the action of gallate" [2].

Caftaric acid, a hydroxycinnamic acid derivative, was found in a range of fluctuation of 50 mg/l ± 4 mg. The samples showed no association with inoculation with the yeast.

Tyrosol has a strong antioxidant effect and offers cell protection. The phenol is the only chemical substance produced during fermentation by the yeast [5,9] and the results of the experiment show a strong connection with the yeast. Fermentation without oxygen forms more tyrosol than with oxygen [10]. It was shown in Figs. 1 and 2 that there is no influence on fermentation time. The values in this yeast experiment vary from 5.4 mg/l to 13.2 mg/l and the values of the individual repetitions are very close together.

Trans-coutaric acid is the main derivative of hydroxycinnamic acid in the berry skin. This component serves as a vehicle for anthocyanins. The values of trans-coutaric acid and cis-coutaric acid are very close together.

Procyanidine B1 (epicatechin ($4\beta \rightarrow 8$) catechol) and procyanidine B2 (-) - epicatechin ($4\beta \rightarrow 8$) - (-) - epicatechin belong to the flavonoids. The different commercial yeasts showed no difference in behavior.

Coffee acid, 3,4-dihydroxycinnamic acid belongs to the group of hydroxycinnamic acids and phenolic acids. Coffee acid is one of the most common secondary plant products. In our experiment, values of 2.3 mg/l ± 0.3 mg/l were determined.

Fertaric acid occurs in grapes as well as in wine. This acid is readily esterified with tartaric acid, and in our experiment fertaric acid showed a fluctuation of 2.1 mg/l ± 0.2 mg/l for all yeasts.

Para-coumaric acid belongs to the hydroxycinnamic acids, which are often found in the presence of vine. The contents depend on the type and maturity of the material. 0.9 mg/l ± 0.3 mg were determined. There is an influence shown by the yeast.

P-coumaric acid is the precursor for 4-ethylphenol, which is formed by the Brettanomyces yeasts (Šucur et al. 2016) in wine. 4-vinylphenol is converted by means of the enzyme cinnamate decarboxylase. 4-vinyphenol is further reduced to 4-ethylphenol by means of the enzyme vinyphenol reductase. This specific reaction for Brettanomyces is often replicated in selection media. Coumaric acid is often added to microbiological media to identify Brettanomyces by the smell.

Epicatechin: the values are at 2.3 mg/ml ± 1 mg/ml. The numbers for the comparable yeasts fluctuate very strongly and this suggests that the influence of the yeast is low here. Epicatechin gallate is a flavanol, a typical polyphenol. Polyphenols are secondary plant metabolites and constituents of the vine and of the wine and thus influence the sensory characteristics of the wine. The yeast affects the polyphenols mainly by the ability to absorb substances through the cell wall.

Ferulic acid belongs to the hydroxycinnamic acids and is often esterified with tartaric acid. It serves as a precursor for 4-ethylguaiacol, an off-aroma formed by *Brettanomyces* [11]. The influence of the yeast on the material is apparent. The data of the repeats are parallel and in a range of 0.8 mg/l ± 0.3 mg/l.

References

[1] S. Cesco, T. Mimmo, G. Tonon, N. Tomasi, R. Pinton, R. Teranzo, G. Neumann, L. Weisskopf, G. Renella, L. Lamdi, P. Nannipieri, Biol. Fertil Soils **48**, 123–149 (2012)

[2] H.H. Dittrich, E. Kerner, Wein-Wissen. **19**, 528–535 (1966)

[3] R. Eder, S. Wendelin, ALVA Tagungsband, 293–296 (2002)

[4] S. Hassan, U. Mathesius, Journal of Experimental Botany **63**(9), 3429–3444 (2012)

DOI: 10.1051/bioconf/20170902016

[5] L.A. Hazelwood, J.M. Daran, A.J. von Maris, J.T. Pronk, J.R. Dickson, Appl. Environ Microbiol **74**, 2259–2266 (2008)
[6] B. Kleine-Gunk, Reservatrol, *Länger jung mit der Rotwein-Medizin: Reservatrol verlängert die Lebenszeit und schützt nachweislich vor Krankheiten*, Trias Verlag (2009)
[7] H. König, G. Unden, J. Fröhlich, *Biology of microorganisms on grapes, in must and in wine*, Springer Verlag, 307–334 (2008)
[8] W.A. Müller, S. Frings, F. Möhrlen, *Tier- und Humanphysiologie, eine Einführung*, 5. Auflage Springer Spektrum Verlag (2015)
[9] Z. Pineiro, E. Cantos-Villar, M. Palma, B. Puertas, J. Agri. Food Chem. **56**, 11683–11689 (2011)
[10] Y. Romboli, S. Mangani, G. Buscioni, L. Granchi, M. Vincenzini, Journal of Microbiology and Biotechnology **31**(7), 1137–1145 (2015)
[11] S. Šucur, N. Čadez, T. Kosmerl, Acta agriculturae Slovenica **107**(2), 453–472 (2016)

Molecular characterization and technological properties of wine yeasts isolated during spontaneous fermentation of *Vitis vinifera* L.cv. Narince grape must grown in ancient wine making area Tokat, Anatolia

Zeynep Dilan Çelik[1,2], Hüseyin Erten[2], Merve Darici[2], and Turgut Cabaroğlu[2,a]

[1] Çukurova University, Department of Biotechnology, Adana, Turkey
[2] Çukurova University, Agriculture Faculty, Food Engineering Dept., Adana, Turkey

Abstract. Narince is a native white grape variety of *Vitis vinifera* L grown in Tokat and produces rich and balanced wines often with a greenish yellow tint and delicate fruity flavour. Fermentation by indigenous yeasts may produce wines with complex oenological properties that are unique to specific region. In this study yeast population during alcoholic fermentation of Narince was investigated. Yeasts were identified by PCR-RFLP analysis of the 5.8 ITS rRNA region and sequence information for the D1/D2 domains of the 26S gene. Eight different species belonging to nine genera were identified as: *Hanseniaspora uvarum*, *Hansenispora guilliermondii*, *Pichia kluyveri*, *Metschnikowia* spp., *Pichia occidentalis*, *Torulaspora delbrueckii*, *Candida zemplinina*, *Lachancea thermotolerance* and *Saccharomyces cerevisiae*. *Hanseniaspora guilliermondii*, *Metschnikowia* spp., *Pichia occidentalis* and *Pichia kluyveri* were identified only in the early stage of fermentation. Selected yeasts tested for their physiological traits, ethanol, SO_2, temperature, pH tolerance, H_2S production, killer and enzymatic activity, fermentation rate, flocculation characteristic, foam, volatile acid and volatile compounds production. Among the yeasts, one,*Lachancea thermotolerance* and four *Saccharomyces cerevisiae* strain showed remarkable technological properties and results were compared with those obtained by using commercial starter culture.

1. Introduction

Vitis vinifera cv. Narince grape is one of the most important native white varieties grown in North-east Anatolia Region of Turkey. Narince produces rich and balanced wines that often have a greenish yellow tint and delicate, fruity aromas. Because of their balanced acidity, these wines are suitable for ageing and acquire a rich and complex bouquet over time [1,2].

One of the important factors affecting wine quality is yeast. Grape must harbour indigenous Non-*Saccharomyces* spp. and *Saccharoyces* yeasts. Many studies state that production of wines by indigenous yeasts isolated from its *terroir* can contribute to the regional character of wines, especially wine flavour [3–5]. On the other hand, winemakers usually prefer to use *Saccharomyces* (*S.*) *cerevisiae* strains unique to the region because those strains adapt easily to regional conditions and can readily dominate must during alcoholic fermentation [6].

Cv. Narince is the commercially important unique white wine grape variety of the Tokat region. Narince vineyards of Tokat region has remarkable historical value dating back to Hittie times civilized 4000 years ago. The wineries produced wines from cv. Narince, unfortunately, use the commercial dry yeasts instead of autochthonous indigenous yeasts. As far as it is known, there is no previous study on the molecular characterization of yeasts isolated from Narince spontaneous wine fermentations and then determination of their technological properties.

The aim of this study was to identify yeasts during spontaneous fermentation of cv. Narince and then to determine technological properties of some selected *S. cerevisiae* and Non-*Saccharomyces* spp. yeasts.

2. Materials and methods

2.1. Chemical analysis of Narince must

Density, brix, titrable acidity, $OD_4 20$, FAN (free amino nitrogen), reducing sugar and pH were measured according to the methods outlined by OIV, 2015 [7]; glucose and fructose were quantified using HPLC (LC-20AD, Shimadzu, Kyoto, Japan) [8,9].

2.2. Sampling and Yeast Isolation

The study was conducted in commercial vineyard located in the Tokat Region, Turkey. White grapes of cv. Narince from Tokat hand harvested, was randomly collected (undamaged and healthy grapes) in sterile plastic bags. From the sampling points approximately 10 kg grapes were collected and transported to the laboratory

Indigenous strains belonging to *Saccharomyces* and non-*Saccharomyces* genera were isolated from Narince grape berries and also from different stage of the

[a] e-mail: tcabar@cu.edu.tr

DOI: 10.1051/bioconf/20170902017

spontaneous fermentation. Yeasts on Narince grape berries were isolated according to Combine (2005) [10]. At the laboratory the samples were crushed under aseptic conditions in the original plastic bags. The juice obtained (1 L) was transferred into sterile bottles and spontaneous fermentation was performed under controlled temperature at 18 °C. During first day after the initiation of fermentation, in the middle and at the end of fermentation, samples of must and wine, diluted in 0.1% peptone-water (decimal dilutions), were inoculated onto plates of YPD, Lysine and modified YPD agar (% 10 ethanol, v/v and 2 g/L photassium metabisulphite) supplemented with chloramphenicol and sodium propionate to inhibit bacteria and filamentous fungi, respectively. Plates were incubated at 28 °C for 48 h and about 30 yeast colonies per plate must were collected, then purified and kept at −40 °C in 50% glycerol.

2.3. DNA extraction and RFLP-PCR analysis

Cell lysis for DNA extraction was performed using the InstaGene Matrix kit (Bio-Rad Laboratories, Hercules, CA) following the manufacturer's instruction. The primers used for amplification of 5.8S ITS rDNA region were ITS1 (5′- TCCGTAGGTGAACCTGCGG-3′) and ITS4 (5′- TCCTCCGCTTATTGATATGC-3′) [11]. The DNA amplifications were carried out in a final volume of 50 μL containing 5 U Taq DNA polymerase (MARKA), 1 X PCR reaction buffer, 25 μM dNTP, 25 mMMgCl$_2$, from each primers 100 μM, DNA (50–100 mg/μL). The PCR reaction was performed on a TC-PLUS thermal cycler (Techne, Germany). The PCR conditions were as follows: initial denaturation at 95 °C for 5 min; 35 cycles of denaturating at 94 °C for 1 min; annealing 55 °C for 2 min; an extension at 72 °C for 2 min; and a final extension step of 10 min at 72 °C. The PCR products were subjected to restriction analysis with the restriction endonucleases Hae III, HinfI and Hha I (Thermo, USA), following the manufacturer's instructions. Amplified products and their restriction fragments were separated on % 1.5 (w/v) agarose gels at 130 V constant voltages for 100 min.

2.4. Partial 26S rRNA gene sequence analysis

Isolates sharing identical restriction patterns were classified into groups and one or two samples were chosen as representative of each group for sequence analysis of the D1/D2 domains of the 26S rRNA gene. Amplification of the D1/D2 domains of 26S rRNA was carried out using NL1 (5′ GCATATCAATAAGCGGAGGAAAAG-3′) and NL4 (5′-GGTCCGTGTTTCAAGACGG-3′) primers (Invitrogen, Milan, Italy), according to Kurtzman and Robnett (1998) [12], and the resulting products were commercially sequenced. The sequences obtained in FASTA format were compared with those deposited at the National Center for Biotechnology Information (NCBI), using BLAST to determine their closest known relatives. In order to confirm the identification of isolates at the species level, the sequences of the D1/D2 domain of the 26S rRNA gene were further investigated. To this purpose, the multisequence alignments among our sequences and those of type strains of their closest relatives were performed using ClustalW (Bioedit v. 7.0.9) [13,14]. The number of nucleotide differences between D1/D2 sequences of our isolates and those of their closest relative were also analysed.

2.5. Determination of oenological properties of strains

The ability to grow at different temperature [15,16], to test the resistance to different concentration of SO_2 and ethanol [17], the H_2S production [18], the killer activity and sensitivity of the strains to K_2 killer toxin [14,20,21], enzymatic activity (BioMerieux), foam production [16,22], fermentation rate [3,23], Helm sedimentation test used to determine strains flocculation capacity [24], and the volatile acidity [7] were determined.

2.5.1. Aroma compounds analysis

In 15 mL screw-capped centrifuge tubes, containing, 4.1 g Ammonium sulphate, 2.7 Ml of wine, 20 μL internal standard (140 μg/ml of 4-methyl −2 pentanol and 2-octanol), 6.3 Ml water, 250 μL of dichloromethane were added. The tube was shaken for 90 min and then centrifuged at 2500 rpm, for 10 min, at 20 °C. Once the phases had been separated, the dichloromethane phases was recovered with a 0.5 mL syringe and transferred to a 0.3 mL viale. The extract was injected into a GC-MS under the conditions listed below. Each sample was extracted in duplicate [25,26]. The extracted were analysed by GC-MS-FID. The gas chromatography (GC) system consisted of an Agilent 6890 chromatograph equipped with a flame ionization detector (FID) (Wilmington, DE, USA), and an Agilent 5973-Network mass selective detector (MSD). Volatile compounds were separated on DB-Wax (30 m length ×0.25 mm i.d. ×0.5 μm thickness; J&W Scientific Folsom, CA, USA) column. 3 μL sample of extract was injected. Injector and FID detectors were set at 250 °C. The flow rate of carrier gas (helium) was 3.3 mL/min. The oven temperature of the DB-Wax column was increased from 40 °C (after 3 min holding) to 90 °C at a rate of 2 °C/min, then at a rate of 3 °C/min to 130 °C and at a rate of 4 °C/min to 240 °C with a final hold, at 240 °C for 12 min. The same oven temperature programs were used for the mass selective detector. The MS (electronic impact ionization)(Agilent 5975B VL MSD) conditions were as follows: ionization energy of 70 eV, mass range m/z of 29–350 a.m.u., scan rate of 1.0 scan s−1, interface temperature of 250 °C, and source temperature of 120 °C. The volatile compounds were identified by comparing their retention index and their mass spectra on the DB-Wax column with those of a commercial spectra database (Wiley 6, NBS 75 k) and of the instrument's internal library created from the previous laboratory studies. Some of the identifications were confirmed by the injection of the chemical standards into the GC-MS system. Retention indices of the compounds were calculated by using an n-alkane series [27,28].

The determination of acetaldehyde and ethyl acetate was carried out by direct injection of 1 μL samples into a gas chromatograph, Agilent 6890 N equipped with FID. Acetaldehyde and ethyl acetate separated using a Chrompack CP-WAX-57CB capillary column (0.25 mm i.d. × 60, m × 0.4 μm film thickness) (Middelburg, The Netherlands). GC settings were as follows: injection temperature: 160 °C; oven temperature: 5 min at 40 °C,

Table 1. General composition of Narince must.

General Composition	Must
Density (g/cm^3, 20 °C)	1.087 ± 0.0
°Brix	20.06 ± 0.3
pH	3.18 ± 0.0
Titratable acidity (g/L)*	5.44 ± 0.0
Reducing sugar (g/L)	189.9 ± 2
OD_{420}	0.06 ± 0.0
FAN (mg/L)	205 ± 0.3
Sugars (g/L)	
Glucose	97.1 ± 1.4
Fructose	92.8 ± 2.0

*As tartaric acid.

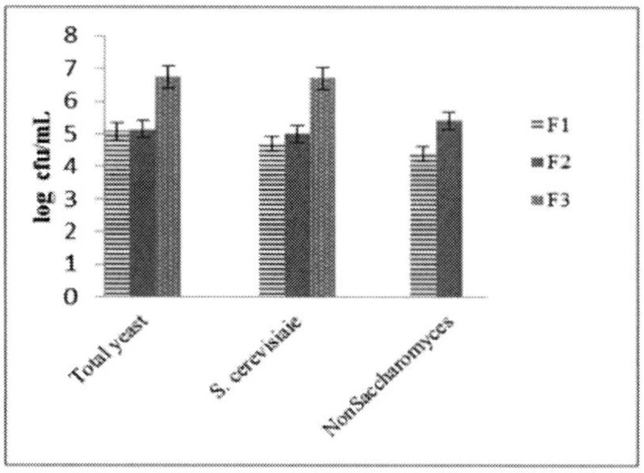

Figure 1. The viable counts from the selective and non-selective agars. F1: beginning of fermentation, F2: middle of fermentation, F3: end of fermentation.

then increased by 4 °C per minute up to 102 °C and 2 °C per minute up to 125 °C and hold for 5 minute and then 3 °C per minute up to 160 °C, 6 °C per minute up to 200 and finally hold 5 minute at 200 °C; carrier gas: He (1.3 mL/min); split rate: 1:50. The quantification was performed by using equation stated reference of Cabaroğlu et al. (2011) [29] with internal standard (4-methyl-2-pentanol) method. Analysis was done in duplicate.

2.6. Statistical analysis

The results of major aroma compounds were compared by the analysis of variance (ANOVA) and principal component analysis (PCA) was carried out using XLSTAT (2015, Addinsoft, New York, USA). PCA tests were used to determine the significant or differences between strains.

3. Results and discussion

Proximate composition of Narince must is given in Table 1.

3.1. Enumeration of yeast population

The viable counts from the selective and non-selective agars are shown in Fig. 1. At the beginning of the spontaneous fermentation a total yeast count on YPD medium was 5 log cfu/mL and they increased to 6.7 log cfu/mL at the middle of fermentation. Initial levels of total non-*Saccharomyces* spp. yeasts were 4.7 log cfu/mL at the beginning and then number increased to 6 cfu/mL at the middle of fermentation and end of fermentation it was not counted.

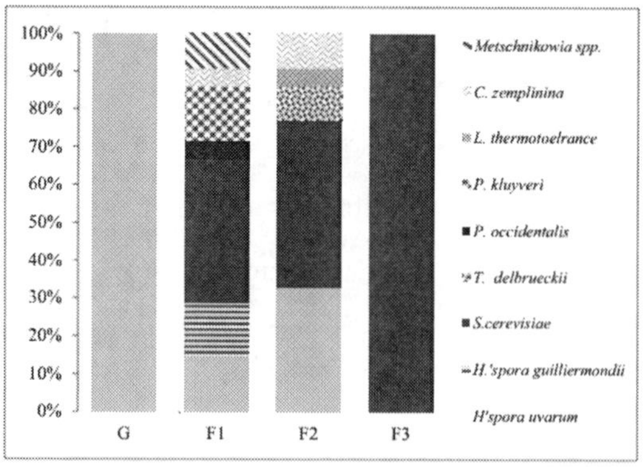

Figure 2. The diversity of yeast at grape and different fermentation stages. G: grape, F: beginning of fermentation, F2: middle of fermentation, F3: end of fermentation, *L. thermotolerance* (*Lachancea thermotolerance*), *T. delbrueckii* (*Torulaspora delbrueckii*), *P. kluyveri* (*Pichia kluyveri*), *C. zemplinina* (*Candida zemplinina*), *P. occidentalis* (*Pichia occidentalis*), *S. cerevisiae* (*Saccharomyces cerevisiae*), *H'spora guilliermondii* (*Hansenispora guilliermondii*), *H'spora uvarum* (*Hanseniaspora uvarum*).

3.2. Strain identification and genetic characterization

The diversity of yeast at grape and different fermentation stages are shown in Fig. 2. *Hanseniaspora uvarum* was the only yeast isolated from Narince grapes before fermentation. During fermentation, however, a diversity of *S. cerevisiae* and non-*Saccharomyces* spp was observed. *S. cerevisiae* and *Hanseniaspora uvarum* were the dominant species at the beginning and *Metschnikowia* spp, *Pichia kluyveri, Hansenispora guilliermondii, Pichia occidentalis and Candida zemplinina* were isolated with less extend at the beginning of fermentation. At the middle of fermentation, *S. cerevisiae, Hanseniaspora uvarum, Candida zemplinina* started to overcome yeasts and others were not isolated from middle of fermentation with the exception of *Lachancea thermotolerance* and *Torulaspora delbrueckii* which were not observed at the beginning. *S. cerevisiae* was the only isolated yeast at the end of fermentation.

The total yeasts diversion is shown in Fig. 3. Five different genera and nine different yeast species were identified. The *S. cerevisiae* yeasts represented most of the total yeast population (49%). *S. cerevisiae*, followed by *Hanseniaspora uvarum* (23%), *Candida zemplinina* and *Pichia kluyveri* (6%), *Torulaspora delbrueckii* (5%), *Hanseniaspora guilliermondii* (4%)*Lachancea thermotolerance* (3%), *Metschnikowia* spp. (3%), and *Pichia occidentalis* (1%).

3.3. Technological properties of yeasts

Eighty two yeast strains were isolated from spontaneous fermentation of Narince must. Strains were grouped according to their phylogenetic mapping (results not shown). Some of the strains belonging to different groups were examined to determine their technological properties according to the Perez-Coello et al., (1999); Labagnara (2016) [3,30]. The results are shown in Table 2. Twenty three strains were eliminated because of high production

Table 2. Technological properties of remaining yeasts.

	884	1088	1120	1333	1378
Resistance to 12% (v/v) ethanol	+	+++	++	++	+++
Resistance to 200 mg/L SO_2	+++	+++	+++	+++	+++
Growth at low temperature 15 °C	++	++	++	++	++
H_2S Production	3:light brown	4:brown	4:brown	3:light brown	4:brown
Killer activity	+	+	+	+	+
Growth at Brix 30°	+++	+++	+++	+++	+++
Foam production (20 °C)	F2	F1	F0	F1	F2
Foam production (15 °C)	F0	F0	F0	F1	F1
Fermentation rate (g CO_2/L.h)	0.45± 0.01	0.99± 0.02	1.7± 0.01	0.8± 0.02	1.6± 0.1
Volatile acidity (g/L)	1	0.72	1.02	1	0.8
Flocculation (%)	69	98	97	86	82
Alkaline phospotase	1	2	2	1	2
Esterase (C4)	3	1	3	3	3
Esterase Lipase (C8)	2	3	4	2	3
Arylamidase	5	5	5	5	4
Valine arylamidase	1	3	4	2	3
Cysteine arylamidase	3	2	3	2	2
Acid phospotase	4	5	2	4	5
Naphthol-AS-BI-phosphohydrolase	3	3	3	2	3
$\beta-$glucosidase	2	2	4	-	3

+: low growth, ++: medium growth, +++: high growth, F0:0-2 mm, F1: 2-4 mm, F2: 4 and higher, 1: very low activity, 2: low activity, 3: medium activity, 4 high activity, 5: very high activity.

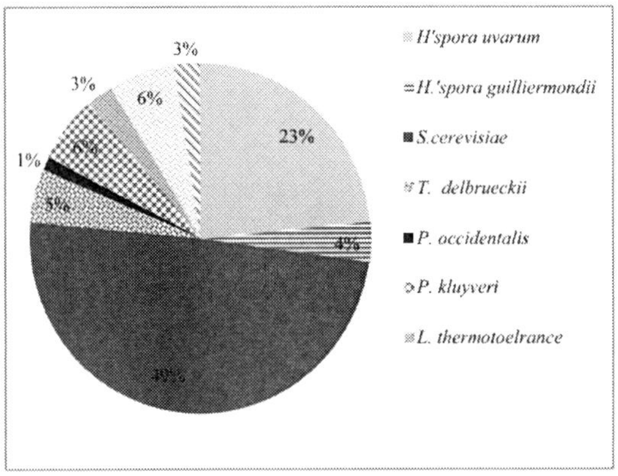

Figure 3. Total Yeast diversion. *L. thermotolerance* (*Lachancea thermotolerance*), *T. delbrueckii* (*Torulaspora delbrueckii*), *P. kluyveri* (*Pichia kluyveri*), *C. zemplinina* (*Candida zemplinina*), *P. occidentalis* (*Pichia occidentalis*), *S. cerevisiae* (*Saccharomyces cerevisiae*), *H'spora guilliermondii* (*Hansenispora guilliermondii*), *H'spora uvarum* (*Hanseniaspora uvarum*).

of the H_2S, weak growth at different levels of ethanol, SO_2 content and temperature. Of the remaining strains were subjected to determine their capacity to initiate fermentation in YPD broth containing 200 ppm of SO_2 and 12% (v/v) alcohol at 15 °C. According to their growth, five strains were eliminated. In the next stage, killer activity, foam production, enzyme profile of the remaining strains was determined and six of them were eliminated. Remaining 4 strains of *S. cerevisiae* and 1 strain of *Lachancea thermotolerance* (Non-*Saccharomyces*) were examined for their volatile acid production, flocculation and seconder metabolites.

Table 2 shows technological properties of studied yeasts. All remaining strains showed high growth rate at 200 mg/L SO_2 and at 30° Brix. Only strain 884 was the more sensitive to the presence of ethanol at 12% (v/v). All strains showed medium growth rate at low temperature (15 °C). All yeasts showed killer activity and their foam production decreased by decreasing temperature. All yeasts were produced more or less H_2S. Fermentation rate of remaining yeasts were found between 0.45–1.7 g CO_2/L.h. Volatile acid production of strains were found between 0.7 g/L-1.02 g/L. The 1120 strain produced high level of volatile acidity. The flocculation capacity was found between 68%–98% and the *S. cerevisiae* strain 884 was the less flocculants one. Enzymatic activities of yeasts depend on strain.

3.4. Seconder metabolites of wines

The results of the analysis of volatile compounds of wines fermented with the 5 selected strains and one commercial strain are shown in Table 3. Principal Component Analysis (PCA) was applied on the data set of the concentration of aroma compounds (Fig. 4).

A total of 30 aroma compounds were identified in Narince wines by GC-MS and GC/FID including 7 higher alcohols, 12 esters, 8 volatile acids, 1 lactone and 1 carbonyl compounds. Major aroma compounds ethyl acetate and acetaldehyde was calculated by GC-FID. The wines fermented with commercial strain, 884, 1088, 1120, 1333 and 1378 contained 328 mg/L, 276 mg/L, 309 mg/L, 381 mg/L, 289 mg/L and 415 mg/L of aroma compounds respectively.

The fermentation of Narince must with the all the isolated yeasts gave different volatile profiles. Among aroma compounds higher alcohols were the most abundance compounds of wines. These compounds are formed during alcoholic fermentation. At concentrations below 300 mg/l, they contribute to the desirable complexity of wine. When their concentration exceeds 400 mg/l, higher alcohols are regarded as a negative quality factor [31,32]. Isoamyl

DOI: 10.1051/bioconf/20170902017

Table 3. Aroma compounds of Narince wines fermented with different autochthonous yeasts.

Aroma Compounds (mg/L)							
Higher Alcohols	C	**884****	**1088****	**1120****	**1333****	**1378****	F
1-propanol	2.36 ± 0.02^b	1.83 ± 0.02^c	2.35 ± 0.04^b	2.85 ± 0.08^a	1.91 ± 0.01^c	1.29 ± 0.05^d	*
Isobutyl alcohol	13.14 ± 0.09^b	9.37 ± 0.13^d	13.50 ± 0.15^b	$9.65 \pm 0.08^{c,d}$	9.82 ± 0.04^c	14.31 ± 0.4^a	*
Isoamyl alcohol	190.27 ± 3.7^b	114.31 ± 2.5^d	188.84 ± 0.01^b	112.08 ± 4.9^d	133.63 ± 0.5^c	253.51 ± 7.5^a	*
2,3-Butanediol	0.56 ± 0.04^b	0.12 ± 0.01^d	0.40 ± 0.01^c	0.61 ± 0.02^b	1.52 ± 0.06^a	0.15 ± 0.01^d	*
Methionol	0.81 ± 0.02^b	0.25 ± 0.01^e	0.34 ± 0.01^d	0.21 ± 0.01^f	0.47 ± 0.01^c	0.89 ± 0.05^a	*
2-Phenylethyl alcohol	63.28 ± 0.7^b	24.99 ± 0.7^e	47.53 ± 0.01^c	30.62 ± 2.9^d	48.63 ± 3^c	102.52 ± 3.5^a	*
Tyrosol	1.69 ± 0.07^a	0.32 ± 0.01^e	0.94 ± 0.05^c	0.5 ± 0.01^d	0.52 ± 0.04^d	1.12 ± 0.07^b	*
Total	272.11	151.19	253.9	156.52	196.5	373.9	
Esters							
Ethyl acetate	30.3 ± 1.7^d	83.39 ± 1.4^b	27.26 ± 0.11^e	184.22 ± 0.05^a	62.39 ± 1.5^c	24.14 ± 0.7^f	*
Isoamyl acetate	1.67 ± 0.1^b	1.37 ± 0.06^c	0.16 ± 0.01^e	0.54 ± 0.03^d	1.68 ± 0.01^b	2.42 ± 0.07^a	*
Hexyl acetate	0.24 ± 0.03^a	0.18 ± 0.01^c	0.05 ± 0^d	nd	0.21 ± 0.01^b	0.21 ± 0.01^b	*
Ethyl hexaonate	0.35 ± 0.01^a	0.33 ± 0.01^a	0.12 ± 0.01^b	0.15 ± 0.0^b	0.32 ± 0.05^a	0.35 ± 0.02^a	*
Ethyl lactate	0.20 ± 0.03^d	0.15 ± 0.01^e	0.22 ± 0.03^d	0.49 ± 0.06^a	0.3 ± 0.01^c	0.33 ± 0.01^b	*
Ethyl octaonate	0.12 ± 0.01^c	0.18 ± 0.01^a	0.09 ± 0^d	nd	0.13 ± 0.01^c	0.16 ± 0.01^b	*
Ethyl-3-hydroxybutanoate	0.22 ± 0.04^a	0.15 ± 0.01^c	$0.16 \pm 0.01^{b,c}$	0.10 ± 0.01^d	0.18 ± 0.01^b	0.18 ± 0.01^b	*
Ethyl decanoate	0.07 ± 0.02^c	0.02 ± 0.01^d	0.40 ± 0.02^a	0.13 ± 0.01^b	0.07 ± 0.00^c	0.06 ± 0.01^c	*
Propanediol diacetate	0.44 ± 0.01^b	0.21 ± 0.01^e	0.34 ± 0.01^c	0.55 ± 0.0^a	0.29 ± 0.01^d	0.15 ± 0.01^f	*
2-Phenylethyl ester	0.27 ± 0.04^b	$0.25 \pm {}^b$	0.27 ± 0.02^b	0.06 ± 0.0^c	0.28 ± 0.01^b	0.86 ± 0.07^a	*
Ethyl-4-hydroxybutanoate	3.06 ± 0.06^c	1.15 ± 0.02^e	2.53 ± 0.3^d	5.17 ± 0.4^a	3.80 ± 0.3^b	2.46 ± 0.09^d	*
Ethyl hydrogen succinate	0.13 ± 0.01^c	0.11 ± 0.01^d	0.07 ± 0.00^e	0.09 ± 0.0^d	0.21 ± 0.01^a	0.19 ± 0.03^b	*
Total	36.8	87.49	31.67	191.50	69.86	31.51	
Volatile acids							
Acetic acid	0.54 ± 0.02^c	0.62 ± 0.01^c	0.37 ± 0.02^d	1.25 ± 0.08^a	1.12 ± 0.5^b	0.60 ± 0.02^c	*
Isobutyric acid	0.23 ± 0.02^d	0.25 ± 0.02^c	0.32 ± 0.01^b	0.2 ± 0.01^e	0.19 ± 0.01^e	0.42 ± 0.02^a	*
Butyric acid	$0.12 \pm 0.01^{b,c}$	0.11 ± 0.01^c	0.12 ± 0.01^c	$0.14 \pm 0.01^{a,b}$	0.15 ± 0.01^a	$0.13 \pm 0.01^{b,c}$	*
Isovaleric acid	0.62 ± 0.01^c	0.27 ± 0.01^e	0.69 ± 0.03^b	0.24 ± 0.0^f	0.34 ± 0.01^d	0.85 ± 0.01^a	*
Hexanoic acid	1.87 ± 0.02^d	1.86 ± 0.02^d	1.93 ± 0.02^c	0.76 ± 0.01^d	3.03 ± 0.2^a	2.31 ± 0.04^b	*
Octanoic acid	2.74 ± 0.1^b	2.42 ± 0.02^c	1.56 ± 0.03^d	0.89 ± 0.03^e	2.95 ± 0.4^a	$2.85 \pm 0.01^{a,b}$	*
Decanoic acid	1.38 ± 0.1^a	0.47 ± 0.01^c	0.07 ± 0.00^d	0.50 ± 0.01^c	0.74 ± 0.02^b	0.57 ± 0.01^c	*
Hexadecanoic acid	0.39 ± 0.01^c	0.15 ± 0.02^e	0.1 ± 0.01^e	0.55 ± 0.03^b	0.28 ± 0.01^d	0.90 ± 0.08^a	*
Total	7.89	6.45	5.16	4.53	8.80	8.63	
Lactones							
gamma-Butyrolactone	0.40 ± 0.04^e	0.49 ± 0.01^d	0.34 ± 0.01^f	0.87 ± 0.03^a	0.67 ± 0.02^c	0.81 ± 0.01^b	*
Total	0.40	0.49	0.34	0.87	0.67	0.81	
Aldehydes							
Acetaldehyde	11.12 ± 0.5^e	21.35 ± 0.7^b	18.43 ± 0.2^c	27.9 ± 0.06^a	14.06 ± 1^d	9.26 ± 0.7^f	*
Total	11.12	21.35	18.43	27.9	14.06	9.26	
TOTAL	**328.30**	**266.97**	**309.5**	**381.32**	**289.89**	**424.11**	

C: wine fermented with commercial yeast strain, ± Standard deviation, nd: not detected, F: significance at which means differ as shown using analysis of variance,*$p < 0.05$ level, **: 1120 is *Lachancea thermotolerance* and others are *S. cerevisiae*

alcohol and 2-phenylethyl alcohol were the most abundant higher alcohols in all wines and they both were found highest in wine with inoculated 1378. Odour threshold values of isoamyl alcohol and 2-phenylethyl alcohol are 30 mg/L and 10 mg/L, respectively. Concentration of these compounds exceeded their threshold values in all Narince wines. 2-phenylethyl alcohol has a floral and rose like aroma and contributes positively to wine aroma [1].

Esters are an important group of volatile compounds produced by yeasts during alcoholic fermentation [31]. Ethyl acetate was the major ester compound produced in Narince wines. Aroma threshold value of ethyl acetate is 7.5 mg/L [31], and it exceeded its odour threshold value in all wines. Ethyl acetate may contribute with pleasant fruity fragrance to the general wine aroma at low concentration, whereas it contributes significantly do defect aroma at a concentration of 150–200 mg/L [32]. The wine inoculated with 1120, the ethyl acetate exceeded this value, which makes this strain unsuitable for making quality wine. Other important volatile esters, isoamyl acetate, 2-phenylethyl acetate, ethyl hexaonate and ethyl octaonate were found above their odour threshold values [31].

As indicated in Table 3, the most abundant volatile acids in Narince wines were octanoic and hexanoic acids. Similar results were found by Selli et al. (2006) [1]. The contribution of volatile acids on Narince wines cannot be considered important because their concentrations were much lower than their odour threshold values [1,31].

γ-Butyrolactone was the only lactone was found in all Narince wines. This compound is found in fermented products and probably some from glutamic acid and related compounds (succinic, 2-oxoglutaric or γ-aminobutyric acids) [1,31].

The acetaldehyde is one of the most important carbonyl compound produced during fermentation and at low levels it contributes to fruity flavours, while high concentrations (200 mg/L) cause flatness in wines [32]. The concentration of acetaldehyde in Narince wines were found between

DOI: 10.1051/bioconf/20170902017

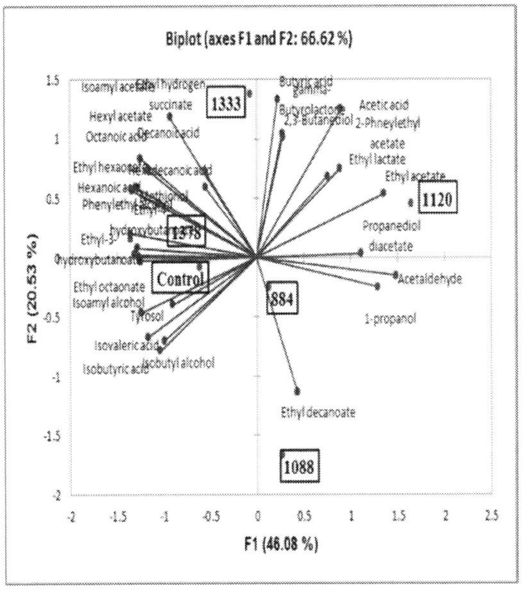

Figure 4. Principal Component Analysis of aroma compounds produced by selected yeasts.

11 mg/L to 27 mg/L. Nevertheless its concentration did not exceed its threshold values of 100 mg/L [31].

The PCA bi-plots (Fig. 4) showed the relationship between better and poorer performing yeast with regards to the aroma compounds. The PCA of aroma compounds showed that most of data explained by the first (F1) and second (F2) factors which are 46.08 and 20.53% respectively. The F1 explained 46% data variation and showed correlation 1120 and 1378. The F2 explained 20.53% of data variation and correlated with 1088 and 1333. The F3 explained 15.4% data variation and correlated with 884 and F4 explained 11% data variation and correlated with the control. As shown in F1, 1088 well differentiated from the other yeasts due to its high content of ethyl decanoate.

4. Conclusion

In this study PCR-RFLP analysis of the 5.8 ITS rRNA region and sequence information for the D1/D2 domains of the 26 S gene were used for yeasts identification. Only one Non-*Saccharomyces* spp. (*Hanseniaspora uvarum*) identified on Narince grape surface. During spontaneous fermentation of Narince musts eight different species belonging to seven genera were identified. Of the 69 isolated native strains of *S. cerevisiae* and Non-*Saccharomyces* from fermenting musts of Narince, only 4 *S. cerevisiae* and 1 *Lachancea thermotoelerance* (Non-*Saccharomyces* spp.) were considered to have good enough technological properties. Moreover 1088 was the most suitable strain to make quality wine with low volatile acid and foam production, high flocculation capacity, high resistance to 12% (v/v) ethanol and producing well balanced aroma compounds.

The authors would like to thank The Scientific and Technical Research Council of Turkey (TUBITAK) for financial support of this research (Project No. 2140173) and DİREN Wine Company for providing the grape samples.

References

[1] S. Selli, A. Canbaş, T. Cabaroğlu, H. Erten, J.P. Lepoutre, Z. Günata, Effect of skin contact on the free and bound aroma compounds of the white *Vitis vinifera* L cv. Narince, Food Control **17**, 75–82 (2006)

[2] M.Ü. Ünal, A. Şener, Effect of harvest year on biochemical properties of Narince grape (*Vitis vinifera* L cv.) polyphenol oxidase, Eur. Food Resc. Tech. **238**, 613–619 (2014)

[3] M.S. Perez-Coello, A.I. Briones Perez, J.F.U. Iranzo, P.J. Martin Alvarez, Characteristics of wines fermented with different *S. cerevisiae* strains isolated from the La Mancha region. Food Microbiol. **16**, 563–573 (1999)

[4] M.L. R. Eder, C. Reynoso, S.C. Lauret, A.L. Rosa, Isolation and identification of the indigenous yeast population during spontaneous fermentation of Isabella (*Vitis labrusca* L.) grape must, Frontiers in Microbiol. **8**, 532 (2017)

[5] K.P. Jemec, N. Cadez, T. Zagorc, V. Bubic, A. Zupec, P. Raspor, Yeast population dynamics in five spontaneous fermentations of Malvasia must, Food Microbiol. **18**, 247–259 (2001)

[6] A. Capece, L. Granchii, S. Guerrini, S. Mangani, R. Romaniello, M. Vincenzini, P. Romano, Diversity of *S. cerevisiae* strains isolated from two Italian wine producing regions. Frontiers in Microbiol. **7**, 1018 (2016)

[7] O.I.V. Compendium of international methods of wine and must analysis. International Organisation of Vine and Wine EN-2015, **1** (2015)

[8] H. Erten, Metabolism of Fructose as an Electron Acceptor by *Leuconostoc mesenteroides*, Process Biochemistry **33**, 735–739 (1998)

[9] H. Cordier, F. Mendes, I. Vasconcelos, J.M. François "A Metabolic and Genomic Study of Engineered *S. cerevisiae* Strains for High Glycerol Production", Metabolic Engineering **9**, 364–378 (2007)

[10] M. Combina, A. Elia, L. Mercado, C. Catania, A. Ganga, C. Martinez, Dynamics of indigenous yeast populations during spontaneous fermentation of wines from Mendoza, Argentina. Int. J. Food Microbial. **99**, 237–243 (2005)

[11] B. Esteve-Zarzosa, A. Gostincar, R. Bobet, F. Uruburu, A. Querol, Selection and molecular characterization of wine yeasts isolated from the 'El Penedes area (Spain), Food Microbial. **17**, 553–562 (2000)

[12] C.P. Kurtzman, C.J. Robnett, Identification and phylogeny of ascomycetous yeasts from analysis of nuclear large subunit (26S) ribosomal DNA partial sequences, Anton. Leeuw **73**, 331–371 (1998)

[13] J.D.Thompson, T.J. Gibson, F. Plewniak, F. Jeanmougin, the CLUSTAL _X Windows interface: flexible strategies for multiple sequence alignment aided by quality analysis tools, Nucleic Acids Res. **25**, 606–615 (1997)

[14] N. Francesca, C. Carvalho, C. Sannino, M.A. Guerreiro, P.M. Almeida, L. Settani, B. Massa, J.P.

Sampaio, G. Moschetti, Yeasts vectored by migratory birds collected in the Mediterranean island of Ustica and description of *Phamffomyces usticensis* f.a. sp. nov., a new species related to the cactus ecoclade, FMS Yeast Res. **14**, 910–921 (2014)
[15] J.A. Regodon, F. Perez, M.E. Valdes, C. De Miguel, H. Suranska, D. Vranova, J. Omelkova, R., Vadkertiova, Monitoring of yeast population isolated during spontaneous fermentation of Moravian wine, Chemical Papers **66**, 861–868 (2012)
[16] E. Nikolaou, E.H. Soufleros, E. Bouloumpasi, N. Tzanetakis, Selection of indigenous *S. cerevisiae* strains according to their oenological characteristics and vinification results, Food Microbiol. **23**, 205–211 (2006)
[17] M. E. Parish, D.E. Carrol, Fermentation characteristics of *S. cerevisiae* isolates from *Vitis rotundifolia* grapes and musts, Am. J. Enol. Vitic **38**, 45–48 (1987)
[18] M. Spiczki, P. Romano, G. Lipani, I. Miklos, Z. Antunovics, Analysis of yeasts derived from natural fermentation on a Tokaj winery, Anton. Leeuw **79**, 97–105 (2001)
[19] S. Orlić, N. Očić, A, Jeromel, K. Huić, S. Redžepović, Selection of indigenous *S. cerevisiae* strains from Kutjevo wine growing area at the laboratory scale, Agriculturae Conspectus Scientificus **70**, 93–97 (2005)
[20] C. J. Jacobs, J. J. Van Vuuren, Effects of Different Killer Yeasts on Wine Fermentations, Am. J. Enol. Vitic **42**, 295–300 (1991)
[21] F. Özçelik, S. Dönmez, Killer yeasts and the determination of killer characters of some yeast". Doğa-Tr. J. of Biology **17**, 1–4 (1993)
[22] H. Surasnka, D. Vranova, J. Omelkova, R. Vadkertiova, Monitoring of yeast population isolated during spontaneous fermentation of Moravian wine, Chemical papers **66**, 861–868 (2012)
[23] J.F.U. Iranzo, A.I.B. Perez, P.M.I. Canas, Study of oenological characteristics and enzymatic activities of wine yeasts, Food Microbiol. **15**, 399–406 (1998)
[24] E.V. Soares and M. Mota, Quantification of yeast flocculation. J. Inst. Brew. **103**, 93–98 (1996)
[25] C. Ortega, R. Lopez, J. Cacho, V. Ferreira, Fast analysis of important wine volatile compounds development and validation of a new method based on gas chromatographic-flame ionization detection analysis of dichloromethane microextracts. J. Chrom. A **92**, 3205–3214 (2001)
[26] M. Darıcı, T. Cabaroğlu, V. Ferreira, L. Lopez, Chemical and sensory characterization of the aroma of Çalkarasırose wine, Austr. J. Grape Wine Res. **20**, 340–346 (2016)
[27] R. Schneider, R. Baumes, C. Bayanove, A. Razungles, Volatile Compounds Involved in the aroma of sweet of fortified wines (Vins Doux Natrurels) from Grenache Noir, J. Agric. Food Chem. **46**, 3230–3237 (1998)
[28] R. Schneider, A. Razungles, C. Augier, R. Baumes, Monoterpenic and Norisoprenoidic Glycoconjugates of *Vitis vinifera* L. cv. Melon B. As Precursors of Odorants in Muscadet Wines, J. Chrom. A **936**, 145–15 (2001)
[29] T. Cabaroğlu, M. Yılmaztekin, Methanol and major volatile compounds of Turkish raki and effect distillate source, J.Inst. Brew. **117**, 98–105 (2011)
[30] T. Labagnara, Wine yeast biodiversity during spontaneous fermentation in response to environmental stress, PhD Thesis, Pisa University, Agricultural and Veterinary Science, Tuscany 147 p. (2016)
[31] J.W. Swiegers, E.J. Bartowsky, P.A. Henschke, I.S. Pretorius, Yeast and bacterial modulation of wine aroma and flavour, Austr. J. Grape Wine Res. **11**, 139–173 (2005)
[32] M. De Benedictis, G. Bleve, F. Grieco, M. Tristezza, M. Tufariello, F. Grieco, An optimized procedure for the enological selection of non-*Saccharomyces* starter cultures, Anton. Leeuw **99**, 189–200 (2011)

La cytométrie appliquée aux mircoorganismes du vin / Application of flow cytometry to wine microorganisms

Cédric Longin, Clément Petitgonnet, Michèle Guilloux-Benatier, Sandrine Rousseaux, et Hervé Alexandre

Institut Universitaire de la Vigne et du Vin, Equipe VAlMiS, UMR PAM A 02.102, Univ. Bourgogne Franche-Comté, AgroSup Dijon, rue Claude Ladrey, BP. 27877, 21078 Dijon, France

Abstract. Flow cytometry (FCM) is a powerful technique allowing detection and enumeration of microbial populations in food and during food process. Thanks to the fluorescent dyes used and specific probes, FCM provides information about cell physiological state and allows enumeration of a microorganism in a mixed culture. Thus, this technique is increasingly used to quantify pathogen, spoilage microorganisms and microorganisms of interest. Since one decade, FCM applications to the wine field increase greatly to determine population and physiological state of microorganisms performing alcoholic and malolactic fermentations. We will describe FCM principles and the available techniques allowing quantification of wine microorganisms by FCM using fluorescent dyes. Different examples of FCM analysis regarding monitoring of alcoholic fermentation, malolactic fermentation (total population, viable population, physiological state), specific detection and quantification of spoilage microorganisms like *Brettanomyces* will be presented.

1. Cytométrie en flux : vue d'ensemble et principe

La cytométrie en flux (CMF) est considérée comme une étude de cellules isolées dans un flux. La CMF est une technique robuste pour analyser des paramètres multiples de cellules individuelles au sein d'un ensemble hétérogène de microorganismes par exemple. Plusieurs revues et livres ont été publiés traitant du principe de cytométrie en flux et donnant de nombreux détails [1–3].

Brièvement, un cytomètre en flux est composé de trois principaux systèmes : fluidique, optique et électronique (Fig. 1).

Le système fluidique permet de séparer et d'aligner les particules grâce au liquide de gaine (focalisation hydrodynamique) [1]. Une fois alignées, les particules passent devant un ou plusieurs lasers.

La lumière émise par les cellules, suite à leur irradiation à l'intérieur de la chambre de flux, est dirigée vers les détecteurs par un système complexe de miroirs et filtres qui composent le système optique. La lumière émise aux petits angles est détectée (FSC). Cette mesure est proportionnelle à la taille des cellules. La lumière est également détectée aux grands angles (SSC). Cette dernière est proportionnelle à la complexité à l'intérieur de la cellule ainsi qu'à la granulosité cellulaire. Un autre dispositif est utilisé pour détecter le rayon lumineux à travers un ensemble de filtres séparant la lumière selon la longueur d'onde (filtres de fluorescences).

La lumière passant par le système optique est convertie en signaux électroniques générés par des photodiodes et des tubes photomultiplicateurs (PMT) puis collectés. De récents cytomètres en flux peuvent avoir jusqu'à 12 PMT et ce nombre ne cesse d'augmenter avec le développement des nouvelles technologies (détection spectrale).

Contexte

Le vin est une matrice complexe contenant de nombreux composés et microorganismes tels que des levures, des bactéries lactiques (BL) et acétiques (BA). Lors de l'élaboration du vin, il est indispensable de connaitre les populations réalisant les fermentations mais également la concentration des microorganismes d'altération présents afin d'avoir une bonne qualité finale. Différentes techniques sont disponibles afin de détecter et/ou quantifier ces microorganismes durant les étapes de fermentation et dans le vin produit.

À ce jour, les laboratoires œnologiques utilisent différentes techniques permettant ces détections et quantifications. Premièrement, l'utilisation de boites de Pétri sélectives permet d'énumérer les microorganismes cultivables présents dans l'échantillon. Cependant la formation de colonies demande des temps d'incubation long (généralement entre 2 et 10 jours) selon les microorganismes. De plus cette méthode ne permet pas la quantification des microorganismes en état viable mais non cultivable (VNC), retrouvés couramment en vin [4–8].

La quantification par microscopie utilisant les cellules de Thoma ou Malassez est une autre méthode, mais celle-ci est longue, fastidieuse, subjective et avec une limite de quantification d'environ 10^4 cellules/mL. Néanmoins, couplée à l'utilisation de marqueurs fluorescents, la microscopie permet de déterminer des fonctions cellulaires spécifiques ou d'identifier les microorganismes présents avec l'utilisation de sondes spécifiques fluorescentes par exemple (Direct Epifluorescence Filter Technique).

D'autres techniques basées sur la réaction en chaine par polymérase (PCR), comme la PCR quantitative peut également permettre de détecter et quantifier les microorganismes. En effet, par l'utilisation d'amorces capables de se fixer spécifiquement à des séquences cibles

DOI: 10.1051/bioconf/20170902018

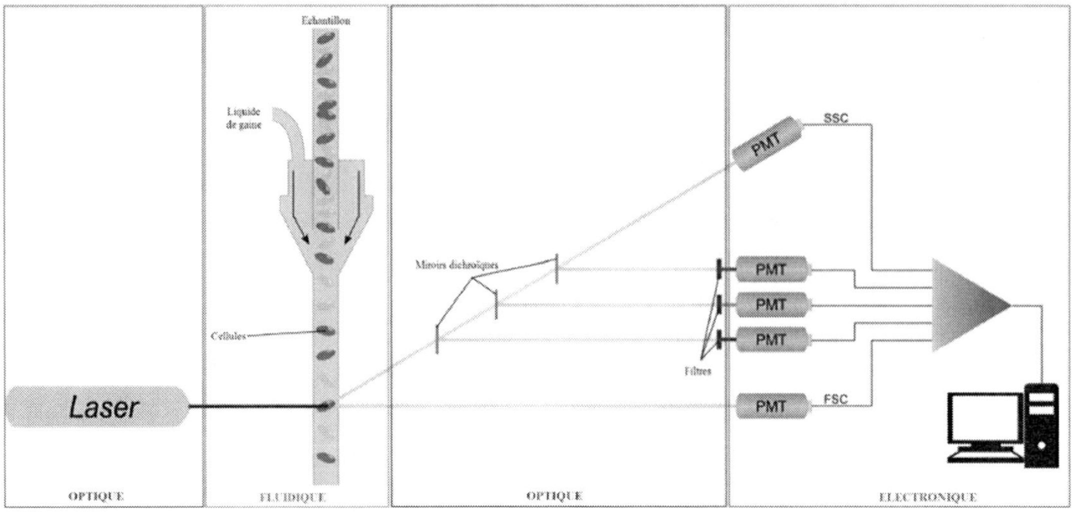

Figure 1. Schéma du principe d'un cytomètre en flux.

d'ADN et l'utilisation d'un intercalant d'ADN (SybrGreen par exemple) ou de sondes spécifiques fluorescentes (sondes TaqMan par exemple), cette technique permet une quantification précise des microorganismes présents. Cependant, cette méthode demande une étape d'extraction d'ADN avant de réaliser la PCR. Celle-ci peut être perturbée par la composition du vin, spécifiquement pour les vins rouges contenant une importante quantité de polyphénols. Ces molécules peuvent s'adsorber au niveau de la paroi cellulaire [9,10] influençant la lyse cellulaire durant l'extraction de l'ADN et pouvant conduire à un faible rendement d'extraction. De plus, un problème lors de l'amplification de l'ADN peut également intervenir par la présence de composés inhibiteurs tels que les tannins, polysaccharides et pigments [11,12].

La méthode CMF permet à la fois une détection et une quantification des microorganismes avec rapidité et une forte spécificité. Premièrement, la CMF permet de suivre l'état physiologique des cellules durant la fermentation alcoolique (FA) ou malolactique (FML). De plus, la CMF peut être utilisée plus spécifiquement, c'est-à-dire couplée à la méthode d'hybridation d'une sonde *in situ* en fluorescence (FISH) afin de quantifier certains microorganismes d'altération par exemple. Dans l'optique d'augmenter voire de généraliser l'utilisation de la CMF appliquée au vin, cet article a pour but de référencer les différentes applications utilisées et envisageables de cette technique en œnologie.

2. Quantification et analyse de l'état physiologique des microorganismes œnologiques par CMF

L'énumération des bactéries et des levures peut être effectuée par CMF (Fig. 2) sans l'utilisation de marqueurs en milieu de culture et certains vins blancs.

Selon le vin étudié, spécifiquement pour le vin rouge, l'énumération peut être plus difficile par la présence d'une concentration importante de composés détectée par le cytomètre en flux, confondu avec les bactéries et même parfois avec les levures. À ce jour, plusieurs

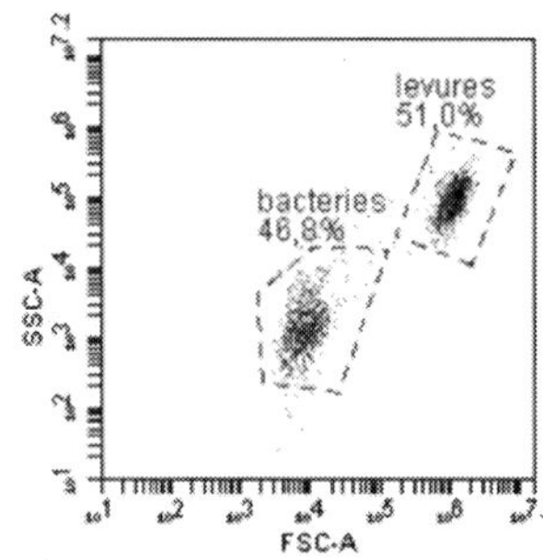

Figure 2. Discrimination des bactéries et des levures par CMF (FSC : taille ; SSC : granulosité) (donnée personnelle).

marqueurs fluorescents comme les marqueurs SYTO, non fluorescents jusqu'à leur liaison avec l'ADN, sont disponibles. Le DAPI et les marqueurs SYBR sont également utilisés. Ceux-ci présentent une fluorescence plus importante lorsqu'ils sont liés à l'ADN.

2.1. Vitalité cellulaire

Afin de réaliser les fermentations durant l'élaboration du vin, les levures et bactéries doivent être en vie et métaboliquement active. Les marqueurs de vitalité permettent de déterminer les fonctions essentielles vitales comme l'activité enzymatique par exemple. La fluorescéine di-acétate (FDA) est un marqueur couramment utilisé dans de nombreuses applications puisque son utilisation est rapide et simple [13] (Figure 3). En effet, le marquage FDA reflète l'activité enzymatique cellulaire à travers l'activité estérase, mettant en évidence l'activité métabolique cellulaire.

2.1.1. Levures

Ainsi une bonne corrélation entre boites de Pétri et CMF avec marquage FDA pour l'énumération de *Saccharomyces cerevisiae* a été mise en évidence [14].

DOI: 10.1051/bioconf/20170902018

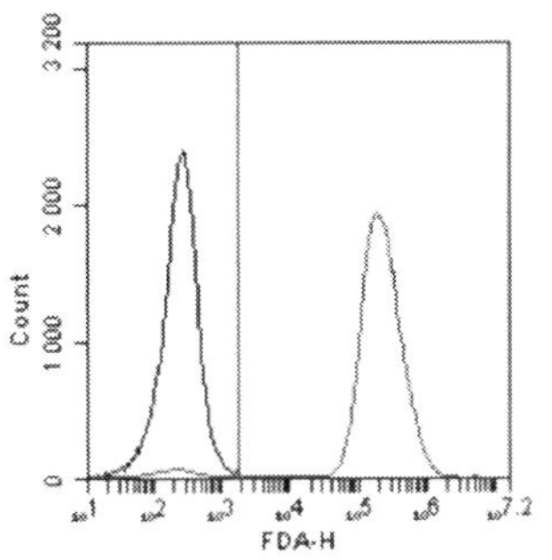

Figure 3. Autofluorescence (histogramme noir) et fluorescence après marquage avec le FDA (histogramme rouge) d'une population de levures lors d'un passage au cytomètre en flux. (donnée personnelle).

Le carboxy-FDA (cFDA) peut également être utilisé car la fluorescéine carboxylée présente une meilleure rétention au sein de la cellule. Le marqueur FUN-1 [7] permet également d'analyser la vitalité cellulaire. Celui-ci pénètre dans les cellules et entraine une fluorescence verte du cytoplasme. Lorsque les levures sont actives métaboliquement, des structures intra-vacuolaires cylindriques se forment et présentent une fluorescence rouge.

2.1.2. Bactéries

Concernant les bactéries, une quantification de *O. oeni* en vin rouge durant la FML a été menée par dénombrement classique (boites de Pétri) et CMF en testant 4 marqueurs différents : Rhodamine 123 ; Calcein acétoxyméthyl ester (Calcein-AM) ; 2′, 7′-bis-carboxyethyl -5(6)-carboxyfluorescéine acétoxyméthyl ester (BCECF-AM) et FDA [14]. Des différences d'intensité de fluorescence selon les marqueurs et les souches testées ont été mises en évidence. La majorité des souches étaient marquées avec la FDA. Cependant, 40 % des *O. oeni* sont fluorescentes après 10–15 min, 30 % après 60 min, et deux souches ne sont pas marquées, traduisant probablement une fuite de la fluorescéine vers le tampon à travers la membrane entrainant ainsi une perte de fluorescence des cellules. Ce problème conduit à une différence significative lors de l'énumération de *O. oeni* en vin Bardolino entre les boites de Pétri et la CMF. En vin blanc, la quantification semble possible avec la FDA [15]. Cependant, une autre étude a mise en évidence une faible corrélation entre les boites de Pétri et la CMF en utilisant la FDA lors de la phase de latence et en fin de croissance en milieu de culture [4].

Pour le marquage FDA, il est important de rappeler que les estérases intracellulaires présentent une activité maximum à pH 7,6, à 30 °C après une incubation de 20 min.

En vin rouge, différents marqueurs de vitalité ont été testés mais aucun ne s'est montré satisfaisant afin de quantifier la vitalité bactérienne au cours de la FML.

2.2. Viabilité cellulaire

Les marqueurs de viabilité déterminent si les cellules sont dans un état physiologique permettant leur survie, comme la mesure de la perméabilité membranaire. L'iodure de propidium (IP) est fréquemment utilisé comme marqueur d'intégrité membranaire [16]. L'IP rentre dans les cellules perméables et se fixe à l'ADN. Les cellules présentent alors une fluorescence rouge. Cependant, l'utilisation de ce marqueur présente un important désavantage. Différents stress comme l'éthanol sont connus pour entrainer une augmentation de la perméabilité membranaire [17] et donc biaiser le résultat, voire rendre fluorescente des cellules vivantes [18].

2.2.1. Levures

Différentes études ont utilisé l'IP pour suivre la sédimentation cellulaire [19] et suivre la population levurienne durant la FA en moût synthétique [6,16,20]. De bonnes corrélations ont été mises en évidence. D'autres études ont été effectuées en moûts blanc et rouge [21,22]. Une sous population a été révélée, appelée «population fragile» qui présentait une perméabilité durant la FA mais cet état était réversible.

2.2.2. Bactéries

La quantification de *O. oeni* a été menée durant la FML. La discrimination entre les bactéries vivantes et mortes en vin rouge est parfois impossible, liée à la présence de débris en grande quantité. Une étude a toutefois démontré que cette quantification était possible par un double marquage [15]. Ces marqueurs, comme ceux cités précédemment, sont non spécifiques, c'est-à-dire que si l'échantillon contient différentes bactéries, il n'est pas possible d'affiner l'analyse à une sous population. Cependant, lors de la FML, il est très probable que cette quantification soit majoritairement celle de bactéries lactiques, voire exclusivement *O. oeni*.

2.3. Double marquage vitalité / viabilité

2.3.1. Levures

Le suivi de la FA de cidres a été réalisé en utilisant les marqueurs CV6, IP et DRAQ5 afin de quantifier la vitalité, la viabilité et la population totale présente. Des cellules doublement marquées ont été ainsi détectées (fonction vitale active mais avec une importante perméabilité membranaire) [23]. Ces résultats ont ensuite été confirmé dans une autre étude [24].

2.3.2. Bactéries

L'analyse de la rétention de la carboxy-fluorescéine ainsi que l'exclusion de l'IP a été réalisée pour *O. oeni* adaptée ou non à l'éthanol en milieu de culture [25]. Les mesures de fluorescence ont été menées suite à un stress éthanol 16 % (v/v) ou sans stress. Les résultats ne montrent aucune différence significative entre les fluorescences en l'absence de stress alors que suite à l'application du stress éthanol, la fluorescence de la fluorescéine diminue de 25 % et celle de l'IP augmente de 80 %. En revanche, lorsque les cellules sont préalablement adaptées à l'éthanol, le stress éthanol 16 % (v/v) entraine un impact cellulaire moins important. Des doubles marquages ont également été réalisés lors d'un suivi de FML [26]. Cependant, dans une étude [25] ils ont démontré que l'intensité de fluorescence

 DOI: 10.1051/bioconf/20170902018

augmente entre 15 min et 60 min contrairement à une étude précédente [14]. Il est donc important d'avoir connaissance de ces potentiels biais qui peuvent intervenir lors d'analyses d'échantillons issus du vin.

2.4. Etat viable mais non cultivable (VNC)

La technique de CMF permet également de mettre en évidence l'état VNC en utilisant un marqueur de vitalité ainsi que le dénombrement sur boites de Pétri.

2.4.1. Levures

Par CMF, le SO_2 a été décrit comme un facteur conduisant à l'entrée en état VNC pour *Brettanomyces bruxellensis* [8] et *S. cerevisiae* [7,27]. Les marqueurs de vitalité sont actifs alors qu'aucune colonie n'est détectée sur milieu sélectif.

2.4.2. Bactéries

Ce même état VNC a été observé chez *O. oeni* par CMF lors de la FML dans un milieu caractérisé par un faible pH, en présence d'éthanol et un déficit en nutriments [23].

2.5. Autres applications

Bien que l'utilisation des marqueurs de viabilité et vitalité soient les principaux utilisés afin de suivre la FA ou la FML dans différentes matrices, d'autres marqueurs fluorescents existent et permettent d'obtenir des informations supplémentaires sur l'état physiologique des cellules.

L'analyse du cycle cellulaire, le potentiel membranaire, la présence d'espèces réactives à l'oxygène, la concentration en lipides intracellulaires, le pH intracellulaire peut être ainsi mesuré [13].

3. Enumeration spécifique des microorganismes du vin

3.1. Utilisation des anticorps spécifiques

L'utilisation d'anticorps spécifiques permet de quantifier spécifiquement *S. cerevisiae* en moût [28]. Afin de réaliser le marquage, une incubation du moût en fermentation avec un premier anticorps polyclonal anti-*Saccharomyces* est effectuée puis un second anticorps couplé à un fluorochrome est utilisé. Il est également possible de suivre la population de *O. oeni* réalisant la FML en même temps que la FA en moût de Chardonnay [28]. Cette méthode a également été développée afin de quantifier *B. bruxellensis* en vin rouge lors de l'élevage [29].

3.2. Utilisation d'une sonde oligonucléotidique

La technique d'hybridation *in situ* en fluorescence (FISH) peut être couplée à la CMF. La FISH est utilisée depuis de nombreuses années [30] mais la visualisation de la fluorescence était principalement effectuée par microscopie avant d'être combinée à la CMF il y a une dizaine d'années [31]. Aujourd'hui, de nombreuses espèces présentent en œnologie peuvent être discriminées au sein d'une population globale par l'utilisation d'une sonde oligonucléotidique [13]. *B. bruxellensis* est une levure dont une quantification précise et rapide est indispensable lors de l'élevage de vin rouge. Plusieurs études se sont donc intéressées à développer une sonde spécifique pour ensuite être appliquée à la CMF [32,33].

4. Futures applications

Comme la microscopie, la CMF est également utilisée dans de nombreux domaines. C'est pourquoi de nombreuses applications sont envisageables en œnologie.

L'utilisation d'un marqueur fluorescent spécifique tel que le 1,6-Diphenyl-1,3,5-hexatriene (DPH) [34] permettrait d'avoir une information sur la fluidité membranaire des levures et des bactéries en vin, et donc sur leur état physiologique à un instant donné.

L'utilisation de sondes peptidiques (PNA) fluorescentes, non chargées donc plus aptes à pénétrer au sein de la cellule, pourrait permettre la quantification spécifique d'une sous population. Des sondes permettant de quantifier les principales bactéries lactiques du vin sont également connues et pourraient être appliquées.

5. Conclusions

La cytométrie en flux est à ce jour encore peu utilisée dans le domaine de l'œnologie. Cependant, de plus en plus d'études sont réalisées afin de développer des outils simples, rapides, fiables avec un faible coût afin de répondre aux différents besoins des vignerons. La matrice vin est très complexe, c'est donc pour cela que la plupart des protocoles développés dans d'autres matrices ne sont pas directement applicables aux échantillons du chai. C'est pourquoi il est nécessaire de mener d'autres études qui conduiront, à terme, à rendre la CMF indispensable lors d'une analyse microbiologique en œnologie.

References

[1] Díaz, M., Herrero, M., García, L.A., Quirós, C.: Application of flow cytometry to industrial microbial bioprocesses. Biochem. Eng. J. **48**, 385–407 (2010)
[2] Longobardi-Givan, A.: Flow Cytometry: First Principles. John Wiley & Sons (2013)
[3] O'Neill, K., Aghaeepour, N., Špidlen, J., Brinkman, R.: Flow Cytometry Bioinformatics. PLoS Comput Biol. **9**, e1003365 (2013)
[4] Bouix, M., Ghorbal, S.: Rapid enumeration of *Oenococcus oeni* during malolactic fermentation by flow cytometry. J. Appl. Microbiol. **114**, 1075–1081 (2013)
[5] Bouix, M., Leveau, J.-Y.: Rapid assessment of yeast viability and yeast vitality during alcoholic fermentation. J. Inst. Brew. **107**, 217–225 (2001)
[6] Branco, P., Monteiro, M., Moura, P., Albergaria, H.: Survival rate of wine-related yeasts during alcoholic fermentation assessed by direct live/dead staining combined with fluorescence in situ hybridization. Int. J. Food Microbiol. **158**, 49–57 (2012)
[7] Salma, M., Rousseaux, S., Sequeira-Le Grand, A., Divol, B., Alexandre, H.: Characterization of the Viable but Nonculturable (VBNC) State in *Saccharomyces cerevisiae*. PLoS ONE. **8** (2013)
[8] Serpaggi, V., Remize, F., Recorbet, G., Gaudot-Dumas, E., Sequeira-Le Grand, A., Alexandre, H.: Characterization of the "viable but nonculturable"

(VBNC) state in the wine spoilage yeast *Brettanomyces*. Food Microbiol. **30**, 438–447 (2012)

[9] Morata, A., Gómez-Cordovés, M.C., Colomo, B., Suárez, J.A.: Cell wall anthocyanin adsorption by different *Saccharomyces* strains during the fermentation of Vitis vinifera L. cv Graciano grapes. Eur. Food Res. Technol. **220**, 341–346 (2004)

[10] Razmkhab, S., Lopez-Toledano, A., Ortega, J.M., Mayen, M., Merida, J., Medina, M.: Adsorption of phenolic compounds and browning products in white wines by yeasts and their cell walls. J. Agric. Food Chem. **50**, 7432–7437 (2002)

[11] Rossen, L., Nørskov, P., Holmstrøm, K., Rasmussen, O.F.: Inhibition of PCR by components of food samples, microbial diagnostic assays and DNA-extraction solutions. Int. J. Food Microbiol. **17**, 37–45 (1992)

[12] Wilson, I.G.: Inhibition and facilitation of nucleic acid amplification. Appl. Environ. Microbiol. **63**, 3741 (1997)

[13] Longin, C., Petitgonnet, C., Guilloux-Benatier, M., Rousseaux, S., Alexandre, H.: Application of flow cytometry to wine microorganisms. Food Microbiol. **62**, 221–231 (2017)

[14] Malacrinò, P., Zapparoli, G., Torriani, S., Dellaglio, F.: Rapid detection of viable yeasts and bacteria in wine by flow cytometry. J. Microbiol. Methods. **45**, 127–134 (2001)

[15] Salma, M., Rousseaux, S., Sequeira-Le Grand, A., Alexandre, H.: Cytofluorometric detection of wine lactic acid bacteria: application of malolactic fermentation to the monitoring. J. Ind. Microbiol. Biotechnol. **40**, 63–73 (2013)

[16] Delobel, P., Pradal, M., Blondin, B., Tesniere, C.: A "fragile cell" sub-population revealed during cytometric assessment of *Saccharomyces cerevisiae* viability in lipid-limited alcoholic fermentation. Lett. Appl. Microbiol. **55**, 338–344 (2012)

[17] Alexandre, H., Rousseaux, I., Charpentier, C.: Relationship between ethanol tolerance, lipid composition and plasma membrane fluidity in *Saccharomyces cerevisiae* and *Kloeckera apiculata*. FEMS Microbiol. Lett. **124**, 17–22 (1994)

[18] Davey, H.M., Hexley, P.: Red but not dead? Membranes of stressed *Saccharomyces cerevisiae* are permeable to propidium iodide. Environ. Microbiol. **13**, 163–171 (2011)

[19] Gerbaux, V., Thomas, J.: Utilisations pratiques de la cytométrie de flux pour le suivi des levures en œnologie. Rev. Fr. Oenologie (2009)

[20] Mannazzu, I., Angelozzi, D., Belviso, S., Budroni, M., Farris, G.A., Goffrini, P., Lodi, T., Marzona, M., Bardi, L.: Behaviour of *Saccharomyces cerevisiae* wine strains during adaptation to unfavourable conditions of fermentation on synthetic medium: Cell lipid composition, membrane integrity, viability and fermentative activity. Int. J. Food Microbiol. **121**, 84–91 (2008)

[21] Chaney, D., Rodriguez, S., Fugelsang, K., Thornton, R.: Managing high-density commercial scale wine fermentations. J. Appl. Microbiol. **100**, 689–698 (2006)

[22] Farthing, J.B., Rodriguez, S.B., Thornton, R.J.: Flow cytometric analysis of *Saccharomyces cerevisiae* populations in high-sugar Chardonnay fermentations. J. Sci. Food Agric. **87**, 527–533 (2007)

[23] Herrero, M., Quirós, C., García, L.A., Díaz, M.: Use of flow cytometry to follow the physiological states of microorganisms in cider fermentation processes. Appl. Environ. Microbiol. **72**, 6725–6733 (2006)

[24] Monthéard, J., Garcier, S., Lombard, E., Cameleyre, X., Guillouet, S., Molina-Jouve, C., Alfenore, S.: Assessment of *Candida shehatae* viability by flow cytometry and fluorescent probes. J. Microbiol. Methods. **91**, 8–13 (2012)

[25] Da Silveira, M.G. da, Romão, M.V.S., Loureiro-Dias, M.C., Rombouts, F.M., Abee, T.: Flow cytometric assessment of membrane integrity of ethanol-stressed *Oenococcus oeni* cells. Appl. Environ. Microbiol. **68**, 6087–6093 (2002)

[26] Amor, K.B., Breeuwer, P., Verbaarschot, P., Rombouts, F.M., Akkermans, A.D.L., Vos, W.M.D., Abee, T.: Multiparametric flow cytometry and cell sorting for the assessment of viable, injured, and dead Bifidobacterium cells during bile salt stress. Appl. Environ. Microbiol. **68**, 5209–5216 (2002)

[27] Divol, B., Du Toit, M., Duckitt, E.: Surviving in the presence of sulphur dioxide: strategies developed by wine yeasts. Appl. Microbiol. Biotechnol. **95**, 601–613 (2012)

[28] Rodriguez, S. b., Thornton, R. j.: Use of flow cytometry with fluorescent antibodies in real-time monitoring of simultaneously inoculated alcoholic-malolactic fermentation of Chardonnay. Lett. Appl. Microbiol. **46**, 38–42 (2008)

[29] Chaillet, L., Martin, G., Genty, V.: Mise au point d'une méthode de détection des *Brettanomyces* par immunocytométrie. AFC (2014)

[30] Wallner, G., Amann, R., Beisker, W.: Optimizing fluorescent in situ hybridization with rRNA-targeted oligonucleotide probes for flow cytometric identification of microorganisms. Cytometry. **14**, 136–143 (1993)

[31] Inácio, J., Behrens, S., Fuchs, B.M., Fonseca, Á., Spencer-Martins, I., Amann, R.: *In situ* accessibility of *Saccharomyces cerevisiae* 26S rRNA to cy3-labeled oligonucleotide probes comprising the D1 and D2 domains. Appl. Environ. Microbiol. **69**, 2899–2905 (2003)

[32] Serpaggi, V., Remize, F., Sequeira-Le-Grand, A., Alexandre, H.: Specific identification and quantification of the spoilage microorganism *Brettanomyces* in wine by flow cytometry: A useful tool for winemakers. Cytometry A. **77A**, 497–499 (2010)

[33] Röder, C., König, H., Fröhlich, J.: Species-specific identification of *Dekkera/Brettanomyces* yeasts by fluorescently labeled DNA probes targeting the 26S rRNA. FEMS Yeast Res. **7**, 1013–1026 (2007)

[34] Müller, S., Ullrich, S., Lösche, A., Loffhagen, N., Babel, W.: Flow cytometric techniques to characterise physiological states of *Acinetobacter calcoaceticus*. J. Microbiol. Methods. **40**, 67–77 (2000)

 DOI: 10.1051/bioconf/20170902019

Diversidad genética de levaduras aisladas a partir de uvas de Vitis vinifera ssp. Sylvestris (Gmelin) Hegi en el área Euroasiática / Genetic diversity of yeasts isolated from Eurasian populations of *Vitis vinifera* ssp. *Sylvestris* (Gmelin) Hegi

Gustavo Cordero-Bueso[1,a], Ileana Vigentini[2], Roberto Foschino[2], David Maghradze[3], y Jesús Manuel Cantoral[1]

[1] Department of Biomedicine, Biotechnology and Public Health, University of Cádiz, Spain
[2] Department of Food, Environmental and Nutritional Sciences, University of Milan, Italy
[3] Institute of Horticulture, Viticulture and Oenology, Agricultural University of Georgia, Tbilisi, Georgia

Abstract. *Vitis vinifera* L. ssp. *Sylvestris* (Gmelin) Hegi is recognized as the dioecious parental generation of today's cultivars. Climatic change and the arrival in Europe of pathogens and pests have led it to be included on the IUCN Red List of Threatened Species in 1997. At best of our knowledge, no studies on microbial populations of grape-berry surfaces have been done. The present work has been focused on the study of yeast occurrence and diversity on grape-berries collected from wild vines. Final outputs have allowed: *i*)to obtain precise information about yeast communities; *ii)* to provide an objective framework for the classification of the broadest range of species according to their extinction risk; *iii)* to select attractive yeast strains for their biotechnological potential, offering new opportunities to winemakers. Sampling plan was performed in Azerbaijan, Georgia, Italy, Romania and Spain. In all, 3180 yeast colonies were isolated and identified as belonging to 50 species, including *Saccharomyces cerevisiae*, by 26S rDNA D1/D2 domains and ITS region sequencing. Isolates of *S. cerevisiae* were also analysed by SSR-PCR obtaining 163 different genotypes. This study highlights the biodiversity potential of pristine environments that still represent a fascinating source to face common problems in winemaking.

1. Introducción

Las poblaciones de vid silvestre euroasiática se extienden desde la región caucásica pasando por la cuenca del Mediterráneo (Turquía, Grecia, Italia, Sur de Francia y Península Ibérica) hasta el macizo del Hindu Kush (Afganistán y Pakistán) y el Magreb (Marruecos, Túnez y Argelia). Sus ejemplares pertenecen al taxón *Vitis vinifera* L. subespecie *sylvestris* Gmelin (Hegi), única especie ancestral en Europa y son parentales dioicos de las variedades de cultivo. Estas últimas son fundamentalmente hermafroditas, aunque pueden encontrarse también ejemplares femeninos, como ocurre en la región Caucásica [1]. La Península Ibérica e Itálica constituyen hoy en día uno de los principales refugios para esta subespecie durante la última glaciación. En España existen pruebas palinológicas del Pleistoceno medio en las turberas de El Padul (Granada), y en la Laguna de Las Madres (Huelva) [2,3].

Según las referencias de Rivera y Walker [4], dentro de la Península Ibérica, las bayas de vid silvestre han contribuido directamente a la alimentación humana desde el Paleolítico. Las plantas domesticadas por el hombre son aquéllas que le sirven para su dieta o son aplicables a sus actividades cotidianas, entre ellas se encontraba esta liana. A partir de los escasos ejemplares hermafroditas aparecidos en la naturaleza como resultado de mutación, se fueron seleccionando variedades de cultivo [5]. Un estudio publicado en 2006 que lleva a cabo los diversos clorotipos y su distribución en 1201 muestras de muy diverso material silvestre y cultivado, procedente de diferentes áreas de la Península Ibérica, Itálica, Oriente Medio y Norte de África refuerza la teoría del origen policéntrico de la domesticación de la vid [6]. El artículo señala, además, que el 70% de los viñedos de la Península Ibérica e Itálica exhiben clorotipos derivados de las poblaciones silvestres de Europa Occidental, por lo que sostiene la idea sobre la existencia de una región de refugio para la vid, entre otras especies botánicas, en la Península Ibérica y cuyos genes son completamente distintos de las variedades cultivadas. Parece ser que el proceso de domesticación fue acelerado y guiado en primer lugar por la influencia cultural, y después, por las aportaciones directas de las actividades que realizaron los colonos fenicios, griegos y púnicos en la cuenca del Mediterráneo occidental. Se puede pensar, por lo tanto, que en las zonas de distribución de la vid silvestre, la introducción en primer lugar del consumo del vino y, posteriormente, de la viticultura, se sobrepusieron al preexistente sustrato de cultura local, caracterizado por una fase de proto-domesticación de la vid [7].

Antes del empleo de las variedades hermafroditas cultivadas, los racimos silvestres constituyeron la materia prima del vino. Probablemente el hombre, durante milenios, llevó a cabo un proceso de frutalización con el fin de aumentar la presencia de este recurso natural en

[a] e-mail: gustavo.cordero@uca.es

sus alrededores en detrimento del uso de la vid silvestre. Con el desarrollo y aumento de las comunicaciones, la acción humana ha ido progresivamente destruyendo los hábitats de esta planta trepadora de forma directa o indirecta, mediante obras públicas (embalses, puentes, trazado de carreteras), expansión de las zonas agrícolas y explotaciones forestales, así como con la limpieza de las cunetas de las carreteras [7], por lo cual a día de hoy sólo quedan algunos reductos de poblaciones en la Península Ibérica, Itálica, Helénica, Anatolia y región Transcaucásica. Arnold et al. [8] concluyen, además, que la vid silvestre en Europa está más afectada que en otros lugares del Mundo debido a que quedan muy pocos refugios para la misma y la tasa de supervivencia es muy baja, pues los intentos de recuperarla mediante trasplantes y reforestación suponen un elevado coste económico y con sin éxito. Así, todo ese creciente impacto ambiental negativo ha llevado a la vid silvestre a figurar como especie amenazada en la lista roja publicada por la Unión Internacional para la Conservación de la Naturaleza (IUCN, 1997) [9].

En la Península Ibérica, los hábitats que albergan todavía un mayor número de parras silvestres son los bosques de ribera (Sierra Norte de Sevilla, Huelva y Parque Nacional de Doñana), ya que las parras son hidrófilas y toman árboles y arbustos como tutor, dado su carácter heliófilo. Los principales tutores corresponden a diversas especies de los géneros: *Acer*, *Alnus*, *Crataegus*, *Ficus*, *Fraxinus*, *Olea*, *Populus*, *Quercus*, *Retama*, *Rubus*, *Ulmus*, *Rubus*, entre otros. Asimismo, aparecen en diversas zonas coluviales de clima húmedo, como es el caso de la costa cantábrica. Hay refugios de vid silvestre en el Norte y en el Sur de la Península Ibérica. La supervivencia de esta subespecie se ve también expuesta a las actividades agrícolas, aves (especialmente alcedinos) y otros factores humanos, además de la filoxera [7]. Según Böhm [10], la vid silvestre que alberga la Península Ibérica también presenta particularidades genéticas diferentes a otras de su misma especie en otras regiones de la cuenca Mediterránea. Este autor denomina a las vides presentes en la península Ibérica (España y Portugal) como vides del "polo ibérico" y a las presentes en la península Itálica como vides del "polo ítalo-etrusco". En Italia se distribuyen por diferentes regiones siendo las principales Piemonte, Lombardía, Cerdeña, Calabria, Basilicata y Sicilia [11].

Aparte de los usos tradicionales citados, la vid silvestre constituye un importante recurso fitogenético que alberga una importantísima diversidad fitogenética, con la que hay que contar para futuros programas de mejora de viníferas y portainjertos, así como para la reforestación de ecosistemas naturales. Actualmente, mediante marcadores moleculares, se abordan estudios sobre la contribución genética de las vides silvestres a las variedades de cultivo características de algunos puntos de la región Mediterránea. Sin embargo, bajo nuestro conocimiento, aún no existen estudios que determinen la ecología de microorganismos asociados a la uva de vides silvestres, siendo de especial relevancia aquellos microorganismos implicados en la fermentación vínica y maloláctica, levaduras, bacterias lácticas o bacterias productoras de vinagre (bacterias acéticas). Por ello, pensamos que preservar la vid salvaje y su consecuente biodiversidad de especies microbianas es de enorme importancia no solo por su interés actual para la industria enológica y vinagrera tan importante en los países del Mediterráneo así como en otros del Mundo, sino también para asegurar la conservación de un *pool* de genes de gran importancia tecnológica para la industria alimentaria en general y mantener estos recursos genéticos microbianos ante la inminente amenaza de extinción de la vid silvestre y crear nuevos estilos de vinos y vinagres que coloquen a España e Italia de nuevo a la cabeza en innovación enológica. En este sentido se hacen necesarios estudios que ayuden a cubrir ese "vacío de conocimiento". La extinción de las pocas poblaciones de vid silvestre en Europa supondría una pérdida irreversible de la biodiversidad, no sólo de la propia liana que nos ocupa, sino también los microorganismos asociados a la misma. El objetivo fundamental evaluar la biodiversidad de microorganismos de interés biotecnológico asociadas a la uva de vides silvestres y proveer datos que refuercen la importancia de la protección de *Vitis vinifera* ssp. *Sylvestris* (Gmelin) Hegi "*in situ*". Además de aportar nuevas cepas de levaduras capaces de hacer frente a problemas comunes entre países dónde la vitivinicultura es una importante fuente económica y social.

2. Material y métodos

La recogida uvas de vid silvestre se llevó a cabo en diferentes muestreos por triplicado en las diferentes regiones de estudio (Tabla 1) durante los meses de Septiembre, Octubre y Noviembre (según latitud y maduración de la uva) en el periodo del 2015 al 2017 la experiencia nos indica que en sistemas abiertos (campo) es importante hacer el muestro a larga escala (mínimo 2 años) para minimizar la influencia de las condiciones climáticas que afectan a las poblaciones de microorganismos presentes en la vid silvestre. Se emplearon bolsas asépticas para la recogida de bayas de uvas como muestra la metodología descrita por Cordero- Bueso et al. (2011). Se realizó un estudio ampelográfico para determinar que las uvas recogidas pertenecía a *V. vinifera* ssp. *Sylvestris* (Gmelin) Hegi siguiendo el protocolo de la O.I.V. También se llevaron a cabo fermentaciones a pequeña escala para aislar microorganismos al inicio, mitad y final del proceso fermentativo tal como se propone en Cordero-Bueso et al. [12]. Además, se aislaron los microorganismos directamente del hollejo de las uvas mediante lavados con cloruro sódico al 0,9% y con la ayuda de un sonicador y mediante rotura de las células vegetales por criogenización y con la ayuda de un bisturí para aislar levaduras endofíticas.

El aislamiento se realizó siguiendo protocolos de Microbiología Clásica. Los medios de cultivo utilizados fueron: medio YPD (2% glucosa, 2% peptona, 1% extracto de levadura, 2% agar) y medio WL-agar (Oxoid). Para su conservación y posteriores estudios de identificación y caracterización se utilizaron los siguientes métodos de conservación: a) en placas Petri, conservándose a 4°C y b) en glicerinados (glicerol 40%), conservadas a –80°C [12].

La identificación de los aislados se hizo mediante la obtención de los diferentes perfiles moleculares del ADN total y mitocondrial previamente extraído mediante protocolos estándares. Las cepas aisladas se identificaron por las técnicas de biología molecular PCR de la

 DOI: 10.1051/bioconf/20170902019

Tabla 1. Puntos de muestreo de uvas de vid silvestre en las diferentes localizaciones de la región Euroasiática.

País	Localización	Provincia	Coordenadas (Latitud, Longitud)	Altitud (m)
Azerbayán	Guruchay 1	Quba	41.38564 N, 48.38264 E	953
	Guruchay 2	Quba	41.32566 N, 48.37580 E	880
Georgia	Tsminda, Gveleti	Gori	42.66244 N, 44.62045 E	1100
	Tsitsamuri, Mtsjeta	Kartli	41.52383 N, 44.43574 E	524
	Zhinvali 03, Dusheti 03	Kartli	42.08784 N, 44.45843 E	986
	Shirikhevi 06, Dusheti	Kartli	41.98121 N, 44.72292 E	978
	Nakhiduri 24, Marnueli	Kartli	41.29143 N, 44.41201 E	429
	Shirikhevi 02, Dusheti	Kartli	41.57850 N, 44.43076 E	1348
	Nakhiduri23, Marneuli	Kartli	41.29140 N, 44.41206 E	437
	Bagichala03, Dusheti	Kartli	42.02226 N, 44.44558 E	644
	Barisakmos, Gadasakmuevi 01	Dusheti	42.08769 N, 44.45849 E	738
Italia	Monte Fenera, Borgosesia	Vercelli	45.71000 N, 08.31600 E	886
	Montalto	Pavia	44.97940 N, 09.20941 E	381
	Ortuabis, Laconi	Oristano	39.85120 N, 09.05184 E	504
	Bau Sa Mela, Nurallao	Cagliari	39.83981 N, 09.12157 E	735
	Santa Sofia, Laconi	Oristano	39.85948 N, 09.12417 E	783
	Ristalu, Aritzo	Nuoro	39.95258 N, 09.19497 E	816
	Fluminimaggiore Nera	Carbonia-Iglesias	39.43689 N, 08.49750 E	62
	Fluminimaggiore Bianca	Carbonia-Iglesias	39.43689 N, 08.48871 E	71
	Gutturu Mannu 1	Cagliari	39.17276 N, 08.91511 E	116
	Gutturu Mannu 2	Cagliari	39.17276 N, 08.90545 E	116
Rumania	Turcul river, Bran	Brasov	45.51731 N, 25.36404 E	763
España	Hueznar, San Nicolás del Puerto	Seville	37.99448 N, 05.66693 W	565
	La Rocina, Doñana	Huelva	37.12380 N, 06.49654 W	8
	El Bosque, Benamahoma	Cádiz	36.77323 N, 05.48759 W	419
	Algeciras, Gran Capitán	Cádiz	36.12112 N, 05.52705 W	406
	La Algaida, Sanlúcar de Barrameda	Cádiz	36.83204 N, 06.31994 W	4
	Guadalupe	Cáceres	39.45132 N, 05.32724 W	637
	Tui	Pontevedra	42.04915 N, 08.64661 W	41

región ITS del ADN ribosómico y RFLP utilizando las endonucleasas de restricción *HaeIII, HinfI y CfoI* [12].

En el caso de las levaduras identificadas como *Saccharomyces cerevisiae*, se aplicó la técnica del análisis de los microsatélites multiplex o SSR con el objetivo de encontrar distintos perfiles a nivel de cepa [12]. Los amplificados se sometieron a electroforesis en geles de agarosa o en secuenciador automático. Y el análisis de las imagenénes en una cámara equipada con un transiluminador UV (BioRad). Se seleccionaron aquéllas que presentaron un perfil distinto y se tomaron al menos dos representantes de la misma especie para ser secuenciadas.

La secuenciación se llevó a cabo a través de los servicios de secuenciación MACROGEN Inc. Korea (`http://www.macrogen.com`). La alineación de sencuencias se realizó con programas de bioinformática online (CLUSTAL W, In-sillico, etc.). Las secuencias fueron comprobadas y depositadas en bases de datos electrónicas de colecciones de microorganismos (CECT, GenBank, NBCI, etc.).

Una vez bien identificados mediante las técnicas señaladas anteriormente, los microorganismos con diferentes perfiles moleculares fueron sometidos caracterización mediante pruebas bioquímicas y físico- químicas para conocer su capacidad de asimilación y fermentación de azúcares, resistencia al etanol y temperatura (curvas de crecimiento en lector de microplaca y fenotipo en placa medio-agar), al anhídrido sulfuroso, stress osmótico (curvas de crecimiento en lector de microplacas) y capacidad enzimática, se atenderá especialmente a las actividades glucosidasa, esterasa, proteasa y pectinasa (uso de kits enzimáticos y medios de cultivos específicos), ya que son de interés en la enología por aportar complejidad a los vinos. Se siguieron protocolos similares a los descritos en Rodríguez et al. [13]. También se obtuvieron las curvas de crecimiento de las cepas de levaduras de géneros *Saccharomyces* y con propiedades enológicas de interés mediante el uso un lector de placas multipocillos (96 pocillos) en mosto estéril natural obtenido a partir de vides silvestres, mosto obtenido de la variedad tempranillo, mosto sintético y en medio YPD como control durante 72 horas a 28°C y obteniendo medidas de lectura de concentración de células por mililitro a una Densidad Óptica de 600 nm e inoculando a un O.D = 0,2–0,4. Una vez recogidos los datos se trataron en Excel para obtener las curvas de crecimiento y se ajustaron de acuerdo al modelo de de cinética de Gompertz descrito por Buchanan et al. [14]. Estas curvas de crecimiento nos proporcionaron información para determinar el momento exacto (horas de crecimiento) de la fase exponencial de estas cepas y conocer el comportamiento en mosto natural de *V. vinifiera* ssp. *Sylvestris* al que están adaptadas, de este modo conoceremos el momento óptimo de inoculación de las cepas posteriores experimentos de fermentación.

Los datos obtenidos en todos los objetivos se analizaron a través de XSTAT. Además, se usaron programas específicos para los análisis filogenético (PAUP v4b10 y Modeltest 3.06), análisis Bayesianos (MRBAYES v3.0b4), análisis de parsimonia heurística y árboles filogenéticos (MAXTREES, PAUP v4b10).

3. Resultados y discusión

El número de cepas muestreadas durante el periodo 2013–2016 asciende a 29 puntos de muestreo (2 en Azerbaijan, 9 en Georgia, 10 en Italia, 1 en Rumanía y 7 en España) en los cuales al menos se tomaron muestras de uva de entre 3 y 5 cepas de vid silvestre. La cantidad de uvas recolectadas fue variable, entre 100 g y 2,5 kg. El grado de maduración de la uva siempre fue óptimo para poder llevar a cabo microvinificaciones como método de autoenriquecimiento con objeto de aislar el número máximo de especies. De todas la muestras tomadas y analizadas se obtuvo un total de 3180 colonias de las cuales, una vez identificadas mediante los métodos moleculares mencionados en la sección de Materiales y Métodos se identificó un total de 50 especies pertenecientes a 10 géneros. La Tabla 2 muestra la frecuencia y distribución de las especies identificadas en la región Euroasiática. Entre los aislados, 5 cepas no fueron identificadas mediante los métodos moleculares conocidos hasta ahora, no obstante fueron caracterizados mediante técnicas clásicas de microscopía y a través de pruebas bioquímicas de fermentación y asimilación de azúcares (datos no mostrados). Adicionalemente, se aislaron levaduras viables pero no cultivables e igualmemente fueron caracterizadas siguiendo los procedimientos mencionados anteriormente. La especie mayoritaria aislada fue *Saccharomyces cerevisiae*, esto es debido al procedimiento experimental empleado de autoenriquecimiento, no obstante, esta especie también fue aislada directamente del hollejo de las uvas recolectadas, aunque en una baja proporción. Todos los aislados pertenecientes al género *Saccharomyces* fueron sometidos a posteriores análisis específicos para obtener diferentes perfiles a nivel de cepa (genotipos). Respecto a levaduras de géneros No-*Saccharomyces*, las especies *Hanseniaspora uvarum* y *Pichia kluyvery* fueron las especies mayoritarias, pero no se aislaron en todas las regiones muestreadas (Tabla 2). Algunas especies fueron exclusivas de cada área, como por ejemplo *Aureobasidium proteae* que solamente fue aislada en Italia, al igual que sucede con especies de los géneros *Cryptococcus*, *Curvibasidium* y algunas de *Hanseniaspora*. Las especies *Clavispora lusitaniae*, *Hanseniaspora meyeri*, *Issatchenkia terricola* y *Xanthophyllomyces dendrorhou* solo se aislaron en Georgia, al igual que sucede con *Saccharomycodes ludwigii* en Azerbayán. Rumanía solo mostró una especie exclusiva entre los aislados, *Pichia fermentans*. En el caso de España cabe descatar que fue la región donde más especies diferentes se aislaron, podría venir motivado por haber muestreado un año más que en el resto de países, sin embargo, también se identificaron numerosas especies no aisladas en otras áreas del Mediterráneo tales como *Candida ethanolica*, *Candida sake*, *Candida stellata*, *Candida zemplinina*, *Lachancea thermotolerans*, *Meyerozyma caribbica*, *Pichia kudriavzevii*, *Rhodosporidium palidugenum*, *Rhodotorula glutinis*, *Scheffersomyces stipitis*, *Schwanniomyces polymorphus*, *Torulaspora delbrueckii* y *Wickerhamomyces anomalus*. Pese a que ninguna de las especies identificadas en los diferentes países donde se realizó el muestreo aparece en todos los países, a excepción de *Saccharomyces cerevisiae*, cabe destacar que la especie *Pichia manshurica* fue aislada en Italia y España pero a la misma latitud (Monte Fenera y Tui).

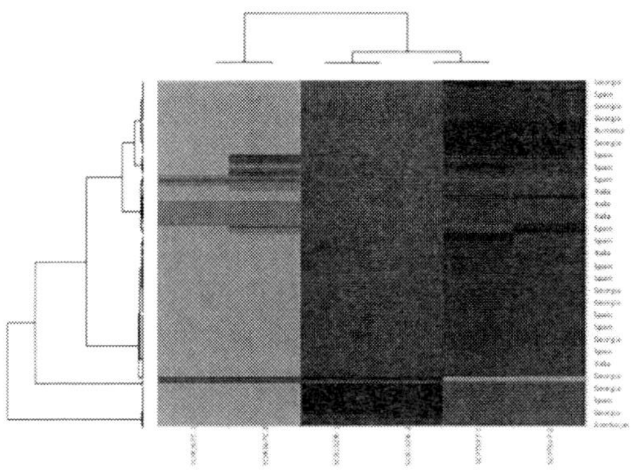

Figura 1. Mapa de calor (Heatmap) de los diferentes genotipos obtenidos mediante PCR-Multiplex de Microsatélites en los diferentes países de la región Euroasiática muestreados.

Aquellos aislados que fueron identificados como pertenecientes al género *Saccharomyces*, dado el interés que suscita en la industria enológica, fueron posteriormente sometidos análisis de PCR Multiplex de microsatélites. Entre los aislados de S. cerevisiae se obtuvieron un total de 169 genotipos diferentes, siendo 80 de España, 56 de Georgia, 24 de Italia, 6 de Azerbaijan y 3 de Rumanía. Se observó cierta relación entre los genotipos de todas la regiones pero sólo para algunos alelos. A continuación se explican los resultados obtenidos.

En la Fig. 1 se muestra el de mapa de calor obtenido a partir del análisis genómico (en base al tamaño en pares de bases del amplificado) por PCR multiplex de microsatélites con las tres parejas de primers. Los alelos se agrupan en filas y las cepas en columnas. Si se observamos individualmente los alelos y las cepas según su origen (Azerbayán, Georgia, Italia, Rumanía y España) vemos claramente que los se dividen en tres grupos que corresponden a los tres genes amplificados.

Según el cluster de la izquierda (dendrograma) se observa que las diferentes cepas de *S. cerevisiae* aisladas de uvas y microvinificaciones de uvas silvestres recogidas en las 5 áreas estudiadas se agrupan de acuerdo al tamaño de los alelos amplificados.

Si nos centrarnos en los patrones rectángulo/cuadrado dentro del mapa, los rectángulos grandes verdes de la izquierda muestran que para los alelos (YOR267) tenemos una coincidencia relativamente alta entre las cepas aisladas en los diferentes países, fijándonos en la parte central del mapa se muestra un rectángulo rojo grande que aúna coincidencias entre los alelos (SC8132) de los diferentes países, pero con una menor frecuencia que en el primer patrón y además se alejan bastante los aislados en Azerbayán (restángulo marrón oscuro) del resto de países. En el caso del tercer patrón de colores más oscuros (marrón-negro) es completamente inverso, en comparación con los dos anteriores, así como el rectándulo verde intenso correspondiente a algunos genotipos aislados en Georgia demuestran la variabilidad y exclusividad de algunos genotipos al área Caucásica. Curiosamente se observa que algunos genotipos encontrados en España comparten varios alelos con aquéllos de los países caucásicos (cuadros verdes, negro y rojo del de la parte inferior del mapa de calor) agrupados en un cuarto cluster (dendrograma de la izquierda). Con estos resultados se

Especies	Azerbayán	Georgia	Italia	Rumanía	España
Aureobasidium proteae			2		
Aureobasidium pullulans			7		18
Candida ethanolica					1
Candida sake					11
Candida stellata					25
Candida zemplinina					4
Clavispora lusitaniae		32			
Cryptococcus flavescens			9		
Cryptococcus wieringae			10		
Curvibasidium cygneicollum			2		
Curvibasidium pallidocorallinum			15		
Hanseniaspora sp			2		
Hanseniaspora clermontiae			9		
Hanseniaspora guilliermondii			18		54
Hanseniaspora meyeri		13			
Hanseniaspora opuntiae			32		120
Hanseniaspora osmophila					
Haseniaspora uvarum		67	158	34	137
Hyphopichia pseudoburtonii			8		
Issatchenkia siamensis		27	5		
Issatchenkia terricola		17			
Lachancea thermotolerans					185
Martiniozyma asiatica			13		
Metschnikowia sp.		21	15		
Metschnikowia fructicola			69	15	
Metschnikowia pulcherrima			36		19
Metschnicowia viticola		18	23		
Meyerozyma caribbica					88
Meyerozyma guilliermondii			30		107
Naganishia diffluens			9		
Pichia fermentans				10	
Pichia kluyveri			43		230
Pichia kudriavzevii					130
Pichia Manshurica			4		24
Rhodosporidium paludigenum					4
Rhodotorula fujisanensis			21		
Rhodotorula glutinis					34
Rhodotorula mucilaginosa		34	1		22
Rhodotorula nothofagi			27		
Saccharomyces cerevisiae	50	296	249	31	517
Saccharomycodes ludwigii	10				
Scheffersomyces stipitis					14
Schwanniomyces polymorphus					2
Torulaspora delbrueckii					67
Wickerhamomyces anomalus					42
Xanthophyllomyces dendrorhou		2			
Zygosaccharomyces sp			2		
Zygosaccharomyces bailii					6
Zygosaccharomyces fermentati					41
Zygosaccharomyces rouxii			2		6
Viables pero no cultivables			30		
No identificada		3			2

podría afirmar que existen tres regiones, áreas o ejes definidos en cuanto a genotipos de levaduras, serían el eje oriental (Azerbaijan y Georgia), el eje central (Rumanía e Italia) y el eje occidental (España, pese a que comparte algunos genotipos con el eje oriental).

4. Conclusiones

Este estudio destaca el potencial de biodiversidad de entornos prístinos que aún representan una fuente fascinante para enfrentar problemas comunes en la elaboración del vino. Se están llevando a cabo estudios de evaluación del potencial biotecnológico de las diferentes especies aisladas, así como estudios para determinar si las cepas no identificadas son posibles nuevas especies.

Referencias

[1] D. Maghradze. O. Failla, R. Bacilieri, S. Imazio, I. Vashkidze, R. Chipashvili, I. Mdinaradze, N. Chkhartishvili, P. This, A. Scienza, *Georgian vitis germplasm : usage, conservation and investigation*

DOI: 10.1051/bioconf/20170902019

(Bulletin oiv, vol. 83, octobre-novembre-décembre n° 956-957-958, 2010)
[2] R. Stevenson, J. Biogeography **12**, 293–314 (1985)
[3] F. Florschütz, J. Menéndez Amor, T.A. Wijmstra, Palaeogeogr. Palaeoclimatol. Palaeoecol **10**, 233–264 (1971)
[4] Rivera-Núñez, Walker, Rev Paleobot. Palynol. **61**, 205–237 (1998)
[5] C. Arnold, A. Schnitzler, A. Douard, R. Peter, F. Gillet, Biodiversity and Conservation **14**, 1507–1523 (2005)
[6] R. Arroyo-García, L. Ruíz-García, L. Bolling, R. Ocete, A. López, C Arnold, A. Ergul, Mol. Ecol. **15**, 3707–3714 (2006)
[7] R. Ocete, C. Arnold, O. Failla, G. Lovicu, B. Biagini, S. Imazio, D. Maghradze, M.A. López, Vitis **50**(2), 97–98 (2011)
[8] Forni. La vite e l'uomo (EDP Ersa, Gorizia, Italia, 2006)
[9] IUCN red list of threatened plants (EDP, K.S. Walter, H.J. Gillett, South Africa, 1997)
[10] J. Böhm, Portugal Viticola, O Grande Livro das Castas (EDP, C. Ferreiras, Lisboa, Portugal, 2007)
[11] B. Biagini, G. De Lorenzis, A. Scienza, O. Failla, S. Imazio, D. Maghradze, D. Act. Horticult. **948**, 211–216 (2012)
[12] G. Cordero-Bueso, T. Arroyo, A. Serrano, J. Tello, I. Aporta, M.D. Vélez, E. Valero, Int. J. Food Microbiol. **145**, 132-139 (2011)
[13] M.E. Rodríguez, J.J. Infante, M. Molina, M. Domínguez, L. Rebordinos, J.M. Cantoral, J. Appl. Microbiol. **108**, 1292–1302 (2010)
[14] R.L. Buchanan, R.C. Whiting, W.C. Damert, Food Microbiol **14**, 313–326 (1997)

Scientific study of $^{13}C/^{12}C$ carbon and $^{18}O/^{16}O$ oxygen stable isotopes biological fractionation in grapes in the Black Sea, Don Basin and the Western Caspian regions

Alexander Kolesnov[1], Margarita Zenina[1], Sergey Tsimbalaev[1], Dmitrii Davlyatshin[2], Mikhail Ganin[2], Nadezhda Anikina[3], Natalya Agafonova[3], Evgeniy Egorov[4], Tatyana Guguchkina[4], Anton Prakh[4], and Mikhail Antonenko[4]

[1] Research Laboratory of Food Quality & Technology (PNIL), Peoples' Friendship University of Russia (RUDN), 117198 Moscow, ul. Mikluho-Maklaya, 8/2, Russian Federation
[2] Laboratory of Quality of the JSC Azbuka Vkusa, 127322 Moscow, ul. Milashenkova 4, Russian Federation
[3] National Research Institute of Viticulture and Winemarking «Magarach», 298600 Yalta, ul. Kirova, 31, Crimea Republic
[4] North Caucasian Regional Research Institute of Horti- and Viticulture, 350901 Krasnodar Region, Krasnodar, ul. 40-letya Pobedy, 39, Russian Federation

Abstract. The report presents the results of a study of carbon and oxygen stable isotopes in carbohydrates and intracellular water of red and white grapes of 2016 wine-growing season in the Crimean peninsula areas, South-west coast of the Greater Caucasus, the Don basin and the Western Caspian region. The mass concentration of reducing sugars in the studied grape samples has been from 17.5 to 25.0 g/100 ml, titrated acids concentration (based on tartaric acid) – from 6.0 to 9.1 g/l, the buffer capacity 34.1–63.2 mg-Eq/l. Red and white wine made from respective grapes contained from 0.5 to 3.6 g/l of residual sugar; from 11.1 to 14.5% ethanol by volume; buffer capacity was 35.2–52.6 mg-Eq/l. It has been found that the $\delta^{13}C_{VPDB}$ values for carbohydrates of red and white grape varieties as a result of biological fractionation of carbon isotopes in the agro-climatic conditions of plant growth for the studied geographical areas are ranging from −26.74 to −20.74‰ (the Crimean peninsula); from −27.31 to −21.58‰ (South West Coast of the Greater Caucasus), from −27.33 to −24.73‰ (Don Basin) and from −26.64 to −23.17‰ (West Caspian). The $\delta^{13}C_{VPDB}$ values for ethanol of the red and white dry wines range from −28.52 to −24.26‰ (the Crimean peninsula); from −29.23 to −24.52‰ (South West Coast of the Greater Caucasus); from −28.97 to −26.22‰ (Don Basin); from −29.14 to −25.22‰ (Western Caspian). Compared with the surface water and groundwater (averages from $\delta^{18}O_{VSMOW}$ −13.90 to −6.38‰) and with precipitation (averages from $\delta^{18}O_{VSMOW}$ − 10.30 to −9.04‰) the $\delta^{18}O_{VSMOW}$ values in intracellular water of grapes are the following: for the Crimean peninsula grapes, from 0.40 to 4.97‰; the South West Coast of the Greater Caucasus, from -2.11 to 6.29‰; the Don Basin, from −2.21 to 6.26‰; the Western Caspian, from −0.24 to 1.44‰. It has been noted that in conditions of water shortage caused by low rainfall or lack of rainfall and irrigation, grapes are prone to the accumulation of "heavy" ^{13}C carbon isotope due to changes in isotopic exchange with the environment through a partial reduction in the intensity of transpiration and photosynthesis.

1. General climatic conditions in the areas of the study

1.1. Black sea

The grapes samples from two main regions of the Black Sea have been used in the study – Crimean peninsula and South West Coast of the Greater Caucasus (Fig. 1). The sampling of the authentic fresh grapes was carried out in 5 viticulture areas of the Crimean Peninsula – Western Maritime-Steppe area (villages Romashkino and Plodovoye), Western Foothills-Maritime area (village Uglovoye), Eastern Mountain-Valleey Maritime zone of the South Coast area (village Morskoye) and East foothill zone of the South Coast area (village Koktebel). The total area of sampling vineyards made 512.0 ha. The main areas on the South West Coast of the Greater Caucasus are the Temryuk, Krymsk and Anapa districts, and also areas close Novorossiysk. The total area of vineyards on South West Coast, in which in a season of 2016 grapes samples have been taken, has made 215.02 ha. Scientific data on the Black Sea areas, including climate, soils and grapes varieties have been reported in the previous publication about results of the study of 2015 [1].

1.2. Don basin

The grapes were studied in two areas – Tsimlyansky (Tsimlyansky Reservoir coast) and Martynovsky districts (Fig. 2). The total area of the surveyed vineyards has made 80.60 ha.

1.3. Western caspian

The main area of viticulture is the The Republic of Dagestan. The list of areas of this region, in which

Figure 1. Black Sea areas.

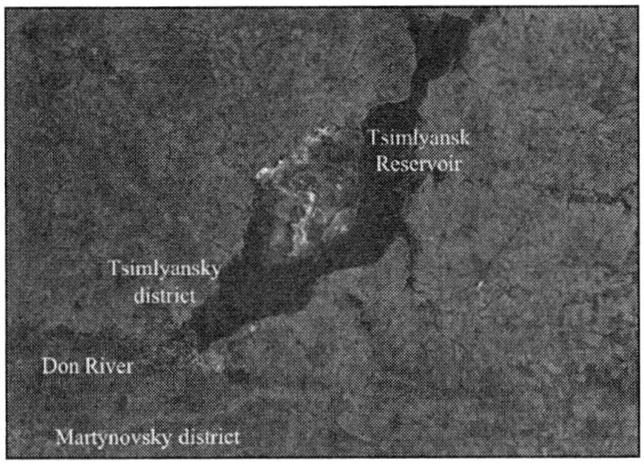

Figure 2. Don Basin areas.

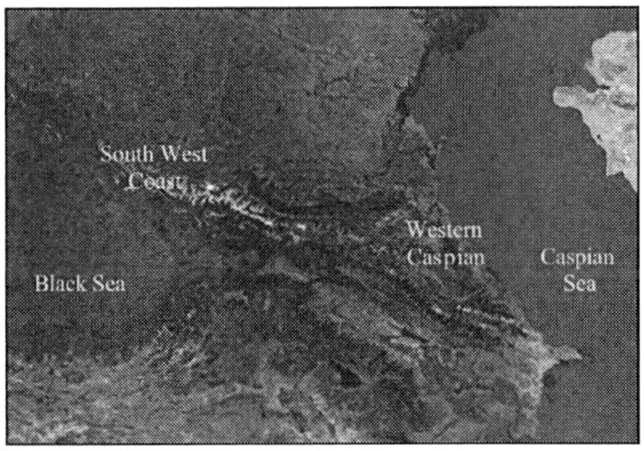

Figure 3. Western Caspian.

grapes samples have been taken, had included the Derbent, Sergokalinsky, Suleyman-Stalsky and Magaramkentsky districts. The total area of the surveyed vineyards has made 375.04 ha.

2. Grape samples

2.1. Varieties

The samples of the 9 white and red grape varieties were used for the study – Aligote, Rkatsiteli, Cabernet Sauvignon, Sauvignon Green (Crimean Peninsula); Cabernet Sauvignon, Chardonnay (South West Coast – Krasnodar region); Bianka, Saperavi Severniy, Tsimlyansky Cherniy (Don Basin); Rkatsiteli, Riesling, Cabernet Sauvignon (Western Caspian – The Republic of Dagestan). The sampling of the authentic fresh grapes was carried out in 15 viticulture areas in shown four regions. The total area of sampling vineyards made 1182.66 ha. The sampling was conducted according to the requirements described in [2].

2.2. Main physical-chemical characteristics

The research program included 43 samples of grapes of the above stated varieties. The fresh musts were obtained under the laboratory conditions. Gentle conditions for the extraction of musts exclude any changes in the natural composition of carbon and oxygen isotopes in the target components of grapes – carbohydrates (sugars) and intracellular water. In order to ensure the microbiological stability the musts samples have been pasteurized and filled in glass containers under aseptic conditions. Before the isotopic measurements the must samples were stored at a temperature from 2 to 4 °C in sealed sterile glass containers without access of air.

36 samples of dry white and red wines (residual sugar - from 0.5 to 3.6 g/l) have been made from the fresh musts under laboratory conditions. The fermentation of fresh musts was carried out using a pure 47 K yeast culture, which was added in an amount of 2% of the must volume. The SO_2 in an amount of 75–100 mg/l was added to musts before fermentation. Prior to the isotopic measurements the dry wines were stored at a temperature from 2 to 4 °C in sealed sterile glass containers without access of air.

The mass concentration of reducing sugars in the studied grape samples has been from 17.5 to 25.0 g/100 ml, titrated acids concentration (based on tartaric acid) - from 6.0 to 9.1 g/l, the buffer capacity 34.1–63.2 mg-Eq/l. Red and white wine made from respective grapes contained from 0.5 to 3.6 g/l of residual sugar; from 11.1 to 14.5% ethanol by volume; buffer capacity was 35.2–52.6 mg-Eq/l. The additional scientific data about buffer capacity measurements in musts and wines are published in [3,4].

3. Stable isotopes compositions of studied grape samples

Scientific study of compositions of stable isotopes of carbon ($^{13}C/^{12}C$) and oxygen ($^{18}O/^{16}O$) was carried out using two types of IRMS/SIRA[1] technique - redox transformation (Flash Combustion - FC-IRMS/SIRA) and isotopic exchange/ equilibration (Equilibration - EQ-IRMS/SIRA). Methodological principles of the FC-IRMS/SIRA and EQ-IRMS-SIRA methods have been described in previous publication [1] and in OIV methods [5,6].

3.1. Materials and methods

3.1.1. Musts and wine samples

The musts and dry wines were used for measurements of the composition of the carbon isotopes $^{13}C/^{12}C$ in carbohydrates (sugars) and ethanol by the FC-IRMS/SIRA method. The composition of oxygen isotopes $^{18}O/^{16}O$ was studied by the EQ-IRMS/ SIRA method directly in musts and dry wines without prior isolation of water from the samples.

[1] IRMS/SIRA = Isotope Ratio Mass Spectrometry/Stable Isotope Ratio Analysis.

Table 1. Composition of stable carbon isotopes ($^{13}C/^{12}C$) in carbohydrates (sugars) of grape varieties and ethanol of dry wines of the selected areas of the Crimean Peninsula (crop of 2016).

Grape variety	Area[1]	$\delta^{13}C_{VPDB}$, ‰	
		carbohydrates (sugars)	ethanol
Aligote	Uglovoye	−23.83 ± 0.21	−25.69 ± 0.05
Cabernet Sauvignon		−23.71 ± 0.16	−25.43 ± 0.05
Rkatsiteli	Morskoye	−26.74 ± 0.02	−28.52 ± 0.01
Cabernet Sauvignon		−23.14 ± 0.03	−24.82 ± 0.04
Rkatsiteli	Plodovoye	−23.88 ± 0.07	-26.52 ± 0.05
Cabernet Sauvignon		−25.08 ± 0.17	−27.29 ± 0.05
Aligote	Koktebel	−24.78 ± 0.06	-26.81 ± 0.03
Cabernet Sauvignon		−20.74 ± 0.03	−24.26 ± 0.04
Cabernet Sauvignon	Romashkino	−21.91 ± 0.08	−24.58 ± 0.03
Sauvignon Green		−22.74 ± 0.04	n.d.[2]

Notes to the Table 1:

[1] For area's description see 1.1.

[2] Not determined.

Table 2. Composition of stable carbon isotopes ($^{13}C/^{12}C$) in carbohydrates (sugars) of grape varieties and ethanol of dry wines of the selected areas of the South West Coast of the Greater Caucasus (crop of 2016).

Grape variety	Area[1]	$\delta^{13}C_{VPDB}$, ‰	
		carbohydrates (sugars)	ethanol
Chardonnay (vineyards 1)	Temryuk district	−25.24 ± 0.37	−28.57 ± 0.04
Chardonnay (vineyards 2)		−25.29 ± 0.35	−27.11 ± 0.08
Chardonnay (vineyards 3)		−27.31 ± 0.40	−29.23 ± 0.06
Cabernet Sauvignon (vineyards 1)		−25.99 ± 0.15	−28.16 ± 0.02
Cabernet Sauvignon (vineyards 2)		−24.60 ± 0.10	−27.32 ± 0.08
Cabernet Sauvignon (vineyards 3)		−21.54 ± 0.23	−28.51 ± 0.10
Chardonnay	Krymsky district	−26.15 ± 0.21	−29.02 ± 0.02
Cabernet Sauvignon		−27.27 ± 0.19	−29.06 ± 0.01
Chardonnay	Anapsky district	−25.83 ± 0.20	−28.88
Cabernet Sauvignon		−26.17 ± 0.25	−28.62 ± 0.16
Chardonnay (vineyards 1/1)	Novoros-siysk	−23.18 ± 0.02	n.d.[2]
Chardonnay (vineyards 1/2)		−23.59 ± 0.11	−25.01 ± 0.02
Chardonnay (vineyards 2/1)		−22.25 ± 0.16	n.d.[2]
Chardonnay (vineyards 2/2)		−21.58 ± 0.13	−24.61 ± 0.01
Chardonnay (vineyards 3/1)		−22.33 ± 0.10	n.d.[2]
Chardonnay (vineyards 3/2)		−23.45 ± 0.07	−24.52 ± 0.10
Chardonnay (vineyards 4/1)		−26.35 ± 0.22	n.d.[2]
Chardonnay (vineyards 4/2)		−23.67 ± 0.16	−28.17 ± 0.05
Cabernet Sauvignon (vineyards 1/1)		−23.84 ± 0.02	−26.69 ± 0.05
Cabernet Sauvignon (vineyards 2/1)		−22.77 ± 0.23	−24.96 ± 0.23
Cabernet Sauvignon (vineyards 3/1)		−23.10 ± 0.17	−25.00 ± 0.08
Cabernet Sauvignon (vineyards 4/1)		−24.75 ± 0.18	−27.91 ± 0.08

Notes to the Table 1:

[1] For area's description see 1.1.

[2] Not determined.

3.1.2. Water samples

To study the composition of stable isotopes of oxygen ($^{18}O/^{16}O$) 19 water samples were taken in all studied areas - two samples from the Crimean Peninsula (water from atmospheric precipitations & underground source), two samples from the Don Basin (water from atmospheric precipitations & underground source), twelve samples form the South West Coast (water from atmospheric precipitations, tap water used for irrigation of vineyards, water from the Anapka river, city tap water and water from underground sources) and three samples from the Western Caspian (water from atmospheric precipitations, city tap water and water from underground sources).

The water samples were stored before the isotopic measurements at a temperature from 2 to 4 °C in sealed sterile glass containers without access of air.

3.1.3. FC-IRMS/SIRA method for $\delta^{13}C_{VPDB}$ analysis

The study of fractionation effect of the stable isotopes of carbon ($^{13}C/^{12}C$) in grape and wine components (carbohydrates (sugars), ethanol) using the FC-IRMS/SIRA method was carried out by using of the following laboratory facilities for stable isotopes measurements:

(*A*) an elemental analyzer FlashEA1112® with oxidation & reduction reactors and autosampler (Thermo Fisher Scientific, Germany);

(*B*) an isotopic interface Conflo III® (Thermo Fisher Scientific, Germany);

(*C*) an IRMS/SIRA isotopic mass spectrometer Delta V Advantage® (Thermo Fisher Scientific, Germany);

Table 3. Composition of stable carbon isotopes ($^{13}C/^{12}C$) in carbohydrates (sugars) of grape varieties and ethanol of dry wines of the selected areas of the Don Basin (crop of 2016).

Grape variety	Area[1]	$\delta^{13}C_{VPDB}$, ‰	
		carbohydrates (sugars)	ethanol
Bianka	Tsimlyan-sky district	−26.48 ± 0.10	−27.74 ± 0.05
Saperavi Severniy		−25.38 ± 0.01	−26.73 ± 0.09
Tsimlyansky Cherniy		−27.33 ± 0.14	−28.97 ± 0.11
Bianka	Martynov- sky district	−24.73 ± 0.20	−26.22 ± 0.09

Notes to the Table 3:

[1] For area's description see 1.2.

Table 4. Composition of stable carbon isotopes ($^{13}C/^{12}C$) in carbohydrates (sugars) of grape varieties and ethanol of dry wines of the selected areas of the Western Caspian (crop of 2016).

Grape variety	Area[1]	$\delta^{13}C_{VPDB}$, ‰	
		carbohydrates (sugars)	ethanol
Rkatsiteli (vineyards 1)	Derbent district	−26.75 ± 0.04	−27.79 ± 0.11
Rkatsiteli (vineyards 2)		−25.92 ± 0.20	−28.07 ± 0.01
Cabernet Sauvignon (vineyards 1)		−26.20 ± 0.01	−28.15 ± 0.06
Cabernet Sauvignon (vineyards 2)		−26.64 ± 0.11	−29.14 ± 0.04
Cabernet Sauvignon	Sergokalin-sky district	−23.17 ± 0.04	−25.22 ± 0.01
Rkatsiteli	Magaram- kent district	−26.20 ± 0.16	−27.77 ± 0.03
Riesling	Suleiman-Stalsky district	−23.66 ± 0.08	−25.78 ± 0.11

Notes to the Table 4:

[1] For area's description see 1.3.

Table 5. Composition of stable oxygen isotopes ($^{18}O/^{16}O$) in intracellular water of grape varieties from selected areas of the Crimean Peninsula (crop of 2016).

Grape variety	Area[1]	$\delta^{18}O_{VSMOW}$, ‰
Aligote	Uglovoye	3.57 ± 0.06
Cabernet Sauvignon		2.34 ± 0.30
Rkatsiteli	Morskoye	0.40 ± 0.10
Cabernet Sauvignon		1.23 ± 0.16
Rkatsiteli	Plodovoye	1.92 ± 0.17
Cabernet Sauvignon		2.77 ± 0.15
Aligote	Koktebel	2.32 ± 0.36
Cabernet Sauvignon		4.11 ± 0.13
Cabernet Sauvignon	Romashkino	4.45 ± 0.03
Sauvignon Green		4.97 ± 0.29

Notes to the Table 5:

[1] For area's description see 1.1.

(*D*) a gas system Sigm-Plus® for the supply of the analytical devices with highly purified gases (Sigm-Plus Ltd., Russian Federation);

(*E*) a PC workstation Optiplex 745® (Dell, USA) for data registration and processing by the high level software Isodat NT 2.5® (Thermo Fisher Scientific, Germany).

The reference substance IAEA-600 (Caffeine, $\delta^{13}C = -27.771$ ‰) was used for the calibration of the working reference gas (WRG) – carbon dioxide 99.9999% (Linde Gas RUS, Russian Federation). The high purity helium 99.9999% (Linde Gas RUS, Russian Federation) was used in the study as carrier gas.

Sample preparation – extraction and purification of carbohydrates (sugars) from musts obtained from fresh grapes – was performed according to the method described in [7]. The isolated and purified preparations of carbohydrates (sugars) were freeze-dried at −50°C and 0.035 mbar pressure in the laboratory lyophilic facility FreeZone[12] (Labconco Corporation, USA). The preparations were stored at a temperature no higher than -20 °C in a sealed glass container. Ethanol was extracted from dry wines by using of the laboratory glass distillation unit according to the conditions for the ethanol extraction as described in [5,8]. Extracted ethanol samples were stored at a temperature no higher than −20 °C in a sealed glass container.

The above mentioned laboratory facilities have been used for isotope measurements. All measurements were carried out taking into account guidelines published in [1,9–11]. Joint hinges of preparations of carbohydrates (sugars) containing from 40 to 100 ng of carbon were placed in tin capsules before carrying out the measurements. The capsules were thoroughly covered with the help of microtweezers, providing its full integrity and eliminating the contact of organic matter with the atmospheric air. The encapsulated hinges (in five replicates for each sample) were placed into the autosampler mounted on the oxidation reactor of the elemental analyzer (*A*). Along with the hinges of carbohydrates (sugars) the encapsulated reference substance was loaded into the autosampler for control of reliability of measurement's results. Ethanol sapmles were injected directly into oxidation reactor of the elemental analyzer (*A*) by using of a microsyringe. The injection volume of each ethanol sample was from 0.08 to 0.1 μl.

The conditions of carring out measurements of isotope composition of carbon in the preparations of carbohydrates (sugars) and ethanol by using the stated above laboratory facilities corresponded to the optimum modes which are developed during the preliminary runs. Quantitative combustion of carbohydrates (sugars) and ethanol was performed in a helium flow (flow rate 100 ml/min) in the oxidation and redox reactors of the elemental analyzer (*A*) at the temperature of 900 and 650 °C in the presence

Table 6. Composition of stable oxygen isotopes ($^{18}O/^{16}O$) in intracellular water of grape varieties from selected areas of the South West Coast of the Greater Caucasus (crop of 2016).

Grape variety	Area[1]	$\delta^{18}O_{VSMOW}$, ‰
Chardonnay (vineyards 1)	Temryuk district	2.28 ± 0.12
Chardonnay (vineyards 2)		2.82 ± 0.30
Chardonnay (vineyards 3)		2.74 ± 0.07
Cabernet Sauvignon (vineyards 1)		−1.87 ± 0.32
Cabernet Sauvignon (vineyards 2)		−0.83 ± 0.14
Cabernet Sauvignon (vineyards 3)		−2.11 ± 0.18
Chardonnay	Krymsky district	1.00 ± 0.23
Cabernet Sauvignon		−1.20 ± 0.06
Chardonnay	Anapsky district	1.55 ± 0.03
Cabernet Sauvignon		−1.98 ± 0.20
Chardonnay (vineyards 1/1)	Novorossiysk	3.99 ± 0.13
Chardonnay (vineyards 1/2)		2.96 ± 0.25
Chardonnay (vineyards 2/1)		6.29 ± 0.16
Chardonnay (vineyards 2/2)		2.71 ± 0.12
Chardonnay (vineyards 3/1)		2.49 ± 0.42
Chardonnay (vineyards 3/2)		2.07 ± 0.09
Chardonnay (vineyards 4/1)		2.61 ± 0.07
Chardonnay (vineyards 4/2)		1.70 ± 0.09
Cabernet Sauvignon (vineyards 1/1)		−0.35 ± 0.12
Cabernet Sauvignon (vineyards 2/1)	Novorossiysk	4.91 ± 0.17
Cabernet Sauvignon (vineyards 3/1)		5.09 ± 0.10
Cabernet Sauvignon (vineyards 4/1)		0.32 ± 0.06

Notes to the Table 6:
[1] For area's description see 1.1.

Table 7. Composition of stable oxygen isotopes ($^{18}O/^{16}O$) in intracellular water of grape varieties from selected areas of the Don Basin (crop of 2016).

Grape variety	Area[1]	$\delta^{18}O_{VSMOW}$, ‰
Bianka	Tsimlyansky district	2.74 ± 0.01
Saperavi Severniy		1.87 ± 0.25
Tsimlyansky Cherniy		−2.21 ± 0.25
Bianka	Martynovsky district	6.26 ± 0.27

Notes to the Table 7:
[1] For area's description see 1.2.

of molecular high purity oxygen 99.999% (Linde Gas RUS, Russian Federation). The separation of elementary gases resulting from the combustion of carbohydrates (sugars) or ethanol in the redox reactor was carried out in automatic mode by the chromatographic purification in the elemental analyzer (*A*). Water contained in the gas flow was removed by the trap column filled with the magnesium perchlorate.

For the quantitative measurement of carbon isotopes $^{13}C/^{12}C$, the isolated carbon dioxide is directed through isotopic interface (*B*) to the IRMS/SIRA mass spectrometer Delta V Advantage® (*C*). Accelerating voltage of the mass spectrometer was 3.07 kV, pressure in the ion source -1.8×10^{-6} kPa, ionization method - electron impact (electron energy 124 eV).

The conducting of a simultaneous correction of measured values of signal strength is an important aspect for control of accuracy of results to exclude the possible influence of isobars $^{12}C^{17}O^{17}O$ and $^{13}C^{17}O^{16}O$ which share is determined at the hardware level by the measurement of a signal of carbon dioxide of mass 45 taking into account the extent of the distribution of the isotopes ^{13}C and ^{17}O in the nature (respectively 1.11 and 0.0375%).

Table 8. Composition of stable oxygen isotopes ($^{18}O/^{16}O$) in intracellular water of grape varieties from selected areas of the Western Caspian (crop of 2016).

Grape variety	Area[1]	$\delta^{18}O_{VSMOW}$, ‰
Rkatsiteli (vineyards 1)	Derbent district	1.12 ± 0.15
Rkatsiteli (vineyards 2)		0.79 ± 0.26
Cabernet Sauvignon (vineyards 1)		−0.02 ± 0.08
Cabernet Sauvignon (vineyards 2)		0.96 ± 0.29
Cabernet Sauvignon	Sergokalinsky district	−0.24 ± 0.02
Rkatsiteli	Magaramkent district	0.96 ± 0.29
Riesling	Suleiman-Stalsky district	1.44 ± 0.42

Notes to the Table 8:
[1] For area's description see 1.3.

3.1.4. EQ-IRMS/SIRA method for $\delta^{18}O_{VSMOW}$ analysis

The study of fractionation effect of the stable isotopes of oxygen ($^{18}O/^{16}O$) in the intracellular water in grapes (musts) and water samples using the EQ-IRMS/SIRA method was carried out by using of the laboratory facilities for stable isotopes measurements:

(*A*) a module used for sample preparation for the analysis (Thermo Fisher Scientific, Germany);

Table 9. Composition of stable oxygen isotopes ($^{18}O/^{16}O$) in water samples from selected areas of the Crimean Peninsula, South West Coast of the Greater Caucasus, Don Basin and Western Caspian (study of 2016).

Sample description	Area[1]	$\delta^{18}O_{VSMOW}$, ‰
Tap water	Western Caspian (Derbent)	−9.42 ± 0.09
	South West Coast of the Greater Caucasus (Temryuk - Taman)	−9.31 ± 0.08
	South West Coast of the Greater Caucasus (Temryuk - Vinograniy)	−9.19 ± 0.08
	South West Coast of the Greater Caucasus (Krymsk)	−13.90 ± 0.01
Water of atmospheric precipitations	Western Caspian (Derbent)	−9.09 ± 0.08
	Don Basin (Tsymlyansky district)	−9.04 ± 0.12
	South West Coast of the Greater Caucasus (Temryuk - vineyard 1)	−10.10 ± 0.01
	South West Coast of the Black Sea (Temryuk - vineyard 2)	−9.62 ± 0.40
	South West Coast of the Greater Caucasus (Temryuk - vineyard 3)	−9.20 ± 0.08
	South West Coast of the Greater Caucasus (Krymsk)	−10.30 ± 0.01
Water from underground sources	Western Caspian (Foothills of the Caucasus)	−8.20 ± 0.09
	Don Basin (Martymovsky district)	−11.34 ± 0.09
	South West Coast of the Greater Caucasus (Temryuk – Taman, vineyard)	−12.78 ± 0.19
	South West Coast of the Greater Caucasus (Temryuk – Taman, winery)	−9.39 ± 0.04
	South West Coast of the Greater Caucasus (Anapsky district)	−7.88 ± 0.15
Water from surface sources	South West Coast of the Greater Caucasus (Anapsky district - river)	−9.55 ± 0.08
	South West Coast of the Black Sea (Anapsky district lake)	−6.38 ± 0.03

Notes to the Table 9:
[1] For area's descriptions see 1.1-1.3.

Table 10. Ranges for $\delta^{13}C_{VPDB}$ and $\delta^{18}O_{VSMOW}$ in carbohydrates (sugars) and intracellular water of white and red grape varieties of the selected areas of the Crimean Peninsula, South West Coast of the Greater Caucasus, Don Basin and Western Caspian (crop of 2016).

Area	$\delta^{13}C_{VPDB}$, ‰ carbohydrates (sugars)	$\delta^{18}O_{VSMOW}$, ‰ intracellular water
Crimean Peninsula	−26.74...−20.74	0.40...4.97
South West Coast of the Greater Caucasus (Krasnodar region)	−27.31...−21.58	−2.11...6.29
Don Basin (Rostov region)	−27.33...−24.73	−2.21...6.26
Western Caspian (The Republic of Dagestan)	−27.67...−23.66	−0.24...1.80

(*B*) a control unit for chromatographic purification and samples injection Gasbench II® (Thermo Fisher Scientific, Germany);
(*C*) an IRMS/SIRA isotopic mass spectrometer Delta V Advantage® (Thermo Fisher Scientific, Germany);
(*D*) a gas system for the supply of the analytical devices with highly purified gases Sigm-Plus® (Sigm-Plus Ltd., Russian Federation);
(*E*) a PC workstation Optiplex 960® (Dell, USA) for data registration and processing by the high level software Isodat 3.0® (Thermo Fisher Scientific, Germany).

The reference substance IAEA-600 (Caffeine, $\delta^{13}C = -27.771$‰) was used for the calibration of the working reference gas (WRG) – carbon dioxide 99.9999% (Linde Gas RUS, Russian Federation). The high purity helium 99.9999% (Linde Gas RUS, Russian Federation) was used in the study as carrier gas. The additional reference substances (all from the IAEA, Austria) with known composition of oxygen isotopes $^{18}O/^{16}O$ were used for adjustment of measurement's results: VSMOW2 (Vienna-Standard Mean Ocean Water, $\delta^{18}O = 0.00$‰);

Table 11. Range for $\delta^{13}C_{VPDB}$ in ethanol of dry white and red wines made from grape varieties of the selected areas of the Crimean Peninsula, South West Coast of the Greater Caucasus, Don Basin and Western Caspian (crop of 2016).

Area	$\delta^{13}C_{VPDB}$, ‰ ethanol of dry wines
Crimean Peninsula	−28.52...−24.26
South West Coast of the Greater Caucasus (Krasnodar region)	−29.06...−24.61
Don Basin (Rostov region)	−28.97...−26.22
Western Caspian (The Republic of Dagestan)	−29.14...−25.22

GISP (Greenland Ice Sheet Precipitation, $\delta^{18}O_{VSMOW} = -24.76$‰); SLAP2 (Standard Light Antarctic Precipitation, $\delta^{18}O_{VSMOW} = -55.50$‰).

High purity gases were used in the study – flushing gas – helium 99.9999%, containing 0.4–0.5% of high purity carbon dioxide (99.9999%), carrier gas - helium 99.9999% (all from Linde Gas RUS, Russian Federation).

Sample preparation for the direct measurement of oxygen isotopes in intracellular water was carried out by isotopic equilibration technique (EQ-IRMS/SIRA method). For isotopic equilibration reaction, 0.500 ml of grape must or 0.200 ml of water was added into a 12 ml round-bottom borosilicate glass vial. The tubes were hermetically sealed by disposable screw cap with Teflon-silicone membrane. 0.200 ml of reference substances VSMOW2, GISP and SLAP2 were added separately in similar tubes.

To carry out measurement, we formed series (groups) of tubes. Each group included the required number of sample tubes and test tubes with reference substances. A batch of tubes was placed in the cells of the sample preparation module (*A*). Results of preliminary experiments on the duration of the isotopic equilibration reactions show that the equilibrium state occurs after at

DOI: 10.1051/bioconf/20170902020

least 18 hours. The optimum temperature for the reaction was 24.0 ± 0.1°C.

At the end of the isotope exchange reaction the gas mixture from each tube was transferred to the control unit Gasbench II® (*B*). Gas mixture was dried over by molecular sieve membrane Nafion® and transferred to the gas chromatograph installed in the control unit Gasbench II® (*B*). Chromatographic separation is carried out in helium flow (2 ml/min), column used – PLOT Fused Silica, stationary phase CP-PoraPLOT Q, length 25 m, diameter 0.32 mm (Varian, USA), column temperature 70 °C. Temperature of samples in the sample module (1) was maintained at a constant level of 24.0 ± 0.1 °C until the end of the measurements.

Quantitative measurements of oxygen isotopes $^{18}O/^{16}O$ were performed according to conditions described above and in [12]. During the measurement cycle the WRG calibrated against a reference substance is introduced into the mass spectrometer – five portions WRG at the beginning of the cycle and three portions WRG at the end. Measurement of each sample was performed in triplicate in series of 10 impulses. For subsequent calculation of the $\delta^{18}O_{VSMOW}$ value (‰), the intensities of 7, 8 and 9th impulses of the carbon dioxide with masses 46 and 44 should be used. Registration and processing of measurements results, as well as control for all devices of the analytical system is done via a PC workstation (*E*).

The results of measurements calculated on the basis of the recorded signals for the masses 44 and 46 are subjected to further adjustment. This adjustment takes into account the linear relationship between the values $\delta^{18}O_{VSMOW}$ of used reference substances VSMOW2, GISP, SLAP2 and the values measured for these references in each batch of samples. The correlation coefficient of the linear dependence must be not lower than 0.99999. Otherwise, the measurement for a particular series of samples, including measurements of reference substance included in this batch, have to be repeated until the above conditions are met.

4. Results and discussion

4.1. Compositions of stable isotopes in grapes, dry wines and waters

The results of measurements of isotope compositions of stable carbon & oxygen isotopes presented as $\delta^{13}C_{VPDB}$ and $\delta^{18}O_{VSMOW}$ values, are shown below in Tables 1-9. The tables contain the average values obtained in a batch of five parallel determinations of carbon isotopes measurements for carbohydrates (sugars) and ethanol of each studied grape variety and dry wines samples, as well as in a batch of three parallel determinations of oxygen isotopes measurements in the intracellular water of each grape variety and water samples from surface, atmospheric and underground sources.

The expanded measurement uncertainty (EMU) of each batch of determinations did not exceed 0.40‰ (for carbon isotopes) and 0.42‰ (for oxygen isotopes). For the calculation of the EMU the coverage factor of 2 was used that corresponds to a confidence level of 95%.

5. Conclusions

Summarizing the data clusters obtained within the research program on the white and red grape's samples taken in the Crimean Peninsula, South West Coast of the Greater Caucasus (Krasnodar region), Don Basin and Western Caspian allows to set the ranges of the natural fractionation of stable carbon and oxygen isotopes in components of this plant within the season of 2016 (Tables 10–11).

The results of our study characterizing the fractionation of carbon isotopes in native carbohydrates (sugars) and also oxygen isotopes in intracellular water as well as in wines correlate very well with the results of our study of 2015 in the Crimean Peninsula [1] and the published data of other researches, which were conducted in other geographical regions with similar agroclimatic conditions, frequent water deficiency and an irregular irrigation (or total absence of that) during the vegetative period. It was shown that the carbohydrates (sugars) of grapes which are grown up in the conditions of water deficiency and high level of active temperatures are enriched by "heavy" ^{13}C carbon isotopes [13–23]. For example, C. van Leeuwen et al. suggested in [15], according to the results of their studies, the following dependence of the range for $\delta^{13}C$ of sugars on the water status during the grape cultivation:

- "no water deficit" ... <-26.0‰,
- "weak water deficit" ... from -24.5 to -26.0‰,
- "moderate to weak water deficit" ... from -23.0 to -24.5‰,
- "moderate to severe water deficit" ... from -21.5 to -23.0‰,
- "severe water deficit" ... >-21.5‰.

In our study the width of the range of $\delta^{13}C_{VPDB}$ values for natural carbohydrates (sugars) in 9 white and red grape varieties amounts to 6.09‰.

Carbohydrates (sugars) of grape varieties from four geographical regions regarding the carbon isotope composition are almost equivalent in accordance with the experimental data from this study, taking into account the quantitative levels of measurement uncertainty. In this part the results of the study correlate well with published data from other scientific researches [18–20].

The research results for water samples show the $\delta^{18}O_{VSMOW}$ values, which are characteristic for the natural fractionation of oxygen isotopes in the atmospheric and geological underground waters (Table 9).

The results of the study can be used in practice for quality assessment, including identification of commercial products offered on national and international foodstuffs markets (e.g. wines from the regions with a Geographical Indication (GI) or Designation of Origin (DO)).

However, in applying the results of this study, including above shown intervals of $\delta^{13}C_{VPDB}$ and $\delta^{18}O_{VSMOW}$ for grapes from the Crimean Peninsula, South West Coast of Greater Caucasus, Don Basin and Western Caspian, for example, when planning a scientific or an applied work, it is necessary to take into account interrelations of the obtained experimental data with agroclimatic conditions of cultivation and harvest of 2016, and also to consider the size of a data matrix which is defined by a quantity of samples, climate peculiarities sampling areas and grape varieties.

We thank Non-profit Self-regulating Organization "Winegrowers & Winemakers" (Krasnodar Region), Union of Winegrowers & Winemakers of Russia (SVVR), winegrowers and wineries from the Crimean Peninsula, Don Basin and Western Caspian regions for supporting of the scientific study, especially for assistance with sampling of authentic grapes.

References

[1] A.Kolesnov, N.Agafonova, BIO Web of Conferences **7**, 02004 (2016)
[2] Г. Валуйко, *Технология виноградных вин* (Таврида, Симферополь, 2001) / G.Valuyko, *Technology of grape wines* (Tavrida, Simferopol, 2001)
[3] V. Gerzhikova, V. Zagorouiko, E. Ostroukhova, N. Anikina, Proceedings of the 27th World congress of vine and wine of the OIV in Bratislava, 85 (2002)
[4] Z. Eremenko, V. Skresanov, A. Shubnyi, N. Anikina, V. Gerzhikova, T. Zhilyakova, Electromagnetic waves, 403–422 (2011)
[5] *Method OIV-MA-AS312-06* (2001)
[6] *Method OIV-MA-AS2-12* (2009)
[7] *BS DD ENV 12140:1997-02-15* (1997)
[8] *Method OIV-MA-AS311-05* (2011)
[9] АКолеснов, И.Филатова, О.малошицкая, М.Зенина, И.Питрюк, Напої. Технологиї та Инноваций **10**, 48–51 (2015) / A.Kolesnov, I.Filatova, O.Maloshitskaya, M.Zenina, I.Pitruyk, Beverages. Technologies & Innovations **10**, 48–51 (2015)
[10] А.Колеснов, И.Филатова, О.малошицкая, м.Зенина, И.аитрюк, Напої. Технологиї та Инноваций **11**, 54–57 (2015) / A.Kolesnov, I.Filatova, O.Maloshitskaya, M.Zenina, I.Pitruyk, Beverages. Technologies & Innovations **11**, 54–57 (2015)
[11] *Finnigan Elemental Analyzer. Operating Manual* (Thermo Electron Corporation, Carbon Measurement, 5-1...5–20, 2010)
[12] *Gasbench II. Operating manual* (Thermo Fisher Scientific, Revision A-1118343: 5-33-5-39, 5-42, 2009)
[13] C. van Leeuwen, J.-P.Gaudillere, O.Tregoat, J. Int. Sci. Vigne Vin **35**, 195–205 (2001)
[14] J.-P.Gaudillere, C.Van Leeuwen, N.Ollat, J. Exp. Bot. **53**, 757–763 (2002)
[15] C. van Leeuwen, O.Tregoat, X.Chone, B.Bois, D.Pernet, J.-P.Gaudillere, J. Int. Sci. Vigne Vin **43**, 121–134 (2009)
[16] S.Gomez-Alonso, E.Garcia-Romero, Aust. J. Grape Wine Res. **16**, 283–289 (2010)
[17] A.Herrero-Langreo, B.Tisseyre, J.-P.Goutouly, Th.Scholasch, C. van Leeuwen, Am. J. Enol. Vitic. **64**(3), 307–315 (2013)
[18] *Code of practice of the European Fruit Juice Association* (2014)
[19] F.Guyon, C. van Leeuwen, L.Gaillard, M.Grand, S.Akoka, G.S.Remaud, N.Sabathie, M.-H.Salagoïty, Anal. Bioanal. Chem. **407**, 9053–9060 (2015)
[20] N.Christoph, BIO Web of Conferences **5**, 02020 (2015)
[21] S.Leonardelli, S.V.Dutra, G.J.Carnieli, F.Spinelli, D. de Siqueira Ferreira, R.Vanderlinde, BIO Web of Conferences **7**, 02001 (2016)
[22] C. van Leeuwen, Ph.Darriet, J. Wine Econ. **11**(1), 150–167 (2016)
[23] F.Camin, M.Boner, L.Bontempo, C.Fauhl-Hassek, S.D.Kelly, J.Riedl, A.Rossmann, Trends Food Sci. Tech. **61**, 176–187 (2017)

Carbon-13 composition of bulk dry wines by irm-EA/MS and irm-^{13}C NMR: An indicator of vine water status

Francois Guyon[1], Cornelis van Leeuwen[2], Laetitia Gaillard[1], Mathilde Grand[3], Serge Akoka[3], Gérald S. Remaud[3], Nathalie Sabathié[1], and Marie-Hélène Salagoïty[1]

[1] Service Commun des Laboratoires, 3 Avenue du Dr. Albert Schweitzer, 33608 Pessac Cedex, France
[2] EGFV, Bordeaux Sciences Agro, INRA, Univ. Bordeaux, 33140 Villenave d'Ornon, France
[3] EBSI Team, CEISAM, UMR CNR6230, 2 rue de la Houssinière, BP. 92208, 44322 Nantes Cedex 3, France

Abstract. Measurements performed on a set of 32 authentic wines (not submitted to any oenological treatment) and their ethanol, recovered by distillation, show high correlation between $\delta^{13}C$ of bulk wine and its ethanol. These measurements were performed by isotope ratio monitoring by mass spectrometry coupled to an elemental analyzer (irm-EA/MS). Then a series of wines produced by vines of which water status was assessed during the growing season with predawn leaf water potential measurements, was studied by irm-EA/MS. As expected $\delta^{13}C$ is correlated to vine water status conditions, as a result of stomatal closure. The ethanol of these specific wines was also analyzed by isotope ratio monitoring and by nuclear magnetic resonance (irm-^{13}C NMR) to determine carbon-13 composition on the two specific sites of the ethanol skeleton. If these measurements confirm the correlation between ^{13}C composition and vine growth conditions, the ^{13}C stereospecific information does not make vine water status assessment more precise.

1. Introduction*

Isotope fractionation in plants is well-known, particularly for ^{18}O and ^{2}H and is a result of the water cycle [1]. In vines, this fractionation is accentuated by transpiration during grape ripening. Carbon isotope fractionation also occurs during plant photosynthesis which selectively uses a light carbon isotope (i.e., $^{12}CO_2$) for molecule synthesis. Previous studies demonstrated that during water deficit periods the plants close their stomata in order to limit transpiration. As a result, the equilibrium with air CO_2 composition, in terms of isotope concentration ($^{12}CO_2$ & $^{13}CO_2$) is modified and the plant need to consume more $^{13}CO_2$ to maintain its photosynthetic activity. As a result, water deficit is characterized by an increase in carbon-13 in photosynthesized compounds [2].

Previous studies demonstrated the impact of water deficit on sugar carbon-13 content at grape level [3]. Another study showed the average $\delta^{13}C$ discrepancy between sugar and its ethanol, resulting from the fermentation process [4]. Because this shift is constant, ethanol $\delta^{13}C$ can be used as an indicator of vine water status during grape ripening (Fig. 1).

Many studies have been performed on grapes but, surprisingly, the final product, the wine, has not been addressed in these studies. The lack of tools to estimate the grape ripening conditions directly on wine is regrettable because water deficit during grape ripening usually provides potentially high quality red wines [5]. Moreover, recent studies correlated this parameter with aging bouquet typicality of red Bordeaux wines [6].

* These results are extracted from the publication in Analytical and Bioanalytical Chemistry, 2015, 407, 9053–9060, with permission.

The aim of this study, previously published elsewhere [7], was to propose a rapid and reliable method directly applicable to wine, able to provide an estimation of grape ripening conditions with regard to vine water status.

2. Methods

2.1. Samples

Two set of samples have been used for this study: the first one based on 34 authentic samples (equally red and white wines) elaborated from grapes in the laboratory according to a previously described protocol [8]. A second set of 28 authentic wines for which the predawn leaf water potential was followed every 2 weeks (from July to end of September) during grape ripening. All samples have been distilled to recover ethanol, allowing analysis to be carried out on the bulk wine and the ethanol, respectively.

2.2. irm-EA/MS measurements

Ψ Measurements have been performed using an elemental analyser (VarioMicroCube, Elementar) coupled to an isotope ratio monitoring by mass spectrometry (Isoprime, Elementar). Masses measured are m/z 44 and 45 corresponding to the CO_2 isotopologues. The values are expressed in ‰ versus Vienna-Pee Dee Belemnite (V-PDB). The provided data correspond to two measurement average if the deviation between the two measurements is lower than 0.3‰.

2.3. irm-^{13}C NMR measurements

A Bruker 400 NMR spectrometer was used to record quantitative ^{13}C spectra at 100.6 MHz (5-mm dual+ probe,

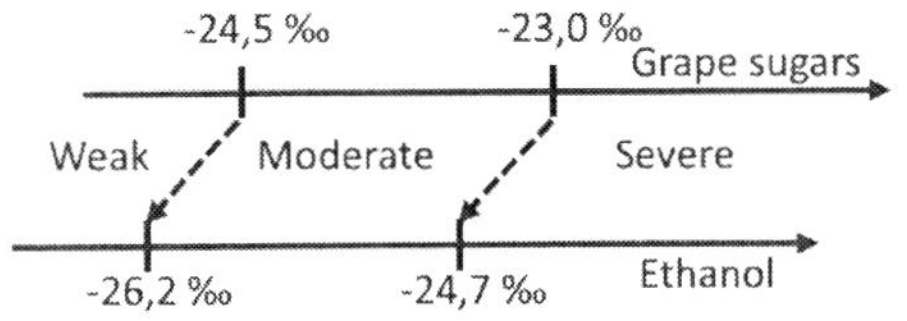

Figure 1. Water deficit thresholds with respect to $\delta^{13}C$ ratio adapted from REF 3.

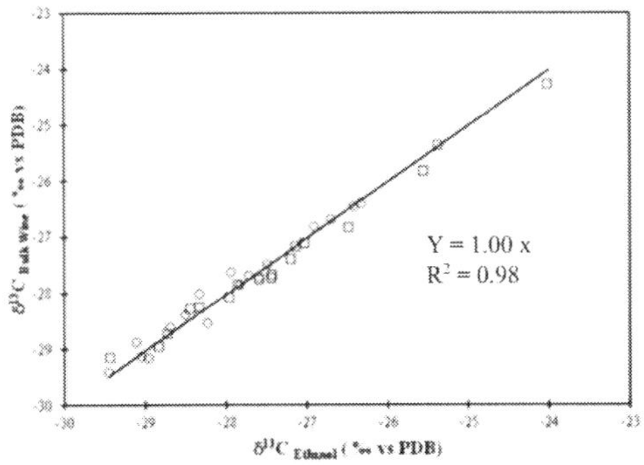

Figure 2. Relation between $\delta^{13}C$ values of authentic wines and their ethanol for red (squares) and white (circles) wines.

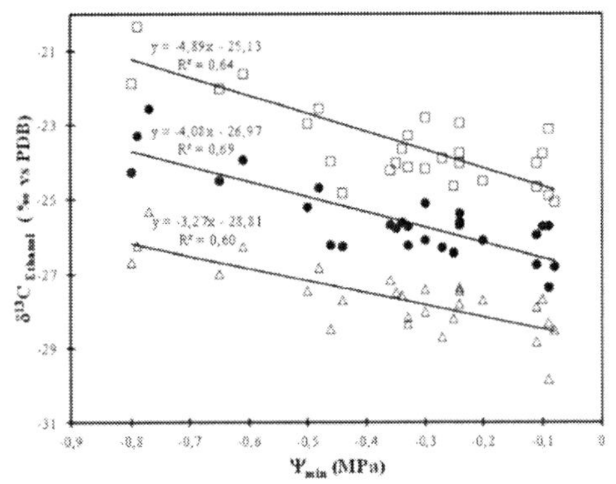

Figure 3. Relation between Ψ_{PDmin} and global ethanol $\delta^{13}C$ ratio (black) determined by irm-EA/MS and with position-specific ^{13}C deviation of ethanol ($\delta^{13}C_{CH2}$: squares, $\delta^{13}C_{CH3}$ triangles) determined by irm-^{13}C NMR.

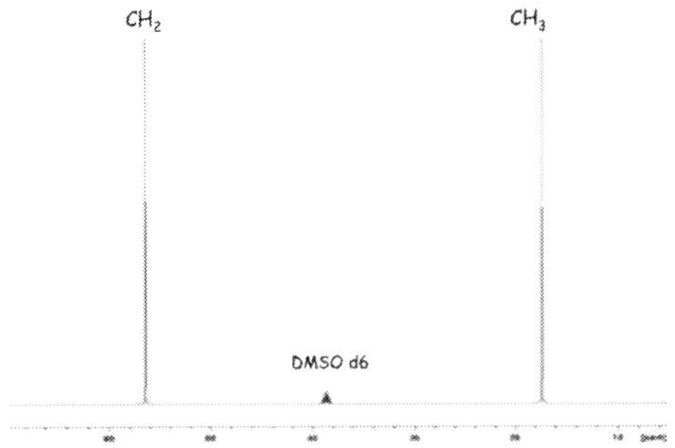

Figure 4. ^{13}C-NMR spectrum of ethanol at natural abundance in DMSO-d_6.

no rotation, 30 °C) Intramolecular ^{13}C composition are described in [9]. Positional ^{13}C distribution on ethanol skeleton is quantified from the ^{13}C mole fraction ($f_i = S_i/S_{tot}$, S_i the ^{13}C signal and S_{tot} the sum of the signals). Considering the statistical mode fraction (F_i), the position-specific relative deviation in ^{13}C abundance for any C atom in position "i" is $d_i = f_i/F_i - 1$. Using the isotope composition of the whole ethanol molecule, d_i is then converted to $\delta^{13}C_i$ (‰).

3. Results

The first step of this work was devoted to determine the relation between ethanol and bulk wine $\delta^{13}C$. A set of 31 authentic samples – without any oenological treatment – have been studied. They corresponds to wine samples elaborated, from grapes, in our laboratory [8]. All these wines, coming from various regions of France were distilled; the recovered ethanol and bulk wine were analyzed by irm-EA/MS to quantify the $\delta^{13}C$. The results, plotted on Fig. 2, show the full correlation between these two measurements. This result is not surprising as a wine is composed, in average, of 84% of water and nearly 16% of organic and inorganic compounds, ethanol corresponding to 11% of the wine.

The second step has been the study of 28 wines for which vine water status during grape ripening was known. Vine water deficit can be characterized by the quantifiable value of the minimum pre-dawn leaf water potential (Ψ_{PDmin}). Ψ_{PDmin} is the maximum level of water deficit experienced by the vine during grape ripening; the more negative this value, the higher the vine water deficit. Carbon-13 isotope ratio was quantified on bulk wine by irm-EA/MS. The results are plotted in Fig. 3 as a function of the minimum pre-dawn leaf water potential. A correlation coefficient of 0.69 is computed revealing the link between wine ethanol $\delta^{13}C$ value and Ψ_{min}, i.e., vine water status during grape ripening.

As photosynthetic is a fractionating process, it was interesting to investigate if it has some impact on the carbon-13 repartition of the ethanol carbonated skeleton and then, determine if a better correlation could be established between stereospecific carbon-13 composition and pre-dawn leaf water potential. These measurements have been performed on the ethanol recuperated by distillation of the wine previously described by irm-^{13}C NMR. A typical ^{13}C NMR spectrum is presented on Fig. 4.

The first peak (18 ppm) corresponds to the methyl carbon (C_I) and the second signal (58 ppm), to the methylene carbon (C_{II}). The quantification for each carbon is possible knowing $\delta^{13}C$ ratio of the ethanol quantified by irm-EA/MS.

The concentration of carbon-13 is significantly different ($\Delta = 4.8 \pm 0.8$‰) between the two position CI and CII but it does not seem to be related to vine water status as the difference appears to be quite homogeneous. Moreover, this carbon-13 content heterogeneity on the ethanol carbonated skeleton is in accordance with previous studies [9,10]. The results plotted on Fig. 3 show the correlation between intramolecular carbon-13 content and minimum pre-dawn leaf water potential. This correlation is slightly better with the C_{II} position which is in agreement with a previous work showing a significant correlation between $\delta^{13}C_{II}$ ratio and the mean of atmospheric temperature during the last three months before harvest [10].

4. Conclusion

The first finding of this work is the similarity between $\delta^{13}C$ of ethanol and bulk wine, as far as the wine has not been supplemented by oenological products.

The second result highlights the correlation between ethanol $\delta^{13}C$ and vine water status. This correlation is also observable for intramolecular ^{13}C distribution. As a result, two techniques can provide information on vine water status conditions during grape ripening.

The coupling irm-EA/MS is preferred as it is less expensive, faster and because the measurements can be performed directly on the wine.

References

[1] H. Craig, Science **133**, 1833–1834 (1961)
[2] GD. Farquhar, JR Ehleringher, KT Hubick, Annu. Rev. Plant Physiol. Plant Mol. Biol. **40**, 503 (1989)
[3] C. van Leeuwen, O. Tregoat, X. Choné, B. Bois, D. Pernet, JP. Gaudillère, J. Int. Sci. Vigne Vin **43**, 121 (2009)
[4] A. Rossmann, HL. Schmidt, F. Reneiro, G. Versini, I. Moussa, MH. Merle, Z. Lebensm. Unters Forsch. **203**, 293 (1996)
[5] C. van Leeuwen, G. Seguin, J. Int. Sci. Vigne Vin **28**, 81 (1994)
[6] M. Picard, C. van Leeuwen, F. Guyon, L. Gaillard, G. de Revel, S. Marchand, Frontiers in Chemistry Submitted (2017)
[7] F. Guyon, C. van Leeuwen, L. Gaillard, M. Grand, S. Akoka, GS. Remaud, N. Sabathié, MH. Salagoity, Anal. Bioanal. Chem. **407**, 9053 (2015)
[8] F. Guyon, C. Douet, S. Colas, MH. Salagoity, B. Medina, J. Agric. Food Chem. **54**, 9918 (2006)
[9] K. Bayle, S. Akoka, GS. Remaud, RJ. Robins, J. Biol. Chem. **290**, 4118 (2015)
[10] A. Gilbert, V. Sylvestre, RJ. Robins, GS. Remaud, Anal. Chem. **81**, 8978 (2009)
[11] A. Gilbert, V. Sylvestre, N. Segebarth, G. Tcherkez, C. Guillou, RJ Robins, S. Akoka, GS. Remaud, Plant Cell Environ. **34**, 1104 (2011)

DOI: 10.1051/bioconf/20170902022

Rapid assessment of mycotoxins in wine by on-line SPE-UHPLC-FLD

Alina-Mihaela Nistor[1], Ştefan-Dragoş Cotan[1], Constantin-Bogdan Nechita[2], Alexandru Tarţian[1], Marius Niculaua[2], and Valeriu V. Cotea[1]

[1] University of Agricultural Sciences and Veterinary Medicine Iaşi, Romania
[2] Reserch Centre for Oenology, Romanian Academy Iaşi branch, Romania

Abstract. According to the latest statistics, grapes are one of the largest fruit crops worldwide. In this regard, it is important to consider all factors influencing quality of grapes and wine. In the last years, scientist focused on the study of mycotoxins that can influence the quality of wine. It is considered that toxins produced by moulds, causing significant economic losses, affect approximately one quarter of the world grape production. If the selective sorting of infected grapes is not done adequately, wine will present a major risk to consumers, mycotoxins being considered by the "International Agency for Cancer Research" a carcinogenic compound. The main mycotoxins monitored in this study come from Aspergillus sp., and are represented by aflatoxins B1, B2, G1, G2 and ochratoxin A. This study purpose is to develop a faster method for the analysis of mycotoxins, in order to increase rapidity and efficiency for the evaluation of the degree of infestation in wine. The purposed method is using an on-line large volume injection coupled to pre-concentration of sample (SPE) which is directly transfer to the ultra-high-pressure liquid chromatography (UHPLC) column for separation and the detection by means of the fluorescence detector (FLD). As the maximum tolerated level for mycotoxins in wines is 2 ppm, this method is able to detect under this limits of quantification with RSD below 2%.

1. Introduction

Worldwide grapes are one of the largest fruit crops, according to the latest FAO-OIV report from 2016. Each year 75 million tons of grapes are produced, which makes them the most widespread fruit [1]. From this total, it is estimated that approximately a quarter is damaged by mould. Such a large proportion of contamination automatically implies economic loss [2,3]. Grape and wine contamination can take place during two stages: during development or after harvest. In the first stage, the crops can be degraded when insects, birds, animals, hailstone, storms, frost, becoming therefore vulnerable to the onset of mould, physically damage it. Also in the development stage, the crops can suffer from stress due to harsh heat and drought. In the second stage, the contamination might happen if the healthy crops are kept in wet and not airy spaces during transportation, storage or usage. A major contamination risk is when the crops are harvested late and it is caught in the rain or this could happen during harvesting [4,5].

Mycotoxins are byproducts of mould metabolism, from where their name comes (Online Etymology Dictionary). Secondary metabolites are filamentary fungi that are highly toxic for humans and animals alike. Diseases caused by mycotoxin intoxication are generally called mycotoxicoses. Their symptoms depend on the type of mycotoxin, the duration of exposure, the age and gender of the infected person, as well as on the person's health. Some common symptoms are anemia, anorexia, bleeding, abdominal pain, convulsions, even depression, excessive nervousness or abnormal behaviour. In the cases of stronger intoxications, mycotoxins can attach to the DNA and cause mutations resulting in infertility and weakly developed or sick children. In the worst cases, they can cause liver and kidney damage, liver cancer, coma and even death. Nowadays there is still no treatment for mycotoxin intoxications, the only known solution being diet and intense hydration [6].

From all the toxins from mould, the most common are: aflatoxins, ochratoxins, citrinin, ergot alkaloids, patulin, zearalenone and trichothecene. These could be identified in food such as cereal, nuts, coffee, cocoa, spices, oleaginous seeds, dried and fresh fruit [7]. Most frequently in samples of grapes and wine, we come across aflatoxins and ochratoxins. It is known that these are actually the most toxic mycotoxins because the "International Agency for Research on Cancer" (IARC) assigned aflatoxins to group I (compounds that can cause cancer in the human body) and ochratoxin A to group 2B (compounds that can possibly cause cancer in the human body) [8]. As well as that, according to assessments performed by Codex Alimentarius, approximately 15% of the total quantity of OTA that is ingested by the human body comes from wine [9].

Aflatoxins were discovered in 1900, in England, when approximately 120,000 birds, fed with *Aspergillus flavus* infested Brazilian nuts, had died. The disease, unknown at the time, was called "Turkey X". They analysed the food sample with thin layer chromatography (TLC) and identified a series of fluorescent compounds that were found responsible for the outbreak [10]. In 1962 the chemical compound derived from *Aspergillus flavus* was called "aflatoxin" [11] – "a" (*Aspergillus*) – "fla" (*flavus*) – toxin.

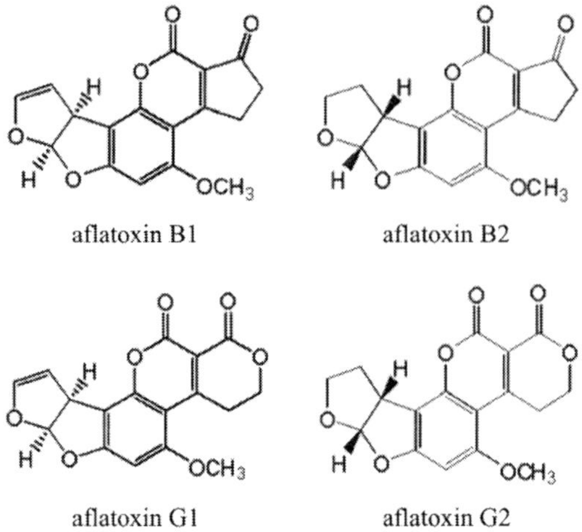

Figure 1. Chemical structures of aflatoxins.

This was analysed with TLC and on a plaque exposed to ultraviolet light it was observed that there are two compounds of different colour: blue and green. They were called "aflatoxin B" (blue) and "aflatoxin G" (green) [11]. The B1, B2, G1 and G2 compounds are the main types of aflatoxins. Besides these, there were identified two metabolites, M1 and M2 (from B1 and B2) [12] and approximately 20 other aflatoxins (P1, Q1, B2a, G2a, etc.), most of which being produced through the biotransformation of the main metabolites [13]. From the total content of aflatoxins, 60%–80% are B1 and B2, G1, G2 do not usually appear without B1.

The main mouldy species that produce aflatoxins are *Aspergillus* (*A. flavus*, *A. parasiticus* and *A. nomicus*) but also *Penicillium*, *Rhizopus*, *Mucor* and *Streptomyces*. It was observer that the incidence of development of the *Aspergillus flavus* species is lower in areas with temperatures smaller than 20 °C, than in the areas with a warm climate with temperatures greater than 25 °C, and in some areas, the infection occurs only when the temperatures rise in association with drought [4]. The B1 aflatoxin generally needs higher temperatures (24–28 °C) than G1 (23 °C) [10].

From a chemical point of view, aflatoxins (Fig. 1) are thermally stable compounds, but they act as an reactive in environments with pH less than 3 and are soluble in water and in organic solvents (methanol, acetonitrile, chloroform) [15].

Ochratoxins were first isolated in 1965 in South Africa. Initially, it was discovered that they come from *Aspergillus* species (*A. melleus*, *A. muricatus*, *A. petrakii*, *A. sulphureus*, *A. albertensis* şi *A. alliaceus*, *A. Negri*, but especially *Aspergillus carbonarius*), but they can also come from *Penicillium verrucosum* and *Penicillium veridicatum* [3,9]. Among the members of this mycotoxin class, ochratoxin A (OTA) is the most known and important due to its high toxicity [3].

Regarding the presence of ochratoxins in oenology, about 10% of the mould developed on grapes that are in the "veraison" stage can produce OTA, while when the grapes are fully maturated 47% of the mould can produce OTA. Regarding the type of wine, it was found that there is a greater incidence of contamination for red wine than for white wine [3]. The wine making technique also has an influence on the OTA quantity, because in the maceration process the quantity of the toxin increases [9].

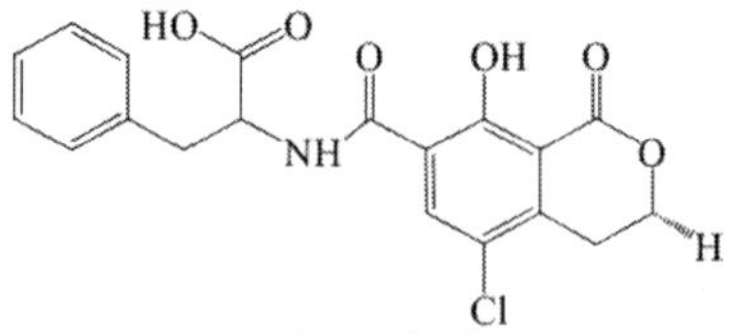

ochratoxin A

Figure 2. Chemical structure of ochratoxin A (OTA).

As a chemical compound, OTA (Fig. 2) is soluble in water and partially soluble in organic solvents (chloroform, ethanol or methanol) and in the presence of light and air, it is relatively unstable.

Taking into account the influence of these kinds of toxins on wine and human health, the current trends are to find a fast and efficient method of detecting these mycotoxins in order to protect the quality of wine. In Europe, the maximum quantity of aflatoxins and OTA allowed in wine and must, according to EU regulations is 2 ppb [16]. Therefore, the method must be very sensitive as well as have thresholds of detection and quantification as low as possible.

2. Experimental

2.1. Chemicals, materials and samples

Standard of ochratoxin A, aflatoxins B1, B2, G1, G2, were acquired LGC standards and formic acid, organic solvents (HPLC grade) methanol and acetonitrile were obtained from Sigma-Aldrich. The ultra-pure water used for mobile phase preparation was purified through a Fistreem Cyclon coupled to a TermoScientific GenPure UV-TOC. Other chemicals and used materials were of analytical grade. Wine samples were from local didactical farm of the University of Agricultural Sciences and Veterinary Medicine "Ion Ionescu de la Brad" Iaşi NE region of Romania. Grapes samples of were proceed by the traditional fermentation method for wines technology. After gravity settler was done seeding with selected yeasts *Saccharomyces cerevisiae*. Alcoholic fermentation is finalized the wine was clarified with bentonite and then bottled. All samples were kept at 12 °C before opening the bottle and analysis.

2.2. Chromatographic LC system

The chromatography equipment was a Shimadzu Nexera system used for method development and validation. Tree pumps module Nexera X2 LC-30AD in binary configuration and one LC-20AT (Prominence series) for washing and maintenance of pressure in columns. All pumps have DGU-20A5R mobile phase degassers. Samples were introduced with Nexera X2 SIL-30 AC autosampler connected to a FCV-32AH valve for large volume injection and bypass for standard injection. Detection of analytes is done with fluorescence detector RF-20A XS (Prominence series) were also used and confirmation true a Nexera X2 SPD-M30A with HS cell of 85 mm. The chromatography and extraction columns were thermostatted in a Nexera X2 CTO-30A column oven and mobile phase flow direction was changed by using an FCV-32AH high pressure column switching valve. The HPLC system was controlled by a CBM-20A communication module. The data evaluation and acquisition of chromatography analyses were performed by Lab-Solution software v 5.75.

2.3. Preparation of stock solution and wine samples

Standard stock solutions of OTA 10 μg/mL, aflatoxins B1, G1, 2 μg/mL and 0.5 μg/mL B2, G2) were utilised. Standard stock solution was stored at −20 °C in the dark. The calibration standard solutions were diluted in the concentration range from 1 to 200 ppb (OTA), 1 to 100 (aflatoxins) using six calibration solutions.

Undiluted and diluted wine samples were filtered through a 0.22 μm nylon filter into glass vials. A 50 μL volume of filtered wine solutions was injected directly into the on-line SPE–LC system. All beer wine samples and working standard solutions were prepared fresh daily and injected in triplicate.

2.4. LC column switching procedure

The simultaneous wine sample extraction and mycotoxin determinations were performed using the on-line SPE– LC system. The SPE column used for mycotoxins extraction from wine samples was a Kinetex 5 μm PFP 100 Å column (50 × 4.6 mm), particles core-shell silica (Phenomenex), with a washing mobile phase of 1% aqueous formic acid pH 2.15 at a flow rate of 1 mL/min and temperature of 40 °C. A 50 μL volume of the filtered wine sample was directly injected in to the guard column where mycotoxins were retained while other polar interferences were washed out to the waste. The valve switch after 2.2 minute and extracted mycotoxins was transferred from the SPE column to the analytical column in the same elution direction.

The chromatographic separation of aflatoxins and OTA from interfering matrix components of the wine was achieved on the fused-core column Kinetex 1.7 μm XB-C18 100 Å column (30 × 3.0 mm) (Phenomenex), using a mixture of 80% acetonitrile (solvent A) and a solution of 1% formic acid in water, pH 2.15 (solvent B) at a flow rate of 2.5 mL/min and a column temperature of 40 °C. A gradient elution was applied on the analytical column, with the initial conditions (solvent A/ B 0.1:99.9) held for 0.2 min. Between 0.2 and 4.5 min of the chromatographic run the percentage of solvent A was linearly increased to 99.9%. Thereafter, the percentage of solvent A was decreased to 0.1% in 90 s, while the valve was switched back to the on-line SPE column position and the analytical column was equilibrated for 1 min to the initial conditions during the next sample pre-treatment step. Fluorescence excitation/emission detection wavelengths were set at 365/455 nm for aflatoxins and 333/460 nm for OTA.

3. Results and discussion

Wine is a complex matrix, which contains water, phenolic compounds, sugars, alcohols, acids, aldehydes, ketones, esters, nitrogen substances, sulphur compounds, mineral substances, stilbenes, bioactive substances and others. The ethanol as a fermentation product can partially affect the extraction of polar mycotoxins from the water because of early elution during the adsorption step. From these reasons, the procedure for extraction of mycotoxins from wine samples should be optimized and is due to cumbersome stapes that the human operator can influence. There is a good compatibility between online extraction (SPE) methodology and separation (LC or UHPLC) steps that must be taken into account. Therefore, the parameters of sample pre-treatment such as conditions of both mobile phases and stationary phases for extraction and separation were optimized before the method applied to wines sample analysis. The main aim of the work was to show a simple and modern analytical tool for the reliable determination of mycotoxins at trace levels in wine.

3.1. Optimization of on-line SPE extraction step

In this part we present the methodology which allowed fast extraction of mycotoxins from wine matrix with preconcentration on a first short column used to simulate SPE conditions. We utilised first a SPE 20 × 2.0 mm 20 μm particles commonly available reversed-phase C18, but the results are not satisfying.

By utilising Kinetex 5μm PFP 100 Å column (50 × 4.6 mm) we have improve the preconcentration of mycotoxins from wine samples with respect to the physicochemical properties (pKa) and lipophilicity (log P) of the analysed mycotoxins. OTA and aflatoxins are weak organic acids. The pKa values of OTA are in the ranges 4.2–4.4 and 7.0–7.3, respectively, for the carboxyl group of the phenylalanine moiety and the phenolic hydroxyl group of the isocoumarin part. The extracted mycotoxins should be in their neutral form to be well retained by hydrophobic interactions. Therefore, for sufficient retention of the mycotoxins of interest on the reversed-phase sorbent, the washing mobile phases were acidified with 1% formic acid to pH 2.15. A compromise between interference removal and the sufficient recovery of mycotoxins must be found during washing solution optimization. Therefore, the influence of the washing mobile phase on the SPE procedure was studied to ensure the fast and quantitative transfer of interfering matrix for disposal and simultaneous retention of the target analytes. The key parameters of extraction are the lipophilicity of the analytes, and the chemistry and extraction capacity of the SPE precolumn. The range of washing mobile phase ratios tested was from 0.1 to 40% of acetonitrile (80%) with 1% aqueous formic acid pH 2.15. When the percentage of pure acetonitrile exceeded 20% the more polar aflatoxins (G2, B2, G1 and B1) can be lost during elution of wine matrix. Therefore, the mobile phase with acetonitrile range from 0.1 to 35% was utilise. After 2.2 min of gradient washing at a flow rate of 2.5 mL/min we switch to first column and connected with the second column that is kept with 1% aqueous formic acid pH 2.15 at 1 mL/min by the third pump (20AT). Since the aim of the experiment is to have a short analyse time, long washing periods were not implied so to improve clean-up efficiency. A second valve switch was performed a soon as the OTA is eluted from second column and is dove in order to minimise the pulse of rotation valves. These on-line SPE conditions produced a

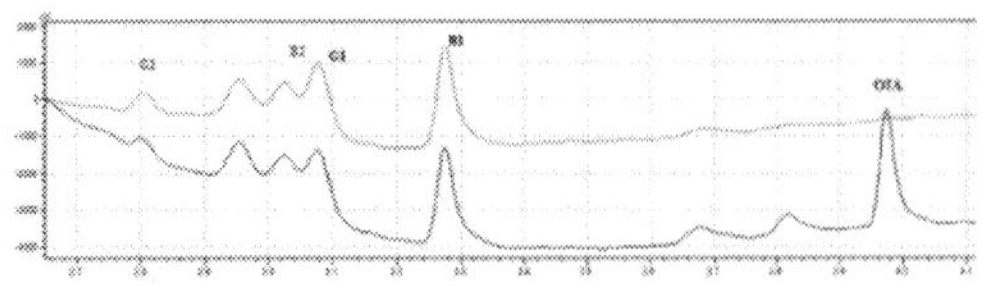

Figure 3. Spectral separation on DAD at 370 nm for aflatoxins and 345 nm for OTA.

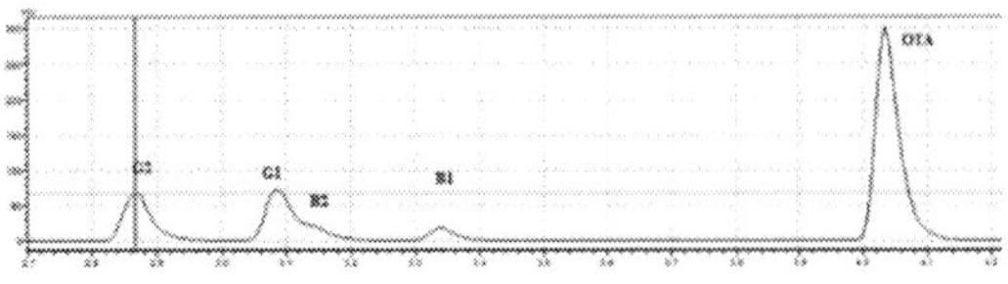

Figure 4. Detection of florescence for all toxins.

satisfactory elution of interfering compounds to the waste and sufficient resolution between peaks. The efficiency of the on-line sample pre-treatment procedure is depicted on the chromatogram in Figs. 3 and 4.

Increasing the volume of the injected sample is an effective way to improve the sensitivity of the on-line SPE–LC method. Nevertheless, the breakthrough volume of the SPE column resulting in analytes elution should not be exceeded. Therefore, compromise must be found between the washing of the residual peaks of the matrix using a larger injected sample volume and efficiency of the clean-up step. Injection volumes larger than 500 μL were not tested in our study because the residual peak of matrix interferences complicated the determination of the aflatoxins peaks (especially B2 and G1). Long isocratic washing step time did not lead to a significant elimination of the ballast matrix peak. Therefore, a compromise between the sensitivity and selectivity of the extraction step was found at 50 μL of undiluted wine which was injected directly into the extraction column.

3.2. Chromatographic separation

Fused-core technology is based on semi-porous particles instead of the traditional totally porous ones. In our study, fused-core particle columns with dimensions of 30 × 3.0 mm and particle size of 1.7 μm with stationary phases Pentafluorophenyl PFP (Phenomenex) is utilise for optimal separation time after on-line SPE extraction. Mobile phases containing mixtures of aqueous acetonitrile (80%) with a 1% formic acid solution in water, pH 2.15 in the range from 0.1 to 99.9% for optimal separation. The column oven temperature was set at 40 °C as a compromise between lowering the back-pressure in the system and stationary phase temperature stability. Other consideration is regarding the spikes that can appear randomly in to the control temperature detection system. Both detectors operate at 30 °C.

In order to optimize the separation conditions in the on-line SPE system, the gradient elution was tuned for better resolution of the aflatoxins peak from the residual peak of interferences or the valve switch. The time of the valve switch was set to 2.2 minute. The gradient elution program started at this moment from 0% of acetonitrile, so the lower content of acetonitrile at the beginning of the separation after the flow switch from the extraction column to the analytical column enabled gradual reabsorption of analytes at the beginning of the chromatogram. Thus, the peak of OTA was eluted over a longer time and well separated from the aflatoxins and wine matrix. A gradient elution was applied on the both analytical columns, with the initial conditions at switch over and its continuation from 35% eluent A (80%). Between 2.2 and 4 min of the chromatographic run, the percentage of solvent A was linearly increased to 99.9%. Thereafter, the percentage of solvent A kept for 90 seconds and was decreased to 0.1% over 60 s, while the valve was switched back to the loading SPE position and the analytical column was equilibrated for 1 min to the initial conditions during for the next sample SPE injection. Under the optimal conditions all five toxins were eluted selectively with some matrix interferences in less than 7 min including the sample pre-treatment step.

3.3. Validation of on-line SPE–HPLC method

The validation and analytical performance of the proposed method was tested in terms of linearity, limits of detection (LOD), limits of quantification (LOQ) and accuracy (recovery). Standard solutions were injected five times into the chromatographic system. Results are summarized in Table 1. Calibration curves were measured for standard solutions at six concentration levels, depending on the fluorescence intensity of each mycotoxin. The linearity for standard were measured in the range of 1 to 200 ppb for OTA also 1 to 100 for aflatoxins G2, B2, G1 and B1. Under the validated conditions, linear relationships with perfect regression coefficients both in standard and matrix calibration curve were found in the range 0.985–0.998 for all toxins. Linear regression parameters determination for standard calibrations are described by the regression equations (slope and intercept). Limits of detection and quantification, calculated as the minimum toxin concentration yielding a signal to noise ratio equal to three and ten times the standard deviation (σ).

LOQs were established as low as 0.15 ppb for OTA and the highest is 5.32 ppb for G2, one cause is the signal registered in from of B2. The obtained LOQ values are low enough to evaluate mycotoxins content. The percentage recovery values were calculated as (measured area of spiked placebo beer sample with standard solution/peak area of standard solution at the same concentration) × 100%. The recovery values were found to be in the range 83.18–109.42% for toxins. Values are relatively close to 100% indicate that there are no interferences or other matrix suppression effects. The on-line SPE–LC system was found to be very sensitive, selective and rugged for the analysis of filtered wine samples without additional sample pretreatment. The values of RSD for retention times, peak symmetry and chromatography resolution were not higher than 5% at the beginning and the end of the study.

3.4. Analysis of wine

In this work, the analysed samples were divided into four categories: dry white wine, wine demi sweet, wine red dry and red wine maturated in barriques. Twelve wine samples are analysed, ten from 2016 year of harvest and tow are from 2015. All samples have OTA levels below the maximum permitted limits of 2 ppb (summarized in Table 2). The results confirmed the presumption that the

Table 1. Analytical characteristics of the validated online SPE–LC method.

	G2	B2	G1	B1	OTA
Standard linear calibration range (ppb)	5–100 ppb	5–100 ppb	1–100 ppb	1–100 ppb	1–200 ppb
Slope	16844.6	273.72	74.096	1028.79	3192.83
Intercept	35153.2	932.83	1610.85	7456.47	154688
Regression coefficient (r^2)	0.98553	0.90239	0.98545	0.98554	0.9982
LOD (ppb)	0.59	0.93	1.61	1.14	0.15
LOQ (ppb)	1.95	3.08	5.32	3.76	0.51
Precision (RSD, %)	2.01	1.17	3.9	2.17	4.07
Accuracy for spike recovery (%)	95.71	94.58	85.18	113.24	109.42

Table 2. Occurrence of ochratoxin A and aflatoxins in the some Romanian wine samples.

Grape sort	year of harvest	wine type	G2 (ppb)	B2 (ppb)	G1 (ppb)	B1 (ppb)	OTA (ppb)
Zghihara	2016	dry white	D	ND	ND	3.33 ± 0.01	D
Agiloté	2016	dry white	D	ND	ND	2.77 ± 0.01	D
Fetească albă	2016	dry white	D	ND	ND	11.09 ± 0.01	D
Fetească regală	2016	dry white	D	ND	ND	D	ND
Sauvignon blanc	2016	dry white	D	ND	ND	D	ND
Traminer	2016	demi sweet	D	ND	ND	18.15 ± 0.02	D
Chadornnay	2016	demi sweet	D	ND	ND	14.27 ± 0.02	D
Fetească neagră	2016	red dry	1.72 ± 0.01	ND	ND	11.71 ± 0.02	D
Merlot	2016	red dry	4.67 ± 0.01	ND	ND	22.95 ± 0.03	D
Cabernet Sauvignon	2016	red dry	2.46 ± 0.01	ND	ND	22.19 ± 0.03	D
Merlot	2015	red barrique	1.88 ± 0.01	ND	ND	28.12 ± 0.03	D
Cabernet Sauvignon	2015	red barrique	D	ND	ND	D	ND

risk of OTA occurrence in wines is very low. Among the samples at 4 wines we find contamination by aflatoxin G2 with concentrations lower than 5 ppb, which is a low level. Other samples have present this aflatoxin but under the detection limit and it was evidence only by DAD detector. For the aflatoxin B2 and G1 we didn't found any amount.

In the case of B1 at a number of samples we found a fairly concentration and in the rest of the samples some quantity is found but cannot be confirmed by spiking or complementary detection technique. All samples analysed are not commercial wines and aim of the study is to evaluate the method and state health of the wines.

4. Conclusion

The method developed is an alternative approach for sample preparation and separation of five mycotoxins in wine. This new method for on-line solid-phase extraction of ochratoxin A and aflatoxins from wine samples is optimized fast and reproducible. Utilization of on-line SPE sample pretreatment with a relatively large volume sample injection was shown to be the method of choice for fast determination of very low concentrations of mycotoxins in wines. The online SPE–UHPLC with dual detection (DAD-FLD) for confirmation showed high sample throughput, low solvent consumption and high sample preconcentration of mycotoxins depending on the injected volume. The combination of sensitive fluorescence detection and a highly efficient chromatography process on both fused-core columns together with sample preconcentration resulted in perfect sensitivity of the developed method. The limit of quantification for OTA at least able to determine maximum permitted limits for OTA occurrence. Aflatoxin B1 is present in all samples in varying amounts.

This work was carried out by the Partnership in priority areas - PN II, developed with the support of MEN - UEFISCDI, project no. 183/2014 (PN-II-PT-PCCA-2013-4-0333) "Technology of capitalization of the bioactive elements from the grape seed waste with usefulness in the food and pharmaceutical industry, plant and environmental protection (Acronym: PROVITIS)".

References

[1] FAO-OIV 2016 `http://www.oiv.int/public/medias/ 4911/fao-oiv-grapes-report-flyer.pdf`
[2] A. Rahmani, S. Jinap, F. Soleimany, Comprehensive reviews in Food Science and Food Safety **8**, 202–251 (2008)
[3] S.M. Jović, A.V. Petrović, N.R. Marković, Proc. Nat. Sci, Matica Srpska Novi Sad **116**, 91–100 (2009)
[4] P.J. Cotty, L.S. Lee, National Cotton Council of America, Las Vegas, NV, 34–36 (1990)
[5] R. Jaime-Garcia, P.J. Cotty, Phytopathology **93**, 1190–1200 (2003)
[6] J.W. Bennett, Klich M., Clin Microbiol Rev. **16**(3) 497–516 (2003)
[7] I. Lhotskál, D. Šatínskýl, L. Havlíkovál, Petr Solich, Anal Bioanal Chem. **408**, 3319–3329, (2016)
[8] IARC, 2016: `http://monographs.iarc.fr/ENG/` Classification
[9] S. Somma, G. Perrone, A.F. Logrieco, Phytopathologia Mediterranea **51**(1), 131–147 (2012)
[10] B.I. Agag, Ass. Univ. Bull. Environ. Res. **7**, 1, March (2004)

DOI: 10.1051/bioconf/20170902022

[11] K. Sargeant, R.B. Carraghan, R. Allcroft, Chem. and Ind., 53–55 (1963)
[12] D.S. Patterson, E.M. Galaney, B.A. Roberts, Food and cosmetic toxicology **16**, 49–50 (1978)
[13] J.G Heathcote., J.R. Hibbert, Elsevier Scientific Publishing Company, Amsterdam, The Netherlands (1978)
[14] M. Weidenborner, Encyclopedia of food mycotoxins, Spronger Publisher Berlin, New York, London (2001)
[15] V.D. Cotea, C. Zănoagă, V.V. Cotea, Tratat de Oenochimie, Editura Academiei Române, II, 103 (2009)
[16] OIV Resolution CST 1/2002

 DOI: 10.1051/bioconf/20170902023

FTIR analysis of ash in wine

Tjaša Jug[1], Saverio Boni[2], and Tatjana Košmerl[3]

[1] Agricultural Chamber Nova Gorica, Pri hrastu 18, Nova Gorica, Slovenia
[2] Steroglass srl, Strada Romano di Sopra, 2/C, 06132 - San Martino in Campo, Perugia (PG), Italy
[3] Biotechnical faculty, University of Ljubljana, Jamnikarjeva 101, Ljubljana, Slovenia

Abstract. In a modern oenological laboratory determination of ash is a limiting factor regarding the speed of analysis, since functional FTIR analysis of ash are not common yet. Therefore 324 samples were analysed by FTIR spectroscopy on TDI Bacchus 3 as well as ISO 17025 accredited reference method (OIV–MA–AS2–04: R2009) in two duplicates. FTIR spectra and reference data were uploaded into software for the calibration development and validation performed. Already the original TDI calibration was ready to use, showing satisfying results. However, we tried to improve it. On 324 spectra, we have looked for optimal wavelengths and mathematical models, which gave the best results. The developed calibration was then confirmed by method validation: repeatability, reproducibility and accuracy. The expanded measurement uncertainty calculated was 0.3 g/L, which I slightly higher than of the reference method used. We have found another advantage over classical method: there is no influence of the air pressure or moisture. Nevertheless, we were able to confirm that our reference method was free of systematic error induced by the influence of the weather. Anyway, we can conclude that analysis of ash in wine by FTIR spectroscopy is a promising, fast and reliable new method.

1. Introduction

Ash determination is an obligatory analysis for certified wines to be placed on the market. It is defined as all the inorganic matter remaining after igniting the residue left from the evaporation of must or wine [1]. Most laboratories, especially those running numerous sample tests, wish faster methods. Ash determination based on empirical equations relating ash content with the measurements of conductivity, seemed promising already more than 40 years ago [2], but they didn't get much practical use. The advantage of conductometric methods is in their precision (1–2%), but the greatest problem concerning its application is the accuracy of results when it comes to such a complex medium with very different chemical composition as wine [2].

IR-spectroscopy is a modern analytical technique, based on measurement of IR spectra. As an indirect method, its results can only be as good as the results used in the calibration. Therefore, FTIR analyses need to overcome the weakness mentioned above anyway by developing a calibration model on few hundred samples.

Actually it is surprising that FTIR determination of ash is not a routine determination yet, since conductivity was one of the first reported parameters [3] with good correlation ($R^2 = 0.95$). Even in the latest review article [4] conductivity of wine is mentioned only once with the reference on the paper from 2004 [3]. We were able to find only one other paper dealing with the conductivity of wine, but even they report it only for spirit drink and not wine, and concluding FTIR data did not provide a good prediction for electrical conductivity [5].

One possible explanation is that the determination of inorganic compounds is more challenging compared to organic compounds owing to the lack of a direct relation between the element involved and the spectra obtained [6].

However, more likely to the authors' opinion the reason is not that the method was not promising, but more likely that there is no widely recognized conductivity's practical use. Even one of the larger inter-laboratory schemes in Europe (BIPEA) does not report its value.

So if ash in wine is theoretically correlated with conductivity and conductivity might be measured by FTIR spectroscopy, it seemed very attractive for us. We have received the basic calibration from the producer – TDI and tried to even improve it to the level of a practical use in a commercial lab.

2. Materials and methods

2.1. Samples

For the development of calibration and its verification samples analysed for the confirmation of the geographical origin for quality wine in Slovenia have been used (324 for development and 162 for verification). Also for method validation, except for its accuracy, samples of inter-laboratory comparison BIPEA have been used.

2.2. FTIR analysis

FTIR analysis was performed on TDI Bacchus 3 FTIR instrument. FTIR spectra and reference data were uploaded into software for the calibration development and optimal bands have been searched.

DOI: 10.1051/bioconf/20170902023

Table 1. Verification of different calibration with 162 samples: A: original calibration, no data, B: 2 wavelengths, first derivative, C: 4 wavelengths, first derivative, D: 4 wavelengths, centred average.

Verification\calibration	A	B	C	D
Number of samples which not achieving the criteria	58	34	38	23
Percent of samples which not achieving the criteria (%)	36	20	24	14
Minimal difference from classical results	−1.0	−0.5	−0.5	−0.6
Maximal difference from classical results	0.7	39.6	22.8	1.1

2.3. Reference method

As a reference, ash was determined in the samples by classical analysis performed according to OIV–MA–AS2–04: R2009 in two duplicates. The method is accredited by Slovenian accreditation (ISO 17025).

2.4. Data analysis

Multivariate analysis has been performed in QL Analyst software of TDI Company. Other calculations such as average, standard deviation, relative standard deviation (RSD) have been performed in LibreOffice 5.2.

3. Results and discussion

3.1. Calibration development

Data evaluation revealed a good correlation, even with the original calibration. Repeatability calculated according to OIV-OENO 10/2005 was 0.05 g/L, which is comparable to the repeatability of the OIV method (0.06 g/L). Reproducibility of 46 control samples was 0.12, which is very good – much lower than MAD. MAD was calculated as an average reproducibility limit of 22 inter-laboratory comparison (BIPEA) divided by $\sqrt{2}$ (MAD = 0.23). Accuracy was confirmed by comparing 12 ILC (BIPEA), where in 2 cases the results were questionable (z-score was 2.5 and 2.7 respectively), while in all other cases z was lower than 2 (average z is 0.6 if we don't take into account questionable results).

Of course, we have tried to improve the accuracy of the method by additional selection of the wavelengths based of the spectra of 324 samples. Already with 2 wavelengths, we have obtained perfect results on 12 ILC samples – all results report z-value less than 2. Even validation set looked optimistic: only 34 samples from 162 (20%) reported difference between reference value and FTIR in absolute terms higher than 0.3 g/L (Table 1, method B).

Nevertheless, new problem has arisen: extreme false positive results, as can be observed from maximal difference. Of course, we have tried to add more wavelengths (Table 1, method C), but the problem with false positive remained. The key improvement of the method was processing of the parameters: if we used first derivative (Table 1, methods B and C), several outliers were produced, but they are not if centred average is used in the calibration model (Table 1, method D).

The calibration based on centred average of 4 chosen wavelengths was chosen for validation.

Table 2. Repeatability of 6 different samples measured in 10 replicates.

Replicate\sample	1	2	3	4	5	6
1	3.18	3.34	3.16	3.10	3.06	2.43
2	3.24	3.33	3.14	3.07	3.09	2.38
3	3.22	3.26	3.08	3.17	3.15	2.33
4	3.24	3.25	3.16	3.17	3.08	2.35
5	3.23	3.34	3.09	3.15	3.04	2.36
6	3.19	3.34	2.99	3.09	3.08	2.38
7	3.17	3.31	3.04	3.09	3.02	2.40
8	3.22	3.41	3.02	3.02	3.08	2.41
9	3.20	3.37	3.05	3.12	3.12	2.40
10	3.26	3.38	3.09	3.05	3.03	2.44
Average	3.22	3.33	3.08	3.10	3.08	2.39
Standard deviation	0.03	0.05	0.06	0.05	0.04	0.03
RSD (%)	1	1	2	2	1	1

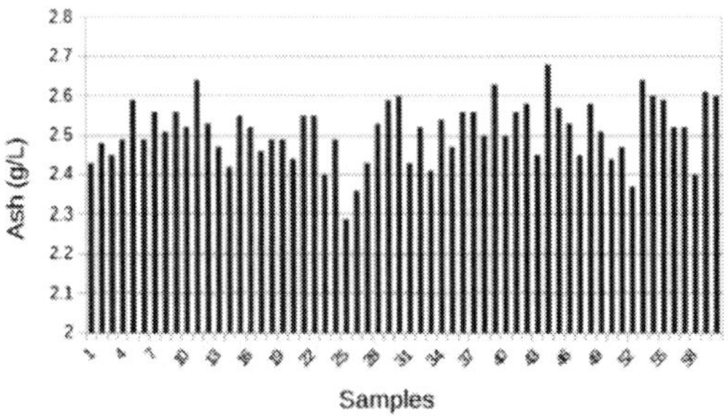

Figure 1. Reproducibility of the method over 2 month's period.

3.2. Method validation

The repeatability has been confirmed on 5 random samples, analysed in 10 replicates. The average of these 5 samples was reported as method repeatability.

Reproducibility has been calculated on one sample, analysed under the same conditions (60 measurements) in the period of 1. 2. 17 until 29. 3. 2017 and can be observed from Fig. 1. The calculated average value is 2.51, standard deviation 0.08 and RSD 3%.

The results of ILC analysis prove that the method is accurate: all ILC give z-score-values lower than 3, only one sample between 2 and 3 (questionable). From the data presented we have calculated measurement uncertainty of bias to be 0.14.

The expanded measurement uncertainty calculated from reproducibility (accounts the same contributions as repeatability) and bias is 0.32 g/L, while expanded measurement uncertainty of reference method is 0.2 g/L.

Linearity doesn't need to be additionally confirmed, since calibration model is straight line with the correlation coefficient $R^2 = 0.94$. Robustness of the method is confirmed through cross-validation performed by the software program itself, showing "performance/ deviation" ratio above 3. Moreover, less than 5% samples were removed from the training set of 324 samples, proving the method selective and specific.

We are sure the method is "ready to use". However, for indirect methods such as FTIR, quality of the developed calibration depends mostly on the quality of the samples, used for the calibration. We believe that our data are of high quality – method accredited according to ISO

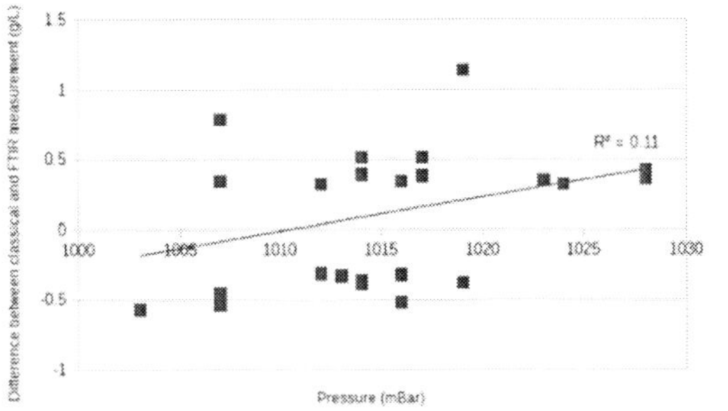

Figure 2. Influence of the air pressure on the difference between FTIR and classical measurement.

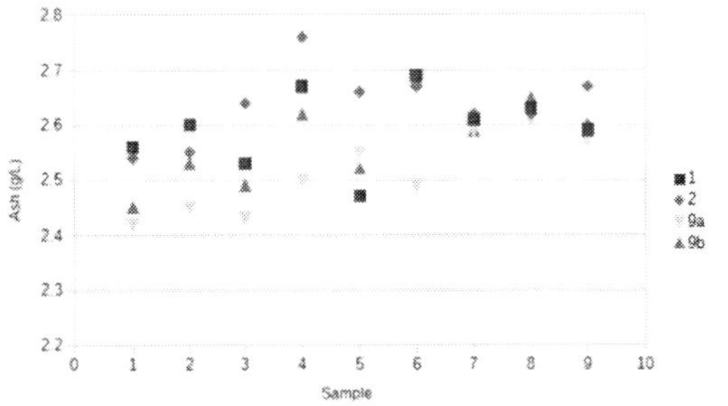

Figure 3. Influence of the order of weighing on the results.

17025, analysis done in duplicates, reproducibility and accuracy of the method are constantly monitored (accuracy at least once a month, reproducibility daily), results of ILC analysis of our reference method are very good (average z of 22 samples is 0.11, no questionable results).

Some doubt has arisen that it might be susceptible to environmental conditions.

3.3. Advantages of the FTIR method

Apart from the obvious – speed, we have found another advantage over the classical method: no influence of environmental factors (moisture, air pressure). The influence might be observable because 100 multiply the weights to achieve results in g/L.

To evaluate the influence of the pressure on reference method we have compared the difference between classical and FTIR measurements, bigger than 0.3 in absolute terms, with average air pressure that day taken for meteorological station Bilje (Fig. 2).

We are not able to prove, which is actually very good information for our developed calibration: the samples used for the calibration do not contain systematic error due to air pressure.

There is another possible source of systematic error induced by the reference method: the samples might moisten during weighing. Again, it would not be a problem if we would not multiply the weights by 100 to achieve g/L. According to AOAC method of ash determination of vinegar, only two samples should be at the same time in the desiccator. There is no such demand in OIV method, so we are using one desiccator for up to 9 samples.

To test this we have analysed one sample in 9 replicates four different ways: 1: weighing samples one by one, so only one sample is in desiccator at the time, 2: 2 samples are in the desiccator at the same time and 9: all samples were in desiccator – 2 sets, a and b, where samples are randomly taken from desiccator. As seen from Fig. 3, weighing 9 samples at the same time has same influence, but it is similar to random error. Even if we cannot prove that FTIR method is better than classical one, because it is not susceptible to environmental factors, we can at least be sure, that we didn't introduce a systematic error that way either.

4. Conclusions

We can conclude that the determination of ash in wine by FTIR spectroscopy is a promising, fast and reliable new method. The best results are achieved if we use four wavelengths and a centred average. The validation of the method proves that it is already fit for purpose with expanded measurement uncertainty 0.3 g/L already near classical method 0.2 g/L.

Apart from the obvious advantage of the method – the speed, we have found another advantage over the classical method: no influence of environmental factors such as air pressure and moisture. We were not able to prove that, however it confirms that calibration data systematic error free.

For the commercial use we still need few hundred samples to be sure, no outliers appear, but might be ready for accreditation next year.

References

[1] *OIV - International Methods of Analysis of Wines and Musts* (2017)
[2] T. Košmerl, D. Bavčar. Zb. Bioteh. Fak. Univ. Ljubl. **81**, 325–334 (2003)
[3] C.-D. Patz, A. Blieke, R. Ristow, H. Dietrich. Anal. Chim. Acta **513**, 81–89 (2004)
[4] C.A.T. dos Santos, R.N. Páscoa, J.A. Lopes. TrAC Trends Anal. Chem. **88**, 100–118 (2017)
[5] B. Ozturk, D. Yucesoy, B. Ozen. Food Anal. Methods **5**, 1435–1442 (2012)
[6] D. Cozzolino, W. Cynkar, N. Shah, P. Smith, Comput. Electron. Agric. **77**, 81–85 (2011)

Comparison of the effect of 8 closures in controlled industrial conditions on the shelf life of a red wine

Jean-Claude Vidal[1], Soline Caillé[2], Alain Samson[1], and Jean-Michel Salmon[1]

[1] UE999 Pech-Rouge, INRA, 11430 Gruissan, France
[2] UMR SPO, INRA, 34000 Montpellier, France

Abstract. Aims: The management of O_2, CO_2 and SO_2 at bottling and the choice of the closure are two key factors of the shelf life of wine in bottles before bringing them to market. The impact of four screw caps, two synthetic and two technical corks was evaluated on a red wine of Merlot/Tannat.

Methods and results: Analytical monitoring (O_2, CO_2, SO_2, aphrometric pressure, L*, a*, b*) was carried out during 538 days of storage at 20 °C. Two sensory analyses at 10 and 17 months completed the study.

The wine was bottled with an average total oxygen content of 2 mg/L. The heterogeneity intra and inter procedure was controlled, including for the dissolved carbon dioxide content.

Conclusion: Unlike closures with highest OTR, the two technical corks and the two screw caps with Saranex seal, harboring the lowest OTR, matched with the wines exhibiting a low total O_2 content at equilibrium (from 4th to 18th month), with more free SO_2 and less changed colour. However this OTR gradient (5 to 67 μg/d) observed through the physicochemical analyses was not necessarily confirmed by both sensory analyses performed.

Significance and impact of study: This study puts into perspective the impact of OTR closure on sensory characteristics evolution of wine consumed during the first two years, especially when the total oxygen at bottling exceeds 1.5 mg/L.

1. Introduction

Oxygen is one of the main factors for wine's evolution. At bottling, oxygen captured in the headspace (HSO) and dissolved in the wine (DO) must be reduced as much as possible. The Oxygen Transmission Rate (OTR) of stopper regulates the transfer of oxygen inside the bottle after bottling. The management of O_2, CO_2 and SO_2 at filling and the choice of stopper are the key factors of the shelf life, manageable by the conditioner before bringing bottles to market.

The oxygen ingresses during and post bottling lead to the decrease of sulfites. In wine, the reaction between O_2 and SO_2 is extremely slow [1]. Sulfites react with the products of wine oxidation and in particular with hydrogen peroxide, product of the oxidation of phenolic compounds [2,3]. The wine becomes more sensitive to oxidation and ages faster. Godden et al. [4] highlighted a critical concentration of free SO_2 of 10 mg/L below which a Semillon wine is perceived as substantially affected by oxidised aroma. For red wines, controlled oxygen ingresses are necessary and variable according to the expected quality before and after bottling especially to avoid the reduction [5,6].

The commercial choice between stopper (natural, technical or synthetic) and screw cap has a direct impact on the volume and the inerting process of the headspace as well as the OTR of closure. The volume and the management technology of headspace (vacuum, gas sparging, snowdrop) explain that the quantity of oxygen trapped in the headspace can vary from 0.4 to 3.6 mg per bottle as measured by Vidal and Moutounet [7]. The bottling line audits outlined by O'Brien et al. [8] confirm this broad range of oxygen amount. Kontoudakis et al. [9] showed that stopper type significantly affected the HSO content. The volume of the headspace of a corked bottle is significantly lower than this of a capped bottle, but on the other hand the cork releases a portion of the oxygen trapped in its own structure due to the compression of the stopper in the bottleneck [10].

Regarding the bottle position, there is no consensus to date on an effect on oxygen mass transfer through the closure and wine aging over time, even if in theory, the oxygen diffusion coefficient through the closure into wine is smaller than into the headspace. Mas et al. [11] concluded that white and red wines were best preserved when bottles were stored horizontally rather than vertically. Puech et al. [12] on rosé and red wines and Skouroumounis et al. [13] on white wines found no significant differences. Godden et al. [4] concluded that for several closures upright storage tended to accelerate loss of SO_2 from a Semillon wine, but in many cases this effect was marginal.

The principal methods for determination of OTR used for wine closures are the coulometric method by Mocon Oxtran and nitrogen flushing of the inner face of the cell [14], the method by permeability meter and pressure difference between the two faces of stopper [15], the luminescence method on corked or capped bottles filled with nitrogen [16] or deoxygenated acid water [17] and

 DOI: 10.1051/bioconf/20170902024

Table 1. Description of the8 closures. [1] PVDC: Polyvinylidene chloride. [2] OTR: Oxygen Transmission Rate in mg/day/closure given by manufacturer. [3] OTR, included release, calculated on 1 year. [4] OTR, included estimated release of 1.5 mg calculated on 1 year.

Code	Type	Length mm	Diameter mm	Weight g	Density kg/m³	OTR[2] mg/d	Method of OTR measurement
B1	Synthetic stopper	42	22.5	4.8	285	0.014[3]	Luminescence
B2	Technical stopper	44	24.5	5.6	270	0.005[4]	Coulometry
B3	Technical stopper	44	24.0	5.7	280	0.005[3]	Colorimetry
B4	Synthetic stopper	42	22.0	7.6	488	0.047	Coulometry
C1	Screw caps + Saranex seal		31.5			0.005	
C2	Screw caps + Saranex seal		31.5			0.005	
C3	Screw cap + seal without PVDC[1]		31.5			0.067	
C4	Screw cap + seal without PVDC		31.5			0.007	

the colorimetric method with indigo carmine [18]. The Mocon Oxtran technology is by far the most commonly used in the packaging industry. But when applied to the bottle/closure system, it cannot mimic exactly the configuration of storage where the closure is wet with wine or in contact with water vapor saturated headspace in case of vertical storage. Moreover, another major drawback of this method is the long time required to reach the steady state of oxygen ingress through the closure when 40 mm long stoppers are tested [19]. These reasons explain why manufacturers use also methods with operating conditions closer to enological reality and which better integrate the desorption of oxygen by the stopper mainly during the first month as luminescence and colorimetric methods [16,20].

However, whatever the used method, the OTR range of natural corks is roughly intermediate between this of screw cap and technical stoppers and this of synthetic stoppers, but with a greater heterogeneity by comparison with industrial stoppers [15,18,21,22].

Logically, many studies have shown the impact of OTR on the quality of wines. Also the higher oxygen permeability it is, the higher are decrease in the SO_2 level, increase of the absorbance at 420 nm and premature emergence of oxidised aroma for white wines [4,13,23–25]. Conversely, screw caps are cited by the majority of the previous articles as closures for which sensorial reduction notes are most frequent due to their low OTR.

Generally speaking, the red wine behaved in a similar way to the white wine, but thanks to its higher phenolic compounds content, it was much more resistant to oxidation [11] but also sensible to reduction when oxygen ingresses post-bottling are insufficient [5,6], in particular with screw caps [26].

Thus, for entry and mid-range wines, synthetic or technical stoppers and aluminum caps usually supersede natural corks.

Based on these findings, an experimental protocol was set up in order to answer the following questions. What is the impact of these closures on the shelf life of a red wine bottled at an industrially achievable Total Oxygen content (TO_2) and intended to be drunk within two years? Are there differences between stoppers and screw caps? Which physicochemical and sensory characteristics are influenced significantly by the oxygen permeability of the closure? To the best of knowledge, this is one of first studies based on the relationship between OTR and the consumptions of oxygen and sulfites under industrial conditions controlled by dissolved gases and sulfites.

2. Material and methods

2.1. Experiment

A 2013 IGP Côtes de Gascogne red wine (70% Merlot + 30% Tannat) was bottled on 26 June 2015 at INRA Pech-Rouge bottling facility (Gruissan, France) in 75 cL flint glass Bordeaux bottles at targeted levels of O_2, CO_2 and SO_2 and with a weak dispersion. Two synthetic stoppers (B1, B4), two technical stoppers (B2, B3), two screw caps with Saranex seal (C1, C2) and two screw caps (C3, C4) with seal without Polyvinylidene chloride (PVDC) were tested (Table 1). The two types of closure and the different length of stoppers (42 or 44 mm) led to different headspace volumes and inerting processes. The target TO_2 in bottle was set at 1.5 mg/bt (2 mg/L), a value reasonably achievable at the industrially level whatever the closure used.

After bottling, bottles were stored upright in the dark in a thermostatically controlled room at 19.9 ± 0.5 °C, with 67.2 ± 15.8 %HR (monitored, but not controlled). Both destructive and non-destructive physicochemical analyses were carried out on several dates spread over 538 days after bottling. An expert jury performed sensory analysis at 10th and 17th months.

Given the small thickness of the seal, the oxygen release of screw caps is negligible as shown by Vidal et al. [17]. For stoppers, as discussed in the introduction, manufacturers generally prefer the luminescence and colorimetric methods of OTR measurement to better quantify the higher release of oxygen by stopper at the beginning of storage which significantly increases the OTR. The coulometry tends to undervalue this phenomenon and gives an OTR of 0.009 mg/d lower than the luminescence method for B1 (0.014 mg/d). The OTR of B2 by coulometry is enhanced by an estimated release of 1.5 mg. For the stoppers B3 and B4, manufacturers have provided OTR value obtained by a single method presented in Table 1.

2.2. Bottles

Cork bottles: OI (Villeurbanne, France), standard 75cL BD CAR II LG, unfilled level 63 mm; screw cap bottles: OI, standard 75cL BD CAR II LG BVS, unfilled level 45 mm.

2.3. Analyse of wine just after bottling

12.9% vol.; sugar 2.6 g/L; TA 3.33 g H_2SO_4/L; VA 0.43 g H_2SO_4/L; pH 3.49; free SO_2 27 ± 0 mg/L; total SO_2

68 ± 0 mg/L; CO_2 325 ± 15 mg/L; L* 67.21; a* 33.89; b* 8.91; A_{420} 2.532; A_{520} 3.379; A_{620} 0.633, Total Phenol Index=49.

2.4. Bottling

The INRA Pech-Rouge bottling line for experimental wines allowed the control and management of dissolved gases on the three elements of the chain:

- a filtration skid (RS IW, Tübingen, Germany) with preparation tank (105 L), prefiltration (1 μm) and final filtration (0.65 μm),
- a single head filler MTB 1/1 (Perrier, Le Cheylar, France) with or without neutral gas flushing of filler tank (46 L) and bottles before filling,
- a single head corking machine Gemini R (Arol, Canelli, Italy) with coupling vacuum and inert gas (N_2 for this study) in several cycles before corking in order to reduce the oxygen amount of the headspace.

This line achieves homogeneous bottling of small volumes of wines with very low variations in TO_2 and dissolved CO_2 [27,28].

Four batches of 100 L were required for the filling of the 450 bottles of the study. The preparation tank was filled by gravity with the starting tank of red wine. The wine was sparged with N_2 gas using a porous injector bolted to the bottom of the preparation tank until DO reached 0.15 mg/L then adjusted to 300 mg/L of dissolved carbon dioxide (DCO_2) by sparging with CO_2 gas.

The circuit was purged with N_2 from the outlet of preparation tank to head filler machine. Wine was forced into the circuit by N_2 to the filler tank through the filtration skid using overpressure of 100 kPa applied to the top of preparation tank. Bottles were blanketed before filling. A slight depression of 8 KPa assisted the filling height adjustment.

Filled cork bottles were sealed by the single head corking machine. 2 combined cycles of N_2 (1 s) followed by vacuum 75 KPa (1 s) were performed. Filled screw cap bottles were crimped by a single head capping machine Galaxy (Costral, Riquewihr, France) without inerting of headspace and cap. The unscrewing torque has been checked for the four kinds of screw cap bottles with Orbis 6 Nm digital torque tester (Mecmesin, Slinfold, England).

Since the DO has been set to a low level for all procedures, the TO_2 target value of 1.5 mg/bt was reached thanks to the management of the headspace according to the type of bottle and its unfilled level.

2.5. Physicochemical analyses

On line O_2 monitoring was performed using a PreSens luminescent probe and PSt3 O_2-sensitive optical spots (PreSens Precision Sensing GmbH, Regensburg, Germany) integrated at four checkpoints on the bottling line and at the top and bottom of the preparation and filler tanks. DCO_2 was monitored by sampling in the preparation tank using a Carbodoseur (Dujardin-Salleron laboratories, Noizay, France).

The following destructive analyses were performed at T0, 1, 4, 8, 12 and 18 months: aphrometric pressure (simplified aphrometer for still wines, Ligapal, Cormontreil, France); free and total SO_2 (potentiometric titration, Titromatic, Crison Instruments, Alella, Spain); ΔEab* (spectrophotometer CM3600d, Konica Minolta, Roissy CDG, France, standard illuminant D65, 10° standard observer). Non-destructive monitoring was conducted every month from T0 to 18 months for unfilled level (set square for wine bottle), calculation of the volume of headspace, dissolved and gaseous O_2 (luminescence with PreSens PSt3 glued spots inside bottles) and dissolved CO_2 (laser spectroscopy, Lsensor CO_2, FT System, Padova, Italy). For destructive and non-destructive chemical analyses, three repetitions were performed by parameter / procedure / date.

Twelve bottles capped with C4 screw cap were stored at 7 °C (C4 7 °C). These bottles were used for SO_2 analyses at 243, 370 and 532 days.

Table 2. Sensory attributes selected and composition of their reference standards. [1] Attributes added at 17 months.

Sensory cluster	Attribute Attribute	Reference standard
Visual	Colour intensity	
Olfactory	Amylic	Isoamyl acetate
	Animal (leather)	4-Ethylphenol
	Cooked red fruits	Red fruits jam
	Dry wood (dust)[1]	Unheated wood powder
	Pastry (vanilla, caramel)	Caramel syrup
	Pepper	Black pepper
Gustatory	Astringency	Grape stem tannin extract
	Bitterness	Caffeine
	Sourness	Tartaric acid
	Sweetness[1]	Grape sugar
	Alcohol[1]	Absolute ethanol

2.6. Sensory analyses

Descriptive quantitative analysis was conducted by an expert sensory panel (22 judges), selected on the basis of their sensory performances and interest (ISO 8586-2, 1994), and trained to descriptive sensory analysis of wines. At the first step of generating vocabulary, the jury selected attributes by consensus, to describe the samples. The Table 2 presents the finally selected attributes. Then the panelists were trained to understand and consistently use attributes and familiarized with the product space. Finally, the judges rated each attribute on an unstructured linear scale from "low" to "high". For olfactory and taste analyses, wines were evaluated in duplicate, in monadic service, according to a random order (Latin square) minimizing carry-over effects, in black tulip-shaped glasses to ensure that visual perceptions did not influence olfactory and taste analyses, between 17 and 19 °C. Following for visual attributes, samples were evaluated in comparative service, in 215 ml wine glasses, with "normal daylight" illumination.

Sensory data were converted into marks from 0 to 10 by Fizz Software version 2.40 A (Biosystemes, Couternon, France).

2.7. Statistical analyses

Statistical analyses were performed using XLSTAT software version 2014 (Addinsoft, Paris, France).

Table 3. Contents of O_2 at bottling and TO_2 at equilibrium. HSO: Headspace Oxygen; DO: Dissolved Oxygen; TO_2: Total Oxygen; mg/bt: mg/bottle. [1] TO_2 at equilibrium: average TO_2 between 90th and 532th day. Averages and standard deviations are based on 3 bottles per procedure.

Type	Code	T0 HSO mg/bt	T0 DO mg/bt	T0 TO_2 mg/bt	TO_2 equilibrium mg/bt[1]
Stopper	B1	1.08 ± 0.00	0.15 ± 0.06	1.22 ± 0.06	0.06 ± 0.02
	B2	1.19 ± 0.12	0.19 ± 0.05	1.38 ± 0.09	0.04 ± 0.01
	B3	1.03 ± 0.07	0.30 ± 0.19	1.34 ± 0.17	0.05 ± 0.01
	B4	0.99 ± 0.07	0.25 ± 0.05	1.24 ± 0.12	0.11 ± 0.06
Screw cap	C1	1.43 ± 0.21	0.19 ± 0.03	1.62 ± 0.18	0.04 ± 0.01
	C2	1.46 ± 0.17	0.18 ± 0.03	1.64 ± 0.17	0.05 ± 0.01
	C3	1.61 ± 0.25	0.15 ± 0.03	1.75 ± 0.28	0.23 ± 0.03
	C4	1.54 ± 0.07	0.20 ± 0.09	1.74 ± 0.13	0.07 ± 0.02
Stoppers averages		1.07 ± 0.06 (10.4 ± 0.6 %v/v)	0.25 ± 0.09	**1.30** ± 0.11	
Screw caps averages		1.51 ± 0.17 (7.9 ± 0.9 %v/v)	0.18 ± 0.05	**1.69** ± 0.19	

For sensory data, after confirming jury good performances (repeatability, consensus and discrimination), results were processed by analysis of variance (2 factors: judge and wine). When significant differences were revealed ($p < 0.05$), mean intensities were compared using the Tukey (HSD) multiple comparison test.

The results of free - total SO_2 and consumption ratios of SO_2/O_2 were processed by analysis of variance (1 factor: closure). When significant differences were revealed ($p < 0.05$), analytical parameters were compared using the Tukey (HSD) multiple comparison test.

The treatment allowed to classify the different wines in several distinguishes groups (A, B, C, D, E). The average values and the several groups are given in this paper.

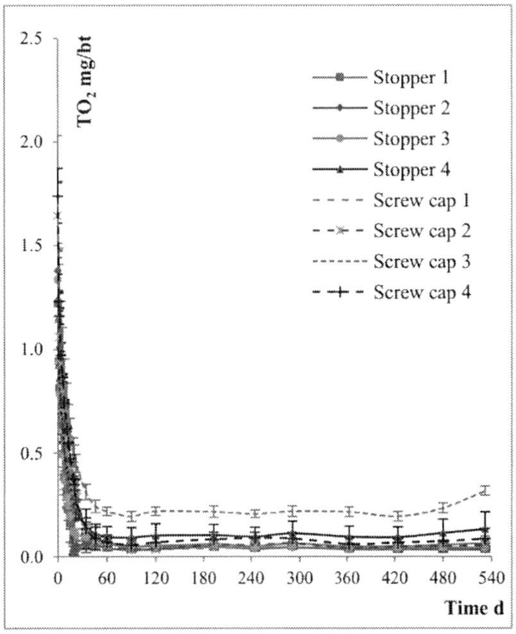

Figure 1. $TO_2 = f(t)$ for each procedure. TO_2 mg/bt: Total oxygen in mg/bottle. Averages and standard deviations are based on 3 bottles per procedure.

3. Results and discussion

3.1. Industrial bottling parameters

The unscrewing torque of capped bottles tested after crimping was on average 16 ± 1 lbf/inch for all four screw caps procedures. The unfilled level which determines the volume of the headspace was on average 63 ± 2 mm for the stoppers and 45 ± 2 mm for screw caps (net of seal thickness) from T0 to 538 d. Aphrometric pressure fluctuated between −160 and 200 kPa from T0 to 538 d. The monitoring of these three parameters was in accordance with usual technical recommendations.

3.2. Dissolved gases

Just after bottling, 86% of TO_2 (TO_2 = HSO + DO) was located in the headspace of bottle. The average TO_2 was 1.5 mg/bt (2 mg/L). Heterogeneity remained limited because the highest intraprocedural standard deviation was 0.28 mg/bt (Table 3) and the maximum intraprocedural deviation was 0.53 mg/bt (between B1 and C3). As already described several times [29–31], both oxygen in the headspace and dissolved in the wine were consumed. This decrease was not linear because 90% of initial TO_2 was consumed after 35 days. From 48th day, TO_2 was less than 0.10 mg/bt, except for B4 and C3. These procedures had the most variable and highest TO_2 content, mainly due to HSO level (higher than 0.1 mg/bt at 532 d), while the DO level was equivalent to other procedures (around 0.01–0.02 mg/bt). Beyond the 90th day, we could consider that all closures reached their steady states (Fig. 1). The TO_2 stabilized at an equilibrium value which was the resultant of O_2 ingresses by the closure and O_2 consumption by the wine, for the same wine and in the same storage conditions. Thus we could link the closure OTR to the average TO_2 between 90th and 532th day (Tables 2 and 3).

The carbon dioxide concentration remained stable up to 532 days, except for the B4 procedure for which the losses represented 18% corresponding to a loss of 60 mg/L, less than the sensory perception threshold.

3.3. Colour

The distance between two colours (ΔEab*) was used to summarise the evolution of the colour of wine between T0 and 538 d.

$$\Delta Eab^* = \sqrt{(L_0 - L_{538d})^2 + (a_0 - a_{538d})^2 + (b_0 - b_{538d})^2} \quad (1)$$

Over time, b* increased and a* decreased, colour tended gradually to tile colour. After 538 d, the ΔEab* varied between 6.6 and 8.3 regardless of the procedure. C4, B1, B3 and C2 were the procedures whose colour has changed the least unlike B4 and C3. However, even if the ΔEab* between T0 and 370 d (538 d) were at least 4.8 (6.6), the maximum interprocedural deviation was 1.7 (ΔEab* = 8.3–6.6 respectively for B4 – C4 at 538 d). The evolution of colour of wine until 538 d is mainly due to the aging of wine. By comparison, the ΔEab* after 18 months storage is less than 1 between wines of Cabernet Sauvignon sealed by natural cork, synthetic closure and screw cap with 16 mL of headspace volume [26].

Meanwhile, the colour of red wine stored at 7 °C changed very little because ΔEab* = 0.6 and 1.1 respectively at 370 and 538 d.

And so until 18 months, the temperature impact on the colour is clearly greater than this of stopper.

3.4. Sulfites

From 370 d, the dispersion stayed or extended between the C1, C2, B2 procedures for which the free SO_2 was at least 11 mg/L and the B4 and C3 procedures for which the free SO_2 toggled below the threshold of 10 mg/L; the B3, C4 and B1 procedures exhibiting an intermediate position (Fig. 2a). The distribution of stoppers is the same for total SO_2 (Fig. 2b).

However it should be mentioned that the impact of the stopper on the free SO_2 conservation reached a maximum of 5 mg/L between the procedures at 538 days (Fig. 2a) that is to say below the value of 7 mg/L of the expanded uncertainty for free SO_2 content of 10 to 30 mg/L [32]. Only the cold treatment had a clear effect on the preservation of free SO_2, as SO_2 consumption reactions slowed (free SO_2 at 538 d = 10/25 mg/ L for C4/C4 7 °C).

3.5. Sulfite versus oxygen consumption

Oxidation of phenolic compounds leads to the production of quinones and hydrogen peroxide. SO_2 reacts with the latter, thus preventing the oxidation of ethanol according to the Fenton reaction, and reduces quinones towards their initial phenolic form. Under ideal experimental conditions, the O_2 :SO_2 molar ratio of the reaction is 1:2 [33], corresponding to a maximum theoretical consumption of 4 mg of SO_2 per mg of consumed O_2. During storage of wine in bottle, a mass ratio below 4 or a molar ratio of 1 :< 2 means that a part of the oxygen which enters inside the bottle does not indirectly reacts with SO_2, but reacts with other constituents of wine [33,34]. Nucleophilic compounds come into competition with sulfites to react

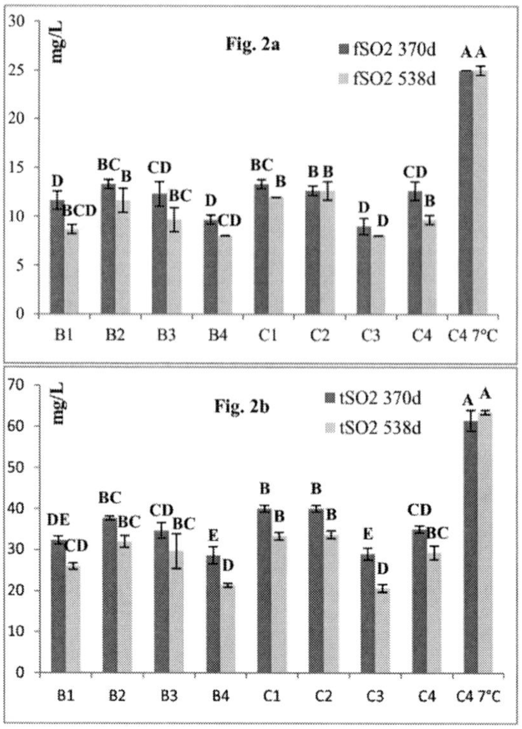

Figures 2a and 2b. Free SO_2, total SO_2 at 370 and 538 days. Stoppers B1 to B4; screw caps C1 to C4. Different labels (A, B, C, D, E) indicate means that significantly differ at $p<0.0001$. Averages and standard deviations are based on 3 bottles per procedure. fSO_2: free SO_2; tSO_2: total SO_2 in mg/L; d:day. Free and total SO_2 at T0 = 27 ± 0/68 ± 0 mg/L.

with the quinones. Waterhouse et al. [35] uses the mass ratios with free SO_2 consumption ($fCSO_2$) and total SO_2 consumed ($tCSO_2$) to evidence this phenomena: more the mass ratio is less than 4 and more the oxidation of other wine constituents is important.

$$TCO_i \frac{mg}{L} = (TO_{2T0} - TO_{2i})\frac{mg}{bt}/0{,}75 + (OTR \times idays) \quad (2)$$

Free Consumed SO_2/TCO =

$$\frac{fCSO_2\frac{mg}{L}}{TCO\frac{mg}{L}} idays = \frac{fSO_2T0 - fSO_2i}{TCO_i} \quad (3)$$

Total Consumed SO_2/TCO =

$$\frac{tCSO_2\frac{mg}{L}}{TCO\frac{mg}{L}} idays = \frac{tSO_2T0 - tfSO_2i}{TCO_i} \quad (4)$$

The ranking of stoppers in descending order of $fCSO_2$/TCO at 538 d (Figure 3) led to the following classification similar to that of OTR: B3 > B2 > C1, C2 > C4 > B1 ⋙ B4 > C3. This classification was identical to the ratio calculated with tSO_2C at 370 and 538 days. The only difference was the slightly inverted order between C1 and C2 at 370 days for the ratio calculated with fSO_2C.

Table 4. Ranking comparison OTR vs physicochemical parameters at 538 days. Stoppers B1 to B4; screw caps C1 to C4. OTR in mg/day/closure; TO_2 at equilibrium in mg/bottle; $fCSO_2$ and TCO in mg/L; d: day.

		OTR and impact of oxygen exposure		
Parameter	**Range**	**low**	**medium**	**high**
OTR mg/d	0.005 to 0.067	B2, B3, C1, C2	< C4 < B1	<< B4 << C3
TO_2 at equilibrium mg/bt	0.04 to 0.23	B2, C1 < B3, C2	< B1 < C4	<< B4 << C3
% losses free SO_2 538d	53 to 70	C2, C1, B2	<B3, C4 < B1	< B4, C3
ΔE* 538d	6.6 to 8.2	C4 < B1 < C2, B3	< C1, B2	< C3 < B4
$fCSO_2$/TCO 538d	3.9 to 0.5	B3, C1, C2, B2	> C4 > B1	>> B4, C3

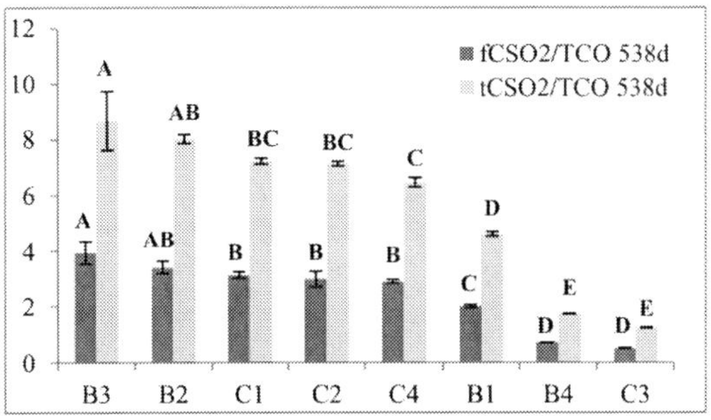

Figure 3. Ratios of free/total Consumed SO_2 / TCO at 538 days. Stoppers B1 to B4; screw caps C1 to C4. f(t)CSO_2/TCO: free (total) Consumed SO_2 / Total Consumed Oxygen. $fCSO_2$, $tCSO_2$ and TCO expressed in mg/L. Ratios ranked in descending order of $fCSO_2$/TCO 538 d. Different labels (A, B, C, D, E) indicate means that significantly differ at $p<0.0001$. Averages and standard deviations are based on 3 bottles per procedure.

Logically between these two dates, a decline was observed for all values, highlighting a more intense oxidation over time. As illustrated in Fig. 3, the ratio $tCSO_2$/TCO was greater than 4 for the 6 least permeable stoppers. This result has already been observed in previous studies on red wine rich in tannins after 12 and 15 months of storage evidencing oxygen-independent SO_2 consumption reactions [6,36].

TO_2 at T0 was between 1.22 and 1.75 mg/bt according the procedures and average TO_2 of corked bottles was 1.30 mg/bt while it was 1.69 mg/bt for capped bottles (Table 3). Therefore the caps procedures started with an average handicap of 0.39 mg/bt compared to the stopper procedures linked to the bottling conditions but independent of the kind of closure. This bias arbitrarily increased the TCO of capped bottles and impacted their sulfites contents without leading the possibility to really quantify it afterwards.

In addition, the OTR of B2 included an estimated (but unmeasured) release of 1.5 mg/stopper. If we considered a release of 2 mg/stopper, mass ratios with fSO_2 and tSO_2 538 d decreased respectively to 3.07 and 7.22, bringing B2 behind C2 and even behind C4 if we considered the handicap on TO_2 at T0 (but excluding the unquantifiable impact on sulfites).

To the more or less accurate assessment of the release of stoppers (in this study, particularly for B2), we must add the high uncertainty on the measurement of free and total SO_2 regardless of the analytical method [32].

Finally, all these sources of variation and uncertainty influenced the ratio values and stoppers ranking.

3.6. Impact of the OTR on physicochemical parameters

TO_2 at equilibrium, fSO_2 losses and ΔEab* are physicochemical parameters which are not statistically related to the OTR, but their evolution is influenced by the diffusion of oxygen through the stopper. Table 4 collects these parameters by comparison to OTR and $fCSO_2$/TCO ratio at 538 d.

Table 4 showed that the impact of oxygen exposure on the red wine followed the rise of closure OTR. But more than a ranking, it rather evidenced an opposition between B2, B3, C1, C2 and B4, C3; C4 and B1 having an intermediate position.

A matter of fact, 2 mg/L of TO_2 at bottling corresponds to one year of oxygen ingresses by B3, B2, C1 or C2 closures and 5 mg/L corresponds to one year of oxygen ingresses by B1 stopper. In addition, previous studies [37,38] on wines stored at ambient temperature show that oxidised characters begin to appear around 12 mg/L (9 mg/75 cL) of oxygen ingresses in the bottle. Thus, given the concerned quantities of oxygen, it is essential to control and manage oxygen at packaging by reducing the targeted amount of total O_2 trapped in bottle (TO_2 at T0) but also its heterogeneity (standard deviation of each lot), so that stopper fully plays its role of oxygen diffuser, especially for wines aimed to be drunk within 18 months.

3.7. Sensory analyses

For each analysis time point, jury performances were checked. The panel repeatability and consensus were good.

At 10 months, wines are significantly discriminated by the attribute *colour intensity* with pvalue <0.0001 (Fig. 4). No olfactory and taste differences were observed between the eight procedures. C2 was significantly different from all other procedures by a lighter colour. Then between the seven other wines, the C3 procedure had a darker colour than B2.

At 17 months, two olfactory descriptors (*animal*, pvalue = 0.030 and *pepper*, pvalue = 0.032) and the visual descriptor (*colour intensity*, pvalue <0.0001) allowed to discriminate wines (Fig. 4). No taste difference was observed. At olfactory level, we could observe that C2 and B2 procedures had significantly more intense animal odour than B4. The B2 procedure also had more peppery odour than B1. As at 10 months, C3 procedure had a significantly darker colour particularly than C2, B2 and C1.

The evolution of wines in 17 months differed according to the closures. The sensory evolution of wine closed by the four screw cap procedures was important, wines became less bitter and the smell of pepper decreased.

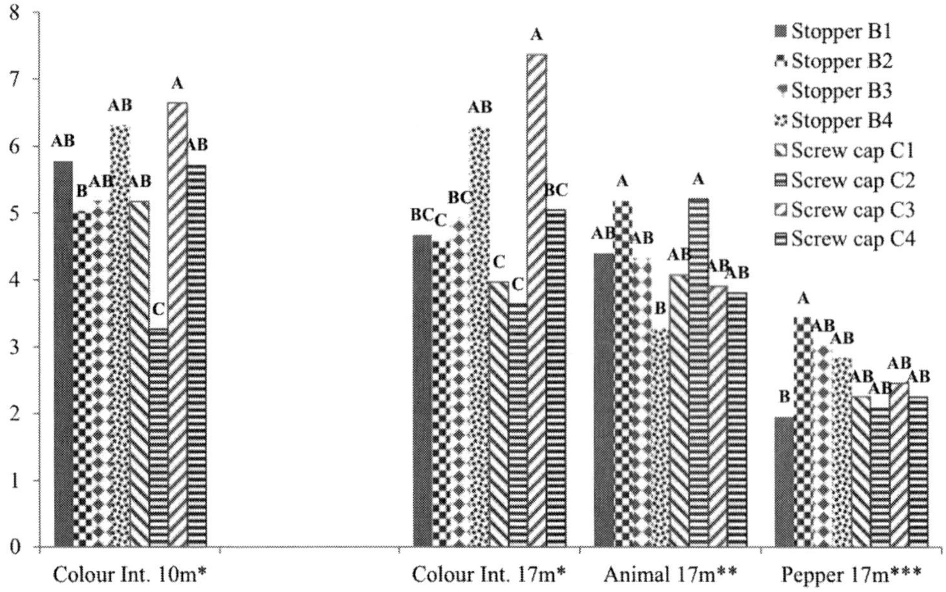

Figure 4. Means for the significant attributes at 10 and 17 months of storage. Colour Int.: Colour Intensity; m: month. Different labels (A, B, C) indicate means that significantly differ at $^*p<0.0001$; $^{**}p<0.030$; $^{***}p<+0.032$.

However, the four stoppers procedures had different developments. For B1 and B4 procedures, animal odour decreased. On the contrary, the animal odour of B3 became more intense. About the B2 procedure, the colour of wine became lighter and the intensities of *cooked red fruit* and *pastry* decreased.

A graduation of OTR was observable on the basis of *animal* and *colour intensity*. Wines with low OTR closures had a more intense animal odour and lighter colour; those with high OTR were darker colour. This observation was made in a previous study on a Grenache wine [5] where it was shown that visual and olfactory differences were observed according to the OTR levels (max. OTR difference: 4.05 mg/year /37.5 cL of wine) but with little impact on taste attributes.

The results obtained by Kwiatokowski et al. [26] with a synthetic cork and a small headspace corroborated our conclusions: a Cabernet Sauvignon wine evolved towards significantly reduced notes, even if these descriptors were not the dominant characteristics of the wines. However, for Ugliano et al. [39], an intake of 1 mg/O_2/year was enough to modify the olfactory characteristics from reduced to fruity; but this conclusion was very dependent on the type of red wine.

4. Conclusion

As regards the physicochemical analyses until 18 months, the important points were:

- the average CO_2 losses were negligible,
- the free SO_2 was always above 10 mg/L, except for B4 and C3 closures,
- HSO of B4 and C3 closures stabilised at higher content, while there was not significant difference of DO between all the closures,
- the evolution of the colour was more and more clear with the time ($\Delta Eab^* \geq 4.8$ as from 370 d), but it was more due to the aging of wine than to the impact of closure,
- the beneficial effect of storage at 7 °C for C4 screw cap was clear on SO_2 conservation and protection of the initial colour.

Finally, the ranking of closures from the overall experiment was similar to this of OTR manufacturers, namely from less to more permeable:

C1, C2, B2, B3 > C4 > B1 ⋙ B4 > C3

C1, C2, B2 and B3 closures were difficult to differentiate, as the oxygen ingress of the first year is around 1.8–1.9 mg for these four closures. The difficulty of finding a link between physicochemical and sensory results mostly came from the fact that between 10 and 18 months, differences of oxygen intakes were low between the stoppers (except for B4 and C3) and wines remained covered by free SO_2. But even with the most permeable closures (C3 and B4), the wines were not systematically characterized by oxidation or aging descriptors.

The study was also conducted on a Colombard and a Rosé de Provence wines under the same conditions of bottling and storage. The physicochemical analyses also highlighted the outlined OTR gradient, but it was mainly on Merlot/Tannat that sensorial analyses were affected by OTR gradient at 17–18 months (data not shown).

The authors thank the CARENE section of UNSCV for its financial support. The authors thank also for their contribution: M. Angènieux, F. Boussuge, F. Dell'Ova, E. Garcia, Y. Sire,

M. Toussaint, M. Veyret from INRA UEPR and E. Picou from INRA UMR SPO.

References

[1] A.L. Waterhouse, V.F. Laurie, Am. J. Enol. Vitic. **57**, 306–313 (2006)
[2] J.C. Danilewicz, J.T. Seccombe, J. Whelan, Am. J. Enol. Vitic. **59**(1), 128–136 (2008)
[3] J.C. Danilewicz, P.J. Wallbridge, Am. J. Enol. Vitic. **61**(1), 166–175 (2010)
[4] P. Godden, L. Francis, J. Field, M. Gishen, A. Coulter, P. Valente, P. Høj, E. Robinson, Aust. J. Grape Wine Res. **7**(2), 64–105 (2001) doi: 10.1111/j.1755-0238.2001.tb00196.x
[5] S. Caillé, A. Samson, J. Wirth, J.B. Diéval, S. Vidal, V. Cheynier, Anal. Chim. Acta **660**(1–2), 35–42 (2009) doi: 10.1016/j.aca.2009.11.049
[6] M. Ugliano, J.B. Diéval, T.E. Siebert, M. Kwiatkowski, O., Aagaard, S. Vidal, E.J. Waters, J. Agric. Food Chem. **60**(35), 8561–8570 (2012) doi: 10.1021/jf3014348
[7] J.C.Vidal, M. Moutounet, OENO One **40**(1), 35–45 (2006) doi: 10.20870/oeno-one.2006.40.1.884
[8] V. O'Brien, C. Colby, with M. Nygaard, Wine Industry Journal **24**(1), 24–29 (2009)
[9] N.Kontoudakis, P. Biosca, R. Canals, F. Fort, J.M. Canals, F. Zamora, Aust. J. Grape Wine Res. **14**(1), 116–122 (2008) doi: 10.1111/j.1755-0238.2008.00013.x
[10] M. Squarzoni, S. Limbo, P. Luciano, Ind. Bevande **XXXII**, 113–116 (2004)
[11] A. Mas, J. Puig, N. Lladoa, F. Zamora, J. Food Sci. **67**(3), 1374–1378 (2002) doi: 10.1111/j.1365-2621.2002.tb10292.x
[12] C. Puech, S. Vidal, J.F. Pegaz, C. Riou, P. Vuchot, Rev. des Œnologues **121**, 13–16 (2006)
[13] G.K. Skouroumounis, M.J. Kwiatkowski, I.L. Francis, H. Oakey, D.L. Capone, B. Duncan, M.A. Sefton, E.J. Waters, J. Aust. J. Grape Wine Res., **11**(2), 369–377 (2005) doi: 10.1111/j.1755-0238.2005.tb00036.x
[14] ASTM-Standard-F1927, ASTM Internationals, West Conshohocken, PA
[15] J. Sanchez, J.M. Aracil, Bulletin de l'O.I.V. **71**, 805–806 (1998)
[16] J.B. Diéval, S. Vidal, O. Aagaard, Packag. Technol. Sci. **24**(7), 375–385 (2011) doi: 10.1002/pts.945
[17] J.C. Vidal, B. Guillemat, C. Chayvialle, Bulletin de l'OIV **84**, 962–964 (2011)
[18] P. Lopés, C. Saucier, C., Y. Glories, J. Agric. Food Chem. **53**(18), 6967–6973 (2005) doi: 10.1021/jf0404849
[19] M. Poças, B. Ferreira, J. Pereira, T. Hogg, Packag. Technol. Sci. **23**, 27–33 (2009) doi: 10.1002/pts.876
[20] P. Lopés, C. Saucier, P.L. Teissedre, Y. Glories, J. Agric. Food Chem. **54**(18), 6741–6746 (2006) doi: 10.1021/jf0614239
[21] T. Karbowiak, R.D. Gougeon, J.B. Alinc, L. Brachais, F. Debeaufort, A. Voilley, D. Chassagne, Crit. Rev. Food Sci. **50**(1), 20–52 (2009) doi: 10.1080/10408390802248585
[22] C. Macku, K. Reed, K. Practical Winery & Vineyard Journal, **winter 2011** (2011)
[23] M. Brajkovich, N. Tibbits, G. Peron, C.M. Lund, S.I. Dykes, P.A. Kilmartin, L. Nicolau, J. Agric. Food Chem. **53**(26), 10006–10011 (2005) doi: 10.1021/jf0512813
[24] P. Chatonnet, D. Labadie, Rev. des Œnologues **106**, 13–20 (2003)
[25] P. Lopés, M.A. Silva, A. Pons, T. Tominaga, V. Lavigne, C. Saucier, P. Darrier, P., P.L. Teisseidre, D. Dubourdieu, J. Agric. Food Chem. **57**(21), 10261–10270 (2009) doi: 10.1021/jf9023257
[26] M.J. Kwiatkowski, G.K. Skouroumounis, K.A. Lattey, E.J. Waters, Aust. J. Grape Wine Res. **13**(2), 81–94 (2007) doi: 10.1111/j.1755-0238.2007.tb00238.x
[27] J.C. Vidal, E. Devic, F. Dell'Ova, *35th World Congress of Vine and Wine*, Izmir, 20 June (2012)
[28] J.C. Vidal, Rev. Fr. Œnol. **272**, 2–5 (2015)
[29] J.C. Vidal, M. Moutounet, Internet journal of viticulture and enology **4/3**, 1–15 (2011) `www.infowine.com`
[30] M. Toussaint, J.C. Vidal, J.M. Salmon, J. Agric. Food Chem. **62**(13), 2946–2955 (2014) doi: 10.1021/jf405392u
[31] C. Dombre, J. Wirth, M. Toussaint, C. Lixon, A. Verbaere, N. Sommerer, J.C. Boulet, S. Caillé, V. Cheynier, P. Rigou, A. Samson, J.M. Salmon, J.C. Vidal, S. Marais, Y. Gerand, P. Roux, M.H. Lemaîstre, A. Bobe, P. Languet, P. Chalier, Rev. des Œnologues **158**, 47–50 (2016)
[32] EURL Œnologues de France, Bilan des résultats des comparaisons Interlaboratoires (CIL), janv. 2006 à déc. 2015. Union des Œnologues de France (2016)
[33] J.C. Danilewicz, Am. J. Enol. Vitic. **67**(1), 13–17 (2016) doi: 10.5344/ajev.2015.15069
[34] G. Han, M. Ugliano, B. Currie, S. Vidal, J.B. Diéval, A.L. Waterhouse, J. Sci. Food Agric. **95**(1), 36–43 (2014) doi: 10.1002/jsfa.6694
[35] A.L. Waterhouse, S. Frost, M. Ugliano, A.R. Cantu, A.R., B.L. Currie, M. Anderson, A.W. Chasey, S. Vidal, J.B. Diéval, H. Heymann, Am. J. Enol. Vitic. **67**(3), 449–459 (2016) doi: 10.5344/ajev.2016.16006
[36] A. Gambuti, T. Siani, L. Picariello, A. Rinaldi, M.T. Lisanti, M. Ugliano, J.B. Diéval, L. Moio, Eur. Food Res. Technol. (2016) doi: 10.1007/s00217-016-2780-3
[37] J.C. Vidal, M. Toussaint, J.M. Salmon, *37th World Congress of Vine and Wine*, Mendoza, **12023** (2014)
[38] M. Ugliano, J. Agric. Food Chem. **61**(26), 6125–6136 (2013) doi: 10.1021/jf400810v
[39] M. Ugliano, J.B. Dieval, S. Begrand, S. Vidal, Rev. des Œnologues **156**, 45–48 (2015)

DOI: 10.1051/bioconf/20170902025

The inhibition of *Saccharomyces cerevisiae* population during alcoholic fermentation of grape must by octanoic, decanoic and dodecanoic acid mixture

Mojmír Baroň, Michal Kumšta, Kamil Prokeš, Lenka Tomášková, and Magdaléna Tomková

Department of Viticulture and Enology, Faculty of Horticulture, Mendel University in Brno, Valticka 337, CZ-691 44 Lednice, Czech Republic

Abstract. The inhibition of alcoholic fermentation by octanoic, decanoic and dodecanoic acid mixture was investigated. Middle chain fatty acids (MCFA) mixture contained 10 grams of C_8:C_{10}:C_{12} in a ratio of 2:7:1 was dissolved in 100 ml of 70% ethanol, and in such form, it was subsequently applied into the fermenting must samples. A flow cytometry test showed that the 10 mg/L dose of MCFA mixture had a toxic effect on *Saccharomyces cerevisiae* (45.9% of viable cells) compared with the control variant (74.35%). In combination with 60 mg/L of SO_2, it had a higher efficiency (3.1% of viable cells) than using a dose of SO_2 alone (13.9%). Direct counting of yeast cells confirmed a higher concentration of dead cells with a higher concentration of MCFA. A dose of 10 mg/L of MCFA mixture caused the highest percentage of dead yeast after 24 hours (about 60%) compared to the control variant without MCFA dosage (about 24%). The results of residues showed that there is no significant quantitative difference between the treated and untreated musts because of fixing of MCFA inside the yeast cells. This method can effectively reduce the cost of production technology for wines with residual sugar and, in general, reduce the dosage of SO_2.

1. Introduction

Alcoholic fermentation is one of the most important processes in wine technology (1). Under appropriate conditions (2), there is an increase in yeast mass and fermentation of the must to a state of exhaustion of the sugar media. In recent years, the popularity of wines with residual sugar has increased, and thus it has become necessary to inhibit and stop fermentation (3).

The issue of early termination of alcoholic fermentation is rather complex. Cooling and filtration lead to an increase in costs and are laborious and generally unavailable, especially for small-scale producers. The individual application of SO_2 is not always fully reliable, and in higher concentrations leads to a reduction in future quality.

Some higher fatty acids (MCFA), particularly long chain (C_{16} and C_{18}), activate alcoholic fermentation. On the other hand, other MCFA with shorter chains, especially hexanoic C_6, octanoic C_8, and decanoic C_10 acids, have fungicidal properties (4; 5; 6). The study of the endogenous MCFA inhibitory effect on alcoholic fermentation and malolactic fermentation were carried out (7). MCFA are produced by yeast during alcoholic fermentation and may contribute to the difficult completion of its general process. In other words, their increased concentration often accompanies problems with the process of alcoholic fermentation (8; 9; 10). The addition of MCFA has an antifungal effect and supplements SO_2. For example, to stop the fermentation of sweet wines, 150 mg/L SO_2 with 9 mg/L of MCFA added has the same efficacy as 250 mg/L of SO_2. MCFA should be added 24 hours before sulphating. Under these conditions, SO_2 causes the majority of the (if not complete) inactivation (11). It was found that the doses sufficient to inhibit varies with the type of yeast tested. For example, *Kluyveromyces marxianus* is less sensitive than *Saccharomyces cerevisiae* (12).

A study comparing the composition of MCFA and their esters in wine treated and untreated with MCFA showed that concentrations remained within the normal range of 2.6 to 12.4 mg/L (MCFA) and 0.2 to 0.81 mg/L (ethylesters) without increase of the concentration of any volatiles (12).

These features, combined with the wholesomeness and increasing popularity of wines with residual sugar, led to the idea that MCFA could serve as a substitute, or rather a complementary method, for the use of SO_2 in wine technology. In 2016, a communication was presented regarding the use of saturated higher fatty acids (MCFA) in oenology in commission II "Oenology" in OIV, and the Czech Republic applied to allow this as new oenological method.

2. Material and methods

2.1. Experimental design

To demonstrate the efficacy of MCFA on yeast inhibition, two inhibition tests were performed. The MCFA solution was applied into the fermenting must of a '*Welschriesling*' variety.

Table 1. Design of experiment.

Determining the efficacy of MCFA in fermenting must	Flow cytometry Flow cytometry
	Direct counting in Bürker chamber
Determination of MCFA residues in treated wines	GC chromatography

2.1.1. *MCFA solution*

The MCFA solution contained C_8:C_{10}:C_{12} in a ratio of 2:7:1. From a total of 10 g of mixture, there was 2 g of C_8, 7g of C_10 and 1 g of C_12. This mixture was dissolved in 100 mL of 70% ethanol. In such a form, different doses of MCFA mixture were subsequently applied into samples of grape must. For GC analysis, the finished wine after the first racking was used.

2.2. Biological material

'Welschriesling' must: Glucose + fructose content 200 g/L, pH 3,0, titratable acids 11.3 g/L, and the must was treated with 20 mg/L SO_2. Mixture of MCFA was added after seven days of fermentation, glucose + fructose content in this time was 50 g/L.

2.3. Flow cytometry test

The first inhibition test was performed using a flow cytometry.

2.3.1. *Determination of live and dead cell concentration*

The cell concentration in the samples was measured with no sample preparation by direct analysis of the sample must using a flow cytometry.

2.3.2. *Determination of yeast viability*

Determination of yeast viability using flow cytometry was performed after the sample preparation. 1 mL of each sample was collected in a micro test-tube, centrifuged and washed 2x with demineralized water. 200 μL of washed cell suspension was added to fluorescein diacetate (FDA) and propidium iodide (PI) at a concentration of 20 μg/mL and 10 μg/mL, respectively. The samples were incubated at room temperature, in the dark, for ten minutes. The experiment was carried out twice; further samples from the first attempt were measured again one hour later. One drop of prepared sample was used for direct microscopic observation using ultraviolet radiation.

2.4. Direct counting in a Bürker chamber

The second inhibition test was performed by direct counting in a Bürker chamber. The fermenting must contain 50 g/L of glucose + fructose was homogenized by stirring. Then it was put into bottles to the volume 0.75 L, to which was added a different dose of MCFA mixture according to Table 2. Each variant was repeated, in the second iteration. The counting was performed 24 and 168 hours after application of the MCFA.

Table 2. Samples with added concentration of MCFA mixture.

	Dose of MCFA mixture [mg/L]
Sample 1	0
Sample 2	4
Sample 3	5
Sample 4	6
Sample 5	7
Sample 6	8
Sample 7	9
Sample 8	10
Sample 9	20
Sample 10	40

2.4.1. *Counting in a Bürker chamber*

The sample was coloured by methylene blue solution to better distinguish the dead and live yeast. On a clean glass of the counting chamber, a sample was pipetted and treated with methylene blue for easy differentiation of live and dead cells. Under the microscope (Olympus CX 31) at 40x magnification, each of the fields of the Bürker counting chamber was monitored. 72 fields were evaluated at 6 columns to 12 squares, 1 square having an area of 1/25 mm^2. The dead yeasts are distinguished from the live ones by the intensity of their colour. The dead cells are coloured dark blue as a result of the disruption of their protective phospholipid layer, thus absorbing a bigger quantity of the pigment. On the contrary, the live cells are coloured blue. Samples were properly diluted so that the number of microorganisms ranged from 10 to 40 cells per square.

2.5. Test of MCFA residues

The purpose of the experiment was testing residues of MCFA in treated wines after application into the fermenting must.

2.5.1. *Experimental design*

- Control sample (only 100 mg/L SO_2)
- Variant with 10 mg/L of MCFA - mixture C_8, C_10 and C_12 (2:7:1).

Each variant had a volume of 350 litres. Subsequently, the finished wine after the first racking (approximately 30 days after must inoculation) was measured with GC-MS (Gas Chromatography/Mass Spectrometry).

The concentration of individual volatile compounds in wine was determined according to the unpublished method of extraction with methyl-t-butylether. 20 mL of wine was pipetted into a 25-mL volumetric flask together with 50 μL of 2-nonanol solution in ethanol; this compound was used as an internal standard (in concentration of 400 mg/L) and 5 mL of a saturated $(NH_4)_2SO_4$ solution. The flask content was thoroughly stirred, and thereafter a volume of 0.75 mL of the extraction solvent (MTBE with added 1% of cyclohexane) was added. After another thorough stirring and separation of individual phases, the upper organic layer (supernatant) was placed into a micro test-tube together with the produced emulsion and centrifuged; the clear organic phase was dried up with anhydrous magnesium sulphate. Extract samples (adjusted in this way) were thereafter used for the GC-MS analysis.

DOI: 10.1051/bioconf/20170902025

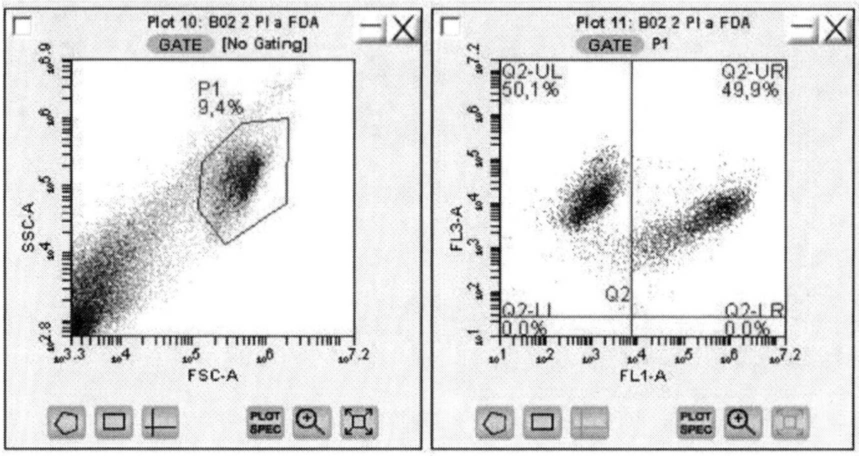

Figure 1. Dot-plot diagrams showing the separation. 1a – left: of yeast from bacteria and impurities. 1b – right: by the FDA and PI labelled yeas.

Table 3. Cell concentration in the experimental must samples.

	MCFA [mg/L]/ 28 hours before test	SO_2 [mg/L]/ 4 hours before test	The number of cells in 1 mL sample
Sample 1 (C)	0	0	8.3×10^6
Sample 2	10	0	9.4×10^6
Sample 3	10	60	8.0×10^6
Sample 4	0	60	11.9×10^6

Instruments: Shimadzu GC-17A, Autosampler: AOC – 5000, Detector: QP-5050A, Software: GCsolution. Separation conditions: column: DB-WAX 30 m × 0.25 mm; 0.25 μm stationary phase (polyethylene glycol). Voltage of the detector 1.5 kV. Individual compounds were identified on the basis of MS spectrum and retention time.

3. Results

3.1. Flow cytometry

Table 3 shows that the lowest cell (both, live or dead) concentration 8×10^6 is in sample 3 treated with 10 mg/L mixture of MCFA 28 hours before the measurement, and 60 mg/L of SO_2 4 hours before the measurement.

In Fig. 1a (below), the total amount of yeast separated by the flow cytometer from bacteria and impurities is marked in red. The value of 9.4% represents the area of yeast. The dot plot diagram in Fig. 1b shows the value of live (49.9%) and dead (50.1%) yeast.

The yeasts were separated from bacterial cells and debris by flow cytometry (see Fig. 1a). Then, based on FSC and SSC parameters, cells labelled FDA were evaluated as "live" while cells labelled PI were evaluated as "dead" (see Fig. 1b). The number of live and dead cells was monitored using flow cytometry.

Submitted samples contained a large number of bacteria and particles of a non-cellular nature, affecting cytometric analysis and the measured data. However, with FSC and SSC parameters, it was possible to locate the yeast population so it could be analysed separately. Bacteria from the impurities could not be sufficiently separated in the submitted samples. For samples 3 and 4, a significant change in the representation of "viable" cells was noted after transfer into the marking culture solution.

Figure 2 shows that the dose of MCFA has an inhibitory effect on *Saccharomyces cerevisiae* compared with the control. In combining SO_2, it has a higher efficiency than using a dose of SO_2 itself. The highest concentration of viable yeast content was in the control (variant C). By contrast, the lowest concentration was in variant 3, where MCFA and SO_2 were added.

3.2. Direct counting in a Bürker chamber

The counting was performed 24 and 168 hours after different doses of application of the MCFA mixture.

Figure 4 shows a higher concentration of dead cells with a higher concentration of MCFA. In the case of variant 8 (concentration 10 mg/L MCFA), the percentage of dead yeast was highest – about 60% (compared to variant 1 – about 24%). From Fig. 5, it is obvious that after more time (168 hours) the number of dead cells were higher than in the variants after 24 hours. The highest concentration of dead cells was measured in variant 10 (40 mg/L of MCFA mixture) – about 94%.

3.3. Determination of MCFA residues in treated wines

To determine MCFA residues after application into the fermenting must to stop alcoholic fermentation, the finished wine after the first racking was measured with GC-MS (Gas Chromatography/Mass Spectrometry).

The results of the GC-MS show that there is no significant quantitative difference in treated and untreated wines. The difference is between C_8 is 0.383 mg/L, C_10 0.226 mg/L and C_12 0.0019 mg/L in treated and untreated wines. These small differences are caused by absorption or assimilation by the yeast. Only a small part is esterified and remains as a residue in the wine.

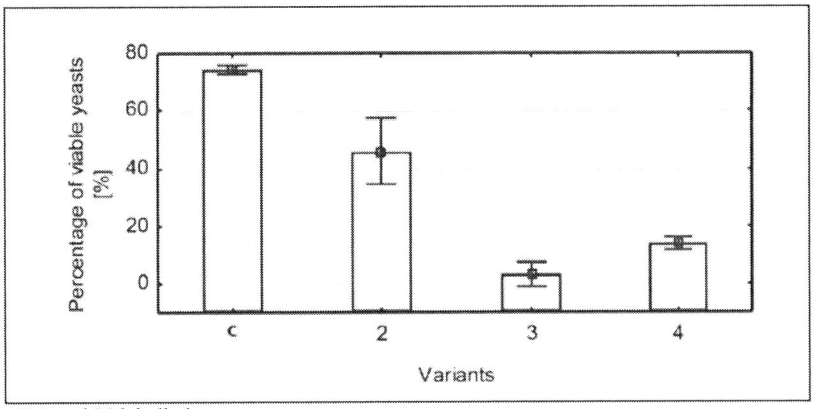

FDA and PI labelled yeast

Figure 2. Percentage of live yeast in the tested must.

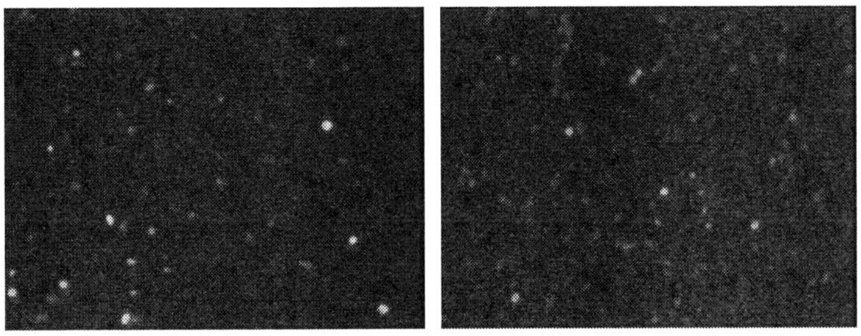

Figure 3. Microscopic view of the lifetime of protoplasts using fluorescein diacetate, propidium iodide and ultraviolet radiation. "Live" yeasts are marked green, and "dead" yeasts are marked red. 3 a) control sample, 3 b) variant 2.

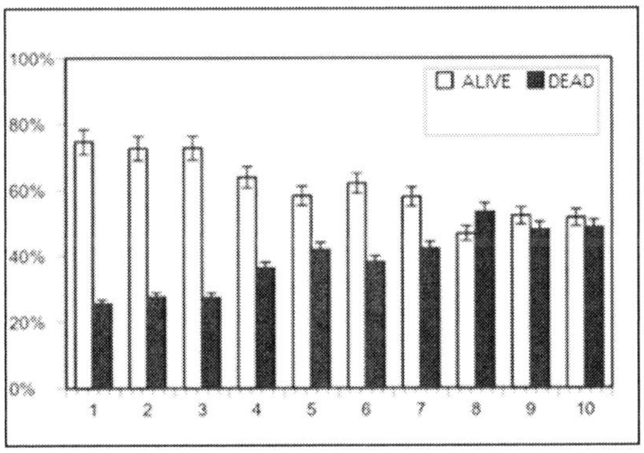

Figure 4. Percentages of live and dead yeast cells after 24 hours of addition of HFA mixture.

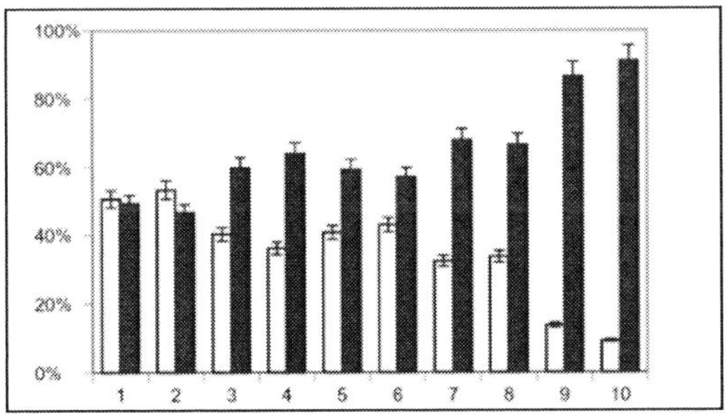

Figure 5. Percentages of live and dead yeast cells after 168 hours of addition of MCFA mixture.

4. Discussion

The results of flow cytometry (Chapter 4.1) show the highest inhibitory effect with the MCFA and SO_2 combination compared to the control variant and must treated with MCFA or SO_2 individually (Fig. 2). These results were achieved at a comparable concentration of yeast cells (dead or live) in a medium and confirm the results of Rib'erau-Gayon., et al., (2006) by providing precise quantification of live and dead cells in the defined medium (Figs. 1 and 3).

Even more accurate results in relation to the used concentration of MCFA mixture were obtained from direct cell counting. Here, 24 hours after application of a MCFA mixture (Fig. 4), there is a much higher concentration of dead cells than the control variant in the case of concentrations of 10, 20 and 40 mg/L MCFA mixture. These results confirm and quantify the results obtained by other authors (e.g., (4; 5; 6)).

Very interesting results have been obtained in the measurement of residues in treated and untreated wines (Sect. 4.3). The yeast basically detoxifies their bodies by esterification of the MCFA (7) which have penetrated into their cells, thus resulting in changes in the sensory characteristics of the wine.

According to our results (Table 4) the quantity of ethylesters MCFA is the same for untreated wine and for wine, to which was applied 10 mg/L mixture of MCFA. This measurement was done after the first racking (12; 13; 14). This shows that a sufficiently low concentration of MCFA has a sufficiently high inhibitory effect on

Table 4. Comparison of treated and untreated wine with MCFA.

T-tests; grouped: Group 1: with MCFA; Group 2: without MCFA											
MCFA	Group 1 [mg/L]	Group 2 [mg/L]	t	sv	p	n1	n2	SD 1	SD 2	F	P
C8	4.62	4.237	0.9051	38	**0.3703**	20	20	1.442	1.405	1.052	0.8911
C10	1.091	0.865	1.48	38	**0.0953**	20	20	0.539	0.36	2.246	0.0591
C12	0.0331	0.0350	0.2643	38	**0.7928**	20	20	0.029	0.020	2.061	0.0909
Ethylesters C8	0.498	0.505	0.132	38	**0.8952**	20	20	0.16	0.179	1.246	0.6291
Ethylesters C10	0.113	0.0953	1.16	38	**0.2523**	20	20	0.054	0.4678	1.331	0.4964

Saccharomyces cerevisiae during fermentation to achieve the stopping of fermentation. The part of the higher fatty acid is eliminated by fixation on the body of the dead yeast (13; 14).

5. Conclusions

Higher fatty acid is not currently used for inhibiting the activity of yeast during alcoholic fermentation. Previously published works demonstrate the properties of individual MCFAs, but do not offer results from wine production conditions and quantification of the effect of these substances on the yeast population after their use. The use of MCFA to stop fermentation is very promising technology. From the results obtained, it is evident that the application of MCFA to the fermentation media causes inhibition of the metabolic activity of the yeast, causing its death and stopping alcoholic fermentation (Figs. 2, 3, 4 and 5). This method can effectively reduce the cost of the production technology of wines with residual sugar and, in general, reduces the dosage of SO_2. This technology is also very simple for use in small wineries, where is no need for energy consuming cooling systems – the higher the temperature, the better the function. The great advantage of this method is its easy application, immediate action, and low cost. MCFA are harmless substances, naturally occurring in wine. Because of their very low concentration of use – less than 10 mg/L (Table 4), they have no effect on the final quality of the wine, primarily because the part of added MCFA is fixed on the yeast bodies and removed with each racking.

References

[1] Hui, Y. H., et al. Handbook of Food Products Manufacturing: Principles, Bakery, Beverages, Cereals,. New Jersey : y John Wiley & Sons, Inc., 2007. ISBN: 978-0-470-04964-8
[2] Fugelsang, K. C. and Edwards, Charles G. Wine microbiology. 2 New York : Springer, 2007. ISBN: 978-0-387-33349-6
[3] Mitchell, Richard and Hall, Colin Michael. Wine Marketing: A Practical Guide. Boston: Elsevier, 2008. ISBN: 9780750654203
[4] Callul, M., et al. HPLC analysis of fatty-acids in wine. American Journal of Enology and Viticulture. 1991, **3**, pp. 268–273
[5] Guilloux-Benatier, M., Le Fur, Y. and Feuillat, M. Influence of fatty acids on the growth of wine micoorganisms Saccharomyces cerevisiae and Oenococcus oeni. Journal of Industrial Microbiology & Biotechnology. 1998, **20**, pp. 144–149
[6] Viegas, C. A., et al. The H+-ATPase in the plasma membrane of Saccharomyes cerevisiae is activated during growth latency in octanoic acd - supplemented medium accomanying he decrease in intracellular pH and cell viability. Applied and Environmental Microbiology. 1998, **64**, pp. 779–783
[7] Bardi, L., Cocito, C. and Marzona, M. Saccharomyces cerevisiae cell faty acid composition and release during fermentation without aeration and in absence of exogenous lipids. International Journal of Food Microbiology. 1999, **47**, pp. 133–140
[8] Pina, C., et al. Ethanol tolerance of five non-Saccharomyces wine yeasts in comparison with a strain of Saccharomyces cerevisiae - influence of different culture conditions. Food Microbiology. 2004, **21**, pp. 439–447
[9] Sajbidor, J., Malik, F. and Kissantalova, H. The relationship between the viability of selected dried wine yeast and their lipid-composition and fatty-acid profile. Food Botechnology. 1992, **6**, pp. 187–196
[10] Torija, M. J., et al. Effects of fermentation temperature and Saccharomyces species on the cell fatty acid composition and presence of volatile compounds in wine. International Journal of Food Microbiology. 2003, **85**, pp. 127–136
[11] Rib'erau-Gayon., P., et al. *Handbook of Enology Volume 1 The Microbiology of Wine and Vinifications.* **2.** s.l.: John Wiley & Sons, 2006. ISBN 0-470-01034-7
[12] Viegas, C. A., Sacorreia, I. and Novais, J. M. Synergistic inhibition of the growth of Saccharomyces

 DOI: 10.1051/bioconf/20170902025

bayanus by ethanol and octanoc or decanoic acids. Biotechnology Letters. 1985, **7**, pp. 611–614

[13] Bábiková, P., et al. *Increasing the efficiency of sulphur dioxide in wine by using of saturated higher fatty acids.* 1. s.l.: Acta Universitatis Agriculturae et Silviculturae Mendelianae Brunensis, 2012. pp. 17–22, **60**. ISSN 1211-8516

[14] Baroň, Mojmír. *Effectiveness of saturated higher fatty acids for inhibition of yeasts and malolactic activities in wine.* 1. s.l.: Acta Universitatis Agriculturae et Silviculturae Mendelianae Brunensis, 2014. pp. 23–26, **62**. ISSN 1211-8516

[15] Correia, I. S., et al. Leakage induced by ethanol, octanoic and decanoic acids in Saccharomyces cerevisiae. Yeast. 1989, **5**, pp. 123–127

[16] Lafon-Lafourcade, S., Geneix, C. and Riberau-Gayon, P. Inhibition of alcoholic fermentation of grape must by fatty-acids produced by yeast and their elimination by yeast ghosts. Applied and Enviromental Microbiology. 1984, **61**(2), pp. 1246–1249

[17] Viegas , C. A., et al. Inhibition of yeast growth by octanoic and decanoic acids produced during ethanolic fermentation. Applied and Environmental Microbiology. 1989, **55**(1), pp. 21–28

[18] Viegas , C. A. and Sacorreia, I. Activation of plasma-membrane ATOase of Saccharomyces cerevisiae by octanoic-acid. Journl of General Microbiology. 1991, **137**, pp. 645–651

[19] Viegas, C. A., et al. Regulation of the expression of the H+-ATPase genes PMA1 and PMA2 during growth and effect of octanoic-acid in Saccharomyces cerevisiae. *Biochemica Et Biophysica Acta-Gene Structure and Expression*. 1994, **1217**, 1, pp. 74–80

[20] Comitini, F. and Ciani, M. The inhibitory activity of wine yeast starters on malolactic bacteria. Annals of Microbiology. 2007, **57**(1), pp. 61–66

[21] Geneix, C., Lafon-Lafourcade, S. and Ribereau-Gayon, P. The effect of fatty acids on the viability of Saccharomyces cerevisiae. *Comptes Rendus De L'Academie Des Sciences*, Série Iii-Sciences De La Vie-Life Sciences. 1983, **19**, pp. 943–947

DOI: 10.1051/bioconf/20170902026

La misura del colore come discriminante nel controllo qualità di vini rossi bulgari

Colour measurement as a discriminating factor in quality control of Bulgarian red wines

Dora Marchi[1], Donato Lanati[1], Desislava Baicheva[2], Alberto Larosa[2], Martina Barisone[1], Chiara Robutti[1], e Daniela Borsa[1]

[1] ENOSIS s.r.l., Strada per Cuccaro, 19, 15043 Fubine, AL, Italia
[2] Edoardo Miroglio Wine Cellar, Nova Zagora Elenovo 8943, Bulgaria

Abstract. Colour is one of the main organoleptic characteristics of wine having a significant impact on consumers' choices, as such it is considered an important factor in assessing the quality of red wines. Wine colour depends on complex extraction and polymerization reactions'; therefore it could evolve over time. This study is based on few samples of Pinot Noir wines, that achieve both good quality levels and a strong market performance, Mavrud, an ancient variety from Thrace, and Rubin, a Syrah and Nebbiolo crossbreed. The selected wines differ for their anthocyanic profile and polyphenolic content. Furthermore, the quality of the selected national varieties is comparable to the most famous international ones. A comparison was made based on chemical and organoleptic characteristics as well as their performance on the market. The chromatic indices, intensity and hue, the anthocyanic profile and their evolution in time, were determined via the polymerization indices. To allow for an objective assessment, the colour of wines was defined using CIELab colour space, which permits an objective identification of any colour perceivable by the human eye. Finally, analytical results were compared to the organoleptic characteristic of the selected wines.

1. Introduzione

Il mercato enologico bulgaro è ancora relativamente poco conosciuto ma in costante sviluppo ed è uno dei più competitivi settori dell'industria alimentare della Bulgaria. A partire dagli anni '90 la Bulgaria ha avviato un processo di privatizzazione dell'industria enologica che si può dire concluso nel 1999. Molta parte della produzione è esportata nei paesi europei (Gran Bretagna, Germania, Ucraina, Polonia e Olanda) ma anche in Giappone e negli USA. La Gran Bretagna, in particolare, importa vini a basso costo da destinare alla grande distribuzione. Da alcuni anni la produzione enologica è stata indirizzata a vini di qualità con un incremento del profitto per gli operatori [1]. Il consumo interno ha invece subito un declino passando da 16.4 L. pro capite nel 1989 a 6.5 L nel 1999 [2].

La Bulgaria, dal punto di vista enologico, è divisa in 5 regioni: la piana del Danubio a nord, la regione del Mar Nero a est, la regione sub-Balcanica, la Struma Valley a sud ovest e la Tracia a sud. Quest'ultima è la regione con la migliore produzione viticola favorita da buone condizioni climatiche per la protezione offerta dai Monti Balcani.

In Bulgaria sono coltivate diverse varietà di uve con una prevalenza delle varietà a bacca colorata (52%) seguite da varietà bianche (32%) e uve da tavola (9%) [2]. La produzione enologica comprende vini autoctoni ed altri ottenuti da vitigni internazionali. Per quanto riguarda i vini rossi, oggetto di questo studio, possiamo ricordare tra i vitigni internazionali: Cabernet sauvignon, Merlot e Pinot Noir, mentre gli autoctoni più conosciuti sono il Mavrud, Dimyat, Gamza e Rubin. Sebbene la qualità dei vini sia in costante crescita, solo un quinto dei vini rossi prodotti possono essere considerati di qualità anche se la produzione migliore ha un positivo e significativo impatto sul prezzo di mercato. In particolare è stato rilevato che i vini bianchi giovani e i rossi da invecchiamento incontrano maggiormente il favore dei consumatori e hanno prezzi più elevati [2].

La relazione tra qualità di un prodotto e prezzo di mercato è stata teorizzata da lungo tempo applicando il "modello edonico" [3]. Rosen, in particolare, ha messo in evidenza che i fattori che influenzano la definizione del prezzo di un prodotto dipendono dall'importanza che il consumatore dà ad ognuno di essi. Nel caso del vino le variabili di interesse possono essere suddivise in 2 gruppi: i fattori intrinseci (varietà dell'uva, colore, origine, qualità tecnologica, e caratteristiche organolettiche) e fattori estrinseci che sono essenzialmente legati alla reputazione dell'azienda e del prodotto. Nella reputazione collettiva riveste molta importanza la regione di origine [4].

Il primo parametro sensoriale che attrae l'attenzione del consumatore è il colore del vino [5] che viene quindi fortemente correlato alla qualità. Esso è determinato da molteplici fattori: la composizione fenolica delle uve di origine (profilo e contenuto antocianico), i composti polimeri formati durante la vinificazione, subendo inoltre una continua evoluzione durante l'affinamento e l'invecchiamento. I cambiamenti sono dovuti a complessi processi di polimerizzazione in cui sono coinvolti gli antociani, l'acetaldeide, i tannini ed altri composti fenolici [6]. Dopo un anno di conservazione gli antociani monomeri iniziali sono in genere dimezzati [7]. I nuovi pigmenti sono molto più stabili [8]. La componente gialla del colore dei vini invecchiati aumenta all'incremento del grado di polimerizzazione e di ossidazione, inoltre l'intensità del colore e la tonalità dipendono dal pH e dalla presenza di SO_2.

La valutazione del colore effettuata mediante analisi organolettica dipende molto dalla competenza e dalla sensibilità del degustatore, quindi per avere una migliore e più oggettiva descrizione è necessario avvalersi di parametri analitici che descrivano il colore sotto i diversi aspetti elencati. Lo scopo del presente lavoro è quello di descrivere il colore di vini rossi prodotti in Bulgaria e commercializzati da diverse aziende utilizzando analisi spettrofotometriche e HPLC.

2. Materiali e Metodi

2.1. Vini analizzati

Per il presente studio sono stati scelti dei vini rossi del commercio di aziende e annate diverse come riportato in Tabella 1. In particolare sono stati scelti 12 campioni di Pinot Noir, vitigno internazionale che negli ultimi anni ha avuto un buon successo commerciale; 8 campioni di Rubin e 10 campioni di Mavrud. Alcuni di essi, indicati in grassetto in tabella, sono stati segnalati tra i 50 migliori prodotti bulgari da DiVino Top 50 nel 2016.

La varietà Rubin è un incrocio di Nebbiolo e Syrah, approvato come nuova varietà dalla Commissione di Stato per le varietà nel 1961. I suoi grappoli sono caratterizzati da un intenso colore rosso.

La varietà Mavrud è coltivata in Bulgaria fin dai tempi antichi. Le sue caratteristiche possono variare molto in funzione del territorio di coltivazione e spesso le uve sono sottoposte ad un periodo di appassimento prima della vinificazione. I vini ottenuti presentano un colore rosso, di intensità variabile, talora molto intenso. Hanno una forte componente tannica per questo richiedono un periodo di invecchiamento, in acciaio o in legno.

2.2. Metodi

Sui vini sono stati determinati i principali parametri analitici mediante analizzatore Foss. Nelle tabelle sottoriportate sono indicati il contenuto di alcol, l'acidità titolabile e il pH.

Il contenuto di antociani totali e flavonoidi totali sono stati determinati mediante analisi spettrofotometrica del vino diluito in etanolo cloridrico [9].

Il colore dei vini è stato valutato attraverso gli indici di colore secondo Glories [10], intensità (IC) e tonalità colorante (TC) mediante determinazione delle assorbanze a 420 nm (giallo), 520 nm (rosso) e 620 nm (blu).

Per rendere le misure del colore dei parametri oggettivi si è scelto di determinare le caratteristiche cromatiche dei vini secondo CIElab. Tale metodica, prevista dalla risoluzione OIV ENO 1/2006, utilizza le componenti tricromatiche X, Y e Z secondo la Commission Internationale de l'Eclairage (CIE, 1976) in

Tabella 1. Elenco dei vini analizzati, varietà e annata e codice Azienda.

Codice	Varietà	Anno	Azienda
P1	Pinot Nero	2012	A
P2	Pinot Nero	2013	A
P3	Pinot Nero	2013	B
P4	Pinot Nero	2013	C
P5	Pinot Nero	2014	D
P6	Pinot Nero	2014	E
P7	Pinot Nero	2014	F
P8	Pinot Nero	2014	G
P9	Pinot Nero	2014	H
P10	Pinot Nero	2015	H
P11	Pinot Nero	2015	I
P12	Pinot Nero	2015	A
R1	**Rubin**	2012	A
R2	Rubin	2012	L
R3	**Rubin**	2013	M
R4	**Rubin**	2014	C
R5	Rubin	2014	N
R6	Rubin	2015	O
M1	Mavrud	2011	P
M2	Mavrud	2012	A
M3	Mavrud	2012	Q
M4	Mavrud	2013	R
M5	Mavrud	2013	S
M6	**Mavrud**	2013	T
M7	Mavrud	2013	U
M8	**Mavrud**	2013	V
M9	Mavrud	2013	M
M10	Mavrud	2015	A

seguito implementato con il sistema CIElab. Il colore di ogni vino viene pertanto definito dai parametri: chiarezza (L*), la componente di colore rosso/verde (a*), la componente di colore giallo/blu (b*) e tramite le grandezze da esse derivate: il Croma (C*) e il tono (h*). Il colore dei vini analizzati è stato rappresentato attraverso i parametri a*/b* posti in un diagramma cartesiano.

I parametri CIElab sono stati ottenuti con le misure spettrofotometriche dell'assorbanza dei vini realizzate contro acqua su un percorso ottico di 2 mm alle λ 450, 520, 570 e 620. Le misure sono state trasformate in trasmittanza e utilizzate per il calcolo secondo il metodo matematico proposto da Peréz-Caballero [11].

Poiché è noto che le reazioni di co-pigmentazione e polimerizzazione portano a composti fenolici non più sensibili alla solforosa del vino si è determinato il dTAT e il dTAT % (frazione di E_{520} dei pigmenti non sensibili alla SO_2) sia al pH del vino che a pH=0 [12].

Per una più completa descrizione dei vini sono stati determinati inoltre i profili antocianinici via HPLC [12].

Tabella 2. Alcol, acidità e polifenoli dei vini Pinot Noir.

Codice	Alcol % v/v	A.T. (g/L)	pH	Ant-tot. (mg/L)	Flav. Tot. (mg/L)
P1	13,93	5	3,59	63	1356
P2	13,68	4,8	3,63	87	1047
P3	13,4	5	3,55	78	1219
P4	14,62	5,8	3,48	86	1924
P5	12,42	5,8	3,64	70	1008
P6	13,63	4,8	3,78	153	1968
P7	13,6	6	3,61	113	1817
P8	13,46	5,2	3,96	125	1620
P9	13,12	5,2	3,67	77	981
P10	13,47	4,6	3,75	87	1411
P11	13,9	5,2	3,58	110	1921
P12	13,66	4,8	3,68	95	1699

Tabella 3. Parametri Cielab, dei vini Pinot Noir.

Codice	L*	a*	b*	h* (Gradi)	C*
P1	78,75	18,25	16,67	42,41	24,71
P2	83,90	13,70	12,78	42,99	18,74
P3	82,68	16,04	11,60	35,87	19,79
P4	66,84	31,34	18,73	30,87	36,51
P5	77,74	19,06	14,65	37,53	24,04
P6	58,24	35,99	20,75	29,97	41,54
P7	69,44	25,86	13,90	28,27	29,36
P8	75,23	20,04	15,02	36,85	25,04
P9	82,67	16,43	8,35	26,95	18,44
P10	84,70	13,51	8,74	32,90	16,09
P11	91,58	5,88	21,23	74,51	22,03
P12	83,46	11,83	7,93	33,81	14,24

Tabella 4. Intensità e tonalità e % di dTAT dei vini Pinot Noir.

Codice	dTAT% pH vino	dTAT % pH=0	I.C.	T.C.
P1	65,01	26,89	3,93	1,14
P2	67,23	13,05	3,16	1,17
P3	63,57	22,15	3,20	1,06
P4	62,72	35,75	6,91	0,89
P5	81,77	21,07	4,19	1,06
P6	61,05	31,16	8,77	0,94
P7	64,10	27,83	5,59	0,98
P8	66,19	18,75	4,61	1,10
P9	55,07	18,53	2,94	0,96
P10	52,17	12,78	2,79	1,04
P11	59,20	7,62	2,88	1,08
P12	69,77	29,87	2,52	1,22

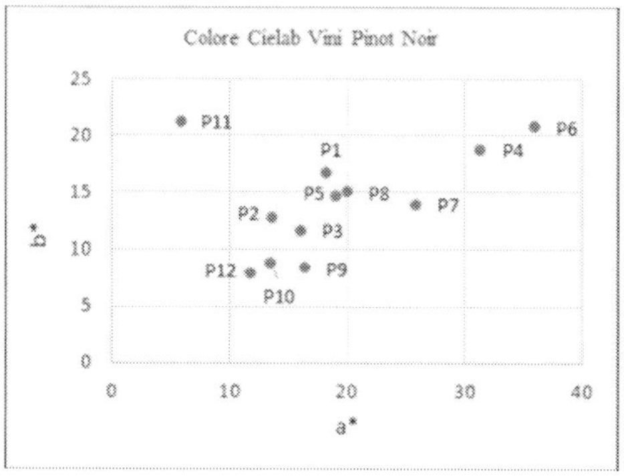

Figura 1. Grafico Colore Cielab Vini Pinot Noir.

3. Risultati

3.1. Vini Pinot Noir

I vini analizzati sono piuttosto eterogenei come si poteva prevedere considerato che sono stati prodotti da aziende diverse e differenti sono anche le annate di produzione. I risultati relativi alle analisi del contenuto alcolico, acidità e pH e del tenore di antociani e flavonoidi dei vini Pinot Noir sono riportati nella Tabella 2. Tutti i campioni presentano un elevato tenore alcolico, più contenuto per il campione P5, mentre il campione P4 raggiunge i 14,62 % v/v.

Gli antociani espressi come malvidina -3-glucoside variano da 63 mg/L per P1 a 153 mg/L per P6.

In Tabella 3 sono riportati i dati relativi alla determinazione del colore dei vini utilizzando il metodo CIElab, mentre in Tabella 4 sono riportati i valori del dTAT% sia a pH vino che a pH0 e gli indici di Sudraud.

Il colore è in genere poco intenso, quasi tutti i vini, infatti, hanno un valore di L* superiore ad 80 o di poco inferiore come si osserva per i vini P1, P5 e P8. In P4, P6 e P7 è minore di 70 pur mantenendosi su valori piuttosto alti. Alle stesse conclusioni si giunge anche osservando i valori C*.

I valori di L* e C* inoltre sembrano in buon accordo con gli indici di colore di Sudraud. Infatti i vini P6 con IC di 8.77, P4 con 6.91 e P7 con 5.59 sono quelli con C* più elevata, quindi colore più intenso.

In Fig. 2 sono stati riportati i valori delle componenti del colore a* e b* posti su diagramma cartesiano.

La componente rossa del colore (a*) ha raggiunto un valore apprezzabile (35,99) solo in P6, in P4 (31,34) e in parte anche in P7 (25,86). Di più difficile valutazione è la componente gialla del colore (b*) i cui valori sono in relazione all'intensità del colore e vanno, comunque confrontati con a*.

Si prestano meglio alla valutazione della componente gialla i valori di h* che non dipendono dall'intensità del colore ma da a* e b*. I valori di h* indicano chiaramente che il colore dei vini possedeva una sensibile tonalità gialla. Essi, pertanto, presentavano più (P1, P2, P3, P5, P8, P10, P11 e P12) o meno (P4, P6, P7 e P9) tonalità aranciate. Anche i valori di h* sembrano ben correlati con l'indice tonalità di Sudraud (T.C.). Infatti i valori più bassi di questo rapporto sono stati registrati per P4, P6, P7 e P9.

È evidente che i valori degli indici CIE L*a*b*, C*, h* solo in parte sono conseguenza del tenore in antociani totali dei vini. Su questi indici, infatti pesano sensibilmente anche i valori del pH e i contenuti in SO_2. Come esempio vale il caso del vino P4 che possiede uno dei più alti valori di C* pur con un contenuto in antociani totali di 86 mg/L, mentre i bassi valori di C* dei vini P8 e P11 vanno imputati agli elevati tenori in SO_2 e nel caso del P8 anche all'alto pH.

I valori di dTAT al pH del vino indichino che il suo colore sia da attribuire al contributo di pigmenti non sensibili alla SO_2, i valori di dTAT a pH = 0 indicano invece chiaramente che la maggior parte dei pigmenti dei vini è ancora sensibile alla SO_2 e alle variazioni di pH. Si tratta pertanto di pigmenti (antociani monomeri e pigmenti polimeri decolorabili da parte della SO_2) che subiranno nel tempo una evoluzione, probabilmente, con accentuazione della tonalità aranciata dei vini in quanto

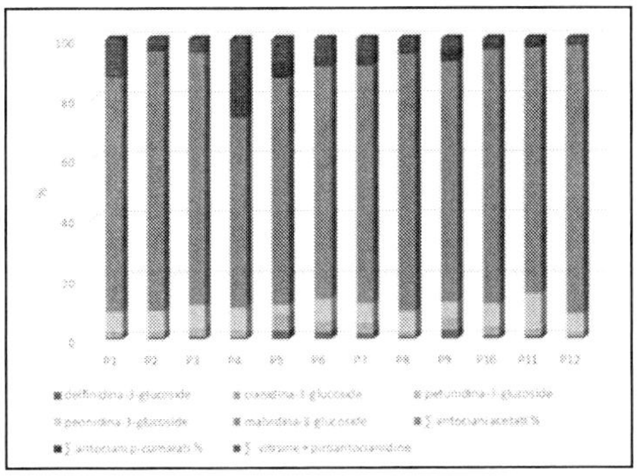

Figura 2. Profilo antocianico Vini Pinot Noir.

Tabella 5. Alcol, acidità e polifenoli dei vini Rubin.

Codice	Alcol % v/v	A.T. (g/L)	pH	Ant-tot. (mg/L)	Flav. Tot. (mg/L)
R1	14,94	6,1	3,58	185	3160
R2	15,72	5,6	3,84	146	3155
R3	15,46	5,3	3,96	159	3248
R4	13,98	4,7	4,00	182	2073
R5	13,88	5,2	3,57	221	2147
R6	14,79	4,8	3,70	281	2212

le reazioni di polimerizzazione saranno condizionate dalla forte prevalenza dei tannini presenti rispetto agli antociani monomeri e polimeri.

Solo il campione P4 presenta valori di dTAT che fanno pensare ad una evoluzione positiva del colore.

In Fig. 2 sono riportati i profili antocianici dei vini Pinot Noir oggetto dello studio. In tutti i campioni si osserva una prevalenza netta di Malvidina-3-glucoside che è l'antociano prevalente già nelle uve. La % di peonidina-3-glucoside pur rimanendo importante è inferiore a quella delle uve a causa della sua facilità di ossidazione. In alcuni campioni (P5, P6 e P9) si osserva una % di antociani acilati che fa ipotizzare che il vino non sia esclusivamente ottenuto da uve Pinot Noir.

3.2. Vini Rubin

Sono stati analizzati 6 campioni di vino Rubin di provenienza e annate diverse: 2 R1 e R2 del 2012, 1 R3 del 2013, 2 R4 e R5 del 2014 e R6 del 2015. Anche i vini Rubin presentano composizione di base poco omogenea. Hanno grado alcolico elevato che varia da 13,88 (R5) a 15,72 (R2), differiscono anche per l'acidità totale che varia da 4,7 g/L di ac. tartarico (R4) a 6,1 (R1) e nel pH da 3,57 a 4,00. Il contenuto in flavonoidi varia da 2073 (R4) a 3248 mg/L di (+)-catechina (R3), mentre quello degli antociani totali come malvidina-3-glucoside da 146 (R2) a 281 mg/L (R6) (Tabella 5).

Il loro colore risulta piuttosto intenso, meno per i vini R4 e R6. Infatti L* varia intorno a 50 in tutti i campioni escluso R4 (65,90) e R6 (58,75). Alle stesse conclusioni si perviene dall'esame di C* (50 in tutti i campioni, 30,50 in R4 e 39,28 in R6), (Tabella 6). I dati di L* e C* sembrano in buon accordo con quelli dell'intensità colorante di Sudraud (6,35 in R4, 8,43 in R6 e da 11,65 a 13,11 negli altri), (Tabella 7). La componente rossa del colore (a*) ha raggiunto valori elevati, variabili da 37,27 a

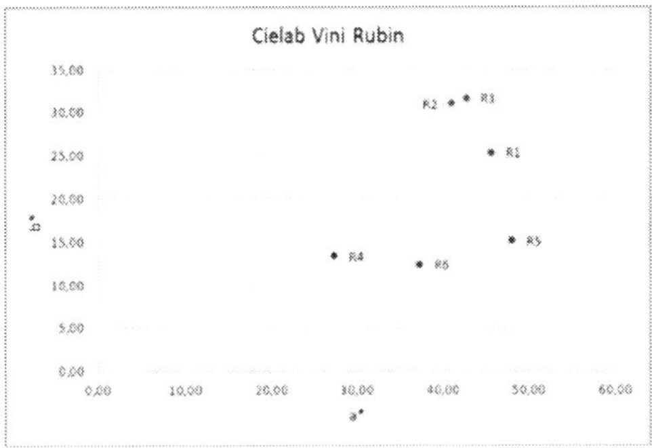

Figura 3. Grafico Colore Cielab Vini Rubin.

Tabella 6. Parametri Cielab, dei vini Rubin.

Codice	L*	a*	b*	h* (gradi)	C*
R1	48,03	45,74	25,33	28,97	52,28
R2	51,45	40,99	31,00	37,10	51,39
R3	50,06	42,86	31,63	36,43	53,26
R4	65,90	27,41	13,36	25,99	30,50
R5	49,68	48,14	15,16	17,48	50,47
R6	58,75	37,27	12,42	18,43	39,28

Tabella 7. Intensità e tonalità e % di dTAT dei vini Rubin.

Codice	dTAT% pH vino	dTAT% pH = 0	I.C.	T.C.
R1	64,17	53,01	12,71	0,83
R2	60,12	52,32	12,22	1,00
R3	61,01	53,95	13,11	0,93
R4	38,44	20,78	6,35	0,93
R5	n.d.	n.d.	11,66	0,71
R6	35,93	16,89	8,43	0,80

48,14 in tutti i campioni escluso R4 (27,41). D'altra parte i bassi valori della componente gialla del colore (b*) fanno pensare che i vini fossero dotati di un colore rosso rubino (Fig. 3).

I valori di h* indicano, più chiaramente di a* e b*, che il colore dei vini R1, R4 e soprattutto R2 e R3 fosse rosso rubino più evoluto di R5 e R6.

I valori delle tonalità di Sudraud (T.C.) da una parte confermano quanto dedotto dall'esame di h*, riguardo all'evoluzione delle tonalità rubino, dall'altra indicano la presenza di tonalità brune nei campioni dal R1 all' R4. I valori più bassi di questo rapporto sono stati registrati nei campioni R5 e R6 nei quali la componente rosso-porpora sembrerebbe dominante. Anche per questi vini è risultato evidente che i valori degli indici CIE L*a*b*, C*, h* solo in parte sono conseguenza del tenore in antociani totali. Su questi indici, infatti pesano sensibilmente anche i valori del pH e i contenuti in SO_2.

I valori di dTAT al pH del vino e a pH = 0 (Tabella 7) indicano che i pigmenti dei campioni R1, R2, R3 non sono più decolorabili da parte della SO_2 e che il loro colore non è più sensibile alle varizioni di pH. I campioni R4 e R6 non presentano ancora simili condizioni di stabilità: è prevedibile che il loro colore possa subire una evoluzione nel tempo.

In Fig. 4 sono riportati i profili antocianici dei vini Rubin. Anche in questo caso si osserva una certa eterogeneità di composizione, la malvidina-3 glucoside

DOI: 10.1051/bioconf/20170902026

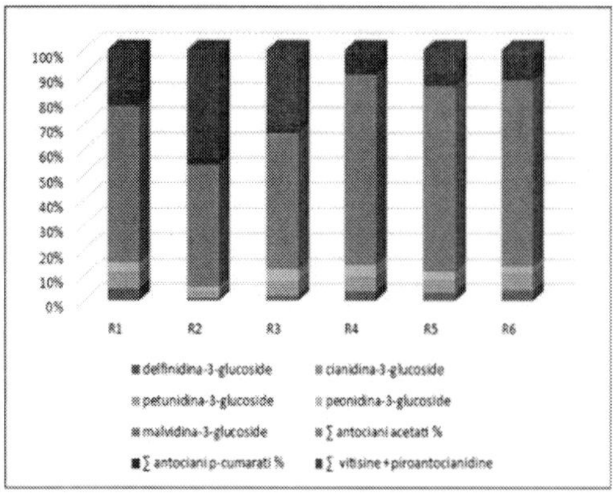

Figura 4. Profilo antocianico Vini Rubin.

Tabella 8. Alcol, acidità e polifenoli dei vini Rubin.

Codice	Alcol % v/v	A.T. (g/L)	pH	Ant-tot. (mg/L)	Flav. Tot. (mg/L)
M1	15,12	5,5	3,58	270	2718
M2	14,83	6,7	3,42	234	2212
M3	12,24	5,6	3,56	102	1454
M4	13,14	5,8	3,58	88	1033
M5	15,14	5,4	3,65	318	2155
M6	15,83	6,0	3,43	186	1191
M7	13,93	4,8	3,75	215	2076
M8	15,32	6,5	3,64	199	3390
M9	14,66	5,0	3,78	119	1906
M10	12,88	5,8	3,50	195	2580

resta in tutti i casi l'antociano prevalente come avviene per le uve, anche se i campioni R2 e R3 hanno una % più bassa rispetto agli altri 4 campioni. La % di petunidina-3-glucoside mantiene una % importante se confrontata con altri vitigni, ma in linea col profilo delle uve.

3.3. Vini Mavrud

Si tratta di vini eterogenei per annata, 1 campione del 2011 (M1), 2 del 2012 (M2 e M3), 6 campioni del 2013 (M4-M9) e 1 del 2015 (M10). Il grado alcolico varia da 12,24 (M3) a 15,14 (M5); l'acidità totale da 4,8 g/L di ac. tartarico (M7) a 6,7 (M2). I pH variano da 3,42 (M2) a 3,78 (M9). Per quanto riguarda la componente polifenolica il contenuto in flavonoidi è molto variabile da 1033 mg/L di (+)-catechina (M4) a 3390 (M8); mentre gli antociani totali calcolati come malvidina-3-glucoside vanno da 88 mg/L (M4) a 318 (M5) (Tabella 8).

Come si può osservare dai dati riportati in Tabella 9 il colore di questi vini risulta piuttosto intenso, meno per i campioni M3, M4 e M10. Infatti L* è minore di 50 o oscilla intorno a questo valore in tutti i vini, a parte i tre sopra citati in cui varia da circa 65 (M3 e M10) a 78,7 (M4). Alle stesse conclusioni si perviene dall'esame di C* che oscilla intorno a 50 eccetto per i campioni M3, M4 e M10.

I dati di L* e C* sembrano in buon accordo con quelli dell'intensità colorante di Sudraud: 7,24 in M3, 4,25 in M4 e 7,50 in M10; negli altri da 10,33 in M9 a 19,20 in M1. La componente rossa del colore (a*) prevale nettamente sulla componente gialla (b*) in tutti i campioni (valori di a* circa doppi rispetto a b*). Si tratta di vini dal colore rosso

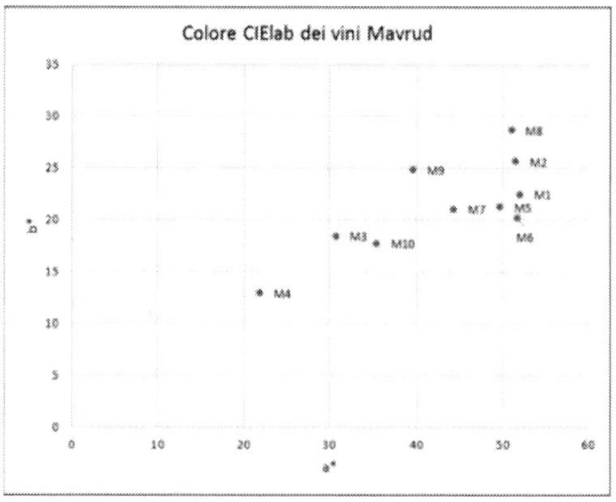

Figura 5. Grafico Colore Cielab Vini Mavrud.

Tabella 9. Parametri Cielab, dei vini Mavrud.

Codice	L*	a*	b*	h*(Gradi)	C*
M1	32,98	52,01	22,45	23,34	56,65
M2	43,11	51,51	25,61	26,44	57,53
M3	65,25	30,77	18,44	30,93	35,88
M4	78,57	21,90	12,92	30,53	25,42
M5	35,17	49,70	21,24	23,14	54,05
M6	43,01	51,65	20,16	21,32	55,45
M7	52,58	44,38	21,01	25,34	49,10
M8	42,58	51,03	28,64	29,30	58,52
M9	54,65	39,66	24,81	32,03	46,78
M10	65,32	35,45	17,70	26,54	39,62

Tabella 10. Intensità e tonalità e % di dTAT dei vini Mavrud.

Codice	dTAT% pH vino	dTAT% pH=0	I.C.	T.C.
M1	61,61	54,01	19,20	0,76
M2	62,91	47,31	13,63	0,77
M3	59,18	50,21	7,24	0,94
M4	62,83	32,09	4,25	0,96
M5	50,38	39,52	17,83	0,73
M6	n.d.	n.d.	14,38	0,71
M7	49,94	38,63	10,88	0,79
M8	60,00	55,63	15,46	0,81
M9	n.d.	n.d.	10,33	0,91
M10	n.d.	n.d.	7,50	0,85

rubino, nella maggior parte dei casi intenso. Anche i valori di h* indicano chiaramente che nel colore dei vini prevale la componente rossa (l'angolo che il vettore C* forma con l'asse a* è, infatti, sensibilmente minore di 45°). Anche i valori di h* sembrano ben correlati con l'indice tonalità di Sudraud (T.C.). Infatti valori che superano 0,9 (inizio delle tonalità brune) sono stati registrati solo nei campioni M3 e M4. In M1, M2, M5, M6, M7 e M8 le tonalità brune sono risultate poco rappresentate. Anche per questi vini si rileva una dipendenza del colore dal tenore in SO_2 e dal pH oltre che dal tenore in antociani totali.

I valori di dTAT al pH del vino e a pH = 0, escluso il campione M4 sono risultati piuttosto simili. Essi mostrano che i pigmenti sono sufficientemente evoluti (non sono decolorati dalla SO_2 e il loro colore varia poco al variare del pH). Considerato che nella maggior parte dei casi il colore è rosso rubino con scarse tonalità brune, si tratta del conseguimento di una buona stabilità dei pigmenti.

DOI: 10.1051/bioconf/20170902026

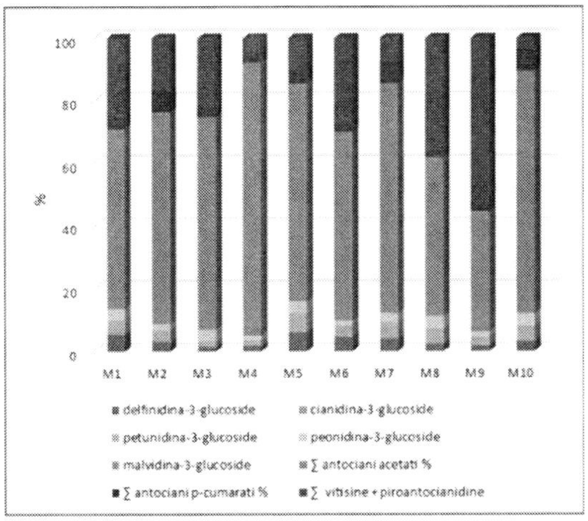

Figura 6. Profilo antocianico Vini Mavrud.

Il profilo antocianico percentuale dei vini Mavrud è riportato nella Fig. 6. In tutti i campioni prevale la malvidina-3-glucoside anche se la sua percentuale è molto diversa nei campioni M8 e M9 rispetto agli altri. Sono diverse tra i campioni anche le % degli antociani acetati e *p*-cumarati. Anche questo parametro conferma l'eterogeneità dei campioni.

4. Conclusioni

Risulta evidente dalle considerazioni sopra riportate per i diversi vini considerati, che la qualità degli stessi, le caratteristiche del loro colore e la valutazione dei fattori che lo influenzano richiede l'utilizzo di metodiche analitiche. Esse infatti possono descrivere il colore ma anche aiutare gli operatori del settore a comprendere quali fenomeni siano avvenuti ed anche quale evoluzione possa avere il vino. La descrizione del colore fatta con le analisi organolettiche su questi stessi vini (dati non riportati) ha permesso di classificarne il colore come rosso granato in alcuni casi o rosso rubino in altri, ed anche di valutarne l'intensità attraverso l'uso di scale astrutturate, ma non di comprenderne l'origine o l'evoluzione. L'uso dei parametri di colore della metodica CIElab può inoltre essere utilizzato per determinare le differenze tra il colore dei vini con precisi parametri matematici o studiarne l'evoluzione nel tempo. La valutazione della qualità dei vini si avvarrà in questo modo di parametri oggettivi.

Riferimenti

[1] B. Geogriev, I.M.E., *Institute of Market Economics* (2007)
[2] N. Noev, `www.vdqs.fed-eco.org`. *Oenometrics IX conference* (2002)
[3] F. Waugh, Journal of Farm economics **10**(2), 185–196 (1928)
[4] L.A. Panzone, British Food Journal **113**, 1060–1078 (2011)
[5] S. Tsanova-Savova, S. Dimov and F. Ribarova, J. Food Comp. And Anal. **15**, 647–654 (2002) l'uso di scale astrutturate C. F. Timbelake and P. Bridle, Am. J. Enol. Vitic. **27**, 105 (1976)
[6] T.C. Somers, Phytochemistry **10**, 2175–2186 (1971)
[7] N. Mateus and V. De Freitas J. Agric. Food Chem. **49**, 5217–5222 (2001)
[8] R. Di Stefano e M.C. Cravero, L'Enotecnico **25**, 81–89 (1989)
[9] Y. Glories, Conn. Vigne Vin. **18**, 253–271
[10] V. Peréz-Caballero, F. Ayala, J.F. Echávarri e A.I. Neguerela, Am.J.Enol. Vitic. **54**(1), 59–62 (2003)
[11] R. Di Stefano, M.C. Cravero, N. Gentilini, L'Enotecnico **25**(5), 83–89 (1989)

DOI: 10.1051/bioconf/20170902027

Timing of leaf removal modifies chemical and phenolic composition of Sauvignon Blanc wine

Douglas André Wurz[1,a], Betina Pereira de Bem[1], Ricardo Allebrandt[1], José Luiz Marcon Filho[1], Alberto Fontanella Brighenti[2], Marcus Outemane[1], Leo Rufato[1], and Aike Anneliese Kretzschmar[1]

[1] Santa Catarina State University (UDESC), College of Agriculture and Life Science, Av. Luiz de Camões, 2090, CEP 88520-000, Lages, SC, Brazil
[2] Santa Catarina State Agricultural Research and Rural Extension Agency (EPAGRI), João Araújo Lima Street, 102, CEP 88520-000, São Joaquim, SC, Brazil

Abstract. The objective of this work was to evaluate the effect of different defoliation times on the chemical and phenolic composition of Sauvignon Blanc wine in high altitude regions of Santa Catarina - Brazil. The experiment was conducted in the 2015 and 2016 seasons, in a vineyard located in the city of São Joaquim – Santa Catarina State. The defoliation was carried out during phenological stages: full bloom, berries pepper-corn size, pea-sized berries, veraison, 15 days after veraison and control without leaf removal. During harvest, severity of *Botrytis cinerea* was assessed by visual evaluation through a diagrammatic scale. For the wine samples from each season, total acidity (meq L^{-1}), pH, total polyphenol content (mg L^{-1} of gallic acid) and color (Abs 420nm) were performed. For the wines of the 2015 harvest, phenolic compounds were analyzed: gallic acid, catechin, vanillic acid, P-coumaric acid and rutin. The chemical composition of Sauvignon Blanc wine was affected as a result of leaf removal timing, defoliation carried out in full bloom, berries pepper-corn size and pea-sized berries reduced total acidity and increased pH of the wines in both seasons. For wine color, in both seasons, it was found a greater yellow coloration in wines originated from grapes where the plants were not defoliated. The increase of yellow coloration is due to the oxidation of catechins and epicatechins, which is related to the increase of botrytis bunch rot. In the 2015 season there were no influence of leaf removal timing in total polyphenols, however, in the 2016, the defoliation performed in full bloom resulted in lower polyphenols contents. Wines from non-defoliated plants presented higher values of catechin, rutin, P-coumaric acid and gallic acid, related to yellow coloration of wines and greater susceptibility to darkening and oxidation. For vanillic acid, wines from non-defoliated plants have the lowest values, this compound is important in wine preservation. Leaf removal is an important management technique in high altitude regions of Santa Catarina - Brazil, it improves wine chemical composition, and if carried out during the stages of full bloom, pepper corn berries and pea-sized berries can reduce the severity of Botrytis bunch rot; and avoid the formation of phenolic compounds related to the processes of oxidation and darkening.

1. Introduction

Highland regions of Santa Catarina State, Brazil, have been acquiring great importance due to the high potential for wine production from *Vitis vinifera* L. cultivars [1–3]. Located between latitudes 26°and 28°S and altitudes between 950 m and 1400 m a.s.l. the knowledge and characterization of this region shows potential to produce quality wines, but due to highlands of Santa Catarina State are being in an emergent wine grown region fundamental investigations into vineyard practices are necessary [4].

In modern viticulture, the management of vine growth plays a fundamental role in the grapevine development, and it is widely recognized as a determining factor in the composition of wines [5]. Management of vine growth for wine grape cultivation have been developed to optimize sunlight interception, photosynthetic capacity and canopy microclimate for the purpose of improve fruit production and wine quality, especially in vigorous varieties with dense canopies [6].

Basal leaf removal may be employed to improve air and light exposure around clusters, as well as to open access to disease- and pest-control chemicals [7]. It is a vineyard practice carried out in all wine-growing regions, which is usually applied between the setting and véraison phenological stages of the grapevine [8,9]. However, leaf removal before setting has become more common [10]. An effect generally associated with basal leaf is a reduce incidence of bunch rot [11]. Besides that, anthocyanin levels and skin color generally are significantly enhanced, particularly in cool climate. Basal leaf removal may increase sugar and polyphenols, improving grape quality at harvest [12]. Whereas, an effect generally associated with basal leaf removal is a drop in titratable acidity [12,13].

To improve our understanding about vineyard practices, the present study aims to investigate the effect of different leaf removal time on the botrytis bunch rot disease development and chemical-phenolic composition

[a] e-mail: douglaswurz@hotmail.com

 DOI: 10.1051/bioconf/20170902027

Table 1. Chemical parameters (titratable acidity, pH and color) of Sauvignon Blanc wines under effect of different leaf removal times in the high-altitude regions of southern Brazil. Vintages 2015 and 2016.

Leaf Removal Time	Acidity (meq L^{-1})		pH		Color (Abs 420 nm)	
	2015	2016	2015	2016	2015	2016
Full bloom	132.5 a	115.8 a	3.00 d	3.01 b	0.06 a	0.05 a
Berries pepper-corn size	131.7 a	113.6 a	2.88 b	2.98 b	0.06 a	0.06 b
Berries pea size	136.6 b	114.3 a	2.88 b	2.99 b	0.05 a	0.06 b
Veraison	145.2 c	124.4 b	2.85 a	2.90 a	0.05 a	0.07 b
15 days after veraison	137.1 b	135.6 c	2.93 c	2.88 a	0.06 a	0.10 c
Control (no leaf removal)	157.5 d	142.4 c	2.93 c	2.86 a	0.08 b	0.10 c
CV (%)	1.0	4.0	0.5	1.2	7.2	9.3

*Means followed by different letters in the same column differ by Skott Knott test ($p < 0.05$).

of Sauvignon Blanc wines in order to guide appropriate management for winemaking in regions 900 m a.s.l. of Southern Brazil.

2. Materials and methods

Experiment was undertaken in a commercial Vitis vinifera L. cv. Sauvignon blanc vineyards located in São Joaquim county (28°17'39" S, 49°55'56" W and 1.230 m above sea level), State of Santa Catarina (SC), southern Brazil, during the year 2014–2015 and 2015–2016 growing seasons.

The climate of the region has been classified as "Cool, Cool nights and Humid", Heliothermic Index of 1.714, average annual rainfall of 1.621 mm and average annual relative humidity of 80% [14]. Soil type is cambisol, which is characterized as having high clay (492 g kg^{-1}) and organic matter (71 g kg^{-1}) [15].

Sauvignon Blanc vines (grafted on '1103 Paulsen') were planted in 2004 in a northwest to southeast row orientation, with a 3.0 m (row) × 1.5 m (vine) spacing. Vines were trained on a vertical shoot position in a double cordon spur-pruning at 1.2 m height and covered with anti-hail net.

The treatments consisted of leaf removal at cluster-zone in five different phenological stages according to Baillod & Baggiollini [16] scale: full bloom, berries pepper-corn size, berries pea size, veraison, 15 days after veraison and control (no leaf removal). Leaf removal was performed manually, removing three basal leaves, including leaf opposite to cluster, exposing completely the grapes. Vineyards practices (pruning, shoot topping and control diseases) were maintained as recommended to commercial growers in this region.

During harvest, 60 kg of grape per treatment was hand-picked to winemaking process. The winemaking was performed in research scale following protocols described by Pszczolkowski & Lecco et al. [17]

Wine samples were immediately submitted to the following analysis: a. titratable acidity (meq L^{-1}); b. pH. These analyses were determined by protocols of OIV [18]; c. Color, determined by the spectrophotometry method with absorbance in 420 nm wavelength described by Rizzon [19] and d. Total polyphenols (mg L^{-1}), determined by the spectrophotometry method described by Singleton and Rossi [20], using the Folin-Ciocalteu reagent, with absorbance in 760 nm wavelength. Phenolic content was determined using a gallic acid calibration curve.

Table 2. Botrytis bunch rot severity in the Sauvignon Blanc's cluster under effect of different leaf removal times in the high-altitude regions of southern Brazil. Vintages 2015 and 2016.

Leaf removal time	Severity (%)	
	2015	2016
Full bloom	7a	13 a
Berries pepper-corn size	7a	14 a
Berries pea size	19 b	21 a
Veraison	32 c	29 b
15 days after veraison	33 c	35 b
Control (no leaf removal)	42 d	44 d
CV (%)	23.9	20.7

*Means followed by different letters in the same column differ by Skott Knott test ($p < 0.05$).

Table 3. Total polyphenol of Sauvignon Blanc wines under effect of different leaf removal times in the high-altitude regions of southern Brazil. Vintages 2015 and 2016.

Leaf removal time	Total polyphenol (mg L^{-1})	
	2015	2016
Full Bloom	221.3 a	267.5 a
Berries pepper-corn size	223.3 a	299.2 b
Berries pea size	235.6 a	320.5 b
Veraison	216.4 a	314.6 b
15 days after veraison	229.6 a	311.6 b
Control (no leaf removal)	245.4 a	321.7 b
CV (%)	5.1	8.6

*Means followed by different letters in the same column differ by Skott Knott test ($p < 0.05$).

Individual polyphenols in the wine were determined using high-performance liquid chromatography (UV-HPLC), as described previously [21]. Wine samples were analyzed in triplicate (two injections from the triplicate bottles).

For the severity evaluation of botrytis bunch rot, the evaluations began at the onset of the first symptom, at intervals of 15 days, under natural infection condition. Thirty clusters per treatment were randomly selected, and the evaluations were carried out through a diagrammatic scale described by Hill et al. [23], based on twelve levels of disease severity: 1, 5, 10, 15, 20, 30, 30, 40, 50, 60, 70, 80 and 90%.

The experimental design was a randomized blocks with four blocks and five vines per plot. Data was analyzed by the F test and Skott Knott's multiple range tests ($p < 0.05$).

Table 4. Concentration of polyphenols (Gallic acid, vanillic acid, p-cumaric acid and rutin) obtained by HPLC in the final wines of Sauvignon Blanc under effect of different leaf removal times in the high-altitude regions of southern Brazil. Vintage 2015.

Leaf removal time	Gallic Acid (mg L^{-1})	Catequin (mg L^{-1})	Vanillic Acid (mg L^{-1})	P-cumaric acid (mg L^{-1})	Rutin (mg L^{-1})
Full bloom	9.23 b	4.96 b	0.67 d	0.32 c	0.84 b
Berries pepper-corn size	7.05 a	4.64 b	0.54 c	0.12 a	0.49 a
Berries pea size	5.88 a	4.71 b	0.53 c	0.09 a	0.46 a
Veraison	6.34 a	4.38 b	0.42 b	0.19 b	0.87 b
15 days after veraison	8.88 b	4.57 b	0.44 b	0.16 b	0.49 a
Control (no leaf removal)	8.77 b	6.07 a	0.32 a	0.31 c	1.22 c
CV (%)	10.1	10.8	13.2	10.4	9.2

*Means followed by different letters in the same column differ by Skott Knott test ($p < 0.05$).

3. Results and discussion

We observed influence of leaf removal timing on the total acidity, pH and color of Sauvignon Blanc wines (Table 1). At both harvests, leaf removal carried out until berries pepper-corn size phenological stages resulted wines with the lowest values of total acidity, while wines from plants without leaf removal show the highest total titratable acidity. Early leaf removal associated with high solar radiation provides a reduction in berries acidity due to the greater degradation of malic acid [24,25].

For pH of the wine, it was observed that earlier leaf removal, before véraison, showed elevation of pH (Table 1). As reported by other authors [24–26], leaf removal resulting in an increase in pH, as a consequence of direct exposure of the cluster to solar radiation.

To color (Abs 420 nm) was observed higher values in wines from non-defoliated vines at both vintages (Table 1). This behavior could be related to more botrytis bunch rot severity observed in the Sauvignon Blanc vines without leaf removal (Table 2). Similar results were observed by Meneguzzo et al. [27] who verified that color index of Gewurztraminer wines increased with the increase of the bunch rot.

In the Sauvignon Blanc clusters, it was observed that early leaf removal carried out up to berries pea size phenological stage resulted a reduction in the severity of botrytis bunch rot, while a significant increase in the severity of botrytis was observed in plants not submitted to leaf removal management (Table 2). Viticulture has a need to acquire more sustainable forms of cultivation (3), and consequently reducing the use of chemical methods of disease control. Vineyards practice that result in greater sun exposure and ventilation in the cluster zone such leaf removal are efficient tools in disease control in grapevine cluster [11,28].

For the concentration of total polyphenols in wines, no difference was observed among leaf removal treatments, except for the defoliation performed at full bloom phenologial stage in the 2016 harvest (Table 3). Although there was no influence of the leaf removal time on the total polyphenol concentration in the 2015 vintage, it was verified that leaf removal influenced the concentration of individual polyphenols (Table 4).

Wines from vines with leaf removal performed at full bloom, 15 days after véraison and control show higher values of gallic acid (Table 3). Also it was observed that plants with no leaf removal resulted in wines with the highest content of catechin and rutin, and lowest value for vanillic acid (Table 3). In white wines the catechins are the main flavonoids. These compounds are responsible for browning and some bitterness in wines [29,30].

Vine not submitted to the leaf removal management as well as the leaf removal performed in the full bloom phenological stage present the highest contents of the P-coumaric acid (Table 3). The importance of these acids are related to phenomena of oxidative browning that the musts or white wines can suffer. These compounds, rich in hydroxyl groups, are the first phenolic substances to be oxidized by the phenoloxidase enzymes in the respective quinones. These quinones are involved in reactions that lead to the appearance of compounds, with colorings ranging from yellow to brown, in musts.

According to the results observed in this study, higher values of polyphenol were found in wines from vines not submitted to the leaf removal management; however, such compounds are responsible for phenomena of darkening of white wines, oxidation, bitterness, thus having a negative effect to the wine quality.

A thorough knowledge of the various polyphenolic structures present in the grape and the mechanisms of its evolution during winemaking is an indispensable basis to development of the technological processes adapted both the vineyard and the winery [31]. Therefore, the results observed in this work demonstrate that the management of basal leaf removal must be performed according to winery standards due its impact to oenological performance of grapes.

4. Conclusion

Leaf removal is an important vineyard practice in high altitude regions of Santa Catarina – Brazil because it improves the chemical composition of wine and if it was carried out until berries pea size phonological stage can reduce the severity of botrytis bunch rot, besides avoid the formation of phenolic compounds related to the processes of oxidation and darkening.

References

[1] E. F. Gris, V.M. Burin, E. Brighenti, H. Vieira and M.T. Bordignon-Luiz. Phenology and ripening of Vitis vinifera L. grape varieties in São Joaquim, southern Brazil: a new South American wine growing region. Cie. Inv. Agr. **37,** 61–75 (2010)
[2] L.I. Malinovski, L.J. Welter, A.F. Brighenti, H.J. Vieira, M.P. Guerra and A.L. Da Silva, A.L. 2012. Highlands of Santa Catarina/Brazil: a region with

high potential for wine production. Acta Hort. **931**, 433–440 (2012)
[3] A.F. Brighenti, E. Brighenti, V. Bonin, and L. Rufato. Caracterização fenológica e exigência térmica de diferentes variedades de uvas viníferas em São Joaquim, Santa Catarina – Brasil. Ciên. Rural. **43**, 1162–1167 (2013)
[4] J.L. Marcon Filho. *Sistemas de condução na produção de uvas viníferas e composição química e aromática de vinhos em região de altitude de Santa Catarina* (Thesis, Lages, 188, 2016)
[5] M. Bogicevic, V. Maras, M. Mugosa, V. Kodzulovic, J. Raicevic, S. Sucur, O. Faila. The effects of early leaf removal and cluster thinning treatments on berry growth and grape composition in cultivars Vranac and Cabernet Sauvignon. Chemical and Biological Technologies in Agriculture **2**, 1–8 (2015)
[6] S. Jogaiah, D.P. Oulkar, A.N. Vijapure, S.R. Maske, A.K. Sharma, R.G. Somkuvar. Influence of canopy management practices on fruit composition of wine grape cultivars grown in semi-arid tropical region of India. African Journal of Agricultural Research **8**, 3462–3472 (2013)
[7] A. Miele, F. Mandelli. Manejo dossel vegetativo e seu efeito nos componentes de produção da videira Merlot. Revista Brasileira de Fruticultura **34**, 964–973 (2012)
[8] D.C. Percival, K.H. Fisher, J.A. Sullivan. Use of fruit zone leaf removal with Vitis vinifera L. cv Riesling grapevines. II. Effect on fruit composition, yeld, and occurrence of bunch rot (*Botrytis cinerea* Pers.:Fr). American Journal of Enology and Viticulture **45**, 133–140 (1994)
[9] M.P. Diago, M. Vilanova, J. Tardaguila. Effects of timing of manual and mechanical early defoliation on the aroma of Vitis vinifera L. Tempranillo wine. American Journal of Enology and Viticulture **61**, 382–391 (2010)
[10] B.L. Komm, M.M. Moyer. Effect of early fruit-zone leaf removal on canopy development and fruit quality in Riesling and Sauvignon Blanc. American Journal of Enology nd Viticulture **66**, 424–434 (2015)
[11] D. Molitor, M. Behr, S. Fischer, L. Hoffamann, D. Evers. Timing of cluster-zone leaf removal and its impact on canopy morphology, cluster architecture and bunch rot susceptibility of grapes. Journal International des Sciences de la Vigne et du Vin **45**, 149–159 (2011)
[12] S. Poni, L. Casalini, F. Bernizzoni, S. Civardi, C. Intrieri. Effects of early defoliation on shoot photosynthesis, yield components, and grape quality. American Journal of enology and Viticulture **57**, 397–407 (2006)
[13] M.P. Diago, Z.N. Ayestarán, A. Garrido, J. Tardaguila. Phenolic composition of Tempranillo wines following early defoliation of the vines. Journal Science Food Agriculture **92**, 925–934 (2012)
[14] J. Tonietto, A. Carbonneau. A multicriteria climatic classification system for grape-growing worlwide. Agricultural and Forest Meteorology **124**, 81–97 (2004)
[15] SOLOS do Estado de Santa Catarina (Rio de Janeiro, Embrapa Solos, 726, 2004)
[16] M. Baillod, M. Baggiolini. Les stades repéres de la vigne. Revue suisse de viticulture arboriculture horticulture **25**, 7–9 (1993)
[17] P. Pszczolkowski, C.C. Lecco. *Manual de vinificácion: guia práctica para la elaborácion de vinos* (Santiago, Universidade Católica do Chile, 2011)
[18] OIV - Organization Internationale de la Vigne et du Vin. Compendium of International Methods of Wine and Must Analysis (OIV, Paris, 2009)
[19] L.A. Rizzon. *Metodologia para análise do vinho.* (Brasília, Embrapa Iformações Tecnológicas, 120, 2010)
[20] V.L. Singleton, J.A. Rossi. Colorymetry of total phenolics with phosphomolibdic-phosphotungstic acid reagent. Am. J. Enol. Vitic. **16**, 144–158 (1965)
[21] E. Cadahía, B. Fernandez de Simon, M. Sanz, P. Poveda, J. Colio. Chemical and chromatic characteristics of Tempranillo, Cabernet Sauvignon and Merlot wines from Do Navarra aged in Spain and French oak barrels. Food Chemistry **115**, 639–649 (2009)
[22] G.N. Hill, R.M. Beresford, K.J. Evans. Tools for accurate assessment of botrytis bunch rot (Botrytis cinerea) on wine grapes. New Zealand Plant Protection **63**, 174–181 (2010)
[23] D.S. Intrigliolo, E. Llacer, J. Revert, M.D. Esteves, M.D. Clement, D. Palau, I, Gómez. Early defoliation reduces cluster compactness and improves grape composition in Mandó, an autochthonous cultivar of Vitis vinifera form southeastern Spain. Scientia Horticulture **167**, 71–75 (2014)
[24] D. Risco, D. Pérez, A. Yeves, J.R. Castel, D.S. Intrigliolo. Early defoliation in a temperate warm and semi-arid Tempranillo vineyard: vine performace and grape composition. Autralian Journal Grape Wine Research **20**, 111–122 (2014)
[25] S. Poni, F. Bernizzoni, S. Civardi. The effect of early leaf removal on whole-canopy gas exchange and vine performace of Vitis vinifera L. Sangiovese. Vitis **47**, 1–6 (2008)
[26] J. Meneguzzo, L.A. Rizzon, A. Miele, M.A. Ayub. Efeito de Botrytis cinerea na composição do vinho Gewurztraminer. Ciência e Tecnologia Alimentos **26**, 527–532 (2006)
[27] D. Molitor, N. Baron, T. Sauerwein, C. André, A. Kicherer, J. Doring, M. Stoll, M. Beyer, L. Hoffmann, D. Evers. Postponing first shoot topping reduces grape cluster compacteness and delays bunch rot epidemic. American Journal of Enology and Viticulture **66**, 164–176 (2014)
[28] B.W. Zoecklein, K.C. Fugelsang, B.H. Gump, F.S. Nury. *Wine Analisys and production* (The Chapman e Hall Enology Library, International Thompson Publishing, 1995)
[29] M.O. Downey. Analysis of tannins in seeds and skins of Shiraz grapes throught berry development. Australian Journal of Grape and Wine Research **9**, 15–27 (2003)
[30] V. Cheynier, M. Moutounet, Los compuestos fenólicos. In C. Flanzy. *Enología: fundamentos científicos y tecnológicos* (Madri, AMV Ediciones, 114–136, 2000)

DOI: 10.1051/bioconf/20170902028

Methanol in wine

Greg Hodson[1], Eric Wilkes[2], Sara Azevedo[1], and Tony Battaglene[1]

[1] FIVS, 18 rue d'Aguesseau, 75008 Paris, France
[2] Australian Wine Research Institute, Hartley Grove, Urrbrae SA 5064, Australia

Abstract. This paper examines the origins of methanol in grape wine and the quantities typically found in it, as well as in other foods such as unpasteurised fruit juices. The toxicology of methanol and the associated regulatory limits established by competent authorities in various parts of the world are also considered. It is concluded that such limits are not driven by public health considerations and thus authorities are requested to consider the need for methanol analyses to be performed and reported on certificates of analysis as a condition of market entry for wine. Where methanol limits are still deemed to be necessary to achieve policy objectives, authorities are encouraged to establish them in the light of the levels of methanol typically found in grape wines produced by the full array of internationally permitted winemaking practices, and to consider harmonising their limits with those that have already been established by other governments or recommended by appropriate intergovernmental organisations.

1. Introduction

The origin and significance of methanol in wine, and the associated establishment of regulatory limits for its presence there, are causes of much confusion and misunderstanding in international trade. This paper, produced by the FIVS Scientific and Technical Committee, reviews the topic in some detail, providing reference materials to assist with further study. It concludes that the levels of methanol commonly found in grape wines are broadly similar to those that may be found in many freshly squeezed and unpasteurised fruit juices if they are stored for a period of time after squeezing. It is further demonstrated, from a comparison of regulatory limits for methanol in wine with food safety risk assessments that have been conducted by reputable bodies, that the limits themselves do not serve any real food safety purpose. This is because many litres of wine per day or even per hour would need to be consumed (even if the product contained the highest content of methanol permissible in regulations) to even approach intake levels of any known toxicological concern [1].

2. Chemical properties and other information for methanol

Methanol is chemically characterized as follows [2]:

2.1. Chemical Formula, Synonyms, CAS Registry number

- Chemical formula: CH_3OH
- Synonyms: Methyl alcohol, Carbinol, Wood alcohol
- CAS Registry Number: 67-56-1.

2.2. Physico-chemical properties

- Physical appearance: Methanol is a colourless liquid with characteristic odour.
- Melting Point: –98 °C
- Boiling Point: 65 °C
- Solubility in water: Miscible.

3. Origin of Methanol in wine

3.1. Action of pectinase enzymes

3.1.1. Action of endogenous pectinase enzymes on pectin in grapes

Methanol is produced before and during alcoholic fermentation from the hydrolysis of pectins by pectinase enzymes (such as pectin methylesterase) which are naturally present in the fruit. More methanol is produced when must is fermented on grape skins; hence there is generally more in red wines than in rosé or white wines (see Sect. 4 below).

3.1.2. Use of exogenous pectinase enzymes in winemaking

Exogenous pectinase enzymes are permitted for use in winemaking (generally as clarifying agents) in at least the following countries: Australia, Canada, the 28 Member States of the European Union, Japan, the Republic of Georgia, New Zealand, South Africa, and the United States. Their use is also deemed to be an acceptable winemaking practice by the International Organisation for Vine and Wine (OIV) [3]. As with the activity of pectinases naturally present in grapes, the use of exogenous pectinases as a winemaking practice will have the effect of increasing the levels of methanol in the resulting wine.

3.2. Treatment of wine with Dimethyldicarbonate

Dimethyldicarbonate (DMDC) is an effective pre- bottling sterilant, accepted for use in winemaking in Argentina,

Australia, Cambodia, Canada, Chile, the 28 Member States of the European Union, the Republic of Georgia, Hong Kong China, Myanmar, New Zealand, Russia, Singapore, South Africa, Thailand, Turkey and the United States, whose use is generally limited in international regulations and recommendations to a maximum treatment of 200 mg/L of wine [4]. For other alcoholic beverages and mixtures of alcoholic and other beverages with an alcoholic strength by volume of less than 15%, the limit on usage is often set at 250 mg/L. The use of DMDC can be important in stabilizing lower alcohol products from additional fermentation in the bottle, and also allows a reduction in the quantity of sulphur dioxide that is used where the oxygen in wine is kept below 1 mg/L. DMDC breaks down rapidly in wine, producing carbon dioxide and leaving methanol at very low levels not harmful to health and other innocuous products in the wine. Methanol at a level of about 100 mg/L is created in wine from a DMDC treatment at the typical maximum treatment level of 200 mg/L [5].

4. Typical levels of methanol in wine

It was noted above that the presence of low levels of methanol in wine is expected due to the action of pectinase enzymes that are naturally present in the grapes. A study of the literature indicates the following information concerning the typical levels of methanol that may be found in wine (these levels generally do not account for any additional amount that may result from a DMDC treatment):

- Red wines will tend to contain more methanol (between 120 and 250 mg/L of the total wine volume) than white wines (between 40 and 120 mg/L of the total wine volume), because of the longer exposure to grape skins during the fermentation [6].
- Wines made from grapes that have been exposed to *Botrytis cinerea* (e.g. late harvest wines, such as Sauternes or Tokay) also have higher methanol levels than standard grape wines (as much as 364 mg/L of the total wine volume) [7].
- Wines made from non *Vitis vinifera* grapes tend to contain more methanol than wine from pure *Vitis vinifera* [8].

4.1. Case study: Typical levels of methanol in Australian wine

A recent survey looked at 150 wines from across Australia to determine typical levels of methanol in commercial wine [9]. The sample set consisted of 90 red and 60 white wines from multiple varieties and vintages. All wines were analysed using a GC-FID in the Australian Wine Research Institute's ISO 17025 accredited laboratory. No evidence of DMDC treatment (a source of methanol) was found for any of the wines tested.

Typical ranges for methanol found in Australian wines were; 60–280 mg/L in reds (mean 170 mg/L) and 40–120 mg/L in whites (mean 58 mg/L). All wines tested had some methanol content. The main driver for higher methanol levels appeared to be skin contact during processing. Variety or vintage had no significant impact.

4.1.1. Typical values

Results for red and white wines were significantly different. Red wines typically contained higher levels of methanol across a larger range of content, reflecting greater variation in skin contact times. All wines were found to be within Australian and OIV guidelines (Fig. 1).

4.1.2. Impact of variety

No significant differences of methanol content were found based on grape variety. The only difference found was between red and white wines, reflecting the differences in processing for the different wine styles (Fig. 2).

4.1.3. Impact of vintage

No significant impact of year of production on the methanol concentration was found (Fig. 3).

5. Typical levels of methanol in fruit juices

In comparison with the quantities found in wine, non-alcoholic fruit juices naturally contain an average of 140 mg/L of methanol, in a range from 12 mg/L to as much as 640 mg/L. This is quite similar to the average methanol level in white wine and within the range of values seen for red wines [10–12].

6. Toxicology of methanol

Methanol is a toxic chemical. Acute methanol poisoning symptoms resemble those of ordinary alcoholic intoxication followed by the presence of severe upper abdominal pain, visual disturbance sometimes proceeding to incurable blindness and prolonged coma, which may terminate in death from respiratory failure. The fatal dose varies but is usually from 100 mL to 200 mL of methanol. Permanent blindness has been claimed to have been caused by as little as 10 mL. The toxicological properties of methanol are mediated by those of ethanol, and patients suffering from methanol poisoning may sometimes be treated by administration of ethanol [13, 14].

According to a paper produced by the United Nations World Health Organization International Programme on Chemical Safety, it is difficult to perform a risk assessment on methanol because the chemical itself and its primary metabolites of concern are naturally present in humans – a fact that pre- supposes a certain level of these substances poses no toxic risk [15]. In this paper, the authors propose a level of methanol intake that is of insignificant risk to humans, in that it causes no increase in the most toxic metabolite (formic acid) above the normal background levels found in humans. That intake level is a single oral exposure to methanol of 20 mg/kg body weight. This represents 1400 mg for an individual weighing 70 kg. To receive this amount of methanol from a wine containing 400 mg/L methanol (the maximum methanol limit recommended by OIV for red wines – see Section 7), that individual would need to consume 3.5 litres (almost five 750 mL bottles) of wine in a single oral exposure.

According to the European Union Scientific Committee on Food, a healthy individual can metabolize 1500 mg

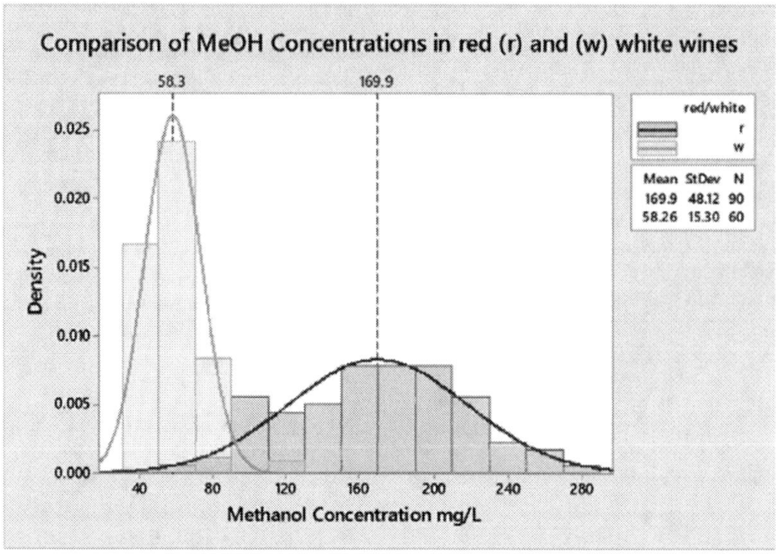

Figure 1.

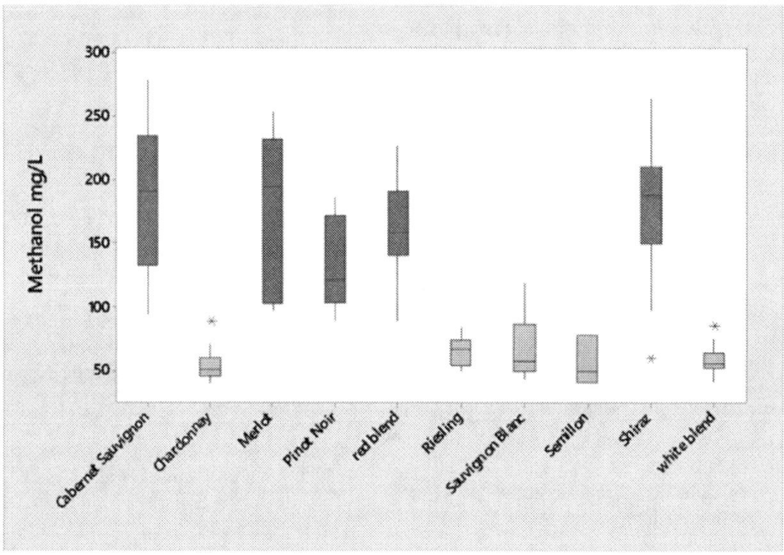

Figure 2. Comparison of MeOH concentrations in varietal wines

per hour of methanol without showing any ill effects [16]. Note that 1500mg of methanol is approximately the amount of methanol contained in 3.75 litres (five 750 mL bottles) of a wine with a methanol content of 400mg/L (the maximum methanol limit recommended by OIV for red wines – see Sect. 7).This volume of wine would have to be consumed in one hour to ingest more methanol than a healthy individual could metabolise with no ill effects.

The Food and Drug Administration of the United States of America (FDA) reported that the No Observed Adverse Effect Level (NOAEL) in humans for methanol is 71 to 84 mg/kg body weight (bw)/day. Because this NOAEL was derived from studies in humans, the FDA developed an Acceptable Daily Intake (ADI) of 7.1 to 8.4 mg/kg bw/day by using a safety factor of 10 [17]. Note that an individual weighing 70 kg would have to consume about 1.25 litres of wine (almost two 750 mL bottles) a day with a methanol content of 400 mg/L (the maximum methanol limit recommended by OIV for red wines - see Sect. 7) to reach the low end of this ADI range.

7. Regulatory limits established for methanol levels in wine

Table 1 summarizes the limits set for methanol in wine in different global markets. It is clear that most economies and regions have established limits that take account of the levels of methanol typically found in wine, and make additional allowances for the possible use in production of exogenous pectinase enzymes and of DMDC, as mentioned earlier in this paper [18].

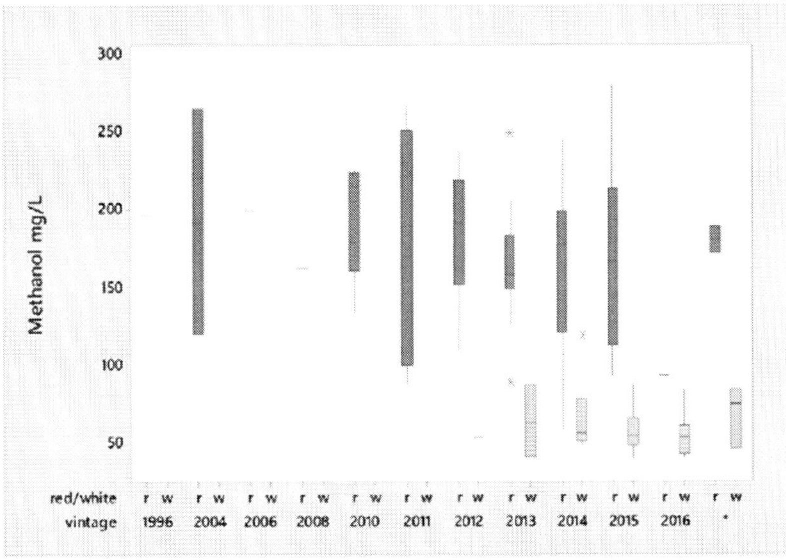

Figure 3. Comparison of MeOH concentrations in vintage wines.

Table 1.

Methanol limits based on volume of wine	
Argentina	350 mg/L
Chile, China, Vietnam, OIV	400 mg/L of wine (Red) 250 mg/L of wine (White and Rosé)
India	2.0 (expressed in terms of g/L of absolute alcohol)
Canadian Provinces	400 mg/L of wine (Ontario) 420 mg/L of wine (Quebec)
Japan, Korea, USA	0.1% (or about 1000 mg/L) of wine
Republic of Georgia	0.05 mg/kg
South Africa	300 mg/L of wine
Switzerland	300 mg/L of wine (Red) 150 mg/L of wine (White and Rosé)
Thailand	420 mg/dm^3
Taiwan	2 g/L
Turkey	10 mg/kg (about 10mg/L) of wine
Limits based on alcohol content of wine	
Australia, Mexico, New Zealand, Singapore	3 g/L of ethanol
Chinese Taipei	2 g/L of ethanol
Indonesia	0.1% based on ethanol

8. Purpose of methanol limits for wine

Comparing the toxicological information in Sect. 6 with the limits that have been established for methanol in wine (Sect. 7), it seems clear that the purpose of these limits is not to protect public health. This is because an individual weighing 70 kg would have to drink 1.25 litres of red wine a day containing the OIV recommended limit amount of methanol (400mg/L) in order to reach the lowest ADI value derived by the FDA (and including a safety factor of 10). According to the paper produced by the World Health Organization International Programme on Chemical Safety, a single oral exposure to 3.5 litres of this wine would not cause an increase in background levels of formic acid (the most toxic metabolite of methanol) in that consumer. Finally, the European Commission Scientific Committee on Food data suggests that the individual would have to consume 3.75 litres of the same wine in an hour to ingest more methanol than the 1500 mg that individuals can metabolize without ill effects.

If the methanol limits that have been established for wine do not represent a public health protection measure, it may be asked what purpose they actually serve. In practice, it seems possible that at one time they were intended to serve as an index of appropriate fruit handling in harvest and subsequent processing. In Sect. 3.1.1., the action of endogenous pectinases in forming methanol in wine was reviewed. Such enzymes will be liberated, and methanol generated, whenever fruit is damaged during its growth, its journey from the vineyard to the winery and in subsequent processing. It is recorded that fungal diseases that attack grapes in the vineyard most likely result in higher levels of methanol in the resulting juice and wine, and in Sect. 4 it was noted that late harvest wines (where fungal pressures will be more pronounced as the grapes remain longer in the vineyard) tend to have higher methanol contents [19]. Therefore, methanol content in wine may be expected to be higher where the fruit has been damaged either by moulds in the vineyard or by handling at harvest and during transportation to the winery.

The possibility of methanol limits historically serving as indices of appropriate fruit handling in wine production seems to be supported by the fact that many authorities differentiate between red wines on the one hand, and white wines and rosés on the other, establishing different methanol limits in each case. If the limits were purely set for toxicological reasons, there would be no need to do this; a single limit for methanol in wine would be appropriate regardless of whether it was white, rosé or red.

In Sect. 3 it was mentioned that the use of exogenous pectinase enzymes in wine production and also of DMDC as a pre-bottling sterilant are widely permitted, and that both of these processes, if used, will contribute somewhat to the total methanol content of wine (in recognition of which, governments and other appropriate intergovernmental bodies have increased their permissible limits for methanol in wine). It is clear, therefore, that there will no longer necessarily be such a clear relationship between a wine's content of methanol and the fruit handling practices in the harvest and production of that wine.

As has been seen, even wine containing the maximum allowable methanol content by regulation never has a high enough concentration to give rise to public health concerns.

9. Conclusions

Methanol is produced quite naturally in wine by the action of endogenous pectinase enzymes on the grape pectins. The possible use of exogenous pectinase enzymes in the winery, as well as the possible treatment of the wine with dimethyldicarbonate just prior to bottling, will have the effect of increasing the methanol content of the wine to varying but small degrees. Many competent authorities around the world have chosen to establish limits for the methanol content of wine, and many have chosen to establish different limits for red wines compared with white and rosé. Yet in each case, when compared with toxicological data, it is clear that the limits do not really serve to protect public health, because the methanol content in wine does not become sufficiently high to even approach a safety concern.

At least two possible future directions are suggested by the observations in this paper:

First, there is no public health necessity for competent authorities to require methanol analysis of wine as a condition of market entry, since no additional consumer safety will be obtained by such measures. It is true that there have been rare wine adulteration incidents involving methanol in wine historically, but an extensive requirement for methanol analysis at the border seems an excessive and costly approach to mitigate the small possibility of such an event in international trade.

Second, if authorities determine that it is necessary to retain these limits, international trade could be greatly facilitated if they would consider taking the following steps:

- Establishing limits that take account of the levels of methanol typically present in grape wines, and that also take account of the full range of permitted winemaking practices that may be applied during their production.
- Given the wide variety of numerical values for existing methanol limits in wine, and of methods chosen for the expression of those limits (as revealed in Table 1), considering the harmonisation of limits with those established by other competent authorities and/or appropriate intergovernmental bodies.

References

[1] A. L. Waterhouse, G. L. Sachs and D. W. Jeffery, Understanding Wine Chemistry, 54 (2016)
[2] *Martindale: Extra Pharmacopoeia 25th Ed,* 85–86
[3] Data on wine production and pectinase usage. www.FIVS-Abridge.org (accessed on August 5, 2016).
[4] Data on wine production and DMDC usage. www.FIVS-Abridge.org (accessed on August 5, 2016).
[5] EFSA Panel on Food additives and Nutrient Sources added to Food (ANS), *Scientific opinion on the re-evaluation of dimethyl dicarbonate (DMDC, E 242) as a food additive*, **13**(12), 4319, 11 (2015)
[6] W. R. Sponholz, Chemie des Weines. 385–411 (1989)
[7] B. Zoecklein, Wine Analysis and Production, 107 (1995)
[8] P Ribereau-Gayon, *Handbook of Enology: The Chemistry of Wine Stabilization and Treatments* 2nd Edition, **2**, 53 (2006)
[9] AWRI / Wine Australia Report, Typical Methanol Levels in Australian Wines (2016)
[10] C. Hou, Y. Lin, Y. Wang, C. Jiang, M. Wu, Journal of Food Composition and Analysis, *Effect of storage conditions on methanol content of fruit and vegetable juices*, **21**(5), 410–415 (2008)
[11] *The Complete Technology Book on Alcoholic and Non-alcoholic Beverages (Fruit Juices, Whisky, Beer, Rum and Wine),* 712 (2008)
[12] J. Clary, *The Toxicology of Methanol*, 48 (2013)
[13] Public Health England, *Methanol: Toxicological Overview*, 3 (2015)
[14] Public Health England, *Methanol: Incident Management,* 8 (2016)
[15] International Programme on Chemical Safety, Environmental Health Criteria, *Methanol* 196 (1997) http://www.inchem.org/documents/ehc/ehc/ehc196.htm (accessed August 5, 2016)
[16] European Commission Scientific Committee on Food, *Opinion of the Scientific Committee on Food on the use of dimethyl dicarbonate (DMDC) in wines*, (2001)
[17] *Federal Register 58*, no 204, 6088–6091, (1993)
[18] www.FIVS-Abridge.org (accessed on August 5, 2016)
[19] M. Amerine, M. Joslyn, *Table Wines: The Technology of Their Production*, 435 (1970)

 DOI: 10.1051/bioconf/20170902029

Phenolic content and antioxidant activity of Armenian cultivated and wild grapes

Kristine Margaryan[1,2], Gagik Melyan[3], Diana Vardanyan[1], Hakob Devejyan[2], and Rouben Aroutiounian[1,2]

[1] Department of Genetics and Cytology, Yerevan State University, 1 Alex Manoogian, 0025 Yerevan, Armenia
[2] Research Group of Plant Genetics and Immunology, Institute of Molecular Biology of National Academy of Sciences RA, 7 Hasratyan, 0014 Yerevan, Armenia
[3] Scientific Center of Viticulture, Fruit-Growing and Wine-Making of the Armenian National Agrarian University, 1139 Merdzavan, Armenia

Abstract. Nowadays the mobilization and conservation of Armenian grapevine genetic resources is becoming global concern and has crucial importance. Armenia is regarded as homeland of viticulture and earliest 'wine culture'. Being studied enough by the methods of ampelography, Armenian grapevine diversity needs to be investigated in accordance with the modern European requirements. In recent years, grape phenolics have been a theme of major scientific and applied interest. These metabolites contribute to grapes and wine sensory properties, such as color, flavor, astringency, and determines the strong antioxidant capacity. The purpose of the presented research was the evaluation of total phenolic content and antioxidant property of forty Armenian aboriginal varieties, interspecific and intraspecific hybrids and wild species with different genetic background and geographic origin. The realized research has revealed a notable difference among the cultivated varieties and wild species in the total phenolic content and antioxidant activity, increased the scientific knowledge about the aboriginal varieties and wild genotypes. Obtained results will support the importance of preserving the biological diversity and favor the reintroduction of grape cultivars and wild genotypes thanks to the present valorization.

1. Introduction

The awareness of the necessity for the protection and characterization of genetic diversity of the cultivated grapevine and its wild ancestor *Vitis vinifera* L. subsp. *sylvestris* is very high. Measures and actions to safeguard the existing germplasm diversity are increasing in all wine growing countries. The conservation and sustainable use of grapevine genetic resources depends on comprehensive description, characterization and the effective management of germplasm collections. This ensures their survival and use in breeding, research and viticulture.

Worldwide, *Vitis* genus consists from about 60 grapevine species but, among them, *V. vinifera* L. is the most widely cultivated. In Armenia, viticulture is particularly rich with numerous autochthonous genotypes and wild species and grapevine biodiversity completes a part of the territory, tradition and history. These events are the result of both natural and human-induced selections that, through the centuries, have led to a strong relationship between cultivars and environment [1–3]. In fact, grapevine organoleptic and nutraceutical features are strongly associated with the geographical areas where vines grow [4]. This phenomenon is due to evolutionary mechanisms according which plants develop specific biosynthetic pathways, able to produce secondary metabolites, in order to protect vegetal tissues from biotic and abiotic stresses and to promote their propagation [5].

Therefore, the metabolic profile can be used as a biochemical marker for the characterization and the identification of grapevine varieties. In *V. vinifera* berries phenolic compounds are the most abundant among the secondary metabolites. These molecules are characterized by a basal structure of an aromatic ring with one or more hydroxyl groups and they are classified into several classes (simple phenols, phenolic acids, coumarins), according to their additional chemical subunits and number, type and orientation of their substituents. In particular, plant phenolic substances presenting more than one benzenic ring are usually known as polyphenols (flavonoids, stilbenes, lignans, tannins) [6,7]. Phenolic compounds are predominantly accumulated in berries skins and seeds.

For this reason, grape residue extract has become popular in recent years as a nutritional supplement. However, although the literature abounds with reports about phenolic compounds of grape seeds or skins, there are very few reports comparing distributions of phenolic compounds between seeds and skins among different wild species and cultivated varieties.

Knowledge of the phenolic compound distribution between seed and skin in a berry will contribute to a more comprehensive assessment of the berry biological activities. In the literature, several studies demonstrated that phenolic compounds play an important role in regulating the oxidative status of plant cells [8,9]. In fact, the free-radical scavenging activity of these molecules was clearly confirmed and associated with specific structural elements as the number, position and chemistry of the substituents, presence of double bonds, methylations and degree of polymerization [10]. For these reasons, a lot of

DOI: 10.1051/bioconf/20170902029

research teams also investigated the antiradical effect of secondary metabolites. Among them, vine and wine gained special interest, since the antioxidant activity of some of its constituents, such as the proanthocyanidins, was found to be 20 times greater than vitamin E and 50 times greater than vitamin C [11].

This work represents the part of a national project whose objective was the study and the conservation of autochthonous, rare and neglected grapevines and wild species, in order to increase knowledge about local genotypes and to preserve Armenian grape biodiversity from extinction, essentially due to globalization, and the agronomic cultural heritage. It has been well known that the grape nutritional qualities are affected by environmental, cultural, and post-harvesting conditions, but the genotype is the main determinating factor, leading to the variation [12,13]. The knowledge of health-beneficial nutrition distribution among cultivated and wild grape species is very important for improving grape nutritional properties by breeding. European grapes commercially spreads around the world and their phenolic compounds and antiradical activities have been well studied but Armenian grape germplasm have yet to be fully explored for their bioactive constituents and nutritional activities.

The goal of the presented research was the evaluation of total phenolic content and antioxidant property of forty Armenian aboriginal varieties, interspecific and intraspecific hybrids and wild species with different genetic background and geographic origin. The realized research is planned to improve the assessment of phenolic compounds and to better understand their distributions in grape seeds and skin.

2. Material and methods

2.1. Plant material

The present research includes forty aboriginal varieties, interspecific and intraspecific hybrids and wild species. Grapes were harvested at their technological maturity, from four province of Armenia - Syuniq as southernmost, Tavush as northeast, Vayots Dzor as southeastern and Armavir as western part of Armenia during 2016 vintage. General descriptions of the grape varieties and wild species are given in Table 1. Grapes free from visible blemish or disease were selected. Three separate batches of grapes from different sites were used to prepare triplicate samples. For quantitative analysis, 100 to 150 grape berries, randomly selected from each grape accession, were collected for extraction. All data collected for each grape variety and wild species were reported as mean ± SD for at least three replications

2.2. Sample preparation

The selected berries were finger pressed to remove juice and pulp. Seeds and skin were separated, washed several times with distilled water, and moisture was absorbed on blotting paper. The skins and seeds were then extracted in 20 ml of an ethanol: water: hydrochloric acid (70:29:1) solution for 24 hours at room temperature. The extracts were filtered before the total phenolics spectrophotometric determination.

2.3. Analysis of total polyphenols

The Folin–Ciocalteu method was used for the determination of the total polyphenols as suggested by L. Rustioni et al., 2014 [14]. In brief, an aliquot (0.5 ml) of the appropriate diluted skin and seed extracts were added to a 10 mL volumetric flask, containing 2.5 ml of distilled water. Then, 0.5 ml of Folin-Ciocalteu reagent was added and the contents mixed. After 3-5 min, 2 ml of 10% Na_2CO3 solution was added and made up to a total volume of 10 ml distilled water. After keeping the samples for 90 min at room temperature extracts absorbance was read at 700 nm against distilled water as the blank. The total polyphenols were expressed as catechin (mg· L^{-1}) concentration and calculated applying the formula:

$$Catechin\ (mg \cdot L^{-1}) = 186.5 \times E_{700} \times d \qquad (1)$$

where E_{700} = absorbance at 700 nm; d = dilution. Then data were converted in mg/kg of grape, based on the fresh berries weights. All samples were prepared in triplicate.

2.4. Free radical-scavenging activity on DPPH

Free radical-scavenging ability of grape skin and seed extracts was determined by use of a stable 2,2-diphenyl-2-picrylhydrazyl radical (DPPH). The DPPH assay was based on the slightly modified method of Brand-Williams et al. [15]. Briefly, 100 μl of adequately diluted sample was added to 3.9 ml methanolic solution of DPPH (0.0025 g/100 ml CH_3OH). The mixture was vigorously shaken and then incubated for 30 min in the dark. The reduction of the DPPH radical, after addition of the grape skin and seed extract, was determined by measuring the absorbance of the mixture at 517 nm (SpectrumLab Gold S54T). All determinations were performed in triplicate. The affinity of the test material to quench DPPH radicals was calculated using the following equation:

$$\%IC = \{(A_{Blank} - A_E)/A_{Blank}\} * 100 \qquad (2)$$

where % IC is the Inhibitory Concentration of the radical scavenging activity, A_{Blank} is the absorbance of the control sample and A_E is the absorbance of the grape extract in the presence of DPPH molecules.

Statistical methods: All data were expressed as mean ± standard deviation (SD) of three replications for each grape skin and seed extracts tested. The data obtained were analyzed statistically by the one-way analysis of variance (ANOVA) and Multiple Range Test (STATGRAPHICS Plus).

3. Results and discussion

3.1. Total phenolic content

Spectrophotometric analysis was carried out to get insight into the phenolic content of grape skin and seed extracts of forty Armenian aboriginal varieties, interspecific and intraspecific hybrids and wild species.

The content of total phenols found in cultivated grape varieties traditionally grown in Armenia and wild species is presented in Table 2. According to the obtained results, the total phenolic content determined in aboriginal sample skins varied between 1296.6 mg/kg of grape and 272.5 mg/kg of of grape. "Garan dmak" presented the

Table 1. Species and pedigree Armenian grape genotypes.

Grape Genotypes	*Pedigree*	*Parents*	*Colour*
Tchilar	*V. vinifera*	aboriginal	white
Garan dmak	*V. vinifera*	aboriginal	white
Mskhali	*V. vinifera*	aboriginal	white
Arevik	*V. vinifera*	aboriginal	white
Qrdi khaghogh	*V. vinifera*	aboriginal	yellow
Mormor	*V. vinifera*	aboriginal	yellow
Hakobi vordi	*V. vinifera*	aboriginal	rose
Krhagi karchevan	*V. vinifera*	aboriginal	red
Kakhet	*V. vinifera*	aboriginal	black
Meghri vahravar (Shireyi)	*V. vinifera*	aboriginal	black
Sali	*V. vinifera*	aboriginal	black
Sev krop	*V. vinifera*	aboriginal	black
Vardananq	*V. vinifera*	Madeleine Angevine × Yanvarskiy chyorniy	black
Narekaci	*V. vinifera*	Ichikmar × Parkent	red
Ayvazyani Vardabuyr	*V. vinifera*	Italia × Muscat Hamburg	white
Itsaptuk	*V. vinifera*	Katta Kurgan × Kishmish khishrau	white
Parvana	*V. vinifera*	Kata kurgan × Kishmish khishrau	white
Hayastan	*V. vinifera*	Nimrang × Italia	white
Hayastani quyry	*V. vinifera*	Nimrang × Italia	white
Tzeni	*V. vinifera*	Nimrang × Italia	white
Shahumyani	*V. vinifera*	Khushaine beliy O.P.	white
Anahit	*V. vinifera*	Nimrang × Muscat Hamburg	white
Berqanush	V.*vinifera*	C - 484 (Madeleine Angevine × Shasla muskatnaya) × No979/2	white
Arevshat	V.*vinifera*	Katta kurgan × Parkent	rose
Arazi	V.*vinifera*	C-484 (Madlen anjevin × Shasla muscatnaia) × Seneka	rose
Aygezard	*V. vinifera*	Katta kurgan × Parkent	rose
Armenia	*V. vinifera*	Khushayne beliy × Sateni cherniy	black
Hayreniq	*V. vinifera*	Italia × Muscat Hamburg	black
Charenci	*V. vinifera* x*V.amurensis*	Seyanets C 1262 (Amurensis × Zhemchug Saba) × Karmrahyut	black
Nali	*V. vinifera*	*unknown*	white
Megri vahravar (semiwild)	*V. vinifera*	*unknown*	black
Wild grape Megri vahravar	*Vitis vinifera ssp. sylvestris*		black
Wild grape 1-8 accessions	*Vitis vinifera ssp. sylvestris*		black

highest total phenolic content (1296.6 mg/kg of grape), followed by "Kakhet", "Krhagi karchevan", "Sev crop" and "Sali" grape varieties. The total phenolic content of "Garan dmak", "Kakhet" and "Krhagi karchevan" grape varieties was significantly higher from other aboriginal varieties ($p < 0.05$). "Garan Dmak" and "Kakhet" are common to Ararat valley and are used to make white and black wines respectively. "Krhagi karchevan" is a rare and long-neglected grape variety, grows in old vineyards of southernmost of Armenia (Syuniq Region, Meghri).

However, significant differences in skin total phenolic content were not found among "Tchilar", "Mskhali" and "Megri vahravar" and among "Sali" and "Sev crop" grape varieties ($p > 0.05$). Significantly lower skin total phenolic content was found in "Mormor" grape variety (272.5 mg/kg of grape) mainly preserved in old vineyards in southeastern of Armenia (Vayots Dzor Region). "Mormor" is characterized by biggest, grey berries and used to make wines with unique taste. Significant quantitative differences were identified also by comparing the total phenolic spectra of aboriginal grape seed extracts. "Kakhet" variety had the highest total phenolic content in seeds (620.33 mg/kg of grape), followed by "Tchilar" (504.68 mg/kg of grape) and "Hakobi vordi" (493.86 mg/kg of grape) varieties, while the lowest appeared in the "Qrdi khaghogh" (85.87 mg/kg of grape). Significant differences in seed total phenolic content were not found among "Arevik", "Mormor", "Meghri vahravar, Shireyi" and among "Sali" and "Sev crop" grape varieties ($p > 0.05$).

In accordance with the presented results, the skin total phenolic content determined in interspecific and intraspecific hybrids varied between 1355.6 mg/kg of grape and 154.0 mg/kg of of grape. "Berqanush" presented the highest total phenolic content (1355.6 mg/kg of grape), followed by "Vardananq" (953.8 mg/kg of grape), "Meghri vahravar" (semiwild) (887.00 mg/kg of grape)", "Charenci" (802.8 mg/kg of grape) hybrids. The parents of "Berqanush" hybrid are C - 484 (Madeleine Angevine x Shasla muskatnaya) x No979/2 (Spitak Araqseni x Chyorniy sladkiy). "Berqanush" is wine variety and common to Ararat valley and Vayots Dzor Region. However, significant differences in skin total phenolic content of interspecific and intraspecific hybrids were not found among "Itsaptuk", "Armenia", "Nali" varieties.

The total phenolic compounds in grape seeds of interspecific and intraspecific hybrids varied significantly. Among the grape cultivars studied "Meghri vahravar" (semiwild)" had the highest total phenolic contents in seeds (725.2 mg/kg of grape), followed by "Berqanush" hybrid (534.5 mg/kg of grape), while the lowest appeared in the "Shahumyani" and "Aygezard" hybrids.

Owing to recent prospections in Armenian woods and river floodplains, many forms of wild grapevine

Table 2. Content of total phenols in Armenian grape genotypes.

Grapevine genotypes	Skin phenolic content (mg/kg of grape FW)	Seed phenolic content (mg/kg of grape FW)
	Aboriginal grapes	
Tchilar	494.13 ± 17.72bc	504.68 ± 16.78g
Garan dmak	1296.63 ± 15.53f	418.43 ± 13.10e
Mskhali	494.97 ± 19.22bc	480.40 ± 7.27ef
Arevik	699.9 ± 19.92d	243.07 ± 9.40cd
Qrdi khaghogh	552.27 ± 16.09c	85.87 ± 5.10a
Mormor	272.47 ± 15.12a	295.38 ± 12.78d
Hakobi vordi	372.13 ± 15.94ab	493.86 ± 31.08g
Krhagi karchevan	1244.87 ± 16.62f	185.33 ± 17.55bc
Kakhet	1251.50 ± 15.01f	620.33 ± 10.4h
Meghri vahravar (Shireyi)	506.73 ± 12.42bc	297.11 ± 11.03d
Sali	1005.77 ± 12.32e	150.34 ± 11.03ab
Sev krop	1105.94 ± 12.57e	127.03 ± 4.70ab
	Interspecific and intraspecific hybrids	
Vardananq	953.80 ± 11.47jk	274.51 ± 10.1g
Narekaci	681.53 ± 14.73h	362.15 ± 16.00h
Ayvazyani Vardabuyr	333.77 ± 16.67e	173.32 ± 9.37def
Itsaptuk	569.03 ± 13.98g	283.20 ± 12.78g
Parvana	279.57 ± 14.84de	200.27 ± 13.31f
Hayastan	264.77 ± 5.65cde	165.54 ± 9.18def
Hayastani quyry	186.53 ± 11.17ab	247.58 ± 16.92g
Tzeni	234.57 ± 14.03bcd	149.01 ± 1.03cde
Shahumyani	154.00 ± 17.51a	75.08 ± 14.19a
Anahit	436.80 ± 11.41f	288.44 ± 31.06g
Berqanush	1355.57 ± 17.13k	534.79 ± 17.76i
Arevshat	210.70 ± 8.30abcd	102.30 ± 8.13abc
Arazi	204.67 ± 13.01abc	161.85 ± 12.56def
Aygezard	160.60 ± 16.14a	85.87 ± 5.10ab
Armenia	512.57 ± 13.80g	137.58 ± 12.69cde
Hayreniq	712.77 ± 12.05h	394.61 ± 18.42h
Charenci	802.80 ± 13.32i	176.56 ± 12.00ef
Nali	554.57 ± 13.14g	127.72 ± 12.59bcd
Megri vahravar (semiwild)	887.00 ± 12.10j	725.24 ± 17.72j
	Wild grapes	
Wild grape Megri vahravar	1160.17 ± 17.44e	1263.29 ± 10.82d
Wild grape 1	2911.50 ± 11.38g	2385.29 ± 14.72f
Wild grape 2	527.03 ± 16.18b	648.89 ± 17.82c
Wild grape 3	944.94 ± 18.54d	489.27 ± 12.63b
Wild grape 4	328.20 ± 18.97a	325.51 ± 14.63a
Wild grape 5	339.47 ± 18.52a	372.98 ± 14.42a
Wild grape 6	838.67 ± 10.59c	349.86 ± 12.70a
Wild grape 7	911.67 ± 15.91cd	597.76 ± 14.07c
Wild grape 8	2118.30 ± 15.77f	1819.19 ± 17.47e

Values represent means of triplicate determination ± S.D.

Data were analyzed by ANOVA and within each column different letters indicate statistically different values according to Multiple Range test at $p \leq 0.05$.

Vitis vinifera ssp. *sylvestris* (approx. 3000 accessions) were discovered, which need to be maintained and characterized. The presented study just estimated the content of total phenolics and antioxidant capacity of wild species with different genetic background and geographic origin, when the quantitative significant differences by comparing the total phenolic spectra of wild grape skin and seeds extracts were identified. The total phenolics in berry skins of nine wild grapes ranged between 2911.50 mg/kg of grape and 328.20 mg/kg of grape with significant differences among the genotypes. We revealed that the Wild grape 1 contains significant high content of skin and seed total phenolics (2911.50 mg/kg of grape, 2385.29 50 mg/kg of grape) respectively, followed Wild grape 8 (2118.30 mg/kg of grape, 1819.19 mg/kg of grape) and Wild grape Megri vahravar (1160.17 mg/kg of grape, 1263.29 19 mg/kg of grape) respectively.

The obtained results showed that the total phenolic content of cultivated varieties and wild species was quite variable. In average, the content of total phenolics determined in wild species (2035,8 mg/kg of grape) was significantly higher in comparison with aboriginal varieties (1106,6 mg/kg of grape) and hybrids (630,2 mg/kg of grape), $p \leq 0.05$. The variability found in total phenolic content between different grape genotypes confirmed the hypothesis that genetic and environmental factors are key influencers of a cultivar's phenolic content.

DOI: 10.1051/bioconf/20170902029

3.2. DPPH radical-scavenging activity

Analysis of DPPH radical-scavenging activity is widely used to evaluate the ability of free radicals to react with antioxidants. The measurement of the consumption of DPPH radical allows to determine exclusively the intrinsic ability of a substance to donate hydrogen atoms or electrons to this reactive species in a homogenous system. The method is based on the reduction of DPPH solution in the presence of a hydrogen donating antioxidant due to the formation of non radical form DPPH-H by the reaction blue-violet colour changes gradually to yellow and a decrease in absorbance at 517 nm is monitored during the reaction [16]. Free radicals cause the oxidation of biomolecules and leads to the cell injury and death. Free radical scavenging capacity was the predominant mechanism in the antioxidant activity of samples, indicating that phenolic antioxidants are potent free radical terminators.

In our study, all the extracts of cultivated and wild grape genotypes exhibited appreciable scavenging activity and ranging from 43.3 to 92.2% for aboriginal grape cultivars, from 32,4 to 80.9% for interspecific and intraspecific hybrids and from 40.61 to 65.48% for wild species. A regression analysis was used to correlate the results of total phenolic content and antioxidant activity detected in cultivated and wild genotypes (Figs. 1, 2, 3). For all analyzed aboriginal genotypes, hybrids and wild species the significant correlation was detected ($R^2 = 0.887$, $R^2 = 0.654$ and $R^2 = 0.776$, $p \leq 0.001$) respectively.

Extracts of all the aboriginal grape varieties demonstrated DPPH radical scavenging activity, which suggests that, they all contained antioxidants. The highest antioxidative activity was revealed by extracts of "Garan dmak" and "Êakhet" cultivars, characterized by the highest concentration of total phenolics. The weakest antioxidative activity was found for the extract obtained from "Mormor" cultivar, demonstrated lowest content of total phenolics.

Among interspecific and intraspecific hybrids the highest capacity to scavenge DPPH free radicals was determined for an extract from "Vardananq", "Berqanush" and "Hayreniq" hybrids, demonstrated the highest concentration of total phenolics. Strong correlations suggested that the phenolic content of these varieties contributed significantly to the antioxidant capacities of grape extracts. Figure 3 showed that all the wild genotypes have a significant inhibitory activity against the DPPH radicals, but the higher inhibition of DPPH radical observed for Wild grape 1 and Wild grape 8.

The obtained results demonstrated that scavenging activity was different between the cultivated grapes and wild species, which was due to the difference in their phenolic contents. One of the aims of the presented research was to determine the correlation between total phenolics and antioxidant activity in cultivated and wild grapes. Some researchers reported a strong correlation between antioxidant activity and total phenolics whereas others also indicated that total phenolics may be less significantly correlated with the antioxidant properties.

According to the results of the present screening study, we can emphasize on the great variability of total phenolic content and on the notable scavenging activity of studied grape genotypes.

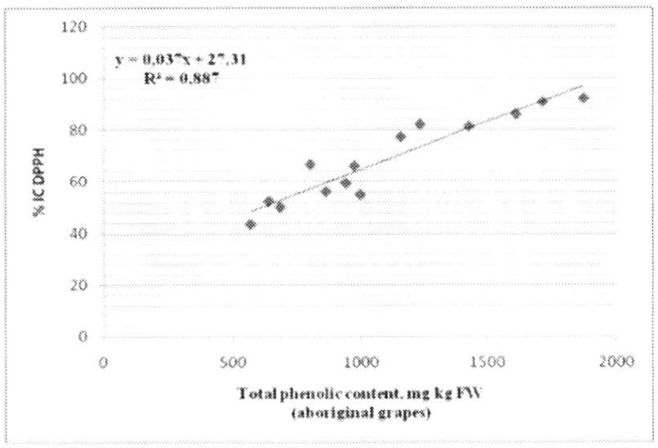

Figure 1. Linear regression between total phenolics and antioxidant activity (DPPH) of Armenian aboriginal grapes.

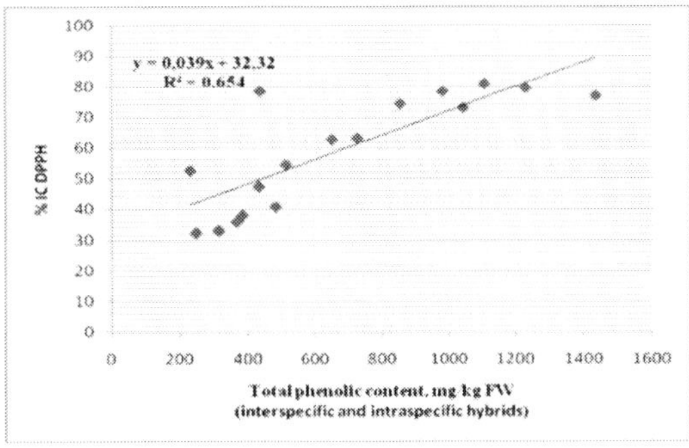

Figure 2. Linear regression between total phenolics and antioxidant activity (DPPH) of Armenian interspecific and intraspecific hybrids.

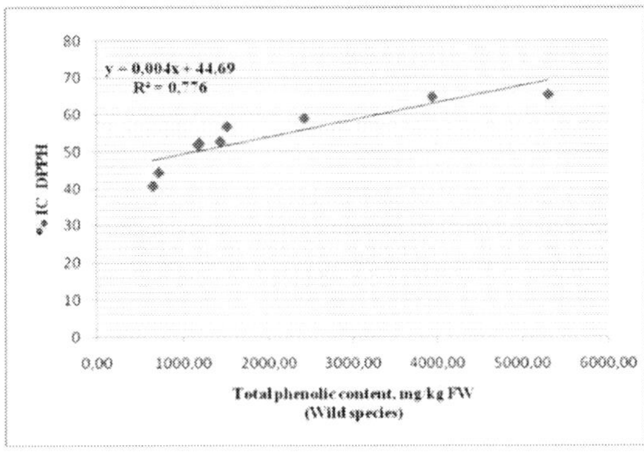

Figure 3. Linear regression between total phenolics and antioxidant activity (DPPH) of wild species.

Presented data indicates the marked antioxidant activity of Armenian cultivated and wild grape extracts strongly related with high content of total phenolics, acting as reductones by donating the electrons and reacting with free radicals to convert them into more stable product.

4. Conclusions

Knowledge of the phenolic compounds in grape berry seed and skin will contribute to more comprehensive assessment of their biological activities. The present research was aimed at characterizing forty Armenian *V. Vinifera* spp. *Sativa* and *V. Vinifera* spp. *Sylvestris* through the determination of their total phenolics and antioxidant activity.

The realized research has revealed a notable difference among the cultivated varieties and wild species in the total phenolic content and antioxidant activity decreasing in following order: wild species > aboriginal varieties > interspecific and intraspecific hybrids. High levels of correlation between the total phenolics and antioxidant activity for all studied grape genotypes were revealed.

Obtained results will support the importance of preserving the genetic diversity and favor the reintroduction of grape cultivars and wild genotypes thanks to the present valorization. Finally, presented research will contribute to the sustainable development of agriculture in Armenia through establishing the basis for the improvement and modernization of the grape and wine production.

The present research was realized as a part of the project "The role of biodiversity of Armenian cultivated and wild grapevine in environmental balance", 15T-1E232.

References

[1] F. Mannini, I International Symposium on Grapevine Growing, Commerce and Research **652**, 87–95 (2003)
[2] Z. Liang, C. L. Owens, G. Y. Zhong and L. Cheng, Food Chem. **129**(3), 940–950 (2011)
[3] Z. Liang, L. Cheng, G. Y. Zhong and R. H. Liu, PLoS ONE **9**(8), e105146 (2014)
[4] A. De la Cerda-Carrasco, R. López-Solís, H. Nuñez-Kalasic, Á. Peña-Neira and E. Obreque-Slier, J. Sci. Food Agric. **52**(14), 4360–4367 (2014)
[5] C. Manach, A. Scalbert, C. Morand, C. Rémésy and L. Jiménez, Am. J. Clin. Nutr. **79**(5), 727–747 (2004)
[6] S. H. Nile, S. H. Kim, E. Y. Ko and S. W. Park, Biomed.Res. Int. **2013**, 718065 (2013)
[7] L. M. Ramirez-Lopez, W. McGlynn, C. L. Goad and C. A. Mireles DeWitt, Food Chem. **149**, 15–24 (2014)
[8] I. M. Huseynova, Biochim. Biophys. Acta **1817**(8), 1516–1523 (2012)
[9] G. Di Marco, A. Gismondi, L. Canuti, M. Scimeca, A. Volpe and A. Canini, Plant Biol. **16**(4), 792–800 (2014)
[10] K. E. Heim, A. R. Tagliaferro, D. J. Bobilya, J. Nutr. Biochem. **13**(10), 572–584 (2002)
[11] J. Shi, J. Yu, J. E. Pohorly and Y. Kakuda, J. Med. Food **6**(4), 291–299 (2003)
[12] A. M. Connor, J. J. Luby, C. B. S. Tong, C. E. Finn, J. F. Hancock, Journal of the American Society for Horticultural Science **1**, 82–88 (2002)
[13] A. R. Proteggente, A. S. Pannala, G. Pagana, L.Van Buren, E. Wagner, S. Wiswman, et al., Free Radical Research **36**, 217–233 (2002)
[14] L. Rustioni, D. Maghradze, C.F. Popescu, G. Cola, E. Abashidze, R. Aroutiounian et al., Vitis **53**(4), 219–226 (2014)
[15] W. Brand-Williams, M. E. Cuvelier, C. Berset, Food Science and Technology-Lebensmittel-Wissenschaft and Technologie **28**(1), 25–30 (1995)
[16] X. J. Duan, W. W. Zhang, X. M. Li, B. G. Wang, Food Chemistry **95**(1), 37–43 (2006)

 DOI: 10.1051/bioconf/20170902030

Procédés de traitements des effluents de cave adaptés au secteur des vins biologiques : état des lieux et perspectives / Processes for wineries wastewater adapted to the sector of the biological wines: Prospect and inventory of fixtures

Joël Rochard

IFV (Institut Français de la Vigne et du Vin), 17 rue Jean Chandon Moët, 51200 Epernay, France

Résumé. Au-delà des itinéraires viticoles et œnologiques, la viticulture biologique doit également s'intéresser aux différents aspects de la viticulture durable concernant notamment le traitement des effluents de cave, avec une approche écologique. Parallèlement à la performance d'épuration, le traitement des effluents doit intégrer différentes orientations écologiques : faible consommation d'énergie et limitation des déchets (boues) de plus en plus difficile à gérer par voie agronomique. En complément, en liaison avec le concept d'éco-oenotourisme, une intégration harmonieuse de dispositif peut être envisagée, qui associe à la fois une limitation des nuisances olfactives et sonores, une valorisation paysagère et de la biodiversité ainsi qu'éventuellement une réutilisation des eaux traitées pour l'irrigation. Les techniques de lits ou massifs plantés ont prouvé leur efficacité dans le domaine de l'épuration des effluents d'origine domestique. Leur application aux effluents de cave qui fait l'objet de différentes recherches depuis plusieurs années, permet de valoriser l'environnement esthétique et écologique de la cave. L'objectif de la communication et de présenter le principe de l'épuration par lit planté de roseaux avec les différentes mises en œuvre envisageable au niveau des caves.

Abstract. Beyond the wine and oenological routes, the biological viticulture must also be interested in the various aspects of the durable viticulture relating to the liquid waste processing in particular of cellar, with an ecological approach. Parallel to performance of purification the liquid waste processing must integrate various ecological orientations: low energy consumption and limitation of waste (muds) increasingly difficult to manage by agronomic way. In complement, in connection with the concept of eco-wine tourism, a harmonious integration of device can be considered, which associates at the same time a limitation of smell pollutions and sound, a landscape and biodiversity valorization and possibly a re-use of the water treated for the irrigation. The techniques of beds or planted solid masses proved their effectiveness in the field of the purification of the effluents of domestic origin. Their application to the effluents of cellar which has been the object of various research for several years, makes it possible to develop the esthetic and ecological environment of the cellar. The objective of the communication and to present the principle of purification by planted bed of reeds with the various implementations possible to the level of the cellars.

1. Introduction

Issu d'un mouvement idéologique né au début du 20ème siècle, l'agriculture et la viticulture biologiques s'appuient sur un lien étroit entre agriculture et nature, privilégiant les équilibres biologiques entre la plante cultivée et son environnement, le respect des rythmes naturels et excluant les produits de synthèse. Au-delà des itinéraires viticoles et œnologiques, la viticulture biologique doit également s'intéresser aux différents aspects de la viticulture durable concernant notamment le traitement des effluents de cave, avec une approche écologique.

Les rejets issus des pressoirs et des caves sont susceptibles de perturber l'équilibre biologique des rivières en particulier pendant la période des vendanges. En effet, les éléments organiques issus des activités vinicoles génèrent, dans un milieu aquatique, le développement de micro-organismes qui puisent l'oxygène dissous au détriment de la faune piscicole.

Les effluents de cave (0,5 à 5 litres/litre de vin) sont de nature organique (DCO de 5 à 30 g/litre) et sont rejetés majoritairement pendant la période de vendanges (2 à 8 semaines). Le traitement a généralement pour objectif, selon la réglementation locale de réduire la pollution un niveau de 125 à 300 mg de DCO par litre.

La lutte contre la pollution dans le domaine vinicole repose sur deux démarches complémentaires. En amont, une adaptation du processus d'élaboration doit être mise en œuvre pour réduire la charge polluante et assurer une gestion optimale de l'eau. En aval, le traitement des effluents de cave réalisé individuellement ou collectivement, peut être envisagé avec plusieurs techniques : évaporation, épandage, dispositifs biologiques [1,2].

Jusqu'à présent, les procédés de traitement les plus utilisés étaient basés sur des développements technologiques de procédés aérobies et dans une moindre

 DOI: 10.1051/bioconf/20170902030

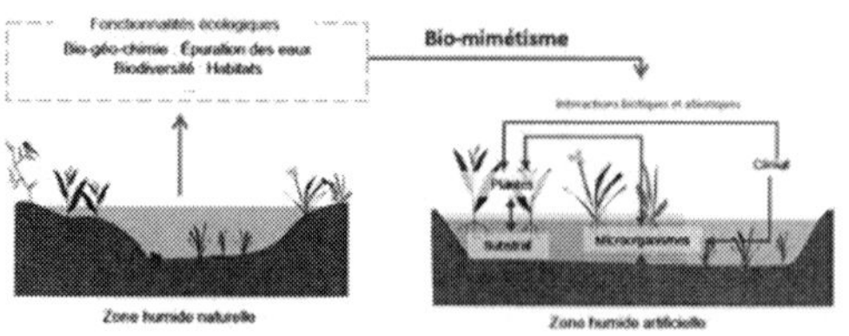

Figure 1. Comparaison d'une zone humide naturelle par rapport un dispositif artificiel. *Source Blueset.*

Figure 2. Rôle épurateur des zones humides. Source: Les zones humides un patrimoine à préserver agence de l'eau RMC/ Région Rhône-Alpes.

mesure anaérobies. Les recherches actuelles ont pour objectif d'intégrer les orientations de développement durable dans le fonctionnement du dispositif de traitement. Le traitement des effluents doit intégrer différentes orientations : faible consommation d'énergie et limitation des déchets (boues) de plus en plus difficile à gérer par voie agronomique. En complément, en liaison avec le concept d'éco-oenotourisme, une intégration harmonieuse de dispositif peut être envisagée, qui associe à la fois une limitation des nuisances olfactives et sonores, une valorisation paysagère et éventuellement de la biodiversité. Bien évidemment, la gestion optimale de l'eau en amont s'impose afin de faciliter le traitement et dans une perspective de raréfaction de la ressource dans de nombreuses régions en liaison avec les changements climatiques.

L'épuration par le sol est un mécanisme utilisé depuis longtemps, notamment la technique d'épandage des effluents. L'épandage des effluents de cave est souvent utilisé sur des terrains agricoles ou de manière plus intensive sur des zones plantées avec des espèces à fort potentiel de développement végétatif (saule, bambou, eucalyptus).

Une autre approche consiste à utiliser le principe d'épuration naturelle des zones humides lié à des plantes dotées d'un fort potentiel racinaire, adaptées à des alternances de conditions sèches et humides [3]. Certaines de ces plantes (roseaux/*Phragmita Australis*) assurent parallèlement un transfert d'oxygène dans le sol par l'intermédiaire de la tige (Fig. 1).

Les filtres plantés de roseaux (ou «constructed wetlands») se classent parmi les filières de traitement biologique basées sur le principe de l'infiltration – percolation. Ce procédé reproduit la dynamique autorégulatrice d'un véritable écosystème : il comprend les interactions vivantes entre différentes espèces de bactéries, les racines des macrophytes, le substrat, l'air, le soleil, l'eau (Fig. 2).

Les techniques de lits ou massifs plantés ont prouvé leur efficacité dans le domaine de l'épuration des effluents d'origine domestique. Leur application aux effluents de cave a fait l'objet de différentes recherches depuis plusieurs années.

Figure 3. Inflorescence et vue en coupe d'un roseau (*Phragmita Australis*).

2. Principe du dispositif par lit planté de roseaux

Le processus reproduit d'une certaine manière le processus naturel d'épuration de l'eau dans les marais dans lesquels les eaux usées sont naturellement préfiltrées et débarrassées des particules solides, puis subissent des traitements naturels physiques, chimiques et surtout biologiques favorisés par des plantes aquatiques ayant au final pour effet de dégrader les matières organiques, de transférer vers les feuilles les composés métalliques, de filtrer et de réduire sensiblement les germes pathogènes des eaux usées.

La présence des végétaux induit de façon indirecte un certain nombre de mécanismes favorisant l'épuration : maintien de la structure du massif, apport d'oxygène dans le milieu filtrant et développement de la flore bactérienne.

Le massif filtrant est installé dans un bassin d'une profondeur généralement comprise entre 50 et 70 centimètres étanchéifié le plus souvent par une géo-membrane pour éviter que l'eau non encore traitée ne s'infiltre dans la nappe phréatique.

Pour le traitement d'eaux usées domestiques, le dimensionnement standard est de 1.5 m^2 par habitant soit environ 60 g DCO/m^2.j. Concernant les effluents vinicoles, compte tenu de la forte variabilité de la caractéristique des effluents selon les caves, le calcul de surface est à établir au cas par cas en fonction du type de mise en œuvre de ce procédé.

Les plantes aquatiques, et particulièrement les roseaux (Fig. 3) possèdent un tissu particulier permettant le transfert d'oxygène depuis les parties aériennes (tiges et feuilles), vers les parties souterraines : il est libéré au niveau des jeunes racines dans le film aqueux entourant le «chevelu racinaire» (Fig. 4). Les bactéries épuratrices présentes à proximité de ces racines sont ainsi alimentées en oxygène.

L'exploitation des capacités épuratoires des massifs filtrants plantés peut être mise en œuvre de deux façons:

- Soit par **infiltration de l'eau verticalement** dans des sols plantés de macrophytes et généralement drainés (filtre à flux vertical ou filtre vertical).

Le lit à flux vertical (Figs. 4 et 5) est un sol artificiel constitué de plusieurs couches de matériaux granulaires superposées dans lequel se développent les rhizomes. Pour favoriser l'oxygénation du filtre, les eaux à traiter sont injectées de manière séquentielle dans un réseau d'épandage placé à la surface du massif. Le réseau

DOI: 10.1051/bioconf/20170902030

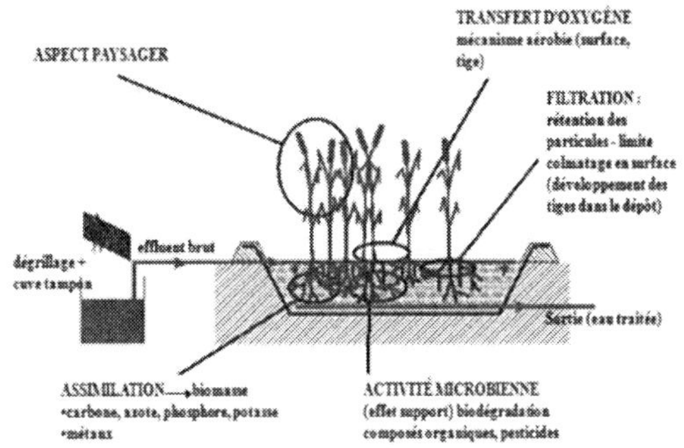

Figure 4. Principe d'un lit planté vertical d'après J. Rochard.

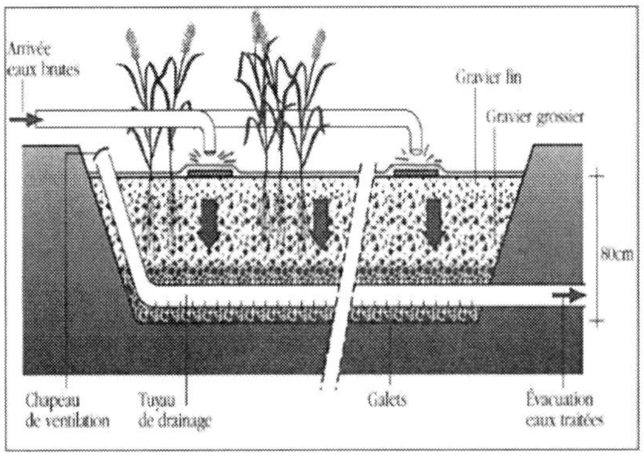

Figure 5. Principe du lit planté vertical.

d'alimentation se trouvant en charge à chaque bâchée, l'effluent est réparti de façon régulière, évitant ainsi la formation de zones de saturation.

Les effluents percolent par gravité jusqu'à des drains au fond du bassin et sont ainsi évacués en partie basse du système. Les tiges, par leurs oscillations, sous l'effet du vent maintiennent à leur base un anneau libre qui facilite la circulation hydraulique dans le massif et réduit le colmatage notamment lorsque les effluents sont très chargés en matière en suspension.

Le temps de séjour assez court et l'alimentation séquentielle préviennent la saturation, permettent l'aération du massif et favorisent les phénomènes de dégradation aérobies.

- Soit par **circulation de l'eau** dans la rhizosphère des macrophytes par cheminement horizontal sous la surface du sol (filtre horizontal).

Le lit à flux horizontal (Fig. 6) est un sol artificiel dont les granulométries sont échelonnées en barrières filtrantes selon un vecteur horizontal. Les eaux à traiter, injectées à l'une des extrémités du lit filtrant, pénètrent horizontalement dans la structure puis sont évacuées par drainage à l'autre extrémité.

L'alimentation est, en général, réalisée en continu, de manière à saturer en permanence les matériaux.

Seule une faible aération de surface, complétée par le transfert d'oxygène au travers de la tige des roseaux intervient. Ce dispositif exige de traiter des effluents avec une faible teneur en matière suspension. Le risque de colmatage impose de procéder à un dégrillage très fin.

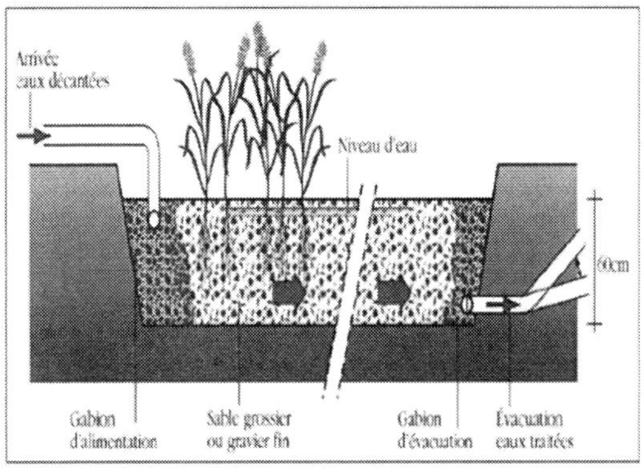

Figure 6. Principe du lit planté horizontal.

Figure 7. Un exemple de valorisation écologique (biodiversité, paysages) du traitement des effluents de cave par lit plante de roseaux (dispositif Blueset en cours d'installation à la coopérative de Buzet dans le sud-ouest de la France).

Parallèlement à leur fonction épuratoire, l'implantation d'un lit planté de roseaux peut s'intégrer dans une démarche paysagère et de biodiversité dans l'environnement de la cave et servir de support à une démarche d'éco-oenotourisme (Fig. 7).

3. Application à la filière vitivinicole

3.1. Traitement des boues de procédés biologiques issus des dispositifs d'épuration biologique

Les boues peuvent faire l'objet d'un traitement par l'intermédiaire d'un lit spécifique. Comparativement à un séchage sur lit de sable, les plantes permettent une dégradation et une stabilisation des boues qui réduit le volume final de produit et ses nuisances olfactives. Ce phénomène est probablement accentué par la présence du couvert végétal qui limite le déplacement d'air et la dispersion des odeurs.

Les boues sont prélevées dans le bassin de stockage après traitement et vidange de l'eau épurée, et déposées par couches successives en surface des lits, selon les cycles d'alimentation et de repos.

Figure 8. Exemple de traitement des boues par un dispositif de lits plantés dotés de système d'aération naturelle par le dispositif Sint au Château Mont Redon à Châteauneuf-du-Pape.

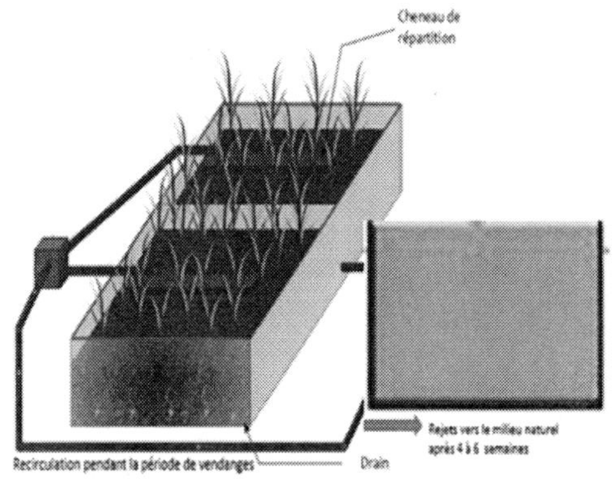

Figure 9. Dispositif de traitement des effluents sur lit planté avec période de recyclage pendant les vendanges. Schéma adapté de `www.plantepure.fr`.

Au final elles sont extraites par curage et valorisées par épandage sur terres agricoles ou compostage (Fig. 8).

3.2. Traitement des effluents par recirculation

Le potentiel de biodégradation limité d'un lit planté (environ 1 gramme de DCO/litre), peut être surmonté par une recirculation de l'effluent à partir d'un bassin tampon de stockage fermé sans aération (Fig. 8). Ainsi, l'effluent est progressivement épuré au cours des percolations successives sur le dispositif (32m3 d'effluent pour une surface de 27 m^2 de lit planté de roseaux).

Des essais réalisés dans une petite cave de la région Bordelaise ont permis de montrer la faisabilité de ce dispositif.

Selon le niveau de DCO de départ, l'épuration peut être obtenue dans un délai de 4 à 6 semaines (Figs. 9–10).

3.3. Traitement de finition

La plupart des dispositifs de lits plantés de roseaux assurent le traitement d'effluent dont la teneur en DCO est proche de 1 gramme par litre pour atteindre les normes de rejets qui varient selon les régions de 125 à 300 milligrammes par litre [4]. Le lit planté est généralement disposé en aval d'un bassin aérobie ou éventuellement d'un dispositif anaérobie, associant selon les cas une épuration de 80 à 95 % pour atteindre un niveau proche de 1 à 1,5 gramme de DCO par litre. Dans ce cas, parallèlement au traitement de finition, le lit planté peut assurer une dégradation des boues du dispositif biologique amont (Figs. 11 et 12).

Cette approche classique de traitement par lit planté en finition impose de maintenir un bassin d'aération en amont, consommateur d'énergie est source potentielle de nuisance olfactive et visuelle d'où la recherche de

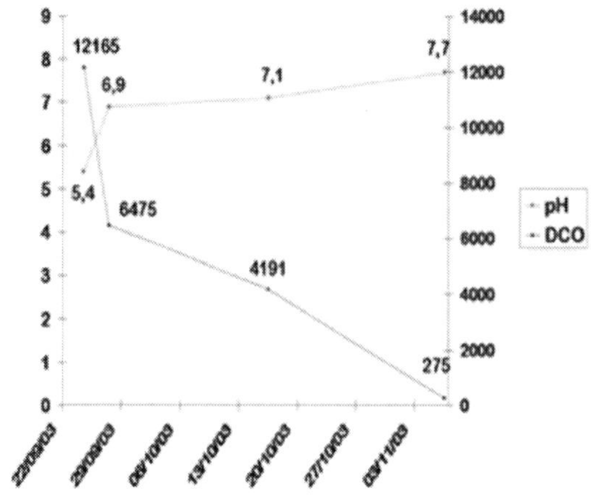

Figure 10. Cinétique d'épuration d'un dispositif de lit planté par recirculation (Source : S. Kerner et J. Rochard).

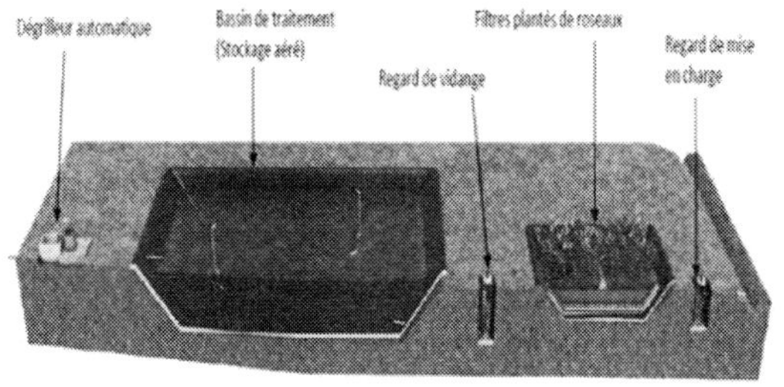

Figure 11. Schéma d'un traitement associant un premier étage par stockage aéré et une finition par filtres plantés de roseaux. (*Source : Agroenvironnement/ Syntea*).

Figure 12. Exemple de traitement de finition par lit planté de roseaux après un premier étage aérobie (Cave Spier en Afrique du Sud) [5].

Figure 13. Une zéolithe, ou zéolite, est une roche formé d'un squelette microporeux d'aluminosilicate, dont les espaces vides permettent d'optimiser l'adsorption et les processus de biodégradation du système racinaire des roseaux (Source Zeofito®).

procédés susceptibles de traiter directement les effluents en associant bien évidemment des mesures en amont du processus d'élaboration pour limiter la charge et la concentration polluante.

3.4. Lit planté sur support de zéolithes

Une piste d'optimisation du procédé consiste à utiliser un matériau très adsorbant comparativement au sable ou gravier des filtres traditionnels (Fig. 13). Une expérimentation a été menée dans une cave de Barolo (Rochard et al.) en Italie avec un filtre composé de zéolite qui possède des propriétés intéressantes d'adsorption et d'échange cationique [6].

Les mesures réalisées pendant une campagne de vendange et de vinification ont montré la possibilité de traiter des effluents d'une teneur moyenne de 3 grammes de DCO avec une concentration moyenne de l'effluent traité proche de 100 mg/litre (Fig. 14).

À la suite de cette expérimentation de nombreuses caves, notamment en Italie, se sont équipées de ce dispositif en traitement direct sans bassin d'aération avec au préalable un dispositif de dégrillage/filtration et selon les cas éventuellement à une neutralisation (Fig. 15).

3.5. Lit planté intensifié par aération forcée

Le procédé Rhizosph'air®) est un traitement végétalisé planté (roseaux, iris) et intensifié par aération forcée. Il vise à combiner, par insufflation d'air, la rusticité d'un premier étage de filtre alimenté en eaux brutes et une gestion intégrée des boues (Fig. 16). Parallèlement, l'apport d'oxygène, permet d'adapter le traitement aux variations de charges et à différents niveaux de rejets. L'aération forcée fonctionne à basse pression (faible hauteur d'eau dans le filtre), ce qui conduit, selon le concepteur, à une faible consommation énergétique (inférieure à 0,3 kWh/m^3 d'eaux traitées), comparativement aux procédés intensifs pour lesquels, pour un niveau de traitement équivalent, la consommation énergétique liée

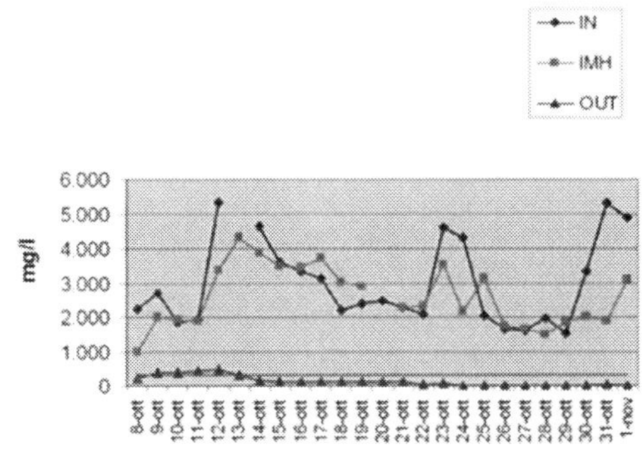

Figure 14. Teneur en DCO des effluents entrée après décantation (IMH) et sortie du dispositif avec massif de zéolitite Zeofito (traitement direct), (Rochard et coll.).

Figure 15. Traitement des effluents par lit planté de roseaux sur support de zéolithes avec le dispositif Zeofito® (cave Baroli dans le Piémont italien).

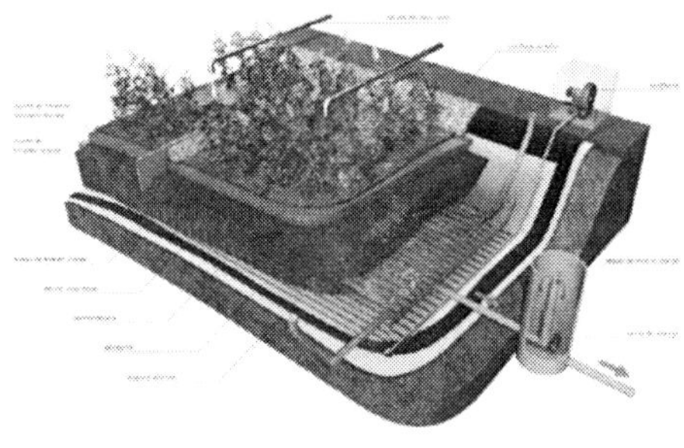

Figure 16. Schéma du traitement des effluents par aération forcée (Rhizosph'air®).

à l'aération est supérieure (généralement supérieure à 0,6 kWh/m^3 d'eaux traitées).

4. Conclusion

Les lits plantés qui s'inspirent des écosystèmes de milieux humides, s'intègrent dans la diversité des dispositifs de traitement des effluents de cave et de pulvérisation.

 DOI: 10.1051/bioconf/20170902030

La conception rustique, la simplicité de gestion, la faible consommation énergétique, la valorisation paysagère sont autant d'arguments qui intéressent les professionnels désireux de développer des démarches durables vis-à-vis des effluents de cave. Au-delà de la dimension esthétique, il est possible d'envisager de recréer des zones humides artificielles qui valorisent la biodiversité locale.

Le traitement de finition ou la gestion des boues des dispositifs issus du traitement des effluents domestiques est largement développé. Des développements plus novateurs (recyclage, traitement direct sur zéolitite, dispositif d'aération forcée) offrent des perspectives intéressantes pour l'avenir.

Comme tout système écologique, le choix de ce type de traitement impose une gestion optimale de l'eau et des sous-produits (bourbes, lies, solution détartrage, terres de filtration, etc.). Parallèlement, si le fonctionnement de ces systèmes est simple et rustique, la réalisation doit être confiée à des spécialistes, avec si possible une bonne expérience du secteur vinicole, pour optimiser la conception et dimensionnement (en intégrant notamment la pointe d'activité pendant les vendanges).

References

[1] J. Rochard, *Traité de viticulture et d'œnologie durables*, éditions Avenir œnologie (2005)
[2] J. Rochard, F. Jourjon, Y. Racault, *Effluents vinicoles, gestion et traitements*, éditions Féret (2001)
[3] J. Rochard, Innovation environnementale dans la gestion des effluents de cave : application des lits plantés de roseaux ? 32ème Congrès Mondial de la Vigne et du Vin, Zagreb, Croatie, 28 Juin – 3 juillet (2009)
[4] Meunier et al., *treatment of winery effluent by a process combinig an activated sludge reactor with reed bed filters*, winery waste and ecologic impacts management (2009)
[5] A.R. Muldizi, *Winery and distillery wastewater treatment by constructed wetland with shorter retention time*, 5th international specialized conference on sustainable viticulture: winery waste and ecologic impacts management (2009)
[6] J. Rochard, A. Oldano, D. Marengo, Zeofito™, *the active phytopurification system*, 5th international specialized conference on sustainable viticulture: winery waste and ecologic impacts management (2009)

Influence of supplementation with inactivated dry yeasts during the tirage of sparkling wines (Cava) on its composition, its foaming properties and its sensorial quality

Laura Medina-Trujillo[1], Elena González-Royo[1], Nathalie Sieczkowski[2], José M. Heras[2], Francesca Fort[1], Joan Miquel Canal[1], and Fernando Zamora[1]

[1] Departamento de Bioquímica y Biotecnología, Facultad de Enología de Tarragona, Universidad Rovira i Virgili, C/ Marcel.lí Domingo, 1, 43007 Tarragona, España
[2] Lallemand Bio S.L. C/ Galileu 303, 1ªplanta. 08028 Barcelona, España

Abstract. Different sparkling wines were elaborated with a base wine of Macabeo from the AOC Cava using the traditional method with supplementation or not of inactivated dry yeasts of *Saccharomyces cerevisiae* (Optimumwhite™) or *Torulaspora delburueckii* (TD291). After 9 months, the samples were analyzed and tasted. The supplementation with both IDY (*S. cerevisiae* and *T. delbrueckii*) produced Cavas with better foaming properties than the Cava control. The tasters could significantly distinguish between the Cavas supplemented with both IDY from the Cava control. Moreover, taster clearly preferred the Cava elaborated with supplementation of IDY of *T. delburueckii* than the control one.

1. Introduction

There are different effervescent beverages including carbonated drinks, beer and, of course, sparkling wines. Their effervescence and their foam are very different. Thus, gasified beverages generally have a very tumultuous effervescence, with large bubbles and relatively short foam persistence. On the opposite side we find beer, with a weak effervescence, very fine bubbles and extremely persistent foam. Well, sparkling wines have intermediate properties, which is very important to guarantee their organoleptic characteristics [1].

It should be taken into account that effervescence and foam are not only visual attributes linked to the beauty of sparkling wines but have also significantly influence the sensations perceived on the palate [2]. A carbonated beverage with a tumultuous effervescence and large bubbles leads to a feeling of aggressiveness in the mouth. In the same way, it will quickly lose its effervescence, given the rapid evolution of gas observed in this type of product. On the contrary, a quality sparkling wine, will present in the mouth a pleasant tingling accompanied by the sensation of creaminess that generates its foam. For these reasons, the effervescence must last the time necessary for the consumer to enjoy their tasting.

For all that reasons, the sparkling winemakers are especially interested in understanding what the sparkling properties depend on and how they can improve them [3,4]. The sparkling wines made by the traditional method develop the second fermentation in the bottle and are maintained in aging with the lees to enrich the wine in proteins, mannoproteins and polysaccharides with the aim of obtaining better effervescence and foam [5,6]. Long contact with the lees gives a greater aromatic and gustatory complexity and a better integration of the carbonic gas [7,8].

The yeasts' autolysis is the main responsible for these changes [9]. However, long aging is costly and for this reason there is interest in techniques that allow obtaining better foam in shorter time. Supplementation with inactivated dry yeasts (IDY) as additional source of proteins, mannoproteins and polysaccharides has been proposed for this purpose [2,10]. However, there are no scientific studies that support its usefulness. For this reason, the aim of this study was to determine if the supplementation with specific IDYs during the tirage really improves the foaming properties of sparkling wines.

2. Materials and methods

The elaboration of the sparkling wines (Cava) was made by the traditional method with a Macabeo base wine. 36 bottles were prepared. Twelve of them were added with 30 mg/L of a specific glutathione-rich IDY from a *Saccharomyces cerevisiae* yeast (Optimumwhite™, Lallemand Inc.). Another 12 bottles were added with 30 mg/L of an experimental IDY of *Torulaspora delbrueckii* TD291 (Lallemand Inc.). The remaining 12 bottles were considered as control. The sparkling wines were disgorged, analyzed and tasted 9 months later.

The analytical methods recommended by the OIV were used to determine the ethanol content, titratable acidity, pH and volatile acidity [11]. Glycerol and sugars (D-glucose and D-fructose) were measured using enzymatic kits (R-Biopharm AG., Darmstadt, Germany). The foaming properties were determined by Mosalux method [12]. Proteins were analyzed by HRSEC-DAD [13]. The polysaccharides were determined by HRSEC-RID [14]. The sensorial analysis was performed with 12 trained oenologists by triangular test.

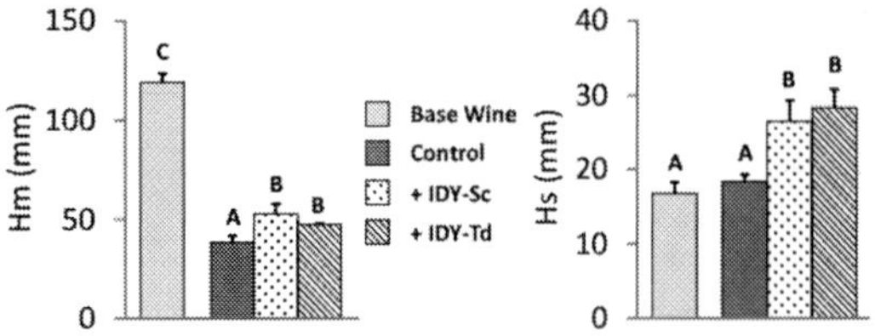

Figure 1. Hm: maximum height of the foam; Hs: stable height of the foam. + IDY-Sc: supplemented with inactive dry yeasts of S. cerevisiae; + IDY-Td: supplemented with inactive dry yeast from T. delbrueckii. Different letters indicate the existence of statistically significant differences (p <0.05).

All data are expressed as the arithmetic average ± the standard deviation from three replicates. One-factor analysis of variance (ANOVA) and Tukey's test were carried out with SPSS software (SPSS Inc., Chicago, Illinois, USA).

3. Results and discussion

Table 1 shows the influence of supplementation with inactive dry yeasts during the second fermentation in the general parameters of sparkling wines.

All the sparkling wines have higher alcoholic strength, pH, volatile acidity and glycerol concentration, and lower titratable acidity than the original base wine. These differences were expected and attributable to the natural changes produced by the second fermentation. No differences on ethanol, titratable acidity, pH and volatile acidity were found between the different sparkling wines. However, glycerol levels were significant higher in the Cavas supplemented with both IDY.

Figure 1 shows the foam parameters of the original base wine and the different sparkling wines.

As expected, the maximum foam height (Hm) of all Cavas was significantly lower than that of their base wine. This drastic reduction of Hm, already described in previous studies [15, 16], has been attributed to the increase of the ethanol produced during the second fermentation and the absorption of proteins by the bentonite used with adjuvant [17].

The supplementation with both IDY (*S. cerevisiae* and *T. delbrueckii*) produced Cavas with better foaming properties than the Cava control. Thus, both parameters, Hm and Hs, were significantly higher in sparkling wines supplemented with both IDY than in their corresponding control. It should be noted that the increase of both Hm and Hs was similar in the Cavas supplemented with IDY of *S. cerevisiae* and those supplemented with IDY of *T. delbrueckii*.

In contrast, no significant differences were found in protein and polysaccharide composition between Cavas supplemented with IDY and the Cava control (Data not shown). In this sense, it should be noted that some previous studies have shown that supplementation with IDY of the sparkling wines led to an increase of the concentration in proteins [2] while in others this effect was not detected [18]. In our case, no differences were observed between the Cava control and the other two supplemented with the two LSI, probably because the protein decrease generated in the Cavas was higher than the sensitivity limit of the analytical method.

Table 1.

Parameter	Base Wine	Sparkling Wines		
		Control	+ IDY-Sc	+ IDY-Td
Glucose + Fructose (g/l)	< 0,5	< 0,5	< 0,5	< 0,5
Ethanol (%)	10,7 ± o,11 **A**	12,0 ± 0,1 **B**	11,9 ± 0,33 **B**	12,0 ± 0,20 **B**
TA (g/l)	5,68 ± 0,01 **A**	5,25 ± 0,06 **B**	5,30 ± 0,04 **B**	5,20 ± 0,04 **B**
pH	2,81 ± 0,01 **B**	3,03 ± 0,01 **A**	3,02 ± 0,01 **A**	3,02 ± 0,01 **A**
VA (g/L)	0,18 ± 0,01 **A**	0,22 ± 0,02 **B**	0,22 ± 0,01 **B**	0,23 ± 0,02 **B**
Glycerol (g/L)	4,70 ± 0,30 **A**	5,80 ± 0,36 **B**	6,65 ± 0,22 **B**	6,84 ± 0,54 **B**

+ LSI-Sc: supplemented with inactive dry yeasts of *S. cerevisiae* ; + LSI-Td: supplemented with inactive dry yeasts of *T. delbrueckii* ; TA: titratable acidity expressed as g tartaric acid / L; VA: volatile acidity expressed as g acetic acid / L. Different letters indicate the existence of statistically significant differences (p <0.05).

Table 2.

Triangular Test	Positive Identifications	*P*	Preferences		
			Control	*+ IDY-Sc*	*+ IDY-Td*
Control vs + IDY-Sc	9/12	0.01	5	4	–
Control vs + IDY-Td	11/12	<0,001	3	–	8

+ IDY-Sc: supplemented with inactive dry yeasts of *S. cerevisiae*; + IDY-Td: supplemented with inactive dry yeast from *T. delbrueckii*.

Table 2 shows the results of the comparative sensorial analysis (triangular test) between sparkling wines supplemented with inactive dry yeasts and control.

Nine of the 12 tasters could distinguish between the Cava supplemented with IDY of S. cerevisiae and the Cava control. This result is significant (p <0.01) and demonstrates that both sparkling wines were different from the sensory point of view.

Among the tasters who were able to differentiate which was the different sample, 5 of them preferred the Cava control and 4 the Cava supplemented with IDY of *S. cerevisiae*, thus it exists a technical tie in their preferences since the tasters did not selected one of the samples clearly

The results were much clearer when comparing Cava control with that supplemented with IDY of *T. delbrueckii*. Specifically 11 of the 12 tasters were able to distinguish between both Cavas, which allows to affirm that the samples were different in a very significant way (p <0.001). In addition 8 of the 11 tasters who were able to distinguish between samples preferred the Cava supplemented with IDY of *T. delbrueckii* which clearly indicates that this Cava was better appreciated from the sensory point of view.

It can be concluded therefore that the supplementation with specific IDY is an interesting tool to improve the foam and the general quality of the sparkling wines.

This study has been funded by CDTI (CIEN-VINySOST2014 project).

References

[1] F. Zamora, Enólogos **23**, 28–32 (2003)
[2] G. Vanrell, P. Cabanillas, S. Albet, JM. Canals, L. Arola, F. Zamora, F. 2002. Rev. Fr. OEnol. **196**, 30–36 (2002)

DOI: 10.1051/bioconf/20170902031

[3] A. Maujean, Rev. Fran. Œnol., **120**, 11–17 (1989)
[4] G. Vanrell, M., Esteruelas, JM. Canals, F. Zamora, P. Poinsaut, N. Sieczkowski, D. Leboeuf, Rev. Oenol. **114**, 28–30 (2005)
[5] BEN. Todd, GH. Fleet, PA. Henschke, P. A. Am. J. Enol. Vitic. **51**, 65–72 (2000)
[6] Y.P. Núñez, AV. Carrascosa, R. González, MC. Polo, A. Martínez-Rodríguez, J. Agric. Food Chem. **54**, 7898–7903 (2006)
[7] F. Brissonet, A. Maujean, Am. J. Enol. Vitic. **44**, 297–301 (1993)
[8] C. Luguera, V. Moreno-Arribas, E. Pueyo, B. Bartolomé, MC. Polo, Food Chemistry **63**, 465–471 (1998)
[9] MJ. Leroy, M. Charpentier, B. Duteurtre, M. Feuillat, C. Charpentier, Am. J. Enol. Vitic. **41**, 21–28 (1990)
[10] L. Martínez-Lapuente, Z. Guadalupe, B. Ayestarán, M. Ortega-Heras, S. Pérez-Magariño, J. Agric. Food Chem. **61**, 12362–12373 (2013)
[11] Organisation Internationale de la Vigne et du Vin. Methods of analysis of wines and must (2014)
[12] A. Maujean, P. Poinsaut, H. Dantan, F. Brissonet, E. Cossiez, Bulletin de l'OIV **711–712**, 405–426 (1990)
[13] J.M. Canals, L. Arola, F. Zamora, Am. J. Enol. Vitic. **49**, 383–388 (1998)
[14] B. Ayestarán, Z. Guadalupe, D. León, 2004. Anal Chim Acta **513**, 29–39 (2004)
[15] E. Pueyo, PJ. Martín-Alvarez, MC. Polo, Am. J. Enol. Vitic. **46**, 518–524 (1995)
[16] M. Esteruelas, E. González-Royo, N. Kontoudakis, A. Orte, A. Cantos, JM. Canals, F. Zamora, F. J. Sci. Food Agric. **95**, 2071–2080 (2015)
[17] G. Vanrell, R. Canals, M. Esteruelas, F. Fort, JM. Canals, F. Zamora, Food Chem. **104**, 148–155 (2007)
[18] S. Pérez-Magariño, L. Martínez-Lapuente, M. Bueno-Herrera, B. Ayestarán, J. Agric. Food Chem. **63**, 5670–5681 (2015)

The influence of the different techniques of maceration on the aromatic and phenolic profile of the Busuioacă de Bohotin wine

Alexandru C. Tartian[1], Valeriu V. Cotea[1,a], Marius Niculaua[2], Cătălin-Ioan Zamfir[1], Cintia Lucia Colibaba[1], and Ana-Maria Moroşanu[1]

[1]Faculty of Horticulture, University of Agricultural Studies and Veterinary Medicine Iasi, Romania
[2]Research Centre for Oenology, Romanian Academy - Iasi Branch, Romamia

Abstract. The main objective of the present study is to evaluate the influence of the maceration techniques on the volatile and phenolic compounds content of the wines obtained from an autochthonous aromatic variety from Romania. For this study, a pink biotype of the Busuioacă de Bohotin variety from the Huşi vineyard was used. The wine samples were obtained by applying five types of maceration: Classic maceration (V1; S1), Thermo-maceration (V2; S2), Microwave maceration (V3; S3), Ultrasound maceration (V4; S4) and Cryo-maceration (V5; S5). The results of the aromatic profile of the wine samples obtained from Busuioacă de Bohotin variety, indicated that the wine samples obtained through cryo-maceration and classic maceration contained a large variety of volatile compounds, namely: alcohols (1-propanol; 2-methyl-1-propanol, 3-methyl-1-butanol, 1-hexanol, phenethyl alcohol e.g.), esters (ethyl octanoate; ethyl caprylate; ethyl lactate; ethyl benzoate; ethyl palmitate e.g.) terpenoid compounds (α-terpineol, linalool, hotrienol, citronellol, nerol, geraniol) and so an. Among alcohols, phenethyl alcohol, which is responsible for floral aroma of roses and is typical for the Busuioacă de Bohotin wine, was found in large quantities in wine samples V1 (4.3 mmol/L) and S5 (5.53 mmol/L). Terpene compounds and their oxides were identified with elevated levels for linalool (*cis*- and *trans*-) and also nerol- as well as linalool-oxide. Analysing the phenolic compounds, it was observed that the total polyphenol index and Folin-Ciocâlteu index expressed higher values for the wine samples V1 and V3. Concerning the total anthocyanins and tannins content, the wine samples obtained had low amounts of this compounds.

1. Introduction

The volatile and phenolic compounds of wines depends by various factors such as grape variety, terroir, vintage and applied technological process. Among winemaking techniques, maceration plays a primordial roll for the extraction of aromatic and phenolic compounds in wines.

Maceration is responsible for all the specific characteristics of sight, smell and taste where the volatile compounds (primary aromas and aroma precursors) and phenolic compounds (anthocyanins and tannins) are extracted from skins, seeds and sometimes from stems of the grape [1].

This technological operation implies the contact between the juice and the solid parts of the grapes for a specific time. In the evolution process of maceration, four processes of physical-chemical nature are distinguished: the extraction and dissolution of different substances, the diffusion of the substances extracted, refitting the substances extracted on certain molecules from the medium and modification or even the destruction of the substances extracted [2].

Maceration process is influenced by numerous factors such as temperature, duration, alcohol content, SO_2 where the extraction of the volatile and phenolic compounds varies according to variety, maturation conditions and other factors. Consequently, the maceration should be modulated and fractioned according to the grape variety and quality and also to the style of wine desired.

With the evolution of the technology various methods were developed that are available to the winemaker to adjust extraction levels during maceration. These are based on the destruction of the cells' walls favouring the dissolution of chemical compounds of interests [1].

Thermo-vinification or thermo-maceration is a method of maceration known for a long time and implies high-temperature extraction prior to fermentation. Various techniques of thermo-maceration were developed on the basis of evolution of the industrial technologies. The extraction principle consists in heating the whole grapes or the crushed and destemmed grapes to high temperatures (65–80°C), promoting the diffusion of phenolics from the skins, especially anthocyanins.

For the extraction process of the phenolic and volatile compounds from the grape, non-conventional techniques can also be used such as microwave and ultrasound maceration, that recognised as outstanding energy sources to promote extraction, cutting down extraction times and increasing yields as well as often the quality of extract [3].

Microwave extraction principle consists in the absorption of the energy by the water present within the cells thus, promoting internal superheating [4]. The heated water evaporates and generates tremendous pressure on the cell wall, resulting thereby the disruption of the cell wall and migration of the active constituents [5].

[a] e-mail: vvcotea@yahoo.com

Microwave maceration is based on heating the pomace by exposing it to microwave radiation in the electromagnetic spectrum. This method has been highlighted to be efficient in terms of rate and effectiveness of polyphenolic extraction from the grapes [6–11].

The use of ultrasounds technology in food engineering has become very popular. The continuous interests in this is due to numerous advantages given by this method (reduces the process time, saves energy and improves shelf life and quality of food products) and the large applications offered such as: peeling, disruption of cell, the production of enzymes without bacterial cells, extraction, acceleration of microbial fermentation, mixing, homogenization, extraction etc. [11–13].

In the oenology department, various studies analyse this method, for: extraction of oil and polyphenols from grape seeds [14], effects on cold maceration and fermentation [15], extraction of anthocyanins [16], extraction of antioxidant bioactive compounds from winery wastes and by-products [17], extraction of volatile compounds [18], acceleration of wine's aging [19].

The extraction application of ultrasounds is based on cavitation phenomenon which consists in the formation of cavities when the sound wave passes through an elastic medium, inducing a longitudinal displacement of particles and acting like a piston, therefore producing the disruption of the cell tissue.

Cryo-maceration is another technique of maceration which involves lowering the temperature of the pomace or the whole grapes to at least −18°C and maintaining this temperature for a specific period of time. Temperature is achieved by using a freezer or cryogenic gases (CO_2 or N_2). During this process, intracellular water from the grape is frozen and the ice crystals formed cause the rupture of the cell tissue and migration of the chemical compounds of interest. (aroma precursors, anthocyanins, tannins) [20,21]. Generally, the wines obtained by cold temperature gain a more aromatic complexity [1], cryo-maceration favouring the extraction of varietal aromas such as monoterpenes while the phenolic compounds are limited [22,23].

Busuioacă de Bohotin variety originates either from ancient Greece, as a close relative of Tămâioasă românească, another aromatic Romanian grape variety, either from France, from the very similar Muscat violet de Frontignan [24].

The wines obtained from Busuioacă de Bohotin variety have a flavour reminiscent of the scent of roses, basil, coriander, wild strawberries where the colour can range from pink with violet hues to pale orange (salmon, onion skin).

The aim of the present study is to evaluate the influence of different techniques of maceration on the aromatic and phenolic profile of the wines obtained from Busuioacă de Bohotin grape variety.

2. Materials and methods

2.1. Grape samples

Busuioacă de Bohotin grape variety has two biotypes which differ mainly by the colour of the grapes' skin: dark violet and pink. The pink biotype presents a low pigmentation degree, being commonly used in combination with the other.

For this reason, a less known biotype of this variety was used in this study in order to explore the aromatic and phenolic profile of the wines obtained under different conditions of maceration.

The experiment was done over the course of two years, during which ten wine samples were obtained from the pink biotype of Busuioacă de Bohotin variety. The pink biotype grapes were harvested from Hu?i vineyard at technological maturity in the years 2014 (A_0) and 2015 (A_1). The main physico-chemical characteristics of the grapes are presented in Table 1 below.

Table 1. The main physic-chemical parameters of the grapes.

Sample	Total acidity g/L tartaric acid	Sugars g/L
A_0	5.7	248.7
A_1	5.41	230

2.2. Wine samples

The wines samples analysed in this study were obtained by applying five maceration techniques:

2.2.1. *Classical maceration (V_1, S_1):*

the grapes were crushed, destemmed and the resulting pomace was introduced in maceration vessels.

2.2.2. *Thermo-maceration (V_2, S_2):*

the juice resulted from the crushing and destemming operations was pumped into a metallic recipient and heated to a temperature of 70°C for 15 minutes. After reaching this temperature, the juice was transferred into a maceration vessel and mixed with the remainder pomace.

2.2.3. *Microwave maceration (V_3, S_3):*

the obtained pomace was microwaved at 750 W for 15 minutes and was manually homogenized at 3 minutes intervals. The end temperature of the pomace was 60°C.

2.2.4. *Ultrasound maceration (V_4, S_4):*

it was performed in an ultrasonic bath where the pomace was sonicated at a frequency of 35 KHz for 30 minutes.

2.2.5. *Cryo-maceration (V_5, S_5):*

the whole grapes were frozen slowly to a temperature of −20°C for 24 hours in order to promote the formation of large ice crystals and thus favouring the break of the skin grape tissue. The frozen grapes were manually destemmed and crushed.

The remainder maceration time after all the above mentioned processes was of seven days for all the variants. The temperature of the pomace was maintained at 20°C and yeast - Fermactive Rouge Primeur® and pectolytic enzyme was added – Zymovarietal Aroma $G^{®}$.

After 7 days of maceration the pomace was pressed using a hydraulic press and the alcoholic fermentation ended in glass vessels. After alcoholic fermentation, the wines were racked, filtered, SO_2 was added (1 ml/L) and bottled.

The two groups of obtained wine samples were registered as follows: **V** – for the wine samples obtained in **2014** and **S** – the wine samples obtained in **2015**. The the wine samples were analysed from a physico-chemical point of view according to the O.I.V. methods and the following parameters were determined: volatile acidity – (**a**), total acidity – (**b**), alcohol content – (**c**), total dry extract – (**d**), pH – (**e**) and conductivity – (**f**) in Table 2.

Table 2. Main physico-chemical parameters of the wines obtained through different maceration techniques.

Sample	a	b	c	d	e	f
V_1	0.35	5.84	14.31	22.9	3.80	2.37
V_2	0.33	5.94	14.01	23.5	3.78	2.45
V_3	0.37	5.84	14.41	25.3	3.84	2.40
V_4	0.31	5.45	13.91	21.6	3.77	2.28
V_5	0.33	5.40	14.23	21.9	3.78	2.30
S_1	0.37	5.87	15.20	27.1	3.34	1509
S_2	0.37	5.82	15.10	26.8	3.34	1450
S_3	0.62	6.01	15.24	26.8	3.30	1476
S_4	0.37	4.79	14.78	25	3.35	1513
S_5	0.31	4.37	14.55	22.7	3.31	1493

Volatile acidity g/L – (**a**), total acidity g/L- (**b**), alcohol content %vol.– (**c**), total dry extract g/L – (**d**), pH – (**e**) and conductivity μ S/cm – (**f**).

2.3. Gas-chromatographic analysis of volatile compounds

In order to conduct the determination of the volatiles compounds, all samples were analysed using a gas chromatograph with a mass spectrometer system (Shimadzu HS 20trap-GC 2010plus-MS8040TQ). The system uses helium as carrier gas and the adsorption material in trap as stationary phase is a resin of 2,6-diphenylene oxide (TENAX).

6 mL of wine were added to mixture of 0.8 g from the following salts: Na Cl, Na_2SO_4 and KH_2PO_4 in a vial for analysis of vapours (headspace fraction) using the method of salting out. All samples were heated to 70°C and shacken at 25 rpm for 5 minutes. The vials were pressurised to 60 KPa with helium to remove condense. The volatile fraction was transferred into a Tenax trap at −10 °C by a transfer line at 150 °C. Before injection for half a minute the trap is dry purged and then the volatiles are transferred at 280 °C with a split of 1/50 in to the analytic column.

The separation column used is a Phenomenex® ZB WAXplus 60 m × 0.25 mm ID ×0.15 μm which is contained in a temperature - controlled oven. Temperature program for separation is: the total injection time 2 minute, temperature remains constant for 3 minutes at 82°C, then starts increasing with 3°C/min. up to 135°C, maintains for 1 minute, then 7.5°C/min. up to 160°C, maintains for 1 minute and finally increases with 27°C/min. to 240°C, remaining constant for 8 minutes (total temperature program 37 minutes). Separation is carried-out at a constant linear velocity of 35 cm/s (column flow 1.5 mL/min).

The 8040 triple quadrupole is utilised in Q3 (third quadrupole) scan mode from 30–400 Da at 0.1 s event time (5000 scan/sec). The ion source interface is maintained at 230°C and source at 200°C.

The results are processed qualitatively comparing them to different MS databases available: NIST 14, Wiley 10, FFNSC and SZTERP. For the quantitative results, all samples are processed by internal standard method relative to 4-methyl-pentan-2-ol.

2.4. Folin- Ciocalteu index and total polyphenols index

Folin-Ciocâlteu Index (FCI) describes the phenolic compounds with reducing proprieties and is based on the oxidation in a basic medium of the phenolic compounds from wine by the Folin–Ciocâlteu reagent.

The blue coloration produced has a maximum absorption around 750 nm (visible light) and it is proportional to the total amount of phenolic compounds (OIV-MA-AS2-10).

Total Polyphenols Index (PTI) or D280 index expresses the total content of the phenolic compounds of wine (anthocyanins, tannins, flavones etc.) and consists on the strong absorption of the ultraviolet light by the benzenic cores characteristic for the phenolic compounds, with a maximum absorption at 280 nm.

2.5. Total content of anthocyanins by pH variation method

This method is based on the existence of a balance between the coloured and uncoloured forms of the anthocyanins in acidic medium. The absorption differences obtained between the two pH solutions (0.6 and 3.5) at a wavelength of 520 nm is proportional to the content of anthocyanins.

2.6. Tannins by MCP (methyl cellulose precipitable) method

The assay is based on the interactions between the tannins and a polymer, resulting in the formation of insoluble polymer tannin complexes, which precipitates.

The polymer used is methyl cellulose and the values of tannin content of the wines analysed are obtained by subtracting the absorbance values at 280 nm (A280) of the both solutions, with and without precipitation.

The FCI and D280 indexes, total anthocyanins and tannins were determined using a Analytic Jena S 200 spectrophotometer.

2.7. Evaluation of chromatic characteristics

In order to evaluate the chromatic features of the analysed samples a spectrophotometer Analytic Jena S was used.

CIE La*b* is the most complete colour model used to describe the colours of the visible spectrum because it allows the specification of the largest number of shades of colour that a person could perceive with a* b* values. Therefore, the range of colours based on this model is described as the widest range of reproducible colours of the visible spectrum.

Basically, the CIE Lab 76 colour or space system is based on a sequential or continuous Cartesian representation of three orthogonal axes: *L*, *a*and *b*. *L* represents the clarity, *a* represents the green/red shades of colour and *b* represents the blue/yellow shades of colour.

3. Results and discussions

Analysing the values of the main physico-chemical parameters (Table 2) of the samples obtained through different technique of maceration were highlighted some differences.

In terms of total acidity (**b**), it was revealed that the samples obtained by ultrasound maceration and cryo-maceration expressed the lowest values regardless of the year they were produced. The most pronounced differences were observed in 2015, where the minimum value of total acidity was registered in the cryo-maceration wine sample (4.37 g/L tartaric acid) followed closely by the ultrasound sample (4.79 g/L). This situation can be caused by the favouring of the precipitation of tartaric acid determined by the used techniques of maceration.

pH (**e**) or the true acidity plays an important role in the wines, enhancing their microbiological and physico-chemical stability. Regarding the wine samples obtained in the two years of experiment, the pH values did not vary greatly indifferent of the technique of maceration applied or the total acidity values. Thus, the pH values varied between 3.77 and 3.80 for 2014 wine samples and between 3.30 and 3.35 in 2015. This situation can be due to the tampon effect of the wine and the presence of highly ionized acids.

In terms of alcohol content (**c**), the use of different techniques of maceration didn't have a big influence on the wine samples obtained.

Thereby, in 2014 the alcohol content of the wines samples varied between a minimum of 13.91% for the ultrasound maceration sample (V4) and a maximum of 14.41% for the microwave maceration (V3) sample while in 2015 this parameter varied between 14.55% (S5-cryo-maceration sample) and 15.24% (S3 -microwave maceration). If the vintage of the wine samples is taken into account, the wine samples obtained in 2015 had bigger values of alcoholic concentration in contrast with 2014 wine samples and it is mainly because of the higher values of the sugars of the grapes harvested in 2015.

The volatile acidity (**a**) of the wine is considered to be a highly important physico-chemical parameter that describes the general quality of the wine and it is given mainly by the presence of the acetic acid. In the case of the wine samples obtained in this study, this parameter ranged in general between 0.31 (g/L acetic acid) and 0.37 (g/L). The only difference was identified in 2015 for the microwave maceration sample with a value of 0.62 (g/L) and can be caused probably by the contact with a larger amount of oxygen during vinification.

In order to highlight the influence of the technique of maceration upon the degree of extraction of the non-volatile organic and inorganic substances, the total dry extract (**d**) was analysed. Thus, the most extractive methods of maceration have been revealed to be microwave maceration sample (V3) with 25.3 g/L in 2014 and classic maceration sample (S1) with 27.1 g/L in 2015.

At the opposite pole, the cryo-maceration (22.7 g/L) and ultrasound maceration samples (21.6 g/L) obtained the lowest values for total dry extract.

The conductivity (**f**) of a wine depends on the concentration and the mobility of the ionic species (anionic and cationic) present in the solution such as: acids, bases, salts, nitrogen compounds (proteins, enzymes, amino acids) etc. Concerning the wine samples studied, in 2014, the values of the conductivity are above 2 μS/cm while the samples from 2015 expressed lower values, been situated between 1450 and 1509 μS/cm.

Table 3. Folin Ciocâlteu Index, PTI Index, Tannins and Anthocyanins.

Sample	FCI	PTI	Anthocyanins (mg/L)	Tannins (epicatechin mg /L)
V1	15.79	12.25	2.84	n.d.
V2	8.95	7.83	1.37	n.d.
V3	12.85	7.80	1.12	n.d.
V4	4.98	3.89	1.12	n.d.
V5	10.21	8.46	0.57	n.d.
S1	25.29	21.38	n.d.	104.85
S2	27.77	21.79	n.d.	63.61
S3	32.53	24.28	n.d.	76.66
S4	22.15	17.44	n.d.	78.55
S5	22.36	17.97	n.d.	42

n.d- Non-detectable.

This variation of the wine samples conductivity is depending only on the year of production and is not influenced by the technique of maceration applied, being in direct correlation with the values of pH. Analysing the values presented in the Table 3 it can be observed that are some differences regarding the phenolic compounds of wine samples obtained by different techniques of maceration.

Thereby, in the year 2014 the wine samples obtained by ultrasound maceration contain the lowest amounts of phenolic compounds with values of 3.89 for PTI and 4.98 for FCI while the highest values where identified at classic maceration technique with 12.25 (PTI) and 15.79 (FCI). In 2015, the most extractive technique of maceration in terms of phenolic compounds indexes has been pointed out to be microwave maceration with values of 24.28 for PTI and 32.53 for FCI while at the opposite pole stands the cryo-maceration (17.97 PTI and 22.36 FCI) followed closely by the ultrasound maceration sample.

Comparing the wine samples obtained in this two years of experiment it was highlighted clearly that the wine samples obtained in 2015 contained more phenolic compounds than the 2014 samples. This fact can be explained by the higher alcohol content of the 2015 wine samples and his strong effect on the extraction of the phenolic compounds from the grapes during maceration.

Concerning the anthocyanins and tannins content of the wine samples analysed were observed the presence in lower amounts and even the absence of this two compounds. Thus, regarding the anthocyanins content, were identified only in 2014 wine samples with values that ranged between a maximum of 2.84 mg/L (V1) and a minimum of 0.57 mg/L (V5). The small amounts of anthocyanins identified in this wine samples can be primarily due to the low pigmentation level of the pink biotype of Busuioacă de Bohotin variety and secondly because of the irreversible reactions of anthocyanins by oxidation, poly-condensation with tannins or even by combination with amino acids and other nitrate compounds.

Concerning the tannin content, they were identified only in the 2015 wine samples with values that varied between 42 mg/L epicatechin (S5) and 104.85 mg/L (S1).

 DOI: 10.1051/bioconf/20170902032

Table 4. Chromatic parameters of the analysed samples.

Samples	Clarity (L)	Chromaticity	
		a*	b*
V1	96.2	2.69	11.21
V2	95.6	2.47	11.85
V3	95.9	2.16	10.70
V4	95.7	0.74	7.42
V5	97.4	0.54	6.78
S1	93.6	3.96	17.57
S2	93.6	3.72	18.27
S3	92.7	4.88	19.74
S4	96.0	2.02	10.27
S5	95.1	2.57	13.44

The presence of the tannins just in the 2015 wine samples can be explained by the influence of the high alcoholic concentration during the maceration of these samples, thus promoting a better extraction of this compounds from the seeds and skins of grapes.

In the case of 2014 wine samples, the absence of the tannins can be due to a possible hydrolysis of the tannins and the formation of a larger amounts of tannins with low polymerization degree that can react with other molecules such as proteins, amino acids and therefore resulting the precipitation of these compounds.

Observing the main chromatic parameters (Table 4) it can be stated that wine samples with high levels of clarity (L) were obtained, where the yellow shades (b) prevailed the red shades (a). These results are primarily due to the low content of anthocyanins and tannins.

Regarding the influence of the technique of maceration upon the chromatic characteristics of the wine samples, the ultrasound and cryo-maceration methods had an impact. Thus, the wines samples obtained by these techniques had the lowest values of red (+a) and yellow (+b), also presenting the highest values of clarity (L).

For identifying the volatile compounds from the wine samples, the GC-MS method was used and the results were expressed quantitatively in mmol/L according to the standard solution 4-methyl-pentan-2-ol.

Alcohols are formed during alcoholic fermentation process whether from sugars or from amino acids by the yeast metabolism, contributing to the wine's aromatic complexity. Alcohols like 1-propanol, 2-methyl-1-propanol (isobutyl alcohol) and 3-methyl-1-butanol (isoamyl alcohol) are the most representative fusel alcohols which are found in the largest quantities in wines and imprints in general a fermented, alcoholic and ethereal odour.

Referring to the wine samples from this study, these compounds were found in higher quantities in the samples where were applied microwave maceration (2014) and cryo-maceration (2015) techniques (Table 5).

Another significant part of the alcohols they also have the higher alcohols with six carbon atoms which derive from grape polyunsaturated fatty acids (linoleic acids) and give the herbaceous smell of wines and they could hypothetically contribute to judge wine origin and affiliation [26]. From this group of compounds, they were identified 1-hexanol, *cis*-3-hexen-1-ol, *cis* and *trans*-2-hexen-1-ol where 1-hexanol was the only compound that was identified in the both years of the wine samples studied and in the largest amounts.

Regarding to the influence of the technique of maceration upon this compound it was observed that the highest quantity of 1-hexanol was identified in the cryo-macerated wines samples (3.06 mmol/L in 2014 and 6.8 mmol/L in 2015).

Of the benzene-derived higher alcohols, phenyl ethyl alcohol is the most important and is also the only fusel alcohol described with positive terms such as sweet, perfumed, and dry rose [26]. In this study, this compound was identified in the largest amounts in the wine samples where were applied classic maceration (4.30 mmol/L-2014) and cryo-maceration (5.53 mmol/L - 2015).

Benzyl alcohol is another higher alcohol compound which has a floral, rose odour and it was found in large quantities in the classic-maceration (0.20 mmol/L - 2014) and in cryo-maceration (0.26 mmol/L - 2015) wine samples.

Esters occur in wines as a major volatile compound mainly from the enzymic esterification during fermentation and generally contribute to the fruity wine odours. Among the esters identified in this study (Table 6), ethyl caprylate, ethyl caprate, ethyl laurate, ethyl myristate and ethyl palmitate were found in higher concentrations. These compounds have very pleasant odours of wax and honey and the concentration of these esters tend to decrease over time as wine ages due to spontaneous hydrolysis.

In the wines samples analysed were identified bigger amounts of ethyl caprylate (16.84 mmol/L) at thermo-maceration sample (S2), higher concentrations of ethyl myristate (1.97 mmol/L) and ethyl palmitate (4.32 mmol/L) for the cryo-maceration sample (S5) while the classic maceration sample (S1) expressed high quantities of ethyl caprate (4.88 mmol/L) and ethyl laurate (2.71 mmol/L).

Ethyl lactate is an ester responsible for the buttery, creamy odours with hints of fruit and coconut which was found in large amounts in the cryo-maceration wine samples (2.95 mmol/L-2014; 6.18 mmol/L-2015).

Esters like 2-phenylethyl acetate responsible for the apple and honey like aromas, diethyl succinate who gives fruity, apple odours and ethyl benzoate responsible for the minty odours were found in higher concentration in wine samples where cryo-maceration and classic maceration were applied. The cryo-maceration wine samples illustrated also the presence in larger amounts of the ethyl 3-hydroxybutanoate who contributes to the marshmallow-like aromas. It was observed that the wine samples obtained by cryo-maceration and classic-maceration had bigger amounts of esters, thus, it can be stated that the techniques of maceration applied in this study had an important influence on the amounts of this compounds formed during maceration.

It was also highlighted that the wine samples obtained in 2015 had bigger concentration of esters comparing with the wine samples obtained in 2014.

The terpenoid compounds (Table 7) are known as the substances that are present in grapes (skins mostly) and are partially responsible for the typical varietal character of the grape cultivar (especially for the Muscat varieties) and consequently to the wine produced.

Table 5. Alcohols identified in the analysed wine samples.

Alcohols (mmol/L)	V1	V2	V3	V4	V5	S1	S2	S3	S4	S5
1-propanol	0.54	0.43	0.58	0.49	0.52	0.87	0.99	0.77	0.003	1.36
2-methyl-1-propanol	0.01	0.67	0.69	0.64	0.63	1.04	1.03	0.73	0.002	1.41
3-methyl-1-butanol	3.89	3.39	3.58	2.51	2.51	6.17	6.83	5.27	0.021	9.50
1-pentanol	0.17	0.13	0.03	0.03	0.14	0.21	0.12	0.09	0.002	0.13
4-methyl-1-pentanol	0.15	0.13	0.01	0.01	0.10	0.04	0.07	0.05	nd	0.07
3-methyl-1-pentanol	0.16	0.26	0.03	0.03	0.15	0.11	0.11	0.09	0.002	0.19
1-hexanol	2.91	2.61	1.90	2.50	3.06	2.99	4.31	1.99	0.004	6.80
cis-3-hexen-1-ol	0.49	0.25	0.64	0.33	0.48	n.d.	n.d.	n.d.	n.d.	n.d.
trans-2-hexen-1-ol	n.d.	n.d.	n.d.	n.d.	0.05	0.03	n.d.	n.d.	n.d.	0.04
1-octen-3-ol	0.05	0.15	n.d.	n.d.	n.d.	0.07	0.12	0.05	n.d.	0.13
1-heptanol	0.40	0.15	0.05	0.07	0.22	0.56	0.48	0.32	0.001	0.58
2-ethyl-1-hexanol	0.09	0.06	0.02	0.02	0.08	0.07	0.05	0.05	n.d.	0.08
2,6-dimethyl-4-heptanol	0.04	n.d.	0.02	n.d.	0.06	n.d.	n.d.	n.d.	n.d.	0.03
1-octanol	0.07	0.22	0.03	0.02	0.07	0.17	0.14	0.11	n.d.	0.19
benzyl alcohol	0.20	0.20	0.12	0.12	0.12	0.21	0.15	0.12	nd	0.26
phenethyl alcohol	4.30	1.98	2.71	2.18	2.74	3.63	3.45	2.89	0.002	5.53
2,6-dimethyl-3,7-octadiene-2,6-diol	0.21	0.28	0.10	0.14	0.16	n.d.	n.d.	n.d.	n.d.	n.d.
1-tetradecanol	0.32	0.27	n.d.	0.09	0.27	n.d.	n.d.	n.d.	n.d.	n.d.

Table 6. Esters identified in the wine samples analysed.

Esters (mmol/L) (mmol/L)	V1	V2	V3	V4	V5	S1	S2	S3	S4	S5
ethyl pyruvate	0.11	0.10	n.d.	0.02	0.12	n.d.	0.06	0.08	n.d.	0.11
ethyl caprylate	0.17	n.d.	n.d.	0.16	0.20	15.85	16.48	11.74	n.d.	6.84
ethyl lactate	2.22	2.12	1.60	n.d.	2.95	3.51	3.87	3.24	n.d.	6.18
ethyl 3-hidroxybutanoate	0.03	0.17	0.01	n.d.	0.22	n.d.	0.04	n.d.	n.d.	0.01
diethyl malonate	n.d.	n.d.	n.d.	n.d.	0.02	n.d.	0.03	n.d.	n.d.	n.d.
ethyl caprate	0.06	n.d.	n.d.	0.05	0.06	4.88	4.19	2.74	n.d.	1.35
ethyl benzoate	0.29	0.33	0.14	0.14	0.15	0.19	0.18	0.12	n.d.	0.22
diethyl succinate	1.21	0.75	0.90	0.79	1.12	0.58	0.61	0.38	n.d.	1.00
ethyl salicilate	0.06	0.04	0.01	0.02	n.d.	0.03	0.04	n.d.	n.d.	0.01
2-phenylethyl acetate	0.10	0.14	0.05	0.04	0.14	0.03	0.03	0.03	n.d.	0.13
ethyl laurate	0.12	0.09	0.07	0.05	0.05	2.71	2.51	1.88	n.d.	1.09
isopropyl myristate	0.28	0.00	0.04	0.06	n.d.	0.05	0.01	0.05	n.d.	0.06
ethyl myristate	0.28	0.38	0.28	0.12	0.36	1.64	1.21	1.11	n.d.	1.97
ethyl heptanoate	n.d.	n.d.	n.d.	0.03	n.d.	0.20	0.15	0.12	n.d.	0.03
ethyl cinnamate	0.26	0.14	0.06	0.16	n.d.	n.d.	n.d.	0.03	n.d.	n.d.
ethyl pentadecanoate	0.21	0.24	0.12	0.00	0.25	0.32	0.14	0.27	trace	0.55
ethyl palmitate	1.09	1.04	0.61	0.63	1.11	3.09	3.07	2.46	n.d.	4.32
ethyl 9-hexadecenoate	0.34	0.42	0.33	0.41	0.68	0.16	0.10	0.20	trace	0.44
(Z)-ethyl pentadec-9-enoate	n.d.	n.d.	0.08	0.00	0.20	n.d.	n.d.	0.09	n.d.	0.12
ethyl stearate	0.63	n.d.	0.31	0.14	0.47	0.35	0.18	0.15	n.d.	0.27
ethyl oleate	n.d.	0.25	0.17	n.d.	0.48	0.04	0.48	nd	n.d.	n.d.

In the studied wine samples, 21 terpenoid compounds were identified such as: simple hydrocarbons (limonene, α-terpinene), alcohols (linalool, hotrienol, α-terpineol, α -citronellol, citronellol, nerol, geraniol, prenol), oxides of alcohols (*cis*-and *trans*-linalool oxide in pyranic and furanic form, nerol oxide, rose oxide), acids (geranic and neric acids), and even esters (α-terpinyl acetate).

Among the most odoriferous monoterpenes are alcohols such as linalool, nerol, geraniol which have a have a floral odour of rose, α-terpineol responsible for the lily of the valley odour, citronellol (citronella) and hotrienol (linden) were found in the highest amounts in the wine samples obtained through cryo-maceration regardless of the vintage. Also, for this odorants compounds, the lowest values were identified in the wine samples obtained by ultrasound maceration and microwave maceration which seems to have negative effects on the terpenoid compounds.

In the wine samples a large variety of terpenoids oxides such as oxides of linalool, nerol, geraniol and traces of rose oxide was also found.

These oxides of monoterpenes compounds are due to the oxidation reactions that occur prior to fermentation, being triggered during extraction of the juice and maceration [27].

Linalool oxides which are responsible for the camphorous and mineral odour and nerol oxides (green and herbaceous) have a high perception threshold compared to the unoxidized forms of the terpenes and consequently they have little olfactory impact on

Table 7. Terpenoid compounds identified in the wine samples.

Terpenoids compounds (mmol/L)	V1	V2	V3	V4	V5	S1	S2	S3	S4	S5
limonene dioxide	0.07	n.d.	n.d.	n.d.	n.d.	n.d.	n.d.	n.d.	n.d.	n.d.
nerol oxide	0.07	n.d.	0.01	0.01	0.13	0.07	0.11	0.04	n.d.	0.04
trans-linalool oxide (furanoid)	n.d.	n.d.	n.d.	0.10	0.32	0.12	0.10	0.07	n.d.	0.20
α-terpinene	n.d.	n.d.	n.d.	n.d.	0.07	0.08	n.d.	0.60	n.d.	n.d.
cis linalool oxide	0.23	0.31	0.23	n.d.	n.d.	n.d.	0.04	0.04	trace	n.d.
trans linalool oxide	0.14	nd	0.08	0.06	0.15	n.d.	n.d.	n.d.	n.d.	n.d.
linalool	1.48	0.81	0.74	0.90	1.14	2.17	2.44	1.71	0.002	2.24
hotrienol	0.47	0.34	0.35	0.25	0.54	0.13	0.21	0.13	0.001	0.14
α-terpineol	0.62	0.56	0.63	0.59	1.15	0.40	0.43	0.36	0.002	0.51
linalool oxide-pyranoyd	0.29	0.21	0.23	0.14	0.14	0.07	0.07	0.04	n.d.	0.09
α -citronellol	0.08	n.d.	n.d.	n.d.	n.d.	n.d.	n.d.	n.d.	n.d.	n.d.
citronellol	0.53	0.32	0.14	0.43	1.00	0.59	0.50	0.39	0.0007	0.82
nerol	0.38	0.31	0.16	0.34	0.54	0.41	0.36	0.23	0.0002	0.46
geraniol	0.33	0.23	0.27	0.28	0.37	0.20	0.20	0.12	n.d.	0.25
trans-geranylgeraniol	n.d.	n.d.	n.d.	n.d.	0.18	n.d.	n.d.	n.d.	n.d.	0.13
prenol	n.d.	0.10	n.d.	n.d.	0.06	trace	n.d.	n.d.	n.d.	n.d.
rose oxide	n.d.	n.d.	n.d.	n.d.	n.d.	0.02	n.d.	n.d.	n.d.	n.d.
α-terpinyl acetate	n.d.	n.d.	n.d.	n.d.	n.d.	0.04	n.d.	0.01	n.d.	0.03
γ - isogeraniol	n.d.	n.d.	n.d.	n.d.	n.d.	n.d.	n.d.	0.01	n.d.	n.d.
neric acid	0.97	n.d.	0.40	n.d.	0.36	n.d.	n.d.	n.d.	trace	n.d.
geranic acid	n.d.	n.d.	n.d.	n.d.	n.d.	0.16	0.14	0.09	n.d.	n.d.

wines. These oxides seem to occur in lower quantities compared with their homologues. Thus, the linalool oxides express values which are ranging between a minimum of 0.04 mmol/L and a maximum of 0.31 mmol/L (*cis*-linalool).

It was also observed the presence of the oxides of nerol and linalool in the wine samples obtained in 2014.

Geranic acid is a compound resulting from the oxidation of geraniol and it was identified in small amounts in the 2015 wine samples.

Rose oxide is a more odoriferous compound and it is partly responsible for the floral aroma of Gerwurztraminer wines [27]. Referring to the wine samples analysed, only a trace of this compound (0.02 mmol/L) was found, in one sample (S1 – classic maceration).

α-terpinyl acetate is the only ester who can be classified as a terpenoid compound, imprinting herbaceous and spicy odours and it was found in small amounts (0.01 mmol/L - S3; 0.04 mmol/L - S1).

α-terpinene is a terpenoid compound who was identified in very large amounts only in the microwave wine sample (0.6 mmol/L) and it is responsible for the resin-like odours.

4. Conclusions

The detailed physico-chemical analyses of the wine samples obtained by different techniques of maceration revealed the following aspects:

In terms of the main physico-chemical parameters, it was highlighted that the wine samples obtained by microwave and classic maceration expressed higher values of total dry extract while the cryo-maceration and ultrasound maceration were the least extractive techniques.

The pink biotype of the Busuioacă de Bohotin variety had a low content of anthocyanins and tannins, being samples where these compounds could not be detected by the assay applied. The wine samples obtained by cryo-maceration and ultrasound maceration contained the lowest amounts of phenolic compounds, being and the less coloured wine samples produced.

Concerning the volatile profile, it was highlighted that the wines obtained by cryo-maceration technique were the most abundant in compounds such as phenethyl alcohol, nerol, geraniol, hotrienol, citronelol and α-terpineol which have a floral aroma reminiscent of rose essence and gives the typical varietal character of Busuioacă de Bohotin wines.

With respect to the influence of the vintage upon the phenolic and aromatic profile of the wines it was revealed that the 2015 wine samples had in generally higher amounts of phenolic and volatile compounds.

References

[1] P. Ribéreau-Gayon, D. Dubourdieu, B. Donèche, A. Lonvaud, "Handbook of Enology Volume I – The Microbiology of Wine and Vinifications", 2nd Edition, 347 (2006)

[2] V.D. Cotea, -"Tratat de Oenologie" – Vinificaţia şi biochimia vinului **I**, 479 (1985)

[3] A. Casazza, B. Aliakbarian, S. Mantegna, G. Cravotto, P. Perego, Journal of Food Eng. **100**, 50–55 (2010)

[4] L. Wang, C.L. Weller, Trends Food Sci. Technol. **17**, 300–12 (2006)

[5] V. Mandal, Y. Mohan, S. Hemalatha, Microwave Assisted Extraction – An Innovative and Promising Extraction Tool for Medicinal Plant Research, Pharmacognosy Reviews **1**(1), (2007)

[6] S.A. Bittar, S. Périno-Issartier, O. Dangles, F. Chemat, An innovative grape juice enriched in polyphenols by microwave-assisted extraction, Food Chem. **141**(3), 3268–3272 (2013)

[7] A. Liazid, M. Palma, J. Brigui, C.G. Barroso, Investigation on phenolic compounds stability during

microwave-assisted extraction, Journal of Chromatography A **1140**, 29–34 (2007)
[8] L. Anna, Carew, M.A. Sparrow, D.C. Curtin, C. Dugald, Close, G.R. Dambergs, Microwave Maceration of Pinot Noir Grape Must: Sanitation and Extraction Effects and Wine Phenolics Outcomes, Food Bio.Techno. l **7**, 954–963 (2014)
[9] A. Liazid, R.F. Guerrero, E. Cantos, M. Palma, C.G. Barroso, Microwave assisted extraction of anthocyanins from grape skins, Food Chem. **124**, 1238–1243 (2011)
[10] Ş. Tudose-Sandu-Ville, V.V. Cotea, C. Colibaba, B. Nechita, M. Niculaua, M. Codreanu, Phenolic Compounds in Merlot Wines Obtained Through Different Technologies in Iaşi Vineyard, Cercetări Agronomice în Moldova **(XLV 4)**, 152 (2012)
[11] A.C. Vollmer, E.C. Everbach, Halpern, M. Kwakye, Bacterial stress responses tol-megahertz pulsed ultrasound in the presence of microbubbles, Applied Env. Microb. **64**, 3927–3931 (1998)
[12] G.D. Betts, A. Williams, R.M. Oakley, Ultrasonic standing waves, inactivation of foodborne microorganism using power ultrasound, Encyclopedia of Food Microb., 2202–2208 (1999)
[13] A. Demirdoven, T. Baysal, The use of ultrasound and combined technologies in food preservation, Food Reviews Intern. **25**, 1–11 (2009)
[14] C.D. Porto, E. Porretto, D. Decorti, Comparison of ultrasound-assisted extraction with conventional extraction methods of oil and polyphenols from grape (Vitis vinifera L.) seeds, Ultra.Sonochemistry **20**, 1076–1080 (2013)
[15] N.E. Darra, H.N. Rajha, M.A. Ducasse, M.F. Turk, N. Grimi, R.G. Maroun, N. Louka, E. Vorobie, Effect of pulsed electric field treatment during cold maceration and alcoholic fermentation on major red wine qualitative and quantitative parameters, Food Chem. **213**, 352–360 (2006)
[16] S.Y. Leong, D.J. Burritt, O. Indrawati, Evaluation of the anthocyanin release and health-promoting properties of Pinot Noir grape juices after pulsed electric fields, Food Chem. **196**, 833–841 (2016)
[17] F.J. Barba, Z. Zhu, M. Koubaa, A.S. Sant'Ana, V. Orlien, Trends in Green alternative methods for the extraction of antioxidant bioactive compounds from winery wastes and by-products: A review, Food Scien. &Techn. **49**, 96–109 (2016)
[18] S. Cabredo-Pinillos, T. Cedron-Fernandez, M. Gonzalez-Briongos, L. Puente-Pascual, C. Saenz-Barrio, Ultrasound-assisted extraction of volatile compounds from wine samples: Optimisation of the method, Talanta **69**, 1123–1129 (2006)
[19] A.C. Chang, F.C. Chen, The application of 20 kHz ultrasonic waves to accelerate the aging of different wines, Food Chem. **79**, 501–506 (2002)
[20] M. Couasnon, Une nuvelle technique: la maceration pre-fermentaire à froid-Extraction a la nieve carbonique, part I. Revue des oenologues et des techniques vitivinicoles et oenologiques **92**, 26–30 (1999)
[21] A. Parenti, P. Spugnoli, L. Calamai, S. Ferrari, C. Gori, Effects of cold maceration on red wine quality from Tuscan Sangiovese grape, European Food Res. and Techn. **218**, 360–366 (2004)
[22] V. Gerbi, G. Zeppa, C. Manara, J.L. Minati, Experience in cold maceration with Muscat grapes, L'Enotecnico **27**, 57–66 (1991)
[23] S. Radeka, S. Herjavec, Peruri, I. Luki, B. Sladonja, Effect of different maceration treatments on free and bound varietal aroma compounds in wine of *Vitis vinifera* L.cv. Malvazija istarska bijela, *Food Technol.* Biotechnol. **46**, 86–92 (2008)
[24] L.C. Colibaba, V.V. Cotea, F.G. Lacureanu, S. Tudose-Sandu-Ville, L. Rotaru, M. Niculaua, C.E. Luchian, Studies of phenolic and aromatic profile of Busuioacă de Bohotin wines**,** BIO Web of Conf. **5** (2015)
[25] J.M. Oliveira, M. Faria, S. Filomena, F. Barros, M.A Isabel, C6-alcohols as varietal markers for assessment of wine origin, Analytica Chim. Acta **563**, 300–309 (2006)
[26] R.F. Simpson, Some important aroma components of white wine, Food Technol. Aust. **31**, 516–522 (1979)
[27] P. Ribéreau-Gayon, Y. Glories, A. Maujean, D. Dubourdieu, "Handbook of Enology Volume II – The Chemistry of Wine. Stabilization and Treatments", 2nd Edition, 9 (2006)

Charakterisierung der Birnenaromatik von Österreichischem Weißburgunder (Pinot blanc) hinsichtlich Typizität und Qualität / Characterization of the pear-aroma profile and its impact on the quality and typicity of Austrian Pinot blanc wines

Christian Philipp, Eder Phillip, Walter Brandes, Ferdinand Regner, Elsa Patzl-Fischerleitner, and Reinhard Eder

Höhere Bundeslehranstalt und Bundesamt für Wein- und Obstbau, Wienerstraße 74, 3400 Klosterneuburg, Austria

Abstract. Austria is the third largest Pinot blanc producing country in the world. Pinot blanc is cultivated on 1916,5 ha of vineyards, which corresponds to 4.3% of the Austrian wine-growing area and 12.3% of the world production of Pinot blanc (15,493 ha). Only in Germany (4794 ha; 30.9%) and in Italy (3086 ha; 19.9%) more Pinot blanc is cultivated. The bouquet of dry Austrian Pinot Blanc is usually discreet, with a predominance of pear and apple flavors, these often coming together with a walnut aroma and a hint of flower (acacia flowers) and herbs. The taste is delicate and full-bodied; it shows fruity sweetness and a slight acidity, and the aftertaste is moderately lemony. Aged wine often reveals notes of honey and almond. Ethyl trans-2-cis-4-decadienoate is well known as an impact compound in fresh and processed pear products. This substance had not been noticed in wine until now. In the course of this study, the content of ethyl trans-2-cis-4-decadienoate and other pear aromas (isoamyl acetate, methyl trans-geranoate, ethyl hexanoate, ethyl octanoate, ethyl decanoate, ethyl dodecanoate) in different Austrian Pinot blanc wine samples were analyzed directly in wine using HS-SPME-SIM-MS. According to the results the wines contain a relevant quantity of ethyl trans-2-cis-4-decadienoate. The concentrations in the considered wines ranged from >0,036 to 4.04 μg/L. With the use of a BET-3-Alternative-Forced-Choice-Process, the detection detection threshold was calculated for this compound. The threshold value was estimated to be at a concentration of 2 μg/L. These data support the view that this volatile substance is partially relevant for the aroma of the analyzed wines. The aroma substances were selected according to a consumer study and the so-called "pear-specific odour-activity" was calculated through two different systems. The results of the study showed, however, that the perceived pear flavor in Pinot blanc wines is a result of the interplay of some of the analyzed aromas. It has also been confirmed that the pear flavor is important for the typicity and quality of the Austrian Pinot blanc wines.

Zusammenfassung. Österreich ist mit einer Anbaufläche von 1.914 ha das drittgrößte Weißburgunder-produzierende Land der Welt. Diese Fläche entspricht 4,3% der österreichischen Weinbaufläche und 12,3% der Weltanbaufläche von Weißburgunder (15.493 ha). Nur in Deutschland (4.794 ha, 30,9%) und in Italien (3.086 ha; 19,9%) wird mehr Pinot blanc angebaut. Das Aroma von trockenem Weißburgunder ist in der Regel diskret, mit vorherrschenden Birnen- und Apfelaromen, diese kommen oft mit einem Nussaroma und einem Hauch von Blüten (Akazienblüten) und Kräutern zusammen. Der Geschmack ist zart und vollmundig. Gereifte Weine zeigen oft Honig- und Mandelnoten. Ethyl-trans2-cis4-decadienoat ist bekannt als Leitaroma in frischen und verarbeiteten Birnenprodukten. Diese Verbindung wurde bis jetzt nicht in Wein beschrieben. Im Rahmen dieser Studie wurde der Gehalt an Ethyl-trans2-cis4-Decadienoat und anderen Ethyl- und Methylester der cis-trans-Isomere der Decadiensäure sowie andere mit Birne assoziierten Aromen (Isoamylacetat, Methyl-trans-Geranoat, Ethylhexanoat, Ethyloctanoat, Ethyldecanoat und Ethyldodecanoat) in österreichischen Weißburgunderproben direkt in Wein mit HS-SPME-SIM-MS analysiert. Bei den Analysen konnten relevante Quantitäten an Ethyl-trans-2-cis-4-decadienoat gefunden werden. Die Konzentrationen der untersuchten Weine lagen zwischen >0,036 und 4,04 μg/L. Ein vorangegangener Test zur Feststellung des Wahrnehmungsschwellenwertes nach dem BET-3-Alternative-Forced-Choice-Verfahren ergab bei dieser Verbindung einen Wert von 2 μg/L. Somit ist der Aromastoff teilweise relevant für den Charakter der untersuchten Weine. Die Aromen wurden nach einer Konsumentenstudie ausgesucht und die sogenannte "birnenspezifische Odour Activity Value" nach zwei Verfahren berechnet. Im Zuge der sensorischen und analytischen Studien konnte festgestellt werden, dass ein Zusammenspiel einiger analysierten Aromen für den Charakter, die Qualität und die Typizität des Birnenaromas verantwortlich ist. Es hat sich auch bestätigt, dass das Birnenaroma für die Qualität und Typizität der österreichischen Weißburgunder wichtig ist.

1. Einleitung

Auch unter dem Namen Pinot Blanc (CH, FR), Pinot Chardonnay, Weißer Burgunder (DE), Pinot Bianco (IT), Klevner (AT), Rulandské bílé (CZ), Fehér burgundi (HU) und Beli pinot (SL) bekannt, handelt es sich beim Weißburgunder um eine sehr alte autochthone Rebsorte aus Frankreich. Die Burgundersorte ist durch spontane Mutation des Grauen Burgunders entstanden und bereits seit Beginn des 14. Jahrhunderts bekannt [1,2]. Seine erste Erwähnung fand der Weißburgunder im Jahr 1895 im Burgund. Bis zur Weingartenerhebung 1999 wegen seiner Ähnlichkeit zum Chardonnay in Österreich noch gemeinsam kultiviert und vinifiziert, verliert der Weißburgunder vor allem in seinem Ursprungsgebiet zunehmend an Bedeutung [3–5].

Der Weißburgunder wird vorwiegend in Deutschland, Italien, Österreich, und Frankreich angebaut. In Österreich hat der Weißburgunder 2015 eine Anbaufläche von 1.916,5 ha. Die Verbreitung der Sorte liegt somit bei einem Anteil von 6,3% im Hinblick auf die Weißweinfläche von 30.502 ha [5].

Seine Anbaufläche ist seit seiner ersten Erfassung 1999 in Österreich etwa gleichbleibend [5]. In Deutschland hingegen ist ein kontinuierlicher Anstieg der Rebfläche zu beobachten. Im Juli 2015 sind es bereits 4.973 ha (Veränderung seit 2014: plus 179 ha) [7].

Bekannte Weißburgunder entstehen in Österreich vor allem in der Südsteiermark, im Vulkanland Steiermark, im Weinviertel, in Wien und Klosterneuburg, im nördlichen Burgenland (Leithaberg DAC) und auch in der Thermenregion. Der Weißburgunder ist jedoch in allen Weinbaugebieten Österreichs vertreten [5].

1.1. Weinqualität, Typizität und Aroma

Für eine gute Weinqualität des Weißburgunders sind hohe Mostgewichte und eine gute Lage erforderlich. Seine besten Weine sind finessenreich und fruchtig, gelegentlich leicht nussig. Das Aroma eines typischen Österreichischen Weißburgunders lässt Birne, Apfel, Quitte, Banane, Aprikose, Karamell und Zitrusfrüchte erkennen. Gereifte Weine erinnern an frisches Brot und können durch Barriqueausbau und längere Flaschenreife an Dichte und Struktur gewinnen. Der Weiße Burgunder zeigt keine große Intensität, jedoch kommen seine Primäraromen bei jugendlichen Weinen perfekt zum Ausdruck. Sein Geschmackseindruck ist als junger, schlanker Wein eher neutral. Seine dezente Art macht ihn daher zum idealen Cuvéepartner. Der Weißburgunder genießt nicht denselben Ruf wie Grauburgunder oder Chardonnay, kann aber unter optimalen Bedingungen ebenso interessante Weine hervorbringen. Er zeigt bei hoher Reife wesentlich weniger Säure als Chardonnay, welch Eigenschaft häufig als seine Schwachstelle beschrieben wird [1,8].

Einen großen Teil der Weinqualität bestimmen die Aromastoffe. Sie prägen sich dem Verbraucher wesentlich ein und beeinflussen somit die Bereitschaft, den Wein zu trinken oder abzulehnen. In der Weinaroma-Forschung fand in den letzten drei Jahrzehnten eine rasante Entwicklung statt. Weltweit wurden wichtige Erkenntnisse zur Charakterisierung von sortentypischen Aromen (typischen Aromen) erarbeitet und dank leistungsfähiger Analysenmethoden weiterhin neue Aromastoffe entdeckt [9].

Tabelle 1. Verteilung der Weißburgunder-Rebfläche international (abs. und rel.) [6].

Anbauland	Fläche in ha	rel. Anteil an Weltanbaufläche
Deutschland[1]	4794	30,9%
Italien[2]	3086	19,9%
Österreich[2]	1914	12,4%
Frankreich[2]	1292	8,3%
Tschechien[2]	732	4,7%
Russland[2]	695	4,5%
Slowenien[2]	525	3,4%
Slowakei[2]	523	3,4%
Moldavien[2]	350	2,3%
Ukraine[2]	338	2,2%
Gesamt	**15493**	

[1] Daten von 2010
[2] Daten von 2014

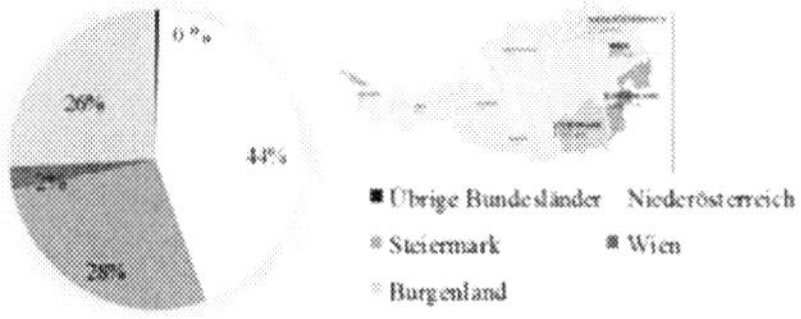

Abbildung 1. Anbaufläche Weißburgunder (Eigendarstellung) [5].

Obwohl das Aroma im Wein von vielen hunderten Einzelverbindungen (größer als 1200 Verbindungen) geprägt ist, sind nur wenige dieser Substanzen für das tatsächlich wahrgenommene Aroma verantwortlich. Die Gesamtkonzentration aller Aromastoffe liegt bei etwa 0,8 bis 1,2 g/l (ca. 1% des Ethanol-Gehaltes), wobei die Gehalte einzelner Substanzen zum Teil in einem Bereich von 10^{-3} bis 10^{-10} g/l liegen [10,11]. Es handelt sich dabei um Verbindungen, die unterschiedlichen Substanzklassen wie Kohlenwasserstoffen, Alkoholen, Aldehyden, Ketonen, Carbonsäuren, Estern, Ethern, Aromaten, Terpenen, Lactonen, Schwefel- und Stickstoffverbindungen zuzuordnen sind [10–12].

Für die Beschreibung des Beitrags zur Intensität von Aromasubstanzen wird der Begriff der Geruchsaktivität definiert. Geruchsaktivitätswert (Odour Activitity Value) ist ein Maß der Bedeutung einer spezifischen Verbindung zu dem Geruch einer Probe. Sie wird als das Verhältnis zwischen der Konzentration der einzelnen Substanz in einer Probe und der Schwellenkonzentration dieser Substanz berechnet [39].

1.2. Birnenaroma von Weinen

Die Birnenaromatik in Wein wird bisher mit den Aromastoffen Isoamylacetat [9], Ethylpentanoat [12], Hexylacetat [13,14], Ethyl-Isobutyrat [15], Methyl-trans-Geranoat [13], Ethyloktanoat [14], Isoamyl-octanoat [13] und Butylbutyrat [16,17] in Verbindung gebracht. Bemerkenswert ist die Tatsache, dass sowohl kurzkettige als auch langkettige Ester mit birnenähnlichem Aroma beschrieben werden.

Ethyl-trans2-cis4-Decadiensäure (t2c4E) ist ein essentieller Bestandteil des typischen Aromas in Birnen und deren Produkten [18]. Neben diesen bereits erwähnten Ester der trans2-cis4-Decadiensäure, welcher in Schalen

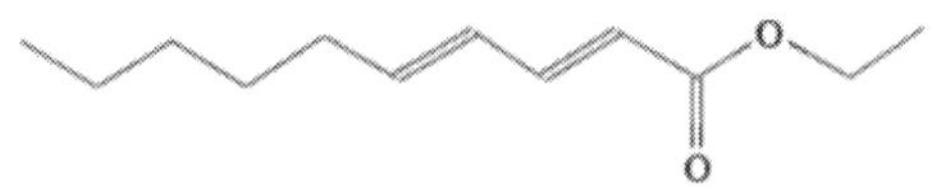

Abbildung 2. Chemische Struktur von 2,4-Decadienoat Ethylester [23].

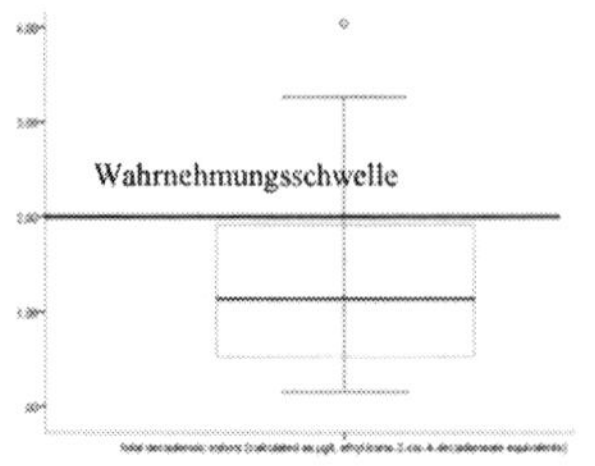

Abbildung 3. Boxplot-Darstellung der Ethyl-trans2-cis4-decadienoat Equivalente in österreichischen Weißburgundern.

und überreifen Birnenfrüchten nachgewiesen werden kann, ist auch Ethyl-trans2-trans4-Decadiensäure ein wertbestimmender Ester von Birnenprodukten. Erst in der Maische überreifer Früchte kann dieser quantifiziert werden [18,20]. T2c4E wird durch die Gärung und Lagerung von Birnenprodukten nicht wesentlich verändert [21].

Im Wein von Trauben aus *Vitis vinifera* wurden diese Ester bisher nicht beschrieben, aber in Trauben von *Vitis lambrusca* Reben (Concord und Niagara) und Elvira (*Vitis lambrusca x Vitis riparia*) wurden Konzentrationen zwischen 10 und 300 μg/L Ethyl-trans2-trans4-decadienoat und t2c4E gefunden [19].

T2c4E hat einen charakteristischen Geschmack nach Birne und ein leicht fruchtiges Aroma. Die langkettige Esterverbindung ist aufgrund ihrer geringen Polarität schwer oder nicht in Wasser löslich. Ihre Aromacharakteristik wird folgend beschrieben: Süße reifer Birne, cremig, dicht und fruchtig, grüne, zum Teil auch fleischige Nuancen [22].

2. Material und methoden

2.1. Proben

Jeweils sechs kommerziell erhältliche Proben (100% Weißburgunder), aus sechs bedeutenden Weißburgunder-Anbaugebieten Österreichs (Steiermark, Wien, Leithaberg, Wagram, Weinviertel und Thermenregion) von insgesamt 36 Winzern wurden für diese sensorischen und analytischen Untersuchungen verwendet. Die 36 Weine wurden vorwiegend für den österreichischen Markt produziert, besitzen alle eine staatliche Prüfnummer und wurden somit als Qualitätswein in Verkehr gebracht. Alle Weine stammen vom Jahrgang 2015. Alle Weine wurden von den Produzenten als gebiets- und sortentypisch beschrieben.

2.2. Sensorische Beurteilungen der Proben

Das sensorische Experten- Panel setzte sich aus insgesamt 14 Männern und 10 Frauen im Alter von 18 bis 91 Jahren zusammen. Alle Mitglieder des Panels konnten eine einschlägige Fach-Expertise in Bezug auf Weinverkostungen vorweisen (staatliche abgelegte Prüfung für amtliche Prüfweinkosten). Die Proben wurden in vier Kommissionen als Duplikat (1. Durchgang mit Herkunftsangabe, 2. Durchgang ohne Herkunftsangabe) auf einer unstrukturierten Skala (0–10 cm) nach folgenden deskriptiven Kriterien bewertet: Fruchtintensität, merkbarer Holzeinsatz, Karamellnoten, Brotnoten, Nussnoten, Kräuternoten, Birnenaromatik und Typizität. Des Weiteren wurde das Gesamturteil der Weine mittels 20 Punkte-Schema (10 – 12 Punkte =einfacher Wein; 12 – 14 Punkte = guter Wein; 14 – 16 Punkte = sehr guter Wein; 16 – 18 Punkte = ausgezeichneter Wein; 18 – 20 = absoluter Spitzenwein) ermittelt. Bis 18 Punkte war die Vergabe von nur ganzen Punkten möglich, ab 18 Punkte waren auch halbe Punkteschritte erlaubt. Die Weine wurden bei 15 °C temperiert. Es wurden jeweils sechs Weine aus sechs unterschiedlichen Regionen gleichzeitig in 23 cl Standardweingläser eingeschenkt.

2.3. Verwendete Chemikalien

Alkohol (99%) wurde von der FA AustrAlco (Österreichische Alkoholhandels-GmbH) bezogen. Die chemischen Standards wurden von der Fa. Merck (Deutschland), Fa. Roth (Österreich), Fa. Riedel-de Haen (Deutschland), Fa. Fluka (USA), Fa. Aldrich (USA) und der Fa. Schuchardt (Deutschland) gekauft und wiesen alle die maximal erhältlicher Konzentration auf (95% – 99.5%). Alle nicht käuflich erhältlichen Standards wurden durch Eigensynthese hergestellt [24,25]. Das für die Aromaanalysen verwendete wasserfreie Kaliumdihydrogenphosphat stammte von der Fa. Zeller (Österreich).

2.4. Konsumentenstudien zur Auswahl der potentiellen Birnenaromen

Diese Studie wurde mit 81 Konsumenten (davon 32 Frauen) aus drei Altersgruppen (18–30 Jahre, 31–50 Jahre, über 50 Jahre) durchgeführt. Dazu wurden 100 Wein-Hauptaromasubstanzen aus folgenden Stoffklassen eingesetzt: Ester-Verbindungen, volatile Phenole, Carbonsäuren, höhere Alkohole, Terpene, Carbonyl-Verbindungen, Norisoprenoide, Lactone und schwefelhaltige Verbindungen. Die Teilnehmer mussten intuitiv entscheiden, ob eine Aromasubstanz (in 10% alkoholischer Lösung) nach Birne riecht. Die Verkostung wurde in 23 cl Standard-Weingläsern durchgeführt. Die Konzentrationen der untersuchten Substanzen sind dem Ergebnisteil zu entnehmen. Eine Verbindung wurde in dieser Studie als potentielles Birnenaroma klassifiziert, wenn die Substanz zumindest von zwei Konsumenten aus jeder Altersgruppe bzw. von mindestens neun Konsumenten gesamt (>10%) als Birnenaroma beschrieben wurde.

2.5. Evaluierung der ausgewählten Aroma-substanzen mittels Expertenverkostung

Zur weiteren Evaluierung und Absicherung der ausgewählten Aromen wurde die Typizität (0–100% typisch) der von den Konsumenten ausgewählten Aromen mittels sechs Experten (langjährige Erfahrung in Weinforschung und Weinverkostung) bewertet. Die Aromen wurden in einer Blindverkostung analog ihrer Typizität bewertet.

2.6. Sensorische Bestimmung der Wahrnehmungsschwelle (detection-threshold)

Die Wahrnehmungsschwelle wurden mittels "best estimated threshold Dreickstest" (BET – 3 AFC) in 10%

alkoholischer Lösung als Duplikat durchgeführt [26]. Für diese Tests wurden insgesamt drei verschiedene Experten-Panels (amtliche Prüfnummern Koster- Panels) verwendet. Eine definierte Menge an Aromasubstanz in 10% alkoholischer Lösung musste in einem blind angelegten Dreieckstest von einer Nullprobe (10% alkoholische Lösung) unterschieden werden. Dieses Verfahren wurde mit insgesamt fünf ansteigenden Aromakonzentrationen wiederholt. Die "Detection-threshold" wurde nach Formel (1) berechnet.

$$BET_{single} = \sqrt{C_a \cdot C_b}$$
$$BET_{average} = \emptyset BET_{single} \quad (1)$$

C_a = Konzentration die gerade noch erkannt wird
C_b = Konzentration die gerade nicht mehr erkannt wird.

2.7. Quantitative Analysen der evaluierten Aromen mittels HS-SPME-GC-MS

5,00 ml der zu analysierenden Proben wurden mit 1 g NaH_2PO_4 und 5 μl einer internen Standardmischung (Ethyl-2-decenoat, 20 μg/L Endkonzentration) und in einem Head-Space-Vial (20 ml) mit einem Magnetrührknochen in den Autosampler (CombiPal Probengeber – CTC Analytics (Schweiz)) gestellt. Als Probeaufgabesystem wurde Head-Space-Solid-Phase-Micro-Extraction (HS SPME) mit einer 65 μ m PDMS/DVB, Fused Silica 24 Ga Faser (Fa. Supelco/ Sigma Aldrich, USA) gewählt. Mit einer 6890 N Serie II Gaschromatograph der Fa. Agilent (USA) und einem massenselektiver Detektor 5975 der Fa. Agilent (USA) wurden die 36 Weine analog nach der von Philipp et al. [45] beschriebenen Methode als Duplikat analysiert.

2.8. Berechnung der "birnenspezifischen Odour-Activity Value"

Die "birnenspezifischen" Odour- Activity Values" wurden für jede "Birnenaroma-Gruppe" (siehe Kapitel 3.6) nach Formel (2) berechnet. Die Ergebnisse dieser Berechnungen werden im Kapitel 3.4 und 3.5 diskutiert und fortan als System 1 bezeichnet.

$$OAV_{Birnengruppe} = \sum_{Aroma\,1}^{Aroma\,x} OAV_{Aroma} \cdot Typizität_{Aroma}. \quad (2)$$

2.9. Statistische Berechnungen der Ergebnisse

Bei den Verkostungsergebnissen der Proben wurde nach erfolgter ANOVA basierend auf Kruskal-Wallis-Friedmann mit LSD-Test (Least significant Difference, auf dem Signifikanzniveau α =0,05) gearbeitet. Hierbei betrachtet man die mindeste Differenz zweier Mittelwerte, die gerade noch signifikant ist.

Für die Evaluierung und Überprüfung des Systems 1 und einiger weiterführender Diskussionen werden die sensorischen Ergebnisse der 36 untersuchten Proben statistisch mit den berechneten "Odour Activity Values" und den analysierten Aromakonzentrationen verglichen. Die Ergebnisse wurden sowohl einer bivariaten Pearson-Korrelationsanalyse auf einem Signifikanzniveau von α =0,05 als auch einer Hauptkomponentenanalyse mit einer Promax Rotation (kappa =4000) und einer Reduktion auf 2 Hauptkomponenten unterzogen. Die Ergebnisse werden als Ladungsdiagramm und Streudiagramm präsentiert. Diese statistischen Aus-wertungen erfolgten mit IBM – SPSS Statistics 22 (Fa. International Business Machines Corporation, USA).

Tabelle 2. Ergebnisse der Konsumentenbefragungen – Auswahl der Aromen(n=81).

Aromasubstanz	μg/L	Anzahl ja Stimmen			
Zugesetzte Konzentration		Gr. 1	Gr. 2	Gr. 3	Gesamt
Ethyl-trans2-cis4-decadienoat	200	22	17	17	56
trans-3-Methyl-2-hexenoic acid (Methyl-trans-Geranoat)	200	16	10	9	35
Isoamylacetat	200	8	20	7	35
Hexylacetat	1000	5	3	0	8
Ethylhexanoat	400	4	7	2	13
Ethyloctanoat	500	2	2	2	6
Isoamyloctanoat	400	8	9	0	17
Butylbutyrat	200	5	1	0	6
Ethylpentanoat	200	0	2	3	5
Ethyldecanoat	400	8	14	8	30
Ethyldodecanoat	400	7	20	13	40

3. Ergebnisse und diskussion

3.1. Ergebnisse der Konsumentenstudie

Eine Studie mit insgesamt 81 Konsumenten (3 Altersgruppen) wurde durchgeführt, um festzustellen, welche der Aromastoffe einen Beitrag zur Birnenaromatik leisten können (Tabelle 2).

Eindeutig als Birnenaroma wurde t2c4E, Methyl-trans-Geranoat, Isoamylacetat, Ethyldodecanoat und Ethyldecanoat erkannt ($\geq$1/3 der befragten Konsumenten erkannten die Substanzen als Birnenaroma). Ebenfalls haben die Aromastoffe Ethyloctanoat, Ethylhexanoat und Isoamyloctanoat die Kriterien erfüllt (Kapitel 2.4). Bemerkenswert ist, dass die in der Literatur erwähnten Aromastoffe Butylbutyrat, Ethylpentanoat und Hexylacetat unter den befragten Konsumenten nicht als birnentypisch beurteilt wurden [12–17]. Genauso interessant ist die Tatsache, dass Ethyldecanoat, Ethyldodecanoat und Ethylhexanoat bis jetzt nicht als Birnenaroma beschrieben wurden.

Um den Nachteil der geringeren Reproduzierbarkeit von Konsumentenstudien zu minimieren, wurde versucht die Probandenanzahl möglichst groß und repräsentativ über alle Altersgruppen und beiden Geschlechter zu verteilen. Außerdem wurde darauf geachtet, dass bei der Verkostung genügend Zeit für die Adaption des Geruchssinnes gegeben war [29,30].

3.2. Ergebnisse der Expertenverkostung und ermittelte Geruchsschwelle für die ausgewählten Aromen

Zur weiteren Absicherung, Diskussion und Bewertung der Typizität der ausgewählten Aromen wurde diesbezüglich mittels sechs Experten eine deskriptive Studie durchgeführt. Die ausgewählten Aromen sind von der

Tabelle 3. Ergebnisse der Expertenbefragung/ Wahrnehmungsschwellen nach BET Verfahren.

Aromasubstanz	Durchschnittliche Typizität [%]						Detection threshold
	T1	T2	T3	T4	T5	Ges.	μg/L
Ethyl-trans2-cis4-decadienoat	28	36			16	80	2
trans-3-Methyl-2-hexenoic acid (Methyl-trans-Geranoat)			48	10		58	8
Isoamylacetat				52		52	30
Ethylhexanoat	10			4	38	52	14
Ethyloctanoat		42		16		58	80
Isoamyloctanoat	2	22	6			30	28
Ethyldodecanoat			26		14	40	70
Ethyldecanoat		18	2		16	36	88

Charakteristik sehr verschieden, daher wurden fünf Hauptgruppen definiert: Überreife Birne (T1); frische Birne (T2); gekochte, verarbeitete Birne (T3); tropische Birnendrops (T4); ölig wachsig birnig (T5).

Vergleicht man die Ergebnisse der Tabelle 2 mit Tabelle 3 fällt zunächst auf, dass in beiden Studien t2c4E die höchste Bewertung erhielt. Isoamyloctanoat wurde bei beiden Verfahren als wenig aromatisch bewertet, aber wenn dann als frische Birne beschrieben. Der Aromastoff Ethylhexanoat erhielt bei den Experten die höchste Bewertung in Bezug auf die Typizität "ölig, wachsig, birnig" wurde aber von den Konsumenten mit nur 13 von 81 Stimmen fast nicht als birnentypisch erkannt. Ethyloctanoat erreichte bei den Experten die höchste Typizität für "frische Birne", wurde aber von den Konsumenten kaum als birnentypisch erkannt.

3.3. Ergebnisse der Aromaanalysen

Neben den relevanten Aromaverbindungen Isoamylacetat, Ethylhexanoat, Ethyloctanoat, Isoamyloctanoat, Ethyldecanoat, Methyl-trans-Geranoat, Ethyldodecanoat und t2c4E wurden analog zu Philipp et al. [45] und weiteren Autoren [24,25] ebenfalls andere Methyl- und Ethylester der Cis-Trans-Isomere der Decadiensäure in 36 österreichische Weine der Sorte Weißburgunder vom Jahrgang 2015 bestimmt.

Die Konzentrationen an Isoamylacetat, Ethylhexanoat, Ethyloctanoat und Ethyldecanoat sowie Ethyldodecanoat entsprechen durchschnittlichen Angaben aus der Literatur [13,31,37].

Trans-3-Methyl-2-hexenoic acid (Methyl-trans-Geranoat) wurde das erste Mal in österreichischen Weinen detektiert. Mit einer Konzentration bis 2,46 μg/L entspricht dieser Wert den Angaben der Literatur. Antalik und Kollegen [13] haben in französischen Rotweinen bis 0,83 μg/L gefunden. Dieser Stoff ist aber vor allem in Gewürztraminer und einigen Süßweinen, wie zum Beispiel Coteaux du Layon aus der Rebsorte Chenin, in hohen Konzentrationen enthalten. Der Nachweis von Methyl-trans-Geranoat in Weißburgunder ist ein weiteres Indiz dafür, dass die Konzentration dieser Verbindung von der Sorte und einigen enzymatischen Aktivitäten abhängt [13,38].

Im Zuge dieser Studie ist die Detektion von Ethyl- und Methylestern der Decadiensäure in Wein

Tabelle 4. Gehalte der Aromastoffe in 36 Weinen der Sorte Weißburgunder.

Substanz	Min [μg/L]	Max [μg/L]	Mittelwert [μg/L]	StdAbw. [μg/L]
Isoamylacetat	242,64	3886,40	1933,19	880,14
Ethylhexanoat	281,92	1367,20	741,68	256,03
Ethyloctanoat	113,94	2207,63	1409,73	410,03
Ethyldecanoat	240,59	1224,97	778,62	238,67
Isoamyloctanoat	0,83	3,65	2,19	0,74
Methyl-trans-Geranoat	n.d.	2,46	0,36	0,45
Ethyldodecanoat	2,02	73,23	35,06	22,84
Ethyl-trans2-cis4-decadienoat				
Ethyl-cis2-trans4-decadienoat*1	>0,036	4,04	1,24	0,81
Methyl-cis2-cis4-decadienoat*1				
Ethyl-cis2-cis4-decadienoat*1				

*1 Berechnet als Ethyl-trans2-cis4-decadienoate Äquivalente [45].

gelungen. Die österreichischen Weißburgunder-Weine beinhalten Konzentrationen von >0,036 bis 4,04 μg/L. Die ebenfalls festgestellte Wahrnehmungsschwelle liegt bei 2 μg/L (Tabelle 3). Daher kann diese Verbindung im geringen Ausmaß partiell zur Aromatik der untersuchten Weine beitragen. Ein Großteil der untersuchten Proben lag in Bezug auf diese Substanz jedoch unter der Wahrnehmungsschwelle. Die Geruchsschwelle, ab der man "Birne" anspricht, liegt wesentlich höher. Sie wurde mit einer individuell hohen Streuung und einer geringen Reproduzierbarkeit von BET_{single} 20–400 μg/L festgestellt. Die hohe Streuung erklärt sich eventuell dadurch, dass t2c4E nicht für alle Menschen sofort als Birne empfunden wird.

3.4. Birnenspezifische - Odour-Activity-Values

Die "birnenspezifischen Odour activity values" ergaben sich aus der mathematischen Zusammenführung der analytischen und sensorischen Aromastudien dieser Arbeit. Die Durchschnittswerte der birnenspezifischen OAVs über die 36, in Philipp et al. [45] diskutierten Weine, sind grafisch als "sensorisch-analytisches Typogramm" in Abbildung 4 dargestellt. Die Aromastoffe Isoamylacetat, Ethylhexanoat und Ethyloctanoat tragen im Durchschnitt mit OAV Werten über 10 zur Birnenaromatik bei [45]. Alle drei Aromastoffe sind von der Charakteristik sehr unterschiedlich.

Isoamylacetat prägt primär den Begriff der Birnendrops. Dieser Aromastoff wird in der Literatur häufig auch als Kaltgäraroma mit einem Charakter von Banane, Drops und Birne beschrieben [9,13,42]. Der Aromastoff Ethylhexanoat wird in der Literatur häufig mit den Begriffen Apfelschale, grüner Apfel [13], süß, fruchtig [42] in Verbindung gebracht. Der starke Beitrag zur Aromagruppe "ölig – wachsig – birnig" kann über die Assoziation der wachsigen Apfel- oder Birnenschale erklärt werden. Ethyloctanoat wird in der gängigen Literatur als Birne [14] oder Ananas, süß und Banane [41] und Ethyldecanoat als fruchtig [13,42] beschrieben. Alle anderen detektierten Aromasubstanzen zeigten im Durschnitt eine "birnenspezifische Odour-Activity" unter

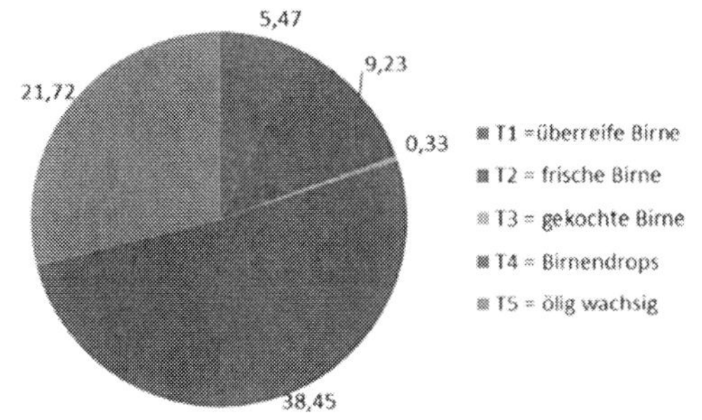

Abbildung 4. Sensorisch-analytisches Typogramm der Birnearomatik im weiteren Sinn (Durchschnitt aus 36 Weinen der Sorte Weißburgunder).

eins. Das bedeutet aber keineswegs, dass sie nicht synergistisch zum empfundenen Aroma beitragen können. In einigen Studien wurde bereits gezeigt, dass das Vorhandensein einiger Aromasubstanzen das Empfinden anderer Aromastoffe beeinflussen kann [40,43]. Ein Blick auf das "sensorisch-analytische Typogramm" der Birnenaromatik zeigt auch, dass mit dieser Abbildung 4 nicht nur das klassisch nach reifer Birne riechende Aroma abgedeckt wird – sondern versucht wird, den Begriff "Birne im Wein" etwas weiter zu fassen. Dies wiederum rechtfertigt den hohen Gesamtwert der "birnen-spezifischen Odour Activity Value" für den Durchschnitt der analysierten österreichischen Weißburgunder von 75,21.

3.5. Diskussion der kalkulierten Odour-Activity-Values mit den Ergebnissen der sensorischen Überprüfung der Weine

Die Hauptkomponentenanalyse (Abb. 5) mit allen analytischen, kalkulierten und sensorisch ermittelten Daten diente zur Diskussion des Systems 1 der "birnenspezifischen Odour Activity Values". Hierfür reicht die Präsentation der Ladungsdiagramme, um gegensätzliche Korrelationen zwischen den Ergebnissen zu veranschaulichen. Je näher die Punkte zueinander liegen, desto größer ist die Korrelation zwischen ihnen. Der dickmarkierte rote Cluster der "birnenspezifischen Odour Activity Value" sollte, sofern das System 1 geeignet ist, die empfundene Birnenaromatik zu beschreiben, neben dem dicken violettmarkierten Deskriptor Birne in Abb.5 liegen (0,093 → keine Korrelation aus Tabelle 5). Tatsächlich zeigte die Hauptkomponentenanalyse, dass die "birnenspezifische Odour Activity Value" stark von den Aromasubstanzen Ethyloctanoat, Isoamyloctanoat und Ethyldecanoat in die entgegengesetzte Richtung verschoben wurde. Das verwendete Experten-Panel sprach hingegen eher auf die Verbindungen Isoamylacetat, Ethyldodecanoat, Methyl-trans-Geranoat und die Decadiensäureester an. Die kalkulierten "birnenspezifischen Odour Activity Values" konnten nicht verwendet werden, um den Deskriptor Birne zu beschreiben. Die kalkulierten "birnen-spezifischen Odour Activity Values" dienten, wie bereits erwähnt, zur grafischen Darstellung eines sehr weit gefassten Begriffes "Birne im Wein".

Die violett präsentierten Deskriptoren stehen für die Verkostungsergebnisse. Interessant ist, dass es offensichtlich drei Gruppen (blaue Kreise, grüne Schrift) gibt. Die eine Gruppe beschreibt vor allem reife Weißburgunder.

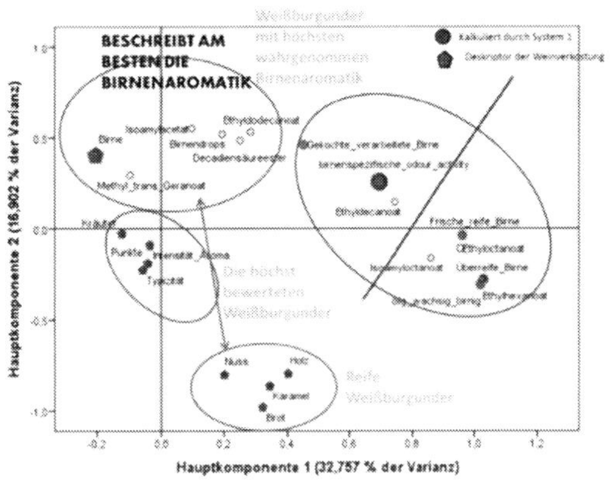

Abbildung 5. Hauptkomponentenanalyse – System 1.

Tabelle 5. Pearson-Korrelationsdaten im System 1.

Wechselwirkung	Pearson Korrealtionsfaktor	Signifikanz
Birne – Birnen Odour Activity	0,093	0,596
Birne – Typizität	**0,514**	0,002*
Birne – Qualität	**0,646**	0,000*
Kräuter – Birne	-0,046	0,799
Kräuter – Typizität	0,327	0,055
Qualität – Typizität	**0,856**	0,000*
Holz – Typizität	0,019	0,913
Holz – Karamel	**0,525**	0,001*
Karamel – Typizität	**0,410**	0,014*

* Signifikant.

Sie sind geprägt von den Deskriptoren Holz, Nuss, Karamell und Brot. Auf der anderen Seite steht, völlig unabhängig davon, der Deskriptor Birne. Es scheint so, dass reife Weißburgunder wenig durch die Aromanote der Birne beschrieben werden. Dazwischen liegt ein Cluster mit den Deskriptoren Gesamtintensität des Aromas, Gesamtpunkte und Typizität. Diese drei Eigenschaften der Weine korrelierten miteinander (Tabelle 5).

Das verwendete Experten-Panel war sich darüber einig, dass die höchst bewerteten Weißburgunder gleichzeitig die typischen Weißburgunder sind. Offensichtlich werden typische österreichische Weißburgunder eher vom Deskriptor Birne (0,514*) und Karamell (0,410*) beschrieben, als von den Deskriptoren Holz (0,019) und Kräuter (0,327). Etwas überraschend ist auch die hohe Korrelation (0,646*) zwischen Birne und Qualität, was sich am ersten Blick nicht durch die Abbildung 5 erklären lässt.

Im nächsten Schritt (Abb. 6) wurden die Störgrößen (Ethyldecanoat, Isoamyloctanoat, Ethyloctanoat und Ethylhexanoat) und zusätzlich die starke Gewichtung von Isoamylacetat aus dem System entfernt und die "birnenspezifische Odour Activity Value" neu berechnet (System 2). Der Durchschnittswert für österreichischen Weißburgunder lag danach bei einer "birnenspezifischen Odour Activity Value" im engeren Sinne (System 2) bei 10,58 (2,16 – 19,91).

Die Pearson-Korrelationsanalyse (Tabelle 6) ergab eine mittlere signifikante Korrelation zwischen der deskriptiven Birne und der "birnenspezifischen Odour Activity Value" (System 2). Diese Korrelation spiegelte sich aber nicht mit den Deskriptoren Typizität und Qualität. Das System

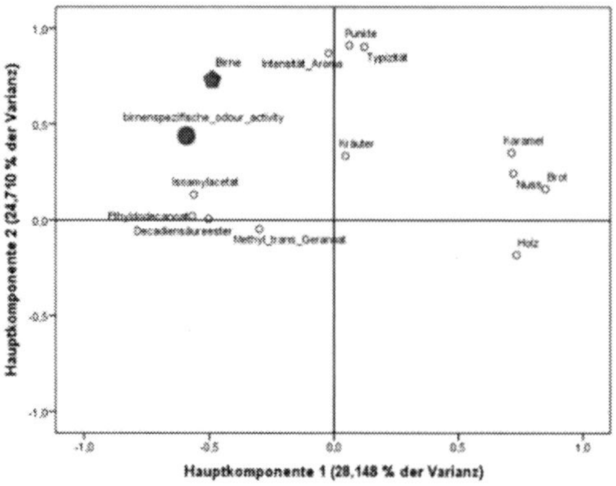

Abbildung 6. Hauptkomponentenanalyse nach Eliminierung der Störgrößen– System 2.

Tabelle 6. Pearson-Korrelationsdaten im System 2.

Wechselwirkung	Pearson Korrelationsfaktor	Signifikanz
Birne – Birnen Odour Activity	**0,491**	0,003*
Birnen Odour Activity – Typizität	0,213	0,219
Birnen Odour Activity – Qualität	0,237	0,171
Birne – Typizität	**0,514**	0,002*
Birne – Qualität	**0,646**	0,000*
Decadiensäureester – Birne	0,055	0,752
Ethyldodecanoat – Birne	0,106	0,543
Isoamylacetat – Birne	0,178	0,307
Methyl-trans-Geranoat – Birne	0,130	0,457

* Signifikant.

2 – eine Wechselwirkung zwischen den Aromastoffen Isoamylacetat, Ethyldodecanoat, Decadiensäureester und Methyl-trans-Geranoat - konnte in eingeschränkten Maße zur Beschreibung der Birnenaromatik der in dieser Studie untersuchten Weine beitragen.

Keine der Reinsubstanzen alleine ergab eine Korrelation zur empfundenen Birnenaromatik – nur im System miteinander funktionierte es. Interessant ist die Tatsache, dass die Konzentrationen von den Decadiensäureestern, Ethyldodecanoat und Methyl-trans-Geranoat in den analysierten Weinen unter der Wahrnehmungsschwelle lagen, dies könnte eventuell ein weiteres Indiz dafür sein, dass auch Aromasubstanzen unter der Wahrnehmungsschwelle, das Empfinden beeinflussen können [40,43].

3.6. Einfluss der Birnenaromatik auf die Qualität und die Typizität

Die Korrelationsanalyse der empfundenen Birnen-aromatik mit der Typizität und Qualität hat eine signifikant-mittelhohe Korrelation ergeben (Tabellen 5 und 6). Die Abbildungen 7 bis 9 zeigen, dass eine Clusterung nach Qualität und Typizität möglich und eine Clusterung nach Intensität der empfundenen Birnenaromatik angedeutet ist. Sie zeigt aber auch, dass eine signifikant höhere Birnenaromatik nicht automatisch eine ausgezeichnete Gesamtbewertung bedeutet und umgekehrt.

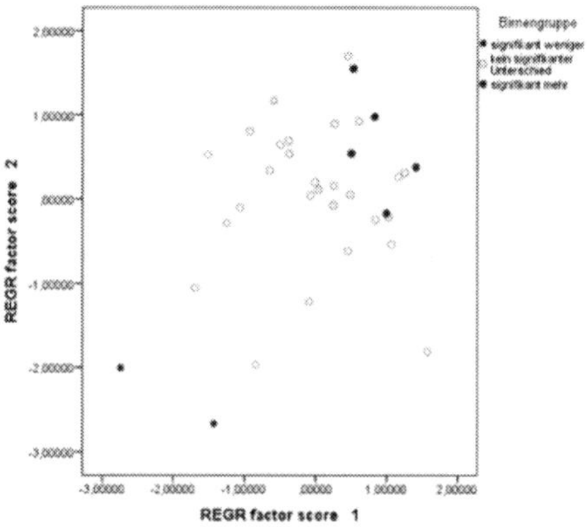

Abbildung 7. Scoreplot nach Birnengruppen (System 2).

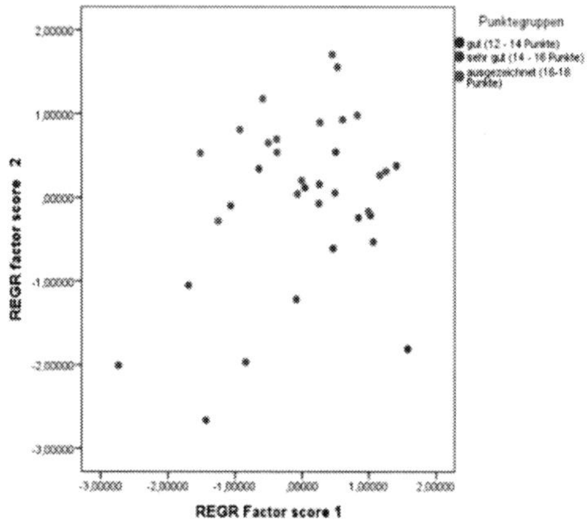

Abbildung 8. Scoreplot nach Gesamtbewertung (System 2).

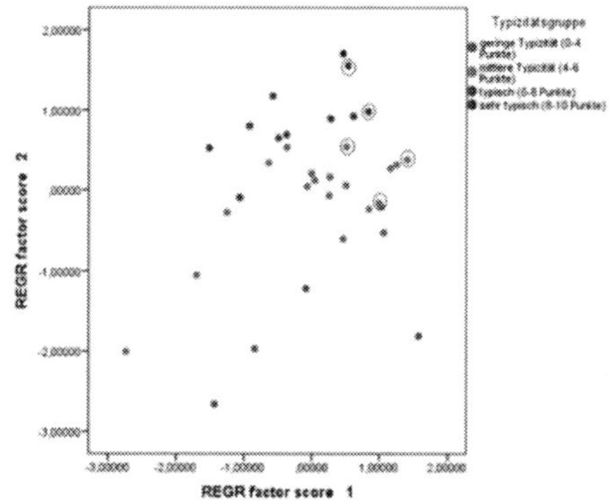

Abbildung 9. Scoreplot nach Typizität (System 2).

Der Deskriptor Typizität ist in diesem Fall gleich zu verstehen wie die Qualität. Das heißt, die Birnenaromatik beeinflusst die Typizität (0,514*) in gewissen Maße, aber es bedeutet nicht, dass Weine mit besonders hoher Typizität auch die höchste Birnenaromatik aufweisen.

Literatur

[1] F. Regner, K. Hanak, C. Eisenheld, *Verzeichnis der österreichischen Rebsorten und deren Klone*, 71–72 (2015)

[2] J. Robinson , J. Harding, J. Jouillamoz, *Wine Grapes*, 806–821 (2012)
[3] Statistik Austria, *Weingartengrunderhebung* (1999)
[4] D. Ehrlich, S. Braun. *Das Rebsorten ABC – Reben und ihre Weine* (2005)
[5] Statistik Austria, *Weingartengrunderhebung* (2015)
[6] Deutsches Weininstitut, *Deutscher Wein: Statistik 2015/2016* (2017)
[7] H.R. Schultz, M. Stoll, *Deutsches Weinbaujahrbuch* (2016)
[8] J. Schmid, F. Manty, B. Linder, R. Ries, E. Rühl, H. Konrad, E. Schönhals, E. Bleser, K. Schaller *Geisenheimer Rebsorten und Klone* (2009)
[9] G. Styger, B. Prior, F. Bauer, J Ind Microbiol Biotchnol **38**, 1145–1159 (2011)
[10] R. Woller, G. Würdig, *Handbuch der Lebensmitteltechnologie - Chemie des Weines* (1989)
[11] A. Rapp, H. Mandery, Experientia **8** (1986)
[12] L. Jakob, *Lexikon der Önologie – Kellerwirtschaft, Weinchemie, Weinrecht* (2012)
[13] G. Antalik, M. C. Perello, G. de Revel, Food Chem **121,** 1236–1245 (2010)
[14] M.G. Álvarez, C. G. Barreiro, B. Cancho-Grande, J. Simal-Gándara, Food Chem **129**, 890–898 (2011)
[15] J. Lozano, J.P. Santos, P. José, M. Aleixandre, I. Sayago, J. Gutiérrez, M.C. Horrillo, IEEE Sensors J. **6**, 173–178 (2006)
[16] E. Mehinagic, G. Royer, R. Symoneaux, F. Jourjon, C. Prost, J Agric Food Chem **54**, 2678–87 (2006)
[17] C. A. de March, S. E. Ryu, G. Sicard, C. Moon, J. Golebiowski, Flavour Frag J **30**, 342-361 (2015)
[18] W.G. Jennings, J Food Sci **26**, 564–568 (1961)
[19] P. Schreier, J.H. Paroschy. Can Insi Food Sci Technol **14**, 112–118 (1981)
[20] M. Gössinger, H. Sämann, R., Baumann, W. Patzl, K. Vogl, Mitt Klosterneuburg **53**, 184-194 (2003)
[21] L. Adam, N. Christoph, G. Versini, Kleinbrennerei **47**, 188–199 (1995)
[22] G. Burdock, *Handbook of Flavor Ingredients* (2009)
[23] Nist, *2,4-decadienoic acid, ethylester (E,Z)* (2016)
[24] W. Brandes, M. Karner, R. Eder, Mitt Klosterneuburg **57**, 63–82 (2007)
[25] W. Brandes, M. Karner, R. Eder, Mitt Klosterneuburg **53**, 103–112 (2003)
[26] M. Czerny, Ma. Christlbauer, Mo. Christlbauer, A. Fischer, M. Granvogl, M. Hammer, C. Hartl, N.M. Hernandez, P. Schieberle, Eur Food Res Technol **228**, 265–273 (2008)
[27] M. Kolb, A. Bahr, S. Hippich, W. Schulz Hydrochimica et Hydrobiologica **21**, 308–311 (1993)
[28] Arbeitsausschuss chemische DIN, *DIN 32645: Nachweis-, Erfassungs-und Bestimmungsgrenze: Ermittlung unter Wiederholbedingungen, Begriffe, Verfahren, Auswertung* (1994)
[29] H.T. Lawless, H. Heymann. *Sensory Evaluation of Food* (2010)
[30] H.T. Lawless, *Laboratory Exercises for Sensory Evaluation* (2013)
[31] R. Lopez, M. Aznar, J. Cacho, V. Ferreira, J Chromatogr A **966**, 167–177 (2002)
[32] S. J. Pérez-Olivero, M. L. Pérez-Pont, J. E. Conde, and J. P. Pérez-Trujillo, J Anayltical Methods in Chemistry, 1–10 (2014)
[33] C.H. Xu, G.S. Chen, Z.H. Xiong, Y.X. Fan, X.C. Wang, Y Liu, Trends Anal Chem **80**, 12–29 (2016)
[34] M.A. Pozo-Bayón, E. Pueyo, P.J. Martín-Álvarez, M.C. Polo, J Chromatogr A **922**, 267–275 (2012)
[35] X. Yang, T. Peppard, J. Agric. Food Chem. **42**, 1925 (1994)
[36] R.R. Villamor, C.F. Ross, Annu Rev Food Sci Techn **4**, 1–20 (2013)
[37] C. Ortega, R. Lopez, J. Cacho, V. Ferreira, J Chromatogr A **923**, 205–214 (2001)
[38] G. Versini, A. Rapp, A. Dalla Serra, U. Pichler, M. Ramponi, (1994). Vitis **33**, 139–142 (1994)
[39] I. L. Francis, J. L. Newton, Aust J Grape Wine Research **11**, 114–126 (2005)
[40] Ferreira, V., & Cacho, J. *Identification of impact odorants of wines.* In M. Victoria Moreno-Arribas, M. Carmen Polo (Eds.), Wine chemistry and biochemistry, 393–415 (2009)
[41] L. Fariña, V. Villar, G. Ares, F. Carrau, E. Dellacass, E. Eduardo, Food Research Internationalm **69**, 244–255 (2015)
[42] J.A. Pino, S. Tolle, R. Gök, P. Winterhalter, Food Chemistry **132**, 1436–1441 (2012)
[43] A. de-la-Fuente-Blanco, M.P. Sáenz-Navajas, V. Ferreira, Food Chemistry **210**, 107–114 (2016)
[44] Kromidas, Stavros, ed. *Handbuch Validierung in der Analytik* (2011)
[45] C. Philipp, P. Eder, W. Brandes, E. Patzl-Fischerleitner, R. Eder, Mitt Klosterneuburg (to be published)

Approaches to outline the aromatic profile of Kyoho wines from South Korea

Enoch Yang, Foteini Drosou, Euthalia G. Dourtoglou, Archodoula Chatzilazarou, and Vassilis G. Dourtoglou[a]

Department of Oenology and Beverage Technology, Faculty of Food Technology and Nutrition, Technological Educational Institute of Athens, 12 Ag. Spyridonos St., Egaleo, 122 10 Athens, Greece

Abstract. Kyoho cultivar is a hybrid of *V. vinifera* and *V. labrusca* and is one of the major varieties cultivate in South Korea. Kyoho grapes are characterized for their intense grape flavor and sour attributes. Korean climatic conditions are different from worldwide wine producing countries. The summer season is hot and dominated by heavy rains. Wines from Kyoho grapes are distinct for their intense grape flavor and the key odors that characterize the wines are aromas of caramel and strawberry notes. The past few years efforts are put to improve the quality and to advance the potential of the domestic wines. The objective of the present study is to provide insight to the main aromatic compounds that form the aroma profile of Kyoho wines. For this aim, Gas Chromatography was coupled with Mass Spectrometry for the separation and identification of the compounds. Analysis allowed to identify the volatile compounds that form the aroma of the wine and to detect the key odors of its characteristic aroma. (2,5-dimethyl-4-hydroxy-3(2H)-furanone) Furaneol®, which is associated with caramel-like flavor, was one of the compounds that were detected along with ((3R)-hydroxy-4,4-dimethyloxolan-2-one) pantolactone.

1. Introduction

The past years the grape industry in Korea has been showing a positive trend in terms of increase area and production volume. Korean grapes are mainly produced for table use and only a small portion is used for wine production. Owing to hot and rainy weathers in the summer and cold conditions in winter, the climatic conditions in Korea are unsuitable for the growth of *V*.vinifera. [1] Thus, Korean wine makers try using table grape from *V. labrusca* cultivars and their hybrids to produce wines, as American grapes show stronger adaptation to the adverse weather conditions [1,2].

Main grapes growing in Korea are, 'Campbell Early' which occupies 74.3% of the total production area, 'Kyoho' which occupies 13.1% and 'Muscat Bailey A' which covers a small portion of 5.9% of the total area cultivated for grape production [1,3]. The majority of Korean domestic wines are produced using the most cultivated variety, that is Campbell Early. Though it has the limitations of containing low levels of sugar, high malic acid concentration and significant foxy aroma content that affects negatively its preference [4]. Kyoho, the second most cultivated grape, is a tetraploid hybrid of *V. labrusca* and *V. vinifera* ('Ishihara Wase' × 'Centennial') [5–7] and it is famous for its large-sized berry and for its intense grape flavor [5]. Several studies have been conducted for the improvement and optimization of the wines produced from *V. labrusca* grapes. Hong & Park [4] analyzed the diversity of indigenous yeasts isolated from Campbell Early grapes. Furthermore, the identified yeast strains were examined for their tolerance to potassium metabisulfite as well as their alcohol and flavor production [4]. Ahn et al. [8], examined the physiochemical differences between *V.vinifera* and *V. labrusca* grapes by analyzing their organic acid contents. Son et al. [9] used targeted analysis of ^{1}H NMR to characterize and differentiate between *V. labrusca* grapes and their wines. In their study, Kyoho wines presented the highest levels of glycerol, 2,3-butanediol, succinate and alcohol compared to other *V. labrusca* wines. Lee et al. [10] evaluated the sensory attributes of a mixture of two different grape varieties, Kyoho and Campbell Early and its chemical attributes were determined and compared to a French Beaujolais wine. Principal component analysis of the descriptive data, showed high intensities of "ripe-fruit aroma", "caramel aroma" and "sour attributes for Kyoho wines.

In this study, the volatile compounds of Kyoho wines were analyzed to determine which groups of compounds consist the aroma profile of the wines and which compounds contribute to the grape/caramel-like odors of the wines.

2. Material and methods

2.1. Chemicals

Dichloromethane, chloroform, sodium chloride and anhydrous sodium sulphate were purchased from Chem Lab (Zedelgem, Belgium). (2,5-dimethyl-4-hydroxy-3(2H)-furanone) Furaneol® was purchased from Sigma-Aldrich (Athens, Greece).

2.2. Samples

For the fulfillment of this study four monovarietal Kyoho wines of two different produces were examined (ES and

[a] e-mail: vdourt@teiath.gr

GS). From each producer two wine samples were provided (ES1, ES2 and GS1, GS2, respectively). The grapes were obtained and produced in Cheonan, Korea.

2.3. Volatile analysis

The extraction of volatile compounds was carried away by Liquid-Liquid extraction (LLE) using dichloromethane as elution sorbent. For the analysis of the volatile compounds of the each examined wine, 50 mL of sample, 10 mL of saturated NaCl solution and 50 mL of dichloromethane were added in laboratory flask. The flask was stirred for 10 min at room temperature. Then, the sample was centrifuged at 3500 rpm for 5 min for the separation of phases. The supernatant was then extracted again using the same volume of solvent for 10 min. The organic layers were then washed with distilled water in a separation funnel. The organic phase was dried over anhydrous sodium sulphate, filtered and then condensed in a vacuum rotary evaporator. Prior to instrumental analysis, the sample was recovered to 50 μL with Chloroform. For the injection 1 μL was used. Each sample was analyzed in duplicate.

The identification of examined compounds was made using their retention time and by comparing their mass spectra with a spectral library of known standard compounds. For the quantification of (2,5-dimethyl-4-hydroxy-3(2H)-furanone) Furaneol® external calibration curves were prepared from standards. For the quantification of ((3R)-hydroxy-4,4-dimethyloxolan-2-one) Pantolactone and (2,5-dimethyl-4-methoxy-3(2H)-furanone) Mesifuran, the area of the corresponding peaks was normalized by that of the corresponding internal standard. The quantification of the compounds was made in Full Scan. The internal standard used for quantification was 3-Octanol (3.0 mg/L). The GC peak area of each compound was obtained from the ion extraction chromatogram by selecting target ions for each one.

2.4. Gas Chromatography/ mass spectrometry analysis

The apparatus used was an Agilent 6890 series GC System (Agilent Technologies, USA), equipped with 5975C VL MSD and a fused silica capillary column, 30m × 0.32 mm i.d. × 0.25 μm film thickness (HP-5MS, Agilent Technologies). 1.0 μL of sample was injected using a split ratio of 100:1. The injector temperature was set at 250°C, the carrier gas was helium at a flow rate of 1 mL/min, and the oven temperature program was 50°C for 2.5 min, increased to 180°C at 2.5°C /min, then to 230°C at 2°C /min, then to 250°C at 6°C /min, then the temperature was held at 250°C for 5 minutes, then increased to 270°C at 5°C/min and lastly it was held for 2 mins at 270°C. The temperature of the transfer line was set at 280°C. The mass spectrometer operated in the ionization mode (EI) at an ionization voltage of 70 eV in the mass range 40–550 amu and a manifold temperature of 270°C. All data was recorded using the Turbomass 5.0 ChemStation software.

3. Results and discussion

3.1. Aroma volatiles composition by GC-MS

A total of 52 compounds were identified across four Kyoho wine samples. The determined compounds are summarized and classified by their functional groups in Table 1. Among the determined compounds, the group with the most compounds identified were higher alcohols and esters. Altogether 16 alcohols, 15 esters, 7 fatty acids, 3 furans, 2 lactones and 3 miscellaneous compounds were detected.

Higher alcohols, also known as fusel alcohols, are, from a quantitative point of view, the most important group of volatile compounds produced by yeast during alcoholic fermentation. [11] Among the higher alcohols in wines, compounds which exhibit the greatest abundance are isoamyl alcohol 2-phenylethanols. 2-phenylethanol is one of the most important phenol-derived higher alcohol and it produces rose-like fragrances. [12–14]. Likewise, Kyoho wines presented significant peaks for the above-mentioned compounds.

Esters are mainly by-products of the alcoholic fermentation and they are considered important to the sensory properties of wines, as they contribute fruity notes. Among the detected esters in the examined wines, major compounds which are known to contribute to the fruity character of a wine are isoamyl acetate, 2-phenyl acetate, ethy hexanoate and ethyl decanoate. [13,15] These compounds were determined in all samples. Iso-butyl acetate which has a strawberry-like odor was detected only in one wine as shown in Table 1.

Fatty acids are mainly formed by yeasts during alcoholic fermentation. Volatile fatty acids tend to have a negative effect on wine aroma, contributing green and cheese, fatty aromas [12,13]. Some of the fatty acids that were identified in the examined wines were octanoic and decanoic acids. Free fatty acids occur only as traces in grapes, but they are biosynthesized during fermentation by yeasts and bacteria. Considering the compounds' unpleasant odors and similar ranges of concentrations across wines, these acids are not considered to be important contributors to the flavor of these wines [10].

Among phenolics, Eugenol was detected which is characterized as clove-like note. At usual concentrations, eugenol adds a general spic note as well [16].

In the examined Kyoho wines two compounds containing the 1-oxacycloalkan-2-one structure were identified: γ-butyrolactone and pantolactone. Concentrations of pantolactone were from 0.063 to 0.145 mg/L. Pantolactones appear to be generated in grapes under carbonic maceration, derived from amino or organic acids, notably glutamic and succinic acids [14]. γ-Butyrolactone, has been suggested that it contributes to the aroma characteristics of Sherry wines, giving pleasant cotton, peachy-like odors [11].

Contrast to major substances, minor classes can contribute greatly to the flavor of the wine differentiating each wine. Such classes are furans. Furans have very sweet aromas and are normally classified in the same group of aromas with give an odor of "burnt" sugar and caramel. [17] Furaneol®, pantolactone and (2,5-dimethyl-4-methoxy-3(2H)-furanone) mesifuran, compounds that contribute to a sweet odor, were detected in all wines, except for mesifuran which was not found in GS2. The quantified concentrations of all three compounds are listed in Table 2.

The quantities of mesifuran determined in the Kyoho wines ranged between 0.046 to 0.066 mg/L. It is reported

Table 1. Volatile compounds detected in Kyoho wines.

Compound	ES1	ES2	GS1	GS2	Odor description
Higher alcohols					
1-Butanol	–	–	–	+	Intoxicant alcohol[b]
Isobutanol	–	–	–	+	Fusel, alcohol[b]
1-Pentanol	–	–	+	+	Fusel alcohol, sweet balsam[c]
Iso-Amyl alcohol	+	+	+	+	Harsh, fruity, banana[a, c]
Prenol	+	+	+	+	Fruity, green lavender[c]
3-methyl-1-pentanol	–	+	+	+	Alcohol, fusel, cognac, green fruity[b, c]
Isohexanol	–	+	+	+	Nutty[c]
1-Hexanol	+	+	+	+	Green grass[a]
3-Hexen-1-ol	+	+	+	+	Green leafy[c]
2-Ethyl-1-hexanol	+	+	+	+	Mushroom, citrus floral[b, c]
1-Heptanol	–	–	–	+	Fragrant plant[b]
Benzyl alcohol	+	+	+	+	Floral, rose[c]
Phenylethyl alcohol	+	+	+	+	Floral, rose, rose water[a, c]
Tyrosol	+	+	+	+	Mild sweet, floral, fruity[c]
Homovanillyl alcohol	+	+	+	+	N.f.
Tryptophol	+	+	+	+	N.f.
Esters					
Ethyl Lactate	+	+	+	+	Solvent[a]
Iso-Butyl acetate	–	–	–	+	Strawberry, fruity, flowery[b]
Ethyl Butanoate	+	+	+	+	Sour fruit, strawberry[b]
Isoamyl Acetate	+	+	+	+	fresh banana
Isoamyl Lactate	+	+	+	+	Sharp tart, buttery, butterscotch[c]
Ethyl Hexanoate	+	+	+	+	Green apple, fruity[b]
Diethyl Succinate	+	+	+	+	Light fruity[b]
Methyl vanillate	+	+	+	+	Warm spicy, vanilla[c]
Ethyl vanillate	+	+	+	+	Phenolic, burnt, smoky[c]
Ethyl octanoate	+	+	+	+	Melon, pineapple[a, b]
Phenethyl acetate	+	+	+	+	Honey, pollen, rose[b]
Ethyl Decanoate	+	+	+	+	Fruity, fatty, pleasant[b]
Succinic acid, ethyl 3-methylbutyl ester	+	+	+	+	N.f.
Ethyl Dodecanoate	+	+	–	–	Fruity, fatty, pleasant[b]
Ethyl palmiate	+	+	+	+	Waxy, fruity, creamy[c]
Acids					
Octanoic acid	+	+	+	+	Rancid, harsh, cheese[c]
Decanoic acid	+	-	+	+	Fatty, unpleasant[c]
Tetradecanoic acid	+	+	+	+	N.f.
Hexadecanoic acid	+	+	+	+	N.f.
Octadecanoic acid	-	+	-	-	N.f.
Homovanillic acid	+	+	+	+	N.f.
Homo-syringic acid	+	+	+	+	N.f.
Phenols					
Phenol, 2,5-di-tert-butyl-	+	+	+	+	N.f.
Eugenol	+	+	+	+	Clove[c]
Phenol, 2,6-dimethoxy-4-(2-propenyl)	+	+	-	-	N.f.
4-((1E)-3-Hydroxy-1-propenyl)-2-methoxyphenol	+	+	+	+	N.f.
Phenol, 2,4-bis(1,1-dimethylethyl)-	+	+	+	+	N.f.
Furans					
2-Furanmethanol	-	-	+	+	Piquancy, bitter[b]
Furaneol®	+	+	+	+	Caramel[b, c]
Mesifuran	+	+	+	-	Sweet, maple, fruity[c]
Lactones					
Pantolactone	+	+	+	+	Cotton candy[c]
γ−Butyrolactone	+	+	+	+	Creamy, peach[d]
Miscellaneous					
6R,9R-3-Oxo-α-ionol	+	+	-	+	Spicy[c]
Limonene	-	-	+	+	Lemon, orange[a]
3-Hydroxy-beta-damascone	+	+	-	+	N.f.

N.f.: not found
[a]Refers to Wang et al. (2016) [13]
[b]Refers to Wang et al. (2017) [19]
[c]Obtained from the Good Scents Company, `http://www.thegoodscentscompany.com` [20]
[d]Refers to Pereira et al. (2014) [21].

Table 2. Concentrations of quantified volatile compounds (mg/L)[a].

	ES1	ES2	GS1	GS2
Furaneol®	0.384	0.459	0.066	0.098
Mesifuran	0.046	0.063	0.066	N.d.
Pantolactone	0.098	0.094	0.063	0.145

N.d.: Not detected

[a] Means are derived from duplicate determinations.

that mesifuran has been detected in substantial amounts in *V. labrusca* hybrids often giving a caramel-like undertone [18].

The determined concentration of Furaneol® ranged between 0.066 to 0.459 mg/L. As stated by Rapp et al. (1980) in [17], Furaneol® can be found in quantities above 1mg/L in wines elaborated with hybrid grapes; concentrations at which it produces a disagreeable strawberry scent. In our case, although the examined wines of our study did not present concentrations above 1mg/L, the wines had a predominant odor of strawberries and caramel-like notes, which can be imparted to the presence of Fureaneol and Mesifuran. Similar sensory attributes were described in the work of Lee et al. [10]. A blend of Kyoho and Campbell Early (*V. labrusca varieties)* wines were evaluated organoleptically; descriptive analysis showed high correlation for grapiness, sweetness and sourness. Likewise, Jackson [16] mentions that it is widely referred for *V. labrusca* grape cultivars to possess intense strawberry odor, which is probably induced by furaneol® and tis methoxy derivatives.

Other wines in which Furaneol® imparts a significant role are rose wines from Provence. Masson & Schneider, [15] identified that the key compounds contributing to the distinct aroma of Provence rose wines are derived by the furanic compounds furaneol® and homofuraneol, the C13 norisoprenoid (β-damascenone) and the two volatile thiols 3-mercaptohexanol and 3-mercaptohexyl acetate. In their study, the wines examined presented high concentrations of furaneol® which indicate that furaneol® play a significant role in the aroma of rose wines. Similarly, other authors in [13] affirm that the key compounds of rose wines are ethyl esters, higher alcohol acetates (particularly isoamyl acetate and 2-phenyl acetate), furaneol® and the polyfunctional thiols 3-mercapto-1-hexanol and 3-mercaptohexyl acetate. In our study polyfunctional thiols were not detected but it could be due to inadequate sensitivity in the extraction procedure as the polyfunctional thiols are present in extremely low concentrations [13]. Besides these compounds, β-damascenone is considered to contribute significantly in the rose wine key odors enhancing their fruity notes [15]. Taking into consideration the fact that Kyoho grapes do not produce deep red wines, but are weak in color as stated in [10], attempts to produce rose wines could be encouraged, since they present similar key odors to the Provence rose wines.

4. Conclusions

In summary, this study aimed to determine the volatile compounds of Kyoho grapes. Specifically, this work involved comprehensive evaluation of 4 Kyoho wines, in which its sweet, strawberry, caramel like odor were identified; these sweet aromas are characteristic of all *V. labrusca* grapes even in their hybrid grapes showing intense grape flavor, roses and strawberry like odors [14], which are imparted by furans like furaneol® and its methoxy derivatives.

This research has been co-funded by the European Union (European Social Fund-ESF) and Greek national funds through the Operational Program "Education and Lifelong Learning" of the National Strategic Reference Framework (NSRF) – Research Funding Program: ARCHIMEDES III. Investing in knowledge society through the European Social Fund.

References

[1] H. K. Yun & K. S. Park, Int. J. Pl. Brd., **1**(1), 22–26 (2007)
[2] M. B. M. de Castilhos, O. L. dos Santos Correa, M. C. Zanus, J. D. G. Maia, S. Gomez-Alonso, E. Garcia-Romero, V. L. Del Bianchi & I. Hermosin-Gutierrez., Food Res. Intr. **76**(3), 697–708 (2015)
[3] H.J. Ahn & H.S. Son, Korean J. Food Sci. Tech. **44**(3), 280–286 (2012)
[4] Y. Hong & H. Park, Foo. Micr. **34**, 297–214 (2013)
[5] Y. Deng, Y. Wu & Y. Li, LWT **38**, 584–590 (2006)
[6] S. Shiraishi, Y. Watanabe, H. Okubo & S. Uemoto, J. Japan. Soc. Hort. Sci. **55**(2), 123–129 (1986)
[7] M. Shiraishi, R. Shinomiya & H. Chijiwa, Sci.Hort. **137**, 107–113 (2012)
[8] H. Ahn & H. Son, Kr. J. Food Sci. Tech. **44**(3), 280–286 (2012)
[9] H. Son, G. Hwang, H. Ahn, W. Park, C. Lee & Y. Hong, Food Res. Intr. **42**, 1483–1491 (2009)
[10] S. Lee, J. Lee, H. Kim, S. Kim & K. Koh, Food Chem. **94**, 385–393 (2006)
[11] M. V. Moreno-Arribas & M. Carmen Polo, *Wine Chemistry and Biochemistry* (2009)
[12] F. Vararu, J. Moreno-Garcia, C. Zamfir, V. V. Cotea & J. Moreno, Food Chem. **197**, 373–381 (2016)
[13] J. Wang, D. L. Capone, K. L. Wilkinson & D. W. Jeffery, Food Chem. **196**, 682–693 (2016)
[14] R. S. Jackson, *Wine Science: Principles and Applications* (2008)
[15] G. Masson & R. Schneider, A.J.E.V. **60**(1), 116–122 (2009)
[16] R. S. Jackson, *Wine Tasting A Professional Handbook* (2009)
[17] V. Ferreira, I. Jarauta, R. Lopez & J. Cacho,J. Chrom. A **1010**, 95–103 (2003)
[18] I. Zabetakis, J. W. Gramshaw & D. S. Robinson, Food Chem. **65**(2), 139–151 (1999)
[19] X. Wang, Y. Tao, Y. Wu, R. An & Z. Yue, Food Chem. **226**, 41–59 (2017)
[20] `http://www.thegoodscentscompany.com/`
[21] V. Pereira, J. Cacho & J. C. Marques, Food Chem. **162**, 122–134 (2014)

DOI: 10.1051/bioconf/20170902035

Profilo aromatico varietale delle uve Mavrud, Bouquet e Rubin Varietal aromatic profile of Mavrud, Bouquet and Rubin grapes

Donato Lanati[1], Dora Marchi[1], Giacomo Mazza[1], Desislava Baicheva[1], e Alberto La Rosa[2]

[1] Enosis s.r.l., Cascina Meraviglia, 15043 Fubine (Alessandria), Italia
[2] Edoardo Miroglio Wine Cellar, 8943 Elenovo (Nova Zagora), Bulgaria

Riassunto Mediante gascromatografia accoppiata alla spettrometria di massa (GC/MS) è stato tracciato il profilo aromatico varietale delle uve Mavrud, Bouquet e Rubin e dei corrispondenti vini ottenuti dalle uve vinificate in purezza. Nelle uve sono stati identificati 33 composti appartenenti alla classe dei benzenoidi, dei terpeni e dei norisoprenoidi; tra i composti più interessanti sono stati trovati il cis e trans-8-OH-linalolo, il p-ment-1-ene-7,8-diolo, il 3-OH-β-damascone, il 3-oxo-α-ionolo, il grassopperchetone e il vomifoliolo. Nelle frazioni volatili libere dei vini sono stati identificati circa 60 composti, per la gran parte composti di fermentazione quali esteri, acetati, alcoli, benzenoidi e lattoni, oltre a terpeni e norisoprenoidi varietali. Nella frazione legata dei vini, ottenuta dopo idrolisi enzimatica sono stati trovati 30 composti varietali.

Abstract. This study defines the aromatic varietal characteristics of Mavrud, Bouquet and Rubin grapes, which are grown in the Thrace region. The varietal profile of the 3 varieties of grapes was obtained through gas chromatography coupled with mass spectrometry (GC/MS). The profile identified 33 compounds belonging the benzenoid, terpene and norisoprenoid classes; some of the most interesting compounds found were the cis- and trans-8-OH-linalool, the p-ment-1-ene-7,8-diol, the 3-oxo-α-ionol, the grassopperketon and the vomifoliol. The grapes were vinified in purity, obtaining the corresponding wines. The analysis of the wines' free volatile fractions led to the identification of approximately 60 volatile compounds, most of them fermentation compounds such as esters, acetates, alcohols, benzenoids and lactones, as well as varietal terpenes and norisoprenoids. In the bound fraction of the wines, which was obtained after enzymatic hydrolysis, 30 varietal compounds were found.

1. Introduzione

La vite viene coltivata in Bulgaria da oltre 3000 anni e già Omero nei suoi poemi parlava dei vini dell'antica Tracia. Con l'arrivo dei Romani si introdussero nuove tecniche di vinificazione e si ampliarono le aree coltivate.

La coltivazione della vite in Bulgaria si concentra in alcune aree privilegiate che godono di condizioni pedoclimatiche favorevoli. Le zone viticole di un certo interesse vanno dalla regione Danubiana nel nord del paese alle coste del Mar Nero, verso il nord-est, per estendersi al centro del paese nell'area sub Balcanica della Rose Valley, fino al sud-ovest della Struma Valley, per giungere a Sud alla più famosa Thracian Valley (Fig. 1).

Il **Mavrud** è il vitigno autoctono più importante della Bulgaria meridionale e viene coltivato sin dai tempi antichi nella regione della Tracia, nel sud della Bulgaria. Lo troviamo ad Asenovgrad nel distretto di Plovdiv a Pazardzhik, Stara Zagora, Nova Zagora e verso il Mar Nero a Bourgas e Pomorie, dove è noto come Katchileva; di recente le coltivazioni si sono sviluppate verso le zone di Melnik, Sakar e l'area sud-orientale ai piedi dei monti Rodopi. Il suo nome deriva dal greco μαροσ (màvros), che significa nero; i grappoli sono molto compatti con una buccia spessa e scura.

E' un vitigno a maturazione tardiva,che predilige zone calde e terreni alluvionali profondi e freschi; le uve raggiungono una piena maturazione verso la seconda metà di ottobre nel distretto di Plovdiv.

L'uva **Rubin** nasce nel 1944 da un incrocio tra Nebbiolo e Syrah realizzato all'Istituto di Viticoltura ed Enologia di Pleven in Bulgaria. Solo dagli anni 50 si diffonde su larga scala nel distretto di Plovdiv e viene riconosciuta come varietà nel 1961. Si coltiva in tutte le regioni viticole della Bulgaria, ma soprattutto nella Pianura Traciana e nella Struma River Valley. Il vitigno si sviluppa bene su terreni collinari con suoli profondi e le uve giungono a maturazione nella metà di settembre.

Le uve **Bouquet o Buket** derivano dall'incrocio tra Mavrud e Pinot noir, creato nel 1951 all'Istituto di Viticoltura ed Enologia di Pleven. Le coltivazioni sono diffuse soprattutto nella regione sub Balcanica a Silven e Nova Zagora e nella regione di Plovdiv. Le viti resistono bene alle basse temperature e le uve, in normali condizioni climatiche, maturano a metà settembre.

2. Materiali e metodi

Tutte le uve analizzate provengono dalla regione della Tracia e precisamente dalla zona di Elenovo village vicino Nova Zagora. Le uve Buket e Rubin sono state raccolte nella prima settimana di settembre, mentre le uve Mavrud nella prima metà di ottobre.

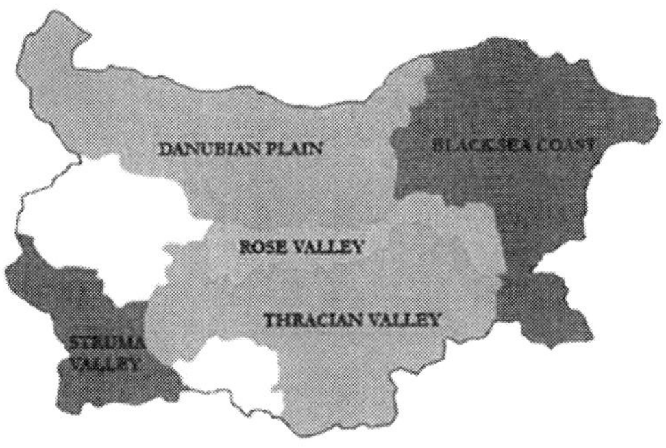

Figura 1. Aree vitivinicole della Bulgaria.

Tabella 1. Analisi delle uve alla raccolta.

UVE 2016		MAVRUD	BUKET	RUBIN
Analisi				
peso medio acino	g	**1.52**	**1.04**	**1,03**
acidità totale	g/L	**5,89**	**6,4**	**5,6**
pH		**3,33**	**3,26**	**3,39**
zuccheri riduttori	g/L	**221**	**216**	**249**
alcool potenziale	%vol	**13,04**	**12,74**	**14,69**
antociani estraibili	mg/kg	**397**	**387**	**488**
antociani totali	mg/kg	**1584**	**798**	**993**
flavonoidi estraibili	mg/kg	**889**	**1041**	**1557**
flavonoidi totali	mg/kg	**3448**	**2177**	**3115**

2.1. Analisi delle uve

Nella Tabella 1 vengono mostrati i dati analitici riferiti alle uve alla raccolta.

L'uva Mavrud analizzata ha un acino medio grande e un contenuto zuccherino di 221 g/L; i dati in Tabella 1 mostrano una concentrazione molto elevata di antociani e flavonoidi, ma con una estraibilità degli stessi non molto alta [1].

L'uva Buket è meno zuccherina, alla raccolta, ma mostra un buon tenore in antociani e flavonoidi estraibili.

Alla raccolta l'uva Rubin possiede un buon contenuto in zuccheri pari a 249 g/L e una ottima quantità di antociani e flavonoidi, soprattutto estraibili.

Tutte le uve sono state vinificate in purezza, ottenendo i vini corrispondenti. I dati del profilo aromatico dei vini Mavrud e Rubin sono stati ottenuti, dopo tre settimane, circa, dalla fermentazione alcolica ma prima della fermentazione malolattica; il vino Buket, ottenuto in macerazione carbonica, nello stesso periodo, aveva già fatto la fermentazione malolattica.

2.2. Estrazione degli aromi delle uve e del vino

I composti aromatici delle uve e dei vini vengono separati su cartuccia C_{18} e analizzati per GC/MS [2–6].

2.3. Gascromatografia-spettrometria di massa

Le analisi sono state effettuate utilizzando un gascromatografo Agilent 6890N accoppiato ad uno spettrometro di massa MSD 5973, equipaggiato con una colonna capillare HP-Innowax in silice fusa di 30 m di lunghezza e di 0,25 mm di diametro; flusso di elio 1 ml/min; iniezione splitless, tempo di iniezione 2 min; temperatura dell'iniettore 230 °C; programma di temperatura: da 60 °C (2 min) a 160 °C a 2 °C/min e da 160 °C a 230 °C a 5 °C/min con 10 min di isoterma finale. La corrente ionizzante era di 70 eV. I composti sono stati identificati comparando i tempi di ritenzione e gli spettri di massa con quelli di composti standards o per confronto con gli spettri di massa riportati in letteratura [7]. La quantificazione dei composti volatili è stata effettuata utilizzando uno standard interno (2-ottanolo) e considerando uguali ad uno tutti i fattori di risposta.

2.4. Estrazione degli antociani dell'uva

Le bucce di 10 acini, vengono separate dalla polpa e dai semi e poste in tampone tartarico a pH=3.2. Dopo circa 4 ore le bucce vengono omogeneizzate e dopo varie operazioni si ottiene un estratto che verrà poi analizzato in HPLC [8]. Nel caso del vino si effettua una iniezione diretta dopo filtrazione su membrana da 0.45 μm.

2.5. Cromatografia in fase Liquida (HPLC-DAD)

Per l'analisi degli antociani monomeri, è stato utilizzato un cromatografo Perkin Elmer Series 200 con rivelatore a diodi (HPLC-DAD), con una colonna LiChrospher 100 RP-18 (5 μm) LiChroCart 250-4 e precolonna LiChroCart 4-4. L'analisi è stata effettuata in gradiente lineare da 94% di A a 6% di B in 40 min, con fase mobile contenente acqua, acido formico e acetonitrile; solvente A 87:10:3 (v/v/v) e solvente B 40:10:50 (v/v/v); flusso 0.8 mL/min, lunghezza d'onda 518 nm. Per l'identificazione degli antociani si valuta il tempo di ritenzione, lo spettro Uv-Vis del picco cromatografico e la λ_{max}.

3. Risultati e discussione

Le uve analizzate, Mavrud, Buket e Rubin, sono classificabili come uve non aromatiche. In questo tipo di uve le molecole, sono presenti sotto forma legata ad uno o più zuccheri. Le principali classi chimiche presenti sono gli alcoli, i benzenoidi, i terpeni e i norisoprenoidi; questi ultimi derivano dalla degradazione dei caroteni durante la maturazione dell'uva e sono tra i composti più importanti per i vini destinati ad un lungo invecchiamento.

Nella Tabella 2 vengono mostrati i profili aromatici delle uve Mavrud, Buket e Rubin, mentre nelle Tabelle 3 e 4 troviamo i composti aromatici dei vini. Infine la Tabella 5 ci mostra il profilo antocianico caratteristico delle tre uve.

3.1. Profilo aromatico delle uve

La concentrazione dei norisoprenoidi nell'uva **Mavrud** è sufficientemente elevata per dotare il vino di un potenziale aromatico adeguato (Tabella 2), condizione indispensabile per garantire al vino un lungo invecchiamento. Tra i composti più importanti troviamo il 3-OH-b-damascone e il grassopperchetone, precursori del b-damascenone,

Tabella 2. Profilo aromatico delle uve.

Uve 2016	Mavrud	Buket	Rubin
COMPOSTI LEGATIμg/L			
ALCOLI E BENZENOIDI			
alcool benzilico	237,8	138,5	156,0
2-fenil etanolo	157,3	144,7	144,8
4-vinil guaiacolo	64,3	141,1	74,1
2,6-dimetossifenolo	38,2	67,7	30,4
4-vinil fenolo	39,3	50,9	46,3
vanillina	39,4	41,7	26,8
metilvanillato	121,7	85,0	71,8
zingerone	19,8	25,6	16,8
siringaldeide	64,3	39,2	37,1
diidroconiferilalcool	400,6	333,5	350,2
3,4,5-trimetossifenolo	70,7	156,0	114,7
TOTALE	**1.253**	**1.224**	**1.069**
TERPENI			
trans-furanlinalolossido	6,3	1,8	1,5
cis-furanlinalolossido	4,8	2,5	1,6
trans-piranlinalolossido	24,7	3,1	2,8
cis-piranlinalolossido	4,4	2,2	2,0
nerolo	3,6	1,7	3,1
geraniolo	14,4	5,2	14,1
2,6-dimetil-3,7-ottadien-2,6-diol	8,0	2,5	2,9
trans-8-OH-linalolo	137,5	35,1	44,4
OH geraniolo	9,9	4,6	2,4
cis-8-OH-linalolo	33,1	11,8	18,1
p-ment-1-ene-7,8-diolo	57,6	64,3	77,1
TOTALE	**304**	**135**	**170**
NORISOPRENOIDI			
3,4-diidro-3-oxoactinidolo I	7,3	3,0	6,3
3,4-diidro-3-oxoactinidolo II	13,1	12,0	14,8
3,4-diidro-3-oxoactinidolo III	7,1	8,4	9,2
3-OH-β-damascone	102,5	77,3	66,9
megastigma-7-en-3,9-diolo	102,6	75,6	8,5
3-oxo-α-ionolo	75,7	73,3	163,3
3-idrossi-7,8-diidro-β-ionolo	41,6	20,0	24,2
3-OH-β-ionone	3,1	21,3	11,1
3-oxo-7,8-diidro-α-ionolo	62,9	45,3	44,2
3-idrossi-7,8-deidro-β-ionolo	52,7	47,7	47,9
grassopperchetone+vomifoliol	623,8	696,8	495,1
TOTALE	**1.093**	**1.081**	**891**

un composto dalla soglia olfattiva molto bassa che si forma nel vino durante l'invecchiamento. La quantità di terpeni presenti è quella tipica di un'uva neutra, con una modesta concentrazione in terpeni monoidrossilati ed una più consistente presenza di dioli terpenici. Elevata la concentrazione in didroconiferilalcool e in alcool benzilico.

Anche le uve Buket (Tabella 2) mostrano una elevata concentrazione in norisoprenoidi; particolarmente alto il tenore in vomifoliolo e grassopperchetone. La quantità di terpeni e fenoli presenti è piuttosto modesta.

Tabella 3. Composti della frazione libera dei vini.

VINI Rossi	Mavrud	Buket	Rubin
COMPOSTI LIBERIμg/L			
ESTERI ED ACETATI			
etil butirrato	187,2	284,7	402,6
2-metil-etil butirrato	6,0	6,6	12,4
3-metil-etil butirrato	7,1	3,6	6,1
etilesanoato	488,2	691,9	827,8
etilottanoato	400,2	881,3	707,9
etildecanoato	190,8	585,2	293,5
isoamil acetato	1158,7	1698,4	4493,4
esil acetato	5,3	20,0	124,4
2-feniletilacetato	89,8	144,4	550,3
TOTALE	**2533**	**4316**	**7418**
ESTERI D'INVECCHIAMENTO			
etil-2-OH-4-metilpentanoato	53,5	47,4	25,1
2-metilbutil+3-metilbutillattato	124,3	147,0	57,0
dietil succinato	850,3	3705,4	2683,2
dietil malato	47,5	25,3	185,0
dietil-2-OH-glutarato	66,1	27,3	225,5
monoetil-2-OH-glutarato γ-latt.	269,1	116,6	480,1
etil-3-fenil-2-OH-propionato	223,4	194,1	126,8
dietilpiroglutammato	34,1	46,0	129,2
TOTALE	**1668**	**4309**	**3912**
ALCOLI E BENZENOIDI			
esanolo	1057,8	959,8	1631,6
trans-3-esenolo	19,8	21,1	66,7
cis-3-esenolo	38,3	127,0	95,5
guaiacolo	15,1	12,7	13,2
alcool benzilico	7,9	27,9	23,8
2-fenil etanolo	35642	16546	29202
4-vinil guaiacolo	201,6	326,3	100,5
4-vinil fenolo	38,0	131,9	30,3
etil cinnammato	0,0	16,3	0,0
siringaldeide	75,5	94,5	60,4
diidroconiferilalcool	387,4	191,4	510,7
etil siringato	95,2	0,0	81,2
acetosiringone	112,1	3,2	79,9
TOTALE	**37691**	**18458**	**31896**
TERPENI - NORISOPRENOIDI			
p-ment-1-ene-7,8-diolo	2,5	2,0	14,3
3-OH-β-damascone	5,7	2,3	33,6
megastigma-7-en-3,9-diolo	81,4	15,0	0,0
3-oxo-α-ionolo	11,9	14,8	225,7
3-idrossi-7,8-diidro-β-ionolo	8,7	2,9	30,3
3-OH-β-ionone	0,0	3,3	36,8
3-oxo-7,8-diidro-α-ionolo	10,2	38,1	0,0
3-idrossi-7,8-deidro-β-ionolo	2,7	29,5	32,7
grassopperchetone+vomifoliolo	19,1	81,6	103,3
TOTALE	**143**	**190**	**477**
LATTONI			
γ-butirolattone	570,3	1049,7	1116,6
trans-β-metil-γ-octalattone	10,4	0,0	17,7
cis-β-metil-γ-octalattone	23,1	0,0	14,8
γ-nonalattone	20,1	17,6	25,3
pantolattone	17,7	9,8	11,7
solerone	24,9	9,6	13,5
sherry lattone 1	54,1	51,4	203,5
sherry lattone 2	99,1	56,4	119,7
TOTALE	**820**	**1195**	**1523**

Tabella 4. Composti della frazione legata dei vini.

VINI Rossi	Mavrud	Buket	Rubin
COMPOSTI LEGATI µg/L			
ALCOLI E BENZENOIDI			
alcool benzilico	21,7	65,2	14,9
2-fenil etanolo	64,0	63,5	31,7
4-vinil guaiacolo	23,9	38,8	41,9
2,6-dimetossifenolo	10,5	76,5	15,7
4-vinil fenolo	39,9	253,7	149,8
diidroconiferilalcool	108,7	192,4	83,1
TOTALE	**269**	**697**	**346**
TERPENI			
trans-furanlinalolossido	7,5	5,7	3,4
cis-furanlinalolossido	5,7	15,9	3,8
trans-piranlinalolossido	12,1	2,9	0,0
nerolo	9,4	7,6	2,4
geraniolo	25,7	13,2	5,1
2,6-dimetil-3,7-ottadien-2,6-diolo	63,9	0,0	0,0
trans-8-OH-linalolo	60,9	21,7	25,3
OH geraniolo	10,3	4,6	0,0
cis-8-OH-linalolo	69,8	6,2	11,0
p-ment-1-ene-7,8-diolo	12,8	12,6	17,0
TOTALE	**278**	**90**	**68**
NORISOPRENOIDI			
3,4-diidro-3-oxoactinidolo I	4,2	7,6	4,8
3,4-diidro-3-oxoactinidolo II	7,4	18,3	14,3
3,4-diidro-3-oxoactinidolo III	5,9	26,9	11,0
3-OH-β-damascone	55,9	104,9	44,2
megastigma-7-en-3,9-diolo	96,6	115,2	18,3
3-oxo-α-ionolo	178,6	367,1	263,2
4-oxo-β-ionolo	55,3	192,6	61,6
3-idrossi-7,8-diidro-β-ionolo	37,8	87,9	34,4
3-OH-β-ionone	3,5	98,2	12,3
3-oxo-7,8-diidro-α-ionolo	103,7	44,1	32,3
3-idrossi-7,8-deidro-β-ionolo	22,3	38,2	14,4
grassopperchetone + vomifoliolo	232,8	122,8	97,8
TOTALE	**804**	**1224**	**609**

Le uve Rubin (Tabella 2) possiedono un profilo aromatico dove prevalgono i norisoprenoidi, mentre appare modesta la concentrazione in terpeni, in linea con un profilo aromatico tipico delle uve neutre. Didroconiferilalcool, alcool benzilico e 2-feniletanolo sono gli alcoli più rappresentativi, mentre sono presenti in modesta quantità buona parte dei fenoli; tuttavia, va ricordato che molti di questi composti, a partire dal 2-feniletanolo si formano in gran quantità durante la fermentazione alcolica, mentre altri composti come il 2,6-dimetossifenolo, la vanillina, la siringaldeide possono raggiungere elevate concentrazioni nella frazione libera del vino, a contatto con il legno.

3.2. Profilo aromatico dei vini

Le uve sono state vinificate in purezza ottenendo i vini corrispondenti; i dati aromatici della frazione libera e di quella legata sono mostrati nelle Tabelle 3 e 4.

Il vino da uve Mavrud possiede una più marcata presenza di benzenoidi rispetto agli altri due vini con note fruttate più delicate e un ottimo livello di invecchiamento chimico; l'elevata concentrazione del 2-fenil etanolo, dalle note di rosa, è un fattore positivo nei vini rossi. Trattandosi di un vino fermentato da poco, nella frazione libera (Tabella 3) non sono ancora pienamente espressi i composti derivanti dall'affinamento in legno; anche i composti varietali derivanti dall'idrolisi dei terpeni e dei norisoprenoidi si trovano in bassa concentrazione e non sono ancora presenti alcuni composti, come i vitispirani, gli actinidoli, il riesling acetale e in piccola parte anche il b-damascenone, derivanti dalla ulteriore evoluzione dei precursori primari, e che, invece, sono presenti nel vino Mavrud dopo invecchiamento. Nella frazione legata del vino (Tabella 4) troviamo ancora una buona concentrazione sia in norisoprenoidi, sia in terpeni.

Il vino Bouquet, analizzato in questo studio, è stato ottenuto con la tecnica della macerazione carbonica e, quindi, destinato ad un consumo in tempi molto brevi. Il vino mostra una buona concentrazione in esteri e acetati di fermentazione ed una più blanda presenza di benzenoidi e norisoprenoidi. Tuttavia, come si può notare nella Tabella 4 i norisoprenoidi della frazione legata sono presenti in concentrazione molto elevata.

Nei vini ottenuti con uve Rubin troviamo, nella frazione libera (Tabella 3), una concentrazione elevata di esteri e acetati di fermentazione che conferiscono al vino delle note fruttate intense. Trattandosi di un vino analizzato a meno di un mese dalla fermentazione alcolica risulta molto bassa la quantità di composti di invecchiamento, sia esteri, sia lattoni, e sono presenti in quantità modesta i composti correlabili all'affinamento in legno come furani, cis e trans-β-metil-γ-octalattone, siringaldeide, etil siringato. Notiamo, inoltre, una consistente presenza di norisoprenoidi, a dimostrazione di una attività idrolitica già in fase avanzata; la frazione legata (Tabella 4) è caratterizzata da un modesto tenore in terpeni e da una buona concentrazione in norisoprenoidi, però, più bassa di quella riscontrata negli altri vini.

I composti della frazione legata dei tre vini analizzati, mostrati in Tabella 4, rappresentano il potenziale aromatico del vino; durante l'invecchiamento, questi composti si idrolizzano nella frazione libera del vino, assicurando una qualità aromatica elevata e il mantenimento delle caratteristiche varietali tipiche della cultivar.

3.3. Profilo antocianico delle uve e dei vini

Le uve rosse delle tre varietà analizzate hanno un profilo antocianico caratteristico, che comprende i 5 principali antociani non acilati e i relativi antociani acetati e cumarati.

In comune, le tre uve, hanno il fatto di possedere dei profili antocianici abbastanza stabili, con una presenza elevata di malvina e di antociani acilati (Tabella 5).

L'uva Mavrud contiene il 55.5% di malvina e un buon 9.5% di peonidina, mentre tra gli antociani acilati, la

Tabella 5. Profilo antocianico delle uve.

	Antociani	Mavrud	Buket	Rubin
ant. nonacilati %	**delfinidina-3-glucoside**	3,8	3,9	7,6
	cianidina-3-glucoside	0,7	0,4	1,8
	petunidina-3-glucoside	5,4	6,0	10,2
	peonidina-3-glucoside	9,4	6,8	7,6
	malvidina-3-glucoside	55,5	61,5	48,4
antociani acilati %	**delfinidina-3-acetilglucoside**	0,0	0,1	0,3
	cianidina-3-acetilglucoside	0,0	0,0	0,1
	petunidina-3-acetilglucoside	0,2	0,3	0,8
	peonidina-3-acetilglucoside	0,3	0,3	0,6
	malvidina-3-acetilglucoside	2,4	3,5	4,0
	delfinidina-3-cumarilglucoside	n.d.	n.d.	n.d.
	cianidina-3-cumarilglucoside	0,1	0,1	0,3
	petunidina-3-cumarilglucoside	1,2	1,2	2,2
	peonidina-3-cumarilglucoside	2,6	1,6	1,8
	malvidina-3-cumarilglucoside	18,3	14,5	14,3
$\sum$ **antociani non acilati %**		74,8	78,5	75,6
$\sum$ **antociani acilati %**		25,2	21,5	24,4
$\sum$ **antociani acetati %**		3,0	4,1	5,8
$\sum$ **antociani p-cumarati %**		22,2	17,4	18,6

quantità di malvina cumarata raggiunge il 18% circa e un tenore complessivo in antociani acilati di oltre il 22%.

L'uva Buket ha un profilo antocianico simile, ma con una maggior % di malvina che raggiunge il 61% dell'intero profilo antocianico; anche in questo caso risulta elevata la percentuale di antociani acilati con una forte evidenza della malvina cumarata.

L'uva Rubin si differenzia dalle altre per il contenuto elevato in petunina e in delfinina, mentre risulta sempre elevato il tenore complessivo in antociani acilati con oltre il 24%; meno accentuato il rapporto tra cumarati e acetati.

Il passaggio da uva a vino porta a modificazioni del profilo antocianico con una costante evoluzione nel tempo, comunque legata al profilo originario [9]. Nei vini analizzati dopo circa un mese dalla vinificazione si vede già, una modificazione nel profilo antocianico, con un aumento della percentuale relativa di malvina. Contemporaneamente diminuiscono le altre antocianine ma in modo più selettivo, mentre più netta appare la diminuzione degli antociani acilati, con una variazione del rapporto tra antociani cumarati e antociani acetati.

Sin dalla fermentazione alcolica, e durante l'affinamento dei vini l'evoluzione del profilo antocianico porta alla formazione di nuovi derivati della malvina, quali le vitisine A e B. Nei vini destinati ad un lungo invecchiamento compaiono anche le piranoantocianine, derivati antocianici che sono il frutto delle interazioni tra fenoli e antociani.

Riferimenti

[1] Di Stefano R., Cravero M.C., Metodi per lo studio dei polifenoli dell'uva, Riv.Vitic. Enol. **44**(2), 37–45 (1991)

[2] Williams, P. J., Strauss, C. R., Wilson, B., Massy-Westropp, R. A., Use of C18 reversed-phase liquid chromatography for the isolation of monoterpene glycosides and nor-isoprenoid precursors from grape juice and wines, J. Chromatogr. **235**, 471–480 (1982)

[3] Di Stefano R., Gentilini N., Ummarino I., Studio dei profili aromatici di varietà a frutto bianco coltivate nella zona del Collio, L'Enologo **38**(12), 95–102 (2000)

[4] Mazza G., Cascio P., Barbieri E., Composti volatili liberi e glicoconiugati presenti nelle foglie e negli acini della Vitis vinifera Cv. Moscato bianco, Riv. Vitic. Enol. (**54**-4), 25–35 (2002)

[5] Mazza G., Raifer W., Lanati D., Il quadro aromatico della cultivar Cabernet Sauvignon coltivata in Alto Adige, L'Enologo **42,** 111–117 (2006)

[6] Cook Papini P., Mazza G., Gatti M. and Bavaresco L., Anthocyanin and aroma profiling of the Albarossa grapevine crossbreed (*Vitis vinifera* L.) and its parent varieties Barbera and Nebbiolo di Dronero, Vitis **49**(3), 121–127 (2010)

[7] Stenhagen, E., Abrahamsson S., Mc Lafferty, F.V. Registry of mass spectral data. John Wiley and Sons, New York, 1974

[8] Di Stefano R. et al., Estrazione degli antociani dalle bucce dell'uva durante la fermentazione. L'Enotecnico **30**(5), 75–83 (1994)

[9] Mazza, G., Marchi, D., Cascio, P., Barbieri, E., Effetto dell'impiego di tannini ed enzimi nella vinificazione di uve Nebbiolo e Sangiovese, Vignevini, Bologna **32**, 62–67 (2005)

International competitiveness of the wine industry: Consistency of its indicators – Competitividad internacional de la industria vitivinícola: Consistencia de sus indicadores[a]

Francesc Cervera y Raúl Compés

Universitat Politècnica de València (UPV), Departamento de Economía y Ciencias Sociales (DECS), Valencia, España

Abstract. In recent years, the concept of competitiveness has gained significant importance in the context of an increasingly globalized economy. In this scenario, international wine markets have also undergone significant changes in its structure and in its performance that require appropriate tools in order to understand it. This paper analyses the current situation of these markets in terms of area, production, consumption and trade flows. It also assesses the competitive position of the Spanish wine industry shown by seven indicators of Revealed Comparative Advantage in relation with the main actors of the international wine market. Finally, it analyses cardinal, ordinal and dichotomous consistency in order to know which indicators are stable with each other -and therefore replaceable-, and which not, and therefore complementary in measuring the international competitiveness of the Spanish wine industry between 2000 and 2015, whose results show the existence of three groups of indicators, one whose indices only contemplate exports, another whose indices compute both exports and imports, and a last one in which two indicators of the two different nature coexist.

Resumen. En los últimos años, el concepto de competitividad ha adquirido una importancia relevante en el marco de la economía crecientemente globalizada. En este escenario, los mercados internacionales del vino también han sufrido cambios importantes en su estructura y en su comportamiento que requieren herramientas adecuadas para su comprensión. El presente trabajo analiza la situación actual de estos mercados en cuanto a superficie, producción, consumo y flujos de comercio. Asimismo, evalúa la posición competitiva de la industria española del vino mediante siete indicadores de Ventaja Comparativa Revelada con respecto a los principales actores en el mercado internacional del vino. Finalmente, realiza un análisis de consistencia cardinal, ordinal y dicotómica con el fin de conocer qué indicadores de los propuestos son estables entre sí -y por tanto sustituibles-, y cuáles no, y por tanto complementarios a la hora de medir la competitividad internacional de la industria vinícola española en el periodo comprendido entre 2000 y 2015, cuyos resultados evidencian la existencia de tres grupos de indicadores, uno con índices que contemplan únicamente las exportaciones, otro con índices que computan tanto exportaciones e importaciones, y un último en el que coexisten dos indicadores de las dos diferentes naturalezas.

1. Introducción

La globalización y creciente segmentación de los mercados del vino han acentuado la necesidad de analizar con más precisión la competitividad de los principales países productores del mundo [1–4]. En unas pocas décadas el negocio vitivinícola ha cambiado radicalmente. Su ámbito ha dejado de ser europeo, y han surgido nuevos y dinámicos exportadores e importadores, estrategias comerciales ambiciosas y transacciones intra–industriales cada vez más importantes [4,5]. El comercio internacional se ha convertido en el principal factor de crecimiento del sector, en particular para algunos países como España o Chile que se han convertido en vini–exportadores [6,7].

Los análisis convencionales muestran que los países europeos han perdido cuota y, aparentemente, competitividad, en todos los mercados -de producción, consumo, y exportaciones- con la globalización vitivinícola [8–10]. No obstante, las metodologías tradicionales de análisis basadas en los indicadores clásicos de comercio internacional resultan insuficientes [11,12]. Es necesario utilizar las nuevas herramientas de competitividad [13,14]. Sin embargo, los diferentes enfoques que existen sobre este concepto dificultan la elección [3]. A partir del índice original de Ventaja Comparativa Revelada (VCR) han surgido otros que incorporan algunas variaciones, sin que se haya estudiado suficientemente el grado de consistencia de la familia de indicadores surgida de él y las consecuencias de emplear uno u otro [15].

Teniendo en cuenta estos antecedentes, el objetivo de esta comunicación es doble: en primer lugar, pretende determinar la competitividad de las once principales economías vitivinícolas del mundo y, en segundo lugar, contrastar la consistencia de los resultados de siete índices de VCR diferentes (Índice de Balassa Simétrico, Variante de Chi-Cuadrado, Índice de Ventaja Comparativa Revelada Aditiva, Índice Simétrico de Ventaja Relativa del Comercio, Índice Simétrico de Ventaja Competitiva Revelada, Variante del Índice de Michael y Índice de Iapadre). De esta forma es posible conocer con más precisión el desempeño de los grandes países

[a] Este trabajo de investigación ha contado con el apoyo del proyecto nacional eMarketwine (CS02016-78775-R).

DOI: 10.1051/bioconf/20170903001

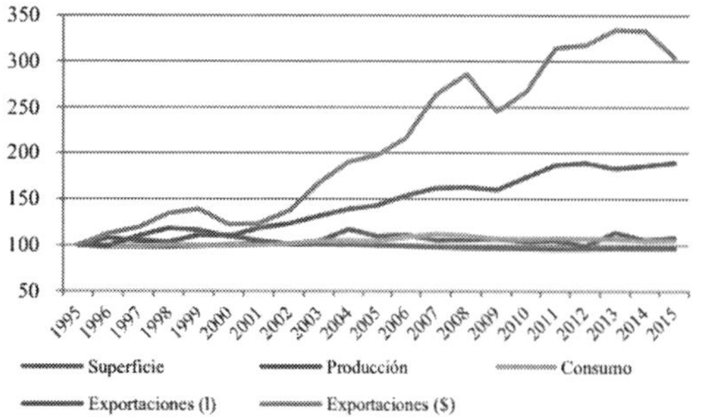

Figura 1. Evolución del comportamiento de las macromagnitudes de la industria vitivinícola mundial periodo 1995–2015 (Índice, 1995 = 100). Elaboración propia con datos de la Organización Internacional del Vino (OIV) y UNCOMTRADE.

productores y exportadores de vino y, adicionalmente, saber qué indicador o indicadores son más fiables y pueden ser sustitutivos entre sí. Tanto esta metodología como los resultados obtenidos a partir de ella pueden ser instrumentos útiles para el diseño de estrategias de marketing de bodegas y países.

2. La globalización de la industria del vino

La industria vinícola se encuentra en un proceso de transformación. En el 2015, los países con mayor contribución a las macromagnitudes mundiales, superficie, producción, consumo y comercialización ya no son únicamente los tradicionales del "Viejo Mundo" del vino o los productores europeos, como pasaba en el siglo XX, sino que además, el "Nuevo Mundo" y otros estados, especialmente China, tienen mayor presencia en el devenir de la industria global del vino.

De esta forma, cogiendo datos desde finales del siglo pasado, se puede observar (Fig. 1) que tanto la superficie como la producción y el consumo se han mantenido estables, aunque con la diferencia de una mayor contribución de nuevos países dentro del mercado, adquiriendo mayor protagonismo a nivel internacional, en detrimento de los países típicamente vinícolas. Mientras tanto, la comercialización de litros de vino se ha duplicado, y más que triplicado en el caso de su valor, entre 1995 y 2015. Con la tendencia a la obertura de nuevos mercados de destino, y con el incremento de la competencia global, se desencadenan cambios importantes en las estrategias empresariales de producción, comercialización, distribución y márquetin [13, 14].

2.1. Cambios en la superficie

El cultivo de la viña en el mundo se encuentra muy concentrado en los países del Viejo Mundo, donde se acapara alrededor del 40% de la superficie mundial. En concreto, España es aglutina la mayor superficie dedicada a la viña con más de un millón de hectáreas en el 2015 – pero un 15% menos respecto al 1995, tendencia seguida por los otros países europeos, Francia e Italia, con caídas alrededor del 14% y 17%, respectivamente en el mismo período de 20 años- (Fig. 2). Estos descensos en Europa, motivados principalmente a las primas a la

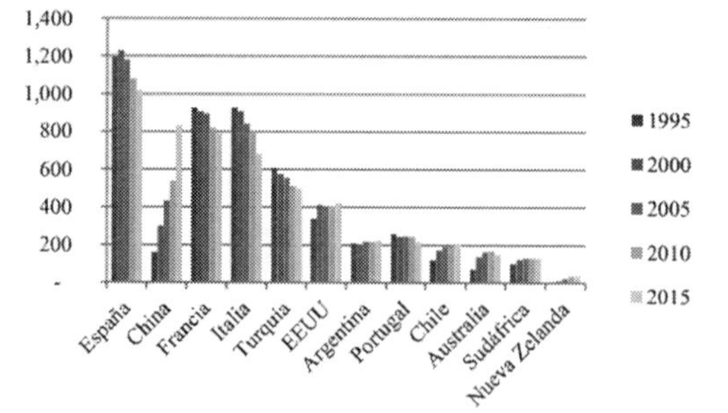

Figura 2. Superficie dedicada a la viña de los principales países productores vinícolas, años 1995, 2000, 2005, 2010 y 2015 (en 1.000 hectáreas). Elaboración propia con datos de la OIV.

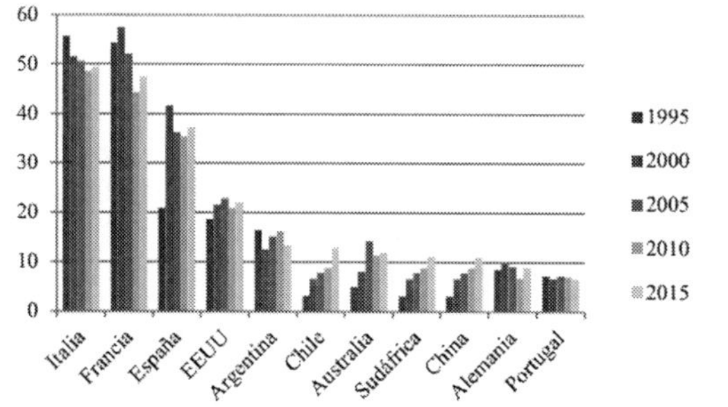

Figura 3. Producción de vino de los principales elaboradores de vino, años 1995, 2000, 2005, 2010 y 2015 (en millones de hectolitros). Elaboración propia con datos de la OIV.

reducción de viña del OCM del vino (2008–2013) -que contribuyó a un decrecimiento de 269.000 hectáreas en los Estados Miembros de la UE- [16], han provocado que la superficie total mundial de viñedo haya experimentado una reducción del 3,5% desde el 1995.

El resto de zonas productoras han registrado aumentos de su superficie de viñedo. En el Nuevo Mundo, aunque ésta representa únicamente el 16%, ha experimentado una crecimiento de más de 300.000 hectáreas. Además, China ha visto cómo la superficie de sus viñas crecía sin descanso, alzándose como el segundo estado con mayor área de este cultivo en 2015.

2.2. Deslocalización de la producción

Al igual que en la superficie, la producción mundial de vino está acaparada por pocos países. Entre Italia, Francia y España producen prácticamente la mitad del vino elaborado en el mundo, aunque con menor incidencia que a principios del siglo XXI (Fig. 3). La tendencia en los países del Nuevo Mundo es la contraria. Sus producciones han crecido con gran incidencia, como el caso de Chile y Sudáfrica, o se han mantenido estables con ciertos altibajos, casos estadounidense o australiano. De nuevo, el caso chino es digno de destacar; su producción ha aumentado sin descanso y ya se encuentra entre los principales elaboradores mundiales de vino.

2.3. Aparición de nuevos consumidores

Aunque los dos principales países consumidores de vino, Estados Unidos y Francia, aúnan casi el 25% del consumo mundial, su tendencia es bien diferente. El primero ha visto

DOI: 10.1051/bioconf/20170903001

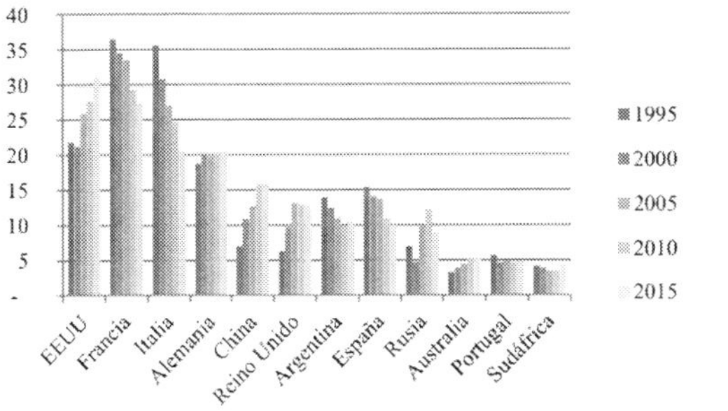

Figura 4. Consumo de vino en los principales países, años 1995, 2000, 2005, 2010 y 2015 (en millones de hectolitros). Elaboración propia con datos de la OIV.

cómo desde principios de siglo el consumo de vino dentro de sus fronteras ha crecido sin descanso. El segundo ha presenciado una reducción del consumo de 9 millones de hectolitros entre el 1995 y 2015, paradójicamente la misma cantidad que ha aumentado el país estadounidense (Fig. 4).

La tendencia francesa en referencia al consumo de vino no es singular. Sus compatriotas europeos, Italia y España, padecieron similar comportamiento y sendos desplomes del vino consumido dentro de sus fronteras. Así pues, aquellos países con mayor tradición de consumo vinícola están experimentado un retroceso en cuanto a la presencia de este producto dentro de las cestas de consumo. También es verdad, y dentro del segmento europeo, que Alemania, Reino Unido y Portugal, los otros grandes consumidores del continente, han conseguido estabilizar el consumo de vino en los últimos 15 años. Por el contrario, y además del caso estadounidense, presenciamos aumentos de consumo de vino de países menos tradicionales y externos a los europeos, cuyas tendencias han sido positivas, como el caso extraordinario de China, el mayor consumo en Australia, o la estabilidad en Sudáfrica. Argentina, por su parte, sigue una trayectoria poco estable en el consumo de vino dentro de sus fronteras.

2.4. La mundialización del mercado del vino

Los mercados de vino se han internacionalizado. Si bien en los años ochenta los mercados del vino eran nacionales o, como mucho, regionales, y estaban copados por países europeos, treinta años después la situación de la geografía económica del vino es muy diferente. Los vinos europeos se comercializan cada vez en más en el exterior, ya sea en mercados tradicionales o en nuevos y exóticos mercados vinícolas, aunque ya no sólo compiten entre sí, sino que lo hacen con vinos procedentes de los Estados Unidos, Chile, Argentina, Sudáfrica, Australia y Nueva Zelanda, entre otros.

2.4.1. Los exportadores de vino

España es el país con mayor número de litros vendidos en los diferentes mercados internacionales. Casi uno de cada cuatro litros que se comercializan internacionalmente proceden de España, 1 cada 5 son italianos y 1 cada 7 de origen francés (Fig. 5). Esto hace que más de la mitad de litros (55%) que podemos encontrar en el mercado internacional de vino sea procedente de estos tres países europeos, con un total para 2015 de más de 104 millones de

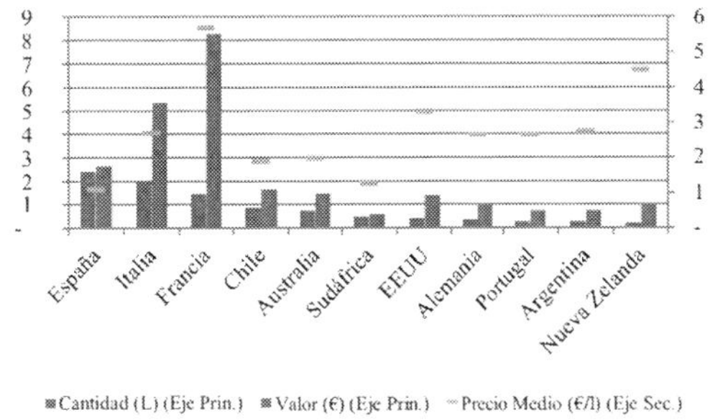

Figura 5. Exportaciones mundiales de vino (en miles de millones de litros y de €, eje principal) de los principales actores en el mercado internacional, y precio medio (€/l, eje secundario), año 2015. Elaboración propia con datos de UNCOMTRADE (S.A.: 220410, 220421 y 220429).

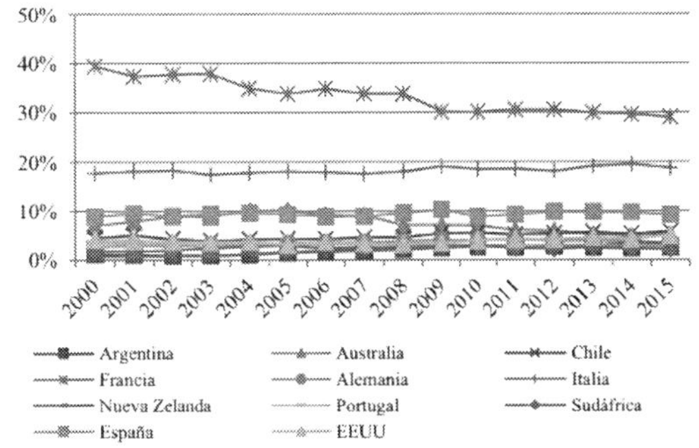

Figura 6. Evolución de la cuota de mercado en valor de las exportaciones de vino de los principales actores en el mercado internacional entre los años 2000 y 2015. Elaboración propia con datos de UNCOMTRADE (S.A.: 220410, 220421 y 220429).

hectolitros. El resto de principales actores, tanto del Viejo como del Nuevo Mundo, se reparten un 35% del total de litros comercializados, con unas horquillas que van desde el 9% de origen chileno hasta el 2% neozelandés en 2015.

En el plano de los ingresos generados por estas ventas en el mismo año de estudio, el escenario es bien distinto. De los más de 28 mil millones de €comercializados en 2015, los vinos de Francia contribuyeron el 29%, los italianos el 19% y los españoles el 9%. El resto ingresaron entre los 1.650 millones de euros de Chile y los 600 millones de euros sudafricanos. Este hecho determina la presencia de unos vinos franceses en los mercados globales con un mayor valor añadido, un precio medio de 5,7 €/l, frente a los 2,7 €/l italianos y los 1,1 €/l españoles, quien posee los vinos más baratos de los once principales exportadores. De entre el resto de competidores, únicamente los vinos neozelandeses, 4,5 €/l, y los estadounidenses, 3,3 €/l, se venden a un precio superior a la media mundial, de 2,7 €/l.

Ahora bien, la hegemonía francesa en el mercado internacional del vino está en declive. La cuota de mercado referente a los ingresos derivados de la venta de sus vinos ha retrocedido del 40% del año 2000 hasta el actual 29%, mientras que el resto de potencias, excluyendo a Australia y Portugal, han alcanzado cuotas mayores respecto al mismo año (Fig. 6).

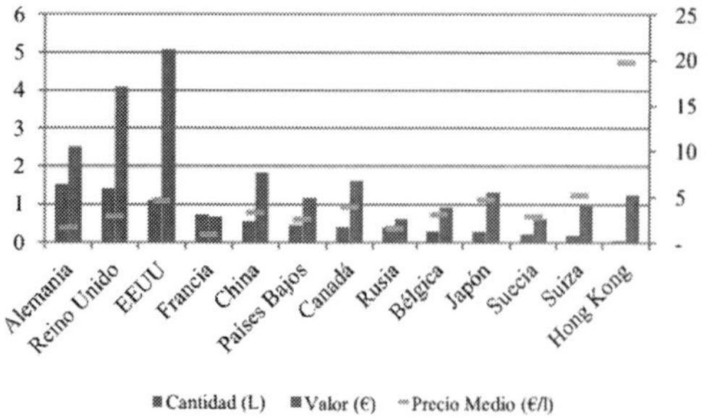

Figura 7. Importaciones mundiales de vino (en miles de millones de litros y de €, eje principal) de los principales mercados de vino, y precio medio (€/l, eje secundario), año 2015. Elaboración propia con datos de UNCOMTRADE (S.A.: 220410, 220421 y 220429).

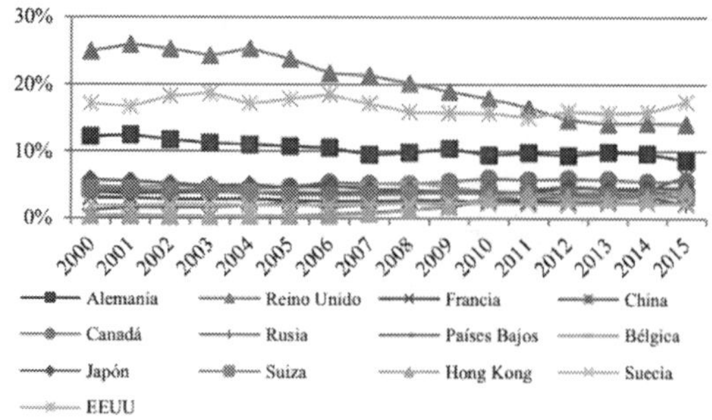

Figura 8. Evolución de la cuota de mercado en valor de las importaciones de vino de los principales mercados entre los años 2000 y 2015. Elaboración propia con datos de UNCOMTRADE (S.A.: 220410, 220421 y 20429).

2.4.2. Los mercados de importación de vino

El mercado europeo aún está considerado el destino principal de los vinos en el mercado internacional. De hecho, entre los trece grandes mercados de importación se encuentran siete Estados Miembros, ocho si ampliamos el criterio al continente europeo. En este sentido, Alemania es el mercado que más vino importa, 15 millones de hectolitros en 2015, por encima del Reino Unido, 14 millones de hectolitros en el mismo ejercicio, y los Estados Unidos, 11 millones de hectolitros. De este modo, 4 de cada 10 litros comercializados acaban en estos tres mercados (Fig. 7), para un total de más de 102 millones de hectolitros.

Según el valor generado por esta comercialización, la lista se invierte. Así es que los Estados Unidos es el principal mercado vinícola, 17% del valor total, seguido del Reino Unido, 14%, y Alemania, 9%, con unos precios medios de importación que van de los 4,6 €/l en Estados Unidos, 2,9 €/l británicos y 1,7 €/l alemanes.

Ahora bien, los grandes mercados europeos del vino, Alemania y, en especial, Reino Unido, han perdido importancia relativa en cuanto al valor de sus mercados de importación. El primero, ha presenciado una tendencia regresiva de su cuota de valor de las importaciones, del 12% en el 2000. El segundo partía de cuotas de alrededor del 25% en los primeros años del siglo, perdiendo en el 2012 su condición de líder de entre los mercados vinícolas en beneficio del mercado estadounidense. Loes Estados Unidos han mantenido una contribución en el valor global del mercado internacional de entre el 15–17% en los últimos 15 años (Fig. 8). El resto de principales mercados, Francia, China, Países Bajos, Canadá, Rusia, Bélgica, Japón, Suecia, Suiza y la Región Administrativa Especial de Hong Kong, representan en conjunto el 35% de las importaciones de vino, el 40% del valor generado, aunque con tendencias claramente contrapuestas. Los mercados europeos han sufrido una caída en la importancia relativa sobre el valor de importación, mientras que los no europeos, especialmente en China y Hong Kong, han aumentado sus cuotas

3. La competitividad internacional de la industria vitivinícola

El concepto de Competitividad no está completamente definido. Nos podemos encontrar diferentes enfoques de lo que la Competitividad mide. Se define como un conjunto de instituciones, políticas y factores que determinan el nivel de productividad de un país [17], la productividad con la que un país utiliza sus recursos humanos, económicos y naturales con el objetivo principal de generar las condiciones para elevar el nivel de vida de sus ciudadanos, habilidad que depende de la productividad, con la forma en que las naciones utilizan su capital y trabajo [18], la capacidad de las empresas, industrias, regiones, naciones o grupos transnacionales de enfrentar la competencia internacional y asegurar la sostenibilidad de una tasa relativamente alta de retornos a los factores de producción, y un nivel relativamente alto de ocupación, que se obtiene a largo plazo mediante un crecimiento de la productividad total [19]. Ser competitivo se entiende como el crecimiento constante de ingresos reales y los estándares de vida de las regiones o los países, con una ocupación disponible para todos los solicitantes de trabajo que lo deseen [20].

Una economía más competitiva es probable que crezca más rápido a lo largo del tiempo [17]. En ese sentido, la mejora de la productividad es importante para que la competitividad tenga más éxito en mercados abiertos a la competencia internacional, ya que aporta una mejora a largo plazo en calidad de vida y en la creación de trabajo y, finalmente, ofrece un mejor aprovechamiento de las ventajas competitivas, que ya no son, por tanto, limitados a la disponibilidad de recursos naturales en la economía y la competencia global [21].

La Competitividad depende de factores macroeconómicos -políticas económicas, industriales, de comercio, monetarias, fiscales y de impuestos, el funcionamiento del mercado de capital, el entorno social y económico, el sistema de normas jurídicas, el sistema educativo, y la ética de trabajo- y microeconómicos -la capacidad productiva empresarial, disponibilidad de medios de producción, gestión empresarial, estrategia de márquetin y relaciones laborales [22].

Ahora bien, el concepto de competitividad no se encuentra plenamente definido y se detectan diferentes tipos de competitividad, que varían en base al tiempo, el lugar y el objeto de estudio [23]. Para entender la Competitividad desde un punto de vista del comercio internacional, podemos encontrar un esfuerzo investigador significante: la Ventaja Absoluta [24], la Ventaja Comparativa [25], las Proporciones de los Factores del Comercio [26], el Desfase Tecnológico o de la Brecha Tecnológica [27], el Ciclo de

Vida del Producto [28] y la Ventaja Comparativa de las Naciones [18].

No obstante, el análisis de la competitividad basándose en estas teorías es bastante complicado debido a las dificultades en la obtención y la medida de los datos requeridos. Por esta razón, los estudios se han centrado en relacionarla con la Ventaja Comparativa.

Ligada al comercio internacional, la Ventaja Comparativa dice que el nivel económico óptimo se consigue cuando los países exportan aquellos bienes en los que tienen una ventaja competitiva e importan aquellos en los que tienen una desventaja comparativa [15]. La determinación de si un país tiene o no ventajas comparativas ayuda a orientar la inversión y el comercio y, por tanto, a tomar mayores ventajas de las diferencias que existen al lado de la demanda y de la oferta internacional de productos y factores de producción [15].

Así pues, la Ventaja Comparativa es la habilidad de una economía determinada para la fabricación de un producto de manera más eficiente que otros países, que se refleja en las direcciones de las especializaciones de exportación e importación, y que se define por la relativa provisión de los medios de producción de bienes a diversos países, la estructura de la demanda, las ventajas de escala, el potencial para la diversificación de productos, etc. [22].

Para medir la Ventaja Comparativa se utiliza el concepto de Coste de Oportunidad [29], aunque debido a las muchas dificultades que se presentan a la hora de su cálculo, sobre todo por su laboriosidad y dificultad [15], se prioriza la utilización de la Ventaja Comparativa Revelada (VCR) [30]. La VCR se plantea si es posible extraer las ventajas comparativas desde los flujos del comercio mundial, atendiendo a que el intercambio real de bienes refleja los costes relativos y también las diferencias que existen entre los países, por factores no necesariamente de mercado [15].

3.1. Indicadores de Competitividad Internacional

Los indicadores de Ventaja Comparativa Revelada que se utilizarán para medir la Competitividad de las industrias vinícolas en el mercado internacional, y que a su vez se utilizarán para comprobar su estabilidad se presentan a continuación. Un VCR positivo, para todos los indicadores, significa la presencia de una industria competitiva en el mercado internacional, siendo los valores negativos reflejo de una industria no competitiva.

3.1.1. *Índice de Balassa Simétrico (I1)*

El Índice de Balassa (1) [30] fue el primer indicador que se propuso como medida de la VCR.

$$IB_{ij} = \left(\frac{X_{ij}}{X_{it}}\right) \Big/ \left(\frac{X_{nj}}{X_{nt}}\right) \tag{1}$$

De donde X representan las exportaciones, i es el país de análisis, j es la industria a analizar, t es el conjunto de industrias globales, y n es el mundo. Ahora bien, para eliminar su asimetría [31] se presenta un índice alternativo (2).

$$I1_{ij} = \frac{IB_{ij} - 1}{IB_{ij} + 1} \tag{2}$$

3.1.2. *Variante de Chi-Cuadrado (I2)*

Otro indicador de VCR es una variante del Chi-Cuadrado [32] (3).

$$\chi^2_{ij} = \frac{[(X_{ij}/X_{it}) - (X_{nj}/X_{nt})]^2}{(X_{nj}/X_{nt})} \tag{3}$$

De donde X representan las exportaciones, i es el país de análisis, j es la industria a analizar, t es el conjunto de industrias globales, y n es el mundo. Para el cálculo del segundo indicador de VCR se incorpora una mejora con el que indicar la especialización o no de un país en una industria [31] (4).

$$I2_{ij} = \begin{cases} (-1)^{*}\chi^2_{ij}\ si\ \left(X_{ij}/X_{it}\right) - \left(X_{nj}/X_{nt}\right) < 0 \\ (1)^{*}\chi^2_{ij}\ si\ \left(X_{ij}/X_{it}\right) - \left(X_{nj}/X_{nt}\right) > 0 \end{cases} \tag{4}$$

3.1.3. *Índice de Ventaja Comparativa Revelada Aditiva (I3)*

Como los dos índices anteriormente propuestos, la Ventaja Comparativa Revelada Aditiva [33] (5) asume únicamente el valor de las exportaciones.

$$I3_{ij} = ARCA_{ij} = \left(\frac{X_{ij}}{X_{ia}}\right) - \left(\frac{X_{rj}}{X_{ra}}\right) \tag{5}$$

De donde X representan las exportaciones, i es el país de análisis, j es la industria a analizar, a es el conjunto de industrias globales menos la industria analizada, y r es el mundo menos el país analizado. Se trata de una modificación del Índice de Balassa caracterizada por la eliminación de la doble contabilización del país/industria.

3.1.4. *Índice Simétrico de Ventaja Relativa del Comercio (I4)*

Los tres índices anteriores han sido criticados por ser parciales, debido a la omisión de las importaciones en su cálculo, esencialmente importante cuando el tamaño del país es significativo [34]. De esta forma se presenta una cuarta opción para el cálculo de la VCR basada en la Ventaja Relativa del Comercio (8), que a su vez es producto de la diferencia entre la Ventaja Relativa de las Exportaciones (6) y la Ventaja Relativa de las Importaciones (3.1.4) [35].

$$RXA_{ij} = \left(\frac{X_{ij}}{X_{ia}}\right) \Big/ \left(\frac{X_{rj}}{X_{ra}}\right) \tag{6}$$

$$RMA_{ij} = \left(\frac{M_{ij}}{M_{ia}}\right) \Big/ \left(\frac{M_{rj}}{M_{ra}}\right) \tag{7}$$

$$RTA_{ij} = RXA_{ij} - RMA_{ij} \tag{8}$$

De donde X representan las exportaciones, M las importaciones, i es el país de análisis, j es la industria a analizar, a es el conjunto de industrias globales menos la industria analizada, y r es el mundo menos el país analizado. Debido a que este indicador no es simétrico, se propone un indicador alternativo para proporcionar esta cualidad (9).

$$I4_{ij} = \frac{RXA_{ij} - 1}{RXA_{ij} + 1} - \frac{RMA_{ij} - 1}{RMA_{ij} + 1} \tag{9}$$

Tabla 1. VCR según los indicadores *I1, I2, I3, I4, I5, I6* e *I7* de los principales actores del mercado internacional del vino, año 2015.

	I1	*I2*	*I3*	*I4*	*I5*	*I6*	*I7*
Argentina	0,759	0,078	0,013	1,656	1,648	0,014	1,008
Australia	0,629	0,023	0,007	0,484	0,473	0,006	0,525
Chile	0,872	0,367	0,028	1,716	1,705	0,029	0,985
Francia	0,808	0,140	0,017	1,068	1,012	0,017	0,915
Alemania	−0,418	−0,001	−0,001	−0,588	−0,555	−0,002	−0,559
Italia	0,736	0,062	0,012	1,186	1,138	0,012	0,835
Nueva Zelanda	0,881	0,435	0,030	0,644	0,639	0,028	0,829
Portugal	0,764	0,083	0,013	0,809	0,802	0,013	0,830
Sudáfrica	0,658	0,029	0,008	1,358	1,350	0,009	0,983
España	0,684	0,037	0,009	1,234	1,206	0,010	0,923
EEUU	−0,315	−0,000	−0,001	−0,450	−0,410	−0,001	−0,357

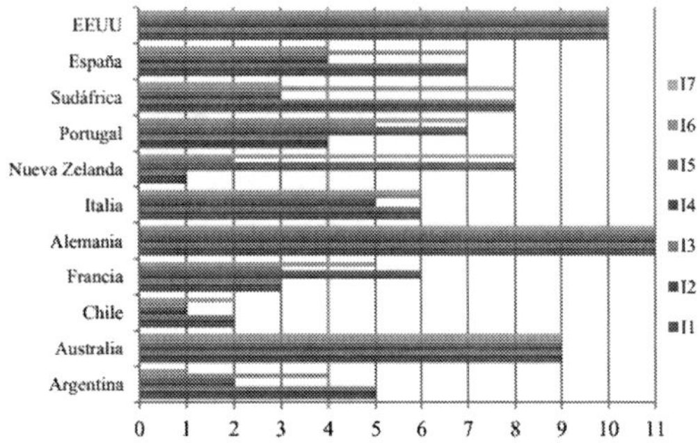

Figura 9. Posición competitiva de los principales países del mercado internacional del vino según *I1, I2, I3, I4, I5, I6* e *I7*, año 2015.

3.1.5. Índice Simétrico de Ventaja Competitiva Revelada (I5)

El índice de Ventaja Competitiva Revelada (12) [36], otro indicador que incorpora en su cálculo las exportaciones y las importaciones, se diferencia respecto al anterior en que no elimina la doble contabilización de la industria/país. Para su cálculo es necesario obtener la Ventaja Competitiva del Suministro (10) y la Ventaja Competitiva de la Demanda (11).

$$RCS_{ij} = IB_{ij} = \left(\frac{X_{ij}}{X_{it}}\right) \Big/ \left(\frac{X_{nj}}{X_{nt}}\right) \quad (10)$$

$$RCD_{ij} = \left(\frac{M_{ij}}{M_{it}}\right) \Big/ \left(\frac{M_{nj}}{M_{nt}}\right) \quad (11)$$

$$RCA_{ij} = RCS_{ij} - RCD_{ij} \quad (12)$$

De donde X representan las exportaciones, M las importaciones, i es el país de análisis, j es la industria a analizar, t es el conjunto de industrias globales, y n es el mundo. De nuevo, debido al problema de la asimetría, se presenta una alternativa (13).

$$I5_{ij} = \frac{RCS_{ij} - 1}{RCS_{ij} + 1} - \frac{RCD_{ij} - 1}{RCD_{ij} + 1} \quad (13)$$

3.1.6. Variante del Índice de Michaely (I6)

Otro cálculo de la RCA que tiene en cuenta las exportaciones y las importaciones es una variante del Índice de Michaely [37] (14).

$$I6_{ij} = IM_{ij} = \frac{X_{ij}}{X_{it}} - \frac{M_{ij}}{M_{it}} \quad (14)$$

De donde X representan las exportaciones, M las importaciones, i es el país de análisis, j es la industria a analizar, t es el conjunto de industrias globales, y n es el mundo.

3.1.7. Índice de Iapadre (I7)

El último índice propuesto, el Índice de Iapadre (15) [38], es una variante del Índice de Contribución en la Balanza Comercial de Lafay [39].

$$I7_{ij} = II_{ij} = \frac{X_{ij} - M_{ij}}{X_{ij} + M_{ij}} - \frac{X_{it} - M_{it}}{X_{it} + M_{it}} \quad (15)$$

De donde X representan las exportaciones, M las importaciones, i es el país de análisis, j es la industria a analizar, t es el conjunto de industrias globales, y n es el mundo.

3.2. La posición competitiva de las industrias vitivinícolas en el mercado internacional

La posición competitiva de las principales industrias vitivinícolas en el mercado internacional es completamente sensible al indicador de VCR utilizado. Los resultados arrojan listas de competitividad diferentes entre los indicadores de la misma en 2015 (Tabla 1).

En primer lugar, según todos los indicadores, los vinos españoles se posicionan en lugares medio-bajos de competitividad, respecto a los competidores. En este sentido, *I4, I5* e *I7* los colocan en el cuatro lugar, mientras que el resto lo ubican el séptimo del ranquin de competitividad internacional de las industrias vitivinícolas (Fig. 9). La industria francesa del vino tiene una competitividad, en referencia con sus competidores, media-alta. Los indicadores *I1, I2, I3* e *I6* establecen que ésta es la tercera industria más competitiva, mientras que *I4* e *I5* indican que su posición es en sexto lugar. Por último, *I7* la posiciona en quinto lugar. Los vinos italianos, para todos los indicadores, son de competitividad media. *I1, I2, I3, I6* e *I7* los colocan en sexto lugar, *I4* e *I5* en quinta. Mientras que la última industria vitivinícola europea competitiva de las analizadas, Portugal, tiene una posición parecida a los vinos españoles, medio-baja. En octavo lugar para los indicadores *I1, I2* e *I3*; séptimo para *I4, I5* e *I7*; y quinto para *I6*.

Dentro del Nuevo Mundo, las industrias más sensibles al indicador para medir el RCA utilizado son Nueza Zelanda y Sudáfrica. El país neozelandés pasa de ser el líder en *I1*, *I2* e *I3*, y segundo en *I6*, a ser el octavo más competitivo en *I4*, *I5* y *I7*. El país sudafricano tiene un desempeño diferente, y casi contrario al caso anterior: octavo para *I1*, *I2*, *I3* e *I6*, y tercero para *I4*, *I5* e *I7*. Chile no sufre tal variabilidad de resultados, o es líder para ciertos indicadores (*I4*, *I5* e *I6*) o segundo (*I1*, *I2*, *I3* e *I7*), al igual que la industria vitivinícola australiana, novena para todos los indicadores propuestos. El último país con unos vinos competitivos en el ámbito internacional, Argentina, se posiciona en unos estándares medio altos. Pasa de ser quinto para *I1*, *I2* e *I3*, a cuarto para *I6*, segundo para *I4* e *I5*, e incluso la mayor industria competitiva para el indicador *I7*.

Las dos únicas industrias vitivinícolas de las principales en el mercado internacional que no son competitivas son Estados Unidos y Alemania, esto es, sus indicadores son negativos para todos los indicadores que se han establecido.

Estas diferencias principalmente recaen en la contabilización de las importaciones a la hora de calcular el RCA. De esta forma, algunos países ven alterada su posición competitiva, con una alta variabilidad, según si se contemplan o no las importaciones. Los resultados establecen que hay países a quienes beneficia su computación (Sudáfrica, Argentina y España), mientras que hay otros a quienes les perjudica (Nueva Zelanda, Francia y Portugal) y otros que no ven alterada con mucha incidencia su posición competitiva (Chile, Italia, Australia, Estados Unidos y Alemania).

No obstante, se determina la necesidad de un análisis más profundo para establecer si existe estabilidad entre los indicadores propuestos, si su consistencia permite que el cálculo de la RCA sea indiferente al indicador escogido o, por el contrario, se tenga que establecer un consenso en el ámbito del estudio de la competitividad a la hora de establecer un único indicador.

4. Consistencia de los indicadores de Competitividad Internacional de la industria vitivinícola

Aunque los indicadores presentados y propuestos en el apartado anterior, *I1*, *I2*, *I3*, *I4*, *I5*, *I6* e *I7*, miden la Ventaja Comparativa Revelada, todos ellos difieren en su sustentación teórica-metodológica. Por esta razón, y por tan de comprobar su estabilidad, se utilizarán tres pruebas que medirán la consistencia dicotómica, la consistencia cardinal y la consistencia ordinal entre ellos [40]. La aplicación de estas pruebas se realizará para la industria española del vino en el periodo que comprende entre el año 2000 y 2015.

4.1. Metodología de cálculo

Desde el punto de vista metodológico, el análisis de consistencia resulta necesario básicamente por dos motivos. En primer lugar, porque si nos encontramos con un conjunto de indicadores muy consistentes entre sí, estos pueden llegar a ser buenos sustitutivos para medir la competitividad de una industria en el mercado internacional, pudiéndose elegir aquél que tenga una mayor ventaja metodológica adicional. En segundo lugar, puesto que si se tiene indicadores alternativos para medir la competitividad poco consistentes o inconsistentes, es la consecuencia de que algún factor que entra en el cómputo de alguno de ellos lo lleve a tal divergencia. Este hecho sugeriría la posible utilización de todos los indicadores bajo una naturaleza complementaria.

Las pruebas de consistencia permiten ver hasta qué nivel dos indicadores dados son coincidentes como medidores de VCR en tres sentidos: (i) establecer en cada año si revela un sector competitivo o no (sentido dicotómico); (ii) revelar un grado de competitividad para cada año que se computa, de acuerdo con el valor específico que asume el indicador (sentido cardinal); y (iii) establecer un ranquin desde aquellos años en los que la competitividad es mayor hasta los que tienen menor competitividad o no son competitivos (sentido ordinal). Si un par de indicadores coinciden en estos tres sentidos, podemos asumir que ambos son perfectamente consistentes. En caso de coincidencias parciales, estas pueden ser mayoría o no. Si no lo son, se dice que los indicadores son inconsistentes.

Las pruebas estadísticas de todas las pruebas permiten cuantificar la consistencia de dos indicadores cualquiera. No obstante, la consistencia entre dos indicadores no es una propiedad que se disfrute independientemente del producto, periodo y país sobre el que se desea medir la competitividad internacional. En cada caso específico se requiere de un análisis de consistencia particular [41].

4.1.1. Consistencia cardinal (CC)

La primera prueba a la que se verán sometidos los indicadores, mediante un análisis de comparación por pares, es la consistencia de los índices como medida cardinal. Para su logro, se utilizará el coeficiente de correlación de Pearson (16, 17) entre los valores que asume la serie de los indicadores *Iv* e *Iw* ($v, w = 1, \ldots, 7; v \neq w$) en el periodo dicho, 2000–2015.

$$CC_{Iv_{ij},Iw_{ij}} = \frac{\mathrm{Cov}(Iv_{ij}, Iw_{ij})}{S_{Iv_{ij}} * S_{Iw_{ij}}} = = \frac{\sum_{t=1}^{T}[(Iv_{ij}(t) - \overline{Iv}_{ij}) * (Iw_{ij}(t) - \overline{Iw}_{ij})]}{\sqrt{\sum_{t=1}^{T}(Iv_{ij}(t) - \overline{Iv}_{ij})^2} * \sqrt{\sum_{t=1}^{T}(Iw_{ij}(t) - \overline{Iw}_{ij})^2}} \quad (16)$$

$$\overline{Iu}_{ij} = \frac{\sum_{t=1}^{T} Iu_{ij}(t)}{T}; \quad u = v, w \quad (17)$$

$Iu_{ij}(t)$ es un indicador para el país i del bien j computado en el año t, representando T la cantidad total de años analizados, para el que la desviación estándar muestra para el periodo viene dado por $S_{Iu_{ij}}$ y $\mathrm{Cov}(Iv_{ij}, Iw_{ij})$ representa la covarianza entre los dos indicadores.

4.1.2. Consistencia ordinal (CO)

La prueba de consistencia ordinal se realiza correlacionando los diversos indicadores ordenándolos por su valor

DOI: 10.1051/bioconf/20170903001

numérico en orden descendente y aplicando la correlación de rangos de Spearman (17).

$$CO_{Iv_{ij},Iw_{ij}} = 1 - \frac{6^{*}\sum_{t=1}^{T}\left(R_{Iv_{ij}}(t) - R_{Iw_{ij}}(t)\right)^{2}}{T(T^{2}-1)} \quad (18)$$

De donde $R_{Iv_{ij}}$ (t) $R_{Iw_{ij}}$ (t) indican la posición en el ranquin determinado por $Iv_{ij}(t)$ y $Iw_{ij}(t)$ en los que se encuentran en el año t. El análisis de este resultado permite ver la consistencia al ordenamiento del grado de especialización que muestra cada indicador [40].

4.1.3. Consistencia dicotómica (CD)

La última prueba a la que los indicadores de RCA serán sometidos es la consistencia dicotómica (19, 20) que indica la coincidencia de mostrar la existencia de competitividad o no.

$$CD_{Iv_{ij},Iw_{ij}} = \frac{\sum_{t=1}^{T} D_{Iv_{ij},Iw_{ij}}(t)}{T} \quad (19)$$

$$D_{Iv_{ij},Iw_{ij}}(t) = \times \begin{cases} 1\, si\; Iv_{ij} > 0\, e\; Iw_{ij} > 0; & o\; Iv_{ij} < 0\, e\; Iw_{ij} < 0 \\ 0\, si\; Iv_{ij} > 0\, e\; Iw_{ij} < 0; & o\; Iv_{ij} < 0\, e\; Iw_{ij} > 0 \end{cases} \quad (20)$$

4.2. Resultados

Debido a que las pruebas de CC y CO están definidas por los coeficientes de correlación de Pearson y Spearman, respectivamente, y que la CD es una proporción de casos favorables (coincidentes) sobre el total de años en que se calcula el indicador, el valor máximo que pueden llegar todos ellos es de la unidad (−1, en caso negativo), siendo esta la coincidencia perfecta (imperfecta en el caso contrario). Por tanto, cuanto más cerca del valor unitario se encuentren CC, CO y CD, mayor será la consistencia cardinal, ordinal y dicotómica, respectivamente (menor consistencia o inconsistencia en el caso negativo).

El análisis de consistencia entre los diferentes indicadores de competitividad internacional propuestos para la industria vitivinícola española en el periodo 2000–2015 muestra resultados que difieren entre qué pareja haya sido analizada (Tabla 2).

La competitividad internacional medido por el indicador *I1*, el Índice de Balassa Simétrico, es bastante consistente con *I2*, la Variante de Chi-Cuadrado, en todas las pruebas de estabilidad. Asimismo, *I1* también es consistente con *I3*, el Índice de Ventaja Comparativa Revelada Aditiva, en las tres pruebas pero esta vez a un nivel menor, tanto en la cardinal como la ordinal. Además, a su vez, entre *I2* e *I3* hay una alta consistencia en los tres test. El hecho de que en este grupo de indicadores sus componentes tengan estabilidad entre sí puede derivar de la naturaleza de ellos, puesto que los tres únicamente tienen en cuenta el valor de las exportaciones en su cálculo, obviando el valor de las importaciones. Por otra parte, estos tres indicadores, *I1*, *I2* e *I3*, llegan a ser inconsistentes, cardinal y ordinal, con tres de los cuatro indicadores de VCR propuestos que incorporar

Tabla 2. Resultados de la Consistencia Cardinal (CC), Consistencia Ordinal (CO) y Consistencia Dicotómica (CD) de los indicadores de VCR elegidos para el caso de la industria vitivinícola española en el periodo 2000–2015.

	CC	*CO*	*CD*
I1vsI2	0.950	0.962	1.000
I1vsI3	0.748	0.788	1.000
I1vsI4	−0.432	−0.479	1.000
I1vsI5	−0.421	−0.479	1.000
I1vsI6	0.479	0.426	1.000
I1vsI7	−0.751	−0.732	1.000
I2vsI3	0.913	0.885	1.000
I2vsI4	−0.314	−0.403	1.000
I2vsI5	−0.305	−0.403	1.000
I2vsI6	0.715	0.535	1.000
I2vsI7	−0.718	−0.706	1.000
I3vsI4	−0.102	−0.203	1.000
I3vsI5	−0.098	−0.203	1.000
I3vsI6	0.928	0.785	1.000
I3vsI7	−0.571	−0.653	1.000
I4vsI5	1.000	1.000	1.000
I4vsI6	0.233	0.271	1.000
I4vsI7	0.588	0.674	1.000
I5vsI6	0.234	0.271	1.000
I5vsI7	0.578	0.674	1.000
I6vsI7	−0.321	−0.262	1.000

las importaciones en su cálculo, estos son *I4*, el Índice Simétrico de Ventaja Comparativa del Comercio, *I5*, el Índice Simétrico de Ventaja Competitiva Revelada, e *I7*, el Índice de Iapadre. Cosa que no ocurre con el *I6*, la Variante del Índice de Michaely, con quien *I1* tiene consistencia, aunque baja en los casos de la cardinal y ordinal, *I2* llega a tener una estabilidad media, de nuevo en las mismas correlaciones, e *I3* tiene casi una consistencia cardinal y ordinal perfecta, cosa que ya ocurre con la dicotómica.

En el caso del grupo de indicadores que incorporan las importaciones en su cálculo, entre *I4*, *I5* e *I7* hay consistencia, llegando a ser perfecta entre los dos primeros, debido a la similitud de cálculo, mientras que en las otras dos combinaciones esta estabilidad es media, en referencia a la cardinal y la ordinal. Ahora bien, entre estos tres indicadores y el otro que contabiliza también las importaciones, *I6*, la consistencia es muy baja en el caso del par con *I4* y con *I5*, llegando a ser inconsistente con *I7*.

De esta forma, podemos identificar principalmente tres grupos de indicadores de VCR de los vinos españoles en el mercado internacional, entre el 2000 y 2015. El primero engloba a los tres índices basados únicamente en las exportaciones (Índice Simétrico de Balassa, Variante de la Chi-Cuadrado e Índice de Ventaja Comparativa Revelada Aditiva). El segundo relaciona tres de los cuatro indicadores propuestos que computan las importaciones además de las exportaciones (Índice Simétrico de Ventaja Relativa del Comercio, Índice Simétrico de Ventaja Competitiva Revelada e Índice de Iapadre). El tercero junta al Índice de Ventaja Comparativa Revelada Aditiva con la Variante del Índice de Michaely (Tabla 3).

De esta forma, se observa que no todos los indicadores propuestos son consistentes y que no serían sustitutos entre sí. Atendiendo a la correlación existente entre los análisis de consistencia cardinal, ordinal y dicotómica entre el Índice Simétrico de Balassa (*I1*), la Variante de

DOI: 10.1051/bioconf/20170903001

Tabla 3. Grupos de indicadores de VCR de la industria española del vino para el periodo 2000–2015 según los resultados de la CC, CO y CD.

	Indicadores de RCA con sólo valor de exportaciones	Indicadores de RCA con valor de exportaciones e importaciones	Ambos grupos de indicadores
Consistencia Alta CC, CO, CD [0.75, 1]	I1-I2 I2-I3 I1-I3	I4 - I5	I3-I6
Consistencia Media CC, CO ,CD [0.5, 0.75[		I4-I7 I5-I7	
Consistencia Baja CC, CO, CD [0, 0.5[		I4-I6 I5-I6	I1-I6 I2-I6
No consistentes CC, CO, CD [-1, 0[		I6-I7	I1-I4, I1-I5, I1-I7, I2-I4, I2-I5, I2-I7, I3-I4, I3-I5, I3-I7

Chi-Cuadrado (*I2*) y el Índice de Ventaja Comparativa Revelada Aditiva (*I3*), todos ellos podrían ser utilizados de forma independiente para determinar la competitividad internacional de los vinos españoles en el mercado global para el periodo 2000–2015, y los tres determinarían la misma posición competitiva para esta industria. Ahora bien, si se quisiera calcular la competitividad teniendo en cuenta también el valor de las importaciones, además de las exportaciones, se podría utilizar cualquier indicador de entre el Índice Simétrico de Ventaja Relativa del Comercio (*I4*), el Índice Simétrico de Ventaja Competitiva Revelada (*I5*) y el Índice de Iapadre (*I7*), puesto que los tres son sustitutivos, debido a los resultados obtenidos referentes a su consistencia cardinal, ordinal y dicotómica.

5. Conclusiones

Los resultados obtenidos demuestran la pertinencia del planteamiento de la investigación. La posición competitiva de las principales industrias vitivinícolas en el mercado internacional es muy sensible al indicador de VCR utilizado. Por ejemplo, cuatro de los siete colocan a la industria española del vino en una situación competitiva media, mientras que el resto la posicionan en un lugar bajo -diferencia que se repite para otros países, con especial intensidad en los casos sudafricano y neozelandés-. Esta situación exige realizar análisis de consistencia -cardinal, ordinal y dicotómica- sobre todos los indicadores, cuyos resultados evidencian una falta de estabilidad, principalmente entre aquellos que contemplan solamente las exportaciones y aquellos que contemplan, además, las importaciones. Por lo tanto no todos los indicadores de VCR son sustitutivos entre sí. De los siete se establecen tres grupos de índices de competitividad cuya consistencia interna es alta: a) el Índice Simétrico de Balassa, la Variante de Chi-Cuadrado y el Índice de Ventaja Comparativa Revelada Aditiva; b) el Índice Simétrico de la Ventaja Relativa del Comercio, el Índice Simétrico de la Ventaja Competitiva Revelada y el Índice de Iapadre y c) el Índice de Michaely de Ventaja Comparativa Revelada Aditiva.

Así es que para medir la competitividad, de los siete índices podemos determinar que: (i) el Índice Simétrico de Balassa (I1) se puede utilizar en detrimento de la Variante de Chi-Cuadrado (I2) y del Índice de Ventaja Comparativo Revelado Aditivo (I3), y viceversa, todos ellos son consistentes y obtienen los mismos resultados; (ii) el Índice Simétrico de la Ventaja Relativa del Comercio (I4) es sustituible por el Índice Simétrico de la Ventaja Competitiva Revelada (I5) o por el Índice de Iapadre (I7), y viceversa, los tres son consistentes y con ellos se obtienen los mismo resultados; (iii) el Índice de Michaely (I6) solamente podría ser sustituido por el Índice de Ventaja Comparativa Revelada Aditiva (I3), aunque la afirmación en el caso contrario es falsa, puesto que I3 también es estable con I1 e I2; (iv) en términos generales, entre los indicadores que contemplan solamente las exportaciones en su cálculo y aquellos que incorporan también las importaciones o hay baja consistencia o son inconsistentes y, por esta razón, el uso o no del flujo comercial de las importaciones a la hora del cálculo de la VCR crea diferentes resultados y, por tanto, o se deberían de contemplar como indicadores complementarios o habría que elegirse un grupo u otro como el único medidor de VCR; (v) utilizar uno u otro depende del tipo de competitividad que se pretenda medir, que ahora se puede definir con mayor precisión; y (vi) respecto a la priorización en el uso de los índices propuestos como indicadores de VCR del comercio internacional, se debería destacar que cada índice presenta mucha susceptibilidad respecto al comportamiento de los flujos de comercio que intervienen en su construcción y, además, los análisis aquí realizados muestran la consistencia entre estos indicadores para el caso de la industria vitivinícola española en el mercado internacional en el periodo 2000–2015 y, por consiguiente, a la hora de elegir uno o diversos indicadores para otra industria u otro país, sería recomendable la realización de pruebas específicas para determinar si se pueden utilizar de forma sustitutiva o complementaria.

Tanto esta metodología como los resultados obtenidos a partir de ella pueden ser instrumentos útiles para el diseño de estrategias de marketing de bodegas, para sustentar las investigaciones de las organizaciones referentes del mundo vitivinícola, como la OIV, y para la elaboración de planes sectoriales de actuación de los sectores vinícolas de países productores, ya que para todos ellos les permiten tener un enfoque y una visión nuevos e innovadores del funcionamiento de los mercados internacionales del vino.

Referencias

[1] R. Castaldi, S. Cholette, M. Hussain, *A Country-Level Analysis of Competitive Advantage in the Wine Industry* (3rd International Wine Business Research Conference, Montpellier, 2006)
[2] European Commission, *Study on the competitiveness of European wines* (Office for Official Publications of the European Communities, Luxembourg, 2014)
[3] F. Cervera, *Consistencia de los indicadores de competitividad de la industria vitivinícola española en el mercado internacional* (Tesina Final Máster, Valencia, 2015)
[4] J. Castillo, E. Villanueva, M. García-Cortijo, Agribusiness, **32**(4), 466–481 (2016)
[5] R. Compés, J. Castillo, *Economía del vino en España y en el mundo* (Cajamar, Almería, 2014)
[6] R. Compés, SEVI. **3**, 440 (2015)
[7] R. Compés, F. Cervera, *Más allá del Viejo y el Nuevo Mundo. La posición de España en los mercados*

DOI: 10.1051/bioconf/20170903001

internacionales del vino (X Conferencia Economía Agraria, Córdoba, 2015)

[8] K. Anderson, *The World's Markets – Globalization at Work* (Edward Elgar, Cheltenham, 2004)

[9] K. Anderson, S. Nelgen, *Wine's Globalization: New Opportunities, New Challenges* (Wine Economics Research Centre, Adelaide, 2011)

[10] A. Mariani, E. Pomarici, V. Boatto, Wine Econ. Pol. **1**(1), 24

[11] K. Almgren, IJHSS, **4**(9), 71–76 (2014)

[12] A. Miecinskiene, V. Stasyte, J. Kazlauskaite, Soc. Behav. Scien. **110**, 116–1175 (2014)

[13] J. Martínez, F. Medina, Rev. Hist. Ind. **52**, 139–174 (2013)

[14] J. Martínez, F. Medina, J. Ramón, Ame. Lat. Hist. Econ. **21**(2), 40-83 (2014)

[15] J. Arias, O. Segura, Rev. Intercamb. Ar. Comer. Agroneg. **4**, 1–10 (2004)

[16] OIV, *Elementos de coyuntura vitivinícola mundial en 2013* (OIV, Paris, 2014)

[17] World Economic Forum, *The global Competitiveness Report 2014-2015* (World Economic Forum, Genève, 2014)

[18] M. Porter, *Competitive Advantage of Nations* (Free Press, New York, 1990)

[19] OECD, *Industrial Structure Statistics* (OECD, Paris, 1996)

[20] European Commission, *European Competitiveness Report 2001* (Office for Official Publications of the European Communities, Luxembourg, 2001)

[21] OECD, *Globalisation and Competitiveness: Relevant Indicators* (OECD, Paris, 1996)

[22] Z. Wysokninska, *Competitiveness in international and global technology trade* (PWN, Warsaw, 2001)

[23] C. Bianco, Cent. Est. Cienc. Desarr. Educ. Sup. **31** (2007)

[24] A. Smith, *The wealth of nations* (The Modern Library, New York, 1776)

[25] D. Ricardo, *The principles of political economy and taxation* (Everyman's Library, London, 1817)

[26] E. Henckscher, B. Ohlin, *Heckscher-Ohlin trade theory* (Cambridge MA; The MIT Press, Cambridge, 1991)

[27] M. Posner, Oxf. Econ. Pap. **13**(3), 323–341 (1961)

[28] R. Vernon, Quart. Journ. Econ. **80**(2), 190–207 (1966)

[29] I. Tsakok, *Agricultural price policy. A practitioner's guide to partial equilibrium analysis* (London Cornell University Press, London, 1990)

[30] B. Balassa, Manch. Sch. Econ. Soc. Stud. **33**, 99–123 (1965)

[31] K. Laursen, *Revealed Comparative Advantage and the alternatives as measures of international specialisation* (Danish Research Unit for Industrial Dynamics, Copenhagen, 1998)

[32] D. Archibugui, M. Pianta, *The technological specialization of advanced countries. A report to the EEC on international science and technology activities* (Kluwer Academic Publishers, Dortrecht, 1992)

[33] A. Hoen, J. Oosterhaven, Ann. Reg. Scien. **40**, 677–691 (2006)

[34] D. Greenaway, C. Milner, *Trade and industrial policy in developing countries: a manual of policy analysis* (The University of Michigan Press, Michigan, 1993)

[35] T. Vollrath, Weltw. Arc. **127**, 264–280 (1991)

[36] T. Vollrath, *Revealed competitive advantage for wheat* (Economic Research Service, Illinois, 1987)

[37] M. Michaely, *Trade, income levels, and dependence* (North Holland, Amsterdam, 1984)

[38] L. Iapardre, IAER. **7**(2), 173–184 (2001)

[39] G. Lafay, *The measurement of revealed comparative advantages* (Chapman & Hall, London, 1992)

[40] R. Ballance, H. Forstner, T. Murray, Rev. Econ. Stat. **69**(1), 157–161 (1987)

[41] E. Depetris, R. García, G. Rossini, Rev. Met. Cuant. Econ. Empr. **9**, 85–105 (2010)

BIO Web of Conferences 9, 03002 (2017) DOI: 10.1051/bioconf/20170903002
40th World Congress of Vine and Wine

Forecasting and evaluation for raisin export in turkey

Hülya Uysal and Selçuk Karabat

Viticulture Research Institute, Manisa, Turkey

Abstract. Turkey is the world's second largest raisin producer after the United States and the leading exporter, accounting to 25% of the world's total raisin production and 40–45% of total raisin exports. The European Union is the top export destination for Turkish raisins receiving an average of 80 percent of all Turkish raisin exports. The main objective of the study is to forecast Turkey's raisin export. A time series modeling approach (Double Exponential Smoothing) has been used to forecast raisin export in Turkey. The forecasts for the export of raisin in Turkey have been made for the next five years using by annual data on 1982–2015 years in this study. The forecast results have shown that the raisin export in Turkey will be around 209,738 tons in 2016.

1. Introduction

Geographically, Turkey has one of the most favorable areas for viticulture. According to FAO 2014 statistics', Turkey has the 6th-largest area of land devoted to vineyards, with 467,000 hectares, and it is the 6th-largest world grape producer with 4.2 million tons annually [1].

Drying grapes make up 35%, table grapes make up approximately 30% of the production, grapes for traditional products such as molasses make up 25% and grapes used for alcoholic beverage production make up approximately 10% of this total. The raisin growing area is mainly located in the Aegean region, especially in Manisa, Turgutlu, Salihli, Akhisar, Menemen, Kemalpaşa, Çal and Çivril [2].

The most important grape variety grown in Turkey is Sultana. This grape is the main variety for raisins. Sultana is a white, thin skinned grape, which produces the best raisins available today. Its small berries are oval and elongated. It does not contain seeds and has high sugar content. One kilogram of raisins requires about four to five kilograms of fresh grapes.

Turkey is the second largest producer of raisin in the world. The global raisin production was 1,139,000 tons for the 2015/16 period. During 2015/16, Turkey's raisin supplies dropped drastically due to damages to the fresh crop caused by severe frost and hail. Especially the late spring frost damaged in April 23 and 24. Turkish sultana raisin production is decreased in 2015/16 to 196,109 tons [3].

Raisin has traded domestic and export markets in Turkey. Despite's Turkey's leading role in the production of raisins, domestic consumption is comparatively low, and most of the production (generally around 80 percent) is exported. Mainly raisins are consumed as a snack food in Turkey. Human consumption is around 20–25,000 MT and the industry consumption (for cakes, biscuits and snacks) is about the same amount which account to a total of 45,000–50,000 MT. The reason for low domestic consumption is that grapes are mostly consumed as fresh in the season and there are so many grape alternatives grown throughout the country. Raisins are only consumed when there are no fresh grapes around [3].

This paper aims to forecast Turkey's raisin export by using the time series modeling approach (Double Exponential Smoothing). Time series models use the past movements of variables in order to predict their future values.

Various models such as regression models Auto regressive integrated moving average (ARIMA) model introduced by Box and Jenkins, (1976) and exponential smoothing methods have been used in the literature to forecast time series data. Double exponential smoothing model is a popular tool in short-term forecasting [4]. Bowerman and O'Connell (1993) also suggested this technique for forecasting [5]. Aydoğan used double exponential smoothing method for forecasting dry beans production in Turkey [6]. Chiken's egg production in Turkey by Çiçekgil and Yazıcıhad been forecasted using time-series modeling techniques Double Exponential Smoothing and ARIMA [7]. Rani and Raza used Double Exponential Smoothing method for forecasting price estimation of major pulses in Pakistan [8]. Another study by Marinah and Abdullah used Double Exponential Smoothing method for forecasting the yearly total paddy production of Kelantan based on the data collected for the period of 1970–2009 [9].

Forecasting provides information for decision makers and it will help in developing the strategic plan. This study provides information for strategic planners, international executives and export/import managers who are concerned with the market for raisins. Forecasting raisin export is important as it is the base for creating achievable foreign trade policy, creating adequate regional development policy, formulating and implementing raisin production strategy etc.

This paper is organized as follows. In the next section, the methodology used for this study and the data are described. The results and discussion are presented

DOI: 10.1051/bioconf/20170903002

Table 1. Typical MAPE Values for Model Evaluation.

MAPE (%)	Evaluation
MAPE ≤ 10%	High Accuracy Forecasting
10% <MAPE ≤ 20%	Good Forecasting
20% <MAPE ≤ 50%	Reasonable Forecasting
MAPE >50%	Inaccurate Forecasting

in following section. Finally, concluding remarks are highlighted.

2. Materials and methods

This study is based on the secondary data to forecast the exports volume of raisin. The data used in this study were obtained from the Turkish Statistical Institute (TIS), FAO and the Ministry of Food Agriculture and Livestock records. The data were used in the analysis includes quantity of export between 1982 and 2015. Export volumes of raisin for five years (2016–2020) have been estimated using Double exponential smoothing method.

The observations' normality was investigated using the Ryan-Jonier (RJ Similar to Sapiro-Wilk) test.

The accuracy of forecast is evaluated based on the estimation of error, thus the smaller the value of MAE, RMSE and MAPE, the better the forecast is. The criterion of MAPE is the decisive factor because it is expressed in easy generic percentage term. Table 1 shows the criteria of MAPE for model evaluation based on Lewis [10, 11]

Double Exponential Smoothing Method

Exponential smoothing is probably the most widely used class of procedures for a wide variety of time series data in order to forecast the future. It weights past observations using exponentially decreasing weights. In other words, recent observations are given relatively more weight in forecasting than the older observations and these weights are automatically calculated by use of smoothing constants. There is no need to assign weights to each previous period.

In exponential smoothing, there are one or more parameters to be determined by the forecaster. These parameters assign the weights which are exponentially decreasing weights as the observations getting older. This is a desired situation because "future events usually depend more on recent data than on data from a long time ago" [12, 13]. This gives the power of adjusting an early forecast with the latest observation. The reason for this is that the future is more dependent upon the recent past than on the distant past [14].

The Double Exponential Smoothing model is a refinement of the popular single exponential smoothing model that includes another component, which takes into account any trend in the data. When, there is an increasing or decreasing trend in the data over time, single exponential smoothing forecasts tend to lag behind observations. Double exponential smoothing is designed to deal with such data by taking their trends into account. The double exponential smoothing model for a time series (yt) is given by the model equation:

$$yt = \mu_t + \beta_t t + \varepsilon_t \quad (1)$$

Table 2. Raisin Exports Amount from Selected Countries (tons) (2012–2016).

	2012	2013	2014	2015	2016
Turkey	247,200	186,900	259,900	201,800	220,000
USA	123,899	159,389	127,279	114,393	125,000
Iran	149,700	130,700	102,200	112,700	120,000
Uzbekist.	20,000	15,700	49,400	67,800	70,000
Chile	67,150	66,000	62,000	67,000	60,000
S. Africa	31,900	35,050	52,900	45,000	45,000
Argentina	29,050	16,300	29,700	35,000	35,000
Other	65,300	78,700	70,000	78,400	74,000
Total	734,199	688,739	753,379	722,093	749,000

Source: [16].

The smoothing equations are,

$$Ft = \alpha y_t + (1 - \alpha)F_{t-1} \quad (2)$$

$$F_t' = \alpha(F_t - F_{t-1}) + (1 - \alpha)F_{t-1}' \quad (3)$$

where μt represents the time-varying mean (level) term, β_t represents the time varying slope term (called "trend" term by exponential smoothers), ε_t is a white noise error term, F_t is the single smoothed series, and F_t' is the double smoothed series. The symbol α stands for the smoothing parameter, between $0 < \alpha \leq 1$.The forecasts from double smoothing are computed using the following formula.

$$\hat{y}_{(t+h)} = F_t + F_t'[(h - 1) + 1/\alpha)]. \quad (4)$$

3. Results and discussion

3.1. Raisin exports

Today dried fruit consumption is widespread. Nearly half of the dried fruits sold are raisins. Raisins may be eaten snack, raw or used in cooking, baking [15]. The global raisin export was 749,000 tons in 2016 [16]. The largest amount of raisin is exported by Turkey. Turkey with 220,000 tons has the highest quantity followed by USA with 125,000 and Iran with 120,000 tons. After these countries, Uzbekistan, Chile, and South Africa are major countries that have the largest quantity of raisins export in the world (Table 2).

Raisin is one of the traditional agricultural export items of Turkey, which exports to approximately 100 countries on five continents depending on changing market conditions. EU countries continue to be the top export destinations for Turkish raisins, accounting for 80 percent of Turkey's raisin exports. The United Kingdom, Germany and the Netherlands continue to be the top three importers of Turkish raisins. Other important raisin import markets from Turkey included Russia, Canada and Japan [3].

3.2. Forecasting exports of raisin, using Double Exponential Smoothing Method

Firstly; time series plot was created to determine the trend in the export of raisin from 1982 to 2015 (Fig. 1). Turkey's raisin export volume has increased 2.4 times from 1982 85,000 tons to 202,000 tons in 2015. The exports volume of raisin show an upward trend from 1982 to 2012 but sudden decline is apparent in 2013. This year raisin

DOI: 10.1051/bioconf/20170903002

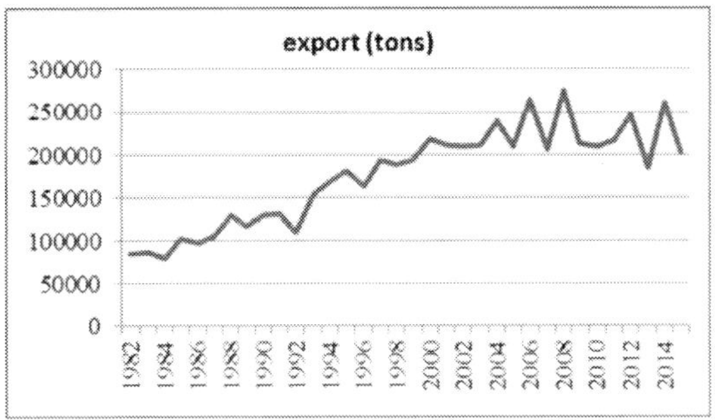

Figure 1. Time series plot of raisin export in Turkey.

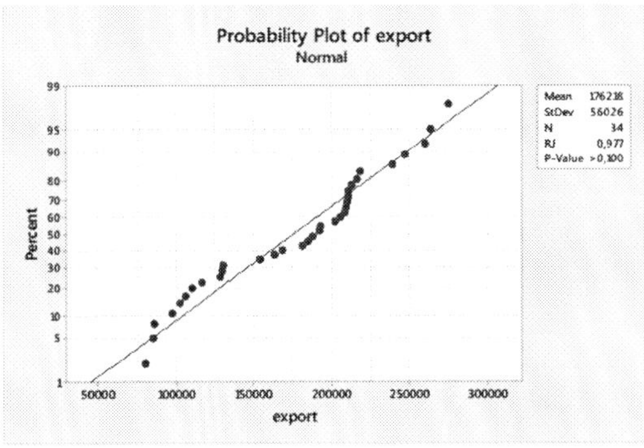

Figure 2. Normality Test results for raisin export.

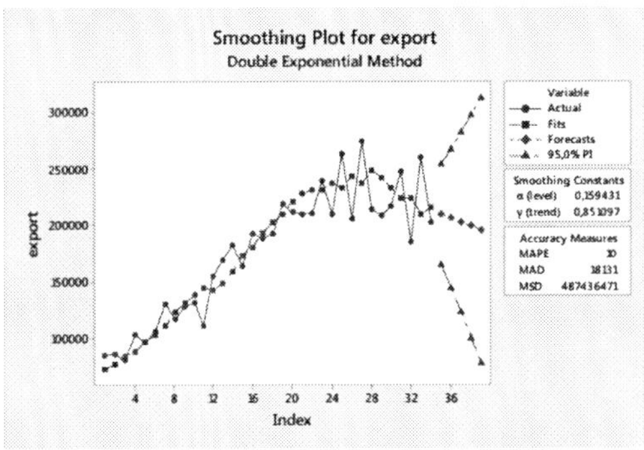

Figure 3. Double Exponential Smoothing forecasting results.

exports decreased to 185,000 tons due to the decrease in production (cold weather and freeze damage on the yields). Export rebound to 260,000 tons in 2014. After a good season in 2014 the freeze and hail in 2015 that resulted in a 22.3% decrease in exports to 202,000 tons in 2015.

The results of the normality tests of series are presented in Fig. 2. The results of the series normality test indicate that the p-value of the series is not significant at the 5% level. As a result, the null hypothesis that the time series have a normal distribution cannot be rejected. Therefore, the raisin export series can be assumed to have a normal distribution as shown in Fig. 2.

The graph of forecasted data which had been calculated using Double exponential smoothing are shown in Fig. 3.

The results of raisin export with 95% forecast interval showed that if the present growth remains the same then

Table 3. Five years 95% forecast of raisin export in Turkey using Double exponential smoothing model.

Period	Forecast	Lower	Upper
2016	209738	165317	254159
2017	206127	144573	267681
2018	202516	122658	282373
2019	198905	100222	297587
2020	195294	77515	313072

the export of raisin will be 209,738 tons in 2016 (Table 3). As shown in the Table 3, the value predicted using in model is near the actual (observed) values for 2016 to 220.000.

It is obvious from the analysis that export of raisin will decline in the coming years. Based on this model, raisin export in the following five years will decrease for approximately 3611 tons.

4. Conclusions

Turkey has good ecological conditions for growing raisin grapes. Despite's Turkey's leading role in the production of raisins, domestic consumption is comparatively low, and most of the production (generally around 80 percent) is exported. Raisin has the largest share in the traditional Turkish agricultural export items. It has strategic geographical position offers significant opportunities for trade.

This paper provided a short-term estimation of raisin export for foreign trade. Double exponential smoothing model was used for this purpose. Time series data of 34 years (1982–2015) were used in this study. The analysis of data showed that Turkey's raisin export is decreasing trend in export market. The forecast results have shown that the raisin export in Turkey will be around 209,738 tons in 2016/17. A forecasted value of dates also shows the situation that 2016 to 2019 Turkey will continue to acting important roles in raisin export in the world.

References

[1] FAO, Food and Agricultural Organization, Statistics Division, 2014, `http://www.fao.org` (2014)
[2] TUİK, (Turkish Statistical Institute), Agricultural Production Statistics (`www.tuik.gov.tr`) (2015)
[3] Gain Report, Turkey Annual Raisin Report, (2015) `https://gain.fas.usda.gov/Recent%20GAIN%20Publications/Raisin%20Annual_Ankara_Turkey_8-12-2015.pdf`
[4] R.G. Brown, *Statistical forecasting for inventory control*. New York, McGraw-Hill, Canadian Prairies, Agron. J. **92**(6), 1047–1053 (1959)
[5] Bowerman, B.L. and R.T. O'Connell, *Forecasting and Time Series: An Applied Approach, Solutions Manual*. 3rd edition. Belmont, CA Duxbury Press, p. 526 (1993)
[6] M, Aydoğan, K.Demiryürek, N.İ. Abacı, Turkish Journal of Agriculture-Food Science and Technology **3**(12), 962–968 (2015)
[7] Z. Çiçekgil, E. Yazıcı, TEAD, **2**(2), 26–34 (2016)
[8] S. Rani and I. Raza, Pakistan J. Agrie. Res. **25**(3) (2012)

DOI: 10.1051/bioconf/20170903002

[9] M. Muhammed and M.H.H. Abdullah, Academia Journal of Agricultural Research **1**(7), 106–113 (2013)

[10] C. D. Lewis, *International and Business Forecasting Methods.* London: Butterworths (1982)

[11] D. Emang, M. Shitan, A.N.A. Ghani, K. M. Noor, Journal of Sustainable Development **3**(3), September (2010)

[12] M. Xie, G. Y. Hong, and C. Wohlin, *Quality and Reliability Engineering International* **13**(6), 347–353 (1997)

[13] S. Çapar, Dokuz Eylül Üniversitesi Sosyal Bilimler Enstitüsü Dergisi **17**, Sayı:3, sayfa: 291–302 (2015)

[14] R. Hyndman, A. Koehler, J. Ord, R. Snyder, *Forecasting with Exponential Smoothing: The State Space Approach Springer*, 375 p. (2008)

[15] `https://en.wikipedia.org/wiki/Dried_fruit` (2017)

[16] `https://apps.fas.usda.gov/psdonline/circulars/raisins.pdf` (2017)

DOI: 10.1051/bioconf/20170903003

Las Denominaciones de Origen Protegidas españolas de vino en los mercados globales / Spanish wine Protected Designations of Origin in global markets

Samuel Esteban Rodríguez

Grupo de Estudios de Desarrollo Territorial de la Universidad de Zaragoza, España

Resumen. España es uno de los principales exportadores de vino del mundo y opera simultáneamente en diversos mercados, sujetos a modelos de competencia diferentes: precio y calidad. En este trabajo se analiza el segmento de vinos amparados bajo la figura de la Denominación de Origen Protegida (DOP), pese a jugar en la actualidad un papel secundario en las exportaciones de vino español. Salvo Jerez, las DOP con un mayor volumen exportado se encuentran en el valle del Ebro, Cataluña, Levante y el centro del país. El vino de calidad español se dirige primordialmente al mercado europeo, especialmente Reino Unido y Alemania. Fuera de Europa, destaca el mercado norteamericano. Existen diferentes modelos entre las DOP españolas, el precio medio del vino exportado es un elemento importante para su identificación. Las denominaciones con precios medios más altos comercializan vino en mercados internacionales donde la nueva cultura del vino se encuentra más asentada; las DOP que comercializan vinos a unos precios más moderados tienen mayor peso en mercados emergentes.

Abstract. Spain is one of the main exporters of wine in the world and operates simultaneously in different markets, subject to different competition models: price and quality. This paper analyzes the protected wines segment under the name of Protected Designation of Origin (PDO), despite playing a secondary role in Spanish wine exports. The PDOs with a higher exported volume are found in the Ebro Valley, Catalonia, Levante, Sherry, and the center of the country. The Spanish quality wine is aimed primarily at the European market, especially the United Kingdom and Germany. Outside Europe, the North American market stands out. There are different models among Spanish PDOs, the average price of exported wine is an important element for their identification. High priced PDOs commercialize wine in international markets where the new wine culture is more settled; The PDOs that market wines at more moderate prices have greater weight in emerging markets.

1. Introducción

En las tres últimas décadas, en España la vitivinicultura ha experimentado trasformaciones en lo que respecta a mercados y procedimientos de elaboración. El desarrollo de la "nueva cultura del vino" ha llevado a éste de su consideración tradicional, como producto alimenticio, a otra centrada en su papel como fuente de placer sensorial. Este cambio ha influido en la reducción del consumo per cápita, pasando entre 1995 y 2012 de 47 a 25 kilos de vino por persona y año [1].

Ligados a esta evolución, se observan dos procesos conectados: por una parte, el desarrollo de la exportación para compensar la reducción del mercado interior; por otra, la mejora de la calidad, para satisfacer las demandas de los consumidores de esta nueva cultura del vino.

Respecto a la primera cuestión, cabe afirmar, que el vino desde tiempos remotos ha sido un importante producto de exportación [2]. Sin embargo, en los últimos años, España ha pasado a ser el primer exportador mundial de vino [3], pudiendo presentarse en este sentido como un caso de éxito. En la campaña de 2014–15, la balanza comercial del sector es claramente positiva, con un saldo entre volumen exportado e importado superior a los 24 millones de hectolitros [4].

En segundo lugar, en lo que concierne a la mejora de la calidad, en España se ha producido un importante aumento del número de Denominaciones de Origen Protegidas (DOP): prácticamente se ha triplicado su número desde la entrada en la Unión Europea (UE) hasta la actualidad. El mercado interior de vinos de calidad con denominación de origen, entre 2000 y 2014, se mantiene en torno a los 6,5 millones de hectolitros [5]. Esta cifra apunta, de nuevo, a cambios en el mercado, pues pese a haber pasado el consumo total de vino dentro de España de 15,34 a 9,85 millones de hectolitros entre 1995 y 2012 [1], el consumo total en el segmento de los vinos de calidad es bastante estable [5]. Entre estos últimos, las fluctuaciones se asocian a otros factores, registrando un mínimo de 5,9 millones de hectolitros en 2008-09, coincidiendo con los primeros años de crisis económica.

A raíz de estos cambios, la estructura del sector se ha reconfigurado, presentando cierta dualidad. Aunque los vinos de calidad, DOP e Indicaciones Geográficas Protegidas (IGP), han ido ganando terreno, se observa, por parte de la industria española, una especialización en la exportación de vino a precios asequibles [6]. En la campaña vitivinícola de 2014–15, el volumen total exportado fue de 24,5 millones de hectolitros [4], sin embargo, fue a un precio medio bajo; por ello, en términos de valor económico del vino comercializado, España continúa detrás de Francia e Italia [3].

Así mismo, analizando los precios medios de las exportaciones de los vinos con DOP, se observa que la ratio de valor económico por volumen comercializado es mayor en el mercado interior que en el exterior [5]. Con estos datos, se deduce, de nuevo, que el vino Español tiende a competir predominantemente en los mercados internacionales en precios. Sin embargo, se encuentran numerosas excepciones, especialmente DOP concretas: casos como Rioja, Ribera del Duero, Priorato o la mayor parte de los pagos vitícolas se encuentran orientados hacia la calidad también en lo que respecta a exportaciones, con precios por encima del promedio.

En consecuencia, cabe afirmar que el sector del vino español opera en múltiples tipos de mercado, observándose una primera diferenciación entre vinos de calidad y el resto. Pese a ello, en los primeros, cuando se comparan regiones productoras, también se pone de relieve la presencia de diferentes modelos: aunque las DOP son el reflejo de la búsqueda de una calidad genuina, vinculada a las características ambientales y sociales del territorio, la figura acoge realidades diversas.

Por un lado, se cuenta con unidades que comercializan pequeños volúmenes de vino; éstas tienden a estar compuestas por un reducido número de viticultores y de bodegas. En el otro extremo se encuentran DOP que llegan a abarcar la extensión de varias provincias y que se encuentran integradas por un elevado número de actores; en éstas últimas el volumen comercializado es muy alto. Cuando se analizan los agregados de precio medio y otros indicadores de competencia, se observan importantes diferencias ligadas a la tipología de cada DOP [7] y también en lo que respecta a destino del vino calificado.

En la campaña vitivinícola de 2014–15 se exporta un 41,20% del vino con DOP español [8]. Sin embargo, algunas regiones productoras, como Jerez o Cava, se encuentran por encima de este valor de referencia; mientras otras, como Penedés o La Mancha, no alcanzan el promedio. El peso de la exportación en cada unidad y el volumen que comercializa en cada destino, permite trazar los flujos comerciales de cada DOP e identificar aquellos de mayor relevancia para el conjunto.

En este trabajo, de forma escalonada, se analiza el papel de las denominaciones de origen protegidas en la exportación de vino español, en el contexto de la nueva cultura del vino. Para ello, en primer lugar, se describe la estructura del conjunto del sector y, posteriormente, se analiza el segmento de los vinos DOP, realizando una comparación interregional y una identificación de flujos de comercio internacional.

2. Marco de referencia: Las exportaciones de vino en España

La evolución del volumen de vino exportado por España entre 1995 y 2012, sigue una tendencia lineal con una pendiente ascendente muy acusada. En la serie de datos sobre exportaciones de vino español de la Organización Internacional de la Viña y el Vino (OIV) [9], se observa

Tabla 1. Evolución de las exportaciones de vino.

Año	Miles de hectolitros
1995	6.260
2000	8.651
2005	14.439
2010	17.156
2012	21.411

Fuente OIV [9].

que, frente a los 6,26 millones de hectolitros exportados en 1995, se exportan 21,41 en 2012 (Tabla 1); es decir, el volumen exportado prácticamente se ha multiplicado por 3,5. Esta progresión continúa en la actualidad: atendiendo a los datos publicados por el ministerio de agricultura, el volumen alcanza los 24,58 millones de hectolitros en la campaña de 2014–15 [4]. Tomando como referencia la cifra de comercio mundial de vino a nivel internacional de 104,30 millones de hectolitros, registrada en 2015 [10], España cuenta con una cuota de mercado del 23,57%, siendo el mayor exportador mundial.

El crecimiento de las exportaciones no se ha visto acompañado por un incremento similar del volumen de vino producido. El rendimiento del viñedo depende de un conjunto de factores numeroso, entre los que destacan su extensión y los de carácter ambiental [11,12]. La producción entre 1995 y 2012 se ha multiplicado por 1,49 [13], sin embargo presenta grandes oscilaciones a lo largo de toda la serie anual, especialmente importantes entre campañas. Se puede mencionar, por ejemplo, el aumento de 8,36 millones de hectolitros producido entre 2002 y 2003 o el descenso de 6,83 millones, registrado entre 2004 y 2005. Pese a estas fluctuaciones de la producción, la tendencia general presenta un suave incremento, aunque no alcanza al registrado por las exportaciones.

Comparando los resultados publicados por el ministerio de agricultura en el balance del vino de 2009–10 [14] y 2014–15 [4], se observa como, aunque en estos 5 años el volumen total exportado ha aumentado 8,7 millones de hectolitros, el peso de los vinos de calidad sobre las exportaciones ha disminuido. En 2009–10 el total exportado con DOP o IGP era de 6,2 millones de hectolitros, lo que suponía un 39,48% de las exportaciones. En la campaña de 2014–15, la suma de los vinos de calidad ha pasado a ser de 7,7 millones de hectolitros; sin embargo, su peso se ha reducido hasta suponer el 31,45% del total [4,14]. A su vez, dentro de los vinos de calidad el porcentaje de vinos DOP, figura de mayor exigencia, también ha disminuido.

La presencia vinos poco diferenciados incide en el bajo precio medio de las exportaciones de vino español. A este fenómeno se añade la venta de vinos a granel, que supone el 45,2% de las exportaciones [15], lo que, de nuevo, reduce el precio. A este respecto, se puede mencionar el ejemplo de los vinos con DOP: entre ellos se exporta a granel un 3,54% del total, aunque este tipo de envase únicamente representa el 0,63% de su valor económico total [8].

En el sector del vino, Francia e Italia se encuentran a la cabeza del ranking mundial de valor económico. Su posición aventaja a la del vino español por varios miles de millones de euros. En *State of the Vitivinicultura World Market* de la OIV [3] aparecen publicados los datos de valor económico del vino exportado por país, siendo de 8,24 miles de millones de euros el del vino francés y de 5,35 el italiano, frente a los 2,64 del español. Realizando una estimación del precio medio a partir de los datos de la mencionada publicación, se obtiene una ratio por hectolitro de 110€para el vino español, 267 para el italiano y 588 para el francés. Estas diferencias reflejan la existencia de diferentes modelos competitivos a nivel internacional, siendo la competencia en precios predominante en el vino español. Pese a ello, el elevado volumen comercializado compensa en parte los bajos precios, situando a España como tercer país a nivel mundial en cuanto a valor económico del vino exportado.

A nivel global se observa un incremento del comercio internacional de vino, con un aumento de la demanda en mercados como Estados Unidos y China [10]. Sin embargo, para España, como miembro de la UE, la exportación de vino al resto de socios juega un papel primordial, siendo el mercado preferente con un 69% del total de las exportaciones [8]. Los principales flujos comerciales se dirigen a Francia, Alemania e Italia, que suman prácticamente el 50% del vino exportado. Pese a ello, los mencionados países tan solo suman el 27,63% del valor económico de la exportación, debido al alto peso que tiene la venta de vino a granel, especialmente en Francia e Italia.

Los flujos entre países, en cuanto a valor económico del vino, están determinados por el volumen comercializado y el precio. El precio medio depende en gran medida de la proporción de vino a granel, que es más barato, y de la proporción de vinos con DOP, cuyo promedio es superior. El resultado de ello es que son, por este orden, Alemania, Reino Unido, Francia y Estados Unidos, los países donde el valor económico total del vino español comercializado es mayor [15].

En las exportaciones, los vinos de calidad diferenciada españoles, DOP e IGP, han crecido en términos absolutos. Sin embargo, el aumento registrado ha sido inferior al experimentado por el resto de vinos, es por ello que el peso final de estos vinos ha descendido. Las Denominaciones de Origen Protegidas, en cuyo segmento es menor la competencia en precios, juegan un papel secundario en las exportaciones. Este tipo de vinos se encuentra orientado hacia mercados donde el reconocimiento de la calidad juega un papel muy importante. La figura de la DOP fue dinamizada notablemente por la aprobación de la Ley de la Viña y el Vino de 2003, creándose numerosas pequeñas denominaciones y aumentando considerablemente su diversidad.

3. Las exportaciones de vino con DOP

En la actualidad, la regulación de nombres tradicionales, elemento de diversificación dentro del sector, puesta en marcha por la ley de 2003, se encuentra regulada por la ley 6/2015, de 12 de mayo, de Denominaciones de Origen e Indicaciones Geográficas Protegidas de ámbito territorial supraautonómico. Sin embargo, los efectos de la regulación se han hecho notar desde 2003, en cuanto a la presencia de diversas tipologías de DOP; entre las 90 que hay actualmente reconocidas, la distribución por nombres tradicionales de denominación, que siguen un orden jerárquico en cuanto a requisitos, es la siguiente: 6 vinos de calidad (VC), 68 denominaciones de origen (DO), 2 denominación de origen calificada (DOca) y 14 vinos de pago (VP).

Tabla 2. Exportaciones por tipo de DOP, 2014–15.

Tipo	Volumen (miles de hl)	Precio medio (€/hl)	Valor económico (millones de €)
VC	0,64	630,94	0,40
DO	3.687,61	267,09	984,94
DOca	1.088,33	434,74	473,14
VP	2,51	912,55	2,29
Total	4.779,09	305,66	1.460,76

Fuente: Ministerio de Agricultura [8].

Las exportaciones dentro de este abanico de tipologías se concentran en la DO y la DOca, que suman el 99,93% del total. La presencia de VC y, especialmente, VP ha cambiado la composición del sector dentro de España; sin embargo, el tamaño de las denominaciones acogidas a estos nombres tradicionales es tan reducido que entre todas no alcanzan el 1% de la exportación.

Los precios medios del vino exportado son más bajos entre las DO que en el resto (Tabla 2), con todo, aglutinan el 67,43% del valor económico final, seguidas por las DOca que suman el 32,39%, aunque en su caso el precio del vino es superior, con un promedio por encima de los 400€por hectolitro. En España, las altas exigencias en cuanto a requisitos, han derivado en que las únicas denominaciones calificadas sean Rioja y Priorato, aunque realmente el 98,46% de la exportación de las DOca se corresponda de manera exclusiva con la DOca Rioja [8].

En la campaña vitivinícola de 2014–15, la exportación se encuentra muy concentrada en un pequeño número de denominaciones: el 16,67% de las DOP exporta el 90,02% del vino; la distribución es bastante similar a la del total de volumen de vino comercializado con DOP, donde un 22,22% de las unidades comercializa el 90,28% del vino [8]. A la cabeza se encuentran Cava y Rioja, sumando entre ambas el 46,36% de la exportación. En este *ranking*, Cava y Rioja son seguidas por Valencia, Cariñena, Jerez, Cataluña, Valdepeñas y La Mancha. El volumen exportado es muy bajo en las DOP del este de España y las islas, destacando únicamente entre las primeras Ribera del Duero, Rías Baixas y Rueda (Fig. 1).

El mapa de volumen exportado es muy similar al del volumen comercializado, aunque no idéntico; esto es debido a la diferente vocación exportadora observada entre denominaciones: destacan Yecla, Utiel-Requena, Cariñena, Valencia, Almansa, Jerez, Calatayud, Cava y Uclés, todas ellas exportan por encima del 60% del vino que comercializan. Por lo general, el porcentaje de vino exportado es menor en los VC, las DOca y los VP; es digno de mención el caso de Rioja pues, pese al elevado volumen de vino exportado, apenas supone un 38,23% de su comercialización total, siendo el mercado español su principal destino.

Existen importantes diferencias entre denominaciones en lo que respecta a exportaciones, sin embargo, a nivel territorial se observa que en las DOP del este del país tienen mayor importancia, destacando el valle del Ebro, Cataluña y Levante. Junto a las 3 regiones mencionadas puede añadirse el caso de la DO Valdepeñas y el de Jerez,

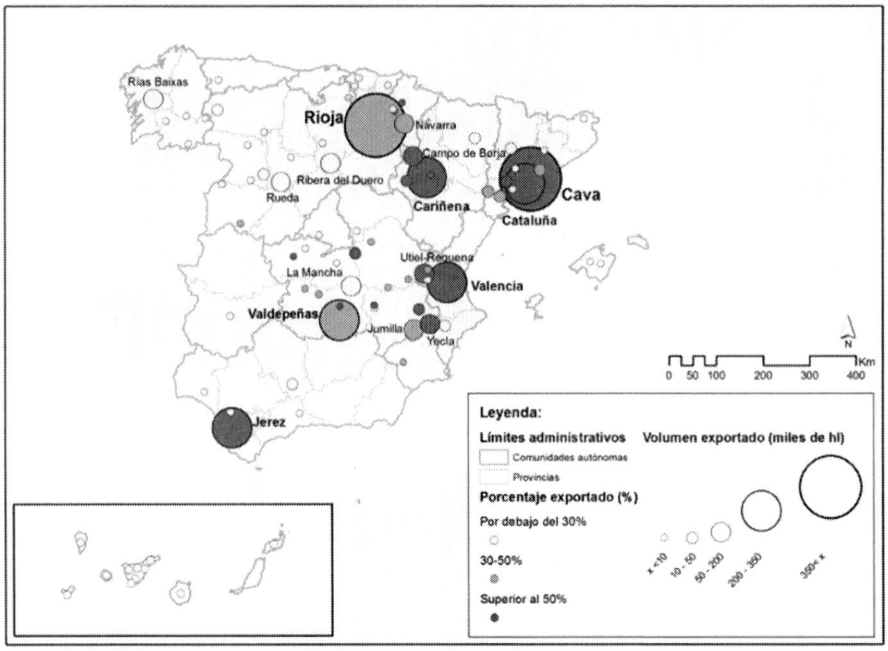

Figure 1. Exportaciones de vino de las DOP, campaña 2014–15. Elaboracin propia a partir de datos del Ministerio de Agricultura [8].

Tabla 3. Exportaciones por tipo de DOP, 2014–15.

Continente	Volumen Exportado (miles de hl)	Porcentaje
Europa	3.520	73,63
América	782	16,35
Asia	418	8,73
Oceanía	24	0,49
Resto del mundo	38	0,79

Fuente: Ministerio de Agricultura [8].

que es una zona con larga tradición exportadora al mercado británico [16]. En el extremo contrario, el caso de las regiones con mayor peso del mercado nacional, destacan las DOP del noroeste de España y las de las islas.

A nivel internacional, en la campaña de 2014–15, el vino de calidad español se dirige en su mayor parte al resto de países de la UE y, de forma secundaria, a los países de Norteamérica, especialmente Estados Unidos. En Asia, China y Japón son los princiales destinos. Europa suma el 73,63% de las exportaciones de vino con DOP; el resto de continentes suma 1,26 millones de hectolitros, es decir, un 26,37% del total exportado, que se reparte fundamentalmente entre América y Asia (Tabla 3, Fig. 2), destacando América, además, por el elevado precio medio del vino que se exporta [15]. En América, el volumen exportado a Estados Unidos es de 456.987 hl, lo que contrasta con los 173.612 hl que suma el conjunto de países iberoamericanos. En Asía destaca China con 207.590 hl [8]; sin embargo, entre las grandes regiones del mundo se sitúa, muy por encima del resto, la UE con un 66,03% de la exportación total.

Dentro de Europa, el de vino de calidad español se dirige fundamentalmente a los mercados de Reino Unido y Alemania, aunque destacan también por sus cifras Bélgica y Holanda. El volumen de vino con DOP procedente de España consumido es mayor en la Europa occidental que en el resto, pese a contar los países del norte y el este de Europa con un consumo incipiente (Fig. 3).

En el sureste europeo, el vino de calidad español tiene una penetración escasa. Descontando la población residente dentro de España, el valor de población de la UE de los 28 a fecha 1 de enero de 2015 era de 462.054.755 personas [17]; poniendo esta cifra en relación con el volumen de vino con DOP exportado a la UE se obtiene una ratio de 0,68 litros per cápita en la campaña de 2014–15. Realizando la misma operación a nivel de país se observa que en países como Bélgica o Suiza el consumo por habitante supera los 2 litros/años y que en Alemania o el Reino Unido supera el litro anual, al igual que en otros países como los bálticos: Lituania, Estonia y Letonia. En Europa, respecto al consumo per cápita, por debajo de la media se encuentran países productores como Francia o Italia, con valores muy bajos, al igual que Austria, Eslovenia, Grecia o Bulgaria. Así mismo, Rusia tiene un consumo por habitante de vino con DOP español escaso, de 0,05 litros por persona y año.

Las denominaciones que suministran la mayor parte del vino a los mercados internacionales son un número reducido, algo similar sucede cuando se analizan los flujos entre DOP y país: solo las ventas de Rioja y Cava a Reino Unido y Alemania suponen el 20,64% del vino con DOP exportado por España. El 50,52% del volumen exportado se encuentra concentrado en 27 flujos DOP-país.

En estos flujos, únicamente se encuentran Estados Unidos, Japón, Suiza y China, como países no comunitarios; las denominaciones que suministran una mayor cantidad de vino a estos mercados son Cava, Rioja, Cariñena y Valencia. Dentro de la UE, se dirigen a Reino Unido, Alemania, Bélgica, Holanda, Francia y

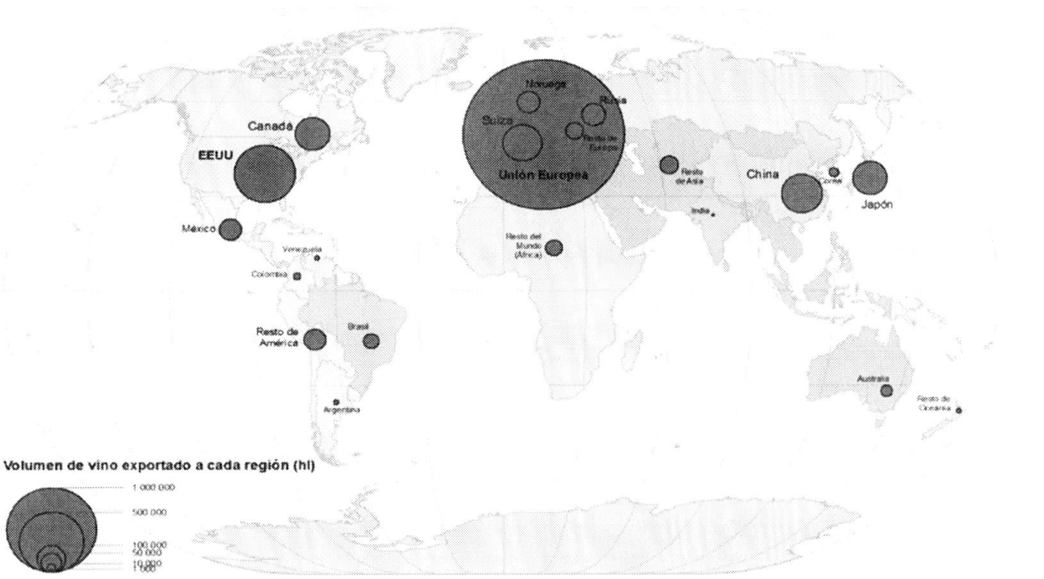

Figure 2. Volumen de vino español con DOP exportado a cada región, campaña 2014–15. Elaboracin propia a partir de datos del Ministerio de Agricultura [8].

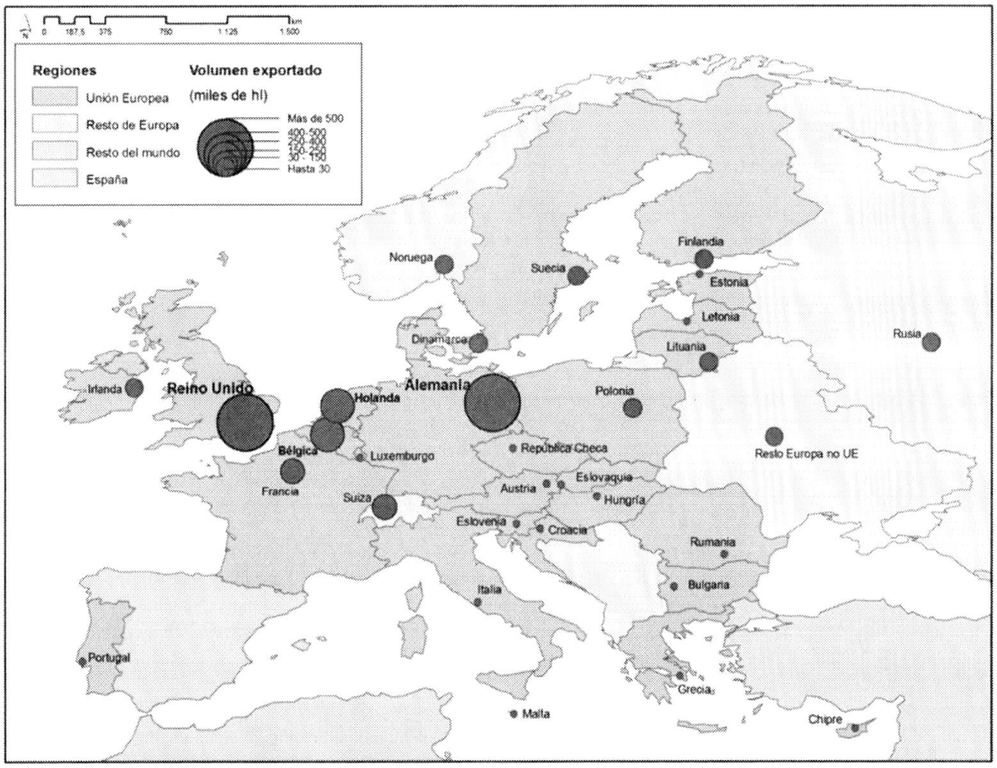

Figure 3. Europa: volumen de vino español con DOP exportado a cada país, campaña 2014–15. Elaboración propia a partir de datos de Ministerio de Agricultura [8].

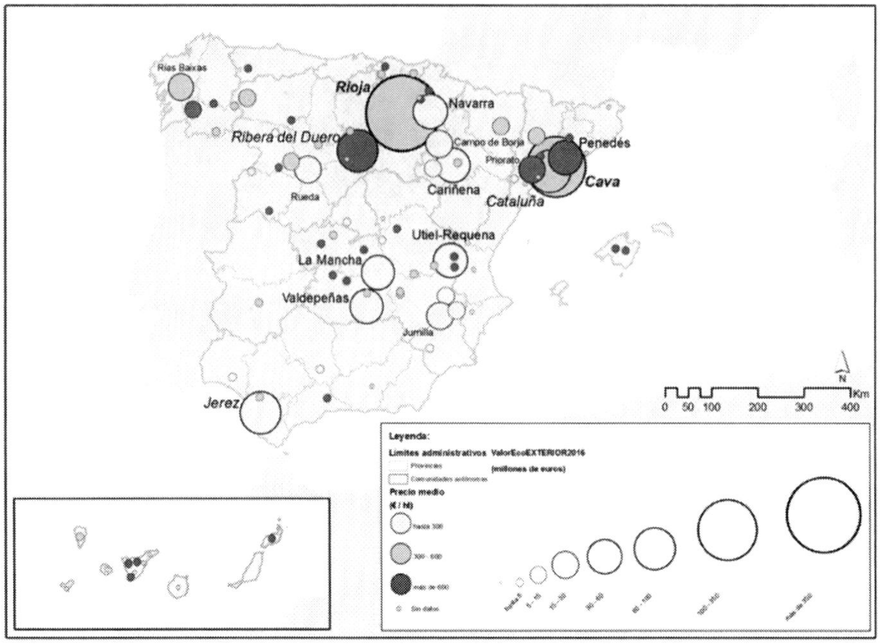

Figure 4. Precio y valor económico del vino exportado por DOP, campaña 2014–15. Elaboración propia a partir de datos de Ministerio de Agricultura [8].

Suecia, teniendo como puntos de partida Rioja, Cava, Jerez, Cariñena, Valencia, Valdepeñas, La Mancha, Utiel-Requena y Cataluña [8].

El precio medio del vino exportado con DOP es de 319 €/hl, frente al precio medio de la exportación de vinos indiferenciados de 97€/hl [15]. Es decir, en los mercados internacionales, el precio medio del vino con DOP es el triple al del total. A su vez, el precio medio del vino no es igual en todas las denominaciones, pues se observan algunas que tienden a competir más en este aspecto y otras que lo hacen en mayor medida en calidad diferenciada; entre estas últimas, el precio medio suele ser más alto. Analizando el volumen exportado, el total es bajo en algunas denominaciones como Ribera del Duero, Penedés o Priorato; sin embargo, combinando volumen exportado y precio se pone de relieve su importancia (Fig. 4).

Las denominaciones con un valor económico total del vino exportado mayor son, de nuevo, Rioja y Cava, a gran distancia del resto; posteriormente se encuentran Cataluña, Jerez, Ribera del Duero, Cariñena, Valdepeñas, Utiel-Requena, La Mancha, Navarra y Penedés. Como puede verse, el orden que mostraban las DOP en cuanto al volumen exportado total se ve sustancialmente alterado, al entrar en juego el precio medio del vino comercializado, perdiendo peso denominaciones como Cariñena o Valdepeñas y subiendo posiciones en el ranking otras como Priorato, Penedés o Rías Baixas. No se dispone para Valencia de datos válidos respecto a valor económico del vino comercializado en la campaña de 2014–15, sin embargo, dado el elevado volumen de vino exportado previsiblemente los valores serán altos también, con una situación similar a otras DOP con características comunes, como por ejemplo Utiel-Requena.

La repercusión económica de la exportación es diferente en función de la región: por un lado, las grandes denominaciones del valle del Ebro, salvo Rioja, las del centro del país y las de Levante y, por otro, las denominaciones que cuentan con precios medios por encima del promedio, Rioja, Cava, Ribera del Duero o Penedés. En el último grupo las cuentas son mayores que en el primero, lo que supone que gran parte del país obtenga, ante niveles similares de vino exportado, un valor económico total menor; esto remarca la importancia de una orientación hacia la calidad. En el segmento de las DOP, que por naturaleza se trata de vinos diferenciados, es muy importante el papel de la calidad, para poder maximizar estos resultados; en este punto puede plantearse la cuestión de si la imagen de los vinos españoles puede estar siendo afectada por los bajos precios medios, pues se relacionan con una menor calidad. Un precio medio bajo permite poder acceder a los grandes mercados indiferenciados de cada país, pero es necesario tener en cuenta que en este tipo de mercados el papel de la competencia en precios es mucho mayor, con un promedio de 95 €/hl para el vino embotellado de origen español indiferenciado [15].

En la campaña de 2014–15, en los mercados internacionales analizados se observa que las DOP que exportan a un precio medio superior a 300 €/hl venden una mayor proporción de vino en Reino Unido, Alemania, Bélgica y EEUU; por el contrario, las DOP que venden a precios más bajos tienen mayor importancia en mercados emergentes como el chino [8]. Este hecho plantea la posibilidad de que el bajo precio medio del vino de algunas denominaciones esté abriendo nuevos mercados al sector de los vinos de calidad españoles. La dualidad que se observa entre las DOP respecto al precio medio, plantea la existencia de diferentes modelos y estrategias comerciales,

 DOI: 10.1051/bioconf/20170903003

con unas denominaciones enfocadas a mercados más elitistas, con una cultura del vino evolucionada, y otras orientas a mercados más indiferenciados y emergentes.

4. Conclusiones

Las exportaciones de vino español han aumentado en los últimos años, tanto en lo que se refiere a vinos indiferenciados como a vinos de calidad. En el segmento de los vinos con denominación de origen el crecimiento entre 2000–01 y 2014–15 ha sido del 29,05%, sin embargo el porcentaje que suponen los vinos de calidad sobre el total exportado se ha visto reducido, pasando en 5 años del 39,48% en 2009–10 al 31,45% en 2014–15. El vino español indiferenciado compite en precios en los mercados internacionales, con un precio medio muy bajo, muy inferior al registrado por el vino de otros países como Francia e Italia. Estos vinos fundamentalmente se exportan a la UE, especialmente a Francia, Alemania e Italia. Así mismo, el comercio con Reino Unido y Estados Unidos también juega un papel muy importante.

El precio medio de los vinos con DOP es superior al promedio de los vinos exportados, sin embargo las Denominaciones de Origen Protegidas juegan un papel secundario en las exportaciones, entre ellas destacan las DO y las DOca. Las DOP con un mayor volumen exportado se encuentran en el valle del Ebro, Cataluña, Levante, Jerez y el centro del país; el resto de áreas cuentan con un peso de la exportación menor. Tanto en cuanto a volumen, como a valor económico de la exportación, destacan Rioja, Cava, Valencia, Cariñena, Jerez, Cataluña, Valdepeñas y La Mancha. A estas denominaciones cabe añadir Utiel-Requena, Penedés, Ribera del Duero, Priorato y Rías Baixas si se considera el valor económico total de las exportaciones.

El mercado al que se dirige un mayor volumen de vino con DOP español es el europeo, especialmente Reino Unido y Alemania, llamando también la atención la penetración del sector en países como Bélgica, Suiza, Dinamarca y Holanda. Es llamativo el contraste que presentan Francia e Italia, ambos países productores de vino, respecto a la importación de vinos indiferenciados y vinos de calidad de origen español, siendo muy alta entre los primeros y muy baja en los segundos. En Europa se observan 3 grupos de países respecto a la importación de vinos de calidad españoles: Por un lado la Europa occidental, que importa una cantidad elevada de vino, un mercado incipiente en la Europa del este y una importación muy reducida en los países del sureste europeo. En América el mercado de los vinos de calidad españoles es escaso en Iberoamérica, concentrándose en el norte del continente: Estados Unidos y Canadá. En Asia destacan el mercado chino y el japonés. El comercio con el resto de países del mundo es bajo.

Se observa un modelo diferente entre los vinos indiferenciados y los vinos con DOP en los mercados internacionales, los primeros tienden a competir en precios y entre los segundos el peso de este modelo competitivo es menor, debido a la calidad diferenciada que protege la propia figura de protección. Sin embargo, entre denominaciones también se observan diferentes estrategias: Las denominaciones con precios medios más altos comercializan vino en mercados internacionales donde la nueva cultura del vino se encuentra más asentada, por el contrario, las DOP que comercializan vinos a unos precios más moderados tienen mayor peso en mercados emergentes como el chino. Entre las primeras destacan Cava, Rioja o Ribera del Duero y entre las segundas se puede citar Valencia, La Mancha o Valdepeñas.

Referencias

[1] Organización Internacional de la Viña y el Vino: *estadísticas de consumo de vino en España, 1995–2012*. Disponible en: `http://www.oiv.int/es/bases-de-datos-y-estadisticas/estadisticas` (consulta 1 de enero de 2017)

[2] J.M. Blázquez, *Hispania: el legado de Roma, en el año de Trajano* "Los productos de la tierra en Hispania". Zaragoza. pp. 95–102 (1998)

[3] Organización Internacional de la Viña y el Vino: *State of the Vitiviniculture World Market*. Abril de 2016 Disponible en: `http://www.oiv.int/public/medias/4710/oiv-noteconjmars2016-en.pdf`> (consulta 6 de marzo de 2017)

[4] Ministerio de Agricultura, Alimentación y Medio Ambiente: *balance del vino 2014–15*. Disponible en: `http://www.mapama.gob.es/es/estadistica/temas/estadisticas-agrarias/agricultura/balance-del-vino/` (consulta 1 de abril de 2017)

[5] Ministerio de Agricultura, Alimentación y Medio Ambiente: *Serie de cifras y datos de las DOP 2000–2014*. Disponible en: `http://www.mapama.gob.es/es/alimentacion/temas/calidad-agroalimentaria/calidad-diferenciada/dop/htm/cifrasydatos.aspx` (consulta 22 de abril de 2017)

[6] J. Martínez, F. Medina. *La competitividad internacional de la industria vinícola española durante la globalización del vino*. Revista de Historia Industrial **22** (52), 139–174 (2013)

[7] S. Esteban, E. Climent, *Los 'mundos de producción' del vino en España: un enfoque cuantitativo aplicado a las denominaciones de origen protegidas*, en de la Riva, J. et al. (eds.). Zaragoza, Universidad de Zaragoza-AGE, 1995–2004 (2015)

[8] Ministerio de Agricultura, Alimentación y Medio Ambiente: cifras y datos de las DOP 2014–15 (2016). Disponible en: `http://www.mapama.gob.es/es/alimentacion/temas/calidad-agroalimentaria/informedops2014-2015-web_tcm7-428808.pdf` (Consulta 22 de abril de 2017)

[9] Organización Internacional de la Viña y el Vino: *estadísticas de exportaciones de España, 1995–2012*. Disponible en: `http://www.oiv.int/es/bases-de-datos-y-estadisticas/estadisticas` (Consulta 20 de abril de 2017)

[10] Organización Internacional de la Viña y el Vino: *World Vitiviniculture Situation, 2016*. Disponible en: `http://www.oiv.int/public/medias/5029/world-vitiviniculture-situation-2016.pdf` (Consulta 20 de abril de 2017)

[11] E. Barco, M.C. Navarro. *Transformación y cambio estructural en la D.O.Ca. Rioja (1983–2005)*. Zubía **16–17**, 121–138 (2005)

[12] C. Rodríguez. *La denominación de origen Valdepeñas en el último tercio del s. XX. Aspectos*

económicos. Cuadernos de estudios manchegos **31**, 267–296 (2007)

[13] Organización Internacional de la Viña y el Vino: *estadísticas de producción de vino de España, 1995-2012* Disponible en: `http://www.oiv.int/es/bases-de-datos-y-estadisticas/estadisticas` (Consulta 20 de abril de 2017)

[14] Ministerio de Agricultura, Alimentación y Medio Ambiente: *balance del vino 2009-10*. Disponible en: `http://www.mapama.gob.es/es/estadistica/temas/estadisticas-agrarias/agricultura/balance-del-vino/` (consulta 1 de abril de 2017)

[15] Ministerio de Agricultura, Alimentación y Medio Ambiente: Dirección General de Producciones y Mercados Subdirección General de Frutas y Hortalizas, Aceite de Oliva y Vitivinicultura: Situación competitiva de los vinos españoles en el mundo y previsiones de evolución (2016) Disponible en: `http://www.mapama.gob.es/es/agricultura/temas/producciones-agricolas/160609analisisexportacionesespanolasdevino_tcm7-424093.pdf` (consulta 18 de abril de 2017)

[16] J. Maldonado, *Cambios de consumo y de gusto de los vinos de Jerez en el Reino Unido y sus consecuencias en la zona de producción entre mediados de los siglos XVIII y XIX*. Historia contemporánea **48**, 117–140 (2014)

[17] Eurostat: Población a 1 de enero (2016). Disponible en: `http://ec.europa.eu/eurostat/tgm/table.do?tab=table&init=1&plugin=1&language=en&pcode=tps00001` (Consulta 15 de abril de 2017)

 DOI: 10.1051/bioconf/20170903004

Wines of Brazil: From the past to the future

Marieta Ferreira[1] and Valdiney Ferreira[2]

[1] PhD UFF, Executive Director of the FGV Publishing and full professor at UFRJ, Rio de Janeiro, Brasil
[2] Executive Director of Vinisa Projetos and visiting professor of the FGV Management Program Rio de Janeiro, Brasil

Abstract. How was the wine industry born in Brazil? What were the most striking facts of the 20th century? What was the nature of the relationship between the wine cooperatives and the wine industry? What were the impacts on the Brazilian market of the entry and exit of multinationals companies? What are the strategies of the Brazilian wines companies that were established essentially aftter he departure of the multinationals, when it was necessary to face the high cost of the Brazilian production and the strong competition of the imported wines? This research aims to present a panorama of the development of the wine industry in Brazil through the testimony of descendants of Italian immigrants who settled in the Serra Gaúcha at the end of the 19th century to devote themselves to vitiviniculture. To this end, we used the methodology of oral history, which involved, in the first place, a consistent and extensive research on the subject to be treated. To this end, we carried out historical and documentary interviews which made it possible to construct a representative panel and gave rise to an excellent tool for reflection.

1. Introduction

In the final decades of the nineteenth century a strong migratory current left the north of Italy for Brazil. The majority settled in São Paulo. However, between 1875 and 1894a large number went to colonize the Serra Gaúcha mountain range in Rio Grande do Sul, where over 60,000 Italians established themselves. They purchased colonial lots where to survive they practiced subsistence agriculture with primary foodstuffs, such as corn, potatoes, and beans, as well as raising domestic animals such as goats, pigs, cows, and mules, amongst others.

By 1879 they were already making the first wine in the Italian colony. This activity soon came to be the one to which the colonists devoted most efforts because it produced the best results. The possibility of practicing vitiviniculture was an important and decisive factor in the decision of Italian immigrants to settle in the Serra Gaúcha.

1886 records show that some immigrants in Campo dos Bugres (Caxias do Sul) had imported and planted European varieties of grape such as *Barbera, Trebbiano, Traminer, Vernaccia*, starting a movement which would strongly contribute to the development of the nascent Brazilian vitiviniculture.

Among the colonies established, Conde d'Eu (Garibaldi), Dona Isabel (Bento Gonçalves), and Campo dos Bugres (Caxias do Sul) are most noteworthy, due to their decisive importance in the development of vitiviniculture in what would be transformed into the most important Brazilian winegrowing region.

The first Brazilian wineries, cooperatives, and the important crises of the sector

Grape production expanded, generating excess which permitted commercial exchanges between neighbors and traders. By the beginning of the twentieth century some families were able to commercially undertake vitiviniculture. Some examples are Carlos Dreher Filho & Cia (1908), Paulo Salton & Irmãos (1910), and Armando Peterlongo (1913).

In parallel, in 1911–13 there occurred the first attempt to create a cooperative movement, when under the leadership of Stefano Paternó various cooperatives were created and disappeared. In 1929, at the peak of one of the important crises of the wine sector which occurred in 1928, 1935, and 1958, this movement would be returned to.

Blamed for these serious crises were the large harvests which, in an environment with a total lack of control of production associated with low consumption in the internal market, created strong oscillations in the price paid for grapes, the formation of elevated stocks of wines, and an accelerated fall in their rices. Moreover, particularly in the 1935 and 1958 crises, this was aggravated by the emergence of new centers of production in the state.

During the 1928 crisis, with government support wine producers created the *Sindicato Vitivinícola do Rio Grande do Sul* to organize the sector. The following year its commercial arm, *Sociedade Vinícola Riograndense*, was created, amongst whose founding partners were 49 of the largest wine producers in the region. Theirs actions provoked numerous conflicts with grape growers, who in a defensive action between 1929 and 1936 resurrected the cooperative movement. More than 50 cooperatives were created at this moment, including Aurora, Aliança, and Garibaldi.

In the 1940s *Sociedade Vinícola Rio Grandense* innovated by strategically establishing bottlers outside Rio Grande do Sul, creating units in Bahia, Paraná, Rio de Janeiro, and São Paulo, as well as the *Porto do Rio Grande* (RS). At its peak it was responsible for 25% of wine production in the state and was fundamental in the

 DOI: 10.1051/bioconf/20170903004

increased sale of Brazilian wines. Its policy of regulating stocks to avoid shortages or excess supply in the market was important in avoiding disastrous oscillations in prices.

Weakened and pressurized by the serious economic crisis of the 1990s, *Sociedade Vinícola Riograndense* closed its industrial units in Brazil and diversified its activities, looking for business opportunities in other sectors. Part of its large wine-growing property of Granja União, which had been created in 1931, was divided into urban lots due to its proximity with the urban perimeter of Flores da Cunha.

In 1997 there occurred the inevitable dismantling of *Sociedade Vinícola Riograndense* as the company closed down and its trading name sold. The brands it owned like Granja União label were sold to the Cordelier Winery.[1]

The arrival of multinationals and their impact on Brazilian viticulture

In 1973 and 1974 various multinationals effectively entered the Brazilian markets, including Martini & Rossi/De Lantier, Heublein, Provifin/Moët&Chandon, Seagram's do Brasil/Maison Forestier, National Distillers/Almadén.

The entrance of these large companies into the national market considerably improved the quality of Brazilian wines. Using their economic strength they propelled the development of winegrowing in the country through important policies for viticulture, the introduction of new equipment and technology, and innovation strategies for operating in the market including: differentiated prices and bonuses for grape growers who produced higher quality grapes; incentives for the conversion of vineyards from the *trellis* or *pergola* vine training system to *espalier*; the planting of fine European grapes which permitted an improvement in the quality of wine; the introduction of advanced technologies which made the production of superior white wine feasible; and aggressive marketing campaigns for the expansion of the consumer market.

This provoked a strong reaction from Brazilian producers who had no other alternative than to improve the quality of their products. Technological investments were made in industrial installations, in viticulture, and in enology courses for members of the families which owned the wineries. Furthermore, practical enologists began to be replaced by enologists with technical and scientific training.

During the 1970s and 1980s, in the wake of the success of the multinationals, many new wineries producing fine wines emerged: Jota Pe/Perini (1972), Luiz Valduga providing wines for Dreher (1973), Monte Lemos/Dal Pizzol (1974), Wizard/Monte Reale (1975), Courmayer (1976), Provino (1978), Cave de Amadeu/Vinicola Geisse(1979), Don Giovanni (1982), Boscato (1983), Adegas Domecq/AlliedDomecq (1985), Casa Valduga (1985), Giacomin (1985), Dom Cândido (1986), Cavalleri (1987), and Cordelier (1987). Some older wineries, such as Salton (1910), Marco Luigi (1946), and Cave Marson (1970) also began to produce fine wines.

One of the most important was Casa Valduga, currently part of Grupo Famiglia Valduga Co. Founded in 1985, in the 1990s the winery effectively and definitely migrated to the fine wines sector, in a development which placed it among the leaders of the Brazilian market. In 1992 it expanded its activities aimed at enotourism with the establishment of restaurants and *pousadas* (rural guest houses). Following this it constructed the first enotourism complex in Brazil.

The opening of the Brazilian market in the 1990s to imported wines provoked an immediate improvement in the quality of products offered, but at the same time it created difficulties for the permanence of the multinationals. As a result, all departed from Brazil or left the fine wines market. Their departure caused a serious crisis in Brazilian viticulture as they left the large grape producers of the Serra Gaúcha without purchasers. The only multinational which remained was Moet & Chandon.

Another consequence was the emergence of small and mid-sized wineries with the aim of producing fine wines to compete with the imported ones.[2] Various of their owners were small or mid-sized grape growers who had traditionally sold their grapes to the large wineries.

To improve the quality of their wines, the education level of their children was increased by sending them to study enology in Brazil and abroad. The principal strategy adopted was the construction of small wineries, retail shops, and hotels to attract rural tourism, commencing the by now sophisticated enotourism in the Serra Gaúcha.

Outstanding in this group was Vinícola Miolo, which between 1990 and 1998 moved from being a small winery to becoming part of the group of the largest wine producers, transforming itself into the largest producer of fine wines in Brazil. Between 1979 and the creation of the winery in 1990, the Miolo family had already been recognized by the large wineries established in the market, such as Salton, Dreher, Martini & Rossi, Maison Forestier, due to the quality of the grapes it produced. Due to difficulties in grape sales, in 1989 Miolo started to produce its own wine.

In search of modernity

Traditional Brazilian wineries and grape growers who managed to resist the hurricane of multinationals and investors had no other alternative but to fight. Resulting from this was the commencement of investments in viticulture and training in enology for family members who stopped being practical and curious and transformed themselves into professionals with scientific backgrounds, as well as improvements in organization and administration.

[1] Ferreira, Valdiney and Ferreira, Marieta. *Vinhos do Brasil; do passado para o futuro.* Rio de Janeiro, FGV Editora, 2016.

[2] We can cite here: Vinicola Miolo (1990/2006), Vinícola Don Laurindo (1991), Cantina Strapazzon (1992), Vinícola Monte Rosário (1993), Famiglia Tasca (1994), Vinícola Salvati & Sirena (1994), Vallontano Vinhos Nobres (1995), Terrasul (1996), Adega Esplendor (1997- closed), Lovara Vinhas e Vinhos (1967/1997), Vinícola Valmarino (1997), Velha Cantina (1997), Cave de Pedra (1997), Lídio Carraro (1998), Vinicola Pizzato (1998), Luís Argenta Vinhos Finos (1999), Vilmar Bettú (end of the 1990s), Vinícola Salvador (1998), Cristófoli Vinhedos e Vinhos Finos (1998), Máximo Boschi (1998), Angheben Adega de Vinhos (1999), Cordilheira de Santana (1999), and Vinho Laurentis (2000).

In 1995, at the initiative of wine producers, Brazil became a member-state of the International Organization of Vine and Wine (OIV).

At the end of 1997 the Brazilian Wine Institute (IBRAVIN) was created in Bento Gonçalves. It represents around 13,000 wine producing or grape growing families, and more than 600 vinification units in the state of Rio Grande do Sul alone. These bodies are organized as grape producers, wine producers, and cooperatives. The fundamental role of the Institute is to unify the viticulture sector and keep it cohesive, guaranteeing the egalitarian participation of grape producers, the producers of wine, and cooperatives in its decision-making.

In1999 the President JK Federal Agrotechnical School (founded in 1959 as the School of Viticulture and Enology and altered in 1964 to the College of Viticulture and Enology – CVE), which offered a technical course in enology, also began to offer third level courses. In 2002 school changed its name again to the Federal Center of Technological Education (CEFET). Almost all current Brazilian enologists have graduated from this school. Another result attributed to the entrance of these enologists to the market is the increase in the number of small and mid-sized wine producing companies.[3]

In the following years, a significant number of recently created wineries, or previously existed ones which had undergone modernization processes, expanded their activities enormously, ranging from investments aimed at intensifying the use of technology to enotourism development programs with the purpose of attracting a new consumer public who would associate wine with history and culture. Some central elements have been consolidated as directives to be followed, while the focus on the quality of wine has come to be an obsession for the new generations who have come to occupy the commanding positions of command in the new business of these families.

These important transformations can be perceived in the statements collected[4] from different generations of the owners of the companies and their families. It is clear in their reports how they have connected past, present, and future, and how family memories are tools for the construction of new identities. In this valorization of memories, elements such as *family, discipline, education, work,* and *professionalization* are selected as references to integrate and guide the construction of future projects and the formatting of family business.

A distant past

In the interviews with the older and newer generations we sought to identify which elements were fundamental in the recovery of the past. In the account of Juarez Valduga, president of Grupo Famiglia Valduga Co., about his time in the seminary where he studied to become a priest, we can observe how important the values of discipline and family are:

"What made me behave like that, I believe, was the fact of my foundations, which are family, simplicity, modesty, and work. It was this which led me to the seminary. In the seminary, there was a lot of discipline, but I had already been disciplined in my family. And I am like that still today."

In another part of his statement Juarez Valduga shows how the notion of family, portrayed in the figures of his parents, had a powerful influence on his behavior as a man, father, and businessman:

"I would say that my parents influenced by behavior. They did not speak much. They did. This attitude of theirs is probably the strongest thing that I have. My father did not need to say that he had to work. He worked. He did not say that he had to be honest. He was extremely honest. He was not a cultivated man. For me he was more wise than cultivated."

Alexandre Miolo, commercial director of Miolo Wine Group and who is from a younger generation, in his account of his daily life in the colony showed how important the values of education and work were:

"All the family worked directly in the colony since they were young. We lived close to where the winery is now, in Linha Leopoldina. Five hundred meters below, along that little road. The houses are still there. My father and uncles now live in the city of Bento and go there occasionally. But it was there that they all lived. We studied there in the morning, or the afternoon, depending on the year. If we studied in the morning, in the afternoon we worked in the colony; if we studied in the afternoon, we worked there in the morning. Afterwards I went to study in Bento and later to Caxias. The work was in the fields. In the vineyard, we did the pruning, harvesting, the phytosanitary treatment. Moreover corn and sugarcane were planted, and everything necessary for our subsistence. The principal business was grapes, which was where we made money. We worked, we studied, and for the reality of the time, we lived well."

As can be seen, the statements, even when from different generations, present common elements, amongst

[3] After 2000 more than 100 wineries producing fine wines were created. Of these only four have since closed down. Among these new wineries are: **2000:** Cordilheira de Sant'Ana, Don Bonifácio, Don Guerino, Ouro Verde, Quinta Ribeiro de Mattos (closed), Vinhedos Hood, Villagio Grando; **2001:** Décima X, Hermann, Larentis, Terragnolo, Villa Francioni (SC); **2002:** Campos de Cima, Cave Ouvidor (closed), Don Miguel, Don Pedrito, Dunamis, Routhier & Darricarrère, Sanjo, Sozo, Terras Altas; **2003:** Copetti & Czarnobay, Don Guerino, Pericó, Peruzzo, Santa Augusta, Villa Bari, ViniBrasil; **2004:** Adolfo Lona, Antonio Dias, Coopernatural, Dezem, Ducos (SF), Monte Azurro, Quinta Santa Maria, Santo Emilio; **2005:** Abreu Garcia, Aracuri, Bella Quinta, Don Abel, Estrada Real, Estrelas do Brasil, Generoso (closed), Pirineus, Villagio Bassetti;**2006:** Almaúnica; **2007:** Batalha, Era dos Ventos, Kranz; **2008:** Arte da Vinha, Bellavista Estate, Élephant Rouge, Quinta da Figueira, Ravanello, Vicari, Villagio Conti, Wine Park; **2009:** Camponogara, Guatambu; **2011:** Leone di Venezia, Villa Cristina; **2014:** Negro Ponte Vigna, Vinha Unna; **2015:** Casa Ágora.

[4] A fundamental source for this paper was the Gaúcho Wine Oral History Project, whose objective was to produce a database of interviews about the origin and trajectory of families descended from Italian immigrants who had settled in the Serra Gaúcha. This source of data offers an important set of information and constitutes an extraordinary instrument of reflection. Carried out in 2015 and 2016, the project has 32 statements related to 17 companies and around 40 hours of recording. The material is published in Ferreira, Valdiney and Ferreira, Marieta. *Vinhos do Brasil- do passado para o futuro*. Op. cit.

 DOI: 10.1051/bioconf/20170903004

which the valorization of work, discipline, the family, and determination are especially important.

The challenges of the present

Another important point illustrated by the selected statements is their visions of the present, in which there stand out elements such as *work, innovation, planning, professionalization*, and the obsessive search for *quality* in wine production.

Adriano Miolo, executive director of the Miolo Wine Group, highlighted at different moments of his interview the importance of these values in the traditional family company which has become one of the largest in the Brazilian wine industry; here he discusses the motivation to study enology as a means of evolving his activities:

"What led me to study enology was the obvious – I had spent my entire life in the vineyards with my father. I imagined I could advance, I mean study, to find out more, but always in viticulture."

Talking of his experience in contacts with multinational companies who came to Brazil in the 1970s, his fixation with education and professionalization is evident:

"The almost four years I spent in Martini & Rossi was my foundation in viticulture and enology. I learned a lot. For a young graduate, to join a top company was really very important. I had done my basic training, but since I wanted to understand more, to grow, I decided to study in Mendoza, because at the time we did not have a third level school of enology in Brazil, just the course in the Agrotechnical School. The closest third level course was in Mendoza. So I went there at the end of 1988 and stayed until 1995. It was a long time, obviously, and of course I completed my education."

Eduardo Valduga, director of Casa Valduga, where he shared the helm with his father Juarez, is from a later generation than Adriano Miolo but following a similar trajectory in his education and professionalization, going to Mendoza to complete his studies before assuming his current function in Grupo Famiglia Valduga Co:

"I went to Novo Hamburgo to Fevale, Faculdade do Vale, but did not like it. So, I decided to follow the family tradition and went to do a course in enology. When I left my other dreams behind, I fought tooth and nail to follow the dream of enology. It awoke a great interest in me, it was an enormous pleasure to study enology. And due to the influence of other Brazilian professionals who were going to Mendoza – there were four Brazilians from other companies making names for themselves in the wine market –, there emerged the idea of also taking advantage of this opportunity."

The obsession with quality is constant in the narrative and practice of these companies. It had long since been perceived that without quality investments they would not go far. Let us look at what Adriano Miolo said about the quality project and use of new technology in the pioneering Miolo, years before the creation of the Miolo Wine Group:

"In 1998 we began to invest in a quality project, starting with the vineyards. We knew that 'trellis' vineyards could not obtain the quality intended. So, the conversion of 'trellis' or 'pergola' vineyards to 'espalier' began and also the importation of shoots. We starting the constructing of a new winery, because since 1989 we had only had a small winery. We introduced new technologies, such as reverse osmosis and vacuum concentration."

His brother Alexandre Miolo also spoke of the investments made in new technology in the implementation of the Quality Project as an important and necessary step to improve the product quality, particularly in the production of white wine:

"We invested in a lot stainless steel equipment such as pneumatic press, tanks with automatic control of temperature. We bought the most modern equipment at the time. When we produced the first sauvignon blanc, the temperature control of the fermentation tank was done with a small hose and very cold water from an artesian well. Can you imagine this? Just as well that the tank only held three thousand liters. It what was possible at that time. Summarizing, I would say that the Quality Project fundamentally consisted of a radical change in viticulture to improve grape quality and in the use of modern equipment in the industrial plant."

This is repeated by Juarez Valduga when he comments on the radical position of his company in relation to the preparation of still red wine during the most difficult harvests that are so common in the Serra Gaúcha:

"I do not make the wine. Even if it is a great success in terms of sales, with which I will make money, but I would lose prestige, so I would prefer not to make it. Because in the future the demanding consumer will penalize me. Some of my products which people ask to return, I prefer not to make."

In addition, in the statement of Jones Valduga, from Domno do Brasil, we can see that themes such professionalization and quality were present in his father's generation. Here they are cited by him, as he is preparing to assume control, as a motive of pride:

"The desire to evolve in business and to begin to make a high quality bottle of wine with modern techniques was always present. My father studied and decide to use his knowledge to grow. Casa Valduga was the first company in Vale dos Vinhedos to produce in stainless steel tanks and to use cold fermentation processes. In 1988, Valduga was the first winery to use stainless steel tanks in Vale dos Vinhedos."

Talking of the strategy of Domno do Brasil to diversify the activities of Grupo Famiglia Valduga Co, beginning to import wine and launching more competitive sparkling wines to sell large volumes, Jones even cites the obsession with quality as a problem:

"The focus is on quality. We cannot simply open an importer and bring any wine. This could create a problem with the position of Valduga's premium products. So, the philosophy was always to do the best in Domno as well. When we started the Domno project it was to produce competitive products that sold well. But we have a problem because we always want to improve a label, use a bit of a better bottle. And in this improvement the products ended up being very good."

The new generations who are assuming control of companies have no doubt that quality is the fundamental prerequisite for survival in the competitive Brazilian market. However, dialogue with their parents, who still participated in the business, was not always easy. The interview with Giovanni Carraro, the youngest son of Lidio Carraro founder of the winery which bears his name, shows this in the changes made in the vineyards to improve the grape quality:

"The preponderant vision was to produce more grapes per plant, and per hectare, as for many this was their only source of revenue. And when the moment for espalier production arrived, in which it was necessary to cut branches, I remember that in the first harvest we had to fight with Lidio. Even when he wanted to do something different, this hurt him a lot. After all it was fruit. It was there on the plant, sometimes it was already becoming ripe, and we cut if off and threw it away. At the time this was inadmissible."

While elements of *work* and *family* appear as relevant when the focus is the past, what changes or remains when the perspective is turned to the present and the future? The statements analyzed continue to place *work* as a fundamental value, but now associated with *professionalization* and *planning* to also achieve a high level of quality in business.

On the horizon future expectations

A third point which can be perceived in our reading of reports refers to the strong belief in the future, where new business strategies become the principal focus. Values such as *differentiation, audacity,planning, professionalization* are presented as the basis core in the organization of reports.

Explaining how administration functions with the different shareholders of the Miolo Wine Group, the executive director Adriano Miolo gives us *insights* about intentions for the future:

"After completing the structuration, the company also organized itself to open its capital. In the future, we expect to open to the Stock Exchange. From the analyses we have, it is not worth doing this now. But to a certain extent we have already opened our capital. The RAR, Benedetti, and Galvão Bueno groups are now shareholders of the company. Of course, the company has a project. It has the biggest vineyard in Brazil. No one has close to the over 1000 hectares we have. There are four industries, more than 500 employees. This company has to have a strong structure, including in terms of capital. And it is organizing for this."

When asked about family projects to continuity the successful trajectory of Grupo Famiglia Valduga, Eduardo Valduga, director of Casa Valduga, demonstrated a concern with the professionalization of administration:

"In the short term, what we are responsible for is organizing the family using the science of administration. Looking for the best practices to organize the company in the form of a company which represents the family organization. Finding the best form to organize the three families as a company. Harmonize and distinguish the corporate bodies and the individuals. Maybe it will be necessary to create other companies, holdings, corporations, councils, administrative bodies, boards. I believe we have this task. Jones can complement this if it wants. For me the long-term is unknown. The family began with a father who created Luis Valduga & Filhos, moved onto brothers who formed Famiglia Valduga Co. Now we are cousins, afterwards will come the second cousins and so forth. This will get ever more complex in the question of family involvement. So, each step of our generation has to be in the search of effective and efficient professionalism."

Daniel Salton, who holds the positions of CEO and Chair of the Board of Vinícola Salton, the oldest family owned winery in Brazil, in explaining the reasons for the longevity (more than 100 years) of his family company, highlights the professionalization of administration:

"Today Salton is no longer just the family, it is Salton with a professional family. We have stopped being amateurs. To be a director it is necessary to have minimum requirements. We changed the statutes, the shareholders' agreement, and professionalized the entire company. No one holds a position in the organization just because they are from the family, they have to be competent."

In commenting on the decision-making process for the wine of the 2016 World Cup, Juliano Carraro mentions the daring posture of Brazilian wineries, which in 2011 ran a serious risk of losing the opportunity to a large Chilean winery. In the end, the Lídio Carraro winery won with its Faces line of wines:

"In 2011 we participated in Soccerex, the International Football Fair, held in Rio de Janeiro, since Brazil would host the next World Cup. The participation of Concha y Toro was certain. It was to be the exclusive sponsor of the fair and there were many rumors that it was making a deal with FIFA to prepare the official wine of the World Cup. [...]. We paid for space inside to show Brazilian wine, because although it was not a FIFA fair, it was a football fair, and everyone from FIFA would be there. There would even be a stand inside. So, we did this and repercussion was excellent. They decided that it would not make sense to make an agreement with a company from outside Brazil, when we had wineries here with great wines."

The past as a strategy for the future

Having presented this general vision of the statements as a whole, we can perceive that some of the elements analyzed gained more relevance over time, while others, even though they still remain, have lost prominence. Nevertheless, the *leitmotiv*, the recurrent elements the value of *work* associated with *quality, planning*, and *professionalization*.

In the statements analyzed we can highlight the narrative capacity presented with the organizational and logical connection of the discourse, emphasizing family union and the importance of Italian immigration as an explanatory factor for the dedication to work. The positive valorization of Italian origins and the connection established with their esteem for work justify why the long distant past and the founding moment are always raised as something which reinforces local identity. Conflicts and disagreements do not appear as important elements, in contrast family ties gain importance. One of the manners of strengthening the identity and memory of leaders, communities, and political groups lies in the care taken with the preservation of their memory. The past is important, both as a landmark of an action that has been strengthening over time, and as a reference to more detailed reflections about the paths intended to be followed.

As we know, oral statements are memories which represent different versions of the past and express

 DOI: 10.1051/bioconf/20170903004

contradictory memories, forgetfulness, distortions, conflicts, and cannot be taken as 'true' and 'objective' reports of the narrated facts; nevertheless, and for this reason, they grant us access to rich material and to information rarely found in other sources.

References

Cabral, Carlos. *Presença do vinho no Brasil – um pouco de história*2ª ed. São Paulo: Cultura (2007)

Cavagnolli, Anelise. *Os parceiros do vinho: a vitivinicultura em Caxias do Sul. 1911–1936.* Dissertação. (Mestrado em História Social). Universidade Federal do Paraná. Curitiba (1989)

Dal Pizzol, Rinaldo, Souza, Sérgio. *Memórias do vinho gaúcho.* Vol. I, II, III. Porto Alegre: Age (2015)

Dardeau, Rogerio. *Vinho fino brasileiro.* Rio de Janeiro: Mauad X (2015)

Ferreira, Marieta de M. & Amado, Janaína (orgs.). *Usos e abusos da história oral.* Rio de Janeiro: Ed. Fundação Getulio Vargas (1996)

Ferreira, Valdiney, Ferreira, Marieta. *Vinhos do Brasil - do passado para o futuro.* Rio de Janeiro: Ed. Fundação Getulio Vargas (2016)

Giron, Loraine Slomp, Bergamaschi, Heloísa Eberle. *Casas de negócios: 125 anos de imigração italiana e o comércio regional.* Caxias do Sul: EDUCS (2001)

Gobbato, Celeste. *Apreciações sobre a vitivinicultura no Rio Grande do Sul: sua situação e seu desenvolvimento.* Tese para cátedra da Escola de Agronomia e Veterinária, Universidade de Porto Alegre. Caxias do Sul: Rossi (1938)

Instituto Brasileiro do Vinho (Ibravin) Available at: www.ibravin.org.br. Accessed on: 02/2016 and 03/2016

SÁ, Mem de. *Aspectos econômicos da colonização italiana no Rio Grande do Sul.* Álbum comemorativo do 75° aniversário da colonização italiana no Rio Grande do Sul. Porto Alegre: Globo (1950)

Sandro, Rogerio dos. *A construção do cooperativismo em Caxias do Sul: Cooperativa Vitivinícola Nova Aliança (1931–2011).* Tese (Doutorado em História). PUC-RS. Porto Alegre (2011)

DOI: 10.1051/bioconf/20170903005

Eficiencia técnica de las denominaciones de origen protegidas en España: Un análisis por tipo de vino comercializado / Technical efficiency of protected designations of origin in Spain: An analysis by type of wine marketed

Fernando Vidal[1], Jesús T. Pastor[2], Juan Aparicio[2], y Lidia Ortiz[2]

[1] Departamento de Economía Agroambiental, UMH, EPSO, Carretera de Beniel km 3,2, Orihuela 03312, España
[2] Centro de Investigación Operativa, UMH, Edificio Torretamarit, Elche 03202, España

Abstract. In the European Union there are two figures of quality protection for agri-food products, according to EC Regulation 1151/2012, which are: Protected Designations of Origin (PDOs) and Protected Geographical Indications. In addition, in the case of wine, traditional nomenclatures coexist in each country. Thus, in the Spanish case we could find within these PDOs the traditional Denominations of Origin, Qualified Origin Denominations, Quality wines with Geographical Indication, Pago Wines, applied to high quality individual vineyards or wine estates and Qualified Pago Wines, of the highest category. The technical and economic efficiency of the different protection figures in the wine sector has been the subject of previous works; however, it has not been examined in depth to see if it is correlated to the type of wine predominantly marketed by each one of them. Bounded Adjusted Measure (BAM) has been used in this case, which allows a better characterization of the units according to their technical efficiency indexes. The data used are the latest provided by the Spanish authorities (2014/2015 campaign). The results seem to indicate that there is not a relationship between greater efficiency and the type of wine marketed.

1. Introducción

La normativa europea relativa a la calidad de los productos agroalimentarios (Reglamento -CE- 1151/2012 del Parlamento Europeo y del Consejo, de 21 de noviembre de 2012), distingue entre dos tipos de figuras de protección; por un lado las Denominaciones de Origen Protegidas (DOPs) y por otro las Indicaciones Geográficas Protegidas (IGP). Según los últimos registros del Ministerio de Agricultura existen en España 90 DOP y 41 IGP.

Adicionalmente, los países de la UE pueden seguir empleando nomenclaturas tradicionales para sus vinos. En nuestro país, éstas vienen marcadas por lo establecido en la Ley 24/2003, de 10 de julio, de la Viña y del Vino. Ésta establece las siguientes categorías dentro de las genéricas DOPs: Denominación de Origen (DO), Denominación de Origen Calificada (DOCa), Vino de Calidad con indicación geográfica (VC), Vino de Pago (VP) y Vino de Pago Calificado (VPCa).

Según los últimos datos disponibles [1], en la campaña 2014/2015 el número se hectáreas inscritas en las diferentes DOPs ascendía a 574.038, el número de viticultores alcanzaba los 123.617, mientras que el número de bodegas inscritas era de 4.120 (el 85% embotelladoras). La comercialización total de esa campaña era de 11.582.429 hl, destacando Rioja (24,2% del total nacional), seguido por Cava (15,7%), Rueda (6,1%) y Ribera del Duero (5,8%). Del total comercializado, el mercado interior representaba un 59% del total y el exterior un 41% (aquí el mejor comportamiento lo tiene Cava, con casi la cuarta parte del total exportado). En cuanto al tipo de vino, el tinto representa más de la mitad del total comercializado (55%), seguido por el vino blanco (19,45%), espumoso (15,88%) y rosado (4,74%). El vino de licor apenas llega al 4,67% y el de aguja menos del 1%.

Si profundizamos en el análisis de la comercialización por tipo de vino, y en el caso del mercado interior, destaca el vino tinto (55,3%), seguido del blanco (25,07%), espumoso (10,10%), rosado (5,44%) y de licor (3,94%). Sin embargo, aunque en el mercado exterior el vino tinto sigue siendo el mayoritario (54,87%), es seguido por el espumoso (24,11%) y ya a mucha distancia por los vinos blancos (11,45%) y de licor (5,59%). Los vinos exportados por España lo son mayoritariamente embotellados (96,46%) frente a un 3,54% a granel. Por destinos, destacan la UE (66,04%), América (16,35%), Asia (8,73%) y Europa no UE (7,60%). Por países, el destino mayoritario es Reino Unido (21,16%), Alemania (17,10%) y EE.UU. (9,56%). Habrá que ver en un futuro no muy lejano el efecto que la salida del Reino Unido de la UE tiene en estos datos y en el comportamiento de su mercado respecto a estos vinos.

Se pretende en la presente comunicación determinar si el tipo de vino comercializado por las diferentes figuras dentro de las DOPs españolas guarda relación con su mayor o menor eficiencia técnica. Ello podría suponer el plantear potenciales cambios en sus enfoques comerciales y en los mercados objetivo.

Tabla 1. Valores medios de las variables analizadas por DOP.

	DO	VP	VC
INPUTS			
1) Superficie (HA, hectáreas)	10.917	59,2	196,2
2) Viticultores (NUM, número)	2.198	1,3	130,6
OUTPUTS			
1) Ventas mercado interior (INT, hl)	114.660	275	1.159
2) Ventas mercado exterior (EXT, hl)	78.000	193	149
TIPO DE VINO*			
1) Tinto	42	10	3
2) Blanco	12	2	1
3) Espumoso	1	–	–
4) Rosado	1	1	1
5) Licor	2	–	–
Total	58	13	5

* = Tipo de vino mayoritariamente comercializado por DOP.

2. Material y métodos

Los datos económicos y productivos empleados en el trabajo, tal y como se ha comentado anteriormente, son los facilitados por el Ministerio de Agricultura español [1]. Las variables consideradas en el trabajo han sido:

- Inputs: 1) Superficie (hectáreas, HA) y 2) Número de viticultores (NUM).
- Outputs: 1) Volumen comercializado en el mercado interior (hectólitros, INT) y 2) Volumen comercializado en el mercado exterior (hl, EXT).

El análisis que aquí se presenta se centra en determinar la eficiencia técnica de las diferentes figuras de protección. Para ello se recurre al Data Envelopment Analysis (DEA), método no paramétrico ampliamente utilizado en el sector agrario. Señalar que en esta aplicación se ha optado por los modelos BAM. La Bounded Adjusted Measure (BAM) es una medida de tipo aditiva que agrega mediante unos determinados pesos las ineficiencias técnicas detectadas por el modelo en cada una de las dimensiones del problema de producción (inputs y outputs), y que verifica una serie de propiedades interesantes, desde un punto de vista económico y matemático: (1) Para cada unidad evaluada, BAM devuelve siempre un valor entre cero y uno, indicándonos con un valor de uno que la unidad es eficiente en el sentido de Pareto-Koopmans (es decir, no dominada); (2) BAM es invariante a cambios en las unidades de medida en inputs y outputs; (3) BAM es invariante a traslaciones de los datos; y, finalmente, (4) BAM satisface la propiedad de monotonía. Adicionalmente, y en comparación con las archiconocidas medidas radiales orientadas, BAM es una medida no orientada que permite considerar todo tipo de ineficiencias técnicas, tanto en inputs como en outputs.

Aunque el DEA se ha aplicado profusamente en temas agrícolas, no ha sido muy habitual su aplicación al análisis de la eficiencia de las DOPs en general y de las vinícolas en particular. Señalar, ya dentro de este sector, la aplicación en España a las DO [2,3], las IGP [4], los VP [5] o la comparación entre ellas [6]. No obstante, ninguno de los estudios anteriores trataba de encontrar si existe una relación entre una mayor o menor eficiencia técnica y el tipo de vino comercializado.

Las DOP se han clasificado (Tabla 1) atendiendo a las diferentes figuras presentes (DO, VP y VC) y al tipo

Tabla 2. Índices de eficiencia (IE) modelo BAM (VRS).

IE = 1
Aylés (VP)[1], Calzadilla (VP)[1], Campo de la Guardia (VP)[1], Dehesa del Carrizal (VP)[1], Dominio de Valdepusa (VP)[1], Lebrija (VC)[1], Rioja (DOCa)[1], Rueda (DO)[2], Cava (DO)[5]
IE = [0,87;0,75] (ordenados de mayor a menor IE)
Pago Florentino (VP)[1], Cariñena (DO)[1], Chacolí de Getaria (DO)[2], Valdepeñas (DO)[1], Navarra (DO)[1], Rias Baixas (DO)[2], Somontano (DO)[1], Valencia (DO)[1], Finca Élez (VP)[1], Los Balagueses (VP)[1], Pago de Otazu (VP)[1], Casa del Blanco (VP)[2], Pago de Arínzano (VP)[1], Guijoso (VP)[1]
IE =] 0,75; 0,6] (ordenados de mayor a menor IE)
Montilla Moriles (DO)[1], El Terrerazo (VP)[1], Chacolí de Bizkaia (DO)[2], Condado de Huelva (DO)[2], Yecla (DO)[1], Pla i Llevant (DO)[1], Binissalem (DO)[1], Chacolí de Álava (DO)[2], Campo de Borja (DO)[1], Empordà (DO)[1], Monterrei (DO)[2], Ribera del Duero (DO)[1], Valdeorras (DO)[2], Ribeiro (DO)[2], Calatayud (DO)[1], Bierzo (DO)[1], Ribeira Sacra (DO)[1], Montsant (DO)[1], Jumilla (DO)[1], Cangas (VC)[1], Priorat (DOCa)[1], Alicante (DO)[1], Cataluña (DO)[2], Toro (DO)[1], Ycoden-Daute-Isora (DO)[1], Costers del Segre (DO)[1], Cigales (DO)[3]
IE =] 0,60; 0,5] (ordenados de mayor a menor IE)
Málaga (DO)[4], Tierra de León (DO)[3], Pla de Bages (DO)[1], Uclés (DO)[1], Almansa (DO)[1], Penedés (DO)[1], Utiel-Requena (DO)[1], Lanzarote (DO)[2], Arribes (DO)[1], La Palma (DO)[2], Arlanza (DO)[1], Sierras de Málaga (DO)[1], Bullas (DO)[1], Islas Canarias (VC)[2], Tarragona (DO)[1], Valle de la Orotava (DO)[2], Manchuela (DO)[1], Valles de Benavente (VC)[3], Sierra de Salamanca (VC)[1], Vinos de Madrid (DO)[1], Conca de Barberá (DO)[1], Ribera del Guadiana (DO)[1], Méntrida[1], Tierra del Vino de Zamora (DO)[1], Ribera del Júcar (DO)[1], La Mancha (DO)[1]*

[1] = Vino tinto; [2] = Vino blanco; [3] = Vino rosado; [4] = Vino de licor; [5] = Vino espumoso * IE = 0,345.

de vino comercializado mayoritariamente (tinto, blanco, rosado, espumoso y de licor). A este respecto señalar que cada DOP se ha incluido únicamente en un único tipo de vino, aquel que supone el mayor porcentaje del volumen total comercializado. De esta manera, se ha podido recoger información de 76 DOP, de las cuales 58 serían DO (dos de ellas Calificadas, Rioja y Priorat), 13 VP y 5 VC. Por tipo de vino, señalar que 55 de estas figuras comercializan mayoritariamente vino tinto, 15 vino blanco, tres rosado, dos vino de licor y una espumoso. El modelo BAM empleado tiene orientación output y se consideran rendimientos a escala variables (VRS).

3. Resultados

Lo primero que puede apreciarse de los resultados (Tabla 2) es que nueve de las DOP analizadas forman parte del grupo de unidades plenamente eficientes (Índice de Eficiencia; IE = 1). De ellas, la mayor parte son VP (5), tres son DO y una es VC. Dentro de las DO, destacar por sus particularidades, la DOCa Rioja, líder en ventas tanto nacionales e internaciones, DO Rueda, con un gran posicionamiento en el mercado del vino blanco y la DO

DOI: 10.1051/bioconf/20170903005

Tabla 3. ANOVA (IE) por DOP y tipo de vino.

Índice de Eficiencia (IE) por DOP		
	Media	Desv. stand.
DO (58)	0,645	0,127
VP (13)	0,855	0,124
VC (3)	0,641	0,205
Total (F = 13,647)*	0,680	0,153
Índice de Eficiencia (IE) por tipo de vino		
	Media	Desv. stand.
Vino Tinto (55)	0,679	0,159
Vino Blanco (15)	0,686	0,127
Vino Rosado (3)	0,573	0,042
Vino de Licor (2)	0,669	0,101
Vino Espumoso (1)	1,000	
Total (F = 1,512)	0,680	0,153

* Diferencias significativas al 99%.

Cava, comercializadora por excelencia del vino espumoso español. Un segundo grupo de DOP, ya ineficientes, serían aquellas cuyo IE se encuentra entre 0,75 y 0,87, compuesto por 14 unidades, de las cuales la mitad son VP y la otra mitad DOs. Los grupos más numerosos son los de mayor ineficiencia, con IE entre 0,5 y 0,75, entre los que se encuentran prácticamente todos los VC (excepto uno) y por el contrario sólo aparece un VP (aunque con un IE de 0,73).

Por tipo de vino, señalar desde un punto de vista cualitativo, que dentro del grupo de las DOP eficientes, la mayor parte (7 de 9) comercializan mayoritariamente vino tinto, una de ellas blanco y la otra espumoso. El comportamiento es similar en el segundo de los grupos, aquel que sin ser eficiente presenta mejores valores de su IE, donde de las 14 unidades que lo integran, once de ellas están más centradas en el vino tinto y tres en el blanco. Vemos, pues, que en estos dos primeros grupos no aparecen ni los vinos rosados ni los de licor. Cuantitativamente, se ha tratado de encontrar si existen diferencias significativas en el comportamiento de las DOP en cuanto a su IE y al tipo de vino comercializado. Para ello se ha recurrido a un análisis de varianza (ANOVA, Tabla 3). Este análisis viene a corroborar lo expuesto en el párrafo anterior; es decir, existen diferencias significativas en el comportamiento de las diferentes DOP atendiendo a su naturaleza jurídica, destacando los VP como figuras más eficientes, mientras que el comportamiento de las DO y los VC es similar. Sin embargo, no se han encontrado diferencias estadísticamente significativas entre las DOPs atendiendo al tipo de vino comercializado.

4. Conclusiones

En los competitivos mercados vinícolas, los diferentes vinos deben competir atendiendo a determinadas características. Una de ellas son las relativas a su origen y a los mecanismos de aseguramiento de la calidad, una calidad diferenciada que en nuestro país (al igual que en la UE) se sustenta fundamentalmente en las DOP. Sin embargo, es bien sabido que dentro de éstas, el consumidor debe enfrentarse a diferentes figuras. En el presente trabajo hemos tratado de analizar cuál de estas figuras es más eficiente técnicamente, y si esta eficiencia depende del tipo de vino producido y comercializado por cada DOP. Así, sí se ha podido constatar que los VP, a priori los que presentan las mayores exigencias productivas y de comercialización dentro de las DOP, son la opción más eficiente.

En cuanto al tipo de vino comercializado, no ha sido posible encontrar diferencias de comportamiento entre ellos, de manera que no puede concluirse que una mayor dedicación a uno u otro tipo de vino (tinto, blanco, rosado, espumoso o de licor) pueda traducirse en una mayor eficiencia técnica. Dado que únicamente se ha considerado una campaña, la 2014/2015, sería interesante en posteriores trabajos ver si este comportamiento se ha mantenido o se mantiene en el tiempo, así como tratar de incorporar nuevas variables que amplíen los análisis e incorporar a los mismos aspectos relacionados con la productividad.

The authors thank the financial support from the Spanish Ministry for Economy and Competitiveness (Ministerio de Economía, Industria y Competitividad), the State Research Agency (Agencia Estatal de Investigacion) and the European Regional Development Fund (Fondo Europeo de Desarrollo Regional) under grant MTM2016-79765-P (AEI/FEDER, UE).

Referencias

[1] Magrama, Disponible en: `http://www.mapama.gob.es/es/alimentacion/temas/calidad-agroalimentaria/calidad-diferenciada/dop/` (2017)

[2] J. Aparicio, F. Borrás, J.T. Pastor, F. Vidal, Eur J Oper Res. **231**(2), 443–451 (2013)

[3] F. Vidal, J.T. Pastor, F. Borrás; D. Pastor, Span J Agric Res. **11**(2), 294–304 (2013)

[4] F. Vidal, J.T. Pastor, J. Aparicio, D.B. López-Lluch, Actas de Horticultura **60**, 34 (2012)

[5] J. Pastor, F. Vidal, F. Borrás, D. Pastor. ¿Son los Vinos de Pago la figura más eficiente entre las DOP españolas?. 38th World Congress of Vine and Wine, Mainz, Germany (2015)

[6] F. Vidal, J. Pastor, J. Aparicio, L. Ortiz. The quality wine sector in Spain. An efficiency analysis. International Conference on Business and Information, Sapporo, Japan (2012)

La necesidad de incorporar las competencias de comunicación en la formación de los enólogos / The need of incorporating the communication skills in the oenologists training

Juan Gómez Benítez[1], Cristina Lasanta Melero[1], y Lucía Benítez Eyzaguirre[2]

[1] Departament of Chemical Engineering and Food Technology Faculty of Sciences University of Cádiz, Spain
[2] Departament of Marketing and Communication, Faculty of Social Sciences and Communication, University of Cádiz, Spain

Abstract. The OIV has recently approved the resolutions OIV-ECO 492-2013 on the definition of the profession of oenologist and OIV-ECO 563-2016 on oenologist training programs. These resolutions have incorporated wine communication activities into the roles and training curriculum of enologists, which require updating the training of professionals in view of their new roles and competencies. These resolutions reflect the reality of a world where communication has taken on a decisive importance in all areas of human activity. The authors of this work have developed since 2012 several teaching innovation experiences in the University of Cadiz to incorporate the communication skills in a transversal way in the training of enologists. This paper presents the results of the last experience developed consisting in the recording and editing of videos of wine presentations by students in two styles: technical and popular. This work incorporates the results of satisfaction surveys of the students that show their high interest in the training in wine communication and the difficulty of working in front of the camera. A list of skills in "Public speaking and presentation skills" and "Responsible commercial communication" is also presented.

1. Introducción

La comunicación ha adquirido actualmente una importancia determinante en todos los ámbitos de la actividad humana. El vino es un producto complejo, por sus diferentes dimensiones hedónicas y características socioculturales, que no puede utilizar un único código comunicativo para su difusión, ya que sus consumidores pueden ser muy distintos, desde el especializado de los diferentes ámbitos del mundo del vino hasta el genérico no conocedor de su elaboración y composición. El consumo del vino tiene dimensiones culturales que deben estar presentes en su comunicación para alcanzar tanto a los diferentes públicos como a las nuevas tendencias de su consumo. Por otra parte, los consumidores deben cada vez estar más informados de las características de los productos que consumen y en este contexto el enólogo, como máximo conocedor del vino que elabora, ha adquirido en una nueva función en la comunicación del vino. Esta realidad se refleja en que existe una diferenciación en la percepción y elección del consumidor de vino según sus conocimientos previos, pues mientras los conocedores prefieren los consejos de los especialistas, los no iniciados prefieren los de los vendedores [1]. Por otra parte Barber et al. [2] informaron de la existencia en el mercado norteamericano de "emerging wine learners" interesados en aprender de vinos de forma rápida a los que interesa un mensaje especializado. Además, como señala Charters [3], los consumidores implicados perciben en el vino un nivel de calidad y complejidad superior que los no implicados. El poder de comunicación del enólogo podría verse incrementado aprovechando los nuevos medios on-line [4]. Pero para desarrollar estas actividades de comunicación el enólogo debe tener una formación específica en este ámbito, pues en caso contrario, los enólogos tienden a utilizar un formato excesivamente técnico, de difícil comprensión por los consumidores no iniciados. Por ello, la OIV ha incorporado las siguientes actividades de comunicación en la resolución OIV-ECO 563-2016 relativa a los programas de formación de los enólogos:

- La comunicación en el sector vitivinícola: estructura y actores claves
- Actividades de Relaciones Públicas
- Interacción a través de Internet y las redes sociales
- Prensa vitivinícola
- Habilidades de oratoria y presentación
- Comunicación comercial responsable.

Por ello, las universidades que imparten formación en enología deberían incorporar en estos currículos formativos algunos de las actividades formativas de los estudios del ámbito de la comunicación, y en particular los grados de Publicidad y Relaciones Públicas y de Comunicación Audiovisual.

En este trabajo se presenta la experiencia de innovación docente realizada en el Grado de Enología de la Universidad de Cádiz durante el curso académico 2016–17 para dotar a sus alumnos de competencias en comunicación. Esta experiencia consistió en la grabación en vídeo de presentaciones de vino por los alumnos en dos estilos: técnico y divulgativo. Este trabajo es

Figura 1. Imagen de la grabación en vídeo con Chroma key.

la continuación de experiencias anteriores desarrolladas desde el año 2012 [5], entre las que cabe resaltar:

- Seminarios con comunicadores de vinos especializados (2012–13)
- Talleres de técnicas de comunicación oral en público, mediatraining, elaboración de un vino y de un documental sobre estudios de enología (2013–14).

Asimismo, en esta comunicación se expone una selección de competencias comunicativas recogidas de grados de comunicación de diversas universidades españolas que podrían ser incorporadas a los estudios de enología.

2. Materiales y métodos

2.1. Formación en comunicación

La actividad se desarrolló con alumnos de la asignatura "Análisis Sensorial" de 4° Curso del Grado en Enología. En primer lugar, los alumnos recibieron varios seminarios de técnicas de comunicación y de actuación delante de la cámara de vídeo por profesores especializados y por profesionales del sector audiovisual. Una vez adquiridos esos conocimientos, los alumnos debieron preparar la descripción de dos vinos en dos formatos diferentes, un formato técnico y un formato divulgativo. Estas presentaciones fueron grabadas en vídeo con chroma key para poder ser editadas posteriormente e incorporarles un fondo de ambiente acorde con el vino presentado.

Los alumnos tuvieron a su disposición un teleprompter para poder leer los textos de las presentaciones durante su grabación, aunque no todos lo usaron. El programa utilizado en la edición fue Adobe Premiere Pro CC 2016. Las presentaciones fueron evaluadas por los profesores en una escala 0-10 con la rúbrica que se presenta en la Tabla 1.

Hay que reseñar que la evaluación de la presentación participó en la nota final de los alumnos en la asignatura en una proporción del 10%.

Para el uso académico de estas grabaciones, se solicitó a los alumnos una autorización firmada. Asimismo, los alumnos respondieron a una encuesta anónima de satisfacción sobre la actividad donde debían expresar su interés y las dificultades encontradas.

2.2. Selección de actividades formativas y competencias

Para la definición de las competencias comunicativas necesarias para los futuros enólogos, se han seleccionado entre las universidades de las primeras posiciones del ranking de las universidades españolas en este ámbito. En particular, se ha recurrido a las que describe la Universidad Pompeu Fabra (Barcelona) por ser las más adecuadas para el desarrollo de la comunicación oral y para la comercialización de productos, gracias a la calidad de la información disponible en su página web [6]. Asimismo, se ha utilizado el programa de formación para la profesión de enólogo de la organización internacional sin ánimo de lucro Wine in Moderation [7].

Tabla 1. Rúbrica de evaluación de las presentaciones de vino por los alumnos.

Formato técnico	Fondo	Extensión	1/3	60%
		Detalle	1/3	
		Corrección	1/3	
	Forma	Claridad	1/3	40%
		Naturalidad	1/3	
		Atractivo	1/3	
Formato Divulgativo	Fondo	Extensión	1/3	40%
		Detalle	1/3	
		Corrección	1/3	
	Forma	Claridad	1/3	60%
		Naturalidad	1/3	
		Atractivo	1/3	
Nota final	Formato técnico			50%
	Formato divulgativo			50%

Tabla 2. Encuesta de satisfacción sobre el aprendizaje en comunicación Valores medios (escala 0 lo peor 10 lo mejor).

Formato técnico	Formato divulgativo	Promedio
8,16 ± 1,34	8,54 ± 1,36	8,35 ± 1,06

3. Resultados y discusión

3.1. Formación en comunicación

Los resultados de la evaluación de las presentaciones se presentan en Tabla 2. Se puede apreciar que las notas obtenidas has sido elevadas, tanto las de formato técnico como las de formato divulgativo. aunque, sorprendentemente, las evaluaciones en formato divulgativo han sido superiores a las obtenidas en formato técnico.

Cabe destacar que la calidad comunicativa de algunas de las presentaciones fue muy elevada, comparables a la de un enólogo profesional con experiencia.

Entre los modelos utilizados por los alumnos para describir los vinos en formato divulgativo, algunos de los cuales alcanzaron un gran poder comunicativo, caben destacar los siguientes:

- Utilización de vivencias personales y familiares
- Historias fabuladas
- Teatralización de escenarios ficticios
- Relatos poéticos

En la Fig. 2 se presenta una imagen de una de las grabaciones de vídeo antes de la edición.

La calidad comunicativa de las imágenes después de su edición e inserción de fondos adecuados al tipo de vino presentado se puede apreciar en las Figs. 3, 4 5 y 6.

Los resultados de la encuesta de satisfacción se presentan en la Tabla 3. Cabe destacar que las valoraciones fueron todas muy elevadas, y que la utilidad del teleprompter fue la menos valorada.

Figura 2. Captura de pantalla antes de la edición del vídeo.

Figura 3. Alumno presentando vino en una bodega.

Figura 4. Alumna presentando vino en un bar.

Figura 5. Alumna presentando vino en un viñedo.

En la Tabla 4 se aprecia que la mayor dificultad que encontraron los alumnos fue el de posar delante de la cámara de vídeo.

Por último, el formato en el que los alumnos se encuentran más cómodos, como cabía prever, fue el técnico, como se aprecia en la Tabla 5, a pesar de que los mayores valores en las evaluaciones lo obtuvieron en el formato divulgativo.

Figura 6. Alumno presentado vino en campiña.

Tabla 3. Encuesta de satisfacción sobre el aprendizaje en comunicación Valores medios (escala 0 lo peor 10 lo mejor).

Utilidad global	9,04 ± 0,98
Interés profesional	9,50 ± 0,78
Interés personal	9,39 ± 1,01
Utilidad de la grabación en vídeo	9,14 ± 1,16
Utilidad del telemprompter	8,18 ± 2,27
Diferenciar lenguajes técnico/divulgativo	8,89 ± 1,23
Interés de ampliar el aprendizaje	9,36 ± 0,93
Interés de una asignatura completa	9,08 ± 0,96

Tabla 4. Dificultades encontradas Valores medios (escala 0 la menor 10 la mayor.

Hablar en público	6,93 ± 1,83
Posar delante de la cámara de vídeo	7,46 ± 2,32
Construir un relato resumido	5,52 ± 2,60

Tabla 5. Preferencias de formato Valores en %.

Técnico	70,8
Divulgativo	25,0
Indiferente	4,2

Conviene resaltar también que estos vídeos convenientemente tratados podrían ser utilizados para dar a conocer los estudios de enología y para promoción personal de los alumnos si son difundidos adecuadamente en las redes sociales.

3.2. Competencias en comunicación

En función de la experiencia y el análisis descrito, se proponen las siguientes competencias comunicativas como esenciales para la formación de los futuros enólogos:

3.2.1. Oratoria y presentación en público [6]

- Reconocer las características del lenguaje oral y aplicarlo al diseño de un discurso.
- Emplear los distintos recursos retóricos para captar la atención del público.
- Identificar y controlar los aspectos básicos de la comunicación no verbal, (gesto, voz, postura, miedo) para pronunciar con éxito un discurso oral.
- Elaborar y estructurar discursos eficaces utilizando las estrategias de la comunicación oral.
- Enfrentarse con seguridad a una comunicación pública, ante un público determinado.

3.2.2. Comunicación comercial responsable [7]

- Juzgar, interpretar, discutir y explicar los datos científicos sobre los beneficios y riesgos de las bebidas alcohólicas y del vino
- Colaborar con otros profesionales de la salud y sociales involucrados en temas de alcohol, salud y aspectos sociales
- Explicar el marco legal para vender, servir y consumir alcohol a los clientes
- Asesorar a los clientes sobre patrones de consumo responsables y moderados
- Comprender las diferencias culturales, los comportamientos y los patrones de consumo

 DOI: 10.1051/bioconf/20170903006

- Educar a los demás y difundir el conocimiento sobre los beneficios del consumo moderado de vino y los riesgos de beber en exceso.

Estas competencias posibilitan el acercamiento a los públicos a través de la elaboración de mensajes más ricos y comprensibles, a través de la inclusión de valores culturales, de cara a una mejor percepción pública de los vinos y la enología.

4. Conclusiones

Los estudiantes de enología han mostrado una gran capacidad para incorporar competencias de comunicación en su formación. Cuando estas competencias son adquiridas y convenientemente desarrolladas los estudiantes de enología y futuros enólogos pueden llegar a crear relatos de gran poder de comunicación sobre vino. La adquisición de estas habilidades será sin duda de gran valor para su currículo personal y para el sector vitivinícola. Por ello, las universidades que imparten enología deben incorporar estas competencias en el currículo formativo de sus alumnos.

Referencias

[1] A. G. Parsons, A. M. Thompson, A. G. Parsons, and A. Thompson, "Wine recommendations?: who do I believe??," Br. Food J. **11**(9), 1003–1015 (2009)

[2] N. Barber, J. Ismail, T. Dodd, N. Barber, and T. Dodd, "Purchase Attributes of Wine Consumers with Low Involvement Purchase Attributes of Wine Consumers with Low Involvement," J. Food Prod. Mark. **4446**(April), 69–86 (2017)

[3] S. Charters, S. Pettigrew, and S. Pettigrew, "Product involvement and the evaluation of wine quality," Qual. Mark. Res. **9**(2), 181–193 (2007)

[4] H. P. Peters, S. Dunwoody, J. Allgaier, Y. Lo, and D. Brossard, "Public communication of science 2.0" **15**(7), 749–753 (2014)

[5] J. Gómez Benítez, L. Benítez Eyzaguirre, P. P. Marín Dueñas, C. Silva Robles, and C. Lasanta Melero, "Una experiencia integral de enseñanza-aprendizaje de la comunicación del vino en la universidad de Cádiz," in *Contenidos innovadores en la universidad actual*, 2014, pp. 349–362

[6] Universidad Pompeu Fabra (Barcelona), "Plan docente del Grado de Publicidad y Relaciones Públicas," 2017. [Online]. Available: `https://gestioacademica.upf.edu/doa/ consultaPublica/look[conpub]BuscarPubGuiaDocAs?entradaPublica=true&idiomaPais=es.ES&_centro=338&_estudio=3382&_anoAcademico=2016`

[7] Wine Moderation in, "Training for Enologist Profession," 2014. [Online]. Available:`https://www.wineinmoderation.eu/es/articles/Training-for-Oenologist-Profession.207/`

DOI: 10.1051/bioconf/20170903007

Technical norms for Geographical Indications and their reflexes for the wine sector

Kelly Lissandra Bruch[1,a], Adriana Carvalho Pinto Vieira[2], Ludimila Cesar Moura Gaspar[3], Cristiane Francisco da Silva[4], and Marcos Vinícius Araujo[1]

[1] Center for Studies and Research in Agribusiness - Universidade Federal do Rio Grande do Sul – CEPAN/UFRGS, Av. Bento Gonçalves, 7712 Porto Alegre, RS, Brasil
[2] Universidade do Extremo Sul de Santa Catarina – UNESC - Av. Universitária, 1105 - Bairro Universitário, 88806-000 Criciúma, Brasil
[3] Ministério da Agricultura, Pecuária e Abastecimento – MAPA - Rua Rodrigues Alves 129, 6°andar, sala 601-A, Praça Mauá, 20048-093 Rio de Janeiro, Brasil
[4] Instituto Federal de Alagoas – IFAL – Rua Mizael Domingos, 75, Centro, 57020600 Maceió, Brasil

Abstract. Does the establishment of technical standards for a geographical indication (GI) help in its sustainable development for the wine sector? Based on this question the present work aims to analyse the technical norms already published by the Brazilian Association of Technical Norms – ABNT, with the objective of examining if these can help in the structuring, recognition and sustainable development of the GI in Brazil. Two standards have already been published: Brazilian Standard ABNT NBR n. 16.479-2016 – Geographical Indication – Terminology; And the Brazilian Standard ABNT NBR n. 16.536-2016 – Geographical Indication – Guidelines for structuring a Geographical Indication for product. A technical standard dealing with the control of GI and a technical standard dealing with the GI management system are being elaborated. Considering the still recent publication of the standards, it was verified at the first moment the impossibility of investigating what would be the effective impacts of the technical norms in the wine-growing GI, as initially proposed. However, there is a gain that would not initially be considered: it was possible to systematize the best practices of GI available in Brazil through the elaboration of standards already published and those in the publication phase.

1. Introduction

Does the establishment of technical standards for the structuring, control and management of a geographical indication (GI) help in its sustainable development? Based on this question the present work aims to analyze the technical norms already published by the Brazilian Association of Technical Norms – ABNT, body that represents Brazil in the scope of ISO, with the objective of analyzing if these can help in the structuring, recognition and sustainable development of the GI in Brazil, notably in the scope of GI viticulture.

Two standards have already been published: Brazilian Standard ABNT NBR n. 16.479-2016, called Geographical Indication – Terminology [1]; And the Brazilian Standard ABNT NBR n. 16.536-2016, called Geographical Indication – guidelines for structuring a Geographical Indication for product [2]. A technical standard dealing with the control of GIs and a technical standard dealing with the GI management system are being elaborated.

The Special Commission for the Study of Geographical Indications, which has met monthly in the last two and a half years for the elaboration of these standards, is composed of representatives of the producers of the GIs, associations that manage the GIs, entities that represent related sectors To products with GI, as well as regulatory bodies such as the Ministry of Agriculture, Livestock and Food Supply and the National Institute of Industrial Property, among others. The purpose of this work is to verify, within the vitivinicultural sector, whether these norms can help in the structuring and recognition of new GIs, as well as in the sustained development of GIs already recognized. As it is a still unsuccessful work, the results are still not conclusive, but they can point to an aid in the structuring of GIs in the recognition phase, as well as in a facilitation in the understanding of the producers on what the GIs would be.

This article is divided into six parts. The first is this introduction. The second is a brief review of the geographical indications in Brazil and the Special Study Commission of Geographical Indications from ABNT (SSC-216). The third part presents the methodology for the construction of technical standards used. The fourth concerns the influence of technical standards in the wine sector. In the fifth part we present the methodology used in the present work. In the sixth and final part, the results and conclusions.

[a] e-mail: kellybruch@gmail.com

DOI: 10.1051/bioconf/20170903007

2. Geographical indications and technical standards

The Trade Related Aspects of Intellectual Property Rights (TRIPS), was adopted as part of the creation of the World Trade Organization (WTO). Among the regulated legal institutes is the geographical indication.

Geographical indications (GI) are well-known institutions in Europe, still not very noticeable in South America. Although the legal protection is relatively recent in the European continent, dating from 1756 the granting of the first legal protection, the use of geographical names goes back to the antiquity. It was consolidated as a reference for the markets, regardless of the formalization allowed by legal protection, whose initial objective was more to protect the producers to avoid the usurpation of a consecrated name of what effectively fetassegur the identification of a product with differentiated characteristics, to look for competitive advantages and to value the originating place of said good [14, 17, 18, 20].

In the Brazilian context, the legal protected identification of the products by their geographic origin occurs only a few decades ago and, despite the great interest that the theme has aroused, the number of officially recognized geographical indications is reduced – currently 59, as well as knowledge about paper of the GI. The Industrial Property Law [4] in its text does not define what a GI is, but it classifies into two species: indication of origin (indicação de procedência – IP, in portuguese) and denomination of origin (denominação de orgem – DO, in portuguese).

According to art. 177, IPL [4], IP is considered as: "the geographical name of a country, city, region or locality of its territory, which has become known as a center for the extraction, production or manufacture of a particular product or the provision of a particular service" [4]. For Locatelli and Souza [9], the characteristic element is the notoriety of the geographical origin related to the product. However, in Brazil, although there is no legal requirement, the authors point out that IP applications are, as a rule, related to products that involve tradition and are linked to the local culture. Nor is there any normative requirement related to the quality or characteristic of the product linked to the geographical environment of origin, if the geographical name is recognized.

The IPL defines DO as: "the geographical name of a country, city, region or locality of its territory, which designates a product or service whose qualities or characteristics are due exclusively and essentially to the geographical environment, including natural and human factors" [4]. For the proof of this institute there is the requirement of a link with the geographical origin, with the proof of a characteristic or quality linked to it. They are differentiated products, or even unique, distinct from the others due to the geographic environment, including natural and human factors, according to Locatelli and Souza [9].

Therefore, according to Vieira; Watanabe; Bruch [18] the concept of GI is related to products with defined geographical origin. By explaining the origin and adding this value to products of the same origin, it translates the quality value and characteristics of the identity and culture of a geographical area into tangible assets. The producers of a region are organized to value these characteristics, mobilizing an intellectual property right: the GI. Therefore, the legal institute makes it possible to preserve characteristics of the product, as well as to value them in the face of consumers, making tangible intangible assets such as reputation, specific environmental factors and human skills, adding to them a certain value [19].

In order to recognize a GI, IPL and international agreements (mainly TRIPS) are observed. In addition, IPL provides that INPI will establish the conditions for registration. The current regulations that governing such conditions are Normative Instruction (IN) No. 25/2013 [7]. This IN defines the quality criteria, rights and duties of national producers that require protection. To apply for registration, associations, institutes or legal entities representing the collectivity are legitimized. Thus, in practice, who actually manages the GI are usually the producers linked to them.

According to IN No. 25/2013, in its art. 6, in order to formalize the registration with the competent public agency (INPI), the application must be accompanied by the geographical name and the corresponding official document of delimitation of the geographical area; description of the product; the instrument that prove the legitimacy of the institution representing the producers or providers; the graphical representation; the payment of the amount corresponding to the INPI; and of the Regulation of Use of the geographical name, which describe the rules regarding the use of the GI.

According to Vieira, Buainain & Bruch [20], in recent years, has shown a greater awareness of the importance of social and environmental issues related to the production and commercialization of agrifood products. There is a perception that public policy actions aimed only at increasing production, despite being important, are no longer sufficient for the increasingly globalized and competitive market. Therefore, the competitiveness of Brazilian agribusiness is based, today, on the valorization of the product, for its quality rating.

In view of the growing tendency to value intangible assets linked to territories, the GI can be considered as a tool to promote territorial development, since they allow the differentiation of products in an increasingly competitive market. According to Nierdele [10] the GI registry is an important strategy instrument for the induction of rural territorial development, as well as stimulating social actors to promote "qualification processes". This occurs, according to the author, by the fact of introducing a new model of food production and consumption, revaluation of traditions, customs, know-how and other immaterial goods associated with a specific territorial identity and geographical origin.

Other benefits also need to be considered. According to Cerdan et al [5] GIs can generate social and cultural benefits represented by the inclusion of producers or disadvantaged regions in the market and environmental benefits related to the preservation of biodiversity and local genetic resources.

In addition, according to Vieira and Pellin [21], complementary activities may emerge after the granting of the GI registration for traditional products. In most cases indications of origin and appellations of origin may establish relationships with other segments that are not directly linked to the product. This consequence can strengthen important activities, generating employment and local income.

DOI: 10.1051/bioconf/20170903007

Locatelli [8] corroborates Pecquer's assertions that it is possible to observe the development and strengthening of activities focused on tourism and gastronomy in many regions that have obtained the recognition of GI for their products, such as in *Vale dos Vinhedos*, Rio Grande do Sul, Brasil. According to the author, GI by stimulating the tradition and culture of a region attract tourists and enable the exploitation of indirect profitable activities.

In research carried out by Pellin & Vieira [13], the authors highlight three advantages: (i) increase in the production of products not directly linked to the GI, but which end up being benefited; (Ii) consumer recognition of the relationship between product and territory; and, (iii) insertion of innovations in the recognized products.

Regarding the first advantage, the research confirms the ability of the GI to stimulate complementary activities in the demarcated region, which often do not have direct relation with the recognized product, however, they benefit. It is what Pecqueur [12] classifies as 'basket of goods and services of the territory'. The consumer, when interested in purchasing products with GI, may also be interested in consuming other products or services from that demarcated region and, thus, stimulating the production and supply of other goods [13].

The second advantage pointed out shows the cultural dimension stimulated by the GI. By linking product with the territory, the consumer ends up strengthening the identity of the demarcated region and the products carry beyond quality – which is one of the requirements of the Regulatory Council of the GI for a product to achieve the sign – a territorial component that differentiates them from other products [13].

Finally, the third advantage suggested proposes a relationship with the production system. From the moment that the rules for production are defined, through the Regulation of Use, the tendency is for the products to have higher quality and innovations to be implemented in the production system, keeping, however, the typicality of the products and their system productive.

According to the research carried out by Pellin & Vieira [13], regarding the contributions of the GI for territorial development, four advantages were highlighted: (i) increase in the flow of tourists; (Ii) increased sales of products; and (iii) increase in the number of establishments in the GI region and increase of land and property price in the demarcated region. All the advantages are related to the economic dimension and that sometimes is worrisome. This is an activity that is stimulated in the majority of GI vitiviniculture experiences in Brazil and in the world. An example of the GI contributing to territorial development is observed in the *Vale dos Vinhedos*, given the considerable increase in tourist flow in the demarcated region that received in 2014, according to the Aprovale, 290 thousand tourists. As a comparison, in 2006 the flow of tourists was only 45 thousand. There are 26 wineries of the IG. The profiles are varied: there are family wineries, with limited elaboration and exclusive sale in their retail, as well as big companies with international presence. There are 43 associates related to the business of tourism between 9 hotels, an international SPA, 15 restaurants, two tourist agencies, and a travel operator; cheese industries, jellies, biscuits, art and handicraft workshops, furniture industry showrooms, among others. Therefore, it is important to remember that the tourism activity produces direct or indirect benefits for several segments generating employment and local income.

However, according to Pellin; Vieira [13], there is a need for a greater effort of approach between the various stakeholders (public sector, private sector and society) so that all involved, as well as the territory, can take advantage of the possible benefits that the IG allows. Partnerships need to be built and strengthened so that they can assist in the strategic development of the GI and receive support from the government sector so that public policies can be developed for further post-grant support of the GI registry [13].

Currently, some GI present quite expressive results, such as the *Vale dos Vinhedos*. But these results can not be 'romanticized' to the point that GI produces positive results automatically. On the contrary, as the *Vale das Uvas Goethe* representant pointed out, it was after three years of recognition of the IP that the work increased (challenges and demands). Maintaining the coherence and adherence of all members, the participation of stakeholders (public power, private sector and society) so that tourism in the region is recognized, is not an easy task [13].

In this sense, an interesting initiative has attracted attention in the scope of the GI. This is the proposal to develop technical standards to systematize good practices for the recognition, management and control of GI. This initiative is being carried out through a partnership signed between the Brazilian Association of Technical Standards – ABNT and the Brazilian Service of Support to Micro and Small Enterprises – SEBRAE, as discussed in the following item. ABNT represents Brazil in the International Organization for Standardization – ISO, internalizing international standardization standards, as well as taking Brazilian initiatives to the international level.

2.1. The Special Commission of Geographical Indication at the Brazilian National Standards Organization

With the motivation to solve problems related to the incorporation of quality, efficiency and best practices, as well as to increase the competitiveness of small and medium-sized enterprises (SMEs), in 2007 ABNT and SEBRAE signed an agreement aimed at promoting the involvement of SMEs with standardization [11]. The success obtained in this partnership resulted in the maintenance of the agreement, which until the moment of this publication is kept in execution.

The scope of this partnership foresees several actions related to the elaboration and implementation of the technical norms by the SME. Among them, actions are taken to collect sectoral demands for the development of technical standards with support to the participation of SMEs in national committee meetings. In addition, training is provided in the application of technical standards, as well as promotion of access to technical standards from subsidized sales.

The technical standards cover the market segments related to products, services, people skills and procedures. The process of elaboration of technical norms is based on the voluntary participation of specialists, being included in this concept the participation of the business owner, entrepreneur, public and private institutions linked to the respective segment, consumers, and others that have

DOI: 10.1051/bioconf/20170903007

interest in the elaboration and application of the standard.

Within the scope of this partnership, SEBRAE realized the possibility of establishing technical norms in projects aimed at the recognition of Brazilian Geographical Indications. Thus, to confirm the need for elaboration, two workshops were held with producers and entities related to the process, promotion, registration, support and research on GI for the collection of demands. These meetings were held in Bento Gonçalves – RS, on April 10, 2014, with 21 participants, and the second was held in Belo Horizonte – BH, on May 19, 2014, with 23 participants [15].

The result of these workshop was the confirmation of the need to elaborate technical standards for the Brazilian Geographical Indication system and the referral to ABNT for the formation of the Technical Committee. Initially, there was alignment of the participants of the two meetings for the elaboration of norms regarding terminology and for management of governance.

The installation meeting of the Commission for Special Studies for Geographical Indication – ABNT / CEE-216 was held on July 22, 2014 in the city of Rio de Janeiro, with the presence of 25 representatives of producers, consumers and governmental regulatory, development and of research. For the scope of the work it was established the elaboration of four technical norms: Terminology, Structuring of Geographical Indication, Management of Geographical Indication and Control of Geographical Indication.

The purpose of creating technical standards for the Geographical Indication system is to contribute guidelines for the processes aimed at the recognition of Geographical Indications in Brazil, regarding the terminology, structuring, management and control of the Brazilian GI. The technical standards are of voluntary application and act in a complementary character to the existing regulations in the INPI for the recognition of a Geographical Indication. It is worth noting that in Brazil there is no specific regulation for the phase after recognition of the GI, both in the issues related to the extinction or annulment of a recognized GI, and in the scope of its control and management.

Under the ABNT-SEBRAE agreement, the participants and representatives of recognized GI benefit from the participation in the meetings. The democratization of the participation of specialists in multidisciplinary and multi-institutional forms for the elaboration of technical norms allows them to be elaborated in an accessible way to technicians, producers and other interested parties.

It is important to emphasize that of this work that has been proposed to elaborate four technical norms, two are already published, notably the Brazilian Standard n. 16479, published on 09/08/2016, under the name Geographical Indication – Terminology, and Brazilian Standard no. 16536, published on 10/25/2016, under the name Geographical Indication – guidelines for structuring the Geographical Indication for Products. Standards for management and control are still under construction.

3. Methodology of the construction of the norm

The technical standards are essentially built in the context of the creation of a Study Group, named Special Commission for the Study of Geographical Indications – (SSC-216) of ABNT. For each theme, a specific committee is created with the participation of multi professionals, representatives of the entire productive chain, of consumers and of neutral parts such as: universities, laboratories, institutes, public agency, stakeholders, among others. One of the most important aspects of the construction of the standard is the principle of transparency and free access of the interested parties in the process of construction of the standard.

The process of constructing a standard uses the universal concept based on the PDCA (Plan, Do, Check and Act) cycle, based on this concept, the construction of the standard was performed with the following steps:

3.1. Plan:

This stage was initiated with the installation meeting of the Special Commission for the Study of Geographical Indications – (SSC-216) of ABNT, which took place on July 22, 2014, at the ABNT headquarters in Rio de Janeiro. This stage was of fundamental importance to ensure the other stages of the construction of the standard. The meeting was organized by ABNT, with record in minutes of all the discussions held during the meeting. In summary, the Planning activities for the beginning of the construction of the standard were:

- Publication and sending of invitations to interested parties;
- Opening of the meeting by ABNT representative;
- Presentation the reasons about the importance of the construction of the norm and justification regarding the relevance of the theme to society;
- Presentation of the participants in the meeting and justification of the absent guests;
- Selection of the coordinator (a) of the study committee by the members present at the meeting;
- Preparation of Work Program;
- Definition of the work schedule and deadlines;
- Definition and prioritization of the themes of the standards to be built.

The work program defined the development of four technical standards relating to:

a) terminology of geographical indication (216: 000. 00-001),
b) guidelines for the structuring of a geographical indication (216: 000.00-002),
c) management of a geographical indication – good practices (216: 000.00-003)
d) traceability and control mechanism (216: 000. 00-004).

All norms have been defined as norms of orientation and not as directional norms as the goal is to compile best practices.

3.2. Make:

This stage refers to the phase of elaboration of the contents of the technical standard already defined in the planning stage. It is one of the most important steps, which require more time, research and technical discussions with the members of the commission.

In this stage the construction of texts of partricipative and collaborative form, that are denominated initially like draft standards, that will form in sequence the text of the norm. It is important to emphasize that the norm is not made by a person but by a diverse group of people who vote and validate all text before the process of public consultation and publication. All should and can give their opinion, justifying their suggestions and proposals for changes, and they must be accepted without distinction by weight or other criteria.

The group of participants, called SSC-216, presents, discusses the text, until they reach a consensus on the final text.

The face-to-face meetings of SSC-216 occured at intervals of 30 to 60 days, according to the schedule elaborated by the commission, validated and published by means of minutes.

The face-to-face meetings are itinerant, allowing stakeholders from different regions to participate in the discussions and construction of the standards texts. At the same time, all members receive the information and referrals from the meetings and can contribute to the elaboration of the rules even if they do not participate in the face-to-face meetings. This can be done either by e-mail or by the system available, called livelink, in which all the minutes are available, as well as all the content discussed in the meeting, each version of the standard projects, as well as complementary materials that help the participants in the construction of the knowledge necessary for the elaboration of the norm.

The membership of the commission usually includes producers of GI, representatives of collective GI groups, representatives of public or private institutions that carry out support and research activities, GI public regulators, GI product consumers.

As a method to expedite the discussions and construction of the texts of the norms concomitantly, the members were invited to divide into subgroups of work, called GT, with each group being responsible for consolidating the texts received to present in person at the meetings. Thus, three (3) WGs were created: WG for Structuring, WG for control and WG for Management. This method facilitated the process of constructing the text in a collaborative way and with greater agility. The terminology standard was developed in conjunction with the large group during all meetings.

Success in this phase occurred systematically and organically, with the participation of an expressive number of suggestions, there was a consensual construction practice with higher productivity in a shorter time.

3.3. Check:

At this stage all construction of text is carried out by the study committee, being submitted to the public appreciation and public statement of every society. It is an important moment, in which the committee expects a significant number of people who can participate with suggestions and refinement of the text of the norm. The proposal is that at the end there is a standard with full and satisfactory application.

The Terminology Standard was submitted to the National Consultation, on February 4, 2016, after 18 face-to-face meetings, attended by 73 people, during the various meetings.

The Structuring rule was submitted to the National Consultation on June 14, 2016, after 19 face-to-face meetings attended by 61 people during the various meetings.

In both cases, the National Consultation was available for 60 days. After that the norm and the comments returned to the commission's analysis.

The national consultation allows interested parties who were not in person in the discussions of the text of the standard to manifest themselves. This way, it is also given the opportunity for everyone to have an opinion and to validate the document before its publication.

3.4. Action:

After the national consultation, all comments and suggestions are consolidated and appreciated by the committee in a face-to-face meeting specifically led by the Commission's coordinator. During the evaluations of the comments received, the committee consents and there may be rejection or acceptance regarding the comments, this is the moment of final decision of the text for publication of the standard.

The publication of the standard is formalized and its access is made available to every company, for payment by the value of the standard, which is calculated based on the number of pages. With this, it is up to the stakeholders to implement the standards in order to qualify their practices. The ABNT has as practice the periodic analysis of the standard. Thus, if there is a need for revision, the process is open so that the commission can carry out the process of continuous improvement.

As a result, after the Terminology standard was submitted to the National Consultation, it returned to the SSC-216, which evaluated all the comments in a total of 22 manifestations and, following several suggestions, recommended the final publication, which occurred on September 8th 2016, under the following name: Brazilian Standard no. 16479, under the name Geographical Indication – Terminology.

The standard of Structuring after being submitted to the National Consultation, returned to the Special Study Commission, which evaluated all the comments, accepting several suggestions, recommended the final publication, which occurred on October 25, 2016, under the following nomenclature Brazilian Standard n. 16536, under the name Geographical Indication – guidelines for structuring the Geographical Indication for Products (Geographical Indication – Guidelines for structuring of Geographical Indications for product).

4. Sustained development of GI for the Brazilian Wine

As in other countries, in Brazil, wine was the first product to have a recognized geographical indication. Brought from European colonization, this ancient culture was established in Brazil and gained specific characteristics of this *terroir*. It developed its own identity and consolidated itself, be it by tradition, landscape, or because it is a product that collaborates and collaborated with the development of producers and regions throughout history in the country.

By recognizing a GI in the wine sector in Brazil, this product is brought to visibility and placed under the eyes of its consumers, so that it knows its processes, its identity and the place where it is produced, as well as the context. These distinctive signs protect local traditional products, the social and economic structure of which they are inserted, thus showing a distinctive origin, with territorial identity and that brings security to the community, in addition to developing markets [6]. The author is still based on three important factors, product, process and place (territory), with intrinsic qualities of place, environmental practices, manufacturing techniques, traditional processes and others.

By establishing in particular places and protecting environmental and cultural resources, it brings with it the notion of *terroir*, the quality of an agricultural product, cultural practices, knowledge of the producers and resources that collaborated for that production [3]. For Brazilian wine, these practices attest to its identity and characteristics of this terroir, either because it is a tropical vitiviniculture, with specific edafocilmatic identity and characteristics, wines from tropical regions and with landscapes and particular *terroirs*, mountains, valleys and savannas, that expression values of important Brazilian biomes.

These benefits are based on attesting the specificity of the product, a particular terroir expressed in the wine, with local identity and tradition, as well as maintaining winemakers in the field, regional and social development, attracting tourists, providing farmers' interaction and giving visibility For those products on the market. Consumers seek these products recognized by GI traditional characteristics, have quality perception and are willing to pay as a premium product [16], which can provide economic growth for producers and region.

However, there is still much to develop for the recognition of GI in Brazil, mainly because it is relatively recent in some areas and regions.

There are particularly big issues related to the effective and long-term structuring of a GI. The formal recognition procedure is only one of these stages, since the effective construction of a trajectory that includes all the elements necessary for the GI to be recognized and have a healthy continuity after this recognition still constitutes one of the great obstacles to the processes of IG.

Of the 59 applications for recognition already granted in Brazil, 41 are for IP and 18 for DO. Of these, many have not yet effectively reached the market, especially for structural issues, in the implementation of the control and management of the GI. Of these, there is one DO for Brazilian wines – *Vale dos Vinhedos*, and sitx IP for Brazilian wines: *Vale dos Vinhedos*, *Pinto Bandeira*, *Farroupilha*, *Monte Belo*, *Vales da Uva Goethe* and *Altos Montes*. In the case of wine-growing GI, the percentage of GI that have effectively placed products on the market is much higher, if an analysis proportional to the number of GI recognized is made. Only one of them has not yet presented products already controlled at the market. In this sense, on the one hand, the participation of GI representatives from wine in the SSC-216 has been fundamental for an adequate construction of this norm, based especially on the cases in which an effective success can be verified.

It is in this sense that the SSC-216 of ABNT developed the first two standards, the first on terminologies and the second on structuring a Geographical Indication for product. The objective is precisely to give support to the structuring of the GI from its inception, besides allowing the consolidation of a vocabulary common to all GI.

It should be emphasized that in Brazil the legislation, already mentioned, is very sigela with respect to the regulation of the recognition of a GI, making it necessary to elaborate a support in which the producers can gather information on the best practices related to its structuring.

The two standards address in their framework tools for the recognition of natural and human resources in the region, as well as knowledge, biodiversity and local histories [1]. That is, the norms provide support so that the place can have instruments that enable it to know what the principles are and how to treat them in the request for recognition of the GI, and thus provide socioeconomic gains and the preservation of history and "know-how " [2], thus collaborating for the sustainable development of the GIs, as a support and instrument for the application for recognition.

The rules that are currently under development, notably related to the control and management of GI, aim to allow, once recognized, an effective continuity in the GI process, avoiding that many are born alive.

5. Methodology

The work was developed through an exploratory analysis of the work being done in the elaboration of the technical standards applied to the GI in the scope of the Special Studies Commission of Geographical Indications – (SSC-216) of ABNT. This research was based on a review of the literature on geographical indications, as well as on all material produced to date by EEC-216, available to all participants through the livelink platform.

From this perspective, the literature were read in order to understand if the publications collaborate for the sustainable development of the Brazilian winemaking and its establishment as natural resources, product particularity, *terroir* and maintenance of practices in winemaking.

6. Final Considerations

Considering the still recent publication of the standards, it was verified at the first moment the impossibility of analyzing what would be the effective impacts of the technical norms in the wine-growing GI, as initially proposed.

However, there is a gain that would not initially be considered: it was possible to systematize the best practices of structuring, managing and controlling GI available in Brazil through the elaboration of standards already published and those in the publication phase. In addition to scientific publications, this material, systematized by a large group of experts, was not available to the main stakeholders until now: GI producers.

In the face of the partnership between ABNT and SEBRAE, this confluence of people with a common purpose was possible, with the norms already published being the most expressive results. Its incorporation and use by those who aim to work for the recognition of delimited regions depends very much on the dissemination of this work.

Currently, three initiatives related to GI viticulture are underway, and has already taken into account the provisions of the technical standards. IP *Campanha Gaúcha* and IP *Vale do São Francisdo* are already in an advanced state of structuring. The project for the DO *Altos de Pinto Bandeira* already begins with this perception.

Certainly, after the completion of these processes, it will be possible to verify more effectively the applicability of these standards to the wine sector, a proposal that is intended for future work.

References

[1] Associação Brasileira de Normas Técnicas. NBR 16479: Indicação Geográfica – Terminologia. Rio de Janeiro. 5 p. (2016)

[2] Associação Brasileira de Normas Técnicas. NBR 16536. Indicação geográfica — Orientações para estruturação de indicação. Rio de Janeiro. 17 p. (2016)

[3] BOWEN, S. Embedding local places in global spaces: Geographical indications as a territorial development strategy. Rural Sociology **75**(2), 209–243 (2010)

[4] Brasil, "Lei N. 9.279 - de 14 de Maio de 1996. Que Regula direitos e obrigações relativos à propriedade industrial" (1996). [Online]. Available: `http://www.planalto.gov.br/ccivil_03/leis/L9279.htm` [Accessed: 04-Mar-2017]

[5] Cerdan, C.M.T., Bruch, K.L., Silva, A.L., Copeti, M., Fávero, K.C., Locatelli, L. Indicação Geográfica de produtos agropecuários: importância histórica e atual. In: Ministério da Agricultura, Pecuária e Abastecimento. Curso de propriedade intelectual & inovação no agronegócio: Módulo II, indicação geográfica / Ministério da Agricultura, Pecuária e Abastecimento; organização Luiz Otávio Pimentel – **4a ed.** – Florianópolis: MAPA, Florianópolis: FUNJAB (2014)

[6] Ilbery, B. et al. Product, Process and Place: An Examination of Food Marketing and Labelling Schemes in Europe and North America. European Urban and Regional Studies, **12**(2), 116–132 (2005)

[7] Instituto Nacional de Propriedade Industrial. Instrução Normativa nº25/2013. Estabelece as condições para o registro de Indicações Geográficas. Brasília (2013)

[8] Locatelli, L. Indicações Geográficas e Desenvolvimento Econômico. In. Barral, Welber; Pimentel, Luiz Otávio (Org). Propriedade intelectual e desenvolvimento. Florianópolis: Fundação Boiteux (2007)

[9] Locatelli, L., Souza, K. A proteção jurídica e o processo de reconhecimento das indicações geográficas no Brasil: aspectos introdutórios. In: Indicações geográficas: desafios e perspectivas nos 20 anos da Lei de Propriedade Industrial. Org. Locatelli, L. Lumen Juris: Rio de Janeiro, p. 3–16 (2016)

[10] Niederle, P. A. Indicações geográficas e processos de qualificação nos mercados agroalimentares". Indicações geográficas: qualidade e origem nos mercados alimentares. Org. Niederle, P. A. Porto Alegre: Editora da UFRGS, p. 23–54 (2013)

[11] Oliveira, M.C. Normalização e MPE: um caminho sem volta. Boletim ABNT, abril, (2013). [Online]. Available: <`http://abnt.org.br/paginampe/biblioteca/files/upload/anexos/pdf/04818c3f3916ca5b8d17e8033feb cd47.pdf`> [Accessed: 27-dez-2016]

[12] Pecqueur, B. Qualité e développement territorial: l' hyphotès e dupannier de biens et de services territorialisés. Economie Rurale, **261**, Paris (2001)

[13] Pellin, V., Vieira, A.C.P. Indicações geográficas no Brasil: uma perspectiva pro-registro.. In: Indicações geográficas: desafios e perspectivas nos 20 anos da Lei de Propriedade Industrial. Org. Locatelli, L. Lumen Juris: Rio de Janeiro, p. 89–112 (2016)

[14] Bruch, K. L. Signos distintivos de origem: entre o velho e novo mundo vitivinícola. **1**. ed. Passo Fundo: Editora IMED. 320p. (2013)

[15] Sebrae. Parceria ABNT / Sebrae Normalização e Pequenos Negócios. São Paulo. (2015). [Online]. Available: <`https://www.google.com.br/url?sa=t&rct=j&q=&esrc=s&source=web&cd=3&cad=rja&uact=8&ved=0ahUKEwj2707rweTRAhXDHJAKHcKFAbUQFggvMAI&url=http%3A%2F%2Fwww.abnt.org.br%2Feventos%2Fapresentacoes%2Fcategory%2F137-workshop-internacional-sobre-normalizacao-e-pequenos-negocios%3Fdownload%3D319%3Aparceria-abnt-sebrae&usg=AFQjCNFlWxqUUFKIW3ToZGwLmZGTBJNMZw&sig2=0LY6svwxnSueC1yfDCYKuA&bvm=bv.145822982,d.Y2I`> [Accessed: 26-dez-2016]

[16] Soam, S. K.; Hussain, M. Commercialization of indigenous health drinks as geographical indications. Journal of Intellectual Property Rights **16**(2), 170–175 (2011)

[17] Vieira, A. C. P.; Pellin, V. O uso do instituto das indicações geográficas como instrumento de promoção do desenvolvimento territorial rural– o caso dos Vales da Uva Goethe – Brasil. 20th APDR Congress, University of Évora (2014)

[18] Vieira, A. C. P.; Watanabe, M.; Bruch, K. L. Perspectivas de desenvolvimento da vitivinicultura em face do reconhecimento da Indicação de Procedência dos Vales da Uva Goethe. Revista Geintec **2**, 327–343 (2012)

[19] Vieira, A.C.P., Zilli, J. C., Bruch, K. L. As políticas públicas como instrumento para o desenvolvimento das indicações geográficas: o caso dos Vales da Uva Goethe em Urussanga, Santa Catarina In: Anais VII Seminário Internacional de Desenvolvimento Regional. Santa Cruz do Sul – RS: UNISC **1** (2015)

[20] Vieira, A.C.P.; Buainain, A.M.; Bruch, K.L. A indicação geográfica como estratégia para minimizar a assimetria de informação In: Propriedade intelectual e inovações na agricultura. Org. Buainain, A.M., Castro, A.C.,

DOI: 10.1051/bioconf/20170903007

Bonacelli, M.B.M. Rio de Janeiro: Ideia **1** (2015)

[21] Vieira, A.C.P.; Pellin, V. Indicações geográficas como políticas públicas de desenvolvimento territorial – o caso dos Vales da Uva Goethe. In: Valdir Roque Dallabrida. (Org.). Indicação geográfica e o desenvolvimento territorial: reflexões sobre o tema e potencialidades no Estado de Santa Catarina. **1a**, ed. São Paulo: Editora LiberArs, p. 273–288 (2015)

DOI: 10.1051/bioconf/20170903008

Brazilian sparkling wine: A successful trajectory

Douglas André Wurz[a], Ricardo Allebrandt, Betina Pereira de Bem, Juliana Reinehr, Adrielen Tamiris Canossa, Luiz Gabriel Dalmolin, Leo Rufato, and Aike Anneliese Kretzschmar

Santa Catarina State University (UDESC), College of Agriculture and Life Science, Av. Luiz de Camões, 2090, CEP 88520-000, Lages, SC, Brazil

Abstract. The objective of this study was to explore the evolution of internal and external commercialization of Brazilian sparkling wines during the period between 1986 and 2015, giving an overview of the current situation and its market trends, and highlighting the importance of Brazil in the world scenario for sparkling wines. This research is based on quantitative data sources provided by different institutions: International Organization of Vine and Wine (OIV), Brazilian Union of Viticulture (UVIBRA), Ministry of Agriculture, Livestock and Supply (MAPA) and Brazilian Agricultural Research Corporation (EMBRAPA). Brazil is a promising country for the sparkling wine market. From 1986–2015, there was an increase of 465.6% in the sales of sparkling wine in the Brazilian domestic market, especially the last ten, which accounted for an increase of 215.7%. In 1998, Brazilian sparkling wines accounted for 70.5% of sparkling wines sold in Brazil. Today, they represent 82.1%of national market, and proved to be one of the most prestigious products from Brazilian viticulture when compared, for example, to non-sparkling wines, which 79% are imported wines. Of the imported sparkling wines consumed in Brazil, six countries account for 98.5% of the market (France: 32.7%; Italy: 24.8% and Spain: 21.9%). The volume exportation of Brazilian sparkling wine has grown significantly in recent years (4,210.2%), with Paraguay, China, Uruguay, the United States, France and the United Kingdom being the major buyers. The evolution of exports shows that Brazilian sparkling wine becomes an alternative to those produced in traditional countries, such as France, Italy and Spain, due to their quality and price. There is an increase in the commercialization of sparkling wines in the Brazilian domestic market, in addition to the growing participation of Brazil in the volume of exports in the world market, placing Brazil in a leading position in the production and sale of sparkling wines.

1. Introduction

The Brazilian viticulture occupies an area of approximately 83,700 hectares, with an annual production varying between 1,300 and 1,400 tons, in which about 40% of the production is destined for processing, and one of the featured products is sparkling wine [1]. According to Mello [2], Rio Grande do Sul and Santa Catarina States are the main producers, representing 95% of all fine wines and sparkling wines market in Brazil.

Vitiviniculture has been an important activity for the economy of some regions, especially those located in the south of the country, where the largest volume of grape, wine and other products is produced. The importance of the wine industry in South Brazil is entirely related to the sustainability of small and medium-sized rural properties, which produce table grapes and grapes for processing, as well as those that act in the winemaking system, in order to contribute to the generation of employment and income [3].

The production of sparkling wines is lower compared to that of fine wines, but the economic impact of this product is very important because of its high added value. For this reason customer awareness of the quality is increasing and winemakers are constantly looking for product improvements and newproducts. In recent years a newmarket strategy in the oenological industry, based on the diversification of wine production, has increased [4].

Brazil is a booming market for the consumption of wines and sparkling wines, although per capita consumption is still very low. While each Brazilian consumes on average about 50 liters of beer and 6 of cachaça a year, wine consumption (fine wines) is around 2 liters per capita per year [5]. The highest sales growth recorded by the Brazilian wineries are the sparkling wines, among them the sparkling Moscatel, according to the Brazilian Wine Institute [6].

Brazilian sparkling wines participation in the market has become stronger in the national scenario. With the commercial opening occurred in the early 1990s, there was an entrance of imported wines from Chile and Argentina, subtracting space from the national products. However, this was not the reality of the segment of sparkling wines.

In the last 20 years, an increase in Brazilian sparkling wine consumption was recorded, as a consequence of the the increase in quality and tipicity, besides having a competitive price in the domestic market and in the foreign market.

Brazilian sparkling wines are the product of excellence for Brazilian winemaking, as recognized by awards in international enological contests and an increase in production and sales [7].

In this context, the objective of this work is to explore the evolution of the commercialization of Brazilian

[a] e-mail: `douglaswurz@hotmail.com`

DOI: 10.1051/bioconf/20170903008

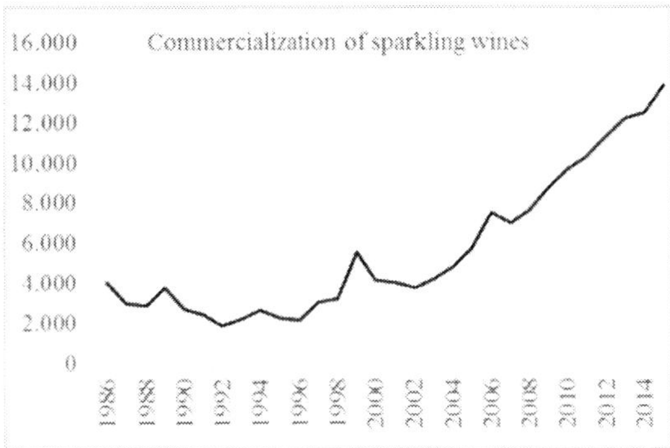

Figure 1. Evolution of the commercialization of sparkling wines in Brazil, in 1000 L, between 1986 and 2014.

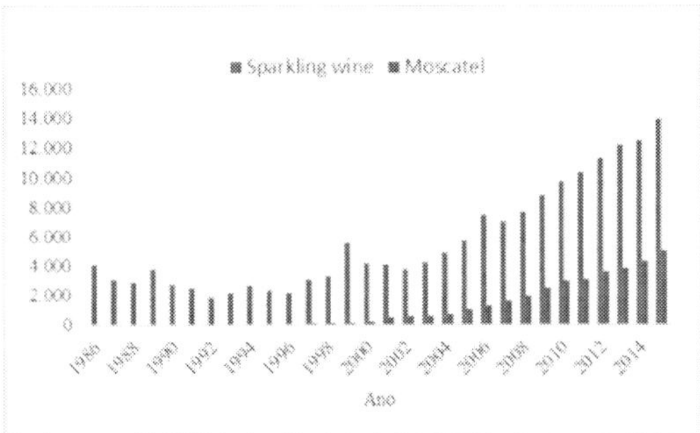

Figure 2. Commercialization volume (in 1000 L) of sparkling wine (Nature, Brut, Demi-Sec) and Moscatel in Brazil from 1986 to 2014.

sparkling wines in the internal and external market during the period 1990–2015, giving an overview of the current situation and its market trends, highlighting the importance of Brazil in the world scenario for sparkling wines.

2. Materials and methods

The study was based on a review of scientific literature and oficial data published by institutions responsible for collecting and disseminating data on the production and marketing of fine wines and sparkling wines in Brazil. For a characterization of the Brazilian sparkling wine market, the following data were researched from the last 30 years: commercialization of national products, importation and exportations.

The institutions assessed were: Instituto Brasileiro do Vinho, IBRAVIN , Embrapa Uva e Vinho, Ministério da Agricultura, Pecuária e Abastecimento, MAPA and União Brasileira de Vitivinicultura, UVIBRA.

All collected data were used to generate graphs with the purpose of showing the panorama and determining trends of the commercialization of wines and sparkling wines in Brazil.

3. Results and discussion

Brazil has a favorable scenario in the sparkling wines market due to the continued growth in the marketing of sparkling wines since 2000 (Fig. 1). From 1986–2015, there was an increase of 465.6% in the sales of sparkling wine in the Brazilian domestic market, especially the last ten, which accounted for an increase of 215.7%.

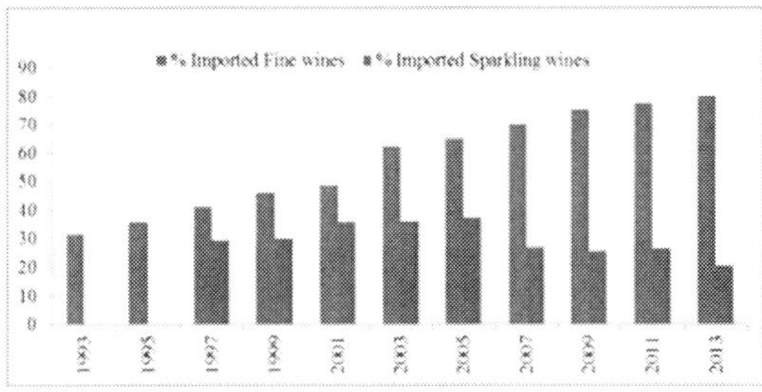

Figure 3. Percentage (%) of imported fine wines and imported sparkling wines marketed in Brazil from 1993 to 2013.

This market is still characterized by seasonality and the consumption of sparkling wine is still lower than the other categories of wines, however, they have been growing sharply in recent years. Its consumption is mostly associated with special occasions and parties, such as the end of the year and other celebrations.

Before 2000, it was observed oscillations of the commercialized volumes of sparkling wines in Brazil. However, from the year 2006 onwards the growth in the marketing of sparkling wines became very significant, going from 7.5 million liters in 2006 to 12.9 million liters in 2014.

The growth of the sparkling wine is pronounced regardless of sparkling categories (Nature, Brut, Demi-Sec or Moscatel). However, the Moscatel style is one of the most promissing categories. Between 2004 and 2014, the volume of muscatel wine sales increased from 0.7 million to 4.3 million, representing a 614% increase in sales (Fig. 2).

One of the determining factors for the oscillations occurred in the commercialization of fine wines in Brazil is related to the quantity of imported wines marketed in the country. In 2014, imported fine wines accounted for 79% of the total volume of wines marketed in the country (Fig. 3). In 1993, imported wines accounted for only 19% of the fine wine market in Brazil.

According to Protas [8], the segment of fine wines faced a strong competition, registering significant growth rates of imported wines, due to the process of opening the Brazilian economy abroad. At the same time, Brazilian consumers started to drink more wines for the last 15–20 years, specially wines from Chile and Argentina [5].

Therefore, over the last 20 years, there has been a considerable increase in the share of imported fine wines marketed in the country (Fig. 4). While for the sparkling segment, there is an inverse trend, that is, over the years, imported sparkling wines have been losing space in the Brazilian market. Much is due to the quality of Brazilian sparkling wine, which over the years has been evolving in quality and conquering the national consumer.

When comparing the marketed values of imported sparkling wines and national sparkling wines, there is an inverse behavior of fine wines. For the sparkling segment, in 2014, only 20.5% of sells come from imported products (Fig. 4). In this context, there is a consolidation of sparkling wines in the Brazilian wine market.

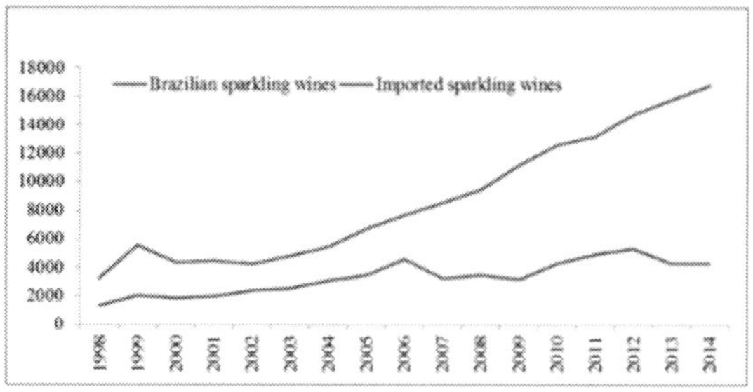

Figure 4. Total Commercialized Value (in 1000 liters) of imported sparkling wines and brazialian sparkling wines from 1998 to 2014.

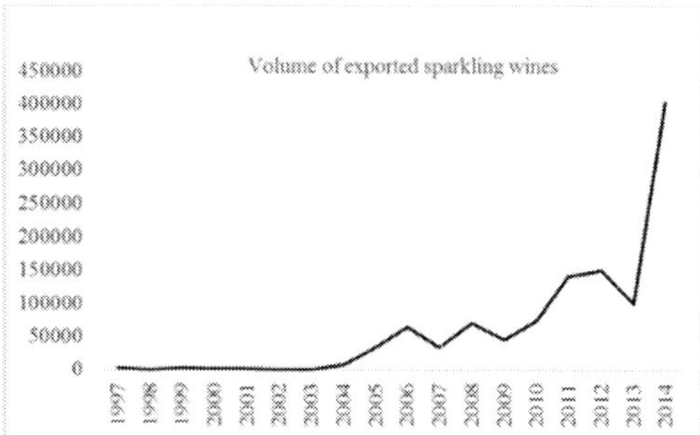

Figure 5. Volume of sparkling wines exported (Liters) by Brazil from 1997 to 2014.

Imported sparkling wines consumed in Brazil come mostly from six countries, which together account for 98.5% of the market: France represents 32.7%, Italy 24.8% and Spain 21.9% [9].

The exported volume of Brazilian sparkling wine has shown a notable growth in recent years, with a growth of 4,210.2% (Fig. 5), and the main buyers are Paraguay, China, Uruguay, the United States, France and the United Kingdom. The evolution of exports shows that Brazilian sparkling wine becomes an alternative to those made in traditional countries, such as France, Italy and Spain, due to their quality and price.

According to Caligari [10], Brazilian wine companies have a limited vision, focused on the present and usually to not predict what may happen in the near future, they have a simplistic way of managing, lack of technical and human resources, especially in the research and marketing. In this sense, the Brazilian Wine Institute - Ibravin, is developing the Wines of Brazil program.

The project Wines of Brazil aims to promote the expansion of the consumption of wines, sparkling wines and other grape derivatives in the Brazilian market. It develops promotional campaigns for consumers and channels, market access projects, fairs, tasting events and training of wine service teams. It is also responsible for the relationship of the sector with buyers, distribution channels, journalists, sommeliers and opinion makers. In the same sense, the Wines of Brasil team works in the dissemination of wine culture in the behavioral, gastronomic and technical fields, using an accessible language to all wine lovers. The program is funded by the Vitis Fund, besides the capital invested by the participating companies, and it is also supported by strategic partners such as SEBRAE and SENAC [11].

4. Conclusion

The evolution of exports shows that Brazilian sparkling wine becomes an alternative to those produced in traditional countries, such as France, Italy and Spain, due to their quality and price. There is an increase in the commercialization of sparkling wines in the Brazilian domestic market, in addition to the growing participation of Brazil in the volume of exports in the world market, placing Brazil in a leading position in the production and sale of sparkling wines.

References

[1] L.M.R. Mello. *Vitivinicultura Brasileira: panorama 2013.* (Comunicado Técnico 156, Embrapa Uva e Vinho, Bento Gonçalves, 6, 2014)
[2] L.M.R. Mello. *Vitivinicultura Brasileira: panorama 2012.* (Comunicado Técnico 137, Embrapa Uva e Vinho, Bento Gonçalves, 5, 2013)
[3] V.N. Duarte. Estudo da Cadeira Produtiva do Vinho em Santa Catarina: características e estágio atual. Evidências **13**, 45–56 (2013)
[4] M.A. Pozo-Bayon, A. Martínez-Rodrígues, E. Pueyo, M.V. Moreno-Arrivas. Chemical and biochemical features involved in sparkling wine production: from a tradicional to an improved winemaking technology. Trend in Food Science Technology **20**, 289–299 (2009)
[5] M. Copello. As regiões produtoras do Brasil. Revista Anuário Vinho do Brasil, São Paulo (2015)
[6] IBRAVIN – Instituto Brasileiro do Vinho Dados estatísticos – Produção de uvas no Brasil (IBRAVIN, Bento Gonçalves, 2015)
[7] M. Gabbardo, F. Battistutta, E.T. Gabbardo, L. Tat, E. Celotti. Aromatic characterization of Brazilian sparkling wines using olfactometry and sensory panel. BIO Web of Conferences **7**, 1–5 (2016)
[8] J.F. Protas, U.A. Camargo, L.M. Mello. A vitivinicultura brasileira: realidade e perspectivas (Embrapa Uva e Vinho, Bento Gonçalves, 2002)
[9] OIV – Organization Internationale de la Vigne et du Vin. State of World Vitiviniculture situation (OIV, Paris, 2014)
[10] N.R.V. Caligari. *Planificacin estrategica del sector vitivinicola de "Rio Grande do Sul"* (Thesis, São Leopoldo, 291, 2000)
[11] IBRAVIN – Instituto Brasileiro do Vinho. Promoção e Comunicação Mercado interno – Vinhos do Brasil (IBRAVIN, Bento Gonçalves, 2017)

DOI: 10.1051/bioconf/20170903009

Diversité des AOC – diversité des labels environnementaux-diversité des consommateurs : quelle stratégie de valorisation pour la filière viticole ? Diversity of the AOC – diversity of the environmental labels diversity of the consumers: Which strategy of valorization for the wine sector?

Frédérique Jourjon, Cécile Coulon, et Ronan Symoneaux

USC 1422 GRAPPE, Univ. Bretagne Loire, Ecole Supérieure d'Agricultures (ESA), INRA, SFR 4207 QUASAV, Angers, France

Résumé. L'objectif du travail est de fournir aux professionnels de la filière viticole des éléments objectifs sur la pertinence d'une stratégie collective de communication environnementale sur les vins d'AOC, de tester la pertinence d'une segmentation des consommateurs selon leur niveau d'implication vis-à-vis du vin et vis-à-vis de l'environnement, de mesurer l'interaction perçue par les consommateurs entre AOC et environnement selon le niveau de notoriété des AOCLe travail a été mené sur 4 années à partir d'enquêtes réalisées sur internet et sur les réseaux sociaux auprès de 3500 consommateurs français. Les résultats obtenus permettent de montrer des différences de perception par les consommateurs sur le lien entre « catégories de vins » et « production respectueuse de l'environnement »Globalement les vins d'AOC sont perçus comme significativement plus respectueux de l'environnement par rapport aux vins d'IGP eux-mêmes mieux perçus que les vins de France. Par ailleurs la réputation d'une AOC peut avoir une influence sur la perception du respect environnemental. Enfin pour les consommateurs, l'impact de la région viticole semble plus important que la réputation d'une AOC quant à « l'image environnementale perçue ». Ce travail permet de donner des éléments d'analyse sur la perception des consommateurs vis-à-vis des liens éventuels entre signes de qualité et production respectueuse de l'environnement. Il semble clair dans le cas de la filière viticole française que l'usage de messages environnementaux comme élément de différenciation et compétitivité pour les vins d'AOC ne répond pas à une attente de la majorité des consommateurs.

Abstract. The objective of this work is to provide for professionals of the wine sector objective elements on the relevance of a strategy of environmental communication in PDO wines and to measure the interaction perceived by the consumers between PDO and environment according to itslevel of notoriety. The work was undertaken between 2012 and 2015 starting from x-ray groups and 3700 surveys conducted into Internet and the social networks. The results got clear differences in perception by the consumers on the link between "categories of wines" and "respectful production for the environment". Overall the wines of PDO are perceived like significantly more respectful for the environment than the better perceived wines of IGP, themselves better percieved than the wines of France. In addition, the reputation of aPDO can affect the perception of the environmental respect For the consumers, the impact of the wine region seems more important than the reputation of a PDO as for "the perceived environmental image" It seems clear in the case of the French wine sector than the use of environmental messages as element of differentiation and competitiveness for the wines of PDO does not answer waiting of the majority of the consumers.

1. Introduction

Depuis plusieurs années, la protection de l'environnement est devenue une priorité pour les pouvoirs publics et pour les filières agricoles entrainant en Europe et dans le monde, une multiplication des démarches environnementales et des dispositifs d'affichage environnemental à l'échelle des produits, tous secteurs et tous produits confondus. Plusieurs études (Ernst et Young, 2009; IPSOS 2010) montrent clairement que les consommateurs français sont de plus en plus sensibles à une information environnementale et qu'ils se déclarent favorables à un affichage environnemental obligatoire. Ces éléments tendent à encourager les filières agricoles à des stratégies de communication environnementale comme facteur clef de différenciation et de compétitivité.

En France, la loi n° 2009–967 stipule que « les consommateurs doivent disposer d'une information environnementale sincère, objective et complète portant sur les caractéristiques globales du couple produit/emballage » pour les produits de grande consommation. Des études montrent qu'il existe une volonté des consommateurs de disposer d'une évaluation de la performance environnementale, notamment sous forme d'un affichage compréhensible et « universel » (Ernst et Young 2009 ; Enquête Ethicity 2011, 2012). Cependant, pour la filière vitivinicole, des études ont montré que la majorité des consommateurs « ne recherchent pas, lorsqu'ils choisissent un vin, d'informations complémentaires sur l'environnement » et « sont peu intéressés par un label environnemental »(Jourjon et Symoneaux 2013a).

A ce jour, aucun travail n'a été réalisé sur l'impact de la notoriété d'un vin sur la perception environnementale par les consommateurs. Cette question est importante dans le cas d'un produit agricole à forte valeur ajoutée, offrant une très large gamme de produits différant par leur label, leur notoriété, leur prix et leur qualité organoleptique ou environnementale. Bien qu'aucune contrainte ne figure dans le cahier des charges des AOC au sujet du respect de l'environnement, ce signe d'appellation pourrait avoir aux yeux des consommateurs une valeur plus ou moins forte selon leur implication dans cette problématique environnementale (Jourjon et Symoneaux 2013a).

Lors de l'achat en supermarché, le consommateur choisit un vin surtout en fonction du vignoble et de l'appellation (Interloire 2013). La majorité des études sur la consommation du vin segmente la population en fonction de la fréquence de consommation (Brugière 2009). Nous avons montré l'importance d'une segmentation en fonction du niveau d'implication dans le monde du vin. Parallèlement, d'autres études se sont attachées à comprendre les implications d'une « conscience environnementale »chez les consommateurs (Barber et al. 2009 ; Najar et Zaiem 2010 ; Vermeir et Verbeke 2006 ; Demarque et al. 2011 ; Xiao et Dunlap 2007 ; Laurent et Kapferer 1985 ; Michaelidou et Dibb 2008 ; Mueller et Remaud 2010 ; Sirieix 2010 ; Stolz et Schmid 2008 ; Zaichkowsky 1985).

Ce travail cherche à étudier la perception du respect environnemental associée par le consommateur à un label de type AOC. Le premier axe de travail tente d'établir si le consommateur est influencé par son niveau d'implication dans l'environnement et dans le monde du vin. Le second axe vise à savoir si, dans l'esprit du consommateur, une AOC respecte plus l'environnement qu'une autre catégorie de vin (IGP, Vin de France). Enfin, le troisième axe de travail se demande si la réputation d'une AOC influence la perception environnementale du consommateur.

2. Matériel et méthodes

Cette étude quantitative poursuit les précédents travaux menés par Jourjon et Symoneaux (2013b). Les résultats présentés s'appuient sur les enquêtes réalisées entre 2012 et 2015 auprès de 3700 consommateurs au total.

Les questionnaires d'enquête s'articulent autour de trois axes :

- l'implication du consommateur dans le monde du vin
- l'implication du consommateur dans l'environnement
- la perception du consommateur du respect porté à l'environnement dans la la production d'un vin de type AOC (Appellation d'Origine Contrôlée), IGP (Indication Géographique Protégée) ou VDT (Vin de France).

Pour les AOC, des questions spécifiques sont intégrées sur leur réputation et leur région viticole.

Les questions relatives à l'implication du consommateur dans le monde du vin sont issues d'études précédentes (Barber et al. 2009 ; Demarque et al. 2011 ; Enquête Ethicity 2011, 2012 ; Ernst et Young 2009 ; Laurent et Kapferer 1985 ; Michaelidou et Dibb 2008 ; Mueller et Remaud 2010 ; Najar et Zaiem 2010 ; Sirieix 2010 ; Stolz et Schmid 2008 ; Vermeir et Verbeke 2006 ; Xiao et Dunlap 2007 ; Zaichkowsky 1985 ; Jourjon et Symoneaux 2013b). Les questions liées à l'implication du consommateur dans l'environnement sont basées sur le modèle proposé par Schlegelmilch et al. (1996).

Selon ce modèle, l'intérêt du consommateur pour l'environnement peut être évalué à travers son comportement d'éco-achat, ses connaissances et attitudes envers l'environnement, son comportement face au recyclage et l'importance portée aux initiatives politiques en matière environnementale.

Pour chaque variable, les questions sont présentées aux enquêtés selon une échelle de Likert de 1 à 5 (1 = « pas du tout d'accord » et 5 = « tout à fait d'accord »). Au retour des questionnaires, la pertinence des questions concernant les implications vis-à-vis du vin et de l'environnement a été vérifiée et validée par ACP. Le troisième axe concernant la perception du respect environnemental de la production du vin est divisé en quatre sous-parties : i) le respect environnemental de chaque AOC (10 questions), ii) la perception de la notoriété de chaque AOC (10 questions), iii) la perception du respect environnemental de chaque catégorie de vin (AOC, IGP, VDT) et iv) la perception du respect environnemental de chaque région viticole (5 questions).

Pour la sous-partie concernant les AOC, deux AOC ont été choisies par région viticole : une très réputée, l'autre moins réputée et correspondant à une appellation générique (ex: pour Bordeaux: « Saint Emilion » et « Bordeaux »). Les AOC et les régions ont été choisies d'après les travaux de Cassagnes (2013) et Gotti (2012) sur la notoriété des AOC françaises en France. Pour éviter tout biais dans les réponses, chaque AOC a fait l'objet d'une question indépendante. L'ordre des questions concernant chacune des AOC a été randomisé avant la diffusion du questionnaire (pour éviter d'avoir à chaque fois l'AOC très réputée avant l'AOC réputée) mais reste le même pour tous les questionnaires.

Les réponses ont été collectées par internet via le logiciel Question Data. Les questionnaires ont été envoyés par mail, en utilisant les bases de données fournies par les Vignerons Indépendants de France (VIF), Familles Rurales et l'Ecole Supérieure d'Agriculture d'Angers (ESA). Il a aussi été diffusé sur les forums « Consommation éthique » et « 60 millions de consommateurs », et dans des réseaux personnels. La segmentation de l'échantillon en fonction de l'implication vin et environnement est obtenue par la somme des notes données aux questions de l'axe 1 et 2. L'échantillon est ensuite divisé entre trois classes, pour avoir trois niveaux d'implication (Faible, Moyen, Fort) : V1, V2, V3 et E1, E2, E3. Les bornes de chaque classe ont été déterminées afin de pouvoir analyser statistiquement chaque classe Les données sont traitées à partir de tris à plat, de tris croisés, d'analyse en composantes principales (ACP) et d'analyse de variance (ANOVA). L'analyse des résultats comprend plusieurs étapes: une vérification de l'influence de l'implication vin et de l'implication environnement sur la notation de la perception du respect environnemental, puis une analyse de la perception du respect environnemental par catégorie de vin, par région et par AOC. Les différentes classes

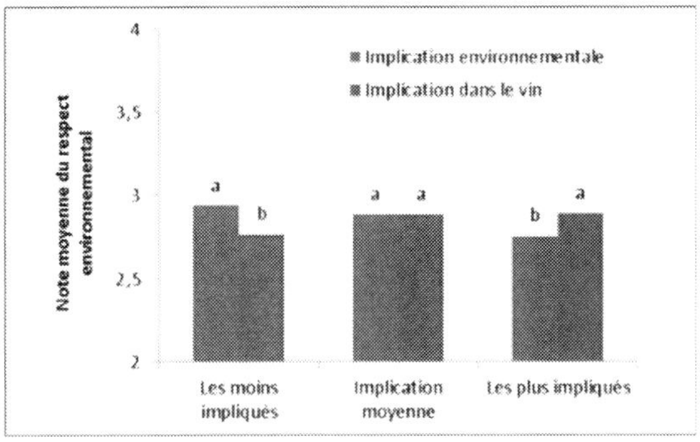

Figure 1. Note moyenne de la perception du respect environnemental du vin pour chaque classe de notre échantillon. Implivin = personne impliquée dans le monde du vin; implienv = personne impliquée dans l'environnement.

établies sont croisées avec les questions du troisième axe de l'enquête pour identifier d'éventuels comportements divergents entre les groupes. Ces traitements sont réalisés avec le logiciel XLSTAT 2014.

3. Résultats

Dans un premier temps, l'influence de l'implication « vin » et « environnement » sur la perception du respect environnemental du vin a été vérifiée pour toutes les catégories de vin confondues (Fig. 1). La segmentation des consommateurs impliqués dans le vin se distingue en deux groupes : **les personnes les moins impliquées (V1) considèrent le vin comme moins respectueux de l'environnement que des personnes plus impliquées** (V2 et V3). Ceci signifie que plus une personne est impliquée dans le monde du vin, plus elle a une perception favorable du respect environnemental de la production de vin. La corrélation est inverse concernant l'implication environnementale : **les personnes les plus impliquées dans l'environnement notent plus sévèrement le respect environnemental du vin que des personnes moins impliquées (E1 et E2).** Ces différences sont significatives, mais il faut être prudent vis-à-vis de l'échelle utilisée, les notes moyennes de chaque classe étant assez proches.

La même analyse a été menée pour chaque catégorie de vin: AOC, IGP et VDT (Fig. 2). L'analyse statistique montre des différences significatives entre chaque catégorie. Les vins de types AOC ont en moyenne une note supérieure à celle des vins IGP, elle-même supérieure à celle des vins de France. Donc en moyenne, **les personnes de l'échantillon considèrent que la production d'un vin AOC respecte plus l'environnement que celle d'un vin de type IGP, elle-même plus respectueuse de l'environnement que celle d'un vin de France**. Ces résultats ne dépendent pas du niveau d'implication « vin » et « environnement » de chaque individu : une personne non connaisseuse du vin (V1) ou non impliquée environnementalement (E1) notera dans le même ordre le respect environnemental des AOC, des IGP et des VDT qu'une personne très impliquée (V3 ou E3).

La Fig. 3 présente la distribution des réponses concernant les résultats présentés sur la Fig. 3 : ce n'est

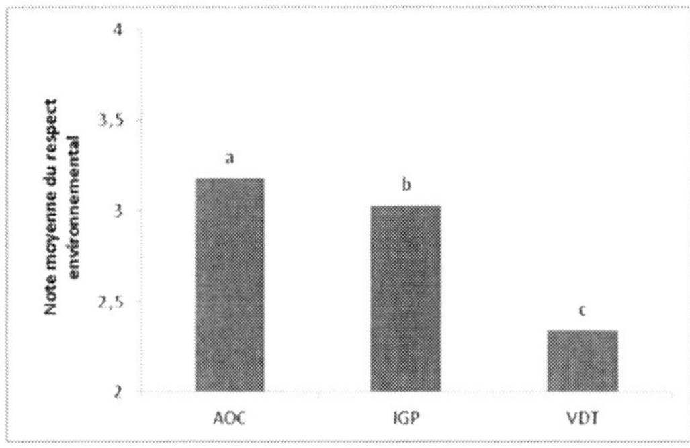

Figure 2. Note moyenne de la perception du respect environnemental de chaque catégorie de vin.

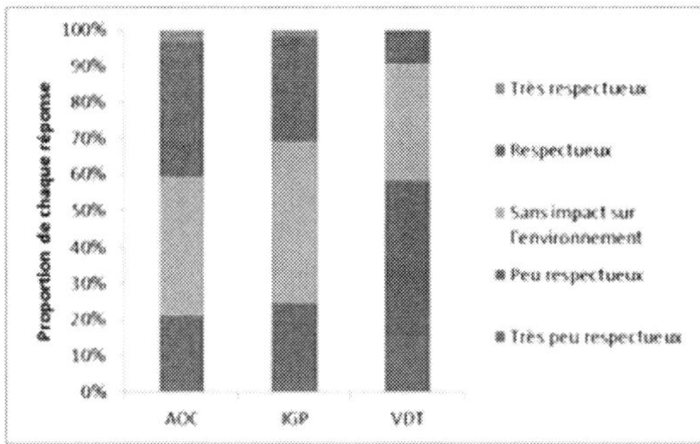

Figure 3. Proportion de chaque type de réponse dans la notation moyenne des catégories de vin.

pas parce que les VDT ont la note moyenne la moins respectueuse que tous les enquêtés ont mis une note basse. En fonction de la catégorie de vin, il y a une proportion de chaque réponse plus ou moins importante: la catégorie AOC a une proportion plus importante de notes « respectueuses » et « très respectueuses » que les catégories IGP et VDT.

En observant uniquement les résultats de la catégorie AOC, nous avons comparé entre elles les 5 régions viticoles retenues. L'implication environnementale a été croisée avec les notes de perception du respect environnemental de la production pour chaque région: en accord avec nos résultats précédents, plus une personne est impliquée dans l'environnement plus sa perception du respect environnemental est faible.

L'Alsace est mieux notée que les autres régions et celle de Bordeaux jugée moins respectueuse de l'environnement par tous les groupes, en particulier par les plus impliqués (E3). Le même type de croisement a été fait entre implication vin et note du respect environnemental: plus une personne est impliquée dans le vin plus sa perception du respect environnemental est forte, sauf dans le cas des régions Vallée du Rhône et Bordeaux. Que ce soit pour l'implication vin ou l'implication environnement, Bordeaux ressort avec l'image de la région la moins respectueuse de l'environnement par toutes les classes de notre échantillon et surtout par les plus impliquées (E3 et V3). La Bourgogne est aussi jugée plus sévèrement mais dans une moindre mesure par ces deux classes, tandis que la Vallée du Rhône obtient une note identique pour les classes V2 et V3.

La comparaison de la perception du respect environnemental entre les 10 AOC du questionnaire fait apparaître deux cas de figure: soit l'AOC à très haute réputation est perçue comme plus respectueuse de l'environnement que l'AOC à moindre réputation (Saint Emilion vs. Bordeaux; Châteauneuf du Pape vs. Vallée du Rhône), soit il n'y a pas de différence significative entre les deux AOC d'une même région, c'est le cas notamment pour les régions Alsace, Vallée de la Loire et Bourgogne. **La perception du respect environnemental d'une AOC ne dépend donc pas uniquement de son niveau de notoriété, la région viticole de l'AOC y contribue aussi fortement** (Figs. 2 et 3).

4. Discussion

L'approche adoptée dans cette étude étant relativement nouvelle, les résultats présentés sont difficilement comparables à d'autres travaux. Ceux qui concernent la « hiérarchisation » des AOC et des régions par rapport à leur réputation rejoignent cependant les constats de Cassagnes (2013) et de Gotti (2012) sur la réputation et la notoriété des AOC et des régions.

La difficulté majeure de cette étude réside dans la représentativité de l'échantillon dans les notions d'implication « vin » et « environnement ». D'une part, il n'existe pas de données nationales permettant de valider la segmentation de notre échantillon et d'autre part on observe de manière récurrente dans ce genre d'étude, une surreprésentation des personnes les plus impliquées dans le vin et dans l'environnement. Deux raisons majeures sont à l'origine de ce constat : la première provient des bases de données utilisées pour la diffusion du questionnaire, qui surreprésentent les « cadres et professions intermédiaires » (45 % de notre échantillon), biaisant ainsi la part de consommateurs à implication vin et environnement plus élevée. La seconde tient au fait que l'échelle de Likert n'a pas été entièrement utilisée par les enquêtés, qui ont répondu sur une partie restreinte de celle-ci. Par exemple, très peu de personnes ont noté 1 ou 2. Nous avons construit les classes d'implication en tenant compte de ce décalage de notation, de manière à obtenir trois niveaux d'implication (Faible, Moyen, Fort) avec des intervalles de notation égaux sur une échelle corrigée. Cette segmentation a permis de donner des effectifs équivalents pour chaque classe.

L'échantillon se caractérise également par une surreprésentation d'individus issus de la région ouest. Nous avons vérifié si cette surreprésentation induisait un biais trop important dans l'analyse de nos résultats, notamment pour la note basse attribuée à la région bordelaise. Pour cela, les analyses ont été refaites en comparant les réponses des individus de l'ouest à celles des autres consommateurs, les différences ne sont pas significatives. Même surreprésentées, les réponses d'habitants de l'Ouest coïncident avec celles des autres régions : le biais reste donc mineur.

Il est cependant important de relativiser les résultats car, à la question « Quelles sont les trois principales raisons pour lesquelles vous choisiriez un vin AOC plutôt qu'un autre type de vin ? *(Cochez trois réponses)* », la réponse « des modes de production moins polluants » n'est sortie qu'à 4 %, montrant que **le respect environnemental ne fait pas partie des trois premières attentes associées à une AOC, qui sont : « tradition », « savoir-faire » et « lieu de production garanti et meilleure qualité gustative ».**

5. Conclusion

Cette étude confirme à nouveau la pertinence de la segmentation des consommateurs selon leur implication dans le monde du vin et/ou dans l'environnement, en particulier pour des questions spécifiques comme la perception des démarches environnementales. Plus une personne est impliquée dans le monde du vin, plus elle considère les processus de production du vin comme respectueux de l'environnement. A l'inverse, plus une personne est impliquée dans l'environnement, moins elle considère les processus de production du vin comme respectueux de l'environnement.

Les consommateurs perçoivent significativement des différences entre les indications de qualité du vin, dans le respect de l'environnement lors des processus de production. L'AOC semble effectivement perçue comme un peu « plus respectueuse de l'environnement » que l'IGP. Dans l'esprit du consommateur, les Vin de France sont moins respectueux de l'environnement.

Au-delà des trois facteurs influençant la perception environnementale (implication dans le vin, dans l'environnement et catégorie de vin), la réputation d'une AOC peut aussi jouer un rôle. Dans le cas de Bordeaux et de la Vallée du Rhône, les AOC à très forte réputation sont ainsi significativement perçues comme plus respectueuses de l'environnement que celles à moyenne réputation. Dans les autres régions étudiées (Val de Loire, Alsace, Bourgogne), aucune différence significative n'est observée entre les deux niveaux AOC. L'impact de la région viticole semble donc plus important que la réputation d'une AOC pour le respect de l'environnement aux yeux des consommateurs.

Dans un contexte de montée en puissance des mouvements bio et de sensibilité sociétale accrue vis-à-vis des questions environnementales, il peut être intéressant pour les bassins de production viticoles d'intégrer cette perception environnementale des consommateurs et d'être attentifs dans leur stratégie de communication aux démarches environnementales mises en œuvre sur un plan national ou dans chaque région.

En France, l'accord signé le 3 mars 2016 entre le Ministre de l'agriculture et le président de l'INAO, intégrant 6 mesures agroécologiques dans l'objectif de contrat des AOC, apparait à ce titre comme un premier élément de réponse pertinent pour les régions et les producteurs engagés dans des démarches d'AOC.

Les auteurs tiennent à remercier les étudiants du domaine d'approfondissement Marchés, Entreprises et Territoires du ViN de l'ESA d'Angers ainsi que du Master International Vintage pour la réalisation et l'administration des questionnaires.

 DOI: 10.1051/bioconf/20170903009

Références

[1] Barber N., Taylor C. & Strick S., 2009. Wine consumers' environmental knowledge and attitudes: influence on willingness to purchase. International Journal of Wine Research **1**(1), 59–72
[2] Brugière F., 2009. Pratiques culturales sur vignes et pratiques œnologiques : connaissances et opinion des Français. Vinifhlor-Infos Vins et cidres **160**, 71–80
[3] Cassagnes J., 2013. Vins de Loire - Une notoriété en hausse auprès des consommateurs français. Accès: www.Viti-net.com [10.03.2014]
[4] Demarque C., Monaco G. L., Apostolidis T. et al., 2011. Socialisation, perspectives temporelles et implication personnelle: une étude dans le champ de l'environnement. Cahiers Internationaux de Psychologie Sociale **92**, 353–369
[5] Enquête Ethicity, 2011. Les Français et la consommation responsable – On se prend en main http://www.blogethicity.net/share/docs/
[6] Enquête Ethicity, 2012. Les Français et la consommation responsable – Le Ernst & Young, 2009. Pistes pour un étiquetage environnemental lisible et efficace – Résultat d'enquête consommateurs, 31 p. plaisir de la simplicité ! http://www.blog-ethicity.net/share/
[7] Gotti L., 2012. Le Top 5 des appellations. 365 jours en Bourgogne. Le blog de Laurent Gotti. www.laurentgotti.com. Accès : [10.03.2014]
[8] Interloire, 2013. Enquête consommateurs - Choisir son vin, les six critères clefs [en ligne] http://www.vitinet.com/vigne_vin/article/choisir-son-vin-les-six-
[9] Jourjon F. & Symoneaux R., 2013a. Comment le niveau d'implication des consommateurs de vin modifie leurs perceptions et attentes d'un affichage environnemental du vin. Bulletin OIV **86**, 992–994
[10] Jourjon F. & Symoneaux R., 2013b. Perception des consommateurs et intérêt d'un étiquetage environnemental pour les productions de vins d'AOP. Revue des Œnologues et des Techniques vitivinicoles et œnologiques **40**(148), 53–56
[11] Laurent G. et Kapferer J. N., 1985. Measuring consumer involvement profiles. Journal of Marketing Research (JMR) **22**(1), 41–53
[12] Michaelidou N. & Dibb S., 2008. Consumer involvement: a new perspective. The Marketing Review **8**(1), 83–99
[13] Mueller S. & Remaud H., 2010. Are Australian wine consumers becoming more environmentally conscious? Robustness of latent preference segments over time. Thèse de doctorat. University of Auckland business School
[14] Najar C. et Zaiem I., 2010. Influence de l'implication durable sur l'intention et le comportement d'achat écologique. Revue Libanaise de Gestion et d'Économie **3**(4), 1–35
[15] Schlegelmilch B., Bohlen G. & Diamantopoulos A., 1996. The link between green purchasing decisions and measures of environmental consciousness. European Journal of Marketing **30**(5), 35–55
[16] Sirieix L., 2010. Consumer perceptions of eco-friendly vs. conventional wines in Australia. Thèse de doctorat. University of Auckland business school
[17] Stolz H. & Schmid O., 2008. Consumer attitudes and expectations of organic wine. Organic wine and viticulture conference, Levizzano, Italy, June 18–20, poster
[18] Vermeir I. & Verbeke W., 2006. Sustainable food consumption: Exploring the consumer "attitude–behavioral intention" gap. Journal of Agricultural and Environmental Ethics **19**(2), 169–194
[19] Xiao C. & Dunlap R. E., 2007. Validating a Comprehensive Model of Environmental Concern Cross-Nationally: A US-Canadian Comparison*. Social Science Quarterly **88**(2), 471–493
[20] Zaichkowsky J., 1985. Measuring the involvement construct. Journal of Consumer Research **12**(3), 341

DOI: 10.1051/bioconf/20170903010

La sostenibilidad como medio para conseguir el posicionamiento de mercado de una bodega / sustainability as a means to achieve the market positioning of a winery

Pilar Gargallo[1], Nieves Garcia-Casarejos[1], y Javier Carroquino[2]

[1] Facultad de Economía y Empresa. Universidad de Zaragoza, España
[2] Departamento de Ingeniería eléctrica. Universidad de Zaragoza, España

Resumen. El cambio climático, el efecto invernadero, el agujero de la capa de ozono, la acidificación del suelo y el agua, el agotamiento de los recursos no renovables y la contaminación de las aguas y de los suelos son factores que han aumentado la conciencia ecológica de la población. El sector vitivinícola es uno de los mejores sensores para detectar los cambios en la evolución del clima y de las temperaturas y tiene plena conciencia de que esos cambios pueden afectar a la calidad de sus vinos. En este sentido, los consumidores de vino no son ajenos al problema medioambiental y un número de ellos cada vez más creciente, prefiere consumir productos obtenidos con métodos de cultivo y elaboración absolutamente compatibles y respetuosos con el medio ambiente. Los consumidores de este tipo de vino están creando una "demanda de virtud" para los productores, que se ven en la necesidad de incorporan medidas que colaboren con el cuidado del medio ambiente. Por tanto, aquellas bodegas que intenten minimizar las consecuencias sociales que provoca la producción de vino con los sistemas actuales, y que utilicen métodos que reduzcan sus emisiones de gases y residuos, y gestionen su agua y su energía de un modo eficiente, podrían acreditarse como proveedores sostenibles de cara a sus clientes, compradores y consumidores y conseguir así un posicionamiento ventajoso para este tipo de agentes. Este trabajo, a través de encuestas, pretende conocer si las bodegas españolas poseen la conciencia medioambiental que demanda una parte de sus consumidores, así como el grado de compromiso que han adquirido en esta cuestión. En caso de existir diferencias significativas entre las bodegas, un segundo objetivo será establecer una clasificación de las mismas según su grado de responsabilidad en función de las medidas adoptadas de mitigación del cambio climático y de mejora de la eficiencia energética, así como, caracterizarlas a partir de sus tipologías geográficas, su nivel de actividad, su percepción de los aspectos ambientales, etc. Para ello, se utilizan conjuntamente las técnicas estadísticas multivariantes del análisis factorial y análisis clúster, que permitirán construir un conjunto de indicadores para describir tipologías de bodegas frente a su responsabilidad con aspectos medioambientales y, posteriormente, a partir de la información proporcionada por estos indicadores, identificar grupos de bodegas similares en su grado de compromiso con esta cuestión.

Abstract. Climate change, the greenhouse effect, the ozone hole, acidification of soil and water, depletion of non-renewable resources and pollution of water and soil are factors that have increased the ecological awareness of the population. The wine sector is one of the best sensors to detect changes in the evolution of climate and temperatures and is fully aware that these changes can affect the quality of wines. In this sense, wine consumers are not outsiders to the environmental problem and a growing number of them, prefer to consume products obtained with methods of cultivation and processing absolutely compatible and respectful with the environment. The consumers of this type of wine are creating a "demand of virtue" for the producers, who are in need of incorporating measures that collaborate with the care of the environment. Therefore, those wineries that try to minimize the social consequences of wine production with current systems, using methods that reduce their emissions of gases and waste, and managing their water and energy efficiently, could be credited as sustainable suppliers for their customers, buyers and consumers and thus achieve an advantageous positioning for this type of agents. This work, through surveys, aims to know if the Spanish wineries have the environmental awareness demanded by a part of their consumers, as well as the degree of commitment they have acquired in this issue. In case of significant differences among the wineries, a second objective will be to establish a classification of them according to their degree of responsibility in terms of the measures taken to mitigate climate change and improve energy efficiency, as well as characterize them from their geographical typologies, their activity level, their perception of environmental aspects, etc. For this purpose, the multivariate statistical techniques of Factor Analysis and Cluster Analysis are used in order to construct a set of indicators that describe wineries typologies in relation to their responsibility with environmental aspects and, subsequently, based on the information provided by these indicators, identify groups of similar wineries in their degree of commitment to this issue.

DOI: 10.1051/bioconf/20170903010

1. Introducción

En los últimos años, determinadas actuaciones llevadas a cabo por algunas empresas han provocado catástrofes medioambientales, disminución de los recursos hídricos y la regresión de la costa, pérdida de la biodiversidad biológica y ecosistemas naturales, aumentos en los procesos de erosión del suelo, ... lo que está contribuyendo al cambio climático y al calentamiento global.

Si no cuidamos de nuestro planeta, las consecuencias pueden ser devastadoras. Por eso es fundamental que las empresas apliquen políticas de responsabilidad social ambiental y, que el medio ambiente se convierta en uno de los puntos de responsabilidad más cuidados por ellas. De hecho, el interés de las empresas por el medio ambiente está muy relacionado con las demandas de los consumidores. Cada vez es más común que sus clientes sean personas con "conciencia ambiental", es decir, que tengan un estilo de vida en el que se preocupen por el medio ambiente y al realizar sus compras cotidianas exijan que los productos o los servicios que requieren se acoplen a su forma de pensar y de vivir.

Se puede hacer negocio y crear empleo al mismo tiempo que se respeta el entorno, es decir, es posible aprovechar los recursos con eficacia, de un modo sostenible y competitivo, con una visión a largo plazo donde el crecimiento económico, la cohesión social y la protección del medio ambiente vayan clara e indisolublemente unidos.

Esta economía verde no va en contra del mercado y tampoco del libre comercio, simplemente trasciende del modo de producción tradicional e insostenible con la incorporación de variables sociales y ambientales que racionalizan el desarrollo y dignifican al ser humano en su interacción con el entorno.

Una de las industrias con mayor trascendencia para lograr el desarrollo y avance socio-económico de la sociedad es la agroindustria, y dentro de ella, el sector vitivinícola es uno de los mejores sensores para detectar los cambios en la evolución del clima y de las temperaturas y tiene plena conciencia de que esos cambios pueden afectar a la calidad de sus vinos. Lo verde, lo ecológico, lo orgánico enlaza con los gustos y demandas de los consumidores. Cada vez más, un número creciente de consumidores prefiere consumir vinos obtenidos con métodos de cultivo y elaboración absolutamente compatibles y respetuosos con el medio ambiente. Este tipo de consumidor está creando una "demanda de virtud" para los productores, que se ven en la necesidad de incorporan medidas que colaboren con el cuidado del medio ambiente. Por tanto, aquellas bodegas que intenten minimizar las consecuencias sociales que provoca la producción de vino con los sistemas actuales, y que utilicen métodos que reduzcan sus emisiones de gases y residuos, y gestionen su agua y su energía de un modo eficiente, podrían acreditarse como proveedores sostenibles de cara a sus clientes, compradores y consumidores. La mejora de la imagen y reputación, la disminución de riesgos ambientales y la disminución de multas y sanciones de estas bodegas, les permitirá adquirir una ventaja competitiva derivada de una correcta gestión ambiental, fuente de diferenciación (Fiore et al., 2017) que les ayudará a mantener o incrementar su margen de beneficios.

Cambio climático y energía son dos caras de la misma moneda: buena parte de las emisiones de gases de efecto invernadero provienen del sector energético en sus diversas formas. Es por ello que la solución al problema pasa por un cambio fundamental en el sistema energético, que en gran medida solo será posible con una mayor participación de las renovables en el mix energético.

El objetivo de este trabajo es conocer la penetración de las energías renovables dentro del sector vitivinícola español, tanto en lo que se refiere a las bodegas como a sus explotaciones vitícolas. Mediante la realización de encuestas se pretende conocer la voluntad de las empresas vitícolas españolas de implementar las energías renovables en sus bodegas y viñedos y caracterizar el sector. Más concretamente, utilizando la técnica estadística del análisis factorial, se pretende construir un conjunto de indicadores que nos permitan describir tipologías de bodegas frente a su percepción acerca de aspectos medioambientales, su grado de aplicación de medidas tendentes a mitigar el cambio climático o mejorar la eficiencia energética, y, sus usos y consumos energéticos. Además, a partir de la información proporcionada por estos indicadores y haciendo uso de la técnica estadística del análisis clúster, se identifican grupos de bodegas similares en cuanto a su interés por las energías renovables y, sus usos y consumos energéticos. Por tanto, una implicación importante de este estudio es la posibilidad de conocer qué aspectos conllevan un mayor conocimiento, seguimiento y compromiso de las distintas cuestiones vinculadas con el cambio climático, el uso de las energías renovables y aplicación de medidas tendentes a mitigar el cambio climático o mejorar la eficiencia energética.

El esquema de presentación del trabajo se organiza del siguiente modo. En la sección 2 se describe brevemente el cuestionario, justificando su organización y la selección de las preguntas que lo componen. En la sección 3 se presenta la muestra de bodegas que participan en el estudio. En la sección 4 se presentan los resultados obtenidos a partir del análisis estadístico tanto unidimensional como multidimensional y en el apartado 5 se proporcionan las conclusiones más relevantes de dicho estudio.

2. El cuestionario

En esta sección se presenta el diseño del cuestionario utilizado para conocer la voluntad de las empresas vitícolas españolas de implementar las energías renovables en sus bodegas y viñedos y caracterizarlas en cuanto a sus tipologías geográficas, sus niveles de actividad, su percepción acerca de aspectos medioambientales, su grado de aplicación de medidas tendentes a mitigar el cambio climático o mejorar la eficiencia energética, y, sus usos y consumos energéticos (LIFE-REWIND, 2016).

Como es habitual en el diseño de un cuestionario, en primer lugar, se realizó un pre-test a distintos propietarios de bodegas de diferentes regiones españolas, identificando así los aspectos clave y las características más relevantes del sector analizado. Este proceso se llevó a cabo mediante entrevistas en profundidad, semiestructuradas y dirigidas.

El cuestionario definitivo comienza con el *Bloque I: Identificación y localización*. Está formado por preguntas que permiten identificar y localizar las bodegas mediante su nombre, el municipio y provincia en los que se ubican, la zona vitivinícola a la que pertenecen, el año de

DOI: 10.1051/bioconf/20170903010

constitución, el accionista mayoritario y la forma jurídica de la empresa.

A continuación se presenta el *Bloque II: Actividad de la empresa*. Este bloque de preguntas tiene como objetivo conocer la actividad de la empresa. Así, se pregunta sobre el número de empleados, volumen de facturación, etc. con el fin de averiguar el tamaño de la bodega. Además, también se pregunta por el porcentaje que representan sobre el total de sus ventas, las distintas calidades diferenciadas del vino producido (D.O.P, I.G.P, etc.) y el porcentaje de las ventas que destinan al mercado exterior. Otra cuestión importante es conocer si realizan otras actividades complementarias y el porcentaje que representan en el total de su facturación. Por último, también interesa averiguar la procedencia de la uva utilizada en la elaboración de su vino, es decir, si es propia, de viticultores con/sin contrato o de cooperativas. Obviamente, es importante saber el número de hectáreas implicadas y si participan o no en su gestión durante el cultivo.

El siguiente conjunto de preguntas constituye el *Bloque III: Política medioambiental de la empresa*. En este bloque se incluyen preguntas que tienen que ver con su responsabilidad medioambiental. Así, a las bodegas se les pregunta si disponen de alguna certificación de vinificación ecológica, si tienen calculada la huella de carbono de su actividad o de sus productos, si realizan auditorías energéticas y si disponen de recursos propios para gestionar la política medioambiental de la empresa.

El *Bloque IV: Actitud ante el cambio climático* está formado por preguntas que pretende analizar la actitud de las bodegas ante los problemas medioambientales. Se desea saber si están convencidas de que el clima ha cambiado y su grado de disposición a reducir las emisiones de CO_2 para mitigar el cambio climático. En particular, interesa conocer qué tipo de medidas se han adoptado ya en este sentido.

Como es sabido que la solución a este problema del cambio climático pasa por un cambio fundamental en el sistema energético, el siguiente grupo de preguntas constituye el *Bloque V: Energías renovables, uso y actitud*. Este bloque incluye cuestiones relacionadas con las energías renovables. Es decir, se pregunta sobre el tipo de energías responsables con el medio ambiente que utilizan las empresas vitivinícolas, si están convencidas de la necesidad de usarlas, su opinión sobre el gasto que supone su implantación, y los aspectos que valoran para adoptar su uso, tales como la fiabilidad, sostenibilidad ambiental, existencia de subvenciones y el impacto en su imagen.

El último bloque de preguntas va destinado a conocer si el uso de energías renovables es viable técnica y económicamente. Por ello, el *Bloque VI: Uso y consumo de energía no renovable* incorpora preguntas sobre el uso y consumo de electricidad, gasóleo y gas. Se pregunta sobre la evolución de su consumo a lo largo del año, el grado de preocupación por los costes energéticos, si se ha revisado la facturación y la potencia contratada con el fin de reducirlas.

3. La muestra de bodegas

En esta sección se presenta el diseño de la muestra de bodegas seleccionadas para participar en el estudio.

Tabla 1. Tamanos muestrales por CC.AA.

	Población de Bodegas	Muestra de Bodegas
Andalucía	287	27
Aragón	144	14
Asturias	19	2
Baleares	57	6
Canarias	84	7
Cantabria	5	1
Castilla-León	597	54
Castilla-Mancha	445	40
Cataluña	603	53
Extremadura	118	11
Galicia	342	30
La Rioja	326	29
Madrid	195	17
Murcia	87	8
Navarra	116	11
País Vasco	261	24
Valencia	208	18
	3,894	384

Para ello, en primer lugar, partimos de un análisis de la población.

En España según la base de datos SABI, que contiene información de empresas españolas y portuguesas, el mapa vitivinícola español está formado por 3,894 bodegas. Con el fin de simplificar al máximo el procedimiento de muestreo pero conservando la representatividad de la muestra, se decidió utilizar un muestreo aleatorio simple y, posteriormente, una estratificación por comunidad autónoma. Así, con una confianza del 95% y un error del 5%, el tamaño de la muestra necesario es de 384 bodegas. En la Tabla 1 se recogen los tamaños por comunidad autónoma.

4. Resultados

En esta sección se presentan los resultados de los datos proporcionados por la encuesta. Más concretamente, en el apartado 4.1 se detalla el análisis estadístico unidimensional para caracterizar el mapa de bodegas. En el apartado 4.2 se presenta el resultado del análisis factorial llevado a cabo para construir un conjunto de indicadores que nos permitan describir tipologías de bodegas frente a su percepción acerca de aspectos medioambientales, su grado de aplicación de medidas tendentes a mitigar el cambio climático o mejorar la eficiencia energética, y, sus usos y consumos energéticos. Por último, en el apartado 4.3 se describen muy brevemente los grupos de bodegas obtenidos mediante un análisis clúster. Este análisis clúster, a partir de la información proporcionada por los indicadores del apartado 4.2, ha identificado 2 grupos de bodegas similares en cuanto a su interés por las energías renovables y, sus usos y consumos energéticos.

4.1. Análisis estadístico unidimensional

En este apartado presentamos esquemáticamente algunos de los resultados más relevantes obtenidos mediante la aplicación de un análisis estadístico unidimensional.

Bloque I: Identificación y localización

- En la muestra están representadas las principales denominaciones de origen del país y todas las Comunidades Autónomas
- Todas las bodegas pertenecen a municipios diferentes
- El 60% son Sociedades Limitadas

Bloque II: Actividad de la empresa

- El 51% de las bodegas solamente obtienen sus ingresos del vino que elaboran
- El 49% restante obtiene ingresos adicionales de:

  81% Turismo y uso recreativo
  19% Transformación y/o venta de otros productos agrícolas
  11% Producción de energía renovable

- Las bodegas destinan en media el 43.66% de sus ventas a la exportación
- El 81.3% de las bodegas dispone de viñedos propios
- El 33.3% tienen algún otro cultivo diferente de la vid

Bloque III: Política medioambiental de la empresa

- El 28% de las bodegas disponen de certificación de vinificación ecológica (agricultura ecológica, vino ecológico, viñedo ecológico, bodega ecológica, ISO 14001, CAAE...)
- El 16% calculan su huella de carbono o realizan auditorías energéticas
- El 58.7% de las explotaciones vitícolas utilizan métodos de agricultura ecológica y de éstas un 50% disponen de alguna certificación
- Recursos destinados por las bodegas a la gestión medioambiental:

  33% No asigna ningún recurso a estas tareas
  7% Dispone de un departamento exclusivo de medioambiente
  46% Disponen de técnicos con otras funciones asignadas

Bloque IV: Actitud ante el cambio climático

- El 93,3% de las bodegas analizadas recicla
- El 69% gestiona de una forma más eficiente sus consumos
- El 64% ha mejorado los aislamientos térmicos
- El 51% ha adquirido maquinaria, instalaciones y vehículos de bajo consumo
- El 35% utiliza nuevos envases
- El 44% ha reducido el peso de sus cajas, botellas, envases, embalajes...,
- El 15% ha reducido las emisiones
- El 13% ha puesto en marcha otras medidas como: aprovechar mejor el calor residual, control inteligente de temperaturas, plantación de árboles...

Bloque V: Energías renovables, uso y actitud

- La utilización de energías renovables no representa más del 10% del consumo de todas las bodegas
- Los tres tipos de energía renovable más habituales son: la biomasa (11%), la energía fotovoltaica (11%) y la energía solar térmica (9,3%)
- A pesar del bajo nivel de uso de las energías renovables, las bodegas están muy convencidas de la necesidad de utilizarlas
- Este gap puede ser debido a que las bodegas consideran muy elevada la inversión inicial que supone implantar energías renovables aunque no los costes de operación y los costes de mantenimiento
- Aspectos que favorecen la adopción de energías renovables por parte de las bodegas:

  La fiabilidad (valorada con 7,24 sobre 10 puntos)
  La sostenibilidad ambiental (valorada con 8,6 sobre10 puntos)
  La existencia de subvenciones (valorada con, 12 sobre 10 puntos)
  El impacto en la imagen (valorada con 7,84 sobre 10 puntos).

4.2. Análisis factorial

En esta sección hemos realizado un análisis factorial con las variables que muestran para cada bodega su grado de acuerdo con que el clima ha cambiado, su disposición a reducir las emisiones de CO_2, su convencimiento del uso de energías renovables, su opinión del gasto que supone la implantación de las energías renovables en cuanto a inversión, costes de operación y de mantenimiento y la importancia que para ella tiene la sostenibilidad ambiental, la fiabilidad, la existencia de subvenciones y el impacto de la imagen en la adopción de energías renovables. A continuación, describimos los 3 factores obtenidos con las variables seleccionadas:

FACTOR GESTIÓN REPUTACIONAL.- Este factor está relacionado positivamente con las siguientes variables: disposición a reducir las emisiones de CO_2, convencimiento del uso de energías renovables, importancia de la sostenibilidad ambiental y del impacto de la imagen en la adopción de las energías renovables. Una bodega que tiene una puntuación alta en el factor de gestión reputacional es una bodega que está muy dispuesta a reducir sus emisiones de CO_2 y muy convencida del uso de las energías renovables ya que valora mucho la sostenibilidad ambiental y el impacto de su imagen.

FACTOR GESTIÓN A CORTO PLAZO.- Este factor está relacionado positivamente con los costes de operación y mantenimiento. De manera que una bodega que tiene una puntuación alta en el factor de gestión a corto plazo es una empresa que piensa que los costes de operación y mantenimiento asociados a la implantación de las energías renovables son muy elevados.

FACTOR GESTIÓN A LARGO PLAZO.- Este factor está relacionado positivamente con la inversión, fiabilidad y la existencia de subvenciones. De manera que una bodega que tiene una puntuación alta en el factor de gestión a largo plazo es una bodega que piensa que la inversión asociada a la implantación de las energías renovables es muy elevada, por eso considera que la fiabilidad y la existencia de subvenciones es absolutamente determinante para implantar las energías renovables.

4.3. Análisis clúster

A partir de estos 3 indicadores, se ha utilizado la técnica multivariante del análisis clúster para realizar una clasificación de las bodegas en dos grupos diferentes.

Grupo 1 (BODEGAS NO SOSTENIBLES) (57,3%): Este grupo está formado por bodegas con puntuaciones bajas en el factor gestión reputacional y en el factor gestión a corto plazo y puntuación alta en el factor gestión a largo plazo. De manera que las bodegas de este grupo no están dispuestas a reducir sus emisiones de CO_2 ni convencidas del uso de las energías renovables puesto que piensan que la inversión asociada a su implantación es excesivamente elevada aunque no los costes de operación y mantenimiento, por eso, no son la sostenibilidad ambiental y el impacto de su imagen sino la fiabilidad y la existencia de subvenciones las cuestiones absolutamente determinantes para que implantasen energías renovables.

Grupo 2 (BODEGAS SOSTENIBLES) (42,7%): Este grupo está formado por bodegas con puntuaciones altas en el factor gestión reputacional y en el factor gestión a corto plazo y puntuación baja en el factor gestión a largo plazo. De manera que las bodegas de este grupo si están dispuestas a reducir sus emisiones de CO_2 y convencidas del uso de las energías renovables puesto que consideran que la inversión asociada a su implantación supone un gasto reducido, a pesar de que sean elevados los costes de operación y mantenimiento. Son bodegas muy preocupadas por su imagen y por la sostenibilidad ambiental.

5. Conclusiones

En este trabajo, mediante la realización de encuestas se ha caracterizado el sector vitivinícola español en cuanto a sus tipologías geográficas, sus niveles de actividad, su percepción acerca de aspectos medioambientales, su grado de aplicación de medidas tendentes a mitigar el cambio climático o mejorar la eficiencia energética y sus usos y consumos energéticos. Asimismo, se han estudiado las motivaciones para la implantación de energías renovables en las bodegas y sus explotaciones vitivinícolas. Por último, mediante un análisis factorial, se han construido un conjunto de indicadores para describir tipologías de bodegas frente a su responsabilidad con aspectos medioambientales y, posteriormente, por medio de un análisis clúster, a partir de la información proporcionada por estos indicadores, se han identificado dos grupos de bodegas similares en su grado de compromiso con el cuidado del medio ambiente.

Para llevar este estudio se ha utilizado un muestreo aleatorio simple estratificado por comunidad autónoma. Las bodegas que componen la muestra están muy centradas en lo que constituye su *core business*, la producción y venta de vino y, además, se observa un alto grado de vinculación entre la propiedad de la bodega y la propiedad de la explotación vitícola.

En este mismo sentido, la penetración de las energías renovables se encuentra todavía en niveles bajos y entre los motivos que aducen las bodegas de la muestra para implantar energías renovables, destaca la sostenibilidad medioambiental, lo que es congruente con su concienciación acerca del cambio climático y su disposición a reducir emisiones. Así, Los consumidores de vino se interesan cada vez más por las cuestiones ecológicas para que el concepto de sostenibilidad ambiental pueda actuar como un canal relacional entre bodegas y consumidores, poniendo en evidencia que la política medioambiental de las empresas influye positivamente en el comportamiento de compra del consumidor, lo que mejora sus oportunidades de mercado (Christofi et al., 2015). Por su parte, Fiore et al. (2017) concluyen que el uso de nuevas tecnologías para penetrar en nuevos mercados muestra una correlación positiva con la adopción de prácticas de producción sostenibles y el enfoque respetuoso con el medio ambiente.

A pesar de que tanto en bodegas como en sus explotaciones vitícolas el salto cualitativo a la incorporación de energías renovables todavía no se ha producido, las condiciones para que esta incorporación se produzca sí que existen, puesto que la principal fuente de energía en el sector vitivinícola sigue siendo la energía eléctrica, cuyos costes cada vez son mayores. Esto podría deberse a la opinión que manifiestan acerca de la elevada inversión inicial que supone la implantación de este tipo de energías en el sector vitivinícola y no tanto a otros motivos, como pueden ser los costes de operaciones y mantenimiento. En este mismo sentido apuntan otros estudios, que señalan las barreras financieras (Meyers et al., 2016) como las más obvias para no implantar energías limpias, debido a los elevados periodos de amortización. Asimismo, estos autores destacan como el interés por el ahorro energético es menor mientras no exista una política interna que permita monitorizar y controlar los ahorros energéticos que se producen. Por tanto, para incentivar la implantación de energías renovables en el sector, sería necesario establecer mecanismos que pudieran cambiar su percepción y elaborar un balance coste-beneficio en el medio-plazo o diseñar fórmulas más acordes con las posibilidades financieras de cada bodega.

No obstante, el análisis clúster ha demostrado que las bodegas españolas se clasifican en dos grandes grupos diferenciados. Las bodegas no sostenibles (57%) que no están dispuestas a reducir sus emisiones de CO_2 ni convencidas del uso de las energías renovables puesto que piensan que la inversión asociada a su implantación es excesivamente elevada, por eso, la fiabilidad y la existencia de subvenciones son las cuestiones absolutamente determinantes para que implantasen energías renovables. Y el grupo de bodegas sostenibles (43%) dispuestas a reducir sus emisiones de CO_2 y convencidas del uso de las energías renovables puesto que consideran que la inversión asociada a su implantación supone un gasto reducido, a pesar de que sean elevados los costes de operación y mantenimiento. Son bodegas muy preocupadas por su imagen y por la sostenibilidad ambiental.

Los autores agradecen la financiación proporcionada por el Proyecto Life REWIND "Renewable Energy in the Wine Industry Profitable small scale renewable energy systems in agri-food industry and rural areas: a demonstration in the wine sector" LIFE13/ENV/ES/000280 Asimismo, esta investigación también ha contado con el apoyo financiero del proyecto ECO2016-77-P (AEI/FEDER, UE).

 DOI: 10.1051/bioconf/20170903010

Referencias

Christofi, M., Leonidou, E., Vrontis, D., 2015. Cause-related marketing, product innovation and extraordinary sustainable leadership: the root towards sustainability. Glob. Bus. Econ. Rev. **17**(1), 93–111

Fiore, M., Silvestri, R., Contò, F., & Pellegrini, G. (2017). Understanding the relationship between green approach and marketing innovations tools in the wine sector. Journal of Cleaner Production **142**, 4085–4091

LIFE-REWIND (2016). *Profitable small scale renewable energy systems in agrifood industry and rural areas: demonstration in the wine sector.* Midterm Report LIFE7ENV7ES7000280: UE

Meyers, S., Schmitt, B., Chester-Jones, M., & Sturm, B. (2016). Energy efficiency, carbon emissions, and measures towards their improvement in the food and beverage sector for six European countries. Energy **104**, 266–283

 DOI: 10.1051/bioconf/20170903011

The perspective of organic wine in Brazil – trends, demands and production

Marcos Vinícius Araujo[a], Marielen Aline Costa da Silva, Daniela Callegaro de Menezes, and Kelly Lissandra Bruch

Center for Studies and Research in Agribusiness – CEPAN / Universidade Federal do Rio Grande do Sul – UFRGS, Av. Bento Gonçalves, 7712 Porto Alegre, RS, Brazil

Abstract. This work aims to analyze the productive reality and propose a panorama for the Brazilian organic wine production, from the perspective of producers, representatives of the sector, market and consumers, thus seeking to highlight the market perspectives and adequacy to the consumers' aspirations. This production that aims to meet a growing demand for sustainable products, which include organic, biodynamic and natural foods, combined with the growing Brazilian tendency to consume better quality wines. To prosecute this study, an exploratory research was developed, interviewing twenty consumers, one intermediary, one representative of the sector and four wineries. The data were analyzed crossing the information collected with these actors and bring out key contents. The results show that production is not aligned with consumer demand that it is still necessary to review the information made available to consumers, in addition to making it clearer, other values could be aggregated to those products or better communicated. On this way, it is possible to conclude that an intermediate product between organic and conventional, which has sustainable practices in its production could meet the current demand, so that in the future it can produce an organic product that corresponds to the consumer's demand.

1. Introduction

From the sensible changes in the consumption habits of Brazilians, the emerging Brazilian economy and the great increase of the middle class [1], the Brazilian consumer started to buy products that had not consumed before due to the greater purchasing power [2], mainly fine and sparkling wine.

This change was perceived in several sectors, such as in the wine market and also organic products. In the Brazilian wine sector, there was a significant increase in the marketing of imported fine wines and national sparkling wines and decrease in table wines between 2010 and 2016, according to Chart 01. The production of grape juice may be one of the causes of the decrease in marketing of table wine, which in Brazil are produced from the same grape variety – *Vitis Rupestris*, *Vitis Labrusca* and hybrids of *vitis sp*.

It is also important to note that in Brazil only fine wines produced from *Vitis Vinifera* varieties are imported. In Brazil, fine wine is seen as a more elitist product and consumed by a population with more income and education, while table wine is consumed by a smaller income parcel, presents characteristics of the varieties of grapes used and has more affordable prices [3].

In this sense, consumption of organic products is also increasing in developing countries [1]. These are consumers motivated by their environmental attitudes, which make them consume green products [4], such as organic foods. In Brazil, the organic production has grown more than 20% per year, but production still cannot meet the demand for these products, due, among other factors, to the quantity exported to Europe [5]. On the other hand, the trend in the consumption of organic wines still presents a challenge for the sector, which started in the production of organic grapes for wine processing in the 1990s.

These difficulties are resulting mainly by the climatic resistance of the vines, in particular the *Vitis Vinífera* varieties. Due to the climate, these varieties are more susceptible to diseases and require more chemical interventions [6]. As an alternative to the traditional system, organics it combines traditional methods with modern agricultural practices, crop rotation, diversification, better land use, such as green and animal fertilization, and natural pest control [7].

These practices make organic wine is seen as safe, healthy and high-quality, yet only be sustainable is not enough to be consumed, the consumer of organic food have to like the taste of the product [8]. In this same perspective, it is important to emphasize that there is difference between organic, biodynamic and natural wines, which are often confused by consumers. If organic wine is restricted to the use of synthetic materials in production, biodynamic is an expression of terroir and climate, man only collaborates so that the roots of the plant reach to a depth that can best express characteristics of that ground and plant [9], some practices resemble homeopathy and follow the lunar calendar.

Natural wine, however, is not clearly defined and is a relatively modern term [10], but is regarded as a wine that uses ancient techniques to produce a wine without any additives, it is the pure expression of the grape, with no added water, sugar, tannins or any other [11], even the fermentation is done naturally. Thus, the main difference

[a] e-mail: araujovmarcos@gmail.com

 DOI: 10.1051/bioconf/20170903011

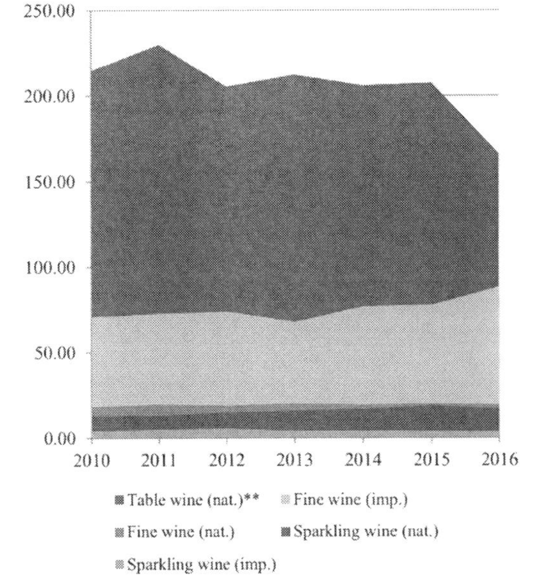

* From the production of the State of Rio Grande do Sul, representing 90% of the national production; ** Produced from *Vitis Labrusca*, *Vitis Rupestris* and hybrids of *vitis sp.*

Source: IBRAVIN-SISDECLARA, adapted by the author.

Chart 01 – Comparison between commercialization of table wines and fine wines in Brazil*

between these three products is that the natural does not use any additives for production, not even natural, it is a pure product. Biodynamic uses only natural processes and follows the lunar calendar and seeks expression of the plant and soil in the product, but uses natural additives. The organic is a production without the use of synthetic additives and with limited use of some additives in winemaking, such as sulfite.

In Brazil, organic wine is regulated by the organic system of agricultural production, subject to certification under Law 10,831, of 23 December 2003, and processes described in Joint Normative Instruction n°18, of 28 May 2009, for processed organic food. This product is certified by accredited organizations at the Brazilian Ministry of Agriculture, Livestock and Supply, which certifies minimum processes, as described above, thus ensuring a safe and reliable product to the consumer.

It is the characteristics of these practices, within the organic production system that attract consumers' attention. More specifically organic wine, these consumers are concerned about the health of the product, the environment and they seek more information, tend to buy in specialized stores, associate this category with a more expensive product, want to know the region produced, take into consideration the price, environmental aspects and the organic system, moreover, they feel like a sacrifice, not a pleasure, which may justify the segment still being small, besides not being willing to change the quality of the conventional wine for an ecological one [14]. The authors further conclude that the perception of genuine and modern flavor could be a strong argument for the product.

In the organic wine segment there is a subjectively perceived status in the consumption process of this product. In this sense, it is important to value the positive social image of organic wine, health effects, due to the absence of chemical residues, positive environmental effect, its origin, production system and wine color [15].

Attracted by traceability, concerned about the future of the planet and consuming products that are better for health, the search for organic wine increases in Europe, especially among the female and young people. In addition, they are informed consumers and the lack or unavailability of quality information can be a limitation in the search for this product [16]. In France, conventional wine consumption has not changed much in recent years, unlike organic wine, in 2013, one in three consumers consumed organic wine [17]. On the other hand, from 2015 to 2016 the growth was 10%, in addition to the increase of 227 to 323 organic wine producers in the country, also having an increase of more than 10% in internal and external marketing, in the same period [18].

In Brazil, organic wine continues have little representation for the sector, which according to IBRAVIN – Brazilian Wine Institute – may be linked to difficulties in organic grape production, due to climatic problems and adaptation of vines to the region's climate. Thus, what we have is the production of organic table wines, which go to the disagreement of the consumer's preference for organic wines.

Intended to meet a demand for healthier and ecologically correct products, associated with the market trend, which motivated the Brazilian wineries to convert to organic wine production. Given this scenario, in which the Brazilian organic wine industry is inserted, the difficulties, limitations and promising perspective of the sector, due to the demand of this market niche, this work aims to analyze the productive reality and propose a panorama for the organic wine production in Brazil, based on the vision of producers, representative of the sector, market and consumers, thus seeking to highlight the market prospects and adequacy to the consumers' wishes.

2. Material and methods

A qualitative exploratory research was developed since May 2015, and secondary data collection and interviews with wineries that produces wine and organic grape juice and they had a complete mix of products, such as table wines, grape juice, fine wines and sparkling wines. Four wineries were visited and interviewed between May 2015 and May 2016, in the cities of Bento Gonçalves (Winery 01 and 03), Garibaldi (Winery 02) and São Marcos (Winery 04), located in the State of Rio Grande do Sul, Southern Brazil. The wineries studied represents the most of Brazilian organic wine production.

Were conducted semi-structured interviews were conducted with: 01 (one) intermediary from São Paulo, which sells organic, natural wines and specializes in Brazilian wines, also conventional; 01 (one) representative of the productive sector belonging to IBRAVIN – Brazilian Wine Institute; and 19 (nineteen) consumers of organic foods, from the city of Porto Alegre and 01 (one) from São Paulo. Other secondary data was collected in articles from EMBRAPA (Brazilian Corporation of Agricultural Research), which dedicate research to the development of organic grape production. The interviews occurred first with the wineries, the intermediary, the wine sector, so that

 DOI: 10.1051/bioconf/20170903011

it could elucidate issues from the chain links to interview the consumers.

The interviewees were chosen for convenience, due to the difficulty in finding producers, marketers and consumers of this product, except the wine sector, which was chosen IBRAVIN for being the representative of the Brazilian wine sector. The intermediary was chosen because it is inserted on the dynamic of commercialization of Brazilian wines, also providing organic and natural wines and being close to consumers and producers.

Consumers were interviewed at ecological fairs in Porto Alegre, according to availability to respond to the interview. Other consumers were interviewed for participating in related organizations and indicated by other consumers of organic foods. The criteria of selection were whether consumed organic foods weekly, consumed wine or had consumed some type of organic wine previously.

The interviews were conducted in person and by video-conference, with a consumer from São Paulo and the audio recorded. In addition, secondary data were also collected from advertising materials from wineries, social networks, sites and *in loco*.

3. Results and discussion

The results were analyzed from the information cross-referencing, seeking to show which values consumers expect from organic wine, criteria in the wine buying moment and organic foods. Keywords were taken from the responses of the wineries, the market (intermediary) and the sector, which presented the questions about the product offer, for sampling the current Brazilian overview, following to the future outlooks of Brazilian organic wine, from the perspective of these three actors and consumers.

From the view of these actors, it separated into two topics. The first approach the current overview of Brazilian organic wine, and then seeks to understand the future outlooks of this product, based on the production and commercialization of the product, which includes the second topic.

3.1. Current overview of organic wine

The results show a production still in the process of development, with the production of organic grapes being the most sensitive part of the sector, since the more resistant varieties do not meet the type of product that is in increasing demand in Brazil. On the other hand, the industry is already moving to begin to meet these demands, seeking solutions, in addition to understanding that this is a growing niche market that demands attention. Table 1 presents a deeper understanding of the themes that emerged in interviews with wineries, intermediary of market and representative of sector, from productive realities, availability of information, marketing approaches and consumer profile.

It is possible to perceive that each winery approaches this production and market relation in a different way, starting from different interests. The wineries that understand this production as a market and positioning approach the product more broadly, within their communication and with specific advertising materials, in addition to include it strongly in their positioning, as in the case of the winery 03 and 04. In the case of winery 03, because it has a small structure and little market share, it uses as a positioning and reports that it receives visits from consumers who come in search of the organic and see the other products better. In the case of winery 04, the care with residues and the positioning of safe food, which goes beyond organic products, made it certified and adopted some practices for all products. In this sense, they hay a market positioning based on quality and safety of products.

In the case of winery 01, its positioning is aimed at the large retail and the complete mix of products. This winery stopped producing and offering these products in 2016 due to the difficulties, besides not being tied to its market positioning. It reports production difficulties and certification, which made it unfeasible to provide the product because the organic wine produced today is not well perceived by its consumers. The winery 02 is positioned by cooperativism. Since its members are interested in organic production, due to their concern for their health, the cooperative needed to benefit the production of these. However, as the grape used has better acceptance for juice, the winery followed by this segment, as well as all the other wineries interviewed. This fact made the winery 02 the largest producer of organic grape juice in Brazil, which produces more than 300 thousand liters in 2015, organic wine is not representative and it is marketed only supermarkets in the State of Rio Grande do Sul, Brazil.

On the other hand, the IBRAVIN representative says that it is still difficult to produce grapes organically due to climatic issues and lack of specific techniques, but there are projects to develop cultivars and appropriate techniques. In addition, new producers are emerging and seeking collaboration, mainly with FECOVINHO – Rio Grande do Sul Wine Cooperatives Federation – and EMBRAPA Uva e Vinho – Brazilian Agricultural Research Corporation – Grape and Wine.

In an interview with the market intermediary, the interviewee shows that today the consumer, for the most part, still does not know the organic wine. For him, the important thing is that the product is the most sustainable possible and that from the first contact with this type of product, it will search for new, whether biodynamic, organic or natural. In addition, consumers today perceive as a poor quality product with poor communication quality, either because they present little information on labels and on websites, or because they are not accessible, by price or availability.

3.2. Future outlooks

Following the current scenario, the interviewees presented proposals for organic wine to gain space in this segment. An intermediate product, between the conventional and the organic wine, but certified, would serve the current market of organic wine, provided that it has a treatment with less waste and include in its production more sustainable processes.

In this sense, the sector presents programs that are being developed to make this possible.It is being implemented in the Brazilian wine sector actions such as Integrated Grape Production for Processing (PIUP, abbreviation in Portuguese), which enables the food production with safe, quality and the traceability of

Table 1. Current view of the Brazilian wine sector.

Availability of information	Website, digital social networks, search engine, commercial representatives, advertising materials, health professionals and points of sale.
Actual production	Few producers, few commercial production of organic wine, most are still in conversion, or are doing sustainable work in wine, in the chain, less chemical intervention, tending to the natural. Little certified product. Lack of methodology, technology for the production of fine wines. There are experiments, but few products, is a field to develop. It represents little for the wineries, some 3 to 5%, the one that represents more (winery 03), and 1% for the winery 01.
Challenges	Producing grapes in an organic way due to climatic adversities, better management of the vineyard, demand more work, labor, greater risk, needs better remuneration. Serra Gaúcha is the biggest producer of wines, but it is not propitious to the organic ones due to the climatic adversities.
Product demand	Natural wine is best known to consumers, for which there is already a demand. When the consumer taste the organic wine he look for other wines, whether natural, organic, biodynamic, sustainable wines. (Note: importance of the intermediary in communicating the value of the product to the consumer, since it presents and teaches what the product is and the difference between them).
Trust and communication	Organic label.
Definition	There is still a lot of confusion in the conception (of what is organic wine), even for professionals in the area, who sometimes confuse the terminology, even the use of "nature" for sparkling wine, which has no connection with intervention in grape production, but rather in not using sugar for the expedition liquor.
Availability	Supermarkets, specialized stores, large retail.
Packaging, communication and marketing	Underdeveloped communication, there is usually no communication, website that gives access to these products, they enter or in the same approach of the conventional or gain a communication very similar to organic in general, there being no differentiation of organic wine. Missing information on labels, the site does not show information about that system. In the supermarkets the products are next to the table wines and in some cases in the section of organic foods.
Regulation	High cost to certify, there is no specific legislation for organic wine, only for organic grape and processed organic food.
Market	The consumer still does not identify the product properly, he has difficulty in differentiating the products of the wine in general, in addition, the cost of production is high, soon the product reaches the most expensive consumer. It brought visibility to the other products of the winery.
What consumers are looking for	They look for the product because it is more sustainable, that has the face of Brazil, not because it is organic, biodynamic or natural. They seek quality of life, do not have pesticides, health, have no chemicals, do not affect the environment, health, quality of life and sustainability.
Objectives of production	Meet the demand of the segment, niche market, which is still little explored. Sustainability of the business, mainly for the production of organic grapes and the quality of life that brings grape producers. Health, well-being, quality of life of the cooperative members (in the case of the cooperative), business opportunity. Aggregation of sustainable value for the brand of the winery and its other products. A genuine product, without transgenic, purer, more natural, because it is a world trend. Offer a quality product to the consumer.
Other values	ISO 22000, food safety, controls the use of agrochemical, even in conventional production, grace period, to avoid waste. Rainwater reuse and recycling. Conservation of local diversity, such as animals and water resources.
Consumer profile	Consumers of class A and B, mostly between 28 and 50 years, up to 70 years, few young people, some consumers of class C.
Public policies	Program PAS (Safe Food Program), for grapes for processing, with training of producers to reduce waste, with guarantee of food safety, without residues in the final product.
Product	The sanity of the grape changes the wine standard, final quality, even using NS220, with the restrictions. The product is a result of the quality of the grape and the crop.

Source: Author.

the whole system, thus seeking the lowest possible environmental impact [19]. Another existing practice on the Brazilian wine industrie for the production of grapes for processing is the Safe Food Program (PAS, abbreviation in Portuguese), which aims to manage the risks and dangers of grape production and processing, ensure product quality and consumer-safe products [20].

Furthermore, in Table 2 it is possible to perceive that there is a collective effort and that shows the importance of the participation of the consumer in this process.

Therefore, consumers are willing to collaborate, which is seen by their participation in organizations, as organic consumer associations and cooperatives. These consumers seek to know better the origin, the company, the producers and the techniques used. It is necessary to provide quality information and easy access, as well as placing the product in adequate and well-marked places of purchase. Moreover, communication and product marketing approach has to follow strategies merged between wine and organic food, with labels designed for this specific product and to meet the information requested by consumers as organic wine is seen as a premium product.

Thus, consumers seeking a product that meets the social, environmental, issues that go beyond organic quality itself, which is the absence of synthetic additives. That is, they are not looking for an organic product in itself, but for practices that involve organic agriculture.

Table 2. Future prospects for the production and consumption of Brazilian organic wine.

Information access	*Work colleagues, organized groups, network of contacts; Internet, websites, groups in digital social networks, web pages and portals; Public agents, technical assistance; Consumer networks, consumer cooperative, direct from producer, visits, point of sale, label, advertising material; Events, news, studies and health professionals.*
Production	The tendency is to produce a product free of pesticides (Agrochemical), or to do within the limits, in the most sustainable way possible; Future challenge of producing more economically way; New producers are migrating to the organic, or more sustainable; The Region of the Campanha Gaúcha, located on the south of the State of Rio Grande do Sul, presents good potential for the production of organic wines, because it has sandy soils and lower humidity.
Challenges	Reach the consumer who is willing to pay a little more; Lack of research; EMBRAPA is developing resistant varieties, such as powdery mildew, mold and other diseases of the vine.
Product demand	The Brazilian organic market is still starting in the wine segment. Is growing, consumers are looking for a more sustainable product, not exactly organic or biodynamic. It is creating a niche.
Trust and communication	*Consumers trust on certification by audit, because the analysis process, participatory certification, consumer participation and producer networks, knowing who produces, direct contact, product being at the fair and being able to talk to the producer, information on the label, product design, people who sell, produce, their networks of contacts, know the origin of the product, the production processes and the organic label.*
Availability	*Direct from producer, ecological fairs, consumer cooperatives, local commerce and supermarkets, specialty shops and organic food restaurants.*
Packaging, communication and marketing	*It lacks an approach, good communication, good visual, it is necessary to qualify and sophisticate the product communication. The consumer cooperative transmits these demands to producers, since the group of consumers can achieve this approximation. It needs to have wine packaging, well presented, with process information, who produces, quality, raw material. Sometimes you have a good product but have not advanced in packaging, availability, scale, regularity in the market, product standard, so you have to establish regularity and constant information.*
What consumers are looking for	*For being sustainable, diversification of production, environmental issue and respect for nature, thinking about future generations, motivate this production, for being a real food, healthy, natural, with special flavor, without chemical additives, pesticide residues and other waste.*
Other values	***Approach all sustainability actions, in addition to the minimum to organic production certification, in the advertising discourse, since consumers take this information into account to decide on a product**.***
Public policies	MAPA's*** Project, published by EMBRAPA, which developed the Integrated Grapes Production for Processing (PIUP), which are within the limits of resource use and agrochemicals. Developing methodology, in addition to dissemination, development and certification, so that it avoids the increase of costs and can add value to the final product.
Product	Thermovinification, which eliminates oxidative enzymes, may be the solution for further standardization of the product.

* The framework addresses the vision of the sector, as actions to be able to supply organic wine, is included the sector and the wineries, in italics the vision of the consumers. ** Suggested by the author from the response of consumers. ***Brazilian Ministry of Agriculture, Livestock and Supply.
Source: Author.

4. Conclusion

It is possible to conclude that production efforts are moving in the sense of meeting consumer demand, but more integration between producers is needed to find methodologies for this crop. On the other hand, it is necessary to think in a specific segment of organic wine, a product positioning between wine and organic food. In this same sense, an intermediate product could meet the current demand of consumers, for a product that has sustainable guidelines and some organic-related practices in order to overcome production difficulties and thus, the wineries can meet the demand for a product with the minimum of possible residues and low production impact, such as a sustainable wine, with some certification process of sustainability.

Moreover, public policies would be important to encourage this production, rethinking value and tax incentives to producers. Regarding the market, it is necessary to rethink the approach, aligning the discourse and practice with what the consumer expects, sufficient and accessible information, a product of quality, availability and regularity in production. Companies also need to open up more to consumers, let them participate in the company, with access to information, making transparent the difficulties faced in providing this product and the benefits it brings.

Finally, as an exploratory research requires new research with a more conclusive focus, based on a more solid database, with a larger base of Brazilians consumers of natural, organic or sustainable wine. But it is still necessary that the market consolidates better, since there is still a small supply, and it is not possible to establish a specific profile of this consumer.

This research was financed with a scholarship by the Coordination for the Improvement of Higher Education Personnel – CAPES, an agency of the Brazilian Government.

References

[1] Á. M. Ritter, M. Borchardt, G. L. R. Vaccaro, G. M. Pereira, and F. Almeida, "Motivations for promoting the consumption of green products in an emerging country: Exploring attitudes of Brazilian consumers," J. Clean. Prod. **106**, 507–520 (2015)

 DOI: 10.1051/bioconf/20170903011

[2] A. N. Almeida, C. Bragagnolo, and A. L. S. Chagas, "A Demanda por Vinho no Brasil: elasticidades no consumo das famílias e," Rev. Econ. e Sociol. Rural **53**(3), 433–454 (2015)

[3] L. M. R. de Mello, "Tendência de Consumo e Perspectivas do Mercado de Vinhos no Brasil," in *SIMPÓSIO NACIONAL DE FERMENTAÇÕES*, p. 7 (2003)

[4] B. Chekima, S. Azizi Wafa Syed Khalid Wafa, O. Aisat Igau, S. Chekima, and S. Laison Sondoh Jr, "Examining green consumerism motivational drivers: does premium price and demographics matter to green purchasing?," J. Clean. Prod. **112**, 3436–3450 (2016)

[5] SEBRAE, "O mercado para os produtos orgânicos está aquecido," *Agronegócio*. (2015). [Online]. Available: `https://www.sebrae.com.br/sites/PortalSebrae/artigos/o-mercado-para-os-produtos-organicos-esta-aquecido,5f48897d3f94e410VgnVCM1000003b74010aRCRD` [Accessed: 03-Feb-2017]

[6] J. F. da S. Protas, U. A. Camargo, and L. M. R. de Melo, "A vitivinicultura brasileira: realidade e perspectivas," *Embrapa Uva e Vinho*. (2002). [Online]. Available: `http://www.cnpuv.embrapa.br/publica/artigos/vitivinicultura/` [Accessed: 07-Mar-2017]

[7] J. P. Reganold and J. M. Wachter, "Organic agriculture in the twenty-first century," Nat. Plants **2**(February), 1–8 (2016)

[8] M. Laureati, D. Jabes, V. Russo, and E. Pagliarini, "Sustainability and organic production: How information influences consumer's expectation and preference for yogurt," Food Qual. Prefer. **30**(1), 1–8 (2013)

[9] F. Humbert, «Europe: la naissance douloureuse du vin «bio»», *La Revue du Vin de France*. (2012). [Online]. Available: `http://www.larvf.com/,vins-bio-ue-commission-europeenne-reglementation-bruxelles-nicolas-joly-gerard-bertrand-olivier-humbrecht,13182,4248636.asp` [Accessed: 10-Mar-2017]

[10] D. Wilson, "Flavoured Wine – Fad, Fashion or Trend?," BIO Web Conf. **7**, 3008 (2016)

[11] L. Chan, "Natural Wine Is The Hottest Trend In Winemaking Right Now: Here's Everything You Need To Know," *Tech Times* (2016). [Online] Available: `http://www.techtimes.com/articles/133381/20160213/natural-wine-is-the-hottest-trend-in-winemaking-right-now-here-s-everything-you-need-to-know.htm` [Accessed: 10-Mar-2017]

[12] BRASIL, "LEI N. 10.831 – DE 23 DE DEZEMBRO DE 2003. Que dispõe sobre a agricultura orgânica e dá outras providências" (2003) [Online]. Available: `http://www.planalto.gov.br/ccivil_03/leis/2003/L10.831.htm` [Accessed: 04-Sep-2016]

[13] BRASIL, "INSTRUÇÃO NORMATIVA CONJUNTA Nº 18, DE 28 DE MAIO DE 2009, Que regulamenta o processamento, armazenamento e transporte de produtos orgânicos," *Diário oficial da União*. (2009) [Online]. Available: `http://www.inmetro.gov.br/barreirastecnicas/pontofocal/..%5Cpontofocal%5Ctextos%5Cregulamentos%5CBRA_342.pdf` [Accessed: 05-May-2017]

[14] L. Lockshin and A. M. Corsi, "Consumer behaviour for wine 2.0: A review since 2003 and future directions," Wine Econ. Policy, **1**(1), 2–23 (2012)

[15] S. Mann, A. Ferjani, and L. Reissig, "What matters to consumers of organic wine?," Br. Food J. **114**(2), 272–284 (2012)

[16] IPSOS-SUDVINBIO, «Les européens et le vin bio. chiffres et tendances», Montpellier (2015)

[17] IPSOS-SUDVINBIO, "Viticulture bio: une filière en pleine croissance," Lattes (2014)

[18] Agencebio, «Croissance Historique de la bio en france le secteur confirme son envol en 2016», *Dossier de presse* (2016) [Online]. Available: `http://www.agencebio.org/sites/default/files/upload/dp_bio_conf_septembre_200916.pdf` [Accessed: 09-Jun-2017]

[19] S. V. da Silveira, L. da R. Garrido, and A. Hoffmann, *Produção integrada de uva para processamento: bases para a adoção da produção integrada*, 1st ed. Brasília: Embrapa (2015)

[20] IBRAVIN, "PAS: Garantia de qualidade do campo à mesa," *Informativo Saca-rolhas*, Bento Gonçalves, p. 16 (2013)

DOI: 10.1051/bioconf/20170903012

Producers' approaches about good agricultural practices in Manisa and İzmir

Selcuk Karabat[1] and Başak Aydın[2]

[1] Viticulture Research Institute, Manisa, Turkey
[2] Atatürk Soil Water and Agricultural Meteorology Research Institute, Kırklareli, Turkey

Abstract. This study includes the agricultural enterprises which apply and do not apply good agricultural practices in Manisa and İzmir provinces. The basic purposes of this research are; to make a comparison between the enterprises apply and do not apply good agricultural practices, to determine the awareness of the producers to good agricultural practices, to specify the tendency of the farmers towards good agricultural practices and to present the factors affecting to good agriculture. In the scope of the study, 25 grape producers in Manisa and 20 mandarin producers in Izmir that apply good agricultural practices have been participated in a survey. Besides, the same survey was also conducted with the same number of producers which do not apply good agricultural practices. To analyze the data, simple calculations such as average, percentage, frequency and through the use of the crosstabs were used and the social-economic status of the producers and some technical and economical specifications of the enterprises were determined. The tendency of the producers towards performing good agricultural practices and the factors affecting to these tendencies have been observed. The tendency of the producers were determined by using Logit Regression Analysis.

1. Introduction and objectives

Today, society's greatest need is to provide safe food items. Rapid increase in the world population, environmental pollution due to developing technology, economic weakness and inadequate education deepen nutritional problems and make it difficult to provide safe food.

Emerging food technology and consumer consciousness are also increasing efforts to improve product quality. The production of foods with the basic requirements for the lives of the consumers in line with current technological requirements is an important service for providing healthy nutrition. Efforts to ensure food safety and quality assurance are also of great importance for consumer and community health.

Good Agricultural Practices (GAP) is one of the well-known standards developed by the Food and Agriculture Organization (FAO) and presently GAP certification is a must for export in most of the developed countries. GAP aims at attaining social, economic and environmental sustainability by following certain criteria right from the planting of crop to the marketing [1]. Turkey with more than 4 million tons of grapes is the 6th largest producer of grapes in the world. Turkey also has the distinction of achieving the high production in mandarin in the world. More than 40 per cent of the total grape production is produced in Manisa-Turkey. Also 15 per cent of the total mandarin production is produced in İzmir-Turkey [2].

This study includes the agricultural enterprises which apply and do not apply good agricultural practices (GAP) in Manisa and İzmir provinces. The main objectives of the research; The economic comparison of farmers applying and not applying GAP in the area of research, determination of farmers' awareness of GAP, the determination of farmers' tendencies towards GAP and the factors affecting GAP.

2. Data and methodology

2.1. Material

The main material of the study was obtained from the surveys conducted with producers in İzmir and Manisa provinces. In addition, it has been used in previous domestic and foreign studies and statistics related to the research topic.

2.2. Method

According to the year 2013 data; 25 grape producers engaged in GAP in Manisa province; A total of 20 mandarin producers were surveyed in İzmir. A survey was conducted with the same number of non-GAP farmers in the study.

2.2.1. Method used in analysis of data

In the analysis of the data, using simple calculations such as average, percentage, and cross tables; some socio-economic characteristics of the surveyed producers and some technical and economic characteristics of the enterprises have been determined. Factors influencing the tendency of producers to make GAP were identified.

 DOI: 10.1051/bioconf/20170903012

Table 1. Some characteristics of GAP and Non-GAP producers.

	GAP Farmers	Non-GAP Farmers
Age of farmers (year)	49,49	51,29
Education period (year)	9,64	7,00
Number of family members	3,91	4,56
Experience with agriculture (year)	27,49	32,64

GAP tendencies of producers were determined using Logit regression analysis.

Logit Regression is a nonlinear regression model specially designed for binary dependent variable. In the literature, Logit regression is also called "Logistic Regression" [3]. If the dependent variable in the model is expressed in two categories, then the model "Dual Logistic Regression Model" is called "Multiple Logistic Regression Model" if it is expressed in more than one category [4]. In the binary logistic regression model, the observed value of the dependent variable is 1 if the event occurs, 0 if the event does not occur, to express two possible states [5]. Logistic regression was developed by statistician David Cox in 1958. The binary logistic model is used to estimate the probability of a binary response based on one or more predictor (or independent) variables (features). It allows one to say that the presence of a risk factor increases the probability of a given outcome by a specific percentage. In this study, farmers' GAP status was taken as a dependent variable and 0 for those who did not 1 for GAP farmers.

3. Findings and results

The average age of the GAP producers was 49.49 and the average age of non-GAP producers was 51.29. GAP producers average year of education 9.64, Non-GAP producers education year average is 7 years. The average number of family members of GAP producers was 3.91, and the average number of family members of non-GAP producers was 4.56. The average agricultural experience of GAP producers was 27.49 and the average agricultural experience of non-GAP producers was 32.64 years. GAP producers and non-GAP producers have also been statistically different in terms of education, number of family members and agricultural experience.

The average land size of GAP producers is 61.57 daa, and the average land size of non-GAP producers is 25.24 daa. The share of land belonging to GAP producers in total land was determined as 97.97%, and the share of land belonging to non-GAP producers in total was determined as 95.76%. As a result of the t-test to determine whether the difference in land size between GAP and Non-GAP producers is statistically significant, ($P = 0.004$) between producers' land sizes at 1% significance level.

Logistic regression analysis results

Factors affecting GAP status of producers were analyzed using logistic regression model. In the binary logistic regression model used in the study, GAP (1) and non-GAP (0) were used as dependent variables.

The independent variables of the model are age, education, number of family members, agricultural experience, non-agricultural income, living place, attitude towards innovations, and number of tractor, land size, total revenue, and total cattle.

Table 2. Estimation results of the logistic regression model.

Variables	Coefficient	Std Dv	P Value
Constant	−1,507	3,052	,621
Age	,042	,048	,378
Education	,060	,111	,589
Num of family member	−,436	,242	,072***
Agr. Experience	−,021	,041	,610
Non Ag. Job	,246	,742	,741
Living place	1,523	,638	,017**
Mem. of an org.	,038	,316	,904
Attitude towards innovations	−1,206	,570	,034**
Number of tractor	,273	,814	,738
Land size	,019	,033	,568
Total revenue	,000	,000	,727
Number of cattle	−,053	,039	,173
Nagelkerke R Square = 0,52			
−2 Log likelihood = 79,74[a]			
$X^2 = 9{,}374$ $p = 0{,}312$ (Hosmer and Lemeshow test)			
$X^2 = 42{,}632$ $P = 0{,}000$ (Omnibus test)			

The GAP approaches of producers were estimated using the logistic regression model and the results obtained are given in Table 2. The coefficients of the variables used in the model to determine the GAP approaches of the producers were quite different and statistically significant ($p = 0{,}000$). The independent variables account for 52% of the dependent variable. As a result of the Hosmer Lemeshow test, the chi square value was calculated as 9,374. p value was obtained as $p = 0{,}312 > 0{,}05$, indicating that the model is appropriate.

The age, education, non-agricultural income, membership of an organization, number of tractor, land size and total revenue are variables included in the model affects the GAP in the positive direction, the agricultural experience and the number of cattle in the negative direction. These variables are statistically insignificant ($p > 0{,}10$).

The number of family members of the producers affected GAP status ($p = 0.072$) at the 10% significance level on the negative side.

The producers' living place affects the GAP status positively at the 5% significance level ($p = 0.017$). The GAP tendencies are increasing as the place where the producers are residing moves from the villages to the small cities. This may be a sign that residence in the county or city provides easy access to the most effective pesticides retailers and governmental agricultural organization units among the consulting resources. As the place of residence of a non-GAP producer moves towards the city, the probability of participating the GAP increases 4.587 times.

Producers' approaches to innovations have a positive impact on good agricultural practice at a 5% significance level ($p = 0.034$). GAP approaches are increasing as producers increase their positive attitudes towards agricultural investment and technology. This situation is an expected result. As the positive attitude for innovations of a non-GAP farmer increases, the GAP participating probability increases by 0.299.

DOI: 10.1051/bioconf/20170903012

References

[1] FAO report, Development of a Framework for Good Agricultural Practices, Rome (2003)

[2] TUİK, (Turkish Statistical Institute) Agricultural Production Statistics. (`www.tuik.gov.tr`) (2015)

[3] Stock, J. H., Watson, M W. Introduction to Econometrics, Pearson Addison Wesley, Boston (2007)

[4] Leech, N. L., Barrett, K. C., Morgan, G. A. SPSS For Intermediate Statistics: Use and Interpretation, Lawrance Erlbaum Associates Publishers, Manwah New Jersey (2004)

[5] Walker, S. H., Duncan, D.B. "Estimation of the Probability of an Event as a Function of Several Independent Variables", Biometrika **54**, 167–179 (1967)

Une prospective pour la filière française des vins biologiques / Prospective study for french Organic wines industry

Patrick Aigrain[1], Françoise Brugière[1], et Hervé Hannin[2]

[1] FranceAgriMer, France
[2] Montpellier SupAgro, France

Résumé. La viticulture biologique française est une filière dynamique. Si certaines évolutions de la demande de bio sont similaires à celles observées pour le vin conventionnel (rosé, Bag-in-box, ...), d'autres comme la demande de vins sans sulfites ou de vins «vegan» interpellent plus spécifiquement ce segment. La filière des vins bio se trouve à la fois devoir respecter la règlementation viticole et celle de la production en bio, que ce soit par le cahier des charges de production de la matière première «raisin» que de sa transformation en vin. Elle doit satisfaire des consommateurs exigeants et fédérer des producteurs passionnés parfois jusqu'à l'excès. Face à cette forte dynamique de développement soutenue par une demande qui contourne partiellement les circuits classiques de distribution, la filière française des vins bio (France Vins Bio) a souhaité mener un exercice de prospective conduit par FranceAgriMer et Montpellier SupAgro. La présente communication va présenter, après un rappel des grands principes et de la méthode de prospective utilisée, les thématiques explorées, les principales hypothèses combinées et l'essentiel des scénarios sur lesquels elles débouchent.

Abstract. The French organic viticulture is a dynamic sector. If certain evolutions of the request for organic are similar to that are observed for the conventional wine (rosé, Bag-in-box, ...), others, as the demand for wines without sulphites or vegan wines challenge this segment more specifically. The sector of the organic wines is supposed, at the same time, to respect the wine regulation and organic regulation like an agricultural product but also as a processed product. It must satisfy the demanding consumers and federate producers impassioned sometimes until excess. Vis-a-vis this strong dynamics of development supported by a request which partially circumvents the classical circuits of distribution, the French sector of the organic wines (Organic France Wines) wished to carry out an exercise of futurology led by FranceAgriMer and Montpellier SupAgro. The present communication will present, after a recall of the great principles and method of futurology used, the various topics examined, the principal combined assumptions and the main part of the scenarios to which they lead.

1. Une filière viticole biologique qui s'interroge

La viticulture biologique française est une filière dynamique: entre 2007 et 2015, les surfaces certifiées Bio[1] ont augmenté de 14 600 à 57 600 hectares auxquels s'ajoutent près de 11 000 ha en conversion, soit, au total 8,7 % du vignoble national. Les objectifs du Grenelle de l'environnement, qui visait 6 % de la surface agricole utile, ou S.A.U., en bio en 2012, ont été atteints et dépassés dès 2010 pour la viticulture.

1,5 million d'hl de vins bio français ont été mis en marché en 2015 dont 54 % sur le marché intérieur. (estimation AgenceBio).

Sur la même période 2007–2015, les sommes dépensées par les ménages pour leurs achats de vins bio sont passées de 249 à 670 millions d'euros[2] dont 41 % en achat directement au producteur. En France, de tous les produits agricoles Bio, le vin est celui qui est le plus commercialisé en circuit court.

Interpellée par cette forte dynamique de développement, elle-même soutenue par une demande qui contourne partiellement les circuits classiques de distribution, la filière française des vins biologiques à travers France Vin Bio (association interprofessionnelle des vins biologiques, qui fédère les associations interprofessionnelles des régions Nouvelle-Aquitaine, Occitanie, Centre-Val-de-Loire, Grand-Est, et Bourgogne-Franche-Comté), a souhaité mener un exercice de prospective.

2. Méthode de prospective utilisée

L'objet de cet exercice conduit par FranceAgriMer et Montpellier SupAgro, est d'élaborer des scénarios pour l'avenir de la filière française des vins biologiques qui puissent servir aux décideurs de la filière, en vue de l'élaboration de stratégies gagnantes.

La méthodologie spécifique à la prospective n'a pas pour objet de prédire l'avenir mais de permettre d'anticiper différentes situations que l'on pourrait rencontrer, sans préjuger de leur caractère probable, souhaitable ou au contraire redoutable. Ces situations et les chemins qui y conduisent sont inscrits dans des scénarios qu'il s'agit de

[1] Source Agence Bio.
[2] Source AgenceBio AND I.

présenter aux commanditaires et décideurs. L'élaboration de ces scénarios incombe à des experts qui sont identifiés et réunis, et dont les compétences diversifiées sont croisées dans un contexte de liberté de parole et d'efforts constants de compréhension mutuelle.

Le travail a pour base la méthode SYSPAHMM (Système, Processus, Agrégat d'Hypothèses, Micro et Macroscénarios) formalisée par Michel et Clementina Sebillotte à l'INRA. Il s'appuie ensuite sur un ajout méthodologique (dit «bouturage-greffage») permettant de valoriser le travail mené à l'échelle de l'ensemble de la viti-viniculture française par l'INRA pour s'adapter à l'échelle inférieure de la filière vins bio.

Cette méthode repose sur la représentation commune du système étudié et des problématiques porteuses d'enjeux pour l'avenir ; celles-ci sont l'objet d'exposés d'experts puis le support de la rédaction d'hypothèses dont l'expression en recto et en verso permet d'envisager différents futurs possibles. 476 hypothèses ont ainsi été rédigées par la cellule d'animation de la prospective. Un travail de regroupement d'hypothèses proches, voire similaires (type doublon), a été poursuivi jusqu'à la réduction de la liste à 210 hypothèses, puis une sélection des hypothèses les plus importantes a été opérée (par un vote). Le réservoir d'hypothèses constitué par les 104 hypothèses de la prospective «Vignes et vins de France» publiée par l'INRA en 2003[3] a également été mobilisé. Ainsi, **46 hypothèses**, jugées les plus déterminantes pour l'avenir de la filière, ont, *in fine*, été sélectionnées. Ces hypothèses ne sont pas uniquement tendancielles, elles incluent également des signaux faibles actuellement d'ampleur limitée mais jugés potentiellement porteurs de modifications sensibles de la trajectoire de la filière française des vins biologiques à l'avenir. Les relations d'influences / dépendances entre ces hypothèses considérées deux à deux ont été établies et utilisées pour regrouper les hypothèses les plus liées entre elles en agrégats ; ceux-ci constituent la trame des scénarios interdisciplinaires qui tiennent ainsi compte des liens entre les hypothèses et en valorisent les libellés recto et verso. A l'issue de plusieurs étapes de réduction de la complexité, **5 scénarios contrastés** d'évolution de la filière française des vins biologiques à l'horizon 2035 ont été écrits.

3. Contenu des scénarios

Les cinq scénarios pour la filière, présentés ci-après dans leur format résumé, ont en commun d'aborder des familles de questions. Ces questions peuvent être structurées en quatre catégories de thématiques :

- Celles relatives au **contexte général global** dans lesquels évolue la filière des vins biologiques, en abordant notamment les questions de l'intensité du changement climatique et de ses conséquences, de l'évolution de la situation économique et financière mondiale (contexte macroéconomique global), et des politiques publiques en matière de santé et de prévention de la consommation excessive d'alcool.
- Celles relatives au **contexte de la filière vins** : transgénèse, nouveaux cépages, IGP/AOP/terroir, définition internationale du vin, contexte culturel d'usage, marchés export, étiquetage/normes internationales, ...
- Celles relatives au **contexte spécifique à la filière de la viticulture et des vins biologiques** : recherche & impasses techniques, formation & enseignement, «chapelles», vins natures, organisation de la filière, distribution, promotion, communication, image, consommation, segmentation, coopératives, conversion & déconversion, installation & transmission d'exploitations.
- Enfin celles relatives à **l'agriculture et aux produits biologiques** : politiques publiques bio (conversions, aménités, eau, ...), bio & normes environnementales, réglementation bio, ...

In fine, chacune de ces cinq histoires correspond à une combinatoire unique de réponses apportées à ces familles communes de questions.

4. Les cinq scénarios

Scénario 1 - La filière vin bio essaye de survivre

Synthèse : Confrontée à une situation macro-économique très difficile et au renforcement de la lutte anti-alcool, bloquée par les impasses techniques et victime des effets du changement climatique, mais jugeant les rares innovations techniques transférables contraires à ses principes philosophiques, qui l'empêchent de plus de profiter de la flexibilité réglementaire offerte pour faciliter l'adaptation de ses acteurs économiques, supplantée par d'autres labels, la filière vin bio voit son marché s'effondrer. Sa survie dépend de l'installation de néo-ruraux et de la conversion d'exploitants conventionnels.

Scénario 2 - La filière réduite au segment premium «vin bio & santé»

Synthèse : Dans un contexte macro-économique très difficile, avec une réglementation bio de plus en plus contraignante, bloquée par les impasses techniques mais jugeant les rares innovations techniques transférables contraires à ses principes philosophiques, la filière vin bio profite des conséquences du changement climatique qui jouent en sa faveur pour rééquilibrer son différentiel de productivité avec le conventionnel. Supplanté par d'autres labels, le vin bio tire parti de son élément de différenciation en matière de problématiques de santé humaine pour occuper et travailler ce seul segment du marché.

Scénario 3 – Le phénix bio

Synthèse : Dans un contexte particulièrement hostile, combinant une crise économique généralisée, une forte pression anti-alcool des autorités de santé, un impact sévère du changement climatique et une restriction continue du nombre de molécules utilisables, la plupart des producteurs d'IG et de vins bio font converger leurs pratiques et leurs cahiers des charges ouvrant la voie à la création d'une catégorie IGB (Indications géographiques bio). Quelques-uns refusent l'assimilation bio-IG, tentent de jouer la carte de la surenchère en s'interdisant toutes les innovations techniques et se marginalisent. Ce mouvement exigeant pourra sortir de sa marginalité par l'émergence

[3] SEBILLOTTE M. et all. Prospective Vignes et Vins – Scénarios et défis pour la recherche et les acteurs – INRA 2003.

de refondateurs qui mettent en avant un contre modèle politique.

Scénario 4 - La filière vin bio gère sa rente de situation

Synthèse : Devenue référente environnementale, identifiée comme voie de résistance aux chocs climatiques, reconnue dans les instances du vin, soutenue par les Pouvoirs Publics, la filière «vin bio» contrôle son développement de manière restrictive pour protéger sa rente de situation en intégrant avec parcimonie les avancées de la recherche et en restreignant la surface de son vignoble...

Scénario 5 – Croissance quantitative assumée

Synthèse : La vitiviniculture bio sort de sa niche pour s'imposer comme le contre modèle agricole en lieu et place de l'IG dans un contexte où le vin est dorénavant majoritairement produit agroindustriel comme les autres. Respectueux de l'environnement et de la santé des consommateurs, apte à atténuer les effets du changement climatique, et pour cela soutenue par les pouvoir publics, la filière vin bio accepte les pratiques physiques et l'intégration de l'innovation technique sans nier ses principes philosophiques essentiels, et relève le défi d'une croissance quantitative conséquente.

Enjeux, conséquences, attitudes stratégiques

Pour chaque scénario, il est possible de dégager les enjeux et conséquences pour la filière française des vins biologiques, l'interprofession, telle ou telle famille professionnelle :

si le scénario se produit, alors,

- qu'est-ce qu'on gagne (en quoi est-ce une opportunité) ?
- qu'est-ce qu'on perd (en quoi est-ce un risque ou une menace) ?

et quelles sont les conséquences ?

Ces 5 visions scénarisées du futur de la filière française des vins biologiques, ne s'apparentent en rien à de la prévision. Elles proposent simplement des futurs possibles et crédibles sans considération de probabilité pour leur occurrence. Il appartiendra au Conseil d'Administration de France Vin Bio de se prononcer sur les attitudes à adopter vis-à-vis des scénarios en choisissant parmi cinq attitudes possibles :

- Proactivité positive : agir dès aujourd'hui pour favoriser l'advenue du scénario.
- Proactivité négative : agir dès aujourd'hui pour défavoriser l'advenue du scénario.
- Réactivité anticipée : se préparer dès aujourd'hui à l'advenue du scénario.
- Veille : ce scénario doit être placé sous surveillance, pour savoir si son advenue se dessine au fur et à mesure du temps.
- Aucune attitude : ce scénario ne présente pas d'intérêt particulier.

C'est à partir de ce positionnement que pourra s'engager une démarche stratégique.

5. Déclinaison régionale de la réflexion prospective

A la demande de la Direction régionale de l'agriculture, un exercice de prospective avait été réalisé en 2011 pour le bassin viticole Languedoc-Roussillon et avait abouti à la description de quatre futurs possible pour la viticulture régionale.

Une nouvelle démarche est actuellement entreprise pour tenter de construire des scénarios pour la viticulture bio du Languedoc-Roussillon en utilisant ces deux prospectives «Filière française des vins biologiques» d'une part et «Vignes et vins du Languedoc Roussillon».

On the effects of search attributes on price variability: An empirical investigation on quality wines

Antonio Seccia[1], Domenico Carlucci[2], Fabio Gaetano Santeramo[1], Tiziana Sarnari[3], and Gianluca Nardone[1]

[1] Department of Agricultural, Food and Environmental Sciences, University of Foggia, Foggia, Italy
[2] Department of Agro-Environmental and Territorial Sciences, University of Bari, Bari, Italy
[3] Ismea, Roma, Italy

Abstract. The choice of a bottle of wine is affected by the presence of attributes that are searched by consumers and can be evaluated before the purchase. The aim of the paper is to analyze the effect of some search attributes on wine price variability applying the Hedonic Price Model. It allows explaining how the price of wine varies depending on its main quality attributes. The analysis has been based on a sample of wines made in Puglia, Italian region characterized by a tradition in wine production and consumption. Data have been collected from a wine guidebook considering the years 2008–2013. The study provided a measure of the market value of some search attributes for wines produced in Puglia. Attributes as alcoholic content, age and score given by experts, influence price variability allowing wines to obtain a premium price, such as the most known Protected Designation of Origin (PDO) and some Protected Geographical Indication (PGI). The name of the variety seems not to have high influence with the exception of less known and locally grown varieties. Results may be of interest for marketers and policy makers of wine industry. Managerial implications could refer to the importance of differentiation strategies aimed to market segmentation and to the pricing strategy. Policymakers could also find interesting hints about the influence of the different appellations and the importance of minor autochthonous grape varieties.

1. Introduction

Wine is a highly differentiated product made from grapes of different varieties grown under various pedoclimatic conditions which change a lot across geographical areas and years. This wide heterogeneity is reflected on the wine price that, such as for other products, is associated with the quality as perceived by consumers. However, in most cases, purchasers compare different bottles of wine with no past consumption experience and so without really knowing the characteristics of the product. Consequently, a relevant role in the choice process is played by attributes whose presence can be verified before the purchase, known as "search attributes", reading the label on the bottle and looking for wine's description and evaluation, referred to a rating system, provided by experts.

The aim of this work is to analyze and quantify the effects that search attributes of wine such as color, alcoholic content, variety, age, area of production and sensory characteristics, can have on price. For this purpose, a "hedonic price model" has been estimated. This model relates the price of a generic good to its quality attributes. The concept is that any product embodies a bundle of characteristics that define its quality. The price of each attribute is implicit but the sum of implicit prices of all attributes determines the whole price of the product. Statistical analysis helps to measure consumers' evaluations of the different product attributes.

The analysis has been conducted on a sample of wines produced in Puglia, which is the third largest wine producing region in Italy, with an average production of more than 6 million hectolitres, equal to almost 13% of the national production, from a vineyard surface area of almost 86,000 hectares. Nearly 19% is certified as Protected Geographical Indication (PGI), whereas the incidence of wines with Protected Designation of Origin (PDO) is 4% and the remaining production (77%) is referred to table wine [1]. In Puglia, there are 32 wines with PDO and 6 wines with PGI. Red wines from autochthonous varieties (*Primitivo*, *Negramaro* and *Nero di Troia*) are the main exported products and the primary export market is Europe (Germany, Austria, the Netherlands and Denmark) but the most profitable market is United States (6–7% of the total export by value from the region). Export share towards Canada and Japan is increasing, whereas Apulian wine is still less known in new consumers countries as Brazil, Russia, India and China.

This study aims to contribute to a better understanding of the influence of search attributes on wine price by using an approach that takes into account both the demand and supply sides. This kind of analysis has important practical implications related to the possibility of quantifying the individual effect of each quality attribute on the overall price of wine. In fact, as Oczkowski [2] pointed out, if the benefit associated with a particular quality attribute (implicit price) could be compared with the relative costs incurred, producers could make better strategic choices. Thus, the results of this study may be useful in understanding the evolutionary dynamics of Apulian wine market and in addressing marketing strategies as companies face an expanding market that is also characterized by increasing competitive pressure and rapidly changing consumer preferences. Moreover, results could be useful to policymakers for decisions regarding

DOI: 10.1051/bioconf/20170903014

the effectiveness of appellations as Protected Designation of Origin (PDO) and Protected Geographical Indication (PGI) and the relevance that minor autochthonous grape varieties have in characterizing the identity of a territory.

The remainder of this paper is organized as follows: Sect. 2 briefly presents an overview of the literature about hedonic price with a focus on wine; Sect. 3 gives a detailed description of the applied methodology; Sect. 4 discusses the results; Sect. 5 summarizes the main findings and highlights some implications.

2. Literature review

According to Lancaster's theory of demand [3], the utility that a consumer can derive from a product depends on the characteristics embedded in it and, under the assumption of perfect competition, the theory suggests that consumers' willingness to pay depends on the bundle of several quality attributes that are independently valued by consumers at the time of purchase. So, the observed market price of a product is the sum of implicit prices paid for each quality attribute [4]; implicit prices can be estimated by employing a hedonic price model which is a regression model capable of expressing the observable price of any particular product as a function of its characteristics (directly or indirectly observable). This theoretical model is based on the assumption of a general economic equilibrium in a perfectly competitive market; therefore, consumers maximize utility by choosing available products under budget constraints and firms maximize profits given the available technology and factor prices [4]. As a consequence, being related to both supply and demand conditions, implicit prices cannot be considered merely as indicators of consumer preferences [2,4,5]. Moreover, in a situation in which there is imperfect competition, implicit prices are also affected by the choices of producers who take into account their own market power, price elasticity of demand for each attribute, and the costs required to incorporate each attribute in the final product [6].

In literature, a large number of studies have adopted this approach for analysis applied to the market of wine because this product is characterized by high degree of differentiation associated with price variability more than other food products [7,8]. For estimating implicit prices, many authors have focused on some wine characteristics that consumers evaluate when making a purchasing decision: the importance of the area of production [5,9,10], the reputation of the winery [11,12], grape varieties [9,13–15], colour [16] and sensory quality ratings [8,9,17]. In addition, this approach has been adopted to assess the influence in pricing structure of product packaging characteristics [18], different price segments [17] and retail formats [19].

The estimation of a hedonic price function deals with some methodological issues. First, a sufficiently large sample size is needed to conduct the estimate and, regarding that, in previous works, wine guidebooks have been used as source of data [2,5,9,20–23]. Further, a crucial aspect is the choice of wine attributes to include in the function as regressors, which is influenced by both data availability and specific objectives of the analysis. Generally, attributes directly valued by consumers before purchasing wine are considered the most suitable for this methodology: colour, alcohol content, area of origin (country, region, sub-region), vintage, variety. Many works have considered such attributes which have shown high significance [2,5,7,9,14,20–27]. In addition, some authors have proved that, other characteristics being equal, brand and certification of origin play a significant role on the price variability [5,9,20,21,23]. Finally, a hedonic price function should include a variable referred to the sensory characteristics of wine. In fact, a positive judgement by consumers on a wine quality will probably lead to repeat purchases and to attract new customers with the result of a price increase. However, it is not easy to find one or more variables that objectively measure the organoleptic quality of a wine. Such issue has been taken into account considering a scoring system for the evaluation of wines by a panel of experts. Some authors have found that scores reported in wine guides do have a significant impact on prices, stating that wine experts judges, playing a role as opinion leaders, have a heavy influence on consumers, mainly when choosing premium wines [2,5,9,20]. On the other hand, other authors have observed a low significance of experts' scores and tried to explain such result with the high degree of subjectivity in wine sensory evaluation that could not reflect preferences of consumers [7,21–23].

The present paper contributes to the existing literature focusing on wines produced in a specific region, the third for production in Italy, and using an approach that considers prices suggested by wineries, as dependent variable, and information readable on the label, as independent variables, so taking into account both the demand and supply sides.

3. Methodology

3.1. Hedonic price model

In this study a hedonic price equation has been estimated with the aim of analyzing the relationship between the price and the main search attributes of wine. Almost every considered attribute can be easily recognized by consumers at the time of purchase by reading label information, particularly, alcoholic content, vintage, colour, area of production (as designation of origin and geographical indication) and variety. Since the price of a wine also largely depends on its organoleptic characteristics, as shown in previous researches [2,5,9,20], it has been considered appropriate to include, as an attribute, the evaluation of the sensory characteristics of the wine made by a panel of experts.

3.2. Data collection and data-set

Data were collected from the annual wine guidebook "Guida dei Vini di Puglia" published by the newspaper "La Gazzetta del Mezzogiorno" considering the editions of 6 years from 2008 to 2013. This guide includes more than 150 wineries located in Puglia region, reporting, for each of them, a description of three wines: the most expensive, the winemaker choice and a new product. The description provides information readable on the label as well as the suggested retail price (Euro/bottle 0.75 liter) and a rating (ranging from 1 to 4 stars) based on the organoleptic evaluation provided by a team of experts from the Italian Sommelier Association (AIS), the Italian Association of Oenologists and the National Wine Tester Organization (ONAV). The number of only three wines

Table 1. Characteristics of the Sample (Price in Euro/bottle 0.75lt).

	Number	Min Price	Max Price	Average Price	St. Dev.
	2280	1,90	80,00	10,57	6,80
Alcohol content					
<12	51	2,46	16,20	5,83	2,81
12–12,9°	464	2,30	25,00	6,89	2,52
13–13,9°	1022	1,90	42,00	9,68	4,80
14–14,9°	593	3,50	55,00	13,03	7,12
15–15,9°	106	5,00	78,00	16,92	9,43
>16°	44	4,50	80,00	26,90	16,17
Age					
1 year	915	2,46	35,00	7,70	3,42
2 years	608	2,30	45,00	10,83	6,10
3 years	407	1,90	80,00	13,30	8,48
4 years	208	4,00	52,00	13,30	6,52
5 years	84	5,50	43,50	13,13	6,93
6 years	36	6,00	50,00	17,61	10,52
>6 years	22	5,50	78,00	24,93	18,45
Colour					
Red	1514	1,90	80,00	12,01	7,62
White	463	2,50	25,00	8,02	3,52
Rosè	303	2,46	25,00	7,29	2,85
Score					
1 star	286	1,90	25,00	7,70	3,37
2 stars	1024	2,46	42,00	9,17	4,51
3 stars	843	3,00	80,00	11,85	7,56
4 stars	127	4,00	75,00	19,80	11,42
Appellation of Origin					
PGI	1539	2,3	78	9,31	5
PDO	741	1,9	80	10,29	5,45
Protected Geographical Indication					
Salento	882	2,50	78,00	11,07	7,25
Puglia	501	2,30	30,00	9,53	5,08
Daunia	64	3,50	20,00	8,54	4,58
Tarantino	34	3,00	20,00	9,27	4,07
Murgia	31	3,00	18,00	8,27	3,32
Valle d'Itria	27	4,50	35,00	9,19	5,68
Protected Designation of Origin					
Primitivo	151	3,50	48,00	13,41	6,68
Salice Salentino	145	4,20	35,00	10,17	5,42
Castel del Monte	119	2,46	55,00	11,67	9,00
Gioia del Colle	81	5,00	80,00	14,93	12,28
Brindisi	36	1,90	16,00	7,59	3,11
Copertino	33	4,80	35,00	8,84	5,34
Locorotondo	26	4,00	14,00	6,97	2,31
San Severo	19	3,00	10,00	5,60	2,45
Moscato di Trani	17	5,00	16,00	11,97	3,15
Others PDOs	114	3,25	35,00	10,66	5,27
Variety					
Blend of Varieties	730	1,90	78,00	10,50	6,14
Single varietal	1550	2,46	80,00	10,84	7,06
Primitivo	516	3,00	80,00	12,91	8,13
Negroamaro	376	3,00	52,00	9,91	6,63
Other National	287	2,46	30,00	8,44	3,99
Minor local	263	3,00	55,00	11,79	7,44
International	108	3,00	22,50	7,76	4,10

for each winery could represent a limit for the sample but the considered guidebook is the most comprehensive for the Apulian wineries. The collected data-set contains 2280 observations coming from 6 editions of the guidebook. The data-set has been considered as cross-section under the hypothesis that prices are not affected by inflation.

A preliminary analysis of the data-set has been carried out by calculating descriptive statistics - such as the number of cases, minimum, maximum, average and standard deviation of price - regarding both the total sample and specific sub-samples distinguished according to a particular quality attribute (Table 1).

In the sample, wines' great price variability is noteworthy, ranging from a minimum of 1.90 Euro/bottle to a maximum of 80.0 Euro/bottle with an average of 10.57 Euro/bottle.

According to the alcohol content (%V/V), 6 subsamples have been derived (less than 12°; 12°–12.9°; 13°–13.9°; 14°–14.9°; 15–15.9°; 16° or more). It is interesting to note that the average price of the wines included in each subsample regularly increases as the alcohol content increases. In fact, it is 5.83 Euro/bottle for wines with alcohol content of less than 12° while it is 26.90 Euro/bottle for wines with alcohol content higher than 16°.

The influence of age on the price of wine is showed considering the increasing trend of the average price in the 7 subsamples distinguished by the age of wine; it ranges from 7.70 Euro/bottle for the first group to 24.93 Euro/bottle for the group of wines of more than 6 years.

Considering the colour, the average price of red wines is higher than white and rosé (12.01 Euro/bottle compared respectively to 8.02 and 7.29 Euro/bottle); however, standard deviation of red wines is much higher showing greater price variability.

The prices of wines also show a fairly clear relationship with the score assigned by experts in accordance with the sensory characteristics. In fact, the average price of wines that have received the minimum score of one star is 7.7 Euro/bottle, rising to 9.17 Euro/bottle for those in the next group, to 11.85 Euro/bottle for those of the three stars group, and, finally, reaching 19.8 Euro/bottle in the top group.

In the sample, there are 741 wines with the appellation of origin PDO (Protected Designation of Origin) and 1539 with the appellation of origin PGI (Protected Geographical Indication). The difference between the average prices of the two groups is quite less than expected: in the former group is 10.29 Euro/bottle whereas for the latter is 9.31 Euro/bottle; however, the standard deviation for PDOs reveals higher price variability. With reference to the PGI appellations, 6 subsamples have been derived reflecting the number of PGIs in Puglia which are: *Salento*, *Puglia*, *Daunia*, *Tarantino*, *Murgia*, and *Valle d'Itria*. *Salento* and *Puglia* concentrate almost 90% of the PGIs observations, with the former accounting for more than 57%, and showing the highest average prices (11.7 and 9.53 Euro/bottle respectively). Wines with PDO appellations have been split up into 10 subsamples considering the frequency of observations in the group: *Primitivo di Manduria, Salice Salentino, Castel del Monte, Gioia del Colle, Brindisi, Copertino, Locorotondo, San Severo, Moscato di Trani, Other PDOs*. *Primitivo di Manduria* has the highest number of observations among PDOs, 151 cases corresponding to 20%, followed by *Salice Salentino, Castel del Monte, Gioia del Colle* and *Brindisi*. The average price, not considering the group *other PDOs*, ranges from 5.6 Euro/bottle for *San Severo* to 14.9 Euro/bottle for *Gioia del Colle*, which shows a great price variability.

The sample includes 1550 (68%) wines made from single varietal grapes and 730 (32%) from varietal blends (using two or more grape varieties). In the first group, the most observed varieties are *Primitivo* with 516 cases (23%) and *Negramaro* with 376 cases (16,5%) which are autochthonous of Puglia and among the 20 most widespread cultivars in Italy. They are followed by wines from varieties widespread in Italy (*Montepulciano, Sangiovese, Falanghina, Malvasia, Moscato*, etc.), from other Apulian autochthonous varieties (*Nero di Troia, Fiano Minutolo, Bombino Bianco, Primitivo di Gioia, Bombino Nero, Verdeca, Aglianico*, etc.) and from international varieties (*Chardonnay, Merlot, Cabernet Sauvignon, Sauvignon, Sirah*, etc.). Price comparison shows a small difference in the maximum price between blends and varietal wines, 80 Euro/bottle for the latter and 78 Euro/bottle for the former.

3.3. Empirical model

Information included in the above-described data-set has been used for the specification of the following hedonic price equation:

$$lnPrice = \alpha + \beta\ Alcoholic\_content + \gamma\ Age + \delta\ Score + \eta_i\ Colour_i + \theta_n Variety_n + \lambda_j\ Appellation_j + \varepsilon.$$

The variables included in the empirical model are listed and briefly described in Table 2.

The price of the bottle has been used as dependent variable (*Price*) in the empirical hedonic price equation, and it is a continuous variable ranging from the lowest value 1.9 Euro/bottle to the maximum 80.0 Euro/bottle. Three explanatory variables, alcoholic content (*Alcoholic_content*), age (*Age*) and score (*Score*) are continuous variables as well, while the other explanatory variables, being categorical, have been entered as dummy variables. Alcoholic content (*Alcoholic_content*) ranges from the minimum of 10.5° to the maximum of 19.0° with an average of 13.4°; age (*Age*) ranges from 1 year to 14 years with a medium age of 2; score (*Score*) is referred to the evaluation of experts who have ranked wines using a scale from 1 to 4 stars.

The remaining explicative variables, being categorical, have been transformed into groups of dichotomous variables or dummies. The colour (*Colour*) has been coded as 3 dummy variables: red, white and rosé; the first has been considered as the baseline variable. The appellation of origin (*Appellation*) has coded as 16 dummy variables: 6 for each Apulian PGI, 9 for each of the most common Apulian PDOs and 1 for the remaining PDOs which have been considered together as a baseline dummy variable. Finally, 6 dummy variables have been considered for *Variety*: the first and the second are referred to wines produced from the main Apulian varieties, *Negramaro* and *Primitivo*; the third to wines made from other autochthonous grapes (*Nero di Troia, Sussumaniello, Ottavianello, Verdeca, Bombino Nero, Aleatico, Bianco di Alessano, Fiano, Greco*, etc.); the fourth to wines from national varieties (*Montepulciano, Malvasia Bianca, Moscato*, ecc.); the fifth to wines from international varieties (*Chardonnay, Merlot, Cabernet Sauvignon, Sauvignon, Sirah, Pinot Noir*); the sixth to wines obtained by blended varieties which has been considered as the baseline variable. The justification to create three distinct groups for varietal wines from, respectively, minor autochthonous varieties, other national varieties and international varieties, each of them as

DOI: 10.1051/bioconf/20170903014

Table 2. Variables of the Empirical Model.

Variables	Tipology	Description
Dependent variable		
Price	Continuous variable	Price of wine expressed in Euro/bottle 0.75 lt
Regressors		
Alcohol_content	Continuous variable	Alcohol content expressed in % Vol
Age	Continuous variable	Age of wine expressed in years
Score	Continuous variable	Score expressed in number of stars
Colour	Categorical variable	
	Dummy	Red = 1;otherwise = 0 (baseline)
	Dummy	White = 1; otherwise = 0
	Dummy	Rosè = 1; otherwise = 0
Appellation	Categorical variable	
	Dummy	Other PDOs = 1; otherwise = 0 (baseline)
	Dummy	Salento = 1; otherwise = 0
	Dummy	Puglia = 1; otherwise = 0
	Dummy	Daunia = 1; otherwise = 0
	Dummy	Tarantino = 1; otherwise = 0
	Dummy	Murgia = 1; otherwise = 0
	Dummy	Valle d'Itria = 1; otherwise = 0
	Dummy	Primitivo di Manduria = 1; otherwise = 0
	Dummy	Salice Salentino = 1; otherwise = 0
	Dummy	Castel del Monte = 1;otherwise = 0
	Dummy	Gioia del Colle = 1; otherwise = 0
	Dummy	Brindisi = 1; otherwise = 0
	Dummy	Copertino = 1; otherwise = 0
	Dummy	Locorotondo = 1; otherwise = 0
	Dummy	San Severo = 1; otherwise = 0
	Dummy	Moscato di Trani = 1; otherwise = 0
Variety	Categorical variable	
	Dummy	Blend = 1; otherwise = 0 (baseline)
	Dummy	Primitivo = 1; otherwise = 0
	Dummy	Negramaro = 1; otherwise = 0
	Dummy	Other National = 1; otherwise = 0
	Dummy	Minor Autochtonous = 1; otherwise = 0
	Dummy	International = 1; otherwise = 0

categorical variable, lies in the fact that the number of observations for each single variety in the sample is too small to provide a sufficiently robust estimation.

Finally, regarding the functional form of the equation, the literature does not clearly suggest among linear, semi-logarithmic and logarithmic forms. Nevertheless, for this work a semi-logarithmic functional form has been chosen, taking into account the easy interpretation of its coefficients and its flexibility.

4. Results

Estimation results of the hedonic price Equation using the method of Ordinary Least Square (OLS) are summarized in Table 3, which also includes the most important performance indicators of the empirical model. In particular, it is relevant to highlight a good overall significance (F-statistic equal to 61.01 with a P-value lower than 0.01) and a good capability to explain the variability of the data-set (adjusted R-squared equal to 0.56). It is possible to observe that the three continuous explicative variables have high statistical significance, even if with different magnitude of the coefficients.

The first, *Alcohol_content* has a coefficient equal to 0.24. Taking into account the log-linear regression model, the coefficient of a continuous variable such as alcoholic content, explains the percentage change in price due to a unit variance of the explicative variable, all other characteristics being equal. So, one point increase in alcoholic content of wine is worth about + 24.0% in its price.

The variable *Score*, referred to the number of stars assigned to the wines by experts for the organoleptic characteristics, is significant as well. As a continuous variable, the regression outcome means that wines earn a 13% premium per unit of score, all other characteristics being equal.

Results show high correlation between price and the third continuous variable, *Age*, whose coefficient is 0.1 explaining that, as expected, wine aging has a positive effect on wine pricing, in this case the increase is 10% per year.

Considering the first of the three categorical variables, *Colour*, only the dummy referred to white wines is statistically significant. Since in the adopted model the coefficient of a dichotomous explanatory variable expresses the percentage change in price due to the presence of a given quality attribute (all other characteristics being equal), it follows that white wines show a difference of 6% in prices compared to red wines used as baseline.

DOI: 10.1051/bioconf/20170903014

Table 3. Estimation Results.

Variable	Coefficient	StandardError	TStatistic	PValue	Significance
Costant	−1, 8079	0,1782	−10, 1411	<0,0001	***
Alcoholic Content	0,2458	0,0135	18,1368	<0,0001	***
Colour					
Rosè	0,0092	0,0274	0,3358	0,737	
White	0,0666	0,0303	2,1975	0,0281	**
Score	0,1349	0,0111	12,1226	<0,0001	***
Age	0,1026	0,0078	13,001	<0,0001	***
Variety					
Primitivo	−0, 1016	0,0316	−3, 2136	0,0013	***
Negroamaro	−0, 0272	0,0248	−1, 0947	0,2737	
International	−0, 1313	0,0441	−2, 9763	0,0029	***
Minor autochtonous	0,1565	0,0315	4,9582	<0,0001	***
Others national	−0, 0049	0,0287	−0, 1715	0,8638	
Appellation of Origin					
Salento-PGI	0,221	0,0478	4,6186	<0,0001	***
Puglia-PGI	0,2106	0,0499	4,2206	<0,0001	***
Daunia-PGI	0,1753	0,072	2,4325	0,0151	**
Valle d'Itria-PGI	0,2942	0,082	3,5885	0,0003	***
Primitivo di Manduria-PDO	0,2025	0,0607	3,3353	0,0009	***
Salice Salentino-PDO	0,0947	0,054	1,753	0,0797	*
Castel del Monte-PDO	0,2365	0,0567	4,1691	<0,0001	***
Gioia del Colle-PDO	0,2362	0,0644	3,6631	0,0003	***
Locorotondo-PDO	0,1981	0,0769	2,5735	0,0101	**
Moscato di Trani-PDO	0,2047	0,1017	2,0126	0,0443	**
Colline Joniche Tarantine-PDO	0,5068	0,1276	3,9705	<0,0001	***
Cacc' e mitte di Lucera-PDO	0,3954	0,1276	3,0972	0,002	***
Dependent Variable = lnPrice					
F Statistic F (28, 2251) = 61,01907			P-value (F) = 1,3E-251		
R^2 = 0,568074			Adjusted R2 = 0,561457		
Log-likelihood = −979,0143			Akaike criterion 2016,029		
Significance: *** 1% ; ** 5% ; * 10%			Hannan-Quinn 2076,659		

Among dummies for the categorical variable *Variety*, only three - Primitivo, International and Minor autochtonous - are statistically significant. In the case of Primitivo and International the coefficients are negative, meaning that consumers are willing to pay less (−10% in the first case and −13% in the second) in comparison with wines made from blends of grapes, used as baseline. Wines from minor Apulian autochthonous varieties do have a higher value, equal to 15%, all other characteristics being equal.

The last categorical variable, *Appellation*, which is referred to the area of production, contributes to explain price variability displaying statistical significance for four PGIs and six PDOs. All coefficients of the dummy variables are positive and have to be interpreted as price premium compared to the dummy Other PDOs which has been considered as baseline, all other characteristics being equal. So, PGI *Valle d'Itria* presents the highest coefficient, meaning that wines with such appellation get a 29% premium price compared to the baseline appellations, followed by PDO *Castel del Monte* (+23%), PDO *Gioia del Colle* (23%), PGI *Salento* (22%), PGI *Puglia* (+21%), PDO Moscato di Trani (+20%), PDO *Primitivo di Manduria* (20%), PDO *Locorotondo* (+19%), PGI *Daunia* (17%) and PDO *Salice Salentino* (9%).

5. Conclusions

Wine is a highly differentiated product and some of the main attributes which affect market competition are defined as search attributes since they can be known prior to purchase. In fact, consumers can easily look for wines that satisfy their own preferences for characteristics as colour, alcohol content, variety, age, area of production and sensory characteristics. The hedonic price model applied to a sample of wines produced in Puglia provided a measure of the market value of these attributes that can be used to investigate some important features of this industry and to provide some insights on the wineries strategies. The study shows that Apulian wines prices evidence a high variability which mainly depends on attributes that can be valued by consumers before the purchase.

A first evidence is the positive influence of alcoholic content on wine price, that has been already found in a study about wines from three Italian regions: Piedmont, Liguria and Valle d'Aosta [13]. It can be explained with the traditional pattern of wine consumption in Italy, but particularly in Southern areas, where alcohol content is still considered an important signal of quality. Moreover, it needs to be outlined that weather conditions of Puglia influence the production of grapes with high sugar content and so high alcoholic strength of wines.

The empirical results about the effect of aging conformed to a priori expectations, as in Costanigro et al. [17], showing that the age of the wine, similarly to the alcoholic content, is an attribute that heavily affects consumers perception of wine quality. However, wine ageing implies higher costs for wineries related to storage and to the delay in selling.

DOI: 10.1051/bioconf/20170903014

Organoleptic quality can be known by consumers with reference to the evaluation provided by experts through a scoring system easily readable on guidebooks, magazines, web sites and other media. It has a clear correlation with wine prices since purchasers evidence a willingness to pay more for wines with sensory characteristics objectively judged superior. Findings are consistent with previous researches and suggest that consumers value such information [5,7,9,13,21,22,24,25,28,29].

The role played by the Geographical Indications and the Designations of Origin gives interesting results since significant dummies variables have positive coefficients. Purchasers recognize a premium price to the well known PDO as *Castel del Monte, Gioia del Colle, Moscato di Trani, Primitivo di Manduria, Locorotondo* and*Salice Salentino* and it is worth to note that three PGI (*Salento, Puglia* and *Daunia)* have a higher influence on price compared to the group Other PDOs, used as baseline. There is evidence that PDO certification, that is at the top of the Italian pyramid of quality, in some cases has less influence on price than PGI certification, characterized by looser roots in the area of origin. Indeed, PDO certification is characterized by more stringent rules about both grapes production and winemaking process and needs proper coordination mechanisms for the collective governing body (the Consortium) jointed to adequate management. On the other hand, the production of PGI wines takes advantage of more flexibility in all production stages, allowing the blending of more varieties and making possible to better react to changes in market trends. For this reason many Apulian producers prefer to use the certification PGI for their top quality production in order to compete in national and international markets. The influence of the region of origin on price has been showed by Schamel [16], Schamel and Anderson [9] and Troncoso [22] and, particularly, some studies showed a strong relationship between Denomination of Origin certification and price [22,30–32]. On the other hand, Nerlove [24] found that the origin of wine has low significant influence on price and Steiner [14] found a low valuation of French wines with geographical appellation in United Kingdom.

The study of the influence of the grape variety on the purchase price does not always lead to similar results according to the literature. Some authors found a sensitivity of the price to varieties, positive or negative depending on variety [23] or just positive [22], while other studies proved a weak correlation [14]. In the present study the estimation model has highlighted the positive influence of the Apulian minor autochthonous varieties on price variability, compared to wines made from blends of varieties. The influence of blends on price is also highlighted by San Martin et al. [23] who, however, refer to blends including Malbec and Cabernet Sauvignon, which are not minor varieties. In the present paper, the price premium associated with wines from minor autochthonous varieties can be related to the feature of such wines to convey a strong identity, both sensory and emotional, clearly opposed to the so-called international style. So, probably, these wines better satisfy the requests from consumers characterized by curiosity, with a medium-high and high level of wine knowledge and oriented to new taste experiences. However, an Italian study about wines from the Italian regions Piedmont, Liguria and Valle d'Aosta evidences different results showing negative coefficients for wines from autochthonous varieties [13]. It is important to take into account that the production of varietal wines from autochthonous grapes requires great investments aimed to varietal conversion and to solve some important technical problems. In fact, they are characterized by low productivity, strongly linked to microclimatic and pedological conditions, and by some critical aspects in oenological process.

Finally, although some limitations in the sample of data provided by the considered guidebook, as the number of only three wines per winery, the results of this study may contribute to understand the Apulian wine market. Further research could be carried out considering a greater number of wines and comparing the benefit associated with a quality attribute (implicit price) with the relative costs incurred.

References

[1] Ismea 2016
[2] E. Oczkowski, "A Hedonic Price Function for Australian Premium Table Wine", Australian Journal of Agricultural Economics **38**(1), 93–110 (1994)
[3] K.J. Lancaster, "A new approach to consumer theory", Journal of Political Economy **74**(2), 132–157 (1966)
[4] S. Rosen, "Hedonic prices and implicit markets: Product differentiation in pure competition", Journal of Political Economy **82**(1), 34–55 (1974)
[5] G. Schamel, "Geography Versus Brands in a Global Wine Market", Agribusiness **22**(3), 363–374 (2006)
[6] D. Hassan, & S. Monier-Dilhan, National brands and store brands: Competition through public quality labels. Agribusiness **22**(1), 21–30 (2006)
[7] S. Lecocq, and M. Visser, "What Determines Wine Prices: Objective vs Sensory Characteristics", Journal of Wine Economics **1**(1), 42–56 (2006)
[8] E. Oczkowski, "Hedonic Wine Price Functions and Measurement Error", The Economic Record **77**(239), 374–382 (2001)
[9] G. Schamel and K. Anderson, "Wine Quality and Varietal, Regional and Winery Reputation: Hedonic Prices for Australia and New Zealand", The Economic Record **79**(246), 357–369 (2003)
[10] L.A. Panzone and O.M. Simões, "The importance of regional and local origin in the choice of wine: Hedonic models of Portuguese wines in Portugal", Journal of Wine Research **20**(1), 27–44 (2009)
[11] S. Landon and C. Smith "Quality Expectations, Reputation, and Price", Southern Economic Journal **64**(3), 628–647 (1998)
[12] B. Ling, and L. Lockshin, "Components of wine prices for Australian wine: How winery reputation, wine quality, region, vintage, and winery size contribute to the price of varietal wines", Australasian Marketing Journal **11**(3), 19–32 (2003)
[13] P. Pavese and R. Zanola "Autochthon vs. blended wines: Do objective and sensory characteristics matter?" WP n.**18**, Periodico mensile on-line POLIS Working Papers (2008)
[14] B.E. Steiner "Australian wines in the British wine market: A hedonic price analysis", Agribusiness **20**(3), 287–307 (2004)

[15] D. Carlucci and A. Seccia, "Vini varietali e vitigni autoctoni: evidenze empiriche del mercato italiano", Rivista di Economia Agraria **65**(1), 43–62 (2010)
[16] G. Schamel, "Individual and Collective Reputation Indicators of Wine Quality Cies Discussion Paper 0009 (2000)
[17] M. Costanigro, J.J. McCluskey and R.C. Mittelhammer, "Segmenting the wine market based on price: hedonic regression when different prices mean different products", Journal of Agricultural Economics **58**(3), 454–466 (2007)
[18] S. Mueller Loose and G. Szolnoki, "Market price differentials for food packaging characteristics", Food Quality and Preference **25**(2), 171–182 (2012)
[19] E. Brentari, R, Levaggi and P. Zuccolotto, "Pricing strategies for Italian red wine" Food Quality and Preference **22**(8), 725–732 (2011)
[20] A. Coppola, V, Sodano and F. Verneau, "Explaining price variability in the Italian market for high quality wines", Rivista di Economia Agroalimentare **5**(1), 127–139 (2000)
[21] J.W. Haeger and K. Storchmann, "Prices of America Pinot Noir wines: climate, craftsmanship, critics", Agricultural economics **35**(1), 67–78 (2006)
[22] J.L. Troncoso, and M. Aguirre, "Price determinants of Chilean wines in the U.S. market: a hedonic approach", Spanish Journal of Agricultural Research **4**(2), 124–129 (2006)
[23] G. San Martin, J.L. Troncoso and B. Brümmer, "Determinants of Argentinean Wine Prices in the U.S.", Journal of Wine Economics **3**(1), 72–84 (2008)
[24] M. Nerlove, "Hedonic price functions and measurement of preferences: the case of Swedish wine consumers", European Economic Review **39**(9), 1697–1716 (1995)
[25] P. Combris, S, Lecocq and M. Visser, "Estimation of a hedonic price equation for Bordeaux wine: Does quality matter?", The Economic Journal **107**(441), 390–402 (1997)
[26] J.J. Fogarty, "The return to Australian fine wine", European Review of Agricultural Economics **33**(4), 542–561 (2006)
[27] O. Ashenfelter, "Predicting the quality and prices of Bordeaux wine", The Economic Journal **118**(529), 174–184 (2008)
[28] A. Di Vittorio and V. Ginsburgh, "Pricing Red Wines of Médoc: Vintages from 1949 to 1989 at Christie's auctions". Journal de la Société Statistique de Paris **137** (1995)
[29] T. Lima "Price and Quality in the California Wine Industry: An Empirical Investigation" Journal of Wine Economics **1**(1), 176–190 (2006)
[30] L. Benfratello, M. Piacenza, and S. Sacchetto, "Taste or reputation: What drives market prices in the wine industry? Estimation of a hedonic model for Italian premium wines", Appl. Econ. **41**(17), 2197–2209 (2009)
[31] C. Chambolle, E. and Giraud-Héraud, "Certification of Origin as a non-tariff barrier", Review of International Economics **13**(3), 461–471 (2005)
[32] R. Veale, and P. Quester "Consumer Sensory Evaluations of Wine Quality: The Respective Influence of Price and Country of Origin" Journal of Wine Economics **3**(1), 10–29 (2008)

 DOI: 10.1051/bioconf/20170903015

Behaviour of Millenial wine consumers in southern Serbia

Vladimir Radovanović, Jelena Petrović, and Blaga Radovanović

University of Niš, Faculty of Natural Sciences and Mathematics, Višegradska 33, 18000 Niš, Serbia

Abstract. Consumer behaviour when purchasing wine is the result of a complementary operation of the large number of different factors, which may include economic, geographic, social, psychological, and other. Discovering consumer preferences for wine and their buying behavior would allow the application of an appropriate marketing strategy to increase the sales of wine. Special attention was given to one of the most promising new demographic segments that likes to buy – Millennial generation. The main purpose of this research is to find out if the wine attributes of the Millennials are different from other generations in southern Serbia. Results of this research are shown that when choosing wine, the quality of wine has the biggest impact on consumers, rather than others wine attributes. The paper also asserts that, the wine industry should particularly focus on marketing to Millennial age group, as this segment has a high willingness to experiment.

1. Introduction

Wine is a product that has a high proportion of attributes that can only be assessed during consumption, experience attributes in other words, with the consequent ability of consumers to assess quality prior to purchase being severely impaired and resulting in consumers falling back on extrinsic cues to assess quality. Consumer behavior is the main field and source for Customer Relationship Management (CRM) programs [1]. A number of studies have been undertaken to investigate the interplay between the various wine attributes, such as product brand, region of origin, various packaging elements, grape variety, as well as other factors, such as price, consumption situation, or occasion, consumers' level of involvement with and knowledge of wine, consumption frequency, wine-related lifestyle segmentation, and so forth [2].

Consumer behavior when purchasing wine is the result of a complementary operation of the large number of different factors, which may include economic, geographic, social, psychological, and other. Literature relating to the behaviour of wine consumers focuses on the following two areas: factors affecting the behaviour of wine consumers and the factors that influence the decision-making process of consumers who buy wine [3]. A key consideration in exploring the wine consumer decision-making process and consumers' evaluation of alternatives is that wine attributes represent intrinsic and extrinsic cues for the consumer [4]. Sanchez and Gill [5] illustrate how consumers have preferences according to the bundle of benefits they are seeking.

Market segmentation or the process of dividing a market into meaningful, relatively similar and identifiable segments or groups is used in multiple industries [6].

Segmentation assists marketing researchers in analyzing the needs of a specific customer segment and it allows marketing campaigns to be focused on these identified needs [7]. Every generation has it's own characteristics. The changes in consumption expenditure can best be understood by taking a generation approach [8]. Different generations and demographic consumer groups are exposed to: (a) different social and economic opportunities and barriers, (b) different types of technology activities, (c) different social perceptions and different community norms, and (d) different life experiences and events [9].

One of the premiere studies on market segmentation was that of Bruwer and House [10], in which he focused on the geographic segment of consumers in the South Australian wine industry. A recent qualitative study conducted in the United States, in which segmentation focused on when and why consumers drink wine [7]. Regarding demographic segmentation, most consumer industries segment this category by the four generations: of Traditionalists (born between 1900 and 1945); Baby boomers (born between 1946 and 1964); Generation X-ers (born between 1965 and 1977); and Millennials or Y generation or Echo boomers (born between 1977 and 2000) [11]. Atkin and Thach [12] were demonstrated how the marketing strategies of Millennials differ from older consumers.

Determining the specific factors that influence Millennials and their purchasing attitudes and patterns has become an important focus of consumer research as their potential spending power, the ability to be trendsetters, adoption to new products and potential for becoming a lifetime customer [7]. A majority of Millennials (56%), Generation Xers (52%) and Boomers (54%) think technology helps people use their time more efficiently, Three-fourths (75%) of Millennials have created a social networking profile compared with 50% of Generation X-ers, 30% of Boomers and 6% of members of the Silent generation [11].

In order to implement an appropriate marketing strategy, which would contribute to increased market share of domestic wine producers on the wine market, it is necessary to conduct appropriate marketing research of wine consumers in southern Serbia. At the same time, it is necessary to analyse the influence of wine attributes on

 DOI: 10.1051/bioconf/20170903015

Table 1. Demographic characteristic of the wine consumers.

Age (year)	Sample numbers	Frequency (%)
18–28	830	48.7
29–38	313	18.4
39–50	262	15.4
51–65	300	17.6

making a decision to purchase and consume wine, in order to understand the preferences of consumers, as well as their behavior.

This paper relies on the conducted marketing research in southern Serbia of wine consumers in order to identify the wine attributes that has a dominant influence on the wine purchasing decision. The main purpose of this research is to find out if the wine attributes of the Millennials are different from other generations in southern Serbia. The paper also asserts that, the wine industry should particularly focus on marketing to Millennial age group, as this segment has a high willingness to experiment.

2. Research methodology

2.1. Samples

The survey was carried out in the southern Serbia. Consumers were surveyed by personal interview, i.e. the "face to face" technique. 1,720 wine consumers were surveyed. After elimination of questionnaires with incomplete answers, we got a sample of 1,705 questionnaires.

2.2. Statistical analysis

Data collected by surveying wine consumers was entered in the SPSS database, and all further analyses were performed using the Statistical Package for the Social Sciences, SPSS. Data analysis relied on statistical methods, descriptive statistics, Chi-test and cluster analysis.

3. Results and discussion

With a view to help Serbia's producers of wine in selling their products on the domestic market, where is large competition high quality imported wine, were carried out this investigation. The first major question was whether the respondent likes wine, and what are the reasons to decide for certain types of wine.

The Table 1 shows the classification of consumers according to their age in the test region of Serbia. Millennials generation is the largest group of wine consumers with a total of 1143 (67.1%). This segment group is divided into two subgroups: Generation Y_1 from 1977 to 1986 (18.4%). and Generation Y_2 from 1987 to 1997 (48.7%):

The largest number of samples of Millennials, Generation Y_2 (830 consumers), who used in this research, were students in University of Niš (Serbia). The age wine consumers from 39 to 50 (Generation X-ers, 1965–1976) participated in 15.4%, and from 51 to 65 year (Baby Boomers generation, 1950–1964) in 17.6% (Fig. 1):

Table 2 shows that 50.76% of consumers in Generation Y_2 make a decision to buy wine based on its quality

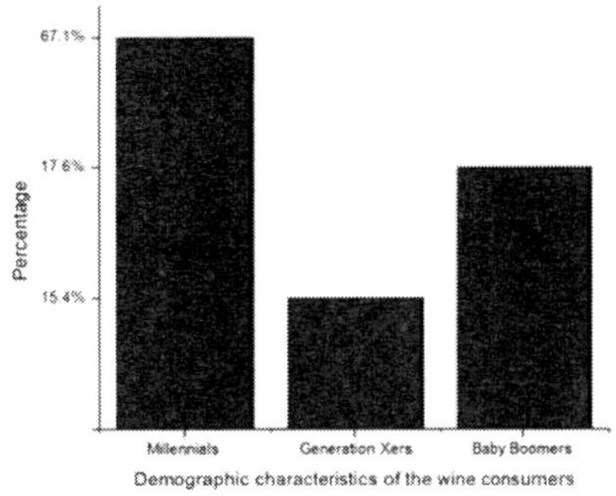

Figure 1. Percentage of Demographic characteristic of the wine consumers.

Table 2. Choice of wine attributes, as a relevant factor when making a decision about purchasing wine (A-attribute, B-Brend, Q-Quality; O-Origin, P-Price, PL- Packaging and Label).

CROSSTAB							
			Age				Σ (%)
			18–28	29–38	39–50	51–65	
A	Q	Count	345	115	87	134	681
		% within A	50.7	16.9	12.8	19.7	100
		% within age	41.6	36.7	35.1	42.7	39.9
		% of Σ	20.2	6.7	5.1	7.9	39.9
	B	Count	118	74	47	36	275
		% within A	42.9	26.9	17.1	13.1	100
		% within age	14.2	23.6	19.0	11.5	16.1
		% of Σ	6.9	4.3	2.8	2.1	16.1
	P	Count	249	79	58	104	490
		% within A	50.8	16.1	11.8	21.2	100
		% within age	30.0	25.2	23.4	33.1	28.7
		% of Σ	14.6	4.6	3.4	6.1	28.7
	PL	Count	69	18	25	22	134
		% within A	51.5	13.4	18.7	16.4	100
		% within age	8.3	5.8	10.1	7.0	7.9
		% of Σ	4.0	1.1	1.5	1.3	7.9
	O	Count	49	27	31	18	125
		% within A	39.2	21.6	24.8	14.4	100
		% within age	5.9	8.6	12.5	5.7	7.3
		% of Σ	2.9	1.6	1.8	1.1	7.3
Σ		Count	830	313	248	314	1705
		% within A	48.7	18.4	14.5	18.4	100
		% within age	100	100	100	100	100
		% of Σ	48.7	18.4	14.5	18.4	100

Source: Prepared by the authors (SPSS Statistics 19).

(chemical and sensory properties) of wine. This generation also their price is very important (50.8%) when deciding to buy a wine and only 39.3% based on country of origin.

Generation Y_1 can also be decided based on the reputation of wine ie brand wine, knowing that reputation is correlated with the quality of wines. Then, Generation X in deciding the purchase of wine to account for the origin of the wine (24.8%), i.e. whether it is domestic or imported wine.

It can be concluded that quality and price it is a very important attribute in deciding the purchase of wine in all generations. Investigation is shows that 39.9% of all respondents make a decision to buy wine based on the quality.

DOI: 10.1051/bioconf/20170903015

Table 3. Impact of age of consumers on purchasing of domestic+ or imported wine.

CHI-SQUARE TESTS			
	Value	**df**	**Asymp. Sig (2 sid.)**
Pearson Chi-Square	39.473[a]	3	.000
Likelihood Ratio	37.335	3	.000
Linear-by-Linear Association	.007	1	.935
N of Valid Cases	1705		
a. 0 cells (.0%) have expected count less than 5. The minimum expected count is 40.15.			

Source: Prepared by the authors (SPSS Statistics 19).

Table 4. Purchasing wine of certain origin by age (D-Domestic, I-Imported, W-wine).

CROSSTAB							
			Age				Σ (%)
			18–28	**29–38**	**39–50**	**51–65**	
W	D	Count	708	258	178	285	1429
		% within W	49.5	18.1	12.5	19.9	100
		% within age	85.3	82.4	71.8	90.8	83.8
		% of Σ	41.5	15.1	10.4	16.7	83.8
	I	Count	122	55	70	29	276
		% within W	44.2	19.9	25.4	10.5	100
		% within age	14.7	17.6	28.2	9.2	16.2
		% of Σ	7.2	3.2	4.1	1.7	16.2
Σ		Count	830	313	248	314	1705
		% within W	48.7	18.4	14.5	18.4	100
		% within age	100	100	100	100	100
		% of Σ	48.7	18.4	14.5	18.4	100

Source: Prepared by the authors (SPSS Statistics 19).

The lowest percentages of respondents who make a decision to buy wine on the basis of the country of origin are those older than 51 years (5.7%).

The lowest percentage of respondents who make a decision to buy wine based on the packaging and label are those aged 29 to 38 years (5.8%) and 39 to 50 years (10.1%).

On the basis of the value of Sig., Table 3 shows that age has a significant influence in the choice of wine attributes, as a factor in making a decision about buying wine, because its value is less than 0.05.

Regardless of the fact whether they buy wine of domestic or imported origin highest percentage of consumers (39.9%) make a decision to buy wine based on its quality.

In addition, the authors carried out an analysis of the impact of age of consumers on the purchase of wine of a certain origin (Table 4):

Respondents of more than 51 years of age mostly buy domestic wine (90.8%), while those 27 from 39 years of age (Generation Y) usually buy wine of imported origin (28.2%).

Wine attributes that affect the consumers to buy wine of domestic origin are: 39.8% of respondents make this decision based on the quality of wine, 31.3% based on the price of wine, 15.4% based on the reputation or brand, 7.6% based on 20 the packaging and label, and only 5.9% based on the country of origin (Table 5).

Consumers who buy wine of imported origin start from the following attributes: quality (40.6% of respondents),

Table 5. Purchasing wine of domestic or imported, based on wine attributes (A-attribute, B-Brend, D-Domestic, Q-Quality, I-Imported, O-Origin, P-Price, PL-Packaging and Label, W-wine).

CROSSTAB								
			Wine attributes					**Σ(%)**
			Q	**B**	**P**	**PL**	**O**	
W	D	Count	569	220	447	108	85	1429
		% of W	39.8	15.4	31.3	7.6	5.9	100
		% of A	83.6	80.0	91.2	80.6	68.0	83.8
		% of Σ	33.4	12.9	26.2	6.3	5.0	83.8
	I	Count	112	55	43	26	40	276
		% of W	40.6	19.9	15.6	9.4	14.5	100
		% of A	16.4	20.0	8.8	19.4	32.0	16.2
		% Σ	6.6	3.2	2.5	1.5	2.3	16.2
Σ		Count	681	275	490	134	125	1705
		% of W	39.9	16.1	28.7	7.9	7.3	100
		% of A	100	100	100	100	100	100.0
		% of Σ	39.9	16.1	28.7	7.9	7.3	100.0

Source: Prepared by the authors (SPSS Statistics 19).

Table 6. Impact of wine attributes on the purchase of wine of domestic or foreign origin.

CHI-SQUARE TESTS			
	Value	df	Asymp. Sig. (2-sided)
Pearson Chi-Square	46.879[a]	4	.000
Likelihood Ratio	45.865	4	.000
Linear-by-Linear Association	2.417	1	.120
N of Valid Cases	1705		
a. 0 cells (.0%) have expected count less than 5. The minimum expected count is 20.23			

Source: Prepared by the authors (SPSS Statistics 19).

brand (19.9%), price of wine (15.6%), country of origin (14.5%), and packaging and label of wine 23 (19.4%):

Regardless of the fact whether they buy wine of domestic or imported origin, the highest percentage of consumers make a decision to buy wine based on its quality of wine and brand of wine.

However, a larger number of Serbian consumers, because of the low purchasing power, make a decision to buy wine of domestic origin.

On the basis of the value of Sig. which is given in Table 6, we can conclude that the attributes of wine have significant impact on the choice of wine of domestic or imported origin.

4. Conclusions

Results of marketing research have shown that when choosing wine, the quality of wine has the biggest impact on consumers, rather than others wine attributes. The results of the research suggest that age Serbia's consumers have a significant impact, when making a decision about purchasing wine.

Based on Chi-test, the authors have been proven that the age affect the decision on the purchase of destined wine.

Starting from the obtained research results, the conclusion offers a recommendation for the implementation of successful strategies of the marketing mix instruments.

The research was supported by the Ministry of Education and Science of the Serbia, No. project TR 31020 and TR 34012.

References

[1] P. Arsil, E. li, J. Bruwer, J Inter Food & Agribus. Marketing **26**(2), 107 (2014)
[2] J. Bruwer, E. Li, S. Lang, S. Chaumont, J Foodservice Bus. Res. **15**(3), 300 (2012)
[3] P. Combris, S. Lecocq, M. Visser, The Economic J. **107**(441), 390 (1997)
[4] S. Geraghty, A. Torres, A. Int J. Wine Business Res. **21**(2), 143 (2009)
[5] M. Sanchez, J.M. Gil, Int. J Wine Marketing **10**(1), 25 (1998)
[6] W. Zikmund, Exploring marketing research, 8th ed. Chicago: Dryden Press Publ., 1999
[7] E. Tach, J. Olsen, Agribusiness **22**(3), 307 (2006)
[8] G. Ordun, Canadian Social Science **11/4**, 40 (2015)
[9] M. Hume, J. World Business **45/4**, 385 (2010)
[10] J. Bruwer, M. House, The Australian & New Zeland Grape grower & winemaker **479**, 56 (2003)
[11] L. Lancaster, D. Stillman, When generation sollide. New York: Harper Collins Publishing, 2002
[12] T. Atkins, L. Tach, Wine Economics and Policy **1**(1), 54 (2012)

DOI: 10.1051/bioconf/20170903016

Eine Frage des Alters – Deutsche Weintrinker und ihr Konsumverhalten mit dem Fokus auf die jüngere Generation / A question of age – German wine drinkers and their consumption behaviour with focus on the younger generation

Gergely Szolnoki and Simone Loose

Geisenheim University, Institut für Betriebswirtschaft und Marktforschung, Von-Lade-Str. 1, 65366 Geisenheim, Germany

Abstract. Deutschland ist mit einer durchschnittlichen Produktionsmenge von durchschnittlich 9,5 Millionen hl sowie mit 15 Millionen hl importiertem Wein einer der wichtigsten Weinmärkte der Welt. Dass der Wein ein beliebtes alkoholisches Getränke in Deutschland ist, zeig der Pro-Kopf-Verbrauch von Still- (21 l) und von Schaumwein (3 l). Was ist die Rolle und was für eine Bedeutung haben jüngere Konsumenten auf dem deutschen Weinmarkt? Unterscheiden sich ihre Geschmackspräferenz und Konsumverhalten von den älteren Konsumenten? Um den deutschen Weinmarkt auf die nächste Generation vorzubereiten, ist es von Bedeutung, die Altersgruppe jüngerer Konsumenten besser zu verstehen. Dafür wurde 2016 eine repräsentative Befragung durchgeführt und die 2078 Befragten in vier Kategorien unterteilt: Jüngste Konsumenten zwischen 16–29 Jahren, die noch in Ausbildung sind oder gerade mit dem Arbeiten anfingen; Konsumenten zwischen 30–49 Jahren, die bereits einen festen Job und ein regelmäßiges Gehalt haben; Konsumenten zwischen 50–65 Jahren, die auf der Höhe ihrer Karriere stehen und ein fundiertes Weinwissen besitzen; und Konsumenten über 65 Jahre, die bereits in Rente sind und das höchste Weinwissen aufweisen. Nach der Analyse der Präferenzen, des Konsumentenverhaltens und der wirtschaftlichen Bedeutung der Gruppen kann festgestellt werden, dass sich das erste Alterssegment von den anderen Konsumenten in allen Bereichen signifikant unterscheidet. Die jüngste Generation trinkt am seltensten Wein und hat niedrigeres Involvement. Aufgrund der in der Kindheit etablierten Präferenzen, bevorzugen sie eher Weißweine mit höherem Restzuckergehalt. Bedingt durch den eher unregelmäßigen Weinkonsum, das niedrige Involvement und durch die niedrige Zahlungsbereitschaft, hat diese Altersgruppe zurzeit eine eher marginale wirtschaftliche Bedeutung auf dem deutschen Weinmarkt. Jedoch als zukünftiges Rückgrat des deutschen Weinmarktes besitzen junge Konsumenten ein großes Potential.

With a total wine production of in average 9.5 million hl and an import of another 15 million hl Germany is one of the most important wine markets of the world. Wine is a popular beverage in Germany with a per capita consumption of 21 litres still wine and 3 litres sparkling wine. Wine has a high penetration in Germany where 60% of the population 16 years and above consume wine at least once a year. What is the role and importance of younger consumers for the German wine market? Do their taste preferences and wine buying behaviour differ from older consumers? To prepare the German wine market for the next generation of wine consumers it is important to better understand this age group. A representative survey was conducted in Germany in 2016. The sample of 2.078 participants was divided into four age groups. The youngest consumers between 16–29 years are mainly still in education or just started working; consumers between 30–49 years mostly in their first permanent job and with a regular salary; wine drinkers between 50–65 years at the peak of their career and established wine experience; retired wine consumers above 65 years with the highest experience in wine. Taste preferences, wine consumption behaviour as well as the economic importance of these age segments were analysed. Particularly the youngest age group under 30 years differs significantly from older consumers in almost all indicators analysed. Consumers in this age group drink wine least frequently and have a lower involvement in wine than older consumers. Because of the taste preferences acquired during childhood they select white and rather sweet wines over red and dry wines. Because of their rather infrequent wine consumption, low wine involvement and low willingness to pay for wine the youngest wine consumers currently have the lowest economic importance for the German wine market. Nevertheless as the future backbone of the German wine demand they represent enormous potential to German wine marketers.

1. Einleitung und Methode

Deutschland ist mit einer durchschnittlichen Produktionsmenge von 9,5 Millionen hl sowie mit 15 Millionen hl importiertem Wein einer der wichtigsten Weinmärkte der Welt. Dass Wein ein beliebtes alkoholisches Getränk in Deutschland ist, zeigt der Pro-Kopf-Verbrauch von Still- (21 l) und von Schaumwein (3 l). Wein ist auch ein weit verbreitetes Getränk. 60% der Bevölkerung im Alter von 16 Jahren und darüber konsumieren mindestens einmal pro Jahr Wein.

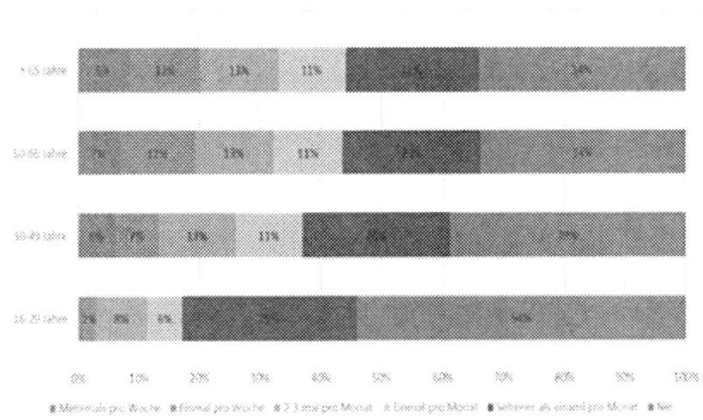

Abb 1. Häufigkeit des Weinkonsums nach Alter.

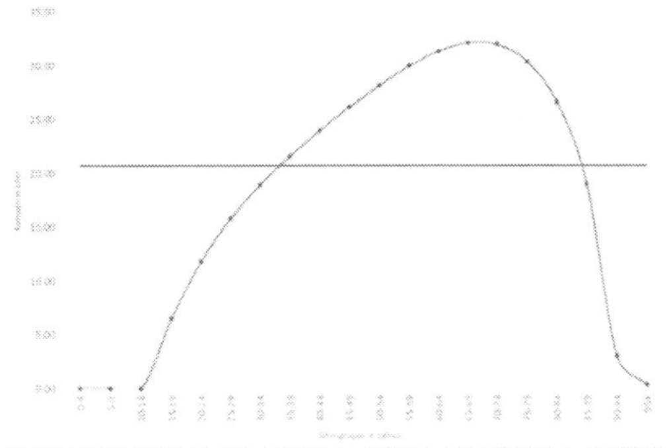

Abb 2. Weinkonsumvolumen nach Alter.

Welche Rolle spielen jüngere Konsumenten auf dem deutschen Weinmarkt? Unterscheiden sich ihre Geschmackspräferenzen und ihr Konsumverhalten von dem älterer Konsumenten? Um den deutschen Weinmarkt auf die nächste Generation vorzubereiten, ist es wichtig, die jüngsten Konsumenten besser zu verstehen. Dafür wurde 2016 eine repräsentative Befragung durchgeführt und die 2078 Befragten in vier Kategorien unterteilt: Jüngste Konsumenten zwischen 16–29 Jahren, die noch in Ausbildung sind oder gerade mit dem Arbeiten anfingen; Konsumenten zwischen 30–49 Jahren, die bereits einen festen Job und ein regelmäßiges Gehalt haben; Konsumenten zwischen 50–65 Jahren, die auf der Höhe ihrer Karriere stehen und ein fundiertes Weinwissen besitzen; und Konsumenten über 65 Jahre, die bereits in Rente sind und das höchste Weinwissen aufweisen.

Tabelle 1. Präferenz für Herkunft, Weinart und Ge-schmacksrichtung – gewichtet nach Volumen.

	16–29 Jahre	30–49 Jahre	50–65 Jahre	>65 Jahre
Deutscher Wein	56%	47%	57%	65%
Importierter Wein	44%	53%	43%	35%
Weißwein	51%	42%	39%	47%
Roséwein	12%	14%	12%	10%
Rotwein	37%	44%	49%	44%
Trocken	49%	52%	55%	53%
Halbtrocken	30%	31%	33%	35%
Lieblich/süß	21%	17%	12%	12%

2. Ergebnisse

2.1. Weinkonsum

Wenngleich jüngere Deutsche alkoholische Getränke konsumieren, spielt Wein dabei eine geringe Rolle. Mehr als die Hälfte dieser Altersgruppe trinkt überhaupt keinen Wein, während 30% seltener als einmal pro Monat Wein konsumieren (Abb. 1).

Nur 17% trinken regelmäßig Wein, was nur einem Bruchteil der anderen Altersgruppen entspricht. Es gibt scheinbar eine Altersgrenze zwischen den jüngsten Konsumenten und den anderen drei Altersgruppen. Ab dem Alter 30+ nimmt die Wahrscheinlichkeit Wein zu trinken signifikant zu und die Zahl der Häufig-Konsumenten steigt auf über 30% an.

Zwischen 2006 und 2014 wurden im Rahmen von Studien, repräsentativ für die deutsche Bevölkerung ab 16 Jahren, mehr als 10.000 Personen zu ihrem Weinkonsum befragt. Das Ergebnis ist die wellenförmige Kurve in Abb. 2. Das konsumierte Weinvolumen steigt stetig bis zum Alter von 69 Jahren an und nimmt ab 80 Jahre rapide ab. Die Abbildung veranschaulicht auch den Zusammenhang zwischen Alter und Konsumvolumen. Die Altersgruppen unterscheiden sich nicht in der Häufigkeit ihres Bierkonsums, während der Konsum von Weinmischgetränken wie Hugo oder Aperol Spritz mit steigendem Alter stetig abnimmt.

Die Gründe keinen Wein zu konsumieren sind vielschichtig. Junge Konsumenten nennen vor allem fehlende familiäre Prägung oder regionale Weintradition als auch die Tatsache, dass nur sehr wenige ihrer Freunde Wein trinken. Neben diesen kulturellen Gründen scheinen auch der Preis und die fehlende Erfahrung eine Barriere für den Weinkonsum zu sein.

2.2. Präferenzen

Sowohl in der Konsumhäufigkeit wie bei der Auswahl des Weins gibt es signifikante Unterschiede zwischen jüngeren und älteren Konsumenten. Tabelle 1 zeigt die Unterschiede zwischen den Altersgruppen hinsichtlich der Weinherkunft, Weinart und Geschmacksrichtung. Bezüglich der Herkunft zeigen die erste und dritte Altersgruppe (16 – 29 Jahre und 50 – 56 Jahre) ein ähnliches Konsumverhalten; beide konsumieren etwas mehr deutschen als importierten Wein. Konsumenten zwischen 30–49 Jahren sind dem Konsum ausländischer Weine gegenüber generell etwas aufgeschlossener (53%), während Konsumenten über 65 Jahre eine stärkere Präferenz für deutsche Weine aufweisen. Bezüglich der Weinart unterscheiden sich die Altersgruppen kaum. Interessanterweise tendieren sowohl die jüngsten als auch die ältesten Konsumenten eher zu Weißwein. Die Präferenz für Rotwein ist am höchsten ausgeprägt in den mittleren Altersgruppen und am niedrigsten bei den jüngsten Konsumenten. Dies ist angesichts unserer genetisch bedingten Abneigung gegenüber bitterem Geschmack nicht überraschend; dies muss erst durch Erfahrung überwunden werden.

Ebenso genetisch bedingt ist die Präferenz für Süßes bei den jüngeren deutschen Konsumenten. Die Präferenz für trockene Weine steigt mit zunehmendem Alter stetig

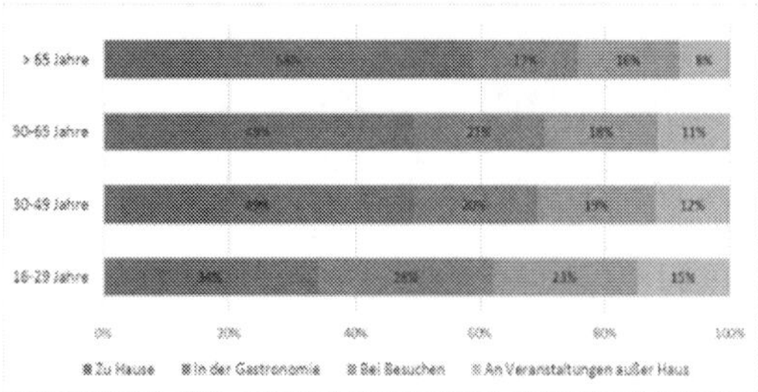

Abb 3. Konsumort nach Alter – gewichtet nach Volumen.

an. Interessanterweise ist die Präferenz für halbtrockene Weine über die Altersgruppen hinweg konstant.

2.3. Einkaufs- und Konsumort

Auch bei der Frage nach den Konsumorten von Wein macht die Altersgrenze von 30 Jahren den größten Unterschied (Abb. 3). Ältere Weintrinker konsumieren nicht nur mehr Wein, sie trinken auch erheblich weniger Wein außer Haus.

Über alle älteren Altersgruppen hinweg wird ungefähr jedes zweite Glas zu Hause getrunken. Für die jüngste Altersgruppe ist Wein eher ein soziales Getränk, das entweder beim Besuch von Verwandten oder Freunden getrunken wird oder außer Haus im Rahmen eines Restaurantbesuchs oder einer Veranstaltung. Die Wahrscheinlichkeit, dass sie in alltäglichen Situationen Wein trinken ist geringer; nur ein Drittel konsumiert Wein zu Hause. Ihr Weinkonsum konzentriert sich auf soziale Gelegenheiten, bevorzugt mit Personen der gleichen Altersgruppe, während es die älteren Weintrinker genießen, zu Hause bei einem Glas Wein zu entspannen.

Die Altersgruppen unterscheiden sich auch hinsichtlich der *Einkaufsstätten* (Abb. 4). Die jüngste Altersgruppe nutzt am stärksten den Discount. Dies ist im Wesentlichen auf ihr begrenztes Budget zurückzuführen, da sie sich noch in der Ausbildung befinden oder gerade erst eine Arbeit aufgenommen haben. Der Discountanteil sinkt stetig mit steigendem Alter. Supermärkte werden von allen Altersgruppen zu ca. einem Drittel genutzt. Ab-Hof Verkauf und Weinfachhandel spielen bei den Kunden mit zunehmendem Alter eine immer größere Rolle. Dies ist sowohl auf ihr höheres Einkommen als auch ihr stärkeres Involvement / größeres Weinwissen zurückzuführen. Tatsächlich sind Involvement, Weininteresse und Weinwissen die treibenden Kräfte für die meisten von uns beobachteten Unterschiede und sind ihrerseits wiederum stark altersabhängig. Die älteste Gruppe – die heutigen Rentner - weist den höchsten Anteil an ab-Hof Käufen auf. Dies ist teilweise darauf zurückzuführen, dass ihr Konsumverhalten in einer Zeit geprägt wurde, als heimische Weine den Markt dominierten.

Auf den ersten Blick scheint der Online-Einkauf von Wein für deutsche Konsumenten keine große Rolle zu spielen. Abhängig von der Altersgruppe variiert der Mengenanteil zwischen 1% und 3%. Betrachtet man jedoch den Wert der online gekauften Weine, beläuft sich der Marktanteil auf 5%. Aktuell nutzt die Gruppe der 30-49-jährigen den Online-Einkauf von Wein am stärksten. Ungefähr einer von fünf Befragten gab an sich vorstellen zu können, in naher Zukunft Wein online zu

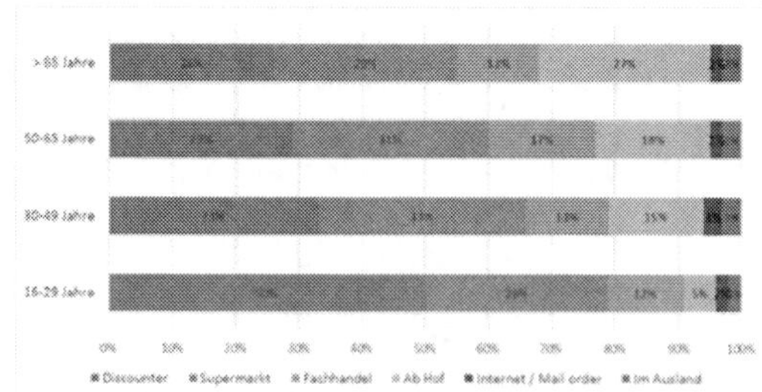

Abb 4. Nutzung von Vertriebskanälen nach Alter – gewichtet nach Volumen.

Tabelle 2. Gruppengröße, Konsumvolumen, Wertanteil, Ausgaben und Durchschnittspreis nach Alter.

	16–29 Jahre	**30–49 Jahre**	**50–65 Jahre**	**>65 Jahre**
Größe der Altersgruppe (%)	15	33	27	26
Pro-Kopf-Verbrauch (l)/ Jahr	18	33	37	39
Pro-Kopf-Ausgaben für Wein (€)/ Jahr	69	153	180	190
Durchschnittspreis für Wein (€/l)	3,8	4,6	4,9	4,8
Wertanteil der Altersgruppe (%)	3	30	32	35

kaufen, außer in der ältesten Gruppe: Hier lehnen 92% den Online-Einkauf ab. Angesichts der Tatsache, dass der Weinkonsum junger einheimischer Konsumenten steigen wird (Abbildung 1), kann man nur vermuten, wie stark die Bedeutung des E-Commerce für den Weinhandel künftig zunehmen wird.

2.4. Wirtschaftliche Bedeutung der Altersgruppen

Um die wirtschaftliche Bedeutung der verschiedenen Altersgruppen für den Weinhandel einzuschätzen, wurden ihr Weinkonsum, die Ausgaben für Wein und die Durchschnittspreise analysiert und quantifiziert (Tabelle 2). Aus zwei Gründen ist das jüngste Segment mit einem Anteil von nur 15% der Weinkonsumenten das kleinste. Erstens wachsen durch den demografischen Wandel weniger junge Deutsche nach. Zweitens ist es, wie schon oben gezeigt, am unwahrscheinlichsten, dass diese Gruppe Wein konsumiert.

Ein ähnliches Muster zeigt sich, wenn man die jährlichen Ausgaben für Wein analysiert. Jüngere Weintrinker geben lediglich 69 €/ Jahr für Wein aus. Dies hat zum einen mit dem geringeren Konsumvolumen und zum anderen mit dem geringeren Durchschnittspreis zu tun. Die jüngsten Konsumenten geben durchschnittlich 1 €weniger / Liter Wein aus, was auf ihr geringeres Involvement aber auch auf ihr begrenztes Einkommen zurückgeführt werden könnte. Mit zunehmendem Alter steigen die Ausgaben an, vor allem weil das Konsumvolumen und der Durchschnittpreis, der für Wein ausgegeben wird, zunehmen.

Mithilfe der Variablen Gruppengröße, Pro-Kopf-Verbrauch und jährliche Ausgaben kann man den Wertanteil der verschiedenen Altersgruppen berechnen. Dieser Anteil beläuft sich in der jüngsten Konsumentengruppe auf magere 3%. Dies bedeutet, dass Weintrinker zwischen 16-29 Jahren 3% des gesamten

Umsatzes der deutschen Weinwirtschaft generieren, was ca. 7,3 Milliarden Euro pro Jahr (nur Stillwein) entspricht. Die anderen drei Altersgruppen teilen den großen Rest (des Kuchens) fast gleichmäßig unter sich auf.

3. Fazit

Die Ergebnisse dieser repräsentativen Umfrage zeigen, dass der Weinkonsum in Deutschland mit dem Alter zunimmt. Insbesondere die jüngste Altersgruppe unter 30 Jahren unterscheidet sich signifikant von den älteren Konsumenten in fast allen untersuchten Aspekten. Sie konsumieren seltener und zeigen eine geringere Begeisterung für Wein. Aufgrund der in der Jugend erworbenen Geschmackspräferenzen trinken sie eher Weiß- als Rot- und eher süße als trockene Weine. Sie konsumieren eher bei sozialen Anlässe als zu Hause und kaufen vorwiegend im Discount ein. All dies führt zu einer eher marginalen wirtschaftlichen Bedeutung, betrachtet man ihren geringen Verbrauch und Mengenanteil am gesamtdeutschen Weinmarkt.

Obwohl diese Gruppe heute von geringer Bedeutung scheint, wäre es ein Fehler, sie zu vernachlässigen. Mit zunehmendem Alter werden sie sich mehr am Weinkonsum beteiligen und langsam in die nächste Altersgruppe übergehen. Verpasst die Weinbranche die Chance sie auf diesem Weg zum häufigeren und stärkeren Weinkonsum zu begleiten und zu unterstützen, wird sie künftig einem Generationenproblem gegenüberstehen. Die jüngere Generation möchte unterhalten werden. Daher muss die Weinbranche den besten Weg finden, die Aufmerksamkeit und das Interesse dieser Konsumenten zu wecken. Glücklicherweise stehen dafür eine Vielzahl möglicher Instrumente zur Verfügung – angefangen vom Verpackungsdesign über innovative Produktentwicklungen bis hin zu Veranstaltungen, Seminaren und Messen.

Literaturverzeichnis

Szolnoki, G., Loose, S. (2016), The younger generation. Meininger's Wine Business International **6**(6), 34–36

Small and medium sized wine producers dealing with business to business e-commerce: Towards a relational model for enhancing value (co-) creation

Andrea Saba[1], Silvia Rolandi[1], and Marco Pilloni[2]

[1] Scuola Superiore di Studi Universitari e di Perfezionamento Sant'Anna, Pisa, Italy
[2] Studio Psicologia del Lavoro, Cagliari, Italy

Abstract. The internet has produced a forward-looking shift in the way we communicate, think and run economic activities around the globe. The revolution in Information and Communication Technology (ICT) has generated an innovating process in the agri-food sector that is far from being unwind. The wine industry is a case in point. Indeed, the industry has attempted to address the multifaceted challenges posed by electronic commerce since always. Business to business electronic commerce has been slower to develop than its counterpart, i.e. business to consumer E-commerce. However business to business e-commerce is starting to attract an increased interest, by showing a significant business value in terms of potential. Against this background, the slow development of B2B e-commerce has been linked to regulatory and economic reasons. In order to deal with such stumbling blocks, the paper suggests a shift towards an open platform that is aimed at facilitating transactions and interactions among businesses through a continuous problem-solving process in which trust and value is co-created by producers and customers through relational contracting.

1. Introduction

The internet has produced a forward-looking shift in the way we communicate, think and run economic activities around the globe. The Internet has allowed us to build a virtual place in which items are traded as well as orders and payments are made. As a result, the spatial and temporal concept of commercial interchanges is still evolving. The wine industry is a case in point. Indeed, the industry has attempted to address the multifaceted challenges posed by electronic commerce since always. Many business experiences have been accumulated in selling products directly to consumers, i.e. business to consumer E-commerce. Conversely, business to business commerce has been slower to develop than its counterpart.

The slow development of B2B e-commerce – in particular in relation to small and medium sized businesses – has been linked to regulatory and economic reasons. Through an investigation of the related stumbling blocks, the paper aims at exploring the implications of an intermediary-oriented e-marketplace that facilitates transactions and interactions in a problem-solving process in which trust and value is co-created by producers and customers through a relational contracting. This is analysed in the framework of further developing emerging trading opportunities for small and medium sized wine producers.

The structure of the paper is threefold. Part I examines what the main aspects of e-marketplaces and their role within the contest of ICT. Part II recognises a need of rethinking the role of e-marketplace as intermediary in facilitating and building trust among businesses. Part III explores a (preliminary) model of relational intermediary–oriented e-marketplace. By filling the gap generated by the distance and the resulting lack of personal interaction among businesses, this model may represent a consistent opportunity to build trust and create value through e-marketplaces.

2. B2B e-commerce in wine business: present and past

ICT has changed the way we do trade at all levels, including Business to Consumers (B2C) and Business to Business (B2B) transactions [1]. A relatively vast body of literature has investigated the role played by the e-commerce in the agri-food supply chain, thus enabling new commercial and trading patterns in the agri-food sector [2]. The potential of ICT in B2B e-commerce has been widely recognised in the agri-food sector; indeed, the revolution in ICT has generated an innovating process that is far from being unwind [3]. While B2B e-commerce is starting to attract an increased interest – by showing a significant business value in terms of potential – its slow development and diffusion requires a renovated attention in (re-) considering the stumbling blocks that are linked to both regulatory and economic reasons [4].

The wine industry is a case in point. Indeed, the industry has attempted to address the multifaceted challenges posed by e-commerce since always. The wine sector has been deeply involved with high expectations regarding the potential growth of the sector as results of adopting ICT. Producers had made many attempts in adjusting their habits as well as increasing their confidence in the digital instrument, by taking into consideration possible advantages and disadvantages in terms of legal and economic factors.

In this framework, the development of e-marketplace has been considered in the literature as a corner-stone in the global process of improving ICT for building more efficient business models [5]. E-marketplaces have played (and are continuing to play) an important role in boosting the e-business world.

In the scientific literature, several scholars have devoted huge efforts in exploring definitions for electronic marketplace [6]. This consistently reflects the evolution that has involved e-marketplaces over the years [7]. By focusing on their unique characteristic of being able to bring multiple buyers and sellers together in one virtual market space, the present paper consider an e-marketplace as an open platforms that is aimed at facilitating transactions and interactions among businesses [8]. In doing this, e-marketplace may enable both economy of scale and economy of scope.

2.1. E-commerce and its (legal) geographical boundaries

At first sight e-marketplace – and generally speaking, the entire e-commerce sector – may give the perception of eliminating the usual geographic boundaries. Indeed, it seems that the Internet may be considered as a unique and exceptional *unicum* of buyers and sellers that act in the same and infinite space. One of the consequences of this representation is the idea associated to the possible elimination of intermediaries, thus resulting in cutting costs – particularly for small and medium-sized businesses that may concretely access potentially unlimited markets with a higher profit margin.

The real world, however, suggests something different. When dealing with e-commerce, geographical boundaries exist. A real product, in our case wine, is physically shipped from one country to another one after finalising the purchase over the Internet. Therefore, the online transaction will be subject to all the possible issues that usually affect international trades. This implies that import and export regulations as well as food laws have to be considered. Consequently, the regulatory framework assumes a main relevance to determine the expansion and the success of the wine e-commerce.

Before the discussion moves to the regulatory issues, one preliminary aspect is worth to be further explored. The wine is a product to which not only import/export regulation applies, but also a (legal) awareness has been paid as a result of being destined to human consumption. Therefore, the regulatory burden rises considerably if compared to other items, such as clothes for example, due to the health issues of drinking wine. The health concerns are considered both in terms of consequent diseases and direct [9], and indirect effects [10].

The complex and indissoluble twine of legislations has been listed in relation to different areas, accordingly to the relevant literature [11]. Examples are specifically provided in relation to the EU legislation on wine export from non-EU countries to EU countries.

a. *Taxation systems*. It should be taken into consideration when deciding to export wine, in particular different rules regulating the tax system may considerably differ from country to country. In Europe, for example, the VAT rate can change among Member States as a uniform VAT system does not exist [12];
b. *Tariffs*. These include a wide range of expenses that need to be managed when dealing with certain countries [13]. In Europe, a specific database, called the 'TARIC', is provided in order to allow interested companies to find out the amount of the tariffs [14];
c. *Licensing*. In order to export products in certain countries, producers should be aware of the existing regulation about licensing, i.e. a specific licence may be required to import wine [15];
d. *Labelling*. While the content of the labels may change over time, a producer that is willing to export to different countries should be confident with labelling legislation, thus considering different labels in compliance with the legal provisions. This part includes also Geographical Indications and other certification schemes, such as the organic certification process [16]. The EU regulatory framework is particularly stringent, i.e. Regulation 1169/2011 in combination with the implementing rules provided under the Regulation 607/2009 [17];
e. *Controls*. As previously mentioned, wine represents a good that could affect the consumer health; therefore, countries may establish different control system considering the fact that substances might be admitted in one country and not in others. In Europe, the Commission Regulation 555/2008 lays down implementing rules regarding, *inter alia*, certification and analysis report requirements [18]. Furthermore, Regulation 606/2009 specifies the oenological practices authorized in the EU [19].

2.2. Stumbling blocks

The wine sector has been traditionally influenced to a great extent by a number of long-established trading patterns and common habits in building business relationship. In particular, intermediaries have played an essential role in aggregating needs of buyers and producers, thus consolidating long-term relationship, personal commitment and trust among them [20]. This clashes with the risk and uncertainty involved in an environment in which all these patterns are less perceivable [21]. As illustrated in the literature, such patterns result in a higher perceived information asymmetry among parties in an e-business in which the transactions are carried out in an environment with only *virtual* relations – that is, an e-environment. More easily, it may give room for opportunistic behavior. Information asymmetry affects all the points that characterise the electronic transaction: product, buyer-and-seller relationships and market environment [22].

In the literature, three key areas of concern have been identified in taking the decision to adopt an electronic marketplace: (1) market blocks; (2) organizational blocks; and, (3) socio-cultural blocks [23]. Such areas of concern represent stumbling blocks in the choice of small and medium-sized producers of using an electronic marketplace in their own business to business relationships. Such blocks has been further strengthen by a generalized lack of knowledge and skills in adopting an ICT tool, while the information asymmetry increases the resulting perceived risk [24].

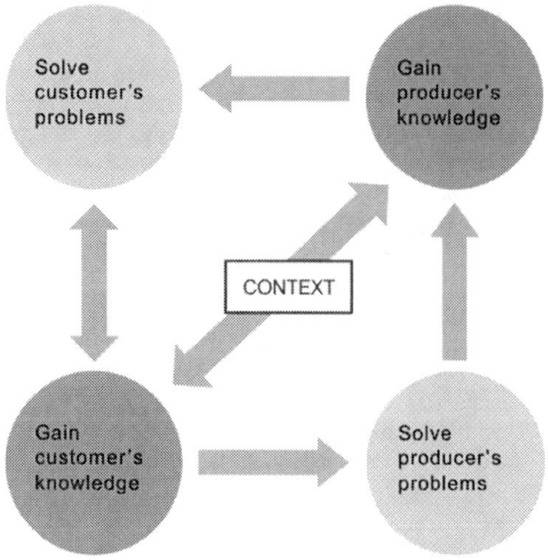

Figure 1. Intermediary reinforcing cycle [28].

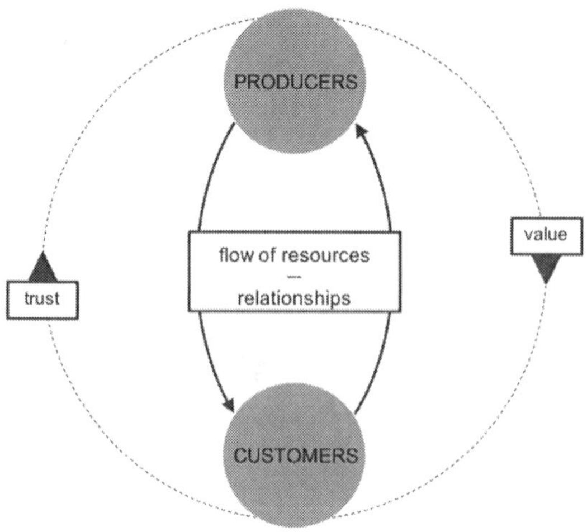

Figure 2. Relational intermediary-oriented model.

3. Rethinking the role of intermediaries

While ICT has had the ambition of carrying out a disintermediation of (nearly) all the economic sectors, they are boosting intermediation in a different form that does not match with traditional categories [25]. An intermediary is valuable in consideration of its being-in-between nature [26]. The proposal, here, consists in a call for an intermediary oriented e-marketplace. Intermediaries are best situated to aggregate the needs of different and eventually heterogeneous groups of customers in purchasing products. Conversely, producers have a range of needs that include negotiation, transaction settling, pre- and post-sale assistance, advertising and logistics. An intermediary may provide an augmented value for producers by bundling such needs. This results in an intermediary that gathers knowledge while serving the interrelated needs of customers and producers [27]. Indeed, the intermediary-oriented e-marketplace becomes the best-informed player in business to business e-relationship. Being able to solve customer problems, it will be able to solve producer problems.

As illustrated in Fig. 1, this creates a mutually reinforcing cycle in which intermediaries give context to the information gained. This creates a competence based framework in which intermediaries may be considered as facilitator and 'complementors' to the producer [29]. Such a cycle will continue its problem-solving process by following the evolutionary movement of ICT. In this scenario, an e-marketplace acts as an intermediary in facilitating electronic transactions and enhancing the intangible asset of trust among businesses [30]. Indeed, such an intermediary-oriented e-marketplace may reduce uncertainties and risks through information sharing and problem solving; thus, limiting information asymmetry and the resulting transaction costs.

4. Towards a relational e-marketplace: A way to enhance value (co-)creation?

A shift is suggested towards an open platform that is aimed at facilitating transactions and interactions among businesses through a continuous problem-solving process in which trust and value is co-created by producers and customers through a relational contracting – as illustrated in Fig. 2.

While a comprehensive model around the many implications that are involved is too pretentious for this paper, the aim, here, is to contribute to the current discussion by building a common (yet preliminary) understanding of a relational intermediary-oriented model of e-marketplace. In order to discuss the resulting implications, this section will firstly investigate the role played by relational contracting in the model and, secondly, the conceptual framework of value co-creation within which long-term relationships and trust are built.

Relational contracting usually emerges under conditions of uncertainty and complexity, in which a relational contract relies on the trust between parties that enter a contractual relationship in order to work together for a mutual benefit outcome [31]. Such contractual arrangement is particularly common in business contexts, where flexible relational contracts are essential [32]. An extensive body of literature exists in the United States on the introduction of a degree of flexibility in contracts and the use of informal sanction to maintain cooperation among business actors [33]. Based on a qualitative research on business executives, Macaulay found that commercial long-term relationship rarely involves, or are based on, formal written contracts [34]. The primary motivation is performing contractual obligations consists in preserving reputation of being good business partners on which others may count for future deals.

The relational approach maintains that the contract should be considered as embedded in a specific socio-economic context. In such perspective, contracting does not occur in a 'social vacuum', but in a 'relational web' [35]. The temporal dimension of the relationship plays a key role in such understanding [36]. Accordingly to a commentator, 'emphasis on discrete transactions abstracted from the ongoing relationships in which they occur distorts the character of the transactions and of the relationships themselves' [37]. The introduction of a relational dimension into the contractual mechanism has been widely discussed in academia (particularly in the United States) during the last fifty years; indeed, it produced extensive scientific research in the law area and economics studies. Arguably, the understanding of a relational dimension resulted in multiple changes in the way in which legal scholars deal with contractual arrangements, by contributed to some extent to adjust the classical contractual perspective [38]. Among others, the empirical studies and, more recently, the law and economics perspective have been set as a response to the changes in the understanding of law as a scientific area [39].

It is essential to note one of the aspects that plays a key role in such an approach: the contract should be understood as a relational event affected by external influences and internal values that produce ongoing adjustments and assure its progression [40]. Indeed, the value of solidarity and reciprocity play a key role in a relational approach, also representing a reaction to the individualism involved in classical contracting [41].

In order to build long-lasting relationships and trust among producers and customers involved in electronic transition, both parties need to interact in the value generation process. Such a long-lasting relationship and trust are created through customer retention [42]. This is achieved by 'accessing customers in the target market, continuing customer dialogues, and choosing efficient ways in treating them such as personalized treatment' [43]. However, the lack of direct interaction in e-commerce is considered as an element that may undermine the real possibility of converting customer retention in a long-term relationship [44].

In relationship building, a number of interactive activities has been identified in literature: (1) transactional interaction, in which the focus is on the transaction as a central economic activity and in particular the time frame is considered the key aspect; (2) value-added interaction, in which customers are attracted by a relational adjustment and amendment of the products to the needs of the customers; and, (3) collaborative interaction, in which customers and producers are committed in a close collaboration for mutual long-term benefits [45]. While substantial differences exist among the activities, these processes are considered as related parts that contribute to relationship building [46].

Customer and producer are closely intertwined during the interaction process. In this line, 'co-creation is regarded as a function of interaction and value creation as direct interactions between value-creating resources and a beneficiary' [47]. According to the literature, value creation process need to be considered through an interactive perspective under which is possible to investigate how the contribution of different resources are integrated in the process [48]. By way of interaction, both customers and producers share (and gain new) information, receive advice and build trust [49].

Within this framework, the literature has identified interrelated components as useful for a business to build a value co-creation capability: producers need to listen to and engage with their customers, thus becoming able to respond externally through the provision of tools for value co-creation and internally by establishing investment efforts in co-creation structures [50]. Such components are complementary in relationship building and value co-creation [51].

5. Conclusion

While e-marketplaces are considered a corner-stone in developing more efficient business models and boosting the e-business world, a number of stumbling blocks – both economic and regulatory – are identified as reasons of its slow development in business to business relationships. Against this background, the wine industry is a case in point. At first sight e-marketplace may give the perception of eliminating the usual geographic boundaries. However, such an e-environment and the relating lack of personal interaction increase the risks and uncertainty involved. In this line, information asymmetries between producers and customers have been recognised as affecting all the point in the electronic transaction.

A shift has been suggested towards an open platform that facilitates transactions and interactions among businesses through a continuous problem-solving process in which trust and value is co-created by producers and customers through a relational contracting. In a relational intermediary-oriented model, e-marketplace acts as an intermediary in facilitating electronic transactions and enhancing the intangible asset of trust among businesses. Such a model may reduce uncertainties and risks through information sharing and problem solving; thus, limiting information asymmetry and the resulting transaction costs.

References

[1] See, among others, Z. Andreopoulou and others, *Typology for e-business activities in the agricultural sector*, International Journal of Business Information Systems, 3 (2008)

[2] For a comprehensive overview, see Y. Zeng and others, *E-commerce in Agri-food Sector: A Systematic Literature Review*, The International Food and Agribusiness Management Review (2017). A fascinating contribution has been recently provided by the research findings achieved by the EU Project 'E-Trust - Building trust for quality assurance in emerging markets for food chains', funded under the EU Sixth Framework Programme. See M. Fritz and others, *Commercio elettronico per la dinamica delle catene agro-alimentari internazionali: un'analisi del potenziale*, Working Paper DEIAgraWP-08-006

[3] See, *inter alia*, M. Fritz, T. Hausen, G. Schiefer, *Developments and Development Directions of Electronic Trade Platforms in US and European Agri-Food Markets: Impact on Sector Organization*, International Food and Agribusiness Management Review, 7 (2004)

[4] Among others, see J. Henderson and others, *Adoption of Internet strategies by agribusiness firm*, International Food and Agribusiness Management Review, 8 (2005); and, N. Leroux, S. Wortman, D. Mathias, *Dominant factors impacting the development of business-to-business (B2B) e-commerce in agriculture*, International Food and Agribusiness Management Review, 4 (2001)

[5] See M. Canavari, M. Fritz, G. Schiefer, *Food Supply Networks. Trust and E-business* (CABI, Wallingford, 2016). See also T. Skjøtt-Larsen, H. Kotzab, M. Grieger, *Electronic marketplaces and supply chain relationships*, Industrial Marketing Management, 32 (2003)

[6] See, among other, T.Y. Eng, *The role of e-marketplaces in supply chain management*, Industrial Marketing Management, 33 (2004); M. Grieger, *Electronic marketplaces: a literature review and a call for supply chain management research*, European Journal of Operational Research, 144 (2003); M. Grieger, *An empirical study of business process across Internet-based electronic marketplaces. A supply-chain management perspective*, Business Process Management Journal, 10 (2004); C. Standing, S. Standing, *The relationship between electronic marketplace strategy and structure*. In: C. Standing, (ed.), *Electronic Markets - Benefits, Costs and Risks* (Palgrave Macmillan, Basingstoke, 2009)

[7] See M. Canavari, M. Fritz, G. Schiefer, *Food Supply Networks. Trust and E-business* (CABI, Wallingford, 2016)

[8] See, among others, H.H. Holzmuller, J. Schluchter, *Delphy studies about the future of B2B marketplaces in Germany*, Electronic Commerce Research and Applications, 1 (2002)

[9] An example regarding direct effects may be linked to an improper content of the product that can cause side effects to the consumers, such as intoxication

[10] An example regarding the indirect effects is the addiction that might arise from an exaggerate consumption of wine

[11] The identification of different areas with subdivisions is very well presented in the contribution given by S.A. Hincliffe, *Trademarks, GIs, and Commercial Aspects of Wine Distribution Agreements*, Journal of Food Law & Policy, 10 (2014)

[12] A complete list of VAT rates applicable can be found at `http://ec.europa.eu/taxation_customs/resources/documents/taxation/vat/how_vat_works/rates/vat_rates_en.pdf`. Accessed 10 April 2017

[13] An example is represented by Europe

[14] For more information, see the database at `http://ec.europa.eu/taxation_customs/dds2/taric/taric_consultation.jsp`. Accessed 10 April 2017

[15] A number of specific examples exists that are applicable to EU Member States. It is also worth mentioning the ongoing debate in the United States regarding the fact whether it is legal or not to sell wine, over the Internet, without a licence to the States where a licence is required. See J.A. Tanford, *E-commerce in Wine, Journal of Law, Economics and Policy*, 3 (2007); and, V. H. Jones, *Put a Cork in it: the Use of H.R. 1161 to end Direct Wine Shipping throughout the States once and for All*, Journal of Food Law & Policy, 8 (2012)

[16] See S.A. Hinchliffe, *Trademarks, GIs, and Commercial Aspects of Wine Distribution Agreements*, Journal of Food Law and Policy, 10 (2014)

[17] See Regulation (EU) No 1169/2011 of the European Parliament and of the Council of 25 October 2011 on the provision of food information to consumers, amending Regulations (EC) No 1924/2006 and (EC) No 1925/2006 of the European Parliament and of the Council, and repealing Commission Directive 87/250/EEC, Council Directive 90/496/EEC, Commission Directive 1999/10/EC, Directive 2000/13/EC of the European Parliament and of the Council, Commission Directives 2002/67/EC and 2008/5/EC and Commission Regulation (EC) No 608/2004, OJ L 304 [2011]; and, Commission Regulation (EC) No 606/2009 of 10 July 2009 laying down certain detailed rules for implementing Council Regulation (EC) No 479/2008 as regards the categories of grapevine products, oenological practices and the applicable restrictions, OJ L 193 [2009]

[18] See article 41, Commission Regulation (EC) No 555/2008 of 27 June 2008 laying down detailed rules for implementing Council Regulation (EC) No 479/2008 on the common organisation of the market in wine as regards support programmes, trade with third countries, production potential and on controls in the wine sector, OJ L 170 [2008]

[19] See Commission Regulation (EC) No 606/2009 of 10 July 2009 laying down certain detailed rules for implementing Council Regulation (EC) No 479/2008 as regards the categories of grapevine products, oenological practices and the applicable restrictions, OJ L 193 [2009]

[20] Among others, see E. Pignatti and others, *Nuove dinamiche nel commercio dei prodotti agroalimentari: resistenze all'adozione dell'e-commerce nelle relazioni B2B*, Working Paper DEIAgraWP-08-005 (2008)

[21] See M. Fritz, M. Canavari, *Management of perceived e-business risks in food-supply networks: e-trust as prerequisite for supply-chain system innovation*, Agribusiness, 24 (2008)

[22] See M. Canavari, M. Fritz, G. Schiefer, *Food Supply Networks. Trust and E-business* (CABI, Wallingford, 2016)

[23] See, *inter alia*, E. Pignatti and others, *Nuove dinamiche nel commercio dei prodotti agroalimentari: resistenze all'adozione dell'e-commerce nelle relazioni B2B*, Working Paper DEIAgraWP-08-005 (2008)

[24] See M. Canavari, M. Fritz, G. Schiefer, *Food Supply Networks. Trust and E-business* (CABI, Wallingford, 2016)

[25] See P. Anderson, E. Anderson, *The New E-Commerce Intermediaries*, MIT Sloan Management Review (Summer, 2002)

[26] Compare with S. Wang, N. Archer, *Business-to-business collaboration through electronic marketplaces: an exploratory study*, Journal of Purchasing & Supply Management, 13 (2007)

[27] Among others, see M. Kourgiantakis, E. Petrakis, *Modelling B2B E-marketplace: the role of intermediaries*, Work supported by the Greek Ministry of Development, General Secretariat of Research and Technology under the Operational Programme Competitiveness (PENED, 2007)

[28] This figure is based on the research carried out in P. Anderson, E. Anderson, *The New E-Commerce Intermediaries*, MIT Sloan Management Review (Summer, 2002)

DOI: 10.1051/bioconf/20170903017

[29] See P. Anderson, E. Anderson, *The New E-Commerce Intermediaries*, MIT Sloan Management Review (Summer, 2002)
[30] Compare with M. Canavari and others, *Prioritizing trust factors in an agrifood B2B e-commerce environment*. In: G. Schiefer, M. Fritz, U. Rickert, (eds), *Trust and Risk in Business Networks* (University of Bonn – ILB Press, Bonn, 2006)
[31] For a comprehensive discussion of relational contract, see, among others, the topical research in C.J. Goetz, R.E. Scott, *Enforcing Promises: An Examination of the Basis of Contract*, Yale Law Journal, 89 (1980); C.J. Goetz, R.E. Scott, *Principles of Relational Contracts*, Virginia Law Review, 67 (1961); and, I.R. Macneil, *Relational Contract: What We Do and Do Not Know*, Wisconsin Law Review (1985)
[32] Among others, see J.M. Feinman, *Relational Contract Theory in Context*, Northwestern University Law Review, 94 (2000); and, M. Diathesopoulos, *Relational contract theory and management contracts: A paradigm for the application of the Theory of the Norms*, MPRA Paper No. 24028 (2010)
[33] On this, see S.Baker, A. Choi, *Contract's role in relational contract*, Virginia Law Review, 101 (2015). For example, see R. Ellickson, *Order Without Law: How Neighbors Settle Disputes* (Havard University Press, Cambridge, 1991); L. Bernstein, *Opting Out of the Legal System: Extralegal Contractual Relations in the Diamond Industry, Journal of Legal Studies*, 21 (1992); L. Bernstein, *Private Commercial Law in the Cotton Industry: Creating Cooperation Through Rules, Norms, and Institutions*, Michigan Law Review, 99 (2001)
[34] See S. Macaulay, *Non-Contractual Relations in Business: A Preliminary Study*, American Sociological Review, 28 (1963)
[35] See I. Macneil, *Values in Contracts: Internal and External*, Northwestern University Law Review, 78 (1983)
[36] A fascinating analysis has been carried out in the Italian literature, see particularly M. Granieri, *Il tempo e il contratto. Itinerario storico-comparativo sui contratti di durata* (Giuffrè Editore, Milano, 2007)
[37] See G.A.G. Gottlieb, *Relationism: Legal Theory for a Relational Society*, University of Chicago Law Review, 50 (1983)
[38] Among others, see E. Mertz, *Afterword: Tapping the Promise of Relational Contract Theory – Real Legal Language and a New Legal Realism*, Northwestern University Law Review 94 (1999)
[39] A true pioneer may be considered K.N. Llewellyin which in 1925 argued that law is "as a thing in a flux" and its development depends on "non-legal- factors" See K.N. Llewellyin, *The effects of legal institution upon economics*, American Economic Review, 14 (1925). For a comprehensive analysis, see C. Farelli, *L'eredità del realismo giuridico americano*, Materiali per una storia della cultura giuridica, 1 (2006). Also, compare with M. Granieri, *Il tempo e il contratto. Itinerario storico-comparativo sui contratti di durata* (Giuffrè Editore, Milano, 2007)
[40] See I.R. Macneil, *The Many Futures of Contracts*, Southern California Law Review, 47 (1974)
[41] See M. Granieri, *Il tempo e il contratto. Itinerario storico-comparativo sui contratti di durata* (Giuffrè Editore, Milano, 2007)
[42] *Mutatis mutandis*, it is interesting to compare with the research findings in J. Riegelsberger, A. Sasse, *Trustbuilders and trustbusters. The role of trust cues in interfaces to e-commerce applications*, Proceedings of the 1st IFIP Conference on towards the e-Society: ecommerce, e-business, e-government: 17–30. IFIP Conference, 3–5 October 2001, Zurich (Switzerland)
[43] See G. Bhalla, *Collaboration and Co-creation. New Platforms for Marketing and Innovation* (Springer, Heidelberg, 2011)
[44] See G. Bhalla, *Collaboration and Co-creation. New Platforms for Marketing and Innovation* (Springer, Heidelberg, 2011)
[45] See M.D. Hutt, T.W. Speh, *Business Marketing Management: B2B* (Cengage, Boston, 2013)
[46] See G. Bhalla, *Collaboration and Co-creation. New Platforms for Marketing and Innovation* (Springer, Heidelberg, 2011)
[47] See L. Chen, J. Marsden, Z. Zhang, *Theory and Analysis of Company-Sponsored Value Co-Creation*, Journal of Management Information Systems, 29 (2012)
[48] See, among others, S.L. Vargo, R.F. Lusch, *Evolving to a New Dominant Logic for Marketing*, Journal of Marketing, 68 (2004)
[49] See, among others, A. Walter, T. Ritter, *The influence of adaptations, trust, and commitment on value-creating functions of customer relationships*, Journal of Business & Industrial Marketing, 18 (2003)
[50] See G. Bhalla, *Collaboration and Co-creation. New Platforms for Marketing and Innovation* (Springer, Heidelberg, 2011)
[51] Mutatis mutandis, see M. Fritz, *E-commerce partnering due diligence: a methodology for trust in e-commerce in food networks*, Acta Agriculturae Scandinavica, Section C - Food Economics, 4 (2007)

DOI: 10.1051/bioconf/20170903018

Evolution of grapevine surfaces in Romania after accession to European Union – period 2007–2016

Arina Oana Antoce[1,a,b] and Laura Lidia Calugaru[2]

[1] University of Agronomic Sciences and Veterinary Medicine of Bucharest, Faculty of Horticulture, Department of Bioengineering of Horti-Viticultural Systems, 59 Marasti Ave., Sector 1, 011464 Bucharest, Romania
[2] Romanian National Office of Vine and Wine Products (ONVPV), 49 Şos. Iancului, Sector 2, 021719 Bucharest, Romania, office@onvpv.ro

Abstract. The paper focuses on the evolution of the surfaces planted in Romania with the main wine grape varieties and also attempts to describe the impact of the funds received for the vine and wine sector after the country's accession to European Union in 2007. The evolution of planted surfaces and production is followed for the period 2007–2016, based on the data recorded in the digital database called National Register for Vine Plantations. The study presents the evolution of the surfaces covered with the most widely spread autochthonous and international varieties, for both white and red wines, which are, in decreasing order, Fetească regală, Fetească albă, Merlot, Riesling, Aligote, Sauvignon blanc, Cabernet Sauvignon, Muscat Ottonel, Fetească neagră, Pinot noir and Chardonnay. The largest newly planted surfaces were recorded in 2011 for most of these varieties, but a high impact of the support received from the EU was visible throughout the period 2008–2012. Later on, from 2013–2016, the rate of planting started to decrease.

1. Introduction

The vine and wine sector is one of the most developed sectors in Romanian agriculture, which largely benefited from the country accession to European Union in 2007. The National Support Programme of Romania for 2009–2013 reported full absorption of the funds allocated for viticulture, with important investments in new plantations and in some new technologies.

The programme is followed by another similar one, planned for 2014–2018, which is ongoing at present.The funds allocation was made on the basis of the European legislation dealing with the common organisation of agricultural markets – (EC) No 1234/2007 and (EU) No 1308/2013 [1,2] and common organization of the market in wine – (EC) No 1493/1999 and (EC) No 479/2008 [3,4] and the respective implementing regulations [5–7]. For example, in the period 2009–2013 the investments in the vine and wine sector in Romania were included among the measures for the rural development and not under special measures dedicated only to viti-vinicultural sector. For the new period of 2014–2018, however, Romania included such investments under the specific measures for vine and wine sector.

The allocation took into account the priorities of the country, concentrating mostly to the restructuring and conversion of vineyards and planting wine grape varieties meant for the production of quality wines with Controlled Denomination of Origin or at least with Geographical Indication. The present study is an attempt to evaluate from several viewpoints the impact of the European funds allocated to Romania with regard to the development of the vine and wine sector during the period 2007–2016. The main points of the paper refer to the evolution of the surfaces planted with the main wine grape varieties cultivated in Romania from 2005 to 2016, evolution of vine surfaces in accordance with the destination of production in each wine region and the new plantations established by using existent planting rights or allocated from the national reserve.

2. Materials and methods

Data were obtained from several sources, such as Statistical yearbooks of Romania issued by National Institute of Statistics [8], Ministry of Agriculture and Rural Development of Romania [9], Romanian Payment and Intervention Agency for Agriculture (APIA) [10], but mostly from reports of the Romanian National Office of Vine and Wine Products (ONVPV) [11] and Romanian National Organisation of Vine and Wine (ONVV) [12] submitted to the European Commission or to OIV. An example of such a report, which is not available for public, is "Agriculture and Rural Development ISAMM CM" sent by ONVV to the European Commission in March 2015 regarding the inventory and measurement of the area planted and also the transitional planting rights regime.

Due to the complexity of the field, both European regulations and Romanian legislation for the wine and vine sector [11] were consulted. Some regulations referred only to wine market common organisation, while others referred generally to the common organisation of agricultural markets, thus, the measures for vine and wine were based on specific provisions of either one or the other type of regulation. Basically, in accordance to

[a] e-mail: `arina.antoce@horticultura-bucuresti.ro`
[b] e-mail: `aantoce@yahoo.com`

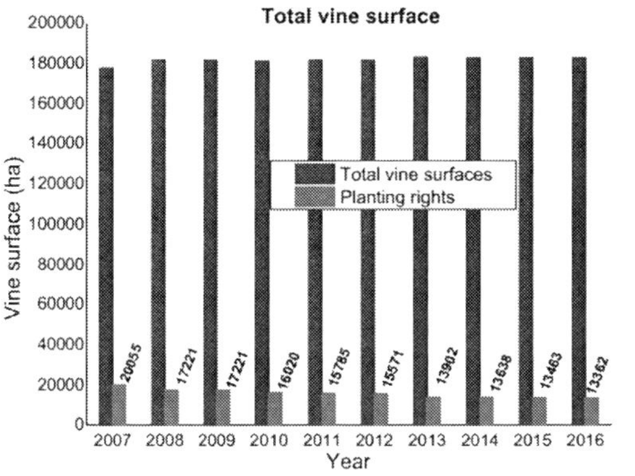

Figure 1. Vine surfaces evolution in Romania after accession to EU to present time.

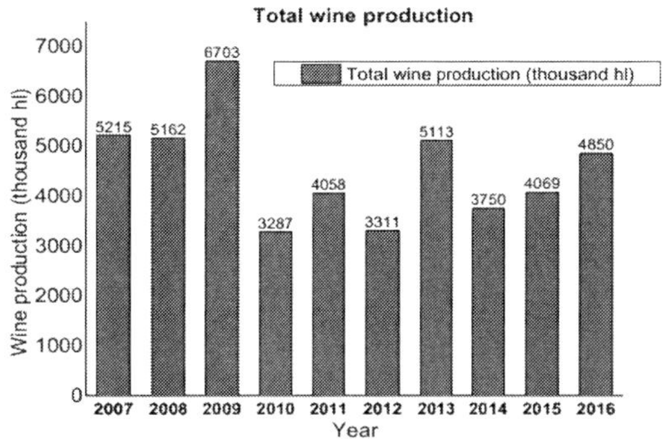

Figure 2. Wine production evolution in Romania from UE accession to present time.

(EC) No 1234/2007, the regulations regarding common organisation of agricultural markets did not cover those parts of the CMOs which were subject to policy reforms, certain elements for wine being part of these reforms.The data regarding grape varieties and planted surfaces were extracted from the Registry of the Vineyard Plantations, a national IT system developed based on an Oracle database, in which the declarations of producers are introduced since 2003. Until 2009 corrections and updates of the registry were allowed, in accordance with the documents from the vineyards owners. A functional registry of the vine plantations was a prerequisite for accession of Romania to EU. At present it is the main evaluation and control tool for the viti-viniculture sector. A new enhanced software application for the Registry was launched this year, comprising the recording of compulsory declarations (harvest, yield, stock) as mentioned in EU legislation [4,13], along with required information for market research.

Based on the data from the Registry of the Vineyard Plantations the surface evolution of autochthonous and international varieties from 2007 to 2016 was examined in order to evaluate how did the National Support Plan helped Romania achieve its objectives assumed as a Member State, in the field of wine and vine.

The main grape varieties monitored and included in the reports for the European Commission are: Fetească albă, Fetească regală, Merlot, Riesling, Aligote, Sauvignon blanc, Cabernet Sauvignon, Muscat Ottonel, Fetească neagră, Pinot noir, Chardonay.

3. Results and discussion

3.1. The evolution of the Romanian vine and wine sector in the period of 2007–2016

The total vine surface in Romania, as presented in Fig. 1, has not fluctuated much since 2007, only a very slight increase being recorded in recent years due to the new plantings.

The total surface covered with vine in Romania reported in 2015 was 182,849 ha, of which 96,492 ha *Vitis vinifera* vine varieties and 86,357 ha of interspecific hybrids (hybrid vine varieties temporarily accepted in culture).

To this surface 13,463 ha are added as planting rights in various stages of usage (planting rights allocated to producers but not yet used – 1,739 ha, replanting rights currently held by the producers – 9,536 ha and planting rights still existing in the reserve – 2,188 ha), leading to a total viticultural potential of 196,312 ha (ISAMM-CM 2015).

The values of 2016 (not yet officially reported at the moment of paper writing) are not very much different from those of 2015, the total vine surface being 182,763 ha to which planting rights are added in surface of 13,362 ha.

The planting rights reserve allocated to Romania upon accession to EU (in amount of 2,830.5 ha) are included in the present national reserve.

Basically, between 2007 and 2016 the total vine surface remained relatively constant, even though the influx of EU funds made a clear difference in the varieties planted and vine renewal. From 2007 the planting rights from the national reserve decreased slowly, with a more important decrease in 2013, when due to the infusion of the EU funds more vine was planted. In Romania the vine planting is only permitted if a planting right is used, either a new planting right, or a replanting right or a right allocated from the national reserve. This is set in Art. 63 of Regulation (UE) nr. 1308/2013 [2] which repeals previous regulations forbidding the planting of grapevine until December 31st, 2015.

The destinations of the vine surfaces for the wine production were in 2015 (ISAMM-CM 2015) as follows:

- 31,532 ha (17.25%) - surfaces for wine with protected denomination of origin (DOC)
- 18,600 ha (10.17%) - surfaces for wine with protected geographic indication (IG)
- 132,717 ha (72.58%) - surfaces for wine without protected DOC/IG.

The wine produced and meant for commerce is only based on the *Vitis vinifera* varieties; thus, when considered in view of the total surfaces covered with vine, the Romanian wine production is lower than expected. The evolution of the wine produced and commercialized in Romania from 2007 to 2016 is presented in Fig. 2, with the resulted categories of wines included in Fig. 3.

We can see that in the period 2010–2013 the wine production decreased, but this is explained by the fact that in this period old vineyards were grubbed and new

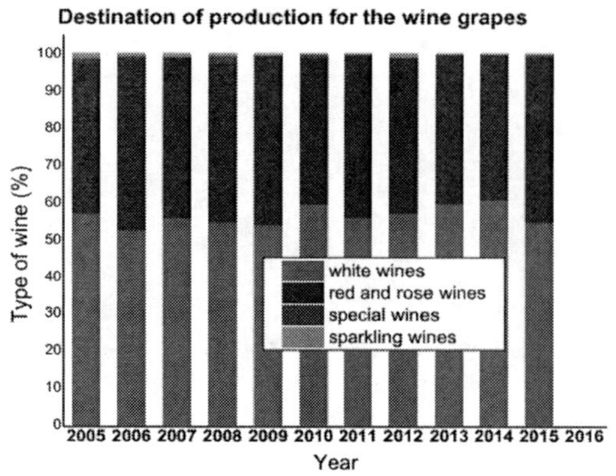

Figure 3. Destination of the wine grape production in Romania from 2005 to 2015.

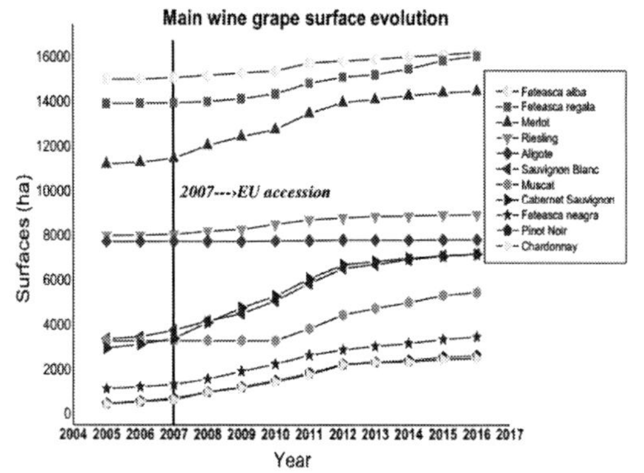

Figure 4. Evolution of surfaces cultivated with the main wine grape varieties in Romania from 2005 to 2016.

vineyards, which need time to enter production, were planted with European funds.

3.2. Evolution of wine grape varieties from 2005 to 2016

More than 140 *Vitis vinifera* and interspecific grape varieties are cultivated in Romania, but not all cover important surfaces. Out of these only about 40 wine grape varieties are found on surfaces larger than 100 ha. The most important varieties cultivated in Romania on significant surfaces are included in Table 1 and their evolution in time is plotted in Fig. 4.

Starting with 2007, the year of Romania's accession to EU, all these main varieties had their surfaces increased. With the exception of Riesling and Aligote, which registered only slight surface increases during the entire reported period, all the other main varieties recorded clear growth, with a sharp increase between 2008 and 2012, when the European funds were absorbed the most efficiently (Fig. 4).

Due to the infusion of European funds, the relative distribution of the main varieties in 2016 differs compared with 2007 (Fig. 5), with a relative increase of the international varieties Cabernet Sauvignon, Chardonnay, Sauvignon blanc, Pinot noir and Muscat, but also with an increase for the indigenous Feteasca neagra. While

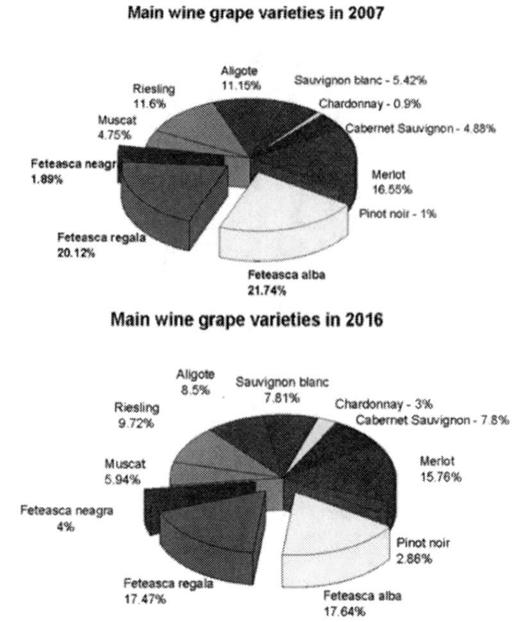

Figure 5. Relative distribution of the main wine grape varieties in Romania at the accession to EU (2007) and at present (2016). The percentages are calculated as reported to the total surface of these varieties.

still being the most widespread varieties in Romania, the white autochthonous varieties Feteasca alba and Feteasca regala recorded a relative decrease. Also a reduction was observed for surfaces with Riesling, Aligote and Merlot.

3.3. Evolution of vine surfaces in accordance to the wine regions and destination of production

In accordance to the geographic position and some other particularities of climate and soil, in Romania are defined eight wine regions. More information about Romanian wine regions, varieties and wines can be found in our previous publications [14, 15].

As mentioned before, the vine surfaces are distributed in the most favourable regions of the country, the largest wine regions being the 1-Hills of Moldova followed by the 2-Hills of Vallachia and Oltenia (Fig. 6).

The new investments in vineyard took into account the necessary increase in wine quality, thus, the new plantations were mostly destined for the DOC wine production. As a consequence, in all the regions were DOC wines are approved, the DOC surfaces increased in 2016 as compared to 2007, by 1.1 to 5.5 times, in accordance to the wine region (Table 2).

Thus, it can be seen that after EU accession the surfaces meant for DOC wine production significantly increased in all wine regions, with the exception of the region of Danube Terraces, where there is no denomination of origin approved.

However, the overall quantities of red wines obtained did not increase as much as expected (Table 3), the percentage of red wines remaining in the range of 40–45% (Fig. 7), even though the total wine production fluctuated largely in accordance to each year's climatic conditions.

Table 1. Evolution of the main wine grape varieties cultivated in Romania from 2005 to 2016 (from ONVPV).

Grape variety	Vine surface (ha)											
	2005	**2006**	**2007**	**2008**	**2009**	**2010**	**2011**	**2012**	**2013**	**2014**	**2015**	**2016**
Feteasca alba	15004	15007	15075	15180	15287	15375	15733	15828	15901	16005	16111	16210
Feteasca regala	13915	13934	13950	14005	14134	14353	14828	15104	15206	15469	15850	16052
Merlot	11216	11294	11472	12044	12440	12758	13480	13964	14106	14278	14411	14480
Riesling	7997	8001	8045	8180	8271	8497	8698	8787	8848	8872	8922	8936
Aligote	7727	7727	7728	7728	7728	7753	7769	7793	7793	7794	7805	7811
Sauvignon blanc	3375	3465	3757	4170	4499	5078	5851	6528	6696	6919	7096	7176
Cabernet Sauvignon	2956	3125	3387	4100	4768	5287	6060	6692	6837	6985	7087	7170
Muscat	3267	3278	3293	3296	3296	3296	3830	4447	4741	5000	5323	5457
Feteasca neagra	1130	1195	1307	1568	1924	2245	2643	2885	3043	3170	3360	3467
Pinot noir	466	554	695	970	1196	1501	1841	2241	2333	2407	2567	2627
Chardonnay	434	544	626	970	1158	1432	1777	2223	2317	2344	2454	2511

Table 2. The surface with vineyards distributed in accordance to the Romanian wine regions in 2007 (time of EU accession) and in 2016 (from ONVPV Communication 2007 and Vine Plantations Registry, ONVPV, 2015–2016).

Wine region	Total surface (ha) in 2007 (EU accession)	*DOC surface in 2007 (ha)*	*Surfaces with IG and without DOC/IG in* 2007 (ha)	Total surface (ha) *in 2016*	*DOC surface in 2016 (ha)*	*IG surface in 2016 (ha)*	*Surfaces without DOC/IG in 2016 (ha)*	DOC surface increase in 2016 as compared to 2007	
								times	%
1. Transylvanian plateau	6604.84	*1432.57*	*5172.27*	6953.13	*3919.99*	*80.62*	*2952.52*	2.74	173.63
2. Hills of Moldova	70504.87	*7419.38*	*63085.49*	69156.73	*8267.69*	*11901.23*	*48987.81*	**1.11**	11.43
3. Hills of Vallachia and Oltenia	49492.97	*3911.92*	*45581.05*	52829.18	*8300.78*	*1596.43*	*42931.97*	2.12	112.19
4. Hills of Banat	2722.83	*838.07*	*1884.76*	2962.43	*1774.72*	*22.94*	*1164.77*	2.12	111.76
5. Hills of Crisana and Maramures	7942.99	*356.82*	*7586.17*	9614.48	*1957.98*	*710.55*	*6945.95*	**5.49**	448.73
6. Hills of Dobrogea	15748.56	*6924.77*	*8823.79*	17008.88	*8333.3*	*3239.57*	*5436.01*	1.20	20.340
7. Danube Terraces	11822.0	*0*	*11822.0*	11250.6	*0*	*417.57*	*10833.03*	0	0
8. Sands and other Favourable Lands of the south of the country	13261.68	*0*	*13261.68*	12941.35	*245.82*	*296.23*	*12399.3*	N/A	N/A
Total:	**178100.74**	*20883.53*	*157217.21*	**182716.78**	*32800.28*	*18265.14*	*131651.36*	1.57	57.06

Table 3. Amounts and types of wine produced in each year.

Type of wine	Wine quantity (1000 hl)										
	2005	**2006**	**2007**	**2008**	**2009**	**2010**	**2011**	**2012**	**2013**	**2014**	**2015**
White	1518.7	2860.7	2768.0	289.0	3637.0	1976.3	2276.7	1913.8	2976.1	2275.6	1998.1
Red	1083.5	2521.0	2153.0	2264.8	3066.0	1311.0	1781.5	1396.8		1474.2	1629.5
Special	32.3	28.1	7.3	103.0	0.0	26.06	0.30	17.0	12.1	0.1	14.7
Sparkling	32.0	33.3	36.0	30.1	23.1	5.4	9.9	25.9	4.6	4.9	6.1
Total	2666.5	5443.1	4964.3	2686.9	6726.1	3318.7	4068.4	3353.5	2992.8	3754.9	3648.3

3.4. Evolution of investments and new plantations after the EU Accession

The infusion of funds after the EU accession greatly benefited the Romanian vine and wine sector.

In viticulture, most funds were allocated for restructuring and conversion of the vine plantations, as seen in Table 4. A total of 288.7 million Euro were invested in the wine and vine sector between 2007–2014, around 98% going each year into the replacement of old plantations.

The new plantations were approved and financed only with compliance with the grape varieties zoning regulations of Romania (Ministry Order 225 of 2006), as a prerequisite of quality wine production in each specific wine region.

The new surfaces planted with wine grapes are shown in Figs. 8–11, grouped in surfaces covered with the main white and black varieties, autochthonous and international, respectively.

Also, in Fig. 12 the total new vine plantations are reported, from 2005 to 2016.

Until 2013 the funds allocated to the wine and vine sector were absorbed entirely. The maximum impact of the European funds allocated to Romania after 2007 was seen in the year 2011 (Fig. 12), when the plantings with the white and black international varieties were given priority,

DOI: 10.1051/bioconf/20170903018

Table 4. Funds allocated in Romanian viticulture between 2007 and 2016 (Romanian Payment and Intervention Agency for Agriculture, 2017).

Vintage year	Vineyards restructuring conversion		Harvest insurance must	Use of concentrated (rectified) markets	Promotion on third	Investments from EU	Total allocated	TOTAL used	
	euro	% of allocation	euro	euro	euro	euro	euro	euro	%
2006 / 2007	6,893,687.59	81.1	–	–	–	–	8,500,000.00	6,893,687.59	81.1
2007 / 2008	35,050,227.75	100.1	–	–	–	–	35,000,000.00	35,050,227.75	100.1
2008 / 2009	41,692,714.49	99.0	323,423.24	29,494.27	54.40	–	42,100,000.00	42,045,686.40	99.9
2009 / 2010	41,548,576.08	98.7	360,658.94	81,283.31	109.48	–	42,100,000.00	41,990,627.81	99.7
2010 / 2011	41,182,059.13	97.8	269,524.10	76,458.11	571.96	–	42,100,000.00	41,528,613.30	98.6
2011 / 2012	40,990,098.37	97.4	306,211.10	56,044.92	747.64	–	42,100,000.00	41,353,102.03	98.2
2012 / 2013	41,183,602.36	97.8	230,808.69	–	663.12	–	42,100,000.00	41,415,074.17	98.4
2013 / 2014	47,084,204.96	98.7	253,321.27	–	362.47	–	47,700,000.00	47,337,888.70	99.2
2014 / 2015	18,995,454.62	39.8	228,617.54	–	699.41	–	47,700,000.00	19,224,771.57	40.3
2015 / 2016	10,940,798.41	22.9	186,734.05	–	479.33	70,473.87	47,700,000.00	11,198,485.66	23.5
TOTAL 2007–2016	**325,561,423.76**	**82.0**	**2,159,298.93**	**243,280.61**	**3,687.81**	**70,473.87**	**397,100,000.00**	**328,038,164.98**	82.6
	99.24%		**0.66%**	**0.07%**	**0.00%**	**0.02%**		**100%**	

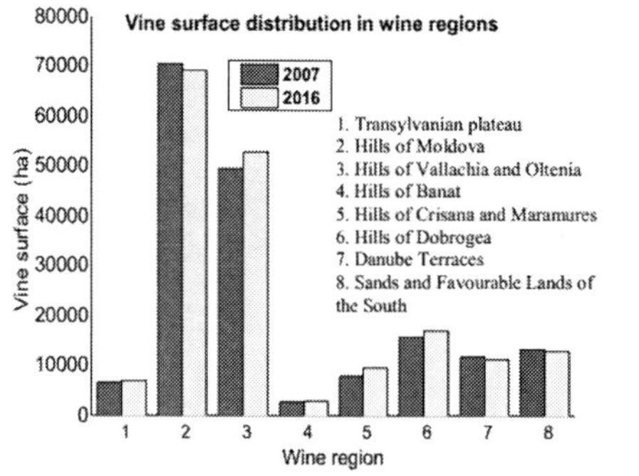

Figure 6. The vine surface distribution in wine regions at EU accession (2007) and at present (2016).

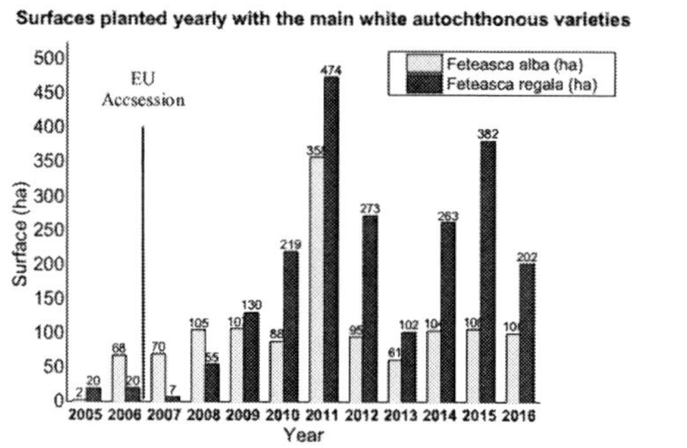

Figure 8. The surfaces planted yearly with the main white autochthonous varieties - period 2005–2016.

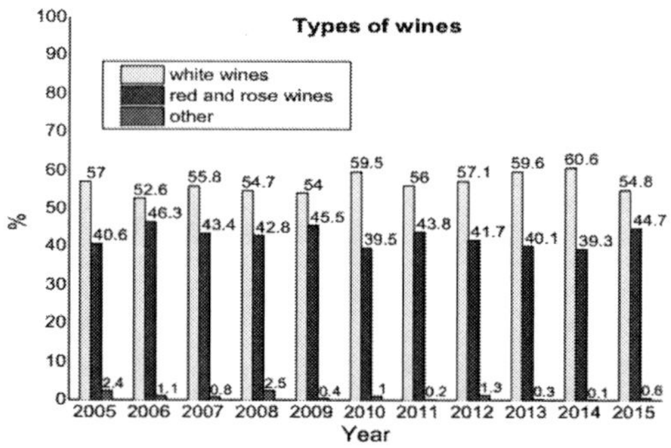

Figure 7. The wine assortment in accordance to the main types: white, red/rosé and other.

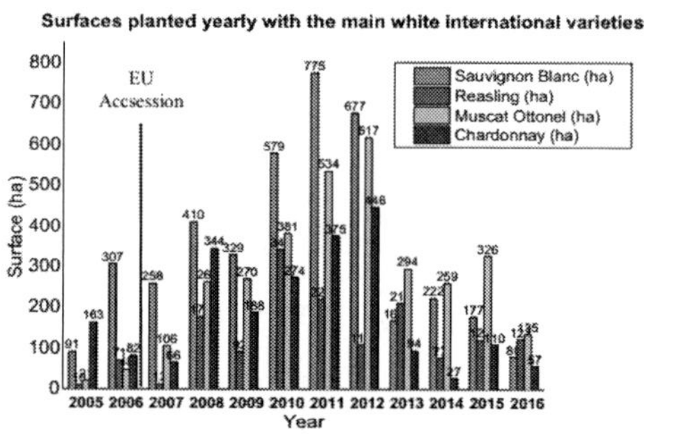

Figure 9. The surfaces planted yearly with the main white international varieties - period 2005–2016.

Sauvignon Blanc (Fig. 9), Cabernet Sauvignon and Merlot (Fig. 11) being also at the top of the producers' list.

The black autochthonous variety Fetească Neagră also had a significant evolution, being one of the most preferred varieties for the new plantings (Fig. 10).

The autochthonous white varieties, Feteasca alba and Feteasca regala were not neglected either (Fig. 8).

Starting with 2013 a sharp decrease is seen in the investments in new vine plantations, due mainly to some legislative measures which led to a temporary blockage. Thus, for the year 2016, out of the available 47.7 million Euro only 12 million were absorbed, the new plantings reaching only 895 ha.

There are reasons to believe that Romania will continue the success story in its viticulture domain,

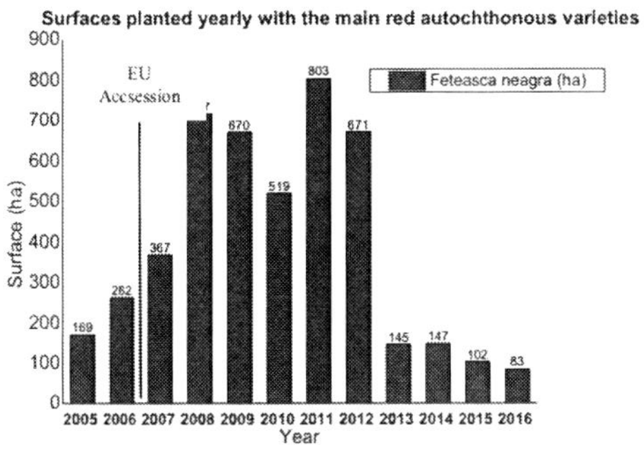

Figure 10. The surfaces planted yearly with the main black autochthonous varieties - period 2005–2016.

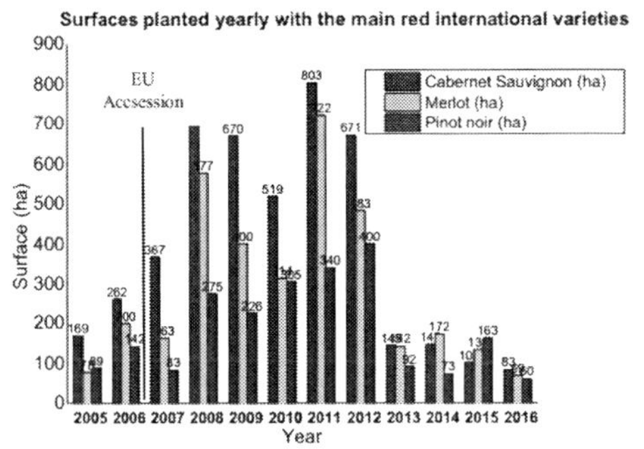

Figure 11. The surfaces planted yearly with the main black international varieties - period 2005–2016.

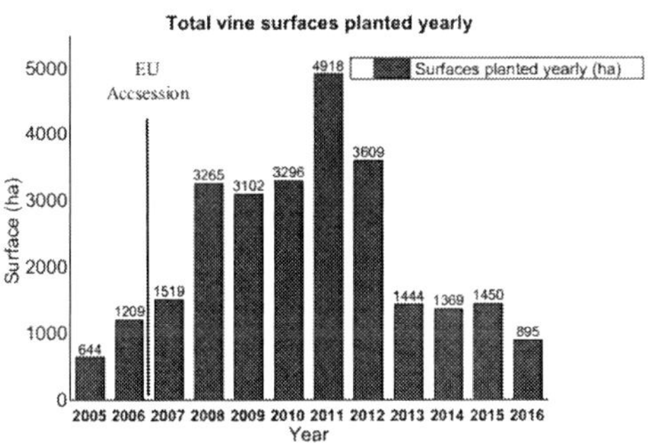

Figure 12. The total vine surfaces planted yearly - period 2005–2016.

with the new National Viticultural Programme 2014–2018, for which 238.5 million Euro are allocated, that is 47.7 million Euro per year.

4. Conclusions

The Romanian vitiviniculture was one of the areas of the economy which greatly benefited from the accession to EU. The most important changes were seen in the renewal of the vine plantations, as almost 90% of the funds allocated through the National Support Plan 2009–2013 for viticulture were spent in restructuring and conversion. During this period the investments in equipment for viticulture and winemaking were actually not covered in the viticulture programme, but in the rural and development programmes for agriculture, thus, only few such investments were made due to high competition from other agricultural sectors. Even so, the entire funds distributed to Romania were very useful, several years in a row during this first programme the funds allocated in viticulture being totally absorbed. The second programme, for 2014–2018, is ongoing, with some changes regarding the objectives, including the investments in equipment. The absorption slowed down, however, due to some legislation inconsistencies and delays in application. The restructuring and conversion helped renew the existing plantations, which were not anymore adapted to market demand – but also helped with the establishing of new plantations, based on using individual planting rights or from the national reserve, so that the total surface of vineyard existent in the country could be kept at the level negotiated at the time of the accession to EU. Grapevine varieties currently in low demand were replaced with valuable autochthonous varieties or international varieties. An increase was observed for the autochthonous Feteasca Neagră, as well as for the international varieties such as Cabernet Sauvignon, Chardonnay and Sauvignon Blanc.

References

[1] COUNCIL REGULATION (EC) No 1234/2007 of 22 October 2007 establishing a common organisation of agricultural markets and on specific provisions for certain agricultural products (Single CMO Regulation) (http://eur-lex.europa.eu/legal-content/en/TXT/?uri=CELEX:32007R1234)
[2] REGULATION (EU) No 1308/2013 OF THE EUROPEAN PARLIAMENT AND OF THE COUNCIL of 17 December 2013, establishing a common organisation of the markets in agricultural products and repealing Council Regulations (EEC) No 922/72, (EEC) No 234/79, (EC) No 1037/2001 and (EC) No 1234/2007. (http://eur-lex.europa.eu/legal-content/EN/TXT/?qid=1491852068735&uri=CELEX:32013R1308)
[3] COUNCIL REGULATION (EC) No 1493/1999 of 17 May 1999 on the common organisation of the market in wine (http://eur-lex.europa.eu/legal-content/EN/TXT/?qid=1491852068735&uri=CELEX:31999R1493)
[4] COUNCIL REGULATION (EC) No 479/2008 of 29 April 2008 on the common organisation of the market in wine, amending Regulations (EC) No 1493/1999, (EC) No 1782/2003, (EC) No 1290/2005, (EC) No 3/2008 and repealing Regulations (EEC) No 2392/86 and (EC) No 1493/1999 (http://eur-lex.europa.eu/legal-content/EN/TXT/?qid=1491852068735&uri=CELEX:32008R0479)
[5] Commission Delegated Regulation (EU) 2015/560 of 15 December 2014 supplementing Regulation (EU) No 1308/2013 of the European Parliament and of the Council as regards the scheme of authorisations for vine plantings (http://eur-lex.europa.eu/legal-content/EN/TXT/?qid=1491855210950&uri=CELEX:32015R0560)

[6] Commission Implementing Regulation (EU) *2015/561* of 7 April 2015 laying down rules for the application of Regulation (EU) No 1308/2013 of the European Parliament and of the Council as regards the scheme of authorisations for vine plantings (http://eur-lex.europa.eu/legal-content/EN/TXT/?qid=1491852414527&uri=CELEX:32015R0561)

[7] Commission Implementing Regulation (EU) 2015/1991 of 5 November 2015 amending Regulation (EC) No 555/2008 laying down detailed rules for implementing Council Regulation (EC) No 479/2008 on the common organisation of the market in wine as regards support programmes, trade with third countries, production potential and on controls in the wine sector. (http://eur-lex.europa.eu/legal-content/EN/TXT/?qid=1491852414527&uri=CELEX:32015R1991)

[8] Statistical yearbooks of Romania issued by National Institute of Statistics, Archives: http://www.insse.ro/cms/en/content/statistical-yearbooks-romania

[9] Ministry of Agriculture and Rural Development of Romania, http://www.madr.ro/en/; http://www.madr.ro/en/horticulture/viticulture-enology.html

[10] Romanian Payment and Intervention Agency for Agriculture (APIA), http://www.apia.org.ro/

[11] Romanian National Office of Vine and Wine Products (ONVPV), http://www.onvpv.ro/; http://www.onvpv.ro/ro/content/legislatie

[12] Romanian National Organisation of Vine and Wine (ONVV), http://www.onvv.ro/

[13] Commission Regulation (EC) No 436/2009 of 26 May 2009 laying down detailed rules for the application of Council Regulation (EC) No 479/2008 as regards the vineyard register, compulsory declarations and the gathering of information to monitor the wine market, the documents accompanying consignments of wine products and the wine sector registers to be kept. http://eur-lex.europa.eu/legal-content/en/TXT/?uri=CELEX:32009R0436

[14] Antoce A. O., Nămoloşanu I., Duşa D., Mereanu D., Rebigan C., Nicolau V., Călugăru L., Some considerations regarding the grapevine variety assortment and wine categories in Romania in recent years", Bulletin de l'OIV, **86**, ian-feb-mar 2013, No. 983-984-985, pp 27–45 (2013).

[15] Nămoloşanu I. and Antoce A. O., 2000, Romanian viticulture on the verge of the third millennium, Die Wein Wissenschaft, Vitic. Enol. Sci. **55**(2), 67–72

DOI: 10.1051/bioconf/20170903019

From new clones to flowers – innovative business models in the Hungarian wine economy

Balázs Kismarjai

Eötvös Loránd University, Department of Social and Economic Geography, 1117 Budapest, Pázmány Péter sétány 1/C, Hungary

Abstract. Due to the ovestocked Hungarian wine sector wineries often need to find entirely new ways that enable them to increase their market share. These opportunities are determined by the current market potentials, the geographic location and of course the wine district itself. In this study I examined some examples of outbreak opportunities for producers in this difficult economic situation. Young winemakers from different Hungarian wine districts unite to open wine bars - these are now present in several towns all over the country. Other winemakers have been experimenting with new clones in less recognized wine districts. A small winery from a historic wine district recommends its wines with flowers in a new shop.

1. Introduction

Based on experiences from the past years it is reasonable to say that the Hungarian wine market has been going through significant and dynamic changes. Fortunately for the wine consumers there is a new generation the members of which are looking for quality wines, so it forces the winemakers to dismiss low-quality table wines ordinary in socialist era. Beside the "great veterans" expropriating the wine market after the end of the communist regime new generations have grown up, small family wineries have appeared as well as modern wineries with up-to-date technologies procreated by investors. Due to the EU subsidies the technological gap has narrowed, so smaller producers are capable of building up modern wineries.

Meanwhile, due to the increased wine-making capacity the Hungarian wine market has been greatly overstocked. One can tend to think that the competition on wine market takes place only among wine makers for the sake of consumers. However the competition on the wine market is twofold. On the one hand it is a real competition among wineries to introduce a product to the market, catch and keep the consumers with the wine label and brand. The necessary marketing absorbs significant amount of money not to mention adjunct costs such as being present in bigger supermarkets. On the other hand, there is a sharpening competition among vine varieties and types of wines as well, for the producers have to find a sensitive equilibrium between red, white and rosé wines, as well as between the worldwide known grape varieties and the so called hungaricums[1]. Furthermore, viticulture and wine-making is affected by climate change. Thus, the winemakers have to take into consideration the local climate and other biophysical factors when planting grapes or selecting the appropiate technology [5].

Of course, these two competition factors are correlated, but they have different impacts. The first type of competition primarily affects the producers in the form of increasing prestige and income, while the main beneficiaries of the second competition type are the consumers who can taste more and more new wines.

The newcomers on wine market have to face a serious challenge, namely reaching a market share that ensures their pay-off and selling their products. Because of this they need proper marketing to create a special, unique product or to formulate a special product tie to base on it. To reach these purposes they need to perform market surveys as well. Unfortunately, these processes are not well-rooted in Hungarian wine-business traditions, as the assigned amount of money is missing from the production process itself. Several promising new wineries went under because they did not measure properly their possibilities, like the current market gaps.

Despite the initial difficulties wine producing enterprises can be remunerative in the long run. The ones surviving this period are stable, their survival ratio is the highest among agricultural enterprises [13].

2. Innovation examples

This publication shows three different ways of reaching bigger market shares by winery managers. These innovations point ahead, they prove the long-term thinking of the investors/owners.

2.1. Licence Koch Winery – new clones, new wines

Csaba Koch has been cultivating his domain in the southern part of the Hungarian plain, in the Hajós-Baja wine region since 1991. In this region one can find 130 hectares of vineyards out of the total 140 hectares owned by Koch (the remaining ten hectares are in the

[1] Hungaricum is an animal, a plant or a product made from it, which is related to the Hungarian production culture, habits or traditions. The Hungaricums are accepted by the Hungarians as specific and well-known things and the Hungaricums are connected with the Hungarian roots by the foreigners too [1].

Villány wine region). The soil of the Hajós-Baja wine region is mainly sand and loess and this region is charactirized by the highest sunny hours ratio per year in Hungary (2070 sunny hours/year). Therefore the grown grape varieties are relevant to these facts: Blaufrankisch, Kadarka and the cabernet-type grape varieties dominate among the red grapes and the aromatic, redolent white grape varieties such as Cserszegi fűszeres (fűszeres means 'spicy'), Irsai Oliver, Muscat Ottonel are the most popular [2].

The **Koch Winery** producing about 900.000 bottles of wine yearly and employing 70 employees belongs to bigger Hungarian wineries. Besides the Hungarian market the winery exports its products to several European countries (e.g. France, Germany) and overseas as well (e.g. Brazil, South-Korea). The main traditional grape varieties of the winery are the Cserszegi fúszeres, Cabernet sauvignon, Muscat Ottonel, Blaufrankisch and Chardonnay. They produce 25–26 types of wines that vary from table wines to premium selections.

Alongside with others these facts have led the owners to start experimenting with new varieties. According to Koch that is to say, the wines from the lower part of the product range should not be made from the same grape varieties as the premium wines. On the basis of these ideas the winery has been experimenting with new clones for 12–13 years. Naturally, this kind of experiment requires a scientific background [5]. Cooperation among Koch Winery, Grape Breeding Gene Conservation Department of Vine, Wine Institute of University of Pécs and its leader, Dr Pál Kozma is pre-eminent. By now half a dozen of clones have been planted and are being investigated meanwhile.

The experiment is based on the recognition that due to its asexual propagation the grapevine is unable to accomodate properly. As a result the genetic codes of the old Vinifera varieties are not changing, so the grapes are incapable against pathogens. With weak resistance they have to be sprayed with powerfull and highly toxic chemicals, which makes ecological and organic production of wines impossible. The purpose is to breed clones that are possibly fully resistant to downy mildew, powdery mildew or grey rot by the process of cross-breeding and applying the most up-to-date technology of gene manipulation. Koch believes it is clear that gene manipulation and breeding resistant varieties must be the foundation of tomorrow's viticulture.

Anyway, the Koch winery is devoted to ecological grape production not using insecticides and trying to minimize applying fungicides. The soil between the grape lines is covered by natural vegetation or sward and instead of applying chemicals they leave killing the pests to nature. Also, they place artificial bird nests to the vineyards.

While breeding new clones they have been paying attention to the different soil types as well. The black soil of the region has a very high humus content accompanied with good lime and other nutrients. The water supply of the above mentioned area between the Danube and the Tisza will be one of the key factors in the next decades which also deserves special attention.

Two of the applied clones in wine production have been already used on bigger territories. The Bácska clone is grown on 15 hectares, the Pannonia clone on 12 hectares. The Bácska clone matures into a kind of mid-quality

Picture 1. Vineyard at Borota. Source: `www.facebook.com/kochboraszat`.

grape, with similar taste and aromatic compounds as the Müller-Thurgau and with high load at the same time, so it can be suitable to make high volume base wine. The Pannónia clone is a definitely good quality variety with outstanding resistance. Half of its genetic stock is inherited from Reisling, so the wine made from it reminds the Reisling type. Even though the yield of vine bases is relatively low (100–110 q/ha), the quality is excellent.

The wines made from the new clones have still been under investigation. The purpose is that the wines come onto the market as organic, however three years have to pass before the qualification procedure is carried out. According to Koch's plans, the vintage of 2018 will be introduced to the market as well distinguished brands from the traditional wines of the winery using differend marketing strategy. The rest is up to the customers.

2.2. Junibor – Association of Young Winemakers/DiVino – "it is good to be a young winemaker nowadays!"

The **Junibor** association was founded in 2008 based on ideas of mainly young, second-generation winemakers. The original idea was to create an association which helps the professional development of its members. Study tours, forums and professional cooperation give the possibility to increase professional experience of the members. Today the association has 34 members from 13 out of the 22 Hungarian wine regions. One of their main purpose is to recruit young wine makers from all the wine regions. At the same time admission is limited. The potential member has to be younger than 35 years old and one can only be a member until the age of 40. Furthermore, in spite of the fact that the members of this wine maker generation know each other well, the prerequisite of admission is a procedure including an interview, introduction, tasting and voting.

Naturally, one of the tasks of the association is to propagate the culture of responsible, self-conscious wine consumption. This intention got a great support from an outsider idea, namely to distribute the wines made by the members of the association collectively. The owners came up with a determined conception of not allowing wines made by non-members to be put on their wine list.

DOI: 10.1051/bioconf/20170903019

The wine bar rooted in this idea was given the name of **DiVino**. The wine bar which was opened in 2011 in a frequented place next to the Basilica in Budapest, has become a must-see place for wine consumers. The great amount of young dedicated wine makers cooperating with each other is an inimitable possibility and a spiritual capital to start with. Making use of this fact the bar owners have managed to give rise to a franchise with several members. The first franchise-member bar was opened in Budapest too, taking advantage of the popularity of the city among youngsters from the West. The DiVino Gozsdu was opened in the party district where it reached significant share and outstanding guest traffic among other bars and restaurants offering similar high standard services.

Picture 2. An ordinary summer day on DiVino's terrace. Source: `www.facebook.com/DiVinowinebar`.

Beacuse of the Hungarian demographic and ecomomic trends the rural enterprises like DiVino have to face serious risks. In spite of this it seems that DiVino Debrecen (2014) and DiVino Győr (2014) are likely to become successful. Naturally, the size and the volume of the guests do not reach the dimensions of the Budapest DiVino, but the well-selected locations (main squares) can guarantee the sufficient guest traffic. The purpose of the owners is the further extension of the franchise system mainly in Hungary. According to the plans three other DiVino bars are going to be opened not later than in 2017. Kecskemét, Székesfehérvár and the tourist-capital of lake Balaton Balatonfüred, will get premium wines. The international expansion represents long term pursuits, primarily in the Carpathian Basin. Though not in a DiVino bar, wines made by members of the Junibor Association can be purchased in Cluj-Napoca, Romania (Kolozsvár). Further extension needs serious and comprehensive tourism and marketing analyses. Based on them the potential cities could be the ones that have outstanding tourist attractiveness (e.g. Krakow in Poland).

As the necessary part of the extension, marketing is managed by a special professional company. In one respect it helps operate the wine bars with ads, secondly it helps Junibor members to synchronize their advertising activities.

The exceptional wonderful place next to the Basilica allowed the Budapest DiVino being the "mother-bar" of the franchise system to organize the fifth Junibor Wine Festival in Szent István Square in 2016.

The members of the Junibor Association sell their wines here as well as at other traditional wine festivals.

Picture 3. V. Junibor Wine Festival. Source: `www.facebook.com/DiVinowinebar`.

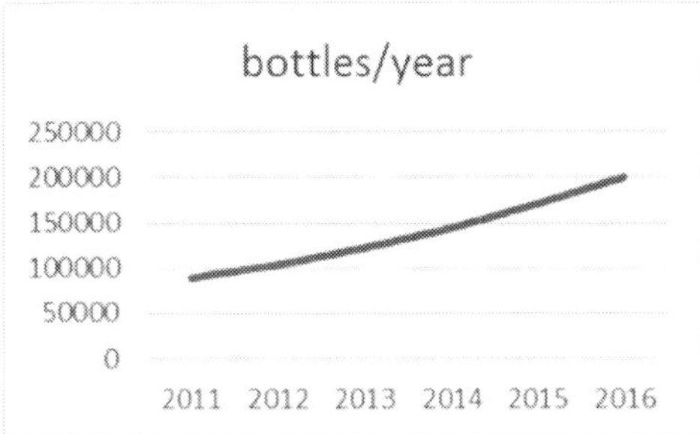

Source: DiVino

Figure 1. Number of sold bottles in the bars of DiVino franchise system. Source: DiVino.

The festival is really successful and the guarenteed quality of wines attracts thousands of visitors to the festival each day of the event (totally eight thousand guests in 2016).

With its quality drinks and foods the DiVino wine bar serves the expert and interested guests at a high level. Today it owns the "Michelin" and "Bib Gourmand" recommendations. These titles naturally increase the traffic rooted in popularity. During the last six years after opening the bar it has been raising its traffic continuously. Last year resulted in 200.000 sold bottles.

2.3. Wines&Roses – wines with flowers

Csaba Miklós, the young wine maker from the wine region of Mór has graduated as a horticulturist, so besides the wines flowers are close to him as well. This background gave him an idea to link these two products and open a shop in Budapest with the name of **Wines&Roses**.

The Mór wine district, which is an old traditional wine producing settlement, is situated to the west of Budapest, about ninety kilometers far from the capital. This wine district covering all in all 724 ha (2011) is one of the smallest out of the total 22 Hungarian wine districts. It primarily produces white wines the typical varieties of which include the local-specific Ezerjó, Sauvignon blanc, Reisling and Pinot blanc. Among the very few red wines Blaufrankisch should be mentioned [2,4,8].

Miklós's winery ("miiklóscsabi's winery") can be found here and he cultivates his vineyards on 3,5 ha. Being a family winery he isn't planning to buy more vineyards.

Picture 4. Mór, Csóka-vineyard. Source: www.mikloscsabi.com.

Picture 5. Wines&Roses Shop. Source: www.facebook.com/wnrbudapest.

They produce about 15.000 bottles of wines yearly and due to the well structured marketing and popularity they can sell most of them. Yet, to increase the popularity of the winery and bridge the seasonal trade Miklós has been looking for new ways. One of these can be matching flowers and wines.

Miklós sells only his own wines in his shop, so he obtains the flowers according to the principles of fair-trade mostly from Kenyan producers (e.g. Credible Blooms, PJ Dave Group, Waridi Farm), in smaller extent from Dutch producers (e.g. Dutch Flower Group, Zandbergen Group). This ensures that seasonal flowers are constantly arriving at the shop through the shortest way. Besides the vase-durability the all-year-round prices are guaranteed in his shop.

The shop has not had a long history yet. The owner keeping to his conseption has been awaiting for customers for a year now. The flowers are sold not only in a traditional manner (packed in cellophane) as imprinted in the last decades, but in cut-bouquets or even in trusses. The purpose is to make buyers feel like decorating and placing flowers in their houses creatively. At the same time Miklós recommends his wines, as his motto "passionate weekdays" matches the flavours of wines and flowers.

3. Summary

As the examples detailed above show, competition forces the wine makers to do their best. In order to achieve the operation costs and reach some profit, the wine makers are continuously coming up with new ideas. One can hope that these ideas will work in the long run and serve as examples of good practices for other wine makers considering similar innovations. Hungarian wine makers can only be successful as opposed to the New World's wine dumping with innovative and striking ideas and producing quality, unique and specific grapes and wines [5]. The relation between scientific and economic domains has an increasingly important role as well. Cooperation, technological transfer from science to production and the direct involvement of researchers into the economical processes could provide kind of an advantage that should not be lost [5].

References

[1] A. Andrásfalvy, A kertészeti hungarikumok sajátos minősége, MTA Társ. Kut. 25 (2003)
[2] G.B. Bányai, D. Ercsey, G. Mészáros, I. Tompa, Nagy Magyar boratlasz (2012)
[3] E.P. Botos, K.S. Moulton, Wine prod. and distr., OIV Congr. (2007)
[4] Sz. Frey, L. Száraz, Az ezerjó házája (2010)
[5] E. Giuliani, A. Morrison, R. Rabellotti ed, Innovation and technological catch-up (2011)
[6] I-né Hajdú ed, Borpiac (2005)
[7] G. Harsányi, A hazai borágazat versenyképessége (2007)
[8] Á. Kiss, Az ezerjó hazájában (1990)
[9] J. Lehota, Market orientation and comp. OIV Congr. (2007)
[10] E. Nagy-Kovács, A szőlő- és bortermékpálya jövedelmezősége (2011)
[11] I. Szűcs, M. Farkasné Fekete ed, Hatékonyság a mezőgazdaságban (2008)
[12] G. Totth, Gazd. **2**, 140 (2016)
[13] G. Valkó, Á. Kincses, Gazd. **1**, 3 (2014)

DOI: 10.1051/bioconf/20170903020

How intrinsic values influence wines prices

Péter Gál

Corvinus University of Budapest, 1093 Budapest, Fővám tér 8., Hungary

Abstract. The use of hedonic price indices is quite common in the wine economics literature, yet they mainly include scores of organoleptic tests and some dummy variables representing varieties and quality signs as geographical indications. This study focuses on the relation between the composition and the price of wines on the example of Hungarian wines. In Hungary, the wine law renders chemical analysis compulsory for all wines released to the market. The study includes five main compounds: actual alcoholic strength, total sugars, total acidity, sugar free extract and pH value and is based on hedonic price indices calculated on a sample of 2,453 wines. Results of several regressions – using different model specifications – consistently show that actual alcoholic strength, sugar content, sugar free extract and pH value are related with the price. Some characteristics have an optimal level, while in other cases the relation is linear.

1. Introduction

The use of hedonic price indices is quite common in the literature of wine economics as these models explain differences in prices with variables describing the products' intrinsic values.

The article is structured as follows. Section 2 presents an overview of the empirical literature, followed by a demonstration of methodology and data used in Sect. 3. Results are discussed and a graphical illustration of optimal values is provided in Sect. 4.

2. Literature on hedonic price indices

Rosen's [1] basic model regards goods as an aggregate of their characteristics. Therefore, differences in prices reflect differences in the set of features. Later, several models were developed for the explanation of wine prices [2–8]. These studies mainly include scores of organoleptic tests and some dummy variables representing varieties and quality signs as geographical indications or the name of the producer. These models reveal the relation of these factors with prices, however do not consider the role of wine character.

Meanwhile, Lecocq and Visser [9] used data describing the weather of the vintage year for Bordeaux *en primeur* sales to confirm that weather data explain the changes in prices. However, in case of en primeur sales, when wines are far from being ready to be marketed and are to be aged for several additional months, weather data represent the most important information on quality for buyers – in addition to the name of the producer.

Combris et al. [10] developed several hedonic price index models on a sample of Burgundy wines. The most important results showed that some elements of the character (acidity, concentration and full-bodiedness) are significantly related to the price.

3. Sample and method

This study is based on hedonic price indices calculated on a sample of 2,453 wines. The prices were observed in the Hungarian off-trade sector (main wine shops and supermarkets). If a wine was observed on multiple sites, the lowest price was included in the dataset. The scope of the study extended only to wines, other grapevine products (such as sparkling wine) were excluded. Tokaj wine specialities were also excluded due to their relatively high concentration of nearly every compound in concern.

Creating a hedonic model using variables describing the characteristics of wines is quite rare as the availability of such data is highly limited. However, the Hungarian wine law imposes analytical tests on all wine products to be marketed. These tests include the measurement of the following compounds and indicators:

- total and actual alcohol content,
- free and bound sulphites,
- pH value,
- acidity (titratable, volatile),
- density,
- extract content (total and sugar free),
- sugar content.

This study focuses on the most important compounds and indicators such as alcohol content, acidity, extract, sugar and pH. Data were provided by the national wine authority.

Descriptive statistics and measurement units are shown on Table 1. All prices were re-calculated for a unit of 0.75 l bottle.

The hedonic price indices calculated here are easily describable as follows:

$$P = ß_0 + ß_q\, Q + \Sigma ß_i\, C_i + u \qquad (1)$$

where Q represents the actual quantity (lot size) of the wine and C_i represents the variable describing compound i.

DOI: 10.1051/bioconf/20170903020

Table 1. Descriptive statistics. N=2,453.

Variable	Mean	Std. Dev.	Min	Max
Price (HUF per 0.75 l)	2,205.78	2,301.02	194.85	53,820
Quantity or lot size (litre)	20,435.5	39,687.56	250	607,568
Total alcohol (%vol)	13.04	1.39	9.77	23.63
Actual alcohol (%vol)	12.59	1.15	7.14	16.45
Sugar (g/l)	8.46	22.01	0	251
Sugar free extract (g/l)	24.81	4.92	15.6	71.8
Acidity (g/l, in tartaric acid)	5.64	0.78	3.7	13.3
pH	3.48	0.16	2.88	4.01

Table 2. Coefficients of models A-C and U.

Model	A	B	C	U
Dependent variable	price	ln price	ln price	ln price
Quantity (lot size)	−0.0028 (0.003)	−0.000005 (0.000)	−0.000005 (0.000)	
(ln)				−0.2076 (0.000)
Actual alcohol			0.3472 (0.000)	
(quadratic)	37.2301 (0.000)	0.0139 (0.000)		0.0126 (0.000)
Sugar	−25.0372 (0.000)			
(quadratic)	0.5061 (0.000)	0.0001 (0.000)	0.0001 (0.000)	0.00005 (0.000)
Sugar free extract		0.0228 (0.000)	0.0237 (0.000)	0.0188 (0.000)
Acidity	185.1998 (0.000)			
pH		−0.5268 (0.000)	−0.5058 (0.000)	−0.2766 (0.000)
constant	−4797 (0.000)	6.5271 (0.000)	4.2733 (0.000)	7.7233 (0.000)
adj. R^2	0.3849	0.5076	0.5147	0.5902
VIF	3.02	1.49	1.50	1.52

Table 3. Coefficients of models D-E.

Model	D	E
Dependent variable	ln price	ln price
Quantity (lot size / ln)	−0.2282 (0.000)	−0.2196 (0.000)
Actual alcohol (ln)	3.1771 (0.000)	3.1426 (0.000)
Sugar (ln)	0.0322 (0.000)	
(for white wines – ln)		0.0624 (0.000)
(for red wines – ln)		−0.0580 (0.000)
Sugar free extract (ln)	0.8659 (0.001)	0.8463 (0.000)
pH (ln)	−1.6855 (0.000)	−1.0908 (0.001)
constant	0.7224 (0.150)	0.0620 (0.899)
R^2	0.5390	0.5505
VIF	1.41	1.42

However, the basic model could not be applied using all right-hand side variables as some of them are interdependent. To avoid endogeneity, some pair of variables could not be present in the same model (for example: total alcohol vs. actual alcohol or pH vs. acidity).

Moreover, considering logarithms of some variables (price, in particular) resulted in better model fit, while including quadratic tags allowed to calculate optimal values. All in all, several regression models were run.

4. Results and discussion

Tables 2 and 3 summarise the results of the various regression models calculated.

As some of the variables observed are functions of each other (e.g. total alcohol content is virtually the sum of actual alcohol and sugar content), endogeneity would mean a major problem if all compounds were included in the same model. Therefore, four sets of models were calculated using whether total alcohol content or actual alcohol and sugar content and pH or acidity. As models using actual alcohol and sugar content with pH fitted the best, they are the sole to be presented here (except for the lin-lin specification, where the inclusion of acidity resulted in a better model).

As all models consistently show, the quantity and pH of wines are negatively related to the price. Other compounds (alcohol, sugar, acidity and sugar free extract) affect positively the price.

As comparing models A-C and D show, using a log-log specification resulted in the best fit.

However, all these models were possible to be improved by two ways.

On one hand, introducing quadratic tags resulted in a better (the best available) fit and the possibility to calculate optimum (model U), while on the other hand, differentiating between red and white wines helped to reveal the different role of sugar in red and white wines (model E).

As model U shows, the relation between prices and some compounds (actual alcohol and sugar) is not of linear nature. As the quadratic tag is significant, these compounds have an optimal level form the point of view of the price.

Meanwhile, model E confirms the different role of sugar in the case of red and white wines: while the more sugar, the more the price is for whites, the more sugar, the less the price is for reds. The effect is almost of the same extent, but to a different direction.

Estimations presented above showed that the variance of wine prices is due to differences of their chemical composition to a large extent (almost 60% in case of the best model).

It is up to further research to reveal the rest of the factors that influence wine prices.

References

[1] Rosen, S. J. Pol. Econ. **82**(1), 34–55 (1974)
[2] Oczkowski, E. Aus. J. of Agr. Econ. **38**(1), 93–110 (1994)
[3] Combris, P., Lange, C., & Issanchou, S. J. Wine Econ. **1**(01), 75–88 (2006)
[4] Galletto, L., Rossetto, L. Wine Econ. and Pol. **4**(01), 60–68 (2015)
[5] Landon, S., & Smith, C. E. J. of Cons. Pol. **20**(3), 289–323 (1997)
[6] San Martín, G. J., Troncoso, J. L., & Brümmer, B. J. of Wine Econ. **3**(01), 72–84 (2008)
[7] Schamel, G. J. of Wine Econ. **4**(01), 62–80 (2009)
[8] Schamel, G., & Anderson, K. Econ. Rec. **79**(246), 357–369 (2003)
[9] Lecocq, S., & Visser, M. J. of Wine Econ. **1**(02), 114–124 (2006)
[10] Combris, P., Lecocq, S., & Visser, M. Appl. Econ. **32**(8) 961–967 (2000)

Wine and culinary tourism: Preferences of experiential consumers

Guenter H. Schamel

Free University of Bozen-Bolzano, Faculty of Economics and Management, 39100 Bozen-Bolzano, Italy

Abstract. We explore the key components of consumer demand for the development of a successful wine and culinary tourism segment. In particular, we investigate the demand preferences that are important to consumers interested in a wine and culinary related hotel stay in South Tyrol. Conceptually, we utilize the 4E-Model of experiential consumption by Pine & Gilmore [1,2] and propose that the four realms of an experience (i.e., entertainment, education, escapist, and aesthetic relate to the principal components of consumer preferences. We survey potential tourists to gain a better understanding of their demand preferences for culinary and wine related hotel stays. Using an exploratory factor analysis, we identify the principal components of consumer demand preferences. The most preferred demand feature of a culinary wine experience relates to informative entertainment. The second feature relates to social-cultural activities that are educating. The third are escapist wine and food-specific activities and the forth relates to the aesthetics of accommodation traits and style in the culinary and wine domain. Moreover, we study the current supply attributes of wine and culinary related hotel offer in South Tyrol and their pricing using a hedonic model. Attributes that provide an entertainment experience form the basis of any offer in the wine and culinary domain. Attributes that provide an educational experience are highly relevant and add will a significant price premium. Matching demand preferences and supply conditions is crucial in developing a successful culinary and wine related tourism segment.

1. Introduction

In this study, we want to obtain a better understanding of the demand preferences that consumers are looking for in a wine and culinary related vacation, i.e. integrated winery visits, wine seminars, specific wine selections for dinner, etc. Specifically, we study these preferred elements of consumer demand in order to assess efforts to develop a wine and culinary tourism segment and related hotel offers in South Tyrol. In addition, we explore the nature and pricing of the current supply related to such offers. We relate the preferred elements of consumer demand to the key components of experiential consumption and to the current supply in the market in order to determine their implicit prices.

In the literature, we find some relevant studies on food (culinary) or wine tourism features from a demand perspective. Charters and Ali-Knight [3] identify specific wine tourist activities in an effort to define wine tourism. Getz and Brown [4] study the demand for long distance wine tourism. Schamel [5] analyses the importance and potential of wine tourism in South Tyrol. He concludes that in addition to oenophiles, relishing culinary tourists are an interesting target group for the further development of a wine and culinary tourism segment since they have only a limited knowledge about wine.

There are also a number of papers examining general aspects related to food in tourism. Mak, Lumbers, and Eves [6]] explore the motivations underlying food consumption in tourism. Quan and Wang [7] propose a structural model integrating peak and consumer experiences in tourism and illustrate it in terms of food consumption. The theory of experiential consumption by Pine and Gilmore [1,2] has inspired numerous studies related to tourism in general (for example, [8]) as well as to wine or food tourism (for example, [9]).

What we do not understand very well, however, are the preferences of tourists when it comes to a combined food (culinary) and wine related experience. Moreover, the literature seems to lack a look at the pricing of experiential consumption related to culinary and wine tourism. We aim to close this gap by providing an examination of the specific elements of wine and culinary related offers that are available in the market, how they relate to the theory of experiential consumption and how they are valued in the market. Consequently, we follow two objectives in this study. First, to examine the demand preferences that are important for tourists in identifying a suitable wine and culinary experience and to relate it to the theory of experiential consumption. Second, to study the attributes of current supply that offer a wine and culinary tourism experience and relate them to actual prices charged in the market, i.e., valuing key components of experiential consumption by estimating implicit prices.

Methodologically, we developed an online survey among potential tourists visiting a wine region and asked them about their preferred features related to a wine and culinary hotel stay. We employ the data obtained in the survey to test the theory of experiential consumption using a principal component analysis. In addition, we gathered a second data set from hotels offering a culinary and wine experience to tourists visiting South Tyrol. This data set characterizes the supply of culinary and wine tourism experiences. We can use it to relate the attributes of supply to the theory of experiential consumption [1,2] and to build a hedonic model in order to estimate implicit prices for them.

We have structured the remainder of the paper as follows. Section 2 provides a brief overview of the literature related to wine and culinary tourism and the theory of experiential consumption. In Sect. 3, we describe the data sets collected and the methodologies applied. In Sect. 4, we discuss the results of our empirical analyses and we conclude and derive some practical implications for our research in Sect. 5.

2. Literature

We find some, but relatively few relevant studies on culinary and wine tourism from a demand perspective. Charters and Ali-Knight [3] identify specific activities of "wine tourists" in an effort trying to define what constitutes them in general. Using an exploratory study, Getz and Brown [4] identify the core wine-product, the general destination appeal and cultural product as the critical features of a wine tourism experience from a consumer perspective. Sidali, Kastenholz and Bianchi [10] develop a theoretical framework for niche marketing food specialties through rural tourism that incorporates the experience economy model.

Quadri-Felitti and Fiore [11] review studies on the evolution of understanding the wine tourism experience and propose a model to advance a theory of experiential consumption (4E = Entertainment, Education, Escapist, Esthetic) and referenced to Pine and Gilmore [1,2]. Within this framework, Charters and Pettigrew [12] ask whether wine consumption is an aesthetic experience.

Conceptually, Oh, Fiore, and Jeoung [8] are among the first to embark on measuring experience economy constructs. Tsai and Wang [13] evaluate a food tourism experience and their value in branding. In this study, we identify and evaluate important demand features for hotel stays that combine a wine and culinary experience, applying the theory of experiential consumption. We also add to the growing literature relating tourism product prices and attributes (for example [14,15]), i.e., evaluating experience economy concepts in terms of observed market prices.

Consequently, this paper will extend the existing literature in two ways. First, it will apply the theory of experiential consumption to combined culinary and wine related tourist experiences, in an effort to identify preferred elements of consumer demand and using an exploratory factor analysis, we reveal the key components of tourism demand preferences in this domain. Second, it will relate the 4E-model of experiential consumption to the hedonic literature on tourism product pricing in an effort to estimate implicit prices for the realms of an experience. In doing so, we can critically assess the development of wine and culinary tourism segments and to create more successful marketing initiatives.

3. Data and methodology

This study combines primary data obtained through an online survey and secondary data compiling the nature of supply related to culinary and wine vacation offers available in South Tyrol. We use the survey to assess the theory of experiential consumption and to relate its key components to hedonic pricing models of tourism products. Together with the survey, the secondary data source identifies the pricing of the key components of experiential consumption related to culinary and wine tourism demand.

Table 1. Survey Sample Statistics.

Gender	Female	66%
	Male	34%
Family Status	Single	38%
	Relationship	50%
	Separated	6%
	No answer	7%
Country Origin	Germany	29%
	Italy	60%
	Austria/Switzerland	7%
	Other	4%
Education Level	<12 Years	11%
	12–13 Years	16%
	>13 Years	22%
	University	51%
Work Status	Self-employed	12%
	Full-Time	36%
	Part-Time	8%
	Students	29%
	Retired	8%
	Other	8%

Source: Own calculations from sample data.

3.1. Survey

We developed an online survey aimed at potential tourists interested in culinary and wine related holiday stays in South Tyrol. We compiled a convenience sample comprised of 182 respondents about their demand preferences when booking a wine and culinary related offer. We translated the survey into two languages for Italian and German speakers to reflect the bilingual character of the destination studied. The usable sample size with complete responses for all relevant items reduced to 134 observations. We list important sample statistics including origin, gender and education in Table 1 and note that the average age of the respondents was 40.7 years (range 16–74 years).

We elaborated 13 selected items to identify important demand features that wine and culinary tourists would like to have included in their vacation. The items consist of the following features: a winery visit, a guided wine tasting, a guided vineyard walk, a guided cellar visit, the availability of a trained hotel sommelier, access to a specific wine information events, other hotel guests with a wine interest, the option to attend a regional cooking class or a wine seminar, a hiking activity with a traditional food experience, a culinary activity (for example a gourmet wine dinner), a dedicated hotel wine bar and exclusive wine cellar, a daily restaurant menu with featured wines, and the accessibility of wine and culinary books and magazines. Respondents had to rank the importance of having these features offered during their stay in South Tyrol using a 7-point Likert scale.

Table 2 lists a ranking of the 13 items characterizing the selected demand features in declining order based

Table 2. Demand Features: Response Ranking.

Demand Feature (7-pt Likert scale)	**Average**	**Std. Dev.**
Guided Wine Tasting	5.78	1.49
Guided Cellar Visit	5.60	1.61
Gourmet Wine Dinner	5.48	1.52
Visit of a Winery	5.39	1.73
Daily Menu with Featured Wines	5.33	1.66
Hiking Activity with Food Experience	5.22	1.67
Trained Hotel Sommelier	4.99	1.71
Guided Vineyard Walk	4.64	1.59
Hotel Wine Bar & Exclusive Cellar	4.57	1.71
Regional Cooking Class with wine	4.45	1.98
Specific Wine Information Event	3.66	1.76
Wine/Culinary Books and Magazines	3.57	1.73
Hotel Guests with a Wine Interest	3.32	1.56

Source: Own calculations from sample data.

on average importance expressed by the respondents. A reliability analysis for these items reveals a Cronbach's Alpha equal to 0.835. We observe that a "Guided Wine Tasting" has the highest average response rating (and the lowest standard deviation) followed by a "Guided Cellar Visit" while other "Hotel Guests with a Wine Interest" receives the lowest average response rating. Note that all these extremes relate to wine tourism demand. "Gourmet Wine Dinners" and "Daily Menus with featured wines" receive the highest average response ratings of features combining both culinary and wine tourism demand.

The survey data is used to perform an exploratory factor or principle component analysis (PCA) in order to group the items (demand features) with the aim to relate the groupings to the theory of experiential consumption of Pine and Gilmore [1,2]. Hence, we aim to associate the preferences related to wine and culinary wine tourism demand with the realms of 4E-Model (i.e., entertainment, education, escapist, and aesthetic).

3.2. Nature of supply

We obtain our second data set from hotels offering a culinary and wine experience to tourists visiting South Tyrol. Recently, the regional tourism sector has started an initiative to promote wine and culinary related hotel offers and to develop a dedicated website using the domain vinumhotels.com. From this domain, we gather data on the featured attributes of culinary and wine tourism experiences on offer and about the prices that are charged. The initiative promotes a vision of supplying international comfort and local hospitality of a well- known wine region with an ancient cultural landscape making the tourism experience unique and exciting. While the domain name focusses on wine, it equally stresses food, culture and wellbeing to offer an authentic experience to the visitor. The domain includes 29 hotels in South Tyrol that emphasize a wine and culinary tourism experience when developing their product offers.

Hence, the data set derived from the website comprises 29 basic offers characterized as culinary and wine tourism with an average price of 88.64 €/person per night. The average hotel has 4-stars. Dinner at the hotel restaurant or half-board is included for about 2/3 of all offers. Using the descriptors found on the website of the initiative, we classify each of the 29 hotel offers according to whether it

Table 3. Rotated Component Matrix [a].

	4-Factor Solution			
Demand Features	Enter-tainment	Edu-cation	Escapist	Esthetic
Visit of a Winery	**.823**			
Guided Wine Tasting	**.775**			
Guided Vineyard Walk	**.754**			
Guided Cellar Visit	**.701**	.300		
Trained Hotel Sommelier	**.644**			.433
Specific Wine Information Event		**.775**		
Hotel Guests with Wine Interest		**.656**		.406
Regional Cooking Class / Wine Seminar		**.615**	.384	
Hiking Activity with Food Experience			**.823**	
Culinary Activity (Gourmet Wine Dinner)			**.806**	
Hotel Wine Bar and Exclusive Cellar				**.706**
Daily Restaurant Menu with Featured Wines	.309			**.673**
Wine/Culinary Books and Magazines	.344	.442		**.512**

Extraction Method: Principal Component Analysis. Rotation Method: Varimax with Kaiser Normalization. [a] Rotation converged in 12 iterations.

includes important features that wine and culinary tourists prefer, i.e., we relate them to the items presented in the survey so that we can group them according to the results of the PCA.

4. Empirical results

4.1. Principal component analysis

We use the items surveyed to perform a PCA on the preferred elements of culinary and wine tourism demand. The analysis reveals a 4-factor solution and Table 3 shows the result of the PCA with factor loadings greater than 0.300. It identifies four key factors of demand that tourists look for in a wine and culinary related hotel stay.

The factor or principle component analysis indicates logical groupings of demand features for wine and culinary related hotel stays. The total variance explained with the model is 65.2% of which the four individual factors explain between 13% and 24%. Factor names should reflect the main demand features from a tourist perspective and they should relate to the 4E-model of experiential consumption.

As suggested in Table 3, we can relate each of item groups to one of the 4E's proposed by the theory of experiential consumption (entertainment, education, escapist, and aesthetic). These relationships then confirm that the 4E-model of experiential consumption applies to wine and culinary tourism as suggested previously by other researchers addressing wine tourism (for example [11,16,17]).

The factor names given in Table 4 reflect the main demand features from a tourist perspective and relate to realms of the 4E-model of experiential consumption. The first factor groups features that refer to "Wine as Infotainment" and associates with *entertainment*. The second factor exemplifies "Social-Cultural Engagement"

Table 4. Factor Names reflecting the 4E Model.

Factor	Wine & Culinary Feature	4E-Realm
1	Wine as Infotainment	Entertainment
2	Social Cultural Engagement	Education
3	Wine and Food Activities	Escapist
4	Accommodation Traits & Style	Esthetic

Source: Own description from sample data.

Table 5. Hedonic Model: Dependent Variable Log (Price/Person/Night).

Variable	Coefficient	Std. Err	t-Stat.	Prob.	€[b]
Constant	2.314***	0.269	8.599	0	—
½ Stars	0.221***	0.033	6.656	0	21.92
Dinner	0.175***	0.057	3.090	0.005	16.95
Education	0.141**	0.059	2.418	0.024	13.42
Escapist	0.125*	0.070	1.793	0.086	11.80
Esthetic	0.138*	0.070	1.959	0.062	13.12
N = 29, Adj. R^2 = 0.719, F-Stat. = 15.31					
Significance Level: *** (1%), ** (5%), * (10%)					

Source: Own calculations. [b]with adjustments according to [18].

and relates to *education*. The third factors concerns "Wine and Food Activities" and suggests being *escapist*. Finally, the forth factor groups features that relate to "Accommodation Traits & Style" and is commonly associated with *aesthetic* dimensions of culinary and wine related demand.

4.2. Hedonic pricing model

As just discussed, the results of the PCA allow us to relate each item group to the 4E-model of experiential consumption (entertainment, education, escapist, and aesthetic). We use this result and the descriptors of the 29 basic culinary and wine related hotel offers found on vinumhotels.com to classify whether a realm of the 4E-model are included in the offer or not. Thus, we classify the attributes of hotel offers according to their experiential dimension based on the PCA and derive four dummy variables representing the 4E's, respectively.

As it turns out, all 29 basic hotel offers include at least one feature classified as "Entertainment" while only 52% of them include a feature related to "Education." Moreover, we can find "Escapist" and "Esthetic" features in 79% and 66% of all hotel offers, respectively. Next, we use the supply data retrieved from the website and the dummy variables classifying the attributes of each hotel offer to estimate a simple hedonic pricing model, which relates both tourism product and experiential attributes to market prices.

Table 5 shows the estimation results for the hedonic model with Log(Price/Person/Night) as the dependent variable. Independent variables included in the model are the hotel star rating, a dummy variable for dinner or half-board included, and dummy variables categorizing the features offered into one of the 4E-realms of experiential consumption. The last column calculates the relative price effect in Euros based on the average room price of 88.64 €/person per night. Reference category is a stay in an average 4-star hotel with a basic offer that includes at least one "Entertainment" feature.

As shown in many previous applications that relate hotel room prices and their attributes (for example [14,15]), the accommodation's star rating is the most significant characteristic also explaining prices for wine and culinary related hotel offers. Another ½ star would make a stay about 22€/person/night more expensive. Including dinner or half-board at the hotel's restaurant is another important attribute adding about 17€/person/night to the price level of a hotel offer in this domain.

In terms of the realms of the 4E-model, educational features are the most significant experiential attribute of a culinary and wine related hotel offer adding about 13.40€/person/night to its price. Experiential attributes related to the escapist and aesthetic realms of the 4E-model are also significant at least at the 10% level. According to our model, they contribute about 11.80€ or 13.10€ per person/night, respectively to the price level of a wine and culinary related hotel offer.

5. Conclusions

Our analysis indicates that wine and culinary tourists are experiential consumers. The empirical results obtained for demand preferences and for hotel offers in this domain for South Tyrol seem plausible and in line with previous research relating to wine tourism features in general and the 4E-model of experiential consumption in particular.

We relate the results of the PCA to current supply conditions for culinary and wine related hotel offers in South Tyrol and show that the tourism sector is developing offers with supply features that both reflect consumer preferences and market prices, respectively. The demand preferences identified in the survey are relevant in order to develop and market successful culinary and wine vacation offers.

Based on the results of our PCA, the 4E-model is a useful vehicle to classify the attributes of hotel offers according to their experiential dimension. Attributes of supply that provide an entertainment experience form the basis of any offer in the wine and culinary domain. Attributes of supply that provide an educational experience are highly relevant and add will a significant price premium. The demand features examined in this domain (e.g., specific wine information events, cooking lessons, wine seminars) require human capital investments, but they can be implemented relatively easily into an existing hotel infrastructure.

The wine and culinary tourism segment can help to foster regional development and the preservation of the culinary and cultural heritage in wine regions. Even though our study is exploratory, the results validate the relevance of experiential consumption related to wine and culinary vacation themes. Future research needs to explore a wider range of potential demand preferences and to exploit more representative sample to generalize the results beyond the local context to other wine regions.

References

[1] B.J. Pine II, J.H. Gilmore (1998): Welcome to the experience economy. Harv. Bus. Rev. **76**(4), 97–105
[2] B.J. Pine, II, J.H Gilmore (1999): The experience economy. Harvard Business Press. Boston, MA
[3] S. Charters, J. Ali-Knight (2002): Who is the wine tourist? Tour. Manage. **23**(3), 311–319

[4] D. Getz, G. Brown (2006). Critical success factors for wine tourism regions: a demand analysis. Tour. Manage. **27**(1), 146–158

[5] G. Schamel (2013): Bedeutung und Potential des Weintourismus in Südtirol. in K. Scherhag (Editor): Weintourismus und Marketing. Lohmar-Köln: Josef Eul Verlag: 139–150

[6] A.H. Mak, M. Lumbers, A. Eves (2012): Globalization and food consumption in tourism. Annals of Tourism Research **39**(1), 171–196

[7] S. Quan, N. Wang (2004): Towards a structural model of the tourist experience: An illustration from food experiences in tourism. Tour. Manage. **25**(3), 297–305

[8] H. Oh, A.M. Fiore, M. Jeoung (2007): Measuring experience economy concepts: Tourism applications. J. Travel Res. **46**(2), 119–132

[9] E. Cohen, L. Ben-Nun (2009): The important dimensions of wine tourism experience from potential visitors' perception. Tour. Hosp. Res. **9**(1), 20–31

[10] K.L. Sidali, E. Kastenholz, R. Bianchi (2015): Food tourism, niche markets and products in rural tourism: combining the intimacy model and theexperience economy as a rural development strategy. J. Sust. Tour. **23**(8/9), 1179–97

[11] D. Quadri-Felitti, A.M Fiore (2012): Experience economy constructs as a framework for understanding wine tourism. J. Vacat. Mark. **18**(1), 3–15

[12] S. Charters, S. Pettigrew (2005): Is wine consumption an aesthetic experience? J. Wine Res. **16**(2), 121–136

[13] C.T.S. Tsai, Y.C. Wang (2017): Experiential value in branding food tourism. J. Destin. Mark. Manage. **6**(1), 56–65

[14] C. Thrane (2005): Hedonic Price Models and Sun-and-Beach Package Tours: The Norwegian Case. J. Travel Res. **43**(3), 302–308

[15] G. Schamel (2012). Weekend vs. midweek stays: Modelling hotel room rates in a small market. Int. J. Hosp. Manage. **31**(4), 1113–1118

[16] B. Pikkemaat, M. Peters, P. Boksberger, M. Secco (2009): The staging of experiences in wine tourism. J. Hosp. Mark. Manage. **18**(2), 237–253

[17] J. Rüdiger, J.H. Hanf, E. Schweikert (2015): Die Erwartungshaltung von Weintouristen in Deutschland. Berichte über Landwirtschaft-Zeitschrift für Agrarpolitik und Landwirtschaft **93**(2)

[18] R. Halvorsen, R. Palmquist (1980): The interpretation of dummy variables in semilogarithmic equations. Am. Econ. Rev. **70**(3), 474–75

DOI: 10.1051/bioconf/20170903022

Contributions de l'œnotourisme pour la Vitiviniculture de l'État de São Paulo (Brésil) / Contributions of enotourism to the vitiviniculture to the State of São Paulo (Brazil)

Adriana Renata Verdi

APTA, Gabinete do Coordenador, CEP: 01037-912 - Praça Ramos de Azevedo, 254, São Paulo, Brasil

Abstract. The Pro-wine project, funded by the Foundation for Research Support of the State of São Paulo - FAPESP, characterized the progress of a group of small wine producers in the State of São Paulo from the insertion in the routes rural tourism and wine tourism. The main wine routes of the state are located in the metropolitan area of São Paulo and express traditional regions of wine production. This work aims to highlight the main contributions of wine tourism to the wine industry of the State of São Paulo, mainly: enhancement of the landscape, strengthening of governance, formalization of the smallholder and improvement of production quality. The work is based on the results of projects financed by FAPESP: "The Modalities of Territorial Governance in the State of São Paulo: coordination, territories, public policies and development" (2011/50837-9) and "Territorial Governance in Brazil: institutional specificities, spatial logics and development policies" (2015/25136-8). The growing demand of tourists increases the potential of the wine business with new investments in the productive structure. The need for expansion of activity vitivinicole and conquest of new markets goes through formalization through the organization of cooperatives and governance between agents.

1. Viticulture dans l'État de São Paulo : principales experiences de l'oenotourisme

Lorsque l'on considère le panorama de la vitiviniculture du Brésil, l'État de São Paulo a 8 mille ha de surface cultivée de raisins, la deuxième position parmi les états brésiliens. Dans la production de 2015, l'État de São Paulo a enregistré 143 mille tonnes, la troisième plus grande participation du total national, au-dessous des États de Rio Grande do Sul et de Pernambuco [1].

Selon l'enquête du Secrétariat de l'Agriculture de l'État de São Paulo, les régions de Campinas, Itapetininga, Jales et Sorocaba ont produit en 2015 plus de 122 mille tonnes, soit 93 % des raisins paulistas.

Les concentrations vitivinicoles traditionnelles de l'État de São Paulo sont situés au voisinage de zones métropolitaines, suivant les grandes lignes de l'expansion urbaine et industrielle (Fig. 1).

Dans ces zones d'alentours de la capitale, la culture de la vigne et la production de vin entrent en conflit avec d'autres usages de l'espace agricole et subissent les effets de la pression exercée par la spéculation immobilière, en particulier la hausse des prix des terrains et la pénurie de main-d'œuvre.

Ainsi, la viabilité socio-économique de l'industrie vitivinicole dans l'entours métropolitain est devenue de plus en plus tributaire du développement d'autres activités, généralement non agricoles, qui fournissent de nouvelles sources de revenus. Dans ce contexte, le tourisme rural et l'oenotourisme acquièrent un rôle clé dans la survie des activités viticoles dans les zones périurbaines et perimetropolitanas. De plus, ils sont aussi des stratégies de développement territorial.

L'oenotourisme implique un ensemble d'activités à la hausse dans le monde entier. En plus de générer de nouveaux emplois et une valeur ajoutée aux produits, le tourisme du vin exige l'amélioration des infrastructures et la promotion de l'approche des ressources culturelles et scéniques de l'espace local.

Depuis les années 90, l'amélioration de la proximité organisée entre les agents locaux des principales concentrations de vin de l'État de São Paulo a fourni un environnement institutionnel propice à la mise en place des projets collectifs de base locale pour le développement de l'oenotourisme et des territoires.

Les concentrations vitivinicoles traditionnelles de Jundiaí et São Roque, situés à une distance de 60 km de la région métropolitaine de São Paulo, sont les principaux cas de ce procédé (Fig. 2).

2. Méthodologie

Compte tenu des menaces fournies par les externalités négatives de la proximité géographique à d'autres utilisations des espaces agricoles, qui sont confrontés à la viticulture et à la vinification dans les régions vitivinícoles traditionnelles de l'État de São Paulo, il est d'observer les contributions de l'enotourisme à la durabilité de ces activités agro-industrielles et au développement territorial.

Selon Bourdon et Pichery, l'oenotourisme ou le tourisme du vin peut être définie comme une gamme complète de services qui nécessitent d'une organisation et

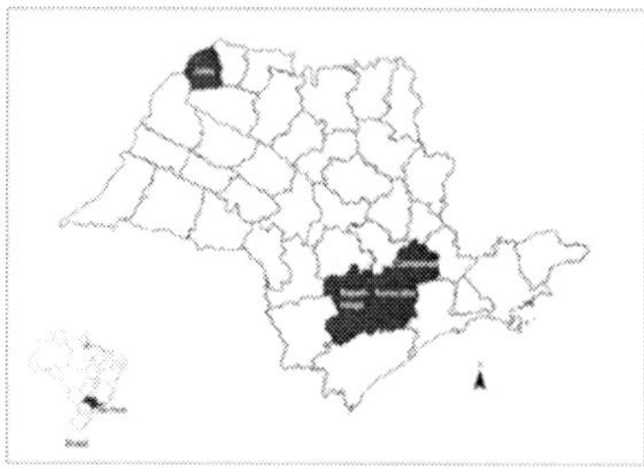

Figure 1. Les principales régions productrices de raisin dans l'État de São Paulo.

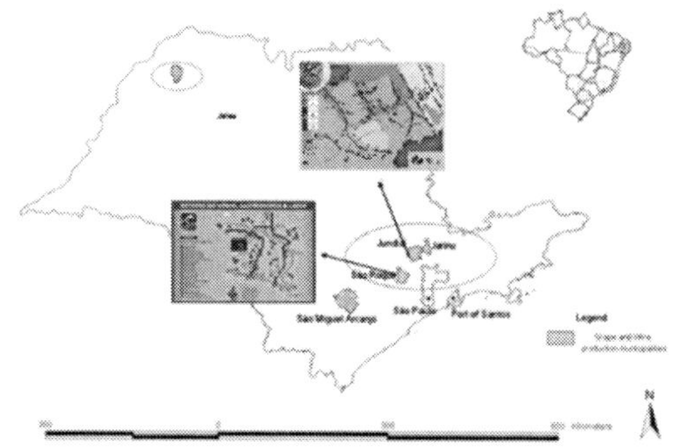

Figure 2. Les principales routes oenotouristiques dans l'État de São Paulo : Jundiai et São Roque.

d'une animation des activités pour le séjour des touristes dans les zones où se développe la culture de la vigne et la production de vins [2].

La littérature internationale sur le sujet a mis en garde contre le fait que les activités liées au tourisme viticole extrapolent le secteur du vin pour incorporer l'exploration d'autres attributs du territoire local. L'oenotourisme a la possibilité d'explorer des ressources territoriales spécifiques, y compris la beauté des paysages, la cuisine et l'architecture typiques.

De ce point de vue, par l'avis de plusieurs experts, Hall reconnaît l'expansion de la base du marché de l'oenotourisme. Selon l'auteur, lors de l'établissement des liens stratégiques plus explicites avec d'autres produits et activités touristiques, l'oenotourisme élargit la gamme de fonctionnement et intègre l'organisation de festivals, la promotion de restaurants, de l'hébergement et de l'appréciation de l'artisanat.

L'articulation stratégique entre les activités touristiques liées au secteur du vin et d'autres ressources spécifiques territoriales locales a été très bien fait par l'oenotourisme en développant des routes et des itinéraires qui intègrent le vin, le patrimoine, la gastronomie et la nature [3].

Selon Ruschmann (2009) « les routes du vin sont les outils qui permettent et organisent la diffusion de l'oenotourisme, la promotion du potentiel des régions où elles se trouvent » [4].

Dans un contexte conceptuel plus large, le concept de l'oenotourisme est lié à la multifonctionnalité de l'agriculture. Sur la base de la réinterprétation du rôle de l'agriculture pour le développement rural et l'évolution du rôle des chefs d'exploitation, chaque fois plus integrés dans un groupe plus large des gestionnaires de paysage, dans un espace qui est de plus en plus espace de la consommation et moins espace de production [5].

Le document analyse les principales routes du vin dans l'État de São Paulo à partir de cette conception large de l'oenotourisme afin d'étudier le rôle de ce pour le développement sectoriel et territorial.

Les stratégies oenotouristiques, prises par la gouvernance des acteurs locaux, sont en mesure de révéler et d'activer des ressources spécifiques qui résultent d'une longue histoire, d'une accumulation de mémoire et d'un apprentissage cognitif collectif. Les caractéristiques spécifiques de l'activation dynamique favorise le développement territorial [6].

Par conséquent, la base de données empiriques est donnée par le cadastre vitivinicole de l'État de São Paulo, le résultat du projet FAPESP « revitalisation de la chaîne vitivinícola pauliste : la compétitivité, la gouvernance et le développement durable », en plus de la surveillance du secteur en participant à d'autres projets FAPESP : Les modalités de gouvernance territoriale dans l'État de São Paulo: coordination, territoires, politiques publiques et développement » (11/50837-9) et « La gouvernance territoriale au Brésil : spécificités institutionnelles, logiques spatiales et politiques de développement » (15/25136-8), ainsi que la participation aux réunions de la chambre sectorielle de la vigne et du vin de l'État de São Paulo, les activités qui assurent un contact permanent avec les leaders du secteur et les gouvernements municipaux.

3. L'oenotourisme dans l'État de São Paulo : les cas de Jundiaí et São Roque

Les principales expériences de l'oenotourisme dans l'État de São Paulo ont émergé dans les régions vitivinicoles traditionnelles, dans l'entourage métropolitain de la capitale. Tant la municipalité de São Roque comme Jundiaí sont à 60 km de la métropole de São Paulo, les principaux axes de l'expansion urbaine et industrielle du pays : la Municipalité de São Roque vers la région métropolitaine de Sorocaba et la Municipalité de Jundiaí vers la région métropolitaine de Campinas. Les deux municipalités sont des pionniers dans la production vitivinicole nationale, ces espaces développent l'activité depuis le début de l'occupation du Brésil.

Malgré les similitudes, le développement de l'oenotourisme à Jundiaí et à São Roque présente des différences dans la motivation, la structure et la gouvernance.

3.1. Le cas de Jundiaí

En Jundiaí, le développement de l'œnotourisme a été plus spontané, parce qu'il a été conjuguée à la survie de la petite production de vin informel depuis le début du processus.

Bien que les premiers registres de la commercialisation du vin à partir de raisins cultivés dans la municipalité de Jundiaí datent de 1669, la viticulture à São Paulo n'acquérir une importance économique qu'à partir de l'interaction entre deux facteurs qui formaient le binôme favorable: l'immigrant italien et le cépage Isabel [7].

Les immigrants italiens qui se sont installés dans la région Jundiaí ont été embauchés comme métayers sur les plantations de café, où ils ont développé des plantations de nourriture pour leur subsistance et vignobles pour la production de vins pour leur propre consommation familiale.

Les gelées de la première moitié du XXe siècle, ainsi que l'importation de la variété Isabel et la crise du café en 1929 ont été les facteurs principaux pour le développement de l'exploitation commerciale de la viticulture à Jundiaí. Dirceu Lino de Mattos, dans leur travail Vignoble et Vignerons de São Roque et Jundiaí, décrit l'environnement dans lequel est passé la transition des cultures: « de la viticulture domestique, de l'arrière-cour, va progressivement à la viticulture sur une petite échelle (...). Le colon, après quelques années de travail dans les fermes, pourrait accumuler des économies (...). Progressivement, l'immigrant libérera son état de colon à devenir propriétaire du terrain et la vigne imposera dans le paysage et dans la vie rurale. L'agriculteur et le vignoble apparaissent comme un seul phénomène résultant de la désintégration de la grande propriété caféière » [8].

Au début des années 70 il y a un changement d'orientation de la viticulture de l'État. La culture des raisins pour la production de vin perd l'espace pour la culture des raisins de table. Les vinicoles de la région Jundiaí utilisent maintenant les raisins du sud, en raison de la baisse des prix. D'autre part, la production de raisin de table a fourni un meilleur rendement financier pour les viticulteurs. Ainsi, la combinaison de ces facteurs a favorisé le remplacement des raisins de vin par les raisins de table, en particulier le cultivar « Niagara ».

La production de vin avec de raisin local a été réduit. En fait, il y avait une réévaluation de la production artisanale et des petites caves informelles, qui sont orientés à la consommation des producteurs eux-mêmes, de la voisinage, du quartier et des districts ruraux de Jundiaí.

En 2008, le projet Pro-vin, financé par la FAPESP a identifié trois principaux embouteilleurs de vin et 95 producteurs artisanaux dans la ville de Jundiaí.

La proximité de la métropole de São Paulo et de Campinas, combinée aux investissements réalisés par les vignerons à la réception des visiteurs, ont promu la vente de vins artisanaux dans les petits établissements vinicoles.

Ce processus s'est développé à partir des années 2000 et a stimulé la gouvernance d'un groupe de petits vignerons dans le district agricole du Caxambu en faveur de l'organisation d'une feuille de route dirigé vers l'œnotourisme.

La proximité organisationnelle de ce groupe de vignerons a commencé en 2003 avec la création de l'Association des Producteurs de Vin Artisanale dans le district du Caxambu - AVA. Les petits agriculteurs, membres de l'Association, ont organisé une feuille de route pour orienter la visite des touristes (Fig. 3).

Tout d'abord, la route était assez rudimentaire et a dû indiquer l'emplacement des membres de AVA.

Ainsi, l'iteneraire a été développé par les petits agriculteurs qui dépendent de l'oenotourisme pour la commercialisation de leurs vins artisanaux.

Cette structuration de l'oenotourisme est une étape importante pour la contribution financière et organisationnelle du groupe, facteurs clés pour la mise en place de la coopérative AVA et l'expansion du marché.

La demande croissante de visiteurs a fourni une plus grande commercialisation du vin et, par conséquent, une

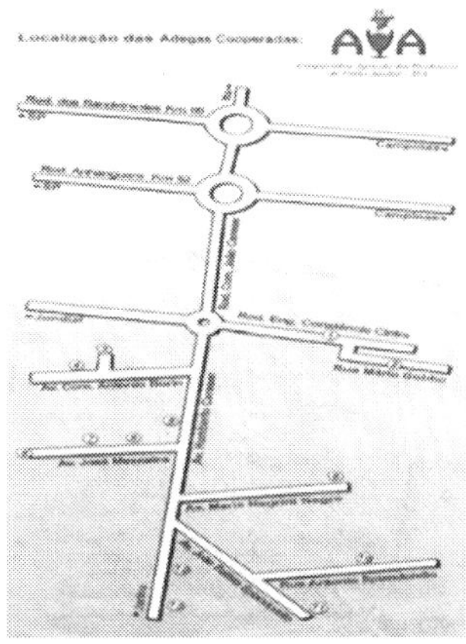

Figure 3. La route des vignobles dans l'AVA, Jundiaí - État de São Paulo. Source : AVA [9].

augmentation du potentiel commercial de l'activité avec de nouveaux investissements dans la structure productive. Compte tenu des conditions de production informelle et de petite production familiale, la nécessité d'une expansion du marché a conduit à la formalisation de l'activité par l'organisation coopérative. En outre, le statut informel des établissements vinicoles est devenu l'objet d'une pression de l'ordre juridique. Entre 2003 et 2004, plusieurs caves ont été signalés au Ministère de l'Agriculture, de l'élevage et de l'approvisionnement - MAPA, ont reçu la surveillance et ont été contraints de se conformer à certaines règles par les autorités juridiques [10].

Pour répondre à l'instruction MAPA n° 34, publiée le 4 Octobre 2006 - portant modalités d'enregistrement du d'établissement producteur de vins et des produits dérivés, organisés en coopérative légalement constitué et l'enregistrement de leurs produits - vingt et un producteurs de l'Association AVA constitué en 2007, la Coopérative AVA. En 2008, la coopérative a remporté l' enregistrement auprès l'Organisation des coopératives de l'État de São Paulo - OCESP. Actuellement, l'AVA compte 16 membres producteurs.

Ainsi émerge une nouvelle catégorie de petits producteurs de vin à fort potentiel entrepreneurial, le plus souvent liés à des stratégies de l'oenotourisme et à la gastronomie locail, héritées de la colonisation européenne.

Les principaux résultats de l'oenotourisme pour les producteurs inscrits à cette feuille de route : le rénforcement du coopérativisme pour formaliser l'activité; les investissements dans les propriétés et les établissements vinicoles pour améliorer l'accueil des visiteurs et le développement de nouvelles activités, en général à attirer les touristes.

Après plus d'une décennie, le feuille de route du vin à Jundiaí est devenu plus complexe en montrant les articulations des activités du vin avec d'autres activités touristiques locales, principalement des restaurants, des bars, des étals de fruits, les sites de tourisme rural, brasserie, cachaçaria, musée du vin, en plus l'exploration des spécificités locales des paysages bucoliques des districts ruraux avec leurs églises au centre.

Ainsi, la configuration actuelle de la route exprime le concept général de l'oenotourisme distingué par Hall,

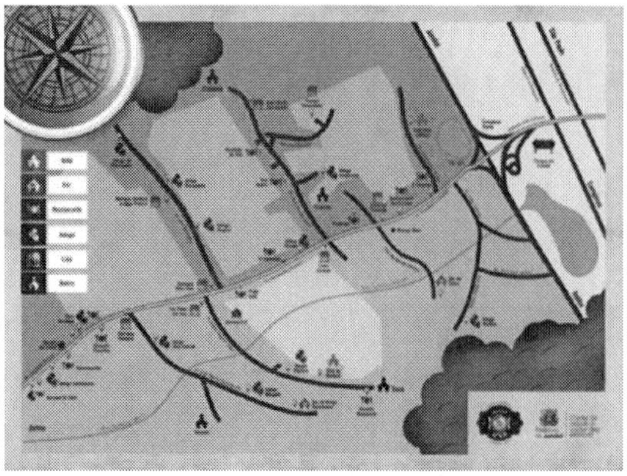

Figure 4. La configuration actuelle de la route de la vigne et du vin à Jundiaí. Source : `http://rotadauva.com.br/`.

Figure 5. Les affiches des fêtes traditionnelles de la vitiviniculture à Jundiaí.

par la réalisation des liens stratégiques développés par le tourisme du vin avec d'autres activités touristiques régionales (Fig. 4).

Dans cette perspective, la route de l'oenotourisme à Jundiaí intègre le vin, le patrimoine et la gastronomie.

L'oenotourisme à Jundiaí a stimulé les événements liés à la culture de la vigne et la production de vin artisanal, le patrimoine immatériel régional.

Les affiches des des fêtes démontrent des liens avec la religiosité et indiquent la tradition des activités vitivinicoles dans la municipalité liée aux événements de l'église (Fig. 5).

Cette tradition est un ressource spécifique bien activé par l'oenotourisme, un fait prouvé par l'affiche du festival du vin artisanal.

Malgré la célébration religieuse célébrer 105 ans et toujours compter sur la présence du vin artisanal local, l'inscription du produit dans l'affiche est plus recente. Ceci est le resultat de la mise en place des stratégie de commercialisation liés au développement de l'oenotourisme local.

3.2. Le cas de São Roque

La municipalité de São Roque, situé à 60 km de la métropole de São Paulo, a développé la vitiviniculture depuis sa fondation en 1657, grâce à l'influence de l'occupation européenne, en particulier la portugaise et l'italienne.

L'occupation européenne a été cruciale pour le développement de la viticulture dans la commune, étant donné que cette population a tenu la tradition de la consommation de vin, en plus de la connaissance sur la culture de la vigne et la production de la boisson.

L'installation de Vermouth Gancia, d'origine italienne, fondée en 1850 à São Roque (SP), a promu la professionnalisation de l'activité vitivinicole dans la municipalité, surtout par l'immigrations des experts. Avec la présence de Gancia, qui a été transformée en S.A. Vinicole Sanroquense - S.A.V.A.S. et de la Cinzano, la municipalité a acquis un laboratoire du Ministère de l'Agriculture - MAPA.

Dans l'âge d'or de la production de vin à partir de raisins cultivées dans la municipalité, il a été développée le cultivar Seibel II. Les producteurs de vin ont organisé le Syndicat de l'Industrie du Vin à São Roque - SINDUSVINHO en 1936.

Les investissements dans la recherche et l'embauche d'experts associés à la mise en œuvre de nouvelles techniques de production raisin et du vin ont été les principaux facteurs de la formation d'agglomération 150 établissements vinicoles. Le pic d'activité a eu lieu dans les années 1960 avec la conquête du titre « Terre du vin pauliste ».

Après la fin des années 70, la vitiviniculture de São Roque commence un processus de déclin, principalement en raison de la spéculation immobilière et les coûts de main-d'œuvre en hausse. Ces facteurs justifient la réduction significative du nombre d'établissements vinicoles locaux au 14 courant.

La vitiviniculture de São Roque prend une nouvelle phase de développement dans les années 90 en raison d'une combinaison de facteurs : la conquête de la dénomination « Station Touristique » ; la reprise du festival du vin par la ville afin de promouvoir les principaux produits agricoles locaux, y compris les artichauts et la reprise des SINDUSVINHO en 2001 avec l'élection d'un conseil d'administration, formé par de nouvelles générations des familles des vignerons pionniers.

La phase de décroissance de l'industrie du vin a commencé à être inversée avec la plus grande proximité organisationnelle entre les principaux acteurs locaux, en particulier embouteilleurs de vin, organisés par le sundicat SINDUSVINHO et l'Administration Municipale, ce qui a fourni un environnement institutionnel propice à la mise en forme des projets collectifs visant au développement de l'oenotourisme.

Les principaux projets collective développés par la gouvernance entre les vignerons et l'administration municipale appréciaient les ressources territoriales spécifiques, en particulier le sauvetage de l'organisation du festival traditionnel du vin et la structuration de la feuille de route des « Vins, Gastronomie et Loisirs ».

Le festival du vin a été repris par une nouvelle orientation stratégique définie par la nouvelle gouvernance. Tout d'abord le festival a été réédité sous le titre « EXPO São Roque : le vin et les artichauts » afin de se concentrer sur un public de haut pouvoir d'achat en offrant des produits à plus forte valeur ajoutée. Plus tard, la stratégie était de changer le profil de la production et de la commercialisation des vins de table à vins fins, associés à une cuisine différente et de qualité, généralement basée sur des éléments culturels portugais et italiens.

En plus du festival, les caves « Goes » et « Quinta do Olivardo »offrent la traditionnelle « pisa-da-uva »,

Figure 6. - La route « Vin, Gastronomie et Loisirs » à São Roque, État de São Paulo. Source : `http://rotadauva.com.br/`.

l'évennement où les visiteurs peuvent apprendre l'histoire de la vitiviniculture locale et vivre l'expérience de la récolte. Cet événement a été une stratégie importante pour attirer les visiteurs.

Actuellement, la ville a 14 embouteilleurs de vins qui sont produits principalement dans la région sud du pays.

En général, ces embouteilleurs de vin appartiennent à plus d'une classe d'entité, comme l'Association Nationale des Embouteilleurs de Vin- ANEV, le syndicat des producteurs de vin affilié à la Fédération des Industries de São Paulo - FIESP et à l'Association Brésilienne des Boissons - ABRABE.

À São Roque on observe que l'oenotourisme, basé sur le patrimoine viticole de la commune, s'est adapté à l'activité de remplissage du vin fait dans l'Etat de Rio Grande do Sul.

Le nom de la feuille de route «Vin, Gastronomie et Loisirs» indique la portée des liens stratégiques des établissements vinicoles avec d'autres activités touristiques du territoire local, et l'activation des ressources spécifiques de la région (Fig. 6).

À l'heure actuelle, la feuille de route du vin de São Roque comprend 36 membres. La configuration actuelle comprend 11 établissements vinicoles; 12 restaurants, dont quatre font partie de la structure interne des quatre caves; une station pour la pêche; un resort pour les chiens; un centre équestre, un parc animalier, une distillerie, une cachaçaria, deux auberges, une vente d'artichaut et un magasin d'art et des objets de décoration.

On observe que l'industrie du vin, en particulier les embouteilleurs de vin de la municipalité ont diversifié l'offre de services aux touristes car il ont investi dans 4 restaurants et ont construit une nouvelle cave adaptée pour répondre aux nouveaux flux de visiteurs.

Selon les données recueillies par le département du tourisme dans les trois Offices placés dans la municipalité de São Roque, la route du vin était la destination de 36% des voitures qui ont visité la ville de Janvier à Octobre 2007. De plus, le festival « Expo São Roque - Artichauts et vin » a enregistré la présence de 54.000 visiteurs pendant trois week-ends de 2006, restant pratiquement le même contingent pendante cet année, quand 55.000 personnes ont été enregistrées.

En ce qui concerne le flux actuel de visiteurs dans la route, en moyenne sont enregistrées 3000 véhicules - environ 9000 personnes par week-end. Dans l'ensemble, on peut dire que la route reçoit 450–500 mille visiteurs par an.

Les principaux résultats de ce processus: la reprise de l'organisation du festival traditionnel du vin en 2005 ; la récupération des plats typiques de la cuisine portugaise et italienne ; l'amélioration de la route des vins, du système de signalisation responsable de l'orientation des touristes; le développement et la promotion des éléments caractéristiques de la vitiviniculture local exprimés par le paysage ; l'investissement dans la beauté et dans l'architecture pittoresque des caves en sous-traitant professionnels de renom.

4. Considérations finales

Dans l'État de São Paulo les principales expériences de l'oenotourisme ont été mises en place dans les concentrations vitivinicoles traditionnelles en organisant des structures de gouvernance locales, soutenues par des différentes bases de production et commercialisation.

Alors que l'oenotourisme à São Roque a mis au point à partir d'un contexte caractérisé par les embouteilleurs de vin qui détient structure de commercialisation axée sur le marché régional et même national, dans le cas de Jundiaí, le contexte de l'oenotourisme se caractérise par la performance des petits producteurs artisanaux qui dépendent de la vente directe sur la propriété pour la survie de l'activité.

Dans cette perspective, le développement de l'œnotourisme a été plus spontanée à Jundiaí parce qu'il était conjugué à la faible production de vin depuis le début du processus. Alors qu'à São Roque, le tourisme du vin, au cours des dernières années, il s'est adapté aux activités de remplissage au début dirigé par quelques établissements vinicoles. Deux réalités différentes en termes de production et de la structure du marché.

La gouvernance constitue une condition sine qua non pour le développement de l'oenotourisme dans l'État de São Paulo. En Jundiaí, la plus grande proximité organisationnelle entre les petits producteurs de l'association AVA, et à São Roque, le renforcement du SINDUSVINHO et l'articulation de ce avec la municipalité ont été des étapes fondamentales pour la structuration la route des vignes et des vins.

Dans le cas de Jundiaí, l'une des contributions les plus importantes de l'oenotourisme est la transformation de l'AVA qui a passé d'une association à une coopérative avec la possibilié de formalization des leurs petits producteurs membres. Cette transformation a permis aux membres l'accès aux politiques publiques, comme ce fut le cas de la conquête de remplissage mobile avec des fonds provenant du Programme Microbassins du Secrétariat de l'Agriculture et de l'Approvisionnement de l'État de São Paulo.

Les actions de l'oenotourisme vont au-delà des activités liés à la vigne et au vin et se tournent vers d'autres ressources du territoire. Les liens stratégiques que l'oenotourisme devient établi avec d'autres produits locaux et d'autres activités ont la possibilité d'activer les ressources spécifiques à fort potentiel de développement territorial.

Dans les deux cas, il a été observé que les actions oenotouristiques ont valorisé les fêtes du raisin et du vin, la célébration de la récolte, les événements culturels, les traditions religieuses et la gastronomie italienne et portugaise. L'oenotourisme se nourrit de ces ressources qui sont devenus une partie de l'ensemble plus large des attractions régionales. Ainsi, la route de la vigne et du vin de Jundiaí couvre l'étals de fruits traditionnels de la région, l'établissement de production de cachaça, les sept restaurants basés sur la cuisine typiquement italienne, et trois d'entre eux font partie de la diversification des attractions des vinicoles aux visiteurs. La route du vin de São Roque est constituée d'un établissement de pêche, un parc animalier, une distillerie et vente d'artichaut, produit typique de l'agriculture locale, ainsi que des restaurants proposant des plats typiques de la cuisine italienne et portugaise. Lorsque l'on considère que les vignerons de São Roque sont plus capitalisé, il y a un plus grand investissement dans la beauté du paysage et de l'architecture des établissements vinicoles en embauchant des professionnels de renom.

Tant dans la route de Jundiaí comme de São Roque, l'oenotourisme a conduit les investissements dans les établissements vinicoles, en particulier dans la zone réservée aux dégustations et ventes de produits, en plus les préoccupations soulevées par l'exposition des vignes et la mise en valeur des traditions des récoltes.

En reconnaissant et en activant des ressources spécifiques dormants et latentes, ces stratégies confirment aussi le rôle de l'oenotourisme comme vecteur de développement territorial.

Références

[1] Instituto Brasileiro de Geografia e Estatística – IBGE. Levantamento Sistemático da Produção Agrícola – LSPA. Pesquisa Mensal de Previsão e Acompanhamento de Safras (2017)

[2] F. Bourdon e M.-C. Pichery, Le territoire viticole en France : de la destruction à la valorisation, LEG, Laboratoire d'Economie et de Gestion, CNRS, Université de Bourgogne (2010)

[3] B. Cambourne, N. Macionis, C.M. Hall, L. Sharples. The Furure of Wine Tourism in: C.M. Hall, L. Sharples, B. Cambourne, N. Macionis. *Wine Tourism Around The World: development, management and markets*, Elsevier (2000)

[4] C.V. Bordini, D. Ruschmann. O Enoturismo. In: A. Panosso Netto (Org) *Segmentação do Mercado-Turístico*. São Paulo: Manole, p. 241-253 (2005)

[5] M.N. Otani, A.R. Verdi, C.E. Fredo, M.L. Maia, M.C.M. Souza. Processo de Consolidação da Vinicultura Artesanal: um estudo de caso no entorno metropolitano de São Paulo e Campinas. Rev. Informações Econômicas, SP **43**(4) (2013)

[6] B. Pecqueur. O Desenvolvimento Territorial: uma nova abordagem dos processos de desenvolvimento para as economias do sul. Rev. Raízes, Raízes, Campina Grande **V24**(01 e 02), 10–22 (2005)

[7] J.R. Inglez de Sousa (Coord.). Uvas para o Brasil. 2. ed. Piracicaba: FEALQ, p. 791 (1996)

[8] R. Viveiros. *O Principado dos paisanos*. 3. ed. Vinhedo: Prefeitura Municipal de Vinhedo (2005)

[9] Associação De Produtores De Vinho Artesanal Do Bairro De Caxambu E Região - AVA. Localização das Adegas Associadas. Folheto Informativo. (s. d.). 2p. (2007)

[10] M.C.M. Souza, M.N. Otani, A. Verdi, A valorização da cultura italiana e o consumo de vinho artesanal. Rev. Informações Econômicas, SP **40**(6) (2010)

 DOI: 10.1051/bioconf/20170903023

The use of wine tourism as a possibility of the marketing with wine cooperatives / Die Umsetzung von Weintourismus als Vermarktungsinstrument bei Winzergenossenschaften

Jens Rüdiger und Jon H. Hanf

Hochschule Geisenheim University, 65366 Geisenheim, Germany

Abstract. Aufgrund des zunehmenden nationalen und internationalen Wettbewerbs, muss sich die deutsche Weinwirtschaft neue Diversifikationsmöglichkeiten erarbeiten. Eine dieser Möglichkeiten ist der Weintourismus. Hier haben auch deutsche Winzergenossenschaften die Möglichkeit, dieses Potential für sich zu nutzen. In dieser Untersuchung wird das in der Literatur noch neue Phänomen beleuchtet, inwieweit sich Winzergenossenschaften im Bereich des Weintourismus engagieren und welche Ziele sie damit verfolgen. Zudem wird eruiert, inwieweit sich die Mitglieder einer Winzergenossenschaft als Multiplikatoren bei der Durchführung von weintouristischen Aktivitäten einbringen. Des Weiteren wird die definitorische Begrifflichkeit des Weintourismus genauer analysiert. In der Arbeit zeigte sich, dass Winzergenossenschaften ebenso wie Weingüter weintouristische Aktivitäten in ihre Unternehmensstruktur integriert haben. Dies erfolgt im Gegensatz zu den meisten Weingütern nicht mit dem Ziel, eine eigene Wertschöpfung zu generieren, sondern dient eher als reines Marketinginstrument. Hierbei wird unterschieden zwischen Genossenschaften mit vermehrtem indirektem Vertrieb, die diese Kommunikationsmaßnahme als Direktkommunikation für die Erschließung neuer Zielgruppen sehen, bzw. Genossenschaften mit vermehrtem direktem Vertrieb, welche die Verkaufsförderung verfolgen. Bei der Umsetzung der jeweiligen weintouristischen Aktivitäten zeigte sich, dass noch Potential für Innovationen besteht. Auch die Möglichkeiten der Einbindung der Mitglieder als Multiplikatoren bei der Durchführung, Kommunikation oder auch Anbindung an die Genossenschaft durch weintouristische Aktivitäten wird von vielen Geschäftsführern unterschätzt. Hinsichtlich der Definition lässt sich festhalten, dass Weintourismus immer - je nach Blickwinkel - die Sichtweise des Anbieters oder Nachfragers annimmt.

Due to increasing national and international competition the German wine industry has to develop new alternatives. One of these alternatives is wine tourism. To exploit this potential market could also be an opportunity for German wine cooperatives. The study at hand illuminates the new literary phenomenon of wine tourism in regards to how deeply wine cooperatives are involved in this topic and which goals they are perusing. This research paper also investigates to what extend the members of wine cooperatives involve themselves as multipliers in the operation of wine tourism activities. Furthermore, the definatory concept of wine tourism will be analyzed more closely. It was discovered that wine cooperatives incorporate wine tourism events into their corporate structure, just as wineries do. In contrast to wineries, wine cooperatives do not do so to generate value added, but simply use it as a marketing tool. It should be differentiated between cooperatives with increasingly high indirect sales and cooperatives that are pursuing sales promotion. They later think of this communication tool as direct communication in order to exploit new target groups. When looking at the implementation of wine tourism activities, it was discovered that there is still potential for more innovative ideas. The managing directors also seem to underestimate the possibility of including cooperative members during the implementation and communication. Additionally, the bonding possibilities within the cooperative through wine tourism activities are often overlooked. Regarding the definition it is to retain that, wine tourism always- depending on the perspective- adopts the viewpoint of either provider or consumer. The definition concepts have in common that regardless of benefit, consumption or direct marketing, it is the wine related experience and a change in location that have to be main focal point.

1. Einleitung

Weintourismus ist erst Mitte der 1990er Jahre in den Blickwinkel der Forschung gerückt [1], auch wenn sich die ersten schwerpunktmäßigen Untersuchungen hauptsächlich auf die "Neue Welt" beziehen [2]. In unserer mitteleuropäischen Gegenwart haben mittlerweile viele Branchenverbände, Organisationen und Institutionen das Thema zum Gegenstand wissenschaftlicher Tagungen gemacht [3]. Auch die Bundesregierung sieht - wie sie in ihrer Antwort auf eine kleine Anfrage über die Potenziale der Winzerbetriebe durch Tourismus deutlich macht - im Weintourismus ein interessantes Marktsegment für das Reiseland Deutschland, der Winzern Diversifizierungsmöglichkeiten und Chancen bietet, um damit ihre Wertschöpfung sowohl im Weinbau als auch im Tourismus zu steigern [4]. So können Weinbaubetriebe durch Tourismusaktivitäten ein zusätzliches Einkommen generieren und ein weiteres ökonomisches Standbein realisieren [5], denn der Verkauf weintouristischer Leistungen steht in engem Zusammenhang mit dem Direktabsatz des Produktes Wein [6]. Generell ist es dabei am Produzenten zu entscheiden, welche möglichen Vertriebsarten er mit seinen weintouristischen Aktivitäten kombiniert: Zum einen kann dies in Form eines Direktvertriebs geschehen, in dem er den Wein unmittelbar an gewerbliche oder private Kunden vertreibt und durch die Einbeziehung der Komponente Weintourismus eine zusätzliche Möglichkeit der Wertschöpfung nutzt [7]. Zum Zweiten besteht die Option, indirekt als Zulieferer von Handel und Gastronomie über den in der Destination durchgeführten Weintourismus zu partizipieren [3,8].

Insgesamt gesehen hat sich der Weintourismus in den letzten Jahren in Deutschland zu einem bedeutenden wirtschaftlichen Faktor in der Weinbranche entwickelt. Im Weinanbaugebiet Franken beispielsweise wird dem weintouristischen Dienstleistungsumfeld eine Verzehnfachung des eigentlichen Weinumsatzes zugeschrieben [9]. Bezogen auf die Akteure der deutschen Weinwirtschaft spielen dabei die Winzergenossenschaften eine bedeutende Rolle, denn mit ihren Mitgliedern verantworten sie ein Drittel der gesamten deutschen Weinproduktion (10). Der Deutsche Raiffeisenverband (DRV) zählte im Weinwirtschaftsjahr 2013/2014 169 Winzergenossenschaften, davon 96 mit eigener Kellerwirtschaft, die beiden Weinbaugebiete Baden und Württemberg stellen dabei fast 65 Prozent der von deutschen Winzergenossenschaften verarbeiteten Anbaufläche [11]. Allerdings sank sowohl die Zahl der deutschen Winzergenossenschaften ebenso wie der Umsatz im Ganzen in den letzten 20 Jahren. Die Winzergenossenschaften verloren fast 10.000 Hektar Anbaufläche und rund 20 Prozent Marktanteil [12]. In diesem Spannungsfeld stellt sich die Frage, inwieweit Winzergenossenschaften vom Trend des Weintourismus, den damit verbundenen Direktabsatzmöglichkeiten und den Möglichkeiten der Diversifizierung partizipieren. Denn laut KAGERMEIER und HARMS [13] profitieren gerade Winzergenossenschaften auch von den zusätzlichen Vertriebsmöglichkeiten des Weintourismus, da sie über einen hohen Grad an regionaler Bekanntheit und Vernetzung verfügen sowie durch ihre Größe und Anzahl der involvierten Winzer großes Potenzial bei der Durchführung aber auch hinsichtlich der Kommunikation von Weinveranstaltungen haben.

Ziel des vorliegenden Beitrags ist es, der Frage nachzugehen, inwieweit sich Winzergenossenschaften im Bereich des Weintourismus engagieren und welche Ziele sie damit verfolgen, wie sie den in der Literatur beschriebenen Direktabsatz fördern und ob dies ein weitergehender Wertschöpfungsbereich für die Winzergenossenschaften darstellt. Zudem soll der Frage nachgegangen werden, inwieweit sich die Mitglieder einer Winzergenossenschaft als Komponente bei der Durchführung von weintouristischen Aktivitäten einbringen.

2. Methodisches Vorgehen

Für eine erste Einschätzung über den Stellenwert von Weintourismus für Winzergenossenschaften wurde die bestehende Fachliteratur im Bereich Tourismus und Weintourismus gesichtet. Die bestehende Literatur setzt sich jedoch mit dem Forschungsbereich Weintourismus und Winzergenossenschaften nur wenig auseinander. Vor allem aber zeigte der Vergleich, dass in der Literatur sehr unterschiedliche und weitläufige Definitionen von "Weintourismus" vorherrschen. Diese dort bestehenden Definitionen werden deshalb zunächst in Kapitel 3 gegenübergestellt und die Kernaussagen der definitorischen Ansätze zusammengefasst.

Nach der begrifflichen Definition wurden im zweiten Schritt die Geschäftsführer der Winzergenossenschaften der beiden Weinanbaugebiete Baden und Württemberg anhand eines standardisierten Online-Fragebogens befragt. Hierfür wurden die Teilnehmer telefonisch kontaktiert und der Link zum Fragebogen elektronisch versandt. Die Wahl des Untersuchungsgebietes ergab sich aus der Größe der verarbeiteten Fläche und der räumlichen und kulturellen Nähe. Dies läßt einen direkten Vergleich der beiden Gebiete zu. Bei der Befragung nahmen 37 der 45 kontaktierten Betriebe teil.

Der Fragebogen untergliederte sich in drei Teile: Die Erhebung allgemeiner Betriebsdaten, Fragen zum Einfluss von Weintourismus und inwieweit die Mitglieder bei weintouristischen Aktivitäten involviert sind. Aufgrund der Flächenabfrage nahmen bei der Befragung Betriebe, die insgesamt rund 65 Prozent der baden-württembergischen Genossenschaftsfläche verarbeiten, teil. Im Wesentlichen wurden geschlossene Fragen verwendet. Wo möglich, wurde eine sechsstufige Likert-Skala (1 = sehr häufig/sehr hoch bis 6 = sehr gering/nie) angewendet. Auf offene Fragen wurde weitgehend verzichtet. Zur Überprüfung der Signifikanz wurde der Wilcoxon-Mann-Whitney-Test (MWU-Test) verwendet. Der MWU-Test ist ein nichtparametrischer Test, der keine großen Stichproben voraussetzt. Liegen mindestens sechs Beobachtungen vor, kann er verwendet werden [14].

Um die Ergebnisse der quantitativen Befragung in das weite Feld des Weintourismus einzuordnen und auch unterschiedliche Blickwinkel einzubinden, wurden vier der schriftlich befragten Geschäftsführer um ein persönliches Interview gebeten. Die Auswahl der Befragten folgte der Größe und unterschiedlichen Mitgliederzahl der Winzergenossenschaften. Für die Befragung wurde ein einheitlicher Interviewleitfaden verwendet, um eine vergleichbare Auswertung sicherzustellen. Die Interviews wurden im Anschluss transkribiert.

3. Überblick aus der Literatur

Der Begriff des Weintourismus wurde erstmals von der AUSTRALIAN TOURIST COMMISSION zu Beginn der 1990er Jahre aufgegriffen [15]. Da sich in der Wissenschaft bzw. auch in der Wahrnehmung der Marktteilnehmer jedoch bis heute keine einheitliche Definition des Weintourismus findet, bietet dies Raum für eine Vielzahl von definitorischen weitgreifenden Ansätzen [16]. Die literarische Recherche zeigt, dass die Begrifflichkeit "Weintourismus" nicht genau abgegrenzt wird, ebenso wenig, inwieweit weintouristische Leistungen als solche definitorisch maßgeblich sind.

In einer der Ersten, sehr nachfrageorientierten Definitionen für Weintourismus von HALL [17] sowie der SOUTH AUSTRALIAN TOURISM COMMISION [18], wird aufgezeigt, dass der Besuch von Weinproben und das Weinerlebnis den primären Motivationsfaktor ausmachen. Diese Definition wurde von vielen Autoren aufgegriffen und erweitert [16]. Die Grundlagendefinition über den primären Motivationsfaktor zeigt aber auch, dass es schwierig ist, einen globalen definitorischen Ansatz zu finden und Weintourismus sehr unterschiedlich ausgeprägt ist. So scheint diese Grundlagendefinition mit dem primären Motivationsfaktoren auf viele Regionen der "Neuen Welt" übertragbar zu sein. In RÜDIGER et al. [19] wird dargelegt, dass dieses Reiseverhalten für den deutschen Markt nur sehr bedingt anwendbar ist, da der Weinreisende, der gezielt Zeit in einer Weinregion verbringt, um mehr über die Weine der entsprechenden Region, die Weinproduktion und die Winzer zu erfahren, nur einen geringen Prozentsatz der weinaffinen Touristen ausmacht. Die potenziellen Nutzer von weintouristischen Angeboten sind weinaffine Touristen, die in einer bestimmten Region ihren Urlaub verbringen und vom Thema Wein am Rande berührt werden möchten. Dies wird auch durch die Definitionen von CARLSON & DOWLING [20], WINEMAKERS FEDERARTION OF AUSTRALIA [21], ASSEMBLÉE DES RÉGIONS EUROPÉ-ENNES VITICOLES (arev) [22] oder DELOITTE [23] unterstützt, die nicht von Weinproben oder dem Weinerlebnis als primären Motivationsfaktor für einen Besuch der Region ausgehen, sondern von einem Zusammenspiel von Wein, Kultur, Gastronomie, Kunst, Bildung und Reiseerlebnisse. Diese sehr nachfrageorientierten Definitionen insgesamt, sehen den Weintourismus aus Sicht des Konsumenten. GETZ [7] und GETZ und BROWN [24] erweitern diese Sichtweise um eine weitere Hauptperspektive, nämlich die der Marketingstrategie der Anbieter. Hier ist der Weintourismus eine Möglichkeit des Direktmarketings für Weinanbieter. So zeigt sich, dass Weintourismus in seiner Definition mehrere Sichtweisen beinhalten sollte. In Abbildung 1 sind die Kernaussagen der am häufigsten in der Literatur verwendeten Definitionsansätze zusammengefasst, sowie in die beiden Blickwinkel der Konsumenten und Anbieter aufgeteilt.

Eine Schnittmenge, den die meisten Definitionen aufgreifen, ist die Reise zum Destinationsziel. Diese Tatsache deckt sich auch mit der Tourismusdefinition der Welttourismusorganisation (WTO) bei der Touristen Personen sind: "die zu Orten außerhalb ihres gewöhnlichen Umfelds reisen…" Damit fallen unter den Begriff Tourismus auch Ausflügler und Tagesgäste. Ausschlaggebend ist, dass eine Ortsveränderung stattgefunden hat [25].

Im Zuge der persönlichen Befragung der Geschäftsführer wurde auch nachgefragt, wie sie den Begriff Weintourismus definieren würden. Hier zeigte sich deutlich, dass die jeweilige subjektive Sichtweise die grundlegende Zusammenfassung der definitorischen Kernaussagen prägt. So definierte Experte 1: "… Weinveranstaltungen, Aktionen im Weinberg, um gezielt Verbraucher in unsere Vinotheken für unser Direktgeschäft zu begeistern…" oder Experte 2: "…eine Weinregion besuchten und dort neben dem reinen Kennenlernen von Weinen auch, sag ich mal, auch Veranstaltungen besuchen die sich mit dem Thema Wein auseinandersetzt, das heißt Weinproben, Vinothekenbesuch oder vielleicht auch mal einen Weinbaulehrpfad besuchen."

In der Folge lässt sich festhalten, dass die Definition des Weintourismus immer - je nach Blickwinkel - die Sichtweise des Anbieters oder Nachfragers annimmt

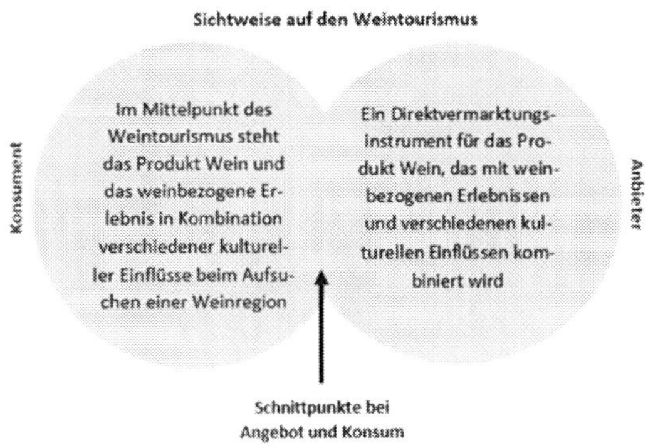

Abb 1. Sichtweisen auf den Weintourismus.

Tabelle 1. Einfluss Weintourismus auf den Direktverkauf von Winzergenossenschaften.

Einfluss	alle (%)	Baden (%)	Württemberg (%)
sehr hoch (1)	37,84	40	35,29
(2)	43,24	35	52,94
(3)	10.81	15	5,88
(4)	8,11	10	5,88
(5)	0	0	0
überhaupt nicht (6)	0	0	0

und die jeweiligen komprimierten Kernaussagen aus Abbildung 1 enthält.

4. Ergebnisse

4.1. Einfluss von Weintourismus auf den Direktverkauf

Um den Bezug von Winzergenossenschaften zum Weintourismus zu eruieren, wurde in der Befragung erhoben, ob weintouristische Aktivitäten einer Winzergenossenschaft Einflüsse auf den Direktverkauf haben (siehe Tabelle 1). Dies wurde mithilfe einer sechsstufigen Likert-Skala (1 = sehr hohen Einfluss bis 6 = keinen Einfluss) erhoben (siehe Tabelle 1). Hier zeigte sich, dass über 90 Prozent einen direkten Zusammenhang sehen (Stufen 1, 2 und 3). Zwischen den beiden Anbaugebieten waren hier keine Unterschiede festzustellen (zweis. MWU-Test: $p = 0{,}8103$).

Nach Aussage des Deutschen Weininstituts (DWI) hat sich Weintourismus in den letzten Jahren zu einem bedeutenden Wirtschaftsfaktor für die deutsche Weinindustrie entwickelt. Dies stützt sich auf die Aussage, dass Menschen eine besonders intensive Beziehung zu deutschem Wein haben, wenn sie selbst in Kontakt mit dem Winzer gekommen sind und die Weinbauregion besucht haben [26]. Um zu überprüfen, ob auch Winzergenossenschaften von diesem Trend profitieren, wurde erfragt, ob Betriebe von vermehrter Tourismus- bez. Weintourismusaktivität in der jeweiligen Region in der Form profitieren, dass Sie neue Käuferschichten oder Konsumenten gewinnen konnten. Von den 37 befragten Winzergenossenschaften bestätigten 89,19%, dass sie wirtschaftlich von diesem Zuwachs profitieren. 2,7% verneinten die Frage und 8,11% haben dies nicht wahrgenommen. Auch hier gibt es keine signifikanten Unterschiede in den Aussagen von Baden und Württemberg.

Um die Antwort bei der jeweiligen Genossenschaft, welche die Frage mit "ja" beantwortet haben, zu konkretisieren, wurden die jeweiligen Teilnehmer zu einer Unterfrage weitergeleitet und befragt, in welcher Form sich dies äußert. Bei der Beantwortung waren Mehrfachnennungen möglich. Dabei bestätigten sich die Aussagen der ersten Frage, dass 90,91% einen erhöhten Direktverkauf feststellten, bei 63,64% steigt die Zulieferung an die Gastronomie, 42,42% nehmen vermehrte Veranstaltungen in der Region wahr, die sie beliefern, und 63,64% haben einen vermehrten Zulauf bei hauseignen Veranstaltungen.

Um die Wichtigkeit des Weintourismus in Bezug auf den Direktverkauf nochmals zu hinterfragen, wurden die Experten persönlich nach dem Stellenwert befragt. Hier zeigte sich, dass für alle Weintourismus ein Direktmarketinginstrument ist - wenngleich auch der Stellenwert, aufgrund der Größe der befragten Betriebe unterschiedlich eingeschätzt wird. So ist es vor allem für kleinere Winzergenossenschaften als Marketinginstrument elementar, wie Experte 1 konstatierte: "... Weintourismus ist für uns überlebensnotwendig, auch als kleine Genossenschaften, sobald der Weintourist zu uns ins Haus kommt, haben wir die Möglichkeit, in den sogenannten Nahkampf überzugehen... sich direkt auszutauschen ist eigentlich das Wichtigste in der Weinvermarktung." Experte 3 ergänzt: "... bei Veranstaltungen. Weinproben und Weinfeste die wir haben, merke ich, wirken sich meistens direkt am Tag danach auf die Vinothekenumsätze aus." Für größere Winzergenossenschaften ist dies, aufgrund des geringeren Direktabsatzes nicht im gleichen Maß ausschlaggebend für den Direktverkauf. Doch auch hier stellt es ein Instrument dar, um sich regional und überregional zu profilieren, wie Experte 4 feststellt:,,. wir haben eine nationale Marke, wir arbeiten hauptsächlich auch mit dem Handel. Weintourismus machen wir, um gezielt Verbraucher in unsere Vinotheken für unser Direktgeschäft zu begeistern, indem der einmal bei uns war, und das sozusagen erlebt hat, der ist natürlich in ganz besonderer Weise mit uns verbunden."

Zur Einschätzung der Gewichtung des Weintourismus in den Winzergenossenschaften, wurde der Anteil der durch Weintourismus generierte Direktabsatz prozentual zum Gesamtumsatz abgefragt. Bei den 22 von den Teilnehmern abgegebenen Angaben zeigte sich, dass bei einem Mittelwert von 12,36 Prozent - bei einer Standartabweichung von 11,85 - der Anteil sehr individuell von der Betriebsgröße abhängt (siehe Tabelle 2). Es zeigt sich, dass vor allem kleinere Winzergenossenschaften, ähnlich wie bei Weingütern [26] auf das Marketinginstrument Weintourismus angewiesen sind - wenngleich festzuhalten bleibt, dass sich unabhängig von der Größe alle Genossenschaften in diesem Bereich engagieren.

4.2. Das weintouristische Angebot von Winzergenossenschaften

Zur Erhebung, welche Art von Angebot die Winzergenossenschaften anbieten, wurden die in der Literatur am häufigsten aufgelisteten klassischen Angebote der Weingüter abgefragt, um aufzuzeigen, ob diese Angebote auch von Winzergenossenschaften umgesetzt werden.

Tabelle 2. Anteil der durch Weintourismus generierte Direktabsatz.

% vom Gesamtumsatz	*M*	*SD*	*min*	*max*
>1000 ha	k.A	k.A	k.A	k.A
600–1000 ha	8,5	9,19	2	15
400–600 ha	11	7,94	5	20
250–400 ha	7,5	2,89	5	10
100–250 ha	12,8	13,59	2	35
<100 ha	16	15,75	3	50
alle	12,36	11,85	2	50

Die Abfrage wurde noch um das Feld "Sonstiges" erweitert, in das die Teilnehmer eigene Veranstaltungen eintragen konnten. Es zeigt sich in der Auswertung (siehe Tabelle 3), dass die klassischen Angebote Weinproben, Weinwanderungen, Kellerführungen, Hoffeste und der Wein-ausschank im Weinberg von den meisten Winzergenossenschaften angeboten wird. Das Feld "Sonstige" wurde von keinem Teilnehmer ausgefüllt. Signifikante Unterschiede der beiden Anbaugebiete gibt es nur im Bereich Kellerführungen, was auf die unterschiedliche Durchschnittsgröße der Genossenschaften in den beiden Anbaugebieten zurückgeführt werden kann. Ausschlaggebend ist hier, dass in Baden sehr viel kleinere Genossenschaften existieren und hier die Produktion weniger Industriecharakter besitzt. Es zeigte sich bei der Abfrage, dass nur ein sehr geringer Teil der Winzergenossenschaften über einen Kinderspielplatz verfügen. Dies ist sicher der Zielgruppe der Weintouristen geschuldet, die im Alter um die 50 Jahre zu suchen sind [19]. Zusätzlich wurde abgefragt, seit welchem Zeitraum die Winzergenossenschaften diese Angebote im Potpourri haben. Hier zeigte sich, dass die Angebote, die in den meisten Betrieben umgesetzt werden, auch den längsten Bestand haben und so traditionell zu den Winzergenossenschaften gehören. Angebote, die bei den Betrieben in den letzten beiden Jahrzehnten realisiert wurden, sind Weinausschank in den Weinbergen, Kulturveranstaltungen (z.B. Lesungen im Weinkeller), thematische Weinproben (z.B. Wein und Schokolade) und thematische Hoffeste (z.B. Advent in der Winzergenossenschaft). Winzergenossenschaften haben hier sowohl die klassischen Angebote der Weingüter ebenso wie neuere Angebote in ihr Programm aufgenommen.

Wie aus der Literatur entnommen, stellt Weintourismus für viele Weingüter einen eigenen Bereich der Wertschöpfung dar. Um einordnen zu können, inwiefern dies auch eine Rolle bei Winzergenossenschaften spielt, wurden die Experten dahingehend befragt. Die Aussagen belegen, dass für viele Weintourismus nur das Marketinginstrument verkörpert, es wird gezielt keine eigne Wertschöpfung durch die Veranstaltung generiert. Alle Befragten trafen in dieser Frage dieselbe Aussage, dass die Veranstaltungen für sie kostendeckend ausgerichtet sind und für sie hauptsächlich "Werbeveranstaltungen" für den Direktverkauf darstellen.

4.3. Einbindungen der Mitglieder in den Weintourismus

Zur Organisation und Umsetzung der Weintourismus-Strategie ist es weiterhin interessant, wie Winzergenossenschaften auf das Potenzial der involvierten Mitglieder

 DOI: 10.1051/bioconf/20170903023

Tabelle 3. Weintouristische Angebote und Zeitraum bei Winzergenossenschaften.

Angebote	alle (%)	Baden (%)	Württemberg (%)
Weinwanderungen	83,78	80	88,24
Kellerführungen	83,78	100	64,71
Weinproben	94,59	95	94,12
Thematische Weinproben	72,97	70	76,47
Hoffeste	86,49	90	82,35
Thematische Hoffeste	67,57	70	64,71
Kulturveranstaltungen und Wein	54,05	50	58,82
Weinausschank in den Weinbergen	78,38	70	88,24
Kinderspielplätze	2,70	0	5,88
Sonstiges	0	0	0

zurückgreifen, um diese in die Aktivitäten miteinzubeziehen. Dazu wurde erhoben, inwieweit in der jeweiligen Winzergenossenschaft weintouristische Aktivitäten von Mitgliedern eigenständig durchgeführt werden. Denn nicht selten existieren in Winzergenossenschaften divergierende Interessen von Mitgliedern und Management, was sich in einer unterschiedlichen Rollenverteilung innerhalb der Genossenschaften abzeichnet, da Mitglied und Geschäftsführung eine Doppelfunktion ausüben (Prinzipal-Agenten-Theorie) [27]. Deshalb war auch die Motivation ein interessanter Ansatz, der durch die Frage "Inwieweit motivieren Sie ihre Mitglieder, eigene weintouristische Veranstaltungen durchzuführen?" erhoben wurde. Die Fragen wurden mithilfe einer sechsstufigen Likert-Skala (1 = sehr häufig bis 6 = nie) erhoben. Die Antworten liefern kein eindeutiges Bild: Bei 48,66 Prozent der befragten Betriebe werden eher häufig Veranstaltungen von Mitgliedern durchgeführt (Stufen 1, 2 und 3), bei 51,34 Prozent nicht (Stufen 4, 5 und 6). Bei der Motivation gaben 67,57 Prozent der Geschäftsführer an, ihre Mitglieder zu motivieren, eigene Veranstaltungen durchzuführen (Stufen 1, 2 und 3). Diese Aussagen unterscheiden sich - weder in der eigenständigen Durchführung $p = 0{,}6892$) noch in der Motivation (zweis. MWU-Test: $p = 0{,}8103$) - auch nicht in den beiden Anbaugebieten. Bei der Befragung der Experten waren die Aussagen ebenso zweigeteilt: Für zwei der Befragten ist die Einbindung der Mitglieder ein elementarer Geschäftsbereich, um diese an die Winzergenossenschaft zu binden und das "Wirgefühl" zu stärken. Zudem bietet es den Mitgliedern die Möglichkeit, ein Zusatzeinkommen zu generieren. Die Einbindung der Mitglieder ist größtenteils auf Veranstaltungen im Außenbereich und der teilweise eigenen Flächen angelegt, um hier eine gewisse Authentizität der Genossenschaft zu vermitteln. Dies wurde auch vom Experten 4 bestätigt: "...Das fördert in beide Richtungen, das fördert die Verifikation des Endverbrauchers, die sehen, dass die Persönlichkeiten, die selbst den Wein erzeugen, also in dem Fall der Anbau. Bei dem Mitglied selbst ist es auch rückwärts Motivation, weil sie mit den Verbrauchern in Kontakt kommen, die Informationen, Verbraucherwünsche auch persönlich erleben. Das ist für den Geschäftsführer, sagen wir mal, sehr angenehm, weil er dann Mitglieder hat, die auch richtig gebildet sind in dem Punkt, wie schwer es ist, was zu vermarkten."

5. Schlussfolgerung

Der deutsche Weinmarkt steht aufgrund des zunehmenden nationalen und internationalen Wettbewerbs unter Zugzwang, sich neue Diversifikationsmöglichkeiten zu erarbeiten. Eine Möglichkeit ist die Partizipation am zunehmenden Trend des Weintourismus. Auch deutsche Winzergenossenschaften, unabhängig von ihrer besonderen Unternehmensstruktur, haben die Möglichkeit diese Entwicklung für sich zu nutzen. Bei der Frage, inwieweit sich Winzergenossenschaften im Bereich des Weintourismus engagieren und welche Motive sie damit verfolgen, konnte aufgezeigt werden, dass Winzergenossenschaften identisch wie Weingüter weintouristische Aktivitäten in ihre Unternehmensstruktur integriert haben. Im Gegensatz zu den meisten Weingütern nicht vor dem Hintergrund, eine eigene Wertschöpfung zu generieren, sondern lediglich mit dem Ziel, als Marketinginstrument den Direktverkauf zu verstärken, bzw. um ihre Reputation auf- und auszubauen. Dabei muss nach Größe, Vermarktungsmenge und Distributionsstruktur der Winzergenossenschaften unterschieden werden: Für kleinere Winzergenossenschaften mit einem bedeutenden Absatz im direkten Vertrieb, stellt Weintourismus eine betriebswirtschaftliche Überlebensstrategie und Verkaufsförderung dar, um einen geregelten Absatz ihrer Produkte zu realisieren. So werden unterschiedliche Ansätze deutlich. Genossenschaften mit vermehrtem indirektem Vertrieb sehen diese Kommunikationsmaßnahme als Direktkommunikation für die Erschließung neuer Verbraucher- oder Zielgruppen. Wie innovativ Winzergenossenschaften bei der Realisierung und Auswahl ihrer Aktivitäten sind, konnte durch die quantitative Erhebung gezeigt werden, denn es dominieren vorrangig traditionelle Angebote. Dies spiegelt zwar zum einen die Erwartung der Weintouristen wieder, wie in [19] aufgezeigt wurde, allerdings beschränkt man sich dabei im Wesentlichen nur auf eine hauptsächliche Zielgruppe, nämlich die der sogenannten "Best Ager". So werden beispielsweise Rahmenangebote wie Kinderspielplätze, die für die Erschließung neuer Zielgruppen elementar sind [28], bei Winzergenossenschaften eher vernachlässigt und sind nur vereinzelt zu finden. Es zeigt sich, dass neue und zusätzliche Angebote bei Winzergenossenschaften nur sehr langsam integriert werden. Die Erwartungshaltung der Weintouristen, die insbesondere auch den Erlebniswert präferiert, wird vernachlässigt. Ein Gegenstand zukünftiger Forschungsarbeiten könnte daher sein, welcher strategische Zusammenhang hinter Weintourismus als Marketingstrategie und der Erwartungshaltung der Weintouristen steht und in wieweit diese und der Weintourismus generell von institutionellen Organisationen gelenkt wird.

Bei der Frage inwieweit Winzergenossenschaften auf das Potenzial der involvierten Mitglieder zurückgreifen, um diese in die Aktivitäten miteinzubeziehen, konnte kein einheitliches Bild eruiert werden - weder in Form der Größe, Mitgliederzahl oder Region. Es zeigte sich bei den Interviews mit den Experten, dass die Integration in den operativen Ablauf der Veranstaltungsdurchführung sehr stark vom persönlichen Engagement der Geschäftsführer

DOI: 10.1051/bioconf/20170903023

abhängig ist. Dabei wird von vielen Genossenschaften das Potenzial, das von Mitgliedern bei der Durchführung, aber auch hinsichtlich der Kommunikation von Weinveranstaltungen eingebracht werden kann (KAGERMEIER und HARMS, 2013), unterschätzt. Gerade die Einbindung der Mitglieder bietet eine Möglichkeit, um diese emotional an die Genossenschaft zu binden. Zudem bietet diese Zusammenarbeit Potential, das Qualitätsdenken der abliefernden Mitglieder anzuheben, da sie als Multiplikatoren in den direkten Kontakt mit dem Konsumenten treten und so für ihr weiterverarbeitetes Produkt den Absatz generieren.

Hinsichtlich der Definition lässt sich festhalten, dass Weintourismus immer - je nach Blickwinkel - die Sichtweise des Anbieters oder Nachfragers annimmt, wobei unabhängig des Konsums oder der Direktvermarktung das weinbezogene Erlebnis und die Ortsveränderung zur ausführenden Destination im Vordergrund stehen. Bezugnehmend auf die Literatur werden alle Aktivitäten, die Weingüter und Winzergenossenschaften durchführen, unter dem Begriff Weintourismus vermarktet wie z.B. Hoffeste oder gastronomische Einrichtungen. Hierbei ist zu differenzieren, ob es die definitorischen Grundzüge enthält, um die Zielgruppe der Touristen anzusprechen oder nur der lokale Konsument im Fokus stehen.

Literaturverzeichnis

[1] R. Mitchell, C.M. Hall, A. Mc Intosh, Wine tourism and consumer behavior, in: Hall, C. M., Sharples, L., Cambourne, B., Macionis, N., Mitchell, R., Macionis, G. (Hrsg.): *Wine Tourism Around the World – Development, management and markets*, Butterworth-Heinemann, Oxford, Seite 115 bis 135 (2000)

[2] V. Strecker, *Weintourismus: Grundlagen, Konzepte & Potenziale im Untersuchungsgebiet Südlicher Oberrhein*, AV Akademikerverlag, Saarbrücken (2011)

[3] A. Dreyer, J. Ratz, J. Berauer, *Weintourismus – Marketing für Weinregionen und Winzer*, ITD-Verlag, Hamburg (2015)

[4] Deutscher Bundestag, *Antwort der Bundesregierung auf die Kleine Anfrage der Fraktion BÜNDNIS 90/DIE GRÜNEN, Potenziale für Winzerbetriebe durch Tourismus*, Deutscher Bundestag 18. Wahlperiode (2015)

[5] K. Saaler, *Tourismus: Wertschöpfung für die Weinbauregionen*, In: Der Deutsche Weinbau, **Nr. 2**, Januar 2011, S. 28–31 (2011)

[6] P. Winkelmann, *Marketing und Vertrieb – Fundamente für die Marktorientierte Unternehmensführung*, München (2013)

[7] D. Getz, *Explore Wine Tourism: Management, Development & Destination*, New York, Sidney, Tokio , S.4–10 (2000)

[8] W. Freyer, *Tourismus-Marketing: Marktorientiertes Management im Mikro- und Makrobereich der Tourismuswirtschaft*, 5. Auflage, München, S. 507 (2007)

[9] E. Brüscher, *Weintourismus – Wichtiger Wirtschaftsfaktor für die Weinwirtschaft*. In: Das Deutsche Weinmagazin, Ausgabe**14**, Juli 2010, S. 24–27 (2010)

[10] M. Iselborn, L. Jungbluth, J.H. Hanf, *Probleme des Qualitätsmanagement bei Winzergenossenschaften. Wirtschafts- und Sozialwissenschaften des Landbaues* e.V. GEWISOLA, Göttingen (2014)

[11] DRV, *Raiffeisen: Statistischer Bericht 2015*. Berlin (2014)

[12] H. Pilz, *Gordischer Knoten: Die deutschen Winzergenossenschaften verlieren Marktanteile. Können sie sich aus der Situation befreien?* Weinwirtschaft. Meininger Verlag. Online verfügbar unter `https://www.meininger.de/de/weinwirtschaft/ausgabe-142014`, zuletzt geprüft am 24.02.2017 (2014)

[13] A. Kagermeier, T. Harms, *Einsatz von Sozial Media im Weintourismusmarketing: Konkret Praxis und ungenutzte Chancen*. In: Scherhag, Knut (Hrsg.): Weintourismus und Marketing, Josef Eul Verlag GmbH, Lohmar/Köln, S. 65–91 (2013)

[14] S. Siegel, *Nichtparametrische statistische Methoden*. Frankfurt am Main (1997)

[15] ACIL Consulting Pty. Ltd, *Review of Wine Tourism and Wine Exports*, Internet, URL: `http://s3.amazonaws.com/zanran_storage/www.aciltasman.com.au/ContentPages/18965828.pdf` (2002)

[16] J. Müller, A. Dreyer, *Weintourismus: Märkte, Marketing, Destinationsmanagement – mit zahlreichen internationalen Analysen*, ITD-Verlag, Hamburg (2010)

[17] CV.M. Hall, G. Johnson, B. Cambourne, N. Macionis, R. Mitchell, L. Sharples, **Wine tourism: an introduction**, in: Hall, C. M., Sharples, L., Cambourne, B., Macionis, N., Mitchell, R., Macionis, G. (Ed.): Wine Tourism Around the World – Development, management and markets, Butterworth-Heinemann, Oxford, Seite 1 bis 23 (2000)

[18] South Australian Tourism Commission, *Wine and tourism: A background research report*, Internet, Url: www.tourism.sa.gov.au/publications/pubs.asp?archive=0&PubCatID=4#Publication102 (2007)

[19] J. Rüdiger, J.H. Hanf, E. Schweickert, *Die Erwartungshaltung von Weintouristen*. Wirtschafts- und Sozialwissenschaften des Landbaues e.V. GEWISOLA, Göttingen (2014)

[20] J. Carlson, R. Dowling, *Wine tourism marketing issues in australia*, International Journal of Wine Marketing **10**, 23–32 (2008)

[21] Winemakers Federartion of Australia, *Wine Tourism Strategic Business Plan 2002–2005*: Embrance the Challenge, Internet, URL: `www.wfa.org.au/PDF/NWTS%Strategic%20Plan. Pdf` (2002)

[22] Assemblée des Régions Européennes Viticoles (arev), *Europäische Charta des Weintourismus: Allgemeine Grundsätze*, Internet, URL: `www.arev.org/sites/default/files/Charte_DE.doc` (2005)

[23] Deloitte, *VINTUR Project - European Enotourism Handbook*, Deloitte Touche Tohmatsu Limited (2005)

[24] D. Getz, G. Brown, *Critical success factors for wine tourism regions: a demand analysis*. In Tourism Management **27**(1), 146–158 (2006)

[25] A. Schulz, W. Berg, M.A. Gardini, T. Kirstges, B. Eisenstein, *Grundlagen des Tourismus – Lehrbuch in 5 Modulen*. Oldenbourg Wissenschaftsverlag (2005)

 DOI: 10.1051/bioconf/20170903023

[26] Das deutsche Weinmagazin, *Wichtiger Wirtschaftsfaktor für die Weinwirtschaft.* In: Das deutsche Weinmagazin **14**/Juli 2010, 24–27 (2010)

[27] L. Ingenwerth, P. Sparer, A. Thein, J.H. Hanf, M. Iselborn, *Analyse der Rollenverteilung in Winzergenossenschaften im Kontext des Qualitätsmanagements.* Wirtschafts- und Sozialwissenschaften des Landbaues e.V. GEWISOLA, Gießen (2015)

[28] F. Allgayer, *Zielgruppen finden und gewinnen – Wie Sie sich in die Welt Ihrer Kunden versetzen,* MI-Fachverlag, Landsberg am Lech (2007)

DOI: 10.1051/bioconf/20170904001

Utilizzo di vitigni di antica coltivazione del Meridione d'Italia per la produzione di vini passiti di pregio / Autochthonous vines from South Italy for the production of fine dessert wines

Donato Antonacci[1], Matteo Velenosi[1], Perniola Rocco[1], Pasquale Crupi[1], Mario Ventura[2], e Carlo Bergamini[1]

[1] Research Centre for Viticolture and Enology (CREA –VE) seat of Turi-Bari (Italy), Via Casamassima, 148 – 70010, Italy
[2] University of Bari, Department of Biology, Via Orabona 4, 70124 Bari, Italy

Abstract. In recent years the wine sector has been characterized by an increased focus on the diversification of wine products.

This change has been welcomed by consumers. In fact, today's consumers are willing to experiment with new products and unusual combinationswhich has stimulated research aimed at the discovery of native grape varieties suitable for for winemaking. In this context, particular attention are receiving vines for the production of dessert wines. Many regions have made efforts in the search for varieties and techniques able to better promote the drying process. At CREA-VE Turi-BA, there is an ongoing recovery and exploitation of indigenous grape varieties in the regions of Southern Italy and today more than 3,000 accessions of wine grapes are preserved.

In the present work we have investigate the possibility of the use of certain accessions of Vitis vinifera varieties (collection CREA-VE Turi-BA) deemed attractive for the production of dessert wines. Accessions presenting a good sugar content and a good titratable acidity were selected; namely, Greco bianco b. (named Q19) and Mantonico bianco b. (named R8) were compared to the cv of Malvasia di Lipari b. Ampelographic and ampelometric analyses, followed by molecular analysis, were carried out for varietal assessment.

A rate of the grapes were fermented immediately after harvest with standard procedure, in comparison to rates vinified after withering. The latter was made in a greenhouse type structure with parts of natural and unforced air, without any additional energy expense. At the end of the process, a loss in weight of the grapes equal to 30–35% was realized. The grapes are then stemmed, crushed, pressed and fermented by selected yeast under controlled temperature.

Upon completion of the fermentation and subsequent stabilization, the wines were analyzed for the most important enological parameters and subjected to evaluation by a panel of expert tasters. Chemical analysis of wines showed that an alcohol content ranging to 11–12% volume was reached with in the standard wine-making; whilst, for the vinification of dried grapes the alcohol content was equal to 16% vol for Mantonico bianco b. and 18% vol. for and Greco bianco.

Organoleptic analysis of the wines showed a greater appreciation for both of dessert wines. The best ratings were given for the olfactory characteristics (floral and fruity aromas) and taste characteristics. In particular, the Greco bianco b. had excellent scores slightly lower than Malvasia di Lipari b.

Therefore, the tested varieties have shown a good potential for the production of dessert wines related to typicality of the grape and the region. Dessert wines produced, have a complex flavor profile and interesting for the high acidic component, which improved the taste perception of wine and attenuated the sensation of sweetness. That feature could allow a good wine aging, too.

1. Introduzione

La tecnica dell'appassimento è conosciuta fin dal tempo dell'antico Egitto (2000–4000 A.C), ma fu sviluppata e perfezionata successivamente dai Greci [3]. L'appassimento da allora è stato perfezionato sempre più nei secoli contemporaneamente all'accrescimento dell'interesse dei consumatori verso i vini passiti. Molti sono i prodotti ricavati dalla disidratazione o dalla sovra maturazione dei grappoli e in questa categoria sono compresi i vini dolci o da dessert e in particolar modo i vini passiti. I passiti sono prodotti in tutto il mondo ed in Europa, in particolar modo, tale produzione è basata sul regolamento CE 1493/1999 e sua successiva modifica tramite il regolamento CE 2165/2005, entrata poi in vigore in Italia con Gazzetta Ufficiale numero 82/2006. Proprio in Italia, sotto la categoria di vini passiti, abbiamo oltre un centinaio di denominazioni di origine [14]. Generalmente, per la produzione di tali prodotti, vengono usati processi di appassimento dei grappoli in post-raccolta, con o senza condizioni artificiali, [2], inoltre, non è permesso l'uso di etanolo per la correzione del naturale grado alcolico raggiunto nel corso della fermentazione. Le differenti varietà utilizzate per fare i passiti, i processi usati per l'appassimento post raccolta, i metodi di produzione in cantina oltre ai processi fermentativi e invecchiamento influenzano notevolmente i risultati qualitativi [10, 11]. L'appassimento è un processo complesso che comporta modifiche alle caratteristiche chimico-fisiche degli acini [1]. I pre-trattamenti chimici [9], le metodiche e le condizioni di appassimento (appassimento al sole, all'ombra o artificiale), così come la varietà, giocano un ruolo fondamentale nella tipologia di vino passito che si ottiene a fine del processo produttivo [12]. Generalmente, in seguito alla raccolta, i grappoli vengono appassiti in stanze apposite dove con il controllo dell'atmosfera (temperatura e umidità) si riesce ad ottenere un calo di peso per perdita di acqua dei grappoli, senza lo sviluppo di muffe o marciumi. Diversamente, nel caso di sviluppo della Botrite Cinerea nobile sui grappoli in appassimento, la qualità del passito migliora [7]. L'appassimento dei grappoli con uso della luce solare rimane uno dei metodi più utilizzati nella zona a clima mediterraneo. In altre regioni, più fredde o temperate, per produrre i passiti si lasciano le uve (raccolte in cassette o su graticci) esiccare in ambienti chiusi, con o senza uso di tecnologie artificiali: per il controllo della temperatura, dell'umidità relativa (RH) e del flusso di aria all'interno dell'ambiente di essicazione [4,5]. Molti sono gli studi riferiti alle modifiche indotte dall'appassimento in ambienti "artificiali" e che prendono in considerazione le componenti del colore, degli aromi e dei polifenoli [6,8]. Differentemente dal precedente metodo, come per i Sauternes e i Tokay, c'è la possibilità di appassire i grappoli direttamente sulla pianta lasciando le uve a sovra maturare, portando a una disidratazione progressiva degli acini. Molte di tali modifiche direttamente in pianta sono ancora poco studiate.

In Italia meridionale troviamo diversi vitigni utilizzati per la produzione di vini passiti, in particolare la Malvasia di Lipari che da un vino DOC prodotto nelle isole Eolie (o Lipari) in provincia di Messina. Il vino ottenuto per disciplinare deve contenere una quantità del vitigno di Malvasia di Lipari massimo del 95% e con una percentuale di Corinto Nero del 5–8%. Si presenta come un prodotto con titolo alcolometrico volumico totale minimo di 18% vol., una quantità di zuccheri residui naturali minimo di 60 g/L e un'acidità totale minima di 4 g/L. Il prodotto presenta caratteristiche aromatiche che sono apprezzate dai consumatori, infatti, proprio per tale motivo è stato utilizzato come confronto alle prove svolte presso il CREA-VE. Proprio, il CREA-VE nella sede di Turi da anni si occupa del recupero del germoplasma di vitis vinifera del meridione e tra questo si è deciso di approfondire lo studio e caratterizzazione di due varietà tipiche del sud: Mantonico b. e Greco b.. Queste due varietà presentano una caratteristica peculiare in quanto sono in grado di mantenere un elevato livello di acidità totale (espressa come g/L di acido tartarico) nel corso della maturazione. Utilizzando un appassimento in ambiente artificiale (serra), senza alcuna spesa energetica, abbiamo valutato le potenzialità di queste varietà all'appassimento confrontandole con una varietà già ampiamente usata per tale scopo come la Malvasia di Lipari.

2. Materiali e metodi

2.1. Caratterizzazione Vitis Vinifera

Le varietà in esame sono state recuperate in alcune accessioni presenti nel sud Italia. L'accertamento varietale è stato realizzato con analisi ampelografica e ampelometrica, seguita da quella molecolare. In particolar modo, le varietà sono state caratterizzate tramite uso del Sequenziatore Sanger "CEQ8000TM genetic analysis system" (Beckmam Coulter) attraverso l'analisi degli SSR. In questo modo sono state identificate le tre varietà: Malvasia di Lipari, Greco b. e Mantonico b.. Il Greco b. è stato denominato Q19, mentre il Mantonico b. è stato denominato R8

2.2. Raccolta e produzione vini "base"

Le varietà in esame sono state monitorate per i parametri di maturazione principali: Contenuto zuccherino misurato con MA871 refractometer (Milwaukee), acidità totale (g/L di acido tartarico) e il pH. Tutte le valutazioni sono state effettuate in accordo al regolamento ECC 2676/90. Le uve sono state tutte raccolte ad inizio ottobre (2015). Una parte delle uve è stata destinata a immediata vinificazione con uso di lievito *Saccharomyces Cerevisiae* AWRI 350 (Maurivin), e fermentati a 16 °C, ottenendo i vini "base". I vini a fermentazione completata, sono stati stoccati in affinamento in una cella frigo a 10 °C fino al momento dell'imbottigliamento. Le restanti uve raccolte, invece, sono state destinate all'appassimento in serra.

2.3. Appassimento e produzione vini passiti

Le uve predisposte all'appassimento sono state disposte su cassette forate in una struttura tipo serra, senza alcuna fornitura di energia. le temperature interne alla serra oscillavano dai 20 °C durante la notte ai 40/50 °C durante il giorno. Il sistema di circolazione dell'aria all'interno avveniva tramite aperture presenti nella serra, mantenendo l'umidità bassa e vicina alle condizioni atmosferiche esterne. Nel realizzare l'appassimento non sono stati usati pretrattamenti chimici, preferendo una raccolta delle uve in condizioni sane, esenti da qualsiasi marciume o attacco di muffa. Le misure del calo peso sono state valutate come percentuale (WL%), tramite la differenza tra il peso fresco delle uve messe ad appassire e le uve in disidratazione: [(peso delle uve esiccate*100)/peso fresco delle uve]-100. L'appassimento ha avuto una durata complessiva di circa 1 mese. Le uve sono state quindi sottoposte a diraspatura, pigiatura, pressatura (1–1.5 atmosfere di pressione) e successiva vinificazione con inoculo selezionato e a temperatura controllata (16 °C). Le uve sono state vinificate a partire da un piede di fermentazione costituito da *Saccharmoyces Cerivisae* e *Saccharomyces Bayanus*. Dopo un giorno dall'inizio della fermentazione, è stato aggiunto un *Saccharomyces Uvarum* selezionato dal CREA-VE di Asti, in quanto molto osmotollerante e in grado quindi di resistere anche ad alte concentrazioni zuccherine. Le fermentazioni sono state monitorate controllando il consumo di zuccheri tramite mostimetro (°Babo). La fermentazione della Malvasia di Lipari a circa 60 g/L di zuccheri residui è stata fermata per ottenere un prodotto conforme al disciplinare di produzione della Malvasia di Lipari DOC delle isole Eolie (Messina). Greco b. e Mantonico b. sono state portare quasi a termine con un residuo zuccherino non superiore ai 20 g/L.

2.4. Analisi sui vini

Gli zuccheri residui, l'acido malico e l'acido lattico sono stati misurati usando il Saturno 150 (Crony instrument). Gli altri parametri come grado alcolico (%vol.), pH, acidità totale sono stati misurati in accordo al regolamento ECC 2676/90.

2.5. Analisi sensoriale sui vini "base" e "passiti"

I vini "base" e i relativi passiti sono stati sottoposti ad analisi sensoriale utilizzando un panel di degustatori esperti. Ai partecipanti al trial è stato fornito un foglio con le caratteristiche da valutare con uno score da 1 a 10 per ciascun parametro preso in considerazione.

Per le caratteristiche visive sono stati valutati:

- Intensità del colore;
- Tonalità del colore (punteggi più bassi tonalità sul verde mentre punteggi più alti per colore giallo carico per i vini base o aranciato per i passiti).

Per le caratteristiche olfattive:

- Intensità globale del bouquet;
- Intensità degli aromi floreali;
- Intensità degli aromi fruttati.

Per le caratteristiche gustative:

- Sapidità;
- Acidità;
- Alcolicità;
- Morbidezza;
- Dolcezza;
- Persistenza.

Infine, è stato attribuito un giudizio complessivo al vino attraverso un punteggio compresa tra 1 e 10 (punteggio più alto migliore qualità complessiva del vino).

I punteggi sono stati normalizzati sulla base dei voti forniti da ciascun giudice e da essi si è ottenuto il grafico spider plot (Fig. 2) presentato nel lavoro. Su tali valori poi è stata fatta analisi statistica per cercare i parametri non significativi, dove non vi è differenza tra vini passiti e vini "base".

2.6. Analisi statistiche

Per le analisi sensoriali è stata svolta l'analisi ANOVA e test di Tukey, utilizzando il pacchetto statistico "R" versione 3.3.3 (2017-03-06).

3. Risultati e discussione

3.1. Uve alla raccolta

Le uve raccolte a inizio ottobre presentano caratteristiche molto diverse, tra il controllo (Malvasia di Lipari) e le due varietà in esame (Greco b. e Mantonico b) come si osserva dalla Tabella 1. Infatti, mentre la Malvasia ha un livello zuccherino più sostenuto (>22° Brix), nelle altre due varietà il contenuto zuccherino accumulato nei grappoli è inferiore ai 20° Brix. Inoltre, le Greco b. e Mantonico b. presentano un alto contenuto acidico (circa 12 g/L per il Greco b). Considerando che tutti gli altri parametri (suolo, forma di allevamento e pratiche di gestione del vigneto) sono gli stessi, si nota dai dati sulle caratteristiche chimiche dei grappoli al momento della raccolta la differente impronta genetica di queste varietà di *Vitis vinifera*. L'acidità totale, come si osserva in Tabella 1, è rimasta elevata anche nell'annata 2015, caratterizzata da tarda primavera ed estate molto calda e secca. I dati termopluviometrici della Puglia sono rintracciabili tramite il sevizio ARPA-Puglia (`http://www.arpa.puglia.it/web/guest/serviziometeo`).

Tabella 1. Analisi grappoli alla raccolta.

Sample	°Brix	pH	AT
Malvasia di Lipari	22.1	3.38	5.5
Greco b. (Q19)	19.2	3.03	11.9
Mantonico b. (R8)	19.7	3.05	9.5

AT: acidità totale espressa come g/L di acido tartarico.

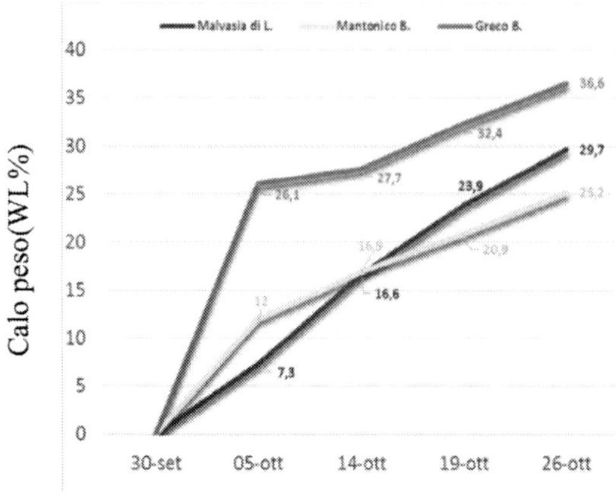

Figura 1. Cinetica di calo peso dei grappoli in serra.

Tabella 2. Analisi delle uve al termine dell'appassimento.

Sample	°Brix	pH	AT
Malvasia di Lipari	38.5	4.44	4.0
Greco b. (Q19)	31.4	4.05	6.3
Mantonico b. (R8)	25.8	3.64	5.6

3.2. Processo di appassimento

In seguito al processo di appassimento durato circa un mese (dal 30 settembre al 26 ottobre) si osserva un calo del peso medio dei grappoli in tunnel di appassimento di circa il 30.5%, anche se le differenze in termini di cinetica di perdita peso tra le varietà è notevole (Fig. 1). In circa 6–7 giorni il Greco b. perde fino al 26.1% di peso nelle condizioni di appassimento in serra e raggiunge oltre il 36% di peso complessivo perso dall'inizio della prova; diversamente il controllo (Malvasia di Lipari) ha una graduale calo peso durante tutto il corso dell'appassimento attestandosi a circa il 30%, mentre il Mantonico b. perde rapidamente nei primi giorni il contenuto di acqua, rallentando poi nell'avanzare del processo di disidratazione degli acini.

In ascisse i giorni in cui si è rilevata la perdita di peso e in ordinate la perdita di peso espressa in WL% (Weight loss).

Tabella 3. Analisi chimiche su vini base e passiti post-imbottigliamento.

Sample	% vol.	pH	AT g/L	Zucch. Residui g/L	Acido malico g/L
Malvasia di Lipari p.	18.0	4.32	6.0	65.2	1.8
Malvasia di Lipari B.	13.7	2.84	7.6	5.9	1.5
Greco b. (Q19) P.	18.0	4.24	8.2	12.8	3.8
Greco b. (Q19) B.	9.6	3.12	11.3	0.3	4.1
Mantonico b. (R8) P.	14.8	3.45	7.3	18.3	2.8
Mantonico b. (R8) B.	11.1	2.78	9.8	0.5	2.8

AT: acidità totale espressa come g/L di acido tartarico.

3.3. Uve post appassimento

Alla fine del processo di appassimento, prima della vinificazione, i grappoli si presentavano completamente sani (nessun ammuffimento o marciume visibile). Il contenuto zuccherino del Greco b. appariva più alto di quello del Mantonico b. a causa della forte perdita di peso e concentrazione degli zuccheri nel corso del processo di passificazione (Tabella 2). Si osserva, inoltre, un decremento di 4–5 g/L nel valore di acidità titolabile per Mantonico b. e Greco b. e di 1.5 g/L nella Malvasia di Lipari, diminuzione che può essere dovuta alla salificazione degli acidi organici. In tal senso, sarà utile svolgere ulteriori studi per capire meglio cosa determina tale fenomeno.

3.4. Analisi chimiche sui vini post-imbottigliamento

Alla fine del processo produttivo dei vini "base" (indicati con B.) ottenuti dall'immediata vinificazione delle uve prima dell'appassimento e dei relativi vini passiti (indicati con P.) si sono ottenuti vini con caratteristiche analitiche ben definite (Tabella 3). I vini passiti e i relativi vini base, hanno lo stesso contenuto di acido malico non avvalorando l'ipotesi che la riduzione dell'acidità che si osserva nei passiti sia dovuta all'ossidazione dell'acido malico come affermato da Rolle L. et al. nel caso di appassimento direttamente sulla vite [13], infatti, troviamo sia nei vini base che nei passiti le stesse quantità di acido malico (g/L) a fine processo produttivo. La quantificazione dell'acido lattico (dati non mostrati) dimostra che la fermentazione malolattica non è iniziata nei passiti e nei vini base, non determinando differenze significative. Nella Malvasia di Lipari si è deciso di produrre un vino seguendo nel dettaglio le caratteristiche chimiche scritte sul disciplinare della Malvasia di Lipari DOC (18% di contenuto alcolico, >4–5 g/L di acidità totale e contenuto di zuccheri residui naturale >60 g/L), differentemente, si è deciso di completare quasi del tutto la fermentazione per gli altri due vitigni per massimizzare la produzione di alcol etilico (Greco b. 18% e Mantonico b. 14.8% vol.). In questo modo si è anche testato il lievito *Saccharomyces Uvarum* fornito da Asti (CREA-VE) per la prova. Il *S.Uvarum* in combinazione con i lieviti del genere *Saccharomyces Cerevisiae* e *Bayanus* è stato in grado di completare quasi del tutto la fermentazione (a 16 °C) in Greco b. (12.8 g/L residui) e Mantonico b. (18.3 g/L residui). Nel Greco le analisi chimiche evidenziano un prodotto con identico grado alcolico volumico alla Malvasia di Lipari, ma con un ridotto contenuto zuccherino e alto contenuto acidico

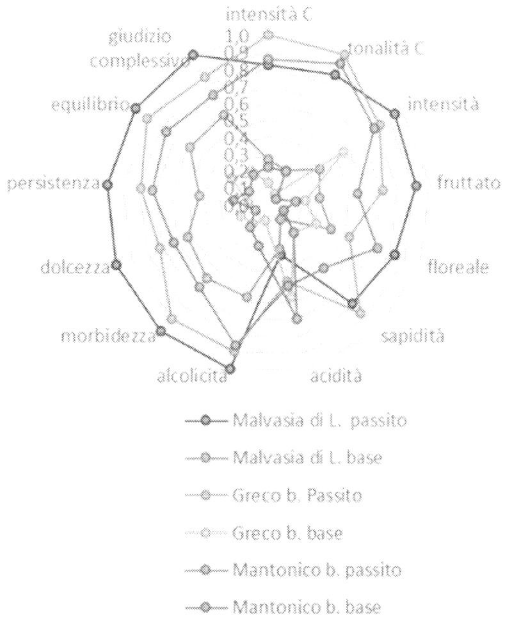

Figura 2. Spider plot delle caratteristiche valutate nel Panel test.

(>8 g/L) dovuto anche alla maggiore presenza dell'acido malico (>3.8 g/L).

3.5. Valutazione del trial sensoriale

A conclusione del processo di stabilizzazione e imbottigliamento del vino è stato svolto un trial sensoriale tramite un panel test di persone esperte e addestrate alla degustazione dei vini. I risultati che sono stati ottenuti sono osservabili in Fig. 2. Come si può notare vi è una forte differenza di punteggio nei descrittori utilizzati per caratterizzare e identificare i vini durante il trial. Nel esaminare i punteggi attribuiti all'intensità complessiva del bouquet aromatico, si osserva che in seguito alla trasformazione delle uve in vino passito si sviluppa una forte complessità aromatica nei vini, infatti, il Mantonico b. base (senza appassimento) che presenta una bassissima intensità aromatica in valori assoluti, migliora con il processo di disidratazione degli acini in serra. Nella Malvasia di Lipari passito le note fruttate e floreali sono poco distinguibili, diversamente, nel Greco b. (vino base e passito) sono più percettibili le note floreali mentre nel Mantonico b. (vino base e passito) sono le note fruttate a prevalere. In futuro svolgeremo un'analisi sulla complessità olfattiva dei principali composti aromatici dei vini base e passiti con il CG-massa, ottenendo indicazioni più dettagliate sulle differenze di questi prodotti. I vini passiti vengono percepiti più sapidi, più morbidi, dolci e anche molto più persistenti dei rispettivi vini base in tutti e tre i vitigni. L'unico parametro che non ha una significativa differenza è quello della sensazione gustativa dell'acidità, anche quando, nel caso della Malvasia, vi è un elevato contenuto zuccherino residuo (>60 g/L) e la stessa cosa avviene per il Mantonico b. e per il Greco b. I commenti che sono stati aggiunti nelle note (dati non riportati), giudicano i vini base come poco bilanciati e eccessivamente aggressivi in bocca per la forte acidità (soprattutto in Mantonico b. e Greco b.), diversamente nei passivi tale acidità viene percepita con una connotazione positiva rendendo i passiti più equilibrati e piacevoli. Sicuramente interessante spunto futuro sarebbe poter invecchiare i passiti prodotti per valutarne la potenzialità all'invecchiamento, considerando l'elevata acidità come un fattore si predisposizione positiva per tale scopo.

4. Conclusioni

Lo studio svolto è un lavoro preliminare di caratterizzazione del Greco b e Mantonico b. presenti nella collezione dell'Azienda sperimentale del CREA-VE di Turi (Ba). Dai dati raccolti dalle prime vinificazioni sperimentali si nota come a parità di condizioni ambientali e climatiche le due varietà in esame presentino un'acidità totale molto sostenuta e un buon contenuto zuccherino, e ciò è evidente anche dal confronto tra tali varietà con la Malvasia di Lipari, una vite aromatica, adatta alla produzione di passiti. L'acidità sostenuta presenta svantaggi quando tali vitigni sono vinificati con metodiche tradizionali, perché i vini appaiono sbilanciati e non apprezzati dal consumatore e anche il valore del giudizio complessivo attribuito dai giudici ne risente (valori bassi). I passiti risultati dalle stesse uve, invece, dopo l'intero processo di appassimento in serra, presentano una percezione sensoriale dell'acidità che valutata sulla base degli score attribuiti dal panel, non è in alcun modo significativamente differente dai relativi vini base. In questo caso la percezione di acidità viene valutata dal panel di esperti come piacevole, non penalizzando il vino passito. In prove future, si sta pensando di arricchire il lavoro di ricerca anche con un periodo di invecchiamento dei passiti prodotti seguendo il disciplinare di produzione della Malvasia di Lipari DOC delle isole Eolie. Con i vini base ottenuti, interessante, sarebbe un loro utilizzo per una prova di spumantizzazione, poiché le caratteristiche chimiche dei vini base, è idonea a tale produzione (basso grado alcolico e acidità sostenuta).

La Malvasia di Lipari passito anche nel trial sensoriale svolto al CREA-VE è effettivamente un prodotto molto apprezzato e superiore negli score agli altri passiti, con forte intensità olfattiva (composti terpenici); anche se, dalla prova sensoriale, emerge anche il Greco b. passito, un prodotto con ottime caratteristiche sulla base dei punteggi attribuiti e nelle note scritte nei commenti (dati non pubblicati). Nel Greco b., infatti, vengono evidenziate note di miele e frutta candita, emergendo come un prodotto che dal procedimento di appassimento in serra viene completamente valorizzato. Inoltre, i grappoli del Greco b. durante la disidratazione in serra hanno mostrato una notevole capacità di perdere peso (circa 26% in una settimana), questa caratteristica può essere interessante per fare differenti prove di appassimento, anche valutando una sua passificazione direttamente in pianta. La velocità nella disidratazione è interessante, allorché permette di ridurre i tempi di mantenimento dei grappoli in pianta durante l'appassimento, riducendo i rischi di tale pratica e i trattamenti chimici per mantenere le uve sane.

In un mercato vitivinicolo in forte sviluppo come quello Pugliese e dell'Italia meridionale in genere, è molto interessante la caratterizzazione e la valutazione delle potenzialità enologiche di queste varietà già conosciute ma con caratteristiche peculiari, che possono essere utilizzate per ottenere prodotti innovativi, in grado di competere con altri vini della stessa categoria già ampiamente affermati come ad esempio i Sauternes francesi o altri passiti Italiani già affermati.

References

[1] Angulo O., Fidelibus M.W. and Heymann H., Grape cultivar and drying method affect sensory characteristics. J Sci Food Agric **87**, 865–870 (2007)

[2] Barbanti D., Mora B., Ferrarini R., Tornielli G.B. and Cipriani M., 2008. Effect of various thermohygrometric conditions on the withering kinetics of grapes used for the production of "Amarone" and "Recioto" wines. J. Food Eng. **85**, 350–358

[3] Belfrage N. e Loftus S., Dried grapes: the classic wines of antiquity, journal of wine research **4**(3), 205–225 (1993)

[4] Bellincontro A., De Santis D., Botondi R., Villa L., Mencarelli F., Different postharvest dehydration rates affect quality characteristics and volatile compounds of Malvasia,Trebbiano and Sangiovese grapes forwineproduction.JSciFoodAgric **84**, 1791–1800 (2004)

[5] Bellincontro A., Nicoletti I., Valentini M., Tomas A., De Santis D., Corradini D., et al., Integration of nondestructive techniques with destructive analyses to study postharvest water stress of winegrapes. Am. J. Enol. Vitic **60**, 57–65 (2009)

[6] Borsa D. e Di Stefano R., Evoluzione dei polifenoli durante l'appassimento di uve a frutto colorato. Riv Vitic Enol **53**, 25–35 (2000)

[7] Doneche B.J. Botrytized wines. In Graham H. Fleet (ed.), wine microbiology and biotechnology (pp. 327–351). Sidney: Taylor & Francis Inc. (1993)

[8] Frangipane M.T., Ceccarelli A., Mencarelli F. and Anelli G-, Study of phenolic compounds in Aleatico grapes dried in a forced air tunnel. ItalJFoodSci **19**, 203–206 (2007)

[9] Gabas A.L., Menegalli F.C. and Telis-Romero J., Effect of chemical pretreatment on the physical properties of dehydrated grapes. Dry Technol **17**, 1215–1226 (1999)

[10] Genovese A., Gambuti A., Piombino P. and Moio L.. Sensory properties and aroma compounds of sweet Fiano wine. Food Chem. **103**, 1228–1236 (2007)

DOI: 10.1051/bioconf/20170904001

[11] Guarrera N., Campisi S. and Asmundo C.N., 2005. Identification of the odorants of two Passito wines by gas chromatography-olfactometryand sensory analysis. Am. J. Enol. Vitic. **56**, 394–399
[12] Pangavhane D.R. and Sawhney R.L., Review of research and development work on solar dryers for grape drying. Energ Convers Manage **43**, 45–61 (2002)
[13] Rolle, L., Torchio, F., Giacosa, S. and Gerbi, V. (2009), Modifications of mechanical characteristics and phenolic composition in berry skins and seeds of Mondeuse winegrapes throughout the on-vine drying process. J. Sci. Food Agric. **89**, 1973–1980
[14] Scienza A., Atlante dei Vini Passiti Italiani, Gribaudo, Savignano (CN), Italy (2006)

Análisis del desarrollo sostenible del territorio en Vale dos Vinhedos, Rio Grande do Sul, Brasil, en la década de 2000 / Analysis of sustainable territorial development in the Vale dos Vinhedos, Rio Grande do Sul, Brazil, in the 2000s

Marielen Aline Costa da Silva, Marcos Vinicius Araújo, y Kelly Lissandra Bruch

CEPAN: Centro de Estudios y Investigación en Agronegocios – Universidad Federal do Rio Grande do Sul, Brasil

Abstract. With a focus on the territories and their development, and in face of the expansion of New Vitiviniculture World, territories such as the "Vale dos Vinhedos", located in the southern region of Brazil, have increasingly awakened the interest of entrepreneurs and researchers, serving as a "model" of human development and progress of wine-related activities. Seeking to understand part of the complexity of the process of territorial development of the "Vale dos Vinhedos" this article proposes to portray the process of development of the municipalities that are encompassed by this territory, at the years 2000 and 2010, by means of the Sustainable Development Index (IDS) - developed by the Inter-American Institute for Cooperation on Agriculture (IICA). The observed data showed that, in general, the indices of sustainable development of each municipality, as well as the aggregate index of the "Vale dos Vinhedos", do not undergo significant transformations from the year 2000 to the year 2010, nor do they express satisfactory development results for the four dimensions analysed. These aspects can be explained by conflicts that are consolidated around the productive and traditional practices and other factors related to the dynamics of the territory

1. Introducción

Entre los años 1950 y 1980, el debate sobre el desarrollo – asociado a la idea de crecimiento económico, vio la expansión de su marco teórico y su crisis suceder en un corto período de tiempo cuando diversos proyectos de desarrollo, que fueron colocados como promesas para la sociedad, se enfrentan a las realidades socioeconómicas que no las legitimaban como alternativas válidas a la sociedad en su conjunto, como se subraya Aguilar et al. [1]. Incluso según los autores, paralelamente las discusiones en torno a la temática desarrollo, emerge en la escena internacional el tema de medio ambiente y su relación con los modelos de crecimiento / desarrollo y con los padrones de producción de consumo asociados con ellos.

Con el avance de los debates sobre desarrollo y sustentabilidad nuevos lentes angulares pasan a permear esas discusiones, como el abordaje territorial. Según Flores y Medeiros [2], con el avance de la globalización el territorio pasó a ser valorizado, siendo persibido como un sistema abierto y dinámico, que sufre y recibe influencias y puede ser abordado en áreas yendo de lo micro a lo macro, dando espacio a la noción de desarrollo local, donde la identidad y las características de cada término pasan a ser un activo capaz de estimular su propio desarrollo, como afirma Azevedo [3].

Dentro de esta perspectiva surge la idea de desarrollo territorial sustentable (DTS). Según Viera [4], el enfoque territorial del desarrollo exige un movimiento efectivo de transferencia de poder del espacio nacional a los espacios regionales y locales, así como necesita un nuevo padrón de inter relacionamiento entre las áreas decisorias de esos diferentes espacios, compatibilizando la gestión y el planeamiento con las especificidades locales de cada uno de ellos, considerando en sí, la complejidad inherente a los sistemas vivos y dinámicos.

Para Vieria y Pellin [5] entre las actividades que vienen siendo implementadas y o estimuladas como estrategias de desarrollo territorial, está la vitivinicultura. Las estructuras productivas locales (EPLs)[1] vitivinícolas además de ser responsables por las características productivas de los llamados territorios del vino, están también ligados a los aspectos culturales e institucionales del ambiente local. En este contexto Flores [7] argumenta que:

> "Los MPLs vitivinícolas constituyen un tipo especial de método... por la posibilidad de estar relacionados al turismo, práctica que recibe la terminología especial de Enoturismo y agregara a ese método características de un cluster de turismo y o territorios turísticos-actividad que, por otra parte, puede actuar creando o modificando territorios y territorialidades".

Según la autora [7], la formación de un cluster de turismo tiende a favorecer la cooperación y la participación entre agentes e instituciones, lo que converge en las

[1] Según Capolari & Volk [6] las EPLś constituyen un tipo particular de cluster, formado por pequeñas y medianas empresas, agrupadas en torno de una profesión o de un negocio, donde se enfatiza el papel desempeñado por los relacionamientos - formales e informales- entre empresas y demás instituciones relacionadas. Las firmas comparten una cultura común e interactúan, como un grupo, con el ambiente sociocultural local.

Cuadro 1. Variables utilizadas en cada dimensión para el cálculo del IDS.

Dimensión			
Social	Ambiental	Económicas	Demográfica
Mortalidad Infantil;	% domicilios con agua corriente;	Indice de Gini;	Tasa de Urbanización;
IDHM Longevidad;	% domicilios con baño y agua corriente;	%desocupados en el sector agropecuario	Tasa de Envejecimiento;
IDHM Educación	% de domicilios com recoleccion de residuos;	- 18 años o mas;	Razon Población masculina femenina;
Jovenes entre 18 a 20 años con secundaria completa;	% de domicilios con agua y desagote sanitario inadecuado	IDHM Renta;	Razon Población rural urbana;
% vulnerable a la pobreza		Renta percapita	

perspectivas del desarrollo territorial sustentable. Para Baida [8] las ventajas de implementaciones de clusters para el desarrollo económico y competitividad territorial, justifican el significativo aumento de políticas para la ploriferacion de clusters a nivel global.

A partir de esta perspectiva, de miradas puestas sobre los territorios, y su desarrollo y frente a la expansión de los EPLś vitivinícolas del Nuevo Mundo, territorios como el Vale dos Vinhedos pasan a despertar, cada vez más, el interés de emprendedores e investigadores. En este sentido, buscando comprender parte de la complejidad del proceso de desarrollo territorial del Vale dos Vinhedos - región de mayor producción de vino de Brasil, localizado en la Serra do Rio Grande do Sul, el presente artículo se propone retratar la situación del territorio en los años 2000 y 2010. Para eso, se optó por el método de Índice de desarrollo Sustentable (IDS) desarrollado por el Instituto Interamericano de Cooperación por la Agricultura (IICA).

2. Metodología

La propuesta metodológica del presente artículo tiene como base el cálculo del Índice de Desarrollo Territorial Sustentable (IDTS), desarrollado por el Instituto Interamericano de Cooperación para la Agricultura (IICA). La metodología consiste en la recolección y sistematización de indicadores representativos de cada una de las dimensiones del desarrollo sustentable y permite la realización de observaciones rápidas, bien como un análisis comparativo de los niveles de desarrollo sustentable en diferentes períodos y locales.

Para este trabajo se utilizo como dimensiones de análisis las dimensiones: sociales, económicas, ambiental y demográfica, siendo que para cada dimensión analizada fueron seleccionados entre 4 y 5 variables representativas, las cuales fueron obtenidas a través del Atlas de Desarrollo Humano Municipal de Brasil (Cuadro1). Segundo Waquill et al. [9] son consideradas múltiples dimensiones, cada una compuesta por un amplio conjunto de variables, con la intención de captar la complejidad y la diversidad de los procesos de desarrollo territorial. De esta forma, la justificación para la elección de las variables en cada dimensión es la búsqueda de la caracterización y distinción de los territorios con gran riqueza de informaciones, sin sobreponerlas en exceso.

En virtud de las diferentes unidades de medida de cada variable, estas fueron transformadas en índices ajustando los valores observados en las variables a escala cuyo valor mínimo es 0 (cero) y el valor máximo es igual a 1 (uno). Según la presente metodología es necesario identificar si las variables presentan una situación en que, al aumentar su valor, favorece o desfavorece el proceso de desarrollo [9].

La operación es echa de la siguiente forma:

- Si es positiva

$$I = x - m/M - m \quad (1)$$

- Si es negativa

$$I = \mathrm{M} - \mathrm{x}/\mathrm{M} - \mathrm{m} \quad (2)$$

Donde:

I: Índice calculado referente a cada variable, en cada período analizado:
x: Valor observado de cada variable en cada período analizado:
m: Valor mínimo considerado:
M: Valor máximo considerado:

Para la determinación de los valores máximos y mínimos de cada variable, se opto por observar la base de datos de 32 municipios pertenecientes al Consejo Regional de Desarrollo (COREDE) de la región Serra Gaúcha.

Así, después de las operaciones descriptas arriba se calculó el IDS de cada dimensión observada, para cada municipio perteneciente al Vale dos Vinhedos (Bento Gonçalves, Garibaldi e Monte Belo do Sul) en los años 2000 y 2010 y finalmente el IDS fue agregado al territorio Vale dos Vinhedos. Cabe resaltar que el IDS varia de 0 a 1, siendo que cuanto más próximo a 1 más cerca del nivel óptimo de desarrollo sustentable está el territorio.

3. Resultados y discusión

A través de la observación de los Ìndices de Desarrollo Sustentable (IDS) del territorio Vale dos Vinhedos se puede persivir que la dimensión social, en el año 2000, el municipio de Bento Gonçalves presento un IDS considerado inestable (0,51) así como Garibaldi y Monte Belo do Sul (0,53 y 0,46 respectivamente). Este resultado se repitió en el año 2010, demostrando una evolución no significativa en los indicadores sociales de los municipios como muestra la tabla 1. Lo mismo ocurre con el análisis del IDS social del territorio agregado1 do Vale dos Vinhedos, cuyo índice se mantuvo inestable durante los años 2000 y 2010. La inestabilidad de la dimensión social en los tres municipios y en el territorio agregado, aparentemente puede estar asociado a las variables vinculadas a la educación, como a la taza de jóvenes con secundaria completa, que quedo muy por debajo de los valores máximos observados en la región de COREDE Serra Gaúcha. Además de la educación, se observo que el porcentaje de vulnerables a la pobreza también se mantuvo distante de los niveles mínimos observados en los demás municipios de la región en ambos años analizados.

En cuanto a la dimensión económica, en los años 2000 los tres municipios, así como el territorio agregado, presentaron IDS inestable (entre 0,4 y 0,6). Ya que en el año 2010 el municipio de Monte Belo do Sul presento IDS económico considerado crítico (0,28) en cuanto Bento Gonçalves tuvo un IDS económico inestable de (0,48) y Garibaldi IDS estable de (0,67). Los datos demuestran

Tabela 1. Ìndices por dimensiones e índices agregados de desarrollo sustentable por municipio do Vale dos Vinhedos en los años 2000 y 2010.

Municipio	Dimenzion	IDS Período /Año 20000	IDS Período /Año 2010
Bento Gonçalves	Social	0,51	0,40
	Económica	0,54	0,48
	Ambiental	0,65	0,49
	Demográfica	0,49	0,73
	IDS (média aritmética)	**0,55**	**0,56**
	IDS (Média harmónica)	**0,53**	**0,50**
Garibaldi	Social	0,53	0,40
	Economica	0,54	0,67
	Ambiental	0,62	0,61
	Demográfica	0,51	0,49
	IDS (média aritmética)	**0,57**	**0,54**
	IDS (Média harmónica)	**0,57**	**0,52**
Monte Belo do Sul	Social	0,46	0,50
	Económica	0,49	0,28
	Ambiental	0,65	0,62
	Demográfica	0,26	0,28
	IDS (média aritmética)	**0,51**	**0,42**
	IDS (Média harmónica)	**0,45**	**0,37**

IDS - Ìndice de Desarrollo Sustentable.

que, a pesar de las tres ciudades tengan la vitivinicultura como una de las principales actividades económicas, apenas las variables de Garibaldi apuntan para una efectiva mejoría del índice económico en un período de 10 años. Tal evolución está vinculada a una expansión de la ganancia *percapita* del IDHM ganancias, que quedaron muy próximos a los mejores valores observados en la región. Vale destacar que Garibaldi tiene en la composición de su base económica grandes empresas del sector metal-mecánica, las cuales fueron responsables por las mayores recetas liquidas y facturaciones en el final de la década del 2000. Como indica el balance económico realizado por el gobierno municipal.

En lo que corresponde a la situación observada en el municipio Monte Belo do Sul- disminución del IDS económico de estado inestable para crítico- queda evidenciado una vez más una diferencia entre los valores máximos (que representan las mejores condiciones económicas, como ganancias y IDHM ganancias) encontrados y los valores percibidios en el municipio en cuestión. Algunos factores que supuestamente pueden explicar esta diferencia son las características relacionadas a los tipos de actividades económicas desarrolladas en Monte Belo do Sul. En cuanto el municipio tiene una economía típicamente agrícola, con aproximadamente el %50 del PBI municipal de la década del 2000 originario de actividades agropecuarias (EMBRAPA 2008). Otros municipios de COREDE Serra Gaúcha, como Caxias do Sul, Bento Gonçalves, Garibaldi poseen como fuente de sus PBIś municipales la recaudación obtenida de las industrias metal-mecánica, procesamiento de alimentos y mueblería, como las ya mencionadas. En este sentido, los indicadores cuantitativos de desarrollo pueden subestimar la realidad del municipio. Un ejemplo es la comparación entre el IDS aquí analizado y el IDESE (Ìndice de Desarrollo Socioeconómico) calculado por la Fundación de Economía y Estadística del estado Rio Grande do SUL (FEE) que demuestran un desarrollo socioeconómico

Tabela 2. Ìndices por dimensiones e índices agregados de desarrollo sustentable por municipio do Vale dos Vinhedos en los años 2000 y 2010.

Período/Año	2000					
Território	Dimension Social	Dimension Económica	Dimension Demográfica	Dimension Ambiental	IDS (média aritmética)	IDS (média harmónica)
IDS Vale dos Vinhedos	0,50	0,53	0,42	0,64	0,54	0,53
Máximo*	0,72	0,8	0,67	0,72	0,63	0,68
Mínimo*	0,28	0,2	0,22	0,19	0,38	0,33
Período/Año	**2010**					
Território	Dimension Social	Dimension Económica	Dimension Demográfica	Dimension Ambiental	IDS (média aritmética)	IDS (média harmónica)
IDS Vale dos Vinhedos	0,43	0,48	0,42	0,65	0,50	0,48
Máximo*	0,57	0,8	0,64	0,85	0,66	0,58
Mínimo*	0,26	0,0	0,23	0,39	0,39	0,43

*Fueron considerados como valores máximos y mínimos extremos los valores observados en los municipios pertencentes a COREDE SERRA, en los años de 2000 e 2010. Apenas El municipio de Pinto Bandeira fuei excluído por no poseer base de datos completa.

medio en el municipio en el año 2010, y no crítico como muestra el presente estudio.

En relación a la dimensión ambiental, los tres municipios muestran índices semejantes, tanto en el año 2000 como en 2010, manteniéndose del lado del desarrollo de la dimensión ambiental estable. Lo mismo ocurrió en el territorio agregado que tuvo un IDS ambiental de 0,64 y 0,65 en 2000 y 2010, respectivamente. En comparación con las demás ciudades pertenecientes al COREDE Sierra Gaúcha, los municipios de Vale dos Vinhedos tuvieron un desempeño inferior, de modo que el mejor IDS ambiental entre las 32 ciudades del Consejo fue de 0,85 en el año 2010. Según informaciones de la Secretaria de Planeamiento, Movilidad y Desarrollo Regional del estado Rio Grande do Sul, el bajo desarrollo de la dimensión ambiental está asociada a el desecho de residuos de la población y de las grandes industrias en los ríos que forman las innumerables cuencas hidrográficas en los municipios con mayor densidad demográfica como Bento Gonçalves, Caxias do Sul, Garibaldi.

En relación a la dimensión demográfica, los municipios de Bento Gonçalves y Garibaldi mantuvieron IDS demográfico considerados críticos en los 2 años observados. Una vez más, este resultado refleja la diferencia entre los mejores valores observados en la región y los valores observados en Monte Belo do Sul. Esta diferencia es explicada por la dinámica del municipio, con baja densidad poblacional, desequilibrio en las relaciones entre población masculina y femenina, bien como en la población rural y urbana. Vale dos Vinhedos, de modo general mantuvo su IDS inestable (0,42) para esa dimensión en ambos años.

Cuando se observan los valores del Ìndice de Desarrollo Sustentable agregado (todas las dimensiones) del Vale dos Vinhedos y de cada municipio integrante del territorio, como muestra la Tabla 2, se percibe que no ocurren mudanzas significativas en ese período de 10 años, siendo que el IDS agregado se mantuvo del lado considerado inestable, con valores de 0,54 en el año 2000 y 0,50 en 2010. Resultado también observado individualmente en cada municipio que constituye el Vale dos Vinhedos. Aún así, vale resaltar que observando los IDS de los municipios pertenecientes a COREDE Serra Gaúcha los mejores resultados tuvieron IDS considerados estables (entre 0,6 y 0,8) tanto en el año 2000 como 2010.

DOI: 10.1051/bioconf/20170904002

4. Consideraciones finales

La presente investigación busco explorar cuestiones relativas al desarrollo territorial sustentable en el Vale dos Vinhedos, importante territorio del vino en el sur de Brasil. Para esto, fueron analizadas cuantitativamente cuatro importantes dimensiones-económica, social, ambiental y demográfica que caracterizan la relación territorio-sustentabilidad-desarrollo, a través de la metodología del Ìndice de Desarrollo Sustentable (IDS), desarrollada por el Instituto Interamericano de Cooperación por la Agricultura (IICA).

De modo general, los resultados observados demuestran que, aun que se halla acompañado un proceso de expansión de la vitivinicultura de la región- con la consolidación del territorio como referencia de enoturismo la conquista de las indicaciones geográficas, la exportación y premiación mundial de sus productos, entre otros, las dimensiones que garantizan un desarrollo sustentable, con énfasis en los sectores endógenos del territorio, no evolucionaran significativamente en el período analizado. Tales aspectos pueden ser explicados por conflictos que se consolidan entorno a las actividades productivas locales (vitivinicultura, metalmecánica, transformación) como la exclusión de la población local a partir del proceso de especialización productiva, bien como factores ligados a la propia dinámica territorial y relaciones político-institucionales.

A pesar de eso, otro factor a ser resaltado en el análisis del desarrollo sustentable del territorio y la limitación de la metodología utilizada. En general la construcción de herramientas que mensuren el proceso de desarrollo está directamente vinculado a los debates y emergencia de diferentes abordajes sobre el tema. En cuanto, muchas veces, metodologías cuantitativas, como el uso de indicadores, no consiguen expresar la realidad, apenas expresan informaciones que simplifican la complejidad de los datos y del territorio. Tal situación puede ser persibída en el correr de este estudio. La utilización de una base para valores máximos y mínimos de municipios con características demográficas económicas tan distintas del territorio Vale dos Vinhedos, a veces, intensifico la limitación de la metodología en describir la realidad de los municipios analizados. Este es el caso del municipio de Monte Belo do Sul, que en la dimensión económica presento indicadores críticos, más un posible análisis cuantitativo de las variables y del objeto de estudio, o sea, el municipio en cuestión, podría tener resultados que no correspondan a la integra realidad local, lo que indicaría graduaciones entre esas metodologías.

Así, se puede concluir que, a pesar de las limitaciones metodológicas, los resultados obtenidos en relación a el vale dos Vinhedos, indican que mismo con la expansión del EPL vitivinícolas, este todavía no alcanzo niveles deseables de desarrollo territorial sustentable. De esta forma, se sugiere que nuevos estudios y otras metodologías sean aplicadas a fin de verificarse de manera más intensa o real proceso de desarrollo del territorio en cuestión. A pesar de eso, se recomienda que otras dimensiones y mas variables sean abordadas en nuevos análisis.

Referencias

[1] M. D. S Aguiar, O. T. Souza, E. E. Filippi, L. Beroldt. Aná. **20**, 84 (2009)
[2] S. S. Flores, R. M. Medeiros, V Enc. Gr. Pesq. Agr. Desen. Reg. e Transf. Socioesp. (2009)
[3] B. Azevedo. Desen. Quest. 99 (2003)
[4] P. F. Vieira. Desenv. Territ. Sust. **4** (2006)
[5] A. C. P. Vieira, V. Pellin. *2ºSem. Nac.l de Planej. e Desenv* (2014)
[6] R. Capolari, Volk, P. SEBRAE (2004)
[7] S.S. Flores, S. S. UFRGS, 153 (2011)
[8] J. A. I. Baidal, J. F. V. Rebollo, A. Fernández, *Clusters de Turism: contrib. do Vale dos Vinhedos econ. Sul do Brasil* (2016)
[9] P. D. Waquil, S. Schneider, E.E. Filippi, M. A. Conterato, S. Specht. *PGDR/UFRGS* (2005)

DOI: 10.1051/bioconf/20170904003

La Cultura del Vino, motor del desarrollo sostenible de las regiones vitivinícolas / The Wine Culture, a driver of the sustainable development of wine regions

Mª Luisa González-SanJosé[1,a], Vicente Gómez-Miguel[2], y Vicente Sotés[2]

[1] Universidad de Burgos (UBU), Facultad de Ciencias, 09001 Burgos, Spain
[2] Universidad Politécnica de Madrid (UPM), ETSIAAB, 28040 Madrid, Spain

Resumen. Existen bastantes evidencias para vincular la cultura del vino con el entorno y el desarrollo sostenible. En primer lugar, destaca que la cultura del vino es el reflejo del aprovechamiento vitivinícola del medio y, por tanto, refleja su efecto sobre el desarrollo social y económico de las regiones, lo que es plausible a través del devenir histórico de las mismas. Más recientemente, el aprovechamiento de la cultura del vino para su explotación a través de las experiencias enoturísticas, ha supuesto una revalorización de muchos territorios, impulsando el desarrollo social y económico. La cultura del vino no es sólo el eje temático de la actividad enoturística, sino que es su base y pilar esencial, por tanto, se relaciona directamente con la sostenibilidad ambiental, económica y social que subyace tras el enoturismo. Además, es importante tener en cuenta que gran parte de la importancia de la cultura del vino se centra en las señas de entidad asociadas a cada territorio, que aportan características distintivas y singulares, contribuyendo a la sostenibilidad cultural. En España, ACEVIN y el Rutas del Vino de España son claros ejemplos de como la cultura del vino es un motor del desarrollo sostenible de las regiones vitivinícolas.

Abstract. Based on definitions of sustainability and wine culture, the purpose of this article is to establish the relationships and contributions of the wine culture, mainly through the enotourism, to the creation and maintenance of sustainable environments in the wine regions. The contents can be summarized as follows: sustainability is a model of development in which the present actions should not compromise the future of future generations, and is linked to economic and social development which has to be respectful with the environment; wine tourism can be defined as the development of tourism, leisure and free time activities dedicated to the discovery and enjoy the cultural and oenological aspects of the vineyards, wines and of the wine regions; the main pillars of wine tourism are the wine culture, the territory and the tourism; and quality enoturism activities should be sustainable, authentic and competitive. In this way, wine culture, and its use through wine tourism experiences, is clearly correlated to social (socially equitable), economic (economically feasible), environmental (environmentally sound) and cultural aspect of the sustainability of the winegrowing regions and territories. In Spain, ACEVIN is an association that clearly illustrates the role of wine culture in the sustainable develop of the wine regions.

1. Contextualización del trabajo: conceptos de sostenibilidad y de cultura del vino

El término sostenibilidad se ha venido haciendo, desde la década de los ochenta del siglo pasado, cada vez más habitual y, actualmente, es un término que se une frecuentemente a muchos aspectos cotidianos (gasto energético y de agua, gestión de residuos, reciclaje, ...). Además, el desarrollo sostenible ha tomado gran relevancia al hacerlo las evidencias del cambio climático y sus posibles consecuencias.

El diccionario de la Real Academia Española (RAE) define la sostenibilidad como "la cualidad de sostenible", y otorga, en segunda acepción, al adjetivo sostenible la siguiente definición "especialmente en ecología y economía, que se puede mantener durante largo tiempo sin agotar los recursos o causar grave daño al medio ambiente" citando como ejemplos desarrollo y economía sostenible. Esta definición concuerda con el concepto de sostenibilidad como modelo de desarrollo en el que las acciones presentes no deben comprometer el futuro de las generaciones venideras, y que se vincula con un desarrollo económico y social respetuoso con el medio ambiente. No obstante, la sostenibilidad encierra ciertos aspectos filosóficos, y la apuesta por el desarrollo sostenible implica la puesta en valor de determinados principios éticos [1], y se considera que la Carta de la Tierra (2000) [2] es el documento que recoge de modo más completo los valores y principios éticos y globales para el desarrollo de un mundo sostenible.

Todas las actividades humanas pueden enfocarse bajo el prisma de la sostenibilidad, y por ello, en función de los contextos se consideran diversas parcelas o tipos de sostenibilidad, destacando o siendo las más habituales la sostenibilidad ambiental, la económica y la social.

[a] e-mail: marglez@ubu.es

La primera se centra en la protección de los sistemas biológicos, en la preservación de los recursos naturales y en el fomento de la conciencia ecológica, apostando por un desarrollo humano fuertemente respetuoso con el medio ambiente. La segunda se focaliza en la generación de riqueza de modo equitativo, así como en la producción y consumo equilibrados, teniendo siempre presente la explotación respetuosa del medio para no comprometer el futuro. La tercera se refiere a la generación del bien social, de la igualdad de oportunidades, del acceso a la educación, la sanidad, etc. y la inculcación de valores de respeto a la naturaleza y del uso armonioso de la misma. Menos conocidas, pero no por ello menos importantes son la sostenibilidad política y la cultural. La distribución del poder político y económico, junto con la generación de estructuras democráticas y de marcos jurídicos de respeto de los derechos humanos y del medio ambiente, fomentando las relaciones solidarias entre los países, naciones, comunidades, etc. son las bases de la sostenibilidad política, mientras que la sostenibilidad cultural se enfoca hacia el respeto, defensa y protección de las diferentes entidades culturales, haciendo suya la parte de la Declaración Universal de la UNESCO sobre Diversidad Cultural (Unesco, 2001), en la que se hace hincapié en que la diversidad cultural es para el hombre tan importante como la diversidad biológica para los seres vivos.

Independientemente de los contextos, los niveles o las escalas temporales, la sostenibilidad es un todo al que contribuyen numerosos aspectos, todos ellos interrelacionados y con fuerte incidencia entre sí. Por tanto, cualquier acción de desarrollo sostenible debe enfrentar consideraciones multidireccionales que permitan un desarrollo equilibrado, considerando como tal un adecuado equilibrio entre el desarrollo socio-económico y cultural de los pueblos y la conservación del medioambiente o entorno natural con el que coexisten.

No existe una única definición de cultura (se han compilado hasta 164 definiciones) y de modo análogo no existe una definición estándar, consensuada y universal de la Cultura del Vino. No obstante, extrapolando o aplicando "al mundo del vino" algunas de las definiciones de cultura, se podría llegar a definiciones más o menos similares. Sin ningún ánimo de polimerizar al respecto, se propone una definición personal, partiendo de dos puntos base, la idea de que la cultura es un concepto multiparamétrico y de dos de las acepciones de la RAE para el vocablo "cultura" que son: "2. f. Conjunto de conocimientos que permiten a alguien desarrollar su juicio crítico. 3. f. Conjunto de modos de vida y costumbres, conocimientos y grado de desarrollo artístico, científico, industrial, en una época, grupo social, etc.". Así, se considera que la cultura del vino es el conjunto de conocimientos, costumbres, modos de vida, expresiones artísticas, científicas e industriales, vinculadas al mundo del vino en sus múltiples facetas, desarrolladas a lo largo de la historia de la humanidad. Es decir supone conocimientos de los vinos, cómo son, dónde, cómo y desde cuándo se hacen, su evolución histórica, sus propiedades (ventajas e inconvenientes), etc., pero también tener conocimientos del medio (suelo, clima, paisaje, etc.), de la planta (variedades de vid, sistemas de cultivo, enfermedades, resistencias, ...), de las limitaciones de las zonas de producción, etc., además de conocimientos de los usos y costumbres, de las tradiciones vitivinícolas, gastronómicas, folclóricas y artísticas (literatura, pintura, escultura, arquitectura, ...) de los diferentes grupos sociales y épocas. Todo ello sin olvidar su vínculo con el desarrollo social y económico y el papel medioambiental del viñedo.

Análogamente, otros autores han dicho que la cultura del vino se debe entender como el conjunto de las manifestaciones que se producen en un territorio determinado y que tienen como eje el que-hacer del vino [3] (Elías Pastor, 2006).

Al respecto de la cultura, en la Declaración de México de la Unesco (1982) puede leerse un fragmento que dice así "*...A través de la cultura el hombre se expresa, toma conciencia de sí mismo, se reconoce como un proyecto inacabado, pone en cuestión sus propias realizaciones, busca incansablemente nuevas significaciones y crea obras que lo trascienden. ...*". Por semejanza podría decirse que, a través de la cultura del vino, considerando como sujeto no sólo al vino sino también lo relacionado con él como las construcciones vitivinícolas (desde las de las viñas a las bodegas), los tratados de cultivo y elaboración, las obras artísticas (literarias, pictóricas, escultóricas, incluso a través de las etiquetas y las botellas, ...), etc., el hombre y la mujer se expresan, toman conciencia de sí, de su entorno, de la historia y otros. Además, seguramente todas las creaciones (vinos, bodegas, recetas, esculturas, espacios enoturísticos, etc.) son el fruto de la reflexión y puesta en cuestión de las propias realizaciones, así como de la búsqueda de nuevas significaciones y, por último, todos ellos transcienden a sus creadores.

2. La cultura del vino como motor del desarrollo sostenible

La cultura del vino contribuye al desarrollo sostenible de los pueblos y regiones vitivinícolas de modo multidimensional, tal y como es requerido, como ya se ha comentado, por la propia naturaleza de la sostenibilidad. Así, además de contribuir a la sostenibilidad cultural, como es obvio, tiene, ha tenido y tendrá una fuerte relación con desarrollo socio-económico (sostenibilidad económica y social), con la explotación cuidada del medio (sostenibilidad del medio), y incluso contribuye a la sostenibilidad política. Todas estas relaciones se comentan más detalladamente en los apartados siguientes.

2.1. Contribución de la cultura del vino al desarrollo del turismo sostenible

Una de las potencialidades de la cultura del vino para el desarrollo sostenible de las regiones es el enoturismo, entendido como una estrategia o modelo de explotación económica y de gestión de la cultura del vino. Es conocido que ya griegos y romanos gustaban de las visitas a las bodegas, pero el enoturismo en el sentido moderno se pued decir que nació en el siglo XIX, como actividad turística ejercida por las clases sociales más altas. No será hasta el siglo XX, cuando las visitas a las bodegas, actividad base de las actividades enoturísticas de la época, se popularizase, y hasta finales de siglo e incluso hasta los inicios del XXI, no se piensa en el enoturismo como un modo de desarrollo económico de los pueblos y regiones.

Existen bastantes evidencias para vincular el turismo enológico con el entorno y el desarrollo sostenible. De forma resumida el enoturismo supone un modo de ocio placentero, de desarrollo de experiencias lúdicas, que conjuga aspectos de la propia expresión humana, como el placer de los sentidos al degustar los vinos, con aspectos relacionados con la cultura, como la adquisición de conocimientos sobre los vinos, y con otros vinculados a los ecosistemas, como los viñedos, y con los propios actores (viticultores, bodegueros, . . .). Además, todo ello es reflejo de como las sociedades interaccionan con el medio para su aprovechamiento y desarrollo social y económico. A lo largo del devenir evolutivo del ser humano, éste se ha adaptado a entornos muy diversos a través del desarrollado de conocimientos y habilidades que le han permitido cubrir las necesidades surgidas en cada momento. Análogamente, en las diferentes épocas y regiones vitivinícolas, el hombre ha aplicado sus conocimientos para conseguir adaptar el cultivo de la vid al medio, haciendo de ello un modo de subsistencia y de desarrollo. A lo largo de los siglos, los conocimientos se han ido modificando, se han incorporado las innovaciones tecnológicas que han ido surgiendo, los vinos se han ido adaptando a las demandas del consumidor, los técnicas y prácticas de cultivo del viñedo se van adaptando al cambio climático, a la eficiencia energética, al respeto ambiental (control de uso de fertilizantes y plaguicidas, etc.), incluyendo los estudios de impacto que el cultivo de la vid puede suponer para en las transformaciones de la naturaleza y del paisaje. Todo ello, constituye parte del desarrollo cultural de las regiones vitivinícolas, en las que la cultura del vino ha marcado su devenir histórico y lo que son hoy en día. Por todo ello, se puede afirmar que la cultura del vino, y su uso a través de las experiencias enoturísticas, está claramente relacionada con la sostenibilidad social, económica, ambiental y cultural de los territorios y por tanto con el desarrollo sostenible de los mismos.

La Carta Europea del Enoturismo [4] indica que el enoturismo es un sistema multidimensional resultado de la interacción de los subsistemas Territorio, Turismo y Cultura del vino, y dice que "se entiende por enoturismo el desarrollo de las actividades turísticas y de ocio y tiempo libre dedicadas al descubrimiento y disfrute cultural y enológico de la viña, el vino y su territorio". La Carta Europea del Enoturismo es un documento de consenso que contiene los fundamentos y objetivos para el desarrollo estratégico del enoturismo y para la defensa y promoción del patrimonio histórico y natural del mundo del vino. Este documento señala que los principios del desarrollo sostenible son una de las bases principales del enoturismo, y todos aquellos que quieran adherirse a actividades reguladas por este documento debe aceptar y comprometerse al cumplimiento de este principio. Por tanto, bajo este prisma, el enoturismo adopta como principio básico el turismo sostenible.

El concepto de turismo sostenible nace hacia finales de los años ochenta (del siglo XX) en contraposición con el turismo de "masas" que termina provocando, en los territorios que lo soportan, una pérdida de calidad de los servicios y, en definitiva, del atractivo. Estos efectos, en general, se derivan de un crecimiento no controlado y que no tiene en cuenta ningún tipo de consideración medioambiental, social o cultural. La Organización Mundial del Turismo en 1999 [5] señaló que el turismo sostenible se sustenta en los siguientes principios: a) los recursos naturales, culturales y otros se conservan para su uso continuado en el futuro, al tiempo que producen beneficios a la sociedad actual; b) el desarrollo turístico se planifica y gestiona de forma que no cause serios problemas ambientales o socioculturales; c) la calidad ambiental se mantiene y mejora; d) se mantiene un alto nivel de satisfacción de los visitantes y el destino retiene su potencial comercial y su prestigio; e) los beneficios del turismo se reparten ampliamente a toda la sociedad.

Atendiendo estos principios, las estrategias de desarrollo enoturístico deben ser plurianuales y deben llevar asociado un programa de actividades que favorezcan y hagan viable el desarrollo socio-económico de las empresas y entidades adheridas, y en general de todo el territorio. Para ello, se establecen dos compromisos esenciales, uno para los agentes del territorio y otro para los agentes turísticos, que respectivamente son: "El territorio o todas aquellas entidades, tanto de carácter local, regional o nacional que tengan competencias en la planificación y gestión de un territorio deben comprometerse en la realización de un diagnóstico de las necesidades del territorio (amenazas y oportunidades) con el fin de definir la orientación enoturística más adecuada para el territorio en su conjunto" y "las empresas que realicen servicios turísticos deben comprometerse a llevar a cabo el diagnóstico de la actividad, adecuando la oferta a las expectativas de los clientes, y tomando medidas para la valorización del patrimonio local, así como para preservar el ambiente e impulsar el desarrollo sostenible en la gestión de la empresa" [4].

El Vademecum del Enoturismo Europeo (VEE) [6], documento a partir del cual se normaliza la metodología para la puesta en marcha, desarrollo y reconocimiento de las Rutas Europeas del Vino, establece el Sistema de Gestión de la Calidad del Enoturismo a nivel europeo. Este sistema, al igual que la Carta, otorga gran importancia a la sostenibilidad, centrando gran parte de ella en el aspecto ambiental, ya que el medioambiente es parte del atractivo enoturístico, constituyendo un valioso patrimonio y capital que hay que proteger. Así, el VEE insiste en que al planificar una Ruta del Vino debe considerarse la incidencia que se ejercerá sobre el medio y sobre todo sobre los recursos no renovables, de tal manera que la afluencia turística no supere en ningún caso la capacidad de carga del territorio, evitando así que se produzcan daños irreparables.

2.2. Contribución de la cultura del vino al desarrollo de la sostenibilidad medio ambiental

La contribución de la cultura del vino al desarrollo de sostenibilidad ambiental se desarrolla a lo largo de dos ejes principales, todo lo que tiene que ver con la sostenibilidad en la explotación agrícola-industrial del medio, y lo relacionado con la explotación sostenible del patrimonio cultural del vino, desde los paisajes a las expresiones artísticas, a través de los diferentes productos turísticos, pero especialmente, en Europa, a través de las Rutas Europeas del Vino.

El VEE considera que la planificación y ordenación del territorio es una herramienta clave para conseguir un desarrollo sostenible de una Ruta del Vino. Una adecuada planificación y ordenación del territorio permite establecer claramente el tipo de actividades soportables en cada territorio, la carga turística y, en definitiva, el aprovechamiento racional del territorio. Por otra parte, el VEE plantea la conveniencia de desarrollar sistemas de gestión medioambiental, insistiendo sobre todo en medidas para el control y gestión de residuos, incluyendo el reciclaje, para así evitar el deterioro del territorio, así como los posibles peligros tanto para la sociedad como para el ecosistema. Además, se indica que se debe trabajar en la reducción de otros tipos de contaminación como la acústica y la lumínica, e incluso la visual (ej. eliminación de excesivo cableado, retirada de elementos urbanos deteriorados, control publicitario, etc.) y se señala la conveniencia de trabajar para mejorar la estética y calidad del paisaje, incluido el urbano. En este proceso deben participar de modo intenso las empresas que además deben contribuir a la creación de una cultura medioambiental sostenible. Para el control del desarrollo de estas actividades el VEE propone la creación de sistemas indicadores de la "calidad" medioambiental.

La relación la explotación agrícola-industrial del medio para el cultivo de la vid se relaciona, directamente y en primera estancia, con la sostenibilidad ambiental por la relación que tienen las prácticas vitivinícolas y enológicas con el medio natural y, por este motivo, está relacionado con todos los aspectos derivados del uso sostenible del territorio, así como con la elaboración sostenible del vino. Por otra parte, el enoturismo se relaciona con la sostenibilidad ambiental a través de los beneficios indirectos, externalidades positivas o efectos positivos (no retributivos) a terceros. Probablemente las externalidades positivas del viñedo más ampliamente reconocidas son la fijación de dióxido de carbono, el efecto cortafuegos, la protección contra la erosión, el efecto corredor y refugio de fauna y, en determinadas ocasiones, la preservación de especies vegetales autóctonas (como pueden ser determinadas variedades de uva). Todos estos factores forman parte de las denominadas externalidades medioambientales [7] y son las que habitualmente se tienen en cuenta en los estudios ambientales del viñedo. Toda externalidad ambiental tiene un determinado valor y éste incide directamente en el valor ambiental del viñedo, parámetro que no es fácil de determinar, ni de valorar. Por ello, se han desarrollado diversos modelos "multicriteria" para poder estimar el valor ambiental de un viñedo en unidades monetarias, esto permite, entre otras cosas, contar con un indicador de su importancia en el bienestar de la sociedad, proporcionar un parámetro ante la restitución por posibles daños, compararlo con otros componentes del bienestar, mejorar la sensibilización de la sociedad sobre su importancia real, y servir a la administración para priorizar las actuaciones de gestión y protección.

Respecto al cultivo sostenible de la vid y la elaboración sostenible de vino, la OIV en sendas resoluciones (CST1/2004 y CST1/2008, respectivamente) define la vitivinicultura sostenible como una "estrategia global a escala de los sistemas de producción y procesamiento de las uvas, asociando a la vez la sostenibilidad económica de las estructuras y los territorios, obteniendo productos de calidad, considerando las exigencias de precisión de la vitivinicultura sostenible, los riesgos vinculados al medio ambiente, la seguridad de los productos y la salud de los consumidores, y valorando los aspectos patrimoniales, históricos, culturales, ecológicos y paisajísticos".

Conseguir una vitivinicultura ambientalmente sostenible implica tener en cuenta diversos factores que pueden dividirse en aquellos generales, aplicables tanto a la gestión del viñedo como de la bodega, como son la gestión del agua y de la energía o el manejo de los residuos, y otros específicos de cada parte del sistema productivo, es decir unos específicamente vinculados a la gestión del viñedo como son el uso fertilizantes y fitosanitarios, y otros específicos de la gestión de la bodega como es el tratamiento de efluentes. Además, en el caso de nuevas plantaciones debe considerarse la elección del emplazamiento de los viñedos, la adecuada selección varietal y del portainjerto, los posibles tratamientos del suelo y, en general, el efecto de las nuevas plantaciones sobre el territorio, incluyendo el efecto sobre la biodiversidad. En este sentido, entre los aspectos organizativos indicados en la resolución CTS1/2008 de la OIV sobre la vitivinicultura sostenible, se indica que es imprescindible "la identificación de zonas de protección y de interés ambiental y paisajístico, y la implementación, en su caso, de medidas de mejora en lo que se refiere a viñedos, edificios e instalaciones".

En general, el viñedo, si se maneja adecuadamente, juega un papel primordial en la protección del suelo contra la erosión [8]. Este hecho se pone en especial evidencia en los territorios en los que de no existir viñedo sería difícil la implantación de otros cultivos, ni prácticamente se instalaría otra vegetación que protegiera al suelo de la degradación, es decir, en zonas catalogadas como de baja capacidad de uso y consideradas zonas de protección, ya que cualquier acción regresiva sobre la cubierta vegetal produciría la pérdida del suelo (ej. zona vitivinícola de Requena (España) [7]).

2.3. Contribución de la Cultura del Vino al desarrollo de la sostenibilidad económica y social

La viña y los productos de la transformación de sus frutos, esencialmente el vino, ha condicionado fuertemente el desarrollo de las sociedades "vitivinícolas" que, han usado estos productos tanto para el consumo propio como para la comercialización, y que, en general, han alcanzado niveles socioeconómicos más elevados en aquellas zonas donde los vinos han tenido mayor prestigio y, por tanto, en las que los viñedos, la uva y el vino han alcanzado mayor valor añadido. Lamentablemente, no todas las regiones vitivinícolas gozan, ni han gozado, del mismo estatus socio-económico, si bien es cierto que, en general, en las zonas donde se ha cultivado y se cultiva la vid la renta per cápita suele ser superior que en las zonas circundantes donde no se produce o se ha producido vino.

Un reflejo del peso de las actividades vitivinícolas en el bienestar socioeconómico de las sociedades europeas fue el ataque de la filoxera a finales del siglo XIX. Este insecto, importado de América, casi supuso la desaparición de las vides europeas y produjo graves daños económicos y sociales en muchas regiones. En algunas de ellas, los daños fueron tales que, en esas regiones, nunca más se ha vuelto a recuperar la actividad vitivinícola.

Más allá de peso socioeconómico de la propia viticultura y de la enología, la explotación de la cultura del vino a través del enoturismo se asocia directamente con la sostenibilidad económica y social porque es un modo de contribuir a la generación de riqueza y a su distribución entre todos los agentes vinculados al vino (desde bodegueros a restauradores, gestores de museos, artesanos de la región, etc.). Por lo tanto, es una forma de contribuir al desarrollo del bien social de las propias sociedades que ocupan los territorios vitivinícolas, a la vez que contribuye a inculcar el respeto a la naturaleza a través de un uso armonioso de la misma.

Actualmente, las actividades enoturísticas engloban actividades que se desarrollan fuera de las bodegas, y que por tanto hacen que se involucren otros miembros de la región con actividades económicas diversificadas, desde la restauración (hoteles, restaurantes y bares), productores de alimentos y artesanía típica, organizadores de manifestaciones culturales y artísticas (exposiciones, museos etnográficos, conciertos, etc.), monitores de actividades de naturaleza (ornitología, senderismo, micología, piragüismos, etc.) Además, el desarrollo del enoturismo puede conseguir otros beneficios sociales importantes ya que puede contribuir a contrarrestar la despoblación del medio rural, y con ello conseguir la mejora de servicios como acceso a internet, mejora de las vías de comunicación, etc., todo ello con una clara repercusión en el desarrollo de las regiones y, por tanto, contribuyendo al desarrollo en igualdad de oportunidades de sus habitantes. Así, algunos autores han señalado que el turismo del vino impulsa el desarrollo socioeconómico integral y sostenible del territorio generando empleo y riqueza y mejorando la calidad de vida de la población local, además la actividad enoturística genera nuevas vías de ingresos ayudando a desarrollar la economía local [9]. Por otra parte, las mejoras asociadas al desarrollo del proyecto enoturístico aumentan el atractivo turístico del territorio, lo que puede atraer nuevas inversiones con la consiguiente generación de empleo, recaudación de impuestos, etc. La multitud de actividades enoturísticas que pueden ofertarse hacen de este campo un sector atractivo para los emprendedores. No son pocos los jóvenes emprendedores que han considerado y consideran el potencial de este sector para desarrollar iniciativas que les han permitido y permiten contrarrestar la falta de empleabilidad por terceros [10]. Por otra parte, un estudio de 2012 indica que, a pesar del aumento del enoturismo observado en los últimos años, aún queda mucho por explorar y por explotar [10] y por tanto la cultura del vino podrá seguir contribuyendo en el futuro al desarrollo socio-económico de las regiones vitivinícolas.

Concordando con lo comentado previamente, varios estudios realizados hasta la fecha ponen de manifiesto que el enoturismo tiene un valor estratégico con gran efecto socio-económico sobre todo en las regiones de interior, no obstante, los efectos no siempre se han cuantificado económica, son variables entre territorios y, en muchos casos, es necesario seguir trabajando para que éste sea un modelo realmente efectivo para la generación de riqueza y de desarrollo [11–15], entreotros.

La calidad y la competitividad de un destino enoturístico depende, en primera estancia y directamente, del grado en que el servicio o el producto cubren las expectativas del cliente, el enoturista. Por otra parte, en segundo lugar, la competitividad depende también del adecuado uso de los recursos desde el punto de vista económico, es decir de la rentabilidad a corto, medio y largo plazo, así como de los beneficios sociales que pueda generar. En este sentido, el VEE señala que "existe una relación directa y clara entre el concepto de sostenibilidad y la competitividad. La sostenibilidad de un destino enoturístico no se puede entender sin la competitividad y viceversa". Además, indica que "la mayor ventaja competitiva se consigue cuando el funcionamiento de todo el sistema enoturístico es único y existe una mutua dependencia y refuerzo entre sus componentes que lo hace más difícil de imitar". Desde este punto de vista, la competitividad también se nutre de la autenticidad y de las peculiaridades intrínsecas a la cultura del vino de cada uno de los territorios.

La instauración en 2009 del Día Europeo del Enoturismo (segundo domingo de noviembre), actividad que comenzó celebrándose de modo conjunto en todas las ciudades europeas asocias (más de 700) a la Red Europea de Ciudades del Vino (RECEVIN) y que actualmente se celebra también en ciudades de otros continentes como Argentina, Estados Unidos y Uruguay, pone de manifiesto el valor económico y social de este sector, y sobre todo su reconocimiento institucional. RECEVIN indica que el "día europeo del enoturismo" es una actividad directamente relacionada con la valorización y promoción de los territorios como activos económicos rentables, de tal modo que se consiga posicionarlos competitivamente en el mercado global de los territorios-producto. Además, esta iniciativa favorece el establecimiento de canales de distribución de mercado bien definidos, reforzando las estrategias colaborativas y de sensibilización intra e intersectoriales, y estimula la innovación comercial del turismo del vino. La promoción conjunta de todos los eventos organizados en todos los territorios genera un impacto de imagen y difusión mayor y más fuerte.

Es interesante tener en cuenta que el establecimiento del día europeo del enoturismo también ha favorecido la homogeneización de los estándares de calidad de las diferentes rutas del vino europeo, lo que tiene un claro y directo beneficio para el consumidor-cliente de este tipo de servicios.

Son escasos los trabajos publicados que han valorado la percepción de las comunidades sobre el efecto del enoturismo en su economía y calidad de vida. Recientemente, se ha publicado un trabajo realizado en una nueva comunidad vitícola situada al norte de California, la región de Piedmont, indicándose que la población percibe que el enoturismo ha favorecido el cuidado de los aspectos culturales y naturales de la comunidad, pero no consideran que haya ejercido ninguna influencia, ni positiva ni negativa, sobre el valor de las propiedades o sobre la calidad de vida, aunque si se señalan una percepción al alza de los impuestos y de los precios de los productos de consumo habitual. Además, indican que el enoturismo produjo que aumentara la variedad de locales comerciales y de las actividades culturales. No obstante, no se detectó un gran entusiasmo de los habitantes locales respecto a la implantación de las bodegas y la llegada de los enoturistas [16]. Este hecho se considera que podría deberse a que se trata de una región sin ninguna tradición vitivinícola, donde las bodegas son una actividad nueva con la que la sociedad de la zona no se ha visto involucrada directamente. La percepción podría ser diferente si los empresarios de las bodegas hubieran interaccionado adecuadamente

 DOI: 10.1051/bioconf/20170904003

con la comunidad, buscando sinergias con los habitantes de la zona.

2.4. Contribución de la Cultura del Vino al desarrollo de la sostenibilidad cultural

No cabe duda que, en torno al vino, los pueblos mediterráneos generaron hábitos y costumbres peculiares, en definitiva, modos de vida distintivos. Así, se dice que, entre otros productos, el cultivo de la vid y el consumo de vino fueron señas de identidad de los pueblos mediterráneos diferenciándolos de los pueblos del norte de Europa que, entre otras muchas cosas, se caracterizaban porque consumían "cerveza". Entorno al vino se generó cierta sofisticación, y son numerosas las presentaciones artísticas, escultóricas, pictóricas, literarias, etc., en las que se incorporan elementos vitivinícolas; así como ha sido constante la vinculación del vino con los dioses (Dionisio, Baco, Osiris y Gestín, entre otros [17], teniendo un papel simbólico importante en muchas religiones.

A lo largo de la historia, la riqueza de las zonas vitivinícolas se refleja en los palacios y viviendas señoriales, en las iglesias y monasterios, en la literatura, escultura y pintura, en la arquitectura de las propias bodegas, así como en el desarrollo de oficios y empresas auxiliares, como toneleros, curtidores, viveristas, etc. Hecho que, adaptado a los tiempos modernos, seguimos observando en fenómenos como bodegas de arquitectura singular construidas por arquitectos de gran renombre, desarrollo de museos del vino, empresas de servicios auxiliares incluyendo desde las elaboradoras de botellas o productos enológicos a las que construyen las depósitos, prensas y demás equipos enológicos, o las implicadas en todo lo relacionado con el viñedo (maquinaria, postes de conducción, sistemas regadío y fertilización, uso de fitosanitarios, etc.), entre otras muchas. Además, se deben tener en cuenta todas aquellas empresas vinculadas con la restauración, y aquellas vinculadas con aspectos más científicos como las editoriales especializadas, los centros de investigación, los encargados de mantener las colecciones varietales y la biodiversidad, etc.

Todo este bagaje histórico ha dejado en los países de la cuenca mediterránea un gran patrimonio histórico, artístico, arquitectónico, etnográfico, gastronómico, folclórico, etc., sin olvidar el patrimonio biológico (como las variedades de uva y los patrones) y enológico (como las formas y modos de elaborar vinos de distintos tipos y los diferentes tipos de equipos y maquinarias asociadas, etc.). En definitiva, un gran patrimonio cultural entorno al vino, que es lo que hace posible desarrollar un turismo enológico de calidad, rico y enriquecedor, diverso y sorprendente, que puede cubrir las expectativas de los diferentes tipos de turistas interesados en él. En todas las actividades turísticas que se pueden desarrollar entorno al vino, se difunde el conocimiento de la cultura del vino, lo que favorece el desarrollo del respeto hacia la misma, y por tanto su defensa y protección, ya que permite identificarla como una cultura con entidad propia. Con todo ello, claramente, se está contribuyendo a la sostenibilidad cultural de los pueblos y regiones vitivinícolas.

Otro aspecto a considerar respecto a la relación de la cultura del vino con la sostenibilidad cultural, es todo lo relacionado con el valor patrimonial del paisaje vitícola. Los paisajes de vitivinícolas son paisajes culturales (unidad en la que se integran las actividades humanas y el medio natural estableciendo una interacción dinámica que se manifiesta en hechos y rasgos físicos, testimonios del transcurso de una sociedad sobre un determinado territorio), del tipo evolutivos, porque fueron generados en el tiempo por transformaciones colectivas, y activos, porque mantienen su actividad y su transformación continúa. Además, muchos de ellos han sido reconocidos y galardonados y reconocidos como Patrimonio Inmaterial de la Humanidad.

2.5. Contribución de la Cultura del Vino al desarrollo de la sostenibilidad politica

La relación del enoturismo con la sostenibilidad política puede ser una faceta no muy evidente y claramente no es la faceta primordial. No obstante, al analizar conjuntamente tanto la definición de sostenibilidad política y las premisas de la Carta Europea del Enoturismo y del Vademecum Europeo del Enoturismo, es más fácil encontrar estos vínculos. Así, por ejemplo, la sostenibilidad política hace referencia a fomentar relaciones solidarias entre los países, naciones, comunidades, etc., hecho que coincide con la cooperación intensa y leal que se exige a todos los agentes implicados en el desarrollo del proyecto enoturístico que se adhieran a la Carta Europea de Enoturismo. Por otra parte, el VEE regula la metodología de puesta en marcha, desarrollo y reconocimiento de las Rutas Europeas del Vino, incluyendo el funcionamiento de los órganos de gestión, marcando las funciones, modos de elección, etc., asegurando la representación de todas las tipologías de socios. Por tanto, aunque sea de modo indirecto, las actividades de enoturismo que podríamos denominar "regladas" en el sentido de estar acogidas a una figura de reconocimiento internacional como las Rutas del Vino Europeas, contribuyendo en alguna medida a la distribución del poder político y económico, junto con la generación de estructuras democráticas a las que también hace referencia la sostenibilidad política.

Más allá de lo relacionado directamente con el enoturismo, la organización de vitivinicultores en cooperativas agrícolas, los órganos de gestión de Consejos Reguladores, las interprofesionales de los sectores vinculados al vino, etc., son otras muestras de la contribución de la cultura del vino a la generación de la cultura de la sostenibilidad política.

2.6. Posibles efectos no sostenibles del enoturismo

Tal y como se ha venido comentando, el enoturismo debiera ser una actividad que contribuyera al desarrollo sostenible del territorio, no obstante, esto requiere de la colaboración de todos los agentes implicados, y de no ser así los principios de sostenibilidad pueden verse comprometidos. De hecho, y especialmente en los inicios de esta actividad, hubo quienes no estaban de acuerdo con el enoturismo considerándolo incluso una intromisión en las actividades autóctonas de los territorios y no como un negocio próspero y rentable, beneficioso para las sociedades que los ocupan. Gran parte de las reticencias derivan o derivaban de que las iniciativas más innovadoras fueron impulsadas de forma particular por las empresas, mientras que no ha habido o está siendo más lenta la innovación enoturística promovida por la iniciativa pública que es la que se considera puede contribuir más al beneficio de todo el territorio.

Elías Pastor comentaba en 2006 que la idiosincrasia propia del medio rural, las ataduras culturales y las tradiciones endogámicas de las sociedades rurales, unidas a la falta de concienciación y confianza, han supuesto un freno importante, en muchos casos, para el desarrollo de los proyectos enoturísticos o de otros que pudieran servirles de soporte. Por tanto, este autor destaca que sin la mentalización de todos los agentes sociales afectados sería difícil el desarrollo de este tipo particular de turismo. Diez años más tarde, la situación del sector enoturístico parece haber evolucionado bastante positivamente en todo el mundo y son cada vez más los enoturistas, así como lo es también la oferta de destinos. Hasta China, uno de los últimos países emergentes en el mundo del vino, tiene oferta enoturística con visitas a bodegas y viñedos, uso de las instalaciones para celebraciones, etc., [18], siendo varias las regiones vitivinícolas chinas que ofrecen actividades enoturísticas. A nivel mundial, Estados Unidos de América, Francia, Italia, Australia, Nueva Zelanda y Sudáfrica, son países en las que la actividad enoturística se ha desarrollado o ha aumentado intensamente en los últimos años, situándose a la cabeza en la lista de países por número registrado de visitas/actividades realizadas por enoturista.

3. Efectos del enoturismo en el desarrollo de las regiones vitivinícolas españolas

España, país eminentemente turístico, el enoturismo ha crecido notablemente en este siglo y comienza a ser un sector de importancia en algunas regiones del interior.

Se puede decir que las primeras actividades de turismo enológico, en sentido moderno, surgen en España en los 90 (del siglo pasado). No obstante, desde finales del siglo XIX grupos sociales de élite ya realizaban, por ocio, visitas a bodegas, destacando aquellas realizadas en las zonas vitícolas de Jerez, Rioja y Penedés (cava). Será a partir de los 60, y más intensamente desde los 70 cuando las visitas a las bodegas se comiencen a realizar de modo más popular. Las primeras iniciativas enoturísticas generalmente se enfocaron como acciones de promoción de los vinos, buscando impulsar las ventas y fomentar el reconocimiento de la imagen de la bodega y la calidad de sus vinos [4]. Sin embargo, en poco tiempo, fueron bastantes las empresas que vieron en el enoturismo un modo de diversificación del negocio. Además, más recientemente ciertas bodegas ven en esta actividad un modo de diferenciarse y de tener la oportunidad de ofertar experiencias únicas que fidelicen clientes ya no sólo por la calidad o por el gusto de los vinos, sino porque quedan "atrapados" por el trato, la historia, los paseos, etc., en definitiva, por las vivencias y experiencias ofrecidas por la bodega y todo su entorno. Por otra parte, los territorios también toman conciencia de las oportunidades que brindan las actividades enoturísticas para su desarrollo económico y tanto los gestores, como los habitantes, se vuelcan en el aprovechamiento del patrimonio cultural enológico disponible poniendo en marcha proyectos de distinta naturaleza. En este sentido, es importante adoptar las estrategias más adecuadas en cada caso y a cada territorio aprovechando los puntos fuertes de cada uno de ellos [19].

Este fue el contexto en el que nació ACEVIN, la Asociación Española de Pequeñas y Medianas Ciudades Vitivinícolas, cuya estrategia se centra en valorizar el potencial endógeno de las zonas y ciudades donde el cultivo y elaboración del vino han sido y son la base productiva en torno a las cuales se han configurado históricamente. Además, buscan que ese potencial sea el motor de desarrollo de otras actividades como las turísticas, generando empleo y fomentando la promoción económica y social de los socios. Pueden ser miembros, todos los municipios, mancomunidades o asociaciones de municipios pertenecientes al ámbito geográfico de una Denominación de Origen vitivinícola, que serán socios ordinarios, y todas aquellas personas jurídicas, públicas o privadas, que de alguna forma estén vinculadas a la viticultura o puedan aportar esfuerzos y puntos de vista que contribuyan de manera favorable a la consecución de los fines que se propone la asociación, que serán socios colaboradores.

ACEVIN forma parte de RECEVIN y ambas asociaciones comparten objetivos y estrategias. Por ello, entre los principales proyectos de ACEVIN destaca el de la creación de las Rutas del Vino de España, producto turístico innovador, temático, cultural, gastronómico y de interior. Se basa en la integración de los recursos y servicios turísticos de interés de una zona vitivinícola, planteados desde la autenticidad y la vocación vivencial. Es un producto que se asienta sobre una estrategia de desarrollo socioeconómico integral del territorio, de cooperación público-privada y de valorización de la identidad y cultura vitivinícola del destino.

Rutas del Vino de España nació en 2001 con el apoyo de la Secretaria General de Turismo, en el marco del Programa "Calidad en los Productos Turísticos", integrado dentro del Plan Integral de Calidad del Turismo Español (PICTE 2000–2006). Las fases iniciales se centraron en la definición y desarrollo de las normas de regulación de la calidad de este producto turístico, para posteriormente seguir trabajando hasta conseguir en lo que hoy es una marca turística de calidad que es referente nacional cuando hablamos de enoturismo o turismo del vino. Rutas del Vino de España "constituyen la oferta enoturística de mayor calidad en nuestro país, ofreciendo la posibilidad de disfrutar de la cultura del vino desde perspectivas muy diferentes; visitar bodegas y viñedos, descubrir las costumbres, la artesanía y el patrimonio de las comarcas vitivinícolas españolas, con la posibilidad de alojarte en casas rurales o magníficos hoteles, algunos de ellos auténticas obras de arte proyectadas por renombrados arquitectos, o disfrutar de la práctica de todo tipo de actividades de ocio en pleno contacto con la naturaleza" (`www.acevin.es`).

Actualmente hay 26 Rutas del Vino certificadas conforme a la marca Rutas del Vino de España. Estas rutas están distribuidas por todo el territorio peninsular, y hay otros proyectos en marcha en los que se está trabajando para alcanzar esta certificación. De momento, unas 2.000 empresas, 600 de las cuales son bodegas y unos 550 municipios forman parte de Rutas del Vino de España. Este producto enoturístico cuenta con el apoyo promocional gubernamental en el exterior a través de los canales de TURESPAÑA, de la web `www.spain.info`, y de las oficinas españolas de turismo, entre otros.

ACEVIN, a través del observatorio turístico asociados a las Rutas del Vino, se encarga de llevar a cabo el seguimiento de la evolución de las actividades enoturísticas que se desarrollan en los centros asociados. El análisis de los resultados recopilados durante el seguimiento se refleja en informes que se hacen públicos de forma resumida en la página web de ACEVIN y en versión completa en la de Rutas del Vino de España. Actualmente (primavera de 2017) están disponibles los informes del primer semestre de 2016 y el del 2015. En ambos casos se reflejan nuevas subidas en el número de visitas, habiéndose superado ampliamente los 2 millones de visitantes (2,25 en 2015), así como se ha detectado un aumento del gasto medio y de las pernoctas en 2016. Respecto a datos de años anteriores, destaca que en el primer semestre del año 2016 se ha notado un aumento del 13% de compra de vino en destino. Todos estos datos indican que las Rutas del Vino de España están incentivando el desarrollo socioeconómico de todas las zonas donde se sitúan, observándose un reparto de beneficios entre todos los sectores implicados en la ruta, bodegas, hoteles, restaurantes, etc. como se desprende del aumento de las pernoctaciones, el gasto medio (incluida la gastronomía y artesanía) y la compra de vino en destino (que incluye la compra en tiendas especializadas y otros negocios y no sólo en las propias bodegas).

El seguimiento de la evolución del enoturismo entre los asociados a las Rutas del Vino permite diseñar acciones de mejora que redunden en el aumento de la satisfacción del enoturista. Así por ejemplo, los resultados de los últimos análisis han hecho que ACEVIN y Rutas del Vino de España decidan impulsar la labor comercial de paquetes turísticos a través de las Agencias Homologadas Rutas del Vino de España, un nuevo programa que supone que ahora en la web de las Rutas del Vino (`www.wineroutesofspain.com`) se dispone de una nueva sección que ofrece información de las agencias especializadas en turismo del vino y gastronómico homologadas por ACEVIN y que ofertan paquetes de viajes temáticos en alguna de las 26 Rutas del Vino de España en funcionamiento. Además, la sección permite solicitar presupuestos de viaje a medida.

ACEVIN apuesta por las bases del VEE y ha participado en diversos programas vinculados con la sostenibilidad como el proyecto "Hacia un modelo de Competitividad, Sostenibilidad y Gobernanza en el Medio Rural: Territorio Enoturístico Socialmente Responsable" (2009–11, financiación del Ministerio de Medio Ambiente y Medio Rural y Marino). Por otra parte, es una asociación fuertemente unida a las regiones vitivinícolas con DO y por ello recientemente ha establecido con la CERCV (Conferencia Española de Consejos Reguladores Vitivinícolas) una alianza de colaboración para potenciar el enoturismo y el vino con Denominación de Origen entre los más jóvenes, comenzando así una colaboración activa, que va a suponer sobre todo acciones de promoción conjunta para aumentar la visibilidad (mutua) en el ámbito online y en redes sociales. Estas acciones son claro ejemplo de estrategias de sinergismo que son las esperables entre todos los agentes sociales que participen del desarrollo del enoturismo sostenible.

Los esfuerzos dirigidos a cumplir con los compromisos del desarrollo sostenible son claros en todos los socios de la Rutas y poco a poco los frutos de estos esfuerzos se van traduciendo en reconocimientos y consecuciones de menciones especiales, de las que periódicamente se hacen eco revistas especializadas, contribuyendo a la divulgación del enoturismo. Sirva de ejemplo la reciente adhesión de la Ruta del Vino de Rioja Alavesa al programa de certificación Bioshere Responsable Tourism, que es el programa de certificación sostenible para la industria turística más prestigioso a nivel internacional.

Las Rutas del Vino de España son un potente producto enoturístico, sin embargo, no engloban a todos los agentes y empresas que ofertan actividades vinculadas a la cultura del vino. Las razones de esta situación son variadas, destacando dos de ellas. La primera es que existe una limitación "legal" desde el punto de vista de que sólo pueden formar parte del proyecto Rutas del Vino las actividades desarrolladas en un territorio perteneciente a una DO; y la segunda es que la asociación al proyecto Rutas del Vino es de carácter voluntario, no existiendo ninguna obligatoriedad de asociarse para poder ofrecer actividades enoturísticas, ni siquiera si estas se desarrollan dentro de una DO en la que exista una Ruta certificada. Cada entidad, municipio, empresa, etc. decide según sus posibilidades e intereses si se integra o no en los proyectos Rutas del Vino de España.

Otro proyecto basado en el asociacionismo de diferentes agentes involucrados en el mundo del vino y de la gastronomía, es la Asociación Española de Enoturismo y Turismo Gastronómico (AEETG) que se constituyó en 2009. Tal y como ellos indican son una asociación empresarial sin ánimo de lucro que se ha propuesto como objetivo principal posicionar a España como referente del turismo del vino y de la gastronomía, para lo que centran sus esfuerzos en la "comunicación" que consideran que es una herramienta imprescindible para posicionarse adecuadamente en el mercado global y para conseguir la promoción del enoturismo y del turismo gastronómico español tanto a nivel nacional como internacional.

La aparición de la AEETG coincidió en el tiempo con la creación del proyecto del Gobierno de España "Saborear España", basado en el turismo gastronómico que nace para reforzar la imagen internacional de España a través de la gastronomía. Este proyecto pretende crear una simbiosis entre vino, gastronomía y turismo, como elementos culturales que identifican a España como destino turístico de calidad en los mercados internacionales [18].

El crecimiento del sector enoturístico ha hecho que en los últimos años se vaya detectando una demanda, cada vez mayor, de profesionales multidisciplinares como lo es la propia cultura del vino. Elías Pastor ya en el 2006 indicaba que era necesario establecer planes de formación de profesionales que pudieran cubrir adecuadamente las demandas del sector enoturismo y, en general, capaces de explotar y gestionar adecuadamente la cultura del vino de cada región contribuyendo al desarrollo de calidad de las mismas. Actualmente, son varios los estudios oficiales que contribuyen a la formación de este tipo de profesionales, siendo pionero a nivel universitario el Máster Oficial en Cultura del Vino de la Universidad de Burgos y que comenzó a impartirse en el curso 2011–12. Egresados de este máster regenta hoy en día empresas turísticas especializadas en enoturismo, trabajan en algunas de las empresas vitivinícolas que más han apostado por el enoturismo, o trabajan como comunicadores en revistas especializadas, blogs, etc. contribuyendo a la difusión de la cultura del vino.

Referencias

[1] H. Komiyama, K. Takeuchi, Sustainability Sci. **1**, 1–6 (2006)
[2] *Carta de la Tierra*, (2000) Disponible en: `http://cartadelatierra.org/`
[3] L.V Elías Pastor, *El turismo del vino. Otra experiencia de ocio* (ED Universidad de Deusto, Bilbao, 2009)
[4] *Carta Europea del Enorutismo*, RECEVIN, (2006) `http://www.recevin.net/downloads/Charte_ES.pdf`
[5] Organización Mundial del Turismo, *Guía para las Administraciones Locales: Desarrollo Turístico Sostenible* (1999)
[6] *Vademecum del Turismo Europeo*, RECEVIN, (2006) `http://www.recevin.net/downloads/VADEMECUM_ENOTURISMO_EUROPEO_ES.pdf`
[7] M.B. Peris-Martínez. Revista digital Ojeando la agenda (2013)
[8] V.D. Gómez-Miguel, En: *Innovación Vitivinícola en la Ribera del Duero: Sostenibilidad* (ED. Consejo Regulador Ribera del Duero, Roa, 2012)
[9] G.M.V. De la Torre, E. Morales-Fernández, M.S. Castro-Freire, Revista Digital Turismo y Desarrollo **12**(5), 15 (2012)
[10] B. Hernando Serrano, Revista digital Acenología **29.04.14** (2014)
[11] M. Marzo Navarro, M. Pedraja Iglesias, Int. J. Contemp. Hospitality Manag. **24**, 312–334 (2012)
[12] A. Abel-Duarte, Y. Liu, J. Sustainable Tourism **20**, 991–1009 (2012)
[13] A.D. Alonso, Y. Liu. Pasos **10**, 345–355 (2012)
[14] P. Lacoste, S. Navarrete, S. IDESIA **32**, 5–14 (2014)
[15] T. López-Guzmán, J. Rodríguez García, S. Sánchez Cañizares, M.J. Luján García, M.J. Int. J. Wine Busin. Research **23**, 374–386 (2010)
[16] O. Simoes. Pasos **6**, 269–279 (2008)
[17] S. Xu, C. Barbieri, D. Anderson, Y.F. Leung, S. Rozier-Rich, S. Tourism Management **55**, 276–286 (2016)
[18] M.P Alonso Abad. Rev. Acenologia **28.02.14** (2014)
[19] L. Hao, X. Li, K. Cao. BIO Web of Conferences **5**, 01021–01024. K. (2015)
[20] M. Gómez, B. González-Díaz, A. Molina. J. Destination Marketing & Management **4**, 258–267 (2015)

Les possibilités d'utilisation ciblée des ressources de la Responsabilité Sociale des Entreprises dans le secteur vinier dans le but d'améliorer la durabilité et la spécificité culturelle des régions viticoles / Opportunities of purposeful use of corporative social responsibility resources in the Wine Sector, to enhance the stability and cultural uniqueness of winery regions

Ivana Murdjeva

Sofia City Province, Sofia, Bulgaria

Résumé. Ces dernières années, la responsabilité corporative sociale (RCS) marque une certaine croissance dans l'intérêt des entreprises envers l'intégration dans des causes d'intérêt social dans leurs pratiques commerciales. L'on note un intérêt accru envers les sujets liés à l'éducation, l'écologie, la santé, l'amélioration de la sensibilisation des consommateurs, l'amélioration des conditions de travail et, tout dernièrement, aux sujets et aux initiatives visant la conservation et la protection du patrimoine culturel matériel et immatériel. Diverses entreprises et produc-teurs, dans leur plupart de petite et moyenne envergure, dirigent leurs ressources vers un tra-vail avec des groupes et des communautés locales, afin d'améliorer le milieu et de créer une motivation pour la conservation durable des emplois et l'augmentation de l'engagement pro-fessionnel et des compétences des personnes employées dans la production. Les secteurs publics ayant contact avec la culture, l'art, la préservation de l'identité et des tra-ditions manquent encore de popularité, y compris les productions spécifiques aux régions, la préservation du milieu et la conservation des phénomènes de la nature, la préservation de la culture des communautés locales et leur intégration dans un échange d'ordre global.

Abstract. Corporative Social Responsibility (CSR) has marked a growth with respect to the interest of companies to integrate some socially important causes into their business practices during last years. A strengthened interest to some topics related to education, ecology, health, improvement of users' information capacity, improvement of labor conditions and, ultimately, also to topics and incentives related to safe-keeping and protection of cultural material and intangible patrimony is noticed. Single companies and manufacturers, small and median mainly, are orienting their resources to the collaboration with local societies and groups, in order to improve the environment and to create a motivation of durable preservation of the number or places of work and enhancement of professional commitments and competence of the persons employed in the manufacturing process. Public Sectors relating to culture, arts, preservation of identity and traditions are still less popular, including, but not limited to specific regional productions, landscape and natural phenomena preservation, conservation of the culture of local communities and its integration into an overall exchange.

1. Analyse de la situation

La base de cette analyse repose sur des observations de longue date de l'état et du développement des régions viticoles qui, en ce moment, disposent d'une structure insuffisamment développée, une faible stabilité du marché et une faible compétitivité. L'attention envers ces régions est provoquée par leur dynamique peu typique de développement, et elles représentent un terrain d'intérêt d'analyse des possibilités d'utilisation des ressources de la Responsabilité sociale des entreprises (RSE), afin d'augmenter la stabilité, le développement des régions et le rétablissement des régions et leur réputation de marché.

Pour en arriver à ses conclusions, cette étude a eu recours à des observations sur la structure et le développement des régions viticoles en Bulgarie, car nous considérons que les résultats peuvent être appliqués avec succès sur d'autres domaines, correspondant aux caractéristiques de base au moment de l'étude :

- L'éloignement
- Une faible structure interne
- Une partie insignifiante de marché
- Une faible compétitivité
- Une perte de réputation
- Un faible intérêt des fabricants et une faible identification et la confiance de la part des consommateurs.

 DOI: 10.1051/bioconf/20170904004

L'actualité de la recherche est soutenue par la tendance établie au cours des dernières années d'accroissement d'intérêt du marché envers les régions viticoles plus ou moins connus, envers la conservation et la promotion des cépages locaux et la manutention des traditions spécifiques et des méthodes de production de vin. Il va sans dire que dans la définition de la portée de l'étude il y eut des débats de plus en plus soutenus concernant le soutien et la préservation des industries locales dans un marché global à travers le développement et la mise à jour du système de protection d'origine.

Le territoire de la Bulgarie est conditionnellement divisé en cinq régions viticoles, chacune avec un sol spécifique et une structure climatique – la vallée de la Thrace, la vallée du Danube, la vallée de la rivière Strouma, la région de la mer Noire et la région de la Vallée des Roses. À l'heure actuelle, la plaine de Thrace et la vallée du Danube existent comme des indications géographiques protégées, la vallée de la rivière Strouma comme une appellation d'origine, la région de la mer Noire, divisée en deux appellations d'origine : le sud-ouest et la côte de la mer du Nord.

La région de la Vallée des Roses se trouve hors du système de protection et d'utilisation pratique dans l'étiquetage des vins. Sur la base de cette division en zones, Bulgarie a identifié et protégé encore 51 appellations d'origine, inégalement éparses sur l'ensemble du territoire. La plupart de ces nom ont connu leur apogée de développement lors et autour des 80-s du XX siècle, sous les conditions complètement différentes nationale et de conjoncture mondiale et à l'heure qu'il est ils se trouvent dans la phase inférieure de leur développement, couvrant des parts négligeables du marché. Seules les zones avec une indication géographique protégée – la Plaine de Thrace et la Vallée du Danube enregistrent un développement et une augmentation de leur part de marché.

Chacune des régions décrites montre une structure socio-économique différente, une distribution de l'exploitations de viticulture et de la structure des producteurs de vin du point de vue de leur volume, résilience, ressources, marché ciblé et priorités du point de vue de la RSE. Les zones ont une structure interne faible et un manque de structures organisationnelles qui puisse entreprendre un contrôle professionnel et une protection de l'appellation. En raison de la variété des entités dans la zone (différence dans la structure de production, capacité, durabilité, ciblage de marché, nomenclature des produits), la structure est très difficile à établir rassemblant des éléments qui puissent servir de base pour une stratégie commune pour la mise en œuvre des mécanismes de la RSE, la protection et le développement de la région.

Il y a une absence de conditions de base pour faciliter l'intégration des principes de la RSE dans la structure de commercialisation des entreprises, telles que : présence de groupes ou réseaux de producteurs qui puissent mener des activités de décentralisation ou délocation, réduction des coûts de production et de distribution, etc. qui soient introduits dans les entreprises de gestion unifiée et une uniformité de standardisation des produits. Une partie importante des entreprises indique une faible connaissance des mécanismes de la RSE et une non/application de ces mécanismes dans leurs stratégies de marché. Dans ces cas de différences substantielles des sujets dans une région et compte tenu de nouvelles approches de la RSE en tant qu'une approche organisationnelle au sein de l'entreprise, autour de laquelle tous les intervenants sont coordonnés afin d'atteindre leurs objectifs communs, la réalisation de la stratégie territoriale pour la RSE fait face à des difficultés importantes telles que :

- Une structure organisationnelle différente des entreprises
- Une différence de volume de production, de stabilité et d'emploi dans les entreprises
- Une différence dans l'approche de marketing
- Une différence d'approche dans la région (présence de producteurs qui n'ont pas de rapport envers la zone, développant un portefeuille en fonction des marques d'origine unique).

Dans un plan positif, l'on devrait souligner que malgré le manque de connaissances des principes de la RSE, ces dernières années la responsabilité sociale marque une croissance dans l'intérêt des entreprises. Elles sont plus en plus informées des causes socialement significatives et appliquent, bien que partiellement, des mécanismes pour l'intégration de leurs pratiques commerciales. L'on note l'intérêt dans les sujets liés à l'éducation, l'écologie, la santé, l'amélioration de la sensibilisation des consommateurs, l'amélioration des conditions de travail. L'intérêt envers les sujets et les initiatives relatives au stockage et à la protection du patrimoine culturel matériel et immatériel est toutefois plus faible. Diverses entreprises et fabricants, notamment les petites et moyennes, dirigent leurs ressources dans le domaine du travail avec les communautés et groupes locaux, afin d'améliorer l'environnement et de créer une motivation pour la conservation permanente des emplois et l'augmentation de l'engagement professionnel et les compétences des personnes engagées dans la production.

Il va sans dire, si au début de l'activité des sociétés l'on pourrait envisager une certaine « fatigue » de l'idée. L'on enregistre la présence de davantage de suggestions pour engager les entreprises dans les causes publiques, parmi elles l'on trouve souvent certaines ayant un caractère imitatif copiant des idées réussies, sans aucune garantie de stabilité et d'utilité du résultat final. Les entreprises courent un risque de s'orienter de façon difficile et de trouver des idées originales, il est de plus en plus difficile de faire confiance aux nouvelles formations et de s'engager dans des causes non vérifiées. C'est alors qu'apparait le risque que certains secteurs publics restent définitivement hors de leur portée d'intérêt en raison de l'activité plus faible ou plus difficile pour les formations.

Les secteurs qui sont liées à la culture, l'art, la préservation de l'identité et les traditions, y compris les industries spécifiques dans les régions, la préservation du paysage et la conservation des phénomènes naturels, la préservation de la culture des communautés locales et de l'intégration dans un inter change général sont toujours peu populaires.

2. Stratégie territoriale pour la mise en œuvre des mécanismes de la RSE – avantages et défis

Les producteurs de vin, laissant de côté les plus grandes structures d'entreprise, dans un certain sens, se trouvent

DOI: 10.1051/bioconf/20170904004

dans un processus d'activation de l'intérêt envers la RSE, en tant que le cadre d'une stratégie de marketing et un facteur de l'augmentation des compétitivités individuelles. Les raisons s'en trouve dans toute une série de facteurs, mais surtout dans la spécificité du produit – le vin, formant l'identité de la base d'origine, de tradition et la réputation de la région et du producteur. Nous dirions que les producteurs de vin ont une position privilégiée pour surveiller le processus de mise en œuvre des mécanismes de la RSE et de comparer les résultats avec une certaine vigilance, même s'ils ne tombent pas sous la « fatigue » de l'idée.

Comme règle générale, les pratiques commerciales chez les fabricants de vins sont liées à la zone de production. Ils prennent en compte la corrélation directe entre la qualité du produit et le développement de la région – publique, infrastructures, éducation, santé, préservation des communautés autochtones, de la culture et de l'art. Chaque cause abstraite et séparée du territoire de production n'aurait pas de sens pour les producteurs de vin, du point de vue du champ d'application de la RSE en tant que stratégie intégrée dans la gestion de l'entreprise. Les principaux éléments de la RSE : marketing, lié à une cause, promotion d'une cause, entreprise de marketing social, le bénévolat et la philanthropie ne fonctionnerait pas ou auraient un moindre effet s'ils ne sont pas activement et spécifiquement ciblés sur la zone de production. La spécificité de l'engagement avec la zone de production rend le producteurs de vin un conducteur naturel des causes publiques, liées à la région et même plus encore avec la préservation du patrimoine culturel, de l'art et de culture, la préservation des caractéristiques spécifiques des régions, la préservation du paysage et de ses pratiques agro-écologiques associées. En ce sens, les producteurs de vin ont été précédemment visés en ce qui concerne les causes qui soutiennent ou soutiendraient la zone.

Des études récentes dans le domaine de la RSE et l'efficacité de la mise en œuvre de ses mécanismes, en vue de l'efficacité marketing de l'entreprise, indiquent qu'il est plus approprié que les éléments de la RSE soient utilisés de façon intégrée avec les bonnes pratiques de fabrication, adoptés par le fabricant. En règle générale, les bonnes pratiques de fabrication appliqués par les producteurs de vin, sont directement ou indirectement liées aux spécificités de la région, formant la qualité du produit (caractéristiques climatiques et du sol, pratiques traditionnelles appliquées pour produire du vin de cépages et pour la vinification, la tradition culinaire locale, les coutumes, la spécificité des produits agricoles cultivés dans la région, etc.). Cette pré condition fait une fois de plus de sorte que les producteurs de vin sont conduits dans un groupe cible d'acteurs liée en permanence à la zone de production.

Il y a aussi un fait incontestable : que l'application de stratégies territoriales pour la RSE est plus efficace que l'application des mécanismes éloignés des objectifs directs des entreprises. Les stratégies territoriales ont l'avantage de prendre en compte les besoins spécifiques d'un groupe d'entités référencées dans une région donnée et d'offrir des solutions et des objectifs autour desquels la réalisation d'une unification serait possible. La stratégie territoriale offre davantage de possibilités de ralliement aux bonnes pratiques de fabrication adoptées par les producteurs dans les régions. Et encore, la stratégie territoriale pour la RSE est une base appropriée pour la formation de réseaux ou de groupes d'entreprises appliquant une gestion uniforme et/ou des normes de qualité.

En ce qui concerne les producteurs de vin opérant dans des zones viticoles périphériques, avec un petit marché d'action territoriale, la stratégie de la RSE aurait été une base appropriée pour l'édification d'une structure interne efficace de la région et la récupération de son importance sur le marché et de son importance comme un facteur majeur dans la formation de la valeur ajoutée du produit. En ce sens, par le biais de la perception de la stratégie territoriale de la RSE au profit de la région et des entités y opérant ces derniers auraient trouvé la solution aux problèmes suivants :

- Renforcement du caractère pragmatique de l'interaction entre les particularités et les besoins de la région et la pratique et les objectifs économiques des producteurs.
- Surmonter le manque de motivation chez les producteurs d'investir dans des stratégies de commercialisation liées à la région, sous condition que cette dernière est en dehors du chiffre d'affaires du commerce et/ou ayant un faible marketing.
- Création d'une base solide pour construire ou restaurer l'infrastructure interne de la région.
- Démarrage du processus de rétablissement de la réputation du marché et de l'identité de la région.
- Création des conditions pour l'unification des modèles organisationnels et des bonnes pratiques de fabrication chez les producteurs - condition nécessaire sine qua non pour la mise en œuvre et l'application de normes communes régionales pour la qualité du produit.

En bref, la stratégie territoriale pour la RSE répond de façon efficace aux questions suivantes: ce que nous pouvons et ce que nous voulons faire ensemble dans la région où nous opérons, afin qu'elle nous assure une distinction et un secteur de marché correspondant aux objectifs individuels de chacun des acteurs ? Quelles sont les ressources de la région, que nous pouvons utiliser avec un consentement mutuel, afin qu'ils puissent être utilisés, stockés et mis au point ? Quel type d'investissements dans des activités non économiques pourrait nous proposer la région, et ont-ils le potentiel réel pour en fait reconstruire l'espace de façon convenable pour la communication de marketing ?

Il est d'une importance cruciale pour le succès de la stratégie territoriale de la RSE d'y avoir, à part les exigences spécifiques de la région, d'intégrer tout aussi les intérêts des sujets, permettant la réalisation de l'accord, bien entendu, sous forme de réduction pour atteindre l'accord général. La stratégie territoriale dans laquelle les entreprises de la région ne reconnaissent pas leurs objectifs spécifiques n'aurait été ni efficace, ni réussie.

En revanche, les opposants de la stratégie territoriale pour la RSE, auront indiqué que l'objectif principal de la stratégie – rétablir la réputation de la région grâce à un contrôle commun sur l'origine et la qualité du produit obtenu dans ses limites, pourrait être résolu avec davantage de succès par l'adoption de législations régionales sectorielles.

L'état actuel des structures législatives se tournent de plus en plus rarement vers la législation sectorielle régionale, adoptant une norme de validité universelle correspondant mieux à la société moderne générale

DOI: 10.1051/bioconf/20170904004

(droit communautaire). Même les lois régissant la production des produits sous la protection de l'origine (des indications géographiques et des appellations d'origine) ne peuvent être définies comme des lois régionales. Leur vérification doit être effectuée à niveau national, avec l'introduction du système E-Bacchus ainsi qu'au niveau communautaire. Les régions, par l'intermédiaire de ses autorités administratives ne saurait invoquer des dérogations à la règle générale en faveur de la spécificité locale au micro-niveau. Une approche similaire pourrait être destructive pour la norme commune réglementant le marché commun et permettant par exception une déviation dans les cas de partialité explicitement déclarés. En ce sens, la législation sectorielle régionale n'est pas la réponse cherchée et ne peut déplacer ou rendre nulle une stratégie territoriale pour la RSE.

Certainement, la législation régionale peut créer les conditions qui encouragent les producteurs à appliquer les mécanismes de la RSE, en étant conformes à la nature volontaire de responsabilité sociale. Il serait important, en ce sens, d'avoir eu des dispositions liées à la philanthropie, au mécénat, aux partenariats publics-privés, etc.

3. Rétablissement de l'importance économique de la région. Édification d'une image positive de la zone. Rôle de la stratégie territoriale pour la RSE dans le processus

Comme mentionné dans cette étude, la même est dirigée vers les régions vitivinicoles périphériques ayant perdu leur part de marché et la réputation commerciale. Il fut indiqué que ces zones sont intéressantes à étudier car elles montrent une dynamique de développement atypique pour les régions viticoles. Dans la plupart des cas, les régions observées ont passé leur pic de développement, n'ont pas retenu leur part de marché et ont cessé d'être économiquement efficaces. Les raisons pourraient être disposées dans un large éventail de relations socio-économique, et que dans le cas examiné, elles pourraient se concentrer dans un changement radical des relations entre le marché et système politique qui les gère. Les régions en Bulgarie ont eu leur pic de développement dans les années 1970 et 1980 du XXème siècle, leur production ayant suivi les règles introduites dans un système de marché qui diffère complètement du marché libéral. Après la réforme politique de la fin du siècle passé, la structure organisationnelle des producteurs de vin est radicalement changée. Les régions sont soumises au défi d'un marché déjà établi, dont ils ont la souplesse nécessaire et la possibilité d'adaptation au marketing. Ainsi, ils perdent leur valeur marchande et leur réputation. La motivation des producteurs pour produire dans la région, dans les limites liées à la préservation et la protection de l'origine et le nom, est perdue.

Le rétablissement de la valeur économique de la région et la construction d'une image positive à ce point n'est plus une restauration de l'image de la région du temps de son apogée. Les producteurs sont placés devant la tâche de parvenir à un accord sur la construction d'une nouvelle image de marketing de la région et de promouvoir sa nouvelle identité. Souvent, la situation socio-économique dans les régions est telle qu'il faut une nouvelle édification de la communauté locale, qui aura une attitude motivée pour préserver la spécificité du territoire. La restauration de l'importance économique de la région doit passer par les étapes suivantes :

1. Identification et développement de la communauté locale.
2. Identification des ressources non économiques de la zone.
3. Identification des besoins de la région du point de vue de la communauté locale.
4. Construction d'une infrastructure organisationnelle interne de la région.
5. Analyse des approches possibles de marché pour la zone. Choix d'une stratégie de marketing appropriée et commune.
6. Identification et acceptation du catalogue de produits, représentant au degré complet la toute nouvelle apparence du marché de la zone.

L'identification des ressources non économiques de la région et leur lien avec les besoins immédiats de la communauté locale, représente la partie la plus intéressante et complexe de la restauration de l'importance du marché de la région et l'imposition de sa nouvelle image.

Il est nécessaire de faire la précision suivante. L'étude se concentre sur les domaines traditionnels de la production de vin, c'est-à-dire les zones ayant une histoire, décrite sur la base de la spécificité du sol et du climat et de la distinction du produit. La perte de leur valeur marchande est en raison de l'appauvrissement ou la perte des ressources de la région et, en résultat de circonstances externes à la zone. En ce sens, nous ne discuterons pas la mise en place d'une zone « nouvelle » au moyen de méthodes de marketing, sinon un « avivement » d'une région traditionnelle existante. Cette précision est importante du point de vue de la possible utilisation spéculative des conclusions susmentionnées concernant de nouvelles zones pour lesquelles les approches devraient être complètement différentes.

L'identification des ressources non économiques d'une région définie est une tâche qui pourrait prendre un temps considérable si elle est mal positionnée dans la relation des activités en séquence. Pratiquement, elle devrait être une des premières tâches posées ensemble ou immédiatement après l'identification de la communauté locale, en tant qu'ensemble de caractères sociaux spécifiques et de relations sociales. Il est possible que la communauté locale indique elle-même ou attire l'attention sur des ressources considérées par elle-même comme importantes. L'identification de la portée exacte des ressources non économiques dans la région est une condition préalable pour la création de conditions de base pour surmonter les problèmes énumérés de haute importance pour la communauté locale :

- Structuration organisationnelle de la communauté locale
- Réduction de la migration et du dépeuplement de la région
- Solution des problèmes des emplois à long terme dans le domaine
- Augmentation de la qualification de la main-d'œuvre dans la région

 DOI: 10.1051/bioconf/20170904004

- Amélioration de l'infrastructure locale dans le domaine de l'éducation et de la culture
- Création d'un engagement personnel envers la zone externe aux personnes étrangères à l'entreprise (les parties intéressées externes).

De leur côté, les ressources non économiques de la zone peuvent se placer dans un très large spectre, ce qui, en plus d'élargir la capacité des producteurs de la région, peut être déterminé en tant que complication principale à leur évaluation correcte et l'établissement des priorités. Les ressources non économiques peuvent être groupées de la façon suivante :

- Education
- Culture de masse
- Possibilités de tourisme alternatif
- Mise de l'accent sur les reliques culturelles et archéologiques et les sites
- Conservation de la tradition culturelle immatérielle
- Richesse locale des variétés spécifiques locales
- Spécificité du paysage, présence de phénomènes naturels uniques présentant de l'intérêt pour la conservation du milieu naturel.

Chacun des groupes pourrait être soumis à un examen individuel. Chacun de ces groupes peut être une base fiable sur laquelle l'on regrouperait les objectifs des sujets de la zone ayant une vision commune pour rétablir sa réputation de marché et sa part de marché. C'est pourquoi chaque groupe peut se trouver à la base d'une stratégie territoriale pour la RSE, en utilisant tout élément: un marketing, lié à une cause, la promotion d'une cause, l'entreprise de marketing social, le bénévolat et la philanthropie.

Il a été constaté que, vers le moment de l'étude, l'intérêt des producteurs envers les différents groupes de ressources non économiques est inégal. L'intérêt accru est dédié au domaine de l'éducation, la protection de l'environnement, l'investissement dans le développement des ressources humaines, la santé, l'amélioration de la sensibilisation des consommateurs, l'amélioration des conditions de travail. L'intérêt, allant de faible à notable, est dédié aux initiatives relatives à la protection du patrimoine culturel, l'éducation liée aux arts et à la culture visuelle.

Les conclusions de cette étude indiquent qu'une stratégie territoriale de la RSE dans le secteur vitivinicole ne serait efficace que si l'on pariait sur des ressources non économiques peu développées, telles que la protection du patrimoine culturel, l'éducation liée aux arts et à culture visuelle, s'il aurait été intégré avec succès parmi les besoins identifiés de la communauté locale.

4. Infrastructures culturelles dans les régions

Selon les données des statistiques nationales il y a des 210 musées sur le territoire du pays qui ont enregistré 4 763 472 visiteurs par an et qui ont un total de 7 539 761 pièces de musée. 36, du nombre total des musées du pays, sont des galeries d'art régionales.

En tant qu'institutions culturelles, les musées gèrent une ressource considérable de possibilités, liées en particulier à l'intégration des causes d'importance sociale, organisation d'événements divers en tant qu'hôtes,

Régions	Musées
Au total	201
Blagoevgrad	7
Burgas	12
Varna	9
Véliko Tarnovo	10
Vidin	2
Vratza	3
Gabrovo	9
Dobritch	7
Kardjali	1
Kyustendil	5
Lovetch	6
Montana	6
Pazardjik	9
Pernik	5
Pléven	5
Plovdiv	14
Razgrad	3
Russé	4
Silistra	5
Sliven	5
Smolyan	3
Sofia	7
Sofia (capitale)	27
Stara Zagora	11
Targoviste	4
Haskovo	5
Shumén	12
Yambol	5

création et mise en œuvre de différentes activités, des programmes visant à accroître la sensibilisation et l'éducation. La coopération avec les musées et le soutien de l'activité du musée jouissent en général d'un haut degré d'approbation sociale. La tradition nationale perçoit les musées en tant que sites pour l'accumulation de la mémoire nationale et de l'identité. Les musées régionaux conservent la mémoire locale, et les galeries, l'identité visuelle et la tradition de la région.

L'intégration des activités d'appui de structures culturelles locales dans une stratégie territoriale pour la RSE transformerait la stratégie dans un acte de coopération établie non seulement entre les producteurs dans les régions, sinon entre les producteurs et la communauté locale. En pratique, cela signifie la construction ou la reconstruction des infrastructures internes dans la région et le positionnement des producteurs de vin dans la catégorie des entités responsables à échelle régionale, ayant non seulement une attitude portée uniquement sur les bénéfices des sociétés de l'entreprise, mais aussi au développement de la région comme un ensemble de relations publiques dans divers domaines.

La construction d'un nouvel aspect dans ce cas passe par la récupération de la mémoire sur la tradition et la culture dans le domaine de la reconstruction de leur image à travers le prisme des nouvelles relations sociales.

Ayant recours à l'héritage culturel en tant que cause et le fondement d'une stratégie territoriale pour la RSE, les entreprises réalisent de la valeur ajoutée à leurs stratégies

DOI: 10.1051/bioconf/20170904004

de commercialisation, les habillant dans des causes de valeur sociale ayant un effet à long terme.

5. Conclusions

Les données les plus récentes du système E-Bacchus, pour l'Union européenne, contiennent 1291 appellations d'origine pour la production de vin.

L'illusion est que chacun de ces domaines ainsi décrites ont un niveau similaire de développement des termes des besoins du public, de la partie commerciale et de la réputation. Naturellement, il est normal que la RSE ait un profil différent dans les diverses régions. Par exemple, en Bulgarie l'utilisation des ressources de la RSE à l'appui des musées et de galeries et de leurs activités sur la préservation du patrimoine culturel contribuerait à renforcer la stabilité du secteur vitivinicole et augmenterait la confiance des communautés locales et amènerait effectivement à la rédaction du nouveau look de zones en stagnation. Dans la pratique, cela signifierait une utilisation efficace, appropriée et socialement bénéfique des ressources de la RSE pour la construction de la nouvelle présentation de la région et le rétablissement de sa réputation de marché.

Les ressources de la RSE dans le secteur vitivinicole sont toujours utilisées de façon insuffisante et peu effective pour améliorer la viabilité des régions, en particulier dans les régions périphériques avec une insuffisante stabilité du marché. Leurs études et leur compilation dans une base de données pour élaborer une stratégie locale contribuerait à l'utilisation rapide des bonnes pratiques et l'activation d'une ressource d'entreprise latente.

DOI: 10.1051/bioconf/20170904005

Women have better olfactory perception for wine aromas

Douglas André Wurz[a], Ricardo Allebrandt, Betina Pereira de Bem, Bruno Bonin, Juliana Reinehr, Adrielen Tamiris Canossa, Leo Rufato, and Aike Anneliese Kretzschmar

Santa Catarina State University (UDESC), College of Agriculture and Life Science, Av. Luiz de Camões, 2090, CEP 88520-000, Lages, SC, Brazil

Abstract. The objective of this work was to verify the influence of the gender on the olfactory perception of aromas found in the wines, as well as to identify the aromatic groups most perceived by men and women. Twenty different aromas of different aromatic classes described in the wines were used: fruity, spices, wood, herbaceous, floral, buttery, defects. The different aromatic groups were packed in Erlenmeyer glasses wrapped with aluminum paper in order to avoid the visualization of the aromas by the participants. Fifty people, 25 men and 25 women, aged between 21 and 65 years, were ramdomly separated in groups of 10 people to participate of the evaluation. The influence of the gender on the ability to identify aromas was verified. Women matched 56.8% of the aromas, while men matched 44.6%. In relation to the aromatic class, a greater index of the feminine gender in all the aromatic classes was verified, being spices the group of aromas that women most perceived, with 80.6% of hits, followed by the floral aromas with 50% accuracy. For men, the aromatic class with the highest index of accuracy was also the spices, however, with a success rate of 58.4%, followed by the herbaceous group with 38.2% of correct answers. Both females and males obtained high scores for the group of wine defects (acetic acid and ethyl acetate), 85.2% and 81.0%, respectively, overcoming the other aromatic classes. Buttery aromas were the ones least recognized by women, with 30.8% of hits, whereas the least perceived aroma for men were the floral ones, with no hits observed in any group of participants. The results found in this study show that there are differences in olfactory perception between men and women, and this factor, in addition to the wine service temperature, wine glass type, olfactory memory, must also be considered in sensory analysis. Female gender has a greater ability to identify aromas in relation to the male gender, since women have a greater number of cells in the olfactory bulb, which is the brain region responsible for smell detection.

1. Introduction

Sensory analysis can be defined as a scientific discipline used to evoke, measure, analyze and interpret reactions of aliment and material characteristics in the way they are perceived by the olfactory, gustatory, tactile, auditory and visual sensory systems that evaluate the attributes of aliments [1]. A sensory quality can be determined by the interaction between man and aliment and implicate in cultural, ethnic and sociological aspects [2]. This quality is associated to the stimuli from aliment and the physiological and social conditions of the individuals that are evaluating the product in an environmental context in which the product and the individual are located [3].

In the sensorial analysis, the olfactory examination is one of the main stages in the wine tasting, because through it we can detect aromas that allow to determine with precision many characteristics about the wine that is being tasted, for example: the variety, the producing region, the age of wine or its defects. However, several factors can affect wine's olfactory sensations, such as the serving temperature, wine glass type, and taster's olfactory memory.

Olfactory sense allows to identify the odor and the aroma of the product, that are perceived by the olfactory organ when volatile substances are aspirated [4].

Studies have pointed out that women have a keener olfactory sense than men, and this fact is correlated with the presence of estrogen. In the gestation and menstrual period women present hyperosmia, that is a moderate accentuation of olfactory sensitivity [5].

It is known that there are differences in olfactory perception between male and female genders, however, few researches have correlated the aromas found in wines according to gender. In this context, the aim of this study was to verify the influence of the gender on the olfactory perception of aromas found in wines, as well as to identify the aromatic groups most perceived by men and women.

2. Materials and methods

The test was performed with twenty different aromas from different aromatic groups described in wines as: fruity, spices, herbaceous, floral, yeast/buttery and defective aromas.

The different aromatic groups were prepared 30 minutes before starting the olfactory evaluation and were placed in Erlenmeyer glasses wrapped with aluminum foil to the evaluators do not visualized the product that brought the aroma.

Fifty people were selected to the evaluation: 25 men and 25 women, aged between 21 and 65 years, separated into groups of 10 people. The evaluation occurred individually, in a specially room prepared for the olfactory examination, free of odors and other external interferences

[a] e-mail: douglaswurz@hotmail.com

Table 1. Influence of the gender on the olfactory perception of different aromas from wines and sparkling wines.

Gender	Hits (%)	Erros (%)
Female and Male	50.1	49.9
Female	56.8	43.2
Male	44.6	55.4

Table 2. Percentage of hits and errors of male and female genders submitted to olfactory perception of different aromatic groups found in wines.

Grupo de Aromas	Female and Male	
	% Hits	% Erros
Fruity	40.1	59.9
Spices	67.4	32.6
Herbaceous	42.0	58.0
Floral	25.0	75.0
Defective	82.6	17.4
Yeast/Buttery	23.9	76.1

that could influence the aroma evaluation. The tasters were instructed to smell the samples in a time of twenty seconds for each sample. As soon as they smelled, they described the odors perceived in a datasheet besides a code from each sample.

After the evaluation the data were computed and a descriptive analysis was processed with the hits obtained by men and women in relation to the aromas and the aromatic groups.

3. Results and discussion

It was observed the influence of the gender in the perception of aromas found in wines and sparkling wines. According to Table 1, it was verified that the female gender presented a mean of 56.8% of correct answers, whereas for the male gender the average of hits was only 44.6%. Evaluating both genders together was observed an average of 50.1% of hits from the aromas evaluated.

It was possible to observe the effect of genders also in the olfactory perception of the different aromatic groups (fruity, spices, herbaceous, floral, yeast/buttery and defective aromas).

Analyzing both male and female genders, it was verified that the aromas related to the defect of the wines (acetic acid and ethyl acetate) were the most noticeable, with 82.6% of hits, followed by the aromatic group of spices and herbaceous, with 67.4% and 42% of hits, respectively. The buttery aromas, that come from the yeasts, were the least identified by the tasters (Table 2).

However, according to Tables 3 and 4, it was verified that for the female gender is easier to identify the fruity, spicy, herbaceous, floral, defective and buttery aromatic groups in relation to male gender. The aromatic groups of defective and spices were the ones that obtained the highest percentages of correct answers in female gender, with 85.2%% and 80.6% of hits, respectively. The buttery aromas were the least identified by the female gender, with only 19.5% of correct answers. According to Minim [3], the sensory quality of a product can vary from person to person and can be influenced by specific factors of the aliment, such as appearance, taste, texture, form, method of preparation, cost and seasonality. In addition to these

Table 3. Percentage of correct answers and errors for the female gender submitted to olfactory perception of different aromatic groups from wines.

Aromatic Group	Female	
	% Hits	% Errors
Fruity	46.1	53.9
Spices	80.6	19.4
Herbaceous	48.5	51.5
Floral	50.0	50.0
Defective	85.2	14.8
Yeast/Buttery	30.8	69.2

Table 4. Percentage of correct answers and errors for the male gender submitted to olfactory perception of different aromatic groups from wines.

Aromatic Group	Male	
	% Hits	% Errors
Fruity	35.9	64.1
Spices	58.4	41.6
Herbaceous	38.2	61.8
Floral	0.0	100.0
Defective	81.0	19.0
Yeast/Buttery	19.5	80.5

aspects, there are still influences related to specific factors of the taster, such as age, sex, educational level, culture, physiological state and financial condition.

Analyzing only the male gender (Table 4), it was verified that the aromatic group of spices was the most identified, with 58.4% of hits, but it is 22.2% less of correct answers than did the female gender for the same aromatic group. For the male gender, the floral aromatic group was not identified by any participant.

Analyzing the aromatic groups evaluated in this work, it was verified that the aromas related to the defects of the wines are the most noticeable, followed by the aromatic group of spices and herbaceous. These data are in agreement with those found in a study carried out by Nascimento et al. (2014), who verified that the aromas of acetic acid, clove and cinnamon had the highest success rates of hits. This result can be justified by the wide use of these spices in Brazilian cuisine, as well as the intense aromas that come from them [6].

The olfactory is a primordial sense of enormous significance to human beings and animals. It is known that a subtle alteration in the structure of chemical odors can drastically alters the perceived smell. Olfactory sensory neurons are responsible to detect odors and express different receptors, causing different signals in the brain and, thus, make possible the perceptions of different specific odors [7].

Female gender has a greater ability to identify aromas in relation to the male gender, since women have a higher number of cells in the olfactory bulb, which is the brain region responsible for smell detection. These cells, when they smell, attach to the olfactory bulb neurons, which in turn trigger memory-related brain cells and conscious interpretation.

DOI: 10.1051/bioconf/20170904005

4. Conclusion

The results found in this study show that there are differences in olfactory perception between men and women, and this factor, in addition to the wine service temperature, wine glass type, olfactory memory, must also be considered in sensory analysis. The most perceptive aromatic group was defective aromas and spices and the more difficult to olfactory sense was yeast/buttery aromas, for the both genders. Female gender has a greater ability to identify aromas in relation to the male gender, especially defective, spices and floral aromatic groups.

References

[1] L.V. Teixeira. Análise sensorial na indústria de alimentos. Revista do Instituto de Laticínios Cândido Tostes **64**, 12–21 (2009)

[2] D.R.S. Simões, J.N. Waszczynskyj, G. Wosiacki. Aromas em maçãs, suco e sidra: uma revisão. Boletim do Centro de Pesquisa e Processamento de Alimentos **27**, 153–172 (2009)

[3] V.P.R. Minin. *Análise sensorial: estudos com consumidores* (UFV, Editora UFV, 225, 2006)

[4] W.M. Araujo, N.P. Montebello, R.B.A. Botelho, L.A. Borgo. *Alquimia dos alimentos*. (2 ed. Brasília, Editora Senac, 500 (2011)

[5] T. Hummel, R. Gollisch, G. Wildt, G. Kobal. Changes in olfactory perception during the menstrual cycle. Cellular and Molecular life Sciences **47**, 715 (1991)

[6] K.O. Nascimento, D.F. Ribeiro, E. Batista. Reconhecimento de aromas e aplicação de testes afetivos como forma de aprendizado. E-xacta **7**, 139–145 (2014)

[7] R.F.R. Kranc. Olfactory receptors and the mechanism of odor perception. Polish Annals of Medicine **20**, 51–55 (2013)

The effect of cold maceration on the phenolic composition of red wines

Mustafa Bayram[1], Burcu Öztürk[2], and R. Ertan Anlı[3]

[1] Gaziosmanpaşa University, Department of Food Engineering, 60250 Tokat, Turkey
[2] Namık Kemal University, Şarköy Vocational School, Wine Production Technologies Programme, 59800 Şarköy/Tekirdağ, Turkey
[3] Ankara University Department of Food Engineering, 06000 Ankara, Turkey

Abstract. In this study, it was aimed to reveal the effect of cold maceration on the phenolic composition according to classical maceration in Cabernet sauvignon wines. For this purpose; total phenol, total anthocyanin and biologically active eight different phenolic compounds were determined by HPLC in Cabernet sauvignon wines produced with two different maceration techniques during the 2009 and 2010 harvest years. According to the results obtained; wines produced in 2009 and 2010 with classical maceration showed the highest total phenol content with 1529.09 mg GAE/L and 2051.81 mg GAE/L, respectively. The highest total anthocyanin content in Cabernet sauvignon wines were determined in classic maceration wine (149.95 mg/L), cold maceration wine (279.50 mg/L) in 2009 and 2010. The amount of catechin in classic and cold maceration wines was determined as 58.30 mg/L; 51.30 mg/L in 2009 and 140.98 mg/L; 86.45 mg/L in 2010, respectively. According the results, catechin was the highest individual phenolic compound among gallic acid and epicatechin in all maceration types in 2009 and 2010. (–) - Epicatechin and gallic acid were the most present phenolic compounds respectively after (+) - catechin. Malvidine 3-glycoside content was higher than cyanidin 3–5 diglucoside content in all kinds of different maceration types in 2009 and 2010.
Keywords: red wine, maceration, phenolic, anthocyanins

1. Introduction

The researches on positive effects of wine on human health are increasing in the leading wine producing countries of the world (Lin 2000, Pervaiz 2001). In addition to alcohol, sugar, organic acids, ester compounds, cations, anions, amino acids, nitrogenous compounds, color compounds, enzymes, polypeptides, polysaccharides and colloidal substances, wine contains phenolic compounds which have considerable effects on human health (Ribereau-Gayon et al., 2000). Among these compounds, phenolic acids, flavanoids, antociyanins, flavones having antioxidant features, have proven to lower thromboses formation and prevent hazards caused by the free radicals (Martinez and Moreno 2000). Phenolic compounds pass into the wine through seed and peel during maceration and affect on the color, flavor and taste of wine (Ribereau-Gayon et al., 2000). In this study, it was aimed to reveal the effect of cold maceration on the phenolic composition according to classical maceration in C.Sauvignon wines.

2. Material and method

Grapes from Cabernet sauvignon variety were manually harvested at optimum maturity in the 2009 and 2010 vintage in Tokat province and transported to the Diren Winery. Grapes were destemmed and crushed in a commercial grape destemmer-crusher. They were then homogeneously transferred into three stainless-steel tanks for the maceration and added with 30 mg/kg sulfur dioxide. Wines were produced by two different macerations (classic maceration, cold maceration). In classical maceration; The must and marc were left to maceration-fermentation at 24 ± 2°C for 10 days. After maceration-fermentation, they were pressed with a 10000 L hydraulic Bucher press and the fermentation were completed at ~ 24°C temperature in controlled stainless steel tanks. In pre-cold maceration; the maceration temperature was adjusted at 4–6°C for 96 hours. The fermentation was then carried out by adding 20 g / hL yeast to the medium. After maceration-fermentation, they were pressed with a 10000 L hydraulic Bucher press and the fermentation were completed at ~ 24°C in temperature controlled stainless steel tanks.

Saccharomyces cerevisiae as yeast strain (Oenoferm Rouge, Erbslöh Gerseheim, France) and *Oenococcus oeni* as malolactic bacteria (Geisenheim, Germany) were used in C.sauvignon wines.

2.1. Determination of total phenolic content (TP)

Spectrophotometric determination of the TP content was done with the Folin–Ciocalteu micro method as adapted for wine analysis (Waterhouse, 2009) using gallic acid as the standard. The calibration curve of absorbance concentration of standard was used to quantify phenolic content. Results were expressed as mg gallic acid equivalents per litre of wine (mg GAE/L).

 DOI: 10.1051/bioconf/20170904006

Table 1. Gradient elution program for phenolic compounds.

Time(min)	A Concentration % (v/v)	B Concentration % (v/v)
0	0	100
3	5	95
18	20	80
25	20	80
30	25	75
35	30	70
40	40	60
55	50	50
65	60	40
67	0	100
68	0	100

2.2. Determination of total anthocyanin content (TA)

TA contents of the samples were determined using the pH-differential method described by Giusti and Wrolstad (1976). The obtained wines were diluted with buffer to give an absorbance reading between 0.4 and 0.6 U. The pH values of the diluted wines were 1.0 (0.025 M potassium chloride buffer) and 4.5 (0.4 M sodium acetate buffer), respectively. Absorbance was measured, by using a Shimadzu spectrophotometer at 520 nm. Results were expressed as mg malvidin 3-glicoside equivalent per litre of wine (mg /L).

2.3. Determination of some individual phenolic compunds

Catechin, epicatechin, gallic acid and other standards were purchased from Sigma-Aldrich (Steinheim, Germany). Detection and quantification of phenolic compounds were carried out with a CBM-20A Prominence System controller, a SIL-10 AXL Autosampler, a 1C–20 AT Prominence pump, a DGU-20 A5 Prominence degasser, a CTO-10A column heater and a diode array detector with wavelengths set at 280 nm. Intersil ODS-3 column ($5\,\mu m - 25 \times 4.6\,mm$) were used for detection of phenolic compounds. The flow rate was 1 ml/min, injection volume was $10\,\mu l$ and the column temperature was set at 30°C. Gradient elution of two solvents was used: Solvent A consisted of: acetic acid (Sigma-Aldrich,Germany)–water (2:98 v/v), solvent B: methanol (Sigma-Aldrich,Germany) and the gradient programme used is given Table 1. The wine samples, standard solutions and mobile phases were filtered by a 0.45-μm pour size membrane filter. The amount of phenolic compounds in the extracts was calculated as $\mu g/l$ wine using external calibration curves, which were obtained for each phenolic standard (Özkan ve Göktürk Baydar 2006).

2.4. Determination of some individual anthociyanin compounds

Anthocyanins of the samples were determined using the OIV method (2003) with some modifications. Malvidine 3-glicoside were purchased from Extrasynthese (France), cyanidine 3–5 diglicoside were purchased from Sigma-Aldrich (Germany). The chromatographic separation was carried out in a C_{18} column with water/Formic acid/Acetonitrile (87:10:3, solvent A) and water/formic acid/acetonitrile (40:10:50, solvent B) as the mobile phase. Gradient elution program is given Table 2.

Table 2. Gradient elution program for anthocyanin compounds.

Time(min)	A Concentration % (v/v)	B Concentration % (v/v)
0	94	6
15	70	30
30	50	50
35	40	60
41	94	6

Table 3. Total phenolic content of C. sauvignon wines in different maceration types (2009).

	Classic	Cold
2009	1529.09± 12.334bA	1501.81± 5.056aA
2010	2051.81± 19.285bB	1715.45± 19.285aB

(n:3). Results were calculated as mg/L gallic acid equivalent. Different letters in the same column and row indicate significant differences between years and different maceration types respectively (P<0.05).

2.5. Statistical analyses

All data were analyzed by ANOVA using Statistical Analysis System. The Least Significant Difference (LSD) was used to compare the significance among means. Significant levels were chosen at 5% ($p<0.05$).

3. Results and discussion

3.1. Total phenolic content

The highest total phenolic content was identified in Cabernet sauvignon wines with the classic maceration (1529.09 mg/L) samples in 2009 and in classic maceration (2051,81 mg/L) samples in 2010 (Table 3, Table 4).

Classic maceration wines had the highest total phenolic content for both years. Long maceration times have a positive effect on the level of total phenolic content (Anlı, 2004). The results indicated that wines produced from classical maceration methods had significantly higher total phenolic content than wines produced from cold maceration methods which is consistent with the research of Salinas et al. (2005). Also Joscelyne (2009) sign that a low maceration temperatures leads to a decrease in phenols.

3.2. Total anthocyanin content

The highest total anthocyanin content in Cabernet sauvignon wines were determined in classic maceration wine (149.95 mg/L), cold maceration wine (279.50 mg/L) in 2009 and 2010 respectively. But the amount of anthocyanin in wines obtained from Cabernet sauvignon grapes by applying classical and cold maceration did not show any statistical difference in same year (Table 4). Sacchi et al. (2005) sign that cold maceration (low temperature extraction in the absence of alcohol), produced little difference in anthocyanin and tannin levels, or sometimes, produced even less anthocyanins, color intensity, and flavonols than non cold-maceration wines (Pinot noir).

Table 4. Anthocyanin content of C. sauvignon wines in different maceration types (2009).

	Classic	Cold
2009	149.95± 5.626aA	134.76± 2.648aA
2010	268.26± 12.511aB	279.50± 7.158aB

(n:3), Results were expressed as mg malvidin 3-glicoside equivalent per litre of wine (mg /L). Different letters in the same column and row indicate significant differences between years and different maceration types respectively (P<0.05). (P<0.05).

Table 5. Individual phenolic compounds of C. sauvignon wines in different maceration types (2009).

	Classic	Cold
Gallic acid	18.68± 0.487b	16.64± 0,771a
Catechin	58.30± 1.052b	51.28± 0,455a
Vanilic acid	0.98± 0.022a	0.67± 0,022a
Caffeic acid	0.32± 0.068a	0.60± 0,055a
Epictechin	25.97± 1.652b	18.06± 0,081a
p-coumaric acid	1.28± 0.074a	1.04± 0,059a
Ferulic acid	0.73± 0,061a	0.61± 0,024a
Hydroxycinnamic acid	5.15± 0,221b	2.82± 0,745a
Quercetin	<LOQ	<LOQ

(n:3), Results were calculated as mg/L. Different letters in the same row indicate significant differences between different maceration types (P<0.05).

Table 6. Individual phenolic compounds of C. sauvignon wines in different maceration types (2010).

	Classic	Cold
Gallic acid	62.57± 0.442b	40.99± 1.420a
Catechin	140.98± 1.198b	86.45± 2.263a
Vanilic acid	3.16± 0,056b	1.75± 0.157a
Caffeic acid	<LOQ	<LOQ
Epictechin	59.51± 0.177b	40.89± 2.355a
p-coumaric acid	2.63± 0,452b	1.48± 0.028a
Ferulic acid	0.84± 0,009a	0.91± 0.160a
Hydroxycinnamic acid	1.86± 0,160a	1.16± 0.518a
Quercetin	0.92± 0,297a	0.63± 0.076a

(n:3), Results were calculated as mg/L. Different letters in the same row indicate significant differences between different maceration types (P<0.05).

3.3. Individual phenolic compounds

Individual phenolic compounds of C. sauvignon wines in different maceration types according to years are given in Table 5 and Table 6.

Catechin was the highest individual phenolic compound in all maceration types in 2009 and 2010 (Tables 5, 6). Epicatechin and gallic acid were the most present phenolic compounds respectively after (+)– catechin. In classical maceration wines the amounts of gallic acid, catechin and epicatechin were found higher than cold maceration wines for both year. The differences between catechin, gallic and epicatechin amounts in classical and cold maceration wines were determined to be statistically significant.

3.4. Individual anthociyanin compounds

The amounts of malvidin 3-glucoside in classic maceration and cold maceration wines were determined as 75.26 mg/L, 66.96 mg/L in 2009, 95.82 mg/L, 99.45 mg/L in 2010, respectively (Tables 7, 8). Malvidine 3-glycoside content is higher than cyanidine 3–5 diglucoside content in all kinds of different maceration types in 2009 and 2010. The differences between the amounts of malvidin 3-glycoside in classical and cold maceration wines were statistically insignificant for both years.

Table 7. Individual phenolic compounds of C. sauvignon wines in different maceration types (2009).

	Classic	Cold
Malvidin 3-glycoside	75,26± 3,016a	66,96± 1,979a
Cyanide 3-5diglycoside	0,53± 0,033a	0,53± 0,000a

(n:3), Results were calculated as mg/L. Different letters in the same row indicate significant differences between maceration types respectively (P<0.05).

Table 8. Individual phenolic compounds of C. sauvignon wines in different maceration types (2010).

	Classic	Cold
Malvidin 3-glycoside	95,82± 3,697a	99,45± 0,632a
Cyanide 3-5diglycoside	1,25± 0,025a	0,96± 0,009a

(n:3), Results were calculated as mg/L. Different letters in the same row indicate significant differences between maceration types respectively (P<0.05).

4. Conclusion

Total phenolics, total anthocyanins, phenolic composition and anthocyanin composition changes during wine process treated by different maceration methods like classic maceration, cold maceration. The results indicate that maceration methods are important factors to control quality of wine made from C.sauvignon grapes. The anthocyanin and total phenolic content of C. sauvignon wines were not increased by the cold pre-fermentative maceration. On the other hand individual phenolic compounds of classical maceration wines were found higher than pre-fermentative maceration wines. In conculison, the maceration types play a role on the individual phenolic composition and anthocyanin content of C. sauvignon red wines. On the other hand, according to the research results "classsic maseration" seems to be more suitable application with the richness of individual phenolic compounds compare to the other process.

References

[1] E. Lin Northwestern Wellness News. II.I: 1–4 (2000)
[2] S. Pervaiz, Leukemia and Lxmphoma **40**(5–6), 491–498 (2001)
[3] P. Ribéreau-Gayon, D. Dubourdieu, B. Doneche, A. Lanvoud, John Wiley and Sons Ltd., England (2000)
[4] J. Martinez, J.J. Moreno, Biochemical Pharmacology **59**, 865–870 (2000)
[5] A.L. Waterhouse, Determination of Total Phenolics, in Current Protocols in Food Analytical Chemistry, I1.1.1–I1.1.8. (2003)
[6] M.M. Giusti, R.E. Wrolstad, In: Wrolstad, R.E, Editor. Current protocols in food analytical

DOI: 10.1051/bioconf/20170904006

chemistry. New York: John & Wiley, Inc. P F1.2.1–F1.2.13. (2001)

[7] G. Özkan, N. Göktürk Baydar, Akdeniz Üniversitesi Ziraat Fakültesi Dergisi **19**(2), 229–234 (2006)

[8] OIV. Resolution Oeno 22/2003 Hplc-Determination Of Nine Major Anthocyanins In Red And Rosé Wine (2003)

[9] R.E. Anlı, Gıda **29**(6), 451–455 (2004)

[10] M.R. Salinas, J. Garija, F. Pardo, A. Zalacain, G.L. Alonso, J.Sci. Food and Agric. **85**, 1527–1536 (2005)

[11] V.L. Joscelyne, Consequences of extended maceration for red wine colour and phenolics. Adelaide University, Doctore Thesis (2009)

[12] K. Sacchi, LF. Bisson, D. Adams, Am. J. Enol. Vitic **56**, 197–206 (2005)

Influence of harvest date on volatile and sensory profile of vine leaves infusions from two Portuguese red grape varieties

António M. Jordão[1,a], Ana C. Correia[1], Fernando M. Nunes[2], and Fernanda Cosme[3]

[1] Polytechnic Institute of Viseu (CI&DETS), Agrarian Higher School, Estrada de Nelas, Quinta da Alagoa, Ranhados, 3500–606 Viseu, Portugal
[2] CQ-VR, Chemistry Research Centre, University of Trás-os-Montes and Alto Douro, School of Life Science and Environment, Chemistry Department, 5001–801 Vila Real, Portugal
[3] CQ-VR, Chemistry Research Centre, University of Trás-os-Montes and Alto Douro, School of Life Science and Environment, Biology and Environment Department, Edifício de Enologia, 5001–801 Vila Real, Portugal

Abstract. The main object of the present study was to investigate the volatile composition and sensorial profile of vine leaves infusions prepared from vine leaves collected 30 and 60 days after grape harvest from two Portuguese red grape varieties (Touriga Nacional and Tinta Roriz). Twelve volatile compounds were detected in vine leaves infusions (6-methyl-5-hepten-2-one, 1-undecene, Z-citral, cedrenol, DL-limonene, 2,4-hexadien-2-ol, benzaldehyde, heptane-1,2,4,6-tetraene, cyclohexene 1-methyl-4-(1-methylethyl), 1,6-octadien-3-ol, 3,7-dimethyl, 5,9-undecadien-2-one-6,10-dimethyl-(5E) and α-cedrol). In addition, it was observed that the volatile compounds present in vine leaves infusions were dependent on the harvest time, with significantly major volatile content in vine leaves collected after 30 days of harvesting. Thus, it was more important the harvesting date for the volatile profile of vine leaves infusion than the vine leave grape cultivar. Similar tendency was also obtained for the sensory profile, with significant differences for astringency and aroma intensity attributes of vine leaves infusions according to the harvesting date.

1. Introduction

Vine leaves infusions may be a potential alternative not fully exploited for valorization of vine leaves after grape harvest that are often left on the vine itself or used as organic material for fertilization, animal feed or energy production. Vine leaves production, commercialization, and quality assessment (including sensory and volatile properties of vine leaves infusions) are not so developed as for wine and table grapes. Efficient discrimination tools can aid producers in optimizing and control the production process, namely harvesting time concerning to the leaves infusion quality. Although several authors have reported diverse biological properties of *Vitis vinifera* L. skins [1], seeds [2] and leaves [3,4], the volatile composition and sensory profile, especially from grapevine leaf infusions is particularly scarce. In addition, it is also particularly scarce the knowledge about the potential influence of vine leaves harvest date on the qualitative characteristics of the respective infusions obtained.

The potential use of vine leaves as herbal infusions is dependent on their composition and sensory characteristics. These properties are depending, namely on the volatile composition of the vine leaves used and the extraction process of these volatile compounds during the infusion preparation.

Thus, the main object of the present study was to investigate the volatile composition and sensory profile of vine leaves infusion prepared from vine leaves collected at two different points after grape harvest from two *Vitis vinifera* Portuguese red grape varieties.

[a] e-mail: antoniojordao@esav.ipv.pt

2. Material and methods

2.1. Vine leaves infusion samples

Vine leaves, from two Portuguese red vine varieties (*Vitis vinifera*), Touriga Nacional and Tinta Roriz, were sampled 30 and 60 days after grapes harvest from 8-year-old grapevines in an experimental vineyard of northeastern Portugal.

Leaves were inspected visually before sampling and only intact and healthy leaves were sampled. Visual appearance of vine leaves in vineyard 30 and 60 days after grape harvest is show in Fig. 1. Samples of 40 leaves were picked randomly from 20 different vine plants from each vine variety. All vine leaves samples were dried until 10% of humidity in the dark and then they were crushed and homogenized and then kept in a dry atmosphere until analysis.

In order to reproduce extraction conditions that are similar to those in the infusions, vine leaves samples were placed in 500 mL of water at a concentration of 5 g/L until the boil is achieved. At the end of this process and after 10 minutes, vine leaves infusions were filtered prior to the volatile analysis and also before the sensory evaluation. All essays were done in duplicate.

DOI: 10.1051/bioconf/20170904007

Figure 1. Visual appearance of vine leaves collected 30 (A) and 60 (B) days after grape harvest.

2.2. SPME-GC-MS methodology

For volatile characterization, a pre-conditioned (1 hour at 300°C) SPME fiber (75 μm carboxen-polydithylosiloxane) was exposed to the headspace of the vine infusion sample during 30 minutes. Following adsorption process, the fiber was immediately inserted in the gas chromatography (GC) injection port. The samples analysis was performed with a trace GC ultra gas chromatograph (Thermo Finningan), equipped with a DB-1 (0.50 μm $\times$ 0.25 mm) fused silica column. GC conditions used were: programmed heating from 40°C during 4 min and heating to 220°C at 2.5°C/min, and 10°C/min to 280°C during 5 min. The injector was maintained at 280°C. Helium was the carrier gas at 1.0 mL/min. The injection was made in the split less mode.

The GC was fitted with an ion trap mass spectrometer (MS), model Polaris Q (Thermo Finningan), using the following the MS conditions: ionization energy 70 eV, electronic impact ion source temperature 220°C, scan mode: full scan, scan event time 0.69 s, mass range 30–800 amu; the MS was operated in positive ion mode. The identification of the compounds were performed by comparing to those published in the literature [5] and comparing the mass spectra to those of the Wiley 6 mass spectra library. All analyses were done in duplicate.

2.3. Sensory evaluation

All vine leaves infusions were performed in a sensory analysis room with individual booths for each expert and according to standardized procedures. All evaluations were conducted in the afternoon (between 16:00 and 17:00) by sixteen expert judges with experience in sensory evaluation of leaves infusions. Vine leaves infusions samples were presented to the panel in white tea cups marked with three digit numbers and in a randomized order. The sensory attributes used were the following: visual (color intensity and hue), aroma (intensity), taste (astringency, bitterness and sweetness) and global appreciation. The judges were asked to evaluate the samples on a 1–10 quality point scale (1 being less intense and 10 more intense) for each attribute.

2.4. Statistical analysis

An analysis of variance and comparison of treatment means (ANOVA, one-way) were carried out using the SPSS software version 23.0 (Chicago, Illinois, USA). Differences between means were tested using Duncan test ($p<0.05$). Principal component analysis (PCA) was used to analyze the relations between sensory characteristics of different vine leaves infusions from the two red grape varieties studied and also collected at two different harvest time.

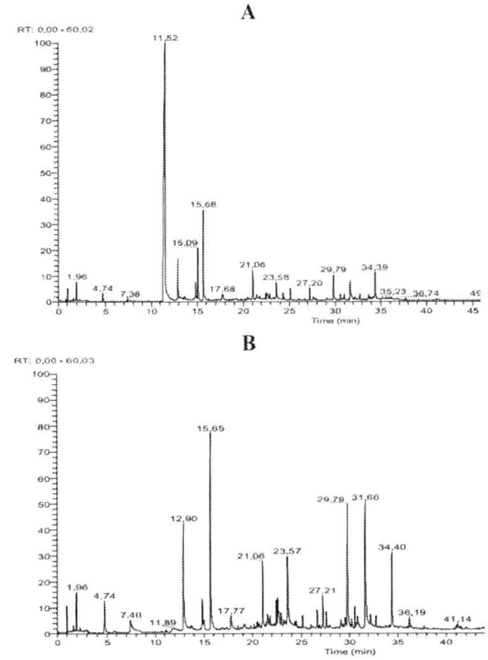

Figure 2. Chromatographic profiles found in vine leaves infusions obtained from Touriga Nacional leaves collected 30 (A) and 60 (B) days after grape harvest.

3. Results and discussion

3.1. Volatile composition of vine leaves infusions

Figure 2 shows an example of a GC chromatographic profile obtained in vine leaves infusions acquired from vine leaves collected 30 and 60 days after grape harvest. By the analysis of Fig. 2, it is clear that it was possible to detect a higher number of different volatile compounds in the infusions prepared from vine leaves collected 60 days after grape harvested than in infusions prepared from vine leaves collected 30 days after grape harvest.

The individual volatile composition of vine leave infusions obtained from two different Portuguese red grape varieties (Touriga Nacional and Tinta Roriz) and harvested after 30 and 60 days of the grape harvest is shows in Table 1. Twelve different compounds were identified in vine leaves infusions.

For the infusions obtained from Touriga Nacional vine leaves collected 30 days after grape harvest a higher abundance of volatiles (total average peak area of 943.2×10^5) in comparison with that prepared from Tinta Roriz (total average peak area of 597.7×10^5) collected at the same harvest time was detected (Table 1). However, they were not significantly different for the leaves collected 60 days after grape harvest, where it was observed a similar total area (total average peak area of 225.2×10^5 and 211.5×10^5 for Touriga Nacional and Tinta Roriz, respectively).

DOI: 10.1051/bioconf/20170904007

Table 1. Average peak area ($\times 10^5$) and standard deviation of volatile compounds determined by SPME-GC-MS from vine leaves infusion of Touriga Nacional and Tinta Roriz leaves collected in two different dates.

Compounds *(sensorial attributes)*	Vine leaves infusions			
	Touriga Nacional		Tinta Roriz	
	30 d (I)	60 d (II)	30 d	60 d
Heptane-1,2,4,6-tetraene	nd	15.4a±1.5 (6.8%)	12.8a±8.7 (2.1%)	14.9a±0.05 (7.0%)
6-Methyl-5-hepten-2-one (*wood, currant*)	45.1b±1.0 (4.7%) (III)	35.9ab±3.5 (16%)	44.0ab±0.3 (7.4%)	35.4ab±0.2 (17%)
1,6-Octadien-3-ol,3,7-Dimethyl (*mushroom*)	105.9a±39.2 (11%)	62.8b±0.1 (28%)	47.8b±0.2 (8.0%)	48.5b±4.3 (23%)
1-Undecene (*burnt*)	nd	1.0a±0.1 (0.4%)	18.7b±1.3 (3.1%)	11.6a±0.7 (5.5%)
Z-Citral (*lemon*)	nd	24.8ab±8.2 (11%)	nd	35.5b±1.0 (17%)
5,9 -Undecadien-2-one-6,10-dimethyl-(5E)	32.6a±9.6 (3.5%)	39.6a±0.1 (18%)	36.4a±0.03 (6.1%)	28.7a±0.5 (14%)
Cyclohexene, 1-methyl-4-(1-methylethyl)	nd	nd	5.3a±0.04 (0.9%)	31.1b±0.5 (15%)
Cedrenol (*cedar wood*)	42.2d±3.9 (4.5%)	45.8d±5.8 (20%)	nd	5.8ab±1.2 (2.7%)
DL-Limonene (*citrus*)	33.6a±3.7 (3.6%)	nd	nd	nd
2,4-Hexadien-2-ol (*geranium, leaves*)	nd	nd	7.7b±0.06 (1.3%)	nd
α-Cedrol (*amphora*)	nd	nd	10.7±3.1 (1.8%)	nd
Benzaldehyde (*almond*)	683.9b±19.8 (73%)	nd	414.4ab±50.6 (69%)	nd
Total average peak area	**943.2b**	**225.2a**	**597.7c**	**211.5a**

(I) 30 days after grape harvest; (II) 60 days after grape harvest; (III) percentage area of each compound; data given as area units of chromatograph peaks; nd = not detected; in brackets are shown some of the sensory attributes for the volatile compounds analyzed. Different letters indicates the existence of statistical differences for each line (p<0.05) by Ducan Test.

Concerning to the individual compounds, the main individual volatile compound present in vine leaves infusions was benzaldehyde (aromatic notes of almond) being responsible for 69% and 73% of the total chromatographic area for Tinta Roriz and Touriga Nacional, respectively collected after 30 days of grape harvesting. However, benzaldehyde was absent in both vine leaves infusions for the leaves harvested 60 days after grape harvest. Opposite tendency was found for z-citral (aromatic note of lemon) that was quantified only in infusions prepared from vine leaves harvested at 60 days in both red grape varieties. This result support that the headspace composition of the vine leaves infusion was different between the two harvesting dates. Previously Weingart et al. [5] by the application of principal component analysis scores plot of vine leaves samples from two different sampling dates obtained two clearly separated clusters.

In addition, only three volatile compounds were quantified in all infusions independently of harvest date and vine variety (6-methyl-5-hepten-2-one, 1,6-octadien-3-ol, 3,7-dimethyl and 5,9-undecadien-2-one-6,10-dimethyl-(5E)). DL-limonene (aromatic note of citrus) was only quantified in infusions prepared from Touriga Nacional leaves harvested at 30 days, while 2,4-hexadien-2-ol (aromatic notes of geranium and leaves) and α-cedrol (aromatic note of amphora) were only quantified in infusions prepared from Tinta Roriz leaves harvested at 30 days.

3.2. Sensory profile of vine leaves infusions

Sensory profile for visual, aroma and taste descriptors of the different vine leaves infusions was evaluated and is shown in Fig. 3. The significantly differences were related to the astringency taste and aroma intensity attributes. Thus, the significantly highest astringency taste was obtained for the vine leaves infusions obtained from leaves collected 60 days after grape harvest, independently of the red grape variety studied. Similar tendency was obtained for aroma intensity, where vine leaves infusions from leaves collected 60 days after grape harvest showed the

DOI: 10.1051/bioconf/20170904007

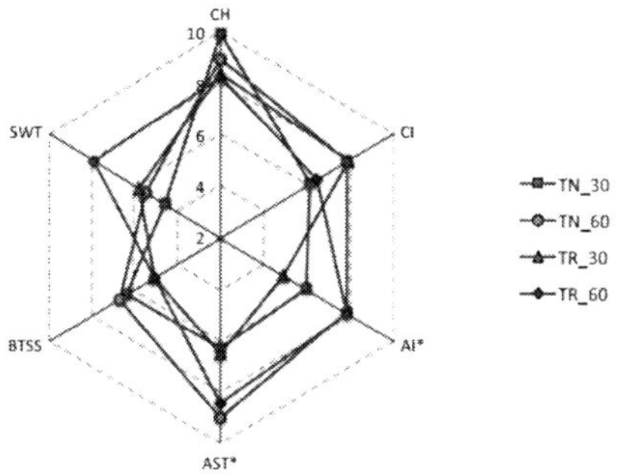

Figure 3. Sensory profile of vine leaves infusions obtained from Touriga Nacional and Tinta Roriz leaves collected 30 and 60 days after grape harvest.
CH = color hue; CI = color intensity; AI = aroma intensity; AST = astringency; BTSS = bitterness; SWT = sweetness. TN_30 = Touriga Nacional infusion from leaves collected 30 days after grape harvest; TN_60 = Touriga Nacional infusion from leaves collected 60 days after grape harvest; TR_30 = Tinta Roriz infusion from leaves collected 30 days after grape harvest; TR_60 = Tinta Roriz infusion from leaves collected 60 days after grape harvest. (*) indicates the existence of statistical differences ($p<0.05$) by Ducan Test.

highest scores for both red grape varieties used. These differences suggest that sensory profile, in particular some taste and aroma attributes, could be determined by the harvest date of vine leaves. For bitterness and sweetness taste and color hue and intensity visual attributes, panel test didn't detect significantly differences between the vine leaves infusions. However, for sweetness taste attribute, there was a slight differentiation (without statistical significance) between the vine leaves infusions obtained from Tinta Roriz leaves collected 60 days after grape harvest in relation to the other vine leaves infusions. Finally, for global appreciation scores (data not show) panel test didn't show significantly differences between the vine leaves infusions.

To get a deeper insight into the relationship between vine grape varieties, sample vine leaves date and their sensory characteristics, a principal component analysis (PCA) was performed on the sensory characteristics of vine leaves infusions (Fig. 4).

The first two principal components (PCs) explained 80.4% of the total variance observed in the original variables. Figure 4A shows that the first PC (PC1, 51.8% of the variance), was positively correlated with the variables, sweetness taste and global appreciation, and negatively correlated with the variables, color hue and bitterness taste, while the second PC (PC2, 28.6% of the variance) was positively correlated with the variables, color hue and astringency taste.

In Fig. 4B it is possible to visualize the spatial distribution of the samples evaluated concerning to the sensory parameters analyzed. Thus, after a cluster analysis, one group is formed by the infusions obtained from vine leaves collected 60 days after grape harvest; these infusions are positioned in the positive side of PC1 and were independently of the red grape variety studied. In addition, the infusions obtained from vine leaves collected 30 days after grape harvest from Touriga Nacional and Tinta Roriz are positioned in second and forth quadrant, respectively.

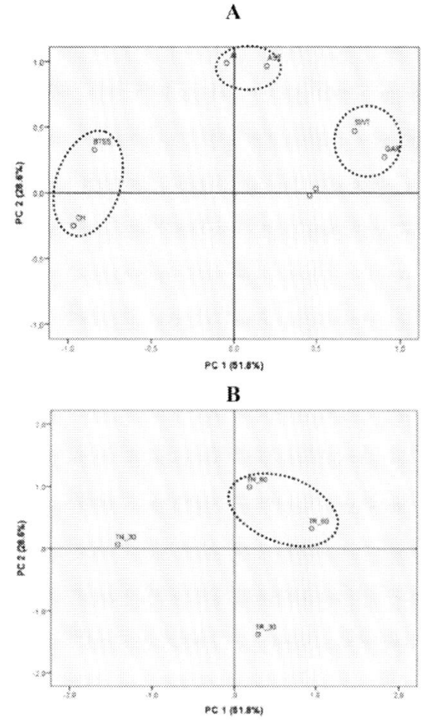

Figure 4. Projection of principal component analysis data from the sensory characteristics of vine leaves infusion samples analyzed: loading plots (A) and scores-plot (B).
CH = color hue; CI = color intensity; AI = aroma intensity; AST = astringency; BTSS = bitterness; SWT = sweetness; GAP = global appreciation TN_30 = Touriga Nacional infusion from leaves collected 30 days after grape harvest; TN_60 = Touriga Nacional infusion from leaves collected 60 days after grape harvest; TR_30 = Tinta Roriz infusion from leaves collected 30 days after grape harvest; TR_60 = Tinta Roriz infusion from leaves collected 60 days after grape harvest.

4. Conclusions

In this work, it was possible to detect a tendency for a more clear influence of the harvesting date on the volatile composition and sensory profile of vine leaves infusions than the vine leave grape variety. Regarding to specifically for the volatile composition, the highest values quantified in vine leaves infusions from leaves collected 30 days after grape harvest were determined namely by the highest content of benzaldehyde. In addition, it was also clear that infusions from leaves collected 60 days after grape harvest showed the highest sensory scores for both red grape varieties used, in particular for aroma intensity and astringency taste.

These results could be important to contribute the better harvest time to collects vine leaves for infusion production. However further research, including more detailed chemical analysis (especially by the analysis of other group of compounds, such as phenolic compounds), more samples points after grape harvest and using other grape varieties it will be necessary to improve the knowledge about the quality of grape vine leaves for infusion production.

DOI: 10.1051/bioconf/20170904007

Authors thanks to FCT (projects UID/Multi/04016/2016 and PEst-OE/QUI/UI0616/2014), Chemistry Research Centre in Vila Real (CQ-VR) and IPV-CI&DETS in Viseu for their financial support.

References

[1] A. El-Din, A. Bekhit, V.J. Cheng, M. McConnell, J.H. Zhao, R. Sedcole, R. Harrison, Food Chem. **129**, 837 (2011)

[2] G.K. Jayaprakasha, T. Selvi, K.K. Sakariah, Food Res. Int. **36**, 117 (2003)

[3] F. Fernandes, E. Ramalhosa, P. Pires, J. Verdial, P. Valentão, P. Andrade, A. Bento, Ind. Crops Prod. **43**, 434 (2013)

[4] N. Ceyhan, D. Keskin, Z. Zorlu, A. Ugur, J. Pure Appl. Microbio. **6**, 1303 (2012)

[5] G. Weingart, B. Kluger, A. Forneck, R. Krska, R. Schuhmacher, Phytochem. Anal. **23**, 345 (2012)

Antioxidant activity of wine assessed by different *in vitro* methods

Chiara Di Lorenzo[1], Mihaela Badea[2], Francesca Colombo[1], Francesca Orgiu[1], Gianfranco Frigerio[1], Raul F. Pastor[3], and Patrizia Restani[1]

[1] Dipartimento di Scienze Farmacologiche e Biomolecolari, Università degli Studi di Milano, via Balzaretti 9, 20133 Milan, Italy
[2] Faculty of Medicine, Transilvania University of Brasov, Brasov, Romania
[3] Cuarta Cátedra de Medicina Interna, Hospital de Clínicas "José de San Martín", Universidad de Buenos Aires, Av Córdoba 2351, Buenos Aires, Argentina

Abstract. Epidemiological studies have suggested that a diet rich in antioxidant compounds could help in counteracting the effects of reactive oxygen species, reducing the risk factors for chronic diseases. The moderate consumption of wine, especially red wine, has been associated with the reduction in mortalities from cardiovascular diseases. One of the possible reasons for the protective effect of wine can be identified in the high content of polyphenols (mainly flavonoids), which have significant antioxidant activity. Even though several *in vitro* tests have been developed for the measure of the antioxidant property, no method has showed a satisfactory correlation with the *in vivo* situation. On these bases, the aim of this study was the application and comparison of different *in vitro* methods to assess the antioxidant activity of red, rosé and white wines. The methods were: 1) Folin-Cocalteau's assay for the quantification of total polyphenol content; 2) the DPPH (1,1-diphenyl-2-picrylhydrazyl) spectrophotometric assay and the Trolox Equivalent Antioxidant Capacity (TEAC) spectrophotometric assay for measuring the antioxidant activity of samples; 3) High Performance Thin Layer Chromatography for separation of phenolic substances and assessment of the associated antioxidant activity; 4) electrochemical detection by using a biosensor. Although all the approaches show some limitations, this battery of tests offers a more reliable body of data on the antioxidant activity of vine derivatives.

1. Introduction

Several epidemiological studies have shown that a nutritionally unbalanced diet contributes to the risk factors for chronic pathological conditions, such as cardiovascular diseases, cancer and diabetes [1].

The production of reactive oxygen species (ROS) is often involved in the initial steps of those conditions, as they can cause the damage of biological structures such as proteins, lipids or DNA. Human metabolism counts on an antioxidant defensive system involving different enzymes to prevent these effects; however, the defences can be overwhelmed in certain conditions (e.g. unhealthy lifestyle and diet) and the progression of oxidative reactions may occur. The inclusion in the diet of foods rich in antioxidant compounds could help in counteracting the toxic effects of ROS. Particularly, over the last decade, the moderate consumption of wine, especially red wine, has been associated with the reduction in mortalities mainly from cardiovascular diseases, slowing LDL oxidation, inhibiting platelet aggregation and stimulating nitric oxide production [2]. Most positive effects of wine are due to the high content of polyphenols (mainly flavonoids), which have shown significant antioxidant properties.

Among them, flavanols [3-O-glycosidesofmyricetin, quercetin,kaempferolandisorhamnetin], flavonols [(+)-catechin, (-)-epicatechin, and (-)-epicatechin gallate] and, in red wines, anthocyanins [3-O-monoglucosidesandthe3-O-acylatedmonoglucosidesofdelphinidin,cyanidin,petunidin,peonidinandmalvidin] are the main classes of compounds identified and associated with the beneficial effects [3].

On these bases, several studies have been focused on the quantification of antioxidant properties of wines, taking into consideration their polyphenol content. The methods used to assess the antioxidant activity differ from each other in terms of chemical bases and reaction conditions. Since a single assay does not accurately reflect all classes of antioxidant molecules, it is recommended the use of two or more assays in order to better describe the antioxidant property of a specific foodstuff in the human body.

Considering the importance of polyphenols in wine, the aims of this study were: 1) the evaluation of the total phenolic content (TPC) in different wine samples, and 2) the set up and application of *in vitro* methods for a fast screening of antioxidant activity. The methods will include spectrophotometric and chromatographic assays and a novel method, based on the use of an electrochemical biosensor. Biosensors are gaining an increasing role in food analysis; they can be defined as a sub-group of chemical systems, in which the analytical device includes a biological sensor coupled with a chemical or physical transducer. The biosensor used in this study was based on the electrochemical measure of potential to determine concentration of analytes or to characterize the chemical reactivity of a compound. Differential Pulse Voltammetry

Table 1. Wine samples included in the study.

Samples analyzed and origin	CODE
Red wine Bibeo, Bitonto, Bari	BRe
Red wine Spiavento, San Severo, Foggia	SRe
Rosé wine Bibeo, Bitonto	BRo
Rosé wine Aro, Sassari	ARo
White wine Spiavento, San Severo	SW
White wine Tuglie, Lecce	TW

(DPV) has been used for quantification, since it is suitable to measure the redox properties of chemical compounds having low molecular weights. Applying a potential, a redox reaction occurs on working electrode surface; electrons involved in the reaction modify the current applied in the cell, and this modification is elaborated by a signal transducer.

Results obtained with biosensors were compared with data from spectrophotometric and chromatographic techniques.

2. Materials and methods

Fast screening methods of the phenolic fraction of wine samples included: 1) Folin-Ciocalteau's assay for the quantification of total polyphenol content; 2) 1,1-diphenyl-2-picrylhydrazyl (DPPH) and Trolox Equivalent Antioxidant Capacity (TEAC) spectrophotometric assays for the assessment of the antioxidant activity of samples; 3) HPTLC (High Performance Thin Layer Chromatography) for the separation and semi-quantitative evaluation of antioxidant property of active compounds. The antioxidant capacity of samples was also evaluated using the electrochemical biosensor described above.

2.1. Samples

The wine samples included in the study were from different regions from Southern Italy. The abbreviations used in this study are listed in Table 1.

Samples of wine Bibeo were kindly offered by Francesca Tassiello (Azienda vinicola Pere Rosse, Bitonto, Bari). Other samples were obtained from the market. Before of the analyses, samples were degassed, filtered on a 0.45 μm filter (Millipore, Billerica, MA, USA) and maintained at $-20\,°C$ till the use.

2.2. Spectrophotometric assays

Three spectrophotometric assays were used in this study.

2.2.1. *Folin-Ciocalteu's assay*

Total polyphenol content (TPC) was determined according to Singleton and Rossi [4].

Samples were suitable diluted with water for HPLC (VWR, France), and aliquots of 300 μL were mixed in test tubes with: 1.5 mL of Folin-Ciocalteau's reagent (Sigma Aldrich, Germany) diluted 10 times, and 1.2 mL of 7.5% sodium carbonate (Sigma Aldrich, Germany). After 30 minutes, the absorbance was measured at 765 nm in a UV-visible spectrophotometer (Varian Cary 50 SCAN, Palo Alto, California, USA).

Blank was prepared using 300 μL of water and treated as described above. Results were expressed as gallic acid (GA) equivalents (mg/mL).

2.2.2. *Antioxidant activity by DPPH assay*

The antioxidant activity (AOA) of wine samples was evaluated spectrophotometrically, as a measure of radical scavenging activity, using 1,1-diphenyl-2-picryl-hydrazyl free radical (DPPH) [5,6]. Aliquots of 1 mL of DPPH (Sigma Aldrich, Germany) in methanol (5 mg/100 mL) were mixed with 0.5 mL of each sample suitably diluted. The absorbance was measured after 30 minutes at 517 nm. Results were expressed as mg/mL of gallic acid equivalents (GA).

2.2.3. *Antioxidant activity by ABTS assay*

The ABTS assay was performed as described by Re et al. with some modifications [7]. ABTS radical cation solution was prepared by mixing 2.45 mM of potassium persulfate and 7 mM ABTS (1:1 v/v); the mixture was then maintained 12–16 h in the dark at room temperature. Before use, the $ABTS^{+\bullet}$ solution was diluted with ethanol to reach an absorbance of 0.7 ± 0.02 at 734 nm. Aliquots of 1.5 mL of $ABTS^{+\bullet}$ solution were mixed with 150 μL of each sample suitably diluted. The absorbance was measured after 6 minutes at 734 nm. Results were expressed as mg/mL of Trolox equivalents.

2.3. Fast chromatographic methods: High Performance Thin Layer Chromatography (HPTLC)

High Performance Thin Layer Chromatography (HPTLC) is a fast and suitable method for the screening of different classes of molecules, allowing the fingerprint characterization of complex products [9]. Furthermore, HPTLC technique can assess some biological properties, which can be directly associated with any specific compound. Among the possible applications, the semi-quantitative measure of antioxidant activity is here described.

Particularly, in this study, HPTLC technique was applied to associate antioxidant activity with wine flavonoids separated by chromatography.

2.3.1. *Chromatographic separation of flavonoids and measure of associated antioxidant activity in red, rosé and white wines*

Aliquots of 5 μL of standard solutions (200 μg/mL) of the main grape flavonoids (quercetin-3-O-galactoside, quercetin-3-O-glu, quercetin-3-glucuronide, kaemferol-3-glu, hyperoside, caftaric acid, epicatechin, epicatechin gallate), and resveratrol were applied on silica-gel plates 254F (10×20 cm, Merck, Darmstadt, Germany) by using a semi-automatic applicator. For wine samples, volumes of 5 μL were loaded onto the plate. The chromatographic run was performed by using a mobile phase (10 mL) containing acetone:toluene:formic acid 4.5:4.5:1 (v/v/v) [8].

At the end of the chromatographic run, the plate was exposed at 254 nm (not shown), derivatized with a DPPH (Sigma Aldrich, Germany) methanolic solution (0.05%)

Figure 1. Edel meter with the sensor insert on the right.

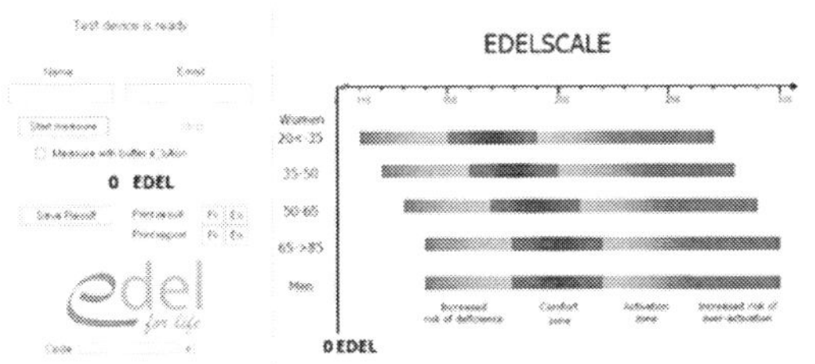

Figure 2. Preview of EDEL program; EDEL values are shown on the left.

and dried for 1 min at room temperature in an extractor hood. The dried plate was wrapped with aluminium foil for 30 min and exposed at UV (366 nm) or at visible light. The images were achieved by using a specific software (VisionCats, CAMAG, Muttenz, Switzerland). The band discoloration from violet to yellow was proportional to the antioxidant activity of each phenolic compound.

2.4. Electrochemical biosensor

Antioxidant activity of samples was also measured by a new approach based on an electrochemical biosensor, EDEL meter (Edel Therapeutics, Lausanne, Switzerland) an analytical device, which includes a biological detector coupled with a chemical transducer and specific software (Fig. 1).

Gallic acid was used as a reference standard. Working solutions were prepared in phosphate buffer (pH 7.4). Each sample (standard solutions or diluted wine samples) was transferred in an aluminium-wrapped becker under magnetic stirring. Then, the electrode was inserted in the EDEL meter and immersed in the solutions. After 10 second, the software was launched and measurements, expressed as EDEL units, were recorded each 30s. An example of EDEL scale is reported in Fig. 2. For each sample, the measure was stopped when EDEL values were stable, reaching the plateau status.

The experiments were repeated three times. For each standard concentration, EDEL units were plotted versus time. Then, EDEL values corresponding to the linear part of the curve were used for quantification. The equation (y=mx+q) was calculated in order to obtain the slope value (m). At the end of the experiments, a calibration curve was calculated plotting each standard concentration vs the slope values. The calibration line correlated the concentration of standard compounds with slope values. Finally, the same protocol was used for wine samples; the antioxidant activity was measured using gallic acid

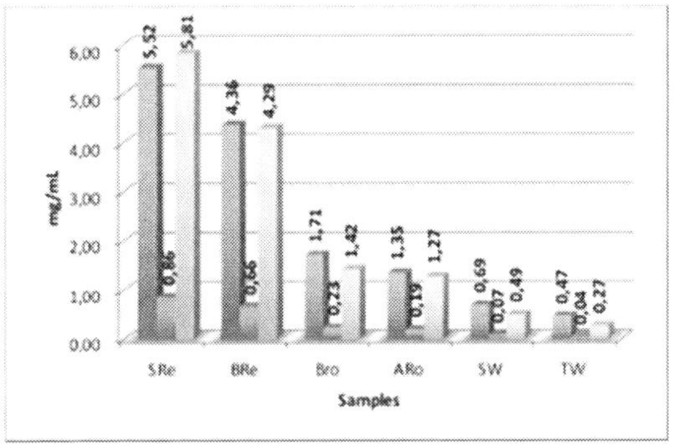

Figure 3. Wine total polyphenol content (mg/mL of gallic acid equivalents, blue bars), and antioxidant activity measured with DPPH (mg/mL of gallic acid equivalents, red bars) and ABTS (mg/mL of trolox equivalents, green bars) assays. For abbreviations of wine samples see Table 1.

solutions to prepare the calibration curve. Data were expressed as mg gallic acid/L.

3. Results and discussion

3.1. Spectrophotometric assays

Total polyphenol content (TPC) and the antioxidant activity (AOA) quantified using DPPH and ABTS assays are illustrated in Fig. 3.

As expected for the known contribution of anthocyanins, the total phenolic content (TPC) of red wines was higher than that measured in rosé and white wines. The TPC in red wines was eight times and three fold higher than TPC in rosé and white wines, respectively.

The red wines (SRe and BRe) and the rosé wine Bibeo (BRo) showed the highest phenol content: 5.52 ± 0.01, 4.36 ± 0.36, and 1.71 ± 0.09 mg/mL, respectively. The lowest polyphenol content was found in white wines Tuglie and Spiavento, and in rosé wine Aro (0.47 ± 0.03, 0.69 ± 0.02, and 1.35 ± 0.10 mg/mL, respectively). These results were in agreement with data published in the scientific literature [9, 10].

Although the Folin Ciocalteau's assay lacks of specificity (sugars, ascorbic acid, aromatic amines, and other molecules could interfere), it is a useful test to obtain preliminary information about the total phenolic content of a sample.

In wine samples, antioxidant activity (AOA) measured using DPPH assay ranged between 0.86 ± 0.03 (red wine Spiavento) and 0.04 ± 0.002 mg/mL (white wine Tuglie), expressed as gallic acid equivalents. The AOA quantified using ABTS assay, and expressed as Trolox equivalents, ranged between 5.81 ± 0.66 (red wine Spiavento) and 0.27 ± 0.02 mg/mL (white wine Tuglie). Wine samples with the highest content of phenol compounds showed also the highest antioxidant activity. In addition, total phenolic content and the antioxidant activity, measured with both ABTS and DPPH assays, were significantly correlated (R^2>0.99), as shown in Fig. 4 and Fig. 5.

The high correlation obtained supports the fact that antioxidant activity is mainly due to phenolic compounds.

Differences between DPPH and ABTS results depend on the chemical reagent involved. Both assays are based on

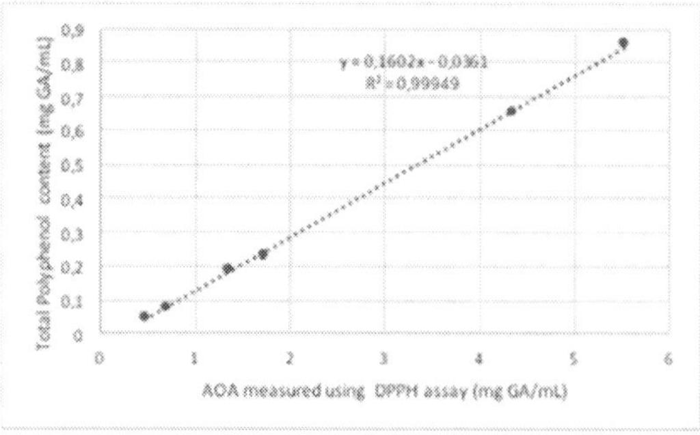

Figure 4. Correlation between total polyphenol content and AOA measured by DPPH assay.

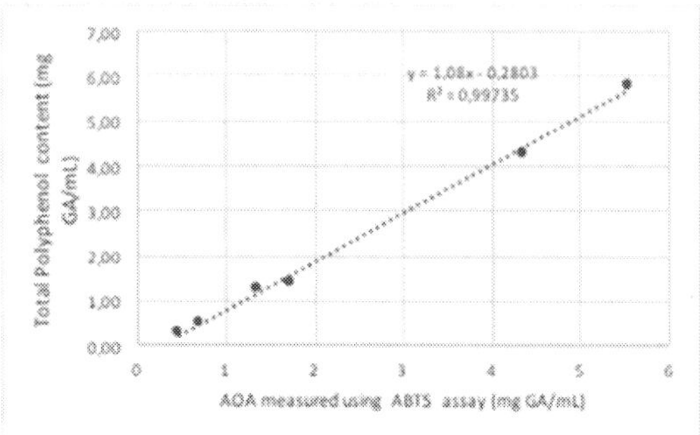

Figure 5. Correlation between total polyphenol content and AOA measured by using ABTS assay.

an electron transfer and involve the reduction of a coloured oxidant. The ABTS assay is based on the generation of a blue/green $ABTS^{+\bullet}$ that can be reduced by antioxidants, whereas the DPPH assay is based on the reduction of the purple DPPH radical to 1,1-diphenyl-2-picryl hydrazine. Although different, both assays are strongly correlated with TPC and can be considered useful tools in evaluating the antioxidant capacity of foods [11].

3.2. High performance thin layer chromatography (HPTLC)

The innovative approach of HPTLC technique allowed the semi-quantitative measure of antioxidant activity associated with a certain number of active compounds.

3.2.1. Flavonoid pattern and antioxidant activity of wine samples

HPTLC technique allowed a parallel evaluation of antioxidant activity and flavonoid distribution in wine samples.

In order to perform a semi-quantitative determination of the antioxidant activity, the plate was exposed at visible light after derivatization with the DPPH solution. To obtain a complete evaluation of flavonoid distribution, the plate was also exposed at 366 nm.

Figures 6 and 7 show the flavonoid distribution and the associated antioxidant activity of the wine samples included in this study.

After derivatization with DPPH solution and exposure of the plate at 366 nm (Fig. 7), a band corresponding to

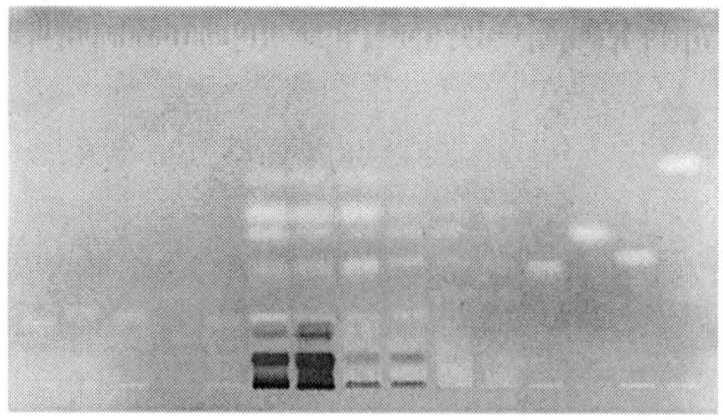

Q1 Q2 Q3 K Hy S_{Re} B_{Re} B_{Ro} A_{Ro} S_W T_W Ca EC ECG R

Figure 6. HPTLC patterns of wine samples at visible light after derivatization with DPPH solution. Standard flavonoids are run in parallel.

Q1 = Quercetin-3-O-galactoside	Aro = Rosé - Aro
Q2 = Quercetin-3-O-glucoside	SW = White - Spiavento
Q3 = Quercetin-3-O-glucuronide	TW = White - Tuglie
K = Kaempferol-3-O-glucoside	Ca = Caftaric acid
Hy = Hyperoside	EC = Epicatechin
Sre = Red - Spiavento	ECG = Epicatecingallate
BRe = Red - Bibeo	R = Resveratrol
BRo = Rosé - Bibeo	

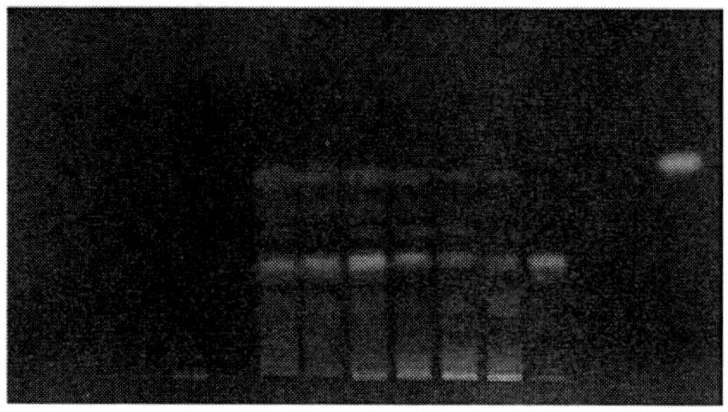

Q1 Q2 Q3 K Hy S_{Re} B_{Re} B_{Ro} A_{Ro} S_W T_W Ca EC ECG R

Figure 7. HPTLC patterns of wine samples at 366 nm after derivatization with DPPH solution. Standard flavonoids are run in parallel. For abbreviation see Fig. 6.

caftaric acid was visible in all samples; this compound was the most abundant and characteristic phenol compound in Rosé wine Bibeo and red wines (SRe and BRe), where the highest antioxidant activity was also observed (Fig. 6). At visible light, hyperoside and epicatechin are clearly detectable in red and rosé wines, both showing also high antioxidant activity. White wines showed the lowest level of flavonoids.

Table 2 shows the relative abundance of antioxidant compounds in wine samples evaluated by HPTLC, in comparison with the quantitative determination of total antioxidant activity measured by spectrophotometric methods.

3.3. Electrochemical biosensor

Figure 8 shows the trend of Edel units versus time, for each gallic acid concentration used.

The preliminary data obtained with this method showed a high variability in the antioxidant activity of samples, in particular for the red wines SRe and BRe:

Table 2. Comparison between the different analytical approaches for the antioxidant activity evaluation.

Samples (Code)	HPTLC (relative abundance)	DPPH assay (mg gallic acid /mL)	ABTS assay (mg trolox /mL)
SRe	++++	0.86	5.81
BRe	++++	0.66	4.29
BRo	+++	0.28	1.42
Aro	++	0.19	1.27
SW	+	0.07	0.49
TW	+	0.04	0.27

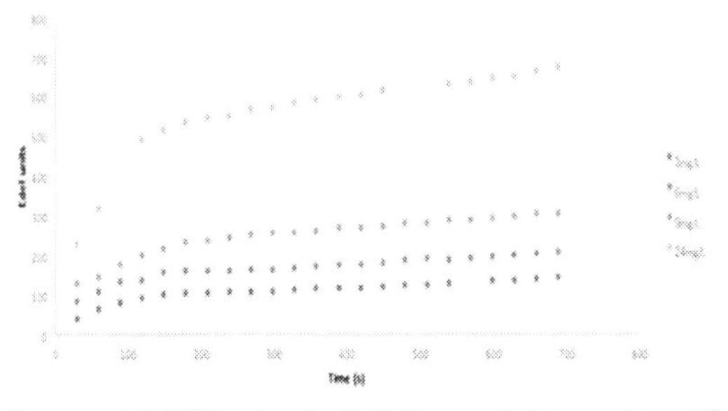

Figure 8. Increase of Edel units for different gallic acid concentrations in function of time.

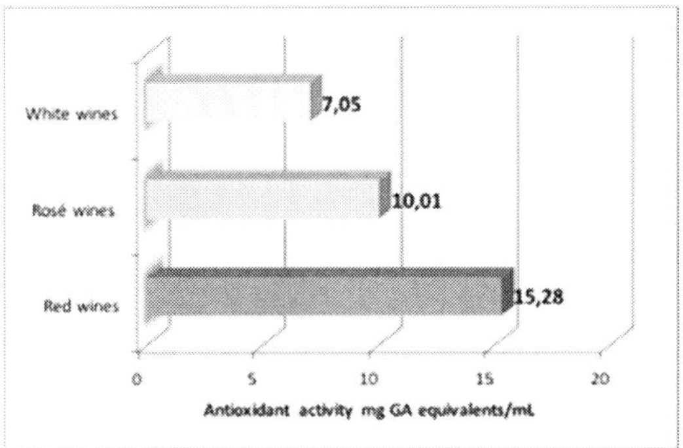

Figure 9. Mean antioxidant activity in the three classes of wines.

22.54 and 8.01 mg gallic acid equivalents/mL, respectively. On the other hand, results for white (5.34 and 8.76 mg GA/mL) and rosé wines (10.26 and 9.76 mg GA/mL) were well correlated with data obtained by spectrophotometric assays. Since biosensors have being only recently applied to the measure of antioxidant activity, no literature reference was at disposal for comparison.

However, considering the mean AOA values for the different types of wine (red, rosé and white) a good correlation can be found with DPPH and ABTS analysis (Fig. 9).

It should be noticed that this method is still in development and a higher number of samples is required to validate the results obtained.

4. Conclusions

In the present study, different analytical approaches have been applied to assess the antioxidant activity in wine.

The Folin-Ciocalteu's method allowed to obtain a fast quantification of total polyphenol content and the data produced showed to be highly correlated with antioxidant activity measured by two different spectrophotometric assays. As expected, red wines showed the highest antioxidant activity, followed by rosé and white wines.

Although not specific, spectrophotometric methods were useful to obtain preliminary information on the phenolic composition of wines and the associated antioxidant activity. The HPTLC showed interesting results not only for the screening of active compounds, but also for the evaluation of their specific antioxidant activity; the most usual *in vitro* assays determine only the total activity.

Notably, rosé wines (mainly BRo) showed an interesting profile for both the total polyphenol content and the antioxidant activity, mainly due to the presence of caftaric acid and epicatechin.

A new analytical approach, based on the use of an electrochemical biosensor was also used. Although with some limitations, probably due to the complexity of the matrix and to possible interferences with the biological nature of sensor, preliminary data have shown that it could be a promising tool to assess antioxidant activity of wines. In addition, if compared with other methods, it does not require specific reagents and the portability makes this device suitable for controls performed in wine cellars.

All analytical methods presented in this study showed a similar and comparable trend, so that they could offer a reliable body of data in the evaluation of antioxidant activity in vine derivatives.

References

[1] M.S. Fernàndez-Pachòn, D. Villano, M.C. Garcìa-Parilla, A. M. Troncoso, Anal. Chim. Acta **513**, 113 (2004)
[2] G.J. Soleas, G. Tomlison, E.P. Diamandis, D.M. Goldberg, J. Agric. Food Chem. **45**, 3995 (1997)
[3] C. Stockley, P.L. Teissedre, M. Boban, C. Di Lorenzo, P. Restani, Food Funct. **3**, 995 (2012)
[4] V. L. Singleton, J.A. Rossi, Am. J. Enol. Viticult. **16**, 144 (1965)
[5] W. Brand-Williams, M.E. Cuvelier, C. Berset, Food Sci. Technol. **28**, 25 (1995)
[6] L.P. Leong, G. Shui. Food Chem. **76**, 69 (2002)
[7] R. Re, N. Pellegrini, A. proteggente, A. Pannala, M. Yang, C. Rice-Evans, Free Radical Biol. Med. **26**, 1231 (1999)
[8] E. Reich, A. Schibli, 2006. High Performance Thin-Layer Chromatography for the Analysis of Medicinal Plants (CIC Edizione internazionale, 2006)

DOI: 10.1051/bioconf/20170904008

[9] E. Büyüktuncel, E. Porgalı, C. Çolak, Food and Nutrition Sciences **5**, 1660-1667 (2014)
[10] M.S. Fernández-Pachón, D. Villaño, M.C. Garcìa-Parrilla, A.M. Troncoso, Analytica Chimica Acta **513**, 113–118 (2004)
[11] A. Floegel, D. Kim, S. Chung, S. I. Koo, O. K. Chun, Journal of Food Composition and Analysis **24**, 1043–1048 (2011)

DOI: 10.1051/bioconf/20170904009

Potential of lactic acid bacteria as suppressors of wine allergies

Hatice Kalkan Yıldırım and Ezgi Dündar

Ege University, Department of Food Engineering, 35100 Bornova, Izmir, Turkey

Abstract. Allergens causes some symptoms as all asthma, allergic conjunctivitis, and allergic rhinitis. These symptoms are seen twice as many in women than in men. The major wine allergens reported in wines are endochitinase 4A and lipid-transfer protein (LTP). This review deal with possibilities of using lactic acid bacteria as suppressors of wine allergies. Phenolic compounds present in wines have not only antioxidant properties causing radical scavenging but also some special properties reported in many *in vitro* studies as regulating functions in inflammatory cells as mast cells. So what is the role of lactic acid bacteria in these cases? Lactic acid bacteria are used during malolactic fermentation step of wine production with purpose of malic acid reduction. During this bioconversion complex phenolic compounds could be hydrolysed by bacterial enzymes to their aglycone forms. Obtained aglycons could pass through the intestinal epithelium of human and allowed reduction of IgE antibody production by affecting Th1/ Th2 ratio. Considering different contents and quantities of phenols in different grape varieties and consequently in different wines more studies are required in order to determine which lactic acid bacteria and strains could be effective in suppressing wine allergens.

1. Introduction

Allergy term is used to describe an inappropriate damaging immune response caused by some substance including foods. The reactions obtained could be antibody- or cell-mediated. Considering clinical disease patterns, the nature of this reactions are classified as Type 1, Type 2, Type 3 and Type 4. Type 1 depends on an interaction between antigen and IgE antibody attached to mast cells. Type 2 is a cytotoxic reaction between tissue- or cell-bound antigen and IgM or IgG antibody. Type 3 is an immune-complex reaction between circulating antigen and IgG antibody. Type 4 is a cellular immune response mediated by sensitized lymphocytes. Antigens that elicit an IgE-mediated type I allergic reaction are called allergens [1].

Even the exact mechanism of activation of these antigens are not understood, the fact that antigens could penetrate mucosal tissues due to their small size and solubility allowed , T-cell activation and Th2-type response reactions. Antibody is synthesized on allergen exposure and subsequently distributed throughout the body where it promptly binds to its high affinity receptor FceR which is densely expressed on the surface of mast cells. Elevated levels reflect recent (within hours) degranulation of mast cells and can be useful where the diagnosis of anaphylaxis is uncertain [1–4].

The food represents the largest antigenic / allergic component that the human body faces during the average life span [5]. Food sensitivities are divided into two groups: primary food sensitivities and secondary ones. IgE-mediated food allergy is in the primary food sensitivity group and can cause anaphylaxis in the high level and lead to immunologic disturbances [6]. Allergens causes some symptoms as all asthma, allergic conjunctivitis, and allergic rhinitis [7]. There are many studies in which grape and wine allergy have been reported [8–10]. Severe anaphylaxis in a patient was determined after consumption of grapes, raisins and wine, caused by lipid transfer protein (LTP) – induced IgE mediated allergy. After application of specific oral tolerance induction (SOTI) which includes allergic food at certain doses, the anaphylaxis risk was reduced [11]. In another study, in patient with IgE-mediated wine allergies was determined allergies to different foods (chicory, peach, strawberry), also [12]. The most important grape and wine allergens are endochitinases 4A and B, lipid transfer protein (LTP) and thaumatin-like proteins [13]. Pathogen-associated proteins (PR) (thaumatin, endochitinase), which are produced in response to pathogens in the grapes, are thought to protect their presence throughout the wine making process because they resist acid hydrolysis and proteases [14].

This review deal with possibilities of using lactic acid bacteria as suppressors of wine allergies.

2. The importance of phenolic compounds

Phenolic compounds present in wines have not only antioxidant properties causing radical scavenging but also some special properties reported in many *in vitro* studies as regulating functions in inflammatory cells as mast cells. Most food polyphenols are found as ester glycosides and polymers forms. Polyphenols are converted to aglycone forms by intestinal enzymes or by colonic microflora before being absorbed by the body [15]. The β-glucosidase enzyme required fo this process may be present in the food itself or may be exogenously added by intestinal enzymes or colonic microflora as mentioned above [16].

There are some studies reporting that various flavonoids have potent anti-allergic activities and treatment with flavonoids results in the down-regulation of serum IgE levels [17,18]. In a study conducted by creating wine medium with water or alcohol and water with Cabernet sauvignon wines were mixed with plasma. As a result, the non-hydrolyzed (+)-catechin in the plasma was undetectable, and the conjugated forms of catechins were predominant. Also reported that β-glucuronidase and arylsulfatase enzymes added to the plasma pool subsequently increased three fold the amount of extracted (+)- catechin [19]. In a study done with mice, it was suggested that after mast cells began to secrete histamine, free radicals bound to ascorbic acid in the cell and ascorbic acid protected mast cells from oxidation [20]. It has been suggested that flavonoids may protect mast cells such as ascorbic in the cell from oxidative damage [21]. In another study were investigated the effects of catechins in tea on the degranulation of mast cells. As a result, depending on reactive oxygen species (ROS) found in cell and the dose of given catechin, it possess an ability to be effective in intracellular oxidation and degranulation of mast cells [22]. In a study investigating the role of Japanese Koshu grape fermented with *Lactobacillus plantarum* on Type 1 allergic reactions, ovalbumin (OVA) was used to induce an allergic reaction in mice. It has been found that fermented grape marcs (at doses of 100 mg/kg fermented grape marcs / orally for 15 days) significantly reduce the IgE antibody response against the OVA. On the other hand, there was not found any contribution to the suppression of IgE production of non-fermented grape marcs. As a result, they indicated that grape seeds fermented with *Lactobacillus plantarum* had anti-allergic effects [23]. At the other state of the same study concerning, have been suggested that resveratrol originating from the red grape variety was not effective in preventing allergic reactions but Koshu white grape marcs fermented with lactic acid bacteria were effective in preventing allergic reactions [23]. Additionally, Negroamoro grape fermented with *Lactobacillus plantarum* was found to be effective in degranulation of mast cells in mouse mesophyll leukemia cells (RBL-2H3. This result was connected to the fact that some special phenols are released during fermentation by lactic acid bacteria [24,25].

3. Lactic acid bacteria and malolactic fermentation

Wine production involves conversion of grape sugars to alcohol and other metabolites by yeast, and malolactic fermentation, which occurs simultaneously or sequentially with alcohol fermentation. Lactic acid bacteria convert L-malic acid to L-lactic acid by decarboxylation and affects the phenolic composition of wine, increasing the contents of total polyphenols and anthocyanins in red wine [26,27]. During this bioconversion complex phenolic compounds could be hydrolysed by bacterial enzymes to their aglycone forms. Obtained aglycones could pass through the intestinal epithelium of human and allowed reduction of IgE antibody production by affecting Th1/Th2 ratio (Fig. 1) [23].

Many factors have an influence on phenolic contents in wines. These factors are grape variety, harvesting time, climatic conditions, soil characteristics, winemaking procedures and postproduction methods. The growth and metabolism of the lactic acid bacteria in the wine is shaped

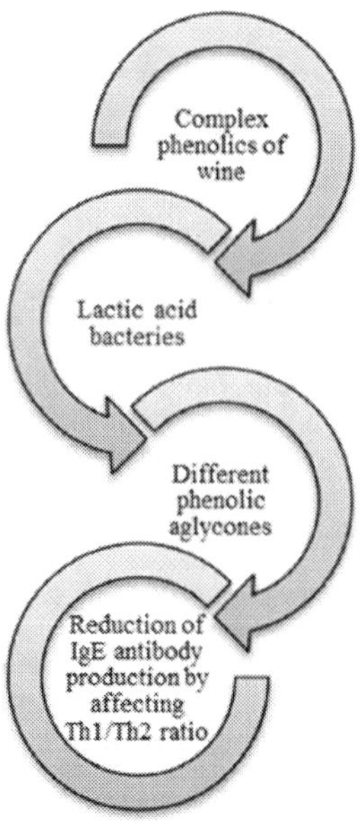

Figure 1. Effects of bioconversion of phenolic compounds with lactic acid bacteria.

by the presence of the fermentable substrate in the wine [28–30].

There are known three main genus of LAB connected with grape, grape marc and wine: *Lactobacillus*, *Oenococcus* and *Pediococcus*. *Onenococcus oeni* has been reported to be predominat species in malolactic fermentation at pH<3.5 [31]. Lactic acid bacteria can hydrolase glycosides, reduce aldehydes to alcohols, and perform many more bioconversion in wine [26]. Many studies have reported that lactic acid bacteria convert glycosylated phenolic compounds into aglycone form through the β-glycosidase enzyme. In a study conducted by creating a synthetic wine medium with aroma components extracted from Muscat wines were used four commercial *O. oeni* strains (EQ 54, Lalvin OSU, Uvaferm Alpha, and Lalvin 31) during the malolactic fermentation and the amounts of free and bound aroma compounds were evaluated. All strains showed a decrease in glycosylated compounds at pH 4. However Uvaferm Alpha and Lalvin 31 strains demonstrate an ability to hydrolyze aromatic components decreased at pH 3.2. The results emphasized that the individual characteristics of the strains play an important role in the degree of hydrolysis of aroma compounds and therefore in changes of the chemical structure of the aglycone obtained [32]. In another study conducted by creating a synthetic wine medium of Chardonnay the effects of *O.oeni* Lalvin EQ54 on phenols were investigated. It has been demonstrated that glycosylated compounds were converted into the form of volatile aglycone forms (3-hydroxydamascone, alpha-terpineol, vanillin, methyl vanillate, 4-hydroxybenzoic acid and tyrosol) by *O. oeni*'s β-glycosidase enzymes. In addition, they have shown that increase the aglycone forms in the Chardonnay glycosidic extract of purely added α-1-rhamnopyranosidase and α-1 arabinofuranosidase enzymes. The results obtained indicate that *O. oeni* possess an enough β-glycoside enzymes to disintegrate the

DOI: 10.1051/bioconf/20170904009

glycosides in wines [33]. In a study conducted by using of synthetic wine medium with different concentration and types of phenolic compounds was determined the effect of *O. oeni* CECT 4100. It has been demonstrated that *Oenococcus oeni* don't cause any changes in phenolic compounds after malolactic fermentation [34].

Lactic acid bacteria and their fermented products are thought to have positive effects on the immune system [35]. In a study done with *Lactobacillus paracasei* KW3110 were determined some positive effects as lowering the level of IgE and balancing effects over Th1/Th2 [36]. In a study conducted in Japan, patients with cedar pollinosis allergy were treated with citrus juice fermented by *Lactobacillus plantarum* YIT 0132. As a results allergic symptoms were alleviates [37].

4. Concluding remarks

In study many studies done at "in vitro" conditions were demonstrated the capabilities of flavonoids to regulate cellular events in inflammatory cells, such as mast cells. The main problem determined was the insufficient absorption of flavonoids and their bioavailability at "in vivo" conditions since flavonoids with the glycoside group could not pass as the intact form through the barrier of the intestinal epithelium in human adults. In case of conditions at which flavonoids are hydrolyzed by β-glucosidase to form the aglycone, the suppression of wine allergens could become possible. Owning such enzymes lactic acid bacteria have a potential to inhibit degranulation of mast cells at the reaction site. Considering different contents and quantities of phenols in different grape varieties and consequently in different wines more studies are required in order to determine which lactic acid bacteria and strains could be effective in suppressing wine allergens.

References

[1] N.F.J. Adkinson, B.S. Bochner, A.W. Burks, W.W. Busse, S.T. Holgate, R.F.J. Lemanske, R.E. O'Hehir, *Middleton's allergy principles and practice*, Elsevier Health Sciences (2013)

[2] D.B. Corry, F. Kheradmand, *Induction and regulation of the IgE response*, Nature **402** (1999)

[3] K.D. Stone, C. Prussin, D.D. Metcalfe, *IgE, mast cells, basophils, and eosinophils*, Allergy Clin. Immunol **125**, 1–16 (2011)

[4] K.A. Packard, M.M. Khan, *Effects of histamine on Th1 / Th2 cytokine balance*, Int. Immunopharmacol. **3**, 909–920 (2003)

[5] H.A. Sampson, *Food allergy*, Curr. Opin. Immunol. **2**, 542–547 (1990)

[6] S.L. Taylor, S.L. Hefle, *Food allergies and other food sensitivities*, Food Technol. Then Chicago **55**, 68–84 (2001)

[7] S.G.O. Johansson, T. Bieber, R. Dahl, P.S. Friedmann, B.Q. Lanier, R.F. Lockey, C. Motala, J.A.O. Martell, T.A.E. Platts-Mills, J. Ring, F. Thien, P.V. Cauwenberge, H.C. Williams, *Revised nomenclature for allergy for global use: Report of the Nomenclature Review Committee of the World Allergy Organization, October 2003*, J. Allergy Clin. Immunol. **113**(5), 832–836 (2004)

[8] D.E. Clayton, W. Busse, *Anaphylaxis to wine*, Clin. Allergy. **10**, 341–343 (1980)

[9] S.G. Schäd, J. Tracka, S. Vieths, S. Scheurer, A. Conti, E.-B. Bröcker, A. Trautmann, *Wine anaphylaxis in a German patient?: IgE-mediated allergy against a lipid transfer protein of grapes*, Int. Arch. Allergy Immunol **136**, 159–164 (2005)

[10] M. Sbornik, J. Rakoski, M. Mempel, M. Ollert, J. Ring, *IgE-mediated type-I-allergy against red wine and grapes*, Allergy **62**, 1339–1340 (2007)

[11] S.G. Schäd, J. Trcka, I. Lauer, S. Scheurer, A. Trautmann, *Wine allergy in a wine-growing district*, World Allergy Organ. J. **3**, 1–5 (2010)

[12] F. Borghesan, D. Basso, F.C. Bianchi, E. Favero, M. Plebani, *Allergy to wine*, Allergy **59**, 1135–1136 (2004).

[13] E.A. Pastorello, L. Farioli, V. Pravettoni, C. Ortolani, D. Fortunato, M.G. Giuffrida, L.P. Garoffo, A.M. Calamari, O. Brenna, A. Conti, *Identification of grape and wine allergens as an endochitinase 4 , a lipid-transfer protein , and a thaumatin*, J. Allergy Clin. Immunol **111**, 350–359 (2003)

[14] H.J.M. Linthorst, L.C. Van Loon, *Pathogenesis – related proteins of plants*, CRC. Crit. Rev. Plant Sci. **10**(2), 123–150 (1991)

[15] C. Manach, A. Scalbert, C. Morand, C. Rémésy, L. Jiménez, *Polyphenols: food sources and bioavailability*, Am. J. Clin. Nutr. **79**, 727–47 (2004)

[16] A. Scalbert, G. Williamson, Chocolate: *Modern science investigates an ancient medicine, dietary intake and bioavailability of polyphenols*, J. Nutr. **130**, 2073–2085 (2000)

[17] S. Yano, D. Umeda, N. Maeda, Y. Fujimura, K. Yamada, H. Tachibana, *Dietary apigenin suppresses IgE and inflammatory cytokines production in C57BL/6N mice*, J. Agric. Food Chem. **54**, 5203–5207 (2006)

[18] S. Yano, D. Umeda, T. Yamashita, Y. Ninomiya, M. Sumida, Y. Fujimura, K. Yamada, H. Tachibana, *Dietary flavones suppresses IgE and Th2 cytokines in OVA-immunized BALB/c mice*, Eur. J. Nutr. **46**, 257–263 (2007)

[19] J.RC. Bell, J.L. Donovan, R. Wong, A.L. Waterhouse, J.B. German, R.L. Walzem, S.E. Kasim-Karakas, *(+) -Catechin in human plasma after ingestion of a single serving of reconstituted red wine*, Am. J. Clin. Nutr. **71**, 103-8 (2000)

[20] J. Ortner, *The Oxidation of Endegenous Ascorbic Acid During Histamine Secretion by rat peritoneal mast cells*, Exp. Cell Res. **129** (1980)

[21] E. Middleton J.R, C. Kandaswami, T.C. Theoharides, *The effects of plant flavonoids on mammalian cells: implications for inflammation, heart disease* , Am. Soc. Pharmacol. Exp. Ther. **52**, 673–751 (2000)

[22] H. Nishikawa, S. Kitani, *Tea catechins have dual effect on mast cell degranulation induced by compound 48/80*, Int. Immunopharmacol. **8**, 1207–1215 (2008)

[23] T. Tominaga, K. Kawaguchi, M. Kanesaka, H. Kawauchi, E. Jirillo, Y. Kumazawa, *Suppression of type-I allergic responses by oral administration of grape marc fermented with Lactobacillus plantarum Suppression of type-I allergic responses by oral–administration of grape marc fermented with–Lactobacillus plantarum*, Immunopharmacol. Immunotoxicol. **32**(4), 593–599 (2010)

DOI: 10.1051/bioconf/20170904009

[24] M. Kaneko, M. Kanesaka, M. Yoneyama, T. Tominaga, E. Jirillo, Y. Kumazawa, *Inhibitory effects of fermented grape marc from Vitis vinifera Negroamaro on antigen-induced degranulation*,Immunopharmacol. Immunotoxicol. **32**(3), 454–461 (2010)

[25] A. Singh, S. Holvoet, A. Mercenier, *Dietary polyphenols in the prevention and treatment of allergic diseases*, Clin. Exp. Allergy **41**, 1346–1359 (2011)

[26] A.L. Waterhouse, G.L. Sacks, D.W. Jeffery *Understanding Wine Chemistry*. John Wiley & Sons (2016)

[27] M.V. Moreno-Arribas, M.C. Polo, *Wine Chemistry and Biochemistry*, New York: Springer (2009)

[28] G. Unden, J. Fröhlich, *Biology of microorganisms on grapes, in must and in wine*, H. König (Ed), Heidelberg: Springer (2009)

[29] J.M Landete, H. Rodríguez, B. De Las Rivas, R. Muñoz, *High-added-value antioxidants obtained from the degradation of wine phenolics by Lactobacillus plantarum*, J. Food Prot. **70**(11), 2670–2675 (2007)

[30] P. van Rensburg, I.S. Pretorius, *Enzymes in winemaking: harnessing natural catalysts for efficient biotransformations – A review*, South African J. Enol. Vitic. **21**, special issue (2000)

[31] D. Wibowo, R. Eschenbruch, C.R. Davis, G.H. Fleet, T.H. Lee, *Occurrence and growth of lactic acid bacteria in wine: A Review*, Am. J. Enol. Vitic. **36**, 302–313 (1985)

[32] M. Ugliano, A. Genovese, L. Moio, *Hydrolysis of wine aroma precursors during malolactic fermentation with four commercial starter cultures of Oenococcus oeni*, J. Agric. Food Chem. **51**, 5073–5078 (2003)

[33] N.D. Incecco, E. Bartowsky, S. Kassara, A. Lante, P. Spettoli, P. Henschke, *Release of glycosidically bound flavour compounds of Chardonnay by Oenococcus oeni during malolactic fermentation*, Food Microbiol. **21**, 257–265 (2004)

[34] C. Reguant, A. Bordons, L. Arola, N. Rozès, *Influence of phenolic compounds on the physiology of Oenococcus oeni from wine*, J. Appl. Microbiol. **88**, 1065–1071 (2000)

[35] I.I. Metchnikoff, *The prolongation of life optimistic studies classics in longevity and aging series*, Springer Publishing Company (2004)

[36] D. Fujiwara, S. Inoue, H. Wakabayashi, T. Fujii, *The anti-allergic effects of lactic acid bacteria are strain dependent and mediated by effects on both Th1 / Th2 cytokine expression*, Int. Arch. Allergy Immunol. **135**(3), 205–215 (2004)

[37] N. Harima-Mizusawa, T. Iino, N. Onodera-masuoka, N. Kato-Nagaoka, J. Kiyoshima-Shibata, A. Gomi, H. Shibahara-Sone, M. Kano, K. Shida, M. Sakai, K. Miyazaki, F. Ishikawa, *Beneficial effects of citrus juice fermented with Lactobacillus plantarum YIT 0132 on Japanese cedar pollinosis*, Biosci. Microbiota, Food Heal. **33**, 147–155 (2014)

 DOI: 10.1051/bioconf/20170904010

Production of ready to drink red and rosé wines from new seedless grapevine crossbreeds

Donato Antonacci, Matteo Velenosi, Perniola Rocco, Teodora Basile, Lucia Rosaria Forleo, Antonio Domenico Marsico, Carlo Bergamini, and Maria Francesca Cardone

Research Centre for Viticolture and Enology (CREA –VE) seat of Turi-Bari (Italy), Via Casamassima, 148–70010, Italy

Abstract. Monomeric and polymeric flavan-3-ols (proanthocyanidins) content in grapes is higher in seeds compared to berry skins. Monomeric flavan-3-ols are more astringent, however, they can combine with other monomer, with anthocyanins and with mannoproteins released by yeast and therefore lose their harsh features in wines. Proanthocyanidins extracted during fermentation and maceration processes in red wines, are important for the organoleptic characteristics of the product and for its aging. There is a difference between skins and seeds proanthocyanidins, with the latter being perceived as more harsh and astringent. One of the most important purposes of refinement and aging of red wines very rich in polyphenols is the slow loss of bitterness. Instead, for wines ready to drink seeds tannins can give bitter overtones, therefore reducing their quality since consumers generally prefer a reduced astringency and attenuated bitterness. This paper investigates the possibility of employ some new seedless grapes crossings of Vitis vinifera L., obtained in recent breeding programs carried out at the CREA-VE of Turi, for the production of improved red and rosé wines made with traditionally red winemaking.

1. Introduction

Among the most important polyphenols found in grape anthocyanins are coloured pigments that characterize red grapes varieties which are mainly present in berry skin [1] and, just in traces, in the subepidermical region of the most coloured varieties. Another class of polyphenols, oligomers of flavan-3-ols called proanthocyanidins, are located part in the skin, others in seeds and in high quantity in grapes stems. [2] There is a difference among the structure of different proanthocyanidins based on their localization in different parts of the berry. Proanthocyanidins stored in seeds are mainly composed of: (+)-epicatechin, (−)-epicatechin and (−)-epicatechin-3-gallate (ECG) [3] whereas in the skin four monomers are usually found: (+)-catechin, (–)-epicatechin, (–)-epicatechin-3-gallate (ECG) and (–)-epigallocatechin (EGC) [4]. Moreover, the mean degree of polymerization (mDP) is usually different between seeds and skin: it is lower in seeds and higher in skin. [5] The lower polymerization of proanthocyanidins in seeds is responsible of important modifications of the perceived astringency of red wines. [6] Indeed, proanthocyanidins influence the bitter perception of wines which increases with the decrease of mDP and the galloylation degree, [7] the hydroxylation of B ring and the stereochemistry of single subunits. [8] The presence of anthocyanins reduces the perception of astringency, [8] indeed during aging the reaction of anthocyanins and proanthocyanidins might explain the change from "hard" to "soft" tannins [7] During fermentation and aging of wine the interaction of proanthocyanidins and anthocyans reduces the astringency, the influence of this interaction depends on the amount of these two classes of polyphenols in wine. Proanthocyanidins also play a role in the stabilization of wine colour due to reactions with anthocyanins. [9] In red winemaking during fermentation and maceration these important compounds pass from grape to wine. The specific winemaking conditions (temperature, SO2, ration grape skin to berry volume and presence of seeds) influence the wine final composition. [10] During fermentation, the mechanical action, punch downs, pumpovers and run-off of replacements, together with the effect of temperature and alcohol, dissolve the protective cuticle present in grape seeds, resulting in the passage in the must/wine of the substances contained in them. Depending of the desired final wine composition different operations are followed during winemaking. Ready to drink wines, which are commercialized after few months from harvesting, are required to be less astringent and with an intense and recognizable aromatic composition, therefore it is necessary to reduce the harsh astringency conferred by proanthocyanidins of the seeds. Wines such as "Beaujolais Nouveau" in France or "Novello" in Italy are produced with carbonic maceration, which differs from classical red winemaking and gives rise to ready to drink wines. In previous works seeds have been manually removed [7] in order to decrease the bitter and astringent compounds present in seeds. This paper investigates the possibility of enhancing some new seedless grapes crossings of Vitis vinifera L., obtained in recent breeding programs carried out at the CREA-VE of Turi, through the production of wines made with traditionally red winemaking. From the crossings available, were selected those who had a black-blue staining of the epidermis, an average berry weight preferably less than 4 g and a titratable acidity greater than 5 g/L. In addition, in the 2016 harvest, characterized

DOI: 10.1051/bioconf/20170904010

by strong development of rot, the grapes of the vines that in mid-late October were completely healthy were picked, with a good chance to possess resistance / tolerance to the above adversity. The crossings used in the trial were 6 in 2015 (early maturation genotypes) and 6 in 2016 (late maturation genotypes). The genetic diversity of the varieties was tested with ampelographic analysis and analysis of molecular markers SSR. The grapes were fermented immediately after collection comprising the steps of disarticulation of the grapes, crushing, activation with selected yeast, maceration (fermentation at controlled temperature), separation of the wine from the pomace. After subsequent stabilization, the wines were analyzed for the most important oenological parameters. The wines are described as fruity and quite balanced, with separate colour intensity as a function of the grapes used in the experimentation and, indeed, they show different winemaking potential.

2. Experimental section

2.1. Harvest and wine production

During ripening the main parameters of grape maturation were checked (pH, total acidity and sugar content), together with the presence and number of seeds. Berries without any trace of seeds belong to the 1 apirenic class (seedless), while berries with small hint of seeds which remain soft and herbaceous even late in ripening belong to the 2/3 apirenic class. Finally to 4 apirenic class belong varieties that have the fully formed, woody and taught seeds. For both vintages grapes have been harvested the same day in order to perform the winemakings at the same time. It was employed the same winemaking technique inoculating the same commercial yeast (Awri350 – Maurivin company). Three manually fullings every 7–8 hours were performed during the first 3/4 of the consumption of sugars during the fermentation process, and then two fullings every 10–12 hours until the total consumption of the sugars (Babo°) were performed. Finally musts were manually pressed in a wine press and stored at 10 °C after two-three days until bottling.

2.2. Chemicals

Formic acid, hydrochloric acid and water (HPLC grade) were purchased from J.T. Baker (Deventer–Holland). Ethanol RPE was purchased from Carlo Erba (Milano, Italy). The anthocyanins delphinidin-3-O-glucoside chloride, cyanidin-3-O-glucoside chloride, peonidin-3-O-glucoside chloride, and malvidin-3-O-glucoside chloride, the flavonols, quercetin-3-O-glucoside and kaempferol-3-O-glucoside, and the flavan-3-ols, procyanidin B2, (+)-catechin and (−)-epicatechin were purchased from Extrasynthese (Genay, France) and used as HPLC reference standards.

2.3. Grape must and wine composition

Total acidity, pH and alcohol were all determined on wine according to EEC regulation 2676/90. °Brix was measured from fresh berries by refractometer (Atago Pocket Refractometer). The spectrophotometric analyses were performed on an Agilent 8453 single-beam UV/VIS spectrophotometer. The Folin-Ciocalteu index for total polyphenols, the total flavonoids and the total anthocyanins amounts were measured following the Di Stefano et al. methods. [11] Due to the limited amount of grape available from the plants of the collection, two wine samples were analyzed for each variety.

2.4. HPLC–DAD analysis

An HPLC 1100 equipped with a DAD detector (Agilent Technologies, Palo Alto, Calif., USA) was used. The samples were injected into a Phenomenex Synergi Hydro-RP 80A C18 column (250 × 4.6 mm, 4 μm) with a pre-column. Separation was carried out at 22 °C, the flow rate was 1.0 mL/min and the injection volume 20 μL. The following gradient system was used with water / formic acid (90:10, v/v) (solvent A) and methanol / water / formic acid (50:40:10, v/v) (solvent B): linear gradients from 85 to 55% A in 15 min; from 55 to 30% A in 30 min; from 30 to 10% A in 10 min; from 10 to 0.1% A in 5 min; from 0.1 to 85% A in 10 min; equilibration time 5 min. The DAD detection was set at 520, 360, and 280 nm. [12] Each wine was filtered through a 0.45 μm syringe cellulose filter prior HPLC analysis. For 2015 vintage wines an anthocyanins extraction was moreover performed, adapted from the method of De Rosso et al. 2012. [12] Compounds were identified according to the retention time and the UV–Vis spectral features described in the literature. The results were expressed in terms of mg/L of malvidin-3-O-glucoside (for anthocyanins), (+)-catechin (for flavan-3-ols) and quercetin-3-O-glucoside (for flavonols).

3. Results and discussion

In 2015 all grapes were harvested that the beginning of October in order to perform a simultaneous winemaking. The analysis on grapes showed that all already accumulated a quantity of sugar over 18° Brix and a titratable acidity higher than 5 g/L. At the end of fermentation the pH was between 3.18 and 3.60 and total acidity was between 5.4 and 6.4 (Table 1), suggesting that grapes reached a perfect timing of maturation, with the only exception of S17/039 that probably was still not completely ripen. In 2016 the harvest was performed at the end of October from new genotypes obtained from late ripening varieties in order to compare both early and late ripening grapevine crossbreads. Wines obtained from these grapes shown a pH between 3.32 and 3.71 and a total acidity not below 4.8 g/l of tartaric acid (S04/132). Just for the N07/110 grapevine crossbread, which was characterized by a high production of grapes (15–20 bunches per vine), the pH was lower (3.19) and at the end the fermentation it showed the lowest amount of alcohol (7.8 %vol). The diversity between the different crossbreedings determine a different accumulation of sugar and other compounds and even a difference in the apirenic class, resulting in different wines, even if it was employed the same winemaking technique. Indeed, N11/030 and N11/008 are obtained from the same parents but have a different number of seeds: totally missing in N11/030 but present, still herbaceous and not well formed in N11/030 (Table 1).

Table 2 shows how alcohol (%vol) remained below 11.5 %vol in all the analyzed wines; it is possible that due to pedoclimatic conditions and type of cultivation of the vines (*tendone*), the plants were able just to produce

 DOI: 10.1051/bioconf/20170904010

Table 1. Grape.

Vintage	Sample	AC	Berry weight (g)	Grape weight (g)	Brix °a	pH[a]	TA[a]
2015	S18/ 009	2/3	4	–	20.6	3.34	6.1
	S17/ 039	2/3	3.1	–	18.3	3.18	4.9
	S20/ 071	2/3	2.5	–	21.8	3.44	7.3
	N06/ 124	1	2	–	20.1	3.67	5.1
	N20/ 137	1	3.2	–	18.8	3.36	6.5
	N22/ 115	1	3	–	20.4	3.51	5.5
2016	S04/ 142	1	2.2	272.3	19.9	3.83	6.2
	S23/ 115	1	4.4	483.8	18.6	3.46	6.2
	N11/ 030	1	3	110.7	20.5	3.80	6.4
	N07/ 110	1	3.6	593.3	15.9	3.45	5.1
	N11/ 008	2/3	4.1	274.7	18.9	3.82	5.2
	S04/ 132	4	3.7	350.5	17.4	3.67	6.3

[a] Calculated on must. AC: apirenic class; TA: total acidity.

Table 2. Chemical composition of wine.

Sample	Apirenic class	pH	Total acidity	%Vol
S18/009	2/3	3.41±0.03	6.4±0.1	9.3
S17/039	2/3	2.94±0.01	7.5±0.1	9.0
S20/071	2/3	3.18±0.00	6.2±0.1	10.9
N06/124	1	3.39±0.01	5.5±0.1	11.1
N20/137	1	3.39±0.01	5.6±0.1	10.4
N22/115	1	3.60±0.03	5.4±0.1	10.5
S04/142	1	3.52±0.01	5.5±0.1	10.5
S23/115	1	3.32±0.01	6.6±0.4	10.1
N11/030	1	3.71±0.01	5.9±0.1	11.2
N07/110	1	3.19±0.00	6.1±0.1	7.8
N11/008	2/3	3.70±0.00	5.1±0.1	9.9
S04/132	4	3.55±0.02	4.8±0.2	9.4
ANOVA		***	***	

Total acidity is expressed as g/l of tartaric acid. ***: Values are significantly different (p<0.001).

Table 3. Colour.

Sample	C.I.	T	%420	%520	%620
S18/009	0.62±0.00	0.75±0.01	36.9	49.4	13.8
S17/039	0.51±0.01	0.65±0.00	34.8	53.1	12.2
S20/071	0.76±0.01	0.70±0.00	35.6	51.3	13.1
N06/124	0.80±0.02	0.65±0.01	34.3	52.9	12.8
N20/137	0.30±0.01	1.01±0.02	44.6	44.0	11.4
N22/115	0.50±0.01	1.07±0.01	44.1	41.3	14.6
S04/142	0.20±0.02	1.46±0.07	52.8	36.4	10.7
S23/115	0.25±0.00	0.93±0.00	44.0	47.3	8.7
N11/030	0.60±0.05	1.45±0.05	50.6	34.8	14.6
N07/110	0.20±0.02	1.09±0.04	44.9	41.3	13.8
N11/008	0.31±0.01	1.05±0.01	45.9	43.8	10.3
S04/132	0.25±0.01	1.00±0.01	45.3	45.5	9.2
ANOVA	***	***			

C.I. (Colour intensity); T (Tint). ***: Values are significantly different (p<0.001).

an amount of sugar that did not exceeded 21° Brix. This was confirmed by the "Saturno 150" analyzer (Crony Instruments) analysis of the residual sugar in the wines at the end of the fermentation, which were all completely dry (<1 g/L of residual sugar).

The colour intensity (C.I.) is different among 2015 and 2016 wines (Table 3). In 2016 the C.I. is 0.3 on average and only N11/030 wine shows the highest value (0.60), instead in 2015 the C.I. is 0.58 on average, with N20/137 as the less coloured wine (0.30). For all wines there was no influence of the "blue" component i.e. 620 nm absorbance, as it was expected since they are not young wines. In 2015 the 520 nm component is especially important if compared with the other two (420 and 620 nm), instead in 2016 the 420 nm is the most important component, with the exception of S23/115 for whom 520 nm component is greater. However, 420 and 520 nm are the two most important components in rosé wines, indeed, even after a traditional red winemaking the wines obtained did not showed a great red colour with a few exception.

Table 4 shows that N06/124 and N22/115 (both wines of 2015) have the highest content of total polyphenols and total flavonoids but only N06/124 has a high content of total anthocyanins. For N22/115 wine a large contribution to the polyphenolic compounds pool comes from the hydroxycinnamic and benzoic acids (data not shown). Generally, for some wines with a high total polyphenolic content we found that a considerable part of this parameter is linked to the presence of benzoic and hydroxycinnamic acids instead of flavonoids (HPLC analysis data not shown), especially for S20/071, N22/115, N11/030 and S04/132 wines. Concerning 2016 vintage wines N11/008 shows the highest content of both total polyphenols and total anthocyanins.

Two 2015 (S20/071 and N06/124) and one 2016 (N11/008) wines show the highest total anthocyanins amount. Anyway, the HPLC analysis of anthocyanins showed different results for wines from the two vintages. Four wines of 2015, namely S18/009, S17/039, S20/071, N20/137 and N22/115 are poor of the main anthocyanins usually found in wines and showed peaks at the end of the chromatogram which are characteristic of polymerized anthocyanins. Indeed, the colour of these wines was orange-red, suggesting the presence of oxidized forms.

Wine N06/124 was an exception showing an anthocyanin profile similar to 2016 vintage wines. The HPLC analysis performed on wines of the 2016 vintage shows how the principal anthocyanin is malvidin-3-O-glucosyde and its coumaroylated product (Table 5). All mono glucosylated forms, even if in small amount in certain varieties, are present with the exception of cyanidin in wines N11/030, N07/110, N11 /008 and N06/124 (2015). Malvidin has two methoxy groups, for that reason it is more resistant to oxidation, it looks like that in our winemaking conditions it was the main anthocyanin present in wine and also after few months after bottling, as it was detected by HPLC in good quantity in 2016 wines and is still the highest anthocyanins in even after one year

Table 4. Polyphenolic indexes.

Sample	Total polyphenols	Total flavonoids	Total anthocyanins
S18/009	567.28±15.09	475.4±50.4	84.0±7.3
S17/039	776.51±18.52	780.4±5.6	64.1±4.4
S20/071	797.23±28.83	605.1±26.6	104.5±2.6
N06/124	1394.53±18.98	1287.5±28.0	134.2±4.6
N20/137	641.39±9.13	530.7±4.7	35.7±5.8
N22/115	1173.18±13.93	800.2±11.2	56.1±7.8
S04/142	332.21±7.02	228.8±4.2	32.8±1.5
S23/115	528.22±6.24	416.0±0.0	52.3±3.0
N11/030	397.67±0.05	186.2±39.2	35.8±3.9
N07/110	329.24±9.01	264.4±54.6	19.2±2.8
N11/008	666.48±13.13	556.6±2.8	149.7±3.2
S04/132	608.07±23.05	454.6±12.6	80.9±1.9
ANOVA	***	***	***

***: Values are significantly different ($p<0.001$). Total polyphenols and total flavonoids are expressed as mg/l of (+) catechin; total anthocyanins are expressed as mg/l of malvidin-3-O-glucoside.

(2015 wines). There are some varieties that present higher amount of other anthocyanins: S23/115 has the highest amount of peonidin-3-O-glucoside; N11/008 of petunidin-3-O-glucoside and S20/071 of cyanidin-3-O-glucoside. Therefore would be interest to further investigate this aspect in order to understand if this is due to a specific and characteristic production of these anthocyanins in the hybrid grapes. Acetylated and coumaroylated forms are also present, unfortunately it was not possible to identify all the detected compound due to a lack of available standards.

In wines from 2015 vintage the HPLC analysis generally showed few anthocyanins in small amounts and polymeric forms of anthocyanins, eluted together at the end of the chromatogram. It is possible that a part of the anthocyanins did polymerized with other organic compounds or polyphenols or have been oxidized, resulting in a decrease of their content by time. Since the number and amount of compounds found did not justify the total anthocyanin parameter and the colour observed (both measured using the 520 nm absorbance) we also hypnotize the presence of cofactors which link with anthocyanins resulted in an increase of perceived colour (hypercromic effect) as reported by other authors. [13] Copigmentation is a phenomenon in which pigments and other non-coloured organic components form molecular associations or complexes, resulting in an enhancement in the absorbance. Copigmentation has not previously been taken into account in traditional wine colour measures, in the relationship between colour and pigment analysis, or in spectrophotometric assays for anthocyanin content. A confirm to this hypothesis might come from wine N06/124 (2015 vintage).

As reported in previous works alcohol content inhibits the formation of cofactor-anthocyanis pigment [13] indeed for N06/124 sample, which is the 2015 wine with the highest alcohol content (11.1%) the HPLC analysis shows an anthocyanins profile compatible with the total anthocyanins value.

In 2015 is the opposite: flavan3ols content is higher in "2/3 seeds" than "1 seeds". Unfortunately it was not possible to identify all the detected compound due to a lack of available standards. Therefore we plan to evaluate both the proanthocyanidin and flavan reactive to vanillin content in order to measure the index of condensation (V/L) and perform more detailed analyses using HPLC-MS instruments.

Among the detected flavan-3-ols, all wines showed the presence of both (+)− catechin and (−)-epicatechin. Procyanidin B2 was only found in wines from 2015 vintage and in wines and the only two wines of 2016 vintage produced from not completely seedless grapes. There is not a clear link between the presence of galloylated forms and the apirenic classes of the grape employed. In 2015 and 2016 wines (+)-catechin is higher in all wines produced from grapes belonging to 2/3 and 4 apirenic classes for both the vintages (Table 6), and moreover it was high in a wine produced from a seedless variety, namely N06/124.

Instead for (–)-epicatechin there was a higher amount in the seedless 2016 vintage wines, a smaller quantity in the 2/3 class (N11/008) and almost nothing in the 4 apirenic class (S04/132). In 2015 vintage the wines from seedless varieties have a content of epicatechin in average 62.3% lower if compared to the 2/3 varieties. It is possible that the climate influenced the proportion of accumulation of catechin and epicatechin in seedless, 2/3 and 4 grape varieties, therefore would be interesting to investigate how the metabolic pathway might have changed depending on the presence or absence of seeds. For procyanidin B2 is interesting to notice how in 2016 this compound is only present in wines produced by 2/3 and 4 apirenic classes grapes. Differently in 2015 procyanidin B2 is present in good quantity in the seedless wines, even if it is higher in 2/3 varieties, involving the possibility that ageing of wine increases the presence of oligomeric and polymeric forms. The astringency and bitterness of red wines cannot be predicted from their component composition. There is considerable confusion as to the relative roles of monomeric and polymeric phenols in both astringency and bitterness, and there is growing evidence that the role of monomeric forms is much greater than has previously been considered. An informal sensory analysis has been already performed, and a future planned sensorial trial of these wines will be conducted with a panel of trained testers in order to understand the perception of some important characteristics of these wines especially concerning taste, body, bitterness and astringency.

We were also able to identify the main flavonols which were quercetin glucoside and glucuronide, followed by

Table 5. HPLC anthocyanin profile.

Sample	DG	CG	PeG	PG	MG	PC	MC
S18/009	1.36±0.07	0.12±0.01	0.18±0.01	0.23±0.013	7.02±0.34	0.24±0.0	0.80±0.04
S17/039	0.16±0.01	0.14±0.01	0.35±0.02	0.203±0.01	4.01±0.19	n.f.	0.404±0.02
S20/071	0.62±0.030	0.35±0.01	0.47±0.02	0.74±0.04	6.33±0.31	0.37±0.02	0.50±0.02
N06/124	0.60±0.03	n.f.	2.83±0.14	1.89±0.09	37.11±1.80	0.43±0.02	2.93±0.14
N20/137	n.f.	0.09±0.00	n.f.	0.14±0.01	0.25±0.01	n.f.	n.f.
N22/115	0.22±0.01	n.f.	n.f.	0.14±0.01	0.17±0.01	n.f.	0.11±0.01
S04/142	0.38±0.02	0.07±0.00	0.89±0.043	1.26±0.06	21.55±1.04	0.37±0.02	1.95±0.09
S23/115	0.282±0.01	0.12±0.01	0.97±0.05	5.78±0.28	23.58±1.14	1.83±0.09	3.10±0.15
N11/030	0.35±0.02	n.f.	0.45±0.02	0.17±0.01	18.50±0.90	0.18±0.01	1.09±0.05
N07/110	0.06±0.00	n.f.	0.30±0.01	0.57±0.03	10.78±0.52	0.35±0.02	1.34±0.06
N11/008	1.46±0.07	n.f.	5.10±0.25	3.18±0.16	144.64±7.01	2.89±0.14	9.29±0.45
S04/132	0.87±0.04	0.20±0.01	2.7±0.13	5.407±0.27	69.73±3.38	1.02±0.05	4.02±0.19
ANOVA	***	***	***	***	***	***	***

***: Values are significantly different ($p<0.001$).Compounds are expressed as mg/l of malvidin -3-O-glucoside; DG: delphinidin -3-O- glucoside; CG: cyanidin -3-O- glucoside; PeG: petunidin -3-O- glucoside; PG: peonidin -3-O- glucoside; MG: malvidin -3-O-glucoside; PC: peonidin -3-O- (6-p-coumaroyl) glucoside; MC: malvidin -3-O- (6-p-coumaroyl) glucoside.

Table 6. HPLC flavan-3-ols profile.

Sample	Catechin	B2	Epicatechin	ECG
S18/009	9.106±0.513	5.074±0.286	6.290±0.354	n.f.
S17/039	4.807±0.271	6.057±0.341	5.521±0.311	1.489±0.084
S20/071	2.468±0.139	4.262±0.240	3.908±0.220	2.147±0.121
N06/124	3.928±0.221	1.148±0.065	2.118±0.119	n.f.
N20/137	1.238±0.070	1.245±0.070	2.423±0.136	0.486±0.027
N22/115	1.324±0.075	2.344±0.132	1.629±0.092	2.631±0.148
S04/142	0.916±0.052	n.f.	25.197±1.419	2.416±0.136
S23/115	1.785±0.101	n.f.	3.263±0.184	1.559±0.088
N11/030	1.657±0.093	n.f.	14.579±0.821	2.418±0.136
N07/110	0.859±0.048	n.f.	5.288±0.298	9.128±0.514
N11/008	2.260±0.127	0.318±0.134	1.171±0.492	n.f.
S04/132	2.023±0.114	0.380±0.021	0.635±0.036	0.972±0.055
ANOVA	***	***	***	***

***: Values are significantly different ($p<0.001$). Compounds are expressed as mg/l of (+)-catechin; B1: procyanidin B1; ECG: epicatechin gallate.

kaempherol glucoside and glucuronide. In all samples Quercetin glucuronide is the main flavonol found, withthe exception of wine S18/009 in which the main compound was kaempherol glucuronide (Table 7). This class of flavonoids anyway was present in small amounts in all wines.

4. Conclusions

The exploratory analyses performed on wines obtained with red winemaking of new seedless grapevine cross-breeds showed how a long aging is not suited for these wines since there is a loss of a large part of the anthocyanins mainly due to polymerization phenomena. These grapes seems optimal for the production of wines to be consumed during the same year of production, exactly as young red wines have to be without the problem of an excess of polyphenolic compounds responsible for the astringent perception which are mainly present in seeds.

An informal sensorial evaluation of these wines has already been performed and a very interesting bouquet of aromas has been found in some wines. In N22/115 wine of the 2015 vintage (hot and dry summer) it was possible to clearly recognize a smell resembling marmalade of red fruits. In another one, namely N06/124 (from seedless grape), the sensorial perception of the wine reminded another particular variety of Central Italy, the Lacrima di Morro d'Alba which is a fruity red wine, which has a strong structure due to its peculiar content of astringent polyphenols. The N06/124 wine possesses the pleasant sensory quality of this famous wine without its astringency. Instead, other wines from 2015 vintage did not show any particular characteristic even if all winemakings, both in 2015 and 2016, were performed using the same commercial yeasts and the same treatments (SO2 addition, etc.). In 2016 wines (colder and rainy summer), instead, the perfumes are less intense but some varieties show light and pleasant bouquets: N07/110 (cherries), N11/008 (fruity and pleasant) and S23/115 (red fruit in alcohol).

Due to the high variability of the new grapevine crossbreeds each grape could be used in order to produce different types of wines e.g. rosé wine, sparkling wine and so on. The present results combined with the analysis of grape and wines produced from the same grapes in the next vintage will clarify the possible final destination of the winemaking product of each variety. Concerning the wines already produced and analyzed, it is possible to understand what would be the most suitable final product. N07/110 wine at a first sight seems to be not interesting (low alcohol, low anthocyanins content and a pale rosé colour), nevertheless it is one of the most interesting wines, because from these grapes it is possible to produce less alcoholic

Table 7. HPLC flavonols profile.

Sample	Qglucur	QGlu	Kglucur	KGlu
S18/009	2.133±0.120	0.961±0.054	3.697±0.208	0.023±0.001
S17/039	7.218±0.407	0.371±0.021	0.357±0.020	n.f.
S20/071	4.027±0.227	1.996±0.112	3.627±0.204	n.f.
N06/124	5.338±0.301	0.569±0.032	1.064±0.060	n.f.
N20/137	4.016±0.226	1.989±0.112	3.616±0.204	n.f.
N22/115	3.189±0.180	0.637±0.036	0.244±0.014	n.f.
S04/142	0.823±0.046	0.450±0.025	1.017±0.057	n.f.
S23/115	2.999±0.169	0.368±0.021	6.330±0.357	n.f.
N11/030	1.318±0.074	0.118±0.007	2.310±0.130	n.f.
N07/110	1.116±0.063	0.612±0.034	1.620±0.091	0.122±0.007
N11/008	5.083±0.286	1.546±0.087	0.133±0.008	0.341±0.019
S04/132	6.893±0.388	0.685±0.039	0.143±0.008	0.290±0.016
ANOVA	***	***	***	***

***: Values are significantly different ($p<0.001$). Compounds are expressed as mg/l of quercetin-3-O-glucoside. Qglucur: quercetin-3-O-glucuronide; QGlu: quercetin-3-O-glucoside; Kglucur: kaempferol-3-O-glucuronide; KGlu: kaempferol-3-O-glucoside.

sparkling wines with a higher residual sugar at the end of the second fermentation (like Moscato d'Asti). In other cases, from some new grapevine crossbreeds it is possible to obtain good quality light rosé wines characterized by few of total polyphenols, flavonoids and anthocyanins but with a pleasant sensorial bouquet. letting the skin-must contact during the maceration process continue until the end of the fermentation, without worries of an excessive extraction of astringent polyphenols compounds or too much anthocyanins (see S23/115 or N20/137). For other wines, such as N06/124 and N11/008, which already showed interesting sensorial features, a good content of polyphenols and a more intense red colour would be possible to obtain red traditional wine (more intense colour) but with characteristics similar to Beaujolais Nouveau or Novello wines with an adequate management of the vineyard (decreasing the number of bunches for each vine) and type of training system (e.g. trellis).

Interestingly even new grapevine crossbreads obtained from the same varieties are different, e.g. N06/124 and N07/110 wines (same grape parents and same wine-making) show a very different content of polyphenolic compounds, and in other cases the same parents produced very different apirenic classes (S04/132 and S04/142). Therefore, due to these strong differences even for grapes coming from same crossbreading, we will investigate other novel grapevine crossbreads varieties created by our research group and present in the experimental field of CREA-VE in Turi (Bari).

This exploratory analysis allowed the identification of interesting seedless grapes to be employed in order to obtain novel red or rosé wines with improved features. The research will go on to further characterize these and other new grapevine crossbreads in the CREA-VE Turi (Bari) experimental field and the resulting wines, especially concerning the aromatic composition and acidic profile.

5. Future perspectives

- Evaluate both the proanthocyanidin and flavan reactive to vanillin content (Index of condensation (V/L)).
- Study of the actual acidic profile of the wines (benzoic and hydroxycinnamic acids) with HPLC.
- Investigate the presence and/or contribution of anthocyanins-cofactor pigments (% copigmentation) and the degree of anthocyanins polymerization (% polymerization) to total wine.

References

[1] P. Ribéreau-Gayon, Y. Glories, A. Maujean, D. Dubourdieu *Handbook of enology, vol 2 Wiley, Chichester*, 232–234 (2000)

[2] A. Lea, P. Bridle, C. Timberlake, V.L. Singleton, Am J Enol Vitic **38**, 277–281 (1979)

[3] J.A. Kennedy, M.A. Matthew, A.L. Waterhouse, Am J Enol Vitic **53**(4), 268–274 (2002)

[4] J.M. Souquet, V. Cheynier, F. Brossaud, M. Moutounet, Phytochemistry **43**, 509–512 (1996)

[5] M. Moutonet, H. Fulcrand, P. Sarni-Manchado, J.M. Souquet, V. Atanalsova, B. Labarbe, C. Maury, S. Vidal, V. Cheynier, *13 Symposium International d'œnologie,* Montpellier, 737–747 (2002)

[6] R. Gawel, Aust J Grape Wine Res **4**, 74–95 (1998)

[7] M. Guaita, M. Petrozziello, L. Panero, C.Tsolakis, S. Motta, A. Bosso, Eur Food Res Technol, doi: 10.1007/s00217-017-2842-1 (2017)

[8] Ma Wen, Guo AnQue, Zhang YuLin, Wang Hua Ye, Li Hua, Trends Food Sci Tech **40**(1), 6–19 (2014)

[9] V. Cheynier, M. Duenas, E. Salas, C. Maury, J.M. Souquet, P. Sarni-Manchado, H. Fulcrand Am J Enol Vitic **57**, 298–305 (2006)

[10] K.L. Sacchi, L.F. Bisson, D.O. Adams, Am J Enol Vitic **56**, 197–206 (2005)

[11] R. Di Stefano, M.C. Cravero, R. Gentilini, L'Enotecnico **XXV**(5), 83–89 (1989)

[12] M. De Rosso, L. Tonidandel, R. Larcher, G. Nicolini, V. Ruggeri, A. Dalla Vedova, F. De Marchi, M. Gardiman, R. Flamini, Analytica Chimica Acta **732**, 120–129 (2012)

[13] R. Boulton, Am. J. Enol Vitic **52**(2), 67–87 (2001)

DOI: 10.1051/bioconf/20170904011

L'histoire du vin comme patrimoine et atout économique – The history of wine as a heritage and an economic asset

Serge Wolikow – Professeur émérite

Université de Bourgogne, MSH Dijon, Chaire Unesco, Culture et tradition du vin, France

Résumé. La connaissance historique du vin est aujourd'hui un enjeu scientifique, culturel mais aussi économique. La précocité du commerce international et de la consommation du vin dans de nombreuses régions du monde permet d'avoir une réflexion en longue durée – sur plusieurs siècles – concernant la globalisation. Au 20e siècle l'instauration de normes internationales partagées, notamment grâce à l'activité de l'OIV, donne aujourd'hui encore à réfléchir sur un développement économique international appuyé sur des normes partagées. Le traitement numérique et la mise à disposition du Bulletin de l'OIV fondés sur un travail universitaire réalisé à Dijon est un outil mis à la disposition de tous pour la réflexion historique et l'étude comparative. Aujourd'hui les territoires vitivinicoles sur tous les continents ont compris que l'histoire du vin est technique mais aussi sociale et culturelle. A l'heure du développement mondial de l'œnotourisme il est temps de réfléchir et partager les expériences et les résultats de la mise en perspectives historique dans les actions de marketing et de promotion. Cela soulève des problèmes éthiques et déontologiques qui méritent d'être évoqués lorsqu'il s'agit de l'histoire du vin, le produit pour lequel a été pensé de garantir sa qualité en direction du consommateur.

Abstract. The historical knowledge of wine is today a scientific, cultural but also economic question. Early international trade and wine consumption in many parts of the world allows for a long-term reflection on globalization over several centuries. In the 20th century, the establishment of shared international standards, in particular through the activities of the OIV, continues to reflect on international economic development based on common standards. The digital processing and the availability of the OIV Bulletin based on a university work carried out in Dijon is a tool made available to all for historical reflection and comparative study. Today, wine regions on all continents have understood that the history of wine is technical, but also social and cultural. At the time of the global development of wine tourism it is time to reflect and share the experiences and the results of the historical perspective in the actions of marketing and promotion. This raises ethical issues that must be explained when it comes to the history of wine, the product for which it was thought to guarantee its quality to the consumer.

Affirmer pour commencer que la connaissance historique du vin est aujourd'hui un enjeu scientifique, culturel mais aussi économique peut sembler étonnant dans un temps où les incertitudes du présent semblent parfois l'emporter particulièrement au plan international. Il me semble, tout au contraire, que l'histoire de la viticulture et du vin peut contribuer à la compréhension de la situation actuelle. La publication sous forme numérique du bulletin de l'OIV depuis 1927 dans le cadre d'un projet mené en partenariat entre l'OIV, l'Université de Bourgogne, appuyée par la Maison des Sciences de l'Homme de Dijon et la Chaire Unesco, culture et tradition du vin nous donne l'occasion d'aborder cette question de manière concrète. Avant de présenter cette réalisation je voudrai insister sur l'intérêt d'une culture historique quand on souhaite comprendre le monde du vin au 21e siècle

Face aux évolutions économiques internationales de la production, de la circulation et de la consommation du vin, son histoire constitue un terrain important de l'analyse pour prendre la mesure des changements. De même, face à l'impact des progrès scientifiques et techniques sur l'élaboration d'un produit dont l'existence est millénaire, quelle est la part de l'innovation mais aussi de la continuité ? L'histoire est également au cœur des représentations et des pratiques culturelles associées à la production et à la consommation du vin, y compris aujourd'hui son marketing. En somme, loin d'être anecdotique, l'approche historique est inévitable lorsqu'on vent comprendre la filière vitivinicole dans sa globalité.

Evoquer l'histoire du vin est une pratique fréquente parmi les acteurs de la filière. Qu'ils soient au niveau de la production ou de la vente, ils mobilisent assez fréquemment des références anciennes censées attester de la qualité du produit et de l'expérience des fabricants. Mais cette évocation du passé, souvent renforcée par des préoccupations juridiques fondée sur l'ancienneté de la viticulture pour justifier d'une appellation, débouche sur un usage restrictif et instrumental de l'histoire. Celle-ci devient une sorte de caution permettant de certifier la qualité au nom d'un passé inchangé. L'ancienneté du vin ne justifie en rien une conception fixiste du passé. Cette représentation de l'histoire, comme le maintien d'un ordre ancien qui serait un gage de qualité n'a pas complètement disparu en dépit de la démarche erronée qu'elle suppose. Il faut donc préciser que l'histoire ne peut être assimilée à l'immobilité mais au contraire à la mesure des changements survenus, insensiblement ou brutalement. Il faut également mentionner quelle est l'échelle géographique de l'analyse. Il convient de rappeler que très anciennement la production et la consommation du vin sont inséparables de sa circulation et de son commerce international.

On ne peut faire une histoire du vin seulement localisée, il faut avoir également une approche globale qui associe production et consommation, aux échelles locale et internationale. Parfois, ce facteur a été décisif comme pour le vin effervescent de champagne qui a d'abord connu une commercialisation internationale au 19e s de même que les vins le bordeaux. Plus généralement, l'histoire globale du vin est inséparable des grands événements planétaires comme les deux guerres mondiales, les crises commerciales et financières, du 20e s à nos jours. Si l'histoire du vin est particulièrement liée à l'histoire générale c'est que très tôt le vin a été l'objet d'un commerce international important en valeur notamment parce que les lieux de production étaient géographiquement limités par apport aux zones de consommation. Ainsi, l'histoire du vin est marquée par de grandes fractures historiques qui éclairent l'évolution, la création ou la disparition des vignobles comme les modifications de la consommation. Les événements politiques ont joué un rôle parfois majeur, avec la fermeture ou l'ouverture des marchés, de la fin du 18e siècle à nos jours : depuis les guerres napoléoniennes en Europe, le libre-échange britannique, les deux guerres mondiales, les accords commerciaux et monétaires après 1945, etc.. Dans ce contexte il ne faut ignorer les profondes évolutions techniques et scientifiques qui ont affecté aussi bien la viticulture que la vinification. Jusqu'au milieu du 19e siècle on ne sait pas ce qu'est la fermentation : elle n'est pas maîtrisée. Après Pasteur on va contrôler et maitriser la fermentation parce qu'on la comprend …Pour la viticulture, dès la fin du 19e siècle, on développe les traitements chimiques ce qui transforme la viticulture, la conduite, les rendements. De même la mécanisation transforme au moins en partie le travail viticole. Enfin, il faut évidemment mentionner le bouleversement des moyens de transports et de conditionnement du vin qui entraîne une modification profonde de la commercialisation. C'est d'un côté la généralisation du transport ferroviaire et maritime dès le 19e s et de l'autre les nouveaux modes de conditionnement, containers et bouteilles en verre. Les règles commerciales affectent également la circulation et la commercialisation des vins, d'abord la constitution des marchés nationaux avec des droits de douanes normalisés puis les accords commerciaux internationaux qui influencent fortement le marché du vin. En somme, l'histoire du vin n'échappe pas aux différents moments de la mondialisation économique depuis le 19e siècle.

L'histoire du vin n'est pas linéaire mais marquée par des cycles, avec des phases de croissances, de progression mais aussi de stagnation voire de régression aussi bien dans le domaine des surfaces, de la production que de la consommation. On peut ainsi identifier quatre grandes phases qui structurent cette histoire du vin. La première croissance du 19e siècle est caractérisée par l'expansion des surfaces plantées et de la production. La longue période de crises et récession qui va de 1870 à 1945, est marquée par la rétraction des surfaces, la stagnation de la production et de la consommation, la désorganisation des marchés. Une deuxième période expansion, de 1945 à 1975, est caractérisée par l'augmentation modérée des surfaces, la croissance forte des rendements, de la production et de la consommation. Enfin, à partir des années 1980, une nouvelle mondialisation de la production et de la consommation s'affirme avec une géographie modifiée de la production et du transport mais aussi de la consommation et donc également de la commercialisation marquée par l'émergence de nouveaux pôles d'activité, en Amérique, Asie et Océanie.

La création de l'OIV durant les années 1920 et le lancement en 1928 du Bulletin s'inscrit donc dans la longue phase de difficultés du monde du vin marquée à la fois par la désorganisation du marché, le repli protectionniste mais aussi les fraudes et la concurrence sans règle. La très riche activité de l'OIV est essentielle pour connaître et comprendre l'histoire mondiale du vin. Avec le programme de numérisation du bulletin des origines jusqu'en 1999 avec c'est un projet documentaire de grande ampleur qui a été réalisé en partenariat avec la direction de l'OIV et une équipe scientifique et technique de l'Université de Bourgogne, la plate forme de numérisation et base de donnée de la Maison des Sciences de l'Homme de Dijon. Il a été réalisé selon les normes en vigueur sur le plan du catalogage, de l'indexation et de l'interopérabilité. Ce projet a été mené de 2014 à 2016 : il a permis de numériser et cataloguer 690 bulletins, 14 415 articles, 31 congrès de l'OIV, 2882 auteurs ont été indexé, 681 lieux géographiques et 2408 noms d'organismes. Il est dès maintenant consultable en ligne de plusieurs manières, en particulier sur le site de la Maison des sciences de l'Homme de dijon. `https://pandor.u-bourgogne.fr/ead.html?id=FRMSH021_00019#!{''content'':[''FRMSH021_00019_e0000030'',true,'''']}`

Le bulletin de l'OIV constitue une ressource exceptionnelle pour le travail historique en permettant des approches scientifiques interdisciplinaires. C'est en particulier le cas dans le domaine de l'histoire des sciences et techniques de la vigne et du vin. Les nombreux rapports, compte rendus de recherches et extraits des travaux des congrès qui paraissent dans la revue offre une ressource inédite pour l'étude des réalités historique et des évolutions des pratiques vitivinicoles sur le XXe siècle. Ils permettent de comprendre la place des savoirs scientifiques dans l'élaboration des décisions internationales en matière de viticulture, de vinification et de droit. Le Bulletin de l'OIV se fait aussi l'écho des législations nationales ou des accords internationaux bilatéraux et multilatéraux. Ces données rendent possible une histoire des normes juridiques, commerciales, techniques et sanitaires de la vigne et du vin au plan au plan mondial et permettront l'étude des rapports de force autour des enjeux commerciaux de la vigne et du vin. L'existence de rubriques statistiques, dans tous les bulletins, offre par ailleurs un outil incomparable pour analyser en détail les marchés, les évolutions de consommation, ou les territoires du vin au sens large. Il est possible, également, de procéder à une étude des normes internationales communes d'analyse des vins et

DOI: 10.1051/bioconf/20170904011

de classification des produits, de questionner les approches sanitaires et environnementales mondiales.

La mise de cette documentation exceptionnelle, à la disposition de tous, doit encourager une démarche historique qui s'appuie sur le travail scientifique et une documentation vérifiée. On a de même, pour la France, la mise en ligne des archives des appellations d'origines contrôlées discutées et délimitées par l'Institut National des Appellations d'Origine (INAO) représente une source précieuse pour connaître et comprendre le processus de qualification et DE validation des terroirs. `https://pandor.u-bourgogne.fr/ead.html?id=FRMSH021_00005#!{''content'':[''FRMSH021_00005_e0000019'',true,'''']}`

A l'heure du développement mondial de l'œnotourisme il est temps de réfléchir et partager les expériences comme les résultats de la mise en perspectives historique dans les actions de marketing et de promotion. Le développement de l'histoire, par les acteurs du monde vitivinicole, ne doit pas être réduit à une simple démarche commerciale ou marketing. C'est, en général, bien davantage, souvent les producteurs individuels ou regroupés ont à cœur de raconter à leurs clients et visiteurs leur aventure professionnelle à travers celle des ancêtres, souvent des nouveaux arrivants, parfois au contraire ils sont les descendants de producteurs enracinés depuis des siècles dans un territoire. Tous s'efforcent d'inscrire leur production vitivinicole dans une aventure qui concrétise le travail engagé mais aussi les progrès réalisé. Cette intervention, dans le champ historique est cependant souvent très localisée et difficilement mise en relation avec les évolutions internationales. Aujourd'hui, la qualité du récit historique présenté aux consommateurs et clients passe par un effort collectif permettant aux professionnels de s'appuyer sur des connaissances partagées et mises à disposition d'un grand public en particulier à travers le web. Le développement de musées, de centres de documentations, de livres anciens et nouveaux sur l'histoire du vin en est la preuve. La Chaire Unesco culture et tradition du vin de l'Université de Bourgogne s'attache, en particulier par un réseau mondial universitaire et une revue électronique «territoires du vin», à diffuser des études comparatives sur les vignobles du monde, associant monographies et propositions théoriques pour penser globalement l'histoire di vin. `https://revuesshs.u-bourgogne.fr/territoiresduvin/`.

Au terme de cette communication nous souhaitons insister sur l'intérêt de mener aujourd'hui un travail historique de qualité dans le cadre des démarches qualitatives reconnues comme nécessaire dans tous les domaines de l'activité vitivinicole. Les outils documentaires numériques permettent d'échanger, de faire circuler les connaissances et la réflexion, ne craignons d'y recourir !

 DOI: 10.1051/bioconf/20170904012

SOECO : indicateurs socio-économiques pour la viticulture et les systèmes de culture innovants / SOECO: Socio-economic indicators for viticulture and innovative cultural systems

Olivier Keichinger[1] et Marie Thiollet-Scholtus[2]

[1] Independant researcher 67600, Sélestat, France
[2] INRA - SAD - UR-0055-ASTER, 28, rue de Herrlisheim, 68000 Colmar, France

Abstract. Research on the concept and assessment of sustainability has developed considerably in recent years. If we found a lot of evaluation methods, majority do not treat or more succinctly the social impacts. Then we developed a method of socio-economic evaluation for wine-producing farms in order to know the impact of practices on profitability (gross margin) but also on human capital (painfulness, pesticides risks, safety) and social capital (landscape, dissemination of acquired knowledge). Whether for more sustainability or to cope with new constraints such as climate change, reduced inputs or agronomic dead ends (limited number of active substances allowed), it is necessary to change the systems of wine production . These so-called innovative systems generally aim to reduce the impact on the environment. But here too it is necessary to evaluate their effects on the other domains of sustainability: the social and economic domains to ensure their adoption by winegrowers. Our method thus has a specific module that makes it possible to carry out this comparison between classical and innovative systems within the same vineyard. In this paper we describe the construction of our method and the feasibility test on four Alsatian viticulture farms.

1. Introduction

La recherche sur le concept et l'évaluation de la durabilité s'est considérablement développée ces dernières années. On le constate par le grand nombre de méthodes d'évaluation que l'on trouve dans la littérature. Toutefois une majorité ne traite pas de la même façon les différents piliers de la durabilité. En général, si la dimension environnementale au travers de l'impact sur l'eau, sur les sols ou bien encore vis-à-vis de la biodiversité est bien prise en compte (nombreux critères, méthodes validées et/ou appliquées, ...) (Thiollet-Scholtus and Bockstaller, [2015]), les impacts économiques et surtout sociaux sont abordés de façon plus succincts.

Ce travail s'inscrit dans un projet plus large qui a pour but de réduire les intrants et en particulier les traitements phytosanitaires. Car pour plus de durabilité ou pour faire face aux nouvelles contraintes qui apparaissent comme le changement climatique ou bien encore des impasses agronomiques (nombre limité de substances actives autorisées) il est nécessaire de faire évoluer les systèmes de production viticole. Ces systèmes, dits innovants, ont généralement comme objectif de réduire l'impact sur l'environnement. Mais ici aussi il est nécessaire d'évaluer leurs effets sur les autres piliers de la durabilité : les piliers social et économique pour s'assurer de leur adoption par les viticulteurs.

Dans ce document, nous détaillerons la construction de la méthode d'évaluation socio-économique pour une exploitation viticole afin de connaitre l'impact des pratiques. Nous y décrirons aussi le module qui permet spécifiquement de réaliser la comparaison entre système classique et innovant au sein d'une même exploitation viticole.

2. Les indicateurs socio-économiques à l'échelle de l'exploitation

Maintenant que les objectifs de cet outil ont été défini. Il devra à la fois évaluer les aspects sociaux et économiques d'une exploitation viticole et pouvoir permettre la comparaison entre le système de culture «classique» et les système innovants qui pourraient être mis en place sur tout ou partie de la surface.

Dans un premier temps, nous allons détailler le mode de construction des indicateurs socio-économiques. Leur démarche d'élaboration suit les principales étapes préconisées par (Bockstaller et al., [2008]) et appliquée par d'autres auteurs (Keichinger, [2001]; Pervanchon et al., [2002]; Thiollet-Scholtus and Bockstaller, [2015]; Tixier et al., [2007]) : i) une première phase de choix préalables et d'hypothèses qui vont déterminer l'étape suivante ii) la construction proprement dite, iii) une phase d'évaluation (application sur le terrain). Les objectifs de cet outil ayant été défini dans l'introduction, nous allons passer directement à la partie construction. Pour plus de compréhension, les objectifs peuvent être divisés en sous objectifs : pour l'évaluation SOECO on différenciera l'économique du social et ce dernier sera divisé en capital humain et en capital social.

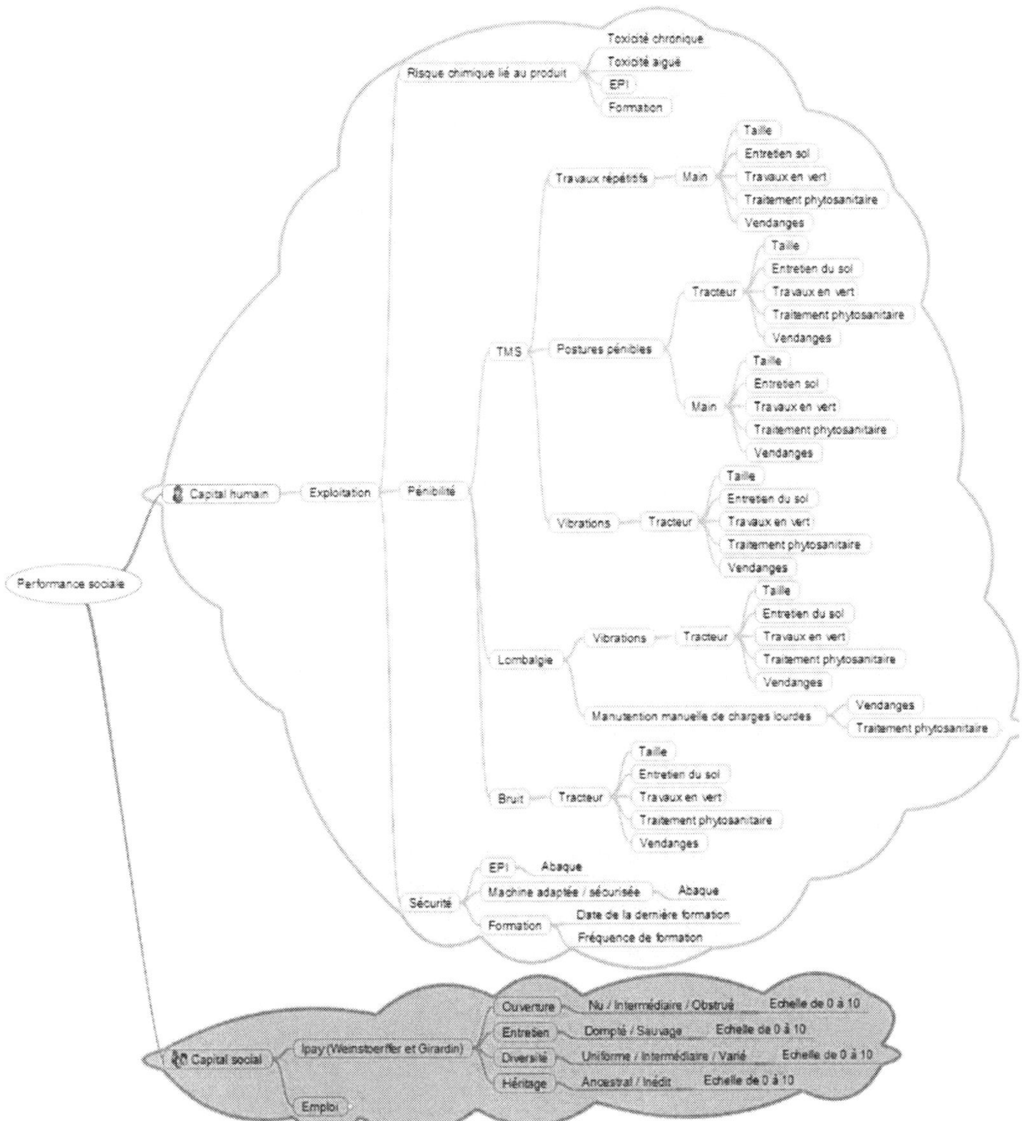

Figure 1. Variables retenues pour l'évaluation social.

2.1. La construction des indicateurs sociaux

Cette étape commence avec la détermination des variables nécessaires à l'évaluation sociale (Cuegniet, 2015). Les performances sociales sont données par la Fig. 1.

Ensuite, le passage à un indicateur nécessite d'agréger les variables (quand il y en a plusieurs) par différents moyens et de déterminer une (ou plusieurs) valeur(s) de référence afin de rendre compréhensible les résultats ainsi obtenus. La construction de l'indicateur «capital humain» ainsi que les sous-objectifs «risque phytosanitaire pour l'opérateur» et «pénibilité au travail» seront décrits ici.

2.1.1. Le risque phytosanitaire pour l'opérateur

Il prend en compte le risque pour l'opérateur lors de la préparation de la bouillie et lors de l'application. Il est difficile d'appréhender le risque pour l'opérateur car certains types de données ne sont pas forcément accessibles pour toutes les substances actives. Par contre on dispose de données extrapolées pour évaluer la toxicité aigüe et chronique : la DL50 et la DJA respectivement.

Pour une exploitation, la ou les personnes responsables des traitements sont exposées au prorata de la quantité épandue. Pour chaque substance active, on pourra réaliser le calcul du «nombre de doses» potentiellement reçu par l'opérateur en divisant la quantité totale épandue par la toxicité (DL50 et DJA). On obtiendra respectivement une indication sur le niveau de toxicité aigüe et chronique à laquelle l'opérateur est potentiellement exposé. Ces valeurs calculées de toxicité aigüe et chronique ne nous renseignent aucunement sur le risque ni même sur la possibilité de faire au mieux en fonction de la pression des ravageurs. C'est pourquoi, pour chaque région on va définir un itinéraire type de traitement. Pour l'Alsace on considère que l'itinéraire type est constitué d'un désherbage sur 40 % de la surface, de 2 traitements anti-mildiou et de 2 traitements anti oïdium. Pour chaque traitement on retiendra le ou les produits ayant la plus faible toxicité et pour éviter les problèmes de résistance, ils seront choisis dans des familles différentes. De plus, on considère que la formation et surtout l'utilisation d'équipement de protection individuelle permet de réduire d'environ 90 % le risque (valeur généralement retenu pour le risque opérateur dans EUROPOEM). Il sera important que l'opérateur soit formé régulièrement à l'usage de ces produits et qu'il se protège soit par des équipements de protection individuelle (pour la préparation de la bouillie, par l'utilisation de lunettes, masque, gants combinaison et bottes et lors de l'épandage par un masque, des gants et une combinaison) soit par une cabine équipée de filtre (lors de l'épandage). Toujours d'après l'EUROPOEM, on considère que l'exposition est plus élevée lors de l'épandage (2/3) que lors de la préparation de la bouillie (1/3). Pour la préparation de la bouillie, la priorité est donnée au port des gants (80 %) et à égale proportion pour les autres protections (5 % chacun). Pour l'épandage c'est la combinaison qui apporte le plus de protection (70 %) suivi du port des gants (20 %) et du masque (10 %), sachant que la cabiné équipée d'un filtre aura le même effet que le port de tous les protections individuelles. Si on considère qu'il faut en moyenne une personne pour gérer 5 ha de vigne on pourra calculer une valeur de référence à partir des cinq traitements annuels et l'utilisation de toutes les protections individuelles citées plus haut. Cette valeur de référence correspondra à la valeur 7 de l'indicateur (niveau acceptable). Pour déterminer la correspondance entre les valeurs de toxicité et les valeurs d'indicateur, on détermine deux autres points : aucun traitement (valeur de toxicité nulle) équivaut à la valeur 10 de l'indicateur et que cinq fois la valeur de toxicité de référence correspond à un indicateur nul. Ces trois points permettent d'établir deux équations reliant valeur de toxicité et valeur d'indicateur.

2.1.2. *La pénibilité au travail*

La pénibilité peut entrainer des maladies professionnelles comme les lombalgies, les troubles musculo-squelettiques ou les effets du bruit. Ces maladies proviennent de travaux répétitifs effectués en vignoble (taille, conduite du tracteur, vendanges, ...) dans des postures quelquefois pénibles (bras levés, accroupi, ...). Une évaluation du risque pour chaque type de maladies professionnelles sera réalisée en listant les différents types de travaux ainsi que le temps passé à les réaliser.

Les troubles musculo-squelettiques peuvent provenir de travaux répétitifs (taille au sécateur), de postures pénibles (position accroupi,...) et des vibrations (conduite du tracteur). Les trois origines de ces troubles étant très différents car n'ayant pas les mêmes causes, ils seront évalués indépendamment puis agrégés. Pour connaitre le temps passé pour chacune des origines des troubles muculo-squelettiques, un questionnaire a été réalisé avec la liste de tous les travaux de la vigne et le temps à l'hectare estimé à leur réalisation. Ce temps est annualisé sur l'exploitation puis ramené à une durée moyenne par semaine et par travailleur. Cette valeur est ensuite comparé à des seuils pour obtenir un indicateur en trois classes : vert (risque faible), orange (risque moyen) et rouge (risque élevé). Les seuils sont de 10 et 20 heures par semaines pour les travaux répétitifs et de 2 et 10 heures pour les postures pénibles et les vibrations. L'évaluation des trois origines des TMS donne trois résultats qu'il faut encore agréger. Les résultats se présentant sous la forme d'un code couleur, on ne peut les agréger que par des règles de décision. Pour cela nous utiliserons la méthode des tableaux de bord (Girardin, 2005). On donnera le même poids aux variables et on aura comme règle d'agrégation : i) le résultat final est «vert» si les trois variables sont «vert» ii) le résultat final est «orange» si une des trois variables est «orange» (les deux autres sont «vert») et iii) le résultat final est «rouge» si une au moins une des trois variables est «rouge» ou si deux au moins des variables sont «orange».

La lombalgie provient de la manutention d'objet lourd et des vibrations et secousses lors de la conduite du tracteur. N'ayant pas la même cause, ces deux origines de la lombalgie seront évaluées indépendamment puis agrégées comme pour les troubles musculo-squelettique. Pour la manutention, on somme les heures passées à porter des charges lourdes lors des travaux de la vigne donc principalement lors des traitements phytosanitaires (pulvérisateur à dos) ou lors des vendanges avec des seuils de 10 et 20 h/semaine. Pour les vibrations, on reprendra les valeurs calculées précédemment ainsi que la méthode d'agrégation pour obtenir le «risque» lombalgie.

Le bruit provient principalement de l'usage d'engin viticole et sera fonction du niveau sonore de ceux-ci et du temps d'utilisation. Il est prévu un questionnaire sur les tracteurs et sur les moyens mis en œuvre pour réduire l'effet du bruit (casque, cabine insonorisé, entretien). Comme précédemment, on va croiser pour chaque tracteur les heures travaillées au cours l'année avec le facteur de niveau sonore de ceux-ci puis sommer le résultat obtenu pour chaque tracteur. Enfin, on obtiendra un indicateur en trois classes en utilisant les seuils de 10 et 20 h/semaine.

La pénibilité est issue de l'agrégation des résultats obtenus pour les troubles musculo-squelettiques, la lombalgie et le bruit calculés précédemment. La méthode est celle des tableaux de bord avec les mêmes règles de décision. Le résultat final de la pénibilité se présentera sous forme d'un indicateur à trois couleurs, vert (les travaux de la vigne ont peu ou pas d'effet sur la santé physique), orange (les effets restent modérés mais il faut par différents moyens essayer de réduire la pénibilité) et rouge (les travaux risquent d'entrainer des problèmes physiques et il est impératif de diminuer la pénibilité).

2.1.3. *L'indicateur «capital humain»*

Pour déterminer le capital humain, il reste à agréger les trois variables qui le constitue : risque phytosanitaire, pénibilité et sécurité. La sécurité issue d'un questionnaire sur l'utilisation de protection individuelle, de sécurité du machinisme et de formations se présente sous la forme d'un indicateur entre 0 et 10. L'agrégation se fera à l'aide de règles de décision (Tableau 1) associée à la logique floue (Ricou et al., [2014]). La pénibilité étant évaluée par trois classes de couleur, on déterminera le degré d'appartenance aux classes floues de la façon suivante : vert (F), orange (50 % F et 50 % D) et rouge (D). Pour le risque phytosanitaire comme pour la sécurité, la valeur 10 de l'indicateur est considérée comme favorable et 0

 DOI: 10.1051/bioconf/20170904012

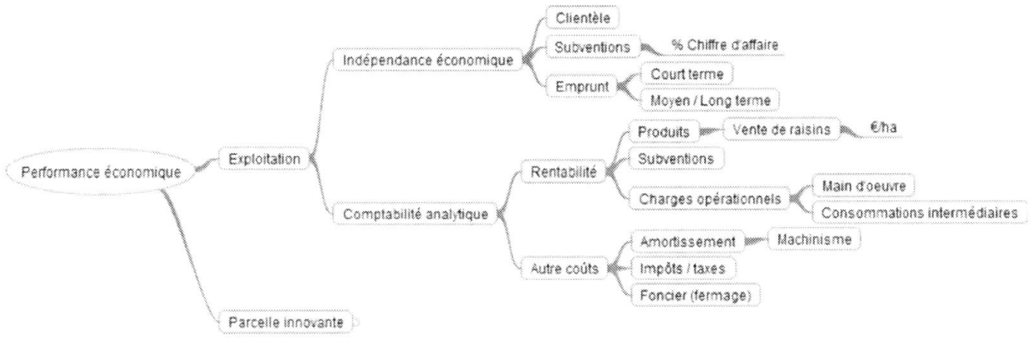

Figure 2. Variables retenues pour l'évaluation économique.

Tableau 1. Conclusions des règles de décision permettant l'évaluation du capital humain.

Risque phytosanitaire	Pénibilité	Sécurité	Conclusions des règles de décision
F	F	F	10
F	F	D	7
F	D	F	7
F	D	D	3
D	F	F	7
D	F	D	3
D	D	F	3
D	D	D	0

comme défavorable. Pour les valeurs intermédiaires le degré d'appartenance à la classe floue favorable est donné par l'équation 0.1* indicateur.

2.2. La comptabilité analytique

Cette étape commence avec la détermination des variables nécessaires à l'évaluation économique (Cuegniet, 2015). Les performances économiques sont données par la Fig. 2.

Elle sera réalisée en deux temps, tout d'abord ce qui a trait directement à la conduite parcellaire (pour pouvoir comparer éventuellement avec des parcelles innovantes) puis ce qui se joue au niveau de l'exploitation. Pour cela nous reprendrons en partie ce qui avait été proposé dans MASC (Sadok et al., 2009) avec une évaluation de la rentabilité à partir du calcul de la marge semi-nette :

Marge semi-nette = PB + AD – CO

Avec : - PB : Produit brut (en €/ha) = quantité récolté * prix de vente

- AD : Aides directes (en €/ha)

- CO : Charges opérationnelles (en €/ha) qui prennent en compte l'achat de semences (enherbement), l'amortissement sur la plantation, les phytosanitaires, les fertilisants, le carburant et la main d'œuvre.

Cette valeur (en €/ha) sera ensuite transformé en indicateur avec un code couleur (noir, rouge, orange et vert) avec comme seuil respectivement 1000, 2000 et 3000 €/ha.

Puis les impôts, l'amortissement du machinisme ou le foncier sont intégrés à cette marge semi-nette pour estimer le bénéfice net à l'hectare. Un indicateur identique à la marge semi-nette sera produit avec comme seuil 0, 1000 et 2000 €/ha.

3. Les indicateurs socio-économiques et les systèmes innovants

La réduction d'intrants passe par la mise en œuvre de systèmes innovants. Si la conception de tels systèmes est important, leur évaluation l'est tout autant afin de connaitre le degré d'atteinte de l'objectif. Pour la partie capital humain, on retrouve la pénibilité, le risque phytosanitaire et la sécurité. S'y ajoute une partie sur la gain ou les pertes que peut engendrer l'innovation dans différents domaines. Cette partie est subjective car il consiste en un questionnaire où le ressenti peut être important. Toutefois, nous avons fait le choix de donner du poids au ressenti car l'adoption ou non de ces innovation par le viticulteur dépendra fortement de celui-ci.

3.1. Le capital humain

3.1.1. L'évaluation

Pour la pénibilité, le risque phytosanitaire et la sécurité, l'évaluation se fera de la même façon que pour l'exploitation mais avec les données propre au système innovant.

Pour le questionnaire, à chaque question le viticulteur attribue une note entre −5 et +5 respectivement selon lui pour une perte maximale ou un gain important (et 0 si pas de différence). Les sept questions portent sur la satisfaction (et motivation) qu'il a à mettre en place ce système innovant, sur l'acquisition de connaissances, sur l'amélioration (ou non) de l'organisation du travail et de la sécurité au travail, sur l'augmentation (ou la diminution) du risque chimique, de la pénibilité et du stress. Pour évaluer ces réponses et donc le ressenti du viticulteur, on fait tout d'abord la moyenne de ces notes puis on transforme cette valeur en indicateur (code à quatre couleurs) à l'aide des limites des classes données dans le Tableau 2.

3.1.2. La comparaison

On pourrait comme pour l'indicateur capital humain, agréger ses différentes composantes pour obtenir un indicateur qui nous renseignerait sur l'impact de ce système innovant. Or en fonction du lieu, du climat, de l'historique ou d'autres paramètres, l'innovation pourrait

DOI: 10.1051/bioconf/20170904012

Tableau 2. Limites des classes permettant la comparaison innovant - classique pour l'indicateur capital humain.

	Limites des classes			
Classes	Pénibilité (%)	Indicateur phytosanitaire (valeur d'indicateur)	Sécurité (valeur d'indicateur)	Indicateur "ressenti" (note)
Vert	≤ 5	≥ 0	≥ 0	≥ 0
Orange	]5 ; 15]	]0 ; –0.5]	]0 ; –0.5]	]0 ; –0.5]
Rouge	]15 ; 25]	]–0.5 ; –1]	]–0.5 ; –1]	]-0.5 ; –1]
Noir	> 25	≤ –1	≤ –1	≤ –1

apparaitre comme peu innovant (voir régressif) par rapport à ce que l'on peut faire par ailleurs. C'est pourquoi il est plus judicieux de comparer cette innovation à ce que faisait le viticulteur auparavant (ou qu'il fait encore sur ses autres parcelles). L'innovation étant testée sur de petites parcelles, l'effet de ce changement de pratique n'aura que peu d'effet sur l'indicateur social à l'échelle de l'exploitation. Mais en ne peut pas non plus calculer d'indicateurs sociaux à la parcelle, car certaines données ne sont pertinentes et/ou calculables qu'à l'échelle de l'exploitation. Donc nous allons supposer que la totalité de l'exploitation se trouve en «innovant» et que l'on comparera avec la totalité de la même exploitation en «classique». Cette comparaison se fera pour chaque sous-objectif du capital humain afin de bien identifier les postes en progrès ou non.

Pour la pénibilité, on pourrait comparer les résultats finaux entre innovant et classique mais du fait de sa construction (mise en classe des valeurs), un changement même relativement important risque de passer inaperçu. C'est pourquoi il est préférable de comparer ici aussi toutes les composantes qui constituent la pénibilité (travaux répétitifs, postures pénibles, vibrations, manutention et bruit). Pour chacune de ces composantes, on calcule l'évolution de «l'innovant» par rapport au «classique» (en %) et ce résultat est réparti en 4 classes de couleur (Tableau 2).

Pour le risque phytosanitaire et la sécurité, on compare la valeur d'indicateur (innovant - classique) et l'on classe le résultat selon les limites données dans le Tableau 2.

Pour la comparaison entre innovant et classique pour le capital humain, il suffit d'agréger les classes obtenues pour la comparaison de pénibilité, risque phytosanitaire et sécurité en y ajoutant aussi l'indicateur ressenti. Les règles de décision étant les suivantes avec comme phrase logique, «pour le que résultat final soit ... il faut que l'on ait... parmi les composantes du capital humain» tel que :

- ... vert ... 4 vert
- ... orange ... 1 orange (et 3 vert)
- ... rouge ... 2 orange (et 2 vert) ou 1 rouge (et 3 vert)
- ... noir ... au moins 1 rouge et 1 orange ou 2 rouge ou 3 orange.

3.2. La comptabilité analytique

Contrairement à l'indicateur «capital humain», on peut cette fois faire une comparaison entre innovant et classique à l'échelle de la parcelle (une marge semi-nette étant calculé à cette échelle dans l'indicateur économique). La seule contrainte étant d'avoir des conditions pédoclimatiques proches ainsi qu'un cépage identique. C'est l'écart (en %) de la marge semi-nette de la parcelle innovante par rapport à la parcelle classique qui sera transformée en indicateur (Tableau 3). Une baisse de moins de 15 % (indicateur vert et orange) est considérée comme acceptable du fait de changement de pratique qui nécessite souvent un temps d'adaptation de la part du viticulteur mais aussi de la vigne.

Tableau 3. Limites des classes permettant la comparaison innovant – classique pour l'indicateur économique.

Indicateur	Limites des classes
Vert	≥ 95 %
Orange	[85; 95[
Rouge	[75; 85[
Noir	<75 %

4. Test sur des exploitations agricoles

Dans le cadre du projet PEPSVI, cette méthode a été appliquée sur quatre exploitations viticoles alsaciennes pour tester sa faisabilité et valider les seuils de référence nécessaires à l'agrégation. Le temps nécessaire au recueil de l'information est raisonnable puisqu'il se situe autour de 3 heures. Le questionnaire ainsi que le calculateur ont été élaboré sur Excel ce qui permettra par la suite aux viticulteurs de faire par eux-mêmes leur évaluation socio-économique.

5. Conclusion

Si l'évaluation socio-économique par la méthode SOECO peut apparaitre complexe dans sa construction, sa mise en œuvre sur des exploitations viticoles est simple et rapide comme le préconise le cahier de charges des indicateurs. La méthode est originale à plusieurs titres. Tout d'abord, l'utilisation de divers mode d'agrégation permet à la fois d'éviter une trop grande perte d'information et de rendre les résultats compréhensibles par tous. Ensuite, elle permet de tester l'effet de systèmes viticoles innovants sur la composante socio-économique de la durabilité en tenant compte de données mesurables (temps de travail, programme de traitement, coût, ...) mais aussi du ressenti du viticulteur, ce dernier point étant très important dans l'appropriation de l'innovation, dans sa réussite et dans sa pérennité.

L'analyse des résultats de l'évaluation socio-économique réalisée en Alsace pour un millésime est difficile du fait du type d'exploitations retenu (un lycée agricole et un institut de recherche). Il est donc important d'élargir la gamme d'exploitation viticole et d'adapter la méthode à d'autres vignobles.

Références

Bockstaller, C., L. Guichard, D. Makowski, A. Aveline, P. Girardin, and S. Plantureux, 2008, Agri-environmental indicators to assess cropping and farming systems. A review **28**, 139–149

Keichinger, O. 2001, Evaluation de l'impact des pratiques agricoles d'exploitations de grandes cultures sur la valeur cynégétique à l'aide d'indicateurs agro-écologiques, Institut National Polytechnique de Lorraine, Colmar

Pervanchon, F., C. Bockstaller, and P. Girardin. 2002. Assessment of energy use in arable farming systems by means of an agro- ecological indicator: the energy indicator. **72**, 149–172

Ricou, C., C. Schneller, B. Amiaud, S. Plantureux, and C. Bockstaller, 2014, A vegetation-based indicator to assess the pollination value of field margin flora. Ecol. indic. **45**, 320–331

Sadok, W., F. Angevin, J. E. Bergez, C. Bockstaller, B. Colomb, L. Guichard, R. Reau, and T. Dore, 2009, MASC, a qualitative multi-attribute decision model for ex ante assessment of the sustainability of cropping systems. Agron. Sustain. Dev. **29**, 447–461

Thiollet-Scholtus, M., and C. Bockstaller, 2015a, Using indicators to assess the environmental impacts of wine growing activity: The INDIGO (R) method **62**, 13–25

Thiollet-Scholtus, M., and C. Bockstaller, 2015b, Using indicators to assess the environmental impacts of wine growing activity: the INDIGO® method. Eur. J. Agron. **62**, 13–25

Tixier, P., E. Malezieux, M. Dorel, C. Bockstaller, and P. Girardin, 2007, Rpest – An indicator linked to a crop model to assess the dynamics of the risk of pesticide water pollution application to banana-based cropping systems **26**, 71–81

Contribution des savoir-faire viticoles au développement durable et à la démarche de responsabilité Sociétale et environnementale, dans le cadre d'une gouvernance patrimoniale des terroirs / Contribution of wine know-how to the sustainable development and the approach of social responsibility and environmental, within the framework of a patrimonial governance of the terroirs

Joël Rochard

IFV (Institut Français de la Vigne et du Vin), 17 rue Jean Chandon Moët, 51200 Epernay, France

Résumé. Les savoirs viticoles constituent, au-delà des atouts locaux, un patrimoine humain irremplaçable. Le génie des hommes a permis à de nombreux écosystèmes terrestres, y compris parfois les plus extrêmes, d'être valorisés afin d'exprimer la quintessence du vin, expression des terroirs, ainsi que des particularités historiques et culturelles locales. Au-delà de la dimension territoriale, les savoir-faire viticoles sont également porteurs de dynamique locale, de préservation et de valorisation. Par ailleurs, ceux-ci peuvent alimenter la créativité scientifique et technologique dans une période de post modernité, en proie à des doutes et parfois des impasses, en quête de valeurs traditionnelles. La communication, à partir d'exemples, souligne la contribution des savoirs historiques locaux à une culture et une gouvernance, indispensable à la valorisation des terroirs et des vins, mais également au développement de démarche opérationnelle du développement durable en lien notamment avec la responsabilité sociale et environnementale de la filière viticole, vis-à-vis de notre planète et des générations futures.

Abstract. The wine knowledge constitutes, beyond the local assets, an irreplaceable human inheritance. Human engineering allowed many terrestrial ecosystems, including sometimes most extreme, to be developed in order to express the quintessence of the wine, expression of the terroirs, as well as historical and cultural characteristics local. Beyond territorial dimension, wine know-how is also carrying local dynamics, safeguarding and valorization. In addition, those can feed the scientific and technological creativity during one time of post modernity, in the grip of doubts and sometimes of the dead ends, in search of traditional values. The communication, starting from examples, underlines the contribution of the local historical knowledge to a culture and a governance, essential to the valorization of the terroir and wines, but also to the development of approach operational of sustainable development in link in particular with the social responsibility and environmental of the wine sector, with respect to our planet and of the future generations.

1. Introduction

La fin des années 1970 et le début des années 1980 marquent la fin de la période des «30 glorieuses», caractérisée par l'intensification des méthodes de production. Elle a permis de réduire la pénibilité des travaux agricoles et viticoles, de régulariser les rendements et de proposer aux consommateurs des denrées abondantes, variées et bon marché. Mais si elle a apporté une réponse à la fois sociale et économique au problème posé par la population dans un contexte de sortie guerre, cette mutation s'est accompagnée dans le même temps de multiples effets collatéraux sur l'environnement. Dès les années 1970, l'absence de prise en compte des équilibres naturels, couplée à la première crise pétrolière, déclencha l'émergence de nouveaux courants de pensée écologiste. En 1972, le rapport Meadows commandé par le Club de Rome, «Les limites de la croissance», prévoyait un effondrement brutal des économies au cours du 21e siècle suite aux effets conjugués de l'évolution démographique, des pressions environnementales et de l'épuisement des ressources.

Depuis la Conférence des Nations Unies de Stockholm en 1972, le problème du rapport entre le développement économique et la détérioration de l'environnement s'est inscrit au centre des préoccupations internationales.

En 1983, L'ONU créa la Commission Mondiale pour l'Environnement et le Développement qui élabora le concept de développement durable : «*Répondre aux besoins du présent sans compromettre la capacité des générations futures de satisfaire leurs propres besoins*». En 1992, 172 états se sont réunis à la conférence des Nation Unies à Rio, au cours de laquelle, cinq accords importants furent adoptés, visant à infléchir l'orientation expansionniste du développement (Fig. 1).

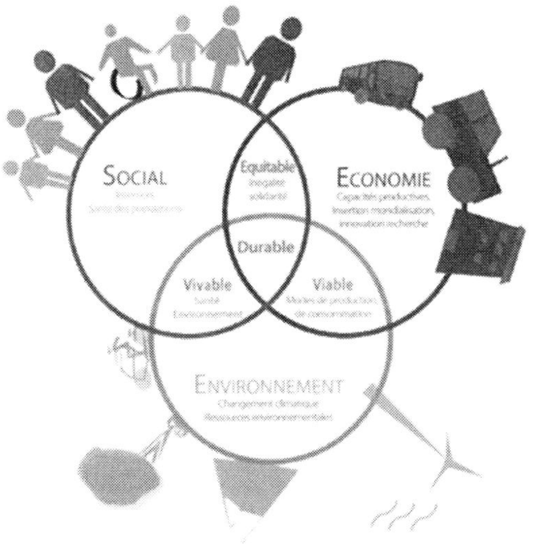

Figure 1. Les trois piliers du développement durable (développement durable source: `www.ledepartement66.fr`).

2. Emergence de la responsabilité sociétale en lien avec l'environnement

La notion de développement durable a mis l'accent sur l'impact de l'homme, par son activité domestique, agricole ou industrielle, sur le devenir des générations futures. Ainsi, la biodiversité, l'utilisation des énergies fossiles, la production de déchet et de gaz à effet de serre, s'immiscent progressivement dans les itinéraires viticoles. Parallèlement la crédibilité des démarches s'appuie de plus en plus sur des notions d'indicateurs en lien avec des méthodologies de type Analyse de Cycle de Vie (Fig. 2).

Au-delà des aspects environnementaux, un peu plus tardivement, une approche sociale puis sociétale du développement durable s'est également intégrée dans la réflexion stratégique de la filière viticole et a été formalisé en France par l'application viticole la norme ISO 26 000.

«L'inégalité a un prix, elle est la cause et la conséquence de la faillite du système politique et elle alimente, dans notre système économique, une instabilité et une inefficacité qui l'aggravent à leur tour. C'est ce cercle vicieux qui nous plonge dans l'abîme. (..) Puisqu'il est flagrant que notre système économique ne peut rien pour la plupart des citoyens et que nos gouvernements sont globalement sous la coupe des intérêts privés, la confiance dans la démocratie va s'éroder. Et puisque nous comprenons peu à peu que nombre de nos pays ne sont plus ceux de l'égalité des chances et du fair play, c'est sans doute notre sentiment de la justice qui est menacé.» Joseph E. Stieglitz, Le prix de l'inégalité.

Le concept de RSE n'apparaît qu'à partir des années 1960 dans la littérature consacrée aux entreprises. Depuis les années 1980, sous l'impulsion notamment des ONG, les concepts de finance éthique, commerce équitable et développement durable sont entrés dans le débat des instances politiques. Des recherches sur la responsabilité sociale et environnementale se sont concrétisées par la mise au point de référentiels internationaux, des guides de conduite des entreprises ou des certificats, normes,

Figure 2. Le développement durable, une approche intégrative. (Source `www.ocealia-groupe.fr`).

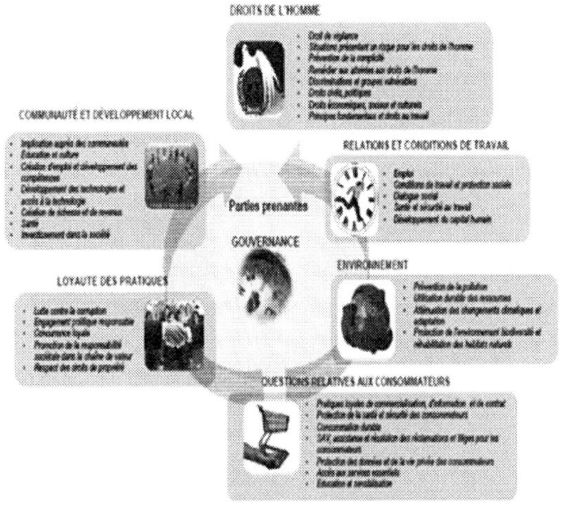

Figure 3. Les 7 points centraux de la démarche RSE/ISO 26 000 (Source `www.luxor-lighting.com`).

label validé par des audits sociaux et environnementaux. C'est dans ce contexte qu'a été formalisée la norme internationale ISO 26 000 (lignes directrices relatives à la responsabilité sociétale).

Pour la Commission européenne, la RSE est un «concept dans lequel les entreprises intègrent les préoccupations sociales, environnementales et économiques dans leurs activités et dans leurs interactions avec leurs parties prenantes sur une base volontaire».

La responsabilité sociétale et environnementale vise en premier lieu les besoins élémentaires des hommes: conditions de travail, nourriture, santé, bien-être, etc. Au-delà, une démarche RSE suppose également une implication de l'entreprise ou de l'organisation vis-à-vis des communautés locales mais également à l'échelle planétaire, en lien notamment avec l'éducation et la culture. Cette vision intègre en particulier la préservation et la valorisation d'un patrimoine commun issu le plus souvent de l'organisation d'une communauté et de la diffusion au cours du temps d'un savoir-faire collectif, autant de sujet qui sont ancrés dans l'histoire de la vigne et du vin.

DOI: 10.1051/bioconf/20170904013

Responsabilité sociétale de la filière vin : un guide AFNOR à la disposition de tous les acteurs

Le guide AFNOR, réalisé par une quarantaine de professionnels de la filière vin notamment, aborde tous les aspects de la profession. Impulsé Y. Chabin au sein de l'Interprofession des Vins Pays d'Oc IGP, il a été élaboré collectivement, sous l'égide d'AFNOR, par une quarantaine d'organismes professionnels.

Le guide (AC X30-032), d'application volontaire, dresse une liste de recommandations pour les entreprises souhaitant initier ou améliorer leur stratégie et leurs actions en matière de RSE. Il concrétise les enjeux de la responsabilité sociétale pour la filière vin en s'appuyant sur le document le plus légitime sur ce sujet: la norme internationale ISO 26000 ainsi que les documents et les retours d'expérience associés.

Des préconisations concrètes et opérationnelles sont proposées pour intégrer la RSE dans le projet d'entreprise. Sur le volet social par exemple, les enjeux des conditions de travail sont traités précisément pour favoriser le développement du capital humain. Pour ceux de la sous-traitance, les propositions ont pour objectif d'avoir des pratiques loyales et respectueuses. Sur le volet environnemental, tous les professionnels engagés ont notamment défini des repères pour prévenir les pollutions, utiliser les ressources durablement et maintenir la biodiversité. Le lien est aussi fait avec la protection et l'information du consommateur sur les produits et les métiers de la filière (Fig. 4).

Figure 4. Accompagnement des démarches RSE par l'AFNOR `http://www.afnor.org/actualites/responsabilite-societale-de-la-filiere-vin-guide/`.

Les savoirs viticoles constituent, au-delà des atouts locaux, un patrimoine humain irremplaçable. Le génie humain a permis à de nombreux écosystèmes terrestres, y compris parfois les plus extrêmes, d'être valorisés afin d'exprimer la quintessence du vin, expression des terroirs, ainsi que des particularités historiques et culturelles locales.

Depuis l'invention de l'agriculture au cours du néolithique, il y a 10 000 ans au Moyen-Orient, l'homme a su s'adapter à l'essor démographique par des savoir-faire initialement empiriques, mais validés au fil des générations. Très tôt les échanges entre les peuples ont permis une confrontation des savoirs et ont contribué à dynamiser la créativité locale. Grâce notamment aux civilisations égyptiennes, helléniques et romaines, relayées au moyen âge par les monastères, chaque région viticole européenne a su trouver des formes singulières de culture de la vigne et d'élaboration des vins, adaptées au contexte physique et historique local. Ce savoir-faire s'est transmis par la migration des hommes et des idées au pays du Nouveau Monde, avec des idées nouvelles, qui a développ une viticulture à la fois moderne mais également fortement imprégnée des savoir-faire ancestraux (Fig. 5).

Figure 5. Fresque égyptiennes de la tombe de Khaemouaset, grand-prêtre sous le règne d'Amenhotep 1 vers – 1500 avant J.-C., qui témoigne du savoir-faire des Égyptiens depuis la vendange jusqu'à la mise en amphores et au transport.

Au-delà des enjeux patrimoniaux, ces savoirs sont souvent porteurs d'une identité locale, de nature à fédérer les énergies collectives autour de valeurs culturelles, indispensable à la valorisation qualitative des vins, mais également au développement opérationnel de projets durables. Bien évidemment, la protection et la valorisation des paysages s'intègrent dans cette vision patrimoniale, mais également écologique des terroirs. Le classement d'une quinzaine de sites viticoles au sein du patrimoine mondial Unesco culturels ou/et naturels et la labellisation paysagère de nombreuses autres régions, témoignent de la gouvernance locale et régionale autour des paysages. Parallèlement aux aspects strictement patrimoniaux, le paysage, dans une stratégie collective d'organisation de l'espace, répond à des enjeux environnementaux notamment pour limiter les risques de ruissellement et d'érosion pour les vignobles de coteaux Il est également est au cœur de l'agroécologie qui vise à valoriser la biodiversité locale pour limiter les parasites.

3. Préservations et valorisation des savoir-faire : de la réflexion à l'action

La modernité, qui s'est imposé notamment après la seconde guerre mondiale, a permis parallèlement à l'autosuffisance alimentaire, de limiter la pénibilité du travail dans les exploitations agricoles et viticoles. En contrepartie, elle a souvent occulté les savoir-faire ancestraux, considérés comme inutiles ou parfois contraignant vis-à-vis d'une pensée «chimique ou technologiste». Ainsi

DOI: 10.1051/bioconf/20170904013

l'histoire de l'humanité, associée au développement des sciences, s'inscrivait jusqu'à présent dans un contexte expansionniste. Toute avancée scientifique offrait de nouvelles perspectives de développement, qui semblaient sans limite. Force est de constater maintenant que les capacités d'adaptation de notre planète ne sont pas infinies.

Ainsi depuis la fin du XXe siècle, la période de postmodernité est en proie à des doutes et parfois des impasses, avec l'émergence d'un passéisme, valeur refuge vis-à-vis d'un avenir incertain. Au-delà des impacts locaux, nos terroirs et chacun d'entre nous, seront probablement demain soumis à un effet de boomerang planétaire, directement lié à l'effet cumulatif de nos comportements individuels. Le développement sans limites des technologies est remis en cause, ce qui suppose une mixité entre des innovations contextualisée par rapport à leurs effets locaux et planétaires sur le long terme et une tradition porteuse de savoir-faire ancestraux souvent validés par le temps. Par ailleurs, cette modernité, porteuse de valeurs «d'individus roi, égocentriques», relayées par le marketing et les médias, s'est également traduite par l'émergence d'un individualisme, destructeur des pratiques collectives, ce qui suppose de créer à nouveau du lien entre les acteurs au sein des territoires viticoles. Fort heureusement, la filière viticole, par son attachement culturel et historique au terroir et aux valeurs ancestrales, a su conserver des savoir-faire et traditions qu'il convient de préserver et de valoriser notamment dans le cadre d'une vision de Responsabilité Sociétale et Environnementale, vis-à-vis des acteurs locaux mais plus globalement de l'ensemble de l'humanité, à l'image du classement d'une quinzaine de sites viticoles au titre du patrimoine mondial Unesco.

«L'Occident a longtemps cru que la modernité était le triomphe de la raison, la destruction des traditions, des appartenances, des croyances, la colonisation du vécu sur le calcul. Mais, aujourd'hui, toutes les catégories qui avaient été soumises à l'élite éclairée, travailleurs et colonisés, femmes et enfants, se sont révoltées et refusent d'appeler moderne un monde qui ne reconnaît pas à la fois leur expérience particulière et leurs accès à l'universel. De sorte que ceux qui s'identifient à la raison apparaissent désormais comme les défenseurs d'un pouvoir.» Alain TOURAINE, Critique de la modernité, 1992.

3.1. Festivités autour du vin : Exemple de la Fête des vignerons de Vevey en Suisse

Créer du lien entre les acteurs des territoires viticoles participe à une réflexion voir une dynamique collective autour du patrimoine commun, mais également dans la valorisation économique et progressivement écologique des terroirs. La fête de la Saint-Vincent dans les régions septentrionales françaises, l'élection de la «Reine des Vins» en Alsace, et dans les zones germaniques, l'organisation de festivités en lien avec l'œnotourisme, contribuent à dynamiser la vie collective dans les zones viticoles.

La Fête des vignerons de Vevey a été inscrite en 2016 sur la Liste représentative du patrimoine culturel immatériel de l'humanité. À Vevey, il existe une tradition qui célèbre le travail des vignerons de la communauté et à laquelle participent non seulement des professionnels mais aussi des membres du public et des artistes locaux.

Figure 6. Paysages suisse de Vevey en bordure du lac Léman, dont la fête des vignerons a été classée par l'UNESCO sur la liste du patrimoine immatériel de l'humanité.

D'abord simple parade, la Fête des Vignerons compte aujourd'hui 15 représentations sur trois semaines et plus de 5 000 figurants. Organisée tous les vingt ans, chaque fête repose sur la recréation de thèmes traditionnels tels que le travail de la terre et des vignes, le cycle des saisons et la fraternité. Des cortèges ont lieu pendant toute la durée de la fête, ainsi que des remises de prix aux meilleurs vignerons. La fête comporte aussi de la musique et des chansons, comme le Ranz des vaches (chant traditionnel des gardiens de troupeaux alpins), et des processions jusqu'à la ville voisine de La Tour-de-Peilz. On peut également y rencontrer des figurants déguisés et des marmousets (figurines de bois fixées sur des bâtons). La Fête des Vignerons repose essentiellement sur le travail de bénévoles, qui la préparent pendant plusieurs années. La transmission se fait par le biais des familles et de l'association à but non lucratif de la Confrérie des Vignerons de Vevey, qui coordonne la fête et dont les membres sont issus du grand public. Cette tradition encourage l'esprit de communauté, contribue à la vie artistique et procure un sentiment de continuité culturelle, tout en stimulant le savoir-faire des artisans vignerons (Fig. 6).

3.2. Terroir et savoir-faire monastique : exemple des climats de Bourgogne

L'écosystème culturel que constitue un terroir dépasse largement sa dimension physique. Le vigneron est l'une des composantes essentielles. Le savoir-faire acquis de génération en génération, l'aptitude à surmonter les défis techniques, la dynamique individuelle et collective, mais également la capacité de construire et pérenniser une image ainsi que la volonté de s'impliquer dans une démarche durable sont, parallèlement à l'approche spatiale, les clés de la valorisation du produit et de la transmission aux générations futures d'un outil de production viable. A l'image de l'homme dont le prédéterminisme génétique est évident, mais pour lequel la réussite personnelle et professionnelle est largement façonnée par ses influences culturelles, le vin est le fruit d'une symbiose intime entre nature et domestication.

UNESCO
Organisation des Nations Unies pour l'éducation, la science et la culture
Patrimoine culturel immatériel

Figure 7. Logo du patrimoine culturel immatériel de l'UNESCO.

Figure 8. Composantes du terroir (source P. Prévost, C. Mathieu, F. Gautier-Pelissier et coll., 2014).

«*Le terroir exprime un patrimoine commun dans un processus de valorisation des ressources locales. Il offre la possibilité de porter, autour d'un (ou de plusieurs) produit(s) typique(s), des pratiques sociales, techniques et économiques procédant d'une vision et d'une vocation. En s'inscrivant dans une profondeur historique et en s'appuyant sur la construction par les acteurs de règles de normalisation et de gouvernance adaptées, le terroir est un système dynamique. Il peut constituer une voie de gestion durable des ressources et de développement adaptée à la communauté qui s'y rattache. Dans cette perspective, le terroir peut aussi être le motif d'un questionnement éthique renouvelé, spécialement à l'endroit de la négociation des règles de normalisation et de gouvernance.*» Philippe Prévost, Mathieu Capitaine, François Gautier-Pelissier et coll., Le terroir, un concept pour l'action dans le développement des territoires, 2014 (Fig. 8).

Les monastères ont contribué à développer la vigne dans la plupart des régions européennes grâce notamment aux connaissances viticoles des moines sur l'art de cultiver la vigne (choix des plants, taille, terroir) et d'élaborer le vin. Historiquement la chute de l'empire romain fut suivie de plusieurs siècles de trouble et d'invasion pendant lesquels les dynasties franques des Mérovingiens (486–751) et des Carolingiens (751–987) ne réussit à imposer que de brèves périodes de calme, laissant l'église chrétienne s'affirmer comme le principal support de stabilité. Monastères et abbayes devinrent alors des centres d'érudition et des pôles artistiques tout en régissant de vastes domaines agricoles et viticoles.

Figure 9. Les moines bourguignons cultivant la vigne, au moyen âge. Extrait de l'ouvrage de M. Figuier, Les merveilles de l'industrie, 1873.

«*Nécessité culturelle et sociale, nécessité religieuse, nécessité économique, car le vin revendu fournit une part conséquente du fonctionnement financier des monastères, chaque abbaye en Europe devient donc un pôle viticole majeur. Les moines assurent aussi la transmission des livres agricoles des Romains, reprenant et améliorant ainsi les techniques culturales que les Anciens avaient mises au point. C'est aux moines que l'on doit l'amélioration du pressoir, la mise en valeur des meilleurs crus et terroirs, l'arrêt du complantage, l'édification de murets, pour dépierrer les sols et protéger les vignes de l'assaut du bétail. L'apport agronomique de ces hommes de Dieu est considérable.*» Jean-Baptiste Noé `http://www.jbnoe.fr/`.

Les climats du vignoble de Bourgogne, classés «Patrimoine Mondial» par l'Unesco, expriment avec raffinement le génie monastique du Moyen Âge. Sans disposer des techniques d'analyse moderne, les moines, dans l'ombre des monastères, avaient su, avec curiosité, intelligence, courage et passion, déterminer la relation entre la quintessence d'un vin et le micro-terroir qui en est à l'origine. Ainsi les parcelles bourguignonnes se distinguent les unes des autres par leurs conditions naturelles spécifiques (géologie, exposition, cépage...) qui ont été façonnées par le travail humain et peu à peu identifiées par rapport au vin qu'elles produisent. Ce paysage culturel est composé de plusieurs éléments: des parcelles viticoles et les unités de production associées, des villages et la ville de Beaune (Figs. 9 et 10).

Au-delà du zonage qualitatif empirique, les moines avaient une expertise de dégustateurs de raisin et de vin, une connaissance subtile des caractéristiques climatiques de l'année avec le recul des millésimes précédents, leur permettant ainsi de définir avec précision l'optimum de

Figure 10. Les climats de Bourgogne figurent sur la liste du patrimoine mondial Unesco depuis 2015.

vendange Cette date historique de récolte est souvent un bon indicateur des particularités climatiques de l'année., Répertoriée avec rigueur par les moines, elles constituent une mémoire qui permet aux historiens de reconstituer le climat depuis plusieurs siècles.

Comme le précise V. Daux, le ban des vendanges constitue, «*des archives continentales d'un grand intérêt du fait de leur résolution annuelle, de leur datation précise et des relations fortes et statistiquement significatives qui les lient à la température moyenne de la période de maturation du raisin*».

Dans un ouvrage remarquable «Histoire du climat depuis l'an mil», Emmanuel Leroy Ladurie a retracé l'évolution des données climatiques à partir d'une série d'indicateurs, parmi lesquels les dates de vendange des régions septentrionales. Sa démarche s'appuyait sur des données historiques répertoriées dans différentes régions (Argenteuil, Bourgogne, Champagne, Jura, Suisse). Ces données étaient calculées à partir de la date des bans de vendange en nombre de jour à partir du 1er septembre. E. Leroy Ladurie a souligné la variabilité, au cours du temps des dates de récolte, en liaison avec des modifications climatiques.

Cette approche permet de reconstituer l'histoire climatique à partir d'une corrélation entre les dates de vendange et les conditions climatiques de la période végétative de la vigne. L'étude d'années atypiques telle 1816 (vendange très tardive) qui faisait suite à l'explosion du volcan Tambora en Indonésie, à l'origine d'un refroidissement significatif dans l'hémisphère nord souligne la pertinence de la démarche. Cette approche permet de conforter la probabilité d'un lien entre l'avancée progressive de la date de récolte depuis deux à trois décennies et des changements climatiques probablement à relier à des effets anthropiques. D'un point de vue pratique cette étude historique, souligne qu'une élévation de 1 °C correspond en moyenne à une avancée d'environ 10 jours des dates de vendanges.

«*Ce serait cependant trop restreindre la portée du ban des vendanges, que de lui attribuer uniquement la facilité de percevoir des redevances. Il avait aussi pour dessein plus noble, la recherche de la qualité, car les «experts» envoyés dans les vignes pour fixer la date du ban, avaient pour mission essentielle de s'assurer que les raisins avaient atteint la parfaite maturité à l'ouverture du ban.*» Pierre Vital, Les vieilles vignes de notre France, 1956.

3.3. Valorisation du savoir-faire de l'élaboration en amphores : exemple géorgien de l'élaboration en kvevris

Depuis quelques années des amphores ou jarres, connaissent un certain développement dans les chais. Celles-ci, en terre cuite, en béton ou céramique, sont valorisées par des propriétés spécifiques concernant notamment la micro-oxygénation. Comme le souligne C. Caillaud, «*À partir du Néolithique, l'usage de la céramique pour la réception du moût, sa vinification, son stockage et son transport semble se généraliser dans les civilisations orientales et méditerranéennes en même temps que se répand la vitiviniculture et la consommation du vin, comme l'attestent les très nombreux témoignages archéologiques. (...) Les contenants en céramique, dolia et amphores, vont subir la concurrence des contenants en bois en Italie, peut-être dès le Ier s. ap. J.-C., d'abord modestement puis de plus en plus fortement à partir des IIe-IIIe s. ap. J.-C., même si l'on connaît quelques productions d'amphores italiennes datant du IVe-Ve s. ap. J.-C. Il semble, en tout cas, qu'à la fin de l'Antiquité, l'usage des contenants céramiques disparaît de manière quasi définitive de la péninsule italique, en attendant que des vignerons éclairés redécouvrent les bienfaits des jarres en terre cuite.*»

La Géorgie, berceau de la viticulture, a su maintenir cette production traditionnelle notamment au sein des domaines familiaux. La méthode géorgienne de vinification à l'ancienne dans des kvevris a été inscrite en 2013 sur la liste représentative du patrimoine culturel immatériel de l'humanité. La vinification en «kvevri» type de jarre (de l'arabe «djara» signifiant vase d'argile a large bouche) se pratique toujours dans les communautés rurale. L'élaboration traditionnelle consiste à fouler les raisins puis à les transférer dans un Kvevri scellé et enfoui dans le sol afin de laisser le mélange fermenter pendant cinq à six mois avant sa consommation. L'Unesco précise: «*Le vin joue un rôle important dans la vie quotidienne des Géorgiens et dans la célébration des rituels et des événements laïques ou religieux. La cave à vin est encore considérée comme le lieu le plus sacré du foyer. La tradition de la vinification en kvevri définit le mode de vie des communautés locales et constitue une part indissociable de leur identité culturelle et de leur héritage, les vignes et le vin étant évoqués dons les traditions orales et les chansons géorgiennes» (Fig. 11).*

A partir de cette élaboration traditionnelle, des études ont été menées pour comprendre les mécanismes physiques en fonction de la forme et du type de récipient, notamment liés à la circulation du vin à l'intérieur de ces contenants comparativement à des cuves classiques. Ces études ont pu montrer l'intérêt de cette configuration traditionnelle en particulier vis-à-vis du processus de clarification naturelle.

3.4. De la tradition écologie d'une cave : exemple de la cave bioclimatique du domaine Léon Boesch en Alsace

La cave est un lieu de calme et de mystère où s'épanouit le vin, en quête de fraîcheur, d'obscurité. Ces paysages souterrains traditionnels témoignent d'un labeur au quotidien, d'un savoir-faire ancestral, d'une géologie singulière et parfois d'une histoire des architectures de surface, bâties avec les matériaux qui sont extraits des

DOI: 10.1051/bioconf/20170904013

Figure 11. Photos de Géorgie qui souligne par les jarres (Kveris) au premier plan la tradition viticole ancestrale, associée à la religion orthodoxe (Source Fotolia).

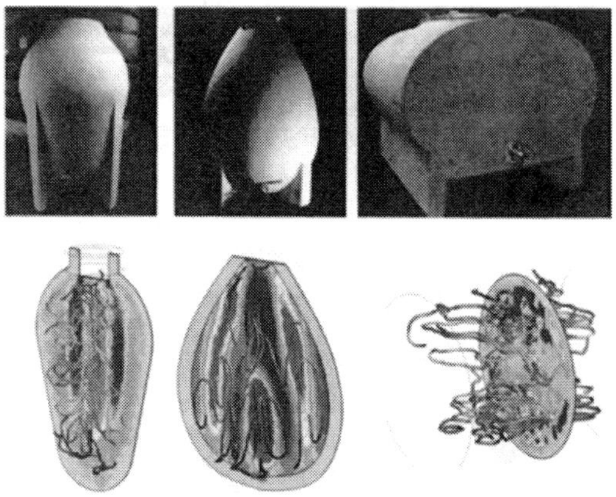

Figure 12. Simulation numérique par le modèle Aquilon des cuves, la vitesse verticale est représentée dans le sens de la gravité (en rouge : positive, en bleu : négative) (Source : R. Guillaument, J.P. Caltagirone, Revue française d'œnologie, 2016).

profondeurs. Les caves ou chais regroupent selon les régions des locaux aux fonctions multiples: local de pressurage, cellier de fermentation et de conservation des vins, cave de conservation des fûts ou des bouteilles, bâtiment d'embouteillage et de stockage.

La conception des chais traditionnels associe tout à la fois une maîtrise naturelle de la température et de l'hygrométrie et un transfert optimal de la vendange des vins et des bouteilles. Ainsi autrefois, le vigneron privilégiait la gravité avec en surface les locaux consacrés au pressurage et à l'embouteillage, les chais de vinification partiellement ou totalement enterrés et la cave en profondeur (Fig. 13).

La modernité a permis de s'affranchir de ces contraintes naturelles grâce à la climatisation, aux pompes et aux dispositifs de transfert du marc, des moûts et des vins mais avec une dépense énergétique croissante, et très souvent un impact sur l'effet de serre ainsi qu'une augmentation de la consommation d'eau qui doit être optimisée notamment dans la perspective de changement climatique. Les nouvelles conceptions écologiques s'inspirent des configurations traditionnelles de nos caves, qui utilisent notamment l'effet d'inertie

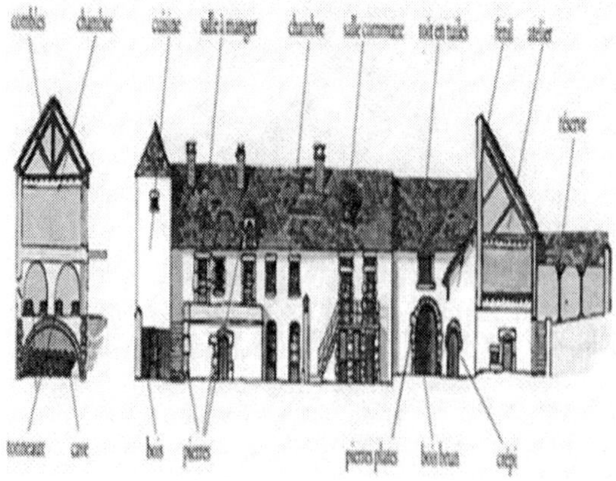

Figure 13. Maison vigneronne traditionnelle de la Côte-d'Or. La conception, qui valorisait l'inertie thermique du sous-sol, le choix de l'exposition permettait d'obtenir une parfaite articulation entre les contraintes de la zone d'habitation et de production des vins vis-à-vis notamment de la régulation température (Source : Encyclopédie touristique des vins de France, édition Hachette, 1994).

Figure 14. Domaine Léon Boesch en Alsace : exemple de conception d'une nouvelle cave bioclimatique qui s'inspire de l'architecture traditionnelle.

thermique du sous-sol, la végétalisation et une disposition optimale en fonction de l'exposition. Parallèlement, de nouvelles technologies ou conceptions telles que la géothermie, le solaire, les toits et murs végétaux, ainsi que des dispositifs de traitement des effluents écologiques, s'intègrent progressivement dans l'architecture et la conception des nouvelles caves.

Conçu par l'architecte Matthieu Winter, le domaine viticole alsacien Léon Boesch est construit à partir d'une ossature bois, mode de construction traditionnel en Alsace. La construction implique notamment des fondations en pierres sèches, une toiture végétalisée, une isolation des murs en bottes de paille, etc. Les matières premières viennent des alentours du village viticole. Les fondations sont constituées de roches calcaires d'un mètre d'épaisseur permettant de garantir l'hygrométrie et la «respiration» de la cave (Fig. 14).

4. Conclusion

Les savoirs faire viticoles constituent, au-delà des atouts locaux, un patrimoine humain universel irremplaçable. Le génie et le courage les hommes ont permis à de nombreux écosystèmes terrestres, y compris parfois les plus extrêmes, d'être valorisés afin d'exprimer la quintessence du vin, expression des terroirs, en lien avec les particularités géologiques, climatiques, historiques et culturelles locales.

Très tôt les échanges entre les peuples ont permis une confrontation des savoirs et ont contribué à dynamiser la créativité locale. En Eurasie, grâce notamment aux civilisations égyptiennes, helléniques et romaines, relayées au moyen âge par les monastères, chaque région viticole a su trouver des formes singulières de culture de la vigne et d'élaboration des vins, adaptées au contexte physique et historique local. Ce savoir-faire s'est transmis par la migration des hommes et des idées au pays du Nouveau Monde qui a su avec des idées nouvelles, développer une viticulture à la fois moderne mais également fortement imprégnée des savoir-faire ancestraux.

Au-delà des enjeux patrimoniaux, ces savoirs sont souvent porteurs d'une identité locale, de nature à fédérer les énergies collectives autour de valeurs culturelles. La fête des vignerons de Vevey, classée par l'UNESCO patrimoine mondial immatériel, témoigne de cette dynamique, support d'une gouvernance, indispensable à la valorisation qualitative des vins, mais également au développement opérationnel de projets durables. Au-delà de la dimension territoriale, les savoir-faire viticoles sont également porteurs de dynamique locale, de préservation et de valorisation. À titre d'exemple, la conception traditionnelle des caves, qui s'appuie notamment sur l'inertie thermique du sous-sol, alimente la vision écologique de leur conception. Il en est de même de l'élaboration traditionnelle en jarres de terre cuite, avec notamment l'exemple des KVEVRIS classés en Géorgie au titre du patrimoine immatériel de l'UNESCO. Ce savoir-faire ancestral sert de support à la conception de cuves ovoïdes qui comportent des propriétés thermiques et de circulation interne du vin spécifiques par rapport aux contenants traditionnels. Parfois ces pratiques ancestrales, à l'image des dates du ban des vendanges, répertoriées notamment par les moines, permettent d'alimenter la réflexion, sur l'évolution du climat passé, pour mieux appréhender son évolution dans un contexte de changements climatiques. Le classement immatériel des climats de Bourgogne, intègre également cette aptitude des moines à déterminer et à répertorier avec précision l'optimum de la date de vendange.

«Après un temps de confiance irrationnelle dans le progrès et dans la capacité humaine, une partie de la société est en train d'entrer dans une phase de plus grande prise de conscience. (...) Le changement climatique est un problème global aux graves répercussions environnementales, sociales, économiques, distributives ainsi que politiques, et constitue l'un des principaux défis actuels pour l'humanité. Les pires conséquences retomberont probablement au cours des prochaines décennies sur les pays en développement.» Encycliques Pape François, 5 juillet 2013.

Au final, les savoirs historiques locaux contribuent à une culture locale et une gouvernance, indispensable à la valorisation des terroirs et des vins, mais également au développement de démarches opérationnelles durable en lien notamment avec une responsabilité sociale et environnementale de la filière viticole, vis-à-vis des territoires mais également de notre planète et des générations futures. Ainsi le secteur viticole, probablement plus que tout autre filière agricole, se doit d'intégrer ce capital historique que constituent les savoir-faire locaux dans une vision de Responsabilité Sociétale et Environnementale universaliste.

Références

[1] Ph. Prévost, M. Capitaine, F. Gautier-Pelissier et coll., *Le terroir, un concept pour l'action dans le développement des territoires*, Vertigo revue électronique en sciences de l'environnement **14**, Numéro 1, mai, www.vertigo.revues.org (2014)

[2] R. Guillaument, J.P. Caltagirone, *Comment définir la cuve la mieux adaptée à ses besoins pour optimiser sa production et obtenir l'équilibre souhaité ?* Revue française d'œnologie, Novembre-Décembre, N° 279 (2016)

[3] Ch. Caillaud, *Le phénomène du «vin en amphore» dans l'Italie d'aujourd'hui*, Territoires du vin n°6 - Territoires du vin d'Italie, 20 août, disponible sur Internet http://revuesshs.u-bourgogne.fr/territoiresduvin/document.php?id=1880 ISSN 1760-5296 (2014)

[4] E. Joseph Stieglitz, *Le prix de l'inégalité*, Editions Les Liens qui Libèrent (2012)

[5] A. Touraine, *Critique de la modernité*, Éditions Fayard (1992)

[6] J.B. Noé, *Les moines, la vigne, le vin*, www.jbnoe.fr/Les-moines-la-vigne-le-vin

[7] V. Daux, *La reconstruction climatique à partir des dates de vendanges*, Revue de la BNF, **3**(36), 26–33 (2010)

[8] E. Leroy-Ladurie, *Histoire du climat depuis l'an mil*, éditions Champs Flammarion (1983)

[9] P. Vital, *Les vieilles vignes de notre France*, réédition Wallada (1984)

AUTHOR INDEX

AUTHOR INDEX

AUTHOR INDEX

AUTHOR INDEX

EDP Sciences – Web of Conferences
17, Avenue du Hoggar
Parc d'Activité de Courtabouf
BP 112
F-91944 Les Ulis Cedex A
France

ISBN 978-1-5108-4435-3

CURRAN ASSOCIATES INC.
proceedings.com